Graduate Texts in Physics

Graduate Texts in Physics

Graduate Texts in Physics publishes core learning/teaching material for graduate- and advanced-level undergraduate courses on topics of current and emerging fields within physics, both pure and applied. These textbooks serve students at the MS- or PhD-level and their instructors as comprehensive sources of principles, definitions, derivations, experiments and applications (as relevant) for their mastery and teaching, respectively. International in scope and relevance, the textbooks correspond to course syllabi sufficiently to serve as required reading. Their didactic style, comprehensiveness and coverage of fundamental material also make them suitable as introductions or references for scientists entering, or requiring timely knowledge of, a research field.

More information about this series at http://www.springer.com/series/8431

Chihiro Hamaguchi

Basic Semiconductor Physics

Third Edition

Springer

Chihiro Hamaguchi
Osaka University
Suita, Osaka
Japan

ISSN 1868-4513 ISSN 1868-4521 (electronic)
Graduate Texts in Physics
ISBN 978-3-319-88329-8 ISBN 978-3-319-66860-4 (eBook)
DOI 10.1007/978-3-319-66860-4

Softcover reprint of the hardcover 3rd edition 2017

Printed on acid-free paper

This Springer imprint is published by Springer Nature
The registered company is Springer International Publishing AG
The registered company address is: Gewerbestrasse 11, 6330 Cham, Switzerland

Preface to the Third Edition

The first edition of *Basic Semiconductor Physics* was published in 2001 and the second edition in 2010. After the publication of the first edition, many typographical errors have been pointed out and the corrected version was published in 2006. The publisher and my friends persuade me to revise the book adding new chapters, keeping the subject at the appropriate level. When I started writing the first edition, I decided not to include the physics of semiconductor devices such as p-n junction diode, bipolar transistor, and metal oxide semiconductor field-effect transistor (MOSFET). This decision is kept in the new (third) edition.

Although many books on semiconductor physics and technology have been published, the basic physics of semiconductor laser is not properly described. When the readers of my book understand the characteristics of two-dimensional electron gas and strain effect of semiconductors, they feel easy to understand double heterostructure lasers and strained quantum well lasers, but it is easier if they study some more detailed discussion on the laser action. Another subject is physics of low-dimensional semiconductors. *Basic Semiconductor Physics*, second edition, deals with two-dimensional electron gas and zero dimensional or quantum dot structure. The physics of quantum dot includes very important physics of artificial atoms and gives a good information of few electron systems. These subjects are included in the second edition published in 2010. Some new topics are also included in the second edition such as electron motion in an external field discussing the derivation of effective mass (Sect. 3.5), the physics of quantum dots (Sect. 8.8), and new Chap. 9 devoted to the discussion on the physics of semiconductor laser, where Einstein coefficients A and B, spontaneous and stimulated emission, luminescence, double heterostructure, and quantum well lasers are discussed. The strain effect of the quantum well laser is described in detail because it is well known that the effect is very important to understand the modes (TE and TM modes) of quantum well laser oscillations.

This new edition (third edition) is revised in various aspects of semiconductor physics as described below.

1. This third edition deals very in detail with the most important subjects of semiconductor physics such as the energy band calculations, transport of carriers, and semiconductor lasers. In Chap. 1, energy band calculations of various semiconductors with spin-orbit interaction are discussed in the full band zone using the local pseudopotential method, nonlocal pseudopotential method, and $\boldsymbol{k} \cdot \boldsymbol{p}$ perturbation method. Once the readers understand the methods, numerical calculations are straight forward because all the matrix elements are given in this textbook.
2. Spin-orbit interaction plays an important role in the electronic states, especially the valence band splitting. The effect is described in Appendix and the matrix elements of the spin-orbit interaction are properly included in the energy band calculations.
3. In Chap. 6, we present the numerical calculations of relaxation times and mobilities limited by the scattering processes as a function of electron energy and temperature, including scatterings by various kinds of phonons (lattice vibrations), impurity density, electron density, and so on. In addition, determination of the deformation potentials is discussed by comparing the calculations with the experimental data. The results help our understanding of electron transport and help to understand semiconductor device physics.
4. Recently, LED's and LD's based on nitrides such as GaN and GaInN are playing the most important role in the optical devices. However, the energy band structures of the nitrides are not well determined. In Chap. 9, energy band calculations of the nitrides, GaN, InN, and AlN are discussed by using the pseudopotential method. In addition, new method to calculate the energy bands of ternary alloys of nitrides such as GaInN, AlGaN, and AlInN is discussed in which the bowing of the band gaps is well explained by introducing only one additional parameter.
5. Entirely new sets of about 70 problems (exercises) and detailed treatments of the answers are provided for better understanding. In order to avoid misleading, most of the curves in this book have been carefully computed and plotted. Calculations use SI units throughout.

In addition to these subjects, errors in the second edition are corrected properly. Especially, actual calculations of electron mobilities are very important and provide detailed information on the transport in semiconductors. Therefore, calculations of the relaxation times and mobilities due to various scattering processes are carried out. These results will provide readers to manage calculations of the electron mobilities in various semiconductors. Also, the physical properties of semiconductors are tabulated in the text.

The author would like to express his special thanks to Professor Nobuya Mori, Osaka University, for his contribution to the energy band calculations of nitrides and for many stimulated discussions on the basic physics. The author is thankful to his wife Wakiko for her patience during the work of this textbook and also his parents Masaru and Saiye Hamaguchi for their support to his higher education. The author expresses his gratitude to Professor Dr. Klaus von Klitzing for inviting him to Max Planck Institute, Stuttgart, to finish the full band $\boldsymbol{k} \cdot \boldsymbol{p}$ perturbation theory, providing him a chance to discuss this subject with Professor Dr. Manuel Cardona. Finally, we want to thank Dr. Claus E. Ascheron and the staff of the Springer Verlag for their help and for the valuable suggestions for clarification of this book.

Osaka, Japan
March 2017

Chihiro Hamaguchi

Preface to the Second Edition

When the first edition of *Basic Semiconductor Physics* was published in 2001, there were already many books, review papers, and scientific journals dealing with various aspects of semiconductor physics. Since many of them were dealing with special aspects of newly observed phenomena or with very fundamental physics, it was very difficult to understand the advanced physics of semiconductors without the detailed knowledge of semiconductor physics. For this purpose, the author published the first edition for the readers who are involved with semiconductor research and development. *Basic Semiconductor Physics* deals with details of energy band structures, effective mass equation, and $\boldsymbol{k} \cdot \boldsymbol{p}$ perturbation, and then describes very important phenomena in semiconductors such as optical, transport, magnetoresistance, and quantum phenomena. Some of my friends wrote to me that the textbook is not only basic but advanced, and that the title of the book does not reflect the contents. However, I am still convinced that the title is appropriate because the advanced physics of semiconductor may be understood with the knowledge of the fundamental physics. In addition, new and advanced phenomena observed in semiconductors at an early time are becoming well known and thus classified in basic physics.

After the publication of the first edition, many typographical errors have been pointed out and the corrected version was published in 2006. The publisher and my friends persuade me to revise the book adding new chapters, keeping the subject at the appropriate level. When I started writing the first edition, I decided not to include physics of semiconductor devices such as p-n junction diode, bipolar transistor, and metal oxide semiconductor field-effect transistor (MOSFET). This is because the large numbers of books dealing with the subjects are available and a big or bulky volume is not accepted by readers. On the other hand, many researchers are involved with optoelectronic devices such as LED (Light Emitting Diode) and LD (Laser Diode) because memory devices such as DVD and blue ray disks are becoming important for writing and reading memory devices. In such devices, semiconductor laser diodes are used. In addition, the communication system based on the optical fiber plays a very important role in network, where laser diode is the key device. Although many books on semiconductor physics and technology have been published, the basic physics of semiconductor laser is not properly described. When the readers of my book understand the characteristics of two-dimensional electron gas and strain effect of semiconductors, they feel easy to understand double heterostructure lasers and strained quantum well lasers, but it is easier if they study some more detailed discussion on the laser action. Another subject is physics of low-dimensional semiconductors. *Basic Semiconductor Physics* deals with two-dimensional electron gas but zero-dimensional or quantum dot structure is not included. The physics of quantum dot includes very important physics of artificial atoms, and gives a good information of few electron systems.

In this revised version I included three main topics. The first one is Sect. 3.5, where electron motion in an external field is discussed with the derivation of effective mass. The most important relation for transport equation is the velocity (group velocity) of an electron in a periodic crystal. In this section, the expectation value of the velocity operator is evaluated and shown to be proportional to the gradient of the electron energy with respect to the wave vector. Then the classical motion of equation is proved to be valid for an electron in a crystal when we use the effective mass. In Sect. 8.8 the physics of quantum dots is discussed in connection with the charging energy (addition energy) required to add an extra electron in a quantum dot. The treatment is very important to understand Coulomb interaction of many electron system. In this section the exact diagonalization method based on Slater determinants is discussed in detail. Chapter 9 is devoted to the discussion on the physics of semiconductor laser, where Einstein coefficients A and B, spontaneous and stimulated emission, luminescence, double heterostructure, and quantum well lasers are discussed. The strain effect of the quantum well laser is described in detail because it is well known that the effect is very important to understand the modes (TE and TM modes) of quantum well laser oscillations.

I would like to express my special thanks to Professor Nobuya Mori for helping me to clarify the subject and providing me his calculated results used in Chap. 9, and also to my colleagues at Sharp Corporation with whom I have had many stimulated discussions on the basic physics of semiconductor lasers. It was very sad that Professor Tatsuya Ezaki of Hiroshima University died very recently, who made the detailed analysis of quantum dot physics for his Ph.D thesis (see Sect. 8.8).

Finally, I want to thank Dr. Claus E. Ascheron and the staff of the Springer Verlag for their help and for the valuable suggestions for clarification of this book.

Osaka
September 2009

Chihiro Hamaguchi

Preface to the First Edition

More than 50 years have passed since the invention of the transistor in December 1947. The study of semiconductors was initiated in the 1930s but we had to wait for 30 years (till the 1960s) to understand the physics of semiconductors. When the transistor was invented, it was still unclear whether germanium had a direct gap or indirect gap. The author started to study semiconductor physics in 1960 and the physics was very difficult for a beginner to understand. The best textbook of semiconductors at that time was "*Electrons and Holes in Semiconductors*" by W. Shockley, but it required a detailed knowledge of solid state physics to understand the detail of the book. In that period, junction transistors and Si bipolar transistors were being produced on a commercial basis, and industrialization of semiconductor technology was progressing very rapidly. Later, semiconductor devices were integrated and applied to computers successfully, resulting in a remarkable demand for semiconductor memories in addition to processors in the late 1970–1980s. Now, we know that semiconductors play the most important role in information technology as the key devices and we cannot talk about the age of information technology without semiconductor devices.

On the other hand, the physical properties of semiconductors such as the electrical and optical properties were investigated in detail in the 1950s, leading to the understanding of the energy band structures. Cyclotron resonance experiments and their detailed analysis first reported in 1955 were the most important contribution to the understanding of the energy band structures of semiconductors. From this work, it was revealed that the valence bands consist of degenerate heavy-hole and light-hole bands. Another important contribution comes from energy band calculations. Energy band calculations based on the empirical pseudopotential method and the $\boldsymbol{k} \cdot \boldsymbol{p}$ perturbation method reported in 1966 enabled us to understand the fundamental properties of semiconductors. In this period, high-field transport and current instabilities due to the Gunn effect and the acoustoelectric effect attracted great interest. In addition, modulation spectroscopy and light scattering were developed and provided detailed information of the optical properties of semiconductors. These contributions enabled us to understand the physical properties of bulk semiconductors almost completely.

At the same time, late in the 1960s and early 1970s, Leo Esaki and his co-workers developed a new crystal growth method, molecular beam epitaxy, and initiated studies of semiconductor heterostructures such as quantum wells and superlattices. This led to a new age of semiconductor research which demonstrated phenomena predicted from quantum mechanics. This approach is completely different from the past research in that new crystals and new structures are being created in the laboratory. This field is therefore called "band gap engineering". It should be noted here that such a research was not carried out up to fabricate devices for real applications but to investigate new physics. The proposal of modulation doping in the late 1970s and the invention of the high electron mobility transistor (HEMT) in 1980 triggered a wide variety of research work related to this field. Later, HEMTs have been widely used in such applications as the receivers for satellite broadcasting. Although the commercial market for LSI memories based on Si technologies is huge, metal semiconductor field-effect transistors (MESFETs) based on GaAs have become key devices for mobile phones (cellular phones) in the 1990s and it is believed that their industrialization will play a very important role in the twenty-first century.

Klaus von Klitzing et al. discovered the quantum Hall effect (later called the integer quantum Hall effect) in the two-dimensional electron gas system of a Si MOSFET in 1980, and this discovery changed semiconductor research dramatically. The discovery of the fractional quantum Hall effect followed the integer quantum Hall effect and many papers on these subjects have been reported at important international conferences. At the same time, attempts to fabricate microstructures such as quantum wires and metal rings were carried out by using semiconductor microfabrication technologies and led to the discovery of new phenomena. These are the Aharonov–Bohm effect, ballistic transport, electron interference, quasi-one-dimensional transport, quantum dots, and so on. The samples used for these studies have a size between the microscopic and macroscopic regions, which is thus called the "mesoscopic region". The research in cmesoscopic structures is still progressing.

The above overview is based on the private view of the author and very incomplete. Those who are interested in semiconductor physics and in device applications of new phenomena require a deep understanding of semiconductor physics. The situation is quite different for the author who had to grope his own way in semiconductor physics in the 1960s, while the former are requested to begin their own work after understanding the established semiconductor physics. There have been published various textbooks in the field of semiconductors but only few cover the field from the fundamentals to new phenomena. The author has published several textbooks in Japanese but they do not cover such a wide range of semiconductor physics. In order to supplement the textbooks, he has used printed texts for graduate students in the last 10 years, revising and including new parts.

This textbook is not intended to give an introduction to semiconductors. Such introductions to semiconductors are given in courses on solid state physics and semiconductor devices at many universities in the world. It is clear from the contents of this textbook that electron statistics in semiconductors, pn junctions, pnp or npn bipolar transistors, MOSFETs, and so on are not dealt with. This textbook is written for graduate students or researchers who have finished the introductory courses. Readers can understand such device-oriented subjects easily after reading this textbook. A large part of this book has been used in lectures several times for the solid state physics and semiconductor physics courses for graduate students at the Electronic Engineering Department of Osaka University and then revised. In order to understand semiconductor physics, it is essential to learn energy band structures. For this reason, various methods for energy band calculations and cyclotron resonance are described in detail. As far as this book is concerned, many of the subjects have been carried out as research projects in our laboratory. Therefore, many figures used in the textbook are those reported by us in scientific journals and from new data obtained recently by carrying out experiments so that digital processing is possible. It should be noted that the author does not intend to disregard the priorities of the outstanding papers written by many scientists. Important data and their analysis are referred to in detail in the text, and readers who are interested in the original papers are advised to read the references. This book was planned from the beginning to be prepared by LATEX and the figures are prepared in EPS files. Figures may be prepared by using a scanner but the quality is not satisfactory compared to the figures drawn by software such as PowerPoint. This is the main reason why we used our own data much more than those from other groups. Numerical calculations such as energy band structures were carried out in BASIC and FORTRAN. Theoretical curves were calculated using Mathematica and equations of simple mathematical functions were drawn by using SMA4 Windows. The final forms of the figures were then prepared using PowerPoint and transformed into EPS files. However, some complicated figures used in Chap. 8 were scanned and then edited using PowerPoint.

The author would like to remind readers that this book is not written for those interested in the theoretical study of semiconductor physics. He believes that it is a good guide for experimental physicists. Most of the subjects are understood within the framework of the one-electron approximation and the book requires an understanding of the Schrödinger equation and perturbation theory. All the equations are written using SI units throughout, so that readers can easily estimate the values. In order to understand solid state physics, it is essential to use basic theory such as the Dirac delta function, Dirac identity, Fourier transform, and so on. These are explained in the appendices. In addition, a brief introduction to group theoretical analysis of strain tensors, random phase approximations, boson operators, and the density matrix is given in the appendices. With this background, the reader is expected to understand all the equations derived in the text book.

The author is indebted to many graduate students for discussions and the use of their theses. There is not enough space to list all the names of the students. He is also very thankful to Prof. Dr. Nobuya Mori for his critical reading of the manuscript and valuable comments. He thanks Dr. Masato Morifuji for his careful reading of the text. Dr. Hideki Momose helped the author to prepare the LaTeX format and Prof. Dr. Nobuya Mori revised it. Also, thanks are due to Mr. Hitoshi Kubo, who took the Raman scattering data in digital format. He is also very thankful to Prof. Dr. Laurence Eaves and Prof. Dr. Klaus von Klitzing for their encouragement from the early stage of the preparation of the manuscript. A large part of the last chapter, Chap. 8, was prepared during his stay at the Technical University of Vienna and he would like to thank Prof. Dr. Erich Gornik for providing this opportunity and for many discussions. Critical reading and comments from Prof. L. Eaves, Prof. K. von Klitzing, Prof. G. Bauer and Prof. P. Vogl are greatly appreciated. Most of the book was prepared at home and the author wants to thank his wife Wakiko for her patience.

Osaka, Japan
March 2001

Chihiro Hamaguchi

Contents

Chapter 1
Energy Band Structures of Semiconductors

Abstract The physical properties of semiconductors can be understood with the help of the energy band structures. This chapter is devoted to energy band calculations and interpretation of the band structures. Bloch theorem is the starting point for the energy band calculations. Bloch functions in periodic potentials is derived here and a periodic function is shown to be expressed in terms of Fourier expansion by means of reciprocal wave vectors. Brillouin zones are then introduced to understand energy band structures of semiconductors. The basic results obtained here are used throughout the text. Nearly free electron approximation is shown as the simplest example to understand the energy band gap (forbidden gap) of semiconductors and the overall features of the energy band structure. The energy band calculation is carried out first by obtaining free-electron bands (empty lattice bands) which are based on the assumption of vanishing magnitude of crystal potentials and of keeping the crystal periodicity. Next we show that the energy band structures are calculated with a good approximation by the local pseudopotential method with several Fourier components of crystal potential. The nonlocal pseudopotential method, where the nonlocal properties of core electrons are taken into account, is discussed with the spin–orbit interaction. Also $\boldsymbol{k} \cdot \boldsymbol{p}$ perturbation method for energy band calculation is described in detail. The method is extended to obtain the full band structures of the elementary and compound semiconductors. Another method "tight binding approximation" will be discussed in connection with the energy band calculation of superlattices in Chap. 8.

1.1 Free-Electron Model

It is well known that the physical properties of semiconductors are understood with the help of energy band structures. The energy states or energy band structures of electrons in crystals reflect the periodic potential of the crystals and they can be calculated when we know the exact shape and the magnitude of the crystal potentials. The shape and the magnitude of the potential are not determined directly from any experimental methods, and thus we have to calculate or estimate the energy bands by using the assumed potentials. Many different approaches to calculations of energy

C. Hamaguchi, *Basic Semiconductor Physics*, Graduate Texts in Physics,
DOI 10.1007/978-3-319-66860-4_1

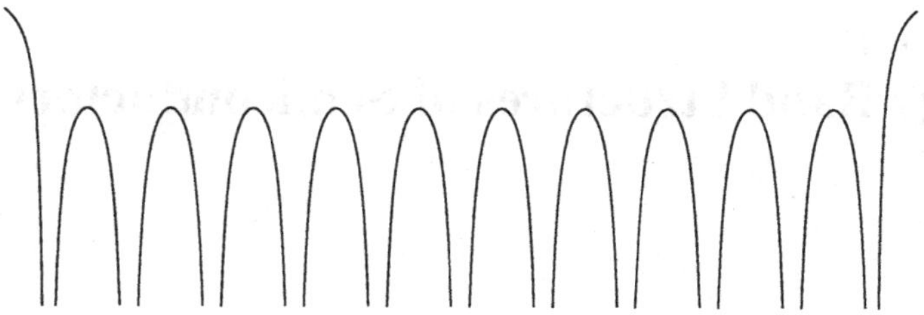

Fig. 1.1 One-dimensional crystal with periodic potential

bands have been reported, but in this textbook we will deal with several methods, which are not so difficult to understand. We begin with the most simplified method to calculate electronic states in a model crystal.

For simplicity we consider a one-dimensional crystal with a periodic potential as shown in Fig. 1.1, and assume that each atom provides one free electron and that the atom has a charge of $+e$, forming an ion.

The ion provides potential energy $V(r) = -\mathrm{e}^2/4\pi\epsilon_0 r$, where r is the distance from the central position of the ion. Therefore, the one-dimensional crystal has a potential energy consisting of the superposition of that of each atom, as shown in Fig. 1.1. From the figure we find the potential energy of the walls is higher than the inside potential and thus the electrons are confined between the walls. However, we have to note that the above results are derived from a very simplified assumption and the potential distribution is obtained without electrons. In a crystal there are many electrons and thus electron–electron interactions play a very important role in the potential energy distribution. Electron–electron interaction will be discussed in the case of plasmon scattering in Chap. 2 and in calculating the electronic states in quantum dots in Sect. 8. In the discussion of the energy band structure we will not deal with the electron–electron interactions and consider a simplified case where we calculate the electronic states for a single electron and then put many electrons in the energy states by taking the Pauli exclusion principle into account.

The large conductivity in metals is understood to arise from the fact that many free electrons exist in the conduction band. Therefore such electrons have an energy higher than the potential maxima and lower than the confining wall potentials. In the extreme case we can make an approximation that the electrons are confined in a square potential well, as shown in Fig. 1.2, where we assume the potential is infinite at $x = 0$ and $x = L$. In such a case the electron energy may be obtained by solving the one-dimensional Schrödinger equation

$$\left[-\frac{\hbar^2}{2m}\frac{\mathrm{d}^2}{\mathrm{d}x^2} + V(x)\right]\Psi(x) = \mathcal{E}\Psi(x)\,, \tag{1.1}$$

and the solutions are given by the following relations:

Fig. 1.2 Simplified quantum well model and electronic states

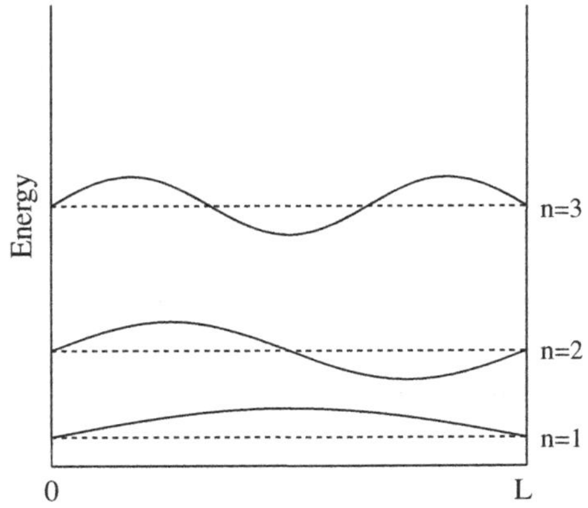

$$\begin{aligned}\Psi(x) &= A\sin(k_n) = A\sin\left(\frac{\pi}{L}\cdot n\right),\\ \mathcal{E} &= \frac{\hbar^2}{2m}k_n^2 = \frac{\hbar^2}{2m}\cdot\left(\frac{\pi n}{L}\right)^2,\\ k_n &= \frac{\pi}{L}\cdot n \qquad (n = 1, 2, 3, \ldots)\end{aligned} \tag{1.2}$$

Eigenfunctions and their energy level for $n = 1, 2, 3, \ldots$ are shown in Fig. 1.2. It is very easy to extend this one-dimensional model to the three-dimensional model, which will not be given here. We have to note that energy band structures are well understood by introducing periodic boundary conditions and the Bloch theorem.

1.2 Bloch Theorem

When we introduce a translational vector $\boldsymbol{T}$, the crystal potential has the periodicity $V(\boldsymbol{r}) = V(\boldsymbol{r}+\boldsymbol{T})$, and thus the squared wave function of the electron $|\Psi(\boldsymbol{r})|^2$ has the same periodicity. The amplitude of the wave function $\Psi(\boldsymbol{r})$ has an ambiguity of a phase factor $\exp(\mathrm{i}\boldsymbol{k}\cdot\boldsymbol{r})$. The cyclic boundary condition in the case of a one-dimensional crystal requires the condition that the wave function including the phase factor is the same at x and at $x + L$, and thus $\Psi(x) = \Psi(x + L)$, where L is the length of the crystal. The results are summarized as follows [1]:

$$\Psi(\boldsymbol{r}) = \exp(\mathrm{i}\boldsymbol{k}\cdot\boldsymbol{r})u_k(\boldsymbol{r}), \tag{1.3}$$

$$u_k(\boldsymbol{r}+\boldsymbol{T}) = u_k(\boldsymbol{r}), \tag{1.4}$$

$$\boldsymbol{k} = \frac{2\pi}{N}(n_x\boldsymbol{a}^* + n_y\boldsymbol{b}^* + n_z\boldsymbol{c}^*) , \tag{1.5}$$

where $\boldsymbol{k}$ is called the electron wave vector and $\boldsymbol{T} = n_1\boldsymbol{a} + n_2\boldsymbol{b} + n_3\boldsymbol{c}$ is the translational vector defined by using the fundamental vectors $\boldsymbol{a}, \boldsymbol{b}, \boldsymbol{c}$ with $n_1, n_2, n_3 = 0, \pm1, \pm2, \ldots$. The function $\Psi(\boldsymbol{r})$ is called the Bloch function and the function $u(\boldsymbol{r})$ is the periodic function of the translational vector,

$$u(\boldsymbol{r} + \boldsymbol{T}) = u(\boldsymbol{r}) . \tag{1.6}$$

The wave vector $\boldsymbol{k}$ is expressed in terms of the reciprocal lattice [1],

$$\boldsymbol{a}^* = \frac{\boldsymbol{b} \times \boldsymbol{c}}{\boldsymbol{a} \cdot (\boldsymbol{b} \times \boldsymbol{c})} , \tag{1.7}$$

$$\boldsymbol{b}^* = \frac{\boldsymbol{c} \times \boldsymbol{a}}{\boldsymbol{a} \cdot (\boldsymbol{b} \times \boldsymbol{c})} , \tag{1.8}$$

$$\boldsymbol{c}^* = \frac{\boldsymbol{a} \times \boldsymbol{b}}{\boldsymbol{a} \cdot (\boldsymbol{b} \times \boldsymbol{c})} , \tag{1.9}$$

which satisfy the following relations [1]:

$$\boldsymbol{a}^* \cdot \boldsymbol{a} = \boldsymbol{b}^* \cdot \boldsymbol{b} = \boldsymbol{c}^* \cdot \boldsymbol{c} = 1 , \tag{1.10}$$

$$\boldsymbol{a}^* \cdot \boldsymbol{b} = \boldsymbol{a}^* \cdot \boldsymbol{c} = \cdots = \boldsymbol{c}^* \cdot \boldsymbol{a} = \boldsymbol{c}^* \cdot \boldsymbol{b} = 0 . \tag{1.11}$$

The reciprocal lattice vector is defined by

$$\boldsymbol{G}_n = 2\pi(n_1\boldsymbol{a}^* + n_2\boldsymbol{b}^* + n_3\boldsymbol{c}^*) , \quad \text{where } n_1, n_2, n_3 \text{are integers.} \tag{1.12}$$

Periodic functions with the lattice vectors $\boldsymbol{a}, \boldsymbol{b}, \boldsymbol{c}$ are Fourier expanded with the reciprocal lattice vectors,

$$u_k(\boldsymbol{r}) = \sum_m A(\boldsymbol{G}_m) \exp(-\mathrm{i}\boldsymbol{G}_m \cdot \boldsymbol{r}) , \tag{1.13}$$

$$V(\boldsymbol{r}) = \sum_n V(\boldsymbol{G}_n) \exp(-\mathrm{i}\boldsymbol{G}_n \cdot \boldsymbol{r}) , \tag{1.14}$$

where $A(\boldsymbol{G}_m)$ and $V(\boldsymbol{G}_n)$ are Fourier coefficients. The coefficients are obtained by the inverse Fourier transformation,

$$A(\boldsymbol{G}_i) = \frac{1}{\Omega} \int_\Omega \exp(+\mathrm{i}\boldsymbol{G}_i \cdot \boldsymbol{r}) u_k(\boldsymbol{r}) \mathrm{d}^3\boldsymbol{r} , \tag{1.15}$$

$$V(\boldsymbol{G}_j) = \frac{1}{\Omega} \int_\Omega \exp(+\mathrm{i}\boldsymbol{G}_j \cdot \boldsymbol{r}) V(\boldsymbol{r}) \mathrm{d}^3\boldsymbol{r} , \tag{1.16}$$

where Ω is the volume of the unit cell of the crystal. From the definition of the reciprocal lattice vector (1.12), we can easily prove the following important relation (see Appendix A.2),

$$\frac{1}{\Omega}\int_{\Omega} \exp\left[\mathrm{i}\,(\boldsymbol{G}_m - \boldsymbol{G}_n)\cdot\boldsymbol{r}\right]\mathrm{d}^3\boldsymbol{r} = \delta_{mn}\,. \tag{1.17}$$

Here we have used the Kronecker delta function defined by

$$\delta_{mn} = \begin{cases} 1 & \text{for } m = n \\ 0 & \text{for } m \neq n \end{cases}\,. \tag{1.18}$$

1.3 Nearly Free Electron Approximation

For simplicity we begin with the one-dimensional case. From (1.16) we obtain

$$V(G_n) = \frac{1}{a}\int_0^a V(x)\exp(\mathrm{i}G_n x)\mathrm{d}x\,, \tag{1.19}$$

which gives the following zeroth-order Fourier coefficient, $V(0)$, when we put $G_n = 0$ in the above equation

$$V(0) = \frac{1}{a}\int_0^a V(x)\mathrm{d}x\,. \tag{1.20}$$

The coefficient $V(0)$ gives the average of the potential energy. In the case of three-dimensional crystals, the coefficient

$$V(0) = \frac{1}{\Omega}\int_{\Omega} V(\boldsymbol{r})\mathrm{d}^3\boldsymbol{r} \tag{1.21}$$

also gives the average value of the potential energy $V(\boldsymbol{r})$ in the unit cell Ω. In the following we measure the energy from $V(0)$ and thus we put $V(0) = 0$. The electronic states of an electron in the periodic potential $V(\boldsymbol{r})$ are given by solving the Schrödinder equation,

$$\left[-\frac{\hbar^2}{2m}\nabla^2 + V(\boldsymbol{r})\right]\Psi(\boldsymbol{r}) = \mathcal{E}(\boldsymbol{k})\Psi(\boldsymbol{r})\,. \tag{1.22}$$

Putting (1.13) into (1.3), we obtain

$$\Psi(\boldsymbol{r}) = \frac{1}{\sqrt{\Omega}} \exp(\mathrm{i}\boldsymbol{k} \cdot \boldsymbol{r}) \sum_n A(\boldsymbol{G}_n) \exp(-\mathrm{i}\boldsymbol{G}_n \cdot \boldsymbol{r}),$$
$$= \frac{1}{\sqrt{\Omega}} \sum_n A(\boldsymbol{G}_n) \exp\left[\mathrm{i}\,(\boldsymbol{k} - \boldsymbol{G}_n) \cdot \boldsymbol{r}\right], \tag{1.23}$$

where the factor $1/\sqrt{\Omega}$ is introduced to normalize the wave function $\Psi(\boldsymbol{r})$ in the unit cell. Putting (1.14) and (1.23) into (1.22), the following result is obtained:

$$\frac{1}{\sqrt{\Omega}} \sum_n \left[\frac{\hbar^2}{2m} (\boldsymbol{k} - \boldsymbol{G}_n)^2 - \mathcal{E}(\boldsymbol{k}) + \sum_m V(\boldsymbol{G}_m) \exp(-\mathrm{i}\boldsymbol{G}_m \cdot \boldsymbol{r}) \right]$$
$$\times A(\boldsymbol{G}_n) \exp\left[\mathrm{i}(\boldsymbol{k} - \boldsymbol{G}_n) \cdot \boldsymbol{r}\right] = 0. \tag{1.24}$$

Multiplying $(1/\sqrt{\Omega}) \exp[-\mathrm{i}(\boldsymbol{k} - \boldsymbol{G}_l) \cdot \boldsymbol{r}]$ to the both sides of the above equation and integrating in the unit cell with the help of (1.17), we find that the first and the second terms are not 0 for $n = l$, and that the third term is not 0 for $-(\boldsymbol{G}_m + \boldsymbol{G}_n) = -\boldsymbol{G}_l$ (or $\boldsymbol{G}_l - \boldsymbol{G}_n = \boldsymbol{G}_m$). Therefore, the integral is not 0 only in the case of $m = l - n$, and we obtain the following result:

$$\left[\frac{\hbar^2}{2m} (\boldsymbol{k} - \boldsymbol{G}_l)^2 - \mathcal{E}(\boldsymbol{k}) \right] A(\boldsymbol{G}_l) + \sum_n V(\boldsymbol{G}_l - \boldsymbol{G}_n) A(\boldsymbol{G}_n) = 0. \tag{1.25}$$

In the free-electron approximation of Sect. 1.1, we assumed the potential is given by the square well shown in Fig. 1.2, and thus the Fourier coefficients are $V(\boldsymbol{G}_m) = 0$ ($m \neq 0$). As stated above, we take the energy basis at $V(0)$ and put $V(\boldsymbol{r}) = 0$. Then (1.22) gives the following solution:

$$\mathcal{E}(\boldsymbol{k}) = \frac{\hbar^2 k^2}{2m}, \tag{1.26}$$

$$\Psi(\boldsymbol{r}) = \frac{1}{\sqrt{\Omega}} A(0) \exp(\mathrm{i}\boldsymbol{k} \cdot \boldsymbol{r}). \tag{1.27}$$

In the nearly free electron approximation, the potential energy is assumed to be very close to the square well shown in Fig. 1.2 and the Fourier coefficients $V(\boldsymbol{G}_l)$ are assumed to be negligible except for $V(0)$. Therefore, we replace the energy $\mathcal{E}(\boldsymbol{k})$ by (1.26) in (1.25) and only the term including $A(0)$ is kept in the second term. This assumption results in

$$\left[\frac{\hbar^2}{2m} (\boldsymbol{k} - \boldsymbol{G}_l)^2 - \frac{\hbar^2 k^2}{2m} \right] A(\boldsymbol{G}_l) + V(\boldsymbol{G}_l) A(0) = 0. \tag{1.28}$$

From this equation we obtain

$$A(\boldsymbol{G}_l) = \frac{V(\boldsymbol{G}_l)A(0)}{(\hbar^2/2m)\left[k^2 - (\boldsymbol{k} - \boldsymbol{G}_l)^2\right]} . \tag{1.29}$$

In the nearly free electron approximation, the electron wave function may be approximated by (1.27). In other words, the terms $A(\boldsymbol{G}_l)$ are very small except $\boldsymbol{G}_l = 0$. From the result given by (1.29), however, we find that the term $A(\boldsymbol{G}_l)$ is very large when $(\boldsymbol{k} - \boldsymbol{G}_l)^2 \approx k^2$. The condition is shown by the following equation:

$$(\boldsymbol{k} - \boldsymbol{G}_l)^2 = k^2 , \tag{1.30}$$

which gives rise to Bragg's law (law of Bragg reflection) and determines the Brillouin zones of crystals. When the electron wave vector ranges close to the value given by (1.30), we keep the term $A(\boldsymbol{G}_l)$ in addition to the term $A(0)$, and neglect the other terms. Then we obtain the following relations from (1.25):

$$\left[\frac{\hbar^2}{2m}k^2 - \mathcal{E}(\boldsymbol{k})\right] A(0) + V(-\boldsymbol{G}_l)A(\boldsymbol{G}_l) = 0 , \tag{1.31a}$$

$$\left[\frac{\hbar^2}{2m}(\boldsymbol{k} - \boldsymbol{G}_l)^2 - \mathcal{E}(\boldsymbol{k})\right] A(\boldsymbol{G}_l) + V(\boldsymbol{G}_l)A(0) = 0 . \tag{1.31b}$$

The solutions of the above equations are obtained under the condition that the coefficients $A(0)$ and $A(\boldsymbol{G}_l)$ are both not equal to 0 at the same time. The condition is satisfied when the determinant of (1.31a) and (1.31b) is 0, which is given by

$$\begin{bmatrix} (\hbar^2/2m)k^2 - \mathcal{E}(\boldsymbol{k}) & V(-\boldsymbol{G}_l) \\ V(\boldsymbol{G}_l) & (\hbar^2/2m)(\boldsymbol{k} - \boldsymbol{G}_l)^2 - \mathcal{E}(\boldsymbol{k}) \end{bmatrix} = 0 . \tag{1.32}$$

From this we obtain

$$\mathcal{E}(\boldsymbol{k}) = \frac{1}{2}\left[\frac{\hbar^2}{2m}\left\{k^2 + (\boldsymbol{k} - \boldsymbol{G}_l)^2\right\} \pm\sqrt{\left(\frac{\hbar^2}{2m}\right)^2 \left\{k^2 - (\boldsymbol{k} - \boldsymbol{G}_l)^2\right\}^2 + 4|V(\boldsymbol{G}_l)|^2}\,\right], \tag{1.33}$$

where the relation $V(-\boldsymbol{G}_l) = V^*(\boldsymbol{G}_l)$ is used. When $k^2 = (\boldsymbol{k} - \boldsymbol{G}_l)^2$ and thus $2\boldsymbol{k} \cdot \boldsymbol{G}_l = G_l^2$, we find

$$\mathcal{E}(\boldsymbol{k}) = \frac{\hbar^2 k^2}{2m} \pm |V(\boldsymbol{G}_l)| , \tag{1.34}$$

which means that there exists an energy gap of $2|V(\boldsymbol{G}_l)|$.

Here we will apply the results to a one-dimensional crystal. Replacing $\boldsymbol{G}_l$ by $\boldsymbol{G}_n$ in (1.33) and using the relation $\boldsymbol{G}_n = 2\pi n/a (n = 0, \pm 1, \pm 2, \pm 3, \cdots)$, we have the following equation:

$$\mathcal{E}(k) = \frac{1}{2}\left[\frac{\hbar^2}{2m}\left\{k^2 + \left(k - \frac{2\pi n}{a}\right)^2\right\}\right.$$
$$\left. \pm\sqrt{\left(\frac{\hbar^2}{2m}\right)^2\left\{k^2 - \left(k - \frac{2\pi n}{a}\right)^2\right\}^2 + 4|V(G_n)|^2}\right]. \tag{1.35}$$

Therefore, $\mathcal{E}(k) \cong \hbar^2k^2/2m$ is satisfied, except in the region close to the condition $k^2 = G_n^2 = (k - 2\pi n/a)^2$ or $k = n\pi/a$. This result gives the choice of $\pm$ in (1.35). Taking account of the sign of the square root, in the region $k < (k - G_n)^2$ we should choose the minus sign and in the region $k > (k - G_n)^2$ we have to choose the plus sign in (1.35). Therefore, in the region $k \approx n\pi/a > 0$, we find we obtain the following relations:

$$k \le \frac{n\pi}{a}: \quad \mathcal{E}(k) = \frac{\hbar^2k^2}{2m} - |V(G_n)|, \tag{1.36}$$

$$k \ge \frac{n\pi}{a}: \quad \mathcal{E}(k) = \frac{\hbar^2k^2}{2m} + |V(G_n)|. \tag{1.37}$$

Using the above relations and plotting $\mathcal{E}(\boldsymbol{k})$ as a function of k, we obtain the results shown in Fig. 1.3a. Such a plot of energy in the whole region of the k vector shown in Fig. 1.3a is called the "extended zone representation". In such a one-dimensional crystal model with N atoms, however, the electron system has N degrees of freedom and thus the wave vector of the electron may take N values in the range $-\pi/a < k \le \pi/a$, corresponding to the first Brillouin zone. When we take this fact into account, the energy can be shown in the first Brillouin zone $-\pi/a < k \le \pi/a$. This may be understood from the fact that the wave vectors $\boldsymbol{k}$ and $\boldsymbol{k} + \boldsymbol{G}_m$ are equivalent because of the equivalence of the wave functions with these two wave vectors from

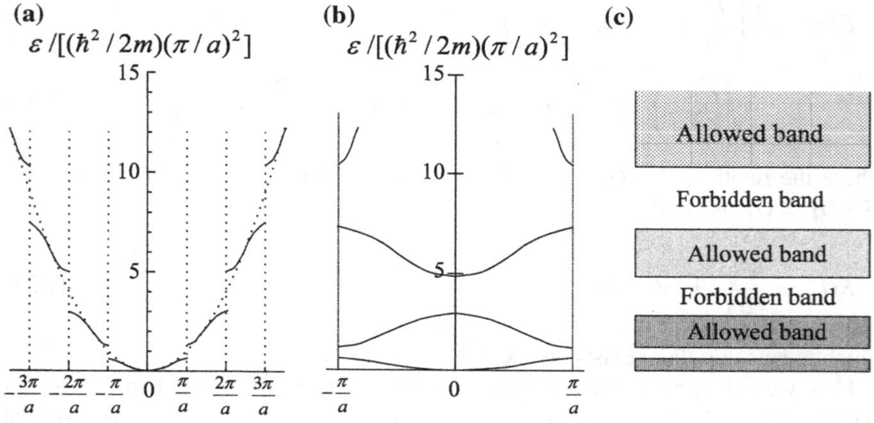

Fig. 1.3 Energy band structure of one-dimensional crystal obtained from the nearly free electron approximation. **a** extended zone representation, **b** reduced zone representation, and **c** energy bands in real space. Energy in units of $(\hbar^2/2m)(\pi/a)^2$

the result shown by (1.23) (see Sect. 1.4). Using this result we easily find that the region $-2\pi/a < k \leq -\pi/a$ in Fig. 1.3a is moved in the region $0 \leq k \leq \pi/a$ of the first Brillouin zone by adding $G = 2\pi/a$ and that $\pi/a \leq k \leq 2\pi/a$ is moved into $-\pi/a \leq k \leq 0$ by adding $G = -2\pi/a$. The region $-2\pi/a < k \leq -\pi/a$ and $\pi/a \leq k \leq 2\pi/a$ is called the second Brillouin zone. The 3rd Brillouin zone, 4th Brillouin zone, ... are defined in the same manner and they can be reduced to the first Brillouin zone. The energy plotted in the first Brillouin zone is shown in Fig. 1.3b and this is called the "reduced zone representation". Usually the energy band structure is shown in the reduced zone scheme. Figure 1.3c shows the allowed energy region with the shaded portion and the region is called the "allowed band", while electrons cannot occupy the region in between the allowed bands, which is called the "forbidden band" or "energy band gap", where the horizontal axis corresponds to the coordinate of real space.

1.4 Reduced Zone Scheme

The Bloch function in a crystal is given by

$$\psi(\boldsymbol{r}) = \frac{1}{\sqrt{\Omega}} \sum_l A(\boldsymbol{G}_l) e^{i(\boldsymbol{k}-\boldsymbol{G}_l)\cdot\boldsymbol{r}} . \tag{1.38}$$

Let us examine the phase between two Bloch functions with $\boldsymbol{k}$-vectors $\boldsymbol{k} - \boldsymbol{G}_l$ and $\boldsymbol{k}$. The phase difference at $\boldsymbol{r}$ between $\exp[i\boldsymbol{k}\cdot\boldsymbol{r}]$ and $\exp[i(\boldsymbol{k} - \boldsymbol{G}_l)\cdot\boldsymbol{r}]$ is $\boldsymbol{G}_l \cdot \boldsymbol{r}$. The phase difference of the Bloch functions at a point displaced by the translational vector $\boldsymbol{T}$, $\boldsymbol{r} + \boldsymbol{T}$, is easily obtained in the following way. Since we have the relations

$$\boldsymbol{T} = n_1\boldsymbol{a} + n_2\boldsymbol{b} + n_3\boldsymbol{c} , \tag{1.39}$$

$$\boldsymbol{G}_l = 2\pi(m_1\boldsymbol{a}^* + m_2\boldsymbol{b}^* + m_3\boldsymbol{c}^*) , \tag{1.40}$$

$$\boldsymbol{G}_l \cdot \boldsymbol{T} = 2\pi(m_1n_1 + m_2n_2 + m_3n_3) = 2\pi n , \tag{1.41}$$

and we obtain

$$\begin{aligned} \boldsymbol{k}\cdot(\boldsymbol{r}+\boldsymbol{T}) - (\boldsymbol{k}-\boldsymbol{G}_l)\cdot(\boldsymbol{r}+\boldsymbol{T}) &= \boldsymbol{G}_l\cdot\boldsymbol{r} + \boldsymbol{G}_l\cdot\boldsymbol{T} \\ &= \boldsymbol{G}_l\cdot\boldsymbol{r} + 2\pi n, \end{aligned} \tag{1.42}$$

where n is an integer and the relation $\boldsymbol{a}^* \cdot \boldsymbol{a} = 1$ is used.

From the above considerations we find the following results. The phase differences between the two functions $\exp[\mathrm{i}\boldsymbol{k}\cdot\boldsymbol{r}]$ and $\exp[\mathrm{i}(\boldsymbol{k}-\boldsymbol{G}_l)\cdot\boldsymbol{r}]$ at two different positions $\boldsymbol{r}$ and $\boldsymbol{r}+\boldsymbol{T}$ differs by the amount $2\pi n$ and thus the Bloch function $\psi(\boldsymbol{r})$ behaves in the same way at the position displaced by the translational vector $\boldsymbol{T}$. In other words, we can conclude that electrons with $\boldsymbol{k}$ and $\boldsymbol{k}-\boldsymbol{G}_l$ are equivalent. Therefore, we can reduce the electronic state of an electron with wave vector $\boldsymbol{k}$ into the state $\boldsymbol{k}-\boldsymbol{G}_l$, and represent the electronic states in the first Brillouin zone. This procedure is called the **reduced zone scheme** and the energy band representation in the reduced zone scheme. On the other hand, the energy band representation over the whole $\boldsymbol{k}$ region is called the **extended zone scheme**.

1.5 Free–Electron Bands (Empty–Lattice Bands)

1.5.1 First Brillouin Zone

In order to calculate energy band structures of a semiconductor the following procedures are required to carry out the calculations. These are

1. Calculate the first Brillouin zone.
2. Calculate the energy band structures in the limit of zero potential energy. This procedure is to obtain the free-electron bands or empty-lattice bands and plot the energy as a function of wave vector $\boldsymbol{k}$ in the reduced zone scheme.
3. Then calculate the energy bands using an appropriate method.

In the energy band calculation the most important procedure is to obtain the empty-lattice bands, which are calculated by assuming zero lattice potential $V(\boldsymbol{r}) = 0$ and keeping the lattice periodicity. In other words we assume the wave functions are given by the free-electron model with the wave vectors of the electrons in the periodic potential. Such an energy band structure is called empty-lattice bands or free-electron bands and thus the band structure exhibits the characteristics of the lattice periodicity.

Here we will show an example of empty-lattice bands in the case of the face-centered cubic (fcc) lattice. First, we calculate the Brillouin zone of the fcc lattice. Figure 1.4a shows the fcc structure. The diamond structure is obtained by displacing the lattice atoms by the amount $(a/4, a/4, a/4)$, which is shown in Fig. 1.4b. Therefore, the diamond structure belongs to the fcc structure. Diamond (C), Si and Ge have this diamond structure. On the other hand, the displaced lattice atoms are different from the original atoms, and the structure is called the zinc-blende structure, which is shown in Fig. 1.4c. Crystals such as GaAs, GaP, AlAs, InAs, InSb belong to the zinc-blende structure. The fundamental vectors and volume v of a fcc lattice are defined by

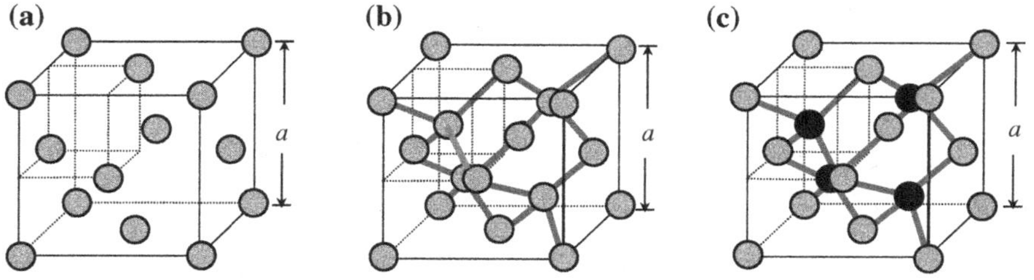

Fig. 1.4 **a** face-centered cubic lattice, **b** the diamond crystal lattice is obtained by displacing the lattice atoms of **a** by $(a/4, a/4, a/4)$, **c** when the displaced lattice atoms are different from the original lattice atoms, the crystal structure is called the zinc-blende crystal structure

$$\begin{aligned}
\boldsymbol{a} &= \frac{a}{2}(\boldsymbol{e}_x + \boldsymbol{e}_y)\,, \quad \boldsymbol{b} = \frac{a}{2}(\boldsymbol{e}_y + \boldsymbol{e}_z)\,, \quad \boldsymbol{c} = \frac{a}{2}(\boldsymbol{e}_z + \boldsymbol{e}_x)\,, \\
v &= \boldsymbol{a} \cdot (\boldsymbol{b} \times \boldsymbol{c}) \\
&= \left(\frac{a}{2}\right)^3 (\boldsymbol{e}_x + \boldsymbol{e}_y) \cdot [(\boldsymbol{e}_y + \boldsymbol{e}_z) \times (\boldsymbol{e}_z + \boldsymbol{e}_x)] = 2\left(\frac{a}{2}\right)^3 = \frac{1}{4}a^3\,, \qquad (1.43)
\end{aligned}$$

where $\boldsymbol{e}$ is the unit vector. The reciprocal vectors of the fcc structure are obtained as follows:

$$\begin{aligned}
\boldsymbol{a}^* &= \frac{\boldsymbol{b} \times \boldsymbol{c}}{v} = \left(\frac{a}{2}\right)^2 \frac{(\boldsymbol{e}_y + \boldsymbol{e}_z) \times (\boldsymbol{e}_z + \boldsymbol{e}_x)}{v} = \left(\frac{a}{2}\right)^2 \frac{(\boldsymbol{e}_x - \boldsymbol{e}_z + \boldsymbol{e}_y)}{a^3/4} \\
&= \frac{1}{a}(\boldsymbol{e}_x + \boldsymbol{e}_y - \boldsymbol{e}_z)\,, \qquad (1.44) \\
\boldsymbol{b}^* &= \frac{1}{a}(-\boldsymbol{e}_x + \boldsymbol{e}_y + \boldsymbol{e}_z)\,, \qquad (1.45) \\
\boldsymbol{c}^* &= \frac{1}{a}(\boldsymbol{e}_x - \boldsymbol{e}_y + \boldsymbol{e}_z)\,. \qquad (1.46)
\end{aligned}$$

From these results we find that the reciprocal lattices of the fcc lattice form body-centered cubic lattices. Therefore, the reciprocal lattice vectors $\boldsymbol{G}$ of the fcc lattice are given by

$$\boldsymbol{G} = 2\pi(n_1\boldsymbol{a}^* + n_2\boldsymbol{b}^* + n_3\boldsymbol{c}^*)\,. \qquad (1.47)$$

The Brillouin zone of the fcc lattice is defined by

$$k^2 = (\boldsymbol{k} - \boldsymbol{G}_l)^2 \qquad (1.48)$$

or

$$2\boldsymbol{k} \cdot \boldsymbol{G}_l = G_l^2\,. \qquad (1.49)$$

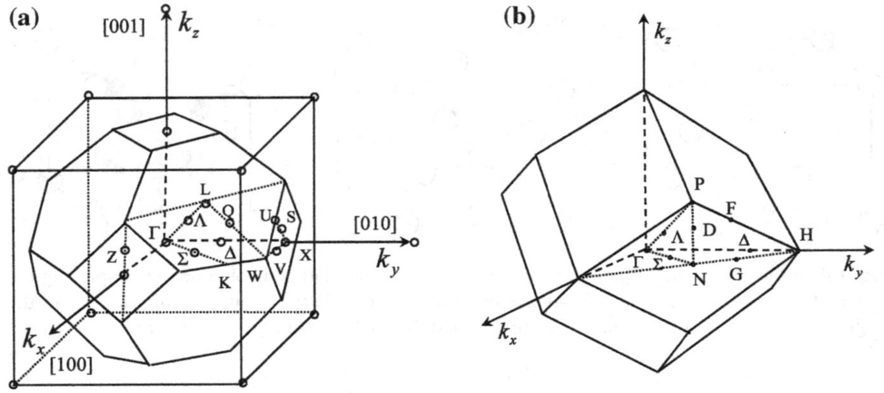

Fig. 1.5 The first Brillouin zone of **a** face-centered cubic lattice and **b** body-centered cubic lattice

Using the above equation the first Brillouin zone is easily calculated, as shown in Fig. 1.5a. For comparison the first Brillouin zone of body-centered cubic lattices, which form face-centered cubic lattices, is shown in Fig. 1.5b.

Since the lattice potential is 0 in the empty-lattice model, the energy of an electron is given by the free-electron model:

$$\mathcal{E}(\boldsymbol{k}) = \frac{\hbar^2}{2m}k^2 . \tag{1.50}$$

We plot the energy $\mathcal{E}(\boldsymbol{k})$ versus wave vector $\boldsymbol{k}$ curves in the reduced zone scheme, using the relation

$$\boldsymbol{k}' = \boldsymbol{k} - \boldsymbol{G} , \tag{1.51}$$

and choose $\boldsymbol{k}'$ in the first Brillouin zone. Then the empty lattice bands are given by

$$\mathcal{E}(\boldsymbol{k}') = \frac{\hbar^2}{2m}\left(\boldsymbol{k}' + \boldsymbol{G}\right)^2 . \tag{1.52}$$

1.5.2 Reciprocal Lattice Vectors of fcc Crystal

In the next Sect. 1.6, we discuss detailed treatment of pseudopotential method for energy band calculations and show how to program the energy band calculation. For this purpose we evaluate the matrix elements of the pseudopotential Hamiltonian of a face centered cubic (fcc) crystal. First, we calculate the reciprocal lattice vectors. Inserting (1.44) $\sim$ (1.46) into (1.47), we obtain the following relations for the reciprocal vectors

$$G = \frac{2\pi}{a}\left[(n_1 - n_2 + n_3)\boldsymbol{e}_x + (n_1 + n_2 - n_3)\boldsymbol{e}_y + (-n_1 + n_2 + n_3)\boldsymbol{e}_z\right] \tag{1.53}$$

and thus the x, y, z components of $\boldsymbol{G}$, and G^2 are

$$G_x = \frac{2\pi}{a}(n_1 - n_2 + n_3) , \tag{1.54}$$

$$G_y = \frac{2\pi}{a}(n_1 + n_2 - n_3) , \tag{1.55}$$

$$G_z = \frac{2\pi}{a}(-n_1 + n_2 + n_3) , \tag{1.56}$$

$$G^2 = \left[G_x^2 + G_y^2 + G_z^2\right] \tag{1.57}$$

$$\equiv \left(\frac{2\pi}{a}\right)^2 \left[(n_1 - n_2 + n_3)^2 + (n_1 + n_2 - n_3)^2 + (-n_1 + n_2 + n_3)^2\right] . \tag{1.58}$$

Using these relations the reciprocal wave vectors of a face–centered cubic lattice are easily evaluated, by putting $n_1,\ n_2,\ n_3 = \pm 0,\ \pm 1,\ \pm 2,\ \pm 4,\ \pm 5,\ \ldots$. It is very convenient to introduce dimensionless lattice vectors $\boldsymbol{K}$ defined by,

$$\boldsymbol{K} = \left(\frac{a}{2\pi}\right)\boldsymbol{G} . \tag{1.59}$$

The calculated reciprocal lattice vectors $\boldsymbol{K}$ are listed in Table 1.1, where $G_x,\ G_y,\ G_z$ are obtained by multiplying $\boldsymbol{K}$ by $(2\pi/a)$ and they are tabulated for $K^2 = K_x^2 + K_y^2 + K_z^2 = 0 \sim 27$.

1.5.3 Free Electron Bands

Free electron bands (empty lattice bands) are easily calculated using the results of Table 1.1, which are shown in Fig. 1.6 in the range of electron energy less than 200 in units of $[\hbar^2/2ma^2]$.

For better understanding we list several reciprocal lattice vectors from the lowest orders, which are given by (see (1.53) and Table 1.1)

Table 1.1 Reciprocal lattice vectors of a face-centered cubic lattice calculated from (1.53), where $K_x = (a/2\pi)G_x$, $K_y = (a/2\pi)G_y$, $K_z = (a/2\pi)G_z$. Here the vectors are obtained for $K^2 = K_x^2 + K_y^2 + K_z^2$ from 0 to 27

$[\boldsymbol{K}]$	Permutations								K^2
[000]	[000]								0
[111]	$[\bar{1}\bar{1}\bar{1}]$	$[\bar{1}\bar{1}1]$	$[1\bar{1}\bar{1}]$	$[\bar{1}1\bar{1}]$	$[1\bar{1}1]$	$[\bar{1}11]$	$[11\bar{1}]$	[111]	3
[200]	$[0\bar{2}0]$	$[\bar{2}00]$	$[00\bar{2}]$	[002]	[200]	[020]			4
[220]	$[\bar{2}\bar{2}0]$	$[0\bar{2}\bar{2}]$	$[\bar{2}0\bar{2}]$	$[0\bar{2}2]$	$[\bar{2}02]$	$[2\bar{2}0]$	$[\bar{2}20]$	$[20\bar{2}]$	8
	$[02\bar{2}]$	[202]	[022]	[220]					
[311]	$[\bar{1}\bar{3}\bar{1}]$	$[\bar{3}\bar{1}\bar{1}]$	$[\bar{1}\bar{3}1]$	$[\bar{3}\bar{1}1]$	$[\bar{1}\bar{1}\bar{3}]$	$[1\bar{3}\bar{1}]$	$[1\bar{3}1]$	$[\bar{3}1\bar{1}]$	11
	$[\bar{3}11]$	$[\bar{1}\bar{1}3]$	$[1\bar{1}\bar{3}]$	$[\bar{1}1\bar{3}]$	$[1\bar{1}3]$	$[\bar{1}13]$	$[11\bar{3}]$	$[3\bar{1}\bar{1}]$	
	$[3\bar{1}1]$	$[\bar{1}3\bar{1}]$	$[\bar{1}31]$	[113]	$[31\bar{1}]$	$[13\bar{1}]$	[311]	[131]	
[222]	$[\bar{2}\bar{2}\bar{2}]$	$[\bar{2}\bar{2}2]$	$[2\bar{2}\bar{2}]$	$[\bar{2}2\bar{2}]$	$[2\bar{2}2]$	$[\bar{2}22]$	$[22\bar{2}]$	[222]	12
[400]	$[0\bar{4}0]$	$[\bar{4}00]$	$[00\bar{4}]$	[004]	[400]	[040]			16
[331]	$[\bar{3}\bar{3}\bar{1}]$	$[\bar{3}\bar{3}1]$	$[\bar{1}\bar{3}\bar{3}]$	$[\bar{3}\bar{1}\bar{3}]$	$[\bar{1}\bar{3}3]$	$[\bar{3}\bar{1}3]$	$[1\bar{3}\bar{3}]$	$[\bar{3}1\bar{3}]$	19
	$[1\bar{3}3]$	$[\bar{3}13]$	$[3\bar{3}\bar{1}]$	$[3\bar{3}1]$	$[\bar{3}3\bar{1}]$	$[\bar{3}31]$	$[3\bar{1}\bar{3}]$	$[\bar{1}3\bar{3}]$	
	$[3\bar{1}3]$	$[\bar{1}33]$	$[31\bar{3}]$	$[13\bar{3}]$	[133]	$[33\bar{1}]$	[331]	[313]	
[420]	$[\bar{2}\bar{4}0]$	$[\bar{4}\bar{2}0]$	$[0\bar{4}\bar{2}]$	$[\bar{4}0\bar{2}]$	$[0\bar{4}2]$	$[\bar{4}02]$	$[0\bar{2}\bar{4}]$	$[\bar{2}0\bar{4}]$	20
	$[2\bar{4}0]$	$[\bar{4}20]$	$[0\bar{2}4]$	$[\bar{2}04]$	$[20\bar{4}]$	$[02\bar{4}]$	$[4\bar{2}0]$	$[\bar{2}40]$	
	[204]	[024]	$[40\bar{2}]$	$[04\bar{2}]$	[402]	[042]	[420]	[240]	
[422]	$[\bar{2}\bar{4}\bar{2}]$	[422]	$[\bar{2}\bar{4}2]$	$[\bar{4}\bar{2}2]$	$[\bar{2}\bar{2}\bar{4}]$	$[\bar{2}\bar{2}4]$	$[2\bar{4}\bar{2}]$	$[2\bar{4}2]$	24
	$[\bar{4}2\bar{2}]$	$[\bar{4}22]$	$[2\bar{2}\bar{4}]$	$[\bar{2}2\bar{4}]$	$[2\bar{2}4]$	$[\bar{2}24]$	$[4\bar{2}\bar{2}]$	$[4\bar{2}2]$	
	$[\bar{2}4\bar{2}]$	$[\bar{2}42]$	$[22\bar{4}]$	[224]	$[42\bar{2}]$	$[24\bar{2}]$	[422]	[242]	
[511]	$[51\bar{1}]$	$[\bar{1}\bar{5}\bar{1}]$	$[\bar{1}\bar{5}1]$	$[\bar{5}\bar{1}\bar{1}]$	$[\bar{5}\bar{1}1]$	[115]	$[1\bar{5}\bar{1}]$	$[1\bar{5}1]$	27
	$[\bar{5}1\bar{1}]$	$[\bar{5}11]$	$[\bar{1}\bar{1}\bar{5}]$	$[\bar{1}\bar{1}5]$	$[1\bar{1}\bar{5}]$	$[5\bar{1}\bar{1}]$	$[\bar{1}1\bar{5}]$	$[5\bar{1}1]$	
	$[1\bar{1}5]$	$[\bar{1}15]$	$[11\bar{5}]$	$[\bar{1}5\bar{1}]$	$[\bar{1}51]$	[511]	$[15\bar{1}]$	[151]	
[333]	[333]	$[\bar{3}\bar{3}3]$	$[3\bar{3}\bar{3}]$	$[\bar{3}3\bar{3}]$	$[\bar{3}33]$	$[3\bar{3}3]$	$[33\bar{3}]$	$[\bar{3}\bar{3}\bar{3}]$	27

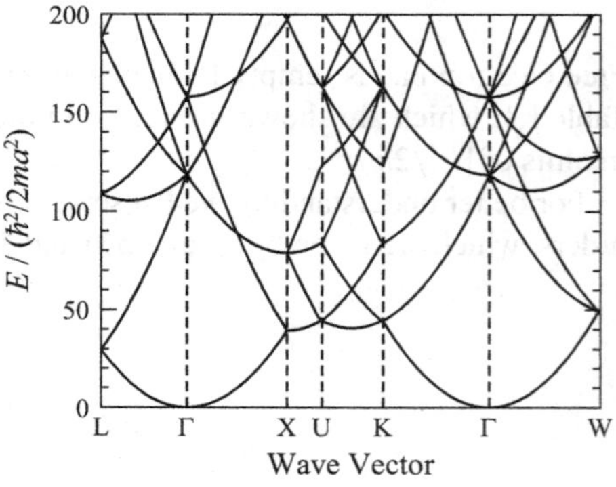

Fig. 1.6 Free electron bands (empty lattice bands) of a face-centered lattice are plotted as a function of electron wave vector along the direction shown in Fig. 1.22, where the energy range is $0 \sim 200$ in units of $[\hbar^2/2ma^2]$

$$\boldsymbol{G}_0 = \frac{2\pi}{a}[0, 0, 0], \tag{1.60a}$$

$$\boldsymbol{G}_3 = \frac{2\pi}{a}[\pm 1, \pm 1, \pm 1], \tag{1.60b}$$

$$\boldsymbol{G}_4 = \frac{2\pi}{a}[\pm 2, 0, 0], \tag{1.60c}$$

$$\boldsymbol{G}_8 = \frac{2\pi}{a}[\pm 2, \pm 2, 0], \tag{1.60d}$$

$$\boldsymbol{G}_{11} = \frac{2\pi}{a}(\pm 3, \pm 1, \pm 1). \tag{1.60e}$$

Putting these values in (1.52), the empty-lattice bands (free-electron bands) are easily calculated. In the following we use $\boldsymbol{k}$ instead of $\boldsymbol{k}'$ and take account of $\boldsymbol{k}$ in the first Brillouin zone. As an example we calculate the energy bands along the direction $\langle 100 \rangle$ in the $\boldsymbol{k}$-space shown in Fig. 1.5. In other words, we calculate the energy band structures $\mathcal{E}$ versus $\boldsymbol{k}$ from the Γ point to the X point. Since $k_y = k_z = 0$ in this direction, putting the reciprocal lattice vectors into (1.50), the energy is given by the following equations:

$$\boldsymbol{G}_0 : \ \mathcal{E} = k_x^2, \tag{1.61a}$$

$$\begin{aligned} \boldsymbol{G}_3 : \ \mathcal{E} &= (k_x \pm 1)^2 + (\pm 1)^2 + (\pm 1)^2, \\ &= \begin{cases} (k_x - 1)^2 + 2 & \text{(4-fold degeneracy)} \\ (k_x + 1)^2 + 2 & \text{(4-fold degeneracy)} \end{cases}, \end{aligned} \tag{1.61b}$$

$$\boldsymbol{G}_4 : \ \mathcal{E} = \begin{cases} k_x^2 + 4 & \text{(4-fold degeneracy)} \\ (k_x - 2)^2 & \text{(single state)} \\ (k_x + 2)^2 & \text{(single state)} \end{cases}, \tag{1.61c}$$

where the energy is measured in the units $\hbar^2(2\pi/a)^2/2m$ and the wave vector $\boldsymbol{k}$ in the units $2\pi/a$. When we plot these relations, we obtain the curves shown in Fig. 1.7.

Next we calculate the $\mathcal{E}$ versus $\boldsymbol{k}$ curves in the $\langle 111 \rangle$ direction of $\boldsymbol{k}$-space, or along the direction from the Γ point to the L point. The results are

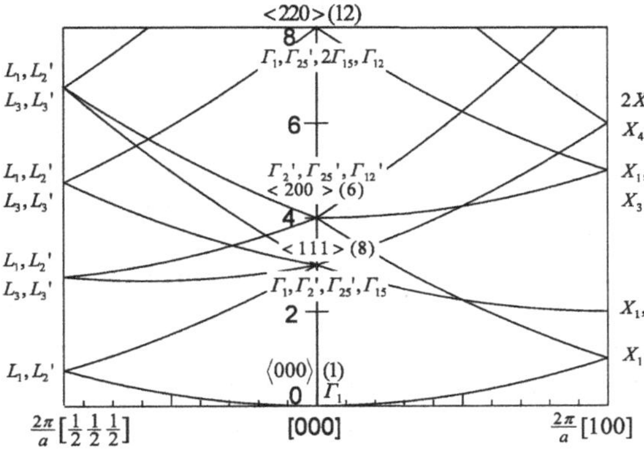

Fig. 1.7 Empty–lattice bands (free-electron bands) of a face-centered cubic lattice. $\langle 000 \rangle$, $\langle 111 \rangle$, $\langle 200 \rangle$, and $\langle 220 \rangle$ represent the reciprocal lattice vectors G_0, G_3, G_4, and G_8, respectively, and the numbers in ( ) show the degeneracy of the wave functions

$$G_0: \quad \mathcal{E} = k_x^2 + k_y^2 + k_z^2 \equiv k_{111}^2, \tag{1.62a}$$

$$G_3: \quad \mathcal{E} = (k_x \pm 1)^2 + (k_y \pm 1)^2 + (k_z \pm 1)^2, \tag{1.62b}$$

$$G_4: \quad \mathcal{E} = \begin{cases} (k_x \pm 2)^2 + k_y^2 + k_z^2 & \text{(2-fold degeneracy)} \\ k_x^2 + (k_y \pm 2)^2 + k_z^2 & \text{(2-fold degeneracy)} \\ k_x^2 + k_y^2 + (k_z \pm 2)^2 & \text{(2-fold degeneracy)} \end{cases}, \tag{1.62c}$$

where $k_x^2 + k_y^2 + k_z^2 = k_{111}^2$ and $k_x = k_y = k_z = k_{111}/\sqrt{3}$. Using these results we obtain the energy bands, $\mathcal{E} - -\boldsymbol{k}$ curves in the direction $\langle 111 \rangle$, which are shown in the left half of Fig. 1.7. In Fig. 1.7, the notation of the point group for O_h is used to represent the symmetry properties of the Brillouin zone edge. Note here that the energy $\mathcal{E}$ is expressed in units of $\hbar^2(2\pi/a)^2/2m$.

1.6 Pseudopotential Method

In this section we will concern with the energy band calculations based on the pseudopotential method. First we introduce local pseudopotential theory in which the nonlocality of the core states are ignored, and we will show how to calculate the energy band structures of the diamond and zinc blende semiconductors by using small number of the pseudopotentials. In the later section we will discuss the nonlocal pseudopotential theory in which the core potential of the occupied states is taken into account.

1.6.1 Local Pseudopotential Theory

The electronic states in a crystal are obtained by solving the following non-relativistic Schrödinger equation in the one-electron approximation:

$$\left[-\frac{\hbar^2}{2m}\nabla^2 + V(\boldsymbol{r}) \right] \Psi_n(\boldsymbol{r}) = \mathcal{E}_n(\boldsymbol{k})\Psi_n(\boldsymbol{r}). \tag{1.63}$$

However, it is possible only when we know the crystal potential $V(\boldsymbol{r})$. In the following we will express the wavefunction $\Psi_n(\boldsymbol{r})$ by the ket vector as $|\Psi_n(\boldsymbol{r})\rangle$ and show how the Schrödinger equation is solved to a good approximation by using empirical parameters, known as pseudopotentials, and the orthogonality of the wave functions [2–7]. The idea of the pseudopotential method is based on the assumption that the real crystal potential $V(\boldsymbol{r})$ is given by the sum of the attractive core potential and the weak repulsive potential (to keep the valence electrons out of the core). The addition of the repulsive potential to the core potential cancels the real potential, resulting in a weak net potential (pseudopotential). The introduction of the pseudopotential enables us to treat valence electrons as nearly free electron approximation or to solve Schödinger equation with a small number of Fourier components of the pseudopotential.

First, we assume the electron wave functions of the core states and their energies are given by $|\phi_j\rangle$ and $\mathcal{E}_j$, respectively. We then have

$$H|\phi_j\rangle = [H_0 + V_{\mathrm{c}}(\boldsymbol{r})]\,|\phi_j\rangle = \mathcal{E}_j|\phi_j\rangle\,, \tag{1.64}$$

where $V_{\mathrm{c}}(\boldsymbol{r})$ is the attractive core potential, and

$$H_0 = -\frac{\hbar^2}{2m}\nabla^2\,. \tag{1.65}$$

The true wave function $|\Psi\rangle$ of an electron is then expressed as the sum of a smooth wave function $|\chi_n(\boldsymbol{r})\rangle$ of a valence electron (subscript n is the band index) and a sum over occupied core states $|\phi_j\rangle$;

$$|\Psi\rangle = |\chi_n\rangle + \sum_j b_j|\phi_j\rangle\,. \tag{1.66}$$

Since the true wave function is orthogonal to the core states, the expansion coefficient $b_{j'}$ is determined by the orthogonality $\langle\phi_j|\Psi\rangle = 0$ as follows.

$$\begin{aligned}\langle\phi_{j'}|\Psi\rangle &= \langle\phi_{j'}|\chi_n\rangle + \sum_j\langle\phi_{j'}|b_j\phi_j\rangle\\ &= \langle\phi_{j'}|\chi_n\rangle + b_{j'} = 0\,,\end{aligned} \tag{1.67}$$

which gives $b_{j'} = -\langle\phi_{j'}|\chi_n\rangle$ and thus we obtain

$$|\Psi(\boldsymbol{k},\boldsymbol{r})\rangle = |\chi_n(\boldsymbol{k},\boldsymbol{r})\rangle - \sum_j\langle\phi_j|\chi_n\rangle|\phi_j\rangle\,. \tag{1.68}$$

We have to note here that $|\chi_n\rangle$ is defined as a smooth wave function for a valence electron and called as the pseudo-wave-function. As in the case of nearly free electron approximation, we calculate energy band structures by using plane waves for $|\chi_n(\boldsymbol{k},\boldsymbol{r})\rangle$ and in this scheme (1.68) is called the OPW (orthogonalized plane wave). Substituting (1.68) into (1.63) we find

$$H|\chi_n\rangle - \sum_j\langle\phi_j|\chi_n\rangle H|\phi_j\rangle = \mathcal{E}_n(\boldsymbol{k})\left\{|\chi_n\rangle - \sum_j\langle\phi_j|\chi_n\rangle|\phi_j\rangle\right\}, \tag{1.69}$$

and then we obtain the following relation:

$$H|\chi_n\rangle + \sum_j[\mathcal{E}_n(\boldsymbol{k}) - \mathcal{E}_j]|\phi_j\rangle\langle\phi_j|\chi_n\rangle = \mathcal{E}_n(\boldsymbol{k})|\chi_n\rangle\,. \tag{1.70}$$

We introduce a new parameter according to the definition of Cohen and Chelikowsky [2]

$$V_r(\boldsymbol{r}) = \sum_j [\mathcal{E}_n(\boldsymbol{k}) - \mathcal{E}_j] |\phi_j\rangle\langle\phi_j| , \tag{1.71}$$

or

$$V_r(\boldsymbol{r})|\chi_n\rangle = \sum_j [\mathcal{E}_n(\boldsymbol{k}) - \mathcal{E}_j] |\phi_j\rangle\langle\phi_j|\chi_n\rangle . \tag{1.72}$$

This term acts like a short-ranged non-Hermitian repulsive potential. Using this definition we obtain the following equation.

$$[H + V_r(\boldsymbol{r})]|\chi_n\rangle = \mathcal{E}_n(\boldsymbol{k})|\chi_n\rangle . \tag{1.73}$$

If H is separated into a kinetic energy $H_0 = -(\hbar^2/2m)\nabla^2$ and attractive core potential $V_c(\boldsymbol{r})$, then (1.73) becomes

$$\left[-\frac{\hbar^2}{2m}\nabla^2 + V_c(\boldsymbol{r}) + V_r(\boldsymbol{r})\right] |\chi_n\rangle = \mathcal{E}_n(\boldsymbol{k})|\chi_n\rangle , \tag{1.74}$$

where $\mathcal{E}_n(\boldsymbol{k})$ is the energy of the band we are interested in. There exists the following inequality between the energies of the core states, $\mathcal{E}_j$, and the energies of the valence and conduction bands, $\mathcal{E}_n(\boldsymbol{k})$:

$$\mathcal{E}_n(\boldsymbol{k}) > \mathcal{E}_j , \tag{1.75}$$

and thus we find from (1.72) that

$$V_r(\boldsymbol{r}) > 0 . \tag{1.76}$$

We may rewrite (1.74) as

$$[H_0 + V_{ps}(\boldsymbol{r})]|\chi_n\rangle = \mathcal{E}_n(\boldsymbol{k})|\chi_n\rangle , \tag{1.77}$$

$$V_{ps}(\boldsymbol{r}) = V_c(\boldsymbol{r}) + V_r(\boldsymbol{r}) , \tag{1.78}$$

and it may be possible to make V_{ps} small enough, since the attractive core potential $V_c(\boldsymbol{r}) < 0$ and the repulsive potential $V_r(\boldsymbol{r}) > 0$ cancel each other. The new potential $V_{ps}(\boldsymbol{r})$ is called the **pseudopotential**. Since the pseudopotential V_{ps} is the sum of the attractive long–range potential V_c and a short–range repulsive potential V_r, V_{ps} becomes weak long–range attractive regions away from the core and weakly repulsive or attractive regions near the core (see Fig. 1.8 [2]).

The pseudopotential $V_{ps}(\boldsymbol{r})$ is also periodic, and we can expand it as the Fourier series

$$V_{ps}(\boldsymbol{r}) = \sum_j V_{ps}(\boldsymbol{G}_j) e^{-i\boldsymbol{G}_j \cdot \boldsymbol{r}} , \tag{1.79}$$

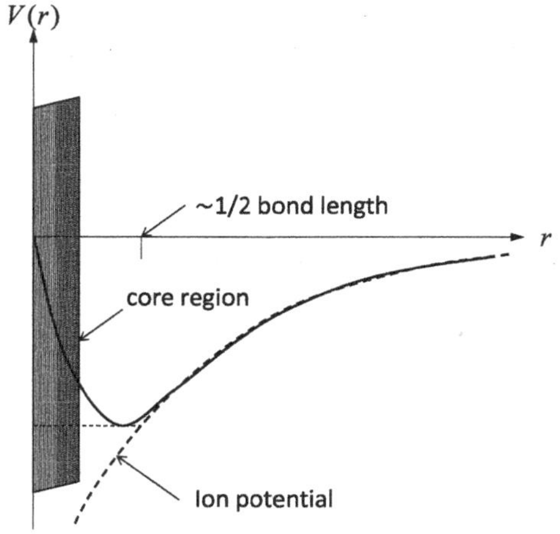

Fig. 1.8 Schematic plot of pseudopotential in real space (after Cohen and Chelikowsky [2])

where the Fourier coefficients $V_{\mathrm{ps}}(\boldsymbol{G}_j)$ are given by

$$V_{\mathrm{ps}}(\boldsymbol{G}_j) = \frac{1}{\sqrt{\Omega}} \int_{\Omega} V_{\mathrm{ps}}(\boldsymbol{r}) \mathrm{e}^{\mathrm{i}\boldsymbol{G}_j \cdot \boldsymbol{r}} \mathrm{d}^3\boldsymbol{r} \,. \tag{1.80}$$

For the reason stated above the potential $V_{\mathrm{ps}}(\boldsymbol{r})$ may be chosen as small as possible, and thus we choose $V_{\mathrm{ps}}(\boldsymbol{G}_j)$ so that the potential $V_{\mathrm{ps}}(\boldsymbol{G}_j)$ is expressed with a small number of the Fourier coefficients $V_{\mathrm{ps}}(\boldsymbol{G}_j)$; in other words, we may keep several values of $V_{\mathrm{ps}}(\boldsymbol{G}_j)$ and neglect the other values because of their smallness. We should note that $|V_{\mathrm{ps}}(\boldsymbol{r})|$ is smaller than $|V(\boldsymbol{r})|$, but it does not mean that $V(\boldsymbol{r})$ converges with only a small number of its Fourier coefficients. The **empirical pseudopotential method** is based on the approximation that the Fourier coefficients of $V_{\mathrm{ps}}(\boldsymbol{r})$ are empirically chosen so that the shape of the critical points and their energies show good agreement with experimental observation.

Energy band calculations based on the empirical pseudopotential method take into account as few the pseudopotentials $V_{\mathrm{ps}}(\boldsymbol{G}_j)$ as possible and use the Bloch functions of the free-electron bands for the wave functions $|\chi_n\rangle$. The energy bands are obtained by solving

$$\left[-\frac{\hbar^2}{2m} \nabla^2 + V_{\mathrm{ps}}(\boldsymbol{r}) \right] |\chi_n(\boldsymbol{r})\rangle = \mathcal{E}_n |\chi_n\rangle(\boldsymbol{r}) \,, \tag{1.81}$$

$$|\chi_n(\boldsymbol{r})\rangle = \frac{1}{\sqrt{\Omega}} \sum_j \mathrm{e}^{\mathrm{i}(\boldsymbol{k}+\boldsymbol{G}_j)\cdot\boldsymbol{r}} \,, \tag{1.82}$$

$$V_{\mathrm{ps}}(\boldsymbol{r}) = \sum_{j'} V_{\mathrm{ps}}(\boldsymbol{G}_{j'}) \mathrm{e}^{\mathrm{i}\boldsymbol{G}_{j'}\cdot\boldsymbol{r}} \,. \tag{1.83}$$

Then the energy band structures are calculated by solving the following equation, where the j-th component is given by dropping the factor $(1/\sqrt{\Omega})\sum_j$:

$$\left[-\frac{\hbar^2}{2m}\nabla^2 + \sum_{j'} V_{\text{ps}}(\boldsymbol{G}_{j'})\mathrm{e}^{\mathrm{i}\boldsymbol{G}_{j'}\cdot\boldsymbol{r}}\right]\mathrm{e}^{\mathrm{i}(\boldsymbol{k}+\boldsymbol{G}_j)\cdot\boldsymbol{r}} = \mathcal{E}_n(\boldsymbol{k})\mathrm{e}^{\mathrm{i}(\boldsymbol{k}+\boldsymbol{G}_j)\cdot\boldsymbol{r}} . \tag{1.84}$$

The eigenvalues and eigen functions of the above equation are easily obtained by solving the following matrix equation. First, we introduce pseudopotential Hamiltonian by

$$H_{\text{ps}} = -\frac{\hbar^2}{2m}\nabla^2 + V_{\text{ps}}(\boldsymbol{r}) , \tag{1.85}$$

and rewrite (1.84) as

$$H_{\text{ps}}|\boldsymbol{k}+\boldsymbol{G}_j\rangle = \mathcal{E}_n(\boldsymbol{k})|\boldsymbol{k}+\boldsymbol{G}_j\rangle , \tag{1.86}$$

$$|\boldsymbol{k}+\boldsymbol{G}_j\rangle = \frac{1}{\sqrt{\Omega}}\mathrm{e}^{\mathrm{i}(\boldsymbol{k}+\boldsymbol{G}_j)\cdot\boldsymbol{r}} . \tag{1.87}$$

Then the solutions are equivalently obtained by solving the determinant

$$||\langle|\boldsymbol{k}+\boldsymbol{G}_i|H_{\text{ps}}|\boldsymbol{k}+\boldsymbol{G}_j\rangle - \mathcal{E}(\boldsymbol{k})\delta_{i,j}|| = 0 , \tag{1.88}$$

where the matrix elements of the Hamiltonian H_{ps} are written as

$$\langle\boldsymbol{k}+\boldsymbol{G}_i|H_{\text{ps}}|\boldsymbol{k}+\boldsymbol{G}_j\rangle = \frac{\hbar^2}{2m}(\boldsymbol{k}+\boldsymbol{G}_i)^2\delta_{\boldsymbol{G}_i,\boldsymbol{G}_j} + V_{\text{ps}}(\boldsymbol{G}_j-\boldsymbol{G}_i) . \tag{1.89}$$

When we know the pseudopotential form factors $V_{\text{ps}}(\boldsymbol{G}_j-\boldsymbol{G}_i)$, the energy band calculations are straightforward by solving the eigen equation (1.88). In the next subsection we will deal with the evaluation of non–vanishing pseudopotential form factors.

1.6.2 Pseudopotential Form Factors

Once we know the Fourier coefficients $V_{\text{ps}}(\boldsymbol{G}_j)$, the solutions of (1.88) are easily calculated with a personal computer. It is very interesting to point out that the calculated energy bands using the reciprocal wave vectors given by (1.60a)–(1.60e)

Table 1.2 Pseudopotentials for several semiconductors in units of Rydberg [Ry] and lattice constants a in [Å] (from [5])

	a[Å]	V_3^S	V_8^S	V_{11}^S	V_3^A	V_4^A	V_{11}^A
Si	5.43	−0.21	+0.04	+0.08	0	0	0
Ge	5.66	−0.23	+0.01	+0.06	0	0	0
Sn	6.49	−0.20	0.00	+0.04	0	0	0
GaP	5.44	−0.22	+0.03	+0.07	+0.12	+0.07	+0.02
GaAs	5.64	−0.23	+0.01	+0.06	+0.07	+0.05	+0.01
AlAs	5.66	−0.221	0.025	0.07	0.08	0.05	−0.004
AlSb	6.13	−0.21	+0.02	+0.06	+0.06	+0.04	+0.02
InP	5.86	−0.23	+0.01	+0.06	+0.07	+0.05	+0.01
GaSb	6.12	−0.22	0.00	+0.05	+0.06	+0.05	+0.01
InAs	6.04	−0.22	0.00	+0.05	+0.08	+0.05	+0.03
InSb	6.48	−0.20	0.00	+0.04	+0.06	+0.05	+0.01
ZnS	5.41	−0.22	+0.03	+0.07	+0.24	+0.14	+0.04
ZnSe	5.65	−0.23	+0.03	+0.06	+0.18	+0.12	+0.03
ZnTe	6.07	−0.22	0.00	+0.05	+0.13	+0.10	+0.01
CdTe	6.41	−0.20	0.00	+0.04	+0.15	+0.09	+0.04

and the free-electron Bloch functions show very reasonable results, where only several pseudopotential parameters derived by Cohen and Bergstresser [5] shown in Table 1.2 are taken into account.

First we explain the pseudopotential parameters. In general a unit cell of a crystal contains a single atom or multi–atoms and thus the pseudopotential is expressed as [6]

$$V_{\rm ps}(\boldsymbol{r}) = \sum_j V(\boldsymbol{G}_j)\mathrm{e}^{-\mathrm{i}\boldsymbol{G}_j\cdot\boldsymbol{r}}\,, \tag{1.90a}$$

$$V(\boldsymbol{G}_j) = \sum_\alpha S_\alpha(\boldsymbol{G}_j)V_\alpha(\boldsymbol{G}_j)\,, \tag{1.90b}$$

$$S_\alpha(\boldsymbol{G}_j) = \frac{1}{N_\alpha}\sum_{\text{cell}\,m} \mathrm{e}^{-\mathrm{i}\boldsymbol{G}_j\cdot\boldsymbol{R}_m^\alpha}\,, \tag{1.90c}$$

$$V_\alpha(\boldsymbol{G}_j) = \frac{1}{\Omega_\alpha}\int \mathrm{e}^{\mathrm{i}\boldsymbol{G}_j\cdot\boldsymbol{r}}V_{\rm ps}^\alpha(\boldsymbol{r})\mathrm{d}^3\boldsymbol{r}\,. \tag{1.90d}$$

Derivation of the above relations is understood by taking account of multi–atoms in the unit cell

$$\sum_j V(\boldsymbol{G}_j)\mathrm{e}^{-\mathrm{i}\boldsymbol{G}_j\cdot\boldsymbol{r}} \to \sum_j\sum_\alpha\sum_{\text{cell}\,m} V_\alpha(\boldsymbol{G}_j)\mathrm{e}^{-\mathrm{i}\boldsymbol{G}_j\cdot(\boldsymbol{r}+\boldsymbol{R}_{\rm m}^\alpha)}\,,$$

where $S_\alpha(\boldsymbol{G})$ is called a **structure factor**, N_α is the number of atomic species α present, $\boldsymbol{R}_m^\alpha$ is the position of the m–th atom of the α–th species, and Ω_α is the atomic volume. Here the crystalline potential is assumed to be a sum of local atomic pseudopotential $V_{\text{ps}}^\alpha(\boldsymbol{r})$.

The diamond-type crystal structure such as Ge and Si contains two atoms A and B ($A = B$) in the unit cell ($N_\alpha = 2$), and the zinc-blende-type crystal structure has two different atoms A and B ($A \neq B$) ($N_\alpha = 2$). When a new vector $\boldsymbol{\tau} = (a/8)(111)$ is defined, the atomic positions of A and B are given by $\boldsymbol{R}^A = -\boldsymbol{\tau}$ and $\boldsymbol{R}^B = +\boldsymbol{\tau}$, respectively. Taking the origin of coordinates to be the center of those two atoms, the structure factors are written as

$$S_{\text{A}}(\boldsymbol{G}_j) = \frac{1}{2}\text{e}^{\text{i}\boldsymbol{G}_j\cdot\boldsymbol{\tau}}, \quad S_{\text{B}}(\boldsymbol{G}_j) = \frac{1}{2}\text{e}^{-\text{i}\boldsymbol{G}_j\cdot\boldsymbol{\tau}} \tag{1.91}$$

and thus the pseudopotential form factor $V(\boldsymbol{G})$ is given by

$$\begin{aligned} V(\boldsymbol{G}_j) &= \frac{1}{2}\left[V_{\text{A}}(\boldsymbol{G}_j)\text{e}^{\text{i}\boldsymbol{G}_j\cdot\boldsymbol{\tau}} + V_{\text{B}}(\boldsymbol{G}_j)\text{e}^{-\text{i}\boldsymbol{G}_j\cdot\boldsymbol{\tau}}\right] \\ &= V^{\text{S}}(\boldsymbol{G}_j)\cos(\boldsymbol{G}_j\cdot\boldsymbol{\tau}) + \text{i}V^{\text{A}}(\boldsymbol{G}_j)\sin(\boldsymbol{G}_j\cdot\boldsymbol{\tau}) \,. \end{aligned} \tag{1.92}$$

Here we introduce following new parameters

$$V^{\text{S}}(\boldsymbol{G}_j) = [V_{\text{A}}(\boldsymbol{G}_j) + V_{\text{B}}(\boldsymbol{G}_j)]/2 \,, \tag{1.93a}$$

$$V^{A}(\boldsymbol{G}_j) = [V_{\text{A}}(\boldsymbol{G}_j) - V_{\text{B}}(\boldsymbol{G}_j)]/2 \,, \tag{1.93b}$$

where V^{S} and V^{A} are called the symmetric and antisymmetric form factors, respectively. The structure factor plays an important role in electronic properties such as energy band structure, diffraction effect and so on. $S^{\text{S}}(\boldsymbol{G}_j) = \cos(\boldsymbol{G}_j\cdot\boldsymbol{\tau})$ and $S^{\text{A}}(\boldsymbol{G}_j) = \sin(\boldsymbol{G}_j\cdot\boldsymbol{\tau})$ are the real part and imaginary part of the structure factor. From the definition of diamond-type crystal we have $V^{\text{A}}(\boldsymbol{G}_j) = 0$ and the structure factor reduces to $\cos(\boldsymbol{G}_j\cdot\boldsymbol{\tau})$. In Table 1.2 the pseudopotentials are defined by using the relations $V^{\text{S}}(\boldsymbol{G}_j) = V_j^{\text{A}}$ and $V^{\text{A}}(\boldsymbol{G}_j) = V_j^{\text{A}}$, where $\boldsymbol{G}_j$ is defined by (1.60a)–(1.60e). As shown in Table 1.2 some of the pseudopotentials $V^{\text{S}}(\boldsymbol{G}_j)$ and $V^{\text{A}}(\boldsymbol{G}_j)$ vanish. This may be understood from the following considerations. The symmetric component of the pseudopotential is written as

$$V^{\text{S}}(G_j)\cos\left(\boldsymbol{G}_j\cdot\boldsymbol{\tau}\right) = V^{\text{S}}(\boldsymbol{G}_j)\cos\left(\frac{a}{8}\left[G_{jx} + G_{jy} + G_{jz}\right]\right) \,. \tag{1.94}$$

Let's examine the pseudopotentials for the reciprocal vectors (1.60a)–(1.60e). The pseudopotentials for smaller values of $\boldsymbol{G}_j$ are evaluated as

$$V^{\mathrm{S}}(\boldsymbol{G}_0)\cos(\boldsymbol{G}_0\cdot\boldsymbol{\tau}) = V^{\mathrm{S}}(\boldsymbol{G}_0)\,,$$
$$V^{\mathrm{S}}(\boldsymbol{G}_3)\cos(\boldsymbol{G}_3\cdot\boldsymbol{\tau}) = V^{\mathrm{S}}(\boldsymbol{G}_3)\cos\left(\frac{\pi}{4}[\pm1\pm1\pm1]\right)\neq 0\,,$$
$$V^{\mathrm{S}}(\boldsymbol{G}_4)\cos(\boldsymbol{G}_4\cdot\boldsymbol{\tau}) = V^{\mathrm{S}}(\boldsymbol{G}_4)\cos\left(\frac{\pi}{4}[\pm2]\right) = 0\,,$$
$$V^{\mathrm{S}}(\boldsymbol{G}_8)\cos(\boldsymbol{G}_8\cdot\boldsymbol{\tau}) = V^{\mathrm{S}}(\boldsymbol{G}_8)\cos\left(\frac{\pi}{4}[\pm2\pm2]\right)\neq 0\,,$$
$$V^{\mathrm{S}}(\boldsymbol{G}_{11})\cos(\boldsymbol{G}_{11}\cdot\boldsymbol{\tau}) = V^{\mathrm{S}}(\boldsymbol{G}_{11})\cos\left(\frac{\pi}{4}[\pm3\pm1\pm1]\right)\neq 0\,.$$

Therefore the symmetric components of the pseudopotentials $V^{\mathrm{S}}(\boldsymbol{G}_0) = V_0^{\mathrm{S}}$, $V^{\mathrm{S}}(\boldsymbol{G}_3) = V_3^{\mathrm{S}}$, $V^{\mathrm{S}}(\boldsymbol{G}_8) = V_8^{\mathrm{S}}$, $V^{\mathrm{S}}(\boldsymbol{G}_{11}) = V_{11}^{\mathrm{S}}$ remain and $V^{\mathrm{S}}(\boldsymbol{G}_4) = V_4^{\mathrm{S}}$ will not contribute. In a similar fashion, there is no contribution from the antisymmetric components of the pseudopotential $V^{\mathrm{A}}(\boldsymbol{G}_0) = V_0^{\mathrm{A}}$ and $V^{\mathrm{A}}(\boldsymbol{G}_8) = V_8^{\mathrm{A}}$. These results give the pseudopotentials for smaller values of $|\boldsymbol{G}_j|$ in Table 1.2, where we find that pseudopotentials of large $|\boldsymbol{G}_j|$ are diminished. Since the energy bands calculated with these pseudopotentials given in Table 1.2 show good agreement with experimental observation, higher order components of the pseudopotentials are usually neglected. The term $V^{\mathrm{S}}(\boldsymbol{G}_0) = V_0^{\mathrm{S}}$ results in a shift of the energy reference and thus we put $V_0^{\mathrm{S}} = 0$.

1.6.3 Nonlocal Pseudopotential Theory

Here we will be concerned with the energy band calculations by the pseudopotential method where the nonlocality of the core potential is considered. The method described above is called local pseudopotential method, where the core potential is assumed to be uniform neglecting the angular orbitals of the core electrons.

The nonlocal pseudopotential method takes account of nonlocal properties of the core electrons. The core potential $V_{\mathrm{c}}(\boldsymbol{r})$ consists of a sum over the occupied core states ϕ_j, and it consists of the various states with the respective angular momentum symmetry as discussed by Cohen and Chelikowsky [2, 6, 7] (see also the references listed there). Therefore the core potential is given by the sum of s–, p–, and d–components of the respective angular momentum quantum number $l = 0, 1, 2, \ldots$

$$V_{\mathrm{c}}(\boldsymbol{r}) = V_s + V_p + V_d + \ldots\,. \tag{1.95}$$

As an example we consider carbon atom C. Its core states are $(1s)^2$ and thus carbon has no p–repulsive potential. The $(2p)$ electrons of the valence states $(2s)^2)(2p)^2$ are affected by the core potential. This repulsive core potential is expected to be stronger because of its closer distance to the core than in Si and Ge. In general, the core potential is energy dependent and the nonlocal (NL) correction term to the local atomic potential term is expressed as the following [2, 4, 7]

$$V_{\mathrm{NL}}^{\alpha}(\boldsymbol{r}, \mathcal{E}) = \sum_{l=0}^{\infty} A_l^{\alpha}(\mathcal{E}) f_l(r) \mathcal{P}_l , \tag{1.96}$$

and

$$f_l^{\alpha}(r) = \begin{cases} 1, & r \le R_m \\ 0, & r \ge R_m \end{cases} , \tag{1.97}$$

where $A_l^{\alpha}(\mathcal{E})$ is an energy-dependent well depth of the α species, R_m is the model radius, which is taken to be the same for all l, and $\mathcal{P}_l$ is projects out the l-th angular momentum component of the wave function.

When we assume a square well for the model potential defined by (1.97), the matrix element of the nonlocal potential is given by

$$V_{\mathrm{NL}}(\boldsymbol{K}, \boldsymbol{K}') = \frac{4\pi}{\Omega_{\alpha}} \sum_{l,\alpha} A_l^{\alpha}(\mathcal{E})(2l+1) \times P_l \left(\cos\left(\theta_{\boldsymbol{K},\boldsymbol{K}'}\right)\right) S^{\alpha}(\boldsymbol{K} - \boldsymbol{K}') F_l^{\alpha}(K, K') , \tag{1.98}$$

where $S^{\alpha}(\boldsymbol{K})$ is the structure factor defined by (1.90c) with $\boldsymbol{K} = \boldsymbol{k} + \boldsymbol{G}$ and $\boldsymbol{K}' = \boldsymbol{k} + \boldsymbol{G}'$, $\theta_{\boldsymbol{K},\boldsymbol{K}'}$ is the angle between $\boldsymbol{K}$ and $\boldsymbol{K}'$, and the sum of α is carried out over the atomic species present.

$$F_l(K, K') = \begin{cases} \dfrac{R_m^3}{2} \left\{ [j_l(KR_m)]^2 - j_{l-1}(KR_m) j_{l+1}(KR_m) \right\} , & K = K' , \\ \dfrac{R_m^2}{K^2 - K'^2} \left[K j_{l+1}(KR_m) j_l(K'R_m) - K' j_{l+1}(K'R_m) j_l(KR_m) \right] , & K \ne K' . \end{cases}$$

$P_l(x)$ is a Legendre polynomial and $j_l(x)$ is a spherical Bessel function, which are given for smaller values of the subscript l:

$$\begin{aligned} &P_0(x) = 1, \ \ P_1(x) = x, \ \ P_2(x) = (1/2)(3x^2 - 1) , \\ &j_0(x) = x^{-1} \sin x , \ \ J_1(x) = x^{-2} \sin x - x^{-1} \cos x , \\ &j_2(x) = (3x^{-3} - x^{-1}) \sin x - 3x^{-2} \cos x , \ \ j_3(x) = 5x^{-1} j_2(x) - j_1(x) . \end{aligned}$$

Energy band calculations require the estimation of energy dependent term $A_l^{\alpha}(\mathcal{E})$ and radii $R_m = R_0$, which are reported by Cohen and Chelikowsky [7]. They make the approximation for $A_0(\mathcal{E})$ for the s state as

$$A_0(\mathcal{E}) = \alpha_0 + \beta_0 \left\{ [\mathcal{E}^0(K) E^0(K')]^{1/2} - E^0(K_{\mathrm{F}}) \right\} , \tag{1.99}$$

where $E^0(K) = \hbar^2 K^2 / 2m$, $K_{\mathrm{F}} = (6\pi^2 Z / \Omega)^{1/3}$ and Z is the valence of the atomic species of interest [6, 7].

Now the energy band calculations with local and nonlocal pseudopotentials are straight forward. The eigenvalues and eigenvectors are obtained by solving the secular equation

$$\det \left| H_{\boldsymbol{G},\boldsymbol{G}'}(\boldsymbol{k}) - \mathcal{E}(\boldsymbol{k})\delta_{\boldsymbol{G},\boldsymbol{G}'} \right| = 0 . \tag{1.100}$$

For the local pseudopotential approximation, we have

$$H^{\mathrm{L}}_{\boldsymbol{G},\boldsymbol{G}'} = \frac{\hbar^2}{2m}(\boldsymbol{k}+\boldsymbol{G})^2 + V_{\mathrm{ps}}(|\boldsymbol{G}-\boldsymbol{G}'|) . \tag{1.101}$$

When we include the nonlocal pseudopotential term we obtain

$$\begin{aligned} H_{\boldsymbol{G},\boldsymbol{G}'} &= H^{\mathrm{L}}_{\boldsymbol{G},\boldsymbol{G}'} + V_{\mathrm{NL}}(\boldsymbol{K},\boldsymbol{K}') \\ &= H^{\mathrm{L}}_{\boldsymbol{G},\boldsymbol{G}'} + \frac{4\pi}{\Omega_\alpha}\sum_{l,\alpha} A^\alpha_l(\mathcal{E})(2l+1)P_l\left(\cos\left(\theta_{K,K'}\right)\right) \\ &\qquad \times S^\alpha(\boldsymbol{G}-\boldsymbol{G}')F^\alpha_l(K,K') , \end{aligned} \tag{1.102}$$

where the sum is carried out over the atomic species α.

Nonlocal pseudopotential parameters reported by Chelikowsky and Cohen [7] are listed in Table 1.3. In their calculations, the model radius of the R_l for the pseudopotential is taken to be the same for all l and $\alpha_0 = 0$ for the cations. Energy band calculations based on the nonlocal pseudopotentials require many parameters, but the calculated results differ only a little compared to the simple local pseudopotential method. We will present only the energy band structure of GaAs calculated by the nonlocal pseudopotential method in Sect. 1.6.6. Before dealing with the energy band calculations by nonlocal pseudopotential, we present energy band structures calculated by th local pseudopotential method for diamond and zinc blende semiconductors without spin–orbit interaction in Sect. 1.6.4. Later in Sect. 1.6.6 we

Table 1.3 Nonlocal pseodopotential parameters for the diamond and zinc blende semiconductors (after Chelikowsky and Cohen [7])

Materials	Pseudopotential form factors [Ry]						Lattice constant [Å]
	V^{S}_3	V^{S}_8	V^{S}_{11}	V^{A}_3	V^{A}_4	V^{A}_{11}	
Si	−0.257	−0.040	0.033				5.43
Ge	−0.221	0.019	0.056				5.65
GaP	−0.230	0.020	0.057	0.100	0.070	0.025	5.45
GaAs	−0.254*	0.014	0.067	0.055	0.038	0.010*	5.65
GaSb	−0.220	0.005	0.045	0.040	0.030	0.000	6.10
InP	−0.235	0.000	0.053	0.080	0.060	0.030	5.86
InAs	−0.230	0.000	0.045	0.055	0.045	0.010	6.05
InSb	−0.200	−0.010	0.044	0.044	0.030	0.015	6.47

The pseudopotential values of GaAs with asterisk differ from the values $V^{\mathrm{S}}_3 = -0.214$ and $V^{\mathrm{A}}_{11} = 0.001$ of Chelikowsky and Cohen [7]

Nonlocal parameters for Si and Ge

Material	α_0 [Ry]	β_0	A_2 [Ry]	R_0 [Å]	R_2 [Å]
Si	0.55	0.32	0	1.06	0
Ge	0	0	0.275	0	1.22

Nonlocal parameters for zinc blende semiconductors
($R_0 = 1.27$ for the cation and 1.06 Å for the anion)
($\alpha_0 = 0$ for the cation)

	Cation			Anion		
Material	α_0 [Ry]	β_0	A_2 [Ry]	α_0 [Ry]	β_0	A_2 [Ry]
GaP	0	0.30	0.40	0.32	0.05	0.45
GaAs	0	0	0.125	0	0	0.625
GaSb	0	0.20	0.20	0	0.30	0.60
InP	0	0.25	0.55	0.30	0.05	0.35
InAs	0	0.35	0.50	0	0.25	1.00
InSb	0	0.45	0.55	0	0.48	0.70

will show calculated results for Ge and GaAs with the spin–orbit interaction for comparison, and finally we present the energy band calculation of GaAs by the nonlocal pseudopotential method.

1.6.4 Energy Band Calculation by Local Pseudopotential Method

In this section we show the calculated results of the energy band structures with the local pseudopotential method by neglecting the spin–orbit interaction. After the discussion of the spin–orbit interaction in Sect. 1.6.5 we will present energy band calculations with the spin–orbit interaction in Sect. 1.6.6. As discussed by Chelikowsky and Cohen [6, 7], the overall feature of the calculated results by the local pseudopotential method shows a good agreement with the results by the nonlocal pseudopotential method, except a small change in the region near some critical points. Instead, the spin–orbit interaction plays a more important role in the energy regions near the critical points. In order to understand the energy band calculation by the pseudopotential, first we will concern with the energy band calculation by the local pseudopotential method.

Since we have only few numbers of the pseudopotential form factors, the energy band calculations are straight forward. However, the accuracy of the calculated energy band structures depends on the number of plane waves used for the pseudopotential Hamiltonian matrix. When the number of plane waves are increased, a large computation time is required to diagonalize the matrix. Therefore we have to limit the number of the plane waves. One of the most popular method is to limit the number of plane waves to form the matrix elements in a reasonable size and the higher energy states are taken into account by using the perturbation method proposed by Löwdin [8] as reported by Brust [9] and Cohen and Bergstresser [5]. Now high performance PC's such as Windows 7 with Intel core i–7 are available and 200 × 200 matrix is solved to give the eigen energies and eigenstates in a reasonable time. In the next section we will deal with $\boldsymbol{k} \cdot \boldsymbol{p}$ perturbation method to calculate energy bands, where 15 eigenstates are used. In this textbook the energy band structures are calculated by the empirical pseudopotential method with 113 plane waves and thus 226 plane waves with spin–up and –down states, and higher energy states up to 169 are treated by Löwdin's perturbation which are believed to be enough number to get accurate energy band structures. The energy bands without the spin–orbit interaction with 59 plane waves and Löwdin's perturbation for the higher states exhibit no noticeable difference with the present results and thus we recommend the readers to use 59 plane waves for the purpose of time saving.

It is very interesting to compare two different results with 15 plane waves and 169 plane waves because these results provide an information of the convergence of the energy band calculation by the pseudopotential method. A beginner for the energy band calculations is recommended to calculate the energy bands of Si for example using 15 plane waves [000], [111] and [200] ($K^2 \leq 4$ in Table 1.1) and disregarding the spin–orbit interaction. The overall features of the calculated energy band structure of Si are quite similar to the result obtained by 169 plane waves as shown in Fig. 1.9,

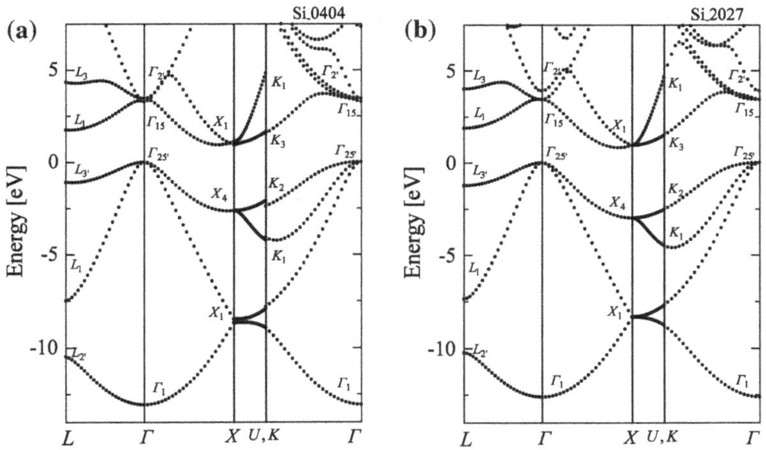

Fig. 1.9 Energy bands of Si calculated by the local empirical pseudopotential method with 15 plane waves (**a**) and 169 plane waves (**b**). The curves of (**b**) are obtained by diagonalizing 113 plane waves $0 \leq \mathcal{E}(\boldsymbol{K}) \leq \mathcal{E}(\boldsymbol{K}) = 20$ (where $\mathcal{E}(\boldsymbol{K}) = (\hbar^2/2m)(2\pi/a)^2)K^2)$ exactly and 56 higher energy states of $20 < \mathcal{E}(\boldsymbol{K}) \leq \mathcal{E}(\boldsymbol{K}) = 27$ by Löwdin's perturbation method. Note the curves of (**a**) exhibit discontinuity at U, K points and a curve of higher conduction band in the region K_1 to Γ is missing

where energy band structures calculated by 15 plane waves are shown by the curves in (a) and the curves in (b) are calculated by 169 plane waves. We find here that overall features are in good agreement but some disagreement exists as follows. The curves obtained by 169 plane waves show smooth continuity at the points U and K (at K_2). The points U and K in the Brillouin zone are equivalent because of the symmetry of the representation as seen in Figs. 1.5 and 1.22 and thus obtained bands are expected to be continuous through the points U and K. In addition to the discontinuity, a higher conduction band obtained by 169 plane waves is missing in the bands of 15 plane waves calculation in the region K_1 to Γ and the conduction bands of Γ_{15} and $\Gamma_{2'}$ are almost degenerate in Fig. 1.9a. This will be discussed in Sect. 1.7, where energy band calculations by $\boldsymbol{k} \cdot \boldsymbol{p}$ perturbation method of 15 states will be discussed.

The energy band calculations carried out by Cohen and Bergstresser [5] reveal that the choice of appropriate values for the pseudopotentials V_3^S, V_8^S, V_{11}^S, V_3^A, V_8^A, V_{11}^A and the neglect of higher-order values give the band structures in good agreement with experimental results. The pseudopotential values determined by Cohen and Bergstresser [5] for typical semiconductors are given in Table 1.2. As discussed by Brust [9], and Cohen and Bergstresser [5], the energy bands are obtained by limited number of plane waves to form pseudopotential Hamiltonian matrix and plane waves with higher free electron energies are taken into account by the perturbation method proposed by Löwdin [8]. Energy band structures calculated by (1.84) with 169 plane waves are shown for Ge and Si in Fig. 1.10, for GaAs, GaP, AlAs and AlSb in Fig. 1.11, for InP, InAs, GaSb and InSb in Fig. 1.12, and for ZnS, ZnSe, ZnTe and CdTe in Fig. 1.13. In the calculations, the pseudopotential matrix of the 113 free-electron states of $0 \leq \mathcal{E}(\boldsymbol{K}) \leq \mathcal{E}(\boldsymbol{K}) = 20$ (where $\mathcal{E}(\boldsymbol{K}) = (\hbar^2/2m)(2\pi/a)^2)K^2)$ are exactly diagonalized. For example, the calculated band gap of GaAs is 1.42 [eV]

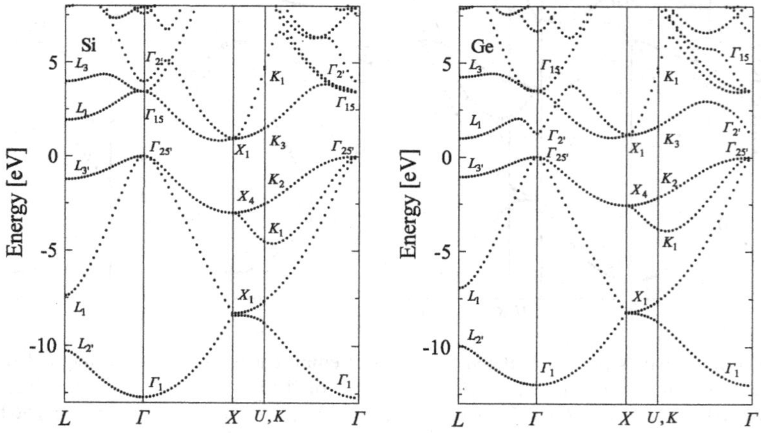

Fig. 1.10 Energy band structures calculated by the empirical pseudopotential method for Si and Ge. The spin–orbit interaction is not included

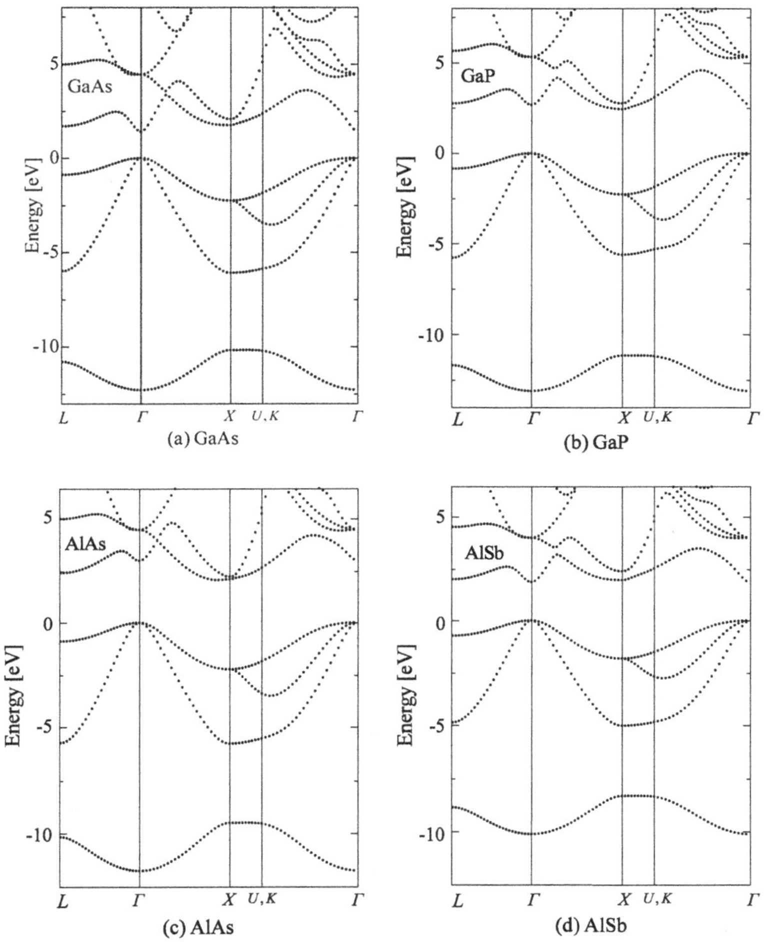

Fig. 1.11 Energy band structures calculated by the empirical pseudopotential method for **a** GaAs, **b** GaP, **c** AlAs, and **d** AlSb. The spin–orbit interaction is not included

which is obtained without spin–orbit interaction (compare the results with spin–orbit interaction shown in Fig. 1.21, where we obtain 1.52 eV for the direct band gap). We have to note here that energy band calculations with 59 plane waves of $0 \leq \mathcal{E}(\boldsymbol{K}) \leq \mathcal{E}(\boldsymbol{K}) = 12$ give quite reasonable results. This is understood from the fact that the next higher levels of the free electron states are $\mathcal{E}(\boldsymbol{K}) = 16$ and well high compared with $\mathcal{E}(\boldsymbol{K}) = 12$. The energy band calculations mentioned above is sometimes called the "local pseudopotential method", and later Chelikowsky and Cohen reported the "nonlocal pseudopotential method" as described in Sect. 1.6.3 in which the spin–orbit interaction is taken into account [6, 7].

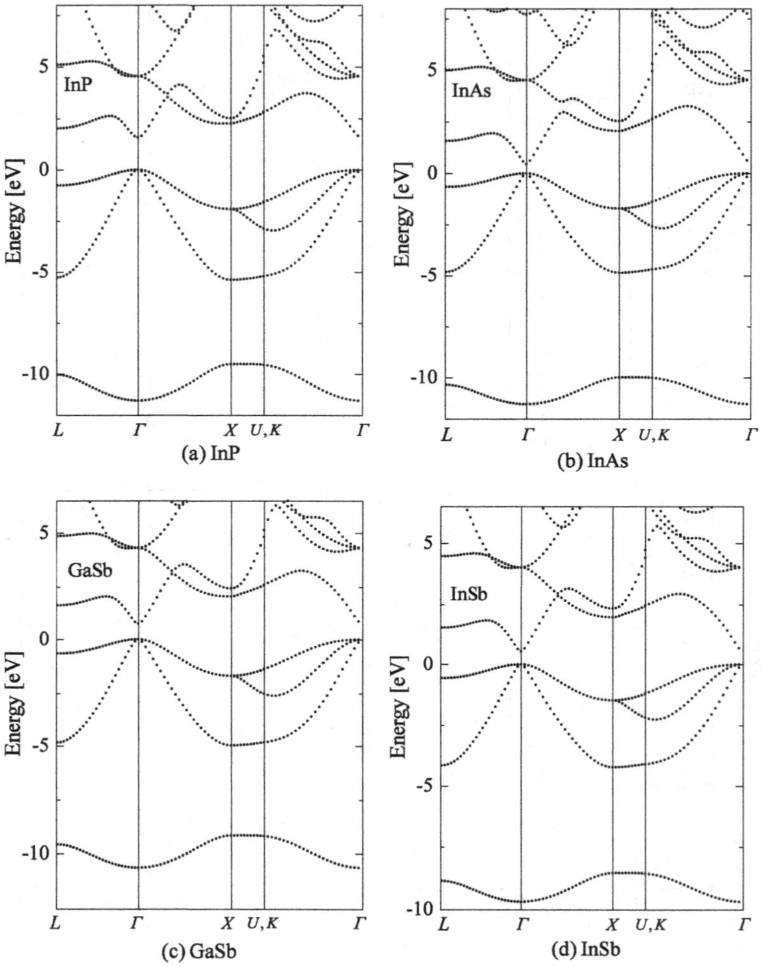

Fig. 1.12 Energy band structures calculated by the empirical pseudopotential method for **a** InP, **b** InAs, **c** GaSb, and **d** InSb. The spin–orbit interaction is not included

1.6.5 Spin–Orbit Interaction

Once the reciprocal vectors are calculated, the free-electron wave functions (1.87) (called as plane waves in this textbook) are easily formulated. Then putting the wave functions into (1.84) we obtain (1.88) which is called eigen-value equation and easily diagonalized to give eigenvalues and eigen functions. The matrix element in (1.84) is written as

$$\langle(\boldsymbol{k}+\boldsymbol{G}_i)|H_{\rm ps}|(\boldsymbol{k}+\boldsymbol{G}_j)\rangle = T(\boldsymbol{k})_{\boldsymbol{G}_i,\boldsymbol{G}_j} + V_{\boldsymbol{G}_i,\boldsymbol{G}_j} + \Delta(\boldsymbol{k})_{\boldsymbol{G}_i,\boldsymbol{G}_j}\,, \tag{1.103}$$

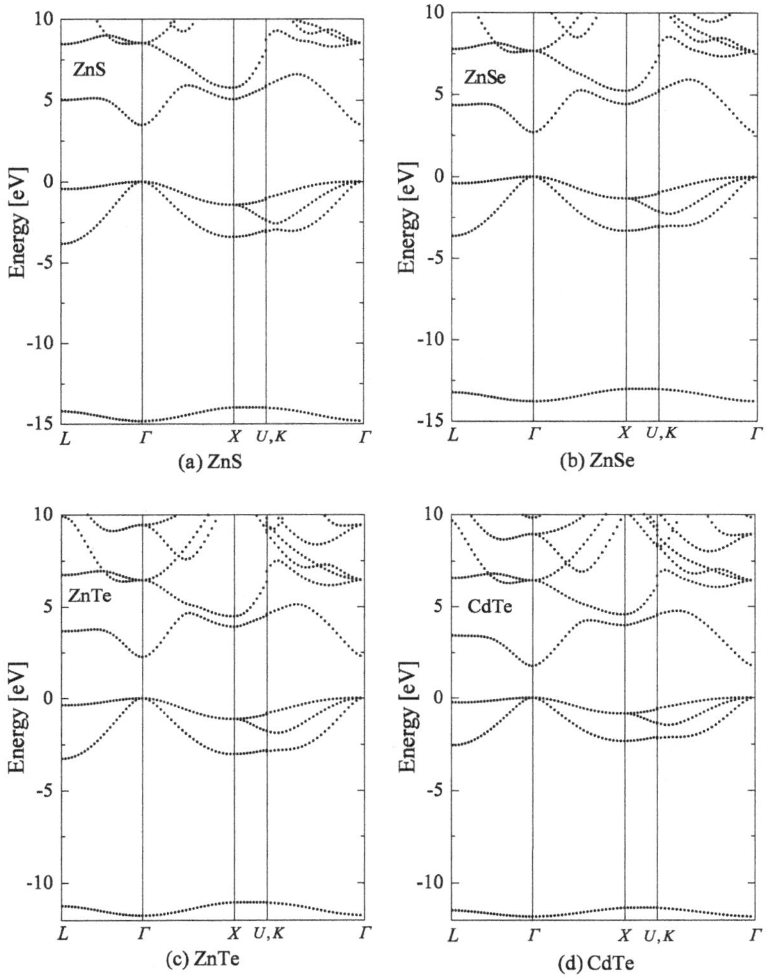

Fig. 1.13 Energy band structures calculated by the empirical pseudopotential method for **a** ZnS, **b** ZnSe, **c** ZnTe, and **d** CdTe. The spin–orbit interaction is not included

where we included spin–orbit interaction by introducing the term $\Delta(\boldsymbol{k})_{G_i,G_j}$. The derivation of spin–orbit interaction term is shown in Appendix H. The three terms of (1.103) are given by

$$T(\boldsymbol{k})_{\boldsymbol{G}_i,\boldsymbol{G}_j} = \frac{\hbar^2}{2m}(\boldsymbol{k}+\boldsymbol{G}_i)^2\delta_{\boldsymbol{G}_i,\boldsymbol{G}_j}\,, \tag{1.104}$$

$$V_{\boldsymbol{G}_i,\boldsymbol{G}_j} = \left[V^{\mathrm{S}}(\boldsymbol{Q})\cos(\boldsymbol{Q}\cdot\boldsymbol{\tau}) + \mathrm{i}V^{\mathrm{A}}(\boldsymbol{Q})\sin(\boldsymbol{Q}\cdot\boldsymbol{\tau})\right]_{\boldsymbol{Q}=\boldsymbol{G}_i-\boldsymbol{G}_j}\,, \tag{1.105}$$

$$\Delta(\boldsymbol{k})_{\boldsymbol{G}_i,\boldsymbol{G}_j} = \mathrm{i}\boldsymbol{\sigma}\cdot\left[\boldsymbol{G}_i\times\boldsymbol{G}_j - \boldsymbol{k}\times(\boldsymbol{G}_i-\boldsymbol{G}_j)\right] \times\left[\lambda^{\mathrm{S}}\cos(\boldsymbol{Q}\cdot\boldsymbol{\tau}) + \mathrm{i}\lambda^{\mathrm{A}}\sin(\boldsymbol{Q}\cdot\boldsymbol{\tau})\right]\,, \tag{1.106}$$

where

$$\boldsymbol{Q} = \boldsymbol{G}_i - \boldsymbol{G}_j \equiv \frac{2\pi}{a}\boldsymbol{K}\,. \tag{1.107}$$

The term $T(\boldsymbol{k})_{\boldsymbol{G}_i,\boldsymbol{G}_j}$ has diagonal elements only and is easily formulated by using the reciprocal wave vectors listed in Table 1.1. The pseudopotential term $V_{\boldsymbol{G}_i,\boldsymbol{G}_j}$ is separated in symmetric parts and antisymmetric parts (for zinc blende crystal) and their matrix elements are evaluated as follows. Using $\boldsymbol{K} = (a/2\pi)\boldsymbol{Q} = (a/2\pi)(\boldsymbol{G}_j - \boldsymbol{G}_i)$, and $\boldsymbol{\tau} = (a/8)[111]$, the symmetric parts of the pseudopotentials are

$$V^{\mathrm{S}}(\boldsymbol{Q})\cos(\boldsymbol{Q}\cdot\boldsymbol{\tau}) = V_3^{\mathrm{S}}\cos\left[\frac{\pi}{4}(K_x+K_y+K_z)\right] \text{ if } |\boldsymbol{K}|^2 = 3 \tag{1.108a}$$

$$= V_8^{\mathrm{S}}\cos\left[\frac{\pi}{4}(K_x+K_y+K_z)\right] \text{ if } |\boldsymbol{K}|^2 = 8 \tag{1.108b}$$

$$= V_{11}^{\mathrm{S}}\cos\left[\frac{\pi}{4}(K_x+K_y+K_z)\right] \text{ if } |\boldsymbol{K}|^2 = 11 \tag{1.108c}$$

$$= 0 \text{ if } |\boldsymbol{K}|^2 > 11\,, \tag{1.108d}$$

and in a similar fashion we obtain the antisymmetric parts of the pseudopotentials for a zinc blende crystal

$$V^{\mathrm{A}}(\boldsymbol{Q})\sin(\boldsymbol{Q}\cdot\boldsymbol{\tau}) = V_3^{\mathrm{A}}\sin\left[\frac{\pi}{4}(K_x+K_y+K_z)\right] \text{ if } |\boldsymbol{K}|^2 = 3 \tag{1.109a}$$

$$= V_4^{\mathrm{A}}\sin\left[\frac{\pi}{4}(K_x+K_y+K_z)\right] \text{ if } |\boldsymbol{K}|^2 = 4 \tag{1.109b}$$

$$= V_{11}^{\mathrm{A}}\sin\left[\frac{\pi}{4}(K_x+K_y+K_z)\right] \text{ if } |\boldsymbol{K}|^2 = 11 \tag{1.109c}$$

$$= 0 \text{ if } |\boldsymbol{K}|^2 > 11\,. \tag{1.109d}$$

The matrix elements of the spin–orbit Hamiltonian are easily evaluated by using the results shown in Appendix H (see also Sect. 1.7.5 for the evaluation of the matrix elements of the spin–orbit Hamiltonian), and the manipulation similar to the pseudopotential term leads to the following relations

$$\lambda^{\mathrm{S}}\cos(\boldsymbol{Q}\cdot\boldsymbol{\tau}) = \lambda^{\mathrm{S}}\cos\left[\frac{\pi}{4}(K_x+K_y+K_z)\right]\,, \tag{1.110}$$

$$\lambda^{\mathrm{A}}\sin(\boldsymbol{Q}\cdot\boldsymbol{\tau}) = \lambda^{\mathrm{A}}\sin\left[\frac{\pi}{4}(K_x+K_y+K_z)\right]\,. \tag{1.111}$$

The expression of spin–orbit interaction shown here is based on the derivation by Melz [10] and known to be mathematically equivalent to the long-wavelength limit of the OPW formulation due to Weisz [11], but the formalism is based on the empirical pseudopotential (local pseudopotential) method. See the paper by Chelikowsky and Cohen [7] for more detailed treatment. Since we are interested in the valence bands and lower lying conduction bands, we may restrict the number of plane waves to calculate the spin–orbit interaction term.

It is well known that the atomic spin–orbit splittings are larger for the heavier elements. Therefore we expect that the spin–orbit splitting at the valence band maximum increases with the heavier elements. For example the spin–orbit splitting of Ge is larger than Si, and it increases in order of the mass for GaP, GaAs, GaSb, InP, InAs, and InSb.

1.6.6 Energy Band Calculations by Nonlocal Pseudopotential Method with Spin–Orbit Interaction

First, we show the calculated results by the local pseudopotential method with spin–orbit interaction for Ge in Fig. 1.14(a) and GaAs in Fig. 1.14(b) with 118 plane waves of spin–up and spin–down with the spin–orbit interaction. The pseudopotential values used in the calculations are local pseudopotentials reported by Cohen and Bergstresser [5] (see Table 1.2), and the spin–orbit interaction parameters are $(2\pi/a)^2\lambda^{\rm S} = 0.0008$, $(2\pi/a)^2\lambda^{\rm A} = 0.0002$, where $\lambda^{\rm S}$ and $\lambda^{\rm A}$ are given by the units $(2\pi/a)^2\lambda^{\rm S}$ and $(2\pi/a)^2\lambda^{\rm A}$ and used as fitting parameters for simplicity throughout the textbook. These parameters are not best fitted but give the spin–orbit splitting energy about 0.340 [eV] for GaAs. There exists only a slight difference in the energies at X, L and other critical points between the present calculations and the results calculated by the nonlocal pseudopotential methods of Chelikowsky and Cohen [7] (the results of the nonlocal pseudopotential method obtained by the present authors' are shown in Fig. 1.15).

Finally we will show the energy band structure of GaAs calculated by using the nonlocal pseudopotential method with the spin–orbit interaction. In the calculation the plane waves for $0 \le K^2 \le \mathcal{E}_1 = 20$ (113 plane waves and thus 226 plane waves with spin–up and –down states) are exactly diagonalized, and $56(= 169 - 113)$ spin–degenerate states for $\mathcal{E}_1 < K^2 \le 27$ are treated by Löwdin's perturbation method [8]. The results are shown in Fig. 1.15, where we used the pseudopotentials $V_3^{\rm S} = -0.254$ (-0.214) and $V_{11}^{\rm A} = 0.010$ (0.001) instead of the parameters shown in the parentheses reported by Chelikowsky and Cohen [7], and the spin–orbit parameters is $\lambda^{\rm S} = 0.00081$ and $\lambda^{\rm A} = 0.000245$. The results are shown in Fig. 1.15 which give the energy gap $\mathcal{E}_{\rm G} = 1.5055$ [eV] and the spin–orbit splitting at the Γ point 0.34018 [eV]. We have to note here that the calculated results depend on the energy cut values $\mathcal{E}_1$ and $\mathcal{E}_2$. When we choose a smaller value for $\mathcal{E}_1$, the convergence is very fast, but the results strongly depend on the value of $\mathcal{E}_2$ and the obtained result is not enough to explain the existing experimental data. On the other hand the results for $\mathcal{E}_1 = 20$ exhibit no recognizable difference between the results with and without the perturbation terms of the plane waves for $\mathcal{E}_1 < K^2 \le 27$.

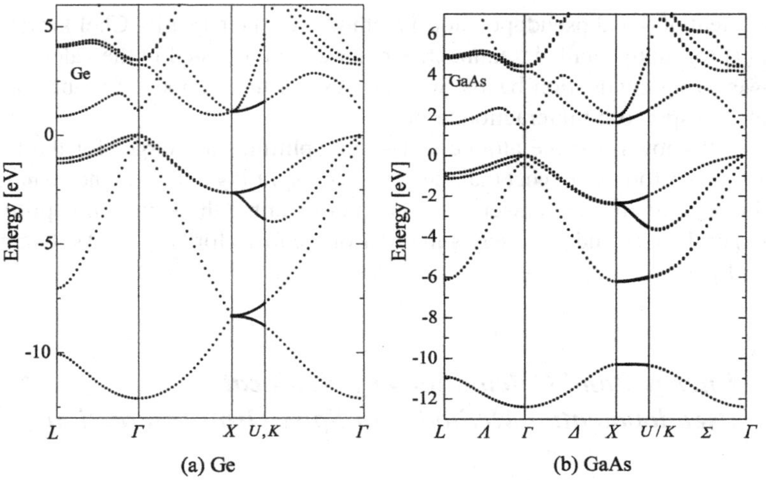

Fig. 1.14 Energy band structures of **a** Ge and **b** GaAs calculated by the local pseudopotential method with the spin–orbit interaction, where 118 plane waves with spin–up and spin–down and the spin–orbit interaction parameters are $(2\pi/a)^2\lambda^{\mathrm{S}} = 0.0008$, $(2\pi/a)^2\lambda^{\mathrm{A}} = 0.0002$. The pseudopotential parameters given in Table 1.2 are used ($(2\pi/a)^2\lambda^{\mathrm{S}} = 0.00097$ for Ge)

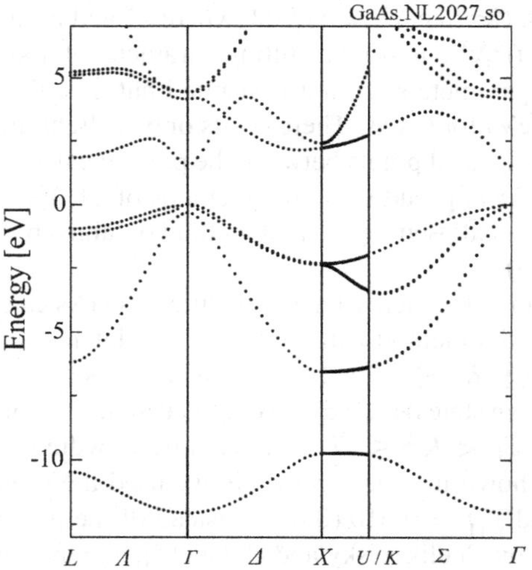

Fig. 1.15 Energy bands of GaAs calculated by the nonlocal pseudopotential method, where the spin–orbit interaction is taken into account. The states for $0 \leq K^2 \leq \mathcal{E}_1 = 20$ (113 plane waves and thus 226 plane waves with spin–up and –down states) are exactly diagonalized and the states for $\mathcal{E}_1 < K^2 \leq 27$ (degenerate 56 waves) are treated by the perturbation method

As reported by Chelikowsky and Cohen [7] the over all features of the energy bands are almost the same as the results calculated by the local pseudopotential method except some critical points. Therefore the results shown in the text are calculated by the local pseudopotential method, unless otherwise mentioned, because we have to adjust more parameters for the nonlocal pseudopotential methods. The local pseudopotential method requires fewer number of the pseudopotential parameters to get results in agreement with the experimental observation, and provides the energy bands and optical properties of various semiconductors which help us to understand the optical characteristics and transport properties based on the full band Monte Carlo simulation.

1.7 $\boldsymbol{k} \cdot \boldsymbol{p}$ Perturbation

1.7.1 $\boldsymbol{k} \cdot \boldsymbol{p}$ Hamiltonian

The $\boldsymbol{k} \cdot \boldsymbol{p}$ perturbation was introduced by Kane [12] in 1956 to analyze the energy band structures of III–V compound semiconductors and led to a successful result. The method was originally used in 1936 to discuss the character table of the symmetry points in the Brillouin zone by Bouckaert, Smoluchowski and Wigner [13]. Later Dresselhaus, Kip and Kittel [14] used the $\boldsymbol{k} \cdot \boldsymbol{p}$ perturbation method to analyze the detailed structure of the valence bands of Ge. This method is described in detail in Sect. 2.1, where the $\boldsymbol{k} \cdot \boldsymbol{p}$ perturbation method is applied to analyze the experimental results of cyclotron resonance in Ge. In this section we will consider the method used by Cardona and Pollak [15] to calculate energy band structures.

We consider the non-relativistic Schrödinger equation for a one-electron system:

$$\left[-\frac{\hbar^2}{2m}\nabla^2 + V(\boldsymbol{r})\right]\Psi(\boldsymbol{r}) = \mathcal{E}\Psi(\boldsymbol{r}), \tag{1.112}$$

where $V(\boldsymbol{r})$ is the crystal potential energy with the lattice periodicity. The solution of (1.112) is given by the Bloch function

$$\Psi(\boldsymbol{r}) = \mathrm{e}^{\mathrm{i}\boldsymbol{k}\cdot\boldsymbol{r}} u_{n,\boldsymbol{k}}(\boldsymbol{r}), \tag{1.113}$$

where $u_{n,\boldsymbol{k}}$ is a function of the lattice periodicity for band index n. Putting this Bloch function into (1.112) and using the following relations

$$\nabla\Psi(\boldsymbol{r}) = \mathrm{i}\boldsymbol{k}\Psi(\boldsymbol{r}) + \mathrm{e}^{\mathrm{i}\boldsymbol{k}\cdot\boldsymbol{r}}\nabla u_{n,\boldsymbol{k}}(\boldsymbol{r}), \tag{1.114a}$$

$$\begin{aligned}\nabla^2\Psi(\boldsymbol{r}) &= -k^2\Psi(\boldsymbol{r}) + 2\mathrm{i}\boldsymbol{k}\mathrm{e}^{\mathrm{i}\boldsymbol{k}\cdot\boldsymbol{r}}\nabla u_{n,\boldsymbol{k}}(\boldsymbol{r}) + \mathrm{e}^{\mathrm{i}\boldsymbol{k}\cdot\boldsymbol{r}}\nabla^2 u_{n,\boldsymbol{k}}(\boldsymbol{r}), \\ &= \mathrm{e}^{\mathrm{i}\boldsymbol{k}\cdot\boldsymbol{r}}(-k^2 + 2\mathrm{i}\boldsymbol{k}\cdot\nabla + \nabla^2)u_{n,\boldsymbol{k}}(\boldsymbol{r}),\end{aligned} \tag{1.114b}$$

we obtain

$$\left[-\frac{\hbar^2}{2m}\nabla^2 + V(\boldsymbol{r}) + \frac{\hbar^2}{2m}k^2 - \mathrm{i}\frac{\hbar^2}{m}(\boldsymbol{k}\cdot\nabla)\right]u_{n,\boldsymbol{k}}(\boldsymbol{r}) = \mathcal{E}_n(\boldsymbol{k})u_{n,\boldsymbol{k}}(\boldsymbol{r}). \tag{1.115}$$

Using the relation $-i\hbar\nabla = \boldsymbol{p}$ for the momentum operator, the above equation may be rewritten as

$$\left[H_0 + \frac{\hbar^2 k^2}{2m} + \frac{\hbar}{m}\boldsymbol{k}\cdot\boldsymbol{p}\right] u_{n,\boldsymbol{k}}(\boldsymbol{r}) = \mathcal{E}_n(\boldsymbol{k})u_{n,\boldsymbol{k}}(\boldsymbol{r})\,, \tag{1.116}$$

where $H_0 = -(\hbar^2/2m)\nabla^2 + V(\boldsymbol{r})$ is Hamiltonian. The terms $\hbar^2k^2/2m$ in [ ] on the left-hand side is a constant (c-number) without any operator and thus the term gives rise to an energy shift of $\hbar^2k^2/2m$ from $\mathcal{E}_n(\boldsymbol{k})$. First solving (1.116) for $\boldsymbol{k} = 0$ and then treating $(\hbar/m)\boldsymbol{k}\cdot\boldsymbol{p}$ as a perturbing term, we obtain eigenstates as a function of $\boldsymbol{k}$ which gives the energy band structure. Therefore, this method is called $\boldsymbol{k}\cdot\boldsymbol{p}$ **perturbation**. The eigenstates for the Hamiltonian H_0 are obtained by using the pseudopotentials, but here we show a simplified method to obtain the eigenstates by solving 2×2 matrices following the method reported by Cardona [16]. Although such a calculation is very simple, obtained eigen energies and eigenvectors are very useful to understand the band structures.

The above $\boldsymbol{k}\cdot\boldsymbol{p}$ Hamiltonian is rewritten by using atomic units as follows:

$$-\frac{\hbar^2}{2m}\nabla^2 = \frac{\hbar^2}{2m}\left(\frac{1}{a_B}\right)^2(-ia_B\nabla)^2 = \mathrm{Ry}\cdot(-ia_B\nabla)^2 \tag{1.117a}$$

$$\frac{\hbar^2k^2}{2m} = \frac{\hbar^2}{2m}\left(\frac{1}{a_B}\right)^2(a_B\boldsymbol{k})^2 = \mathrm{Ry}\cdot(a_B\boldsymbol{k})^2 \tag{1.117b}$$

$$\frac{\hbar}{m}\boldsymbol{k}\cdot\boldsymbol{p} = \frac{\hbar^2}{2m}\left(\frac{1}{a_B}\right)^2\left(\frac{a_B^2}{\hbar}\right)(2\boldsymbol{k}\cdot\boldsymbol{p}) = \mathrm{Ry}\cdot\left(\frac{a_B^2}{\hbar}\right)(2\boldsymbol{k}\cdot\boldsymbol{p})\,, \tag{1.117c}$$

where $a_B = 4\pi\epsilon_0\hbar^2/(me^2) \simeq 0.529$ [Å] is Bohr radius and $\mathrm{Ry} = me^4/(8\epsilon^2h^3c) \simeq 13.6$ [eV] is Rydberg constant. Using the dimensionless notation or the atomic units

$$\boldsymbol{k}\text{ (in [a.u])} = a_B\boldsymbol{k}\,, \qquad \boldsymbol{p}\text{ (in [a.u])} = \frac{a_B}{\hbar}\boldsymbol{p} = -ia_B\nabla\,, \tag{1.118}$$

and then the $(\hbar/m)\boldsymbol{k}\cdot\boldsymbol{p}$ operator of (1.117c) is rewritten as

$$\frac{\hbar}{m}\boldsymbol{k}\cdot\boldsymbol{p} = \mathrm{Ry}\cdot(2\boldsymbol{k}\cdot\boldsymbol{p})\,. \tag{1.119}$$

Finally, the length is expressed in atomic units or normalized by a_B and then the $\boldsymbol{k}\cdot\boldsymbol{p}$ Hamiltonian is rewritten as

$$H_0 + \frac{\hbar}{m}\boldsymbol{k}\cdot\boldsymbol{p} + \frac{\hbar^2k^2}{2m} = -\nabla^2 + 2\boldsymbol{k}\cdot\boldsymbol{p} + k^2\,. \tag{1.120}$$

When we put $\boldsymbol{k} = 0$ in (1.116), we obtain

$$H_0 u_{n,0}(\boldsymbol{r}) = \left[-\frac{\hbar^2}{2m} \nabla^2 + V(\boldsymbol{r}) \right] u_{n,0}(\boldsymbol{r}) = \mathcal{E}_n(0) u_{n,0}(\boldsymbol{r}) , \tag{1.121}$$

which is rewritten in atomic units as

$$\left[-\nabla^2 + V(\boldsymbol{r}) \right] u_{n,0}(\boldsymbol{r}) = \mathcal{E}_n(0) u_{n,0}(\boldsymbol{r}) . \tag{1.122}$$

Since the crystal potential $V(\boldsymbol{r})$ and Bloch function $u_{n,0}(\boldsymbol{r})$ are periodic with the lattice constant, and thus these two functions are expanded by Fourier series. Therefore we may use the pseudopotential theory stated before, and the diagonalization of the Hamiltonian matrix gives the eigenstates and the corresponding eigenvalues. In the $\boldsymbol{k} \cdot \boldsymbol{p}$ perturbation theory we need eigenstates at $\boldsymbol{k} = 0$ (at the Γ point) only. For this purpose we rewrite (1.86) and (1.87) as

$$\left[-\nabla^2 + V_{ps}(|G|^2) \right] |\boldsymbol{G}_j\rangle = \mathcal{E}_n |\boldsymbol{G}_j\rangle , \tag{1.123}$$

$$|\boldsymbol{G}_j\rangle = \frac{1}{\sqrt{\Omega}} \mathrm{e}^{\mathrm{i}\boldsymbol{G}_j \boldsymbol{r}} , \tag{1.124}$$

where $V_{\mathrm{ps}}(\boldsymbol{r})$ is expanded with Fourier coefficients $V_{\mathrm{ps}}(|G|^2)$ (pseudopotential form factors). Simplified solutions of the above equations are very helpful to obtain the eigenstates and to explain the group theoretical representations used in Fig. 1.7. The $\boldsymbol{k} \cdot \boldsymbol{p}$ Hamiltonian for semiconductors with inversion symmetry has off-diagonal terms only. In addition, we classify the matrix elements with the help of group theory, and so the number of the matrix elements are extremely decreased. Cardona and Pollak [15] proposed to calculate the energy band structures using 15 electronic states (wave functions) at the Γ point (at $\boldsymbol{k} = 0$) and obtained very accurate energy band structures of germanium and silicon. Here we will show the energy band calculations of germanium and silicon based on the $\boldsymbol{k} \cdot \boldsymbol{p}$ perturbation method of Cardona and Pollak, where the spin–orbit interaction is not included. First we classify the free electron energy bands shown in Fig. 1.7 with the help of group theory.

Here we summarize important and useful results of group theory without showing the derivation. The most important factors are to use the character table of the crystal. The character table for a face centered cubic lattice which includes diamond and zinc blende crystals is shown in Table 1.4. Although detailed description of group theory is not shown here, the character table is very useful to calculate non-vanishing matrix elements and thus selection rule of optical transition. Another important information about the symmetry properties of quantum states is the basis functions for the representations, which is summarized in Table 1.5.

In a crystalline solid the notation of atomic orbitals is classified by using the spherical harmonics $Y_{lm}(\theta, \phi)(m = -l, \cdots, +l)$ which constitute a basis for the irreducible representation and thus the electronic states are related to the states of an atom as shown below [17],

Table 1.4 Character table of small representations of O_h group

BSW	E	$3C_4^2$	$6C_4$	$6C_2$	$8C_3$	J	$3JC_4^2$	$6JC_4$	$6JC_2$	$8JC_3$
Γ_1	1	1	1	1	1	1	1	1	1	1
Γ_2	1	1	−1	−1	1	1	1	−1	−1	1
Γ_{12}	2	2	0	0	−1	2	2	0	0	−1
$\Gamma_{15'}$	3	−1	1	−1	0	3	−1	1	−1	0
$\Gamma_{25'}$	3	−1	−1	1	0	3	−1	−1	1	0
$\Gamma_{1'}$	1	1	1	1	1	−1	−1	−1	−1	−1
$\Gamma_{2'}$	1	1	−1	−1	1	−1	−1	1	1	−1
$\Gamma_{12'}$	2	2	0	0	−1	−2	−2	0	0	1
Γ_{15}	3	−1	1	−1	0	−3	1	−1	1	0
Γ_{25}	3	−1	−1	1	0	−3	1	1	−1	0

Table 1.5 Basis function of irreducible representation of O_h group at Γ point

Representation	Degeneracy	Basis functions
Γ_1	1	1
Γ_2	1	$x^4(y^2-z^2)+y^4(z^2-x^2)+z^4(x^2-y^2)$
Γ_{12}	2	$z^2-\frac{1}{2}(x^2+y^2),\ (x^2-y^2)$
$\Gamma_{15'}$	3	$xy(x^2-y^2),\ yz(y^2-z^2),\ zx(z^2-x^2)$
$\Gamma_{25'}$	3	$xy,\ yz,\ zx$
$\Gamma_{1'}$	1	$xyz[x^4(y^2-z^2)+y^4(z^2-x^2)+z^4(x^2-y^2)]$
$\Gamma_{2'}$	1	xyz
$\Gamma_{12'}$	2	$xyz[z^2-\frac{1}{2}(x^2+y^2)],\ xyz(x^2-y^2)$
Γ_{15}	3	$x,\ y,\ z$
Γ_{25}	3	$z(x^2-y^2),\ x(y^2-z^2),\ y(z^2-x^2)$

$$
\begin{aligned}
\text{State } s\ (l=0) &= \Gamma_1\,,\\
\text{State } p\ (l=1) &= \Gamma_{15}\,,\\
\text{State } d\ (l=2) &= \Gamma_{25'}+\Gamma_{12}\,,\\
\text{State } f\ (l=3) &= \Gamma_{15}+\Gamma_{25}+\Gamma_{2'}\,,\\
\text{State } g\ (l=4) &= \Gamma_{25'}+\Gamma_{15'}+\Gamma_{12}+\Gamma_1\,,\\
\text{State } h\ (l=5) &= \Gamma_{25}+2\Gamma_{15}+\Gamma_{12'}\,.
\end{aligned}
$$

In a crystalline solid the wave functions of an atom and the next nearest neighbor are hybridized, resulting in bonding and anti-bonding states and thus in energy shift, as discussed below. A concept of atomic orbitals, LCAO (Linear Combination of Atomic Orbitals), is very helpful to understand the energy bands at $\boldsymbol{k}=0$. Here we follow the method of Cardona [16]. First, we consider electronic states in the outer shell of an atomic Ge, where two s–states and two p–states are filled with electrons. Therefore the other two p–states and ten d–states are empty. Since two

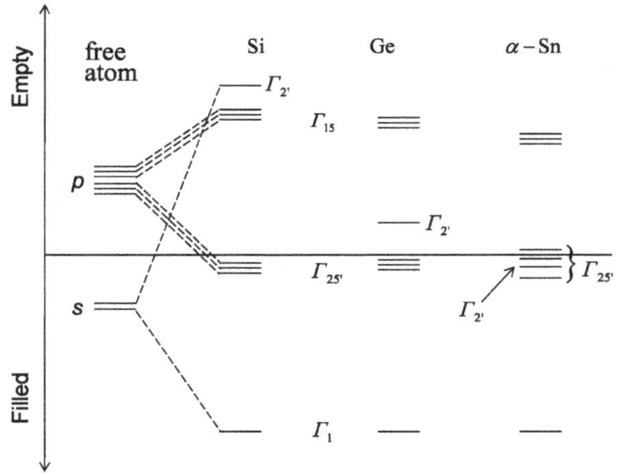

Fig. 1.16 Splitting of the atomic states of Ge, Si and α–Sn under the presence of the crystalline field of the diamond structure at Γ point of the Brillouin zone, where a bar corresponds to a degenerate energy state with spin up and down (after Cardona [16])

atoms are included in a unit cell of diamond structure, the states are doubled when the two atoms are put together. In such a case, the electronic states are modified by linear combinations of the atomic orbitals, the bonding and anti-bonding states, and thus the states split as shown in Fig. 1.16. No mixing by the crystal field occurs between *s*– and *p*–orbitals at $\boldsymbol{k} = 0$. The *s*–orbitals contain a negligible admixture of *f*–orbitals while the *p*–orbitals may contain a significant admixture of *d*–orbitals of the same shell. The ordering of the states are illustrated so as to fit the observed conduction and valence bands. The ordering between the sates is changed by the bonding–anti-bonding splitting. If the splitting is small, as in the case of α–Sn with large lattice constant, the Γ_1 and $\Gamma_{2'}$ are filled and $\Gamma_{25'}$ only partially filled. As the bonding–anti-bonding splitting becomes larger, the $\Gamma_{2'}$–state becomes higher than $\Gamma_{25'}$, resulting in an energy gap. This is the case for Ge and Si as shown in Fig. 1.16.

In order to calculate the energy bands of Ge and Si by the $\boldsymbol{k} \cdot \boldsymbol{p}$ method, we have to choose the eigenstates at $\boldsymbol{k} = 0$ properly, not so many states but enough to lead reasonable results. First we show the free electron bands for lower energy states in Table 1.6, where the free electron energies are estimated for Ge ($a = 5.66$ Å), $\boldsymbol{G}_3$ band is 1.038 a.u. (14.1 eV), and $\boldsymbol{G}_4$ is 1.38 a.u. (18.8 eV). The fourth band $\boldsymbol{G}_8$ is 2.77 a.u. (37.7 eV), which is well higher than the $\boldsymbol{G}_4$ bands and we may limit the plane waves up to $\boldsymbol{G}_4$. This enables us to choose 15 eigenstates with the representations; Γ_1^l, $\Gamma_{25'}^l$, Γ_{15}, $\Gamma_{2'}^l$, Γ_1^u, $\Gamma_{25'}^u$, $\Gamma_{12'}$, and $\Gamma_{2'}^u$. This approximation was used by Cardona and Pollak [15].

Next, we discuss relation between the LCAO and plane waves of free electron bands expressed by a combination of the reciprocal lattice vectors $[G_x, G_y, G_z]$ of the empty lattice bands at Γ point (at $\boldsymbol{k} = 0$ of the Brillouin zone). The plane waves of [000], [111] and [200] are related to the LCAO states as shown by Cardona [16].

Table 1.6 Reciprocal of smaller free electron energies, where dgn means the degeneracy of the states and the free electron energy in [a.u.] is estimated for $a = 5.66$ [Å]

$\boldsymbol{G}$	Vector components	dgn	Representations	Energy in [a.u.]
$\boldsymbol{G}_0$	$= (2\pi/a)[0, 0, 0]$	1	Γ_1^l	0
$\boldsymbol{G}_3$	$= (2\pi/a)[\pm1, \pm1, \pm1]$	8	$\Gamma_{25'}^l + \Gamma_{15} + \Gamma_{2'}^l + \Gamma_1^u$	1.038
$\boldsymbol{G}_4$	$= (2\pi/a)[\pm2, 0, 0]$	6	$\Gamma_{25'}^u + \Gamma_{12'} + \Gamma_{2'}^u$	1.38
$\boldsymbol{G}_8$	$= (2\pi/a)[\pm2, \pm2, 0]$	12	$\Gamma_1 + \Gamma_{12} + 2\Gamma_{15} + \Gamma_{25'}$	2.77
$\boldsymbol{G}_{11}$	$= (2\pi/a)[\pm3, \pm1, \pm1]$	24	$\Gamma_1 + \Gamma_{12} + \Gamma_2 + \Gamma_{12'}$ $\Gamma_{15'} + \Gamma_{25} + 2\Gamma_{25'} + 2\Gamma_{15}$	3.81

$$
\begin{array}{lll}
[000] & \Gamma_1^l; & s\text{-bonding;} \\
[111] & \left\{\begin{array}{l} \Gamma_{25'}^l; \\ \Gamma_{15}; \\ \Gamma_{2'}^l; \\ \Gamma_1^u; \end{array}\right. & \begin{array}{l} p\text{-bonding} \\ p\text{-anti-bonding} \\ s\text{-anti-bonding} \\ s\text{-bonding (next shell);} \end{array} \\
[200] & \left\{\begin{array}{l} \Gamma_{25'}^u; \\ \Gamma_{12'}; \\ \Gamma_{2'}^u; \end{array}\right. & \begin{array}{l} d\text{-bonding} \\ d\text{-bonding} \\ s\text{-anti-bonding (next shell)}, \end{array}
\end{array}
$$

where the superscript l and u denote the lower and the upper of the two states of the same symmetry. For simplicity we use the relation (1.59) to express dimensionless reciprocal lattice vector $\boldsymbol{K}$. The group of reciprocal wave vectors $\boldsymbol{K}[\pm1, \pm1, \pm1](8)$ have 8 components (dimension is 8), and a combinations of the 8 plane waves gives the following representations by using the character table of Table 1.4

$$
\boldsymbol{K}[\pm1, \pm1, \pm1](8) = \Gamma_1^u(1) + \Gamma_{2'}^u(1) + \Gamma_{15}(3) + \Gamma_{25'}^l(3)\,. \tag{1.125}
$$

The representation Γ_1^u has the same character of the lowest valence band Γ_1^l arising from the reciprocal wave vector $\boldsymbol{K}[0, 0, 0]$. Representation Γ_1^u in (1.125) is understood as follows. A summarized combination of orthogonalized plane waves for Γ_1^u is composed from $[\pm1, \pm1, \pm1]$

$$
\begin{aligned}
\Gamma_1^u[\pm1, \pm1, \pm1] = \frac{1}{\sqrt{8}} \{ & [1, 1, 1] - [\bar{1}, 1, 1] - [1, \bar{1}, 1] - [1, 1, \bar{1}] \\
& -[\bar{1}, \bar{1}, 1] - [\bar{1}, 1, \bar{1}] - [1, \bar{1}, \bar{1}] + [\bar{1}, \bar{1}, \bar{1}] \}
\end{aligned} \tag{1.126}
$$

and the character is the same of $\Gamma_1^l[0, 0, 0]$, where we used the notation

$$
[K_x, K_y, K_z] = \exp\left[\mathrm{i}\left(\frac{2\pi}{a}\right)(K_x x + K_y y + K_z z)\right]. \tag{1.127}
$$

In a similar fashion other symmetrized combinations of the orthogonalized plane waves are composed for $\Gamma_{2'}^{u}$, Γ_{15}, and $\Gamma_{25'}^{l}$. The representations of the plane waves are summarized in Table 1.6. In the following several symmetrized combinations of the orthogonalized plane waves $[K_x, K_y, K_z]$ belonging to the Γ–representation are given:

$$\Gamma_{2'}^{l}\,[111] = \frac{1}{\sqrt{8}}\{[111]-[1\bar{1}\bar{1}]-[\bar{1}1\bar{1}]-[\bar{1}\bar{1}1]-[\bar{1}\bar{1}\bar{1}]+[\bar{1}11]+[1\bar{1}1]+[11\bar{1}]\},$$

$$\Gamma_{2'}^{u}\,[200] = \frac{1}{\sqrt{6}}\{[200]+[020]+[002]-[\bar{2}00]-[0\bar{2}0]-[00\bar{2}]\},$$

$$\Gamma_{25'}^{l}(X)\,[111] = \frac{1}{\sqrt{8}}\{[111]-[1\bar{1}\bar{1}]+[\bar{1}1\bar{1}]+[\bar{1}\bar{1}1]+[\bar{1}\bar{1}\bar{1}]-[\bar{1}11]+[1\bar{1}1]+[11\bar{1}]\},$$

$$\Gamma_{25'}^{u}(X)\,[200] = \frac{1}{\sqrt{2}}\{[200]+[\bar{2}00]\},$$

$$\Gamma_{15}(x)\,[111] = \frac{1}{\sqrt{8}}\{[111]-[1\bar{1}\bar{1}]+[\bar{1}1\bar{1}]+[\bar{1}\bar{1}1]-[\bar{1}\bar{1}\bar{1}]+[\bar{1}11]-[1\bar{1}1]-[11\bar{1}]\},$$

$$\Gamma_{12'}(1)\,[200] = \frac{1}{2}\{[020]-[002]-[0\bar{2}0]+[00\bar{2}]\},$$

$$\Gamma_{12'}(2)\,[200] = \frac{1}{\sqrt{12}}\{2[200]-[020]-[002]-2[\bar{2}00]+[0\bar{2}0]+[00\bar{2}]\}.$$

The Y and Z components of $\Gamma_{25'}^{l}$, and y and z components of Γ_{15} are obtained by means of cyclic permutation.

1.7.2 Derivation of the k · p Parameters

Energy band calculation by means of $\boldsymbol{k}\cdot\boldsymbol{p}$ perturbation requires estimation of momentum matrix elements such as $\langle\Gamma_{25'}^{l}|\boldsymbol{p}|\Gamma_{2}^{l}\rangle$ and energy eigenstates at the Γ point of the Brillouin zone ($\boldsymbol{k} = 0$). For this purpose we use pseudopotential method stated in the previous Sect. 1.6. These values are used for the initial data and then adjusted to the values so that the calculated critical points agree with the experiment. Cardona [16] applied the pseudopotential method for the purpose as described below. First we solve (1.123) with (1.124). For simplicity, irreducible representations of linear combinations of orthogonalized plane waves [000], [111] and [200] are considered, and then we find in Table 1.6 that representations Γ_1, $\Gamma_{2'}$ and $\Gamma_{25'}$ appear twice, and designated as upper "u" and lower "l" states. In the case of a diamond crystal such as Ge and Si there exist symmetric pseudopotential terms only, and we obtain eigenvalues and eigenfunctions by diagonalizing 2×2 matrices of Hamiltonian $H_0 = -\nabla^2 + V_{ps}(|G|^2)$ for the Γ_1, $\Gamma_{2'}$ and $\Gamma_{25'}$ states:

$$\begin{array}{cc} \Gamma_1[000] & \Gamma_1[111] \end{array}$$
$$\begin{vmatrix} 0 & 2V_3^{\mathrm{S}} \\ 2V_3^{\mathrm{S}} & 3(2\pi/a)^2 + 3V_8^{\mathrm{S}} \end{vmatrix}, \tag{1.128a}$$

$$\begin{array}{cc} \Gamma_{2'}[111] & \Gamma_{2'}[200] \end{array}$$
$$\begin{vmatrix} 3(2\pi/a)^2 + 3V_3^{\mathrm{S}} & \sqrt{6}(V_3^{\mathrm{S}} + V_{11}^{\mathrm{S}}) \\ \sqrt{6}(V_3^{\mathrm{S}} + V_{11}^{\mathrm{S}}) & 4(2\pi/a)^2 + 4V_8^{\mathrm{S}} \end{vmatrix}, \tag{1.128b}$$

$$\begin{array}{cc} \Gamma_{25'}[111] & \Gamma_{25'}[200] \end{array}$$
$$\begin{vmatrix} 3(2\pi/a)^2 - V_3^{\mathrm{S}} & \sqrt{2}(V_3^{\mathrm{S}} - V_{11}^{\mathrm{S}}) \\ \sqrt{2}(V_3^{\mathrm{S}} - V_{11}^{\mathrm{S}}) & 4(2\pi/a)^2 \end{vmatrix}. \tag{1.128c}$$

In the following we estimate the eigen energies and eigenfunctions using the pseudopotentials listed in Table 1.2 and thus the obtained results differ a little from those reported by Cardona [16] and Cardona and Pollak [15]. The energies of Γ_1 states are −0.171 [Ry] (−2.326 [eV]) and 1.237[Ry] (16.8 [eV] for germanium, and 0.735 [Ry] (10.0 [eV] and −0.2399 [Ry] (−3.263 [eV]) for silicon. The energies of the $\Gamma_{2'}$ states are given by 1.69 [Ry] (22.9 [eV]) and 0.77 [Ry] (10.5 [eV]) for germanium, and 1.81 [Ry] (24.6 [eV]) and 0.977 [Ry] (13.3 [eV]) for silicon. The eigenstates of the lower energies are

$$\Gamma_{2'}^{l} \text{ for Ge}: \quad 0.843\Gamma_{2'}[111] + 0.539\Gamma_{2'}[200], \tag{1.129a}$$
$$\Gamma_{2'}^{l} \text{ for Si},: \quad 0.906\Gamma_{2'}[111] + 0.422\Gamma_{2'}[200]. \tag{1.129b}$$

In a similar fashion the eigenstates of $\Gamma_{25'}$ are obtained by solving 2×2 Hamiltonian matrix. The eigen energies are 1.74 [Ry] (23.6 [eV]) and 0.909 [Ry] (12.37 [eV]) for Ge, and 1.84 [Ry] (25.0 [eV]) and 1.00 [Ry] (13,6 [eV]) for Si.

$$\Gamma_{25'}^{l} \text{ for Ge}: \quad 0.755\Gamma_{25'}[111] + 0.656\Gamma_{25'}[200], \tag{1.130a}$$
$$\Gamma_{25'}^{l} \text{ for Si},: \quad 0.774\Gamma_{25'}[111] + 0.634\Gamma_{25'}[200]. \tag{1.130b}$$

The energies of the single state Γ_{15} is given by $3(2\pi/a)^2 - V_8^{\mathrm{S}}$, and the doubly–degenerate states $\Gamma_{12'}(1)[200]$) and $\Gamma_{12'}(2)[200]$ are given by $4(2\pi/a)^2 - 2V_8^{\mathrm{S}}$, where matrix elements are 0 between the states $\Gamma_{12'}(1)$ and $\Gamma_{12'}(2)$. The calculated energies at $\boldsymbol{k} = 0$ (at Γ point) are listed in Table 1.7, where energies of the first row of the material in the table in units of [Ry] are obtained by solving a single band and 2×2 pseudopotential matrices, and the second row of rel.[Ry] is the relative values with respect to the valence band top of $\Gamma_{25'}^{l}$. The third row of the material shows the eigenvalues calculated by diagonalizing 15×15 pseudopotential matrices. In the 15×15 pseudopotential calculation, 15 reciprocal wave vectors of [000](1), $[\pm 1, \pm 1, \pm 1]$(8), and $[\pm 2, 0, 0]$(6) (numbers in () are degeneracy of the states) are used, and the result reveals that $\Gamma_{25'}^{l}$, $\Gamma_{25'}^{u}$ and Γ_{15} states are triply–degenerate, and $\Gamma_{12'}$ states are doubly–degenerate. We find in Table 1.7 that the values deduced from a simplified method give a very good guide to locate the eigenstates of the valence and conduction bands. The results obtained by the simple method are very close to

Table 1.7 Energy eigenvalues at $\boldsymbol{k} = 0$ (Γ point) calculations by simplified matrices of the pseudopotentials. Values are given in atomic unit Rydberg [Ry], and relative values (rel.[Ry]) are with respect to the energy of the top valence band $\Gamma^l_{25'}$. For comparison eigenvalues obtained by solving 15 × 15 pseudopotential matrix is shown by full [Ry]

At $\boldsymbol{k} = 0$	Waves	Germanium			Silicon		
		[Ry]	rel.[Ry]	full [Ry]	[Ry]	rel.[Ry]	full [Ry]
$\Gamma^l_{25'}$	[111]	0.909	0.00	0.00	1.00	0.00	0.00
$\Gamma^l_{2'}$	[111]	0.769	−0.139	0.0342	0.977	−0.023	0.2396
Γ_{15}	[111]	1.026	0.117	0.2694	1.085	0.086	0.2521
Γ^u_1	[111]	1.237	0.331	0.4805	1.374	0.246	0.5406
Γ^l_1	[000]	−0.171	−1.08	−0.9270	−0.128	−1.28	−0.9611
$\Gamma_{12'}$	[200]	1.36	0.452	0.6044	1.42	0.421	0.5870
$\Gamma^u_{25'}$	[200]	1.74	0.828	0.8938	1.836	0.836	0.9192
$\Gamma^u_{2'}$	[200]	1.687	0.778	0.9396	1.809	0.809	0.9996

the values of "pseudo" in Table 1.8, although the energy levels of $\Gamma_{2'}$ estimated by the simple method are negative and thus lie below the valence band top for both Ge and Si. In the $\boldsymbol{k} \cdot \boldsymbol{p}$ perturbation calculations, however, the parameters are adjusted to fit the data of experimental critical points and thus these estimations will be used for the initial parameters (Table 1.7).

Estimation of the momentum matrix element reported by Cardona [16] is very helpful to understand the energy band structure and thus it is described below in detail. The momentum matrix elements of $\boldsymbol{p}$ between $\Gamma^l_{25'}$ and $\Gamma^l_{2'}$ which is expressed as P are given by using (1.129a) ~ (1.130b)

$$\begin{aligned}\langle\Gamma^l_{25'}(X)|p_x|\Gamma^l_{2'}\rangle &= \langle\Gamma^l_{25'}(Y)|p_y|\Gamma^l_{2'}\rangle = \langle\Gamma^l_{25'}(Z)|p_z|\Gamma^l_{2'}\rangle \\ &= \frac{2\pi}{a}[0.843 \times 0.755 + 0.539 \times 0.656] \\ &= 0.58 = P \quad (\text{: for Ge}),\end{aligned} \tag{1.131}$$

$$\begin{aligned}\langle\Gamma^l_{25'}(X)|p_x|\Gamma^l_{2'}\rangle &= \langle\Gamma^l_{25'}(Y)|p_y|\Gamma^l_{2'}\rangle = \langle\Gamma^l_{25'}(Z)|p_z|\Gamma^l_{2'}\rangle \\ &= \frac{2\pi}{a}[0.906 \times 0.774 + 0.422 \times 0.634] \\ &= 0.59 = P \quad (\text{: for Si}).\end{aligned} \tag{1.132}$$

The momentum matrix elements of $\boldsymbol{p}$ between $\Gamma^l_{25'}$ and Γ_{15} states are

$$\begin{aligned}\langle\Gamma^l_{25'}(Z)|p_x|\Gamma_{15}(y)\rangle &= \langle\Gamma^l_{25'}(X)|p_y|\Gamma_{15}(z)\rangle = \langle\Gamma^l_{25'}(Y)|p_z|\Gamma_{15}(x)\rangle \\ &= \frac{2\pi}{a} \times 0.755 = 0.44 = Q \quad (\text{: for Ge}),\end{aligned} \tag{1.133}$$

$$\begin{aligned}\langle\Gamma^l_{25'}(Z)|p_x|\Gamma_{15}(y)\rangle &= \langle\Gamma^l_{25'}(X)|p_y|\Gamma_{15}(z)\rangle = \langle\Gamma^l_{25'}(Y)|p_z|\Gamma_{15}(x)\rangle \\ &= \frac{2\pi}{a} \times 0.774 = 0.47 = Q \quad (\text{: for Si}).\end{aligned} \tag{1.134}$$

Table 1.8 Energy eigenvalues used for energy band calculations by the $\boldsymbol{k} \cdot \boldsymbol{p}$ perturbation method in units of Rydberg (from reference [15])

At $\boldsymbol{k} = 0$	Waves	Germanium			Silicon		
		$\boldsymbol{k} \cdot \boldsymbol{p}$	OPW[a]	Pseudo	$\boldsymbol{k} \cdot \boldsymbol{p}$	OPW[a]	Pseudo
$\Gamma^{l}_{25'}$	[111]	0.00	0.00	0.00	0.00	0.00	0.00
$\Gamma^{l}_{2'}$	[111]	0.0728[b]	−0.081	−0.007	0.265[b]	0.164	0.23
Γ_{15}	[111]	0.232[b]	0.231	0.272	0.252[b]	0.238	0.28
Γ^{u}_{1}	[111]	0.571	0.571	0.444	0.520	0.692	0.52
Γ^{l}_{1}	[000]	−0.966	−0.929	−0.950	−0.950	−0.863	−0.97
$\Gamma_{12'}$	[200]	0.770	0.770	0.620	0.710	0.696	0.71
$\Gamma^{u}_{25'}$	[200]	1.25[c]		0.890	0.940		0.94
$\Gamma^{u}_{2'}$	[200]	1.35		0.897	0.990		0.99

[a]F. Herman, in *Proceedings of the International Conference on the Physics of Semiconductors, Paris, 1964* (Dunod Cie, Paris, 1964), p. 3
[b]M. Cardona, J. Phys. Chem. Solids **24**, 1543 (1963)
[c]G. Dresselhaus, A.F. Kip, and C. Kittel, Phys. Rev. **98**, 368 (1955); E.O. Kane, J. Phys. Chem. Solids **1**, 82 (1956)

Table 1.9 The values of momentum matrix elements used for the energy band calculations of germanium and silicon by the $\boldsymbol{k} \cdot \boldsymbol{p}$ perturbation (atomic units)

Momentum matrix elements	Germanium			Silicon		
	$\boldsymbol{k} \cdot \boldsymbol{p}$	Pseudo	c.r.[†]	$\boldsymbol{k} \cdot \boldsymbol{p}$	Pseudo	c.r.[†]
$P = 2\mathrm{i}\langle \Gamma^{l}_{25'} \vert \boldsymbol{p} \vert \Gamma^{l}_{2'} \rangle$	1.360	1.24	1.36[a]	1.200	1.27	1.20[b]
$Q = 2\mathrm{i}\langle \Gamma^{l}_{25'} \vert \boldsymbol{p} \vert \Gamma_{15} \rangle$	1.070	0.99	1.07[a]	1.050	1.05	1.05[b]
$R = 2\mathrm{i}\langle \Gamma^{l}_{25'} \vert \boldsymbol{p} \vert \Gamma_{12'} \rangle$	0.8049	0.75	0.92[c]	0.830	0.74	0.68[d]
$P'' = 2\mathrm{i}\langle \Gamma^{l}_{25'} \vert \boldsymbol{p} \vert \Gamma^{u}_{2'} \rangle$	0.1000	0.09		0.100	0.10	
$P' = 2\mathrm{i}\langle \Gamma^{u}_{25'} \vert \boldsymbol{p} \vert \Gamma^{l}_{2'} \rangle$	0.1715	0.0092		−0.090	−0.10	
$Q' = 2\mathrm{i}\langle \Gamma^{u}_{25'} \vert \boldsymbol{p} \vert \Gamma_{15} \rangle$	−0.752	−0.65		−0.807	−0.64	
$R' = 2\mathrm{i}\langle \Gamma^{u}_{25'} \vert \boldsymbol{p} \vert \Gamma_{12'} \rangle$	1.4357	1.13		1.210	1.21	
$P''' = 2\mathrm{i}\langle \Gamma^{u}_{25'} \vert \boldsymbol{p} \vert \Gamma^{u}_{2'} \rangle$	1.6231	1.30		1.32	1.37	
$T = 2\mathrm{i}\langle \Gamma^{u}_{1} \vert \boldsymbol{p} \vert \Gamma_{15} \rangle$	1.2003	1.11		1.080	1.18	
$T' = 2\mathrm{i}\langle \Gamma^{l}_{1} \vert \boldsymbol{p} \vert \Gamma_{15} \rangle$	0.5323	0.41		0.206	0.34	

[†]Values used to analyze the cyclotron resonance experiments
[a]B.W. Levinger and D.R. Frankl, J. Phys. Chem. Solids **20**, 281 (1961)
[b]J.C. Hensel and G. Feher, Phys. Rev. **129**, 1041 (1963)
[c]Calculated from cyclotron resonance data
[d]Calculated from cyclotron resonance data of reference b

The momentum matrix elements for the $\boldsymbol{k} \cdot \boldsymbol{p}$ perturbation should be multiplied by a factor 2, and thus $P = 1.16$ for Ge ($P = 1.18$ for Si) and $Q = 0.88$ for Ge ($Q = 0.94$ for Si), which are very close to the parameters used by Cardona and Pollak [15] in Table 1.9. The matrix elements P play a very important role in the determination of the valence band structure to be dealt with in Chap. 2 and the optical absorption due to the direct transition discussed in Chap. 4.

The 15 states at the Γ point are classified into $\Gamma_1^l(1)$, $\Gamma_1^u(1)$, $\Gamma_{2'}^l(1)$, $\Gamma_{2'}^u(1)$, $\Gamma_{25'}^l(3)$, $\Gamma_{15}(3)$, $\Gamma_{25'}^u(3)$, $\Gamma_{12'}(2)$, where the superscripts l and u correspond to the lower and upper states of the bands, respectively, and number in the parentheses ( ) is the dimension of the representation. It is evident from the character table of the group theory that $\Gamma_{25'}$ is 3-dimensional with three eigenstates. The energy eigenstates are estimated roughly by the pseudopotential method or other approximations [15] as stated above, and shown in Table 1.8. The momentum operator $\boldsymbol{p}$ has the same symmetry as Γ_{15} and thus the matrix elements of $\boldsymbol{k} \cdot \boldsymbol{p}$ for the 15 eigenstates have non-zero components, as shown in the following equations and in Table 1.9.

$$P = 2\mathrm{i}\langle \Gamma_{25'}^l | \boldsymbol{p} | \Gamma_{2'}^l \rangle \,, \tag{1.135a}$$
$$Q = 2\mathrm{i}\langle \Gamma_{25'}^l | \boldsymbol{p} | \Gamma_{15} \rangle \,, \tag{1.135b}$$
$$R = 2\mathrm{i}\langle \Gamma_{25'}^l | \boldsymbol{p} | \Gamma_{12'} \rangle \,, \tag{1.135c}$$
$$P'' = 2\mathrm{i}\langle \Gamma_{25'}^l | \boldsymbol{p} | \Gamma_{2'}^u \rangle, \tag{1.135d}$$
$$P' = 2\mathrm{i}\langle \Gamma_{25'}^u | \boldsymbol{p} | \Gamma_{2'}^l \rangle \,, \tag{1.135e}$$
$$Q' = 2\mathrm{i}\langle \Gamma_{25'}^u | \boldsymbol{p} | \Gamma_{15} \rangle \,, \tag{1.135f}$$
$$R' = 2\mathrm{i}\langle \Gamma_{25'}^u | \boldsymbol{p} | \Gamma_{12'} \rangle \,, \tag{1.135g}$$
$$P''' = 2\mathrm{i}\langle \Gamma_{25'}^u | \boldsymbol{p} | \Gamma_{2'}^u \rangle \,, \tag{1.135h}$$
$$T = 2\mathrm{i}\langle \Gamma_1^u | \boldsymbol{p} | \Gamma_{15} \rangle \,, \tag{1.135i}$$
$$T' = 2\mathrm{i}\langle \Gamma_1^l | \boldsymbol{p} | \Gamma_{15} \rangle \,. \tag{1.135j}$$

The factor 2 of the momentum matrix elements in Table 1.9 and (1.135a) $\sim$ (1.135j) is understood from $\boldsymbol{k} \cdot \boldsymbol{p}$ perturbation Hamiltonian given by (1.119) and (1.120), where energy is in Rydberg [Ry] and the length in unit a_B (Bohr radius) as discussed above. We note here the matrix elements of $\langle \Gamma_{25'} | \boldsymbol{k} \cdot \boldsymbol{p} | \Gamma_{12'} \rangle$ used in the present calculations. Using the character table and the basis function we may deduce the non-vanishing matrix elements for the $\boldsymbol{k} \cdot \boldsymbol{p}$ theory. In the present analysis the matrix elements between $\Gamma_{25'}$ and $\Gamma_{12'}$ are evaluated by using the property of the character table in Table 1.4,

$$\Gamma_{25'} \times \Gamma_{12'} = \Gamma_{15} + \Gamma_{25} \,, \qquad \Gamma_{15} \times \Gamma_{12'} = \Gamma_{25'} + \Gamma_{15'} \,. \tag{1.136}$$

The wave vectors $\boldsymbol{p} = [p_x, p_y, p_z]$ have the same property as $[x, y, z]$, and thus the representation is Γ_{15}. Therefore we find non-zero matrix elements $\langle \Gamma_{25'} | \boldsymbol{p} | \Gamma_{12'} \rangle$ and $\langle \Gamma_{15'} | \boldsymbol{p} | \Gamma_{12'} \rangle$. However, $\Gamma_{15'}$ states belong to the plane waves $[\pm 3, \pm 1, \pm 1]$ and the free electron energy is $11(2\pi/a)^2$ which is much higher than the upper states of $\Gamma_{25'}[200]$ with the free electron energy $4(2\pi/a)^2$, and thus $\Gamma_{15'}$ states may be disregarded. Therefore the following matrix elements for the states $\Gamma_{12'}$ are included in the 15×15 $\boldsymbol{k} \cdot \boldsymbol{p}$ perturbation (see (1.143a) $\sim$ (1.143f) for a detailed treatment).

$$2\mathrm{i}\langle \Gamma_{25'}^l(X) | \boldsymbol{p} | \Gamma_{12'} \rangle = R \,, \tag{1.137a}$$
$$2\mathrm{i}\langle \Gamma_{25'}^u(Y) | \boldsymbol{p} | \Gamma_{12'} \rangle = R' \,. \tag{1.137b}$$

Using the parameters listed in Table 1.9 and following the procedures below, the energy band structures of Ge and Si are easily calculated. First, we calculate the

15×15 matrix elements of the $\boldsymbol{k} \cdot \boldsymbol{p}$ Hamiltonian, and second diagonalize the matrix to obtain the energy eigenvalues and their eigenstates at $\boldsymbol{k}$ of the Brillouin zone. The matrix of the $\boldsymbol{k} \cdot \boldsymbol{p}$ Hamiltonian has 15×15 *complex* elements. When we include the spin–orbit interaction, the matrix of $\boldsymbol{k} \cdot \boldsymbol{p}$ Hamiltonian is given by 30×30 *complex* elements.

1.7.3 15–band k · p Method

It is very important to point out here that the 15×15 matrices for Ge and Si are factorized into smaller matrices when we use group theoretical consideration, as shown by Cardona and Pollak [15]. For simplicity we consider the energy bands along the $\langle 100\rangle$, $\langle 110\rangle$ and $\langle 111\rangle$ directions of the $\boldsymbol{k}$-vector. The direction $\langle 100\rangle$ starts from the Γ point and ends at the X point along the Δ axis, and the $\langle 110\rangle$ direction is from the Γ point to the K point along the Σ axis, while the $\langle 111\rangle$ direction is from the Γ point to the L point along the Λ axis. With the help of the compatibility relation given in Table 1.10 (see [13]) the matrix elements of the $\boldsymbol{k} \cdot \boldsymbol{p}$ Hamiltonian are factorized in several groups of smaller matrices. In the following we use atomic units as stated above, and thus $\hbar^2 k^2/2m$ and $p^2/2m$ are expressed as k^2 and p^2, respectively. In the following we show the factorized matrices of the $\boldsymbol{k} \cdot \boldsymbol{p}$ Hamiltonian in the [100], [110] and [111] directions.

1. [100] **direction**

(a) Δ_5 bands

From Table 1.10 we find that three Γ bands exist, but the bands in the parentheses ( ) are neglected because of their high energy states.

Δ_5 bands: $(\Gamma_{15'})$, $\Gamma^u_{25'}$, $\Gamma^l_{25'}$, Γ_{15}

The matrix elements for these 3 bands are

Table 1.10 Compatibility relations

Γ_1	Γ_2	Γ_{12}	$\Gamma_{15'}$	$\Gamma_{25'}$	$\Gamma_{1'}$	$\Gamma_{2'}$	$\Gamma_{12'}$	Γ_{15}	Γ_{25}
Δ_1	Δ_2	$\Delta_1\Delta_2$	$\Delta_{1'}\Delta_5$	$\Delta_{2'}\Delta_5$	$\Delta_{1'}$	$\Delta_{2'}$	$\Delta_{1'}\Delta_{2'}$	$\Delta_1\Delta_5$	$\Delta_2\Delta_5$
Λ_1	Λ_2	Λ_3	$\Lambda_2\Lambda_3$	$\Lambda_1\Lambda_3$	Λ_2	Λ_1	Λ_3	$\Lambda_1\Lambda_3$	$\Lambda_2\Lambda_3$
Σ_1	Σ_4	$\Sigma_1\Sigma_4$	$\Sigma_2\Sigma_3\Sigma_4$	$\Sigma_1\Sigma_2\Sigma_3$	Σ_2	Σ_3	$\Sigma_2\Sigma_3$	$\Sigma_1\Sigma_3\Sigma_4$	$\Sigma_1\Sigma_2\Sigma_4$
X_1	X_2	X_3	X_4	X_5	$X_{1'}$	$X_{2'}$	$X_{3'}$	$X_{4'}$	$X_{5'}$
Δ_1	Δ_2	$\Delta_{2'}$	$\Delta_{1'}$	Δ_5	$\Delta_{1'}$	$\Delta_{2'}$	Δ_2	Δ_1	Δ_5
Z_1	Z_1	Z_4	Z_4	Z_2Z_3	Z_2	Z_2	Z_3	Z_3	Z_1Z_4
S_1	S_4	S_1	S_4	S_2S_3	S_2	S_3	S_2	S_3	S_1S_4
M_1	M_2	M_3	M_4	M_5	$M_{1'}$	$M_{2'}$	$M_{3'}$	$M_{4'}$	$M_{5'}$
Σ_1	Σ_1	Σ_4	Σ_4	$\Sigma_2\Sigma_3$	Σ_2	Σ_2	Σ_3	Σ_3	$\Sigma_1\Sigma_4$
Z_1	Z_1	Z_3	Z_3	Z_2Z_4	Z_2	Z_2	Z_4	Z_4	Z_1Z_3
T_1	T_2	$T_{2'}$	$T_{1'}$	T_5	$T_{1'}$	$T_{2'}$	T_2	T_1	T_5

$$\begin{array}{ccc} |\Gamma_{25}^{l}\rangle & |\Gamma_{15}\rangle & |\Gamma_{25'}^{u}\rangle \end{array}$$
$$\left|\begin{array}{ccc} k_x^2 & Qk_x & 0 \\ Qk_x & \mathcal{E}(\Gamma_{15}) + k_x^2 & Q'k_x \\ 0 & Q'k_x & \mathcal{E}(\Gamma_{25'}^{u}) + k_x^2 \end{array}\right|. \tag{1.138}$$

(b) Δ_1 bands

Δ_1 bands: Γ_1^l, Γ_1^u, (Γ_{12}), Γ_{15}

The matrix elements for these three bands are

$$\begin{array}{ccc} |\Gamma_{15}\rangle & |\Gamma_1^{u}\rangle & |\Gamma_1^{l}\rangle \end{array}$$
$$\left|\begin{array}{ccc} \mathcal{E}(\Gamma_{15}) + k_x^2 & Tk_x & T'k_x \\ Tk_x & \mathcal{E}(\Gamma_1^u) + k_x^2 & 0 \\ T'k_x & 0 & \mathcal{E}(\Gamma_1^l) + k_x^2 \end{array}\right|. \tag{1.139}$$

(c) $\Delta_{2'}$ bands

$\Delta_{2'}$ bands: $\Gamma_{25'}^l$, $\Gamma_{25'}^u$, $\Gamma_{2'}^l$, $\Gamma_{2'}^u$, $\Gamma_{12'}$

The matrix elements for these five bands are

$$\begin{array}{ccccc} |\Gamma_{2'}^{l}\rangle & |\Gamma_{25'}^{l}\rangle & |\Gamma_{12'}\rangle & |\Gamma_{25'}^{u}\rangle & |\Gamma_{2'}^{u} \end{array}$$
$$\left|\begin{array}{ccccc} \mathcal{E}(\Gamma_{2'}^l) + k_x^2 & Pk_x & 0 & P'k_x & 0 \\ Pk_x & k_x^2 & \sqrt{2}Rk_x & 0 & P''k_x \\ 0 & \sqrt{2}Rk_x & \mathcal{E}(\Gamma_{12'}) + k_x^2 & \sqrt{2}R'k_x & 0 \\ P'k_x & 0 & \sqrt{2}R'k_x & \mathcal{E}(\Gamma_{25'}^u) + k_x^2 & P'''k_x \\ 0 & P''k_x & 0 & P'''k_x & \mathcal{E}(\Gamma_{2'}^u) + k_x^2 \end{array}\right|, \tag{1.140}$$

where the factor $\sqrt{2}$ of $\sqrt{2}Rk_x$ and $\sqrt{2}R'k_x$ arises from the definition of $\Gamma_{12'}$ states ($\Gamma_{12'}(1)$ and $\Gamma_{12'}(2)$) as given by Cardona and Pollak [15] and the $\Gamma_{12'}(2)$ state does not interact with any other state in the [100] direction and behaves like a free electron band (see (1.143a), (1.143b), (1.144)).

2. [110] **direction**

From Table 1.10, Σ_1, Σ_4, Σ_3, Σ_2 are included in this direction.
(a) Σ_1 bands: Γ_1^l, Γ_1^u, (Γ_{12}), $\Gamma_{25'}^l$, $\Gamma_{25'}^u$, Γ_{15} (: 5 bands)
(b) Σ_4 bands: (Γ_2), (Γ_{12}), $(\Gamma_{15'})$, Γ_{15}, (Γ_{25}) (: 1 band)
(c) Σ_3 bands: $(\Gamma_{15'})$, $\Gamma_{25'}^l$, $\Gamma_{25'}^u$, $\Gamma_{2'}^l$, $\Gamma_{2'}^u$, $\Gamma_{12'}$, Γ_{15} (: 6 bands)
(d) Σ_2 bands: $(\Gamma_{15'})$, $\Gamma_{25'}l$, $\Gamma_{25'}^u$, $(\Gamma_{1'})$, $\Gamma_{12'}$, (Γ_{25}) (: 3 bands)
Therefore, for the bands in the [110] direction, Σ bands, 15×15 matrix elements results in irreducible matrix of 6×6, 5×5, 3×3, 1×1.

3. [111] **direction**

From Table 1.10, Λ_1, Λ_3 are included in this direction.
(a) 7 bands of Λ_1 bands: Γ_1^l, Γ_1^u, $\Gamma_{25'}^l$, $\Gamma_{25'}^u$, $\Gamma_{2'}^l$, $\Gamma_{2'}^u$, Γ_{15}
(b) 4 bands of Λ_3 bands: (Γ_{12}), $(\Gamma_{15'})$, $\Gamma_{25'}^l$, $\Gamma_{25'}^u$, $\Gamma_{12'}$, Γ_{15}, (Γ_{25})

and the Λ bands in the [111] direction are classified in 7×7 and 4×4 irreducible matrix.

It should be noted here that above 15×15 $\boldsymbol{k} \cdot \boldsymbol{p}$ matrix is easily extended to include k_y and k_z components. However we have to take account of correct symmetry of the $\Gamma_{12'}(1)$ and $\Gamma_{12'}(2)$ states, which is done by extending the method of Dresselhaus [14] and convert them into the representations of Cardona and Pollak [15] as follows. Using the definition of Dresselhaus et al. and following their procedures, we obtain

$$\langle \Gamma_{25'}(X)|p_x|\gamma_1^-\rangle = R\,, \tag{1.141a}$$
$$\langle \Gamma_{25'}(X)|p_x|\gamma_2^-\rangle = -R\,, \tag{1.141b}$$
$$\langle \Gamma_{25'}(Y)|p_y|\gamma_1^-\rangle = \omega R\,, \tag{1.141c}$$
$$\langle \Gamma_{25'}(Y)|p_y|\gamma_2^-\rangle = -\omega^2 R\,, \tag{1.141d}$$
$$\langle \Gamma_{25'}(Z)|p_z|\gamma_1^-\rangle = \omega^2 R\,, \tag{1.141e}$$
$$\langle \Gamma_{25'}(Z)|p_z|\gamma_2^-\rangle = -\omega R\,, \tag{1.141f}$$

where we have to note that the matrix elements R defined by Dresselhaus, Kip and Kittel [14] (R_{DKK}) and R defined by Cardona and Pollak [15] (R_{CP}) are related by $R_{\text{CP}} = 2R_{\text{DKK}}$. When we choose the eigenstates $\Gamma_{12'}$ defined by Cardona and Pollak [15];

$$\Gamma_{12'}(1) = \frac{1}{\sqrt{2}}(\gamma_1^- - \gamma_2^-)\,, \quad \Gamma_{12'}(2) = \frac{1}{\sqrt{2}}(\gamma_1^- + \gamma_2^-)\,, \tag{1.142}$$

we obtain the following results

$$\langle \Gamma_{25'}|p_x|\Gamma_{12'}(1)\rangle = \sqrt{2}R\,, \tag{1.143a}$$
$$\langle \Gamma_{25'}|p_x|\Gamma_{12'}(2)\rangle = 0\,, \tag{1.143b}$$
$$\langle \Gamma_{25'}|p_y|\Gamma_{12'}(1)\rangle = (\omega + \omega^2)R/\sqrt{2} = -R/\sqrt{2}\,, \tag{1.143c}$$
$$\langle \Gamma_{25'}|p_y|\Gamma_{12'}(2)\rangle = (\omega - \omega^2)R/\sqrt{2} = \omega(1-\omega)R/\sqrt{2} = \mathrm{i}R\sqrt{3/2}\,, \tag{1.143d}$$
$$\langle \Gamma_{25'}|p_z|\Gamma_{12'}(1)\rangle = (\omega^2 + \omega)R/\sqrt{2} = -R/\sqrt{2}\,, \tag{1.143e}$$
$$\langle \Gamma_{25'}|p_z|\Gamma_{12'}(2)\rangle = (\omega^2 - \omega)R/\sqrt{2} = \omega(\omega - 1)R/\sqrt{2} = -\mathrm{i}R\sqrt{3/2}\,, \tag{1.143f}$$

where ω is the solutions of $\omega^3 = 1$ (exclude the solution $\omega = 1$) or the solutions of $\omega^2 + \omega + 1 = 0$. The above results are obtained by using the solution $\omega = (-1 + \mathrm{i}\sqrt{3})/2$. When we choose the solution $\omega = (-1 - \mathrm{i}\sqrt{3})/2$, the sign of the imaginary part is changed, but the energy band calculations give the same result. In addition we have to note that $\omega = 1$, one of the solutions of $\omega^3 = 1$, does not give a correct energy bands. This is because the solution $\omega = 1$ does not represent the correct symmetry of γ_1^- and γ_2^-. Here we show the 15×15 $\boldsymbol{k} \cdot \boldsymbol{p}$ Hamiltonian matrix without the spin–orbit interaction (antisymmetric potential terms for zinc blende crystals are included);

$$
\begin{array}{|cccccc}
|\Gamma^l_{25'}(X)\rangle & |\Gamma^l_{25'}(Y)\rangle & |\Gamma^l_{25'}(Z)\rangle & |\Gamma_{15}(x)\rangle & |\Gamma_{15}(y)\rangle & |\Gamma_{15}(z)\rangle \\
E(\Gamma^l_{25'})+k^2 & 0 & 0 & -\mathrm{i}V_1^- & Qk_z & Qk_y \\
0 & E(\Gamma^l_{25'})+k^2 & 0 & Qk_z & -\mathrm{i}V_1^- & Qk_x \\
0 & 0 & E(\Gamma^l_{25'})+k^2 & Qk_y & Qk_x & -\mathrm{i}V_1^- \\
\mathrm{i}V_1^- & Qk_z & Qk_y & E(\Gamma_{15})+k^2 & 0 & 0 \\
Qk_z & \mathrm{i}V_1^- & Qk_x & 0 & E(\Gamma_{15})+k^2 & 0 \\
Qk_y & Qk_x & \mathrm{i}V_1^- & 0 & 0 & E(\Gamma_{15})+k^2 \\
0 & 0 & 0 & -\mathrm{i}V_4^- & Q'k_z & Q'k_y \\
0 & 0 & 0 & Q'k_z & -\mathrm{i}V_4^- & Q'k_x \\
0 & 0 & 0 & Q'k_y & Q'k_x & -\mathrm{i}V_4^- \\
\sqrt{2}Rk_x & -(R/\sqrt{2})k_y & -(R/\sqrt{2})k_z & 0 & 0 & 0 \\
0 & -\mathrm{i}R\sqrt{3/2}k_y & \mathrm{i}R\sqrt{3/2}k_z & 0 & 0 & 0 \\
Pk_x & Pk_y & Pk_z & 0 & 0 & 0 \\
P''k_x & P''k_y & P''k_z & 0 & 0 & 0 \\
0 & 0 & 0 & Tk_x & Tk_y & Tk_z \\
0 & 0 & 0 & T'k_x & T'k_y & T'k_z
\end{array}
$$

$$
\begin{array}{ccccc}
|\Gamma^u_{25'}(X)\rangle & |\Gamma^u_{25'}(Y)\rangle & |\Gamma^u_{25'}(Z)\rangle & |\Gamma_{12'}(1)\rangle & |\Gamma_{12'}(2)\rangle \\
0 & 0 & 0 & \sqrt{2}Rk_x & 0 \\
0 & 0 & 0 & -(R/\sqrt{2})k_y & \mathrm{i}R\sqrt{3/2}k_y \\
0 & 0 & 0 & -(R/\sqrt{2})k_z & -\mathrm{i}R\sqrt{3/2}k_z \\
\mathrm{i}V_4^- & Q'k_z & Q'k_y & 0 & 0 \\
Q'k_z & \mathrm{i}V_4^- & Q'k_x & 0 & 0 \\
Q'k_y & Q'k_x & \mathrm{i}V_4^- & 0 & 0 \\
E(\Gamma^u_{25'})+k^2 & 0 & 0 & \sqrt{2}R'k_x & 0 \\
0 & E(\Gamma^u_{25'})+k^2 & 0 & -(R'/\sqrt{2})k_y & \mathrm{i}R'\sqrt{3/2}k_y \\
0 & 0 & E(\Gamma^u_{25'})+k^2 & -(R'/\sqrt{2})k_z & -\mathrm{i}R'\sqrt{3/2}k_z \\
\sqrt{2}R'k_x & -(R'/\sqrt{2}k_y & -(R'/\sqrt{2})k_z & E(\Gamma_{12'})+k^2 & 0 \\
0 & -\mathrm{i}R'\sqrt{3/2}k_y & \mathrm{i}R'\sqrt{3/2}k_z & 0 & E(\Gamma_{12'})+k^2 \\
P'k_x & P'k_y & P'k_z & 0 & 0 \\
P'''k_x & P'''k_y & P'''k_z & 0 & 0 \\
0 & 0 & 0 & 0 & 0 \\
0 & 0 & 0 & 0 & 0
\end{array}
$$

$$
\begin{array}{cccc|}
|\Gamma^l_{2'}(xyz)\rangle & |\Gamma^u_{2'}(xyz)\rangle & |\Gamma^u_1\rangle & |\Gamma^l_1\rangle \\
Pk_x & P''k_x & 0 & 0 \\
Pk_y & P''k_y & 0 & 0 \\
Pk_z & P''k_z & 0 & 0 \\
0 & 0 & Tk_x & T'k_x \\
0 & 0 & Tk_y & T'k_y \\
0 & 0 & Tk_z & T'k_z \\
P'k_x & P'''k_x & 0 & 0 \\
P'k_y & P'''k_y & 0 & 0 \\
P'k_z & P'''k_z & 0 & 0 \\
0 & 0 & 0 & 0 \\
0 & 0 & 0 & 0 \\
E(\Gamma^l_{2'})+k^2 & 0 & \mathrm{i}V_2^- & \mathrm{i}V_3^- \\
0 & E(\Gamma^u_{2'})+k^2 & \mathrm{i}V_5^- & \mathrm{i}V_6^- \\
-\mathrm{i}V_2^- & -\mathrm{i}V_5^- & E(\Gamma^u_1)+k^2 & 0 \\
-\mathrm{i}V_3^- & -\mathrm{i}V_6^- & 0 & E(\Gamma^l_1)+k^2
\end{array}
\tag{1.144}
$$

It should be noted here that $\langle \Gamma_{12'}(2)|p_y|\Gamma_{25'}(Y)\rangle$ is given by the complex conjugate of $\langle \Gamma_{25'}(Y)|p_y|\Gamma_{12'}(2)\rangle$. Energy bands without the spin–orbit interaction are easily

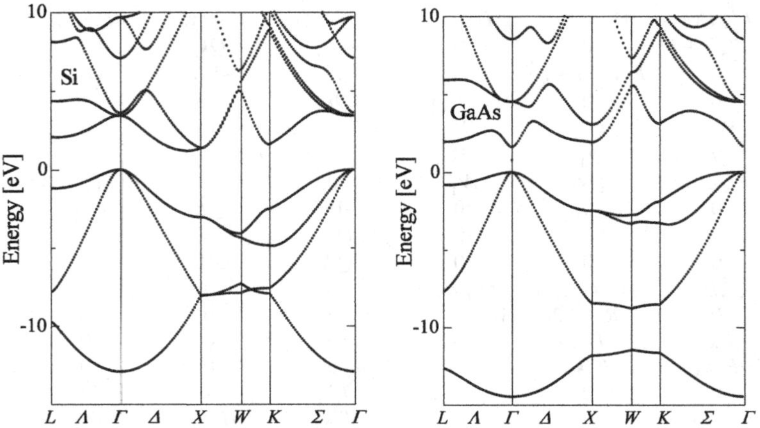

Fig. 1.17 Energy band structure of Si and GaAs calculated by 15 band $\boldsymbol{k} \cdot \boldsymbol{p}$ perturbation without spin–orbit interaction. The results are obtained by solving secular equation (1.144) with the parameters given in Table 1.11

calculated by solving (1.144) in any directions of the Brillouin zone. As an example calculated energy band structure of Si along the L, Γ, X, W, K to Γ points is shown in Fig. 1.17. Also in the figure the energy band structure of GaAs is shown, where the antisymmetric potential is taken account (see 1.7.4).

1.7.4 Antisymmetric Potentials for Zinc Blende Crystals

We have to note here the definition of the matrix elements of the anti-antisymmetric potentials. As discussed in Sect. 1.6, a diamond type crystal has inversion symmetry and thus $V^{\mathrm{A}}(\boldsymbol{G}) = 0$, while a zinc blende type crystal has no inversion symmetry and thus $V^{\mathrm{A}}(\boldsymbol{G}) \neq 0$. In order to extend the $\boldsymbol{k} \cdot \boldsymbol{p}$ Hamiltonian for a zinc blende crystal we have to evaluate the matrix elements of the anti-symmetric potential as shown in (1.144). Here we present how to evaluate approximate values of the matrix elements of the anti-symmetric potential. First, we obtain non-vanishing matrix elements of the 15 plane waves of $[0, 0, 0]$, $[\pm 1, \pm 1, \pm 1]$, and $[\pm 2, 0, 0]$ which are classified as Γ_1^l, $\Gamma_{25'}^l$, Γ_{15}, $\Gamma_{12'}$, $\Gamma_{2'}^l$, Γ_1^u, $\Gamma_{25'}^u$, $\Gamma_{12'}$, and $\Gamma_{2'}^u$ as shown in Table 1.6. In Table 1.4, an inversion operation is expressed by J and representation Γ_i has the inversion symmetry when $J > 0$ but no inversion symmetry when $J < 0$. The product of $\Gamma_i \times \Gamma_j$ has the same symmetry property. Noting the anti-symmetric potential V^- has negative sign for the inversion operation, non-vanishing matrix elements $\langle \Gamma_i | V^- | \Gamma_j \rangle$ are $\langle \Gamma_{15} | V^- | \Gamma_{25'} \rangle$ and $\langle \Gamma_{2'} | V^- | \Gamma_1 \rangle$. When we use the notations of Pollak, Higginbotham, and Cardona [18], these are given by the following relations,

$$V_1^- = \langle \Gamma_{15} | V^- | \Gamma_{25'}^l \rangle = V_4^{\mathrm{A}} , \tag{1.145a}$$
$$V_2^- = \langle \Gamma_{2'}^l | V^- | \Gamma_1^u \rangle = -3V_4^{\mathrm{A}} , \tag{1.145b}$$
$$V_3^- = \langle \Gamma_{2'}^l | V^- | \Gamma_1^l \rangle = 2V_3^{\mathrm{A}} , \tag{1.145c}$$
$$V_4^- = \langle \Gamma_{15} | V^- | \Gamma_{25'}^u \rangle = \sqrt{2} \left(V_3^{\mathrm{A}} - V_{11}^{\mathrm{A}} \right) , \tag{1.145d}$$
$$V_5^- = \langle \Gamma_{2'}^u | V^- | \Gamma_1^u \rangle = -(\sqrt{6}/3) \left[(4V_3^{\mathrm{A}} + 2V_{11}^{\mathrm{A}} \right] , \tag{1.145e}$$
$$V_6^- = \langle \Gamma_{2'}^u | V^- | \Gamma_1^l \rangle = (2\sqrt{6}/3) V_4^{\mathrm{A}} , \tag{1.145f}$$

where the last terms of the above equations are evaluated from the pseudopotentials in Table 1.2 and the relations defined by (1.109a) $\sim$ (1.109d) with linear combinations of the plane waves (1.126) and (1.128 $\sim$ (1.128). As an example we evaluate these terms for GaAs using the pseudopotentials given in Table 1.2. We obtain $V_1^- = 0.05\,(0.12652)$, $V_2^- = -0.15\,(-0.24791)$, $V_3^- = 0.14\,(0.38210)$, $V_4^- = 0.0849\,(0.12297)$, $V_5^- = -0.245\,(-0.34820)$, and $V_6^- = 0.0816\,(0.0)$, where the values in the parentheses are determined form the energy band calculations and summarized in Table 1.11 for several zinc blende type semiconductors. Since the term V_6^- corresponds to interactions between very distant atomic orbitals $\Gamma_{2'}^u$ and Γ_1^l (difference in the free electron energy is $4(2\pi/a)^2 = 1.38$ [a.u.] for GaAs), we may safely assume that $V_6^- = 0.0$ [18].

When we include the spin–orbit interaction, the above equation leads to 30×30 $\boldsymbol{k} \cdot \boldsymbol{p}$ *complex* matrix because each state is doubled with spin–up $\uparrow\rangle$ and spin–down $\downarrow\rangle$. In Chap. 2, we will discuss the second order perturbation of the $\boldsymbol{k} \cdot \boldsymbol{p}$ Hamiltonian and the parameters defined by Dresselhaus et al. [14] and by Luttinger [2.2], where the above results are used to evaluate the contributions from γ_1^- and γ_2^- ($\Gamma_{12'}(1)$ and $\Gamma_{12'}(2)$).

1.7.5 Spin–orbit Interaction Hamiltonian

When we take account of the spin–orbit interaction, the eigenstates are doubled with spin–up and spin–down as discussed above, and then we have to solve 30×30 *complex* matrix. In the following we choose the wave functions $|X \uparrow\rangle$, $|X \downarrow\rangle$, $|Y \uparrow\rangle$, $|Y \downarrow\rangle$, $|Z \uparrow\rangle$, $|Z \downarrow\rangle$ for $\Gamma_{25'}^l$, and $|x \uparrow\rangle$, $|x \downarrow\rangle$, $|y \uparrow\rangle$, $|y \downarrow\rangle$, $|z \uparrow\rangle$, $|z \downarrow\rangle$ for Γ_{15}. This approximation leads to 30×30 *complex* matrix with the spin–orbit interaction. In this subsection we formulate the spin–orbit matrix elements using these eigenstates. Spin–orbit interaction is also discussed in 2.3 to analyze the valence band structure, in addition to describe the effective mass and effective g factor (Landé g factor) of the conduction band. Evaluation of the matrix elements of spin–orbit Hamiltonian is given in detail in 2.3.

We put the spin–orbit interaction term (H.18) of Appendix H into (1.112), and obtain

$$H_{\mathrm{so}} = \frac{\hbar}{4m^2c^2} [\nabla \times \boldsymbol{p}] \cdot \boldsymbol{\sigma} + \frac{\hbar^2}{4m^2c^2} [\nabla \times \boldsymbol{k}] \cdot \boldsymbol{\sigma} , \tag{1.146}$$

which should be added to the terms in the brackets of the left hand side of (1.116). The second $\boldsymbol{k}$–dependent term is very small compared to the first $\boldsymbol{k}$–independent term (see Kane [12]) and thus only the first term is considered in this text. Thus the $\boldsymbol{k}$–independent term of the spin–orbit Hamiltonian is rewritten as

$$H_{\text{so}} \propto \boldsymbol{L} \cdot \boldsymbol{\sigma} = (\boldsymbol{r} \times \boldsymbol{p}) \cdot \boldsymbol{\sigma} = -\mathrm{i}\hbar(\boldsymbol{r} \times \nabla) \cdot \boldsymbol{\sigma} \tag{1.147}$$

and therefore we find that

$$-\mathrm{i}\hbar(\boldsymbol{r} \times \nabla) \cdot \boldsymbol{\sigma} = -\mathrm{i}\hbar\left[\left(y\frac{\partial}{\mathrm{d}z} - z\frac{\partial}{\mathrm{d}y}\right)\sigma_x + \left(z\frac{\partial}{\mathrm{d}x} - x\frac{\partial}{\mathrm{d}z}\right)\sigma_y + \left(x\frac{\partial}{\mathrm{d}y} - y\frac{\partial}{\mathrm{d}x}\right)\sigma_z\right], \tag{1.148}$$

where $\boldsymbol{\sigma}$ is Pauli spin operator[1] and the matrix elements are evaluated by using the basis functions given in Table 1.5.

$$\langle X(\Gamma_{25'}^l) \uparrow | H_{\text{so}} | Y(\Gamma_{25'}^l) \uparrow \rangle = \mathrm{i}\Delta_{25'}^l/3\,, \tag{1.149a}$$
$$\langle X(\Gamma_{25'}^l) \uparrow | H_{\text{so}} | Z(\Gamma_{25'}^l) \downarrow \rangle = \Delta_{25'}^l/3\,, \tag{1.149b}$$
$$\langle Y(\Gamma_{25'}^l) \uparrow | H_{\text{so}} | Z(\Gamma_{25'}^l) \downarrow \rangle = \mathrm{i}\Delta_{25'}^l/3\,. \tag{1.149c}$$

In the same manner the spin–orbit interaction for Γ_{15} is written as

$$\langle x(\Gamma_{15}) \uparrow | H_{\text{so}} | y(\Gamma_{15}) \uparrow \rangle = \mathrm{i}\Delta_{15}/3\,, \tag{1.150a}$$
$$\langle x(\Gamma_{15}) \uparrow | H_{\text{so}} | z(\Gamma_{15}) \downarrow \rangle = \Delta_{15}/3\,, \tag{1.150b}$$
$$\langle y(\Gamma_{15}) \uparrow | H_{\text{so}} | z(\Gamma_{15}) \downarrow \rangle = \mathrm{i}\Delta_{15}/3\,, \tag{1.150c}$$

and the antisymmetric term of the spin–orbit interaction is treated as

$$\langle X(\Gamma_{25'}^l) \uparrow | H_{\text{so}} | y(\Gamma_{15}) \uparrow \rangle = \mathrm{i}\Delta^-/3\,, \tag{1.151a}$$
$$\langle X(\Gamma_{25'}^l) \uparrow | H_{\text{so}} | z(\Gamma_{15}) \uparrow \rangle = +\Delta^-/3\,, \tag{1.151b}$$
$$\langle Y(\Gamma_{25'}^l) \uparrow | H_{\text{so}} | z(\Gamma_{15}) \downarrow \rangle = \mathrm{i}\Delta^-/3\,. \tag{1.151c}$$

The spin–orbit interaction in the valence bands is discussed in Chap. 2. Here we will show the present treatment leads to the same results. The $\Gamma_{25'}^l$ valence bands ($|X\rangle$, $|Y\rangle$, $|X\rangle$) are triply–degenerate at the Γ point ($\boldsymbol{k} = 0$). The degenerate $\Gamma_{25'}^l$ bands split into doubly–degenerate heavy hole and light hole bands, and the spin–orbit–split–off band as discussed in Chap. 2. Here it is shown that the above matrix elements of the spin–orbit Hamiltonian for the valence bands $\Gamma_{25'}^l$ give the same results of the spin–orbit splitting dealt in Chap. 2. The matrix of the spin–orbit Hamiltonian for the valence band $\Gamma_{25'}^l$ is written as

$$\begin{array}{ccc} |X\uparrow\rangle & |Y\uparrow\rangle & |Z\downarrow\rangle \end{array}$$
$$\left|\begin{array}{ccc} 0 & \mathrm{i}\Delta_{25'}^l/3 & \Delta_{25'}^l/3 \\ -\mathrm{i}\Delta_{25'}^l/3 & 0 & \mathrm{i}\Delta_{25'}^l/3 \\ \Delta_{25'}^l/3 & -\mathrm{i}\Delta_{25'}^l/3 & 0 \end{array}\right| \left|\begin{array}{c} |X\uparrow\rangle \\ |Y\uparrow\rangle \\ |Z\downarrow\rangle \end{array}\right|, \tag{1.152}$$

[1] See (2.50) of Chap. 2 for the definition and (H.33c) of Appendix H for the matrix elements.

and diagonalization results in

$$\begin{vmatrix} (1/3)\Delta^l_{25'} & 0 & 0 \\ 0 & (1/3)\Delta^l_{25'} & 0 \\ 0 & 0 & -(2/3)\Delta^l_{25'} \end{vmatrix} \begin{vmatrix} u_{v1} \\ u_{v2} \\ u_{v3} \end{vmatrix} . \tag{1.153}$$

Therefore the spin–orbit splitting of $\Gamma^l_{25'}$ bands is $\Delta_0 = \Delta^l_{25'}$ and the corresponding eigenfunctions are

$$u_{v1} = \frac{\mathrm{i}}{\sqrt{2}}(X - \mathrm{i}Y)\uparrow , \tag{1.154a}$$

$$u_{v2} = \frac{1}{\sqrt{2}}(X\uparrow + Z\downarrow) , \tag{1.154b}$$

$$u_{v3} = \frac{-1}{\sqrt{3}}[(X + \mathrm{i}Y)\uparrow - Z\downarrow] . \tag{1.154c}$$

Using these results we may obtain the matrix elements of the spin–orbit Hamiltonian $H_{\rm so}$ for the $\Gamma_{25'}$ states:

$$\langle \Gamma_{25'}|H_{so}|\Gamma_{25'}\rangle = \frac{\Delta}{3} \begin{array}{c} |X\uparrow\rangle \; |Y\uparrow\rangle \; |Z\downarrow\rangle \; |X\downarrow\rangle \; |Y\downarrow\rangle \; |Z\uparrow\rangle \\ \begin{vmatrix} 0 & \mathrm{i} & 1 & 0 & 0 & 0 \\ -\mathrm{i} & 0 & \mathrm{i} & 0 & 0 & 0 \\ 1 & -\mathrm{i} & 0 & 0 & 0 & 0 \\ 0 & 0 & 0 & 0 & \mathrm{i} & 1 \\ 0 & 0 & 0 & -\mathrm{i} & 0 & \mathrm{i} \\ 0 & 0 & 0 & 1 & -\mathrm{i} & 0 \end{vmatrix} \end{array} , \tag{1.155}$$

and the matrix is separated by two 3×3 matrices and diagonalization of the matrix gives the following eigenvalues and eigen functions,

$$\langle \Gamma_{25'}|H_{so}|\Gamma_{25'}\rangle = \frac{\Delta}{3} \begin{array}{c} |v_{v1}\rangle \; |v_{v2}\rangle \; |v_{v3}\rangle \; |v_{v4}\rangle \; |v_{v5}\rangle \; |v_{v6}\rangle \\ \begin{vmatrix} 1 & 0 & 0 & 0 & 0 & 0 \\ 0 & 1 & 0 & 0 & 0 & 0 \\ 0 & 0 & 1 & 0 & 0 & 0 \\ 0 & 0 & 0 & 1 & 0 & 0 \\ 0 & 0 & 0 & 0 & -2 & 0 \\ 0 & 0 & 0 & 0 & 0 & -2 \end{vmatrix} \end{array} , \tag{1.156}$$

The eigenstates of the fourfold degenerate valence bands are

$$u_{v1} = \frac{1}{\sqrt{2}}[|X\uparrow\rangle + |Z\downarrow\rangle],$$
$$u_{v2} = \frac{1}{\sqrt{2}}[|X\downarrow\rangle + |Z\uparrow\rangle],$$
$$u_{v3} = \frac{\mathrm{i}}{\sqrt{2}}[|(X - \mathrm{i}Y)\downarrow\rangle],$$
$$u_{v4} = \frac{\mathrm{i}}{\sqrt{2}}[|(X - \mathrm{i}Y)\uparrow\rangle],$$

and for twofold degenerate spin–orbit split off bands are

$$u_{v5} = -\frac{1}{\sqrt{3}}[|(X + \mathrm{i}Y)\uparrow\rangle - |Z\downarrow\rangle],$$
$$u_{v6} = -\frac{\mathrm{i}}{\sqrt{3}}[|(X + \mathrm{i}Y)\downarrow\rangle - |Z\uparrow\rangle],$$

where we find that the spin–orbit interaction results in the split of valence bands $\Delta/3$, $\Delta/3$, and $-2\Delta/3$. Note here that the obtained eigen functions differ from those defined by (2.63a) $\sim$ (2.63f) in Chap. 2 because of the different definition of the original (unperturbed) basis functions. Similar relations for the Γ_{15} states and antisymmetric parts are easily evaluated.

1.7.6 30–band k · p Method with the Spin–Orbit Interaction

We have to note here that the full band calculation (energy states at any points of the Brillouin zone) is easily carried out by extending 15×15 $\boldsymbol{k} \cdot \boldsymbol{p}$ matrix of (1.144) to 30×30 $\boldsymbol{k} \cdot \boldsymbol{p}$ matrix with spin–up and spin–down states. The matrices of the spin–orbit interaction Hamiltonian for $\Gamma^u_{25'}$ states, Γ_{15} states, and the antisymmetric parts between $\Gamma^u_{25'}$ and Γ_{15} states are obtained by (1.155). The 30–band $\boldsymbol{k} \cdot \boldsymbol{p}$ Hamiltonian with *complex* elements are solved to obtain 30 eigenstates. The full band calculations based on the 30–band $\boldsymbol{k} \cdot \boldsymbol{p}$ methods have been reported in the literatures [19, 20], where the treatment of the matrix elements R and R_1 are deduced by Voon and Willatzen [19].[2] All the $\boldsymbol{k} \cdot \boldsymbol{p}$ parameters for semiconductors such as Ge and Si with diamond crystal structure and several III-V compound semiconductors such as GaAs and GaP are summarized in Table 1.11. It should be noted again that III-V compound semiconductors with a zinc blende structure have no inversion symmetry and thus we have to include the anti-symmetric terms of the potentials and spin–orbit interaction as discussed by Pollak et al. [18]. A lack of the inversion symmetry results in the antisymmetric potential V^-, and the antisymmetric parts of Δ^- in the spin–orbit Hamiltonian of which matrix elements are defined by (1.151a) $\sim$ (1.151c).

[2]The author is thankful for M. Cardona to remind the work by Voon and Willatzen after his visit to Max Planck Institute at Stuttgart in June, 2013.

Table 1.11 Energy eigenvalues (in Rydberg) and momentum matrix elements (in atomic units) used in the $\boldsymbol{k} \cdot \boldsymbol{p}$ Hamiltonians for Si, Ge, GaAs, GaP, InP and InSb. Matrix elements of the anti-symmetric potentials V^- and anti-symmetric spin–orbit splitting parameter Δ^- (in Rydberg) for GaAs, GaP, InP and InSb are also listed. Values are from references [15, 18, 21–23]

	Si	Ge	GaAs	GaP	InP	InSb
$\Gamma^l_{25'}$	0.00	0.00	0.00	0.00	0.00	0.00
$\Gamma^l_{2'}$	0.265	0.0728	0.0845	0.2566	0.0929	0.022
Γ_{15}	0.252	0.232	0.2596	0.2511	0.2622	0.232
Γ^u_1	0.520	0.571	0.4940	0.5222	0.5057	0.400
$\Gamma_{12'}$	0.710	0.771	0.6063	0.7126	0.5803	0.494
$\Gamma^u_{25'}$	0.940	1.25	0.9002	0.9535	0.8745	0.726
$\Gamma^u_{2'}$	0.990	1.35	0.9849	1.0056	0.9792	0.765
Γ^l_1	−0.950	−0.966	−0.844	−0.9827	−0.8107	−0.846
P	1.200	1.360	1.3225	1.207	1.0876	1.3460
Q	1.050	1.070	1.1599	1.051	1.1346	1.0990
R	0.830	0.8049	0.7635	0.8289	0.8045	0.5914
P''	0.100	0.100	0.2465	0.100	0.1267	0.5324
P'	−0.090	0.1715	0.0438	−0.07863	0.1031	0.0666
Q'	−0.807	−0.752	−0.5511	−0.8046	−0.6585	−0.2120
R'	1.210	1.436	0.9697	1.220	1.1038	1.0760
P'''	1.320	1.623	1.5530	1.333	1.4281	1.2340
T	1.080	1.200	1.1387	1.0852	1.0806	0.9070
T'	0.206	0.5323	0.5323	0.2202	0.3906	0.0210
$\Delta^l_{25'}$	0.0032	0.0213	0.0251	0.00399	0.00823	0.0590
Δ_{15}	0.0036	0.0265	0.0135	0.00459	0.00573	0.0287
V_1	$= \langle \Gamma_{15} \| V^- \| \Gamma^l_{25'} \rangle$		0.12652	0.14924	0.1347	0.0869
V_2	$= \langle \Gamma^l_{2'} \| V^- \| \Gamma^u_1 \rangle$		−0.24791	−0.26885	−0.2003	−0.1558
V_3	$= \langle \Gamma^l_{2'} \| V^- \| \Gamma^l_1 \rangle$		0.38210	0.45687	0.2252	0.2391
V_4	$= \langle \Gamma_{15} \| V^- \| \Gamma^u_{25'} \rangle$		0.12297	0.21044	0.1131	0.0581
V_5	$= \langle \Gamma^u_{2'} \| V^- \| \Gamma^u_1 \rangle$		−0.34820	−0.33021	−0.2601	−0.1252
V_6	$= \langle \Gamma^u_{2'} \| V^- \| \Gamma^l_1 \rangle$		0.0	0.0	0.0	0.0
Δ^-			0.0051	0.00485	0.00682	0.0160

The calculated energy band structures of in the energy range $-15 \sim 10$ [eV] are shown for Ge and GaAs in Fig. 1.18 and for GaP and InP in Fig. 1.19, where we find that the 30–band $\boldsymbol{k} \cdot \boldsymbol{p}$ perturbation method gives a reasonable result although the matrix elements are very few compared to the pseudopotential method. Since the energy bands near the lowest conduction band and the top valence bands are very important to understand the electrical and optical properties of semiconductors, the energy band structures in the vicinity of the conduction minima and the valence band maxima calculated by the 30–band $\boldsymbol{k} \cdot \boldsymbol{p}$ perturbation method are shown in Fig. 1.20

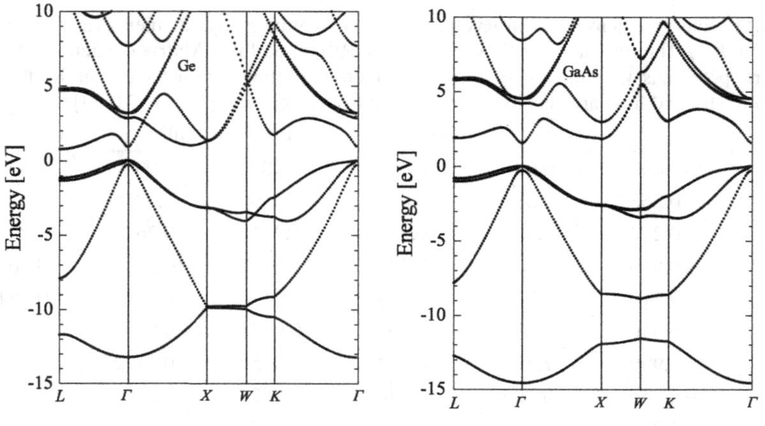

Fig. 1.18 Energy band structure of Ge and GaAs calculated by the 30–band $\boldsymbol{k} \cdot \boldsymbol{p}$ method with spin–orbit interaction in $\boldsymbol{k}$ space along L, Γ, X, W, K, and Γ of the Brillouin zone

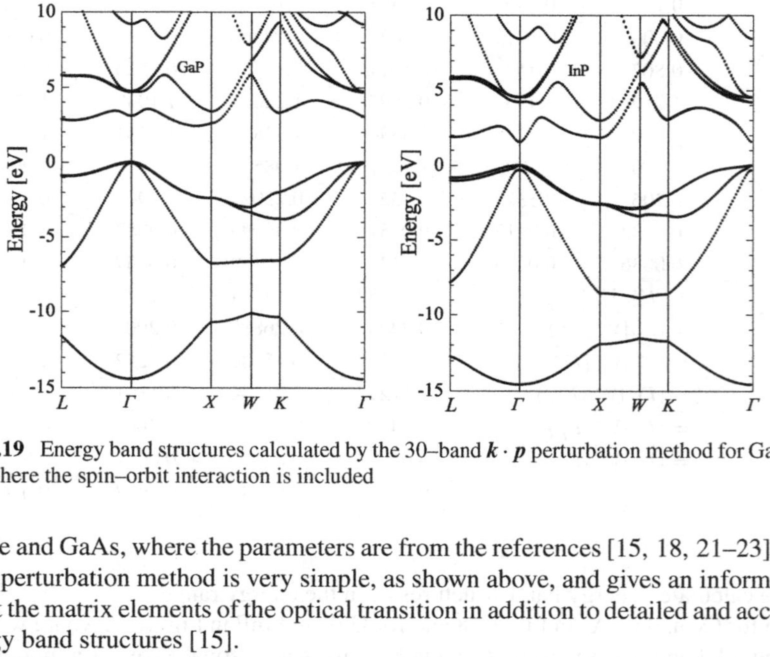

Fig. 1.19 Energy band structures calculated by the 30–band $\boldsymbol{k} \cdot \boldsymbol{p}$ perturbation method for GaP and InP, where the spin–orbit interaction is included

for Ge and GaAs, where the parameters are from the references [15, 18, 21–23]. The $\boldsymbol{k} \cdot \boldsymbol{p}$ perturbation method is very simple, as shown above, and gives an information about the matrix elements of the optical transition in addition to detailed and accurate energy band structures [15].

We have to note here some difference of the calculated band structures between the empirical pseudopotential method and 30–band $\boldsymbol{k} \cdot \boldsymbol{p}$ perturbation method. As seen in the first Brillouin zone of a face centered cubic crystal given in Figs. 1.5a and 1.22 the U point and K point are equivalent and we may expect the same energy eigenvalues at the two points. This feature is understood from the symmetry properties of the free electron band in the region U and K of Fig. 1.6. However,

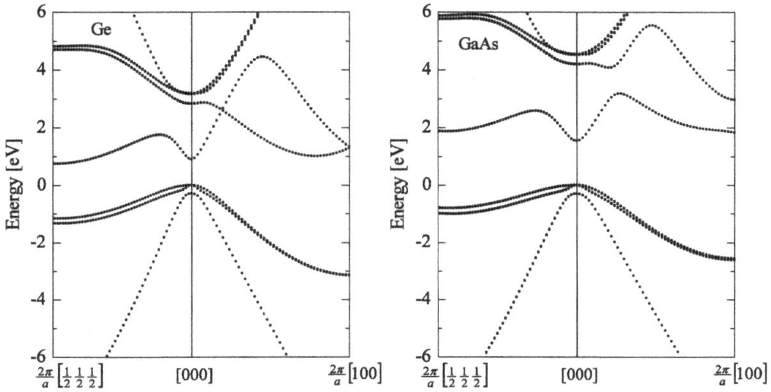

Fig. 1.20 Energy band structures near the conduction and valence bands calculated by the 30–band $\boldsymbol{k} \cdot \boldsymbol{p}$ perturbation method for Ge and GaAs, where the spin–orbit interaction is included

when we plot energy bands obtained by the 30–band $\boldsymbol{k} \cdot \boldsymbol{p}$ method along X point to U point and K point to Γ point, we find a small discontinuity at the points U and K, although the pseudopotential method gives smooth curve in this region. This may be ascribed to the assumption of limited number of eigen sates at the Γ point for the $\boldsymbol{k} \cdot \boldsymbol{p}$ perturbation method as follows. The free electron bands of a face centered cubic lattice are shown in Figs. 1.6 and 1.7, where we find that one of the 12 free electron bands of [220] is merged into one of the 6 free electron bands of [200]. This is clearly seen in Fig. 1.21, where the energy band structures of GaAs calculated by 30–band $\boldsymbol{k} \cdot \boldsymbol{p}$ are shown in Fig. 1.21a and the results obtained from the empirical pseudopotential method with 65 plane waves (130 plane waves with spin–up and spin–down) are shown in Fig. 1.21b. In the calculations of the local pseudopotential method, we used the following parameters replacing the pseudopotentials V_3^{S} and V_{11}^{A} in Table 1.3 by $V_3^{\mathrm{S}} = -0.260$ and $V_{11}^{\mathrm{A}} = 0.015$, and the spin–orbit interaction parameters $\lambda^{\mathrm{S}} = -0.00050$ and $\lambda^{\mathrm{A}} = -0.00012$ in atomic units. Total number of the free electron waves is 113 and the higher lying free electron waves beyond $\mathcal{E} = 16$ are included by Löwdin's perturbation method.These parameters lead to $\mathcal{E}_{\mathrm{G}} = 1.52\,\mathrm{eV}$ and spin–orbit splitting $\Delta = 0.342\,\mathrm{eV}$. In Fig. 1.21a we find the bands obtained from 30–band $\boldsymbol{k} \cdot \boldsymbol{p}$ method are discontinuous at U and K points, while the bands calculated by the empirical pseudopotential method with 65 plane waves are continuous. In addition the second lowest conduction band in the region Γ–X calculated by 30–band $\boldsymbol{k} \cdot \boldsymbol{p}$ method does not appear in the region of between U, K and Γ points. These features are observed in the band structures calculated by the pseudopotential method with 15 plane waves shown in Fig. 1.9, while the energy bands with 65 plane waves show much more smooth (continuous) curves near U and K points.

Although such a small difference exists in the energy band structure calculated by the 30–bands $\boldsymbol{k} \cdot \boldsymbol{p}$ perturbation method, the obtained overall features of the full band structures are very in good agreement with the empirical pseudopotential method and

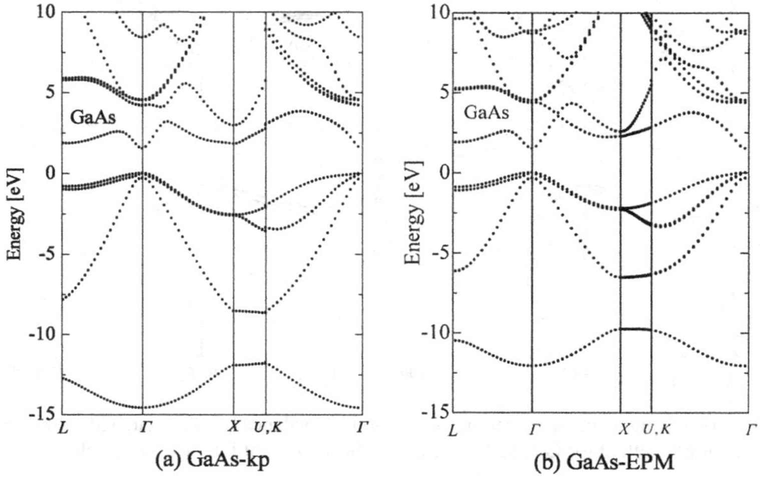

Fig. 1.21 Energy band structures of GaAs along L–Γ, Γ–X, X–U, K, and U, K–Γ points **a** calculated by the 30–band $\boldsymbol{k} \cdot \boldsymbol{p}$ perturbation method and **b** calculated by the empirical pseudopotential method with 65 plane waves (130 plane waves with spin–up and spin–down), where the spin–orbit interaction is included. See the textbook for the used parameters

the bands along L, Γ, X, W, K to Γ are very smooth. As discussed in Sect. 1.8, the density of states are calculated by dividing the polyhedron, the 1/48 volume of the first Brillouin zone, starting the basal plane of L, Γ, X, W, K. Then the calculations of the density of states are straight forward.

From the results of the pseudopotential and $\boldsymbol{k} \cdot \boldsymbol{p}$ perturbation methods we find the important features of the energy band structures of the semiconductors which we deal with in this text. All these semiconductors have valence band maxima at $\boldsymbol{k} = 0$ (Γ point) and most III–V semiconductors (except several compounds such as GaP, AlAs and so on) have the conduction band minimum at $\boldsymbol{k} = 0$ (Γ point), and so are classified as direct gap semiconductors. On the other hand, other semiconductors such as Ge, Si, GaP, AlAs and so on have the conduction band minima at $\boldsymbol{k} \neq 0$, and so are called indirect gap semiconductors. Ge has the conduction band minima at $\boldsymbol{k} = (\pi/a)[111]$ (L point), which are degenerate and consist of equivalent four conduction bands, and thus the conduction bands have a "many–valley structure." On the other hand, Si has the conduction band minima at the Δ point close to the X point and six equivalent conduction band minima. Later we will discuss the optical properties of semiconductors, where we find that the direct and indirect semiconductors are quite different in their optical properties such as absorption and light emission. The electrical properties also exhibit the features of many-valley structures in Ge and Si, and also the Gunn effect of GaAs, which arises from the inter-valley transfer of electrons in a high electric field from the high electron mobility Γ valley to the low mobility L and/or X valleys. From these results we understand that the energy band structures play a very important role in the understanding of the electrical and optical properties of semiconductors. In addition, once we know the procedures for calculating energy band structures it is very easy to extend the method to calculate the energy band structures of superlattices (periodic layers of different semiconductors such as GaAs/AlAs) as treated in Chap. 8. It is also possible to predict the basic features of semiconductors from the results of energy band calculations. In this text the basic physics of semiconductors is treated on the basis of their energy band structures.

1.8 Density of States

As shown in Sect. 4.3, density of states (DOS) is defined as the number of states per unit energy. When the number of states in a small volume of $\boldsymbol{k}$-space in a small range of energy, $[\mathcal{E}, \mathcal{E} + \Delta\mathcal{E}]$ is given by $(1/2\pi)^3 \Delta v(\boldsymbol{k})$, the density of states $J_{\mathrm{DOS}}(\mathcal{E})$ is defined by (spin factor 2 is omitted)

$$J_{\mathrm{DOS}}(\mathcal{E}) = \sum_{\boldsymbol{k}} \frac{1}{(2\pi)^3} \frac{\Delta v(\boldsymbol{k})}{\Delta\mathcal{E}} . \tag{1.157}$$

Here $v(\boldsymbol{k})$ is a small volume of the wave vector $\boldsymbol{k}$ in the first Brillouin zone. The density of states is required to calculate dielectric function (joint density of states) and also to calculate scattering rate of electrons. It is well known that Monte Carlo simulation [24] gives a good description of transport properties at high electric fields. Full band Monte Carlo simulation is used very often, where the calculated energy band structure is used to simulated electron motion in $\boldsymbol{k}$-space, and thus the density of states is required to obtain the scattering rate. In this textbook we will not deal with high field transport (see [25] for a review on high field transport) and thus we will not concern with Monte Carlo simulation. However, it is very important to know how to calculate the density of states defined by (1.157) from the calculated energy band structure. Various methods have been reported to calculate DOS from the energy bands. Brust [9] reported a rigorous analysis of the joint density of states in Ge and Si, but the method is very complicated. Here we will show a simple but accurate method to calculate the density of states. First, let's take a look of the Brillouin zone shown in Fig. 1.22 which is the same as shown in Fig. 1.5a. It is clear from the 48-fold symmetry of the Brillouin zone that 1/48th part of the zone shown in Fig. 1.22 is sufficient to calculate the density of states. In other words, all the other $\boldsymbol{k}$-points in the first Brillouin zone may be obtained by rotation of the 1/48-th of the Brillouin zone. Using a unit length $k_{\mathrm{f}} = (X - \Gamma)/8 = \pi/4a$ with the lattice constant a, 1/48-th part of the first Brillouin zone can be divided into polyhedrons shown in Fig. 1.23. The bottom plane consists of the critical points Γ, X, W, K. The second plane is displaced by k_{f} in the k_z direction with respect to the first pane, and intersects Λ, S, Q. The third plane with the same displacement intersects Λ, U, Q, and the fourth intersects Λ, Q, L. The two neighboring planes

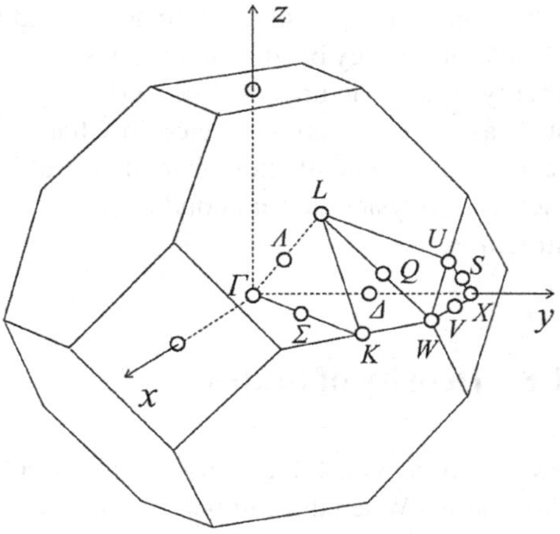

Fig. 1.22 The first Brillouin zone of face centered cubic lattice and the symmetry points. The box defined by the lines is 1/48-th part of the volume of the first Brillouin zone

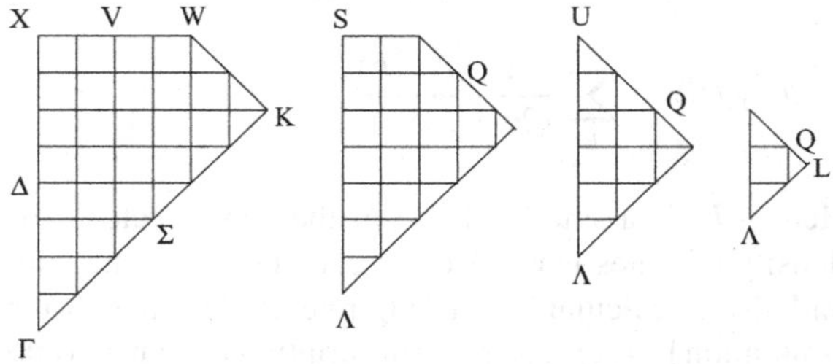

Fig. 1.23 1/48th part of the first Brillouin zone of face centered cubic lattice is discretized into polyhedrons using unit length $k_f = (X - \Gamma)/8 = \pi/4a$, where a is the lattice constant

form polyhedrons of which volume gives the number of states. The discretization shown in Fig. 1.23 is not sufficient to calculate the density of states. Here we have to note the most important idea to use this type of discretization. In order to get smaller volumes of the polyhedrons, use the unit length $k_f/N \to k_f$, where N is 2, 4, 8, ... Then the volumes of new polyhedrons becomes 1/8, 1/64, 1/512, ... of the original polyhedrons and the number of polyhedrons are 8, 64, 512, ... times of the original number. The density of states are easily calculated using this discretization. We calculate the energy eigenstates at the corners of a polygon and tabulate them. Then pick up the minimum $\mathcal{E}_{min}$ and the maximum $\mathcal{E}_{max}$ from the lowest pairs of the tabulated eigenstates. Calculate $\Delta\mathcal{E} = \mathcal{E}_{max} - \mathcal{E}_{min}$ and $\mathcal{E} = (\mathcal{E}_{max} + \mathcal{E}_{min})/2$. This gives us

$$J_{\mathrm{DOS}}(\mathcal{E}) \simeq \frac{1}{(2\pi)^3} \frac{\Delta v_f(\boldsymbol{k})}{\Delta\mathcal{E}}, \tag{1.158}$$

where $\Delta v_f(\boldsymbol{k})$ is the volume of the polyhedron. In a similar fashion we calculate the density of states for the second lowest pairs, third lowest pairs and so on up to a required $\mathcal{E}$ value. The obtained DOS $J_{\mathrm{DOS}}(\mathcal{E})$ is not uniform but scattered in the

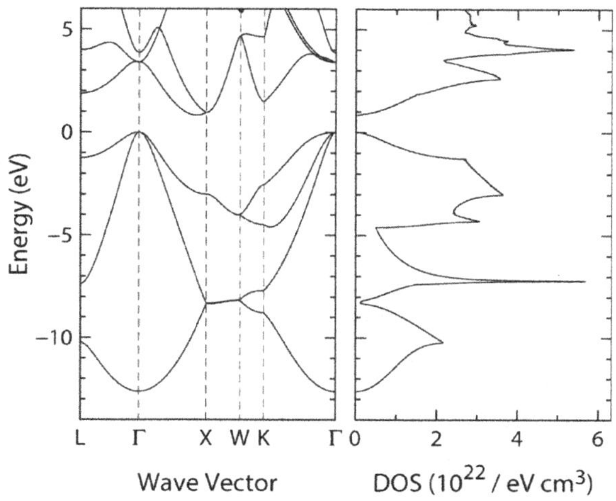

Fig. 1.24 Energy band structures of Si calculated by empirical pseudopotential method with the pseudopotentials listed in Table 1.2 and the calculated density of states (DOS)

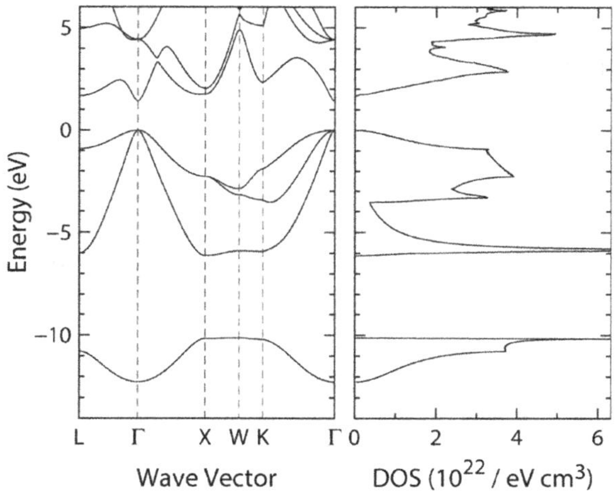

Fig. 1.25 Energy band structures of GaAs calculated by empirical pseudopotential method with the pseudopotentials listed in Table 1.2 and the calculated density of states (DOS)

energy $\mathcal{E}$ and thus we have to rearrange the data in the histogram, 0.1–eV histogram for example. Then smoothing procedure will give a smooth curve of $J_{\text{DOS}}(\mathcal{E})$ as a function of $\mathcal{E}$. Typical examples of the energy bands and the DOS are shown in Fig. 1.24 for Si and in Fig. 1.25 for GaAs, where the energy bands are obtained by the local pseudopotential method using the pseudopotentials given in Table 1.2 and the density of states in the valence bands and the conduction bands are calculated.

1.9 Problems

(1.1) Calculate the reciprocal lattice vectors of a zinc blende type crystal structure and compare with the result of Table. 1.1.

(1.2) Energy band calculations are carried out by using atomic units [a.u]. Give wave vector $k = 2\pi/a_{\mathrm{B}}$, and energy $(\hbar^2/2m)/a_{\mathrm{B}}^2)$ in atomic units, where $a_{\mathrm{B}} = (\epsilon_0 h^2/\pi m e^2) = 0.529177$ [Å] is Bohr radius.

(1.3) Rewrite Equation (1.35), using the atomic units.

(1.4) Evaluate energy bands of two band model based on the nearly free electron approximation, taking free electron bands of $G_n = 0$ and $G_n = 1$, where $G_n = 2n\pi/a$, $a = 0.543$ [Å], and $V(G_1) = V_3^{\mathrm{s}} = -0.21$ [a.u.]. Calculate the energy band $\mathcal{E}(k_x)$ and plot the energy bands together with the free electron bands.

(1.5) Derive fundamental vectors $[\boldsymbol{a}, \boldsymbol{b}, \boldsymbol{c}]$ of (i) simple cubic, (ii) body centered cubic, (iii) face centered cubic and hexagonal closed pack crystals, and calculate their unit cell volume v.

(1.6) Derive the reciprocal lattice vectors $[\boldsymbol{a}^*, \boldsymbol{b}^*, \boldsymbol{c}^*]$ of (i) simple cubic, (ii) body centered cubic, (iii) face centered cubic and hexagonal closed pack crystals, and calculate their unit cell volume v.

(1.7) Derive spin–orbit interaction given by (1.149a)

$$\langle X(\Gamma_{25'}^{l}) \uparrow | H_{\mathrm{so}} | Y(\Gamma_{25'}^{l}) \uparrow \rangle = \mathrm{i}\Delta_{25'}^{l}/3$$

(1.8) In order to calculate full band structure in the first Brillouin zone, we have to the wave vectors at the critical points (symmetry points) and their lengths. Referring Figs. 1.22 and 1.23, evaluate the symmetry points and the lengths between the symmetry points of a face centered cubic crystal.

References

1. C. Kittel, *Introduction to Solid State Physics*, 7th edn. (Wiley, New York, 1996)
2. M.L. Cohen, J.R. Chelikowsky, *Electronic Structure and Optical Properties of Semiconductors* (Springer, Heidelberg, 1989). This article provides a good review of energy band calculations based on the pseudopotential theory. The bibliography is a good guide to find articles of the energy band calculations and optical properties of various semiconductors
3. V. Heine, The pseudopotential concept. Solid State Phys. **24**, 1–36 (1970)
4. M.L. Cohen, V. Heine, The fitting pseudopotentials to experimental data and their subsequent application. Solid State Phys. **24**, 37–248 (1970)
5. M.L. Cohen, T.K. Bergstresser, Phys. Rev. **141**, 789 (1966)
6. J.R. Chelikowsky, M.L. Cohen, Phys. Rev. B **10**, 5095 (1974)
7. J.R. Chelikowsky, M.L. Cohen, Phys. Rev. B **14**, 556 (1976)
8. P.O. Löwdin, J. Chem. Phys. **19**, 1396 (1951)
9. D. Brust, Phys. Rev. A **134**, 1337 (1964)
10. P.J. Melz, J. Phys. Chem. Solids **32**, 209 (1971)
11. G. Weisz, Phys. Rev. **149**, 504 (1966)
12. E.O. Kane, J. Phys. Chem. Solids **1**, 82 (1956); 249 (1957)

13. L.P. Bouckaert, R. Smoluchowski, E. Wigner, Phys. Rev. **50**, 58 (1936)
14. G. Dresselhaus, A.F. Kip, C. Kittel, Phys. Rev. **98**, 368 (1955)
15. M. Cardona, F.H. Pollak, Phys. Rev. **142**, 530 (1966)
16. M. Cardona, Optical properties and band structure of Germanium and Zincblende-type semiconductors, in *Proceedings of the International School of Physics ≪Enrico Fermi≫* (Academic Press, New York, 1972) pp. 514–580
17. F. Bassani, G. Pastori Parravicini, *Electronic States and Optical Transitions in Solids* (Pergamon Press, New York, 1975)
18. F.H. Pollak, C.W. Higginbotham, M. Cardona, J. Phys. Soc. Jpn. **21**(Supplement), 20 (1966)
19. L.C.L.Y. Voon, M. Willatzen, *The $\boldsymbol{k} \cdot \boldsymbol{p}$ Method: Electronic Properties of Semiconductors* (Springer, Berlin, 2009)
20. D.W. Bailey, C.J. Stanton, K. Hess, Phys. Rev. B **42**, 3423 (1990)
21. H. Hazama, Y. Itoh, C. Hamaguchi, J. Phys. Soc. Jpn. **54**, 269 (1985)
22. T. Nakashima, C. Hamaguchi, J. Komeno, M. Ozeki, J. Phys. Soc. Jpn. **54**, 725 (1985)
23. H. Hazama, T. Sugimasa, T. Imachi, C. Hamaguchi, J. Phys. Soc. Jpn. **55**, 1282 (1986)
24. C. Jacoboni, L. Reggiani, Rev. Mod. Phys. **55**, 645 (1983)
25. E.M. Conwell, *High Field Transport in Semiconductors*, vol. 9, Solid State Physics (Academic Press, New York, 1967)

Chapter 2
Cyclotron Resonance and Energy Band Structures

Abstract Cyclotron resonance experiments were successfully used to determine the effective masses of electrons in the conduction band and holes in the valence bands. Electrons are subject to rotational motion in a magnetic field (cyclotron motion) and resonantly absorb the radiation fields (microwaves or infrared radiation) when the cyclotron frequency is equal to the radiation frequency. The resonant condition gives the effective mass of the electron. Analyzing the cyclotron resonance data of Ge and Si, detailed information of the electrons in the conduction band valleys. In addition, the analysis of the hole masses based on the $\boldsymbol{k} \cdot \boldsymbol{p}$ perturbation reveals the detailed valence band structures such as the heavy hole, light hole, and spin–orbit split–off bands. The non-parabolicity of the conduction band is also discussed. Quantum mechanical treatment of the electrons in the conduction band and holes in the valence bands leads us to draw the picture of Landau levels which is used to understand magnetotransport in Chap. 7 and quantum Hall effect in Chap. 8.

2.1 Cyclotron Resonance

Cyclotron resonance has been successfully used to determine the effective masses of electrons in semiconductors. However, the cyclotron resonance has played a very important role in understanding the valence band structures of Ge and Si and later led to an accurate determination of the energy band structures. In this chapter first we will describe the cyclotron resonance experiment and its analysis in order to discuss the anisotropy of the electron effective mass and the many valley structures. Second, we will discuss the analysis of the valence band structures by using the $\boldsymbol{k} \cdot \boldsymbol{p}$ perturbation method, which will reveal the importance of the cyclotron resonance experiments.

When a particle with electronic charge q is placed in a magnetic field, the particle make a circular motion in the plane perpendicular to the magnetic field. This motion is caused by the Lorentz force perpendicular to the magnetic field $\boldsymbol{B}$ and the velocity $\boldsymbol{v}$ of the particle, and is given by the relation

$$\boldsymbol{F}_L = q\boldsymbol{v} \times \boldsymbol{B} . \tag{2.1}$$

C. Hamaguchi, *Basic Semiconductor Physics*, Graduate Texts in Physics,
DOI 10.1007/978-3-319-66860-4_2

When the particle is not scattered, the equation of motion for a particle with mass m is given by

$$m\frac{\mathrm{d}\boldsymbol{v}}{\mathrm{d}t} = q\boldsymbol{v} \times \boldsymbol{B} \,. \tag{2.2}$$

Assuming that the magnetic field is applied in the z direction and putting $\boldsymbol{B} = (0, 0, B_z)$, the equation results in

$$m\frac{\mathrm{d}v_x}{\mathrm{d}t} = qv_y B_z \,,$$
$$m\frac{\mathrm{d}v_y}{\mathrm{d}t} = -qv_x B_z \,,$$

which give rise to

$$\frac{\mathrm{d}^2 v_x}{\mathrm{d}t^2} = -\frac{qv_x B_z}{m} \tag{2.3}$$

for v_x and a similar equation for v_y. Therefore, the solutions for v_x and v_y are

$$v_x = A\cos(\omega_\mathrm{c} t), \qquad v_y = A\sin(\omega_\mathrm{c} t),$$

where

$$\omega_\mathrm{c} = \frac{qB_z}{m} \tag{2.4}$$

is the **cyclotron (angular) frequency** of the particle. From the results we find that the particle in a magnetic field B_z orbits with angular frequency ω_c in the plane perpendicular to $\boldsymbol{B}$ (x, y plane) and the motion is referred to as "cyclotron motion." When an electromagnetic field of frequency $\omega = \omega_\mathrm{c}$ is applied, the particle absorbs the energy of the electromagnetic field resonantly. This is called **cyclotron resonance**. In semiconductors there exist various scattering processes and thus the particle (electron and hole) resonance is modified by the scattering. If we define the scattering time (or relaxation time) by τ, the condition for well-defined resonance is

$$\omega_\mathrm{c}\tau \gg 1 \,. \tag{2.5}$$

In other words, the cyclotron resonance is smeared out and no clear resonance is observed when condition (2.5) is not fulfilled. In order to observe clear cyclotron resonance, therefore, we have to apply a high magnetic field (higher $\boldsymbol{B}$ results in higher ω_c and higher ω) or achieve a condition where less scattering occurs (longer τ is achieved at lower temperature T).

Let us consider the condition for a typical semiconductor with effective mass $m^* = 0.1\,\mathrm{m}$ ($m = 9.1 \times 10^{-31}$ kg is the free electron mass). When a magnetic field $B = 1\,\mathrm{T}$ is applied, the cyclotron frequency is

$$\omega_c = \frac{1.6 \times 10^{-19} \times 1}{0.1 \times 9.1 \times 10^{-31}} \simeq 2 \times 10^{12}\,\mathrm{rad/s}\,,$$

which corresponds to a microwave field of frequency $f = \omega/2\pi \simeq 3 \times 10^{11}$ Hz. In addition, in order to realize the condition given by (2.5),

$$\omega_c\tau = \frac{qB_z}{m^*}\tau = \mu B_z \gg 1\,, \tag{2.6}$$

a semiconductor sample with electron mobility greater than $\mu = 1\mathrm{m}^2/\mathrm{Vs} = 10^4\mathrm{cm}^2/\mathrm{Vs}$ is required. A high magnetic field is achieved by using a superconducting solenoid and a high angular frequency electromagnetic field ($\omega = \omega_c$) by using microwave or infrared light. When we use an electromagnet, a field of 1 T is achieved and thus cyclotron resonance is observed in a sample with electron mobility larger than $10^4\mathrm{cm}^2/\mathrm{Vs}$. The electron mobility becomes higher at lower temperatures, where less phonon scattering occurs and the first observation of the cyclotron resonance was made at 4.2 K. At low temperatures electrons and holes are captured by donors and acceptors, and no free carriers exist for absorbing the microwave field. In the first cyclotron resonance measurements on Ge, light illumination was used to excite the carriers from the valence band to the conduction band, and resonance due to holes in addition to electrons was observed.

Here we will discuss the cyclotron resonance curves. First we consider the case of an electron with isotropic effective mass m^* for simplicity. The equation of motion for the electron is

$$m^*\frac{\mathrm{d}\boldsymbol{v}}{\mathrm{d}t} + \frac{m^*\boldsymbol{v}}{\tau} = q(\boldsymbol{E} + \boldsymbol{v} \times \boldsymbol{B})\,. \tag{2.7}$$

The microwave field is linearly polarized with $\boldsymbol{E} = (E_x, 0, 0)$, and the magnetic field is applied in the z direction as $\boldsymbol{B} = (0, 0, B_z)$. When we write the microwave field with angular frequency ω as $E = E_x \cos(\omega t) = \Re\{E_x \exp(-\mathrm{i}\omega t)\}$, we may use the relation $\mathrm{d}/\mathrm{d}t = -\mathrm{i}\omega$ and the motion of equation is written as

$$m^*\left(-\mathrm{i}\omega + \frac{1}{\tau}\right)v_x = q(E_x + v_y B_z)\,, \tag{2.8}$$

$$m^*\left(-\mathrm{i}\omega + \frac{1}{\tau}\right)v_y = q(0 - v_x B_z)\,. \tag{2.9}$$

From these equations we obtain

$$v_x = \frac{q\tau}{m^*} \cdot \frac{(1 - \mathrm{i}\omega\tau)}{(1 - \mathrm{i}\omega\tau)^2 + \omega_c^2\tau^2}E_x\,, \tag{2.10}$$

$$v_y = -\frac{q\tau}{m^*} \cdot \frac{\omega_c\tau}{(1 - \mathrm{i}\omega\tau)^2 + \omega_c^2\tau^2}E_x\,, \tag{2.11}$$

and the current density in the x direction is deduced from the relation $J_x = nqv_x$. This gives the power absorption P per unit volume given below.

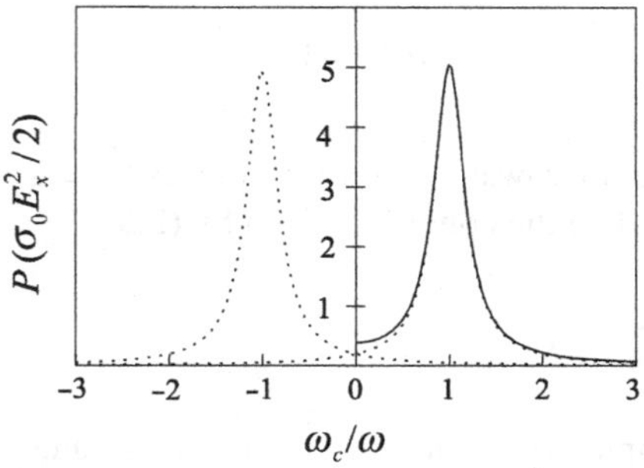

Fig. 2.1 Absorption line shapes of cyclotron resonance. The dotted curve peaked in the right half ($\omega_c > 0$) is the resonance for holes with $q > 0$, and the *dotted curve* peaked in the left half ($\omega_c < 0$) is the resonance for electrons with $q < 0$. These *dotted curves* correspond to the resonance due to a microwave field which is right-circularly polarized and left-circularly polarized, respectively. The solid curve is the absorption for the case of a linearly polarized microwave. The curves are obtained for $\omega\tau = 5$

$$\begin{aligned} P &= \frac{1}{2}\Re(J_x E_x) = \frac{1}{2}\frac{nq^2\tau}{m^*}E_x^2 \cdot \Re\left[\frac{1 - \mathrm{i}\omega\tau}{(1 - \mathrm{i}\omega\tau)^2 + \omega_c^2\tau^2}\right] \\ &\equiv \frac{1}{4}\sigma_0 E_x^2 \cdot \Re\left[\frac{1}{(1 - \mathrm{i}\omega\tau) - \mathrm{i}\omega_c\tau} + \frac{1}{(1 - \mathrm{i}\omega\tau) + \mathrm{i}\omega_c\tau}\right] \\ &= \frac{1}{4}\sigma_0 E_x^2 \left[\frac{1}{(\omega - \omega_c)^2\tau^2 + 1} + \frac{1}{(\omega + \omega_c)^2\tau^2 + 1}\right], \end{aligned} \tag{2.12}$$

where $\sigma_0 = nq^2\tau/m^*$ is the d.c. the conductivity of the semiconductor with carrier density n, and $\omega_c = qB/m^*$ is the cyclotron angular frequency. The two terms on the right-hand side of (2.12) are plotted in Fig. 2.1 as the dotted curves, which are symmetric with respect to the peaked at $\omega = \pm\omega_c$ and are called the Lorentz function

$$F_L(\omega) = \frac{\Gamma/2\pi}{(\omega - \omega_c)^2 + (\Gamma/2)^2}. \tag{2.13}$$

The Lorentz function given by the above equation has a full-width half-maximum of Γ and the integral of the Lorentz function with respect to ω is unity. Therefore, the Lorentz function behaves like a delta function when Γ becomes very small. The two peaks in Fig. 2.1 arise from different signs of the charge q and thus the curves for $\omega_c < 0$ and $\omega_c > 0$ correspond to the resonance for electrons and holes, respectively. When the microwave field is right-circularly polarized and left-circularly polarized, resonance is observed due to holes and electrons, respectively. The resonance curves of (2.12) for a linearly polarized microwave field are plotted by solid curves in Fig. 2.2 for $\omega\tau = 0.5, 1.0, 2.0, 3.0, 4.0$ and 5.0, where a clear resonance is seen for $\omega\tau \gg 1$. In cyclotron resonance experiments a microwave field with fixed frequency is applied and the resonance due to the microwave absorption is detected as a function of the

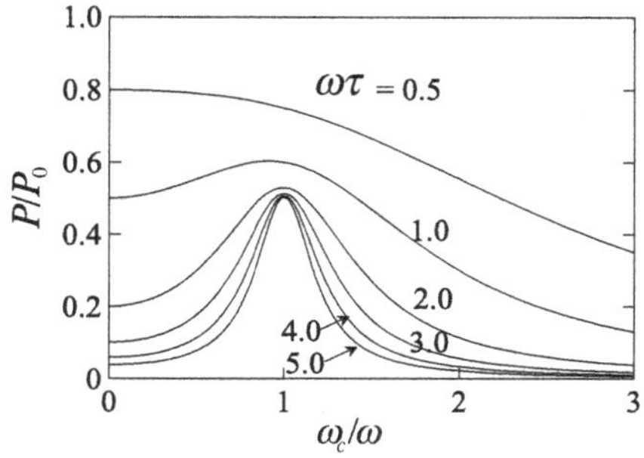

Fig. 2.2 Absorption line shapes of the cyclotron resonance for $\omega\tau = 0.5, 1.0, 2.0, 3.0, 4.0$, and 5.0. The cyclotron resonance is clearly observed in the case of $\omega\tau \gg 1$

magnetic field. Therefore, the horizontal axis of Fig. 2.2 is the magnetic field (ω_c/ω). From (2.12) and the curves in Fig. 2.1 it is evident that the full-width half-maximum is $2/\tau$.

Next, we consider the case of semiconductors with anisotropic effective masses such as in Ge and Si. Defining the effective mass tensor by $\widetilde{m}$, the equation of motion for a carrier with electronic charge q is expressed as

$$\widetilde{m} \cdot \dot{\boldsymbol{v}} + \frac{\widetilde{m} \cdot \boldsymbol{v}}{\tau} = q(\boldsymbol{E} + \boldsymbol{v} \times \boldsymbol{B}) . \tag{2.14}$$

Assuming the electric field and the carrier velocity are given by $\boldsymbol{E} = \boldsymbol{E}\exp(-\mathrm{i}\omega t)$ and $\boldsymbol{v} = \boldsymbol{v}\exp(-\mathrm{i}\omega t)$, respectively, and using the relation $\mathrm{d}/\mathrm{d}t = -\mathrm{i}\omega$, the equation

$$\left(-\mathrm{i}\omega + \frac{1}{\tau}\right)\widetilde{m} \cdot \boldsymbol{v} = q(\boldsymbol{E} + \boldsymbol{v} \times \boldsymbol{B}) \tag{2.15}$$

is derived. In order to solve this equation we introduce the following variables:

$$\omega' = \omega + \mathrm{i}\frac{1}{\tau} . \tag{2.16}$$

For simplicity (valid for Ge and Si) we assume that the effective mass $\widetilde{m}$ has diagonal parts only, i.e.

$$\left[\frac{1}{\widetilde{m}}\right] = \begin{bmatrix} 1/m_1 & 0 & 0 \\ 0 & 1/m_2 & 0 \\ 0 & 0 & 1/m_3 \end{bmatrix} , \tag{2.17}$$

and that the magnetic field is applied in an arbitrary direction:

$$\boldsymbol{B} = (B_x, B_y, B_z) \equiv B(\alpha, \beta, \gamma) .$$

Then (2.15) can be rewritten as

$$\begin{cases} i\omega' m_1 v_x + q(v_y B_z - v_z B_y) = 0, \\ i\omega' m_2 v_x + q(v_z B_x - v_x B_z) = 0, \\ i\omega' m_3 v_x + q(v_x B_y - v_y B_x) = 0, \end{cases}$$

where we put $\boldsymbol{E} = 0$. This assumption of $\boldsymbol{E} = 0$ is based on the fact that the resonance condition is derived in a vanishing electric field. Writing the magnetic field $\boldsymbol{B} = B(\alpha, \beta, \gamma)$, (2.15) becomes

$$\begin{vmatrix} i\omega' m_1 & qB\gamma & -qB\beta \\ -qB\gamma & i\omega' m_2 & qB\alpha \\ qB\beta & -qB\alpha & i\omega' m_3 \end{vmatrix} = 0, \tag{2.18}$$

where $\omega' = (\omega + i/\tau)$ is used. From this equation a solution for ω' is easily obtained. Under the condition of clear resonance ($\omega\tau \gg 1$) we obtain $\omega' \cong \omega$ and thus we may put $\omega' = \omega = \omega_c$. From these considerations the resonance condition is given by

$$\omega_c = \frac{qB}{m^*} = qB\sqrt{\frac{m_1\alpha^2 + m_2\beta^2 + m_3\gamma^2}{m_1 m_2 m_3}}, \tag{2.19}$$

where m^* is the **cyclotron mass.**

Figure 2.3 shows a typical experimental setup for the cyclotron resonance experiment. A microwave field generated by a klystron (with a circuit to control the frequency constant) is guided by a waveguide and fed into a cavity with a semiconductor

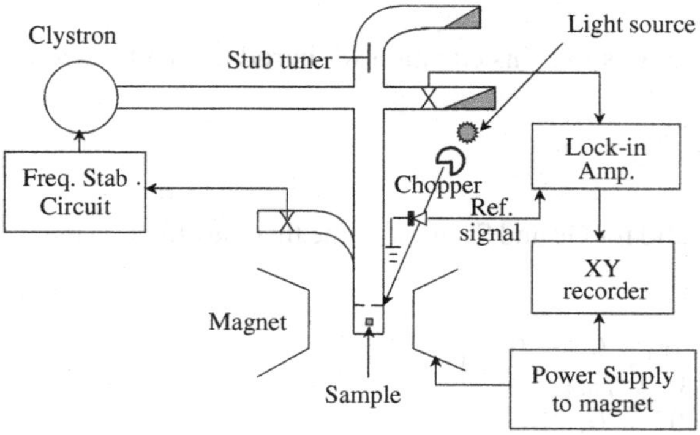

Fig. 2.3 Experimental setup for cyclotron resonance measurements. Microwave field generated by a klystron is guided to a cavity where a sample is inserted. The cavity is installed in a cryostat to be cooled down to 4.2 K. The carriers of the semiconductor sample are frozen out at low temperatures, and electrons and holes are excited by light illumination guided into the cavity by an optical pipe and the light is pulsed by a chopper to provide the reference signal for a lock-in amplifier. The magnetic field is swept slowly and the microwave absorption is recorded as a function of the magnetic field

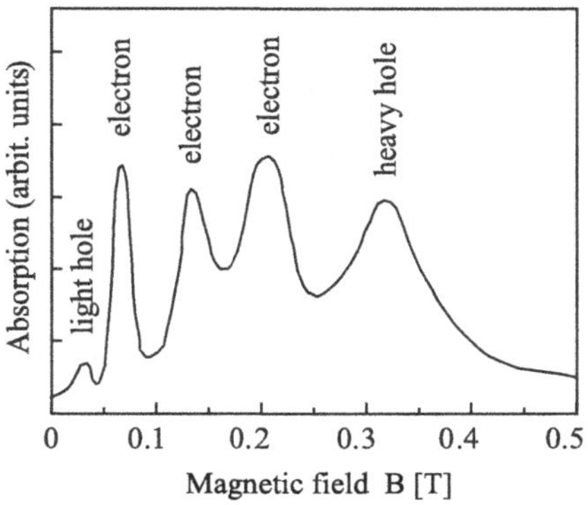

Fig. 2.4 Cyclotron resonance in Ge. $\omega/2\pi = 24\,\text{GHz}$, $T = 4.2\,\text{K}$. The magnetic field is applied in the direction 60° with respect to the $\langle 001 \rangle$ axis in the (110) plane. (From [1])

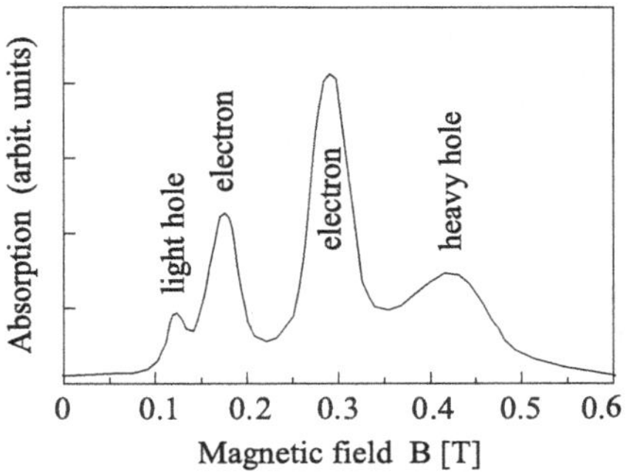

Fig. 2.5 Cyclotron resonance in Si. $\omega/2\pi = 24\,\text{GHz}$, $T = 4.2\,\text{K}$. The magnetic field is applied in the direction 30° with respect to the $\langle 001 \rangle$ axis in the (110) plane. (From [1])

sample. The cavity is installed in a cryostat and the sample temperature is controlled down to 4.2 K. In order to achieve the condition $\omega_c \tau \gg 1$, the sample is cooled down to 4.2 K. At low temperatures the carriers in the semiconductor are frozen out and pulsed light is used to excite photo-carriers (electrons and holes). When the magnetic field is swept, the microwave absorption increases at resonance, which is detected by a microwave diode installed in the waveguide and amplified by a lock-in amplifier. Thus the measured microwave absorption versus magnetic field curve reveals cyclotron resonance. Typical experimental results are shown in Fig. 2.4 for Ge and in Fig. 2.5 for Si [1]. As seen in Fig. 2.4 and Fig. 2.5 signals due to several kinds of electrons and two holes (heavy holes and light holes) are observed. These complicated resonance peaks are discussed here but the detailed interpretation is described later.

We now discuss the cyclotron mass. When we change the direction θ of the magnetic field with respect to the $\langle 001 \rangle$ axis in the plane (110) and calculate the cyclotron mass defined below from the magnetic field B_c for resonance,

$$m^* = \frac{qB_c}{\omega},$$

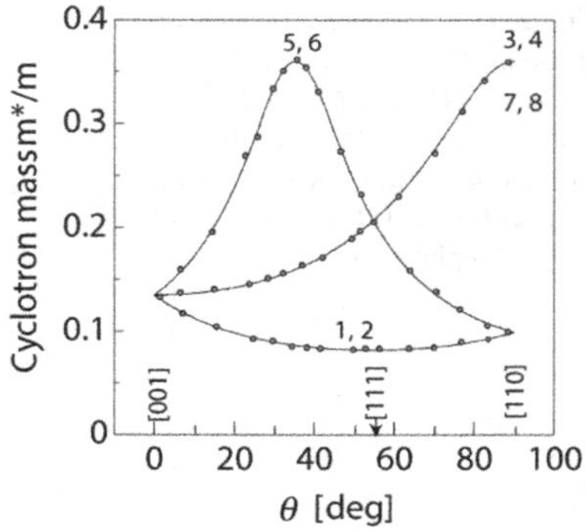

Fig. 2.6 Cyclotron masses of electrons in the conduction bands of Ge as a function of the magnetic field direction θ. The angle θ is the angle between the magnetic field and the $\langle 001 \rangle$ axis in the (110) plane. See the text for the valley indices 1, 2, ..., 8

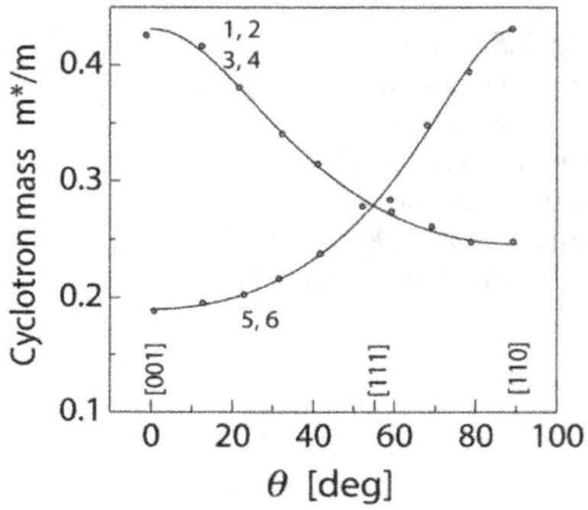

Fig. 2.7 Cyclotron masses of electrons in the conduction bands of Si as a function of the magnetic field direction θ. The angle θ is the angle between the magnetic field and the $\langle 001 \rangle$ axis in the (110) plane. See the text for the valley indices 1, 2, ..., 6

we obtain the cyclotron masses as a function of the direction θ, which are shown in Fig. 2.6 for Ge and in Fig. 2.7 for Si.

The angular dependence of the cyclotron masses shown in Figs. 2.6 and 2.7 is analyzed as follows. From the results of the energy band calculations in Chap. 1, we learned that the conduction band minima of Ge are located at the L point along the $\langle 111 \rangle$ axis and that the conduction band minima of Si are located at the Δ point, 15% from the X point, along the $\langle 100 \rangle$ axis. When we take these axes as the principal axes of a newly defined $\boldsymbol{k}$-vector and put $\boldsymbol{k} = 0$ at the bottom of the conduction band, then the symmetry of the conduction band valleys leads to

$$\mathcal{E} = \frac{\hbar^2}{2m_t}(k_x^2 + k_y^2) + \frac{\hbar^2}{2m_l}k_z^2 \,. \tag{2.20}$$

The constant energy surfaces of the conduction bands are shown in Fig. 2.8 for Ge and in Fig. 2.9 for Si. Since the conduction bands consist of multiple energy surfaces, the conduction bands are called many valleys or many-valley structure. In Si there are 6 valleys, while in Ge there exist 4 valleys (the pairs of valleys along the principal axis are apart from the reciprocal lattice vector $\boldsymbol{G}$ and they are equivalent). The cyclotron

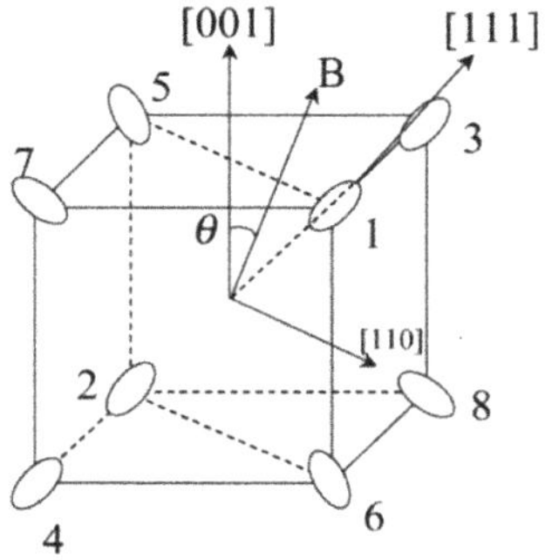

Fig. 2.8 Many-valley structure of the conduction bands in Ge

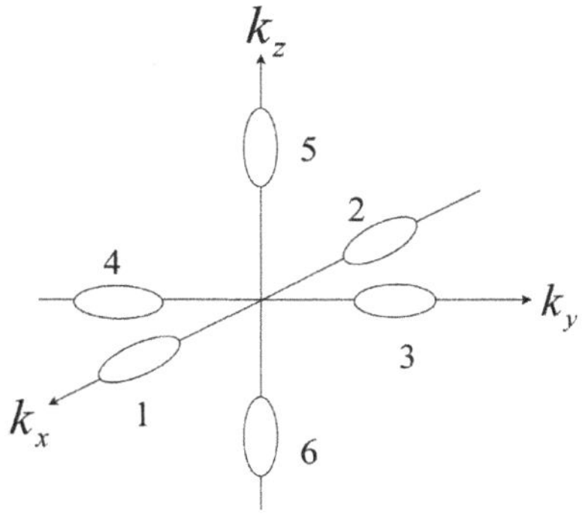

Fig. 2.9 Many-valley structure of the conduction bands in Si

mass in the conduction band with the ellipsoidal energy surface is given by (2.19) with $m_1 = m_2 = m_t$, $m_3 = m_l$ as

$$\frac{1}{m_c^*} = \sqrt{\frac{m_t(1-\gamma^2) + m_l\gamma^2}{m_t^2 m_l}}, \tag{2.21}$$

where γ is the direction cosine between the magnetic field and the principal axis of the ellipsoid (k_z direction), i.e. $B_z = B\gamma$.

Let us assume that the magnetic field is applied in the direction of the angle θ with respect to the $\langle 001 \rangle$ axis in the (110) plane. In this case we have

$$\boldsymbol{B} = B\left(\frac{\sin\theta}{\sqrt{2}}, \frac{\sin\theta}{\sqrt{2}}, \cos\theta\right) \equiv B\boldsymbol{e}_B.$$

First we consider the case of Ge. As stated above, 4 valleys should be taken into account, but here we have 8 valleys (we find pairs along the principal axis are degenerate in the following). The unit vectors of the principal axes are

$$\begin{aligned} \boldsymbol{e}_m : &\pm\tfrac{1}{\sqrt{3}}(1, 1, 1) \quad &&(\text{valleys: } 1, 2), \\ &\pm\tfrac{1}{\sqrt{3}}(-1, 1, 1) &&(\text{valleys: } 3, 4), \\ &\pm\tfrac{1}{\sqrt{3}}(-1, -1, 1) &&(\text{valleys: } 5, 6), \\ &\pm\tfrac{1}{\sqrt{3}}(1, -1, 1) &&(\text{valleys: } 7, 8), \end{aligned} \tag{2.22}$$

and therefore the direction cosine γ is given by

$$\begin{aligned} \gamma &= \boldsymbol{e}_m \cdot \boldsymbol{e}_B \\ &= \pm\tfrac{1}{\sqrt{3}}(\sqrt{2}\sin\theta + \cos\theta); \quad &&(\text{valleys: } 1, 2)\,, \\ &= \pm\tfrac{1}{\sqrt{3}}(\cos\theta) &&(\text{valleys: } 3, 4, 7, 8)\,, \\ &= \pm\tfrac{1}{\sqrt{3}}(-\sqrt{2}\sin\theta + \cos\theta) &&(\text{valleys: } 5, 6)\,. \end{aligned} \tag{2.23}$$

Inserting these relations into (2.21) we obtain three solid curves as shown in Fig. 2.6, where we used the following parameters for the masses:

$$m_{\mathrm{t}} = 0.082\,m\,, \qquad m_{\mathrm{l}} = 1.58\,m\,. \tag{2.24}$$

In Fig. 2.6 good agreement is found between the experimental data and the calculated curves.

Next, we consider the case of Si, where the constant energy surfaces of the conduction band consists of 6 valleys along the $\langle 100 \rangle$ directions as shown in Fig. 2.9. We have already shown from the energy band calculations in Chap. 1 that the conduction band minima are located at $k_x = 0.85(2\pi/a)$ near the X point in the Brillouin zone. When we choose the angle θ between the magnetic field and the $\langle 001 \rangle$ axis in the (110) plane, the direction cosine γ is given by

$$\begin{aligned} \gamma &= \pm\tfrac{1}{\sqrt{2}}\sin\theta \quad (\text{valleys: } 1, 2, 3, 4)\,, \\ \gamma &= \pm\cos\theta \quad\;\; (\text{valleys: } 5, 6)\,. \end{aligned} \tag{2.25}$$

Substituting the above relations in (2.21) and assuming

$$m_{\mathrm{t}} = 0.19\,m\,, \qquad m_{\mathrm{l}} = 0.98\,m\,, \tag{2.26}$$

we obtain the solid curves in Fig. 2.7b, which show good agreement with the experimental observations.

2.2 Analysis of Valence Bands

We see in Figs. 2.4 and 2.5 that there exist two resonance peaks for holes in Ge and Si and that the curves are quite simple compared to the resonance curves for electrons. However, the analysis of the cyclotron resonance for holes is very complicated and requires more detailed analysis based on the $\boldsymbol{k} \cdot \boldsymbol{p}$ perturbation method as stated later. First, we plot the cyclotron masses of holes as a function of the angle between the magnetic field and the $\langle 001 \rangle$ axis in the (110) plane, which are shown

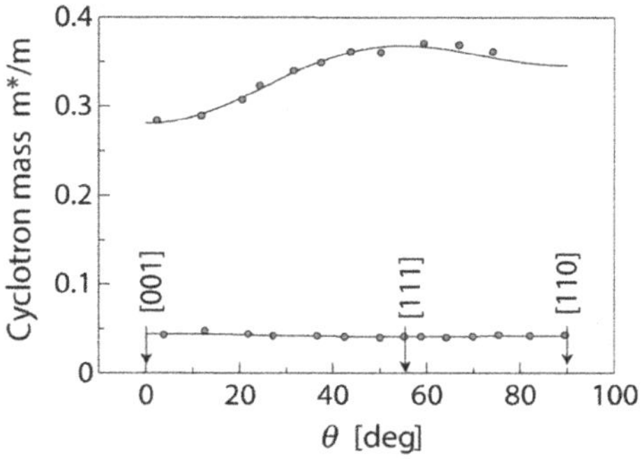

Fig. 2.10 Cyclotron masses of holes in Ge. θ is the angle between the magnetic field and $\langle 001 \rangle$ axis in the (110) plane

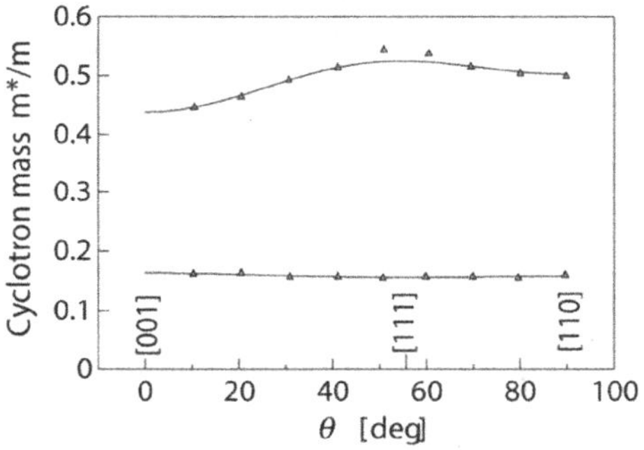

Fig. 2.11 Cyclotron masses of holes in Si. θ is the angle between the magnetic field and $\langle 001 \rangle$ axis in the (110) plane

in Fig. 2.10 for Ge and in Fig. 2.11 for Si. In the figures we find the angular dependence of the hole cyclotron masses exhibits quite different features compared with the electron cyclotron masses. The two hole masses do not cross each other and thus we can conclude that there exist two different holes. We define the heavy holes for the heavy effective-mass carriers and light holes for the light effective-mass carriers. As seen in Figs. 2.10 and 2.11 the heavy hole mass is anisotropic but the light hole mass is isotropic with narrower full-width half maximum. These features are well explained with the help of valence band analysis based on the $\boldsymbol{k} \cdot \boldsymbol{p}$ perturbation method. In the following we will consider valence band analysis by the $\boldsymbol{k} \cdot \boldsymbol{p}$ perturbation method [1].

As described in the energy band calculations of Chap. 1, the valence band top is located at the Γ point ($\boldsymbol{k} = 0$) and the valence bands are triply degenerate when the spin–angular orbit interaction is disregarded. The three valence bands are expressed by the symmetry of $\Gamma_{25'}$, whose basis functions are given by $|X\rangle$, $|Y\rangle$, and $|Z\rangle$. Also, we know that the lowest conduction band at the Γ point has the symmetry $\Gamma_{2'}$ and the basis function is $|xyz\rangle$. For simplicity, first we consider only these four bands in the $\boldsymbol{k} \cdot \boldsymbol{p}$ perturbation method. When the perturbation term is disregarded, the eigenstates and the eigenvalues of these four bands are given by

$$H_0|j\rangle = \mathcal{E}_j|j\rangle \,, \qquad (|j\rangle = |\Gamma_{2'}\rangle,\ |X\rangle,\ |Y\rangle,\ |Z\rangle) \,, \tag{2.27}$$

or the matrix elements of H_0 for the four bands are

$$\begin{array}{cccc} |\Gamma_{2'}\rangle & |X\rangle & |Y\rangle & |Z\rangle \end{array}$$
$$\begin{bmatrix} \mathcal{E}_c & 0 & 0 & 0 \\ 0 & \mathcal{E}_v & 0 & 0 \\ 0 & 0 & \mathcal{E}_v & 0 \\ 0 & 0 & 0 & \mathcal{E}_v \end{bmatrix} . \tag{2.28}$$

Next, we consider how the energy bands are modified by applying the $\boldsymbol{k} \cdot \boldsymbol{p}$ perturbation. This perturbation yields

$$(H_0 + H_1)|j\rangle = \left(\mathcal{E}_j - \frac{\hbar^2 k^2}{2m}\right)|j\rangle \equiv \lambda_j |j\rangle , \tag{2.29}$$

where

$$\lambda_j = \mathcal{E}_j - \frac{\hbar^2 k^2}{2m} . \tag{2.30}$$

The above equation gives the following relation:

$$\begin{aligned} \langle \Gamma_{2'}|H_1|X\rangle = \langle \Gamma_{2'}| - i\frac{\hbar^2}{m}\boldsymbol{k}\cdot\nabla|X\rangle &= \langle xyz| - i\frac{\hbar^2}{m}k_x\frac{\partial}{\partial x}|yz\rangle \\ &\equiv Pk_x . \end{aligned} \tag{2.31}$$

Thus we obtain the following matrix:

$$\begin{array}{cccc} |\Gamma_{2'}\rangle & |X\rangle & |Y\rangle & |Z\rangle \end{array}$$
$$\begin{bmatrix} \mathcal{E}_c - \lambda_j & Pk_x & Pk_y & Pk_z \\ Pk_x & \mathcal{E}_v - \lambda_j & 0 & 0 \\ Pk_y & 0 & \mathcal{E}_v - \lambda_j & 0 \\ Pk_z & 0 & 0 & \mathcal{E}_v - \lambda_j \end{bmatrix} = 0. \tag{2.32}$$

From the above equation we obtain

$$\mathcal{E}_{1,2} = \frac{\mathcal{E}_c + \mathcal{E}_v}{2} \pm \sqrt{\left(\frac{\mathcal{E}_c - \mathcal{E}_v}{2}\right)^2 + P^2k^2} + \frac{\hbar^2 k^2}{2m} , \tag{2.33}$$

$$\mathcal{E}_{3,4} = \mathcal{E}_v + \frac{\hbar^2 k^2}{2m} . \tag{2.34}$$

The results mean that the perturbation term H_1 gives rise to a splitting of the three valence bands: one valence band, $\mathcal{E}_2$, and doubly degenerate valence bands, $\mathcal{E}_3$ and $\mathcal{E}_4$. However, the degenerate valence bands exhibit minima at the Γ point ($\boldsymbol{k} = 0$), which is unreasonable for the valence band characteristic because a light hole and a heavy hole are observed in the cyclotron resonance measurements. The features of the valence bands are well explained when we take into account the spin–orbit interaction, which will be discussed in detail later.

Next, we consider more accurate $\boldsymbol{k} \cdot \boldsymbol{p}$ perturbation. As described in Sect. 1.7, the energy band calculations take into account many bands in addition to $\Gamma_{25'}$ and $\Gamma_{2'}$. Therefore, the unreasonable results obtained above are due to the fact that we disregarded other conduction bands than the conduction band $\Gamma_{2'}$. We will discuss a general treatment of the $\boldsymbol{k} \cdot \boldsymbol{p}$ perturbation method in the following.

The more accurate treatment of the valence bands has been reported by Dresselhaus, Kip, and Kittel [1] and Luttinger and Kohn [2]. Here we will follow the method of Luttinger and Kohn. In the perturbing Hamiltonian,

$$(H_0 + H_1)|j\rangle = (\mathcal{E}_j - \frac{\hbar^2 k^2}{2m})|j\rangle\,, \tag{2.35}$$

we consider all eigenstates in addition to the degenerate valence bands $|\Gamma_{25'}\rangle$. We learned from the above treatment that the closest conduction band $\Gamma_{2'}$ affects the perturbation only slightly. Therefore, we consider the second-order perturbation in order to calculate (2.35) and neglect the first-order perturbation. The eigenstates of the valence bands $\Gamma_{25'}$ at $\boldsymbol{k} = 0$ are defined as

$$|j\rangle = |X\rangle,\ |Y\rangle,\ |Z\rangle\,,$$

and the eigenvalues are given by

$$\mathcal{E}_j = \mathcal{E}_v \equiv \mathcal{E}_0\,.$$

The second-order perturbation of (2.35) for the valence bands $\Gamma_{25'}$ leads to the following result:

$$\mathcal{E} = \frac{\hbar^2 k^2}{2m} + \sum_{i,j'} \frac{\langle j|H_1|i\rangle\langle i|H_1|j'\rangle}{\mathcal{E}_0 - \mathcal{E}_i} + \mathcal{E}_0\,, \tag{2.36}$$

where the eigenstates $|j\rangle$, $|j'\rangle$ are the triply-degenerate valence bands $\Gamma_{25'}$ ($|X\rangle$, $|Y\rangle$, and $|Z\rangle$) with eigenvalue $\mathcal{E}_0$ and the eigenstates $|i\rangle$ correspond to any bands with eigenvalue $\mathcal{E}_i$ other than the valence bands $\Gamma_{25'}$. Since the perturbation term is given by

$$\begin{aligned} H_1 &= \frac{\hbar}{m}\boldsymbol{k} \cdot \boldsymbol{p} = -\mathrm{i}\frac{\hbar^2}{m}\boldsymbol{k} \cdot \nabla = -\mathrm{i}\frac{\hbar^2}{m}\left(k_x\frac{\partial}{\partial x} + k_y\frac{\partial}{\partial y} + k_x\frac{\partial}{\partial z}\right) \\ &= \frac{\hbar}{m}(k_x p_x + k_y p_y + k_z p_z) = \sum_{l=x,y,z} \frac{\hbar}{m} k_l p_l\,, \end{aligned} \tag{2.37}$$

Equation (2.36) is written as

$$\mathcal{E}(\boldsymbol{k}) = \mathcal{E}_0 + \frac{\hbar^2 k^2}{2m} + \frac{\hbar^2}{m^2} \sum_{l,m} \sum_{i,j'} \frac{\pi^l_{ji} \pi^m_{ij'}}{\mathcal{E}_0 - \mathcal{E}_i} k_l k_m$$

$$\equiv \mathcal{E}_0 + \sum_{l,m} \sum_{j'} D^{lm}_{jj'} k_l k_m \,, \tag{2.38}$$

where we define the following relations:

$$\langle i | p_l | j \rangle = \pi^l_{ij} \,, \tag{2.39}$$

$$D^{lm}_{jj'} = \frac{\hbar^2}{2m} \delta_{jj'} \delta_{lm} + \frac{\hbar^2}{m^2} \sum_i \frac{\pi^l_{ji} \pi^m_{ij'}}{\mathcal{E}_0 - \mathcal{E}_i} \,. \tag{2.40}$$

The eigenstates of the valence bands $|j\rangle$ and $|j'\rangle$ of $\Gamma_{25'}$ have the symmetry $|X\rangle = |yz\rangle$, $|Y\rangle = |zx\rangle$, $|Z\rangle = |xy\rangle$ as stated before and therefore we find that there are three combinations of the subscripts j, j', l, m:

$$\begin{matrix} xx & yy & zz & yz,\, zy & xz,\, zx & xy,\, yx \\ 1 & 2 & 3 & 4 & 5 & 6 \end{matrix} \,,$$

and

$$\begin{matrix} XX & YY & ZZ & YZ,\, ZY & XZ,\, ZX & XY,\, YX \\ 1 & 2 & 3 & 4 & 5 & 6 \end{matrix} \,.$$

Using the tensor notation

$$\alpha,\, \beta = 1,\, 2,\, 3,\, 4,\, 5,\, 6, \tag{2.41}$$

the matrix elements are rewritten as

$$D^{lm}_{ij} \equiv D_{\alpha\beta} \,, \tag{2.42}$$

where $D^{lm}_{jj'}$ are fourth-rank tensors. In the case of cubic crystals we have the following components

$$\begin{vmatrix} D_{11} & D_{12} & D_{12} & 0 & 0 & 0 \\ D_{12} & D_{11} & D_{12} & 0 & 0 & 0 \\ D_{12} & D_{12} & D_{11} & 0 & 0 & 0 \\ 0 & 0 & 0 & D_{44} & 0 & 0 \\ 0 & 0 & 0 & 0 & D_{44} & 0 \\ 0 & 0 & 0 & 0 & 0 & D_{44} \end{vmatrix} \,, \tag{2.43}$$

and the non-zero components are

$$
\begin{aligned}
D_{11} &= D^{xx}_{XX} = D^{yy}_{YY} = D^{zz}_{ZZ} = L\,,\\
D_{12} &= D^{yy}_{XX} = D^{ZZ}_{XX} = D^{xx}_{YY} = D^{zz}_{YY} = D^{xx}_{ZZ} = D^{yy}_{ZZ} = M\,,\\
D_{44} &= D^{xy}_{XY} + D^{yx}_{XY} = D^{yz}_{YZ} + D^{zy}_{YZ} = D^{zx}_{ZX} + D^{xz}_{ZX} = N\,.
\end{aligned}
\tag{2.44}
$$

From these results, the matrix elements of the $\boldsymbol{k}\cdot\boldsymbol{p}$ Hamiltonian are given by

$$
\begin{array}{c|ccc|}
 & |X\rangle & |Y\rangle & |Z\rangle \\
\langle X| & Ak_x^2 + B(k_y^2 + k_z^2) & Ck_xk_y & Ck_xk_z \\
\langle Y| & Ck_xk_y & Ak_y^2 + B(k_x^2 + k_z^2) & Ck_yk_z \\
\langle Z| & Ck_xk_z & Ck_yk_z & Ak_z^2 + B(k_x^2 + k_y^2)
\end{array}\,.
\tag{2.45}
$$

The three parameters A, B,and C of Luttinger are written as [3]

$$A = \frac{\hbar^2}{2m} + \frac{\hbar^2}{m^2}\sum_i \frac{\pi^x_{Xi}\pi^x_{iX}}{\mathcal{E}_0 - \mathcal{E}_i}\,, \tag{2.46a}$$

$$B = \frac{\hbar^2}{2m} + \frac{\hbar^2}{m^2}\sum_i \frac{\pi^y_{Xi}\pi^y_{iX}}{\mathcal{E}_0 - \mathcal{E}_i}\,, \tag{2.46b}$$

$$C = \frac{\hbar^2}{m^2}\sum_i \frac{\pi^x_{Xi}\pi^y_{iY} + \pi^y_{Xi}\pi^x_{iY}}{\mathcal{E}_0 - \mathcal{E}_i}\,. \tag{2.46c}$$

The secular equation of (2.45) gives the eigenenergy $\mathcal{E}(\boldsymbol{k})$, while the $\boldsymbol{k}\cdot\boldsymbol{p}$ Hamiltonian of Dresselhaus, Kip, and Kittel [1] is given by replacing A, B, and C with the following parameters, L, M, and N defined by (2.44), respectively

$$L = \frac{\hbar^2}{m^2}\sum_i \frac{\pi^x_{Xi}\pi^x_{iX}}{\mathcal{E}_0 - \mathcal{E}_i}\,, \tag{2.47a}$$

$$M = \frac{\hbar^2}{m^2}\sum_i \frac{\pi^y_{Xi}\pi^y_{iX}}{\mathcal{E}_0 - \mathcal{E}_i}\,, \tag{2.47b}$$

$$N = \frac{\hbar^2}{m^2}\sum_i \frac{\pi^x_{Xi}\pi^y_{iY} + \pi^y_{Xi}\pi^x_{iY}}{\mathcal{E}_0 - \mathcal{E}_i}\,, \tag{2.47c}$$

and the corresponding secular equation gives the eigenenergy $\lambda = \mathcal{E}(\boldsymbol{k}) - \hbar^2k^2/2\,\mathrm{m}$.

The $\boldsymbol{k}\cdot\boldsymbol{p}$ perturbation gives rise to a splitting of the triply degenerate valence bands at $\boldsymbol{k} \neq 0$ and the $\mathcal{E}(\boldsymbol{k})$ curves are given by (2.45), which are shown on the left of Fig. 2.12. Here we have to note that the cyclotron resonance tells us the existence of two hole bands, not triply degenerate bands at $\boldsymbol{k} = 0$. This is explained when we take the spin–orbit interaction into account as shown in the next section and the results are shown on the right of Fig. 2.12.

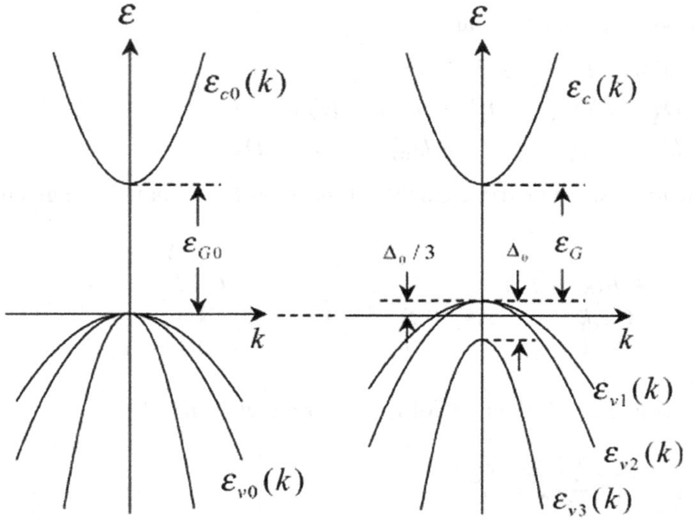

Fig. 2.12 Valence band structure. The *left figure* shows the splitting of the valence bands at $\boldsymbol{k} \neq 0$ due to the interaction between the bands, and the *right figure* shows the valence band structures obtained by taking the spin–orbit interaction into account, where the heavy hole and light hole bands are doubly-degenerate at $\boldsymbol{k} = 0$ and the lowest band is the spin–orbit split-off band due to the spin–orbit interaction (see Sect. 2.3)

2.3 Spin–Orbit Interaction

The interaction between spin angular momentum and orbital angular momentum is due to the relativistic effect of the electrons [1, 3, 4]. In Appendix H the derivation of the spin–orbit interaction Hamiltonian is described in detail.[1] Here we use the interaction Hamiltonian H_{so} given by

$$H_{\mathrm{so}} = \xi(r)\boldsymbol{L} \cdot \boldsymbol{S}, \tag{2.48}$$

$$\xi(r) = \frac{1}{2m^2c^2}\frac{1}{r}\frac{\mathrm{d}V}{\mathrm{d}r}, \tag{2.49}$$

where $\boldsymbol{L}$ and $\boldsymbol{S}$ are the orbital and spin angular momentum, respectively. When we define

$$\boldsymbol{S} = \frac{\hbar}{2}\boldsymbol{\sigma},$$

the Pauli spin matrix is given by

$$\sigma_x = \begin{bmatrix} 0 & 1 \\ 1 & 0 \end{bmatrix}, \quad \sigma_y = \begin{bmatrix} 0 & -\mathrm{i} \\ \mathrm{i} & 0 \end{bmatrix}, \quad \sigma_z = \begin{bmatrix} 1 & 0 \\ 0 & -1 \end{bmatrix}. \tag{2.50}$$

[1] See also Sect. 1.7 of Chap. 1 and (1.146) of p. 51.

Here the spin states $S_z = +\hbar/2$ and $S_z = -\hbar/2$ are sometimes referred to as $\uparrow$ spin and $\downarrow$ spin and their eigenstates are

$$|\uparrow\rangle = |\alpha\rangle = \begin{bmatrix} 1 \\ 0 \end{bmatrix}, \quad |\downarrow\rangle = |\beta\rangle = \begin{bmatrix} 0 \\ 1 \end{bmatrix}. \tag{2.51}$$

Next, we define the angular momentum operators for the $\Gamma_{25'}$ valence bands. As stated in Sect. 2.2 the eigenstates of the valence bands at $\boldsymbol{k} = 0$ are expressed by $|X\rangle$, $|Y\rangle$, $|Z\rangle$. Their angular momentum $\boldsymbol{L} = \boldsymbol{r} \times \boldsymbol{p}$ has a quantum number of unity and the matrix representation of the operator is

$$L_x = \frac{\hbar}{\sqrt{2}} \begin{bmatrix} 0 & 1 & 0 \\ 1 & 0 & 1 \\ 0 & 1 & 0 \end{bmatrix}, \quad L_y = \frac{\hbar}{\sqrt{2}} \begin{bmatrix} 0 & -\mathrm{i} & 0 \\ \mathrm{i} & 0 & -\mathrm{i} \\ 0 & \mathrm{i} & 0 \end{bmatrix}, \quad L_z = \hbar \begin{bmatrix} 1 & 0 & 0 \\ 0 & 0 & 0 \\ 0 & 0 & -1 \end{bmatrix}. \tag{2.52}$$

When we choose the basis of the eigenstates defined by

$$u_+ = \frac{1}{\sqrt{2}}(|X\rangle + \mathrm{i}|Y\rangle) = \begin{bmatrix} 1 \\ 0 \\ 0 \end{bmatrix}, \tag{2.53a}$$

$$u_- = \frac{1}{\sqrt{2}}(|X\rangle - \mathrm{i}|Y\rangle) = \begin{bmatrix} 0 \\ 0 \\ 1 \end{bmatrix}, \tag{2.53b}$$

$$u_z = |Z\rangle = \begin{bmatrix} 0 \\ 1 \\ 0 \end{bmatrix}, \tag{2.53c}$$

the angular momentum operators are given by

$$L_\pm = L_x \pm \mathrm{i}L_y, \tag{2.54}$$

where

$$L_+ = \sqrt{2}\hbar \begin{bmatrix} 0 & 1 & 0 \\ 0 & 0 & 1 \\ 0 & 0 & 0 \end{bmatrix}, \quad L_- = \sqrt{2}\hbar \begin{bmatrix} 0 & 0 & 0 \\ 1 & 0 & 0 \\ 0 & 1 & 0 \end{bmatrix}. \tag{2.55}$$

Similarly, the spin momentum operators are expressed as

$$\sigma_\pm = \sigma_x \pm \mathrm{i}\sigma_y, \tag{2.56}$$

and their matrix representations are given by

$$\sigma_+ = 2 \begin{bmatrix} 0 & 1 \\ 0 & 0 \end{bmatrix}, \quad \sigma_- = 2 \begin{bmatrix} 0 & 0 \\ 1 & 0 \end{bmatrix}. \tag{2.57}$$

Using these results, the Hamiltonian of the spin–orbit interaction becomes

$$\begin{aligned} H_{\text{so}} &= \frac{\Delta}{\hbar}\boldsymbol{L}\cdot\boldsymbol{\sigma} \\ &= \frac{\Delta}{\hbar}\left(L_x\sigma_x + L_y\sigma_y + L_z\sigma_z\right) \\ &= \frac{\Delta}{\hbar}\left[\frac{1}{2}\left(L_+\sigma_- + L_-\sigma_+\right) + L_z\sigma_z\right], \end{aligned} \tag{2.58}$$

where $3\Delta = \Delta_0$ is called *spin–orbit splitting energy* and gives the value of the energy splitting of the valence bands at $\boldsymbol{k} = 0$.

The six states of the valence bands $|u_+\alpha\rangle$, $|u_+\beta\rangle$, $|u_z\alpha\rangle$, $|u_-\beta\rangle$, $|u_-\alpha\rangle$, $|u_z\beta\rangle$ give rise to the matrix elements of the spin–orbit interaction Hamiltonian:

$$\begin{array}{cccccc} u_+\alpha & u_+\beta & u_z\alpha & u_-\beta & u_-\alpha & u_z\beta \end{array}$$
$$\begin{vmatrix} \Delta & 0 & 0 & 0 & 0 & 0 \\ 0 & -\Delta & \sqrt{2}\Delta & 0 & 0 & 0 \\ 0 & \sqrt{2}\Delta & 0 & 0 & 0 & 0 \\ 0 & 0 & 0 & \Delta & 0 & 0 \\ 0 & 0 & 0 & 0 & -\Delta & \sqrt{2}\Delta \\ 0 & 0 & 0 & 0 & \sqrt{2}\Delta & 0 \end{vmatrix}. \tag{2.59}$$

Expression (2.59) is derived as follows:

$$\begin{aligned} L_+\sigma_-|u_+\alpha\rangle &= L_+|u_+\rangle\sigma_-|\alpha\rangle \\ &= \sqrt{2}\hbar\begin{bmatrix} 0&1&0\\0&0&1\\0&0&0 \end{bmatrix}\begin{bmatrix}1\\0\\0\end{bmatrix} \times 2\begin{bmatrix}0&0\\1&0\end{bmatrix}\begin{bmatrix}1\\0\end{bmatrix} = 0 \end{aligned} \tag{2.60a}$$

$$\begin{aligned} L_-\sigma_+|u_+\alpha\rangle &= L_-|u_+\rangle\sigma_+|\alpha\rangle \\ &= \sqrt{2}\hbar\begin{bmatrix} 0&0&0\\1&0&0\\0&1&0 \end{bmatrix}\begin{bmatrix}1\\0\\0\end{bmatrix} \times 2\begin{bmatrix}0&1\\0&0\end{bmatrix}\begin{bmatrix}1\\0\end{bmatrix} = 0 \end{aligned} \tag{2.60b}$$

$$\begin{aligned} L_z\sigma_z|u_+\alpha\rangle &= \hbar\begin{bmatrix} 1&0&0\\0&0&0\\0&0&-1 \end{bmatrix}\begin{bmatrix}1\\0\\0\end{bmatrix}\begin{bmatrix}1&0\\0&-1\end{bmatrix}\begin{bmatrix}1\\0\end{bmatrix} \\ &= \hbar\begin{bmatrix}1\\0\\0\end{bmatrix}\begin{bmatrix}1\\0\end{bmatrix} = \hbar|u_+\rangle|\alpha\rangle = \hbar|u_+\alpha\rangle\,. \end{aligned} \tag{2.60c}$$

Using the above results we find

$$\langle u_+\alpha|L_z\sigma_z|u_+\alpha\rangle = \hbar\,, \tag{2.61}$$

and a similar treatment gives the results of (2.59).

Expression (2.59) is expressed as a 6×6 matrix and many of the non-diagonal elements have the value 0. When we investigate the matrix of (2.59), we find that the matrix is factorized into two 3 × 3 matrices and that they are also factorized into two matrices of dimensions 1 × 1 and 2 × 2. The 2 × 2 matrix is easily diagonalized and (2.59) reduces to

$$\begin{vmatrix} \Delta & 0 & 0 & 0 & 0 & 0 \\ 0 & \Delta & 0 & 0 & 0 & 0 \\ 0 & 0 & -2\Delta & 0 & 0 & 0 \\ 0 & 0 & 0 & \Delta & 0 & 0 \\ 0 & 0 & 0 & 0 & \Delta & 0 \\ 0 & 0 & 0 & 0 & 0 & -2\Delta \end{vmatrix} . \tag{2.62}$$

From the result given in (2.62) we see that the 6-fold degenerate valence bands split into 4-fold valence bands with energy shifted by Δ, consisting of the heavy hole and light hole bands, and into a doubly degenerate spin–orbit split-off band shifted down by 2Δ. These features are shown in Fig. 2.12. It should be noted that the 4-fold degenerate bands at $\boldsymbol{k} = 0$ split into the doubly degenerate heavy hole and light hole bands at $\boldsymbol{k} \neq 0$. The corresponding eigenstates are given by the following equations. The fourfold valence band–edge eigenstates are

$$u_{v1} = \left|\frac{3}{2}, \frac{3}{2}\right\rangle = |u_+\alpha\rangle = \frac{1}{\sqrt{2}}|(X + \mathrm{i}Y) \uparrow\rangle , \tag{2.63a}$$

$$\begin{aligned} u_{v2} &= \left|\frac{3}{2}, \frac{1}{2}\right\rangle = -\frac{1}{\sqrt{3}}|u_+\beta\rangle - \sqrt{\frac{2}{3}}|u_z\alpha\rangle \\ &= -\frac{1}{\sqrt{6}}[|(X + \mathrm{i}Y) \downarrow\rangle + 2|Z \uparrow\rangle] , \end{aligned} \tag{2.63b}$$

$$\begin{aligned} u_{v3} &= \left|\frac{3}{2}, -\frac{1}{2}\right\rangle = -\frac{1}{\sqrt{3}}|u_-\alpha\rangle - \sqrt{\frac{2}{3}}|u_z\beta\rangle \\ &= -\frac{1}{\sqrt{6}}[|(X - \mathrm{i}Y) \uparrow\rangle + 2|Z \downarrow\rangle] , \end{aligned} \tag{2.63c}$$

$$u_{v4} = \left|\frac{3}{2}, -\frac{3}{2}\right\rangle = |u_-\beta\rangle = \frac{1}{\sqrt{2}}|(X - \mathrm{i}Y) \downarrow\rangle , \tag{2.63d}$$

and for twofold spin–spit off bands are

$$\begin{aligned} u_{v5} &= \left|\frac{1}{2}, \frac{1}{2}\right\rangle = -\sqrt{\frac{2}{3}}|u_+\beta\rangle + \sqrt{\frac{1}{3}}|u_z\alpha\rangle \\ &= -\frac{1}{\sqrt{3}}[|(X + \mathrm{i}Y) \downarrow\rangle - |Z \uparrow\rangle] , \end{aligned} \tag{2.63e}$$

$$\begin{aligned} u_{v6} &= \left|\frac{1}{2}, -\frac{1}{2}\right\rangle = -\sqrt{\frac{2}{3}}|u_-\alpha\rangle + \frac{1}{\sqrt{3}}|u_z\beta\rangle \\ &= -\frac{1}{\sqrt{3}}[|(X - \mathrm{i}Y) \uparrow\rangle - |Z \downarrow\rangle] . \end{aligned} \tag{2.63f}$$

The eigenstates with the time reversal symmetry are reported by Luttinger and Kohn [2], which are shown by (2.137b) $\sim$ (2.137g).

Now we discuss the cyclotron masses or the effective masses of the valence bands by deriving the $\boldsymbol{k}$ dependence of the bands. This is done by applying the method of Luttinger and Kohn [2] to the analysis of the valence bands, where we have to use the eigenstates (basis functions) derived by taking the spin–orbit interaction into account. This approach has been adopted by Dresselhaus, Kip and Kittel [1]. They used the eigenfunctions of (2.63a–2.63f) and calculated the matrix elements of the Hamiltonian with the spin–orbit interaction. Defining the matrix element H_{ij} by

$$\begin{aligned}
H_{11} &= Lk_x^2 + M(k_y^2 + k_z^2)\,,\\
H_{22} &= Lk_y^2 + M(k_z^2 + k_x^2)\,,\\
H_{33} &= Lk_z^2 + M(k_x^2 + k_y^2)\,,\\
H_{12} &= Nk_xk_y\,,\\
H_{23} &= Nk_yk_z\,,\\
H_{13} &= Nk_xk_z\,,
\end{aligned}$$

the result is given by the 6 × 6 matrix of (62) in their paper [1]:

$$\begin{vmatrix}
\frac{H_{11}+H_{22}}{2} & -\frac{H_{13}-\mathrm{i}H_{23}}{\sqrt{3}} & -\frac{H_{11}-H_{22}-2\mathrm{i}H_{12}}{2\sqrt{3}} & 0 & -\frac{H_{13}-\mathrm{i}H_{23}}{\sqrt{6}} & -\frac{H_{11}-H_{22}-2\mathrm{i}H_{12}}{\sqrt{6}} \\
-\frac{H_{13}+\mathrm{i}H_{23}}{\sqrt{3}} & \frac{4H_{33}+H_{11}+H_{22}}{6} & 0 & -\frac{H_{11}-H_{22}-2\mathrm{i}H_{12}}{2\sqrt{3}} & -\frac{H_{11}+H_{22}-2H_{33}}{3\sqrt{2}} & \frac{H_{13}-\mathrm{i}H_{23}}{\sqrt{2}} \\
-\frac{H_{11}-H_{22}+2\mathrm{i}H_{12}}{2\sqrt{3}} & 0 & \frac{4H_{33}+H_{11}+H_{22}}{6} & \frac{H_{13}-\mathrm{i}H_{23}}{\sqrt{3}} & \frac{H_{13}+\mathrm{i}H_{23}}{\sqrt{2}} & \frac{H_{11}+H_{22}-2H_{33}}{3\sqrt{2}} \\
0 & -\frac{H_{11}-H_{22}+2\mathrm{i}H_{12}}{2\sqrt{3}} & \frac{H_{13}+\mathrm{i}H_{23}}{\sqrt{3}} & \frac{H_{11}+H_{22}}{2} & \frac{H_{11}-H_{22}+2\mathrm{i}H_{12}}{\sqrt{6}} & -\frac{H_{13}+\mathrm{i}H_{23}}{\sqrt{6}} \\
-\frac{H_{13}+\mathrm{i}H_{23}}{\sqrt{6}} & -\frac{H_{11}+H_{22}-2H_{33}}{3\sqrt{2}} & \frac{H_{13}-\mathrm{i}H_{23}}{\sqrt{2}} & \frac{H_{11}-H_{22}-2\mathrm{i}H_{12}}{\sqrt{6}} & \frac{H_{11}+H_{22}+H_{33}}{3}-\Delta_0 & 0 \\
-\frac{H_{11}-H_{22}+2\mathrm{i}H_{12}}{\sqrt{6}} & \frac{H_{13}+\mathrm{i}H_{23}}{\sqrt{2}} & \frac{H_{11}+H_{22}-2H_{33}}{3\sqrt{2}} & -\frac{H_{13}-\mathrm{i}H_{23}}{\sqrt{6}} & 0 & \frac{H_{11}+H_{22}+H_{33}}{3}-\Delta_0
\end{vmatrix}\,. \tag{2.64}$$

The matrix of (2.64) is approximately divided into two matrices comprising the upper-left 4 × 4 matrix and the lower-right 2 × 2 matrix [1]. The other two blocks of 2 × 4 will give rise to an error of the order of k^4/Δ_0 and thus the above approximation is accurate enough to enable us to discuss the valence bands. The 4 × 4 matrix gives the solutions of

$$\mathcal{E}(\boldsymbol{k}) = Ak^2 \pm \sqrt{B^2k^4 + C^2(k_x^2k_y^2 + k_y^2k_z^2 + k_z^2k_x^2)}\,, \tag{2.65}$$

where

$$\begin{aligned} A &= \frac{1}{3}(L + 2M), \\ B &= \frac{1}{3}(L - M), \\ C &= \frac{1}{3}[N^2 - (L - M)^2], \end{aligned} \tag{2.66}$$

where the coefficients A, B, and C differ from the parameters of (2.46a) $\sim$ (2.46c) defined by Luttinger.

On the other hand, from the 2×2 matrix we obtain the solutions

$$\mathcal{E}(\boldsymbol{k}) = Ak^2 - \Delta_0 . \tag{2.67}$$

Equation (2.65) gives the $\boldsymbol{k}$ dependence of the heavy and light holes in the valence bands and (2.67) leads to the spin–orbit split-off band. It is evident from the above analysis that the splitting is due to the spin–orbit interaction. The constant energy surface of the heavy-hole band is not spherical for $C \neq 0$ but warped as shown in Fig. 2.13a, where the energy contour of the heavy hole band of Si in the (k_x, k_y, k_z) space for $\mathcal{E} = 0.20$ [eV] are shown, and the contour lines in the (001) plane for $k_z = 0$ are plotted in Fig. 2.13b in the region $k_x \leq 0.1,\ k_y \leq 0.1$. Energy contour of the heavy hole band in the region of $k_x, k_y, k_z \simeq 0$ are spherical but for larger values of k_x, k_y and k_z the contour lines are warped. On the other hand, the light-hole band is almost spherical in the wide range of wave vectors as seen in Fig. 2.13c.

As stated above, the valence bands, especially the heavy hole band, are complicated and thus require us to use a different approach for the analysis of the cyclotron masses in the valence bands [1]. We use the cylindrical coordinates (k_H, ρ, ϕ) for the wave vector of the hole, where k_H is the z component of the wave vector (parallel to the applied magnetic field). Using this notation the cyclotron mass is expressed as

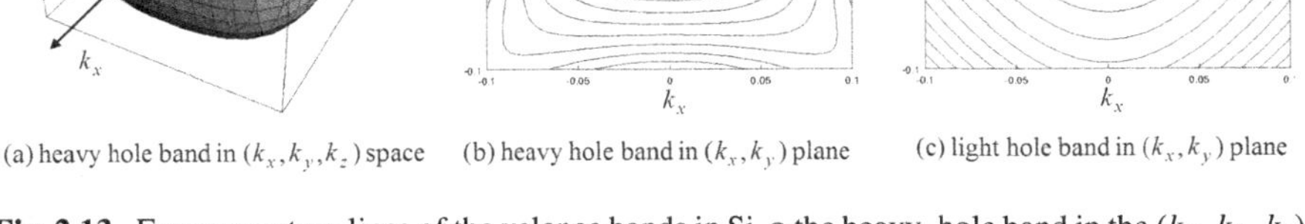

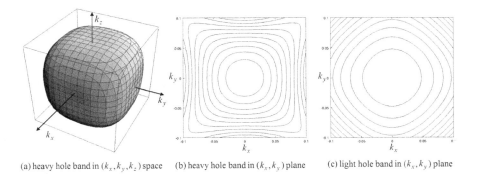

Fig. 2.13 Energy contour lines of the valence bands in Si, **a** the heavy–hole band in the (k_x, k_y, k_z) space for $\mathcal{E} = 0.20$ [eV], **b** the heavy–hole band in the (001) plane $(k_z = 0)$, and **c** the light–hole band in the (001) plane $(k_z = 0)$, where k_x, k_y, k_z are in units of $2\pi/a$

$$m^* = \frac{\hbar^2}{2\pi} \oint \frac{\rho \, \mathrm{d}\phi}{\partial \mathcal{E}/\partial \rho} . \tag{2.68}$$

Assuming the magnetic field is applied in the (001) plane and putting $k_H = 0$, we obtain

$$m^* = \frac{\hbar^2}{\pi} \int_0^{\pi/2} \frac{\mathrm{d}\phi}{A \pm \{B^2 + \frac{1}{4}C^2[1 + g(\phi)]\}^{1/2}} , \tag{2.69}$$

where

$$g(\phi) = -3(3\cos^2\theta - 1)[(\cos^2\theta - 3)\cos^4\phi + 2\cos^2\phi] , \tag{2.70}$$

and θ is the angle with respect to the (001) axis in the (110) plane. The term $g(\phi)$ is expanded as

$$\begin{aligned} m^* &= \frac{\hbar^2}{2} \frac{1}{A \pm \sqrt{B^2 + (C/2)^2}} \\ &\quad \times \left\{ 1 \pm \frac{C^2(1 - 3\cos^2\theta)^2}{64\sqrt{B^2 + (C/2)^2}\left[A \pm \sqrt{B^2 + (C/2)^2}\right]} + \cdots \right\} . \end{aligned} \tag{2.71}$$

Using these results and choosing the parameters A, B and C to fit the experimental data of the cyclotron mass we find the curves shown in Figs. 2.10 and 2.11. From these analyses we derive the valence band parameters A, B and C, which are listed below.

$$\begin{aligned} A &= -(13.0 \pm 0.2)(\hbar^2/2m) , \\ |B| &= (8.9 \pm 0.1)(\hbar^2/2m) , \\ |C| &= (10.3 \pm 0.2)(\hbar^2/2m) , \end{aligned} \tag{2.72}$$

for Ge and

$$\begin{aligned} A &= -(4.1 \pm 0.2)(\hbar^2/2m) , \\ |B| &= (1.6 \pm 0.2)(\hbar^2/2m) , \\ |C| &= (3.3 \pm 0.5)(\hbar^2/2m) , \end{aligned} \tag{2.73}$$

for Si. It may be pointed out here that the parameters L, M and N are easily deduced from (2.66) using the values of A, B and C given above [1]. The effective masses of the heavy and light holes averaged in $\boldsymbol{k}$ space are

$$\frac{m}{m^*_{\mathrm{hh}}} = -A - \sqrt{B^2 + \frac{C^2}{5}} , \tag{2.74}$$

$$\frac{m}{m^*_{\mathrm{lh}}} = -A + \sqrt{B^2 + \frac{C^2}{5}} , \tag{2.75}$$

where A, B, and C are given by dimensionless values in units of $\hbar^2/2\,\mathrm{m}$.[2] The average effective masses of the heavy and light holes are estimated to be 0.336 m (0.520 m) and 0.0434 m (0.159 m) for Ge (Si), respectively.

2.4 Non-parabolicity of the Conduction Band

We are concerned with the valence band structure by taking the spin–orbit interaction in the previous section. Here in this section we will deal with the effect of the spin–orbit interaction on the conduction band, where we consider the conduction band $|\Gamma_{2'}\rangle$ and the valence bands $|\Gamma_{25'}\rangle$ only for simplicity. In this approximation the matrix elements of the $\boldsymbol{k}\cdot\boldsymbol{p}$ Hamiltonian are

$$\begin{vmatrix} \mathcal{E}_{c0}-\lambda & 0 & Pk & 0 \\ 0 & \mathcal{E}_{v0}-\Delta_0/3-\lambda & \sqrt{2}\Delta_0/3 & 0 \\ Pk & \sqrt{2}\Delta_0/3 & \mathcal{E}_{v0}-\lambda & 0 \\ 0 & 0 & 0 & \mathcal{E}_{v0}+\Delta_0/3-\lambda \end{vmatrix} = 0\,, \tag{2.76}$$

where $P=(\hbar/m)\langle\Gamma_{2'}\,|p_x|\,X\rangle$ is defined by (2.31) and $\lambda=\mathcal{E}-k^2$. For simplicity we put $k_x=k_y=0, k_z=k$ without any loss of generality. From the upper-left 3×3 matrix we obtain

$$\left(\lambda-\mathcal{E}_{v0}+\frac{2\Delta_0}{3}\right)\left(\lambda-\mathcal{E}_{v0}-\frac{\Delta_0}{3}\right)\left(\lambda-\mathcal{E}_{c0}\right) - P^2k^2\left(\lambda-\mathcal{E}_{v0}+\frac{\Delta_0}{3}\right)=0\,. \tag{2.77}$$

We may write

$$\mathcal{E}_c=\mathcal{E}_{c0}, \qquad \mathcal{E}_v=\mathcal{E}_{v0}+\frac{\Delta_0}{3}, \qquad \mathcal{E}_c-\mathcal{E}_v-\mathcal{E}_G\,, \tag{2.78}$$

and then (2.77) reduces to

$$\left(\lambda-\mathcal{E}_v+\Delta_0\right)\left(\lambda-\mathcal{E}_v\right)\left(\lambda-\mathcal{E}_c\right)-P^2k^2\left(\lambda-\mathcal{E}_v+\frac{2\Delta_0}{3}\right)=0\,. \tag{2.79}$$

Since we are interested in the effective mass of the conduction band edge, the term $\hbar^2k^2/2\,\mathrm{m}$ is taken to be small and λ in the terms other than $\lambda-\mathcal{E}_c$ is replaced by $\mathcal{E}_c$.

---

[2] In many references the average effective mass is expressed as

$$\frac{m}{m^*}=-A\pm\sqrt{B^2+\frac{C^2}{6}}\,,$$

which should be corrected as above.

This approximation gives rise to

$$\lambda = \mathcal{E}_c(\boldsymbol{k}) - \frac{\hbar^2 k^2}{2m} = \frac{P^2 k^2 (\mathcal{E}_c - \mathcal{E}_v + 2\Delta_0/3)}{(\mathcal{E}_c - \mathcal{E}_v + \Delta_0)(\mathcal{E}_c - \mathcal{E}_v)} + \mathcal{E}_c\,, \tag{2.80}$$

or

$$\mathcal{E}(\boldsymbol{k}) = \frac{\hbar^2 k^2}{2m} + \frac{P^2 k^2}{3}\left[\frac{2}{\mathcal{E}_{\rm G}} + \frac{1}{\mathcal{E}_{\rm G} + \Delta_0}\right] + \mathcal{E}_c\,, \tag{2.81}$$

and therefore the band edge effective mass m_0^* is obtained as

$$\frac{1}{m_0^*} = \frac{1}{m} + \frac{2P^2}{3\hbar^2}\left[\frac{2}{\mathcal{E}_{\rm G}} + \frac{1}{\mathcal{E}_{\rm G} + \Delta_0}\right]. \tag{2.82}$$

We introduce energy parameter $\mathcal{E}_{P0}$ which is often used to analyze the valence band parameters,

$$\mathcal{E}_{P0} = \frac{2}{m} P_0^2\,, \tag{2.83a}$$

$$P_0 = \langle \Gamma_{25'}(X) | p_x | \Gamma_{2'} \rangle\,, \quad P = \frac{\hbar}{m} P_0\,, \tag{2.83b}$$

and then (2.82) is rewritten as

$$\frac{1}{m_0^*} = \frac{1}{m} + \frac{\mathcal{E}_{P0}}{3m}\left[\frac{2}{\mathcal{E}_{\rm G}} + \frac{1}{\mathcal{E}_{\rm G} + \Delta_0}\right] \tag{2.84}$$

This relation is known to explain the observed effective masses quite well in III–V compound semiconductors, as shown below. Here we discuss the case of GaAs as an example. The parameter $\mathcal{E}_{P0} = (2/m)P_0^2$ is rewritten by using the definition of $P_{\rm au} = 2P_0$ used by Cardona and Pollak [5] as $\mathcal{E}_{P0} = P_{\rm au}^2/(2m)(= P_{\rm au}^2$ in atomic units) $= 1.32^2 \times 13.6 = 23.7\,{\rm eV}$ (see Table 1.11),[3] the energy gap $\mathcal{E}_{\rm G} = 1.53\,{\rm eV}$ and the spin–orbit split-off energy $\Delta_0 = 0.34\,{\rm eV}$ give the band-edge effective mass $m_0^* = 0.0643\,{\rm m}$ from (2.84), which is in good agreement with the observed value 0.067 m.

When we put the relation $\lambda = \mathcal{E}_c(\boldsymbol{k}) - k^2$ $(= \mathcal{E}_c(\boldsymbol{k}) - \hbar^2 k^2/2m)$ into (2.79), the $\boldsymbol{k}$ dependence of the energy of the conduction band electrons $\mathcal{E}_c(\boldsymbol{k})$ deviates from the parabolic band

$$\mathcal{E}_c(\boldsymbol{k}) = \frac{\hbar^2 k^2}{2m_0^*}\,. \tag{2.85}$$

[3]Note that P used in Tables 1.9 and 1.11 is $P_{\rm au} = 2\langle \Gamma_{2'} | p_x | X \rangle = 2P_0$ and thus $\mathcal{E}_{P0} = (2/m)P_0^2 = P_{\rm au}^2/2\,{\rm m}$ which is equivalent to $P_{\rm au}^2$ in atomic units.

Such a band is called a *non-parabolic band.* The conduction band of the order of k^4 is obtained by neglecting the term λ^3 which gives the energy of the order of k^6. The solution of the quadratic equation of λ is given by

$$\mathcal{E}(\boldsymbol{k}) = \frac{\hbar^2 k^2}{2m_0^*} - \left(1 - \frac{m_0^*}{m}\right)^2 \left(\frac{\hbar^2 k^2}{2m_0^*}\right)^2 \left[\frac{3\mathcal{E}_G + 4\Delta_0 + 2\Delta_0^2/\mathcal{E}_G}{(\mathcal{E}_G + \Delta_0)(3\mathcal{E}_G + 2\Delta_0)}\right] . \tag{2.86}$$

Another good example of a non-parabolic conduction band is the case of InSb, where $\Delta_0 \gg \mathcal{E}_G$ is satisfied. In this case we may approximate (2.79) by

$$(\lambda - \mathcal{E}_c)(\lambda - \mathcal{E}_v) - \frac{2}{3} P^2 k^2 = 0 , \tag{2.87}$$

and thus the conduction band is expressed by the following equation.

$$\mathcal{E}(\boldsymbol{k}) = \frac{\hbar^2 k^2}{2m} + \frac{\mathcal{E}_c + \mathcal{E}_v}{2} + \sqrt{\frac{(\mathcal{E}_c - \mathcal{E}_v)^2}{4} + \frac{2}{3} P^2 k^2} . \tag{2.88}$$

This approximation is done by considering the conduction band and the top valence band only, and thus it is called the **two-band approximation**. When the energy reference is taken to be the conduction band bottom ($\mathcal{E}_c = 0$), the conduction band non-parabolicity is expressed as

$$\begin{aligned} \mathcal{E}(\boldsymbol{k}) &= \frac{\hbar^2 k^2}{2m} + \frac{\mathcal{E}_G}{2}\left[\sqrt{1 + \left(\frac{8P^2}{3\mathcal{E}_G^2}\frac{2m}{\hbar^2}\right)\frac{\hbar^2 k^2}{2m}} + 1\right] \\ &= \frac{\hbar^2 k^2}{2m} + \frac{\mathcal{E}_G}{2}\left[\sqrt{1 + \frac{4\mathcal{E}_p}{\mathcal{E}_G^2} \cdot \frac{\hbar^2 k^2}{2m}} + 1\right] , \end{aligned} \tag{2.89}$$

where the relation (note the difference between $\mathcal{E}_p$ and $\mathcal{E}_{P0}$)

$$\mathcal{E}_p = \frac{4mP^2}{3\hbar^2} \equiv \frac{2}{3}\mathcal{E}_{P0} \tag{2.90}$$

is used.[4] (2.89) gives the solution for $\hbar^2 k^2/2m$

$$\frac{\hbar^2 k^2}{2m} = \mathcal{E} + \frac{\mathcal{E}_G + \mathcal{E}_p}{2}\left[1 - \sqrt{1 + \frac{4\mathcal{E}_p \mathcal{E}}{(\mathcal{E}_G + \mathcal{E}_p)^2}}\right] . \tag{2.91}$$

---

[4]In Sect. 2.6 we deal with the Luttinger parameters and introduce $\mathcal{E}_{P0} = (2/m)P_0^2$ with the momentum matrix element $P_0 = \langle \Gamma_{2'} | p_x | X \rangle$. The parameter $\mathcal{E}_{P0}$ is often cited for the analysis of the valence band states and of the Luttinger parameters [3, 6]. The values of $\mathcal{E}_{P0}$ for various semiconductors range from 19 to 27 eV. The matrix element introduced by Cardona and Pollak [5] is $P = 2P_0$ and $\mathcal{E}_{P0} = P^2/2m$. Also note the difference between the subscripts of $\mathcal{E}_p$ and of $\mathcal{E}_{P0}$.

When $\mathcal{E}$ is small, the terms of the square root of the above equation may be expanded to give the following result:

$$\frac{\hbar^2 k^2}{2m} = \mathcal{E}\left[\frac{\mathcal{E}_G}{\mathcal{E}_G + \mathcal{E}_p} + \frac{\mathcal{E}_p^2 \mathcal{E}}{(\mathcal{E}_G + \mathcal{E}_p)^3} - \frac{2\mathcal{E}_p^3 \mathcal{E}^2}{(\mathcal{E}_G + \mathcal{E}_p)^5} + \cdots\cdots\right]. \tag{2.92}$$

From the fact that the energy band approaches the parabolic band $\mathcal{E} = \hbar^2 k^2/2m_0^*$ in the limit of $\mathcal{E} \to 0$, the band-edge effective mass is given by

$$m_0^*/m = \frac{\mathcal{E}_G}{\mathcal{E}_G + \mathcal{E}_p}. \tag{2.93}$$

For example, in the case of InSb we have $\mathcal{E}_G = 0.235\,\text{eV}$, $m_0^*/m = 0.0138$, $\Delta_0 = 0.9\,\text{eV}$, and thus the condition $\Delta_0 > \mathcal{E}_G$ is satisfied. From the above result we obtain $\mathcal{E}_p \approx 17\,\text{eV}$, which shows a good agreement with the parameter used for the energy band calculations, $\mathcal{E}_p \approx 20\,\text{eV}$. From (2.92) and (2.93) the dispersion of the conduction band is rewritten as

$$\frac{\hbar^2 k^2}{2m_0^*} = \gamma(\mathcal{E}) = \mathcal{E}[1 + \alpha\mathcal{E} + \beta\mathcal{E}^2 + \cdots\cdots], \tag{2.94}$$

where α and β are defined by

$$\alpha = \frac{\mathcal{E}_p^2}{\mathcal{E}_G(\mathcal{E}_G + \mathcal{E}_p)^2} = \left(1 - \frac{m_0^*}{m}\right)^2 \cdot \frac{1}{\mathcal{E}_G}, \tag{2.95}$$

$$\beta = -\frac{2\mathcal{E}_p^3}{\mathcal{E}_G(\mathcal{E}_G + \mathcal{E}_p)^4}. \tag{2.96}$$

Since $\mathcal{E}_G \ll \mathcal{E}_p$ in general, we can neglect the terms beyond the second term on the right-hand side and we obtain

$$\frac{\hbar^2 k^2}{2m_0^*} \equiv \gamma(\mathcal{E}) = \mathcal{E}\left(1 + \frac{\mathcal{E}}{\mathcal{E}_G}\right), \tag{2.97}$$

which is referred to in the literature very often to express the conduction band non-parabolicity.

2.5 Electron Motion in a Magnetic Field and Landau Levels

2.5.1 Landau Levels

In this section we will deal with the motion of an electron in a magnetic field using the effective-mass equation described in Sect. 3.2. The treatment is based on the method developed by Kubo et al. [7, 8]. The Hamiltonian of an electron in a magnetic field $\boldsymbol{B}$ is written as

$$H = \frac{1}{2m}(\boldsymbol{p} + e\boldsymbol{A})^2 , \tag{2.98}$$

where $\boldsymbol{A}$ is the vector potential defined as

$$\boldsymbol{B} = \nabla \times \boldsymbol{A} . \tag{2.99}$$

Since the vector potential satisfies the relation $\nabla \cdot \boldsymbol{A} = 0$, we find the following relation

$$\boldsymbol{A} \cdot \boldsymbol{p} - \boldsymbol{p} \cdot \boldsymbol{A} = \mathrm{i}\hbar\nabla \cdot \boldsymbol{A} = 0 \tag{2.100}$$

and thus the momentum operator and the vector potential commute with each other. We define the following general momentum operator [7]:

$$\boldsymbol{\pi} = \boldsymbol{p} + e\boldsymbol{A} . \tag{2.101}$$

Then the effective mass Hamiltonian is given by the following equation:

$$H = \mathcal{E}_0(\boldsymbol{\pi}) + U(\boldsymbol{r}) , \tag{2.102}$$

where $\mathcal{E}_0(\boldsymbol{p})$ represents the Bloch state of the electron with **crystal momentum** $\boldsymbol{p} = \hbar\boldsymbol{k}$ and $U(\boldsymbol{r})$ is the perturbing potential. As shown in Chap. 3 the effective-mass equation allows us to express the operator as $\mathcal{E}_0(-\mathrm{i}\hbar\nabla)$ without a magnetic field or $\mathcal{E}_0(-\mathrm{i}\hbar\nabla + e\boldsymbol{A})$ with a magnetic field. The momentum operator satisfies the following commutation relation:

$$\boldsymbol{\pi} \times \boldsymbol{\pi} = -\mathrm{i}\hbar e\boldsymbol{B} . \tag{2.103}$$

When the magnetic field $\boldsymbol{B}$ is applied in the z direction, we obtain

$$\pi_x = p_x + eA_x, \tag{2.104a}$$

$$\pi_y = p_y + eA_y, \tag{2.104b}$$

$$\pi_z = p_z \tag{2.104c}$$

and thus the commutation relations are

$$\left[\pi_x, \pi_y\right] = -\mathrm{i}\hbar eB_z \tag{2.105a}$$

$$\left[\pi_x, x\right] = \left[\pi_y, y\right] = -\mathrm{i}\hbar . \tag{2.105b}$$

Here we introduce following new variables

$$\xi = \frac{1}{eB_z}\pi_y, \quad \eta = -\frac{1}{eB_z}\pi_x , \tag{2.106a}$$

$$X = x - \xi, \quad Y = y - \eta . \tag{2.106b}$$

Using (2.105b) we obtain the following relation,

$$[\xi, \eta] = \frac{\hbar}{\mathrm{i}} \frac{1}{eB_z} \equiv \frac{l^2}{\mathrm{i}} \tag{2.107a}$$

$$[X, Y] = -\frac{\hbar}{\mathrm{i}} \frac{1}{eB_z} \equiv -\frac{l^2}{\mathrm{i}} \tag{2.107b}$$

$$[\xi, X] = [\eta, X] = [\xi, Y] = [\eta, Y] = 0 , \tag{2.107c}$$

where

$$l = \sqrt{\hbar / eB_z} \tag{2.108}$$

is the cyclotron radius of the ground state and is independent of the electron effective mass. From the commutation relations stated above, the combination of the following variables form a complete set of canonical variables:

$$(\xi, \eta), \quad (X, Y), \quad (p_z, z) .$$

The Hamiltonian is then expressed as follows by using the relative coordinates:

$$H = \frac{\hbar^2}{2m^* l^4}(\xi^2 + \eta^2) + \frac{\hbar^2 k_z^2}{2m^*} . \tag{2.109}$$

As an example, we deal first with the case without a perturbing potential ($U = 0$). Since the Hamiltonian $\mathcal{E}_0(\boldsymbol{\pi})$ does not include X and Y, these variables will not change in time. This is consistent with the physical model of the cyclotron motion of the electron, where the coordinates (X, Y) represent the cyclotron motion in the plane x–y perpendicular to the magnetic field $\boldsymbol{B}$. For this reason the coordinates (X, Y) are called the center coordinates of the cyclotron motion. It is evident from (2.106b) that ξ and η represent the relative coordinates of the cyclotron motion (X, Y). We also find the following relation from (2.107b) and the uncertainty principle:

$$\Delta X \Delta Y = 2\pi l^2 = \frac{h}{eB_z} . \tag{2.110}$$

This relation is sometimes referred to as the degeneracy in a magnetic field of the eigenstate $\mathcal{E}_0(\pi_x, \pi_y, p_z)$. Since the energy eigenvalue is independent of (X, Y), the energy eigenvalue is unchanged under the movement of the cyclotron center coordinates (x, y). When the size of a sample in the directions of x and y is assumed to be L_x and L_y, respectively, the allowed cyclotron orbits are $L_x L_y / \pi l^2$. Therefore, this treatment will give the degenerate states of $1/\pi l^2$ per unit area. The degeneracy is different by a factor of 2 from (2.110).

Here we consider a conduction band with a spherical constant energy surface and with a parabolic band, $\mathcal{E}_0(\boldsymbol{k}) = \hbar^2 k^2/2m^*$, where m^* is the scalar effective mass. When we choose the Landau gauge and put $\boldsymbol{A} = (0, B_z x, 0)$, the effective-mass equation in a magnetic field is given by

$$\left[\frac{1}{2m^*}(p_y + eB_z x)^2 + \frac{p_x^2}{2m^*} + \frac{p_z^2}{2m^*}\right]\psi = \mathcal{E}\psi \,. \tag{2.111}$$

Since $\mathcal{E}_0(\boldsymbol{\pi})$, p_z and p_y commute with each other, we may choose the eigenfunction as

$$\psi = \exp(ik_y)\exp(ik_z)F(x) \,. \tag{2.112}$$

Inserting (2.112) into (2.111), the eigenfunction $F(x)$ is found to satisfy the following relation

$$\left[\frac{p_x^2}{2m^*} + \frac{1}{2m^*}\left(2e\hbar k_y B_z x + e^2 B_z^2 x^2\right)\right] F(x) = \mathcal{E}' F(x) \,, \tag{2.113}$$

where

$$\mathcal{E} = \mathcal{E}' + \frac{\hbar^2 k_y^2}{2m^*} + \frac{\hbar^2 k_z^2}{2m^*} \,. \tag{2.114}$$

Expression (2.113) may be rewritten in a very convenient form by using the following relation

$$X = -\frac{\hbar k_y}{eB_z} \tag{2.115}$$

giving rise to the following result

$$\begin{aligned}\left[\frac{p_x^2}{2m^*} + \frac{e^2 B_z^2}{2m^*}(x - X)^2\right] F(x) &= \left(\mathcal{E}' + \frac{\hbar^2 k_y^2}{2m^*}\right) F(x) \\ &= \left(\mathcal{E} - \frac{\hbar^2 k_z^2}{2m^*}\right) F(x) \,.\end{aligned} \tag{2.116}$$

The transform from x to (X, ξ) may be evident from the above equation and we find that the electron motion in a magnetic field is well described by the relative coordinate $\xi = x - X$. The equation is known as the equation for the one-dimensional simple harmonic oscillator with the angular frequency

$$\omega_c = \frac{eB_z}{m^*} \tag{2.117}$$

and (2.116) is rewritten as

$$\left[\frac{p_\xi^2}{2m^*} + \frac{1}{2}m^*\omega_c^2\xi^2\right] F(\xi) = \left(\mathcal{E} - \frac{\hbar^2 k_z^2}{2m^*}\right) F(\xi). \tag{2.118}$$

The energy of the cyclotron motion is then given by

$$\mathcal{E} = \left(N + \frac{1}{2}\right)\hbar\omega_c + \frac{\hbar^2 k_z^2}{2m^*}, \tag{2.119}$$

where N is the quantum number of the Landau level. The eigenstates of (2.118) are given by

$$\begin{aligned}|N, X, p_z\rangle = &\frac{1}{(2^N N!\sqrt{\pi} l)^{1/2}} \exp\left(-\frac{|x-X|^2}{2l^2}\right)\\ &\times \exp\left\{\mathrm{i}\left(\frac{p_z z}{\hbar} - \frac{Xy}{l^2}\right)\right\} H_N\left(\frac{x-X}{l}\right),\end{aligned} \tag{2.120}$$

where $H_N(x)$ are Hermite polynomials. The above results are summarized in Fig. 2.14. The left-hand figure shows the energy bands without a magnetic field. When a magnetic field $\boldsymbol{B}$ is applied in the z direction, the electron motion is quantized in the x, y plane and forms Landau levels as shown in the right-hand figure.

Next, we consider the case of a perturbing potential U. The electron motion in a perturbing potential will be solved by inserting $\mathcal{E}_0 + U$ into the Hamiltonian (2.102). Let us consider the equation of motion for the operator $\boldsymbol{Q}$, which is written as

$$\dot{\boldsymbol{Q}} = \frac{\mathrm{i}}{\hbar}[H, \boldsymbol{Q}]. \tag{2.121}$$

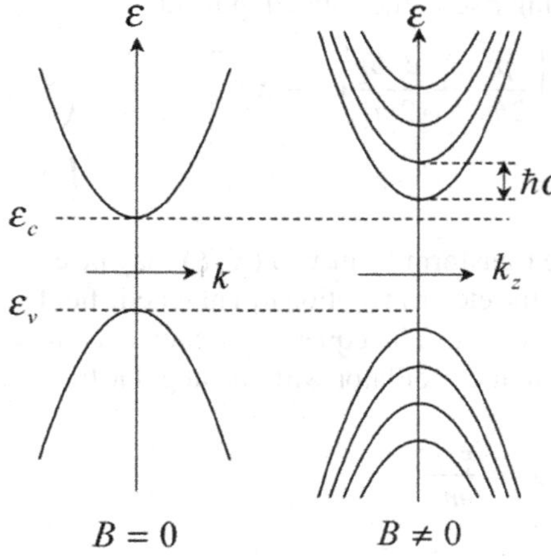

Fig. 2.14 Electronic states in a magnetic field. The *left-hand figure* represents the energy bands without magnetic field. Electrons are quantized in the plane perpendicular to the magnetic field, resulting in the Landau levels as shown in the *right-hand figure*

Therefore, we obtain

$$\dot{\xi} = \frac{\mathrm{i}}{\hbar}[H, \xi] = \frac{\mathrm{i}}{\hbar}\frac{1}{eB_z}[H, \pi_y]$$
$$\dot{\eta} = \frac{\mathrm{i}}{\hbar}[H, \eta] = -\frac{\mathrm{i}}{\hbar}\frac{1}{eB_z}[H, \pi_x] .$$

Using the commutation relations given by (2.105b), we obtain the following equations

$$\dot{\xi} = \frac{\partial \mathcal{E}_0}{\partial \pi_x} - \frac{1}{eB_z}\frac{\partial U}{\partial y} \tag{2.122a}$$
$$\dot{\eta} = \frac{\partial \mathcal{E}_0}{\partial \pi_y} + \frac{1}{eB_z}\frac{\partial U}{\partial x} . \tag{2.122b}$$

In a similar fashion we find the following relations.

$$\dot{X} = \frac{\mathrm{i}}{\hbar}[H, X] = \frac{\mathrm{i}}{\hbar}[U, X] = \frac{1}{eB_z}\frac{\partial U}{\partial y} , \tag{2.123a}$$
$$\dot{Y} = \frac{\mathrm{i}}{\hbar}[U, Y] = -\frac{1}{eB_z}\frac{\partial U}{\partial x} , \tag{2.123b}$$
$$\dot{p}_z = -\frac{\partial U}{\partial z} , \ \dot{z} = \frac{\partial \mathcal{E}_0}{\partial p_z} . \tag{2.123c}$$

These relations are summarized as follows.

$$\dot{\pi}_x = -eB_z\frac{\partial \mathcal{E}_0}{\partial \pi_y} - \frac{\partial U}{\partial x} \tag{2.124a}$$
$$\dot{\pi}_y = +eB_z\frac{\partial \mathcal{E}_0}{\partial \pi_x} - \frac{\partial U}{\partial y} . \tag{2.124b}$$

The velocity $\boldsymbol{v} = (v_x, v_y, v_z)$ is written as

$$v_x \equiv \dot{x} = \dot{\xi} + \dot{X} = \frac{\partial \mathcal{E}_0}{\partial \pi_x} \tag{2.125a}$$
$$v_y \equiv \dot{y} = \dot{\eta} + \dot{Y} = \frac{\partial \mathcal{E}_0}{\partial \pi_y} \tag{2.125b}$$
$$v_z \equiv \dot{z} = \frac{\partial \mathcal{E}_0}{\partial p_z} . \tag{2.125c}$$

When a uniform electric field is applied, the perturbing potential is expressed as

$$\frac{\partial U}{\partial x} = -eE_x , \ \frac{\partial U}{\partial y} = 0 , \tag{2.126}$$

and we obtain the following relation from (2.123a) and (2.123b)

$$\dot{Y} = \frac{E_x}{B_z} , \quad \dot{X} = 0 . \tag{2.127}$$

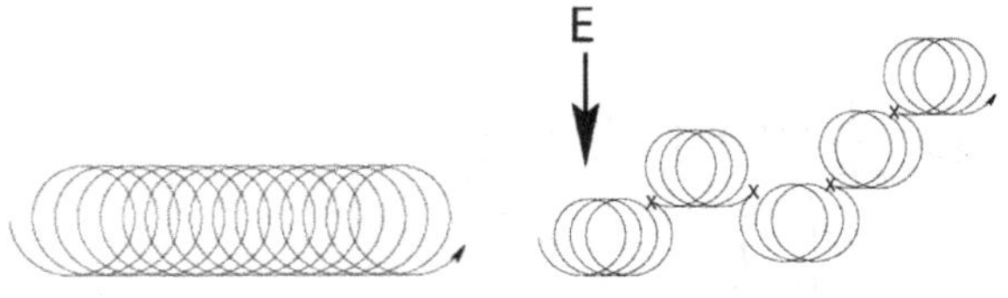

Fig. 2.15 Electron motion in a magnetic field applied in the z direction (perpendicular to the text page) and in a uniform electric field applied parallel to the text page. In the absence of scattering, the electron moves in the direction perpendicular to the magnetic field and the electric field, as shown in the left half of the figure. In the presence of scattering the cyclotron center is scattered and a drift component appears along the direction of the electric field, which is shown in the right half of the figure

This means that the cyclotron center moves with a constant velocity E_x/B_z in the direction perpendicular to the magnetic and electric fields, which is shown in the left half of Fig. 2.15. However, in the presence of scattering centers, the electron is scattered and changes the center of the cyclotron motion, giving rise to a drift motion along the electric field direction, as shown in the right half of Fig. 2.15.

2.5.2 *Density of States and Inter Landau Level Transition*

Now we discuss the density of states of the electrons in the presence of a magnetic field. As explained above, an electron is quantized in the x, y plane and the energy eigenstate is independent of the cyclotron center, resulting in a degeneracy of $p_{\rm d} = 1/2\pi l^2$. Since the wave vector in the z direction is given by $k_z = (2\pi/L_z)n$ ($n = 0, \pm1, \pm2, \pm3, \ldots$), we obtain

$$\sum_{k_z} = \frac{L_z}{2\pi}\int_{-k_0}^{+k_0} {\rm d}k_z = \frac{L_z}{\pi}\int_0^{+k_0} {\rm d}k_z\,, \tag{2.128}$$

where $k_0 = (2\pi/L_z)(N/2)$ and k_z have N degrees of freedom. Equation (2.119) may be rewritten as

$$\mathcal{E}' = \mathcal{E} - (n + \frac{1}{2})\hbar\omega_{\rm c} = \frac{\hbar^2 k_z^2}{2m^*}\,, \tag{2.129}$$

and thus we obtain

$$\sum_{k_z} = \frac{L_z}{\pi}\int_0^{+k_0} {\rm d}k_z = \frac{L_z\sqrt{2m^*}}{2\pi\hbar\sqrt{\mathcal{E}'}}{\rm d}\mathcal{E}'\,. \tag{2.130}$$

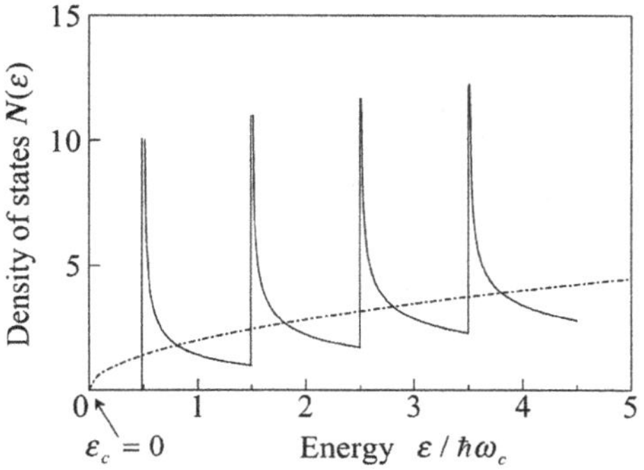

Fig. 2.16 Density of states in a magnetic field. The divergence of the density reflects the one-dimensionality of an electron in a magnetic field and the divergence is removed by taking account of the broadening due to electron scattering. The *dot-dashed curve* represents the density of states for electrons in three dimensions when $B = 0$

Multiplying the degeneracy factor $p_{\rm d}$ to the above equation, the number of electronic states for electrons in a magnetic field is given by

$$N(\mathcal{E}') = \frac{L_x L_y L_z e B_z \sqrt{2m^*}}{(2\pi\hbar)^2} \frac{1}{\sqrt{\mathcal{E}'}} \,. \tag{2.131}$$

Therefore, the density of states $g(\mathcal{E}, B_z)$ per unit volume in the energy range between $\mathcal{E}$ and $\mathcal{E} + d\mathcal{E}$ is defined as

$$g(\mathcal{E}, B_z) = \frac{e B_z \sqrt{2m^*}}{(2\pi\hbar)^2} \sum_n \frac{1}{\sqrt{\mathcal{E} - (n + \frac{1}{2})\hbar\omega_{\rm c}}} \,. \tag{2.132}$$

The density of states in a magnetic filed is shown in Fig. 2.16 as a function of the electron energy.

Next we will discuss the quantum mechanical treatment of cyclotron resonance. In the quantum mechanical picture the cyclotron resonance arises from the electron transition between Landau levels by absorbing photon energy (energy of a radiation field). As described in Sect. 4.2, the interaction Hamiltonian between the electron and the radiation field is given by (4.30), which is written as

$$H_{\rm er} = \frac{e}{m} \boldsymbol{A} \cdot \boldsymbol{p} \,. \tag{2.133}$$

A circularly polarized wave may be expressed as the following by defining the vector potential $\boldsymbol{A} = \boldsymbol{e} \cdot A$:

$$e_{\pm} = \frac{1}{\sqrt{2}} \left(e_x \pm e_y \right) , \qquad p_{\pm} = \frac{1}{\sqrt{2}} \left(p_x \pm p_y \right) , \tag{2.134}$$

where $p_\pm$ are the momentum operators for right and left polarized waves, respectively. When we define the z components of $\boldsymbol{e}$ and $\boldsymbol{P}$ as e_z and p_z, the transition rate calculated for the circularly polarized and linearly polarized waves gives the same result. In order to calculate the transition rate we use the eigenstates of the electron in a magnetic field given by (2.120), and then the transition rate is given by

$$\begin{aligned} w &= \frac{2\pi}{\hbar} \left| \langle N', X', p_z' \,| H_{\mathrm{er}} |\, N, X, p_z \rangle \right|^2 \\ &= 2\pi \frac{e^3 A^2}{m^*} B(N+1) \delta_{k_y, k_y'} \delta_{k_z, k_z'} \delta_{N, N' \pm 1} \delta\left(\hbar\omega - \hbar\omega_{\mathrm{c}}\right) . \end{aligned} \quad (2.135)$$

It is evident from the above equation that the selection rule for the cyclotron resonance is expressed as $\delta k_y = k_y' - k_y = 0$, $\delta k_z = k_z' - k_z = 0$, $\delta N = N' - N = \pm 1$.

2.5.3 Landau Levels of a Non-parabolic Band

In Sect. 2.4 the non-parabolicity of the conduction and valence bands were discussed. The non-parabolicity plays an important role in the magnetotransport and cyclotron resonance. As described above, the Landau levels in a parabolic band are equally spaced in energy. In this section we will deal with Landau levels in a non-parabolic band along with the treatment of Luttinger and Kohn [2], and Bowers and Yafet [9]. The wave functions of the conduction band and valence bands are well described by using Bloch functions at $\boldsymbol{k} = 0$:

$$|n\boldsymbol{k}\rangle = u_{n,0} \mathrm{e}^{\mathrm{i}\boldsymbol{k}\cdot\boldsymbol{r}} , \quad (2.136)$$

where $u_{n,0}$ were derived in Sect. 2.3 by diagonalizing the spin–orbit interaction Hamiltonian and are given by Luttinger and Kohn [2]

$$u_{1,0}(\boldsymbol{r}) = |S \uparrow\rangle , \quad u_{2,0}(\boldsymbol{r}) = |\mathrm{i}S \downarrow\rangle \quad (2.137\mathrm{a})$$

$$u_{3,0}(\boldsymbol{r}) = |(1/\sqrt{2})(X + \mathrm{i}Y) \uparrow\rangle , \quad (2.137\mathrm{b})$$

$$u_{4,0}(\boldsymbol{r}) = |(\mathrm{i}/\sqrt{2})(X - \mathrm{i}Y) \downarrow\rangle , \quad (2.137\mathrm{c})$$

$$u_{5,0}(\boldsymbol{r}) = |(1/\sqrt{6})[(X - \mathrm{i}Y) \uparrow + 2Z \downarrow]\rangle , \quad (2.137\mathrm{d})$$

$$u_{6,0}(\boldsymbol{r}) = |(\mathrm{i}/\sqrt{6})[(X + \mathrm{i}Y) \downarrow - 2Z \uparrow]\rangle , \quad (2.137\mathrm{e})$$

$$u_{7,0}(\boldsymbol{r}) = |(\mathrm{i}/\sqrt{3})[-(X - \mathrm{i}Y) \uparrow + Z \downarrow]\rangle , \quad (2.137\mathrm{f})$$

$$u_{8,0}(\boldsymbol{r}) = |(1/\sqrt{3})[(X + \mathrm{i}Y) \downarrow + Z \uparrow]\rangle , \quad (2.137\mathrm{g})$$

where the time reversal symmetry is taken account.

In a magnetic field applied in the z direction, the vector potential is described as

$$A_x = -By, \quad A_y = 0, \quad A_z = 0 , \quad (2.138)$$

and therefore the Hamiltonian becomes

$$H = H_0 + \frac{s}{m} y p_x + \frac{s^2}{2m} y^2 , \tag{2.139}$$

where s is defined by using the cyclotron radius l as

$$s = \frac{eB}{\hbar} = \frac{1}{l^2} . \tag{2.140}$$

It is easy to find the following relations for the matrix elements of the term which includes the contribution of the magnetic field:

$$\begin{aligned} \langle n\boldsymbol{k}|y p_x|n'\boldsymbol{k}'\rangle &= \int \mathrm{e}^{\mathrm{i}(\boldsymbol{k}'-\boldsymbol{k})\cdot\boldsymbol{r}} u_{n,0}^{*} y (k_x - \mathrm{i}\nabla_x) n_{n',0} \mathrm{d}^3\boldsymbol{r} \\ &= -\mathrm{i}\frac{\partial}{\partial k'_y} \int \mathrm{e}^{\mathrm{i}(\boldsymbol{k}'-\boldsymbol{k})\cdot\boldsymbol{r}} u_{n,0}^{*} (k_x - i\nabla_x) u_{n',0} \mathrm{d}^3\boldsymbol{r} \\ &= -\mathrm{i}\frac{\partial}{\partial k'_y} [(k_x + \pi_{nn'}^x)\delta(\boldsymbol{k}'-\boldsymbol{k})] \\ &= (k_x \delta_{nn'} + \pi_{nn'}^x)\frac{1}{\mathrm{i}}\frac{\partial\delta(\boldsymbol{k}'-\boldsymbol{k})}{\partial k'_y} , \end{aligned} \tag{2.141}$$

In a similar fashion we find

$$\begin{aligned} \langle n\boldsymbol{k}|y^2|n'\boldsymbol{k}'\rangle &= \int y^2 \mathrm{e}^{\mathrm{i}(\boldsymbol{k}'-\boldsymbol{k})\cdot\boldsymbol{r}} u_{n,0}^{*} u_{n',0} \mathrm{d}^3\boldsymbol{r} \\ &= \left(-\mathrm{i}\frac{\partial}{\partial k'_y}\right)^2 \int \mathrm{e}^{\mathrm{i}(\boldsymbol{k}'-\boldsymbol{k})\cdot\boldsymbol{r}} u_{n,0}^{*} u_{n',0} \mathrm{d}^3\boldsymbol{r} \\ &= -\frac{\partial^2\delta(\boldsymbol{k}'-\boldsymbol{k})}{\partial k'^2_y}\delta_{nn'} , \end{aligned} \tag{2.142}$$

and finally we obtain the relation

$$\begin{aligned} \langle n\boldsymbol{k}|H_0|n'\boldsymbol{k}'\rangle &= \langle n\boldsymbol{k}| -\frac{\hbar^2}{2m}\nabla^2 + V(\boldsymbol{r})|n'\boldsymbol{k}'\rangle \\ &= \left[\left(\mathcal{E}_n + \frac{\hbar^2 k^2}{2m}\right)\delta_{nn'} + \frac{\hbar}{m}k_\alpha \pi_{nn'}^{\alpha}\right]\delta(\boldsymbol{k}-\boldsymbol{k}') . \end{aligned} \tag{2.143}$$

First, we will derive approximate solutions for the Landau levels of a non-parabolic conduction band. We define the zeroth-order wave function by a linear combination of Bloch terms

$$\Psi(\boldsymbol{r}) = \sum_j f_j(\boldsymbol{r}) u_{j,0}(\boldsymbol{r}) , \tag{2.144}$$

and then the matrix elements of the Hamiltonian H are derived as follows:

$$\begin{vmatrix}
\mathcal{E}_G-\lambda & 0 & \sqrt{\frac{1}{2}}P\bar{k}_+ & 0 & \sqrt{\frac{1}{6}}P\bar{k}_- & \mathrm{i}\sqrt{\frac{2}{3}}Pk_z & -\mathrm{i}\sqrt{\frac{1}{3}}P\bar{k}_- & \sqrt{\frac{1}{3}}Pk_z \\
0 & \mathcal{E}_G-\lambda & 0 & \sqrt{\frac{1}{2}}P\bar{k}_- & -\sqrt{\frac{2}{3}}Pk_z & \sqrt{\frac{1}{6}}P\bar{k}_+ & \sqrt{\frac{1}{3}}Pk_z & -\mathrm{i}\sqrt{\frac{1}{3}}P\bar{k}_+ \\
\sqrt{\frac{1}{2}}P\bar{k}_+ & 0 & -\lambda & 0 & 0 & 0 & 0 & 0 \\
0 & \sqrt{\frac{1}{2}}P\bar{k}_- & 0 & -\lambda & 0 & 0 & 0 & 0 \\
\sqrt{\frac{1}{6}}P\bar{k}_- & -\sqrt{\frac{2}{3}}Pk_z & 0 & 0 & -\lambda & 0 & 0 & 0 \\
-\mathrm{i}\sqrt{\frac{2}{3}}Pk_z & \sqrt{\frac{1}{6}}P\bar{k}_+ & 0 & 0 & 0 & -\lambda & 0 & 0 \\
\mathrm{i}\sqrt{\frac{1}{3}}P\bar{k}_- & \sqrt{\frac{1}{3}}Pk_z & 0 & 0 & 0 & 0 & -\Delta_0-\lambda & 0 \\
\sqrt{\frac{1}{3}}Pk_z & \mathrm{i}\sqrt{\frac{1}{3}}P\bar{k}_+ & 0 & 0 & 0 & 0 & 0 & -\Delta_0-\lambda
\end{vmatrix}
\begin{vmatrix} f_1 \\ f_2 \\ f_3 \\ f_4 \\ f_5 \\ f_6 \\ f_7 \\ f_8 \end{vmatrix} = 0\,, \tag{2.145}$$

where

$$\begin{aligned}
\lambda &= \mathcal{E} - \frac{\hbar^2\bar{k}^2}{2m} \\
\bar{k}^2 &= \bar{k}_x^2 + \bar{k}_y^2 + k_z^2 \\
\bar{k}_\pm &= \bar{k}_x \pm \bar{k}_y \\
\bar{k}_x &= k_x - \mathrm{i}s\frac{\partial}{\partial k_y} \\
\bar{k}_y &= k_y\,.
\end{aligned} \tag{2.146}$$

Now we have to note the definition of the vector potential. In the previous subsection we defined the vector potential as $(0,\ B_z x,\ 0)$, whereas in this subsection we adopt the gauge of $(-B_z y, 0, 0)$. It is explained in the previous subsection that the energy of the cyclotron motion is independent of the cyclotron center $X = -\hbar k_y/eB_z$. Therefore, the gauge used in this subsection gives rise to the result that the cyclotron energy is independent of k_x. This results allows us to put $k_x = 0$ in the following calculations. Another assumption is made in this subsection that we neglect the contribution of spin–orbit interaction to the $\boldsymbol{k}$ dependence in (2.145). It is found that (2.145) may be easily solved when the term $\hbar^2\bar{k}^2/2m$ is negligible. In this subsection we are interested in the Landau levels at $k_z = 0$ and thus the above assumptions are valid in the analysis [9]. Under these assumptions (2.145) leads to the following two equations, where terms other than f_1 and f_2 are neglected:

$$\left[(\mathcal{E}_G-\lambda)\lambda+\frac{2}{3}P^2\left(k_z^2+k_y^2-s^2\frac{\partial^2}{\partial k_y^2}+\frac{1}{2}s\right)\right.$$
$$\left.+\frac{P^2\lambda}{3(\Delta_0+\lambda)}\left(k_z^2+k_y^2-s^2\frac{\partial^2}{\partial k_y^2}-s\right)\right]f_1=0 \tag{2.147}$$

$$\left[(\mathcal{E}_G-\lambda)\lambda+\frac{2}{3}P^2\left(k_z^2+k_y^2-s^2\frac{\partial^2}{\partial k_y^2}-\frac{1}{2}s\right)\right.$$
$$\left.+\frac{P^2\lambda}{3(\Delta_0+\lambda)}\left(k_z^2+k_y^2-s^2\frac{\partial^2}{\partial k_y^2}+s\right)\right]f_2=0\,. \tag{2.148}$$

When we investigate the above equations, we find that the equations represent simple harmonic oscillator equations with dimensionless variable $k_y/\sqrt{s}$. Therefore we obtain the following solutions:

$$D(\lambda_{n\pm})\equiv\lambda_{n\pm}(\lambda_{n\pm}-\mathcal{E}_{\rm G})(\lambda_{n\pm}+\Delta_0)$$
$$=-P^2\left[k_z^2+s(2n+1)\right]\left[\lambda_{n\pm}+\frac{2}{3}\Delta_0\right]\pm\frac{1}{3}P^2\Delta_0 s=0\,, \tag{2.149}$$

where $\lambda_{n,\pm}$ stands for λ_{k_z}, n (spin quantum number), $\pm$, and $s=eB/\hbar$. It is evident from (2.149) that the latter coincides with the solution for a parabolic band in the absence of a magnetic field, i.e. for $B=0$ (and thus $s=0$). Using the relation $\lambda=\mathcal{E}-\hbar^2k^2/2m$ in (2.149) and neglecting the small spin-splitting term, the Landau levels at $k_x=0$ are approximated as below [10]. [They are easily derived from (2.86).]

$$\mathcal{E}_n=\left(n+\frac{1}{2}\right)\hbar\omega_{\rm c}-\left(1-\frac{m_0^*}{m}\right)^2\left[\frac{3\mathcal{E}_{\rm G}+4\Delta_0+2\Delta_0^2/\mathcal{E}_{\rm G}}{(\mathcal{E}_{\rm G}+\Delta_0)(3\mathcal{E}_{\rm G}+2\Delta_0)}\right]$$
$$\times\left(n+\frac{1}{2}\right)^2(\hbar\omega_{\rm c})^2\,, \tag{2.150}$$

where m_0^* is the band-edge effective mass of the conduction band and is given later (2.152) (or see (2.84)) and $\hbar\omega_{\rm c}=\hbar eB/m_0^*$ as defined before.

2.5.4 *Effective g Factor*

Landau levels of the conduction band electron is expressed as

$$\mathcal{E}_{\pm,n,k_z}(B)=\hbar\omega_{\rm c}\left(n+\frac{1}{2}\right)+\frac{\hbar^2k_z^2}{2m_0^*}\pm\frac{1}{2}\mu_{\rm B}|g_0^*|B\,, \tag{2.151}$$

where m_0^* is the band edge effective mass, $\omega_{\rm c}=\hbar eB/m_0^*$, the quantum numbers $\pm$ refer to the spin, and $\mu_{\rm B}=e\hbar/2m$ is the Bohr magneton. The constant g_0^* is the Landé g factor, referred to as the effective g–factor. The band edge effective mass m_0^*

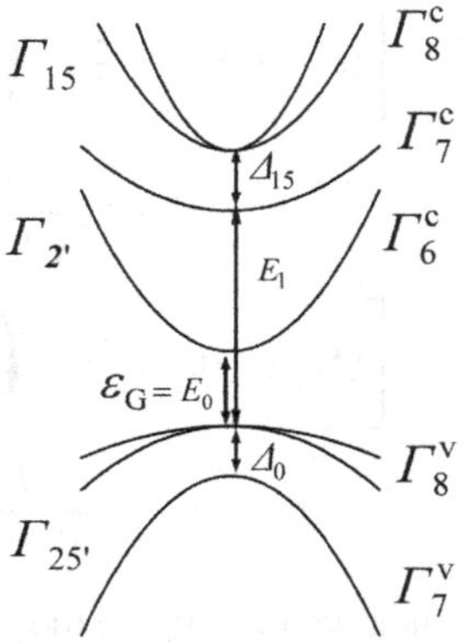

Fig. 2.17 Schematic energy band structure near $\boldsymbol{k} = 0$ in a direct bandgap semiconductor, group theoretical symbols for a diamond type (single group representation) and zinc blende type (double group representation)

and the effective g factor g_0^* are deduced by Roth et al. [11] and cited many papers. [12–15]. Here we use the definition of Herman and Weisbuch [15]:

$$\frac{1}{m_0^*} = \frac{1}{m} + \frac{\mathcal{E}_{P0}}{3m}\left[\frac{2}{\mathcal{E}_G} + \frac{1}{\mathcal{E}_G + \Delta_0}\right], \tag{2.152}$$

$$\frac{g_0^*}{2} = 1 + \frac{1}{m}\frac{1}{i}\sum_u \frac{\langle \Gamma_6^c | p_x | u\rangle\langle u | p_y | \Gamma_6^c\rangle - \langle \Gamma_6^c | p_y | u\rangle\langle u | p_x | \Gamma_6^c\rangle}{\mathcal{E}_c - \mathcal{E}_u} \tag{2.153}$$

$$= 1 + \frac{\mathcal{E}_{P0}}{3}\left(\frac{1}{\mathcal{E}_G + \Delta_0} - \frac{1}{\mathcal{E}_G}\right)$$

$$+\frac{\mathcal{E}_{P1}}{3}\left(\frac{1}{\mathcal{E}(\Gamma_8^c) - \mathcal{E}_G} - \frac{1}{\mathcal{E}(\Gamma_7^c) - \mathcal{E}_G}\right) + C', \tag{2.154}$$

where the factor 2 of the denominator of $g_0^*/2$ is the free electron g factor ($g_0 = 2$) and the energies are define by the magnitude (positive value). The relation of (2.153) and thus (2.154) arises from the spin–orbit interaction. For reference the energy band structure near $\boldsymbol{k} = 0$ with respective energies are shown in Fig. 2.17. $\mathcal{E}_c$ is the energy of the conduction band $|\Gamma_6^c\rangle$ (or $|\Gamma_{2'}\rangle$ in the single group representation). The energies $\mathcal{E}_u$ for the valence bands are $\mathcal{E}(\Gamma_8^v) = -\mathcal{E}_G$ and its spin–split off band $\mathcal{E}(\Gamma_7^v) = -\mathcal{E}_G - \Delta_0$. For the higher conduction band $\mathcal{E}_u$ are $\mathcal{E}(\Gamma_7^c) - \mathcal{E}_G$ and its spin–split off band $\mathcal{E}(\Gamma_8^c) - \mathcal{E}_G = \mathcal{E}(\Gamma_7^c) + \Delta_{15} - \mathcal{E}_G$, and thus the energy denominators are, respectively, $\mathcal{E}_c - \mathcal{E}(\Gamma_7^c) - \mathcal{E}_G$ and $\mathcal{E}_c - \mathcal{E}(\Gamma_8^c) - \mathcal{E}_G = \mathcal{E}_c - \mathcal{E}(\Gamma_7^c) - \Delta_{15} - \mathcal{E}_G$. Putting $\mathcal{E}_c = 0$, we obtain the result given by (2.154). The higher conduction band $\mathcal{E}(\Gamma_7^c)$ and $\mathcal{E}(\Gamma_8^c) = \mathcal{E}(\Gamma_7^c) + \Delta_{15}$ are measured from the valence band top. These higher conduction band energies are obtained as $\mathcal{E}_1$ and $\mathcal{E}_1 + \Delta_{15}$, where $\mathcal{E}_1$ is the optical transition energy between the valence band top Γ_8^v ($\Gamma_{25'}$ band) and Γ_7^c (Γ_{15}) band (called as E_1–edge).

Derivation of (2.154) from (2.153) is straight forward by evaluating the matrix elements using the eigenstates of Γ_7^{v} and Γ_8^{v} valence bands ($\Gamma_{25'}$ bands in single group representation) given by (2.63a) $\sim$ (2.63f) and similar eigenstates for Γ_7^{c} and Γ_8^{c} conduction bands ($\Gamma_{15'}$ bands in single group representation). The diagonalization of spin–orbit Hamiltonian gives the corresponding eigenvalues and eigensates, where different forms of the eigen functions are reported by Luttinger and Kohn [2] (see (2.137b) $\sim$ (2.137g)), where the time reversal symmetry is taken account. Roth et al. [11] defined the eigenstates as given below, which are the same as (2.63a) $\sim$ (2.63f) of this textbook. Here we list the eigenstates derived by Roth et al. [11]: for fourfold valence bands are

$$u_{v1} = \left|\frac{3}{2}, \frac{3}{2}\right\rangle = \frac{1}{\sqrt{2}} |(X + \mathrm{i}Y) \uparrow \rangle \,, \tag{2.155a}$$

$$u_{v2} = \left|\frac{3}{2}, \frac{1}{2}\right\rangle = -\frac{1}{\sqrt{6}} [|(X + \mathrm{i}Y) \downarrow \rangle + 2\,|Z \uparrow \rangle] \,, \tag{2.155b}$$

$$u_{v3} = \left|\frac{3}{2}, -\frac{1}{2}\right\rangle = -\frac{1}{\sqrt{6}} [|(X - \mathrm{i}Y) \uparrow \rangle + 2\,|Z \downarrow \rangle] \,, \tag{2.155c}$$

$$u_{v4} = \left|\frac{3}{2}, -\frac{3}{2}\right\rangle = \frac{1}{\sqrt{2}} |(X - \mathrm{i}Y) \downarrow \rangle \,, \tag{2.155d}$$

and for twofold spin–split off bands are

$$u_{v5} = \left|\frac{1}{2}, \frac{1}{2}\right\rangle = \frac{1}{\sqrt{3}} [|(X + \mathrm{i}Y) \downarrow \rangle - |Z \uparrow \rangle] \,, \tag{2.155e}$$

$$u_{v6} = \left|\frac{1}{2}, -\frac{1}{2}\right\rangle = \frac{1}{\sqrt{3}} [|(X - \mathrm{i}Y) \uparrow \rangle - |Z \downarrow \rangle] \,. \tag{2.155f}$$

The term C' represents the contributions from far–higher lying bands. The energies $\mathcal{E}(\Gamma_7^{\text{c}})$ and $\mathcal{E}(\Gamma_8^{\text{c}})$ are measured from the the valence band top and thus all the energy values are taken to be positive. Here we adopted the notations defined in Sect. 2.6, and $\mathcal{E}_{P0}$ and $\mathcal{E}_{P1}$ are defined as

$$\mathcal{E}_{P0} = \frac{2}{m} |\langle \Gamma_{2'} | p_x | \Gamma_{25'}(X) \rangle|^2 \,, \tag{2.156}$$

$$\mathcal{E}_{P1} = \frac{2}{m} |\langle \Gamma_{2'} | p_x | \Gamma_{15}(x) \rangle|^2 \,. \tag{2.157}$$

The above relations were first derived by Roth et al. [11]. Also similar treatment leads to the band edge effective mass for the 14 band band model and the results are shown in Sect. 2.6. Here we estimate the effective $g-$ factor for GaAs at low temperature, where we choose $\mathcal{E}_{\text{G}} = 1.519$ [eV], $\Delta_0 = 0.341$ [eV], $\mathcal{E}_{P0} = 28.8$ [eV], $\mathcal{E}_1 = 4.488$ [eV], $\Delta_{15} = 0.171$ [eV], $\mathcal{E}_{P1} = \mathcal{E}_{P0}/4 = 7.2$ [eV], and $C' = -0.02$. These parameters give $g_0^* = -0.445$ with good agreement of experiment $g_0^* = -0.44$. The parameters assumption $\mathcal{E}_{P1} = \mathcal{E}_{P0}/4$ and C' are not well established and thus the above values are open for discussion. Litvinenko et al. introduce another mixing term and their parameters give $g_0^* = -0.44$. In this textbook we presented the magnitude of g–factor without discussion of the sign and the reported values for various III-V materials are negative values except some compounds.

The effective g factor of (2.154) leads to (2.159) of 8 band model by neglecting the higher band contributions. When we take account of Γ_6^c conduction band, and Γ_7^v and Γ_8^v valence bands only we obtain

$$\frac{1}{m_0^*} = \frac{1}{m} + \frac{\mathcal{E}_{P0}}{3m}\frac{2\Delta_0 + 3\mathcal{E}_G}{\mathcal{E}_G(\Delta_0 + \mathcal{E}_G)}, \tag{2.158}$$

$$\frac{g_0^*}{2} = 1 + \frac{\mathcal{E}_{P0}}{3}\left(\frac{1}{\mathcal{E}_G + \Delta_0} - \frac{1}{\mathcal{E}_G}\right) = 1 + \left(1 - \frac{m}{m_0^*}\right)\frac{\Delta_0}{2\Delta_0 + 3\mathcal{E}_G}. \tag{2.159}$$

The effective g–factor of InSb is estimated by (2.159) as expected from Kane's two band model. We choose the parameters of InSb at low temperature as, $\mathcal{E}_G = 0.2353$ [eV], $\mathcal{E}_{P0} = 23.1$ [eV], and $\Delta_0 = 0.803$ [eV]. These values give $g_0^* = -48.6$, which shows a reasonable agreement with the measurements $g_0^* \simeq -50$ in InSb [12, 13]. The effective g–factor depends on the electron energy $\mathcal{E}$ and given by Litvinenko et al. [13]

$$\frac{g_0^*}{2} = 1 + \frac{\mathcal{E}_{P0}}{3}\left(\frac{1}{\mathcal{E}_G + \Delta_0 + \mathcal{E}} - \frac{1}{\mathcal{E}_G + \mathcal{E}}\right), \tag{2.160}$$

and shows a good agreement with the experiments.

We have to note here that the above formulations are based on the $\boldsymbol{k} \cdot \boldsymbol{p}$ perturbation theory. Narrow gap semiconductors such as InSb is well expressed by the Kane's two band model (8 band model with spin–orbit interaction), and that more general form with higher bands contributions is useful for more accurate modeling for zinc blende semiconductors such as GaAs, where we have to use 14 band model. The results shown above are obtained by using the energy eigenstate notations of the double group (see also the treatment in Sect. 2.6) and with magnetic field applied in the z direction,

2.5.5 Landau Levels of the Valence Bands

It may be expected that the calculations of Landau levels for valence bands are complicated because of the degeneracy of the bands and the spin–orbit interaction. In this subsection we deal with the Landau levels in the valence bands along with the treatment of Luttinger [3] and the method extended by Pidgeon and Brown [16]. First we investigate the valence bands using (2.44) and (2.45). The non-diagonal term Nk_xk_y is given by

$$D_{XY} = D_{XY}^{xy}k_xk_y + D_{XY}^{yx}k_yk_x. \tag{2.161}$$

We define the antisymmetric term by

$$K = D_{XY}^{xy} - D_{XY}^{yx}, \tag{2.162}$$

and then D_{XY} is rewritten as

$$D_{XY} = N\{k_x k_y\} + \frac{1}{2}K(k_x, k_y), \tag{2.163}$$

where

$$\{k_x k_y\} = \frac{1}{2}(k_x k_y + k_y k_x), \tag{2.164a}$$

$$(k_x, k_y) = k_x k_y - k_y k_x. \tag{2.164b}$$

From the commutation relation of (2.105b)

$$(k_x, k_y) = -\mathrm{i}eB_z \tag{2.165}$$

is derived and therefore we obtain

$$D_{XY} = N\{k_x k_y\} + \frac{K}{2\mathrm{i}}eB_z. \tag{2.166}$$

Similarly we find

$$D_{YX} = N\{k_x k_y\} - \frac{K}{2\mathrm{i}}eB_z. \tag{2.167}$$

The other terms are also expressed in a similar fashion. As a result (2.45) is divided into two terms

$$D = D^{(S)} + D^{(A)}, \tag{2.168}$$

where $D^{(S)}$ is obtained from (2.45) by putting $k_\alpha k_\beta \to \{k_\alpha k_\beta\}$, and $D^{(A)}$ is given by the matrix

$$D^{(A)} = \frac{eK}{2}\begin{vmatrix} 0 & -\mathrm{i}B_z & \mathrm{i}B_y \\ \mathrm{i}B_z & 0 & -\mathrm{i}B_x \\ -\mathrm{i}B_y & \mathrm{i}B_x & 0 \end{vmatrix} \tag{2.169}$$

We define the operators J_x, J_y and J_z, which have the same characteristics as the momentum operators expressed by a 4×4 matrix, i.e.

$$(J_x, J_y) = \mathrm{i}J_z,\ (J_y, J_z) = \mathrm{i}J_x,\ (J_z, J_x) = \mathrm{i}J_y \tag{2.170a}$$

$$J_x^2 + J_y^2 + J_z^2 = \frac{3}{2}\left(\frac{3}{2} + 1\right) = \frac{15}{4}. \tag{2.170b}$$

Using these results we obtain the Luttinger Hamiltonian given by

$$D = \frac{1}{m}\left\{\left(\gamma_1 + \frac{5\gamma_2}{2}\right)\frac{k^2}{2} - \gamma_2(k_x^2 J_x^2 + k_y^2 J_y^2 + k_z^2 + J_z^2)\right.$$
$$- 2\gamma_3(\{k_x k_y\}\{J_x J_y\} + \{k_y k_z\}\{J_y J_z\} + \{k_z k_x\}\{J_z J_x\})$$
$$\left. + e\kappa \boldsymbol{J} \cdot \boldsymbol{B} + eq(J_x^3 B_x + J_y^3 B_y + J_z^3 B_z)\right\}, \tag{2.171}$$

where units are chosen that $\hbar = 1$,

$$\frac{1}{2m}\gamma_1 = -\frac{1}{3}(A + 2B), \quad \frac{1}{m}(3\kappa + 1) = -K, \tag{2.172a}$$
$$\frac{1}{2m}\gamma_2 = -\frac{1}{6}(A - B), \quad \frac{1}{2m}\gamma_3 = -\frac{1}{6}C. \tag{2.172b}$$

The parameter q is introduced for the correction due to the spin–orbit splitting, which is neglected because of the small contribution. The parameters defined above are calledLuttinger parameters. Following the treatment of Luttinger and Kohn [2], the $\boldsymbol{k} \cdot \boldsymbol{p}$ Hamiltonian is expressed as (in atomic units)

$$\sum_j \left\{ D_{jj'}^{lm} k_l k_m + \pi_{jj'}^l k_l + \frac{1}{2}s(\sigma_3)_{jj'} \right.$$
$$\left. + \frac{1}{4c^2}[(\boldsymbol{\sigma} \times \nabla V) \cdot \boldsymbol{p}]_{jj'} + \mathcal{E}_{j'}\delta_{jj'} \right\} f_{j'}(\boldsymbol{r}) = \mathcal{E} f_j(\boldsymbol{r}), \tag{2.173}$$

where

$$D_{jj'}^{lm} \equiv \frac{1}{2}\delta_{jj'}\delta_{lm} + \sum_i \frac{\pi_{ji}^l \pi_{ij'}^m}{\mathcal{E}_0 - \mathcal{E}_i}. \tag{2.174}$$

The summations with respect to j and j' in the above equation are carried out for 2 conduction bands and 6 valence bands, i is for all bands, and l, m are for 1, 2, 3 or for x, y, z. $\mathcal{E}_i$ and $\mathcal{E}_j$ are the average energies for the states i and j, respectively. Equation (2.173) is evaluated by using (2.137g) and (2.144), giving rise to a 8×8 matrix. When the magnetic field is applied in the $(1\bar{1}0)$ plane, the Luttinger matrix D becomes

$$D = D_0 + D_1, \tag{2.175}$$

where D_0 may be solved exactly and D_1 may be solved by a perturbation method [16]. The detail of the analysis is not described here. In the following we put $k_3 = 0$. Then D_0 is divided into two 4×4 matrices:

$$D_0 = \begin{vmatrix} D_a & 0 \\ 0 & D_b \end{vmatrix}. \tag{2.176}$$

In addition we neglect the perturbation term of the conduction band and thus all the terms except the first term of (2.174). This assumption is also adopted by Pidgeon and Brown, where they put $F = 0$. Finally we obtain

$$
\begin{vmatrix}
\mathcal{E}_{\rm G} - \mathcal{E}_a + s(n+1) & \mathrm{i}\sqrt{s}Pa^\dagger & \mathrm{i}\sqrt{\frac{1}{3}s}Pa & \sqrt{\frac{2}{3}s}Pa \\
-\mathrm{i}\sqrt{s}Pa & -s[(\gamma_1+\gamma')(n+\frac{1}{2})+\frac{3}{2}\kappa] - \mathcal{E}_a & -s\sqrt{3}\gamma'' a^2 & -\mathrm{i}s\sqrt{6}\gamma' a^2 \\
-\mathrm{i}\sqrt{\frac{1}{3}s}Pa^\dagger & -s\sqrt{3}\gamma'' a^{\dagger 2} & -s[(\gamma_1-\gamma')(n+\frac{1}{2})-\frac{1}{2}\kappa] - \mathcal{E}_a & \mathrm{i}s\sqrt{2}[\gamma'(n+\frac{1}{2})-\frac{1}{2}\kappa] \\
\sqrt{\frac{2}{3}s}Pa^\dagger & \mathrm{i}s\sqrt{6}\gamma' a^{\dagger 2} & -\mathrm{i}s\sqrt{2}[\gamma'(n+\frac{1}{2})-\frac{1}{2}\kappa] & -s[\gamma_1(n+\frac{1}{2})-\kappa] - \Delta_0 - \mathcal{E}_a
\end{vmatrix}
\begin{vmatrix} f_1 \\ f_3 \\ f_5 \\ f_7 \end{vmatrix} = 0
$$

$$
\begin{vmatrix}
\mathcal{E}_{\rm G} - \mathcal{E}_b + sn & \mathrm{i}\sqrt{\frac{1}{3}s}Pa^\dagger & \mathrm{i}\sqrt{s}Pa & \sqrt{\frac{2}{3}s}Pa^\dagger \\
-\mathrm{i}\sqrt{\frac{1}{3}s}Pa^\dagger & -s[(\gamma_1-\gamma')(n+\frac{1}{2})+\frac{1}{2}\kappa] - \mathcal{E}_b & -s\sqrt{3}\gamma'' a^2 & \mathrm{i}s\sqrt{2}[\gamma'(n+\frac{1}{2})+\frac{1}{2}\kappa] \\
-\mathrm{i}\sqrt{s}Pa^\dagger & -s\sqrt{3}\gamma'' a^{\dagger 2} & -s[(\gamma_1+\gamma')(n+\frac{1}{2})-\frac{3}{2}\kappa] - \mathcal{E}_b & \mathrm{i}s\sqrt{6}\gamma' a^{\dagger 2} \\
\sqrt{\frac{2}{3}s}Pa & -\mathrm{i}s\sqrt{2}[\gamma'(n+\frac{1}{2})+\frac{1}{2}\kappa] & -\mathrm{i}s\sqrt{6}[\gamma' a^2 & -s[\gamma_1(n+\frac{1}{2})+\kappa] - \Delta_0 - \mathcal{E}_b
\end{vmatrix}
\begin{vmatrix} f_2 \\ f_6 \\ f_4 \\ f_8 \end{vmatrix} = 0 .
$$

Here we have used the creation operator $a^\dagger$ and the annihilation operator a defined as

$$
a^\dagger = \frac{1}{\sqrt{2s}}(k_1 + \mathrm{i}k_2) , \tag{2.177a}
$$

$$
a = \frac{1}{\sqrt{2s}}(k_1 - \mathrm{i}k_2) , \tag{2.177b}
$$

$$
n = a^\dagger a \tag{2.177c}
$$

P is the matrix element of the momentum operator defined previously, $P = -\mathrm{i}\langle S|p_z|Z\rangle$, and γ' and γ'' are given by

$$
\gamma' = \gamma_3 + (\gamma_2 - \gamma_3)\left[\frac{1}{2}(3\cos^2\theta - 1)\right]^2 , \tag{2.178a}
$$

$$
\gamma'' = \frac{1}{3}\gamma_3 + \frac{1}{3}\gamma_2 + \frac{1}{6}(\gamma_2 - \gamma_3)\left[\frac{1}{2}(3\cos^2\theta - 1)\right]^2 , \tag{2.178b}
$$

where θ is the angle between the z axis and the magnetic field. The valence band parameters used here, γ_1, γ_2, γ_3 and κ are different from the Luttinger parameters γ_1^L, γ_2^L, γ_3^L and κ^L, and they are given by

$$\gamma_1 = \gamma_1^L - \frac{2P^2}{3\mathcal{E}_G}, \tag{2.179a}$$

$$\gamma_2 = \gamma_2^L - \frac{P^2}{3\mathcal{E}_G}, \tag{2.179b}$$

$$\gamma_3 = \gamma_3^L - \frac{P^2}{3\mathcal{E}_G}, \tag{2.179c}$$

$$\kappa = \kappa^L - \frac{P^2}{3\mathcal{E}_G}. \tag{2.179d}$$

Here we will show the derivation of the relations (2.177a)–(2.177c). Using the generalized momentum defined by

$$\pi_x = p_x + eA_x = p_x - yeB \tag{2.180a}$$

$$\pi_y = p_y + eA_y = p_y \tag{2.180b}$$

we find

$$k_1 = -\mathrm{i}\frac{\partial}{\partial x} + \frac{eA_x}{\hbar}, \tag{2.181a}$$

$$k_2 = -\mathrm{i}\hbar\frac{\partial}{\partial y} + \frac{eA_y}{\hbar}, \tag{2.181b}$$

and thus we obtain the relation

$$\begin{aligned} \hbar^2(k_1, k_2) &= (p_x + eA_x, p_y + eA_y) \\ &= (p_x, p_y) + (p_x, eA_y) + (eA_x, p_y) + e^2(A_xA_y - A_yA_x) \\ &= 0 + (p_x, 0) + (-eBy, p_y) + 0 \\ &= -eB(y, p_y) = -\mathrm{i}\hbar eB, \\ (k_1, k_2) &= \frac{eB}{\mathrm{i}\hbar} \equiv \frac{s}{\mathrm{i}}. \end{aligned} \tag{2.182}$$

When we define new variables by

$$k_1 = \sqrt{s}p, \quad k_2 = \sqrt{s}q, \tag{2.183}$$

p and q are canonical variables which satisfy

$$(p, q) = \frac{1}{\mathrm{i}}. \tag{2.184}$$

We can now define new operators for the creation and annihilation operators $a^\dagger$ and a by

$$a^\dagger = \frac{1}{\sqrt{2}}(p + \mathrm{i}q) = \frac{1}{\sqrt{2s}}(k_1 + \mathrm{i}k_2), \tag{2.185a}$$

$$a = \frac{1}{\sqrt{2}}(p - \mathrm{i}q) = \frac{1}{\sqrt{2s}}(k_1 - \mathrm{i}k_2). \tag{2.185b}$$

The commutation relation for the new operator is

$$(a, a^{\dagger}) = 1 \tag{2.186}$$

and the relations (2.177a)–(2.177c) are derived as follows:

$$\frac{1}{2}(p^2 + q^2)u_n = \left(a^{\dagger}a + \frac{1}{2}\right)u_n = \left(n + \frac{1}{2}\right)u_n\,, \tag{2.187a}$$

$$a^{\dagger}au_n = nu_n\,, \tag{2.187b}$$

$$au_n = \sqrt{n}u_{n-1}\,, \tag{2.187c}$$

$$a^{\dagger}u_n = \sqrt{n+1}u_{n+1}\,, \tag{2.187d}$$

$$n = a^{\dagger}a\,. \tag{2.187e}$$

It is evident from the two eigenvalue equations in matrix form below (2.176) that the solutions of these equations are given by the solutions of a simple harmonic oscillator equation:

$$f_a = \begin{vmatrix} a_1\Phi_n \\ a_3\Phi_{n-1} \\ a_5\Phi_{n+1} \\ a_7\Phi_{n+1} \end{vmatrix}\,, \qquad f_b = \begin{vmatrix} a_2\Phi_n \\ a_6\Phi_{n-1} \\ a_4\Phi_{n+1} \\ a_8\Phi_{n-1} \end{vmatrix}\,. \tag{2.188}$$

Using these relations the eigenvalues for a and b are given by

$$\begin{vmatrix} \mathcal{E}_G - \mathcal{E}_a + s(n+1) & \mathrm{i}\sqrt{sn}P & \sqrt{\frac{1}{3}s(n+1)}P & \sqrt{\frac{2}{3}s(n+1)}P \\ -\mathrm{i}\sqrt{sn}P & -s[(\gamma_1+\gamma')(n-\frac{1}{2})+\frac{3}{2}\kappa]-\mathcal{E}_a & -s\sqrt{3n(n+1)}\gamma'' & -\mathrm{i}s\sqrt{6n(n+1)}\gamma' \\ -\mathrm{i}\sqrt{\frac{1}{3}s(n+1)}P & -s\sqrt{3n(n+1)}\gamma'' & -s[(\gamma_1-\gamma')(n+\frac{3}{2})-\frac{1}{2}\kappa]-\mathcal{E}_a & \mathrm{i}s\sqrt{2}[\gamma'(n+\frac{3}{2})-\frac{1}{2}\kappa] \\ \sqrt{\frac{3}{2}s(n+1)}P & \mathrm{i}s\sqrt{6n(n+1)}\gamma' & -\mathrm{i}s\sqrt{2}[\gamma'(n+\frac{3}{2})-\frac{1}{2}\kappa] & -s[\gamma_1(n+\frac{3}{2})-\kappa]-\Delta_0-\mathcal{E}_a \end{vmatrix} = 0\,, \tag{2.189}$$

and

$$\begin{vmatrix} \mathcal{E}_G - \mathcal{E}_b + sn & \mathrm{i}\sqrt{\frac{1}{3}sn}P & \mathrm{i}\sqrt{s(n+1)}P & \sqrt{\frac{2}{3}sn}P \\ -\mathrm{i}\sqrt{\frac{1}{3}sn}P & -s[(\gamma_1-\gamma')(n-\frac{1}{2})+\frac{1}{2}\kappa]-\mathcal{E}_b & -s\sqrt{3n(n+1)}\gamma'' & -\mathrm{i}s\sqrt{2})[\gamma'(n-\frac{1}{2})+\frac{1}{2}\kappa] \\ -\mathrm{i}\sqrt{s(n+1)}P & -s\sqrt{3n(n+1)}\gamma'' & -s[(\gamma_1+\gamma')(n+\frac{3}{2})-\frac{3}{2}\kappa]-\mathcal{E}_b & \mathrm{i}s\sqrt{6n(n+1)}\gamma' \\ \sqrt{\frac{3}{2}sn}P & \mathrm{i}s\sqrt{2}[\gamma'(n-\frac{1}{2})+\frac{1}{2}\kappa & -\mathrm{i}s\sqrt{6n(n+1)}\gamma' & -s[\gamma_1(n-\frac{1}{2})+\kappa]-\Delta_0-\mathcal{E}_b \end{vmatrix} = 0\,. \tag{2.190}$$

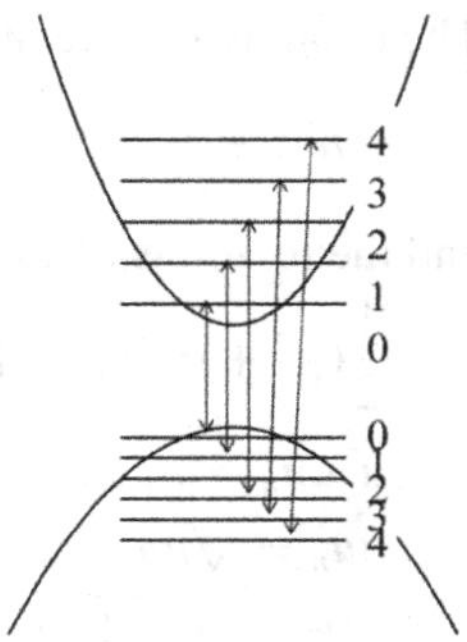

Fig. 2.18 Schematic diagram of the Landau levels at $\boldsymbol{k} = 0$ for a simple two bands model of conduction and valence bands. The allowed transitions are shown by the arrows between the Landau levels with quantum numbers of the conduction and valence bands

These equations are valid for $n \geq 1$. In the case of $n = -1$ we put $a_1 = a_3 = a_2 = a_6 = a_8 = 0$, and in the case of $n = 0$ we put $a_3 = a_6 = a_8 = 0$. The Landau levels of the electron in the conduction band and holes in the valence bands are obtained by solving the above equations. It may also be possible to obtain the selection rule for the transition between the Landau levels from the eigenstates derived here.

2.5.6 *Magneto–optical Absorption*

As described above the Landau levels of the valence bands are complicated and the detailed analysis requires accurate values of Luttinger parameters. One of the methods is the analysis of the cyclotron resonance of holes as discussed in this chapter. Another method is magneto–absorption measurements [11, 16–21]. Here a simple case is described for the purpose of introduction to the magneto–absorption effects. In Fig. 2.18 a schematic diagram of a simplified model of two parabolic bands are shown and the bars are the Landau levels. The Landau levels of the conduction and valence bands are given by

$$\mathcal{E}_{\mathrm{c}} = \mathcal{E}_{\mathrm{c}}(0) + \hbar\omega_{\mathrm{c}}\left(n + \frac{1}{2}\right) + \frac{\hbar^2 k_z^2}{2m_{\mathrm{c}}}, \tag{2.191a}$$

$$\mathcal{E}_{\mathrm{v}} = \mathcal{E}_{\mathrm{v}}(0) - \hbar\omega_{\mathrm{v}}\left(n + \frac{1}{2}\right) - \frac{\hbar^2 k_z^2}{2m_{\mathrm{v}}}, \tag{2.191b}$$

where $\omega_{\mathrm{c}} = eB/m_{\mathrm{c}}$ and $\omega_{\mathrm{v}} = eB/m_{\mathrm{v}}$ are the cyclotron frequencies of the conduction and valence bands, respectively. In Chap. 4 optical properties of semiconductors will be discussed and the absorption coefficient is proportional to the joint density of states. In the absence of magnetic field the absorption coefficient of 3 dimensional case is given by (see (4.56) and (4.58))

$$\alpha_0 = K\frac{4\pi}{(2\pi)^3}\left(\frac{2\mu}{\hbar^2}\right)^{3/2}\sqrt{\hbar\omega - \mathcal{E}_{\mathrm{G}}}\,, \tag{2.192a}$$

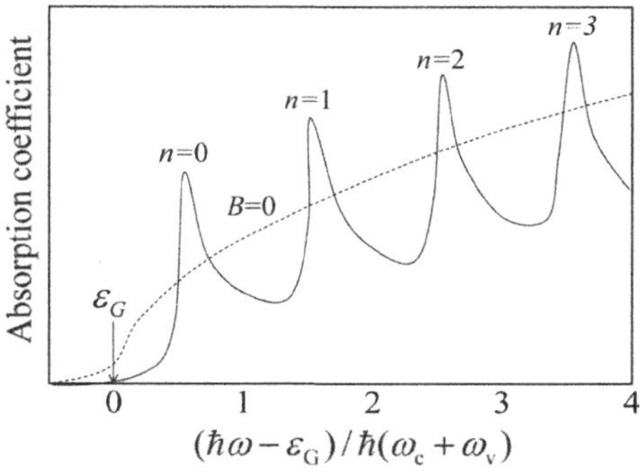

Fig. 2.19 Absorption coefficient as a function of incident photon energy (normalized by $\hbar(\omega_c + \omega_v)$) for the bands shown Fig. 2.18, where broadening effect is taken into account for both $B = 0$ (*dashed line*) and $B \neq 0$. The density of states shown in Fig. 2.16 is obtained without broadening effect

$$K = \frac{\pi e^2}{\epsilon_0 m^2 \omega^2} |\boldsymbol{e} \cdot \boldsymbol{p}_{cv}|^2 \tag{2.192b}$$

$$1/\mu = 1/m_c + 1/m_v, \quad \mathcal{E}_G = \mathcal{E}_c - \mathcal{E}_v . \tag{2.192c}$$

In the presence of a magnetic field the density of states is one dimensional and given by (2.132). Therefore the absorption coefficient is written as

$$\alpha_B = K \frac{eB\sqrt{2\mu}}{(2\pi\hbar)^2} \sum_n [\hbar\omega - \hbar\omega_n]^{-1/2} , \tag{2.193a}$$

$$\hbar\omega_n = \mathcal{E}_G + \left(n + \frac{1}{2}\right)(\hbar\omega_c + \hbar\omega_v) \tag{2.193b}$$

In Fig. 2.19 absorption coefficients (in a.u. units) for $B = 0$ and $B \neq 0$ are plotted, where the broadening effect is taken into account by putting $\hbar\omega \to \hbar\omega + \mathrm{i}\Gamma$, with $\Gamma = 0.05(\hbar\omega_c + \hbar\omega_v)$. In the figure we see the maxima of magneto–absorption appear periodically with the spacing of photon energy $\hbar\omega_c + \hbar\omega_v$. This is not the case observed by experiments as shown below. Since the conduction and valence bands are nonparabolic, and the valence bands consist of the degenerate heavy and light hole bands at the Γ point ($\boldsymbol{k} = 0$) and spin–orbit split–off band and thus the Landau levels are complicated as analyzed above.

Here a comparison between the experimental and theoretical results in InSb reported by Pidgeon and Brown [16] is shown, as one of the best examples. The experiments are carried out by choosing two different configurations of the electric field vector $\boldsymbol{E}$ of the incident radiation (light polarization) and the applied magnetic field $\boldsymbol{B}$, $\boldsymbol{E} \perp \boldsymbol{B} \parallel [100]$ and $\boldsymbol{E} \parallel \boldsymbol{B} \parallel [100]$, and optical absorption data (magneto-absorption data) were obtained at a fixed photon energy by sweeping the magnetic field $\boldsymbol{B}$. In Fig. 2.20 are shown plots of photon energy of the transmission minima as a function of magnetic-field field strength for the principal transmission in the $\boldsymbol{E} \perp \boldsymbol{B}$, with $\boldsymbol{B}$ parallel to the [100] crystal direction, along with the theoretical curves. In Fig. 2.21 are shown similar plots for the configuration $\boldsymbol{E} \parallel \boldsymbol{B} \parallel [100]$.

We see a good agreement between the experimental and assigned theoretical curves in Figs. 2.20 and 2.21. Here the notation for the transitions are defined as

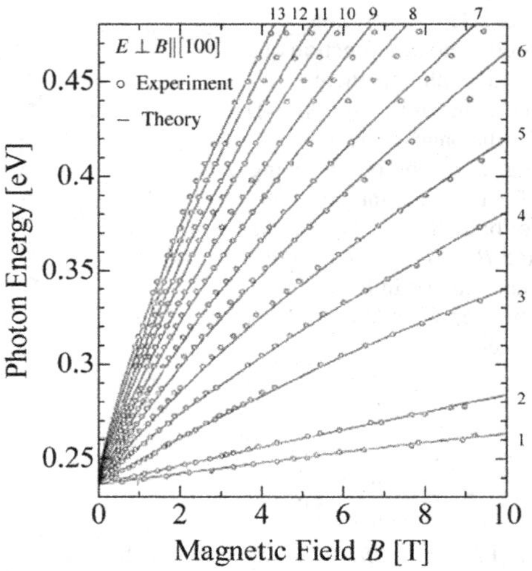

Fig. 2.20 Plot of the photon energy of magneto-absorption maxima (transmission minima) as a function of magnetic field for $\boldsymbol{E} \perp \boldsymbol{B} \parallel [100]$ in n-InSb at $T = 4.2$ K. The solid lines are calculated from (2.189) and (2.190) in the text for a and b series and the notation of the transitions is defined in the text and the transitions are as follows: (after Pidgeon and Brown [16])

1. $a^-(2)a^c(0)$ 6. $b^+(2)b^c(2)$ 11. $b^+(6)b^c(4)$
2. $b^+(0)b^c(0)$ 7. $a^+(4)a^c(2)$ 12. $b^+(5)b^c(5)$
3. $a^+(1)a^c(1)$ 8. $b^+(3)b^c(3)$ 13. $a^+(7)a^c(5)$
4. $b^+(1)b^c(1)$ 9. $a^+(5)a^c(3)$
5. $a^+(3)a^c(1)$ 10. $b^+(4)b^c(4)$

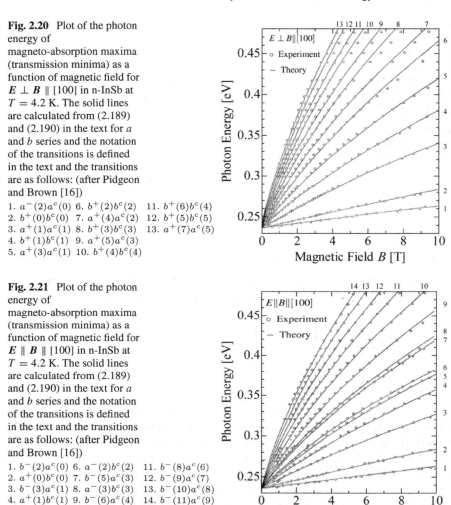

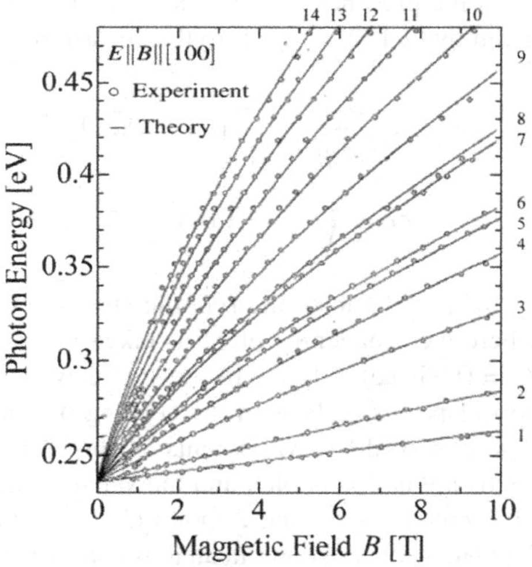

Fig. 2.21 Plot of the photon energy of magneto-absorption maxima (transmission minima) as a function of magnetic field for $\boldsymbol{E} \parallel \boldsymbol{B} \parallel [100]$ in n-InSb at $T = 4.2$ K. The solid lines are calculated from (2.189) and (2.190) in the text for a and b series and the notation of the transitions is defined in the text and the transitions are as follows: (after Pidgeon and Brown [16])

1. $b^-(2)a^c(0)$ 6. $a^-(2)b^c(2)$ 11. $b^-(8)a^c(6)$
2. $a^+(0)b^c(0)$ 7. $b^-(5)a^c(3)$ 12. $b^-(9)a^c(7)$
3. $b^-(3)a^c(1)$ 8. $a^-(3)b^c(3)$ 13. $b^-(10)a^c(8)$
4. $a^+(1)b^c(1)$ 9. $b^-(6)a^c(4)$ 14. $b^-(11)a^c(9)$
5. $b^-(4)a^c(2)$ 10. $b^-(7)a^c(5)$

follows. In the notations $a^{\pm}(n)a^c(n')$ and $b^{\pm}(n)b^c(n')$, a and b mean the a, and b series, and $+$, $-$, and c refer to the light–hole, heavy–hole, and conduction electron, respectively. The orbital quantum number is given in brackets and the selection rules result in allowed transitions for $\Delta n = 0, -2$. For example, the notation $a^-(2)a^c(0)$ means the transition between the light–hole level of a series with quantum number $n = 2$ and the conduction electron level of a series with quantum number $n' = 0$. The non–linearity of the calculated and experimental data in Figs. 2.20 and 2.21 arises from the nonparabolic effects of the conduction and valence bands. Here we find the observed transitions are mostly from the light–hole levels. On the other hand for the configuration $\boldsymbol{E} \parallel \boldsymbol{B} \parallel [100]$ the following transitions are observed.

For the configuration $\boldsymbol{E} \parallel \boldsymbol{B} \parallel [100]$, strongest transitions are from the heavy–hole levels. From the fitting procedure Pidgeon and Brown [16] obtained the following band–edge parameters for InSb:

$$P^2 = 0.403\,\text{au}\,, \tag{2.194a}$$
$$\mathcal{E}_G = 0.2355\,\text{eV}\,, \tag{2.194b}$$
$$\Delta = 0.9\,\text{eV}\,, \tag{2.194c}$$
$$\gamma_1^L = 32.5\,, \tag{2.194d}$$
$$\gamma_2^L = 14.3\,, \tag{2.194e}$$
$$\gamma_3^L = 15.4\,, \tag{2.194f}$$
$$\kappa^L = 13.4\,, \tag{2.194g}$$

where they found that F is very small and put it equal to zero.

2.6 Luttinger Hamiltonian

Here we derive so called "Luttinger Hamiltonian", which will be used for the analysis of the valence band states in a quantum structure and for the analysis of semiconductor quantum well lasers in Chap. 9. First, we calculate several matrix elements of (2.64) defined by Dresselhaus, Kip, and Kittel [1] and relate them to the Luttinger parameters. From (2.64) we obtain the following matrix elements

$$\frac{H_{11}+H_{22}}{2} = \frac{L+M}{2}(k_x^2+k_y^2) + Mk_z^2 \left(\rightarrow \frac{1}{2}P(\boldsymbol{k})\right)\,, \tag{2.195a}$$

$$-\frac{H_{13}-\mathrm{i}H_{23}}{\sqrt{3}} = -\frac{N}{\sqrt{3}}(k_x-\mathrm{i}k_y)k_z\,(\rightarrow L(\boldsymbol{k}))\,, \tag{2.195b}$$

$$-\frac{H_{11}-H_{22}-2\mathrm{i}H_{12}}{2\sqrt{3}} = \frac{(L-M)(k_x^2-k_y^2)-2\mathrm{i}Nk_xk_y}{\sqrt{12}}$$
$$(\rightarrow M(\boldsymbol{k}))\,, \tag{2.195c}$$

$$\frac{H_{33}+H_{11}+H_{22}}{6} = \frac{1}{6}\left[(L+M)(k_x^2+k_y^2)P + 2Mk_z^2\right]$$
$$+\frac{2}{3}M(k_x^2+k_y^2) + Lk_z^2\left(\rightarrow \frac{1}{6}P(\boldsymbol{k}) + \frac{2}{3}Q(\boldsymbol{k})\right)\,, \tag{2.195d}$$

and other elements are also easily calculated. The notations in the parentheses are the definition of Luttinger and Kohn [2], and L, M, and N of Dresselhaus et al. [1] are related to A, B, and C of Luttinger [2, 3] by $A = \hbar^2/2m + L$, $B = \hbar^2/2m + M$, and $C = N$. Using the wave functions defined by (2.63a)–(2.63f) and following the treatment of Luttinger and Kohn [2], (2.64) is rewritten as[5]

---

[5]The matrix elements derived by using the expressions of Dresselhaus et al. are equivalent to the expressions of Luttinger and Kohn except the diagonal elements as discussed before, and the

$$\begin{array}{l|cccccc|} |\frac{3}{2},\frac{3}{2}\rangle & P/2 & L & M & 0 & \mathrm{i}L/\sqrt{2} & -\mathrm{i}\sqrt{2}M \\ |\frac{3}{2},\frac{1}{2}\rangle & L^* & P/6+2Q/3 & 0 & M & -\mathrm{i}(P-2Q)/3\sqrt{2} & \mathrm{i}\sqrt{3/2}L \\ |\frac{3}{2},-\frac{1}{2}\rangle & M^* & 0 & P/6+2Q/3 & -L & -\mathrm{i}\sqrt{3/2}L^* & -\mathrm{i}(P-2Q)/3\sqrt{2} \\ |\frac{3}{2},-\frac{3}{2}\rangle & 0 & M^* & -L^* & P/2 & -\mathrm{i}\sqrt{2}M^* & -\mathrm{i}L^*/\sqrt{2} \\ |\frac{1}{2},\frac{1}{2}\rangle & -\mathrm{i}L^*/\sqrt{2} & \mathrm{i}(P-2Q)/3\sqrt{2} & \mathrm{i}\sqrt{3/2}L & \mathrm{i}\sqrt{2}M & (P+Q)/3-\Delta_0 & 0 \\ |\frac{1}{2},-\frac{1}{2}\rangle & \mathrm{i}\sqrt{2}M^* & -\mathrm{i}\sqrt{3/2}L^* & \mathrm{i}(P-2Q)/3\sqrt{2} & \mathrm{i}L/\sqrt{2} & 0 & (P+Q)/3-\Delta_0 \end{array} \tag{2.196}$$

where

$$P(\boldsymbol{k}) = \frac{\hbar^2}{2m}\left[(A+B)(k_x^2+k_y^2)+2Bk_z^2\right], \tag{2.197a}$$

$$L(\boldsymbol{k}) = -\mathrm{i}\frac{C}{\sqrt{3}}\frac{\hbar^2}{2m}(k_x-\mathrm{i}k_y)k_z, \tag{2.197b}$$

$$Q(\boldsymbol{k}) = \frac{\hbar^2}{2m}\left[B(k_x^2+k_y^2)+Ak_z^2\right], \tag{2.197c}$$

$$M(\boldsymbol{k}) = \frac{1}{\sqrt{12}}\frac{\hbar^2}{2m}\left[(A-B)(k_x^2-k_y^2)-2\mathrm{i}Ck_xk_y\right]. \tag{2.197d}$$

Here A, B, and C are dimensionless constants obtained by dividing the original values by $\hbar^2/2m$ and Δ_0 is spin–orbit splitting at $\boldsymbol{k}=0$. Equation (2.196) is called *Luttinger Hamiltonian* and Luttinger parameters γ_1, γ_2 and γ_3 given by (2.172a)–(2.172b) are rewritten as follows by recovering units of $\hbar = 1$ such as $(1/2m)\gamma_1 \to (\hbar^2/2m)\gamma_1$, and so on,

$$\gamma_1 = -\frac{1}{3}(A+2B), \tag{2.198a}$$

$$\gamma_2 = -\frac{1}{6}(A-B), \tag{2.198b}$$

$$\gamma_3 = -\frac{1}{6}C, \tag{2.198c}$$

or

$$-A = \gamma_1 + 4\gamma_2, \tag{2.199a}$$

$$-B = \gamma_1 - 2\gamma_2, \tag{2.199b}$$

$$-C = 6\gamma_3. \tag{2.199c}$$

It should be noted here that the above parameters are for the electronic states and that the parameters for the holes of the valence bands are obtained by replacing $-A$, $-B$ and $-C$ by A, B and C, respectively.

When the matrix elements are redefined as

$$P' = \frac{1}{3}(P+Q) = -\gamma_1\frac{\hbar^2}{2m}(k_x^2+k_y^2+k_z^2), \tag{2.200a}$$

(Footnote 5 continued)
corresponding secular equation of Luttinger and Kohn Hamiltonian gives eigenvalue $\mathcal{E}(\boldsymbol{k})$, while the secular equation of Dresselhaus, Kip and Kittel gives $\lambda = \mathcal{E}(\boldsymbol{k}) - \hbar^2k^2/2m$.

$$Q' = \frac{1}{6}(P - 2Q) = -\gamma_2 \frac{\hbar^2}{2m}(k_x^2 + k_y^2 - 2k_z^2)\,, \tag{2.200b}$$

$$L = 2\mathrm{i}\sqrt{3}\gamma_3 \frac{\hbar^2}{2m}(k_x - \mathrm{i}k_y)k_z\,, \tag{2.200c}$$

$$M = -\sqrt{3}\frac{\hbar^2}{2m}\left[\gamma_2(k_x^2 - k_y^2) - 2\mathrm{i}\gamma_3 k_x k_y\right]\,, \tag{2.200d}$$

and then (2.196) is rewritten as

$$\begin{vmatrix} P'+Q' & L & M & 0 & \mathrm{i}L/\sqrt{2} & -\mathrm{i}\sqrt{2}M \\ L^* & P'-Q' & 0 & M & -\mathrm{i}\sqrt{2}Q' & \mathrm{i}\sqrt{3/2}L \\ M^* & 0 & P'-Q' & -L & -\mathrm{i}\sqrt{3/2}L^* & -\mathrm{i}\sqrt{2}Q' \\ 0 & M^* & -L^* & P'+Q' & -\mathrm{i}\sqrt{2}M^* & -\mathrm{i}L^*/\sqrt{2} \\ -\mathrm{i}L^*/\sqrt{2} & \mathrm{i}\sqrt{2}Q' & \mathrm{i}\sqrt{3/2}L & \mathrm{i}\sqrt{2}M & P'-\Delta_0 & 0 \\ \mathrm{i}\sqrt{2}M^* & -\mathrm{i}\sqrt{3/2}L^* & \mathrm{i}\sqrt{2}Q' & \mathrm{i}L/\sqrt{2} & 0 & P'-\Delta_0 \end{vmatrix}\,. \tag{2.201}$$

Here we show in Fig. 2.22 the calculated curves of the valence band structures of GaAs by the 6×6 $\boldsymbol{k} \cdot \boldsymbol{p}$ Luttinger Hamiltonian and the full band calculations based on 30×30 $\boldsymbol{k} \cdot \boldsymbol{p}$ Hamiltonian. We find that the 6×6 $\boldsymbol{k} \cdot \boldsymbol{p}$ Hamiltonian gives the dispersion nerar $\boldsymbol{k} \simeq 0$, and that a big deviation appears in the full Brillouin zone. Used parameters for the $\boldsymbol{k} \cdot \boldsymbol{p}$ full band calculations are in Table 1.11 and for the Luttinger parameters are in Table 2.1.

In order to deal with optical properties of quantum structures such as quantum wells, quantum dots and so on, we have to solve the electronic states of the conduction and the valence bands together. In such cases we have to solve 8×8 $\boldsymbol{k} \cdot \boldsymbol{p}$ Hamiltonian adding two electronic states of the conduction band $|\Gamma_{2'} \uparrow\rangle$ and $|\Gamma_{2'} \downarrow\rangle$ to the 6×6 $\boldsymbol{k} \cdot \boldsymbol{p}$ Hamiltonian (2.196) or (2.201). The eight–band theory has been

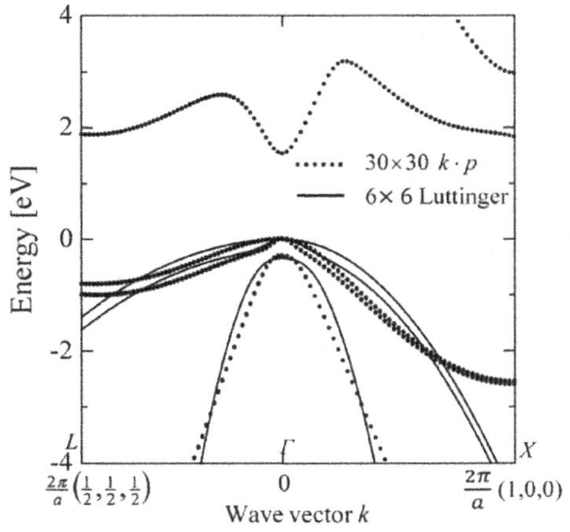

Fig. 2.22 Comparison of the valence band dispersion between the 30–band $\boldsymbol{k} \cdot \boldsymbol{p}$ (*dotted curves*) and 6–band $\boldsymbol{k} \cdot \boldsymbol{p}$ Luttinger Hamiltonian (*solid curves*)

Table 2.1 Energy band parameters of Ge, Si, GaAs, InAs, AlAs, and GaP. The parameters γ_1, γ_2, and γ_3 are the original Luttinger parameters which correspond to the redefined Luttinger parameters γ_1^L, γ_2^L, and γ_3^L, respectively, and the effective mass m^* is normalized by the free electron mass m

Parameters	Ge	Si	GaAs	InAs	AlAs	GaP
$\mathcal{E}_G^{\Gamma}$ [eV]	0.8872	4.185	1.519	0.417	3.099	2.886
$\mathcal{E}_G^X$ [eV]	1.3	1.17	1.981	1.433	2.24	2.35
$\mathcal{E}_G^L$ [eV]	0.82	n.a.	1.815	1.133	2.46	2.72
Δ_0 [eV]	0.297	0.044	0.341	0.39	0.28	0.08
$m_e^*(\Gamma)$	n.a.	n.a.	0.067	0.026	0.15	0.13
$m_l^*(L)$	1.57	n. a.	1.90	0.64	1.32	2.0
$m_t^*(L)$	0.0807	0.19	0.075	0.05	0.15	0.253
$m_l^*(X)$	n.a.	0.916	1.98	1.13	0.97	1.2
$m_t^*(X)$	n.a.	0.19	0.27	0.16	0.22	0.15
γ_1	13.35	4.26	6.98	20.0	3.76	4.05
γ_2	4.25	0.38	2.06	8.5	0.82	0.49
γ_3	5.69	1.56	2.93	9.2	1.42	2.93
$\mathcal{E}_{P0}$ [eV]	26.3	21.6	28.8	21.5	21.1	31.4
F	−27.2	−5.14	−17.8	−43.3	−6.62	−10.78
$F^{\dagger}$			(−1.94)	(−2.90)	(−0.48)	(−2.04)

n.a. means the value is not available

F: Estimated from $F = -(\mathcal{E}_{P0}/3)[2/\mathcal{E}_G + 1/(\mathcal{E}_G + \Delta_0)]$

$F^{\dagger}$: F values in parentheses are recommended by Vurgaftman, Meyer, and Ram-Mohan [27]

reported by Bahder [22], Zhang [23], and Pryor [24]. However, the results in Fig. 2.22 restricts us to use 6×6 Luttinger Hamiltonian or extended 8×8 $\boldsymbol{k} \cdot \boldsymbol{p}$ Hamiltonian for analyzing the periodic quantum dots of superlattices, in which Brillouin zone folding effect plays an important role as discussed in Chap. 8.

2.7 Luttinger Parameters

It should be noted here that the valence band parameters of Luttinger are subject to reinterpretation because they are determined experimentally and thus all the contributions from the bands are included. If we use the parameters to solve $\boldsymbol{k} \cdot \boldsymbol{p}$ 8×8 Luttinger-Hamiltonian, the contribution from the $|\Gamma_{2'}\rangle$ conduction band is included twice. Therefore we have to exclude the contribution form the conduction band. First, we discuss how to determine the valence band parameters of Luttinger using the analysis made by Kane [4], Dresselhaus, Kip and Kittel [1], Groves et al. [25] and Lawaetz [6],

$$F = -\frac{2}{m}\sum_j \frac{|\langle \Gamma_{25'}(X)|p_x|\Gamma_{2'}, j\rangle|^2}{\mathcal{E}(\Gamma_{2'}, j)}, \tag{2.202a}$$

$$G = -\frac{2}{m}\sum_j \frac{|\langle \Gamma_{25'}(X)|p_x|\Gamma_{12'}, j\rangle|^2}{\mathcal{E}(\Gamma_{12'}, j)}, \tag{2.202b}$$

$$H_1 = -\frac{2}{m}\sum_j \frac{|\langle \Gamma_{25'}(X)|p_y|\Gamma_{15}, j\rangle|^2}{\mathcal{E}(\Gamma_{15}, j)}, \tag{2.202c}$$

$$H_2 = -\frac{2}{m}\sum_j \frac{|\langle \Gamma_{25'}(X)|p_y|\Gamma_{25}, j\rangle|^2}{\mathcal{E}(\Gamma_{25}, j)}, \tag{2.202d}$$

where the valence band edge $\Gamma_{25'}$ (Γ_8) at $\boldsymbol{k} = 0$ is taken to be energy zero.[6] The contributions to the parameters L, M, and N of (2.47a) $\sim$ (2.47c) defined defined by Dresselhaus, Kip and Kittel [1] from the above four conduction bands are calculated as follows.

1. Contributions to L (or Lk_x^2) are from $\Gamma_{2'}$ and $\Gamma_{12'}$. The contribution to L (or A) from $|\Gamma_{2'}\rangle$ is evaluated by taking account of the relations $\langle X|p_x|\Gamma_{2'}\rangle = \langle Y|p_y|\Gamma_{2'}\rangle = \langle Z|p_z|\Gamma_{2'}\rangle$, and thus

$$-\frac{2}{m}\frac{\langle X|p_x|\Gamma_{2'}\rangle\langle X|p_x|\Gamma_{2'}\rangle}{\mathcal{E}(\Gamma_{2'})} = -\frac{2}{m}\frac{|\langle X|p_x|\Gamma_{2'}\rangle|^2}{\mathcal{E}(\Gamma_{2'})} = F\,. \tag{2.203}$$

The contribution from $|\Gamma_{12'}(1)\rangle$ and $|\Gamma_{12'}(2)\rangle$ is evaluated by using (1.141a) and (1.141b) or (1.143a) and (1.143b) in Sect. 1.7. In the following we use the notations $|\gamma_1^-\rangle \to |\Gamma_{12'}(1)\rangle$ and $|\gamma_2^-\rangle \to |\Gamma_{12'}(2)\rangle$,

$$\begin{aligned}
&-\frac{2}{m}\sum_j \frac{|\langle \Gamma_{25'}(X)|p_x|\Gamma_{12'}, j\rangle|^2}{\mathcal{E}(\Gamma_{12'}, j)} \\
&\quad = -\frac{2}{m}\frac{|\langle \Gamma_{25'}(X)|p_x|\Gamma_{12'}(1)\rangle|^2}{\mathcal{E}(\Gamma_{12'})} - \frac{2}{m}\frac{|\langle \Gamma_{25'}(X)|p_x|\Gamma_{12'}(2)\rangle|^2}{\mathcal{E}(\Gamma_{12'})} \\
&\quad = -\frac{2}{m}\frac{R*R}{\mathcal{E}(\Gamma_{12'})} - \frac{2}{m}\frac{(-R)*(-R)}{\mathcal{E}(\Gamma_{12'})} = 2G\,.
\end{aligned} \tag{2.204}$$

Therefore the term L is given by

$$L = \frac{\hbar^2}{2m}\left[F + 2G\right]. \tag{2.205}$$

[6] The reason why the upper four conduction bands appear in the matrix elements is understood from the the selection rules $\langle \Gamma_{25'}|\boldsymbol{p}|\Gamma_{l,j}\rangle$, where $\boldsymbol{p}$ is momentum operator and transforms as the representation Γ_{15}, and $\Gamma_{l,j}$ are upper conduction band states. The direct product is given by using the character table of Table 1.4

$$\Gamma_{25'} \times \Gamma_{15} = \Gamma_{12'} + \Gamma_{15} + \Gamma_{2'} + \Gamma_{25}\,,$$

and thus only the conduction band states of the four representations on the right hand perturb the valence band edge.

2. The parameter M (or Mk_y^2) is associated with the states Γ_{15} and Γ_{25} and we easily obtain the following results,

$$H_1 = -\frac{2}{m}\sum_j \frac{|\langle X|p_y|\Gamma_{15}, j\rangle|^2}{\mathcal{E}(\Gamma_{15})} = -\frac{2}{m}\frac{|\langle X|p_y|\Gamma_{15}(z)\rangle|^2}{\mathcal{E}(\Gamma_{15})}, \tag{2.206}$$

$$H_2 = -\frac{2}{m}\sum_j \frac{|\langle X|p_y|\Gamma_{25}, j\rangle|^2}{\mathcal{E}(\Gamma_{25})} = -\frac{2}{m}\frac{|\langle X|p_y|\Gamma_{25}(1)\rangle|^2}{\mathcal{E}(\Gamma_{25})}, \tag{2.207}$$

and then we obtain

$$M = \frac{\hbar^2}{2m}[H_1 + H_2]. \tag{2.208}$$

3. Finally N (or C) is evaluated as follows. Contribution from $\Gamma_{2'}$ to Nk_yk_z:

$$-\frac{2}{m}\frac{\langle Y|p_y|\Gamma_{2'}\rangle\langle\Gamma_{2'}|p_z|Z\rangle}{\mathcal{E}(\Gamma_{2'})} = -\frac{2}{m}\frac{|\langle Y|p_y|\Gamma_{2'}\rangle|^2}{\mathcal{E}(\Gamma_{2'})} = F. \tag{2.209}$$

Contribution from $\Gamma_{12'}$ to Nk_yk_z:

$$\begin{aligned}\sum_j &\langle Y|p_y|\Gamma_{12'}(j)\rangle\langle\Gamma_{12'}(j)|p_z|Z\rangle \\ &= \langle Y|p_y|\Gamma_{12'}(1)\rangle\langle\Gamma_{12'}(1)|p_z|Z\rangle + \langle Y|p_y|\Gamma_{12'}(2)\rangle\langle\Gamma_{12'}(2)|p_z|Z\rangle \\ &= (\omega R)(\omega^{\dagger 2}R) + (-\omega^2 R)(-\omega^\dagger R) = [\omega\cdot\omega^{\dagger 2} + \omega^2\cdot\omega^\dagger)R^2 \\ &= [\omega^\dagger + \omega]R^2 = -R^2,\end{aligned} \tag{2.210}$$

and thus contribution from $\Gamma_{12'}$ to Nk_yk_z is (see subsection 1.7.3 for R)

$$-\frac{2}{m}\frac{|\langle Y|p_y|\Gamma_{12'}|^2}{\mathcal{E}(\Gamma_{12'})} = -G. \tag{2.211}$$

Contribution from Γ_{15} to Nk_yk_z is

$$-\frac{2}{m}\frac{|\langle Z|p_y|\Gamma_{15}(x)\rangle\langle\Gamma_{15}(x)|p_z|Y\rangle|}{\mathcal{E}(\Gamma_{15})} = H_1. \tag{2.212}$$

Contribution of Γ_{25} to Nk_yk_z is evaluated using the following symmetry operation. We define

$$\Gamma_{25,1} = |z(x^2 - y^2)\rangle, \tag{2.213a}$$

$$\Gamma_{25,2} = |x(y^2 - z^2)\rangle, \tag{2.213b}$$

$$\Gamma_{25,3} = |y(z^2 - x^2)\rangle, \tag{2.213c}$$

and the matrix element by

$$\langle X|p_y|\Gamma_{25,1}\rangle = \langle yz|p_y|z(x^2 - y^2)\rangle = U. \tag{2.214}$$

Symmetry operations by $\pi/2$ along x, y and z give the following relations,

$$x-\text{axis}: \quad \langle X|p_z|\Gamma_{25,3}\rangle = -U\,, \tag{2.215}$$

$$y-\text{axis}: \quad \langle Z|p_y|\Gamma_{25,2}\rangle = -U\,, \tag{2.216}$$

$$z-\text{axis}: \quad \langle Y|p_x|\Gamma_{25,1}\rangle = -U\,. \tag{2.217}$$

When we define

$$-\frac{2}{m}\sum_j \frac{\left|\langle X|p_y|\Gamma_{25,j}\rangle\right|^2}{\mathcal{E}(\Gamma_{25,j})} = -\frac{2}{m}\frac{U\cdot U}{\mathcal{E}(\Gamma_{25})} = H_2\,, \tag{2.218}$$

the contribution of Γ_{25} to Nk_yk_x is

$$-\frac{2}{m}\sum_j \frac{\langle X|p_y|\Gamma_{25,j}\rangle\langle\Gamma_{25,j}|p_x|Y\rangle}{\mathcal{E}(\Gamma_{25,j})} = -\frac{2}{m}\frac{U\cdot(-U)}{\mathcal{E}(\Gamma_{25})} = -H_2\,. \tag{2.219}$$

Therefore the total contribution to N (or C) is found to be

$$N = \frac{\hbar^2}{2m}[F - G + H_1 - H_2]\,. \tag{2.220}$$

From these considerations we obtain the following relations for the parameters L, M, N of Dresselhaus, Kip, and Kittel,

$$L = \frac{\hbar^2}{2m}[F + 2G]\,, \tag{2.221}$$

$$M = \frac{\hbar^2}{2m}[H_1 + H_2]\,, \tag{2.222}$$

$$N = \frac{\hbar^2}{2m}[F - G + H_1 - H_2]\,, \tag{2.223}$$

and thus the parameters A, B, and C of Luttinger are given by

$$A = \frac{\hbar^2}{2m}[1 + F + 2G]\,, \tag{2.224}$$

$$B = \frac{\hbar^2}{2m}[1 + H_1 + H_2]\,, \tag{2.225}$$

$$C = \frac{\hbar^2}{2m}[F - G + H_1 - H_2]\,. \tag{2.226}$$

Neglecting the spin–orbit splitting of $|\Gamma_{15}\rangle$ conduction band states, the Luttinger valence band parameters are given by Groves et al. [25] and Lawaetz [6], which are obtained by putting (2.224)–(2.226) into (2.198a)–(2.198c) as follows,

$$\gamma_1 = -\frac{1}{3}(F + 2G + 2H_1 + 2H_2) - 1\,, \tag{2.227a}$$

$$\gamma_2 = -\frac{1}{6}(F + 2G - H_1 - H_2)\,, \tag{2.227b}$$

$$\gamma_3 = -\frac{1}{6}(F - G + H_1 - H_2)\,. \tag{2.227c}$$

These Luttinger parameters are modified by the parameter q when the spin–orbit splitting Δ_{15} of the Γ_{15} conduction band states is taken account [25, 26]. Including the term κ which represents the magnetic filed effect in the Luttinger Hamiltonian, the Luttinger parameters are then given by

$$\gamma_1 = -\frac{1}{3}(F + 2G + 2H_1 + 2H_2) - 1 + \frac{1}{2}q\,, \tag{2.228a}$$

$$\gamma_2 = -\frac{1}{6}(F + 2G - H_1 - H_2) - \frac{1}{2}q\,, \tag{2.228b}$$

$$\gamma_3 = -\frac{1}{6}(F - G + H_1 - H_2) + \frac{1}{2}q\,. \tag{2.228c}$$

$$\kappa = -\frac{1}{6}(F - G - H_1 + H_2) - \frac{1}{3} - \frac{9}{2}q\,, \tag{2.228d}$$

where

$$q \simeq -\frac{2}{9}\frac{H_1}{\mathcal{E}(\Gamma_{15})}\Delta_{15}\,. \tag{2.229}$$

Usually the Luttinger parameters are determined from the experimental data and the energy values of $\mathcal{E}_G = \mathcal{E}_c(\Gamma_{2'}) - \mathcal{E}_v(\Gamma_{25'}) \equiv \mathcal{E}_0$ and $\mathcal{E}_c(\Gamma_{15}) - \mathcal{E}_v(\Gamma_{25'}) = \mathcal{E}_0'$ are known from optical spectroscopies, and contributions from higher lying bands may be neglected because of the larger energy denominators $\mathcal{E}_j$. Then the following two terms are taken into account (note the difference between $\mathcal{E}_{P0}$ and $\mathcal{E}_p$ as stated in (2.90)),

$$\mathcal{E}_{P0} = \frac{2}{m}|\langle\Gamma_{25'}(X)|p_x|\Gamma_{2'}\rangle|^2 = \frac{2}{m}P_0^2\,, \tag{2.230a}$$

$$\mathcal{E}_{P1} = \frac{2}{m}|\langle\Gamma_{25'}(X)|p_y|\Gamma_{15}(z)\rangle|^2\,, \tag{2.230b}$$

and we obtain

$$F = -\frac{\mathcal{E}_{P0}}{\mathcal{E}_G}\,, \qquad H_1 = -\frac{\mathcal{E}_{P1}}{\mathcal{E}_1}\,, \tag{2.231}$$

where the spin–orbit interaction is neglected. When the spin–orbit interaction is included, the matrix element F between the conduction band $|\Gamma_{2'}, j\rangle$ and the valence bands $|\Gamma_{25'}\rangle$ should be modified. Taking account of the lowest conduction band $|\Gamma_{2'}\rangle$ only and of the 6 valence band states $|\frac{3}{2}, \pm\frac{3}{2}\rangle$, $|\frac{3}{2}, \pm\frac{1}{2}\rangle$, and $|\frac{1}{2}, \pm\frac{1}{2}\rangle$, F is evaluated, for the spin–conserved matrix elements, as follows;

$$\begin{aligned} F &= \frac{2}{m}\sum_j \frac{|\langle\Gamma_{25'}(X)|p_x|\Gamma_{2'}\rangle|^2}{\mathcal{E}(\Gamma_{25'}) - \mathcal{E}(\Gamma_{2'}, j)}\,, \\ &= -\frac{\mathcal{E}_{P0}}{3}\left[\frac{2}{\mathcal{E}_G} + \frac{1}{\mathcal{E}_G + \Delta_0}\right]. \end{aligned} \tag{2.232}$$

Since the 8×8 $\boldsymbol{k} \cdot \boldsymbol{p}$ Hamiltonian includes the coupling between the conduction band $|\Gamma_{2'}\rangle$ and the valence bands $|\Gamma_{25'}\rangle$, we have to exclude the contributions from the conduction band $|\Gamma_{2'}\rangle$, the term F. Therefore the Luttinger parameters γ_1, γ_2, and γ_3 should be modified. We redefine the Luttinger parameters as γ_1^L, γ_2^L, and γ_3^L and the new valence band parameters γ_1, γ_2, and γ_3 are given by

$$\gamma_1 = \gamma_1^L - \left(-\frac{1}{3}F\right), \tag{2.233a}$$

$$\gamma_2 = \gamma_2^L - \left(-\frac{1}{6}F\right), \tag{2.233b}$$

$$\gamma_3 = \gamma_3^L - \left(-\frac{1}{6}F\right), \tag{2.233c}$$

and then we obtain

$$\gamma_1 = \gamma_1^L - \frac{1}{3} \cdot \frac{\mathcal{E}_{P0}}{3}\left[\frac{2}{\mathcal{E}_G} + \frac{1}{\mathcal{E}_G + \Delta_0}\right], \tag{2.234a}$$

$$\gamma_2 = \gamma_2^L - \frac{1}{6} \cdot \frac{\mathcal{E}_{P0}}{3}\left[\frac{2}{\mathcal{E}_G} + \frac{1}{\mathcal{E}_G + \Delta_0}\right], \tag{2.234b}$$

$$\gamma_3 = \gamma_3^L - \frac{1}{6} \cdot \frac{\mathcal{E}_{P0}}{3}\left[\frac{2}{\mathcal{E}_G} + \frac{1}{\mathcal{E}_G + \Delta_0}\right], \tag{2.234c}$$

where $\mathcal{E}_G = \mathcal{E}_c - \mathcal{E}_v$ is the energy gap and Δ_0 is the spin–orbit splitting.[7] For more detailed discussion see the [1, 6, 16, 24]. It should be noted here that the hole effective masses of the valence bands are evaluated by using the original Luttinger parameters γ_1^L, γ_2^L, and γ_3^L, while the eigenstates and eigenvalues of the 8×8 $\boldsymbol{k} \cdot \boldsymbol{p}$ Hamiltonian matrix are obtained by using new Luttinger parameters γ_1, γ_2, and γ_3. The new values of the Luttinger parameters are calculated by (2.234a) $\sim$ (2.234c), but the reported values of F have not yet settled. Pigeon and Brown [16] reported that the experimental data of the inter band magneto-absorption are well explained by putting $F = 0$ for InSb. They used $\mathcal{E}_{P0} = 10.96$ [eV], $\mathcal{E}_G = 0.2355$ [eV], $\Delta_0 = 0.9$ [eV], $\gamma_1^L = 32.5$, $\gamma_2^L = 14.3$, and $\gamma_3^L = 15.4$, and they found a good over–all fit of the magneto-absorption spectra in InSb by putting $F = 0$. From their data we obtain $F = -34.5$ ($F/3 = -11.5$) and the correction factor $F/3$ is not negligible. Vurgaftman, Meyer, and Ram-Mohan [27] recommend to use $\mathcal{E}_{P0} = 23.3$ [eV], $\mathcal{E}_G = 0.235$ [eV], $\Delta_0 = 0.81$ [eV], $\gamma_1^L = 34.8$, $\gamma_2^L = 15.5$, $\gamma_3^L = 16.5$ and $F = -0.23$ for InSb. Also they recommend $F = -1.94$ for GaAs which is about $1/10$ of the calculated value from $F = -(\mathcal{E}_{P0}/3)[2/\mathcal{E}_G + 1/(\mathcal{E}_G + \Delta_0)] = -17.8$. The band parameters of several semiconductors are summarized in Table 2.1.

[7] Equations (2.179a) $\sim$ (2.179c) are obtained by neglecting the spin–orbit splitting, or by putting $\Delta_0 = 0$ in (2.234a) $\sim$ (2.234c).

It is well known that the effective masses in III-V semiconductors may be estimated by the relation (2.84), which is rewritten as

$$\frac{m}{m_0^*} = 1 + \frac{\mathcal{E}_{P0}}{3}\left[\frac{2}{\mathcal{E}_{\rm G}} + \frac{1}{\mathcal{E}_{\rm G} + \Delta_0}\right] = 1 - F\,. \tag{2.235}$$

However more accurate values are evaluated by taking account of the contribution from higher lying conduction bands, Γ_8^c and Γ_7^c (which are relate to the single group representation Γ_{15}^c), and expressed by the following relation [6, 15, 28]

$$\frac{m}{m_0^*} = 1 - F - H_1 + F'\,, \tag{2.236a}$$

$$H_1 = -\frac{2}{m}\frac{P_1^2}{3}\left[\frac{2}{\mathcal{E}(\Gamma_8^c)} + \frac{1}{\mathcal{E}(\Gamma_7^c) + \Delta_{15}}\right], \tag{2.236b}$$

where $\mathcal{E}(\Gamma_8^c)$ and $\mathcal{E}(\Gamma_7^c)$ are the energies of the conduction bands (related to the higher lying Γ_{15} conduction bands in the single group expression) measured from the valence band top Γ_8^v and Δ_{15} is the spin–orbit splitting of the bands. The momentum matrix element P_1 is defined by $Q = 2P_1$ in (1.135b) of Chap. 1 and the values are given in Tables 1.9 and 1.11. Observed effective mass of the lowest conduction band is known to be well expressed by the above relation with $F' \simeq -2$. [6, 15, 28]

2.8 Problems

(2.1) Estimate the magnetic field strength to observe cyclotron resonance for microwave frequency 24 GHz for (1) effective mass $m^2 = 0.3\,\mathrm{m}$ (for Si and Ge) and (2) $m^* = 0.067\,\mathrm{m}$ (for GaAs). (3) Discuss the experimental conditions of the cyclotron resonance for the case of (1) and (2).

(2.2) Prove the commutation relation (2.100)

$$\boldsymbol{A}\cdot\boldsymbol{p} - \boldsymbol{p}\cdot\boldsymbol{A} = \mathrm{i}\hbar\nabla\cdot\boldsymbol{A} = 0\,,$$

(2.3) Assume a non–parabolic conduction band and show what kind of dispersion relation of the conduction band is expected when the band gap approaches zero.

(2.4) Figures 2.4 and 2.5 show the cyclotron masses for heavy and light holes in Ge and Si as a function of applied magnetic field. Heavy holes exhibit maximum in the magnetic field direction [1, 1, 1], while the light hole cyclotron mass is minimum. Explain the behaviour by using the constant energy contour of the valence bands in Fig. 2.13.

(2.5) Estimate electron effective mass m_0^* and effective g_0^* factor of InAs using the following parameters. $\mathcal{E}_{\rm G} = 0.417$ [eV], $\Delta_0 = 0.39$ [eV], $\mathcal{E}_{P0} = 21.5$ [eV], $\mathcal{E}_1 = 4.44$ [eV], $\Delta_{15} = 0.16$ [eV], $\mathcal{E}_{P1} = \mathcal{E}_{P0}/4 = 5.375$ [eV], and $C' = -0.02$. Experimental values are $m^* = 0.024\,\mathrm{m}$ and $g = -14.7\sin -17.5$.

(2.6) Calculate the matrix element of spin–orbit interaction $\langle u_{-}\alpha|\boldsymbol{L}\cdot\boldsymbol{\sigma}|u_{-}\alpha\rangle$

(2.7) Calculate the density of states in (1) parabolic conduction band (2.85) and (2) non–parabolic band given by (2.158).

(2.8) When magnetic field is $[0, 0, B]$, show the corresponding vector potential $\boldsymbol{A}$ is given by $[0, Bx, 0]$. This representation is called Landau gauge. List the other representations.

References

1. G. Dresselhaus, A.F. Kip, C. Kittel, Phys. Rev. **98**, 368 (1955)
2. J.M. Luttinger, W. Kohn, Phys. Rev. **97**, 869 (1955)
3. J.M. Luttinger, Phys. Rev. **102**, 1030 (1956)
4. E.O. Kane, J. Phys. Chem. Solids **1**, 82 (1956) and (1957) 249
5. M. Cardona, F.H. Pollak, Phys. Rev. **142**, 530 (1966)
6. P. Lawaetz, Phys. Rev. B **4**, 3460 (1971)
7. R. Kubo, H. Hasegawa, N. Hashitsume, J. Phys. Soc. Jpn. **14**, 56 (1959)
8. R. Kubo, S.J. Miyake, N. Hashitsume, in *Solid State Physics*, vol. 17, ed. by F. Seitz, D. Turnbull (Academic Press, New York, 1965) p. 269
9. R. Bowers, Y. Yafet, Phys. Rev. **115**, 1165 (1959)
10. Q.H.F. Vrehen, J. Phys. Chem. Solids **29**, 129 (1968)
11. L.M. Roth, B. Lax, S. Zwerdling, Phys. Rev. **114**, 90 (1959)
12. W. Zawadzki, Phys. Lett. **4**, 190 (1963)
13. K.L. Litvunenko, L. Nikzat, C.R. Pidgeon, J. Allam, L.F. Cohen, T. Ashley, M. Emeny, W. Zawadzki, B.N. Murdin, Phys. Rev. B **77**, 033204 (2008)
14. W. Zawadzki, P. Pfeffer, R. Bratschitsch, Z. Chen, S.T. Cundiff, B.N. Murdin, C.R. Pidgeon, Phys. Rev. B **78**, 245203 (2008)
15. C. Herman, C. Weisbuch, Phys. Rev. B **15**, 823 (1977)
16. C.R. Pidgeon, R.N. Brown, Phys. Rev. **146**, 575 (1966)
17. E. Burstein, G.S. Picus, Phys. Rev. **105**, 1123 (1957)
18. S. Zwerdling, R.J. Keyes, H.H. Kolm, B. Lax, Phys. Rev. **104**, 1805 (1956)
19. S. Zwerdling, B. Lax, Phys. Rev. **106**, 51 (1957)
20. S. Zwerdling, B. Lax, L.M. Roth, Phys. Rev. **108**, 1402 (1957)
21. S. Zwerdling, B. Lax, L.M. Roth, K.J. Button, Phys. Rev. **114**, 80 (1959)
22. T.B. Bahder, Phys. Rev. B **41**, 11992 (1990)
23. Y. Zhang, Phys. Rev. B **49**, 14352 (1994)
24. C. Pryor, Phys. Rev B **57**, 7190 (1998)
25. S.H. Groves, C.R. Pidgeon, A.W. Ewald, R.J. Wagner, J. Phys. Chem. Solids **30**, 2031 (1970)
26. J.C. Hensel, K. Suzuki, Phys. Rev. Lett. **22**, 838 (1969)
27. I. Vurgaftman, J.R. Meyer, L.R. Ram-Mohan, J. Appl. Phys. **89**, 5815 (2001)
28. H. Hazama, T. Sugimasa, T. Imachi, C. Hamaguchi, J. Phys. Soc. Jpn. **55**, 1282 (1986)

Chapter 3
Wannier Function and Effective Mass Approximation

Abstract Effective–mass equation is very useful to understand the transport and optical properties of semiconductors. In this chapter the effective–mass equation is derived with the help of Wannier function. Using Schrödinger equation based on the effective–mass approximation, we discuss the shallow impurity levels of donors in Ge and Si. Transport properties of electrons and holes are interpreted in terms of the effective mass in the classical mechanics (Newton equation). In this chapter the group velocity (the expectation value of the velocity) is shown to be given by $\langle \boldsymbol{v} \rangle = (1/\hbar)\partial\mathcal{E}/\partial\boldsymbol{k}$ in a periodic crystal potential. In the presence of an external force $\boldsymbol{F}$, an electron is accelerated in $\boldsymbol{k}$ space in the form of $\hbar\partial\boldsymbol{k}/\partial t = \boldsymbol{F}$. The electron motion is then expressed in the classical picture of a particle with the effective mass m^* or $1/m^* = (1/\hbar^2)\partial^2\mathcal{E}/\partial\boldsymbol{k}^2$ and the momentum $\boldsymbol{p} = \hbar\boldsymbol{k} = m^*\langle \boldsymbol{v} \rangle$. The results are used to derive transport properties in Chap. 6.

3.1 Wannier Function

There have been reported several methods to derive the effective–mass equation. In this chapter we use the Wannier function approach to derive the effective–mass equation because of its importance in semiconductor physics. However, in this chapter a simplest case will be dealt with, where only a single band is taken into account. For this purpose we begin with the introduction of the Wannier function showing that it is localized at the lattice point and that the Wannier function is obtained from the Bloch function by the Fourier transform.

When the Bloch function of a energy band n is defined by $b_{kn}(\boldsymbol{r})$, the Wannier function $w_n(\boldsymbol{r} - \boldsymbol{R}_j)$ is derived from the Bloch function by the Fourier transform [1]:

$$w_n(\boldsymbol{r} - \boldsymbol{R}_j) = \frac{1}{\sqrt{N}} \sum_{\boldsymbol{k}} \exp(-\mathrm{i}\boldsymbol{k} \cdot \boldsymbol{R}_j) b_{\boldsymbol{k}n}(\boldsymbol{r}) \,, \tag{3.1}$$

where N is the number of atoms and $\boldsymbol{R}_j$ is the position of jth atom. We have shown in Chap. 1 that the Bloch function extends over the crystal. On the other hand, the

C. Hamaguchi, *Basic Semiconductor Physics*, Graduate Texts in Physics,
DOI 10.1007/978-3-319-66860-4_3

Wannier function is localized at the atom, as shown later. Here we will investigate the features of the Wannier function.

First, we will show the Bloch function can be expanded in Wannier functions:

$$b_{kn}(\boldsymbol{r}) = \frac{1}{\sqrt{N}} \sum_j \exp(\mathrm{i}\boldsymbol{k} \cdot \boldsymbol{R}_j) \cdot w_n(\boldsymbol{r} - \boldsymbol{R}_j) \,. \tag{3.2}$$

We find that this expansion is just the inverse transform of the Wannier function given by (3.1) and it is easily proved as follows. Inserting (3.1) in (3.2) we obtain

$$\begin{aligned} b_{kn}(\boldsymbol{r}) &= \frac{1}{\sqrt{N}} \sum_j \exp(\mathrm{i}\boldsymbol{k} \cdot \boldsymbol{R}_j) \frac{1}{\sqrt{N}} \sum_{k'} \exp(-\mathrm{i}\boldsymbol{k}' \cdot \boldsymbol{R}_j) b_{k'n}(\boldsymbol{r}) \\ &= \frac{1}{N} \sum_{jk'} \exp\left\{\mathrm{i}(\boldsymbol{k} - \boldsymbol{k}') \cdot \boldsymbol{R}_j\right\} b_{k'n}(\boldsymbol{r}) \equiv b_{kn}(\boldsymbol{r}) \,. \end{aligned} \tag{3.3}$$

The final result of the above equation is obtained by using (A.29) of Appendix A.2:

$$\sum_j \exp\left\{\mathrm{i}(\boldsymbol{k} - \boldsymbol{k}') \cdot \boldsymbol{R}_j\right\} = N \delta_{k,k'} \,. \tag{3.4}$$

Next, we show the orthonormal property of Wannier functions that Wannier functions associated with different atoms are orthogonal and that Wannier functions of different energy bands are diagonal.

$$\begin{aligned} &\int w^*_{n'}(\boldsymbol{r} - \boldsymbol{R}_{j'}) w_n(\boldsymbol{r} - \boldsymbol{R}_j) \mathrm{d}^3\boldsymbol{r} \\ &= \frac{1}{N} \sum_{k,k'} \int \mathrm{d}^3\boldsymbol{r} \exp(\mathrm{i}\boldsymbol{k}' \cdot \boldsymbol{R}_{j'} - \mathrm{i}\boldsymbol{k} \cdot \boldsymbol{R}_j) b^*_{k'n'}(\boldsymbol{r}) b_{kn}(\boldsymbol{r}) \\ &= \frac{1}{N} \sum_{k,k'} \exp(\mathrm{i}\boldsymbol{k}' \cdot \boldsymbol{R}_{j'} - \mathrm{i}\boldsymbol{k} \cdot \boldsymbol{R}_j) \int b^*_{k'n'}(\boldsymbol{r}) b_{kn}(\boldsymbol{r}) \mathrm{d}^3\boldsymbol{r} \\ &= \frac{1}{N} \sum_{k} \exp\left\{\mathrm{i}\boldsymbol{k} \cdot (\boldsymbol{R}_{j'} - \boldsymbol{R}_j)\right\} \delta_{n,n'} = \delta_{j,j'} \delta_{n,n'} \,, \end{aligned} \tag{3.5}$$

and thus it is proved that Wannier functions are orthonormal.

Let us examine the localization characteristics of the Wannier function $w_n(\boldsymbol{r} - \boldsymbol{R}_j)$ at each lattice point $\boldsymbol{R}_j$. As shown in Chap. 1 the Bloch functions are not localized in the crystal. For simplicity we assume that the Bloch function is approximated as

$$b_{kn}(\boldsymbol{r}) = u_{0n}(\boldsymbol{r}) \exp(\mathrm{i}\boldsymbol{k} \cdot \boldsymbol{r}) \tag{3.6}$$

and that the term $u_{0n}(\boldsymbol{r})$ is independent of $\boldsymbol{k}$. This approximation is understood to be quite reasonable because the electronic states are well described by the Bloch function at the band edge as shown in the $\boldsymbol{k} \cdot \boldsymbol{p}$ perturbation method of Sect. 1.7. Under this assumption the Wannier function is approximated as

$$w_n(\boldsymbol{r}-\boldsymbol{R}_j)=\frac{1}{\sqrt{N}}u_{0n}(\boldsymbol{r})\sum_{\boldsymbol{k}}\exp\left\{\mathrm{i}\boldsymbol{k}\cdot(\boldsymbol{r}-\boldsymbol{R}_j)\right\}. \tag{3.7}$$

As given by (A.22) and (A.23) in Appendix A.2 the following relation is easily obtained:

$$\sum_{\boldsymbol{k}}\mathrm{e}^{\mathrm{i}\boldsymbol{k}\cdot(\boldsymbol{r}-\boldsymbol{R}_j)}=L^3\delta(\boldsymbol{r}-\boldsymbol{R}_j), \tag{3.8}$$

$$\frac{1}{L^3}\int \mathrm{d}^3\boldsymbol{r}\,\mathrm{e}^{\mathrm{i}(\boldsymbol{k}-\boldsymbol{k}')\cdot\boldsymbol{r}}=\delta_{\boldsymbol{k},\boldsymbol{k}'}, \tag{3.9}$$

where $\delta(\boldsymbol{r}-\boldsymbol{r}')$ is the Dirac delta function and $\delta_{\boldsymbol{k},\boldsymbol{k}'}$ is the Kronecker delta function. Using this result the Wannier function is written as

$$w_n(\boldsymbol{r}-\boldsymbol{R}_j)=\frac{1}{\sqrt{N}}u_{0n}(\boldsymbol{r})L^3\delta(\boldsymbol{r}-\boldsymbol{R}_j), \tag{3.10}$$

and therefore we find that the Wannier function is localized at the lattice point $\boldsymbol{R}_j$.

3.2 Effective-Mass Approximation

Here we will derive of the effective–mass approximation which may be applied to a wide variety of calculations of the electronic properties of semiconductors, for example shallow impurity states, transport, optical properties and so on. Here we assume that the perturbing potential $H_1(\boldsymbol{r})$ varies very slowly compared with the lattice constant a. This assumption allows us to simplify the effective–mass equation based on a single band. The Schrödinger equation for the one-electron Hamiltonian is written as

$$[H_0+H_1(\boldsymbol{r})]\Psi(\boldsymbol{r})=\mathcal{E}\cdot\Psi(\boldsymbol{r}), \tag{3.11}$$

where

$$H_0=\frac{p^2}{2m}+V(\boldsymbol{r}), \tag{3.12}$$

and we will show that the solution of above equation is given by the solution

$$\left[-\frac{\hbar^2}{2m^*}\nabla^2+H_1(\boldsymbol{r})\right]F(\boldsymbol{r})=\mathcal{E}\cdot F(\boldsymbol{r}), \tag{3.13}$$

where m^* is the effective–mass of the electron we are concerned with. In the following we assume the effective–mass is given by a scalar effective–mass for simplicity. Therefore, the energy of the electron $\mathcal{E}$ is assumed to be expressed by the wave vector $\boldsymbol{k}$:

$$\mathcal{E}_0(\boldsymbol{k}) = \frac{\hbar^2 k^2}{2m^*} . \tag{3.14}$$

Equation (3.13) is called the **effective–mass equation** and this approximation is called the **effective–mass approximation**. The reason why we call this the effective–mass approximation is quite clear. The Schrödinger equation (3.11) with the Hamiltonian of (3.12) is expressed in a simple Schrödinger equation by replacing the Hamiltonian of (3.12) by $-(\hbar^2/2m^*)\nabla^2$, where the potential disappears and m is replaced by m^*. The derivation of the effective–mass equation will be shown later.

First, we calculate the eigenfunction $\Psi(\boldsymbol{r})$ of (3.11). Since the Wannier function is expressed by the Fourier transform of the Bloch functions and thus Wannier functions have orthonormality and completeness, any function may be expanded by Wannier functions:

$$\Psi(\boldsymbol{r}) = \sum_n \sum_j F_n(\boldsymbol{R}_j) w_n(\boldsymbol{r} - \boldsymbol{R}_j) . \tag{3.15}$$

In the following we consider a specific band and neglect the band index n of the subscript.

In the absence of perturbation the Schrödinger equation is expressed in terms of the Bloch functions $b_{\boldsymbol{k}n}(\boldsymbol{r})$ as

$$H_0 b_{\boldsymbol{k}}(\boldsymbol{r}) = \mathcal{E}_0(\boldsymbol{k}) b_{\boldsymbol{k}}(\boldsymbol{r}) . \tag{3.16}$$

Inserting

$$\Psi(\boldsymbol{r}) = \sum_{j'} F(\boldsymbol{R}_{j'}) w(\boldsymbol{r} - \boldsymbol{R}_{j'}) \tag{3.17}$$

into (3.11), multiplying $w^*(\boldsymbol{r} - \boldsymbol{R}_j)$ on both sides and integrating over the crystal volume we obtain

$$\begin{aligned} &\int \sum_{j'} w^*(\boldsymbol{r} - \boldsymbol{R}_j) H_0 F(\boldsymbol{R}_{j'}) w(\boldsymbol{r} - \boldsymbol{R}_{j'}) \mathrm{d}^3\boldsymbol{r} \\ &+ \int \sum_{j'} w^*(\boldsymbol{r} - \boldsymbol{R}_j) H_1 F(\boldsymbol{R}_{j'}) w(\boldsymbol{r} - \boldsymbol{R}_{j'}) \mathrm{d}^3\boldsymbol{r} \\ &= \int \mathcal{E} \sum_{j'} w^*(\boldsymbol{r} - \boldsymbol{R}_j) F(\boldsymbol{R}_{j'}) w(\boldsymbol{r} - \boldsymbol{R}_{j'}) \mathrm{d}^3\boldsymbol{r} . \end{aligned} \tag{3.18}$$

When we define the following relations

$$(H_0)_{jj'} = \int w^*(\boldsymbol{r} - \boldsymbol{R}_j) H_0 w(\boldsymbol{r} - \boldsymbol{R}_{j'}) \mathrm{d}^3\boldsymbol{r}\,, \tag{3.19}$$

$$(H_1)_{jj'} = \int w^*(\boldsymbol{r} - \boldsymbol{R}_j) H_1 w(\boldsymbol{r} - \boldsymbol{R}_{j'}) \mathrm{d}^3\boldsymbol{r}\,, \tag{3.20}$$

(3.18) is rewritten as

$$\sum_{j'} (H_0)_{jj'} \cdot F(\boldsymbol{R}_{j'}) + \sum_{j'} (H_1)_{jj'} \cdot F(\boldsymbol{R}_{j'}) = \mathcal{E} \cdot F(\boldsymbol{R}_j)\,. \tag{3.21}$$

The right-hand side of the above equation is rewritten with the help of (3.5). Since we assume that the perturbing potential varies very slowly, the following approximation is possible. From the properties of the Wannier function, $w^*(\boldsymbol{r} - \boldsymbol{R}_j)$ and $w(\boldsymbol{r} - \boldsymbol{R}_{j'})$ are localized at $\boldsymbol{R}_j$ and $\boldsymbol{R}_{j'}$, respectively, and thus only the overlapping part of $w^*(\boldsymbol{r} - \boldsymbol{R}_j)$ and $w(\boldsymbol{r} - \boldsymbol{R}_{j'})$ will contribute to the integration of (3.20). The integration arising from small values of $(\boldsymbol{R}_j - \boldsymbol{R}_{j'})$ or the integration over the nearest neighbor will contribute. In a more simplified approximation only the contribution from $\boldsymbol{R}_j = \boldsymbol{R}_{j'}$ may be taken into account. In this approximation the summation of (3.20) over j' results in

$$\sum_{j'} (H_1)_{jj'} \simeq H_1(\boldsymbol{R}_j) \int w^*(\boldsymbol{r} - \boldsymbol{R}_j) w(\boldsymbol{r} - \boldsymbol{R}_j) \mathrm{d}^3\boldsymbol{r} = H_1(\boldsymbol{R}_j)\,. \tag{3.22}$$

It is evident from (3.20) that $(H_0)_{jj'}$ is a function of $(\boldsymbol{R}_j - \boldsymbol{R}_{j'})$, which is shown by using the Wannier function as follows. H_0 is independent of the translation vector $\boldsymbol{R}$. Therefore, replacing $\boldsymbol{r} - \boldsymbol{R}_j$ with $\boldsymbol{r}$ we find

$$(H_0)_{jj'} = \int w^*(\boldsymbol{r}) H_0 w(\boldsymbol{r} - \boldsymbol{R}_{j'} + \boldsymbol{R}_j) \mathrm{d}^3\boldsymbol{r} \equiv h_0(\boldsymbol{R}_j - \boldsymbol{R}_{j'})\,. \tag{3.23}$$

Using these results (3.21) is rewritten as the following set of equations:

$$\sum_{j'} h_0(\boldsymbol{R}_j - \boldsymbol{R}_{j'}) F(\boldsymbol{R}_{j'}) + H_1(\boldsymbol{R}_j) F(\boldsymbol{R}_j) = \mathcal{E} \cdot F(\boldsymbol{R}_j)\,. \tag{3.24}$$

Exchanging the order of the summation of the first term on the left-hand side of the above equation, $(\boldsymbol{R}_j - \boldsymbol{R}_{j'}) \to \boldsymbol{R}_{j'}$, we obtain the following result:

$$\sum_{j'} h_0(\boldsymbol{R}_{j'}) F(\boldsymbol{R}_j - \boldsymbol{R}_{j'}) + H_1(\boldsymbol{R}_j) F(\boldsymbol{R}_j) = \mathcal{E} \cdot F(\boldsymbol{R}_j)\,. \tag{3.25}$$

Equation (3.12) gives the relation

$$\mathcal{E}_0(\boldsymbol{k}) = \int b_{\boldsymbol{k}}^*(\boldsymbol{r}) H_0 b_{\boldsymbol{k}}(\boldsymbol{r}) \mathrm{d}^3\boldsymbol{r} \tag{3.26}$$

and it may be rewritten as follows by using (3.2), which is an expansion of the Bloch function in Wannier functions:

$$\begin{aligned}\mathcal{E}_0(\boldsymbol{k}) &= \frac{1}{N}\sum_j\sum_{j'}\exp[-\mathrm{i}\boldsymbol{k}\cdot(\boldsymbol{R}_j-\boldsymbol{R}_{j'})](H_0)_{jj'}\\ &= \frac{1}{N}\sum_j\sum_{j'}\exp[-\mathrm{i}\boldsymbol{k}\cdot(\boldsymbol{R}_j-\boldsymbol{R}_{j'})]h_0(\boldsymbol{R}_j-\boldsymbol{R}_{j'})\,. \end{aligned}\tag{3.27}$$

It is very important to point out that $\boldsymbol{R}_j$ and $\boldsymbol{R}_{j'}$ give a combination of the same lattices. Therefore, the summation of $\boldsymbol{R}_j$ with respect to j gives the same value for different $\boldsymbol{R}_{j'}$. In other words, the summation is exactly the same for $j' = 1, 2, 3, \ldots, N$. This will lead us to obtain the equation:

$$\mathcal{E}_0(\boldsymbol{k}) = \frac{1}{N}\sum_j N\exp(-\mathrm{i}\boldsymbol{k}\cdot\boldsymbol{R}_j)h_0(\boldsymbol{R}_j) = \sum_j \exp(-\mathrm{i}\boldsymbol{k}\cdot\boldsymbol{R}_j)h_0(\boldsymbol{R}_j)\,. \tag{3.28}$$

The result may be understood as follows. The energy band $\mathcal{E}_0(\boldsymbol{k})$ is a periodic function in $\boldsymbol{k}$ space and the period is the lattice vectors $\boldsymbol{R}_j$; therefore, $\mathcal{E}_0(\boldsymbol{k})$ can be expanded by the lattice vector $\boldsymbol{R}_j$. In addition, the Fourier coefficient $h_0(\boldsymbol{R}_j)$ of the expansion is given by

$$h_0(\boldsymbol{R}_j) = \frac{1}{N}\sum_{\boldsymbol{k}}\mathcal{E}_0(\boldsymbol{k})\exp(\mathrm{i}\boldsymbol{k}\cdot\boldsymbol{R}_j)\,. \tag{3.29}$$

Next, let us consider a function $F(r - R_j)$ and apply the Taylor expansion to it at the position r, which gives rise to the following result in the case of a one–dimensional crystal:

$$F(r-R_j) = F(r) - R_j\frac{\mathrm{d}}{\mathrm{d}r}F(r) + \frac{1}{2!}(R_j)^2\frac{\mathrm{d}^2}{\mathrm{d}r^2}F(r) - \cdots\,. \tag{3.30}$$

In the case of a three-dimensional crystal the function $F(\boldsymbol{r} - \boldsymbol{R}_j)$ is expanded as

$$\begin{aligned}F(\boldsymbol{r}-\boldsymbol{R}_j) &= F(\boldsymbol{r}) - \boldsymbol{R}_j\cdot\nabla F(\boldsymbol{r}) + \frac{1}{2!}(\boldsymbol{R}_j\cdot\nabla)[\boldsymbol{R}_j\cdot\nabla F(\boldsymbol{r})] - \cdots\\ &= \exp(-\boldsymbol{R}_j\cdot\nabla)F(\boldsymbol{r})\,,\end{aligned}\tag{3.31}$$

and therefore we obtain

$$\sum_{j'} h_0(\boldsymbol{R}_{j'})F(\boldsymbol{r}-\boldsymbol{R}_{j'}) = \sum_{j'} h_0(\boldsymbol{R}_{j'})\exp(-\boldsymbol{R}_{j'}\cdot\nabla)F(\boldsymbol{r})\,. \tag{3.32}$$

On the other hand, when (3.28) is multiplied by $F(\boldsymbol{r})$ on the both sides, we find

$$\mathcal{E}_0(\boldsymbol{k})F(\boldsymbol{r}) = \sum_{j'} h_0(\boldsymbol{R}_{j'})\exp(-\mathrm{i}\boldsymbol{k}\cdot\boldsymbol{R}_{j'})F(\boldsymbol{r})\,. \tag{3.33}$$

Now, let us compare (3.32) with (3.33), we find that a replacement of $\boldsymbol{k}$ with $-\mathrm{i}\nabla$ in (3.33) gives exactly the same result as in the right-hand side of (3.32). Therefore, the following relation may be obtained:

$$\sum_{j'} h_0(\boldsymbol{R}_{j'})F(\boldsymbol{r}-\boldsymbol{R}_{j'}) = \mathcal{E}(-i\nabla)F(\boldsymbol{r})\,. \tag{3.34}$$

Using this result and replacing $\boldsymbol{R}_j$ by $\boldsymbol{r}$, (3.25) becomes

$$\sum_{j'} h_0(\boldsymbol{R}_{j'})F(\boldsymbol{r}-\boldsymbol{R}_{j'}) + H_1(\boldsymbol{R}_j)F(\boldsymbol{r}) = \mathcal{E}\cdot F(\boldsymbol{r}) \tag{3.35}$$

and may be rewritten as

$$\mathcal{E}_0(-\mathrm{i}\nabla)F(\boldsymbol{r}) + H_1(\boldsymbol{r})F(\boldsymbol{r}) = \mathcal{E}\cdot F(\boldsymbol{r})\,. \tag{3.36}$$

Or it is possible to rewrite it in the following form:

$$\left[\mathcal{E}_0(-\mathrm{i}\nabla) + H_1(\boldsymbol{r})\right] F(\boldsymbol{r}) = \mathcal{E}\cdot F(\boldsymbol{r})\,, \tag{3.37}$$

which is called the **effective–mass equation**. This kind of approximation is called the **effective–mass approximation**. We have to note here that the function $F(\boldsymbol{r})$ given by (3.37) is an envelope function and that the Fourier coefficients $F(\boldsymbol{R}_j)$ of the wave function $\Psi(\boldsymbol{r})$ expanded in Wannier functions are given by $F(\boldsymbol{R}_j)$, which may be obtained by replacing $\boldsymbol{r}$ of the envelope function $F(\boldsymbol{r})$ by $\boldsymbol{R}_j$. Now we find that the envelope function $F(\boldsymbol{r})$ should be a slowly varying function of $\boldsymbol{r}$. In other words the coefficient $F(\boldsymbol{R}_j)$ should vary very slowly over the distance of the lattice spacing $(\boldsymbol{R}_{j+1} - \boldsymbol{R}_j)$.

$\mathcal{E}_0(\boldsymbol{k})$ is the electronic energy as a function of the wave vector without the perturbing potential H_1, and thus it gives the energy band of the electron. For simplicity we assume that the energy band is given by a spherical parabolic band with isotropic effective mass m^*:

$$\mathcal{E}_0(\boldsymbol{k}) = \frac{\hbar^2 k^2}{2m^*}\,. \tag{3.38}$$

The effective–mass equation is therefore given by

$$\left[-\frac{\hbar^2}{2m^*}\nabla^2 + H_1(\boldsymbol{r})\right] F(\boldsymbol{r}) = \mathcal{E}\cdot F(\boldsymbol{r})\,. \tag{3.39}$$

When we compare (3.11) with (3.39), we find the following features. The periodic potential $V(\boldsymbol{r})$ of (3.11) disappears and the free electron mass m is replaced by the effective–mass m^*. The effective–mass equation is named after these features. As stated before, various properties of semiconductors are calculated by using the effective–mass equation. This means that the determination of the effective–mass is very important. We have to note that the effective mass equation or effective–mass approximation is derived by different approaches [2] and that one of the methods is shown in the analysis of excitons given in Sect. 4.5.

3.3 Shallow Impurity Levels

One of the best examples of the applications of the effective–mass approximation is the analysis of donor levels in semiconductors. First, we deal with the donor level associated with a conduction band expressed by a scalar effective–mass m^*. Under this assumption the effective–mass equation is given by

$$\left[-\frac{\hbar^2}{2m^*}\nabla^2 - \frac{e^2}{4\pi\kappa\epsilon_0 r}\right] F(\boldsymbol{r}) = \mathcal{E} \cdot F(\boldsymbol{r}) . \tag{3.40}$$

Careful observation of the above equation reveals that it is equivalent to the Schrödinger equation used to derive the electronic states of a hydrogen atom. The above equation is from the Schrödinger equation of the hydrogen atom, replacing the free electron mass m by m^* and the dielectric constant of free space by the dielectric constant of the semiconductor $\kappa\epsilon_0 = \epsilon$. Therefore, the ground state energy $\mathcal{E}$ and the effective Bohr radius a_{I} of the donor are

$$\mathcal{E} = -\frac{m^* e^4}{2(4\pi\epsilon)^2\hbar^2} = -\frac{m^*/m}{(\epsilon/\epsilon_0)^2}\mathcal{E}_{\mathrm{R}} \tag{3.41}$$

$$\mathcal{E}_{\mathrm{R}} = \frac{me^4}{2(4\pi\epsilon_0)^2\hbar^2} \tag{3.42}$$

$$a_{\mathrm{I}} = \frac{4\pi\hbar^2\epsilon}{m^* e^2} = \frac{\epsilon/\epsilon_0}{m^*/m} a_{\mathrm{B}} , \tag{3.43}$$

where $\mathcal{E}_{\mathrm{R}}$ is the ionization energy of the hydrogen atom ($\mathcal{E}_{\mathrm{R}} = 1$ Rydberg $= 13.6$ eV).

Let us try to estimate the ionization energy of a donor by assuming the effective–mass $m^* = 0.25$ m and the relative dielectric constant $\epsilon/\epsilon_0 = 16$. We obtain the ionization energy of donor as 0.013 eV, which is very close to the ionization energy observed in Ge. We have already shown in Chaps. 1 and 2 that the conduction band minima of Ge and Si are located at the L and the Δ points near the X point in the Brillouin zone, respectively, and that they consist of multiple valleys with anisotropic effective–masses. Therefore, the above simplified analysis cannot be applied to the case of donors in Ge and Si. In addition we have to note that the donor levels in many-valley semiconductors are degenerate due to the multiple conduction band minima. First, we neglect the degeneracy to analyze the donor levels of Ge and Si, and later we deal with the degeneracy.

Let us consider a conduction band with an ellipsoidal energy surface and take the z direction to be along the longitudinal axis of the ellipsoid. When we measure the energy with respect to the bottom of the conduction band at $\boldsymbol{k}_0$ and define the effective–masses along the longitudinal axis as m_l and the transverse axis as m_t, we may write the energy band as

$$\mathcal{E}_c(\boldsymbol{k}) = \frac{\hbar^2}{2}\left(\frac{k_x^2 + k_y^2}{m_\mathrm{t}} + \frac{k_z^2}{m_\mathrm{l}}\right). \tag{3.44}$$

Therefore, the effective–mass equation for donor levels associated with the conduction band becomes

$$\left[-\frac{\hbar^2}{2m_\mathrm{t}}\left(\frac{\partial^2}{\partial x^2} + \frac{\partial^2}{\partial y^2}\right) - \frac{\hbar^2}{2m_\mathrm{l}}\frac{\partial^2}{\partial z^2} - \frac{e^2}{4\pi\epsilon(x^2 + y^2 + z^2)^{1/2}}\right]\psi(\boldsymbol{r}) = \mathcal{E}\cdot\psi(\boldsymbol{r}), \tag{3.45}$$

where ϵ is the dielectric constant of the semiconductor. Defining the effective–mass ratio

$$\gamma = \frac{m_\mathrm{t}}{m_\mathrm{l}} \quad (< 1) \tag{3.46}$$

and rewriting (3.45) using cylindrical coordinates we obtain

$$\left[-\frac{\hbar^2}{2m_\mathrm{t}}\left(\frac{1}{r}\frac{\partial}{\partial r}r\frac{\partial}{\partial r} + \frac{1}{r^2}\frac{\partial^2}{\partial\varphi^2} + \gamma\frac{\partial^2}{\partial z^2}\right) - \frac{e^2}{4\pi\epsilon(r^2 + z^2)^{1/2}}\right]\psi(\boldsymbol{r}) = \mathcal{E}\cdot\psi(\boldsymbol{r}). \tag{3.47}$$

The effective–mass Hamiltonian is then written as

$$\mathcal{H} = -\frac{\hbar^2}{2m_\mathrm{t}}\left(\frac{1}{r}\frac{\partial}{\partial r}r\frac{\partial}{\partial r} + \frac{1}{r^2}\frac{\partial^2}{\partial\varphi^2} + \gamma\frac{\partial^2}{\partial z^2}\right) - \frac{e^2}{4\pi\epsilon(r^2 + z^2)^{1/2}}. \tag{3.48}$$

It is impossible to solve (3.47) analytically. Therefore, we adopt the variational principle to obtain approximate solutions. In order to obtain the ground state level of the donor we use the following trial function for the state [3–5]:

$$\psi = (\pi a^2 b)^{-1/2}\exp\left[-\left(\frac{r^2}{a^2} + \frac{z^2}{b^2}\right)^{1/2}\right], \tag{3.49}$$

where $a > b$ and ψ is normalized such that $\langle\psi|\psi\rangle = 1$. We use the following transform of the variables to obtain $\langle\psi|\mathcal{H}|\psi\rangle$:

$$
\begin{aligned}
r &= au \cos v\,, \\
z &= bu \sin v\,, \\
\int_V \mathrm{d}^3\boldsymbol{r} &= 2\pi \int_0^{+\infty} r\mathrm{d}r \int_{-\infty}^{+\infty} \mathrm{d}z \\
&= 2\pi \int_0^{+\infty} \int_{-\pi/2}^{+\pi/2} abu^2 \cos v\, \mathrm{d}y\, \mathrm{d}u\,.
\end{aligned}
$$

Then the ground state of the donor is given by

$$
\begin{aligned}
\mathcal{E} &= \langle \psi^* | \mathcal{H} | \psi \rangle \\
&= -\frac{\hbar^2}{2m_\mathrm{t}} \int_0^{+\infty} \int_{-\pi/2}^{+\pi/2} \left\{ \left(\frac{\cos^2 v}{a^2} + \gamma \frac{\sin^2 v}{b^2} \right) (u^{-1} + 1) \right. \\
&\quad \left. - \left(\frac{2}{a^2} + \frac{\gamma}{b^2} \right) u^{-1} + \frac{2}{a_I (a^2 \cos^2 v + b^2 \sin^2 v)^{1/2}} u^{-1} \right\} \\
&\quad \times e^{-2u} 2u^2 \cos v\, \mathrm{d}v\, \mathrm{d}u \\
&= -\frac{\hbar^2}{2m_\mathrm{t}} \left[-\frac{2}{3a^2} - \frac{\gamma}{3b^2} + \frac{2}{a_I \sqrt{a^2 - b^2}} \sin^{-1} \sqrt{\frac{a^2 - b^2}{a^2}} \right]. \qquad (3.50)
\end{aligned}
$$

The variational principle is used to minimize $\mathcal{E}$ in (3.50) by varying the parameters a and b. To do this we introduce another transform of the variables by

$$
\rho^2 = \frac{a^2 - b^2}{a^2}, \quad (0 < \rho^2 < 1)\,, \qquad (3.51)
$$

and eliminate b to obtain

$$
\mathcal{E} = -\frac{\hbar^2}{2m_\mathrm{t}} \left[-\frac{2(1 - \rho^2) + \gamma}{3a^3(1 - \rho - 2)} + \frac{2 \sin^{-1} \rho}{\rho a a_\mathrm{I}} \right]. \qquad (3.52)
$$

The value of a to minimize the above equation is easily deduced and is given by

$$
a = \frac{\rho[2(1 - \rho^2) + \gamma]}{3(1 - \rho^2) \sin^{-1} \rho} a_\mathrm{I}\,. \qquad (3.53)
$$

Inserting this value of a into (3.50) we obtain the ground state energy:

$$
\mathcal{E} = \frac{3(1 - \rho^2)(\sin^{-1} \rho)^2}{\rho^2 [2(1 - \rho^2) + \gamma]} \mathcal{E}_\mathrm{I}\,. \qquad (3.54)
$$

where

$$
\begin{aligned}
\mathcal{E}_\mathrm{I} &= -\frac{m_\mathrm{t} e^4}{2\hbar^2 (4\pi\epsilon)^2} \\
&= -\frac{m_\mathrm{t}}{m} \left(\frac{\epsilon_0}{\epsilon} \right)^2 \mathcal{E}_\mathrm{R} \quad (\mathcal{E}_\mathrm{R} = 13.6\ \mathrm{eV} = 1\ \mathrm{Rydberg})\,. \qquad (3.55)
\end{aligned}
$$

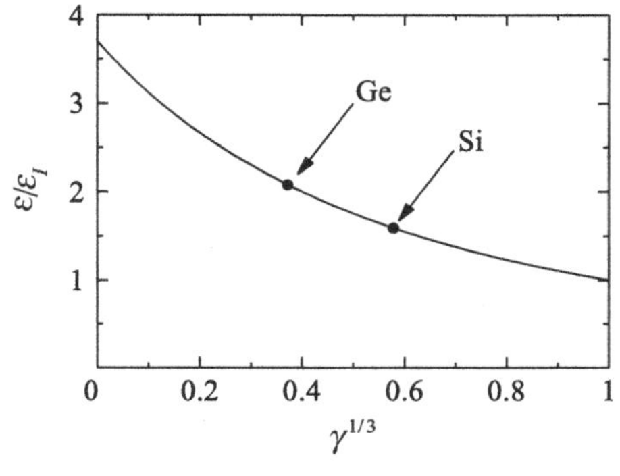

Fig. 3.1 Ionization energy of a donor calculated from the effective–mass equation and by the variational principle plotted as a function of the effective–mass ratio $\gamma = m_t/m_l$

From this expression we find that $\mathcal{E}_I$ is the ground state energy of the donor associated with a conduction band of a semiconductor with the transverse effective–mass m_t and dielectric constant ϵ. The ionization energy of the donor may be obtained from the value of ρ to minimize (3.54) for an arbitrary value of $\gamma = m_t/m_l$. However, the value to minimize (3.54) may not be obtained analytically, and thus we used a numerical analysis. The results are shown in Fig. 3.1.

3.4 Impurity Levels in Ge and Si

The parameters used for the calculations of the impurity levels in Ge and Si are summarized in Table 3.1 and the results obtained by the variational principle are shown in Table 3.2, where the ionization energies are 9.05 meV for Ge and 29.0 meV for Si. The experimentally observed ionization energies (summarized in Table 3.3) are about 10 meV for Ge, in good agreement with the calculation, and about 45 meV for Si, not in good agreement with the calculation. The good agreement is understood from the large value of the effective Bohr radius in Ge and thus the impurity potential varies slowly compared with the lattice constant, leading to a good condition to use the effective–mass approximation. On the other hand, the effective Bohr radius for Si is smaller and the effective mass approximation seems to be invalid for the calculation of the ground states in Si. The discrepancy between the calculation and the experimental values may arise from the neglect of the degeneracy of the conduction band minima in addition to the validity of the effective–mass approximation. We will discuss these two factors in the following.

We may expect that the excited states of an impurity give a good agreement between the calculation and the experiment. This is because the wave functions of the excited states extend over space more than the ground state, and thus the effective–mass approximation becomes more accurate for the excited states. In order to check the validity of the effective–mass equation for the excited states we calculate the excited states using the trial functions shown in Table 3.4.

Table 3.1 Parameters used to calculate the donor levels of Ge and Si. The parameters are given in dimensionless units

	m_t/m	m_l/m	γ	ϵ/ϵ_0
Ge	0.082	1.58	0.0519	16.0
Si	0.190	0.98	0.1939	11.9

Table 3.2 Ground state energies of donors in Ge and Si calculated by the effective–mass approximation and the variational principle

	$-\mathcal{E}_I$ [meV]	$-\mathcal{E}/\mathcal{E}_I$	$(-\mathcal{E})$ [meV]	a_I [Å]	a/a_I	(a) [Å]	b/a_I	(b) [Å]
Ge	4.36	2.08	(9.05)	103.2	0.622	(64.2)	0.221	(22.8)
Si	18.3	1.59	(29.0)	33.14	0.740	(24.5)	0.415	(13.8)

Table 3.3 The ionization energies $\mathcal{E}_0$ (meV) of various donors in Ge and Si determined by experiments

	Li	P	As	Sb	Bi
Si	33	45	49	39	69
Ge		12.0	12.7	9.6	

Table 3.4 Trial functions for calculating the excited states of donors by the variational method

States	Variational functions
1s	$(\pi a^2 b)^{-1/2} \exp\left[-\left(\frac{r^2}{a^2}+\frac{z^2}{b^2}\right)^{1/2}\right]$
2s	$(C_1 + C_2 r^2 + C_3 z^2) \exp\left[-\left(\frac{r^2}{a^2}+\frac{z^2}{b^2}\right)^{1/2}\right]$
$2p_0$	$Cz \exp\left[-\left(\frac{r^2}{a^2}+\frac{z^2}{b^2}\right)^{1/2}\right]$
$2p_{\pm1}$	$Cr \exp\left[-\left(\frac{r^2}{a^2}+\frac{z^2}{b^2}\right)^{1/2}\right] \exp(\pm i\phi)$
$3p_0$	$(C_1 + C_2 r^2 + C_3 z^2) z \exp\left[-\left(\frac{r^2}{a^2}+\frac{z^2}{b^2}\right)^{1/2}\right]$
$3p_{\pm1}$	$(C_1 + C_2 r^2 + C_3 z^2) r \exp\left[-\left(\frac{r^2}{a^2}+\frac{z^2}{b^2}\right)^{1/2}\right] \exp(\pm i\phi)$

Using the parameters given in Table 3.4 and the variational method, the excited states of donors in Ge and Si are calculated and are summarized in Table 3.5. As stated above the discrepancy between theory and experiment is considerable for the ground state in Si. We compare the calculations based on the effective–mass approximation with the experimental data in Si, which is shown in Table 3.6 and

Table 3.5 Ground state and excited states levels of the donors in Ge and Si

States	Si	Ge
1s	-29 ± 1	-9.2 ± 0.2
2p, $m = 0$	-10.9 ± 0.2	-4.5 ± 0.2
2s	-8.8 ± 0.6	
2p, $m = \pm 1$	-5.9 ± 0.1	-1.60 ± 0.03
3p, $m = 0$	-5.7 ± 0.6	-2.35 ± 0.2
3p, $m = \pm 1$	-2.9 ± 0.05	-0.85 ± 0.05

Table 3.6 Experimental values of the ground state and the excited state levels of various donors (P, As and Sb) in Si (meV) and the calculated values by the effective–mass approximation (shown by EMA)

Impurities	Ground state	Excited states			
P	45	10.5	5.5	2.4	0.4
As	53	10.9	5.6	2.4	0.1
Sb	43	11.2	6.5	3.1	
EMA	29	10.9	5.9	2.9	

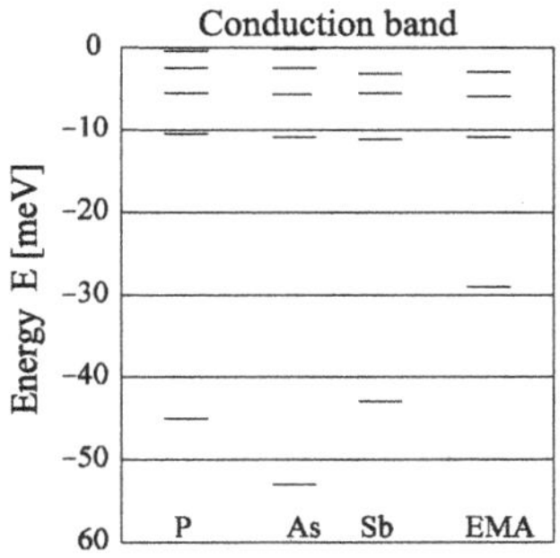

Fig. 3.2 Experimental values of the donor levels of P, As and Sb compared with the calculated values by the effective–mass approximation (EMA)

in Fig. 3.2. Although the ground state level does not show a good agreement, the excited states show reasonable agreement between the experiment and calculation. This suggests us the validity of the effective–mass approximation, giving rise to a good agreement for the excited states.

From these considerations we may conclude that the excited states of donors in Ge and Si are well described by the effective–mass approximation. In order to discuss the ground state we have to take into account the degeneracy of the conduction band minima (many valleys), valley–orbit interaction. and later the method for improving the calculation for the ground state (central cell correction) [6].

3.4.1 Valley–Orbit Interaction

As an example we discuss the valley–orbit interaction in Si, which has six equivalent valleys located at the Δ point in the $\langle 100\rangle$ direction of Brillouin zone. The eigenfunctions of these six valleys are described by

$$\psi^{(i)}(\boldsymbol{r}) = \sum_{j=1}^{6} \alpha^i_j F_j(\boldsymbol{r}) u_j(\boldsymbol{r}) \exp(\mathrm{i}\boldsymbol{k}_{0j} \cdot \boldsymbol{r}), \qquad (i = 1, \ldots, 6) \tag{3.56}$$

where $j = 1, 2 \ldots, 6$ are the indices for the conduction band minima of six valleys at $(\pm k_0, 0, 0)$, $(0, \pm k_0, 0)$ and $(0, 0, \pm k_0)$, and $u_j(\boldsymbol{r}) \exp(\mathrm{i}\boldsymbol{k}_{0j} \cdot \boldsymbol{r})$ and $F_j(\boldsymbol{r})$ are the Bloch function of the jth conduction band and the H atom-like envelope function determined from the effective–mass approximation, respectively. The linear combination of the wave functions is factorized with the help of group theory to give an irreducible representation, and the coefficients of the irreducible representation, α^i_j, are then given by

$$\begin{aligned}
&\alpha^1_j : 1/\sqrt{6}(1, 1, 1, 1, 1, 1) && (A_1)\\
&\left.\begin{aligned}\alpha^2_j &: 1/2(1, 1, -1, -1, 0, 0)\\ \alpha^3_j &: 1/2(1, 1, 0, 0, -1, -1)\end{aligned}\right\} && (E)\\
&\left.\begin{aligned}\alpha^4_j &: 1/\sqrt{2}(1, -1, 0, 0, 0, 0)\\ \alpha^5_j &: 1/\sqrt{2}(0, 0, 1, -1, 0, 0)\\ \alpha^6_j &: 1/\sqrt{2}(0, 0, 0, 0, 1, -1)\end{aligned}\right\} && (T_1)
\end{aligned} \tag{3.57}$$

It may be expected from the above equation that the ground state of a donor in Si splits into three groups due to the valley–orbit interaction: singlet, doublet and triplet. The matrix elements of the Hamiltonian H_{vo} for the valley–orbit interaction are then described as the following for the wave functions $F_j(\boldsymbol{r})$ $(j = 1, 2, \ldots 6)$ by introducing a parameter $\langle F_i(\boldsymbol{r}) \exp[\mathrm{i}(\boldsymbol{k}_{0i} - \boldsymbol{k}_{0j}) \cdot \boldsymbol{r}] | H_{\mathrm{vo}} | F_j(\boldsymbol{r})\rangle = -\Delta$ [7, 8]:

$$\langle i | H_{\mathrm{vo}} | j\rangle = \begin{vmatrix} 0 & -\Delta & -\Delta & -\Delta & -\Delta & -\Delta \\ -\Delta & 0 & -\Delta & -\Delta & -\Delta & -\Delta \\ -\Delta & -\Delta & 0 & -\Delta & -\Delta & -\Delta \\ -\Delta & -\Delta & -\Delta & 0 & -\Delta & -\Delta \\ -\Delta & -\Delta & -\Delta & -\Delta & 0 & -\Delta \\ -\Delta & -\Delta & -\Delta & -\Delta & -\Delta & 0 \end{vmatrix} . \tag{3.58}$$

Diagonalization of (3.58) gives rise to a single state with energy 5Δ and a 5–fold degenerate state with energy $-\Delta$. The single state corresponds to the singlet state given by (3.58), and the 5–fold degenerate state to the doublet and triplet states of (3.58). The splitting of the doublet and triplet states is due to the difference in the valley–orbit interaction between the sets of the valleys. The valley-orbit interaction between the valleys such as $(\pm k_0, 0, 0)$ (g–type) and between the sets such as $(k_0, 0, 0)$ and $(0, k_0, 0)$ (f–type) are expected to be different. When we take account of the difference and assume the interaction for the g–type as $(1 + \delta)\Delta$, the matrix elements are rewritten as

$$\begin{vmatrix} 0 & -(1+\delta)\Delta & -\Delta & -\Delta & -\Delta & -\Delta \\ -(1+\delta)\Delta & 0 & -\Delta & -\Delta & -\Delta & -\Delta \\ -\Delta & -\Delta & 0 & -(1+\delta)\Delta & -\Delta & -\Delta \\ -\Delta & -\Delta & -(1+\delta)\Delta & 0 & -\Delta & -\Delta \\ -\Delta & -\Delta & -\Delta & -\Delta & 0 & -(1+\delta)\Delta \\ -\Delta & -\Delta & -\Delta & -\Delta & -(1+\delta)\Delta & 0 \end{vmatrix}$$
$$= \begin{vmatrix} -5\Delta - \delta\Delta & 0 & 0 & 0 & 0 & 0 \\ 0 & \Delta - \delta\Delta & 0 & 0 & 0 & 0 \\ 0 & 0 & \Delta - \delta\Delta & 0 & 0 & 0 \\ 0 & 0 & 0 & \Delta + \delta\Delta & 0 & 0 \\ 0 & 0 & 0 & 0 & \Delta + \delta\Delta & 0 \\ 0 & 0 & 0 & 0 & 0 & \Delta + \delta\Delta \end{vmatrix} . \quad (3.59)$$

From (3.59) we obtain the following results for the ground states, where we have assumed that the singlet state A_1 is the ground state:

$$\begin{aligned} \mathcal{E}(A_1) &= 0 && \text{(singlet)}, \\ \mathcal{E}(E) &= 6\Delta && \text{(doublet)}, \\ \mathcal{E}(T_1) &= 6\Delta + 2\delta\Delta && \text{(triplet)}, \end{aligned} \quad (3.60)$$

where we find that the valley–orbit interaction results in the splitting of the $1s$ ground state in three groups. From the analysis by infrared spectroscopy of the donors P, As and Sb in Si, the binding energies of the $1s$ states are given by the values listed in Table 3.7 [9]. From the analysis of the intensity of infrared absorption of P donors in Si, the valley–orbit interaction given by (3.60) is determined to be $6\Delta = 13.10$ and $2\delta\Delta = -1.37$ meV, and therefore we can conclude that the $1s$ donor state splits into $1s(A_1)$ (singlet), $1s(T_1)$ (triplet) and $1s(E)$ (doublet) in the order from the lowest to the highest level. The splitting due the valley–orbit interaction is very small and the difference between experiment and theory cannot be explained. This may require more detailed analysis of the donor states. The discrepancy may be explained by taking an exact wave functions for the donor, but it is impossible to go further. Instead we will derive an appropriate method to analyze the optical spectra of donors in Ge and Si in the next subsection.

3.4.2 Central Cell Correction

In the derivation of the effective–mass equation we noted that the validity of the effective–mass approximation requires a larger effective Bohr radius a_I of a donor

Table 3.7 Binding energies of 1*s* states for donors of P, As and Sb in Si (meV). The splitting is due to valley-orbit interaction (see text)

States	P	As	Sb
1s(E)	32.21	31.01	30.23
$1s(T_1)$	33.58	32.42	32.65
$1s(A_1)$	45.31	53.51	42.51

compared with the lattice constant a ($a_I \gg a$). As seen in Table 3.2, the variational parameters a and b are smaller in Si than in Ge. In addition we know that the effective radius of the orbit is larger for the excited states compared to the ground state. These results will explain the better agreement of the ground state energy in Ge than in Si and a reasonable agreement of the excited states in Si between experiment and the effective–mass theory. We have to note that the difference between the donor states in different donor atoms cannot be explained by the effective–mass approximation. The disagreement of the theory may be due to the potential shape near the center of the impurities and it is expected that the potential near the impurity atom is not Coulombic. The exchange interaction of donor electrons with core electrons and their correlations will play an important role in the determination of the ground state. In addition, the screening effect is weakened in the region near the core. Taking all of these effects into account, we have to correct the potential near the core in the effective–mass approximation. Here we will present the "central cell correction" according to the treatment reported by Kohn [6]. The treatment is based on the assumption that the ground state reflects the energy observed from experiments. Then a correction is made for the wave function of the ground state to explain the infrared absorption spectra between the ground state and the excited states. No corrections for the excited states are required because of the good agreement between the experiment and theory.

Since the experimental value of the ground state binding energy $\mathcal{E}_{obs}$ is larger than the result of the effective–mass approximation in general, the wave function is expected to increase steeply near the center of the impurity atom, as shown in Fig. 3.3. Therefore, we assume that

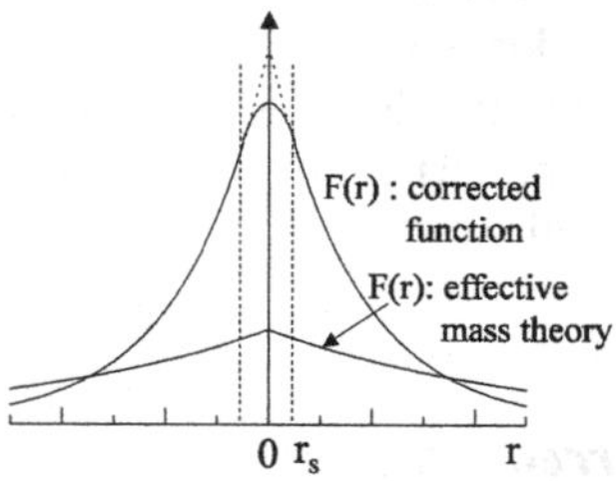

Fig. 3.3 Central cell correction of the donor ground state, where the corrected wave function and the envelope function of the effective–mass equation are shown. The corrected wave function is chosen to give the experimental donor level and extrapolated near the center

$$r \geq r_{\mathrm{s}}: \quad \psi = \frac{1}{6} F(\boldsymbol{r}) \sum_{j=1}^{6} u_j(\boldsymbol{r}) \mathrm{e}^{\mathrm{i}\boldsymbol{k}_j \cdot \boldsymbol{r}}, \tag{3.61}$$

$$\left[-\frac{\hbar^2}{2m^*} \nabla^2 - \frac{e^2}{4\pi\epsilon r} \right] F(\boldsymbol{r}) = \mathcal{E}_{\mathrm{obs}} F(\boldsymbol{r}), \tag{3.62}$$

$$F(\infty) = 0, \tag{3.63}$$

and use the wave function $F(\boldsymbol{r})$, which shows a good agreement with experiment. In other words, we determine the effective Bohr radius of the function

$$F(\boldsymbol{r}) \approx \mathrm{e}^{-r/a^*} \tag{3.64}$$

to give the experimental value of the donor ground state. We find that a^* is determined by the following relation using the values of the effective–mass equation $\mathcal{E}_{\mathrm{effmass}}$ and of experiment $\mathcal{E}_{\mathrm{obs}}$.

$$a^* = \sqrt{\frac{\mathcal{E}_{\mathrm{effmass}}}{\mathcal{E}_{\mathrm{obs}}}} a_{\mathrm{I}}. \tag{3.65}$$

In the region $r < r_{\mathrm{s}}$ the potential reflects the respective type of the donor and thus the corresponding correction to the wave function. However, it may be well approximated by

$$r < r_{\mathrm{s}}; \quad F(\boldsymbol{r}) \simeq F(\boldsymbol{r}_{\mathrm{s}}). \tag{3.66}$$

This procedure is shown in Fig. 3.3 and the process is called the "central cell correction". This method is known to give a good explanation of the infrared absorption spectra and the intensity of the absorption [9].

3.5 Electron Motion Under an External Field

Here we derive equations of electron motion under an external field in a periodic potential. The equations are well known but the derivation so far reported is based on an assumption that group velocity $\boldsymbol{v}_{\mathrm{g}}$ of an electron in a crystal is given by

$$\boldsymbol{v}_{\mathrm{g}} = \frac{1}{\hbar} \frac{\partial \mathcal{E}}{\partial \boldsymbol{k}}. \tag{3.67}$$

The above equation is derived by using the following relations

$$\boldsymbol{v}_{\mathrm{g}} = \frac{\partial \omega}{\partial \boldsymbol{k}}, \qquad \mathcal{E} = \hbar\omega. \tag{3.68}$$

Such a derivation does not deal with electrons in periodic potential correctly, but it is based on an analogy of electron particle with propagating wave as given by (3.68). First, we derive the expectation value of electron velocity in periodic potential, which is shown to coincide with the group velocity given by (3.67). Then we show quantum mechanical treatment of acceleration of an electron in the energy band. The results lead to the similarity of electron motion in classical mechanics and to the definition of effective mass.

3.5.1 Group Velocity

In this subsection we will show that the expectation value of the electron velocity $\langle \boldsymbol{v} \rangle$ is defined by the following relation;

$$\langle \boldsymbol{v} \rangle = \frac{1}{\hbar}\mathrm{grad}_{\boldsymbol{k}}\mathcal{E}(\boldsymbol{k}) = \frac{1}{\hbar}\nabla_{\boldsymbol{k}}\mathcal{E}(\boldsymbol{k})\,. \tag{3.69}$$

The velocity in classical mechanics may be related to the expectation value of the velocity in quantum mechanics. Once we obtain the group velocity, then the electron motion may be described by the equation of motion in classical mechanics.

The expectation value of the electron velocity should be evaluated by using the eigenfunction of an electron. Here we calculate the expectation value by using Bloch function in periodic potential $V(\boldsymbol{r})$. The Bloch function is given by (1.23) of Chap. 1 as

$$\Psi_{\boldsymbol{k}}(\boldsymbol{r}) \equiv |\boldsymbol{k}\rangle = \frac{1}{\sqrt{\Omega}}\sum_{\boldsymbol{G}} A_{\boldsymbol{G}}(\boldsymbol{k})\mathrm{e}^{\mathrm{i}(\boldsymbol{k}+\boldsymbol{G})\cdot\boldsymbol{r}}\,, \tag{3.70}$$

where $\boldsymbol{G}$ is the reciprocal lattice vector and Ω is the volume of the unit cell. In the following we assume that the external field is weak and the electron moves in the same band, and thus we omit the index of the energy bands. Since the momentum operator is given by $\boldsymbol{p} = -\mathrm{i}\hbar\nabla$, the expectation value of the electron velocity $\boldsymbol{v}$ in the state $|\boldsymbol{k}\rangle$ is given by

$$\langle \boldsymbol{k}|\boldsymbol{v}|\boldsymbol{k}\rangle = \frac{1}{m}\langle \boldsymbol{k}|\boldsymbol{p}|\boldsymbol{k}\rangle = -\frac{\mathrm{i}\hbar}{m}\langle \boldsymbol{k}|\nabla|\boldsymbol{k}\rangle = -\frac{\mathrm{i}\hbar}{m}\int \Psi^*_{\boldsymbol{k}}(\boldsymbol{r})\nabla\Psi_{\boldsymbol{k}}(\boldsymbol{r})\mathrm{d}^3\boldsymbol{r}\,. \tag{3.71}$$

The above equation is evaluated by using the following relations derived in Chap. 1

$$\nabla|\boldsymbol{k}\rangle = \nabla\Psi_{\boldsymbol{k}}(\boldsymbol{r}) = \sum_{\boldsymbol{G}} \mathrm{i}(\boldsymbol{k}+\boldsymbol{G})\mathrm{e}^{\mathrm{i}(\boldsymbol{k}+\boldsymbol{G})\cdot\boldsymbol{r}}A_{\boldsymbol{G}}(\boldsymbol{k})\,, \tag{3.72}$$

$$\nabla^2|\boldsymbol{k}\rangle = \nabla^2\Psi_{\boldsymbol{k}}(\boldsymbol{r}) = -\sum(\boldsymbol{k}+\boldsymbol{G})^2\mathrm{e}^{\mathrm{i}(\boldsymbol{k}+\boldsymbol{G})\cdot\boldsymbol{r}}A_{\boldsymbol{G}}(\boldsymbol{k})\,. \tag{3.73}$$

The Bloch function $\Psi_{\boldsymbol{k}}(\boldsymbol{r})$ is normalized in the unit cell Ω as

$$\begin{aligned}\int_{\Omega} \Psi_{\boldsymbol{k}}^{*}(\boldsymbol{r})\Psi_{\boldsymbol{k}}(\boldsymbol{r})\mathrm{d}^3\boldsymbol{r} &= \frac{1}{\Omega}\int_{\Omega}\sum_{\boldsymbol{G}} A_{\boldsymbol{G}}^{*}(\boldsymbol{k})A_{\boldsymbol{G}}(\boldsymbol{k})\mathrm{d}^3\boldsymbol{r} \\ &= \sum_{\boldsymbol{G}} |A_{\boldsymbol{G}}(\boldsymbol{k})|^2 = 1\,. \end{aligned} \tag{3.74}$$

Using (3.72) and the reciprocal lattice vectors $\boldsymbol{G}$ and $\boldsymbol{G}'$, (3.71) is written as

$$\begin{aligned}\langle \boldsymbol{k}|\boldsymbol{v}|\boldsymbol{k}\rangle &= \frac{\hbar}{m}\frac{1}{\Omega}\int \sum_{\boldsymbol{G}',\boldsymbol{G}} A_{\boldsymbol{G}'}^{*}(\boldsymbol{k})\mathrm{e}^{-\mathrm{i}(\boldsymbol{k}+\boldsymbol{G}')\cdot\boldsymbol{r}}(\boldsymbol{k}+\boldsymbol{G})\mathrm{e}^{\mathrm{i}(\boldsymbol{k}+\boldsymbol{G})\cdot\boldsymbol{r}}A_{\boldsymbol{G}}(\boldsymbol{k})\mathrm{d}^3\boldsymbol{r} \\ &= \frac{\hbar}{m}\sum_{\boldsymbol{G}',\boldsymbol{G}}(\boldsymbol{k}+\boldsymbol{G})\frac{1}{\Omega}\int A_{\boldsymbol{G}'}^{*}(\boldsymbol{k})A_{\boldsymbol{G}}(\boldsymbol{k})\mathrm{e}^{\mathrm{i}(\boldsymbol{G}-\boldsymbol{G}')\cdot\boldsymbol{r}}\mathrm{d}^3\boldsymbol{r}\,, \end{aligned} \tag{3.75}$$

where the above integral becomes 0 except the case $\boldsymbol{G} = \boldsymbol{G}'$, and thus we obtain the following equation

$$\langle \boldsymbol{k}|\boldsymbol{v}|\boldsymbol{k}\rangle = \frac{\hbar}{m}\sum_{\boldsymbol{G}}(\boldsymbol{k}+\boldsymbol{G})|A_{\boldsymbol{G}}(\boldsymbol{k})|^2 = \frac{\hbar\boldsymbol{k}}{m} + \frac{\hbar}{m}\sum_{\boldsymbol{G}}\boldsymbol{G}|A_{\boldsymbol{G}}(\boldsymbol{k})|^2\,. \tag{3.76}$$

When the right hand side of (3.76) is shown to be equal to $(1/\hbar)\nabla\mathcal{E}(\boldsymbol{k})$, then the relation (3.69) will be proved.

Using (3.73) Schrödinger equation is written as

$$\begin{aligned}&\frac{\hbar^2}{2m}\sum_{\boldsymbol{G}}(\boldsymbol{k}+\boldsymbol{G})^2 A_{\boldsymbol{G}}(\boldsymbol{k})\mathrm{e}^{\mathrm{i}(\boldsymbol{k}+\boldsymbol{G})\cdot\boldsymbol{r}} + \sum_{\boldsymbol{G}} V(\boldsymbol{r})A_{\boldsymbol{G}}(\boldsymbol{k})\mathrm{e}^{\mathrm{i}(\boldsymbol{k}+\boldsymbol{G})\cdot\boldsymbol{r}} \\ &\quad = \mathcal{E}(\boldsymbol{k})\sum_{\boldsymbol{G}} A_{\boldsymbol{G}}(\boldsymbol{k})\mathrm{e}^{\mathrm{i}(\boldsymbol{k}+\boldsymbol{G})\cdot\boldsymbol{r}}\,. \end{aligned} \tag{3.77}$$

Multiplying both sides of the above equation by $\mathrm{e}^{-\mathrm{i}(\boldsymbol{k}+\boldsymbol{G}')\cdot\boldsymbol{r}}$ and integrating in the unit cell we obtain

$$\begin{aligned}&\frac{\hbar^2}{2m}\sum_{\boldsymbol{G}}(\boldsymbol{k}+\boldsymbol{G})^2\int A_{\boldsymbol{G}}(\boldsymbol{k})\mathrm{e}^{\mathrm{i}(\boldsymbol{G}-\boldsymbol{G}')\cdot\boldsymbol{r}}\mathrm{d}^3\boldsymbol{r} \\ &\quad +\sum_{\boldsymbol{G}}\int V(\boldsymbol{r})A_{\boldsymbol{G}}(\boldsymbol{k})\mathrm{e}^{\mathrm{i}(\boldsymbol{G}-\boldsymbol{G}')\cdot\boldsymbol{r}}\mathrm{d}^3\boldsymbol{r} \\ &\quad = \mathcal{E}(\boldsymbol{k})\int\sum_{\boldsymbol{G}} A_{\boldsymbol{G}}(\boldsymbol{k})\mathrm{e}^{\mathrm{i}(\boldsymbol{G}-\boldsymbol{G}')\cdot\boldsymbol{r}}\mathrm{d}^3\boldsymbol{r}\,. \end{aligned} \tag{3.78}$$

As described in Chap. 1, the periodic potential of a crystal $V(\boldsymbol{r})$ can be expanded in terms of the reciprocal lattice vectors by Fourier expansion theorem, and its expansion coefficient is given by $V(\boldsymbol{G}) = (1/\Omega)\int V(\boldsymbol{r})\mathrm{e}^{\mathrm{i}\boldsymbol{G}\cdot\boldsymbol{r}}$. Therefore we obtain

$$\frac{\hbar^2}{2m}(\boldsymbol{k}+\boldsymbol{G}')^2 A_{\boldsymbol{G}'}(\boldsymbol{k}) + \Omega \sum_{\boldsymbol{G}} V(\boldsymbol{G}-\boldsymbol{G}')A_{\boldsymbol{G}}(\boldsymbol{k}) = \mathcal{E}(\boldsymbol{k})A_{\boldsymbol{G}'}(\boldsymbol{k})\,. \tag{3.79}$$

Since $\boldsymbol{G}$ and $\boldsymbol{G}'$ are the reciprocal lattice vectors, we exchange $\boldsymbol{G}'$ and $\boldsymbol{G}$ in (3.79) and we obtain the following relation

$$\frac{\hbar^2}{2m}(\boldsymbol{k}+\boldsymbol{G})^2 A_{\boldsymbol{G}}(\boldsymbol{k}) + \Omega \sum_{\boldsymbol{G}'} V(\boldsymbol{G}'-\boldsymbol{G})A_{\boldsymbol{G}'}(\boldsymbol{k}) = \mathcal{E}(\boldsymbol{k})A_{\boldsymbol{G}}(\boldsymbol{k})\,. \tag{3.80}$$

The complex conjugate of the above equation is written as

$$\frac{\hbar^2}{2m}(\boldsymbol{k}+\boldsymbol{G})^2 A^*_{\boldsymbol{G}}(\boldsymbol{k}) + \Omega \sum_{\boldsymbol{G}'} V(\boldsymbol{G}'-\boldsymbol{G})A^*_{\boldsymbol{G}'}(\boldsymbol{k}) = \mathcal{E}(\boldsymbol{k})A^*_{\boldsymbol{G}}(\boldsymbol{k})\,, \tag{3.81}$$

First, we multiply both sides of (3.80) by $\nabla_{\boldsymbol{k}} A^*_{\boldsymbol{G}}(\boldsymbol{k})$ and sum up with respect to $\sum_{\boldsymbol{G}}$. Second, we multiply both side of (3.81) by $\nabla_{\boldsymbol{k}} A_{\boldsymbol{G}}(\boldsymbol{k})$ and sum up with respect to $\sum_{\boldsymbol{G}}$. Then we sum up each side of the derived two equations, and find

$$\begin{aligned}&\frac{\hbar^2}{2m}\sum_{\boldsymbol{G}}(\boldsymbol{k}+\boldsymbol{G})\nabla_{\boldsymbol{k}}(A^*_{\boldsymbol{G}}\cdot A_{\boldsymbol{G}})\\ &\quad+\Omega\sum_{\boldsymbol{G}}\sum_{\boldsymbol{G}'}V(\boldsymbol{G}'-\boldsymbol{G})(A_{\boldsymbol{G}'}\nabla_{\boldsymbol{k}}A^*_{\boldsymbol{G}}+A_{\boldsymbol{G}'}\nabla_{\boldsymbol{k}}A_{\boldsymbol{G}})\\ &\quad=\mathcal{E}(\boldsymbol{k})\sum_{\boldsymbol{G}}(A_{\boldsymbol{G}}\nabla_{\boldsymbol{k}}A^*_{\boldsymbol{G}}+A^*_{\boldsymbol{G}}\nabla_{\boldsymbol{k}}A_{\boldsymbol{G}})\equiv 0\,.\end{aligned} \tag{3.82}$$

The equality ($\equiv 0$) of the above equation is proved by using (3.74) as follows,

$$\sum_{\boldsymbol{G}}(A_{\boldsymbol{G}}\nabla_{\boldsymbol{k}}A^*_{\boldsymbol{G}}+A^*_{\boldsymbol{G}}\nabla_{\boldsymbol{k}}A_{\boldsymbol{G}}=\nabla_{\boldsymbol{k}}\left(\sum_{\boldsymbol{G}}|A_{\boldsymbol{G}}|^2\right)=\nabla_{\boldsymbol{k}}(1)\equiv 0\,. \tag{3.83}$$

On the other hand, operating $\nabla_{\boldsymbol{k}}$ on both sides of (3.80) and multiplying $A^*_{\boldsymbol{G}}(\boldsymbol{k})$ from the left, then the summing up with respect to $\sum_{\boldsymbol{G}}$ results in

$$\begin{aligned}&\frac{\hbar^2}{2m}\sum_{\boldsymbol{G}}(\boldsymbol{k}+\boldsymbol{G})|A_{\boldsymbol{G}}|^2+\frac{\hbar^2}{2m}\sum_{\boldsymbol{G}}(\boldsymbol{k}+\boldsymbol{G})^2A^*_{\boldsymbol{G}}\nabla_{\boldsymbol{k}}A_{\boldsymbol{G}}\\ &\quad+\Omega\sum_{\boldsymbol{G}}\sum_{\boldsymbol{G}'}V(\boldsymbol{G}'-\boldsymbol{G})A^*_{\boldsymbol{G}}\nabla_{\boldsymbol{k}}A_{\boldsymbol{G}'}\\ &\quad=\sum_{\boldsymbol{G}}|A_{\boldsymbol{G}}|^2\nabla_{\boldsymbol{k}}\mathcal{E}(\boldsymbol{k})+\mathcal{E}(\boldsymbol{k})\sum_{\boldsymbol{G}}A^*_{\boldsymbol{G}}\nabla_{\boldsymbol{k}}A_{\boldsymbol{G}}\,.\end{aligned} \tag{3.84}$$

In a similar fashion we operate $\nabla_{\boldsymbol{k}}$ on both sides of (3.81) and multiply $\sum_{\boldsymbol{G}} A_{\boldsymbol{G}}$. Then we sum up left hand sides and right hand sides of the derived two equations separately, we obtain the following equation

$$\frac{\hbar^2}{m}\sum_{\boldsymbol{G}}(\boldsymbol{k}+\boldsymbol{G})|A_{\boldsymbol{G}}|^2+\frac{\hbar^2}{2m}\sum_{\boldsymbol{G}}(\boldsymbol{k}+\boldsymbol{G})^2\nabla_k(A^*_{\boldsymbol{G}}A_{\boldsymbol{G}})$$
$$+\Omega\sum_{\boldsymbol{G}}\sum_{\boldsymbol{G}'}V(\boldsymbol{G}'-\boldsymbol{G})(A_{\boldsymbol{G}}\nabla_k A_{\boldsymbol{G}'}+A_{\boldsymbol{G}}\nabla_k A^*_{\boldsymbol{G}'})$$
$$=2\sum_{\boldsymbol{G}}|A_{\boldsymbol{G}}|^2\nabla_k\mathcal{E}(\boldsymbol{k})+\mathcal{E}(\boldsymbol{k})\sum_{\boldsymbol{G}}\nabla_k(A^*_{\boldsymbol{G}}A_{\boldsymbol{G}})\equiv 2\nabla_k\mathcal{E}(\boldsymbol{k})\,. \tag{3.85}$$

The sum of the second and third terms of the left hand side results in 0 from the relation of (3.82). Therefore (3.85) reduces to

$$\frac{\hbar^2}{2m}\sum_{\boldsymbol{G}}(\boldsymbol{k}+\boldsymbol{G})|A_{\boldsymbol{G}}(\boldsymbol{k})|^2=\nabla_k\mathcal{E}(\boldsymbol{k})\,. \tag{3.86}$$

Using above result in (3.76), we obtain the following relations

$$\langle\boldsymbol{k}|\boldsymbol{v}|\boldsymbol{k}\rangle=\frac{1}{\hbar}\nabla_k\mathcal{E}(\boldsymbol{k})\,, \tag{3.87a}$$

or for the matrix element of momentum operator $\boldsymbol{p}=m\boldsymbol{v}$

$$\langle\boldsymbol{k}|\boldsymbol{p}|\boldsymbol{k}\rangle=\frac{m}{\hbar}\nabla_k\mathcal{E}(\boldsymbol{k})\,. \tag{3.87b}$$

The result means that the expectation value of the electron velocity in periodic potential is given by the gradient of electron energy $\mathcal{E}(\boldsymbol{k})$ in a band. When we assume the expectation value of the electron velocity is equivalent to the group velocity, then the classical representation of (3.67) is verified.

3.5.2 Electron Motion Under an External Force

Next we derive equation of electron motion in a crystal in the presence of an external force $\boldsymbol{F}$. In classical mechanics a particle motion, or acceleration, is normally written as

$$\frac{\mathrm{d}\boldsymbol{p}}{\mathrm{d}t}=\boldsymbol{F}\,, \tag{3.88}$$

where $\boldsymbol{p}=m\boldsymbol{v}$ is the momentum of a particle with the mass m and the velocity $\boldsymbol{v}$. The most important case in this textbook is the electron motion in an electric field $\boldsymbol{E}$, and thus the external force is written as

$$\boldsymbol{F}=-e\boldsymbol{E}\,, \tag{3.89}$$

where $-e$ is the electronic charge. We assume that a uniform electric field applied to a crystal is so week that no interband transition occurs and that the wave function of an electron is described by the following time-dependent Schrödinger equation with Hamiltonian H.

$$\mathrm{i}\hbar\frac{\partial}{\partial t}\Psi(\boldsymbol{r},t) = (H - \boldsymbol{F}\cdot\boldsymbol{r})\Psi(\boldsymbol{r},t)\,. \tag{3.90}$$

Now we expand the wave function $\Psi(\boldsymbol{r})$ by Bloch functions $\Psi_{\boldsymbol{k}}(\boldsymbol{r})$. Here we assumed the case where the electric field is so weak that no tunneling occurs, and thus the electron moves in a specific energy band.[1] Also we assume a continuous function with respect to $\boldsymbol{k}$ and the summation $\sum_{\boldsymbol{k}}$ is replaced by integration. Then we obtain

$$\Psi(\boldsymbol{k},t) = \sum_{\boldsymbol{k}} a_{\boldsymbol{k}}(t)\Psi_{\boldsymbol{k}}(\boldsymbol{r}) = \int \alpha(\boldsymbol{k},t)\Psi_{\boldsymbol{k}}(\boldsymbol{r})\mathrm{d}^3\boldsymbol{k}\,. \tag{3.91}$$

Inserting (3.91) into (3.90), and using the relation $H\Psi(\boldsymbol{r}) = \mathcal{E}(\boldsymbol{k})\Psi(\boldsymbol{r})$, we obtain

$$\begin{aligned}\mathrm{i}\hbar &\int \dot{\alpha}(\boldsymbol{k},t)\Psi_{\boldsymbol{k}}(\boldsymbol{r})\mathrm{d}^3\boldsymbol{k} \\ &= \int \alpha(\boldsymbol{k},t)\mathcal{E}(\boldsymbol{k})\Psi_{\boldsymbol{k}}(\boldsymbol{r})\mathrm{d}^3\boldsymbol{k} - \int \alpha(\boldsymbol{k},t)(\boldsymbol{F}\cdot\boldsymbol{r})\Psi_{\boldsymbol{k}}(\boldsymbol{r})\mathrm{d}^3\boldsymbol{k}\,.\end{aligned} \tag{3.92}$$

Multiplying the above equation by $\Psi^*_{\boldsymbol{k}'}(\boldsymbol{r})$, integrating in the unit cell, and using orthonormality relation of the Bloch function $\Psi_{\boldsymbol{k}}(\boldsymbol{r})$, we obtain the following relation.

$$\mathrm{i}\hbar\dot{\alpha}(\boldsymbol{k},t) = \alpha(\boldsymbol{k},t)\mathcal{E}(\boldsymbol{k}) - \int \alpha(\boldsymbol{k},t)\boldsymbol{F}\cdot\int \Psi^*_{\boldsymbol{k}'}(\boldsymbol{r})\boldsymbol{r}\Psi_{\boldsymbol{k}}(\boldsymbol{r})\mathrm{d}^3\boldsymbol{r}\mathrm{d}^3\boldsymbol{k}\,. \tag{3.93}$$

Here we perform the spatial integration of the second term on the right hand side of (3.93). Using the Bloch function $\Psi_{\boldsymbol{k}}(\boldsymbol{r}) = \mathrm{e}^{\mathrm{i}\boldsymbol{k}\cdot\boldsymbol{r}}u_{\boldsymbol{k}}(\boldsymbol{r})$, the spatial integration reduces to

$$\begin{aligned}\boldsymbol{X} &= \int \Psi^*_{\boldsymbol{k}'}(\boldsymbol{r})\boldsymbol{r}\Psi_{\boldsymbol{k}}(\boldsymbol{r})\mathrm{d}^3\boldsymbol{r} = \int \mathrm{e}^{-\mathrm{i}\boldsymbol{k}'\cdot\boldsymbol{r}}u^*_{\boldsymbol{k}'}\boldsymbol{r}\mathrm{e}^{\mathrm{i}\boldsymbol{k}\cdot\boldsymbol{r}}u_{\boldsymbol{k}}(\boldsymbol{r})\mathrm{d}^3\boldsymbol{r} \\ &= \mathrm{i}\int \left(\nabla_{\boldsymbol{k}'}\mathrm{e}^{-\mathrm{i}\boldsymbol{k}'\cdot\boldsymbol{r}}\right)u^*_{\boldsymbol{k}'}(\boldsymbol{r})\mathrm{e}^{\mathrm{i}\boldsymbol{k}\cdot\boldsymbol{r}}u_{\boldsymbol{k}}(\boldsymbol{r})\mathrm{d}^3\boldsymbol{r} \\ &= \mathrm{i}\int \left[\nabla_{\boldsymbol{k}'}\left(\mathrm{e}^{-\mathrm{i}\boldsymbol{k}'\cdot\boldsymbol{r}}u^*_{\boldsymbol{k}'}(\boldsymbol{r})\right) - \mathrm{e}^{-\mathrm{i}\boldsymbol{k}'\cdot\boldsymbol{r}}\nabla_{\boldsymbol{k}'}u^*_{\boldsymbol{k}'}(\boldsymbol{r})\right]\mathrm{e}^{\mathrm{i}\boldsymbol{k}\cdot\boldsymbol{r}}u_{\boldsymbol{k}}(\boldsymbol{r})\mathrm{d}^3\boldsymbol{r} \\ &= \mathrm{i}\nabla_{\boldsymbol{k}}\delta_{\boldsymbol{k},\boldsymbol{k}'} - \mathrm{i}\delta_{\boldsymbol{k},\boldsymbol{k}'}\int u_{\boldsymbol{k}}(\boldsymbol{r})\nabla_{\boldsymbol{k}}u^*_{\boldsymbol{k}}(\boldsymbol{r})\mathrm{d}^3\boldsymbol{r}\,.\end{aligned} \tag{3.94}$$

[1] In a general case (3.91) is defined by replacing $\alpha(\boldsymbol{k},t)$ with $\sum_n \alpha_n(\boldsymbol{k},t)$, where n in the band index.

Inserting this into (3.93), we obtain

$$\begin{aligned} i\hbar\dot{\alpha}(\boldsymbol{k},t) &= \alpha(\boldsymbol{k},t)\mathcal{E}(\boldsymbol{k}) - \int \alpha(\boldsymbol{k},t)\boldsymbol{F}\cdot\boldsymbol{X}\mathrm{d}^3\boldsymbol{k} \\ &= \alpha(\boldsymbol{k},t)\mathcal{E}(\boldsymbol{k}) - i\boldsymbol{F}\cdot\nabla_{\boldsymbol{k}}\alpha(\boldsymbol{k},t) \\ &\quad + i\alpha(\boldsymbol{k},t)\int(\boldsymbol{F}\cdot\nabla_{\boldsymbol{k}}u_{\boldsymbol{k}}^*(\boldsymbol{r}))u_{\boldsymbol{k}}(\boldsymbol{r})\mathrm{d}^3\boldsymbol{r}\,. \end{aligned} \tag{3.95}$$

Complex conjugate of this equation is given by

$$\begin{aligned} -i\hbar\dot{\alpha}^*(\boldsymbol{k},t) &= \alpha^*(\boldsymbol{k},t)\mathcal{E}(\boldsymbol{k}) + i\boldsymbol{F}\cdot\nabla_{\boldsymbol{k}}\alpha^*(\boldsymbol{k},t) \\ &\quad - i\alpha^*(\boldsymbol{k},t)\int(\boldsymbol{F}\cdot\nabla_{\boldsymbol{k}}u_{\boldsymbol{k}}(\boldsymbol{r}))u_{\boldsymbol{k}}^*(\boldsymbol{r})\mathrm{d}^3\boldsymbol{r}\,. \end{aligned} \tag{3.96}$$

Multiplying (3.95) by $\alpha^*(\boldsymbol{k},t)$ and (3.96) by $-\alpha(\boldsymbol{k},t)$, then summation of each sides results in

$$\begin{aligned} i\hbar\frac{\partial}{\partial t}|\alpha(\boldsymbol{k},t)|^2 &= -i\boldsymbol{F}\cdot\nabla_{\boldsymbol{k}}|\alpha(\boldsymbol{k},t)|^2 \\ &\quad + i|\alpha(\boldsymbol{k},t)|^2\int\left[\boldsymbol{F}\cdot\left(u_{\boldsymbol{k}}(\boldsymbol{r})\nabla_{\boldsymbol{k}}u_{\boldsymbol{k}}^*(\boldsymbol{r})\right) + u_{\boldsymbol{k}}^*(\boldsymbol{r})\nabla_{\boldsymbol{k}}u_{\boldsymbol{k}}(\boldsymbol{r})\right]\mathrm{d}^3\boldsymbol{r}\,. \end{aligned} \tag{3.97}$$

Here we find that the second term of the right side gives rise to 0 as shown below. Normalization of the Bloch function is given by

$$\int \Psi_{\boldsymbol{k}}^*(\boldsymbol{r})\Psi_{\boldsymbol{k}}(\boldsymbol{r})\mathrm{d}^3\boldsymbol{r} = \int u_{\boldsymbol{k}}^*(\boldsymbol{r})u_{\boldsymbol{k}}(\boldsymbol{r})\mathrm{d}^3\boldsymbol{r} = 1\,. \tag{3.98}$$

Differentiating this equation with respect to $\boldsymbol{k}$ and we obtain

$$\int (u_{\boldsymbol{k}}(\boldsymbol{r})\nabla_{\boldsymbol{k}}u_{\boldsymbol{k}}^*(\boldsymbol{r}) + u_{\boldsymbol{k}}^*(\boldsymbol{r})\nabla_{\boldsymbol{k}}u_{\boldsymbol{k}}(\boldsymbol{r})\mathrm{d}^3\boldsymbol{r} = 0\,. \tag{3.99}$$

This result reveals that the second term of the right side of (3.97) becomes 0. Therefore, (3.97) results in

$$\hbar\frac{\partial}{\partial t}|\alpha(\boldsymbol{k},t)|^2 = -\boldsymbol{F}\cdot\nabla_{\boldsymbol{k}}|\alpha(\boldsymbol{k},t)|^2\,, \tag{3.100}$$

or

$$\left[\frac{\partial}{\partial t} + \left(\frac{\boldsymbol{F}}{\hbar}\cdot\nabla_{\boldsymbol{k}}\right)\right]|\alpha(\boldsymbol{k},t)|^2 = 0\,. \tag{3.101}$$

A general solution of the above equation is given by an arbitrary function G with $\boldsymbol{k}$ and t,

$$|\alpha(\boldsymbol{k},t)|^2 = G\left(\boldsymbol{k} - \frac{1}{\hbar}\boldsymbol{F}t\right)\,. \tag{3.102}$$

Here $\alpha(\boldsymbol{k}, t)$ is the expansion coefficient of the wave function as given by (3.91) and thus $|\alpha(\boldsymbol{k}, t)|^2$ represents the probability of electrons in the state $\boldsymbol{k}$ at time t. Therefore we choose a function G to exhibit a maximum at time $t = 0$ and $\boldsymbol{k} = \boldsymbol{k}_0$. If the function has a maximum at time Δt and at $\boldsymbol{k}'$ and no scattering occurs, we obtain the following relation.[2]

$$G(\boldsymbol{k}_0) = G\left(\boldsymbol{k}' - \frac{1}{\hbar}\boldsymbol{F}\Delta t\right). \tag{3.103}$$

Therefore we obtain the relation $\boldsymbol{k}_0 = \boldsymbol{k}' - \boldsymbol{F}\Delta t/\hbar$, and thus

$$\Delta \boldsymbol{k} = \boldsymbol{k}' - \boldsymbol{k}_0 = \frac{1}{\hbar}\boldsymbol{F}\Delta t\,, \tag{3.104}$$

In other word, we obtain the following relation

$$\hbar\frac{\mathrm{d}\boldsymbol{k}}{\mathrm{d}t} = \boldsymbol{F}(= -e\boldsymbol{E})\,, \tag{3.105}$$

and (3.88) is proved.

In general, when there exist no scattering events, the electron wave vector $\boldsymbol{k}$ at time t changes from the initial state $\boldsymbol{k}_0$ at time 0 according to

$$\boldsymbol{k} = \boldsymbol{k}_0 + \frac{1}{\hbar}\boldsymbol{F}t\,. \tag{3.106}$$

In this case the state $\boldsymbol{k}$ should be empty because of Pauli exclusion principle. When we include interband transition, $|\alpha(\boldsymbol{k}, t)|^2$ should be replaced by $\sum|\alpha_n(\boldsymbol{k}, t)|^2$ and then (3.102) is valid. In the case of interband transition, the wave vector after the transition should be equal to $\boldsymbol{k}$ given by (3.102). In the calculation of electrical conduction, we can neglect such an interband transition because the energy of an electron is smaller than the band gap energy. If the electron scattering by phonons exists, we have to take account of momentum and energy conservation, which will be discussed in Chap. 6.

3.5.3 Electron Motion and Effective Mass

We have already learned that the electronic states are given by the relation of electron energy as a function of electron wave vector, $\mathcal{E}(\boldsymbol{k})$. In the above subsections we find that an external force induces a change in the electron wave vector $\boldsymbol{k}$, and thus

[2]This assumption is based on the fact that the function G will stay un–changed without scattering and thus the maximum value of G is not changed with time t. Scattering induces a change in the wave function and the maximum value will stay constant before the next scattering.

in the electron energy according the relation $\mathcal{E}(\boldsymbol{k})$. In order to deal with electron transport, therefore, we have to know the energy band. The electron distribution is governed by Fermi–Dirac distribution function and electrons in semiconductors in the thermal equilibrium occupy the states near the conduction band minimum. Such electrons are well described by the relation $\mathcal{E}(\boldsymbol{k}) = \hbar^2 k^2/2m^*$, where m^* is called the effective mass. Here we will show that the effective mass approximation is valid for the description of electron transport.

First, we summarize the results obtained in the above two subsections. The expectation value of the electron velocity or the electron group velocity is given by (3.67)

$$\langle \boldsymbol{v} \rangle \equiv \boldsymbol{v}_{\mathrm{g}} = \frac{1}{\hbar}\frac{\mathrm{d}\mathcal{E}}{\mathrm{d}\boldsymbol{k}}, \tag{3.107}$$

and an external force $\boldsymbol{F}$ produces a change in the wave vector of the electron as

$$\hbar\frac{\mathrm{d}\boldsymbol{k}}{\mathrm{d}t} = \boldsymbol{F}(= -e\boldsymbol{E}). \tag{3.108}$$

The time derivative of (3.107) is given by

$$\frac{\mathrm{d}\langle \boldsymbol{v} \rangle}{\mathrm{d}t} = \frac{1}{\hbar}\frac{\mathrm{d}}{\mathrm{d}t}\frac{\mathrm{d}\mathcal{E}}{\mathrm{d}\boldsymbol{k}} \tag{3.109}$$

Since the acceleration of the electron in the external force is governed by (3.108), and the electron energy is given by a function of the wave vector as $\mathcal{E}(\boldsymbol{k})$, the right hand side of the above equation is rewritten as

$$\frac{\mathrm{d}\langle \boldsymbol{v} \rangle}{\mathrm{d}t} = \frac{1}{\hbar}\frac{\mathrm{d}}{\mathrm{d}t}\frac{\mathrm{d}\mathcal{E}}{\mathrm{d}\boldsymbol{k}} = \frac{1}{\hbar}\frac{\mathrm{d}\boldsymbol{k}}{\mathrm{d}t}\frac{\mathrm{d}^2\mathcal{E}}{\mathrm{d}\boldsymbol{k}^2}. \tag{3.110}$$

Inserting (3.108) into (3.110), we obtain

$$\frac{\mathrm{d}\langle \boldsymbol{v} \rangle}{\mathrm{d}t} = \frac{1}{\hbar^2}\frac{\mathrm{d}^2\mathcal{E}}{\mathrm{d}\boldsymbol{k}^2}\boldsymbol{F}. \tag{3.111}$$

This relation should be compared with Newton's law of motion of a particle with the velocity $\boldsymbol{v}$ and the mass m,

$$\frac{\mathrm{d}\boldsymbol{v}}{\mathrm{d}t} = \frac{1}{m}\boldsymbol{F}. \tag{3.112}$$

Comparing (3.111) with (3.112), we may write (3.111) as

$$\frac{\mathrm{d}\langle \boldsymbol{v} \rangle}{\mathrm{d}t} = \frac{1}{m^*}\boldsymbol{F}, \tag{3.113}$$

$$\frac{1}{m^*} = \frac{1}{\hbar^2}\frac{\mathrm{d}^2\mathcal{E}}{\mathrm{d}\boldsymbol{k}^2}, \tag{3.114}$$

where m^* is equivalent to a particle mass and called as the effective mass. In general the inverse effective mass $1/m^*$ is defined as

$$\left(\frac{1}{m^*}\right)_{ij} = \frac{1}{\hbar^2}\frac{\mathrm{d}^2\mathcal{E}}{\mathrm{d}k_i\mathrm{d}k_j}. \tag{3.115}$$

Since we are interested in the electrons near the band edge (conduction band minimum $\mathcal{E}_c$), the energy band is well approximated by parabolic function of the electron wave vector $\boldsymbol{k}$

$$\mathcal{E} = \frac{\hbar^2}{2m^*}k^2 + \mathcal{E}_c. \tag{3.116}$$

This relation is easily derived from (3.114) by carrying out integration with respect to $\boldsymbol{k}$. We have already used this relation to develop the effective mass approximation. From these considerations we conclude that the effective mass defined by (3.114) can be used to evaluate transport properties of electrons in the form of classical mechanics.

3.6 Problems

(3.1) Effective mass equation is derived in different ways. The other method not delt with this textbook is given by Luttinger and Kohn. See the following paper for the derivation of the effective mass equation used for analysis of shallow impurity states and of the cyclotron resonance: Joaquin M. Luttinger and Walter Kohn. "*Motion of electrons and holes in perturbed periodic fields*." Phys. Rev. **97** (1955) 869–883.

(3.2) An electron with its effective mass m^* is put in a magnetic field B applied in z direction.

(1) Solve the cyclotron motion of the electron and show its angular frequency is give by $\omega_c = eB/m^*$.
(2) Calculate its cyclotron radius R_c.
(3) Apply Bohr's quantization condition $\int p dl = h$ to the orbital motion, where $p = m^* v$ with the velocity v of the orbital motion. Then show this condition leads to the cyclotron radius obtained by solving Schrödinger equation.

References

1. G. Wannier, Phys. Rev. **52**, 191 (1937)
2. J.M. Luttinger, W. Kohn, Phys. Rev. **97**, 869 (1955)

3. W. Kohn, J.M. Luttinger, Phys. Rev. **97**, 1721 (1955)
4. C. Kittel, A.H. Mitchell, Phys. Rev. **96**, 1488 (1954)
5. W. Kohn, J.M. Luttinger, Phys. Rev. **98**, 915 (1955)
6. W. Kohn, in *Solid State Physics*, vol. 5, ed. by F. Seitz and D. Turnbull (Academic Press, New York, 1957) pp. 257–320
7. D.K. Wilson, G. Feher, Phys. Rev. **124**, 1068 (1961)
8. R.A. Faulkner, Phys. Rev. **184**, 713 (1969)
9. R.L. Aggarwal, A.K. Ramdas, Phys. Rev. **140**, A1246 (1965)

Chapter 4
Optical Properties 1

Abstract This chapter deals with fundamental theory of optical properties in semiconductors. First reflection and absorption coefficients are derived by using Maxwell's equations. Then quantum mechanical derivations of direct and indirect optical transition rates are given in addition to the classification of the joint density of states. Optical transitions associated with electron-hole pair, excitons, are also discussed. Dielectric functions are discussed in connection with the critical points of semiconductors. Also we will discuss stress effect on the optical transition such as piezobirefringence and stress-induced change in the energy band structure. The results are used in Chap. 9 to evaluate the strain effect in quantum well lasers.

4.1 Reflection and Absorption

The classical theory of electromagnetic waves is described by Maxwell's equations. Let us define the electric field $\boldsymbol{E}$, electric displacement $\boldsymbol{D}$, magnetic field intensity vector $\boldsymbol{H}$, magnetic field flux (referred to as "magnetic field" in this text) $\boldsymbol{B}$, current density $\boldsymbol{J}$, and charge density ρ; then the macroscopic domain is governed by the classical field equations of Maxwell:

$$\nabla \times \boldsymbol{E} = -\frac{\partial \boldsymbol{B}}{\partial t}, \tag{4.1a}$$

$$\nabla \times \boldsymbol{H} = \boldsymbol{J} = \sigma \boldsymbol{E} + \frac{\partial \boldsymbol{D}}{\partial t}, \tag{4.1b}$$

$$\nabla \cdot \boldsymbol{D} = \rho, \tag{4.1c}$$

$$\nabla \cdot \boldsymbol{B} = 0. \tag{4.1d}$$

In the case of a material with a complex dielectric constant

$$\kappa = \kappa_1 + i\kappa_2 \tag{4.2}$$

we obtain

C. Hamaguchi, *Basic Semiconductor Physics*, Graduate Texts in Physics,
DOI 10.1007/978-3-319-66860-4_4

$$\boldsymbol{D} = \kappa\epsilon_0 \boldsymbol{E} . \tag{4.3}$$

In the following we assume that there exists no excess charge in the domain we are interested in. Then we obtain the following relations from Maxwell's equations. For simplicity we assume that there is no charge to induce the current, i.e. $\sigma = 0$. When charge carriers exist, the following results may be modified by introducing the complex conductivity as described by (4.24), (4.25a) and (4.25b), and also as discussed in Sect. 5.5.

$$\begin{aligned}\nabla \times \nabla \times \boldsymbol{E} &= -\frac{\partial}{\partial t}(\nabla \times \boldsymbol{B}) ,\\ &= -\mu_0 \frac{\partial}{\partial t}\left(\sigma \boldsymbol{E} + \frac{\partial \boldsymbol{D}}{\partial t}\right) ,\\ &= -\mu_0 \kappa \epsilon_0 \frac{\partial^2}{\partial t^2}\boldsymbol{E} ,\end{aligned} \tag{4.4}$$

Since there is no excess charge, we put $\rho = 0$ and find the following result:

$$\begin{aligned}&\nabla \times \nabla \times \boldsymbol{E} = \nabla(\nabla \cdot \boldsymbol{E}) - \nabla^2 \boldsymbol{E},\\ &\nabla \cdot \boldsymbol{E} = 0 ,\\ &\nabla^2 \boldsymbol{E} = \mu_0 \epsilon_0 \kappa \frac{\partial^2 \boldsymbol{E}}{\partial t^2} ,\end{aligned} \tag{4.5}$$

Let us assume a plane wave for the electromagnetic field

$$\boldsymbol{E} \sim \boldsymbol{E} \exp[\mathrm{i}(\boldsymbol{k} \cdot \boldsymbol{r} - \omega t)] . \tag{4.6}$$

Inserting this into (4.5) we obtain

$$(\mathrm{i}k)^2 = (-\mathrm{i}\omega)^2 \mu_0 \epsilon_0 \kappa . \tag{4.7}$$

The phase velocity of the electromagnetic wave is then given by

$$\frac{\omega}{k} = \frac{1}{\sqrt{\epsilon_0 \mu_0}\sqrt{\kappa}} = \frac{c}{\sqrt{\kappa}} = c' , \tag{4.8}$$

where c is the velocity of light in vacuum ($\kappa = 1$)

$$c = \frac{1}{\sqrt{\epsilon_0 \mu_0}} \cong 3.0 \times 10^8\,\mathrm{m/s} \tag{4.9}$$

and c' is the velocity of light in a material with dielectric constant κ

$$c' = \frac{c}{\sqrt{\kappa}} = \frac{c}{n} . \tag{4.10}$$

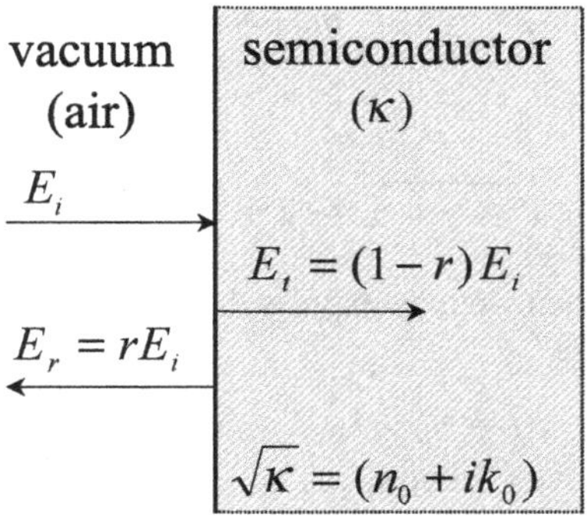

Fig. 4.1 Incident, reflected and transmitted electromagnetic waves (light) at the boundary of a material surface

In the above equation we have used the definition of the **refractive index** $n = \sqrt{\kappa}$. Since we have defined the dielectric constant by a complex number, we have to introduce a complex refractive index. Let us assume that the electromagnetic wave propagates in the z direction with electric field amplitude $\boldsymbol{E}_\perp$:

$$\begin{aligned}\boldsymbol{E}(z,t) &= \boldsymbol{E}_\perp \exp[\mathrm{i}(kz - \omega t)] = \boldsymbol{E}_\perp \exp\left[\mathrm{i}\omega\left(\frac{k}{\omega}z - t\right)\right] \\ &= \boldsymbol{E}_\perp \exp\left[\mathrm{i}\omega\left(\frac{\sqrt{\kappa}}{c}z - t\right)\right]. \end{aligned} \tag{4.11}$$

We also define the incident, reflected and transmitted electric fields of the electromagnetic field at the boundary shown in Fig. 4.1 by

$$\begin{aligned} &\boldsymbol{E}_\mathrm{i} \exp\left[\mathrm{i}\omega\left(\frac{1}{c}z - t\right)\right], \\ &\boldsymbol{E}_\mathrm{r} \exp\left[\mathrm{i}\omega\left(-\frac{1}{c}z - t\right)\right], \\ &\boldsymbol{E}_\mathrm{t} \exp\left[\mathrm{i}\omega\left(\frac{\sqrt{\kappa}}{c}z - t\right)\right]. \end{aligned}$$

Since we assume that there is no excess charge at the boundary, the electric field and its gradient are required to be continuous at z. The reflection coefficient r ($\boldsymbol{E}_\mathrm{r} = r\boldsymbol{E}_\mathrm{i}$) and the transmission coefficient t ($\boldsymbol{E}_\mathrm{t} = t\boldsymbol{E}_\mathrm{i}$) have to satisfy the following relations.

$$1 = r + t \tag{4.12a}$$

$$1 = -r + t\sqrt{\kappa} = -r + (1 - r)\sqrt{\kappa}. \tag{4.12b}$$

From these relations we obtain the reflection coefficient

$$r = \frac{\sqrt{\kappa} - 1}{\sqrt{\kappa} + 1}. \tag{4.13}$$

When we define complex refractive index n^* by

$$\sqrt{\kappa} = n^* \equiv n_0 + \mathrm{i}k_0\,, \tag{4.14}$$

or

$$\sqrt{\kappa_1 + i\kappa_2} = n_0 + \mathrm{i}k_0\,, \tag{4.15}$$

then we may express the real and imaginary parts of the complex dielectric constant by

$$\kappa_1 = n_0^2 - k_0^2, \tag{4.16a}$$

$$\kappa_2 = 2n_0k_0\,, \tag{4.16b}$$

where n_0 and k_0 are called the refractive index and the extinction coefficient, respectively. From these relations, the amplitude reflection coefficient is given by

$$r = \frac{n_0 - 1 + \mathrm{i}k_0}{n_0 + 1 + \mathrm{i}k_0} = |r| \tan\theta\,, \quad \tan\theta = \frac{2k_0}{n_0^2 + k_0^2 - 1}\,. \tag{4.17}$$

The reflection coefficient is generally defined as the power reflection coefficient (the reflection coefficient for the incident energy), and thus we have to calculate the reflection coefficient for E^2, H^2 or the Poynting vector $\boldsymbol{E} \times \boldsymbol{H}$, which is given by

$$\begin{aligned} R = |r|^2 &= \frac{(n_0 - 1)^2 + k_0^2}{(n_0 + 1)^2 + k_0^2} \\ &= \frac{(\kappa_1^2 + \kappa_2^2)^{1/2} - [2\kappa_1 + 2(\kappa_1^2 + \kappa_2^2)^{1/2}]^{1/2} + 1}{(\kappa_1^2 + \kappa_2^2)^{1/2} + [2\kappa_1 + 2(\kappa_1^2 + \kappa_2^2)^{1/2}]^{1/2} + 1}\,. \end{aligned} \tag{4.18}$$

Inserting (4.14) into (4.11),

$$\begin{aligned} \boldsymbol{E}(z,t) &= \boldsymbol{E}_\perp \exp\left[\mathrm{i}\omega\left(\frac{\sqrt{\kappa}}{c}z - t\right)\right] = \boldsymbol{E}_\perp \exp\left[\mathrm{i}\omega\left(\frac{n_0 + \mathrm{i}k_0}{c}z - t\right)\right] \\ &= \boldsymbol{E}_\perp \exp\left(-\frac{\omega k_0}{c}z\right) \exp\left[\mathrm{i}\omega\left(\frac{n_0}{c}z - t\right)\right] \end{aligned} \tag{4.19}$$

is obtained and the attenuation of the electric power or the Poynting vector of the electromagnetic waves is written as

$$I \propto E^2 \propto E_\perp^2 \exp\left(-2\frac{\omega k_0}{c}z\right) \equiv E_\perp^2 \exp(-\alpha z)\,, \tag{4.20}$$

where

$$\alpha = 2\frac{\omega k_0}{c} = \frac{\omega\kappa_2}{n_0 c} \tag{4.21}$$

is called the **absorption coefficient** which gives rise to an attenuation of the incident power by $1/e \simeq 1/2.7 \simeq 0.37$ per unit propagation length.

Next we discuss the relation of the complex conductivity and the complex dielectric constant to the power dissipation (power loss) per unit volume of a material, which is exactly the same as the dielectric loss. Using the complex dielectric coefficient in the electric displacement in (4.1b), it is rewritten as

$$\boldsymbol{D} = \kappa\epsilon_0 \boldsymbol{E} = (\kappa_1 + \mathrm{i}\kappa_2)\epsilon_0 \boldsymbol{E} \,. \tag{4.22}$$

Since we assume a plane wave for the electromagnetic wave, we may put $\partial/\partial t \to -\mathrm{i}\omega$ and then the current density is given by

$$\boldsymbol{J} = \sigma \boldsymbol{E} + \frac{\partial \boldsymbol{D}}{\partial t} = \sigma \boldsymbol{E} - \mathrm{i}\omega(\kappa_1 + \mathrm{i}\kappa_2)\epsilon_0 \boldsymbol{E} \equiv \sigma^* \boldsymbol{E} \,. \tag{4.23}$$

When we write the complex conductivity σ^* as

$$\sigma^* = \sigma_r + \mathrm{i}\sigma_i \tag{4.24}$$

we find the following relations between the complex conductivity and the complex dielectric constant:

$$\sigma_r = \sigma + \omega\kappa_2\epsilon_0, \tag{4.25a}$$

$$\sigma_i = -\omega\kappa_1\epsilon_0 \,. \tag{4.25b}$$

It is well known that the power loss is given by $w = \sigma_r E^2/2$ and thus the power loss per unit volume of a dielectric with $\kappa = \kappa_1 + \mathrm{i}\kappa_2$ is written as

$$w = \frac{1}{2}\omega\kappa_2\epsilon_0 E^2 \,, \tag{4.26}$$

which is known as the "dielectric loss." This relation holds exactly for light absorption in a material and is used later to relate optical transitions and light absorption. In the case of semiconductors where free carriers exist, the real part of the complex dielectric constant κ_1 is related to the imaginary part of the complex conductivity by

$$\kappa_1 = \kappa_l - \frac{\sigma_\mathrm{i}}{\omega\epsilon_0} \,, \tag{4.27}$$

where κ_l is the real part of the dielectric constant due to the crystal lattices and we may expect κ_1 to become zero at a specific frequency (plasma frequency). This relation is often used to discuss the classical theory of plasma oscillation (see Sects. 5.4.1 and 5.5).

4.2 Direct Transition and Absorption Coefficient

In this section we will consider the band-to-band direct transition of an electron from the valence band to the conduction band induced by the incident light. To do this we consider the electron motion induced by the incident light in a perfect crystal. The Hamiltonian of the electron is given by

$$H = \frac{1}{2m}(\boldsymbol{p} + e\boldsymbol{A})^2 + V(\boldsymbol{r}), \tag{4.28}$$

where $\boldsymbol{A}$ is the vector potential of the electromagnetic filed and $V(\boldsymbol{r})$ is the periodic potential of the crystal. The vector potential is expressed by the plane wave:

$$\boldsymbol{A} = \frac{1}{2}A_0\boldsymbol{e}\left[\mathrm{e}^{\mathrm{i}(\boldsymbol{k}_\mathrm{p}\cdot\boldsymbol{r}-\omega t)} + \mathrm{e}^{-\mathrm{i}(\boldsymbol{k}_\mathrm{p}\cdot\boldsymbol{r}-\omega t)}\right], \tag{4.29}$$

where $\boldsymbol{k}_\mathrm{p}$ and $\boldsymbol{e}$ are the wave vector of the electromagnetic field and its unit vector (polarization vector), respectively. In the above equation the vector potential is expressed as a real number by adding its complex conjugate. Using the relation $\boldsymbol{A}\cdot\boldsymbol{p} = \boldsymbol{p}\cdot\boldsymbol{A}$ and neglecting the small term A^2, the Hamiltonian is rewritten as

$$H \simeq \frac{p^2}{2m} + V(\boldsymbol{r}) + \frac{e}{m}\boldsymbol{A}\cdot\boldsymbol{p} \equiv H_0 + H_1\,. \tag{4.30}$$

Assuming $H_1 = (e/m)\boldsymbol{A}\cdot\boldsymbol{p}$ as the perturbation, the transition probability per unit time w_cv for the electron from the initial state $|\mathrm{v}\boldsymbol{k}\rangle$ to the final state $|\mathrm{c}\boldsymbol{k}'\rangle$ is calculated to be

$$\begin{aligned} w_\mathrm{cv} &= \frac{2\pi}{\hbar}|\langle \mathrm{c}\boldsymbol{k}'|\frac{e}{m}\boldsymbol{A}\cdot\boldsymbol{p}|\mathrm{v}\boldsymbol{k}\rangle|^2\delta\left[\mathcal{E}_\mathrm{c}(\boldsymbol{k}') - \mathcal{E}_\mathrm{v}(\boldsymbol{k}) - \hbar\omega\right] \\ &= \frac{\pi e^2}{2\hbar m^2}A_0^2\left|\langle \mathrm{c}\boldsymbol{k}'|\exp(\mathrm{i}\boldsymbol{k}_\mathrm{p}\cdot\boldsymbol{r})\boldsymbol{e}\cdot\boldsymbol{p}|\mathrm{v}\boldsymbol{k}\rangle\right|^2 \\ &\quad\times\delta\left[\mathcal{E}_\mathrm{c}(\boldsymbol{k}') - \mathcal{E}_\mathrm{v}(\boldsymbol{k}) - \hbar\omega\right]. \end{aligned} \tag{4.31}$$

The matrix element of the term which includes the momentum operator $\boldsymbol{p}$ is called the matrix element of the transition and gives the selection rule and the strength of the transition. Let us consider the Bloch function to express the electron state:

$$|j\boldsymbol{k}\rangle = \mathrm{e}^{\mathrm{i}\boldsymbol{k}\cdot\boldsymbol{r}}u_{j\boldsymbol{k}}(\boldsymbol{r}), \tag{4.32}$$

where $j = \mathrm{v}$ and c represents the valence band and conduction band states, respectively. Then the matrix element of the transition is given by

$$\begin{aligned} \boldsymbol{e}\cdot\boldsymbol{p}_\mathrm{cv} &= \frac{1}{V}\int_V \mathrm{e}^{-\mathrm{i}\boldsymbol{k}'\cdot\boldsymbol{r}}u_{\mathrm{c}\boldsymbol{k}'}^*(\boldsymbol{r})\mathrm{e}^{\mathrm{i}\boldsymbol{k}_\mathrm{p}\cdot\boldsymbol{r}}\boldsymbol{e}\cdot\boldsymbol{p}\mathrm{e}^{\mathrm{i}\boldsymbol{k}\cdot\boldsymbol{r}}u_{\mathrm{v}\boldsymbol{k}}(\boldsymbol{r})\mathrm{d}^3\boldsymbol{r} \\ &= \frac{1}{V}\int_V \mathrm{e}^{\mathrm{i}(\boldsymbol{k}_\mathrm{p}+\boldsymbol{k}-\boldsymbol{k}')\cdot\boldsymbol{r}}u_{\mathrm{c}\boldsymbol{k}'}^*(\boldsymbol{r})\boldsymbol{e}\cdot(\boldsymbol{p}+\hbar\boldsymbol{k})u_{\mathrm{v}\boldsymbol{k}}(\boldsymbol{r})\mathrm{d}^3\boldsymbol{r}\,. \end{aligned} \tag{4.33}$$

From the property of the Bloch function, $u(\boldsymbol{r}) = u(\boldsymbol{r}+\boldsymbol{R}_l)$, where $\boldsymbol{R}_l$ is the translation vector, and the matrix element is rewritten as

$$\boldsymbol{e} \cdot \boldsymbol{p}_{cv} = \frac{1}{V} \sum_l \exp\{\mathrm{i}(\boldsymbol{k}_\mathrm{p} + \boldsymbol{k} - \boldsymbol{k}') \cdot \boldsymbol{R}_l\} \times \int_\Omega \mathrm{e}^{\mathrm{i}(\boldsymbol{k}_\mathrm{p}+\boldsymbol{k}-\boldsymbol{k}')\cdot\boldsymbol{r}} u^*_{\mathrm{c}\boldsymbol{k}'}(\boldsymbol{r})\boldsymbol{e} \cdot (\boldsymbol{p} + \hbar\boldsymbol{k})u_{\mathrm{v}\boldsymbol{k}'}(\boldsymbol{r})\mathrm{d}^3\boldsymbol{r}\,, \tag{4.34}$$

where Ω is the volume of the unit cell. Summation with respect to $\boldsymbol{R}_l$ becomes 0 except for

$$\boldsymbol{k}_\mathrm{p} + \boldsymbol{k} - \boldsymbol{k}' = \boldsymbol{G}_m (= m\boldsymbol{G})\,, \tag{4.35}$$

where $\boldsymbol{G}_m$ is the reciprocal lattice vector ($\boldsymbol{G}$ is the smallest reciprocal lattice vector and m is an integer). The wave vector of light (electromagnetic waves) with a wavelength of 1 μm is $|\boldsymbol{k}_\mathrm{p}| = 6.28 \times 10^4\,\mathrm{cm}^{-1}$ and the magnitude of the reciprocal lattice vector for a crystal with a lattice constant of 5 Å is $|\boldsymbol{G}| = 1.06 \times 10^8\,\mathrm{cm}^{-1}$, and thus the inequality $k_\mathrm{p} \ll G$ is fulfilled in general. Therefore, the largest contribution to the integral in (4.34) is due to the term for $\boldsymbol{G}_m = 0$ ($m = 0$). This condition may be understood to be equivalent to the conservation of momentum. From these considerations (4.35) leads to the important relation

$$\boldsymbol{k}' = \boldsymbol{k} \tag{4.36}$$

for the optical transition. That is, electron transitions are allowed between states with the same wave vector $\boldsymbol{k}$ in $\boldsymbol{k}$ space. In other words, when a photon of energy greater than the band gap is incident on a semiconductor, an electron with wave vector $\boldsymbol{k}$ in the valence band is excited into a state with the same wave vector in the conduction band. From this fact the transition is referred to as a **direct transition**.

Since the integral with respect to $\hbar\boldsymbol{k}$ in (4.34) vanishes because of the orthogonality of the Bloch functions, we obtain

$$\boldsymbol{e} \cdot \boldsymbol{p}_{\mathrm{cv}} = \frac{1}{\Omega} \int_\Omega u^*_{\mathrm{c}\boldsymbol{k}'}(\boldsymbol{r})\boldsymbol{e} \cdot \boldsymbol{p} u_{\mathrm{v}\boldsymbol{k}}(\boldsymbol{r})\mathrm{d}^3\boldsymbol{r}\,\delta_{\boldsymbol{k},\boldsymbol{k}'}\,. \tag{4.37}$$

Using the above results, the photon energy absorbed in the material per unit time and unit volume is given by $\hbar\omega w_{\mathrm{cv}}$, which is equivalent to the power dissipation of the electromagnetic waves per unit time and unit volume given by (4.26):

$$\hbar\omega w_{\mathrm{cv}} = \frac{1}{2}\omega\kappa_2\epsilon_0 E_0^2\,. \tag{4.38}$$

From the relation between the electric field and the vector potential, $\boldsymbol{E} = -\partial\boldsymbol{A}/\partial t$, we may put $E_0 = \omega A_0$; thus the imaginary part of the dielectric constant is given by the equation

$$\kappa_2 = \frac{2\hbar}{\epsilon_0 \omega^2 A_0^2} w_{\rm cv} \\ = \frac{\pi e^2}{\epsilon_0 m^2 \omega^2} \sum_{\boldsymbol{k},\boldsymbol{k}'} \left| \boldsymbol{e} \cdot \boldsymbol{p}_{\rm cv} \right|^2 \delta \left[\mathcal{E}_{\rm c}(\boldsymbol{k}') - \mathcal{E}_{\rm v}(\boldsymbol{k}) - \hbar\omega \right] \delta_{\boldsymbol{k}\boldsymbol{k}'} . \tag{4.39}$$

It is evident that the absorption coefficient is obtained by inserting κ_2 into (4.21).

4.3 Joint Density of States

In the previous section we derived the dielectric function for the direct transition, which is given by (4.39). When we assume that the matrix element $\boldsymbol{e} \cdot \boldsymbol{p}_{\rm cv}$ varies very slowly as $\boldsymbol{k}$ or is independent of $\boldsymbol{k}$ (this is a good approximation), the term $|\boldsymbol{e} \cdot \boldsymbol{p}_{\rm cv}|^2$ in (4.39) may be moved out of the summation. The imaginary part of the dielectric constant can then be rewritten as

$$\kappa_2(\omega) = \frac{\pi e^2}{\epsilon_0 m^2 \omega^2} |\boldsymbol{e} \cdot \boldsymbol{p}_{\rm cv}|^2 \sum_{\boldsymbol{k}} \delta[\mathcal{E}_{\rm cv}(\boldsymbol{k}) - \hbar\omega], \tag{4.40}$$

$$\mathcal{E}_{\rm cv}(\boldsymbol{k}) = \mathcal{E}_{\rm c}(\boldsymbol{k}) - \mathcal{E}_{\rm v}(\boldsymbol{k}) . \tag{4.41}$$

The summation with respect to $\boldsymbol{k}$ in (4.40) may be understood as the summation of the pair states of $|{\rm v}\boldsymbol{k}\rangle$ and $|{\rm c}\boldsymbol{k}\rangle$ due to the delta function and called the **joint density of states**. Replacing the summation $\sum$ by an integral in $\boldsymbol{k}$ space, the joint density of states $J_{\rm cv}(\hbar\omega)$ is written as

$$J_{\rm cv}(\hbar\omega) = \sum_{\boldsymbol{k}} \delta[\mathcal{E}_{\rm cv}(\boldsymbol{k}) - \hbar\omega] = \frac{2}{(2\pi)^3} \int {\rm d}^3\boldsymbol{k} \cdot \delta[\mathcal{E}_{\rm cv}(\boldsymbol{k}) - \hbar\omega] , \tag{4.42}$$

where the spin degeneracy factor 2 is taken into account. Integration of the above equation is carried out by the following general method.

Consider two constant energy surfaces in $\boldsymbol{k}$ space, $\mathcal{E} = \hbar\omega$ and $\hbar\omega + {\rm d}(\hbar\omega)$. The density of states in ${\rm d}(\hbar\omega)$ is then obtained as follows.

$$J_{\rm cv}(\hbar\omega) \cdot {\rm d}(\hbar\omega) = \frac{2}{(2\pi)^3} \int_{\hbar\omega = \mathcal{E}_{\rm cv}} \frac{{\rm d}S}{|\nabla_{\boldsymbol{k}} \mathcal{E}_{\rm cv}(\boldsymbol{k})|} \cdot {\rm d}(\hbar\omega) . \tag{4.43}$$

Therefore, the joint density of states $J_{\rm cv}(\hbar\omega)$ is rewritten as

$$J_{\rm cv}(\hbar\omega) = \frac{2}{(2\pi)^3} \int_{\hbar\omega = \mathcal{E}_{\rm cv}} \frac{{\rm d}S}{|\nabla_{\boldsymbol{k}} \mathcal{E}_{\rm cv}(\boldsymbol{k})|} , \tag{4.44}$$

where we have to note that the integral is carried out over the constant energy surface, $\hbar\omega = \mathcal{E}_{\rm cv}(\boldsymbol{k})$, because of the delta function. This general form of the density of states is derived as follows. Referring to Fig. 4.2, let us consider constant energy surfaces

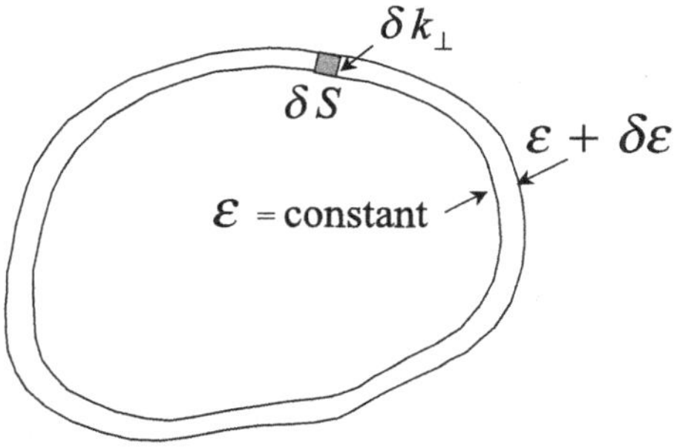

Fig. 4.2 Derivation of the general form of the joint density of states. Consider two constant energy surfaces $\mathcal{E}$ and $\mathcal{E}+\delta\mathcal{E}$ displaced by $\delta\mathcal{E}$. The small element of volume in $\boldsymbol{k}$ space is then defined by the product of the bottom area δS and the distance $\delta k_\perp$ perpendicular to the constant energy surface

displaced by a small amount of energy $\delta\mathcal{E}$ in $\boldsymbol{k}$ space, $\mathcal{E}$ and $\mathcal{E}+\delta\mathcal{E}$. The small element of the volume $\delta V(\boldsymbol{k})$ in $\boldsymbol{k}$ space is defined by the volume contained in the bottom area δS and its height $\delta k_\perp$ between the two constant energy surfaces. The states allowed in a volume element $\mathrm{d}^3\boldsymbol{k}$ in $\boldsymbol{k}$ space per unit volume is given by

$$\frac{2}{(2\pi)^3}\mathrm{d}^3\boldsymbol{k} = \frac{2}{(2\pi)^3}\delta V(\boldsymbol{k})\,. \tag{4.45}$$

The distance between the constant energy surfaces is given by $(\partial k_\perp/\partial\mathcal{E})\delta\mathcal{E} = (\partial\mathcal{E}/\partial k_\perp)^{-1}\delta\mathcal{E}$, where $\partial\mathcal{E}/\partial k_\perp$ is the gradient of $\mathcal{E}$ in the direction normal to the energy surfaces and we find

$$\frac{\partial\mathcal{E}}{\partial k_\perp} = |\nabla_{\boldsymbol{k}}\mathcal{E}| = \sqrt{\left(\frac{\partial\mathcal{E}}{\partial k_x}\right)^2 + \left(\frac{\partial\mathcal{E}}{\partial k_y}\right)^2 + \left(\frac{\partial\mathcal{E}}{\partial k_z}\right)^2}\,, \tag{4.46}$$

or

$$\delta k_\perp = \frac{\delta\mathcal{E}}{|\nabla_{\boldsymbol{k}}\mathcal{E}|}\,. \tag{4.47}$$

From the definition of the volume element $\delta V(\boldsymbol{k}) = \delta S\cdot\delta k_\perp$, the density of states in energy space between $\mathcal{E}$ and $\mathcal{E}+\delta\mathcal{E}$ is given by

$$\rho(\mathcal{E})\mathrm{d}\mathcal{E} = \frac{2}{(2\pi)^3}\int_S \frac{\mathrm{d}S}{|\nabla_{\boldsymbol{k}}\mathcal{E}|}\mathrm{d}\mathcal{E}\,. \tag{4.48}$$

It is evident from the above derivation that the joint density of states is obtained by replacing $\mathcal{E}(\boldsymbol{k})$ by $\mathcal{E}_{\mathrm{cv}}(\boldsymbol{k})$ in the above equation and thus that the derivation of (4.44) is straightforward.

The joint density of states $J_{\mathrm{cv}}(\hbar\omega)$ given by (4.44) diverges when $\nabla_{\boldsymbol{k}}\mathcal{E}_{\mathrm{cv}}(\boldsymbol{k}) = 0$ is satisfied. This leads to a maximum probability for the optical transition. We may expect such behavior at various points in the Brillouin zone. Such a point in the Brillouin zone is called a **critical point** or **singularity** of the joint density of states.

The behavior may be expected to occur under the following two conditions:

$$\nabla_{\boldsymbol{k}} \mathcal{E}_{\mathrm{c}}(\boldsymbol{k}) = \nabla_{\boldsymbol{k}} \mathcal{E}_{\mathrm{v}}(\boldsymbol{k}) = 0, \tag{4.49}$$

$$\nabla_{\boldsymbol{k}} \mathcal{E}_{\mathrm{c}}(\boldsymbol{k}) = \nabla_{\boldsymbol{k}} \mathcal{E}_{\mathrm{v}}(\boldsymbol{k}) \neq 0\,. \tag{4.50}$$

These equations reflect the conditions that the slopes of the two bands are parallel. The former condition means that the slopes are horizontal and will be satisfied at points of Brillouin zone with high symmetry. For example, this condition is satisfied at the Γ point. The second condition is satisfied at various points of Brillouin zone. The critical points behave differently depending on the type of critical point. The properties of the critical points were first investigated by van Hove [1] and thus are called **van Hove singularities**.

In order to discuss the singularities in more detail we expand $\mathcal{E}_{\mathrm{cv}}(\boldsymbol{k})$ at the point $\nabla_{\boldsymbol{k}}(\boldsymbol{k}) = 0$ ($\boldsymbol{k} = \boldsymbol{k}_0$, $\mathcal{E}_{\mathrm{cv}} = \mathcal{E}_{\mathrm{G}}$) in a Taylor series and keep terms up to the second order:

$$\mathcal{E}_{\mathrm{cv}} = \mathcal{E}_{\mathrm{G}} + \sum_{i=1}^{3} \frac{\hbar^2}{2\mu_i}(k_i - k_{i0})^2\,, \tag{4.51}$$

where the first-order term of $\boldsymbol{k}$ disappears due to the singularity condition $\nabla_{\boldsymbol{k}} \mathcal{E}_{\mathrm{cv}} = 0$. The constant μ_i has the dimension of mass and is given by the following equation with the effective mass m_{e}^* of the conduction band and m_{h}^* of the valence band.

$$\frac{1}{\mu_i} = \frac{1}{m_{\mathrm{e},i}^*} + \frac{1}{m_{\mathrm{h},i}^*}\,. \tag{4.52}$$

The mass μ_i is called the reduced mass of the electron and hole. Using this result, the joint density of states J_{cv} at the critical point is then calculated in the following way. Defining the new variable

$$s_i = \frac{\hbar(k_i - k_{i0})}{\sqrt{2|\mu_i|}}\,, \tag{4.53}$$

Equation (4.51) is rewritten as

$$\mathcal{E}_{\mathrm{cv}}(\boldsymbol{s}) = \mathcal{E}_{\mathrm{G}} + \sum_{i=1}^{3} \alpha_i s_i^2\,, \tag{4.54}$$

where $\alpha_i = \pm 1$ is the sign of μ_i, and $\alpha_i = +1$ for $\mu_i > 0$ and $\alpha_i = -1$ for $\mu_i < 0$. Using the definition of the variable we may calculate the joint density of states given by (4.44) (for the M_0 critical point: $\alpha_i > 0$ for all α_i) as

Table 4.1 Joint density of states $J_{\text{cv}}(\hbar\omega)$ for 3-dimensional critical points

Types of critical points				$J_{\text{cv}}(\hbar\omega)$	
Types	μ_1	μ_2	μ_3	$\hbar\omega \le \mathcal{E}_{\text{G}}$	$\hbar\omega \ge \mathcal{E}_{\text{G}}$
M_0	+	+	+	0	$C_1(\hbar\omega - \mathcal{E}_{\text{G}})^{1/2}$
M_1	−	+	+	$C_2 - C_1(\mathcal{E}_{\text{G}} - \hbar\omega)^{1/2}$	C_2
M_2	−	−	+	C_2	$C_2 - C_1(\hbar\omega - \mathcal{E}_{\text{G}})^{1/2}$
M_3	−	−	−	$C_1(\mathcal{E}_{\text{G}} - \hbar\omega)^{1/2}$	0

$C_1 = \dfrac{4\pi}{(2\pi)^3}\left(\dfrac{8\mu_1\mu_2\mu_3}{\hbar^6}\right)^{1/2}$

$$J_{\text{cv}}(\omega) = \frac{2}{(2\pi)^3}\left(\frac{8|\mu_1\mu_2\mu_3|}{\hbar^6}\right)^{1/2}\int_{\mathcal{E}_{\text{cv}}=\hbar\omega}\frac{\mathrm{d}S}{|\nabla_s\mathcal{E}_{\text{cv}}(\boldsymbol{s})|}$$
$$= \frac{2}{(2\pi)^3}\left(\frac{8|\mu_1\mu_2\mu_3|}{\hbar^6}\right)^{1/2}\int_{\mathcal{E}_{\text{cv}}=\hbar\omega}\frac{\mathrm{d}S}{2s}, \tag{4.55}$$

where $s = (s_1^2 + s_2^2 + s_3^2)^{1/2}$. It is straightforward to obtain the density of states for the M_0 critical point. In this case we have $\mathcal{E}_{\text{cv}} - \mathcal{E}_{\text{G}} = s_1^2 + s_2^2 + s_3^2 = s^2$ and then we obtain $\int \mathrm{d}S = 4\pi s^2$ for $\hbar\omega > \mathcal{E}_{\text{G}}$, which leads to the following result for the joint density of states $J_{\text{cv}}(\omega)$:

$$J_{\text{cv}}(\hbar\omega) = \begin{cases} 0 & ; \hbar\omega \le \mathcal{E}_{\text{G}} \\ \dfrac{4\pi}{(2\pi)^3}\left(\dfrac{8\mu_1\mu_2\mu_3}{\hbar^6}\right)^{1/2}\sqrt{\hbar\omega - \mathcal{E}_{\text{G}}} & ; \hbar\omega \ge \mathcal{E}_{\text{G}}. \end{cases} \tag{4.56}$$

It is evident from (4.51) or (4.54) that the singularities or critical points are classified in four categories depending on the combination of the signs of the reduced masses, or in other words we define the critical point M_j ($j = 0, 1, 2$ and 3) by j, the number of negative μ_i. The critical point M_0 has 3 positive reduced masses ($j = 0$: zero negative reduced mass): $\mu_1 > 0$, $\mu_2 > 0$ and $\mu_3 > 0$. The critical points M_1, M_2 and M_3 are for 1, 2, and 3 negative reduced masses, any of μ_1, μ_2 and μ_3. The joint density of states $J_{\text{cv}}(\omega)$ for the critical points M_1, M_2 and M_3 have been calculated and the results are summarized in Table 4.1 (see for example [2, 3]).

Inserting (4.56) into (4.39) the imaginary part of the dielectric constant for the M_0 critical point is given by

$$\kappa_2(\omega) = \frac{e^2}{2\pi\epsilon_0 m^2\omega^2}|\boldsymbol{e}\cdot\boldsymbol{p}_{\text{cv}}|^2\left(\frac{8\mu_1\mu_2\mu_3}{\hbar^6}\right)^{1/2}\sqrt{\hbar\omega - \mathcal{E}_{\text{G}}}, \tag{4.57}$$

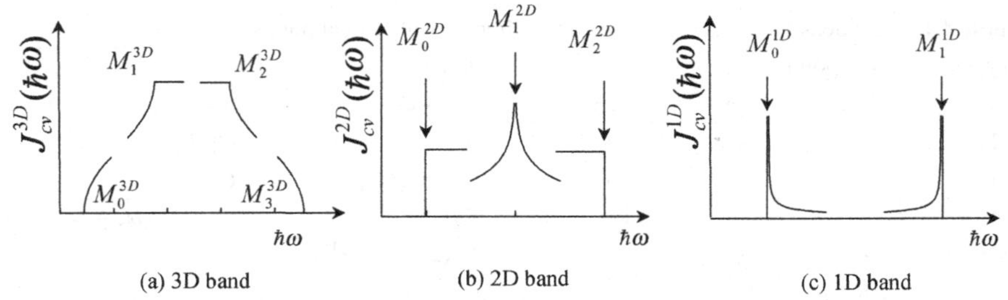

Fig. 4.3 Joint density of states $J_{cv}(\hbar\omega)$ for **a** 3D, **b** 2D and **c** 1D bands

where $\hbar\omega > \mathcal{E}_G$. The absorption coefficient is obtained by inserting this equation for $\kappa_2(\omega)$ into (4.21):

$$\alpha(\omega) = \frac{e^2|\boldsymbol{e}\cdot\boldsymbol{p}_{\rm cv}|^2}{2\pi\epsilon_0 m^2 c n_0 \omega}\left(\frac{8\mu_1\mu_2\mu_3}{\hbar^6}\right)^{1/2}\sqrt{\hbar\omega-\mathcal{E}_{\rm G}} \quad \text{for} \quad \hbar\omega > \mathcal{E}_{\rm G}\,. \tag{4.58}$$

It is also evident from (4.51) that $\mathcal{E}_{\rm cv}$ becomes a maximum at $\boldsymbol{k}_0$ for the M_0 critical point (all of μ_i are positive) and that $\mathcal{E}_{\rm cv}$ becomes a minimum at $\boldsymbol{k}_0$ for the M_3 critical point (all of μ_i are negative). On the other hand, the M_1 (M_2) critical point exhibits a saddle point at $\boldsymbol{k}_0$, where $\mathcal{E}_{\rm cv}$ shows a maximum (minimum) at $\boldsymbol{k}_0$ in one direction and a minimum (maximum) in other directions. When we consider a pair of conduction and valence bands, the lowest energy critical point is M_0 and the highest energy critical point is M_3.

It should be noted that Fig. 4.3a is calculated by using the relations given in Table 4.1 for 3D case. We will present an example of the calculations of DOS (Density of states) for the 3D energy bands obtained by a simple treatment of the tight-binding approximation in the case of the simple cubic lattice. The first Brillouin zone is determined by the reciprocal lattice vectors $\boldsymbol{G}_x = \boldsymbol{G}_y = \boldsymbol{G}_z = \pm 2\pi/a$ and thus the zone edge wave vectors are $k_x = k_y = k_z = \pm\pi/a$. The energy bands are then given by

$$\mathcal{E}_{\rm cv}(\boldsymbol{k}) = (\mathcal{E}_{\rm G} + 3\gamma) - \gamma\left[\cos(ak_x) + \cos(ak_y) + \cos(ak_z)\right] \tag{4.59}$$

The energy band $\mathcal{E}(k_x)$ is shown in Fig. 4.4a. The density of states (DOS) of the energy band is obtained by discretizing the first Brillouin zone as described in Chap. 1, Sect. 1.8, and the results are shown in Fig. 4.4b, where we see M_0 M_1 M_2 and M_3 critical points.

The joint densities of states $J_{\rm cv}$ for 2-dimensional (2D) and 1-dimensional (1D) bands are summarized in Table 4.2 and in Fig. 4.3b–c. It is very important to point out that in Fig. 4.3b–c the joint densities of states $J_{\rm cv}$ at the saddle point for the 2D critical point and at the 1D critical points diverge. The behavior of the 1D critical point has already been shown in Chap. 2 to describe the density of states in a high magnetic field, where an electron is quantized in the perpendicular direction to the

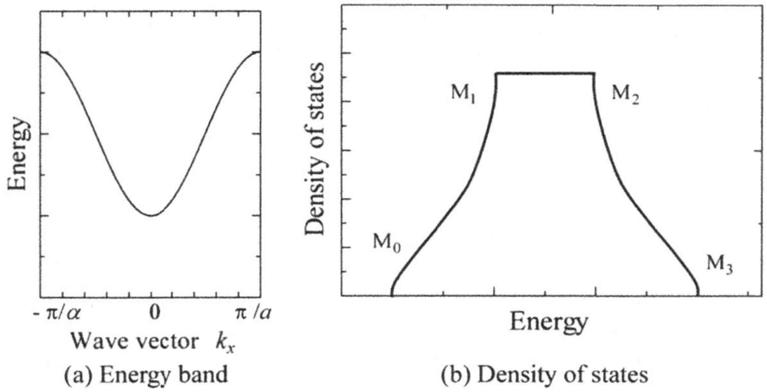

Fig. 4.4 Density of states calculated from (4.59) using numerical calculations. **a** Energy bad structure obtained by the tight-binding approximation for a simple cubic lattice. **b** Density of states (DOS) obtained by numerical calculations from (4.59), where all the four critical points M_0, M_1, M_2, and M_3 are clearly seen

Table 4.2 Joint density of states $J_{cv}(\hbar\omega)$ for 1-dimensional and 2-dimensional critical points

Types of critical points		$J_{cv}(\hbar\omega)$	
Dimension	Types	$\hbar\omega \le \mathcal{E}_G$	$\hbar\omega \ge \mathcal{E}_G$
1D	M_0	0	$A(\hbar\omega - \mathcal{E}_G)^{-1/2}$
1D	M_1	$A(\mathcal{E}_G - \hbar\omega)^{-1/2}$	0
2D	M_0	0	B_1
2D	M_1	$(B_1/\pi)(B_2 - \log\lvert\mathcal{E}_G - \hbar\omega\rvert)$	$(B_1/\pi)(B_2 - \log\lvert\mathcal{E}_G - \hbar\omega\rvert)$
2D	M_2	B_1	0

$A = \frac{2}{4\pi} \cdot \left(\frac{2|\mu|}{\hbar^2}\right)^{1/2}, \quad B_1 = \frac{2}{4\pi} \cdot \left(\frac{4|\mu_1\mu_2|}{\hbar^4}\right)^{1/2},$

$B_2 = \log|2B_3 - (\mathcal{E}_G - \hbar\omega) + 2\sqrt{B_3^2 - (\mathcal{E}_G - \hbar\omega)B_3}|$

magnetic field and the electron can move along the magnetic field, and thus the electron behaves like a 1D carrier.

4.4 Indirect Transition

In Sects. 4.1 and 4.2 we considered the process where an electron in the valence band absorbs one photon and makes a transition to the conduction band vertically in $\boldsymbol{k}$ space, i.e. a direct transition. This process plays the most important role in direct gap semiconductors such as GaAs, InSb and so on. On the other hand, we have shown in Chaps. 1 and 2 that the conduction band minima in Ge and Si are located at the L point and Δ point, respectively, whereas the top of the valence band at the Γ point. Therefore, the fundamental absorption edge (lowest optical transition) is not direct, and thus direct transitions of electrons from the top of the valence band to the lowest conduction band minima is not allowed. Experimental results in Ge and Si reveal a weak transition for the photon energy corresponding to the indirect band gap between the top of the valence band at $\boldsymbol{k} = 0$ and the conduction minima at $\boldsymbol{k} \neq 0$. This process is interpreted as the indirect transition in which an electron in the valence band absorbs a photon and then absorbs or emits a phonon to make the transition to the conduction band minima. This process is caused by a higher-order interaction or second-order perturbation in quantum mechanics, as described below in detail. The higher-order perturbation produces a weaker transition probability compared with the direct transition, and thus the weak absorption is explained. In addition we have to note that the transition is validated through a virtual state for the transition from the initial to the final states, and for this reason the transition is called an indirect transition.

Let us define the Hamiltonian H_{e} for electrons, H_{l} for lattice vibrations, H_{el} for the electron-phonon interaction and H_{er} for the electron-radiation (photon) interaction. Then the total Hamiltonian is written as

$$H = H_{\mathrm{e}} + H_{\mathrm{l}} + H_{\mathrm{el}} + H_{\mathrm{er}} \,. \tag{4.60}$$

We will discuss the Hamiltonian of the lattice vibrations and the electron-phonon interaction in Chap. 6 and will not go into detail here. The Hamiltonian of electrons and phonons is written as $H_0 = H_{\mathrm{e}} + H_{\mathrm{l}}$ and the eigenstates are expressed as

$$|j\rangle = \begin{cases} |\mathrm{c}\boldsymbol{k}, n_{\boldsymbol{q}}^{\alpha}\rangle & \text{(for an electron in the conduction band)} \\ |\mathrm{v}\boldsymbol{k}, n_{\boldsymbol{q}}^{\alpha}\rangle & \text{(for an electron in the valence band)} \end{cases} \,, \tag{4.61}$$

where $\boldsymbol{k}$ is the wave vector of the electron and $n_{\boldsymbol{q}}^{\alpha}$ is the phonon quantum number of mode α and wave vector $\boldsymbol{q}$. Expressing the perturbation Hamiltonian as

$$H' = H_{\mathrm{el}} + H_{\mathrm{er}} \tag{4.62}$$

and keeping the perturbation up to the second order, the transition probability from an initial state $|i\rangle$ to a final state $|f\rangle$ through a virtual state $|m\rangle$ is then given by

$$w_{if} = \frac{2\pi}{\hbar}\left|\langle f|H'|i\rangle + \sum_m \frac{\langle f|H'|m\rangle\langle m|H'|i\rangle}{\mathcal{E}_i - \mathcal{E}_m}\right|^2 \delta(\mathcal{E}_i - \mathcal{E}_f) \,, \tag{4.63}$$

where it is evident that the relation $\langle f|H'|i\rangle = \langle f|H_{\mathrm{er}}|i\rangle + \langle f|H_{\mathrm{el}}|i\rangle = 0$ holds because of the following reasons. The matrix element $\langle f|H_{\mathrm{er}}|i\rangle$ is the same as the element for the direct transition stated in the previous section and the transition is allowed between the same $\boldsymbol{k}$ vectors in $\boldsymbol{k}$ space. In the case of an indirect transition, however, the wave vectors $\boldsymbol{k}$ are different between the initial state $|i\rangle$ at the top of the valence band and the final state $|f\rangle$ at the bottom of the conduction band, and thus we find that $\langle f|H_{\rm er}|i\rangle = 0$. On the other hand, the matrix element $\langle f|H_{\rm el}|i\rangle$ ensures momentum conservation (wave vector conservation), $\delta(\boldsymbol{k}_i \pm \boldsymbol{q} - \boldsymbol{k}_f)$, but energy is not conserved because of the small value of the phonon energy compared with the band gap. Therefore, we find that $\langle f|H_{\rm el}|i\rangle = 0$. From these considerations only the second term in (4.63) will contribute to the indirect transition and so finally we get

$$w_{if} = \frac{2\pi}{\hbar}\left|\sum_m \frac{\langle f|H'|m\rangle\langle m|H'|i\rangle}{\mathcal{E}_i - \mathcal{E}_m}\right|^2 \delta(\mathcal{E}_i - \mathcal{E}_f)\,. \tag{4.64}$$

When we insert (4.62) into (4.64), such terms as $\langle f|H_{\rm el}|m\rangle\langle m|H_{\rm el}|i\rangle$, $\langle f|H_{\rm er}|m\rangle\langle m|H_{\rm er}|i\rangle$ and so on appear. However, these terms will not contribute to the indirect transition for the reason stated above. Finally, we find that the two processes shown in Fig. 4.5 will remain. The first process is: (1) an electron at the top of the valence band A interacts with radiation to absorb a photon and makes a transition to a virtual state D in the conduction band, followed by a transition by phonon absorption or emission to the final state C. The second process is: (2) an electron makes a transition from the top of the valence band A to a virtual state in the valence band B by absorbing or emitting a phonon and then absorbs a photon to end up at the final state C. The second process can be understood in a different way: an electron in the valence band B absorbs a photon, leaving a virtual state of a hole there and making a transition to the final state C, and an electron at the top of the valence band A is then transferred to this virtual state by absorbing or emitting a phonon. These processes are expressed by the following equation:

$$\begin{aligned} w_{if} = \frac{2\pi}{\hbar}\Bigg| &\frac{\langle \mathrm{C}\boldsymbol{k}_f, n_{\boldsymbol{q}}^{\alpha} \pm 1|H_{\rm el}|\mathrm{D}\boldsymbol{k}_i, n_{\boldsymbol{q}}^{\alpha}\rangle\langle \mathrm{D}\boldsymbol{k}_i, n_{\boldsymbol{q}}^{\alpha}|H_{\rm er}|A\boldsymbol{k}_i, n_{\boldsymbol{q}}^{\alpha}\rangle}{\mathcal{E}_i - \mathcal{E}_D} \\ &+ \frac{\langle \mathrm{C}\boldsymbol{k}_f, n_{\boldsymbol{q}}^{\alpha} \pm 1|H_{\rm er}|\mathrm{B}\boldsymbol{k}_f, n_{\boldsymbol{q}}^{\alpha} \pm 1\rangle\langle \mathrm{B}\boldsymbol{k}_f, n_{\boldsymbol{q}}^{\alpha} \pm 1|H_{\rm el}|\mathrm{A}\boldsymbol{k}_i, n_{\boldsymbol{q}}^{\alpha}\rangle}{\mathcal{E}_i - \mathcal{E}_{\rm B}}\Bigg|^2 \\ &\times \delta(\mathcal{E}_i - \mathcal{E}_f)\,, \end{aligned}$$

where the upper sign (+) of $\pm$ represents phonon emission and the lower sign ($-$) of $\pm$ corresponds to phonon absorption. When we neglect the phonon energy term $\hbar\omega_{\boldsymbol{q}}$ because of its smallness compared to the photon energy or the band gap, we find following relations:

$$\begin{aligned} &\boldsymbol{k}_f = \boldsymbol{k}_i \mp \boldsymbol{q}\,, \\ &\mathcal{E}_i - \mathcal{E}_f = \mathcal{E}_{\rm v}(\boldsymbol{k}_i) + \hbar\omega \mp \hbar\omega_{\boldsymbol{q}}^{\alpha} - \mathcal{E}_{\rm c}(\boldsymbol{k}_f)\,, \\ &\mathcal{E}_i - \mathcal{E}_{\rm D} = \mathcal{E}_{\rm v}(\boldsymbol{k}_i) + \hbar\omega - \mathcal{E}_{\rm D}(\boldsymbol{k}_i) \cong \mathcal{E}_{\rm c}(\boldsymbol{k}_f) - \mathcal{E}_{\rm D}(\boldsymbol{k}_i) \cong \mathcal{E}_{\rm c0} - \mathcal{E}_{\rm c1}\,, \\ &\mathcal{E}_i - \mathcal{E}_{\rm B} = \mathcal{E}_{\rm v}(\boldsymbol{k}_i) \mp \hbar\omega_{\boldsymbol{q}}^{\alpha} - \mathcal{E}_{\rm B}(\boldsymbol{k}_f) \cong \mathcal{E}_{\rm v}(\boldsymbol{k}_i) - \mathcal{E}_{\rm B}(\boldsymbol{k}_f) \cong \mathcal{E}_{\rm v0} - \mathcal{E}_{\rm v1}\,. \end{aligned}$$

Assuming that the matrix element is independent of the wave vector $\boldsymbol{k}$, we may approximate

$$w_{if} = \frac{2\pi}{\hbar}\sum_{m,\alpha,\pm} \left|M_{\rm cv}^{m,\alpha,\pm}\right|^2 \delta(\mathcal{E}_i - \mathcal{E}_f) \tag{4.65}$$

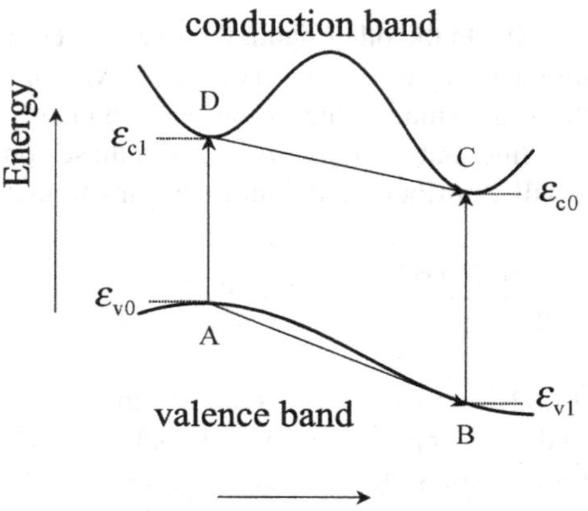

Fig. 4.5 Processes for the indirect transition. (1) An electron at the top of the valence band A interacts with radiation to absorb a photon and makes a transition to a virtual state D in the conduction band, followed by a transition by phonon absorption or emission to the final state C. (2) An electron makes a transition from the top of the valence band A to a virtual state in the valence band B by absorbing or emitting a phonon and then absorbs a photon to end up at the final state C

for the transition probability and

$$\kappa_2(\omega) = \frac{\pi e^2}{\epsilon_0 m^2 \omega^2} \sum_{m,\alpha,\pm} \left|M_{\mathrm{cv}}^{m,\alpha,\pm}\right|^2 \times \sum_{\boldsymbol{k},\boldsymbol{k}'} \delta\left[\mathcal{E}_{\mathrm{c}}(\boldsymbol{k}') - \mathcal{E}_{\mathrm{v}}(\boldsymbol{k}) - \hbar\omega \pm \hbar\omega_{\boldsymbol{q}}^{\alpha}\right] \tag{4.66}$$

for the imaginary part of the dielectric constant. The first summation is carried out for the virtual state ($|m\rangle$), the phonon mode, and its emission (+) and absorption (−). The probability of phonon emission and absorption is proportional to $n_{\boldsymbol{q}}^{\alpha}+1$ and $n_{\boldsymbol{q}}^{\alpha}$, respectively, and the average excited phonon number is given by Bose–Einstein statistics as

$$n_{\boldsymbol{q}}^{\alpha} = \frac{1}{\exp(\hbar\omega_q^{\alpha}/k_{\mathrm{B}}T) - 1} . \tag{4.67}$$

Using these results (4.65) may be written as

$$w_{if} = \frac{2\pi}{\hbar} \sum_{m,\alpha,\pm} \left[\left|A_{\mathrm{cv}}^{m,\pm}\right|^2 \left(n_{\boldsymbol{q}}^{\alpha} + \frac{1}{2} \pm \frac{1}{2}\right)\right] \times \sum_{\boldsymbol{k},\boldsymbol{k}'} \delta\left[\mathcal{E}_{\mathrm{c}}(\boldsymbol{k}') - \mathcal{E}_{\mathrm{v}}(\boldsymbol{k}) - \hbar\omega \pm \hbar\omega_{\boldsymbol{q}}^{\alpha}\right] . \tag{4.68}$$

The second summation with respect to $\boldsymbol{k}$ and $\boldsymbol{k}'$ represents the density of states for the indirect transition, which is calculated as follows.

$$
\begin{aligned}
J_{\mathrm{cv}}^{\mathrm{ind}} &= \sum_{\boldsymbol{k},\boldsymbol{k}'} \delta\left[\mathcal{E}_{\mathrm{c}}(\boldsymbol{k}') - \mathcal{E}_{\mathrm{v}}(\boldsymbol{k}) - \hbar\omega \pm \hbar\omega_{\boldsymbol{q}}^{\alpha}\right] \\
&\sum_{\boldsymbol{k},\boldsymbol{k}'} \delta\left[\mathcal{E}_{\mathrm{c0}} - \mathcal{E}_{\mathrm{v0}} + \frac{\hbar^2}{2}\left(\frac{k_x^2}{m_{\mathrm{h}x}} + \frac{k_y^2}{m_{\mathrm{h}y}} + \frac{k_z^2}{m_{\mathrm{h}z}}\right.\right. \\
&\left.\left. + \frac{k'^2_x}{m_{\mathrm{e}x}} + \frac{k'^2_y}{m_{\mathrm{e}y}} + \frac{k'^2_z}{m_{\mathrm{e}z}}\right) - \hbar\omega \pm \hbar\omega_{\boldsymbol{q}}^{\alpha}\right], \qquad (4.69)
\end{aligned}
$$

where the conduction band near the bottom is approximated by $\mathcal{E}_{\mathrm{c}}(\boldsymbol{k}') = \hbar^2 k'^2/2m_{\mathrm{e}} + \mathcal{E}_{\mathrm{c0}}$ with the electron effective mass m_{e} and the valence band near the top by $\mathcal{E}_{\mathrm{v}}(\boldsymbol{k}) = -\hbar^2 k^2/2m_{\mathrm{h}} + \mathcal{E}_{\mathrm{v0}}$ with the hole effective mass m_{h}. The summation $\sum_{\boldsymbol{k}}$ is replaced by $2/(2\pi)^3 \int \mathrm{d}^3\boldsymbol{k}$ and transformations such as $x = \hbar k_x/\sqrt{2m_{\mathrm{h}x}}$, $x' = \hbar k'_x/\sqrt{2m_{\mathrm{e}x}}$ etc. are used. Then the density of states is given by

$$
\begin{aligned}
J_{\mathrm{cv}}^{\mathrm{ind}} &= \frac{2K}{(2\pi)^6} \int \mathrm{d}x\mathrm{d}y\mathrm{d}z\mathrm{d}x'\mathrm{d}y'\mathrm{d}z' \\
&\times \delta\left(x^2 + y^2 + z^2 + x'^2 + y'^2 + z'^2 + \mathcal{E}_{\mathrm{G}} - \hbar\omega \pm \hbar\omega_{\boldsymbol{q}}^{\alpha}\right), \qquad (4.70)
\end{aligned}
$$

where we have assumed that the spin of the electron is not changed in the transition (no spin-flip transition) and that the spin degeneracy is used for one of the bands. In addition we have used K and $\mathcal{E}_{\mathrm{G}}$ given by

$$
K = \sqrt{\frac{2^6 m_{\mathrm{h}x} m_{\mathrm{h}y} m_{\mathrm{h}z} m_{\mathrm{e}x} m_{\mathrm{e}y} m_{\mathrm{e}z}}{\hbar^{12}}},
$$

$$
\mathcal{E}_{\mathrm{G}} = \mathcal{E}_{\mathrm{c0}} - \mathcal{E}_{\mathrm{v0}}.
$$

In order to carry out the integral of $J_{\mathrm{cv}}^{\mathrm{ind}}$ we use polar coordinates $(x, y, z) = (r, \theta, \phi)$ with $r^2 = s$ and $r'^2 = s'$ ($\int \sin\theta \mathrm{d}\theta \mathrm{d}\phi = 4\pi$), which leads to

$$
J_{\mathrm{cv}}^{\mathrm{ind}} = \frac{2K}{(2\pi)^6} \int (4\pi)^2 \frac{\mathrm{d}s\mathrm{d}s'}{4} \sqrt{ss'}\,\delta(s + s' + \mathcal{E}_{\mathrm{G}} - \hbar\omega \pm \hbar\omega_{\boldsymbol{q}}^{\alpha}).
$$

The integral of the above equation is easily carried out by using the δ function, and we obtain

$$
\begin{aligned}
J_{\mathrm{cv}}^{\mathrm{ind}} &= \frac{4\pi^2}{(2\pi)^6} 2K \int_0^{\hbar\omega \mp \hbar\omega_{\boldsymbol{q}}^{\alpha} - \mathcal{E}_{\mathrm{G}}} \sqrt{s}\sqrt{\hbar\omega \mp \hbar\omega_{\boldsymbol{q}}^{\alpha} - \mathcal{E}_{\mathrm{G}} - s}\ \mathrm{d}s \\
&= \frac{2K}{(2\pi)^4} \cdot \frac{\pi}{8} (\hbar\omega \mp \hbar\omega_{\boldsymbol{q}}^{\alpha} - \mathcal{E}_{\mathrm{G}})^2. \qquad (4.71)
\end{aligned}
$$

Finally, we obtain the dielectric function for the indirect transition given by

$$\kappa_2(\omega) = \frac{\pi e^2}{\epsilon_0 m^2 \omega^2} \cdot \frac{K}{(4\pi)^3} \sum_{m,\alpha,\pm} \left|M_{\rm cv}^{m,\alpha,\pm}\right|^2 (\hbar\omega \mp \hbar\omega_{\boldsymbol{q}}^{\alpha} - \mathcal{E}_{\rm G})^2$$
$$= \frac{\pi e^2}{\epsilon_0 m^2 \omega^2} \cdot \frac{K}{(4\pi)^3} \sum_{m,\alpha} \left[\left|A_{\rm cv}^{m,\alpha,+}\right|^2 \cdot \frac{(\hbar\omega - \hbar\omega_{\boldsymbol{q}}^{\alpha} - \mathcal{E}_{\rm G})^2}{1 - \exp(-\hbar\omega_{\boldsymbol{q}}^{\alpha}/k_{\rm B}T)} + \left|A_{\rm cv}^{m,\alpha,-}\right|^2 \cdot \frac{(\hbar\omega + \hbar\omega_{\boldsymbol{q}}^{\alpha} - \mathcal{E}_{\rm G})^2}{\exp(\hbar\omega_{\boldsymbol{q}}^{\alpha}/k_{\rm B}T) - 1} \right], \tag{4.72}$$

where the first and second terms on the right-hand side are associated with the transition followed by phonon emission and by phonon absorption, respectively.

The virtual state of the indirect transition is determined by the phonon mode and its deformation potential and also by the selection rule for the electron-radiation field interaction. As the simplest case we take into account a virtual state and a phonon mode and calculate the temperature dependence of the absorption coefficient. The denominator of the second term becomes large for the case $k_{\rm B}T \ll \hbar\omega_{\boldsymbol{q}}^{\alpha}$ and the transition followed by phonon absorption disappears at low temperatures. On the other hand, the denominator of the first term becomes 1 at low temperatures. Therefore, the indirect transition will be governed by phonon emission at low temperatures. The feature is shown in Fig. 4.6, where $\sqrt{\kappa_2}$ is plotted as a function of $\hbar\omega - \mathcal{E}_{\rm G}$ for different temperatures, by taking into account the fact that the square root of the imaginary part of the dielectric function and thus of the absorption coefficient is proportional to $\hbar\omega \mp \hbar\omega_{\boldsymbol{q}} - \mathcal{E}_{\rm G}$ ($\sqrt{\kappa_2} \propto \sqrt{\alpha} \propto (\hbar\omega \mp \hbar\omega_{\boldsymbol{q}} - \mathcal{E}_{\rm G})$ (α: absorption coefficient). At lower temperatures the process of phonon absorption decreases and the absorption coefficient becomes very weak. We find that the phonon energy involved in the process is given by the half-width of the lower straight line and that the band gap $\mathcal{E}_{\rm G}$ lies in the middle of the line. The experimental results for Si are shown in Fig. 4.7, where the square root of the absorption coefficient $\sqrt{\alpha}$ is plotted as a function of photon energy and find the feature of the indirect transition shown in Fig. 4.6 [4]. The turning point of the curve shifts to lower photon energy at higher temperatures, which is explained in terms of the temperature dependence of the band gap (the band gap decreases with increasing lattice temperature). The experimental

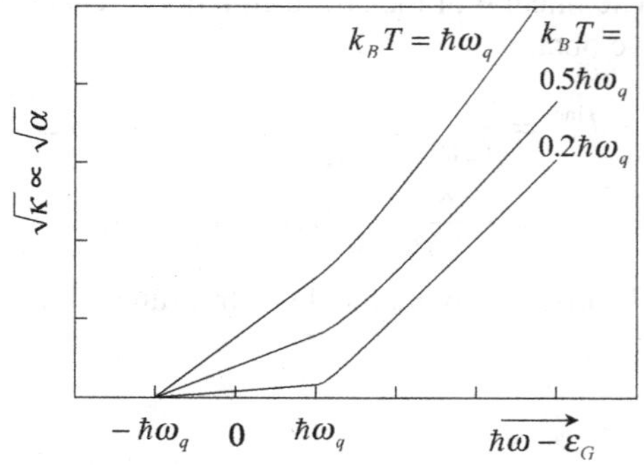

Fig. 4.6 Absorption coefficient at the indirect transition edge, where the square root of the absorption coefficient $\sqrt{\alpha} \propto \sqrt{\kappa}$ is plotted as a function of photon energy minus the band gap $\hbar\omega - \mathcal{E}_{\rm G}$ for different temperatures

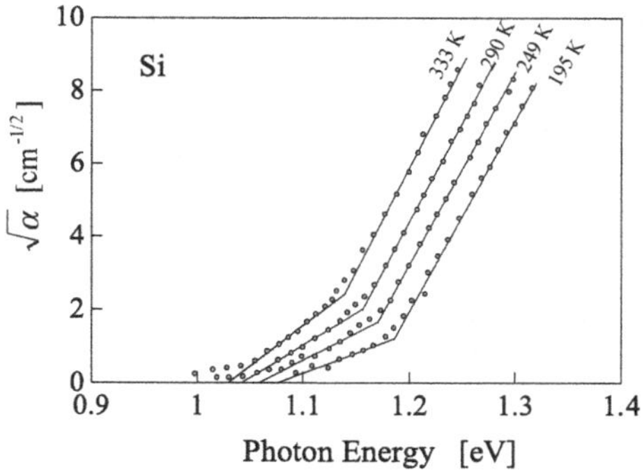

Fig. 4.7 Square root of the absorption coefficient $\sqrt{\alpha}$ plotted as a function of photon energy $\hbar\omega$ for Si, where the transitions due to phonon absorption and emission are well resolved (from [4])

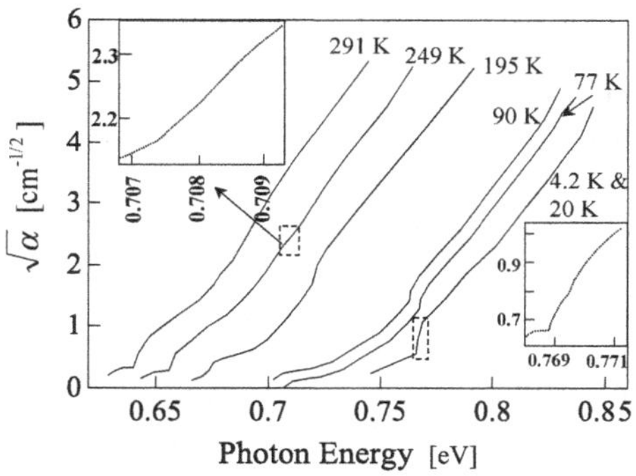

Fig. 4.8 The square root of the absorption coefficient $\sqrt{\alpha}$ of Ge at the indirect transition edge is plotted as a function of photon energy $\hbar\omega$ (from [5])

results for Ge are shown in Fig. 4.8 [5], where turning points are observed in the spectrum for $T = 249$K at about 0.65 and 0.71 eV. These turning points correspond to the transition due to the absorption and emission of one LO phonon, and the transition due to phonon emission disappears at low temperatures. A weak absorption is observed at $T = 4.2$ K below the photon energy around 0.77 eV due to LA phonon emission process, which is believed to be due to a forbidden transition followed by TA phonon emission. It is most important to point out here that the square root of the absorption coefficient is not a straight line but exhibits a hump. This is caused by exciton effects as discussed in Sect. 4.5.

4.5 Exciton

4.5.1 Direct Exciton

In Sect. 4.2 we discussed optical absorption spectra due to direct transitions, where an electron in the valence band (wave vector $\boldsymbol{k}$) is excited by photon absorption into the conduction band (the wave vector at the band is $\boldsymbol{k}' = \boldsymbol{k}_e$) and leaves a hole in the valence band (the wave vector of the hole is $\boldsymbol{k}_{\mathrm{h}} = -\boldsymbol{k}$). The two carriers are subject to the Coulomb interaction, which is disregarded in Sect. 4.2 and also in Sect. 4.4. In this section we will be concerned with the Coulomb interaction between electrons and holes at the critical point M_0, where we assume the bound state is very weak and thus the effective mass approximation is valid for the treatment. The bound state of the electron-hole pair is called an **exciton**. The exciton based on the approximation is called a **Wannier exciton**.

When the electron-hole interaction is neglected, the wave function of the one electron and one hole system is given by

$$\Psi_{ij}(\boldsymbol{r}_{\mathrm{e}}, \boldsymbol{r}_{\mathrm{h}}) = \psi_{i\boldsymbol{k}_{\mathrm{e}}}(\boldsymbol{r}_{\mathrm{e}})\psi_{j\boldsymbol{k}_{\mathrm{h}}}(\boldsymbol{r}_{\mathrm{h}}),$$

where i and j are the band indices, and $\boldsymbol{r}_{\mathrm{e}}$, $\boldsymbol{r}_{\mathrm{h}}$, $\boldsymbol{k}_{\mathrm{e}}$ and $\boldsymbol{k}_{\mathrm{h}}$ are the coordinates and wave vectors of the electron and hole, respectively. In the absence of the Coulomb interaction the Hamiltonians of an electron in a conduction band, H_{e}, and an electron in a valence band, H_{v}, are written as[1]

$$H_{\mathrm{e}}\psi_{\mathrm{c}\boldsymbol{k}_{\mathrm{e}}}(\boldsymbol{r}_{\mathrm{e}}) = \mathcal{E}_{\mathrm{c}}(\boldsymbol{k}_{\mathrm{e}})\psi_{\mathrm{c}\boldsymbol{k}_{\mathrm{e}}}(\boldsymbol{r}_{\mathrm{e}}),$$

$$H_{\mathrm{v}}\psi_{\mathrm{v}\boldsymbol{k}_{\mathrm{v}}}(\boldsymbol{r}_{\mathrm{v}}) = \mathcal{E}_{\mathrm{v}}(\boldsymbol{k}_{\mathrm{v}})\psi_{\mathrm{v}\boldsymbol{k}_{\mathrm{v}}}(\boldsymbol{r}_{\mathrm{v}}).$$

In the following calculations we consider the Hamiltonian H_{h} for a hole in the valence band and use the Bloch function given by $\psi_{\mathrm{v},\boldsymbol{k}_{\mathrm{h}}}(\boldsymbol{r}_{\mathrm{h}}) = \psi_{\mathrm{v},\boldsymbol{k}_{\mathrm{v}}}(\boldsymbol{r}_{\mathrm{v}})$. As stated above, the crystal momentum vectors of an electron and a hole in the valence band have the same magnitude but different signs, and so we define the total wave vector of the electron-hole system by

$$\boldsymbol{K} = \boldsymbol{k}' + (-\boldsymbol{k}) = \boldsymbol{k}_{\mathrm{e}} + \boldsymbol{k}_{\mathrm{h}} \,. \tag{4.73}$$

In the presence of the electron-hole interaction (Coulomb interaction) $V(\boldsymbol{r}_{\mathrm{e}} - \boldsymbol{r}_{\mathrm{h}})$, the Hamiltonian for an exciton (electron-hole pair) is thus written as

$$H = H_{\mathrm{e}} + H_{\mathrm{h}} + V(\boldsymbol{r}_{\mathrm{e}} - \boldsymbol{r}_{\mathrm{h}}) \,, \tag{4.74}$$

and the wave function of the exciton may be expanded by the Bloch functions of the electron and hole:

$$\Psi^{n,\boldsymbol{K}}(\boldsymbol{r}_{\mathrm{e}}, \boldsymbol{r}_{\mathrm{h}}) = \sum_{\mathrm{c},\boldsymbol{k}_{\mathrm{e}},\mathrm{v},\boldsymbol{k}_{\mathrm{h}}} A_{\mathrm{cv}}^{n,\boldsymbol{K}}(\boldsymbol{k}_{\mathrm{e}}, \boldsymbol{k}_{\mathrm{h}})\psi_{\mathrm{c}\boldsymbol{k}_{\mathrm{e}}}(\boldsymbol{r}_{\mathrm{e}})\psi_{\mathrm{v}\boldsymbol{k}_{\mathrm{h}}}(\boldsymbol{r}_{\mathrm{h}}) \,. \tag{4.75}$$

[1]Here we are dealing with the Hamiltonian of an electron in a valence band H_{v} and we express the state by the Bloch function $\psi_{\mathrm{v},\boldsymbol{k}}(\boldsymbol{r}_{\mathrm{v}})$ and its energy eigenvalue by $\mathcal{E}_{\mathrm{v}}(\boldsymbol{k}_{\mathrm{v}})$. When a hole state is considered, we have to use the relations: $\boldsymbol{k}_{\mathrm{h}} = -\boldsymbol{k}_{\mathrm{v}}$, $\mathcal{E}_{\mathrm{h}}(\boldsymbol{k}_{\mathrm{h}}) = -\mathcal{E}_{\mathrm{v}}(-\boldsymbol{k}_{\mathrm{v}})$.

Using the wave function of (4.75) in the Hamiltonian of (4.74), multiplying the complex conjugate of the Bloch functions $\psi^*_{c'\boldsymbol{k}'_e}(\boldsymbol{r}_e)\psi^*_{v'\boldsymbol{k}'_h}(\boldsymbol{r}_h)$, and integrating over $\boldsymbol{r}_e$, $\boldsymbol{r}_h$, we obtain

$$\left[\mathcal{E}_{c'}(\boldsymbol{k}'_e) + \mathcal{E}_{h'}(\boldsymbol{k}'_h) - \mathcal{E}\right] A^{n,\boldsymbol{K}}_{c'v'}(\boldsymbol{k}'_e, \boldsymbol{k}'_h) + \sum_{c,\boldsymbol{k}_e,v,\boldsymbol{k}_h} \langle c'\boldsymbol{k}'_e;\, v'\boldsymbol{k}'_h | V(\boldsymbol{r}_e - \boldsymbol{r}_h) | c\boldsymbol{k}_e;\, v\boldsymbol{k}_h\rangle A^{n,\boldsymbol{K}}_{cv}(\boldsymbol{k}_e, \boldsymbol{k}_h) = 0\,, \tag{4.76}$$

where we have used the orthonormality of the Bloch functions. Since we are interested in the Wannier exciton, which is weakly bound, the electron-hole interaction potential (Coulomb potential) $V(\boldsymbol{r}_e - \boldsymbol{r}_h)$ is assumed to vary very slowly over the distance of the lattice constant, and thus the interaction term may be moved outside of the integral. Taking into account the periodicity of the Bloch functions, the integral may be replaced by a summation of the integral in the unit cell. Therefore, we can derive

$$[\mathcal{E}_c(\boldsymbol{k}_e) + \mathcal{E}_h(\boldsymbol{k}_h) + V(\boldsymbol{r}_e - \boldsymbol{r}_h) - \mathcal{E}]\, A^{n,\boldsymbol{K}}_{cv}(\boldsymbol{k}_e, \boldsymbol{k}_h) = 0\,, \tag{4.77}$$

where c$'$, $\boldsymbol{k}'_e$, ... are written as c, $\boldsymbol{k}_e$, ... for simplicity. Here we assume that the conduction and valence bands are parabolic with scalar masses and that at the top of the valence band $\mathcal{E}_v = 0$, i.e.

$$\mathcal{E}_c(\boldsymbol{k}_e) = \frac{\hbar^2 k^2}{2m_e} + \mathcal{E}_G,$$

$$\mathcal{E}_h(\boldsymbol{k}_h) = \frac{\hbar^2 k_h^2}{2m_h},$$

where $\mathcal{E}_c - \mathcal{E}_v = \mathcal{E}_G$ is the energy gap. The Fourier transform of $A^{n,\boldsymbol{K}}_{cv}$ is given by

$$\Phi^{n,\boldsymbol{K}}_{cv}(\boldsymbol{r}_e, \boldsymbol{r}_h) = \frac{1}{V}\sum_{\boldsymbol{k}_e,\boldsymbol{k}_h} e^{i\boldsymbol{k}_e\cdot\boldsymbol{r}_e} e^{i\boldsymbol{k}_h\cdot\boldsymbol{r}_h} \cdot A^{n,\boldsymbol{K}}_{cv}(\boldsymbol{k}_e, \boldsymbol{k}_h) \tag{4.78}$$

and the Fourier transform of (4.77) is derived in a similar fashion giving rise to

$$[\mathcal{E}_c(-i\nabla_e) + \mathcal{E}_h(-i\nabla_h) + V(\boldsymbol{r}_e - \boldsymbol{r}_h) - \mathcal{E}]\, \Phi^{n,\boldsymbol{K}}_{cv}(\boldsymbol{r}_e, \boldsymbol{r}_h) = 0\,. \tag{4.79}$$

In the derivation of the above equation we used the following relation under the assumption of parabolic bands:

$$\mathcal{E}_c(-i\nabla_e)\Phi^{n,\boldsymbol{K}}_{cv}(\boldsymbol{r}_e, \boldsymbol{r}_h) = \left[-\frac{\hbar^2}{2m_e}\nabla_e^2 + \mathcal{E}_G\right]\Phi^{n,\boldsymbol{K}}_{cv}(\boldsymbol{r}_e, \boldsymbol{r}_h)$$
$$= \left[\frac{\hbar^2 k_e^2}{2m_e} + \mathcal{E}_G\right]\Phi^{n,\boldsymbol{K}}_{cv}(\boldsymbol{r}_e, \boldsymbol{r}_h) = \mathcal{E}_c(\boldsymbol{k}_e)\Phi^{n,\boldsymbol{K}}_{cv}(\boldsymbol{r}_e, \boldsymbol{r}_h)\,,$$

and a similar relation for the hole band. Equation (4.79) is the **effective-mass equation** for exciton. Equation (4.79) is also obtained from the effective-mass equation derived in Sect. 3.2.

The effective-mass equation (4.79) for the exciton may be rewritten as

$$\left[-\frac{\hbar^2}{2m_e}\nabla_e^2 + \mathcal{E}_G - \frac{\hbar^2}{2m_h}\nabla_h^2 - \frac{e^2}{4\pi\epsilon|\boldsymbol{r}_e - \boldsymbol{r}_h|} - \mathcal{E}\right]\Phi_{cv}^{n,\boldsymbol{K}}(\boldsymbol{r}_e, \boldsymbol{r}_h) = 0\,. \tag{4.80}$$

This equation is well known as the two body problem in quantum mechanics and the solution is obtained by separating the variables into center-of-mass motion and relative motion. Then we put

$$\boldsymbol{r} = \boldsymbol{r}_e - \boldsymbol{r}_h, \qquad \boldsymbol{R} = \frac{m_e\boldsymbol{r}_e + m_h\boldsymbol{r}_h}{m_e + m_h},$$

$$\frac{1}{\mu} = \frac{1}{m_e} + \frac{1}{m_h}, \qquad M = m_e + m_h,$$

and (4.80) is rewritten as

$$\left[\left(-\frac{\hbar^2}{2M}\nabla_{\boldsymbol{R}}^2\right) + \left(-\frac{\hbar^2}{2\mu}\nabla_{\boldsymbol{r}}^2 - \frac{e^2}{4\pi\epsilon r}\right)\right]\Phi^{n,\boldsymbol{K}}(\boldsymbol{r}, \boldsymbol{R}) = [\mathcal{E} - \mathcal{E}_G]\cdot\Phi^{n,\boldsymbol{K}}(\boldsymbol{r}, \boldsymbol{R})\,. \tag{4.81}$$

The solution of this equation is given by

$$\Phi^{n,\boldsymbol{K}}(\boldsymbol{r}, \boldsymbol{R}) = \phi_{\boldsymbol{K}}(\boldsymbol{R})\psi_n(\boldsymbol{r}),$$

where $\phi_{\boldsymbol{K}}(\boldsymbol{R})$ and $\psi_n(\boldsymbol{r})$ are respectively given by the separate equations

$$H_{\boldsymbol{R}}\phi_{\boldsymbol{K}}(\boldsymbol{R}) \equiv -\frac{\hbar^2}{2M}\nabla_{\boldsymbol{R}}^2\,\phi_{\boldsymbol{K}}(\boldsymbol{R}) = \mathcal{E}(\boldsymbol{K})\cdot\phi_{\boldsymbol{K}}(\boldsymbol{R}), \tag{4.82}$$

$$H_{\boldsymbol{r}}\psi_n(\boldsymbol{r}) \equiv \left[-\frac{\hbar^2}{2\mu}\nabla_{\boldsymbol{r}}^2 + V(\boldsymbol{r})\right]\psi_n(\boldsymbol{r}) = \mathcal{E}_n\cdot\psi_n(\boldsymbol{r}). \tag{4.83}$$

The total energy of the system (exciton state) $\mathcal{E}$ is then given by

$$\mathcal{E} = \mathcal{E}(\boldsymbol{K}) + \mathcal{E}_n + \mathcal{E}_G\,. \tag{4.84}$$

The energy due to the center-of-mass motion $\mathcal{E}(\boldsymbol{K})$ is calculated by inserting the wave function $\phi_{\boldsymbol{K}}(\boldsymbol{R}) \propto \exp(\mathrm{i}\boldsymbol{K}\cdot\boldsymbol{R})$ into (4.82):

$$\mathcal{E}(\boldsymbol{K}) = \frac{\hbar^2 K^2}{2M}\,. \tag{4.85}$$

On the other hand, the energy due to the relative motion $\mathcal{E}_n$ is obtained from (4.83), which is the same as the bound state energy for shallow donor states derived from the effective mass equation:

$$\mathcal{E}_n = -\frac{\mu}{m}\left(\frac{\epsilon_0}{\epsilon}\right)^2 \cdot \frac{\mathcal{E}_{\mathrm{H}}}{n^2} = -\frac{\mathcal{E}_{\mathrm{ex}}}{n^2} \quad (n = 1, 2, 3, \ldots), \tag{4.86}$$

$$\mathcal{E}_{\mathrm{ex}} = \frac{\mu}{m}\left(\frac{\epsilon_0}{\epsilon}\right)^2 \mathcal{E}_{\mathrm{H}} = \frac{\mu e^4}{2(4\pi\epsilon)^2\hbar^2}, \tag{4.87}$$

where $\mathcal{E}_{\mathrm{H}}$ is the ionization energy for the hydrogen atom. From these results we obtain the total energy of an exciton

$$\mathcal{E} = -\frac{\mathcal{E}_{\mathrm{ex}}}{n^2} + \frac{\hbar^2 K^2}{2M} + \mathcal{E}_{\mathrm{G}}. \tag{4.88}$$

Here we have to note that the energy of an exciton is measured from the bottom of the conduction band ($\mathcal{E}_{\mathrm{c}} = 0$). The above calculations reveal that the exciton states consists of the center-of-mass motion of the electron-hole pair with wave vector $\boldsymbol{K}$ and of the relative motion bound by the Coulomb attraction force, and that the total energy of the exciton is the sum of the energies due to the center-of-mass motion $\mathcal{E}_{\boldsymbol{K}} = \hbar^2 K^2/2M$ and to the motion of relative coordinates $\mathcal{E}_n$.

As an example, we assume parameters close to those of GaAs, such as $\mu/m = 0.05$ and $\epsilon/\epsilon_0 = 13$, and then we obtain $\mathcal{E}_n = -3 \times 10^{-4}\mathcal{E}_{\mathrm{H}}/n^2$ which gives the bound energy of the ground state ($n = 1$) of about 4 meV. The exciton Bohr radius is estimated from (3.43) to be $a_{\mathrm{ex}} = (m/\mu)(\epsilon/\epsilon_0)a_{\mathrm{B}}$ which gives $a_{\mathrm{ex}} \approx 150\,\text{Å}$. Figure 4.9 illustrates the energy states of an exciton, where the energy depends on the wave vector $\boldsymbol{K}$ for the center-of-mass motion. In general, we are concerned with the optical excitation of excitons and thus we can assume $\boldsymbol{K} = 0$.

As shown in the above example, the effective Bohr radius of an exciton is much larger than the lattice constant and thus this condition validates the use of the effective mass approximation. As described above we divide the exciton motion into two motions, relative motion in the space $\boldsymbol{r}$ and center-of-mass motion in the space $\boldsymbol{R}$, and express the wave functions of the effective mass equation by

$$\Phi^{n,\boldsymbol{K}}(\boldsymbol{r}, \boldsymbol{R}) = \frac{1}{\sqrt{V}}\exp(\mathrm{i}\boldsymbol{K}\cdot\boldsymbol{R})\phi_n(\boldsymbol{r}), \tag{4.89}$$

where $\phi_n(\boldsymbol{r})$ is called the envelope function of an exciton. The ground state of the exciton is given in analogy to the hydrogen atom with a shallow donor state as

$$\phi_1(\boldsymbol{r}) = \frac{1}{\sqrt{\pi a_{\mathrm{ex}}^3}}\exp(-r/a_{\mathrm{ex}}), \tag{4.90}$$

where $a_{\mathrm{ex}} = (m/\mu)(\epsilon/\epsilon_0)a_{\mathrm{B}}$ is the effective Bohr radius of the exciton. By the inverse transformation of the wave function (4.89), we obtain the coefficient $A_{\mathrm{cv}}^{n,\boldsymbol{K}}(\boldsymbol{k}_{\mathrm{e}}, \boldsymbol{k}_{\mathrm{h}})$

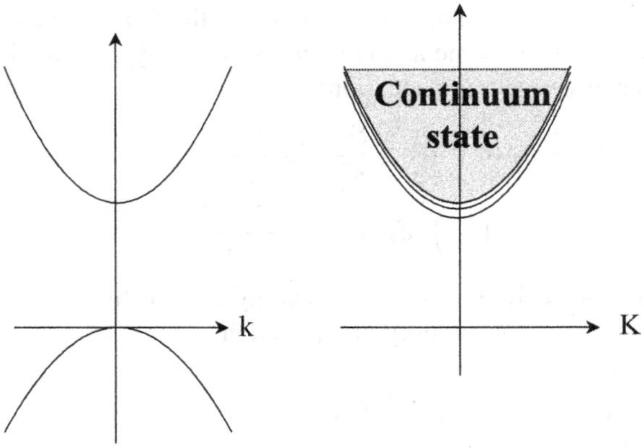

Fig. 4.9 Energy states of exciton as a function of wave vector $\boldsymbol{K}$ for the center-of-mass motion. Discrete states are shown by solid curves and the continuum state by the chequered region

$$\begin{aligned} A_{\mathrm{cv}}^{n,\boldsymbol{K}}(\boldsymbol{k}_{\mathrm{e}}, \boldsymbol{k}_{\mathrm{h}}) &= \frac{1}{V}\int \mathrm{d}^3\boldsymbol{r}_{\mathrm{e}}\mathrm{d}^3\boldsymbol{r}_{\mathrm{h}}\mathrm{e}^{-\mathrm{i}\boldsymbol{k}_{\mathrm{e}}\cdot\boldsymbol{r}_{\mathrm{e}}}\mathrm{e}^{-\mathrm{i}\boldsymbol{k}_{\mathrm{h}}\cdot\boldsymbol{r}_{\mathrm{h}}}\Phi_{\mathrm{cv}}^{n,\boldsymbol{K}}(\boldsymbol{k}_{\mathrm{e}}, \boldsymbol{k}_{\mathrm{h}}) \\ &= \frac{1}{\sqrt{V}}\int \mathrm{d}^3\boldsymbol{r}\mathrm{d}^3\boldsymbol{R}\mathrm{e}^{-\mathrm{i}\boldsymbol{R}\cdot(\boldsymbol{k}_{\mathrm{e}}+\boldsymbol{k}_{\mathrm{h}}-\boldsymbol{K})}\phi_n(\boldsymbol{r})\mathrm{e}^{-\mathrm{i}\boldsymbol{k}^*\cdot\boldsymbol{r}} \\ &= \frac{1}{\sqrt{V}}\int \mathrm{d}^3\boldsymbol{r}\mathrm{e}^{-\mathrm{i}\boldsymbol{k}^*\cdot\boldsymbol{r}}\phi_n(\boldsymbol{r})\delta_{\boldsymbol{K},\boldsymbol{k}_{\mathrm{e}}+\boldsymbol{k}_{\mathrm{h}}}, \end{aligned} \tag{4.91}$$

$$\boldsymbol{k}^* = \frac{m_{\mathrm{h}}\boldsymbol{k}_{\mathrm{e}} - m_{\mathrm{e}}\boldsymbol{k}_{\mathrm{h}}}{M}. \tag{4.92}$$

As defined in (4.73), the wave vector $\boldsymbol{K}$ is the sum of the wave vectors of the electron and hole $(\boldsymbol{k}_{\mathrm{e}} + \boldsymbol{k}_{\mathrm{h}})$. The coefficient $A_{\mathrm{cv}}^{1,\boldsymbol{K}}(\boldsymbol{k}_{\mathrm{e}}, \boldsymbol{k}_{\mathrm{h}})$ for the exciton ground state is obtained by inserting (4.90) into the above equation for $n = 1$:

$$A_{\mathrm{cv}}^{1,\boldsymbol{K}}(\boldsymbol{k}_{\mathrm{e}}, \boldsymbol{k}_{\mathrm{h}}) = \frac{1}{\sqrt{V}}\left(\frac{64\pi}{a_{\mathrm{ex}}^5}\right)^{1/2}\frac{1}{(k^{*2} + 1/a_{\mathrm{ex}}^2)^2}\,\delta_{\boldsymbol{K},\boldsymbol{k}_{\mathrm{e}}+\boldsymbol{k}_{\mathrm{h}}}, \tag{4.93}$$

where $A_{\mathrm{cv}}^{1,\boldsymbol{K}}(\boldsymbol{k}_{\mathrm{e}}, \boldsymbol{k}_{\mathrm{h}})$ is found to decrease very rapidly when $|\boldsymbol{k}_{\mathrm{e}}|$ and $|\boldsymbol{k}_{\mathrm{h}}|$ exceed $1/a_{\mathrm{ex}}$ and thus the wave vectors contributing to the $1s$ exciton are limited to a narrow region of the Brillouin zone $|\boldsymbol{k}| \le 1/a_{\mathrm{ex}}$. In the case of the Wannier exciton, a_{ex} is very large compared to the lattice constant and thus only a narrow region of wave vectors in $\boldsymbol{k}$ space plays an important role in the formation of the exciton. A similar conclusion is drawn for the case of shallow donors.

Next we will discuss light absorption by excitons. Let us assume the initial state to be given by $\Psi_0 = \phi_{\mathrm{c},\boldsymbol{k}_{\mathrm{e}}}\phi_{\mathrm{v},-\boldsymbol{k}_{\mathrm{h}}} = \phi_{\mathrm{c},\boldsymbol{k}_{\mathrm{e}}}\phi_{\mathrm{v},\boldsymbol{k}_{\mathrm{e}}}$ before excitation, with an electron in the valence band, and the transition from the initial state to the excited state of an exciton given by $\Psi^{\lambda,\boldsymbol{K}}$. According to the treatment stated in Sect. 4.2 the transition probability per unit time w_{if} is written as

$$
\begin{aligned}
w_{if} &= \frac{2\pi}{\hbar}\cdot\frac{e^2}{m^2}|A_0|^2\frac{1}{V}\sum_{\lambda}\left|\langle\Psi^{\lambda,\boldsymbol{K}}|\exp(\mathrm{i}\boldsymbol{k}_{\mathrm{p}}\cdot\boldsymbol{r})\boldsymbol{e}\cdot\boldsymbol{p}|\Psi_0\rangle\right|^2 \\
&\quad\times\delta(\mathcal{E}_{\mathrm{G}}+\mathcal{E}_{\lambda}-\hbar\omega) \\
&= \frac{2\pi}{\hbar}\cdot\frac{e^2}{m^2}|A_0|^2\frac{1}{V}\sum_{\boldsymbol{k}_{\mathrm{e}},\lambda}\left|A_{\mathrm{cv}}^{\lambda,\boldsymbol{K}}(\boldsymbol{k}_{\mathrm{e}},-\boldsymbol{k}_{\mathrm{e}})\langle\phi_{\mathrm{c},\boldsymbol{k}_{\mathrm{e}}}|\boldsymbol{e}\cdot\boldsymbol{p}|\phi_{\mathrm{v},\boldsymbol{k}_{\mathrm{e}}}\rangle\right|^2 \\
&\quad\times\delta(\mathcal{E}_{\mathrm{G}}+\mathcal{E}_{\lambda}-\hbar\omega)\,,
\end{aligned}
\tag{4.94}
$$

where the wave vector of light $\boldsymbol{k}_{\mathrm{p}}$ is assumed to be 0 and the momentum conservation law $\boldsymbol{k}_{\mathrm{e}} = -\boldsymbol{k}_{\mathrm{h}}$ is used. The total energy of the exciton $\mathcal{E}_{\lambda}$ includes the continuum state ($\mathcal{E}_n > 0$). Since $A_{\mathrm{cv}}^{\lambda,\boldsymbol{K}}$ has a large value only in a narrow region of $\boldsymbol{k}_{\mathrm{e}} \approx 0$, the term $\langle\phi_{\mathrm{c},\boldsymbol{k}_{\mathrm{e}}}|\boldsymbol{e}\cdot\boldsymbol{p}|\phi_{\mathrm{v},\boldsymbol{k}_{\mathrm{e}}}\rangle$ is regarded as constant in the narrow region of $\boldsymbol{k}$ space we are interested in. Under these assumptions and the condition $\boldsymbol{k}_{\mathrm{e}} = -\boldsymbol{k}_{\mathrm{h}}$ we obtain

$$
A_{\mathrm{cv}}^{\lambda,\boldsymbol{K}}(\boldsymbol{k}_{\mathrm{e}},-\boldsymbol{k}_{\mathrm{h}}) = \frac{1}{V}\int \mathrm{d}^3\boldsymbol{r}_{\mathrm{e}}\mathrm{d}^3\boldsymbol{r}_{\mathrm{h}}\mathrm{e}^{-\mathrm{i}\boldsymbol{k}_{\mathrm{e}}\cdot(\boldsymbol{r}_{\mathrm{e}}-\boldsymbol{r}_{\mathrm{h}})}\Phi_{\mathrm{cv}}^{\lambda,\boldsymbol{K}}(\boldsymbol{r}_{\mathrm{e}},\boldsymbol{r}_{\mathrm{h}})\,. \tag{4.95}
$$

The summation in (4.95) with respect to $\boldsymbol{k}_{\mathrm{e}}$ is non-zero only when $\boldsymbol{r}_{\mathrm{e}} - \boldsymbol{r}_{\mathrm{h}} = 0$, as shown in Appendix A.2. Therefore, we obtain

$$
w_{if} = \frac{2\pi}{\hbar}\cdot\frac{e^2}{m^2}|A_0|^2\sum_{\lambda}|\boldsymbol{e}\cdot\boldsymbol{p}_{\mathrm{cv}}|^2|\phi_{\lambda}(0)|^2\,\delta(\mathcal{E}_{\mathrm{G}}+\mathcal{E}_{\lambda}-\hbar\omega)\,, \tag{4.96}
$$

where $|\boldsymbol{e}\cdot\boldsymbol{p}_{\mathrm{cv}}|$ is the momentum matrix element defined in Sect. 4.2. The imaginary part of the dielectric function $\kappa_2(\omega) = \epsilon_2(\omega)/\epsilon_0$ is then written as

$$
\kappa_2(\omega) = \frac{\pi e^2}{\epsilon_0 m^2\omega^2}|\boldsymbol{e}\cdot\boldsymbol{p}_{\mathrm{cv}}|^2\sum_{\lambda}|\phi_{\lambda}(0)|^2\,\delta(\mathcal{E}_{\mathrm{G}}+\mathcal{E}_{\lambda}-\hbar\omega)\,. \tag{4.97}
$$

First, we consider the excitation of the discrete states of the bound exciton and their transition probability. When we assume parabolic bands with scalar effective masses for simplicity, the relation $\phi_{\lambda}(0) \neq 0$ is fulfilled only for the s state of the bound exciton. The quantum number λ of the exciton states can then be expressed only by the principal quantum number n:

$$
|\phi_n(0)|^2 = \frac{1}{\pi a_{\mathrm{ex}}^3 n^3}\,, \tag{4.98}
$$

$$
\mathcal{E}_{\lambda} = \mathcal{E}_n = -\frac{\mathcal{E}_{\mathrm{ex}}}{n^2}\,, \tag{4.99}
$$

and therefore we find the following relation for the bound states:

$$
\kappa_2(\omega) = \frac{\pi e^2}{\epsilon_0 m^2\omega^2}|\boldsymbol{e}\cdot\boldsymbol{p}_{\mathrm{cv}}|^2\frac{1}{\pi a_{\mathrm{ex}}^3}\sum_{n}\frac{1}{n^3}\,\delta\left(\mathcal{E}_G - \frac{\mathcal{E}_{\mathrm{ex}}}{n^2} - \hbar\omega\right)\,. \tag{4.100}
$$

In the above equation the spin degeneracy is disregarded and $\kappa_2(\omega)$ differs by a factor of 2 from the result in Sect. 4.2. Equation (4.100) reveals that the oscillator strength of exciton absorption spectra is proportional to $1/n^3$ and that the discrete lines converge to $\hbar\omega = \mathcal{E}_G$.

On the other hand, the continuum states of exciton is given by $\phi_\lambda(0) \neq 0$ only for the magnetic quantum number $m = 0$, and its state is determined by the wave vector $\boldsymbol{k}$. Elliott [6] has shown that

$$|\phi_{\boldsymbol{k}}(0)|^2 = \frac{\pi\alpha_0 \exp(\pi\alpha_0)}{\sinh(\pi\alpha_0)}, \tag{4.101}$$

$$\mathcal{E}_\lambda = \mathcal{E}_{\boldsymbol{k}} = \frac{\hbar^2 k^2}{2\mu}, \tag{4.102}$$

where $\alpha_0 = (a_{ex}|\boldsymbol{k}|)^{-1} = [\mathcal{E}_{ex}/(\hbar\omega - \mathcal{E}_G)]^{1/2}$. We obtain the imaginary part of the dielectric function for the continuum state:

$$\kappa_2(\omega) = \frac{\pi e^2}{\epsilon_0 m^2 \omega^2} \frac{2\pi}{(2\pi)^3} \left(\frac{8\mu_1\mu_2\mu_3}{\hbar^6}\right)^{1/2} \mathcal{E}_{ex}^{1/2} |\boldsymbol{e} \cdot \boldsymbol{p}_{cv}|^2 \frac{\pi \exp(\pi\alpha_0)}{\sinh(\pi\alpha_0)}. \tag{4.103}$$

Now, we consider two extreme conditions. The asymptotic solutions of $\kappa_2(\omega)$ for $\alpha_0 \to 0$ ($\hbar\omega \to \infty$) and for $\alpha_0 \to \infty$ ($\hbar\omega \to \mathcal{E}_G$) may be given by, respectively,

$$\kappa_2(\omega)_{\alpha_0 \to 0} = \frac{\pi e^2}{\epsilon_0 m^2 \omega^2} \frac{2\pi}{(2\pi)^3} \left(\frac{8\mu_1\mu_2\mu_3}{\hbar^6}\right)^{1/2} \times |\boldsymbol{e} \cdot \boldsymbol{p}_{cv}|^2 \sqrt{\hbar\omega - \mathcal{E}_G}, \tag{4.104a}$$

$$\kappa_2(\omega)_{\alpha_0 \to \infty} = \frac{\pi e^2}{\epsilon_0 m^2 \omega^2} \frac{1}{2\pi} \left(\frac{8\mu_1\mu_2\mu_3}{\hbar^6}\right)^{1/2} \mathcal{E}_{ex}^{1/2} |\boldsymbol{e} \cdot \boldsymbol{p}_{cv}|^2. \tag{4.104b}$$

In the case of $\alpha_0 \to 0$, $\mathcal{E}_{ex}$ is very small and the relation $\hbar^2 k^2/2\mu \gg \mathcal{E}_{ex}$ holds. Therefore, the energy of the exciton is much larger than the electron-hole interaction potential and thus we may neglect the Coulomb interaction, giving rise to the result obtained for the one-electron approximation. This is evident from the fact that (4.104a) agrees with (4.39) obtained for the M_0 critical point in Sect. 4.3 except for the spin degeneracy factor 2. In the case of $\alpha_0 \to \infty$, on the other hand, (4.104b) may be expected to agree with $\kappa_2(\omega)$ for the discrete excitons. When n is large, the exciton absorption lines will overlap each other and cannot be distinguished each other, leading to a quasi-continuum state. The density of states for this case is estimated from $(\mathrm{d}\mathcal{E}/\mathrm{d}n)^{-1} = n^3/2\mathcal{E}_{ex}$, and then we obtain

$$|\phi_n(0)|^2 \cdot \left(\frac{\mathrm{d}\mathcal{E}}{\mathrm{d}n}\right)^{-1} = \frac{1}{2\pi a_{ex}^3 \mathcal{E}_{ex}} = \frac{1}{2\pi}\left(\frac{8\mu^3}{\hbar^6}\right)^{1/2} \mathcal{E}_{ex}^{1/2}. \tag{4.105}$$

Now we find that the dielectric functions $\kappa_2(\omega)$ obtained from the bound (discrete) states and from the continuum state agree with each other at $\hbar\omega = \mathcal{E}_G$. Figure 4.10

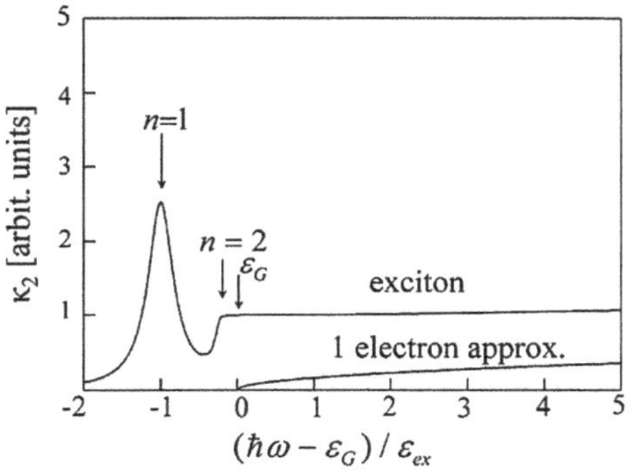

Fig. 4.10 Absorption spectra (κ_2) due to direct excitons and due to the one-electron approximation (without electron-hole interaction) are plotted in the region near the fundamental absorption edge as a function of photon energy, where the spectrum for exciton absorption is calculated by assuming $\Gamma = 0.2\mathcal{E}_{ex}$ and the Lorentz function for the discrete (bound) exciton

shows absorption spectra (spectra of $\alpha(\omega) \propto \kappa_2(\omega)$) calculated by taking account of the electron-hole interaction, where the dielectric function without the electron-hole interaction (one-electron approximation) is also plotted for comparison. We have to note here that the calculations are approximated by using the Lorentz function with the broadening parameter $\Gamma = 0.2\mathcal{E}_{ex}$ and by neglecting the spin degeneracy factor 2. We see exciton absorption associated with $n = 1$ and $n = 2$ in Fig. 4.10.

Here we will discuss the Lorentz function approximation of the dielectric function for the discrete (bound) exciton. We express (4.100) by

$$\kappa_2(\omega) = C \cdot \delta\left(\mathcal{E}_G - \frac{\mathcal{E}_{ex}}{n^2} - \hbar\omega\right).$$

As shown in Appendix A, the Dirac delta function can be approximated by the Lorentz function in the limit of $\Gamma \to 0$. Using the Dirac identity in the limit of $\Gamma \to 0$, we may write

$$\frac{1}{\mathcal{E}_G - \mathcal{E}_{ex}/n^2 - (\hbar\omega + i\Gamma)} = \mathcal{P}\left\{\frac{1}{\mathcal{E}_G - \mathcal{E}_{ex}/n^2 - \hbar\omega}\right\} + i\pi\delta\left(\mathcal{E}_G - \mathcal{E}_{ex}/n^2 - \hbar\omega\right) \quad (4.106)$$

and a comparison of the dielectric function $\kappa_2(\omega)$ with the imaginary part of above equation gives the following relation:

$$\kappa_2(\omega) = \Im\left\{\frac{C/\pi}{\mathcal{E}_G - \mathcal{E}_{ex}/n^2 - (\hbar\omega + i\Gamma)}\right\} = \frac{C\Gamma/\pi}{\left(\mathcal{E}_G - \mathcal{E}_{ex}/n^2 - \hbar\omega\right)^2 + \Gamma^2}$$
$$= C \cdot \delta\left(\mathcal{E}_G - \frac{\mathcal{E}_{ex}}{n^2} - \hbar\omega\right). \quad (4.107)$$

The second relation of the above equation with $C = 1$ is called the Lorentz function. The Lorentz function is characterized by the center located at $\hbar\omega = (\mathcal{E}_G - \mathcal{E}_{ex}/n^2)$ and by its full-width half-maximum Γ. The integral of the Lorentz function with respect to $\hbar\omega$ is equal to unity. In the limit of $\Gamma \to 0$, therefore, the Lorentz function is equivalent to $\delta(\mathcal{E}_G - \mathcal{E}_{ex}/n^2 - \hbar\omega)$. From these considerations it is possible to replace the delta function by the Lorentz function. As shown in Sect. 4.6, the real part of the dielectric function is obtained from the Kramers–Kronig relation of (4.126a):

$$\begin{aligned}\kappa_1(\omega) &= \frac{2}{\pi}\mathcal{P}\int_0^\infty \frac{\omega'\kappa_2(\omega')}{(\omega')^2 - \omega^2}\mathrm{d}\omega' \\ &= \frac{2}{\pi}\mathcal{P}\int_0^\infty \frac{\omega' \cdot C \cdot \delta(\mathcal{E}_G - \mathcal{E}_{ex}/n^2 - \hbar\omega')}{(\omega')^2 - \omega^2}\mathrm{d}\omega' \\ &= \frac{2C}{\pi}\frac{\mathcal{E}_G - \mathcal{E}_{ex}/n^2}{(\mathcal{E}_G - \mathcal{E}_{ex}/n^2)^2 - (\hbar\omega)^2}. \end{aligned} \tag{4.108}$$

The above result may be approximated by

$$\begin{aligned}\kappa_1(\omega) &= \frac{2C}{\pi}\frac{\mathcal{E}_G - \mathcal{E}_{ex}/n^2}{(\mathcal{E}_G - \mathcal{E}_{ex}/n^2)^2 - (\hbar\omega)^2} \\ &= \frac{C/\pi}{\mathcal{E}_G - \mathcal{E}_{ex}/n^2 - \hbar\omega} + \frac{C/\pi}{\mathcal{E}_G - \mathcal{E}_{ex}/n^2 + \hbar\omega} \\ &\simeq \frac{C/\pi}{\mathcal{E}_G - \mathcal{E}_{ex}/n^2 - \hbar\omega}. \end{aligned} \tag{4.109}$$

The final relation of the above equation is obtained from the second relation by assuming that the second term is non-resonant and small compared to the first term, enabling us to neglect the second term. The complex dielectric function of the exciton is therefore approximated by

$$\kappa(\omega) = \kappa_1(\omega) + \mathrm{i}\kappa_2(\omega) = \frac{C/\pi}{\mathcal{E}_G - \mathcal{E}_{ex}/n^2 - (\hbar\omega + \mathrm{i}\Gamma)}. \tag{4.110}$$

The above relation coincides with the result obtained by the second quantization method [7]. The experimental results for the optical absorption in GaAs is shown in Fig. 4.11 [8]. At low temperatures ($T = 186, 90, 21$ K) the exciton absorption peak for $n = 1$ is clearly seen and the overall features are in good agreement with the calculated absorption spectrum shown in Fig. 4.10.

4.5.2 Indirect Exciton

In the above calculations, we dealt with excitons associated with the fundamental absorption edge of the M_0 critical point at the Γ point (direct transition), where the top of the valence band and the bottom of the conduction band are located at $\boldsymbol{k} = 0$. We have also shown that a weak absorption will appear for the indirect transition

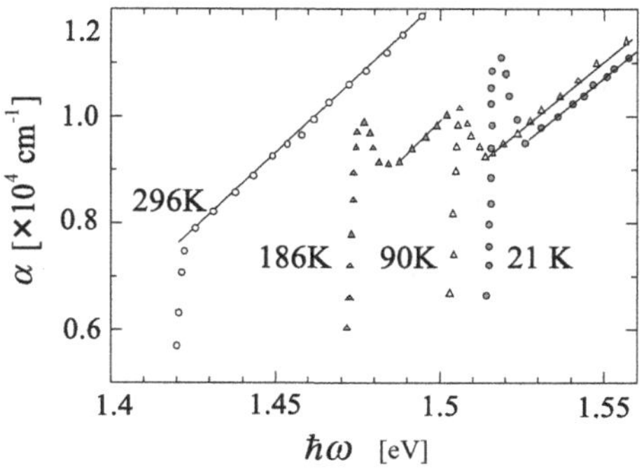

Fig. 4.11 Experimental results of the optical absorption in GaAs at different temperatures. The exciton peak is clearly observed at low temperatures from (M.D. Sturge [8]) and the overall features are in good agreement with the calculated results shown in Fig. 4.10

where the top of the valence band and the bottom of the conduction band are located at different $\boldsymbol{k}$ in the Brillouin zone and is an indication of the existence of the exciton effects in such indirect gap semiconductors. In this subsection we will be concerned with exciton effect in indirect gap semiconductors such as Ge, Si, GaP and so on. The top of the valence band and the bottom of the conduction band are assumed to be located at $\boldsymbol{k}_{\rm v0}$ $(= -\boldsymbol{k}_{\rm h0})$ and $\boldsymbol{k}_{\rm c0}$ $(= \boldsymbol{k}_{\rm e0})$ $(\boldsymbol{k}_{\rm v0} \neq \boldsymbol{k}_{\rm c0})$, respectively, and the bands are approximated by the parabolic functions

$$\mathcal{E}_{\rm c} = \mathcal{E}_{\rm G} + \frac{\hbar^2}{2m_{\rm e}}(\boldsymbol{k}_{\rm c} - \boldsymbol{k}_{\rm c0})^2,$$

$$\mathcal{E}_{\rm h} = -\mathcal{E}_{\rm v} = \frac{\hbar^2}{2m_{\rm h}}(\boldsymbol{k}_{\rm v} - \boldsymbol{k}_{\rm v0})^2.$$

Putting $\boldsymbol{k}'_{\rm e} = \boldsymbol{k}_{\rm c} - \boldsymbol{k}_{\rm c0} = \boldsymbol{k}_{\rm e} - \boldsymbol{k}_{\rm c0}$, $\boldsymbol{k}'_{\rm h} = -(\boldsymbol{k}_{\rm v} - \boldsymbol{k}_{\rm v0}) \equiv \boldsymbol{k}_{\rm h} - \boldsymbol{k}_{\rm h0}$, then the effective mass equation for this exciton is written as

$$\left[-\frac{\hbar^2}{2m_{\rm e}}\nabla_{\rm e}^2 - \frac{\hbar^2}{2m_{\rm h}}\nabla_{\rm h}^2 - \frac{e^2}{4\pi\epsilon r}\right]\Phi^{\lambda,\boldsymbol{K}'}(\boldsymbol{r}_{\rm e},\boldsymbol{r}_{\rm h})$$
$$= \mathcal{E}_\lambda \Phi^{\lambda,\boldsymbol{K}'}(\boldsymbol{r}_{\rm e},\boldsymbol{r}_{\rm h})\,, \tag{4.111}$$

where $\boldsymbol{K}' = \boldsymbol{k}'_{\rm e} + \boldsymbol{k}'_{\rm h} = (\boldsymbol{k}_{\rm e} + \boldsymbol{k}_{\rm h}) - (\boldsymbol{k}_{\rm e0} + \boldsymbol{k}_{\rm h0})$. Following the treatment of the direct exciton, the wave function $\Phi^{\lambda,\boldsymbol{K}'}(\boldsymbol{r}_{\rm e},\boldsymbol{r}_{\rm h})$ may be separated into translational (center-of-mass) motion ($\boldsymbol{R}$) and relative motion ($\boldsymbol{r}$):

$$\Phi^{\lambda,\boldsymbol{K}'} = \frac{1}{\sqrt{V}}\phi_\lambda(\boldsymbol{r})\exp(\mathrm{i}\boldsymbol{K}'\cdot\boldsymbol{R})$$

where we find that $\phi_\lambda(\boldsymbol{r})$ satisfies (4.83) and thus the total energy of the exciton is given for the bound state by

$$\mathcal{E}_\lambda = \mathcal{E}_n = -\mathcal{E}_{\rm ex}/n^2 + \frac{\hbar^2 K'^2}{2M}\,. \tag{4.112}$$

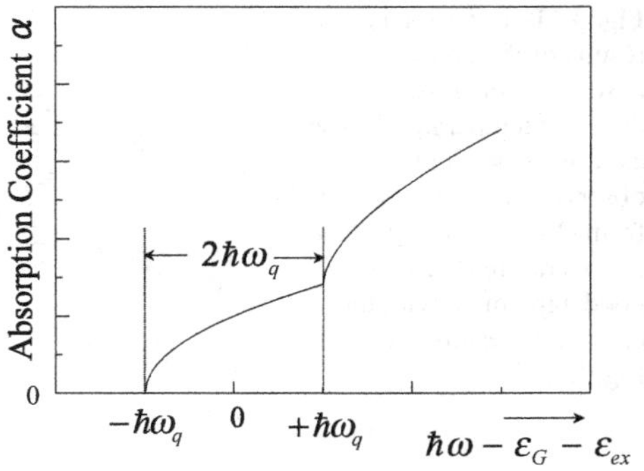

Fig. 4.12 Optical absorption spectra associated with an indirect exciton, where $\alpha(\omega) \propto \kappa_2(\omega)$ is plotted as a function of photon energy. The absorption coefficient increases at photon energies associated with phonon emission and absorption

Therefore, we obtain the dielectric function of the indirect exciton for the bound state as

$$\begin{aligned}\kappa_2(\omega) &= \frac{\pi e^2}{\epsilon_0 m^2 \omega^2} \sum_{m,\alpha,\pm} |\boldsymbol{e} \cdot \boldsymbol{p}_{\text{cv}}^{m,\alpha,\pm}|^2 |\phi_n(0)|^2 \\ &\times \delta\left[\mathcal{E}_{\text{G}} - \frac{\mathcal{E}_{\text{ex}}}{n^2} + \frac{\hbar^2}{2M}\{\boldsymbol{k}_{\text{e}} + \boldsymbol{k}_{\text{h}} - (\boldsymbol{k}_{\text{e0}} + \boldsymbol{k}_{\text{h0}})\}^2 - \hbar\omega \pm \hbar\omega_{\boldsymbol{q}}^{\alpha}\right] \\ &= \frac{\pi e^2}{\epsilon_0 m^2 \omega^2} \cdot \frac{1}{2\pi^2} \sum_{m,\alpha,\pm} |\boldsymbol{e} \cdot \boldsymbol{p}_{\text{cv}}^{m,\alpha,\pm}|^2 \frac{1}{\pi a_{\text{ex}}^3} \cdot \left(\frac{2M}{\hbar^2}\right)^{3/2} \cdot \frac{1}{n^3} \\ &\times \sqrt{\hbar\omega - \mathcal{E}_{\text{G}} \mp \hbar\omega_{\boldsymbol{q}}^{\alpha} + \mathcal{E}_{\text{ex}}/n^2}\,. \end{aligned} \tag{4.113}$$

It is evident from these results that the dielectric function $\kappa_2(\omega)$ for the bound state of the indirect exciton does not exhibit any discrete lines but a continuum spectrum. This feature may be understood by noting that the wave vector $\boldsymbol{K}'$ of the indirect exciton is different from the direct exciton due to the exchange of momentum of the indirect exciton with the lattice vibrations, giving rise to the possibility of the wave vector taking arbitrary values.

Figure 4.12 shows an illustration of (4.113). As stated in Sect. 4.5, the absorption coefficient calculated from the one-electron approximation shows an increase at the photon energies corresponding to phonon absorption and phonon emission. When the Coulomb interaction between the electron and hole is taken into account, $\kappa_2(\omega)$ increases at photon energies $\mathcal{E}_{\text{G}} - \mathcal{E}_{\text{ex}}/n^2 \pm \hbar\omega_{q\alpha}$ and the distance between the two rising points is equal to twice the phonon energy. The most significant difference between the two curves is the photon energy dependence of the absorption coefficients. The dielectric function $\kappa_2(\omega)$ exhibits quadratic behavior with respect to photon energy in the case without Coulomb interaction, whereas $\kappa_2(\omega)$ for the indirect exciton shows a square root dependence of photon energy. These features explain the deviation from a linear relation of the $\sqrt{\alpha}$ versus photon energy plot of Ge in Fig. 4.8 by means of the exciton effect. The experimental observation of indirect excitons in GaP is shown in Fig. 4.13, where the square root of the absorption coef-

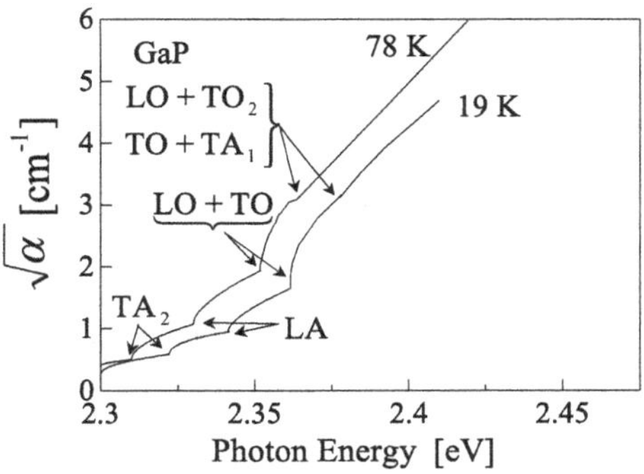

Fig. 4.13 Square root of the absorption coefficient $\alpha^{1/2}$ in GaP is plotted as a function of photon energy. The arrows indicate an increase in the absorption coefficient due to exciton effect (phonon emission) (Gershenzon et al. [9])

ficient is plotted as a function of photon energy. The results are in good agreement with (4.113) [9], where phonon modes involved with the transition are indicated. We note here that a general expression of $\kappa_2(\omega)$ as a function of $\hbar\omega$ is not obtained, but that an asymptotic form is obtained for $\hbar\omega - \mathcal{E}_\mathrm{G} \gg \mathcal{E}_\mathrm{ex}$, which agrees with the result of the one-electron approximation without the exciton effect (see [6]).

4.6 Dielectric Function

The polarization vector $\boldsymbol{P}(\omega)$ in a material due to an external field $\boldsymbol{E}(\omega)$ of angular frequency ω is defined by

$$\boldsymbol{P}(\omega) = \chi(\omega)\epsilon_0 \boldsymbol{E}(\omega)\,, \tag{4.114}$$

where $\chi(\omega)$ is the susceptibility tensor. Since the susceptibility tensor is an analytical function of frequency in real numbers, it is rewritten by using the Cauchy relation as

$$\chi(\omega) = \int_{-\infty}^{+\infty} \frac{\mathrm{d}\omega'}{2\pi\mathrm{i}} \frac{\chi(\omega')}{\omega' - \omega - \mathrm{i}\varepsilon}\,, \tag{4.115}$$

where ε is an infinitely small positive number. In Appendix A we derive the Dirac identity relation given by (A.2)

$$\lim_{\varepsilon \to 0} \frac{1}{\omega - \mathrm{i}\varepsilon} = \mathcal{P}\frac{1}{\omega} + \mathrm{i}\pi\delta(\omega)\,. \tag{4.116}$$

The integration of (4.115) is carried out by using the Dirac identity relation, which gives rise to

$$\chi(\omega) = \mathcal{P}\int_{-\infty}^{\infty} \frac{d\omega'}{2\pi i}\frac{\chi(\omega')}{\omega' - \omega} + \frac{1}{2}\int_{-\infty}^{\infty} d\omega' \chi(\omega')\delta(\omega' - \omega)\,, \tag{4.117}$$

or

$$\chi(\omega) = \mathcal{P}\int_{-\infty}^{\infty} \frac{d\omega'}{\pi i}\frac{\chi(\omega)}{\omega' - \omega}\,. \tag{4.118}$$

Now, we define the complex dielectric function by

$$\chi(\omega) = \chi_1(\omega) + i\chi_2(\omega) \tag{4.119}$$

and then we obtain the following relation from (4.118):

$$\chi_1(\omega) = \mathcal{P}\int_{-\infty}^{\infty} \frac{d\omega'}{\pi}\frac{\chi_2(\omega')}{\omega' - \omega}, \tag{4.120a}$$

$$\chi_2(\omega) = -\mathcal{P}\int_{-\infty}^{\infty} \frac{d\omega'}{\pi}\frac{\chi_1(\omega')}{\omega' - \omega}\,. \tag{4.120b}$$

We divide the integration for χ_1 into the following two regions:

$$\chi_1(\omega) = \mathcal{P}\int_{-\infty}^{0} \frac{d\omega'}{\pi}\frac{\chi_2(\omega')}{\omega' - \omega} + \mathcal{P}\int_{0}^{\infty} \frac{d\omega'}{\pi}\frac{\chi_2(\omega')}{\omega' - \omega}.$$

The reality condition for the susceptibility tensor (dielectric tensor) $\chi(\omega) = \chi^*(-\omega)$ leads to the following relations

$$\chi_1(\omega) = \chi_1(-\omega), \tag{4.121a}$$

$$\chi_2(\omega) = -\chi_2(-\omega)\,. \tag{4.121b}$$

Using these relations we obtain for $\chi_1(\omega)$

$$\begin{aligned}\chi_1(\omega) &= \mathcal{P}\int_{0}^{\infty} \frac{d\omega'}{\pi}\chi_2(\omega')\left[\frac{1}{\omega' + \omega} + \frac{1}{\omega' - \omega}\right] \\ &= \mathcal{P}\int_{0}^{\infty} \frac{d\omega'}{\pi}\chi_2(\omega')\frac{2\omega'}{\omega'^2 - \omega^2}\,.\end{aligned} \tag{4.122}$$

In a similar fashion we calculate $\chi_2(\omega)$ and therefore obtain the following relations:

$$\chi_1(\omega) = \frac{2}{\pi}\mathcal{P}\int_{0}^{\infty} \frac{\omega'\chi_2(\omega')}{\omega'^2 - \omega^2}d\omega'\,, \tag{4.123a}$$

$$\chi_2(\omega) = -\frac{2\omega}{\pi}\mathcal{P}\int_{0}^{\infty} \frac{\chi_1(\omega')}{\omega'^2 - \omega^2}d\omega'\,. \tag{4.123b}$$

The above relations are called the **Kramers–Kronig relations**, and the transform between the real and imaginary part of the susceptibility tensors is called the Kramers–Kronig transform. The dielectric constant $\kappa(\omega)$ is related to the susceptibility by using the definition of electric displacement $\boldsymbol{D}$ through

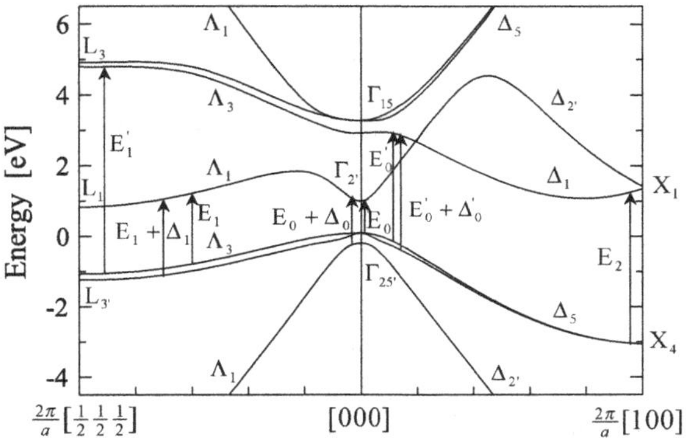

Fig. 4.14 Energy band structure of Ge calculated by the $\boldsymbol{k} \cdot \boldsymbol{p}$ perturbation method and the direct transitions at several important critical points (shown by arrows)

$$\begin{aligned} \boldsymbol{D} &= \epsilon_0 \boldsymbol{E} + \boldsymbol{P} = (1+\chi)\epsilon_0 \boldsymbol{E} \\ &\equiv \kappa\epsilon_0 \boldsymbol{E} \equiv \epsilon \boldsymbol{E}\,, \end{aligned} \tag{4.124}$$

and the following relation holds:

$$\kappa(\omega) = 1 + \chi(\omega)\,. \tag{4.125}$$

Therefore, the Kramers–Kronig relations for the dielectric constant $\kappa(\omega)$ are given by

$$\kappa_1(\omega) = 1 + \frac{2}{\pi}\mathcal{P}\int_0^\infty \frac{\omega'\kappa_2(\omega')}{\omega'^2 - \omega^2}\mathrm{d}\omega' \tag{4.126a}$$

$$\kappa_2(\omega) = -\frac{2\omega}{\pi}\mathcal{P}\int_0^\infty \frac{\kappa_1(\omega')}{\omega'^2 - \omega^2}\mathrm{d}\omega'\,, \tag{4.126b}$$

or, using (4.40), the relations

$$\chi_2(\omega) = \kappa_2(\omega) = \frac{\pi}{\epsilon_0}\left(\frac{e}{m\omega}\right)^2 \sum_{\boldsymbol{k}} |\boldsymbol{e}\cdot\boldsymbol{p}_{\mathrm{cv}}|^2 \delta\left[\mathcal{E}_{\mathrm{cv}}(\boldsymbol{k}) - \hbar\omega\right], \tag{4.127a}$$

$$\begin{aligned} \chi_1(\omega) &= \kappa_1(\omega) - 1 = \frac{2}{\pi}\int_0^\infty \frac{\omega'\kappa_2(\omega')\mathrm{d}\omega'}{\omega'^2 - \omega^2} \\ &= \frac{2}{\epsilon_0}\left(\frac{e\hbar}{m}\right)^2 \sum_{\boldsymbol{k}} \frac{|\boldsymbol{e}\cdot\boldsymbol{p}_{\mathrm{cv}}|^2}{\mathcal{E}_{\mathrm{cv}}(\boldsymbol{k})} \cdot \frac{1}{\mathcal{E}_{\mathrm{cv}}^2(\boldsymbol{k}) - \hbar^2\omega^2} \end{aligned} \tag{4.127b}$$

are obtained.

The energy band structure of Ge calculated by the $\boldsymbol{k} \cdot \boldsymbol{p}$ perturbation method is shown in Fig. 4.14, where the optical transitions at the critical points are indicated by arrows and the notation of the critical points is after Cardona [2, 10]. It should be noted that the representation used for the energy bands is the notation for the single group instead of the double-group representation. The reason is due to the fact that the basis functions used for the $\boldsymbol{k} \cdot \boldsymbol{p}$ perturbation calculations are those for the single-group representation. In the following we will derive the dielectric functions, real and imaginary parts, at the critical points shown in Fig. 4.14. The results are very important for understanding the optical properties of semiconductors. First, we derive the imaginary part of the dielectric function for a critical point, and the real part is calculated by using the Kramers–Kronig relation.

4.6.1 E_0, $E_0 + \Delta_0$ Edge

The critical points E_0 and $E_0 + \Delta_0$ are associated with the direct transition from the valence band and the spin-orbit split-off band to the lowest conduction band at the Γ point and thus they are M_0 type. The imaginary part of the dielectric function $\kappa_2(\omega)$ is calculated in Sects. 4.2 and 4.3. The real part of the dielectric function is calculated from the Kramers–Kronig relations.[2] The imaginary part of the dielectric function for the E_0 edge is

$$\kappa_2(\omega) = \sum_{i=1,2} A|\langle \mathrm{c}|p|\mathrm{v}_i\rangle|^2 \frac{1}{\omega^2}(\omega - \omega_0)^{1/2}, \tag{4.128}$$

$$A = \frac{e^2\hbar^{1/2}}{2\pi\epsilon_0 m^2}\left(\frac{2\mu_i}{\hbar^2}\right)^{3/2}, \tag{4.129}$$

where i represents the heavy and light hole valence bands and the summation is carried out for the two bands. μ_i is the reduced mass of the valence bands and the conduction band, and $\omega_0 = E_0/\hbar \equiv \mathcal{E}_G/\hbar$. Similarly the absorption coefficient for the transition at the $E_0 + \Delta_0$ edge between the spin-orbit split-off and the conduction band is calculated, which is given by replacing the momentum matrix element by $|\langle \mathrm{c}|p|\mathrm{v}_{\mathrm{so}}\rangle|^2 = |\langle \mathrm{c}|p|\mathrm{v}_i\rangle|^2 \equiv P_0^2$ (P_0 is defined in (2.83b) and (2.203d)), the reduced mass by μ_{so} and ω_0 by $\omega_0 s = (\mathcal{E}_G + \Delta_0)/\hbar$.

Using (4.126a) we obtain

$$\kappa_1(\omega) - 1$$
$$= AP_0^2\omega_0^{-3/2}\frac{1}{x_0^2}\left[2 - (1+x_0)^{1/2} - (1-x_0)^{1/2}\right]; \text{ for } x < x_0 \tag{4.130a}$$
$$= AP_0^2\omega_0^{-3/2}\frac{1}{x_0^2}\left[2 - (1+x_0)^{1/2}\right]; \text{ for } x > x_0 \tag{4.130b}$$

where $x_0 = \omega/\omega_0$.

---

[2] In this section we follow the custom adopted in the optical spectroscopy of semiconductors, and the critical points are expressed by symbols such as E_0, E_1 and E_2 instead of the notation for energy $\mathcal{E}_j$ used in this text. In later sections we use $\boldsymbol{E}$ for the electric field, which will not introduce any complication or confusion.

According to the $\boldsymbol{k} \cdot \boldsymbol{p}$ perturbation method described in Sect. 2.4, the effective mass is defined as a function of the momentum matrix element P_0 and the energy gaps. From (2.82) we have

$$\frac{1}{m(\Gamma_{2'})} = \frac{1}{m} + \frac{2P_0^2}{3m\hbar}\left(\frac{2}{\omega_0} + \frac{1}{\omega_0 + \Delta_0}\right) \tag{4.131}$$

for the electron in the conduction band at the Γ point. We take into account the $\Gamma_{2'}$ conduction and the $\Gamma_{25'}$ valence bands and neglect the perturbation due to higher bands. The effective mass in the $\langle 001 \rangle$ direction is obtained from (2.65) and (2.67) as

$$\frac{\hbar^2/2}{m_{\text{hh}}(001)} = -A - |B| \,, \tag{4.132a}$$

$$\frac{\hbar^2/2}{m_{\text{lh}}(001)} = -A + |B| \,, \tag{4.132b}$$

$$\frac{\hbar^2/2}{m_{\text{so}}} = -A - \Delta_0 \,. \tag{4.132c}$$

Using the definitions of Luttinger (2.226), (2.227) and (2.228) in Sect. 2.7 and neglecting contribution from the higher lying conduction bands, we obtain

$$A = \frac{\hbar^2}{2m}[1 + F + 2G] \simeq \frac{\hbar^2}{2m}[1 + F] \,, \tag{4.133a}$$

$$B = \frac{\hbar^2}{2m}[1 + H_1 + H_2] \simeq \frac{\hbar^2}{2m}[1] \,, \tag{4.133b}$$

$$C = \frac{\hbar^2}{2m}[F - G + H_1 - H_2] \simeq \frac{\hbar^2}{2m}[F] \,. \tag{4.133c}$$

and using (2.234)

$$F = -\frac{\mathcal{E}_{P0}}{3}\left[\frac{2}{\mathcal{E}_G} + \frac{1}{\mathcal{E}_G + \Delta_0}\right] , \quad \mathcal{E}_{P0} = \frac{2}{m}P_0^2 \,. \tag{4.134}$$

$$\frac{m}{m_{\text{hh}}(001)} = -2 - F \tag{4.135a}$$

$$\frac{m}{m_{\text{hh}}(001)} = -F \tag{4.135b}$$

$$\frac{m}{m_{\text{so}}(001)} = -(1 + F) \tag{4.135c}$$

$$\frac{m}{m(\Gamma_{2'})} = 1 - F \tag{4.135d}$$

Therefore the reduced masses are given by

$$\frac{m}{\mu_{\rm hh}} = \frac{m}{m(\Gamma_{2'})} + \frac{m}{m_{\rm hh}} = -1 - 2F = -1 + \frac{2\mathcal{E}_{P0}}{3}\left[\frac{2}{\mathcal{E}_G} + \frac{1}{\mathcal{E}_G + \Delta_0}\right]$$
$$\simeq -1 + \frac{2\mathcal{E}_{PO}}{\mathcal{E}_{\rm G}} \tag{4.136a}$$
$$\frac{m}{\mu_{\rm lh}} = \frac{m}{m(\Gamma_{2'})} + \frac{m}{m_{\rm lh}} = 1 - 2F = 1 + \frac{2\mathcal{E}_{P0}}{3}\left[\frac{2}{\mathcal{E}_G} + \frac{1}{\mathcal{E}_G + \Delta_0}\right]$$
$$\simeq 1 + \frac{2\mathcal{E}_{PO}}{\mathcal{E}_{\rm G}} \tag{4.136b}$$
$$\frac{m}{\mu_{\rm so}} = \frac{m}{m(\Gamma_{2'})} + \frac{m}{m_{\rm so}} = -2F = \frac{2\mathcal{E}_{P0}}{3}\left[\frac{2}{\mathcal{E}_G} + \frac{1}{\mathcal{E}_G + \Delta_0}\right]$$
$$\simeq \frac{2\mathcal{E}_{PO}}{\mathcal{E}_{\rm G}} \tag{4.136c}$$

The above results reflect that the electron effective mass $m(\Gamma_{2'})$ is small compared with the effective masses of the heavy, light, and spin-orbit split off bands. As an example we consider the case of GaAs, where $\mathcal{E}_{\rm G} = 1.51$ [eV]. $\Delta_0 = 0.341$ [eV], and $\mathcal{E}_P = 28.8$ [eV]. Assuming $\mathcal{E}_{\rm G} \gg \Delta_0$ and $\mathcal{E}_{PO}/\mathcal{E}_{\rm G} \gg 1$, we obtain the following simple relation:

$$\frac{1}{\mu_{\rm hh}} = \frac{1}{\mu_{\rm lh}} = \frac{1}{\mu_{\rm so}} = \frac{2\mathcal{E}_{PO}}{m\mathcal{E}_{\rm G}} = \frac{2\mathcal{E}_{PO}}{m\hbar\omega_0} \tag{4.137}$$
$$\mu_i \equiv= \frac{m\mathcal{E}_{\rm G}}{2\mathcal{E}_{PO}} = \frac{m\hbar\omega_0}{2\mathcal{E}_{PO}} \tag{4.138}$$

Therefore, the real part of the dielectric function due to the contribution from the E_0 edge and the $E_0 + \Delta_0$ edge is given by

$$\kappa_1(\omega) - 1 = C_0''\left[f(x_0) + \frac{1}{2}\left(\frac{\omega_0}{\omega_{\rm so}}\right)^{3/2} f(x_{0\rm s})\right] \tag{4.139}$$

where $C_0'' \simeq 2AP_0^2\omega_0^{-3/2}$ with $\mu_i = m\hbar\omega_0/(2\mathcal{E}_{PO}) = m\mathcal{E}_{\rm G}/2\mathcal{E}_{P0}$, $x_0 = \omega/\omega_0$ and $x_{0\rm s} = \omega/(\omega_0 + \Delta_0)$. The function $f(x)$ is defined as

$$f(x) = x^{-2}\left[2 - (1+x)^{1/2} - (1-x)^{1/2}\right]. \tag{4.140}$$

4.6.2 E_1 and $E_1 + \Delta_1$ Edge

We have already mentioned that the joint density of states for the optical transition given by (4.44) exhibits a divergence when the condition given by (4.49) or (4.50) is satisfied. The energy band calculations show that the conduction band and the valence band are parallel over a wide range of the wave vector $\boldsymbol{k}$ in the $\langle 111\rangle$ direction of $\boldsymbol{k}$ space, satisfying the condition of (4.50). As discussed in Chap. 2, the conduction band Λ_1 originated from $\Gamma_{2'}$ and the valence bands Λ_3 (doubly degenerate bands split into heavy and light hole bands due to the spin-orbit interaction) originated from $\Gamma_{25'}$ exhibit the parallel behavior, and these features appear in almost all semiconductors, giving rise to a critical point called E_1 and $E_1 + \Delta_1$. Since the critical point is associated with the parallel feature over a wide range in the $\langle 111 \rangle$ direction, the critical point may be approximated as being two-dimensional and the density of states can be expressed by the step function

$$\left.\begin{array}{ll} E_1 & : \kappa_2 = B|\langle \mathrm{c}|p|\mathrm{v}\rangle|^2 \omega^{-2} H(\omega - \omega_1) \\ E_1 + \Delta_1 & : \kappa_2 = B'|\langle \mathrm{c}|p|\mathrm{v}\rangle|^2 \omega^{-2} H(\omega - \omega_1 - \Delta_1) \end{array}\right\} , \tag{4.141}$$

where $B = \sqrt{3}\pi\mu(E_1)e^2/\epsilon_0\hbar^2 m^2 a_0$, $B' = \sqrt{3}\pi\mu(E_1 + \Delta_1)e^2/\epsilon_0\hbar^2 m^2 a_0 \simeq B$, and H is the step function defined as $H(x) = 0$ for $x < 0$ and $H(x) = 1$ for $x > 0$. The detail of the derivation is found in the papers by Cardona [11] and Higginbotham et al. [12]. The lattice constant a_0 appears in the above equation from the integral $\int \mathrm{d}k_z = 2\pi\sqrt{3}/a_0$. The real part of the dielectric function for the E_1 edge and the $E_1 + \Delta_1$ edge is calculated from the Kramers–Kronig relation as

$$\begin{aligned} \kappa_1(\omega) - 1 &= -\frac{B}{\pi}|\langle \mathrm{c}|p|\mathrm{v}\rangle|^2 \omega^{-2} \log\left|\frac{\omega_1^2 - \omega^2}{\omega_1^2}\right| \\ &= \frac{B}{\pi}|\langle \mathrm{c}|p|\mathrm{v}\rangle|^2 \omega_1^{-2}\left(1 + \frac{1}{2}\frac{\omega^2}{\omega_1^2} + \cdots\right) \end{aligned} \tag{4.142}$$

for the E_1 edge, and

$$\begin{aligned} \kappa_1(\omega) - 1 &= -\frac{B'}{\pi}|\langle \mathrm{c}|p|\mathrm{v}\rangle|^2 \omega^{-2} \log\left|\frac{(\omega_1 + \Delta_1)^2 - \omega^2}{(\omega_1 + \Delta_1)^2}\right| \\ &= \frac{B'}{\pi}|\langle \mathrm{c}|p|\mathrm{v}\rangle|^2 (\omega_1 + \Delta_1)^{-2}\left(1 + \frac{1}{2}\frac{\omega^2}{(\omega_1 + \Delta_1)^2} + \cdots\right) \end{aligned} \tag{4.143}$$

for the $E_1 + \Delta_1$ edge.

4.6.3 E_2 Edge

The reflectance R and the imaginary part of the dielectric function $\kappa_2(\omega)$ in semiconductors exhibit a large peak at around $\hbar\omega \simeq 4\,\mathrm{eV}$ and the peak is called the E_2 edge. The origin of the E_2 edge is not well identified, but is believed to be associated with a combination of several critical points. Here the dielectric function of the E_2 edge is treated by the method proposed by Higginbotham, Cardona and Pollak [12]. Since the E_2 edge has a large peak at $\hbar\omega = E_2$, the dielectric function is approximated by using a simple harmonic oscillator model, the Drude model. Then the dielectric function is given by

$$\kappa_1(\omega) - 1 \simeq \frac{C_2''\omega_2^2}{\omega_2^2 - \omega^2} \simeq C_2''(1 + x_2^2) , \tag{4.144}$$

where $\hbar\omega_2 = E_2$ and $x_2 = \omega/\omega_2$.

4.6.4 Exciton

As we discussed in Sect. 4.5, exciton absorption will occur in semiconductors with a direct gap, where excitons consists of electron-hole pairs interacting through the Coulomb potential and the absorption energy is below the fundamental absorption edge by the exciton binding energy (a few meV). For simplicity, we will consider the ground state excitons excited at the E_0 edge, and then the absorption may be approximated by the Lorentz function. The real part of the dielectric function for the exciton is derived in Sect. 4.5 and given by (4.108):

$$\kappa_1(\omega) \simeq \frac{C_{\text{ex}}''\omega_{\text{ex}}^2}{\omega_{\text{ex}}^2 - \omega^2} = \frac{C_{\text{ex}}''}{1 - x_{\text{ex}}^2}, \tag{4.145}$$

where $\hbar\omega_{\text{ex}}$ is the energy of the exciton, the binding energy is $\mathcal{E}_{\text{ex}} = \hbar(\omega_0 - \omega_{\text{ex}})$, and $x_{\text{ex}} = \omega/\omega_{\text{ex}}$.

The above results are summarized to give the real part of the dielectric function below the E_0 edge as

$$\begin{aligned}\kappa_1(\omega) - 1 = &\frac{C_{\text{ex}}''}{1 - x_{\text{ex}}^2} + C_0''\left[f(x_0) + \frac{1}{2}\left(\frac{\omega_0}{\omega_{0\text{s}}}\right)^{3/2} f(x_{0\text{s}})\right], \\ &+ C_1''\left[h(x_1) + \left(\frac{\omega_1}{\omega_{1\text{s}}}\right)^2 h(x_{1\text{s}})\right] + C_2''(1 + x_2^2),\end{aligned} \tag{4.146}$$

where $x_1 = \omega/\omega_1$, $x_{1\text{s}} = \omega/(\omega_1 + \Delta_1)$ and

$$h(x) = 1 + \frac{1}{2}x^2. \tag{4.147}$$

It is well known that the higher-energy edges exhibit only a weak dispersion in the dielectric constant and thus the contribution is approximated by replacing 1 with a constant D in $\kappa_1(\omega) - 1$. This approximation is known to give good agreement with experiment in many semiconductors.

Examples of the comparison between the theory and experiment are shown in Fig. 4.15 for Ge and in Fig. 4.16 for GaAs, where the solid curves are best fitted by using (4.146) by neglecting the contribution from the exciton effect and we find good agreement between the theory and experiment. The parameters used to calculated the theoretical curves in Fig. 4.15 for Ge and in Fig. 4.16 for GaAs are shown in Table 4.3.

It has been shown that the dielectric function is well expressed by taking the contribution from the E_0 edge and the $E_0 + \Delta_0$ edge into account [13]. When we adopt this approximation, the real part of the dielectric function is given by the terms due to the E_0 edge and the $E_0 + \Delta_0$ edge and the constant D due to the contributions from the other edges. Therefore, the dielectric function is well expressed by

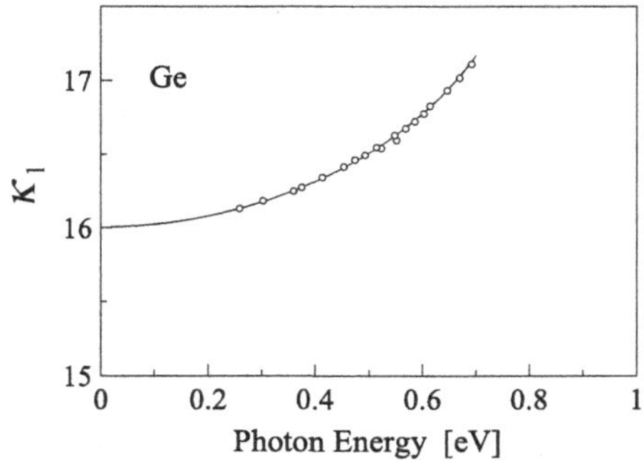

Fig. 4.15 The real part of the dielectric function for Ge, where experimental results are compared with the best fitted curve calculated from (4.146) (without the exciton effect) (from [12])

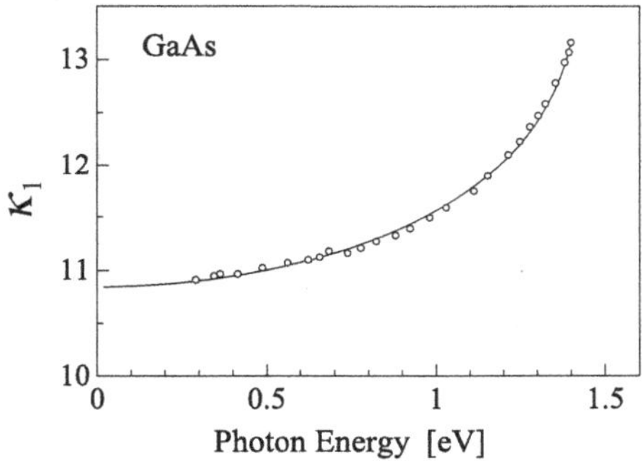

Fig. 4.16 The real part of the dielectric function for GaAs, where experimental results are compared with the best fitted curve calculated from (4.146) (without the exciton effect) (from [12])

Table 4.3 Parameters use to fit the theoretical curves of the dielectric function with the experimental results in Ge and GaAs. All the parameters are for $T = 300\,\mathrm{K}$ and the critical point energies are shown in units of [eV]

	Ge		GaAs	
	Experiment	Calculation	Experiment	Calculation
C_0''	4.85	1.90	7.60	1.53
C_1''	5.60	3.22	2.05	2.50
C_2''	3.10		3.50	
E_0	0.797		1.43	
$E_0 + \Delta_0$	1.087		1.77	
E_1	2.22		2.895	
$E_1 + \Delta_1$	2.42		3.17	
E_2	4.49		4.94	

$$\kappa_1(\omega) = C_0''\left[f\left(\frac{\omega}{\omega_0}\right) + \frac{1}{2}\left(\frac{\omega_0}{\omega_{0s}}\right)^{3/2} f\left(\frac{\omega}{\omega_{0s}}\right)\right] + D\,. \tag{4.148}$$

The difference in C_0'' between the fit and the theory is believed to be due to the exciton effect and neglect of the contribution from the higher-energy critical points.

4.7 Piezobirefringence

4.7.1 Phenomenological Theory of Piezobirefringence

When a stress is applied to a semiconductor crystal, it produces a change in the refractive index in the parallel and perpendicular directions. This effect is called **piezobirefringence**. The plane of polarization of the light makes an angle of 45° with the stress direction. Experimentally, one obtains the phase difference Δ between the components of the light polarized parallel and perpendicular to the stress axis. Assuming a sample thickness t the phase difference per unit length is given by [12]

$$\frac{\Delta}{t} = \frac{2\pi\left(n_{\parallel} - n_{\perp}\right)}{\lambda}, \tag{4.149}$$

where λ is the wavelength of the incident light, and $n_{\parallel}$ and $n_{\perp}$ are the refractive indices of the polarization in the directions parallel and perpendicular to the stress. Normally the piezobirefringence experiment is carried out in the transparent region of the optical spectrum, where the absorption coefficient is small, enabling us to neglect the imaginary part of dielectric function ($\kappa_2 \simeq 0$) and thus to use (4.149)

$$\frac{\Delta}{t} = \frac{2\pi}{\lambda}\left[\frac{(\kappa_1)_{\parallel} - (\kappa_1)_{\perp}}{n_{\parallel} + n_{\perp}}\right] \simeq \frac{\pi}{\lambda n_0}\left[(\kappa_1)_{\parallel} - (\kappa_1)_{\perp}\right], \tag{4.150}$$

where n_0 is the refractive index without stress and relations derived from (4.15) are used such as $n_{\parallel} = \sqrt{(\kappa_1)_{\parallel}}$. Piezobirefringence has a very important property for investigating the stress effect on the energy bands, especially on the valence bands. In this subsection we will deal with a phenomenological theory of piezobirefringence and microscopic theory based on the band structure of semiconductors. It may be easily understood by replacing stress with phonons that piezobirefringence is related to Brillouin scattering and Raman scattering as discussed later.

When a stress T_{kl} is applied to a crystal, the dielectric constant tensor κ_{ij} is changed by $\Delta\kappa_{ij}$. We define the change by

$$\Delta\kappa_{ij} = Q_{ijkl}T_{kl}, \tag{4.151}$$

where Q_{ijkl} is called the piezobirefringence tensor of fourth rank. The fourth-rank tensor is often expressed by $Q_{\alpha\beta}$ ($\alpha, \beta = 1, 2, \ldots, 6$) and the non-zero components are Q_{11}, Q_{12}, Q_{44} for a crystal with cubic symmetry. We consider the case where the stress is applied in the [100] direction ($T_{xx} = T_1 = X$) as an example. The change in the dielectric constant for light polarized parallel to the stress is then given by

$$\Delta\kappa_{\parallel} = \Delta\kappa_{xx} = Q_{xxxx}T_{xx} = Q_{11}X,$$

and for light polarized perpendicular to the stress is given by

$$\Delta\kappa_\perp = \Delta\kappa_{zz} = Q_{zzxx}T_{xx} = Q_{31}T_1 = Q_{12}X.$$

Therefore, piezobirefringence coefficient $\alpha(100)$ is given by

$$\alpha(100) = \frac{\kappa_\parallel - \kappa_\perp}{X} = Q_{11} - Q_{12}\,. \tag{4.152}$$

Similarly, we can obtain the piezobirefringence coefficient for the stress applied in the other directions.

The above treatment is phenomenological and the piezobirefringence coefficients reflect the stress-induced change in the energy band structure and thus depend on the critical points. In general, a stress induces a change in the lattice constant, giving rise to a change in the energy band structure. The change is estimated quantitatively by using the deformation potential, which will be discussed in more detail in Chap. 6 in connection with the electron-phonon interaction. In this section we will consider the general theory of the deformation potential and its application to piezobirefringence theory.

4.7.2 Deformation Potential Theory

Defining the coordinate vectors before and after the deformation by $\boldsymbol{r}$ and $\boldsymbol{r}'$ and the displacement vector by $\boldsymbol{u}$, we find

$$\boldsymbol{u} = \boldsymbol{r} - \boldsymbol{r}' = \boldsymbol{e}\cdot\boldsymbol{r}\,, \tag{4.153}$$

where $\boldsymbol{e}$ is the strain tensor defined by

$$\boldsymbol{e} = e_{ij} = \frac{1}{2}\left(\frac{\partial u_i}{\partial r_j} + \frac{\partial u_j}{\partial r_i}\right)\,. \tag{4.154}$$

Equation (4.153) can be rewritten as

$$r_i' = r_i - \sum_j e_{ij} r_j\,. \tag{4.155}$$

Then the operator $\boldsymbol{\nabla}$ of the old coordinates $\boldsymbol{r}$ is expressed by new coordinates $\boldsymbol{r}'$ as

$$\frac{\partial}{\partial r_i} = \frac{\partial}{\partial r_i'} - \sum_j e_{ij}\frac{\partial}{\partial r_j'}\,, \tag{4.156}$$

and thus we obtain

$$\frac{\partial^2}{\partial r_i^2} = \left(\frac{\partial}{\partial r_i'} - \sum_j e_{ij}\frac{\partial}{\partial r_j'}\right)^2 \simeq \frac{\partial^2}{\partial r_i'^2} - 2\sum_j e_{ij}\frac{\partial^2}{\partial r_i'\partial r_j'} , \tag{4.157}$$

where the above equation takes account of the first order of e_{ij} and neglects the higher-order terms. Then we obtain

$$\nabla^2 = \nabla'^2 - 2\sum_{i,j} e_{ij}\frac{\partial^2}{\partial r_i'\partial_j'} . \tag{4.158}$$

Since the crystal potential is given by

$$V(\boldsymbol{r}) = V(\boldsymbol{r}' + \boldsymbol{e}\cdot\boldsymbol{r}) \simeq V(\boldsymbol{r}' + \boldsymbol{e}\cdot\boldsymbol{r}') , \tag{4.159}$$

it is expanded with respect to $\boldsymbol{e}$ as:

$$V(\boldsymbol{r}) = V_0(\boldsymbol{r}') + \sum_{i,j} V_{ij}(\boldsymbol{r}')e_{ij} , \tag{4.160}$$

where the expansion coefficient $V_{ij}(\boldsymbol{r}')$ is given by[3]

$$V_{ij}(\boldsymbol{r}') = \left.\frac{\partial V(\boldsymbol{r}' + \boldsymbol{e}\cdot\boldsymbol{r}')}{\partial e_{ij}}\right|_{e=0} = -\frac{\partial V(\boldsymbol{r}')}{\partial r_i'}r_j' . \tag{4.161}$$

In the following we neglect the stress effect on the spin-orbit split-off-band, because experiments have shown no significant stress effect on this band. Therefore, the Hamiltonian of an electron is written as

$$H = H_0 + H_{\text{so}} + H_{\text{s}} , \tag{4.162}$$

$$H_0 = -\frac{\hbar^2}{2m} + V_0(\boldsymbol{r}) , \tag{4.163}$$

where H_0 is the Hamiltonian without stress, H_{so} is spin-orbit interaction, and H_{s} is the Hamiltonian for the stress effect. Using (4.158) and (4.160), the stress Hamiltonian is written as:

$$H_{\text{s}} = \frac{\hbar^2}{m}\sum_{ij} e_{ij}\frac{\partial^2}{\partial r_i\partial r_j} + V_{ij}e_{ij} \equiv \sum_{ij} D^{ij}e_{ij} , \tag{4.164}$$

where

[3]From (4.155) we obtain

$$\frac{\partial V}{\partial e_{ij}} = \frac{\partial V}{\partial r_i'}\frac{\partial r_i'}{\partial e_{ij}} = -\frac{\partial V}{\partial r_i'}r_j \simeq -\frac{\partial V}{\partial r_i'}r_j' .$$

$$D^{ij} = \frac{\hbar^2}{m} \frac{\partial^2}{\partial r_i \partial r_j} + V_{ij} \tag{4.165}$$

is called the deformation potential operator.

For simplicity we discuss the stress effect in a cubic crystal. It may be understood that the treatment can be applied to any crystal by taking the crystal symmetry into account. The valence bands are assumed to be 3-fold degenerate p states, which is equivalent to taking into account the $\Gamma_{25'}$ valence bands as shown in Sect. 2.2 to analyze the valence band structure. Following the treatment of Sect. 2.2, the matrix elements of H_s are shown to be

$$\langle k|H_s|l\rangle = \sum_{ij} \langle k|D^{ij}|l\rangle e_{ij} \equiv \sum_{ij} D_{klij} e_{ij} , \tag{4.166}$$

where

$$D_{klij} = \langle k|D^{ij}|l\rangle . \tag{4.167}$$

D_{klij} is a fourth-rank tensor, called the deformation potential tensor, and has elements similar to (2.43). Using (2.53a), we obtain for example

$$\begin{aligned}\langle u_+|H_s|u_+\rangle &= \frac{1}{2}\left(D_{xxxx}e_{xx} + D_{xxyy}e_{yy} + D_{xxzz}e_{zz}\right. \\ &\quad \left. + D_{yyxx}e_{xx} + D_{yyyy}e_{yy} + D_{yyzz}e_{zz}\right) \qquad (4.168)\\ &= \frac{1}{2}(D_{11} + D_{12})(e_{xx} + e_{yy}) + D_{12}e_{zz} \\ &\equiv G_1 . \qquad (4.169)\end{aligned}$$

In a similar fashion other elements are calculated and finally we obtain the following result:

$$\langle k|H_s|l\rangle = \begin{vmatrix} G_1 & G_2 & G_4 \\ G_2^* & G_1 & G_4^* \\ G_4^* & G_4 & G_3 \end{vmatrix} , \tag{4.170}$$

$$\begin{aligned}G_1 &= \frac{1}{2}(D_{11} + D_{12})(e_{xx} + e_{yy}) + D_{12}e_{zz} , \\ G_2 &= \frac{1}{2}(D_{11} - D_{12})(e_{xx} - e_{yy}) - 2\mathrm{i}D_{44}e_{xy} , \\ G_3 &= D_{12}(e_{xx} + e_{yy}) + D_{11}e_{zz} , \\ G_4 &= \sqrt{2}D_{44}(e_{xz} - \mathrm{i}e_{yz}) .\end{aligned}$$

From the above analysis we find that the deformation of the valence bands due to stress or strain is described by three potential constants D_{11}, D_{12} and D_{44}. Another approach to the stress effect has been reported by Picus and Bir [14], who introduced the deformation potential due to the volume deformation ($e_{xx}+e_{yy}+e_{zz}$; hydrostatic term) and to the uniaxial term. The strain Hamiltonian for the $\Gamma_{25'}$ bands at $\boldsymbol{k}=0$ is then written as

$$H_{\mathrm{s}}^{(i)} = -a^{(i)}(e_{xx}+e_{yy}+e_{zz}) - 3b^{(i)}\left[\left(L_x^2-\frac{1}{3}L^2\right)e_{xx}+\text{c.p.}\right] - \frac{6d^{(i)}}{\sqrt{3}}\left[\{L_xL_y\}e_{xy}+\text{c.p.}\right], \tag{4.171}$$

where the index i represents one of the three valence bands, $\boldsymbol{L}$ is a dimensionless operator given by the angular momentum operator for the quantum number $l=1$ of (2.52) divided by $\hbar$, c.p. is the cyclic permutation of x, y, z, and $\{L_xL_y\} = (1/2)(L_xL_y+L_yL_x)$[4] The constant $a^{(i)}$ is the hydrostatic deformation potential for the band i, and $b^{(i)}$ and $\mathrm{d}^{(i)}$ are the deformation potentials for the uniaxial strain (e_{ij}, $i=j$) and the shear strain (e_{ij}, $i\neq j$), respectively. Since experiments measure the change in the energy gap between the $\Gamma_{2'}$ (Γ_1) conduction band and the $\Gamma_{25'}$ (Γ_{15}) valence bands under hydrostatic pressure, the deformation potential $a^{(i)}$ corresponds to the relative change in the energy gap (not the valence bands only). In Chap. 9 we will discuss the strain effect on the quantum well lasers, where the deformation potential a_{c} for the conduction band and a_{v} for the valence bands are defined. When the conduction band-edge shift is expressed by $\mathcal{E}(X)-\mathcal{E}_{\mathrm{c}}(X=0)=a_{\mathrm{c}}(e_{xx}+e_{yy}+e_{zz})$ and the valence band top by $-a_{\mathrm{v}}(e_{xx}+e_{yy}+e_{zz})$, then the change in the energy gaps due to the hydrostatic pressure is given by $(a_{\mathrm{c}}+a_{\mathrm{v}})(e_{xx}+e_{yy}+e_{zz})$. Therefore the total deformation potential is sometimes defined by $a=a_{\mathrm{c}}+a_{\mathrm{v}}$. It should be noted here that the Hamiltonian H_{s} with $D_{\alpha\beta}$ and the Hamiltonian with $a^{(i)}$, $b^{(i)}$, $\mathrm{d}^{(i)}$ give rise to the same results as expected from the analysis given above and that the relations between the two types of deformation potential is easily obtained.

4.7.3 Stress-Induced Change in Energy Band Structure

A typical energy band structure in the vicinity of $\boldsymbol{k}=0$ (Γ) such as Ge and GaAs is shown schematically in Fig. 4.17 for the conduction band and the valence bands. The valence bands without stress consist of doubly degenerate bands $|\frac{3}{2},\frac{3}{2}\rangle$, $|\frac{3}{2},\frac{1}{2}\rangle$ and the spin-orbit split-off band $|\frac{1}{2},\frac{1}{2}\rangle$, as shown on the left of Fig. 4.17. When

[4]The reported matrix elements of the strain Hamiltonian are evaluated by using the angular momentum operators in terms of "spin" matrices corresponding to spin unity according to Picus and Bir [14], where the basis functions are taken as

$$S_+ = \frac{1}{\sqrt{2}}|X+\mathrm{i}Y\rangle, \qquad S_0=|Z\rangle, \qquad S_- = \frac{1}{\sqrt{2}}|X-\mathrm{i}Y\rangle.$$

The corresponding operators are then expressed by the relation (2.52). The matrix elements of the strain Hamiltonian in the text are evaluated by using the basis functions and the corresponding operators.

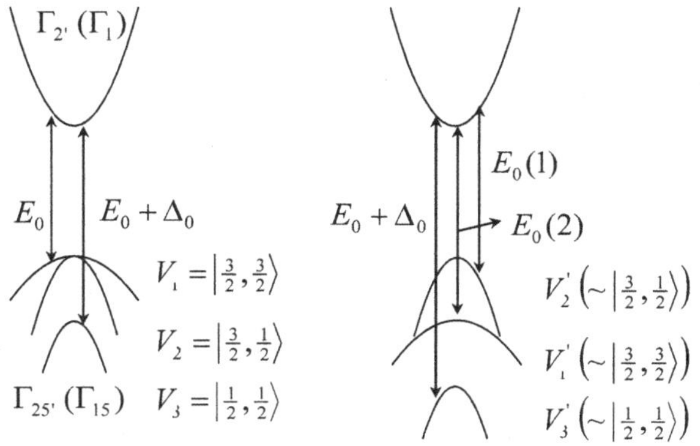

Fig. 4.17 Energy band structure of semiconductors such as Ge and GaAs near $\boldsymbol{k} = 0$ with and without stress. The degenerate valence bands without stress split into three bands under the compressive uniaxial stress ($X < 0$) as shown on the right. The bands V_1', V_2' and V_3' originate from V_1, V_2 and V_3 bands but they are mixed up with each other at $\boldsymbol{k} \neq 0$

stress is applied, the degeneracy is removed, as shown on the right of Fig. 4.17 for the compressive uniaxial stress. Therefore, we may expect a remarkable change in the optical properties of semiconductors by the application of stress. In experiments uniaxial stress along the [001], [111] or [110] direction is applied. Using the results in Appendix C the strain components for the stress X are ($X > 0$ and $X < 0$ correspond to the tensile and compressive stresses, respectively),

[001] stress

$$e_{zz} = s_{11}X\,,$$
$$e_{xx} = e_{yy} = s_{12}X\,,$$
$$e_{xy} = e_{xz} = e_{yz} = 0\,,$$

[111] stress

$$e_{xx} = e_{yy} = e_{zz} = (s_{11} + 2s_{12})\left(\frac{1}{3}X\right)\,,$$
$$e_{xy} = e_{yz} = e_{zx} = \left(\frac{1}{2}s_{44}\right)\left(\frac{1}{3}X\right)\,,$$

[110] stress

$$e_{xx} = e_{yy} = (s_{11} + s_{12})\left(\frac{1}{2}X\right)\,,$$
$$e_{zz} = s_{12}X\,,$$
$$e_{xy} = \left(\frac{1}{2}s_{44}\right)\left(\frac{1}{2}X\right)\,,$$
$$e_{xz} = e_{yz} = 0\,,$$

where $s_{\alpha\beta}$ ($e_{ij} = s_{ijkl}T_{kl}$, $e_\alpha = s_{\alpha\beta}T_\beta$; $i, j = x, y, x$; $\alpha, \beta = 1, 2, \ldots, 6$) is the elastic compliance tensor (see Appendix C). In the following the band index i is dropped out.

The matrix elements of the strain Hamiltonian are easily evaluated using (4.171) and the wave functions (2.63a) $\sim$ (2.63f), which will be used to derive the 6×6 Hamiltonian matrix in Sects. 9.6.2 and 9.6.4. We rewrite the strain Hamiltonian H_s in the following form.

$$\begin{aligned} H_\mathrm{s} &= -a_\mathrm{v}\left(e_{xx}+e_{yy}+e_{zz}\right) - 3b\left[\left(L_x^2 - \frac{1}{3}\boldsymbol{L}^2\right)e_{xx} + \mathrm{c.p.}\right] \\ &= H_\mathrm{s}^\mathrm{hydro} + H_\mathrm{s}^\mathrm{shear}\,, \qquad (4.172\mathrm{a}) \\ H_\mathrm{s}^\mathrm{hydro} &= -a_\mathrm{v}\left(e_{xx}+e_{yy}+e_{zz}\right)\,, \qquad (4.172\mathrm{b}) \\ H_\mathrm{s}^\mathrm{shear} &= -3b\left[\left(L_x^2 - \frac{1}{3}\boldsymbol{L}^2\right)e_{xx} + \mathrm{c.p.}\right]\,, \qquad (4.172\mathrm{c}) \end{aligned}$$

where the deformation potential a_v is defined for the valence bands. When we define the hydrostatic pressure dependence of the conduction band edge by $a_\mathrm{c}(e_{xx}+e_{yy}+e_{zz})$, the total band gap change due to the hydrostatic pressure is given by $(a_\mathrm{c}+a_\mathrm{v})\left(e_{xx}+e_{yy}+e_{zz}\right)$.

Then the matrix elements for H^hydro are given by

$$\begin{aligned} \left\langle\frac{3}{2},\frac{3}{2}\right|H_\mathrm{s}^\mathrm{hydro}\left|\frac{3}{2},\frac{3}{2}\right\rangle = \left\langle\frac{3}{2},\frac{1}{2}\right|H_\mathrm{s}^\mathrm{hydro}\left|\frac{3}{2},\frac{1}{2}\right\rangle = \left\langle\frac{1}{2},\frac{1}{2}\right|H_\mathrm{s}^\mathrm{hydro}\left|\frac{1}{2},\frac{1}{2}\right\rangle \\ = -a_\mathrm{v}\left(e_{xx}+e_{yy}+e_{zz}\right)\,, \qquad (4.173) \end{aligned}$$

and the matrix elements for $H_\mathrm{s}^\mathrm{shear}$ are[5]

$$\left\langle\frac{3}{2},\frac{3}{2}\right|H_\mathrm{s}^\mathrm{shear}\left|\frac{3}{2},\frac{3}{2}\right\rangle = +\frac{b}{2}(e_{xx}+e_{yy}-2e_{zz})\,, \qquad (4.174\mathrm{a})$$

$$\left\langle\frac{3}{2},\frac{1}{2}\right|H_\mathrm{s}^\mathrm{shear}\left|\frac{3}{2},\frac{1}{2}\right\rangle = -\frac{b}{2}(e_{xx}+e_{yy}-2e_{zz})\,, \qquad (4.174\mathrm{b})$$

$$\left\langle\frac{3}{2},\frac{1}{2}\right|H_\mathrm{s}^\mathrm{shear}\left|\frac{1}{2},\frac{1}{2}\right\rangle = -i\frac{1}{\sqrt{2}}b\left(e_{xx}+e_{yy}-2e_{zz}\right)\,., \qquad (4.174\mathrm{c})$$

where we have to note that $\langle 3/2, 3/2|H_\mathrm{s}^\mathrm{shear}|3/2, 1/2\rangle = 0$ because of the diagonal property of the spin directions. We define these non-zero components as

$$\Delta^\mathrm{hydro} = -a_\mathrm{v}(e_{xx}+e_{yy}+e_{zz}) = -a_\mathrm{v}\left(s_{11}+2s_{12}\right)X\,, \qquad (4.175\mathrm{a})$$

$$\Delta^\mathrm{shear} = +\frac{b}{2}\left(e_{xx}+e_{yy}-2e_{zz}\right) = -b\left(s_{11}-s_{12}\right)X\,. \qquad (4.175\mathrm{b})$$

These results give the Hamiltonian matrix

$$\begin{array}{ccc} \left|\frac{3}{2},\frac{3}{2}\right\rangle & \left|\frac{3}{2},\frac{1}{2}\right\rangle & \left|\frac{1}{2},\frac{1}{2}\right\rangle \end{array}$$
$$\left|\begin{array}{ccc} \Delta^\mathrm{hydro}+\Delta^\mathrm{shear} & 0 & 0 \\ 0 & \Delta^\mathrm{hydro}-\Delta^\mathrm{shear} & -\mathrm{i}\sqrt{2}\Delta^\mathrm{shear} \\ 0 & \mathrm{i}\sqrt{2}\Delta^\mathrm{shear} & -\Delta_0+\Delta^\mathrm{hydro} \end{array}\right|\,, \qquad (4.176)$$

---

[5]The relation of (4.174a) is sometimes expressed by using negative sign $(-b/2\cdots)$ as discussed in Chap. 9. The change of expressions, however, is just the change in the sign of the deformation potential b.

where the energy is measured from the top of the doubly degenerate valence band edge. Equation (4.176) will be used to derive 6×6 Hamiltonian matrix (9.161) for a quantum well with strain in Sect. 9.6.4.

In the case of [001] stress, the parameters defined by Pollak and Cardona [15] are given by

$$\delta E_{\mathrm{H}} = -\Delta^{\mathrm{hydro}} = +a(s_{11} + 2s_{12})X\,, \tag{4.177a}$$

$$\delta E_{001} = -2\Delta^{\mathrm{shear}} = +2b(s_{11} - s_{12})X\,, \tag{4.177b}$$

where $a = a_{\mathrm{c}} + a_{\mathrm{v}}$ and thus $\delta E_{\mathrm{H}} = a(s_{11}+2s_{12})X = (\partial \mathcal{E}_g/\partial P)\cdot P$ is the change in the energy gap due to the hydrostatic component of the strain (P: hydrostatic pressure). The term $\delta E_{001} = 2b(s_{11} - s_{12})X$ is the linear splitting of the $\left|\frac{3}{2}, \pm\frac{3}{2}\right\rangle$ and $\left|\frac{3}{2}, \pm\frac{1}{2}\right\rangle$ valence bands, and the mixing between $\left|\frac{3}{2}, \pm\frac{1}{2}\right\rangle$ and $\left|\frac{1}{2}, \pm\frac{1}{2}\right\rangle$ valence bands. Since the stress does not remove the Kramers degeneracy of each state, diagonalization of the matrix of the Hamiltonian $H_{\mathrm{so}} + H_{\mathrm{s}}$ is done by diagonalizing the matrix for positive m_J of the valence states $|J, m_J\rangle$, which gives rise to the following matrix [15]

$$\begin{array}{ccc} \left|\frac{3}{2},\frac{3}{2}\right\rangle & \left|\frac{3}{2},\frac{1}{2}\right\rangle & \left|\frac{1}{2},\frac{1}{2}\right\rangle \end{array}$$
$$\begin{vmatrix} \frac{1}{3}\Delta_0 - \delta E_{\mathrm{H}} - \frac{1}{2}\delta E_{001} & 0 & 0 \\ 0 & \frac{1}{3}\Delta_0 - \delta E_{\mathrm{H}} + \frac{1}{2}\delta E_{001} & \frac{1}{2}\sqrt{2}\delta E_{001} \\ 0 & \frac{1}{2}\sqrt{2}\delta E_{001} & -\frac{2}{3}\Delta_0 - \delta E_{\mathrm{H}} \end{vmatrix}\,, \tag{4.178}$$

where the origin of the energy is set to the top of the triply degenerate valence bands without spin-orbit interaction. Diagonalizing the matrix, the energy splitting and corresponding eigenstates are obtained. Figure 4.17 shows a schematic illustration of the stress induced change of the three valence bands for a compressive stress. Thus the energy band gap shift at $\boldsymbol{k} = 0$ is given by the following equations, where the approximation is valid for $\delta E_{001} \ll \Delta_0$:

$$\Delta(\mathcal{E}_{\mathrm{c}} - \mathcal{E}_{\mathrm{v1}}) = -\frac{1}{3}\Delta_0 + \delta E_{\mathrm{H}} + \frac{1}{2}\delta E_{001}\,, \tag{4.179a}$$

$$\Delta(\mathcal{E}_{\mathrm{c}} - \mathcal{E}_{\mathrm{v2}}) = \frac{1}{6}\Delta_0 + \delta E_{\mathrm{H}} - \frac{1}{4}\delta E_{001} - \frac{1}{2}\left[\Delta_0^2 + \Delta_0\delta E_{001} + (9/4)(\delta E_{001})^2\right]^{1/2} \tag{4.179b}$$

$$\approx -\frac{1}{3}\Delta_0 + \delta E_{\mathrm{H}} - \frac{1}{2}\delta E_{001} - \frac{1}{2}(\delta E_{001})^2/\Delta_0 + \cdots\,, \tag{4.179c}$$

$$\Delta(\mathcal{E}_{\mathrm{c}} - \mathcal{E}_{\mathrm{v3}}) = \frac{1}{6}\Delta_0 + \delta E_{\mathrm{H}} - \frac{1}{4}\delta E_{001} + \frac{1}{2}\left[\Delta_0^2 + \Delta_0\delta E_{001} + (9/4)(\delta E_{001})^2\right]^{1/2} \tag{4.179d}$$

$$\approx +\frac{2}{3}\Delta_0 + \delta E_{\mathrm{H}} + \frac{1}{2}(\delta E_{001})^2/\Delta_0 + \cdots\,. \tag{4.179e}$$

The wave functions under the application of stress are then given by

$$u_{v1,X} = \left|\frac{3}{2},\frac{3}{2}\right\rangle, \tag{4.180a}$$

$$u_{v2,X} = \left|\frac{3}{2},\frac{1}{2}\right\rangle + \frac{1}{\sqrt{2}}\alpha_0\left|\frac{1}{2},\frac{1}{2}\right\rangle, \tag{4.180b}$$

$$u_{v3,X} = \left|\frac{1}{2},\frac{1}{2}\right\rangle - \frac{1}{\sqrt{2}}\alpha_0\left|\frac{3}{2},\frac{1}{2}\right\rangle, \tag{4.180c}$$

where $\alpha_0 = \delta E_{001}/\Delta_0$. Using these results the momentum matrix elements between the valence and conduction bands which determine the strength of the optical transition are calculated as

$$|\langle \mathrm{c}|p_\parallel|\mathrm{v}_1\rangle|^2 = 0, \tag{4.181a}$$

$$|\langle \mathrm{c}|p_\perp|\mathrm{v}_1\rangle|^2 = \frac{1}{2}P^2, \tag{4.181b}$$

$$|\langle \mathrm{c}|p_\parallel|\mathrm{v}_2\rangle|^2 = \frac{2}{3}P^2\left(1+\alpha_0-\frac{3}{4}\alpha_0^2+\cdots\right), \tag{4.181c}$$

$$|\langle \mathrm{c}|p_\perp|\mathrm{v}_2\rangle|^2 = \frac{1}{6}P^2\left(1-2\alpha_0+\frac{3}{2}\alpha_0^2+\cdots\right), \tag{4.181d}$$

$$|\langle \mathrm{c}|p_\parallel|\mathrm{v}_3\rangle|^2 = \frac{1}{3}P^2\left(1-2\alpha_0+\frac{3}{2}\alpha_0^2+\cdots\right), \tag{4.181e}$$

$$|\langle \mathrm{c}|p_\perp|\mathrm{v}_3\rangle|^2 = \frac{1}{3}P^2\left(1+\alpha_0-\frac{3}{4}\alpha_0^2+\cdots\right). \tag{4.181f}$$

In the above equations P is the momentum matrix element between the $|\Gamma_{25'}\rangle$ valence band and the $|\Gamma_{2'}\rangle$ conduction band, which is written as

$$P = \langle c\uparrow|p_x|X\uparrow\rangle = \langle c\uparrow|p_y|Y\uparrow\rangle = \langle c\uparrow|p_z|Z\uparrow\rangle. \tag{4.182}$$

For the cases of [111] and [110] stresses the analysis is similarly straightforward and the results are given in [15].

The change in the real part of the dielectric constant due to stress is calculated from these results and (4.130a). For E_0 we find

$$\Delta\kappa_1(\omega) = \sum_{\mathrm{v}_1,\mathrm{v}_2}\left(\frac{\partial\kappa_1}{\partial M}\Delta M + \frac{\partial\kappa_1}{\partial\omega_0}\Delta\omega_0 + \frac{1}{2}\frac{\partial^2\kappa_1}{\partial\omega_0^2}(\Delta\omega_0)^2 + \frac{\partial^2\kappa_1}{\partial\omega_0\partial M}\Delta\omega_0\Delta M\right), \tag{4.183}$$

where $M = |\langle \mathrm{c}|p_{\parallel,\perp}|\mathrm{v}\rangle|^2$. This may be rewritten as

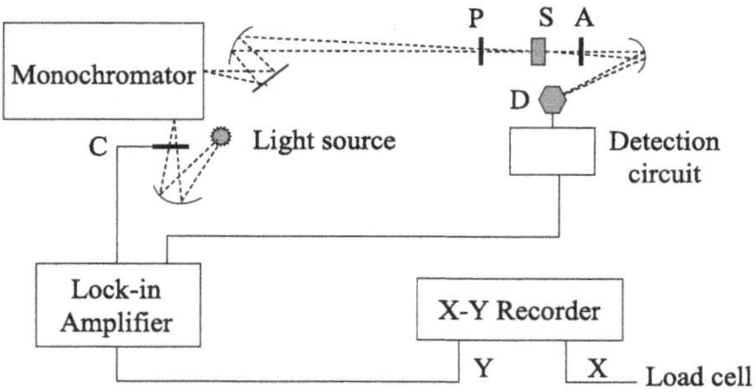

Fig. 4.18 Experimental setup for piezobirefringence measurements. P and A are the polarizer and analyzer, respectively, S is the sample, D is the detector, C is the light chopper, and the load cell is a stress transducer to detect the stress magnitude

$$(\kappa_1)_\parallel - (\kappa_1)_\perp = A\omega_0^{-3/2} P^2 \left\{ \alpha_0 \left[f(x_0) - \frac{\Delta_0}{4\hbar\omega_0} g(x_0) \right] - \frac{3}{4}\alpha_0^2 \left[f(x_0) + \frac{\Delta_0}{2\hbar\omega_0} g(x_0) \right] + \frac{1}{2}\frac{\alpha_0 \alpha_\mathrm{H} \Delta_0}{\hbar\omega_0} \left[g(x_0) + \frac{\Delta_0}{4\hbar\omega_0} l(x_0) \right] \right\} . \quad (4.184)$$

For the $E_0 + \Delta_0$ gap, we obtain

$$(\kappa_1)_\perp - (\kappa_1)_\parallel = A\omega_{0\mathrm{s}}^{-3/2} P^2 \left\{ \left[-\alpha_0 + \frac{3}{4}\alpha_0^2 \right] f(x_{0\mathrm{s}}) - \left[\frac{\alpha_0 \alpha_\mathrm{H} \Delta_0}{2\hbar\omega_{0\mathrm{s}}} \right] g(x_{0\mathrm{s}}) \right\} , \quad (4.185)$$

where

$$f(x) = (1/x^2)[2 - (1+x)^{1/2} - (1-x)^{1/2}] , \quad (4.186a)$$
$$g(x) = (1/x^2)[2 - (1+x)^{-1/2} - (1-x)^{-1/2}] , \quad (4.186b)$$
$$l(x) = (1/x^2)[2 - (1+x)^{-3/2} - (1-x)^{-3/2}] , \quad (4.186c)$$
$$x_0 = \omega/\omega_0, \quad x_{0s} = \omega/\omega_{0s} , \quad \alpha_\mathrm{H} = \delta E_\mathrm{H}/\Delta_0 . \quad (4.186d)$$

When we take account of the E_0, $E_0 + \Delta_0$, E_1, $E_1 + \Delta_1$ and E_2 critical points, the real part of the dielectric constant in the presence of a [100] stress is given by [12]

$$(\kappa_1)_\parallel - (\kappa)_\perp = C_0 X \left\{ -g(x_0) + \frac{4\omega_0}{\Delta_0} \left[f(x_0) - \left(\frac{\omega_0}{\omega_{0s}} \right)^{3/2} f(x_{0s}) \right] \right\} = C_1 X \left\{ 1 + \frac{1}{2}x_1^2 - \left(\frac{\omega_1}{\omega_{1s}} \right)^2 \left(1 + \frac{1}{2}x_{1s}^2 \right) \right\} + C_2 X \left\{ 1 + 2x_2^2 \right\} . \quad (4.187)$$

The piezobirefringence experiment is carried out by using the setup shown in Fig. 4.18. The light from the source is chopped into pulses, dispersed by a monochromator. The polarizer is oriented so that the beam incident on the sample is polarized at 45° to the stress axis. Under these conditions, the light components

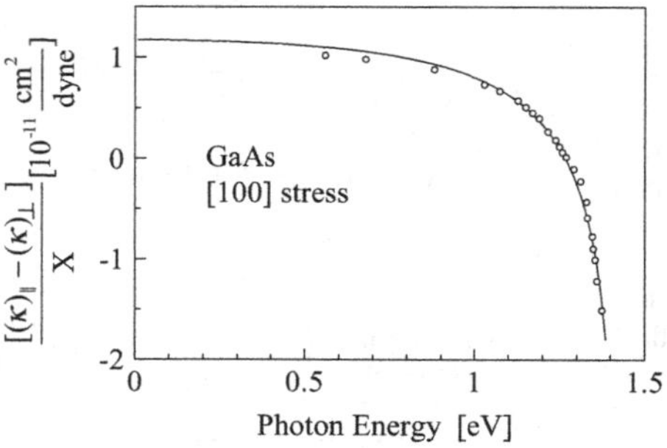

Fig. 4.19 Piezobirefringence coefficient $[(\kappa_1)_{\parallel} - (\kappa_1)_{\perp}]/X$ of GaAs plotted as a function of photon energy [12]. The calculated result from (4.187) is shown by a solid curve fitted to the experimental data. The critical point energies are those listed in Table 4.3 and $C_0 = -0.51$, $C_1 = -10.0$, $C_2 = 3.6\ [10^{-11}\ \text{cm}^2/\text{dyne}]$

polarized parallel and perpendicular to the stress are equal. The signal is obtained as a function of stress for the analyzer with its polarization plane either perpendicular or parallel to that of the polarizer. The signal intensity is given by [12]

$$I = \frac{1}{2} I_0 (1 \pm \cos \Delta), \tag{4.188}$$

where Δ is the phase difference between the parallel and perpendicular components of the transmitted light, and is given by (4.149) or (4.150). The symbol $\pm$ corresponds to the configuration of the analyzer with respect to the polarizer, parallel or perpendicular. The transmitted light intensity oscillates as a cosine function with a frequency depending on the stress, and thus the relation between Δ/t and stress X is obtained. When we plot Δ/t as a function of X, the slope with the help of (4.150) gives $[(\kappa_1)_{\parallel} - (\kappa_1)_{\perp}]/X$. The measured data are plotted as a function of photon energy in Fig. 4.19, where the calculated result from (4.187) is shown by a solid curve fitted to the experimental data. It is evident from Fig. 4.19 that piezobirefringence in GaAs is well explained by the theory stated above. From the analysis we may deduce the deformation potentials.

4.8 Problems

(4.1) Using the relation of (4.18), calculate reflectivity R as a function real part of dielectric constant κ_1 for $\kappa_2 = 0.1, 5.0$ and 10 and show the result in figure.

(4.2) We assume that a conduction band and a valence band are given by

$$\mathcal{E}_{\mathrm{c}} = \frac{\hbar^2 k^2}{2m_{\mathrm{c}}} + \mathcal{E}_{\mathrm{G}}$$
$$\mathcal{E}_{\mathrm{v}} = \frac{\hbar^2 k^2}{2m_{\mathrm{v}}}$$

where $\mathcal{E}_{\mathrm{G}}$ is the band gap, and

$$\mathcal{E}_{\mathrm{cv}} = \frac{\hbar^2}{2\mu}k^2 + \mathcal{E}_{\mathrm{G}}, \quad \frac{1}{\mu} = \frac{1}{m_{\mathrm{c}}} + \frac{1}{m_{\mathrm{v}}}$$

(1) Calculate the joint density of states by using (4.42).
(2) Calculate the joint density of states by using (4.44)

(4.3) Absorption coefficient due to indirect transition is shown in Fig. 4.6. Explain the temperature dependence of the absorption coefficient with respect to the following two features:
(1) Absorption coefficient increases with temperature.
(2) A clear bending point is observed at a low temperature.

(4.4) Express the dielectric functions κ_2 and κ_1 as a function of $x = \hbar\omega/\omega_0$ with parameters A, P_0^2, and $\omega_0 = \mathcal{E}_{\mathrm{G}}/\hbar$.

(4.5) Evaluate the coefficient $AP_0^2\omega_0^{-3/2}$ for GaAs with the parameters $\mathcal{E}_{\mathrm{G}} = 1.51\,\mathrm{eV}$, $\mathcal{E}_{P0} = 28.8$, and the reduced mass $\mu_i = \mu = 0.067\,\mathrm{m}$.

(4.6) Plot $\kappa_2(x)$ and $\kappa_1(x)$ as a function of x using the parameters given above for GaAs for E_0 transition. Since the E_0 critical point reflects the transition from the heavy and light hole bands to the conduction band, we may put $2C_0$ as stated in the text.

(4.7) Derive real part of the dielectric constant of GaAs at the band edge from (4.130a) for E_0 transition. The material constants are $\mathcal{E}_{\mathrm{G}} = 1.51\,\mathrm{eV}$, $\mathcal{E}_{P0} = 28.8$. The reduced mass μ_i is assumed to be the conduction electron effective mass $m_{\mathrm{c}}^* = 0.067\,\mathrm{m}$.
(1) Evaluate κ_1 for the case of simple bands with a conduction band and a heavy hole band.
(2) Evaluate κ_1 for the case with three valence bads with spin-orbit split band.
(3) At higher temperature the energy pap $\mathcal{E}_{\mathrm{G}}$ becomes smaller. Evaluate above two cases for $\mathcal{E}_{\mathrm{G}} = 1.43\,\mathrm{eV}$.

References

1. L. van Hove, Phys. Rev. **89**, 1189 (1953)
2. M. Cardona, in *Modulation Spectroscopy*, ed. by F. Seitz, D. Turnbull, H. Ehrenreich (Academic Press, New York, 1969)
3. B. Batz, in *Semiconductors and Semimetals*, vol. 9, ed. by R.K. Willardson, A.C. Beer (Academic Press, New York, 1972)
4. G.G. MacFarlane, V. Roberts, Phys. Rev. **98**, 1865 (1955)
5. G.G. MacFarlane, T.P. McLean, J.E. Quarrington, V. Roberts, Phys. Rev. **108**, 1377 (1957)
6. R.J. Elliott, Phys. Rev. **108**, 1384 (1954)
7. H. Haug, S.E. Koch, *Quantum Theory of the Optical and Electronic Properties of Semiconductors* (World Scientific, Singapore, 1993) Chaps. 10–11
8. M.D. Sturge, Phys. Rev. **127**, 768 (1962)
9. M. Gershenzon, D.G. Thomas, R.E. Dietz, in *Proceedings of the International Conference on Semiconductor Physics*, Exeter (The Institute of Physics and the Physical Society, London, 1962), p. 752
10. F.H. Pollak, C.W. Higginbotham, M. Cardona, J. Phys. Soc. Jpn. Suppl. **21**, 20 (1966)
11. M. Cardona, Optical properties and band structure of germanium and zincblende-type semiconductors, in *Proceedings of the International School of Physics "Enrico Fermi" on Atomic Structure and Properties of Solids* (Academic Press, New York, 1972), pp. 514–580
12. C.W. Higginbotham, M. Cardona, F.H. Pollak, Phys. Rev. **184**, 821 (1969)
13. P.Y. Yu, M. Cardona, J. Phys. Chem. Solids **34**, 29 (1973)
14. G.E. Picus, G.L. Bir, Sov. Phys.-Solid State **1**, 136 (1959); **1**, 1502 (1960); **6**, 261 (1964); Sov. Phys.-JETP **14**, 1075 (1962)
15. F.H. Pollak, M. Cardona, Phys. Rev. **172**, 816 (1968)

Chapter 5
Optical Properties 2

Abstract In this chapter precision optical spectroscopy is discussed. Modulation spectroscopy has been successfully used to investigate energy band structure of semiconductors. Electroreflectance spectroscopy is described in detail, which is shown to be related to the third derivative form of the dielectric function with respect to the photon energy. Electric field–induced change in the dielectric functions is shown to reflect the type of the critical point. Other optical spectroscopy such as Raman scattering and Brillouin scattering is also described. In addition, resonant Raman scattering and resonant Brillouin scattering are discussed. In the last part of this chapter we deal with coupled modes such as phonon polariton and exciton polariton, and finally free–carrier absorption and plasmon.

5.1 Modulation Spectroscopy

5.1.1 Electro-Optic Effect

Two phenomena are known as the electric field induced change in reflectivity. One is linearly proportional to the electric field and is known as the **Pockels effect** and the other is quadratically proportional to the electric field and is known as the **Kerr effect**. These two phenomena are called the **electro-optic effect** and are observed in the transparent region and thus well below the fundamental absorption edge. For example, the Pockels effect is expressed by

$$(\delta\kappa^{-1})^{ij} = r_{ijk}E_k (\equiv \delta\beta_{ij})$$

where $\boldsymbol{E}$ is the electric field applied to a crystal, $(\delta\kappa^{-1})^{ij}$ is the ij component of the change in the inverse dielectric constant and r_{ijk} is called the Pockels electro-optic constant. In this Section we will not go into detail on this kind of phenomenological treatment but we will consider the magnitude of the change in the dielectric constant near the fundamental absorption edge induced by the applied electric field in a semiconductor.

C. Hamaguchi, *Basic Semiconductor Physics*, Graduate Texts in Physics,
DOI 10.1007/978-3-319-66860-4_5

The application of an electric field $\boldsymbol{E}$ results in the addition of the potential energy term $-e\boldsymbol{E}\cdot\boldsymbol{r}$ to the electron Hamiltonian and thus the translational symmetry is lost in the direction of the electric field. This gives rise to a mixing of the electronic states of the wave vectors $\boldsymbol{k}_0$ and $\boldsymbol{k} = \boldsymbol{k}_0 - e\boldsymbol{E}t/\hbar$ as discussed later. First, we will consider optical transition under an applied d.c. electric field. When the Coulomb interaction is neglected, the effective mass equation for the relative motion of an electron–hole pair is

$$\left[-\frac{\hbar^2}{2\mu}\nabla^2 - e\boldsymbol{E}\cdot\boldsymbol{r}\right]\psi(\boldsymbol{r}) = \mathcal{E}\psi(\boldsymbol{r})\,, \tag{5.1}$$

or

$$\left[-\frac{\hbar^2}{2\mu_x}\frac{d^2}{dx^2} - \frac{\hbar^2}{2\mu_y}\frac{d^2}{dy^2} - \frac{\hbar^2}{2\mu_z}\frac{d^2}{dz^2} \right.$$
$$\left. - e(E_x x + E_y y + E_z z)\right]\psi(x, y, z) = \mathcal{E}\psi(x, y, z)\,, \tag{5.2}$$

where μ_i is the value in the direction of the principal axis of the reduced effective mass tensor. The solutions of the above equation are given by the product of the wave functions $\psi(x)\psi(y)\psi(z)$, where each function satisfies the equations.

$$\left[-\frac{\hbar^2}{2\mu_i}\frac{\mathrm{d}^2}{\mathrm{d}r_i^2} - eE_i r_i - \mathcal{E}_i\right]\psi(r_i) = 0, \tag{5.3a}$$

$$\mathcal{E} = \mathcal{E}_x + \mathcal{E}_y + \mathcal{E}_z\,. \tag{5.3b}$$

Equation (5.3a) may be solved easily by changing the variables:

$$\hbar\theta_i = \left(\frac{e^2E_i^2\hbar^2}{2\mu_i}\right)^{1/3}, \tag{5.4}$$

$$\xi_i = \frac{\mathcal{E}_i + eE_i r_i}{\hbar\theta_i}\,. \tag{5.5}$$

Then (5.3a) can be rewritten as

$$\frac{\mathrm{d}^2\psi(\xi_i)}{\mathrm{d}\xi_i^2} = -\xi_i\psi(\xi_i)\,. \tag{5.6}$$

The solution of this equation is known to be given by the Airy function [1], and we obtain

$$\psi(\xi_i) = C_i\mathrm{Ai}(-\xi_i)\,, \tag{5.7}$$

where C_i is a normalization constant of the function $\psi(\xi)$ and is given by

$$C_i = \frac{\sqrt{e|E_i|}}{\hbar\theta}\,. \tag{5.8}$$

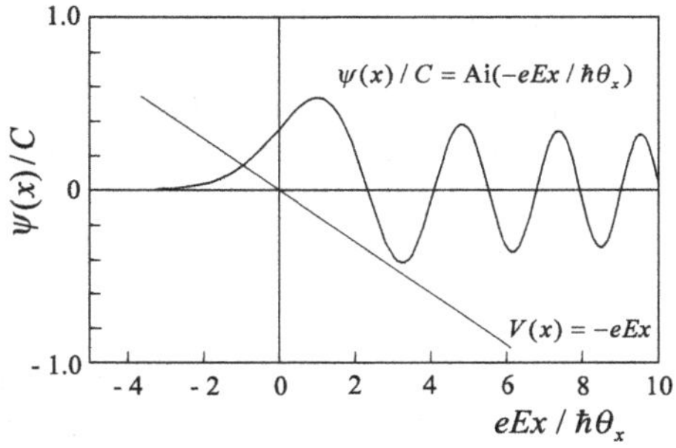

Fig. 5.1 Wave function of an electron–hole pair in a static electric field, with a tilted potential barrier. The wave function penetrates into the region $x < 0$ ($\mathcal{E} < V = -eEx$)

Therefore, the solution of (5.3a) is written as

$$\psi(\xi_x, \xi_y, \xi_z) = C_x C_y C_z \mathrm{Ai}(-\xi_x)\mathrm{Ai}(-\xi_y)\mathrm{Ai}(-\xi_z)\,. \tag{5.9}$$

5.1.2 *Franz–Keldysh Effect*

For simplicity we consider the case where the electric field is applied in the x direction. Since we have $E_y = E_z = 0$, (5.3a) gives rise to

$$\left[-\frac{\hbar^2}{2\mu_x}\frac{d^2}{dx^2} + \frac{\hbar^2 k_y^2}{2\mu_y} + \frac{\hbar^2 k_z^2}{2\mu_z} - eEx\right]\psi(x, y, z) = \mathcal{E}\psi(x, y, x) \tag{5.10}$$

and the solution is given by

$$\psi(x, y, z) = C \cdot \mathrm{Ai}\left(\frac{-eEx - \mathcal{E} + \hbar^2 k_y^2/2\mu_y + \hbar^2 k_z^2/2\mu_z}{\hbar\theta_x}\right) \times \exp\{\mathrm{i}(k_y y + k_z z)\}\,. \tag{5.11}$$

It is evident from this result that the wave function in the y, z directions is periodic, given by a Bloch function, but the localization of the electron–hole pair function occurs in the x direction due to the lack of translational symmetry due to the electric field. As an example, if we assume $k_y = k_z = 0$ and $\mathcal{E} = 0$, then the wave function is written as

$$\psi(x) = C \cdot \mathrm{Ai}(-eEx/\hbar\theta_x)\,. \tag{5.12}$$

Figure 5.1 shows a plot of the wave function as a function of x. Let us consider the one-dimensional motion of a particle with effective mass μ_x. In the classical picture the particle cannot penetrate into the region ($x < 0$) of positive potential ($-eEx > 0$), and thus the point $x = 0$ is the turning point for the motion. In the quantum mechanical picture, however, the particle can penetrate into the region $x < 0$ and the wave function has an exponential tail in that region as shown in Fig. 5.1. As a result, an electron in a valence band is allowed to be excited into a conduction band by absorbing a photon with its energy below the band gap. This effect is known as the **Franz–Keldysh effect** [2, 3] and the detailed treatment is given later.

Next we discuss the effect of the electric field on the interband transition. The wave function $\psi(x, y, z)$ given by (5.9) is the same as the envelope function described in Sect. 4.5 and represents the wave function of the relative motion of the electron–hole pair. Therefore, the imaginary part of the dielectric function $\kappa_2(\omega, \boldsymbol{E}) = \epsilon_2(\omega, \boldsymbol{E})/\epsilon_0$ in the presence of a static electric field may be obtained by inserting (5.9) into (4.97). Replacing the summation with respect to λ in (4.97) by $\int \mathrm{d}\mathcal{E}_x \mathrm{d}\mathcal{E}_y \mathrm{d}\mathcal{E}_z$, we obtain

$$\begin{aligned}\kappa_2(\omega, \boldsymbol{E}) &= \frac{\pi e^2 |\boldsymbol{e}\cdot\boldsymbol{p}_{\mathrm{cv}}|^2}{\epsilon_0 m^2 \omega^2} \frac{e^3 |E_x E_y E_z|}{(\hbar\theta_x \hbar\theta_y \hbar\theta_z)^2} \int \mathrm{d}\mathcal{E}_x \mathrm{d}\mathcal{E}_y \mathrm{d}\mathcal{E}z \\ &\times \left| \mathrm{Ai}\left(-\frac{\mathcal{E}_x}{\hbar\theta_x}\right) \cdot \mathrm{Ai}\left(-\frac{\mathcal{E}_y}{\hbar\theta_y}\right) \cdot \mathrm{Ai}\left(-\frac{\mathcal{E}_z}{\hbar\theta_z}\right) \right|^2 \\ &\times \delta\left[\mathcal{E}_{\mathrm{G}} + \mathcal{E}_x + \mathcal{E}_y + \mathcal{E}_z - \hbar\omega\right] , \end{aligned} \tag{5.13}$$

and the evaluation of this equation for the van Hove critical points is straightforward. For example, we will show the calculations for the case of the M_0 critical point $(\mu_x, \mu_y, \mu_z > 0)$. When we assume that the electric field $\boldsymbol{E}$ is parallel to one of the principal axes, x, $(E_x \neq 0, E_y = E_z = 0)$, the results are the same as stated above. The integral with respect to $\mathrm{d}\mathcal{E}_x \mathrm{d}\mathcal{E}_y$ is given by the joint density of states for the two-dimensional band $J_{\mathrm{cv}}^{\mathrm{2D}}(\hbar\omega - \mathcal{E}_{\mathrm{G}} - \mathcal{E}_x)$, and thus we obtain

$$\begin{aligned}\kappa_2(\omega, \boldsymbol{E}) &= \frac{\pi e^2}{\epsilon_0 m^2 \omega^2} |\boldsymbol{e}\cdot\boldsymbol{p}_{\mathrm{cv}}|^2 \frac{e|E_x|}{(\hbar\theta_x)^2} \\ &\times \int_{-\infty}^{+\infty} J_{cv}^{\mathrm{2D}}(\hbar\omega - \mathcal{E}_G - \mathcal{E}_x) \cdot \left| \mathrm{Ai}\left(-\frac{\mathcal{E}_x}{\hbar\theta_x}\right) \right|^2 \mathrm{d}\mathcal{E}_x . \end{aligned} \tag{5.14}$$

The two-dimensional joint density of states is shown in Table 4.2, which is

$$J_{\mathrm{cv}}^{\mathrm{2D}}(\hbar\omega) = \begin{cases} B_1 = \dfrac{(\mu_y \mu_z)^{1/2}}{\pi\hbar^2}, & \hbar\omega > \mathcal{E}_{\mathrm{G}} \\ 0, & \hbar\omega < \mathcal{E}_{\mathrm{G}} . \end{cases} \tag{5.15}$$

Then (5.14) is given by

$$\begin{aligned}\kappa_2(\omega, \boldsymbol{E}) &= \frac{\pi e^2}{\epsilon_0 m^2 \omega^2} |\boldsymbol{e}\cdot\boldsymbol{p}_{\mathrm{cv}}|^2 \frac{e|E_x||\mu_y \mu_z|^{1/2}}{(\hbar\theta_x)^2 \pi\hbar^2} \int_{-\infty}^{\hbar\omega - \mathcal{E}_{\mathrm{G}}} \left| \mathrm{Ai}\left(-\frac{\mathcal{E}_x}{\hbar\theta_x}\right) \right|^2 \mathrm{d}\mathcal{E}_x \\ &= \frac{e^2}{2\epsilon_0 m^2 \omega^2} |\boldsymbol{e}\cdot\boldsymbol{p}_{\mathrm{cv}}|^2 (\hbar\theta_x)^{1/2} \left(\frac{8|\mu_x \mu_y \mu_z|}{\hbar^6} \right)^{1/2} \\ &\times \left[|\mathrm{Ai}'(-\eta)|^2 + \eta |\mathrm{Ai}(-\eta)|^2 \right] , \end{aligned} \tag{5.16}$$

where

$$\eta = \frac{\hbar\omega - \mathcal{E}_G}{\hbar\theta_x} . \tag{5.17}$$

Here we will discuss the properties of the Airy functions in the limiting case where the photon energy is well above the fundamental absorption edge. In the case of $\eta \gg 0$ ($\eta \to \infty$) and thus in the case of $\hbar\omega \gg \mathcal{E}_G$, (5.16) is approximated by using asymptotic forms of the Airy functions

$$\lim_{z\to\infty} \mathrm{Ai}(-z) = \frac{1}{\sqrt{\pi}} z^{-1/4} \sin\left(\frac{2}{3}z^{3/2} + \frac{\pi}{4}\right) , \tag{5.18a}$$

$$\lim_{z\to\infty} \mathrm{Ai}'(-z) = \frac{1}{\sqrt{\pi}} z^{1/4} \cos\left(\frac{2}{3}z^{3/2} + \frac{\pi}{4}\right) , \tag{5.18b}$$

and then we find

$$|\mathrm{Ai}'(-\eta)|^2 + \eta|\mathrm{Ai}(-\eta)|^2 = \frac{\sqrt{\eta}}{\pi} = \frac{(\hbar\omega - \mathcal{E}_G)^{1/2}}{\pi(\hbar\theta_x)^{1/2}} . \tag{5.19}$$

Insertion of (5.19) into (5.16) gives exactly the same dielectric function as that for the one-electron approximation derived in Chap. 4 (4.39).

On the other hand, we find a very interesting result in the other limiting case where the photon energy is below the fundamental absorption edge. In this case we find $\eta \ll 0$ ($\eta \to -\infty$) and thus $\hbar\omega \ll \mathcal{E}_G$. The asymptotic solutions for the Airy functions are

$$\lim_{z\to\infty} \mathrm{Ai}(z) = \frac{1}{2\sqrt{\pi}} z^{-1/4} \exp\left(-\frac{2}{3}z^{3/2}\right)\left[1 - \frac{3C_1}{2z^{3/2}}\right] , \tag{5.20a}$$

$$\lim_{z\to\infty} \mathrm{Ai}'(z) = \frac{1}{2\sqrt{\pi}} z^{1/4} \exp\left(-\frac{2}{3}z^{3/2}\right)\left[1 + \frac{21C_1}{10z^{3/2}}\right] , \tag{5.20b}$$

where $C_1 = 15/216$ [1]. Therefore, the dielectric function for the incident photon energy below the band gap is

$$\begin{aligned} \kappa_2(\omega, E_x) &= \frac{e^2|\boldsymbol{e}\cdot\boldsymbol{p}_{cv}|^2}{2\epsilon_0 m^2\omega^2}\frac{1}{2\pi}\left(\frac{8|\mu_x\mu_y\mu_z|}{\hbar^6}\right)^{1/2}(\mathcal{E}_G - \hbar\omega)^{1/2} \\ &\quad \times \exp\left[-\frac{4}{3}\left(\frac{\mathcal{E}_G - \hbar\omega}{\hbar\theta_x}\right)^{3/2}\right] \\ &= \frac{1}{2}\kappa_2(\omega)\exp\left[-\frac{4}{3}\left(\frac{\mathcal{E}_G - \hbar\omega}{\hbar\theta_x}\right)^{3/2}\right] , \end{aligned} \tag{5.21}$$

or inserting this into (4.21) the absorption coefficient α is given by the relation

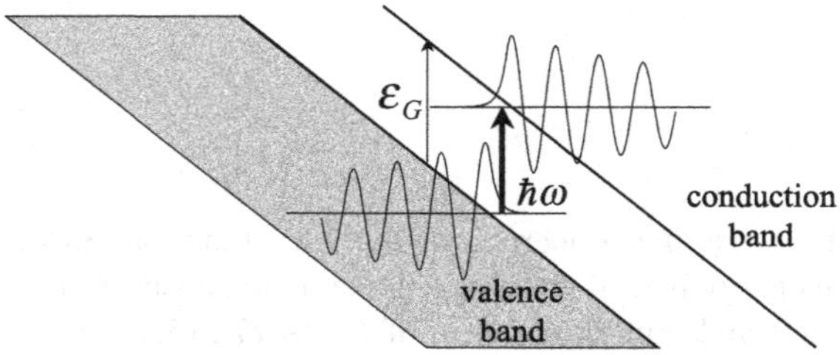

Fig. 5.2 Illustration of Franz–Keldysh effect in a semiconductor. In the presence of a high electric field the wave functions of the electron and the hole penetrate into the band gap and an optical transition is allowed in the photon energy region below the band gap $\hbar\omega < \mathcal{E}_G$

Fig. 5.3 $\kappa_2(\omega, 0)/A$ without electric field and $\kappa_2(\omega, E)/A$ with electric field are plotted as a function of photon energy near the M_0 critical point

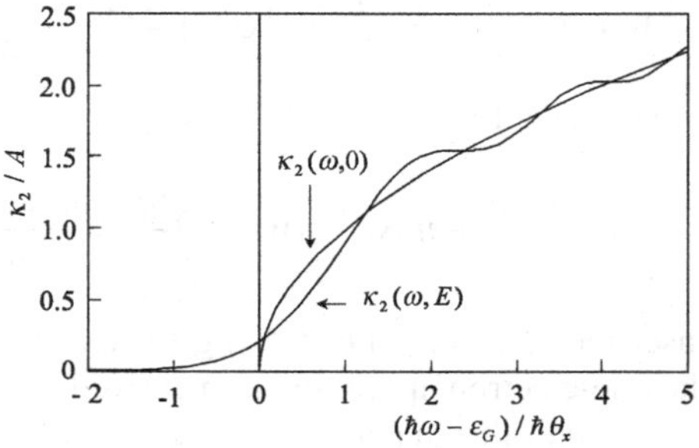

$$\alpha(\omega, E_x) = \frac{e^2|\boldsymbol{e}\cdot\boldsymbol{p}_{cv}|^2}{2\epsilon_0 m^2 c n_0 \omega}\frac{1}{2\pi}\left(\frac{8|\mu_x\mu_y\mu_z|}{\hbar^6}\right)^{1/2}(\mathcal{E}_G - \hbar\omega)^{1/2}$$
$$\times \exp\left[-\frac{4}{3}\left(\frac{\mathcal{E}_G - \hbar\omega}{\hbar\theta_x}\right)^{3/2}\right]$$
$$= \frac{1}{2}\alpha(\omega)\exp\left[-\frac{4}{3}\left(\frac{\mathcal{E}_G - \hbar\omega}{\hbar\theta_x}\right)^{3/2}\right], \tag{5.22}$$

where $\kappa_2(\omega)$ and $\alpha(\omega)$ are the imaginary part of the dielectric constant without and with electric field, respectively. These results indicate that absorption occurs for an incident photon with an energy less than the band gap in the presence of an electric field and that the absorption coefficient has an exponential tail. This effect is predicted independently by Franz and Keldysh and is called the Franz–Keldysh effect. This effect is well explained in terms of the penetration of the wave function in the presence of a high electric field with the help of the illustration of Fig. 5.2, where the electron and hole penetrate into the band gap due to quantum mechanical effects and an optical transition becomes allowed in the region $\hbar\omega < \mathcal{E}_G$. It may be expected that the absorption coefficient shows an exponential decay as the photon energy decreases.

Using the result for the M_0 point given by (5.16), $\kappa_2(\omega, \boldsymbol{E})$ and $\kappa_2(\omega, 0)$ are plotted in Fig. 5.3. In the absence of electric field, $\boldsymbol{E} = 0$, the dielectric constant $\kappa_2(\omega, 0)$ is zero in the region $\hbar\omega < \mathcal{E}_G$. In the presence of an electric field $\boldsymbol{E} \neq 0$, however, the wave function penetrates into the band gap and the dielectric constant $\kappa_2(\omega, \boldsymbol{E})$ is not zero in the region below the fundamental absorption edge ($\hbar\omega < \mathcal{E}_G$), giving rise to an exponential tail. Although $\kappa_2(\omega, 0)$ exhibits an increase as $\sqrt{\hbar\omega - \mathcal{E}_G}$ in the photon energy region above the fundamental absorption edge, $\kappa_2(\omega, \boldsymbol{E})$ exhibits an oscillatory behavior (with a modulation by a weak oscillatory component) and converges to $\kappa_2(\omega, 0)$ in the region well above the fundamental absorption edge.

5.1.3 Modulation Spectroscopy

In the previous section we discussed the change in dielectric constant in the presence of an electric field. The change is pronounced near critical points such that it decreases exponentially below the fundamental absorption edge and oscillates beyond the edge, converging to the dielectric constant without electric field at higher photon energies. When we introduce new variables, (5.16) is written as

$$\kappa_2(\omega, \boldsymbol{E}) = \int_{-\infty}^{\infty} \mathrm{d}\omega' \frac{\omega'^2}{\omega^2} \kappa_2(\omega', 0) \left\{ \frac{1}{|\Omega|} \mathrm{Ai}\left(\frac{\omega' - \omega}{\Omega} \right) \right\}, \tag{5.23}$$

where Ω ($\hbar\Omega$ has the dimensions of energy) is defined by

$$\Omega = \frac{\theta}{2^{2/3}} = \frac{(\theta_x \theta_y \theta_z)^{1/3}}{2^{2/3}} = \left(\frac{e^2 E^2}{8\hbar\mu_E} \right)^{1/3} \tag{5.24}$$

and μ_E is the reduced mass in the direction of the electric field. It should be noted that (5.23) behaves as $\kappa_2(\omega, 0)$ but modulated by the Airy function and that the behavior arises from the mixing of the different states of $\boldsymbol{k}$ induced by the electric field, as pointed out earlier.

The change in the dielectric constant in an electric field is quite small and even in an available high electric field such as ($\leq 10^5$ V/cm) will give rise to a change of a small fraction. Therefore, it is very difficult to observe the exponential tail and oscillatory behavior of the dielectric constant shown in Fig. 5.3. However, applying a periodically oscillating electric field and using a phase-sensitive detector (lock-in amplifier), the change in dielectric constant $\Delta\kappa$

$$\Delta\kappa(\omega, \boldsymbol{E}) = \kappa(\omega, E) - \kappa(\omega, 0) \tag{5.25}$$

can be measured with high accuracy. This method is often referred to as **electroreflectance** or **electroabsorption** and has been applied to study the optical properties of semiconductors in the region from the ultraviolet to the infrared since the 1960s and gives us very important information about the band structure of semiconductors. The method is also called **modulation spectroscopy**, including many other modulation methods. In general modulation spectroscopy is carried out by using the experi-

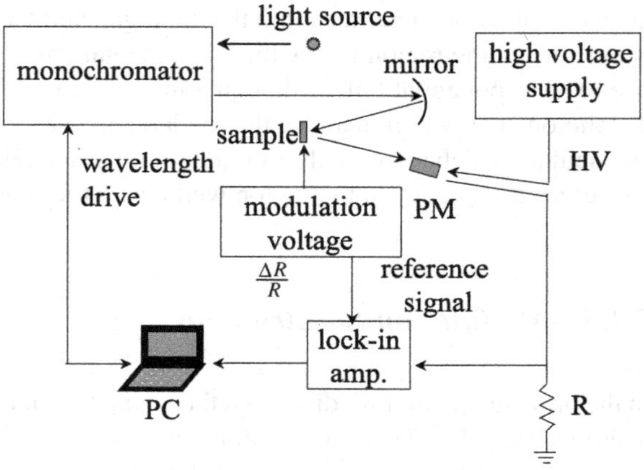

Fig. 5.4 Experimental setup for modulation spectroscopy. Different modulation methods produce different types of modulation spectroscopy

mental setup shown in Fig. 5.4. Modulation methods have been reported so far such as electric field modulation (electroreflectance), laser excitation (photoreflectance), applying high current to a sample or to a sample holder and modulating the sample temperature (thermoreflectance) and stress modulation by using a piezoelectric element (piezoreflectance), and so on.

Let us the dielectric constant of the sample is $\kappa = \kappa_1 + i\kappa_2$ and the modulation is uniform in the modulated sample, then the relation between the change in the reflectance $\Delta R/R$ and the change in the dielectric constants $\Delta\kappa_1$ and $\Delta\kappa_2$ is given by

$$\frac{\Delta R}{R} = \alpha(\kappa_1, \kappa_2)\Delta\kappa_1 + \beta(\kappa_1, \kappa_2)\Delta\kappa_2 \,. \tag{5.26}$$

The above relation is easily calculated to give the following result by using (4.18)

$$R = \frac{(\kappa_1^2 + \kappa_2^2)^{1/2} - [2\kappa_1 + 2(\kappa_1^2 + \kappa_2^2)^{1/2}]^{1/2} + 1}{(\kappa_1^2 + \kappa_2^2)^{1/2} + [2\kappa_1 + 2(\kappa_1^2 + \kappa_2^2)^{1/2}]^{1/2} + 1}, \tag{5.27}$$

and the coefficients are given as follows:

$$\alpha = \frac{2\gamma}{\gamma^2 + \delta^2}, \tag{5.28a}$$

$$\beta = \frac{2\delta}{\gamma^2 + \delta^2}, \tag{5.28b}$$

$$\gamma = \frac{n(n^2 - 3k^2 - n_0)}{n_0}, \tag{5.28c}$$

$$\delta = \frac{k(3n^2 - k^2 - n_0)}{n_0}, \tag{5.28d}$$

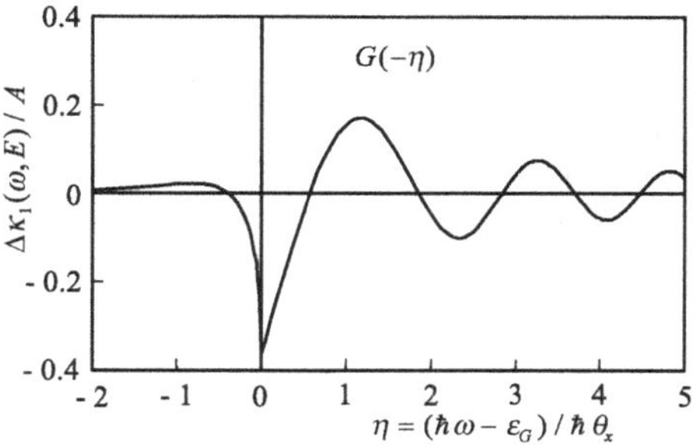

Fig. 5.5 Electric field-induced change in the real part of the dielectric constant for the M_0 critical point $\Delta\kappa_1(\omega, E)$

where n_0 is the refractive index of the non-absorbing medium of incidence and assumed to be $n_0 = 1$ for the air or vacuum. Note here that n and k are the refractive index and the extinction coefficients of a semiconductor as defined in Sect. 4.1 ($n + \mathrm{i}k = \sqrt{\kappa_1 + \mathrm{i}\kappa_2}$). The coefficients α and β are functions of the incident photon energy and are called the Seraphin coefficients [4].

The spectra of the electric field induced change in the dielectric constant $\Delta\kappa(\omega, \boldsymbol{E})$ have been calculated for all the critical points under the assumption of the one-electron approximation and parabolic energy bands [5, 6]. For example, the change for the M_0 critical point is obtained from (4.39) and (5.16) as

$$\begin{aligned}\Delta\kappa_2(\omega, \boldsymbol{E}) &= \kappa_2(\omega, \boldsymbol{E}) - \kappa_2(\omega, 0) \\ &= A \cdot \{\pi[\mathrm{Ai}'^2(-\eta) + \eta \mathrm{Ai}^2(-\eta)] - u(\eta) \cdot \sqrt{\eta}\} \\ &\equiv A \cdot F(-\eta)\,. \end{aligned} \tag{5.29}$$

The real part of the dielectric function is easily derived through the Kramers–Kronig transform given by (4.126a), giving rise to

$$\Delta\kappa_1(\omega, \boldsymbol{E}) = A \cdot G(-\eta)\,, \tag{5.30}$$

where the functions $F(-\eta)$ and $F(-\eta)$ are

$$F(-\eta) = \pi[\mathrm{Ai}'^2(-\eta) + \eta \mathrm{Ai}^2(-\eta)] - \sqrt{\eta} \cdot u(\eta)\,, \tag{5.31a}$$

$$G(-\eta) = \pi[\mathrm{Ai}' \cdot \mathrm{Bi}'(-\eta) + \eta \mathrm{Ai}(-\eta) \cdot \mathrm{Bi}(-\eta)] + \sqrt{-\eta} \cdot u(-\eta)\,, \tag{5.31b}$$

and $u(x)$ is the step function such that $u(x) = 0$ for $x < 0$ and $u(x) = 1$ for $x \geq 0$. The coefficient A is defined as

$$A = \frac{e^2}{2\pi\epsilon_0 m^2 \omega^2} |\boldsymbol{e} \cdot \boldsymbol{p}|^2 \left(\frac{8|\mu_x \mu_y \mu_z|}{\hbar^6}\right)^{1/2} |\hbar\theta|^{1/2}\,. \tag{5.32}$$

These functions are shown in Figs. 5.5 and 5.6.

From these considerations we find that the change in the dielectric functions exhibit spectra rich in structures reflecting the critical points. The component of the modulated dielectric function is quite weak but phase-sensitive detection with the help of a lock-in amplifier provides clear spectra with high signal-to-noise ratio.

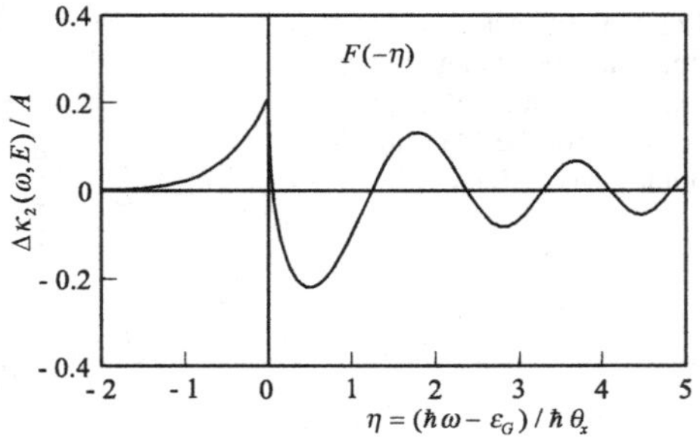

Fig. 5.6 Electric field-induced change in the imaginary part of the dielectric constant for the M_0 critical point $\Delta\kappa_2(\omega, E)$

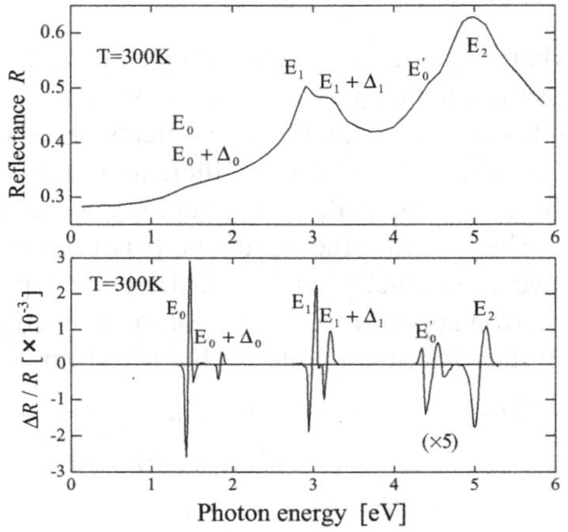

Fig. 5.7 Spectra of reflectance and electroreflectance of GaAs at room temperature (300 K) in the photon energy region from near infrared to ultraviolet. The critical points are not well resolved in the reflectance spectra, whereas the electroreflectance spectra exhibit sharp structures near the critical points and the critical points are easily identified with good accuracy. The reflectance data are from [7]

These features are well displayed in Fig. 5.7. The upper part of the figure shows the reflectivity of GaAs as a function of photon energy and several peaks are noticed such as the E_1, $E_1 + \Delta_1$ and E_2 critical points, but the critical points E_0 and $E_0 + \Delta_0$ are not well resolved. On the other hand, the lower part of the figure shows electroreflectance spectra of GaAs, where all the critical points E_0, $E_0 + \Delta_0$, E_1, $E_1 + \Delta_1$, E_0', and E_2 are well resolved with high signal-to-noise ratio. The electroreflectance spectra are affected by the uniformity and strength of the applied electric field, exciton effects and so on, and thus the spectra are not obtained in the same sample and in the same experiment. However, measurements in the relevant region of photon energy will provide similar spectra to those found in the literature.

5.1.4 Theory of Electroreflectance and Third-Derivative Form of Aspnes

Electroreflectance experiments and their analysis have revealed that the obtained spectra do not exhibit Franz–Keldysh oscillations but very simple structures. From

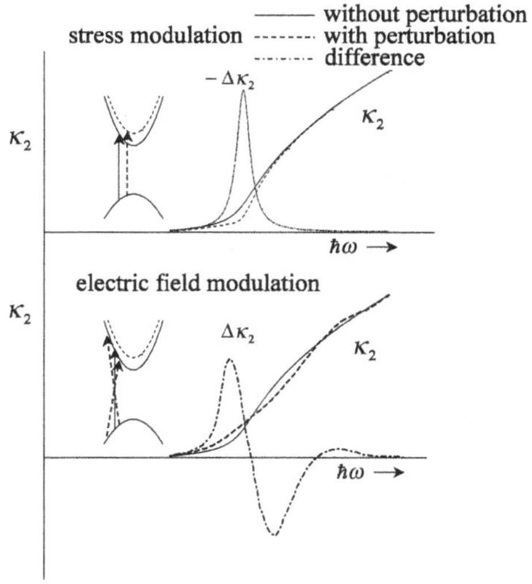

Fig. 5.8 Schematic illustration of the change in the imaginary part of the dielectric constant. The upper figure is the case for a static external force (such as stress modulation) which preserves the periodicity of the lattice, and the change in the dielectric constant is given by the first derivative of the dielectric function with respect to photon energy. On the other hand, electric field modulation breaks the lattice periodicity and the perturbation will give rise to a mixing of the electronic states, resulting in a complicated structure of the change in the dielectric constant. See text for details

the detailed analysis the spectra were found to agree well with the third derivative of the dielectric function with respect to the photon energy. This effect was found by Aspnes [8–11] and is called the **third derivative form of Aspnes**, cited often as **third-derivative modulation spectroscopy**. The principle of third-derivative modulation spectroscopy is illustrated in Fig. 5.8. In the case of static modulation, the periodicity of the lattice is preserved and thus no mixing of the electronic states occurs, resulting only in the displacement of the critical point, as shown in the upper part of Fig. 5.8. Therefore, static modulation gives rise to the first-derivative form of the dielectric function. This is the case for stress modulation, thermo-modulation and so on. For example, let us consider the change in the band gap $\Delta\omega_G(X)$, which is given by

$$\Delta\kappa_2(\omega, X) = \kappa_2(\omega, X) - \kappa_2(\omega, 0) = \frac{d\kappa_2(\omega, 0)}{d\omega}\Delta\omega_G(X) . \tag{5.33}$$

On the other hand, the application of an electric field gives rise to a potential energy term $-e\mathcal{E}x$ and to a lack of periodicity in the electric field direction (x). As a result such transitions as those shown in the lower part of Fig. 5.8 become possible. In order to understand the theory of third-derivative modulation spectroscopy, a qualitative discussion will be given first and then we will present its theoretical analysis.

Equation (4.39) of Sect. 4.3 or (4.128) of Sect. 4.6 may be rewritten as

$$\kappa_2 = C \cdot \frac{1}{\mathcal{E}^2}[\mathcal{E} - (\mathcal{E}_c - \mathcal{E}_v)]^{1/2}, \tag{5.34}$$

where $\mathcal{E}_c$ and $\mathcal{E}_v$ are the bottom of the conduction band and the top of the valence band, respectively, and $\mathcal{E} = \hbar\omega$ is the photon energy. The constant C contains the momentum matrix element, the effective mass and so on and is assumed to be constant in the region near the critical point we are concerned with. The time required for the transition is estimated from the Heisenberg uncertainty principle as

$$\tau = \frac{\hbar}{\mathcal{E} - (\mathcal{E}_c - \mathcal{E}_v)}. \tag{5.35}$$

Under the electric field E_x an electron–hole pair with the reduced mass $\mu_\parallel$ will move a distance x given by

$$x = -\frac{eE_x}{2\mu_\parallel}\tau^2 = -\frac{eE_x}{2\mu_\parallel}\frac{\hbar^2}{[\mathcal{E} - (\mathcal{E}_c - \mathcal{E}_v)]^2}. \tag{5.36}$$

This displacement is caused by the term $-eE_x$ of the potential energy of the Hamiltonian. The term gives rise to a perturbation to the electron in the conduction band and the hole in the valence band, and as a result produces a change in the energy given by

$$\Delta(\mathcal{E}_c - \mathcal{E}_v) = \frac{e^2 E_x^2}{2\mu_\parallel}\frac{\hbar^2}{[\mathcal{E} - (\mathcal{E}_c - \mathcal{E}_v)]^2}. \tag{5.37}$$

The change in the dielectric constant due to the change in energy is estimated from (5.34) as

$$\begin{aligned}\Delta\kappa_2 &= \frac{d\kappa_2}{d(\mathcal{E}_c - \mathcal{E}_v)} \cdot \Delta(\mathcal{E}_c - \mathcal{E}_v) = -C\frac{e^2 E_x^2}{4\mu_\parallel \mathcal{E}^2}\frac{\hbar^2}{[\mathcal{E} - (\mathcal{E}_c - \mathcal{E}_v)]^{5/2}} \\ &= \frac{2\hbar^2 e^2 E_x^2}{3\mu_\parallel \mathcal{E}^2}\frac{\partial^3}{\partial \mathcal{E}^3}(\mathcal{E}^2\kappa_2) = \frac{4}{3\mathcal{E}^2}(\hbar\theta_x)^3\frac{\partial^3}{\partial \mathcal{E}^3}(\mathcal{E}^2\kappa_2).\end{aligned} \tag{5.38}$$

In a similar fashion the change in the real part of dielectric constant is estimated.[1] From this qualitative analysis, we find that the change in the dielectric constant due to an applied electric field is expressed by the third derivative of the dielectric constant with respect to the photon energy. Aspnes [8, 9, 11] has shown from analysis of ellipsometry measurements that the third derivative of the dielectric constant shows good agreement with the electroreflectance data.

Theoretical analysis of third-derivative modulation spectroscopy has been done by Aspnes and the outline of his treatment is given below. Several methods have been reported for the derivation and the details are given in the paper by Aspnes and Rowe [10]. The dielectric constant is expressed as follows in terms of the energy $\mathcal{E}$,

---

[1] Use (5.96) and the calculations are straightforward [12].

broadening constant Γ, and electric field E [10] (the dielectric function is derived from the polarization current under the one-electron approximation and from its Fourier transform):

$$\begin{aligned}
&\kappa_{\mathrm{cv}}(\mathcal{E}, \Gamma, \boldsymbol{E}) \\
&= 1 + \frac{e^2}{\epsilon_0 m^2 \omega^2 \hbar} \int_{\mathrm{B.Z.}} \mathrm{d}^3\boldsymbol{k} \int_0^\infty \mathrm{d}t \left[\boldsymbol{e}\cdot\boldsymbol{p}_{\mathrm{cv}}\left(\boldsymbol{k} - \frac{1}{2}\frac{t}{\hbar}e\boldsymbol{E}\right)\right] \\
&\times \left[\boldsymbol{e}\cdot\boldsymbol{p}_{\mathrm{cv}}\left(\boldsymbol{k} - \frac{1}{2}\frac{t}{\hbar}e\boldsymbol{E}\right)\right] \exp[\mathrm{i}(\mathcal{E} + \mathrm{i}\Gamma)t/\hbar] \\
&\times \exp\left[-\mathrm{i}\int_{-t/2}^{t/2} (\mathrm{d}t'/\hbar)\mathcal{E}_{\mathrm{cv}}(\boldsymbol{k} - e\boldsymbol{E}t'/\hbar)\right].
\end{aligned} \tag{5.39}$$

In the above equation the integral with respect to t' is carried out by expanding the exponential term in the following way under the condition of weak electric field:

$$\exp\left[-\mathrm{i}\int_{-t/2}^{t/2} (\mathrm{d}t'/\hbar)\mathcal{E}_{\mathrm{cv}}(\boldsymbol{k} - e\boldsymbol{E}t'/\hbar)\right] \simeq \frac{t}{\hbar}\mathcal{E}_{\mathrm{cv}} + \frac{1}{3}t^3\Omega^3, \tag{5.40}$$

where

$$(\hbar\Omega)^3 = \frac{1}{3}e^2(\boldsymbol{E}\cdot\nabla_{\boldsymbol{k}})^2\mathcal{E}_{cv} = \frac{e^2E^2\hbar^2}{8\mu_\parallel} \tag{5.41}$$

and the relation between θ defined by (5.4) and Ω is given by (5.24). From this result (5.39) is rewritten as

$$\begin{aligned}
&\kappa_{\mathrm{cv}}(\mathcal{E}, \Gamma, \boldsymbol{E}) \\
&= \frac{\mathrm{i}Q}{\pi\omega^2} \int_{\mathrm{B.Z.}} \mathrm{d}^3\,\boldsymbol{k} \int_0^\infty \frac{\mathrm{d}t}{\hbar} \exp\left[-\frac{1}{3}\mathrm{i}t^3\Omega^3 + \mathrm{i}t(\mathcal{E} + \mathrm{i}\Gamma - \mathcal{E}_{\mathrm{cv}})/\hbar\right],
\end{aligned} \tag{5.42}$$

where the relative dielectric constant of 1 for air is omitted. Q is given by

$$Q = \frac{\pi e^2|\boldsymbol{e}\cdot\boldsymbol{p}_{\mathrm{cv}}|^2}{\epsilon_0 m^2} \cdot \frac{2}{(2\pi)^3}, \tag{5.43}$$

where we have assumed that $\Gamma \gg |\hbar\Omega|$. In addition we assume that the integration may be carried out in the region where $-\mathrm{i}t^3\Omega^3/3$ is not significantly large, and thus we can expand as

$$\exp(-\mathrm{i}t^3\Omega^3/3) = 1 - \mathrm{i}t^3\Omega^3/3 - \cdots. \tag{5.44}$$

Then we obtain

$$\kappa_{cv}(\mathcal{E}, \Gamma, \boldsymbol{E}) \sim \frac{-Q}{\pi\omega^2}\int_{\mathrm{B.Z.}} \mathrm{d}^3\boldsymbol{k}\, \frac{1}{\mathcal{E} + \mathrm{i}\Gamma - \mathcal{E}_{cv}(\boldsymbol{k})} + \frac{2Q}{\pi\omega^2}\int_{\mathrm{B.Z.}} \mathrm{d}^3\boldsymbol{k}\, \frac{(\hbar\Omega)^3}{[\mathcal{E} + \mathrm{i}\Gamma - \mathcal{E}_{\mathrm{cv}}(\boldsymbol{k})]^4} \equiv \kappa_{\mathrm{cv}}(\mathcal{E}, \Gamma, 0) + \Delta\kappa_{\mathrm{cv}}(\mathcal{E}, \Gamma, \boldsymbol{E}). \tag{5.45}$$

In the absence of an electric field, we have $\boldsymbol{E} = 0$ and thus $\hbar\Omega = 0$, which leads to the result

$$\kappa_{cv}(\mathcal{E}, \Gamma, 0) = \frac{-Q}{\pi\omega^2}\int_{\mathrm{B.Z.}} \mathrm{d}^3\boldsymbol{k}\, \frac{1}{\mathcal{E} + \mathrm{i}\Gamma - \mathcal{E}_{\mathrm{cv}}(\boldsymbol{k})}\,. \tag{5.46}$$

It is evident from the Dirac delta function shown in Appendix A that the above relation is exactly the same as (4.40), (4.42) and (4.39) of Sect. 4.43 or (4.128) of Sect. 4.6. From these results we obtain the following relation

$$\Delta\kappa_{\mathrm{cv}}(\mathcal{E}, \Gamma, \boldsymbol{E}) \equiv \frac{1}{3\mathcal{E}^2}\left(\hbar\Omega\frac{\partial}{\partial\mathcal{E}}\right)^3 \mathcal{E}^2\kappa_{\mathrm{cv}}(\mathcal{E}, \Gamma, 0)\,. \tag{5.47}$$

Equation (5.47) agrees with the result obtained from qualitative analysis except for the constant of the prefactor. Therefore, we may conclude that the electroreflectance spectra are given by the third derivative of the dielectric function with respect to photon energy. From this fact (5.47) is referred to as the **third-derivative form of Aspnes**. When we keep the higher-order term in the expansion of (5.44), the experimental data show a better agreement with the theory [13].

The third-derivative modulation spectra valid in the low electric field region are written as

$$\frac{\Delta R}{R} = \Re\left[\sum_j C_j \mathrm{e}^{\mathrm{i}\theta_j}(\mathcal{E} - \mathcal{E}_j + \mathrm{i}\Gamma_j)^{-m_j}\right], \tag{5.48}$$

where E_j is the energy of the jth critical point, C_j is the amplitude constant, θ_j is the phase constant, Γ_j is the broadening constant, and $\Re$ means the real part of the equation. $m_j = 4 - d/2$ depends on the dimension d of the critical point and $m_j = 5/2$, 3, and 7/2 for 3-, 2-, and 1-dimension, respectively.

Figure 5.9 shows electroreflectance spectra of n-GaAs epitaxially grown on an n^+ GaAs substrate, where a semitransparent electrode of Ni on the surface of n-GaAs and a Au electrode on the back of the GaAs substrate were formed and a weak a.c. voltage was applied to the electrodes to produce electric field modulation in the low-field limit. The dotted curve is fitted to the experimental curve by using the third-derivative form of Aspnes. It is evident from Fig. 5.9 that the electroreflectance data are well explained by the Aspnes theory. The electroreflectance measurements and analysis based on the Aspnes theory provide detailed information about the critical points. From the above analysis we obtain $E_1 = 2.926\,\mathrm{eV}$, $E_1 + \Delta_1 = 3.153\,\mathrm{eV}$.

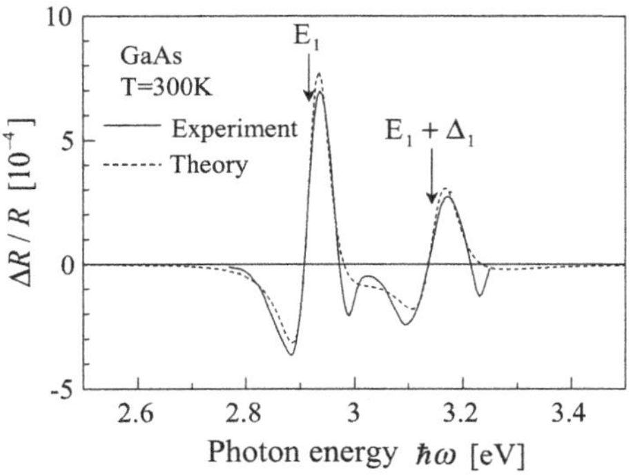

Fig. 5.9 Electroreflectance spectra of GaAs at $T = 300\,\mathrm{K}$. The *solid curve* is experimental and the *dotted curve* is a best fit to the experimental curve using the third-derivative theory of Aspnes. In the analysis the following data were used: $\mathcal{E}_{g1} = E_1 = 2.926\,\mathrm{eV}$, $\Gamma_1 = 46.4\,\mathrm{meV}$, $\theta_1 = 3.275$, $C_1 = 0.00386$, $\mathcal{E}_{g2} = E_1 + \Delta_1 = 3.153\,\mathrm{eV}$, $\Gamma_2 = 59.1\,\mathrm{meV}$, $\theta_2 = 3.061$, $C_2 = 0.003015$

5.2 Raman Scattering

When an electric field $\boldsymbol{E}$ is applied to a medium, a polarization vector $\boldsymbol{P}$ is induced which is related to the field by

$$\boldsymbol{P} = \epsilon_0 \chi \boldsymbol{E}$$

and its components are given by

$$P_j = \epsilon_0 \chi_{jk} E_k \quad \left(\equiv \sum_k \epsilon_0 \chi_{jk} E_k \right), \tag{5.49}$$

where χ_{jk} is the electric susceptibility and is a second-rank tensor. Let us consider an electromagnetic field with frequency ω incident on the medium and express the field by the plane wave

$$\boldsymbol{E}(\boldsymbol{r}, t) = \boldsymbol{E}(\boldsymbol{k}_i, \omega_i) \cos(\boldsymbol{k}_i \cdot \boldsymbol{r} - \omega_i t) . \tag{5.50}$$

Then the polarization is written as

$$\boldsymbol{P}(\boldsymbol{k}_i, \omega_i) \cos(\boldsymbol{k}_i \cdot \boldsymbol{r} - \omega_i t) . \tag{5.51}$$

Therefore, we find the following relation between the amplitudes of the electric field and the polarization:

$$P(k_i, \omega_i) = \epsilon_0 \chi(k_i, \omega_i) E(k_i, \omega_i) . \tag{5.52}$$

The light scattering in which we are interested in this section is induced by a local change of the susceptibility χ. Here we are concerned with the light scattering by lattice vibrations and we express the atomic displacement by

$$u(r, t) = u(q, \omega_q) \cos(q \cdot r - \omega_q t) , \tag{5.53}$$

where q and ω_q are the wave vector and angular frequency of the lattice vibrations. The modes of the lattice vibrations will be described in Chap 6. When the amplitude of the lattice vibrations is small compared to the lattice constant, the change in the susceptibility due to the lattice vibrations is expanded in a Taylor series as

$$\chi(k_i, \omega_\mathrm{i}, u) = \chi^{(0)}(k_i, \omega_\mathrm{i}) + \left(\frac{\partial \chi}{\partial u}\right)_{u=0} u(r, t) + \cdots , \tag{5.54}$$

and thus the susceptibility tensor is expressed as

$$\chi_{jk} = \chi_{jk}^{(0)} + \chi_{jk,l} u_l + \chi_{jk,lm} u_l u_m + \cdots , \tag{5.55}$$

where

$$\chi_{jk,l} = \left(\frac{\partial \chi_{jk}}{\partial u_l}\right)_{u=0} , \quad \chi_{jk,lm} = \left(\frac{\partial^2 \chi_{jk}}{\partial u_l \partial u_m}\right)_{u=0} . \tag{5.56}$$

It is evident from the result that $\chi_{jk,l}$ and $\chi_{jk,lm}$ are the third-rank and fourth-rank tensors, respectively. The first terms on the right hand sides of (5.54) and (5.55) are the susceptibility without perturbation and the terms beyond this are the components induced by the perturbation. Keeping up to the second term on the right-hand side of (5.54) and inserting them into (5.52), we obtain the following result for the polarization:

$$P(r, t, u) = P^{(0)}(r, t) + P^{(\mathrm{ind})}(r, t, u) , \tag{5.57}$$

and the two components on the right-hand side are given by

$$P^{(0)}(r, t) = \chi^{(0)}(k_i, \omega_\mathrm{i}) \epsilon_0 E(k_i, \omega_\mathrm{i}) \cos(k_i \cdot r - \omega_\mathrm{i} t), \tag{5.58}$$

$$P^{(\mathrm{ind})}(r, t, u) = \left(\frac{\partial \chi}{\partial u}\right)_{u=0} u(r, t) \epsilon_0 E(k_i, \omega_\mathrm{i}) \cos(k_i \cdot r - \omega_\mathrm{i} t) . \tag{5.59}$$

The term $P^{(\mathrm{ind})}$ in the above equation is the polarization induced by the lattice vibrations and is related to Raman scattering. It is rewritten as:

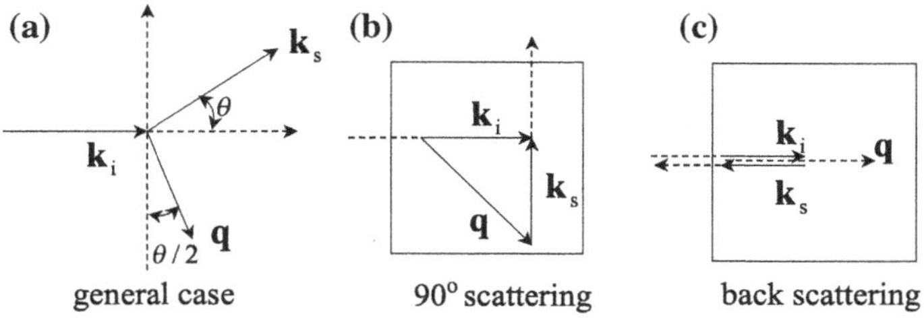

Fig. 5.10 Geometry of the light scattering experiment. **a** General case, **b** right-angle (90°) scattering geometry, and **c** back scattering geometry. $\boldsymbol{k}_\mathrm{i}$ is the wave vector of the incident photon, $\boldsymbol{k}_\mathrm{s}$ is the wave vector of the scattered photon, and $\boldsymbol{q}$ is the phonon wave vector

$$\begin{aligned}
&\boldsymbol{P}^{(\mathrm{ind})}(\boldsymbol{r}, t, \boldsymbol{u}) \\
&= \left(\frac{\partial \chi}{\partial \boldsymbol{u}}\right)_{\boldsymbol{u}=0} \boldsymbol{u}(\boldsymbol{q}, \omega_\mathrm{q}) \cos(\boldsymbol{q} \cdot \boldsymbol{r} - \omega_\mathrm{q} t)\epsilon_0 \boldsymbol{E}(\boldsymbol{k}_i, \omega_\mathrm{i}) \cos(\boldsymbol{k}_i \cdot \boldsymbol{r} - \omega_\mathrm{i} t) \\
&= \frac{1}{2}\epsilon_0 \left(\frac{\partial \chi}{\partial \boldsymbol{u}}\right)_{\boldsymbol{u}=0} \boldsymbol{u}(\boldsymbol{q}, \omega_\mathrm{q}) \boldsymbol{E}(\boldsymbol{k}_i, \omega_\mathrm{i}) \\
&\quad \times \left[\cos\{(\boldsymbol{k}_i + \boldsymbol{q}) \cdot \boldsymbol{r} - (\omega_\mathrm{i} + \omega_\mathrm{q})t\}\right. \\
&\quad \left. + \cos\{(\boldsymbol{k}_i - \boldsymbol{q}) \cdot \boldsymbol{r} - (\omega_\mathrm{i} - \omega_\mathrm{q})t\}\right] .
\end{aligned} \tag{5.60}$$

The above results are understood as follows. The polarization $\boldsymbol{P}^{(\mathrm{ind})}$ consists of the **Stokes shifted wave** with wave vector $\boldsymbol{k}_\mathrm{S} = (\boldsymbol{k}_i - \boldsymbol{q})$ and angular frequency $\omega_\mathrm{S} = (\omega_\mathrm{i} - \omega_\mathrm{q})$ and of the **anti-Stokes wave** with wave vector $\boldsymbol{k}_\mathrm{AS} = (\boldsymbol{k}_i + \boldsymbol{q})$ and angular frequency $\omega_\mathrm{AS} = (\omega_\mathrm{i} + \omega_\mathrm{q})$. In Raman scattering experiments, therefore, we observe two scattered beams, one is a Stokes shifted wave and the other is an anti-Stokes shifted wave. In other words, these two different scattered lines appear on both sides of the laser line shifted by the phonon frequency and the shift is called the Raman shift.

In the scattering event the energy conservation law and momentum conservation law hold:

$$\omega_\mathrm{i} = \omega_\mathrm{s} \pm \omega_\mathrm{q}, \tag{5.61}$$

$$\boldsymbol{k}_\mathrm{i} = \boldsymbol{k}_\mathrm{s} \pm \boldsymbol{q}, \tag{5.62}$$

where the sign $\pm$ means Stokes scattering for $+$ and anti-Stokes scattering for $-$. Since the difference between ω_i and ω_s is small, $|\boldsymbol{k}_i|$ and $|\boldsymbol{k}_s|$ are almost the same. In this case the scattering angle θ in Fig. 5.10a is given by

$$q = 2k_\mathrm{i} \sin(\theta/2) . \tag{5.63}$$

From these discussions we may understand that the intensity of the Raman scattering is determined by $P^{(\mathrm{ind})}$. In the following we will derive the Raman scattering intensity, where the intensity is defined by using the Raman tensor. Here we will discuss the Raman tensor first. The polarization due to the lattice vibrations χ_{ij} is defined by

$$\chi_{ij} = \chi_{ij}^{(0)} + \sum_k \chi_{ij,k} u_k + \sum_{k,l} \chi_{ij,kl} u_k u_l + 0(u^3) , \tag{5.64}$$

where

$$\chi_{ij,k} = \left(\frac{\partial \chi_{ij}}{\partial u_k}\right)_{u=0} , \quad \chi_{ij,kl} = \left(\frac{\partial^2 \chi_{ij}}{\partial u_j \partial u_l}\right)_{u=0} . \tag{5.65}$$

The term proportional to u gives the first-order Raman scattering and the first derivative of χ of the above equation is called the **first-order Raman tensor**. In a similar fashion the second derivative of χ gives **second-order Raman scattering**, which will not be discussed in detail. It is evident that $\chi_{ij,k}$ is the third-rank tensor and its non-zero components are determined from the crystal symmetry. The intensity of the first-order Raman scattering has been derived by Smith [14], and the result for the intensity of unpolarized light is given by

$$I_s = \frac{3\hbar\omega_s^4 L d\Omega}{\rho c^4 \omega_q} |\chi_{zy,x}|^2 (n_q + 1) , \tag{5.66}$$

where L is the sample length along the light wave vector $\boldsymbol{k}_{\mathrm{i}}$, ρ is the crystal density, n_q is the occupation number of phonons (given by Bose–Einstein statistics), and $\mathrm{d}\Omega$ is the detector solid angle. The above equation is for Stokes scattering, and the intensity for anti-Stokes scattering is obtained by replacing $n_q + 1$ by n_q. The result is derived from classical mechanics and its quantum mechanical treatment will be described later.

In standard experiments, Raman scattering is measured by using polarized light, and the electric polarization to produce the scattering is expressed as

$$\begin{aligned} \boldsymbol{P}^{(\mathrm{ind})} &\propto \left(\frac{\partial \chi}{\partial \boldsymbol{u}}\right) \boldsymbol{u}(\boldsymbol{q}, \omega_{\mathrm{q}}) \cdot \boldsymbol{e}^{(i)} = \left(\frac{\partial \chi_{ij}}{\partial u_k}\right) u_k(\boldsymbol{q}, \omega_{\mathrm{q}}) \cdot \boldsymbol{e}^{(i)} \\ &= (\chi_{ij,k}) u_k \cdot e_j^{(i)} \sim R_{ji}^k \cdot e_j^{(i)} , \end{aligned}$$

where $\boldsymbol{e}^{(i)}$ is the unit polarization vector of the incident light. The scattered light intensity of the polarization direction $\boldsymbol{e}^{(s)}$ is given by $\boldsymbol{e}^{(s)} \cdot \boldsymbol{P}^{(\mathrm{ind})}$, and thus we obtain for the scattered intensity

$$I_s \propto \left| \boldsymbol{e}^{(s)} \cdot \left(\frac{\partial \chi}{\partial \boldsymbol{u}}\right) \boldsymbol{u} \cdot \boldsymbol{e}^{(i)} \right|^2 \sim \left| \boldsymbol{e}^{(i)} \cdot R_{ji}^k \cdot \boldsymbol{e}^{(s)} \right|^2 \sim \left| e_j^{(i)} R_{ji}^k e_i^{(s)} \right|^2 , \tag{5.67}$$

where the subscripts i, j, k mean the x, y, z components, and the superscripts (i) and (s) are for incident and scattered light, respectively.

As mentioned before, $\chi_{ij,k}$ is the third-rank tensor, and $\chi_{ij,k}$ multiplied by a displacement vector $\boldsymbol{u} = \boldsymbol{\xi} \cdot u$ ($\boldsymbol{\xi}$ is the unit vector of the displacement) results in the second-rank tensor. In other words

$$\chi_{ij,k} \cdot \boldsymbol{u} = \sum_k \chi_{ij,k}\, u_k \sim \sum_k R^k_{ji} \xi_k$$

is the second-rank tensor and its non-zero components are determined by the crystal symmetry. For the same reason the Raman tensor R^k_{ji} or $\chi_{ij,k}$ is determined by the crystal symmetry. The Raman tensor R^k_{ji} depends on the incident light polarization j, scattered (Raman) light polarization i, and phonon polarization k. The unperturbed susceptibility $\chi^{(0)}_{ij}$ is a second-rank tensor. In the case of Raman scattering the frequencies of the incident and scattered light are different, and thus the Raman tensor R^k_{ji} $(\chi_{ij,k})$ is no longer a second-rank tensor in the strict sense. The Raman tensor determines the selection rule of Raman scattering, and Raman scattering and infrared absorption are complementary. The Raman tensor $\chi_{ij,k} = (\partial \chi_{ij}/\partial \boldsymbol{u})$ reflects the crystal symmetry. As an example, we consider the Raman tensor of a crystal with inversion symmetry. The Raman tensor of a crystal with inversion symmetry is invariant under the operation of inversion, whereas there are two types of phonon modes: one does not change sign under the inversion operation (even parity) and the other changes sign under the inversion operation (odd parity). Since the displacement $\boldsymbol{u}$ of a phonon mode with odd parity changes sign under the inversion operation, the sign of $\chi_{ij,k} = (\partial \chi_{ij}/\partial \boldsymbol{u})$ is changed. In other words, we may conclude that the Raman tensor for the phonon modes with odd parity vanishes. It should also be noted that phonons with odd parity are infrared-active and phonons with even parity are infra-red-inactive.

It is well known that a crystal without inversion symmetry exhibits piezoelectricity. As an example we consider a zinc blende crystal such as GaAs. Piezoelectricity is a phenomenon in which a strain e_{jk} induces an electric field, and the i component of the electric field E_i is related to the strain according to

$$E_i = e_{ijk} e_{jk},$$

where e_{ijk} is a piezoelectric constant and a third-rank tensor. The definition of strain is given in Appendix C. Here we adopt the contraction of the subscripts of the tensor defined in Appendix C: $xx = 1$, $yy = 2$, $zz = 3$, $yz = zy = 4$, $zx = xz = 5$, and $xy = yx = 6$ for jk. Then the piezoelectric constant tensor of a cubic crystal T_d is given by

$$\begin{bmatrix} 0 & 0 & 0 & e_{14} & 0 & 0 \\ 0 & 0 & 0 & 0 & e_{14} & 0 \\ 0 & 0 & 0 & 0 & 0 & e_{14} \end{bmatrix}, \tag{5.68}$$

and its non-zero components consist of one independent constant e_{14}. The non-zero components arise from the components xyz, yzx, zxy. The Raman tensor R^k_{ji} is expected to have the same properties as the piezoelectric tensor. Detailed analysis has been done with the help of group theory. Here we adopt the notation used so far and express the Raman tensor by the 3×3 matrices.

$$R^x_{yz}, R^x_{zy} = \begin{bmatrix} 0 & 0 & 0 \\ 0 & 0 & d \\ 0 & d & 0 \end{bmatrix}, \quad R^y_{zx}, R^y_{xz} = \begin{bmatrix} 0 & 0 & d \\ 0 & 0 & 0 \\ d & 0 & 0 \end{bmatrix}, \quad R^z_{xy}, R^z_{yx} = \begin{bmatrix} 0 & d & 0 \\ d & 0 & 0 \\ 0 & 0 & 0 \end{bmatrix}. \tag{5.69}$$

As described earlier, the Raman tensor ($\sum_k R^k_{ji}$) is a second-rank tensor and it may be possible to express it by an irreducible representation as is the case for the strain tensor. Referring to Appendix C, the Raman tensor shown above belongs to the Γ_4 group. Using (C.15) in Appendix C we obtain the following result:

$$R(\Gamma_1) = \begin{bmatrix} a & 0 & 0 \\ 0 & a & 0 \\ 0 & 0 & a \end{bmatrix}, \tag{5.70}$$

$$R(\Gamma_3) = \begin{bmatrix} b & 0 & 0 \\ 0 & b & 0 \\ 0 & 0 & -2b \end{bmatrix}, \quad \sqrt{3}\begin{bmatrix} b & 0 & 0 \\ 0 & -b & 0 \\ 0 & 0 & 0 \end{bmatrix}, \tag{5.71}$$

$$R(\Gamma_4) = \begin{bmatrix} 0 & 0 & 0 \\ 0 & 0 & d \\ 0 & d & 0 \end{bmatrix}, \quad \begin{bmatrix} 0 & 0 & d \\ 0 & 0 & 0 \\ d & 0 & 0 \end{bmatrix}, \quad \begin{bmatrix} 0 & d & 0 \\ d & 0 & 0 \\ 0 & 0 & 0 \end{bmatrix}. \tag{5.72}$$

The Raman tensor $R(\Gamma_4)$ of the above equation is equivalent to R^z_{xy}, R^x_{yz} and so on of (5.69). The factor $\sqrt{3}$ of (5.71) arises from the normalization of the basis functions shown in Appendix C such as $(1/\sqrt{2})(e_{xx} - e_{yy})$ and $(1/\sqrt{6})(e_{xx} + e_{yy} - 2e_{zz})$. When we use the form of the strain tensor (C.16) in Appendix C, (5.71) is rewritten as

$$R(\Gamma_3) = \begin{bmatrix} b & 0 & 0 \\ 0 & b & 0 \\ 0 & 0 & b \end{bmatrix}, \quad \begin{bmatrix} b & 0 & 0 \\ 0 & b & 0 \\ 0 & 0 & b \end{bmatrix}. \tag{5.73}$$

In the above equation two equivalent tensors are listed because the phonons are doubly degenerate.

A table of Raman tensors have been reported which are obtained with the help of group theory. In this section we show the Raman tensors in Table 5.1 derived by Loudon [15], where the modes with (x), (y), and (z) are both Raman-active and infrared-active with the polarization direction x, y, and z, respectively. It should be noted that all of the Raman-active modes are not observed as the first-order Raman scattering. As shown in the previous example of a crystal with T_d symmetry, optical phonons have Γ_4 symmetry and phonons with Γ_1 (A_1) and Γ_3 (E) symmetry are not optical phonons. These phonon modes are normally observed as the second-order Raman scattering. Also, it should be noted that Table 5.1 obtained under the assumption of infinite phonon wave vectors (wave vector: $\boldsymbol{q} = 0$) and that the tensors are classified with the help of the point group of the crystals. In experiments on Raman scattering, phonons with the wave vector $\boldsymbol{q} \neq 0$ are involved.

Table 5.1 Raman tensors of a cubic crystal and Raman-active phonon modes

	$\begin{bmatrix} a & 0 & 0 \\ 0 & a & 0 \\ 0 & 0 & a \end{bmatrix}$	$\begin{bmatrix} b & 0 & 0 \\ 0 & b & 0 \\ 0 & 0 & b \end{bmatrix}$	$\begin{bmatrix} b & 0 & 0 \\ 0 & b & 0 \\ 0 & 0 & b \end{bmatrix}$	$\begin{bmatrix} 0 & 0 & 0 \\ 0 & 0 & d \\ 0 & d & 0 \end{bmatrix}$	$\begin{bmatrix} 0 & 0 & d \\ 0 & 0 & 0 \\ d & 0 & 0 \end{bmatrix}$	$\begin{bmatrix} 0 & d & 0 \\ d & 0 & 0 \\ 0 & 0 & 0 \end{bmatrix}$
T	A	E	E	F(x)	F(y)	F(z)
T_h	A_g	E_g	E_g	F_g	F_g	F_g
O	A_g	E_g	E_g	F_g	F_g	F_g
T_d	A_1	E	E	$F_2(x)$	$F_2(y)$	$F_2(z)$
O_h	A_{1g}	E_g	E_g	F_{2g}	F_{2g}	F_{2g}

5.2.1 Selection Rule of Raman Scattering

Here we will discuss the first order Raman intensity (single phonon Raman scattering) using the results obtained above. For the purpose of better understanding, we will consider Raman scattering in a crystal with inversion symmetry (O_h group) and a crystal without inversion symmetry (T_d group). From Table 5.1 we find that optical phonons in a crystal with inversion symmetry O_h consist of triply degenerate $F_{2}g$ ($\Gamma_{25'}$) phonons at the Γ point: two transverse optical phonon modes and one longitudinal phonon mode. The scattering intensity is given by

$$I = A\left[\sum_{j,i=x,y,z} e_j^{(i)} R_{ji} e_i^{(s)}\right]^2, \tag{5.74}$$

where A is a constant determined from by the material and the scattered photon frequency, and $e_j^{(i)}$ and $e_i^{(s)}$ are the j and i components of the unit polarization vector of the incident and scattered light. Referring to Fig. 5.10a, we calculate the scattering intensity for the two cases where the polarization of the scattered light is parallel and perpendicular to the scattering plane (x, z plane), respectively. Noting that the summation is carried out for degenerate phonon modes, the intensities for the two cases are given by

$$\begin{aligned} I_{\parallel} &= A\left[\left(e_x^{(i)} R_{xz}\right)^2 + \left(e_y^{(i)} R_{yx}\right)^2 + \left(e_y^{(i)} R_{yz}\right)^2\right] \\ &= A|d|^2\left[\left(e_x^{(i)} \sin\theta\right)^2 + \left(e_y^{(i)}\right)^2 (\cos^2\theta + \sin^2\theta)\right] \\ &= A|d|^2\left[\left(e_x^{(i)} \sin\theta\right)^2 + \left(e_y^{(i)}\right)^2\right], \end{aligned} \tag{5.75a}$$

$$I_{\perp} = A\left[\left(e_x^{(i)} R_{xy}\right)^2\right] = A|d|^2\left[\left(e_x^{(i)}\right)^2\right]. \tag{5.75b}$$

On the other hand, in a crystal without inversion symmetry T_d the doubly degenerate transverse optical (TO) phonons and longitudinal optical (LO) phonons split. In such

Table 5.2 Selection rule for Raman scattering in the T_d group (includes zinc blende crystal such as GaAs) in back scattering and right-angle scattering geometries

Scattering geometry		Selection rule	
		TO phonons	LO phonons
Back scattering	$z(y, y)\bar{z}$; $z(x, x)\bar{z}$	0	0
	$z(x, y)\bar{z}$; $z(y, x)\bar{z}$	0	$\|d_{\rm LO}\|^2$
90° scattering	$z(x, z)x$	$\|d_{\rm TO}\|^2$	0
	$z(y, z)x$	$\|d_{\rm TO}\|^2/2$	$\|d_{\rm LO}\|^2/2$
	$z(x, y)x$	$\|d_{\rm TO}\|^2/2$	$\|d_{\rm LO}\|^2/2$

a case the Raman tensor depends on the phonon polarization direction. Defining the unit polarization vector of the phonons by $\boldsymbol{\xi}$, the scattering intensity is given by

$$I = A\Bigg[\sum_{j,i,k=x,y,z} e_j^{(i)} R_{ji}^k \xi_k e_i^{(s)}\Bigg]^2 . \tag{5.76}$$

For the scattering geometry shown in Fig. 5.10a, the scattering intensity for LO phonons is given by

$$\begin{aligned} I_\parallel(\text{LO}) &= A_{\text{LO}}\left[e_y^{(i)} R_{yz}^x \xi_x e_z^{(s)} + e_y^{(i)} R_{yx}^z \xi_z e_x^{(s)}\right]^2 \\ &= A_{\text{LO}}|d_{\text{LO}}|^2 \left(e_y^{(i)} \sin(3\theta/2)\right)^2 , \end{aligned} \tag{5.77a}$$

$$I_\perp(\text{LO}) = A_{\text{LO}}\left[e_x^{(i)} R_{xy}^z \xi_z e_y^{(s)}\right]^2 = A_{\text{LO}}|d_{\text{LO}}|^2 \left(e_x^{(i)} \sin(\theta/2)\right)^2 , \tag{5.77b}$$

and for TO phonons

$$\begin{aligned} I_\parallel(\text{TO}) &= A_{\text{TO}}\left[\left(e_x^{(i)} R_{xz}^y \xi_y e_z^{(s)}\right)^2 + \left(e_y^{(i)} R_{yz}^x \xi_x e_z^{(s)} + e_y^{(i)} R_{yx}^z \xi_z e_x^{(s)}\right)^2\right] \\ &= A_{\text{TO}}|d_{\text{TO}}|^2 \left[\left(e_x^{(i)} \sin\theta\right)^2 + \left(e_y^{(i)} \cos(3\theta/2)\right)^2\right] , \end{aligned} \tag{5.78a}$$

$$I_\perp(\text{TO}) = A_{\text{TO}}\left[e_x^{(i)} R_{xy}^z \xi_z e_y^{(s)}\right]^2 = A_{\text{TO}}|d_{\text{TO}}|^2 \left(e_x^{(i)} \cos(\theta/2)\right)^2 . \tag{5.78b}$$

When we put $A_{\text{LO}}|d_{\text{LO}}|^2 = A_{\text{TO}}|d_{\text{TO}}|^2 = A|d|^2$, we obtain $I_\parallel(\text{LO}) + I_\parallel(\text{TO}) = I_\parallel$, and $I_\perp(\text{LO}) + I_\perp(\text{TO}) = I_\perp$, which agree with (5.75a) and (5.75b), respectively.

In the cases of the right angle scattering and back scattering geometries shown in Fig. 5.10b and c, the scattered intensities are obtained by putting $\theta = \pi/2$ and $\theta = \pi$, respectively. Raman scattering geometry or selection rule is often defined by $\boldsymbol{k}_i(\boldsymbol{e}^{(i)}, \boldsymbol{e}^{(s)})\boldsymbol{k}_s$, where four vectors are used; two wave vectors $\boldsymbol{k}_i$ and $\boldsymbol{k}_s$ and two polarization vectors $\boldsymbol{e}^{(i)}$ and $\boldsymbol{e}^{(s)}$ for incident and scattered light, respectively. When we use this notation the selection rule of T_d crystal is given by Table 5.2 in the case of back and 90° scattering geometries.

Raman scattering experiments are usually carried out by using a laser in the visible region. The photon energy is higher than the band gap of many semiconductors such as GaAs and Si and thus the light is not transmitted. In such semiconductors the

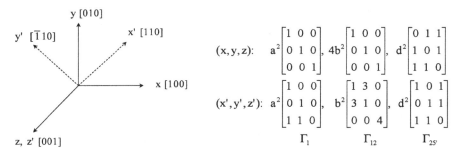

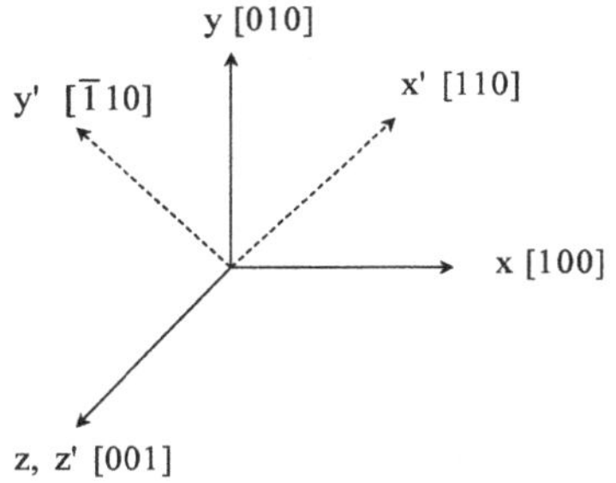

$$(x,y,z):\quad a^2\begin{bmatrix}1&0&0\\0&1&0\\0&0&1\end{bmatrix},\ 4b^2\begin{bmatrix}1&0&0\\0&1&0\\0&0&1\end{bmatrix},\ d^2\begin{bmatrix}0&1&1\\1&0&1\\1&1&0\end{bmatrix}$$

$$(x',y',z'):\quad a^2\begin{bmatrix}1&0&0\\0&1&0\\1&1&0\end{bmatrix},\ b^2\begin{bmatrix}1&3&0\\3&1&0\\0&0&4\end{bmatrix},\ d^2\begin{bmatrix}1&0&1\\0&1&1\\1&1&0\end{bmatrix}$$

$$\Gamma_1 \qquad\qquad \Gamma_{12} \qquad\qquad \Gamma_{25'}$$

Fig. 5.11 Raman scattering intensity for two different configurations of Si. The results on Si, O_h group, may be applied to GaAs, T_d group. The intensity is obtained from Table 5.1

incident photon is absorbed in the surface region and thus Raman measurements with a right-angle (90°) scattering geometry are not possible. In such a case the back-scattering geometry is adopted. From Table 5.2, we find that only LO phonons are detected by Raman scattering. In order to detect TO phonons by Raman scattering we have to use the following configuration. When we define new axes y' and z' along the directions [011] and [0$\bar{1}$1], respectively, the Raman configurations $y'(z', x)\bar{y}'$ and $y'(z', z')\bar{y}'$ will provide a Raman intensity proportional to $|d_{\mathrm{TO}}|^2$ for TO phonons and zero for LO phonons.

We now discuss the experimental results on Raman scattering. There have been numerous papers published so far and it is not the purpose of this textbook to survey all of the data. Instead, only very recent experimental results will be reported, which were obtained for the purpose of this book. First, we will show the experimental results on GaAs, which belongs to the T_d group. As shown in Table 5.1, Raman-active optical phonons are F_2 (Γ_4 or Γ_{15} after the notation of BSW). In addition, we find from Table 5.2 that LO phonons are observed by using the $z(x, y)\bar{z}$ configuration in the (001) plane of GaAs, and that TO phonons are detected by using the $x'(z', y')\bar{x}'$ configuration in the (110) plane of GaAs. Here, $[x, y, z]$ are the coordinates along the crystal axes and $[x', y', z']$ are new coordinates obtained by rotating by 45° around the z axis. Therefore, the (x', z') or (y', z') plane corresponds to the (110) plane. This definition is the same for other cubic crystals. The selection rule and the intensity of Raman scattering in the new coordinates are summarized in Fig. 5.11. Although Fig. 5.11 is for the results on Si, O_h group, the same results may be obtained for GaAs, T_d group, referring to Table 5.1.

Raman scattering experiments have been carried out by utilizing a double monochromator or a triple monochromator and photon counting system or a CCD camera. The data shown here are taken by a new Raman spectrometer "Ramascope" (Raman microscope) made by Renishaw which utilizes a notch filter to remove Rayleigh scattering components and provides high accuracy and convenience. Raman scattering signals of Si observed with the equipment are shown in Fig. 5.12a, where the Raman configuration is $z(x, y)\bar{z}$. Under this configuration Raman scattering due to LO phonons is expected to be observed, while the experimental result shows signals

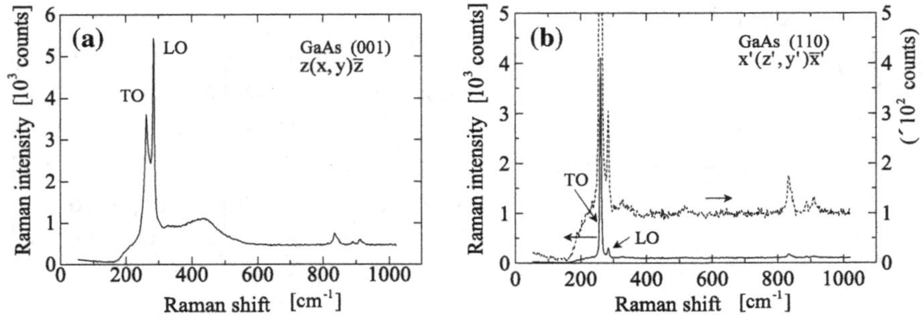

Fig. 5.12 Raman scattering data of GaAs at room temperature. **a** In addition to LO phonon scattering, which is allowed for the $z(x, y)\bar{z}$ configuration, a weak signal due to TO phonon scattering, which is forbidden in this configuration, is observed due to misalignment of the configuration. **b** For the configuration $x'(z', y')\bar{x}'$, TO phonon scattering is allowed but LO phonon scattering is forbidden

due to forbidden TO phonons in addition to LO phonon scattering in the region 200–300 cm^{-1}. The signal may arise from imperfect polarization of light and from the fact that the incident light angle depends on the depth of focus. Raman scattering due to TO phonons may be observed under the Raman configuration $x'(z', y')\bar{x}'$ using the (100) plane, which is determined from Fig. 5.11. The results are shown in Fig. 5.12b, where a strong peak of TO phonons and a weak peak of LO phonons are observed.

Temple and Hathaway [16] have reported detailed results for multi-phonon Raman scattering in Si. Referring to this paper, we discuss how to observe the firs-t and second-order Raman scattering. Experiments are carried out by utilizing the Ramascope of Renishaw in a similar way as used for GaAs. Since Si belongs to the O_h group, the phonon modes of A_1g (Γ_1), E_g (Γ_{12}), F_{2g} ($\Gamma_{25'}$) are found to contribute to Raman scattering from Table 5.1. In the following we adopt the notation of BSW (Bouckaert, Smoluchowski and Wigner [13] in Chap. 1) (in brackets). Since optical phonons belong to $\Gamma_{25'}$, the first-order Raman scattering arises from the Raman tensor belonging to $\Gamma_{25'}$. The experimental results in the (001) plane of Si are shown in Fig. 5.13.

It is evident from Fig. 5.11 that the $z(x', x')\bar{z}$ configuration of Fig. 5.13a enables us to observe all the phonon modes of the representations $\Gamma_1 + \Gamma_{12} + \Gamma_{25'}$. The strong peak at 519 cm^{-1} is due to one-phonon Raman scattering due to the degenerate TO and LO phonons ($\Gamma_{25'}$) at the Γ point. The broad peak in the region 200–450 cm^{-1} is due to the second-order Raman scattering of two acoustic phonons, and the peak slightly above the peak due to the one-phonon Raman scattering arises from the second-order Raman scattering of simultaneously emitted optical and acoustic phonons. The peak near 1000 cm^{-1} is due to second-order Raman scattering which involves two optical phonons. It should be noted here that the energy and momentum conservation rules hold for the second-order Raman scattering and that the two phonons have wave vectors in the reverse direction to each other. The intensity of the second-order Raman scattering reflects the density of states for the phonon mode and thus will

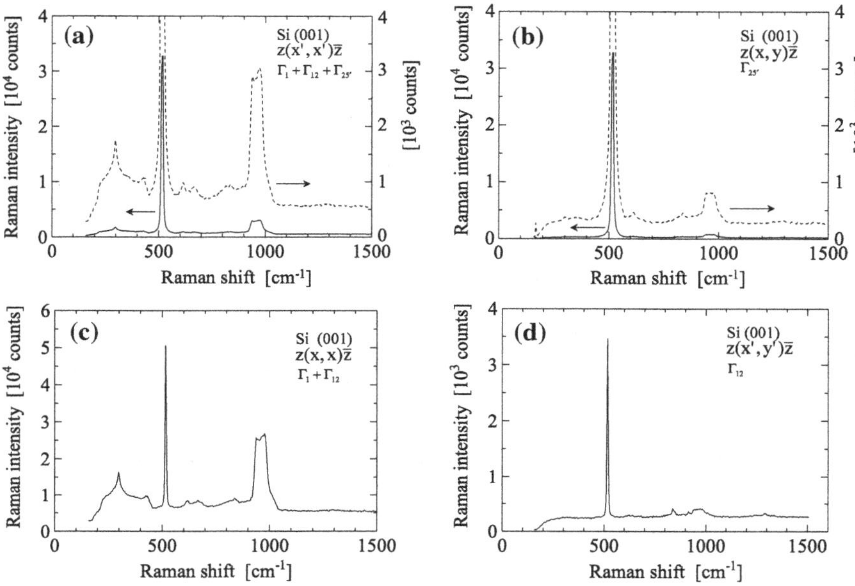

Fig. 5.13 Raman scattering observed in the (001) plane of Si at $T = 300\,\text{K}$. **a** $z(x', x')\bar{z}$ configuration: all the phonon modes of representations $\Gamma_1 + \Gamma_{12} + \Gamma_{25'}$ are observed. The peak at $519\,\text{cm}^{-1}$ is due to first-order Raman scattering by $\Gamma_{25'}$ optical phonons at the Γ point, and the other signals arise from second-order Raman scattering. **b** $z(x, y)\bar{z}$ configuration: only Γ_{25} phonons will provide allowed Raman scattering. **c** $z(x, x)\bar{z}$ configuration: second-order Raman scattering due to phonons of representations $\Gamma_1 + \Gamma_{12}$ is observed. **d** $z(x', y')\bar{z}$ configuration: Raman scattering due to Γ_{12} phonons is expected. According to the slight misalignment second-order Raman scattering is observed in the region $900–1000\,\text{cm}^{-1}$ in addition to the strong peak due to first-order Raman scattering

provide information about the dispersion of the phonon branches. Figure 5.13b shows Raman scattering for the $z(x, y)\bar{z}$ configuration, where Raman scattering due to $\Gamma_{25'}$ phonons is allowed. In the figure, however, we find first-order Raman scattering due to optical phonons at the Γ point and second-order Raman scattering due to two optical phonons. Raman scattering for the $z(x, x)\bar{z}$ configuration is shown in Fig. 5.13c, where phonons of the $\Gamma_1 + \Gamma_{12}$ representations may be observed. In other word, we observe only the second-order Raman scattering due to acoustic and optical phonons except for a weak component due to $\Gamma_{25'}$ phonons. The Raman configuration $z(x', y')\bar{z}$ of Fig. 5.13d enables us to observe the Γ_{12} phonons. The second-order Raman scattering is so weak that the first-order Raman scattering due to optical phonons of $\Gamma_{25'}$ type appears, which may be caused by a slight misalignment of the optical path. Although the intensity is very low, second-order Raman scattering is observed in the region $900–1000\,\text{cm}^{-1}$.

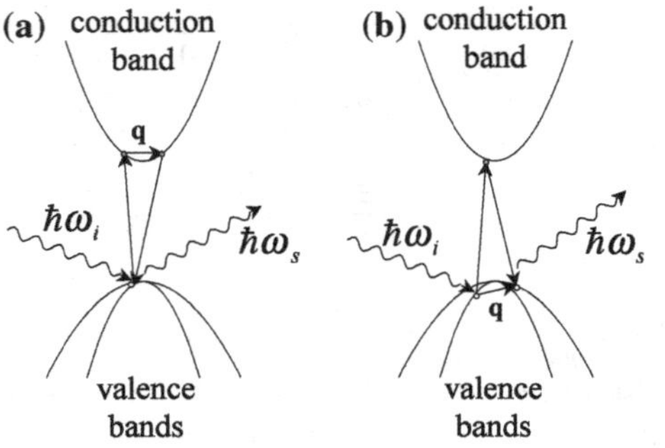

Fig. 5.14 Quantum mechanical representation of Raman scattering. Interband transition induced by an incident photon, electron or hole scattering by a phonon, and emission of a scattered photon

5.2.2 Quantum Mechanical Theory of Raman Scattering

The microscopic interpretation of Raman scattering is as followings. An incident photon creates an excited state of an electron from its ground state, or an electron–hole pair, and the electron or the hole (or electron–hole pair, exciton) interacts with a phonon (phonon scattering) to create a new excited state. Then the electron and the hole recombine to emit a photon. Therefore, we may expect that the scattering intensity increases rapidly when the incident photon energy approaches the fundamental absorption edge. This rapid increase is discussed later and called **resonant Raman scattering**. The processes are schematically illustrated in Fig. 5.14, where the conduction band and the heavy and light hole bands are considered. Figure 5.14a corresponds to the case where an excited electron is scattered by a phonon and (b) to the case of hole scattering by a phonon. In such a quantum mechanical process, three interactions are involved: (1) interaction of an incident photon with an electron in the semiconductor, (2) interaction of an exciton or electron–hole pair with a phonon, (3) recombination (the electron-hole pair with a scattered photon), and therefore the process may be treated by third-order perturbation theory of quantum mechanics. Theoretical analysis has been done by Loudon [17]. Usually quantum mechanical calculations have been carried out with the help of diagrams and various methods have been adopted. Among them two methods adopted by Loudon [17] and by Yu and Cardona [12] are very convenient for us. The process in Fig. 5.14 is shown by diagrams in Fig. 5.15, where (a) is after Loudon and (b) is after Yu and Cardona.

In this textbook we will follow the method of Yu and Cardona. First we define the following rules to draw a diagram:

(a) Excitations such as the photon, electron–hole pair and phonon are called **propagators**, and are drawn by dotted, solid and dashed lines as shown in Fig. 5.16.
(b) The interaction between these excitations is indicated by an intersection connecting two propagators, and the intersection is called a **vertex**. Vertices are

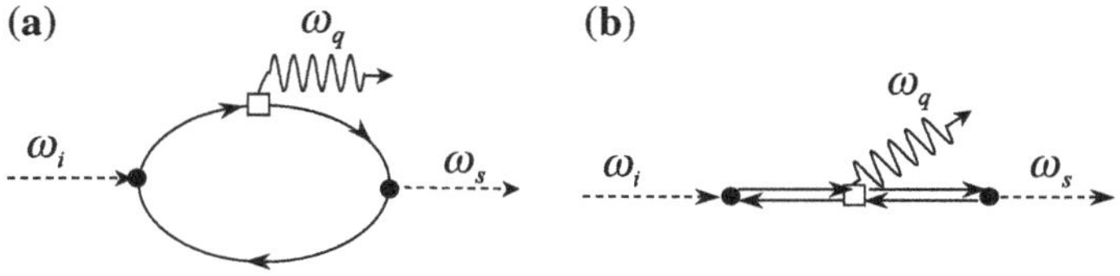

Fig. 5.15 Representations of Raman scattering by diagrams: **a** after Loudon [17] and **b** after Yu and Cardona [12]

Fig. 5.16 Propagator and vertex used to express diagrams

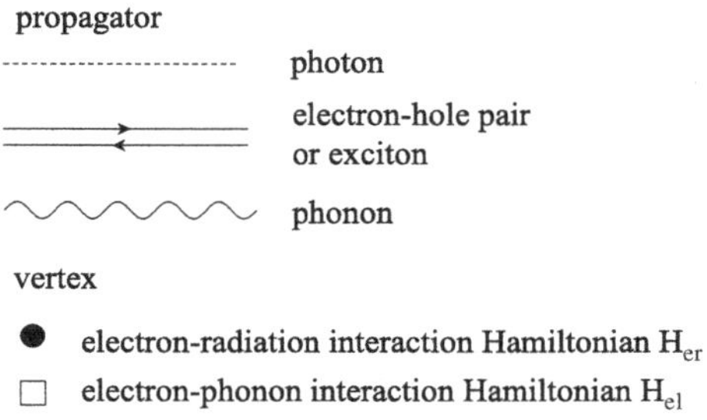

expressed by various notations, but here we use • for electron–photon (radiation field) interaction and □ for the electron–phonon interaction.

(c) Propagators are drawn with arrows. The direction of the arrows represents the creation and annihilation at the interaction. An arrow toward a vertex indicates annihilation and an arrow leaving a vertex indicates creation.
(d) The sequential progress of the interactions are from the left to the right and all the interactions are lined up according to this rule.
(e) Once a diagram is drawn, then draw all the diagrams by changing the order of the vertices.

Using these rules, 6 diagrams for Raman scattering are easily obtained by taking account of 3 vertices, which are shown in Fig. 5.17. Next, we will consider the perturbation calculations of Raman scattering. Defining an initial state by $|\mathrm{i}\rangle$ and the final state by $|\mathrm{f}\rangle$, the scattering rate is calculated by using Fermi Golden Rule as follows. We consider the case of Fig. 5.17a as an example. The Hamiltonian to be solved is

$$H = H_0 + H_{\mathrm{er}} + H_{\mathrm{el}},$$

where $H_0 = H_{\mathrm{e}} + H_{\mathrm{l}}$ is the sum of the electron and lattice vibration Hamiltonians, H_{er} is the interaction Hamiltonian of an electron–photon pair (radiation field), and H_{el} is the interaction Hamiltonian for electron–lattice vibrations. The next step is to calculate the perturbation expansion of the scattering probability using the diagrams, which is shown as follows.

1. From the first vertex we obtain

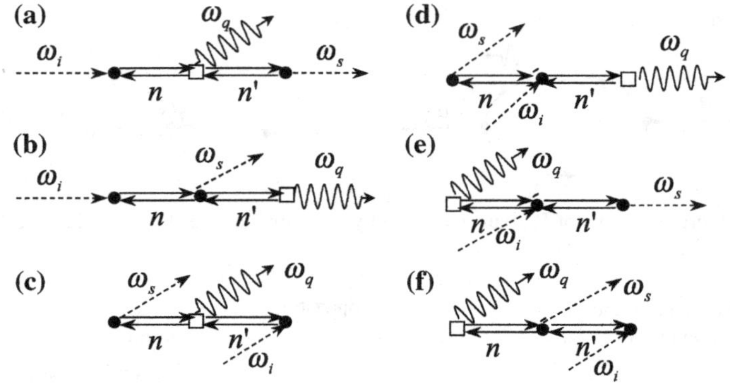

Fig. 5.17 Six diagrams for Raman scattering (Stokes)

$$\sum_n \frac{\langle n|H_{\text{er}}(\omega_{\text{i}})|\text{i}\rangle}{[\hbar\omega_{\text{i}} - (\mathcal{E}_n - \mathcal{E}_{\text{i}})]}, \tag{5.79}$$

where $|\text{i}\rangle$ is the initial state with energy $\mathcal{E}_{\text{i}}$, and $|n\rangle$ is the intermediate state with the energy $\mathcal{E}_n$. The term $\hbar\omega_{\text{i}}$ of the energy denominator has $+$ sign for absorption of the quanta (photon) and $-$ sign for emission. The summation is carried out over all the intermediate states $|n\rangle$.

2. Using this rule, the contribution from the second vertex is carried out and summed as

$$\sum_{n,n'} \frac{\langle n'|H_{\text{el}}(\omega_{\text{q}})|n\rangle\langle n|H_{\text{er}}(\omega_{\text{i}})|\text{i}\rangle}{\left[\hbar\omega_{\text{i}} - (\mathcal{E}_n - \mathcal{E}_{\text{i}}) - \hbar\omega_{\text{q}} - (\mathcal{E}_{n'} - \mathcal{E}_n)\right][\hbar\omega_{\text{i}} - (\mathcal{E}_n - \mathcal{E}_{\text{i}})]}, \tag{5.80}$$

where $|n'\rangle$ is another intermediate state and the minus sign of the term $-\hbar\omega_{\text{q}}$ arises from rule (1), corresponding to the phonon emission. This equation is rewritten as

$$\sum_{n,n'} \frac{\langle n'|H_{\text{el}}(\omega_{\text{q}})|n\rangle\langle n|H_{\text{er}}(\omega_{\text{i}})|\text{i}\rangle}{\left[\hbar\omega_{\text{i}} - \hbar\omega_{\text{q}} - (\mathcal{E}_{n'} - \mathcal{E}_{\text{i}})\right][\hbar\omega_{\text{i}} - (\mathcal{E}_n - \mathcal{E}_{\text{i}})]}. \tag{5.81}$$

In a similar fashion we may take account of other vertices and obtain higher-order perturbations.

3. Note here that the total energy is conserved when we calculate the final vertex term, which is done as follows. The energy denominator for the final (third) vertex is given by

$$\begin{aligned}&\left[\hbar\omega_{\text{i}} - (\mathcal{E}_n - \mathcal{E}_{\text{i}}) - \hbar\omega_{\text{q}} - (\mathcal{E}_{n'} - \mathcal{E}_n) - \hbar\omega_{\text{s}} - (\mathcal{E}_{\text{f}} - \mathcal{E}_{\text{i}})\right]\\ &\quad = \left[\hbar\omega_{\text{i}} - \hbar\omega_{\text{q}} - \hbar\omega_{\text{s}} - (\mathcal{E}_{\text{i}} - \mathcal{E}_{\text{f}})\right].\end{aligned} \tag{5.82}$$

However, the electron returns to the ground state after Raman scattering, and thus the final state $|\mathrm{f}\rangle$ is the same as the initial state $|\mathrm{i}\rangle$. Therefore we obtain the following relation

$$\left[\hbar\omega_{\mathrm{i}} - \hbar\omega_{\mathrm{q}} - \hbar\omega_{\mathrm{s}}\right].$$

The energy conservation rule tells us that the above term should be zero. The calculation of the scattering probability is done by replacing the above term by the Dirac delta function $\delta[\hbar\omega_{\mathrm{i}} - \hbar\omega_{\mathrm{q}} - \hbar\omega_{\mathrm{s}}]$, and the scattering probability is given by

$$\begin{aligned} & w(-\omega_{\mathrm{i}}, \omega_{\mathrm{s}}, \omega_{\mathrm{q}}) \\ & = \frac{2\pi}{\hbar}\left|\sum_{n,n'} \frac{\langle 0|H_{\mathrm{er}}(\omega_{\mathrm{s}})|n'\rangle\langle n'|H_{\mathrm{el}}(\omega_{\mathrm{q}})|n\rangle\langle n|H_{\mathrm{er}}(\omega_{\mathrm{i}})|\mathrm{i}\rangle}{\left[\hbar\omega_{\mathrm{i}} - \hbar\omega_{\mathrm{q}} - (\mathcal{E}_{n'} - \mathcal{E}_{\mathrm{i}})\right][\hbar\omega_{\mathrm{i}} - (\mathcal{E}_{n} - \mathcal{E}_{\mathrm{i}})]}\right|^2 \\ & \times\delta\left[\hbar\omega_{\mathrm{i}} - \hbar\omega_{\mathrm{q}} - \hbar\omega_{\mathrm{s}}\right]. \end{aligned} \tag{5.83}$$

In a similar fashion the calculations for other five diagrams in Fig. 5.17 are straightforward. Summing up all the contributions we obtain the following result, where the initial and final states are assumed to be the ground state $|0\rangle$ and we have used the relation $|\mathrm{i}\rangle = |\mathrm{f}\rangle = |0\rangle$.

$$\begin{aligned} & w(-\omega_{\mathrm{i}}, \omega_{\mathrm{s}}, \omega_{\mathrm{q}}) \\ & = \frac{2\pi}{\hbar}\left|\sum_{n,n'} \frac{\langle 0|H_{\mathrm{er}}(\omega_{\mathrm{s}})|n'\rangle\langle n'|H_{\mathrm{el}}(\omega_{\mathrm{q}})|n\rangle\langle n|H_{\mathrm{er}}(\omega_{\mathrm{i}})|0\rangle}{\left[\hbar\omega_{\mathrm{i}} - \hbar\omega_{\mathrm{q}} - (\mathcal{E}_{n'} - \mathcal{E}_0)\right][\hbar\omega_{\mathrm{i}} - (\mathcal{E}_{n} - \mathcal{E}_0)]}\right. \\ & + \frac{\langle 0|H_{\mathrm{er}}(\omega_{\mathrm{s}})|n'\rangle\langle n'|H_{\mathrm{er}}(\omega_{\mathrm{q}})|n\rangle\langle n|H_{\mathrm{el}}(\omega_{\mathrm{i}})|0\rangle}{[\hbar\omega_{\mathrm{i}} - \hbar\omega_{\mathrm{s}} - (\mathcal{E}_{n'} - \mathcal{E}_0)][\hbar\omega_{\mathrm{i}} - (\mathcal{E}_{n} - \mathcal{E}_0)]} \\ & + \frac{\langle 0|H_{\mathrm{er}}(\omega_{\mathrm{s}})|n'\rangle\langle n'|H_{\mathrm{el}}(\omega_{\mathrm{q}})|n\rangle\langle n|H_{\mathrm{er}}(\omega_{\mathrm{i}})|0\rangle}{\left[-\hbar\omega_{\mathrm{s}} - \hbar\omega_{\mathrm{q}} - (\mathcal{E}_{n'} - \mathcal{E}_0)\right][-\hbar\omega_{\mathrm{s}} - (\mathcal{E}_{n} - \mathcal{E}_0)]} \\ & + \frac{\langle 0|H_{\mathrm{er}}(\omega_{\mathrm{s}})|n'\rangle\langle n'|H_{\mathrm{er}}(\omega_{\mathrm{q}})|n\rangle\langle n|H_{\mathrm{el}}(\omega_{\mathrm{i}})|0\rangle}{[-\hbar\omega_{\mathrm{s}} + \hbar\omega_{\mathrm{i}} - (\mathcal{E}_{n'} - \mathcal{E}_0)][-\hbar\omega_{\mathrm{s}} - (\mathcal{E}_{n} - \mathcal{E}_0)]} \\ & + \frac{\langle 0|H_{\mathrm{er}}(\omega_{\mathrm{s}})|n'\rangle\langle n'|H_{\mathrm{el}}(\omega_{\mathrm{q}})|n\rangle\langle n|H_{\mathrm{er}}(\omega_{\mathrm{i}})|0\rangle}{\left[-\hbar\omega_{\mathrm{q}} + \hbar\omega_{\mathrm{i}} - (\mathcal{E}_{n'} - \mathcal{E}_0)\right]\left[-\hbar\omega_{\mathrm{q}} - (\mathcal{E}_{n} - \mathcal{E}_0)\right]} \\ & \left. + \frac{\langle 0|H_{\mathrm{er}}(\omega_{\mathrm{s}})|n'\rangle\langle n'|H_{\mathrm{el}}(\omega_{\mathrm{q}})|n\rangle\langle n|H_{\mathrm{er}}(\omega_{\mathrm{i}})|0\rangle}{\left[-\hbar\omega_{\mathrm{q}} - \hbar\omega_{\mathrm{s}} - (\mathcal{E}_{n'} - \mathcal{E}_0)\right]\left[-\hbar\omega_{\mathrm{q}} - (\mathcal{E}_{n} - \mathcal{E}_0)\right]}\right|^2 \\ & \times\ \delta[\hbar\omega_{\mathrm{i}} - \hbar\omega_{\mathrm{q}} - \hbar\omega_{\mathrm{s}}]. \end{aligned} \tag{5.84}$$

Loudon deduced the Raman scattering intensity using the above result [15], which is shown below, where the scattering geometry is the same as in thee classical theory used by Smith (5.66):

$$I = \frac{e^4 \omega_s V(n_q + 1) L d\Omega}{4\hbar^3 m^4 d^2 M c^4 \omega_q \omega_i} \left[|R^x_{yz}|^2 + |R^y_{zx}|^2 + |R^z_{xy}|^2 \right] , \tag{5.85}$$

where e and m are the magnitude of the electronic charge and the effective mass of electron, d the lattice constant, V the volume of the crystal, M $(1/M = 1/M_1 + 1/M_2)$ the reduced mass of the lattice atoms, and n_q the phonon occupation number. R^x_{yz} used in (5.85) is the Raman tensor defined previously and given by the following relation for diamond type (O_h) and for zinc blende type crystals(T_d):

$$R^x_{yz}(-\omega_i, \omega_s, \omega_q) = \frac{1}{V} \sum_{\alpha,\beta} \left[\frac{p^z_{0\beta} \Xi^x_{\beta\alpha} p^y_{\alpha 0}}{(\omega_i - \omega_q - \omega_\beta)(\omega_i - \omega_\alpha)} + 5 \text{ other terms} \right] , \tag{5.86}$$

where

$$\hbar\omega_\alpha = \mathcal{E}_\alpha - \mathcal{E}_0, \quad \hbar\omega_\beta = \mathcal{E}_\beta - \mathcal{E}_0$$

and $p^y_{\alpha 0} = \langle \alpha | p_y | 0 \rangle$ is the momentum matrix element for the light polarization y. $\Xi^i_{\alpha\beta}$ is given by

$$\langle \alpha | H_{el} | \beta \rangle = \Xi^i_{\alpha\beta} \frac{\bar{u}_i}{d},$$

where $\boldsymbol{u}$ is the relative displacement of the optical phonons and $\bar{u}_i$ is the amplitude in the x direction of its quantized representation (the quantization of the lattice vibrations will be treated in Chap. 6). $\Xi^i_{\alpha\beta}$ is the deformation potential constant defined by Bir and Picus [18]. According to the analysis of Loudon [15], the scattering efficiency is estimated to be about 10^{-6} to 10^{-7} from (5.85). It is impossible to calculate the scattering efficiency from (5.84) because of the uncertainty of the coefficients involved. Therefore, we pick up the term of the most important contribution from (5.84) and treat the other terms as background, replacing them by a constant. This approximation has been successfully used to analyze resonant Raman scattering and resonant Brillouin scattering.

From these discussions we learn the relation between the macroscopic (classical) theory of the Raman tensor and the quantum mechanical result of the Raman tensor. We have to note here that relations such as $\chi^x_{yz} = \chi_{zy,x}$ hold from the macroscopic theory, but there are no such relations for the quantum mechanical R^x_{yz}. This may be understood from the fact that the subscripts y, z of R^x_{yz} are for the polarization directions of incident (y) and scattered (z) photons and that R^x_{zy} corresponds to the time reversal sequence of Fig. 5.17a, from the right to the left, which should be expressed from the quantum mechanical point of view (time reversal) as

$$R^x_{yz}(-\omega_i, \omega_s, \omega_q) = R^x_{zy}(-\omega_i + \omega_q, \omega_s + \omega_q, -\omega_q)\,. \tag{5.87}$$

The phonon energy is much smaller than the photon energy in Raman scattering and we may safely put $\omega_q \simeq 0$ to obtain

$$R^x_{yz}(-\omega_i, \omega_i, 0) = R^x_{zy}(-\omega_i, \omega_i, 0)\,. \tag{5.88}$$

Therefore, we find that R^x_{yz} has the same property as $\chi_{yz,x}$.

5.2.3 *Resonant Raman Scattering*

As shown in Fig. 5.14, Raman scattering involves the creation of an electron–hole pair as the intermediate state. Therefore, when the incident photon energy approaches the fundamental absorption edge, the efficiency of creating electron–hole pairs increases and the Raman scattering efficiency is enhanced dramatically. The resonant enhancement is called **resonant Raman scattering**. This enhancement is clear from the scattering efficiency obtained in Sect. 5.2.2. The strongest contribution in (5.84) comes from the diagram shown in Fig. 5.17a, which is evident from the following result. From (5.86) we put

$$R^x_{yz} = \frac{1}{V}\sum_{\alpha,\beta} \frac{p^z_{0\beta}\Xi^x_{\beta\alpha}p^y_{\alpha 0}}{(\omega_i - \omega_q - \omega_\beta)(\omega_i - \omega_\alpha)}\,, \tag{5.89}$$

and assuming that the bands are isotropic and parabolic and that the momentum matrix element is independent of the wave vector, we find

$$\begin{aligned} R^x_{yz} &= \frac{2}{(2\pi)^3} p^z_{0\beta}\Xi^x_{\beta\alpha}p^y_{\alpha 0} \\ &\times \int_{\mathrm{B.Z.}} \frac{4\pi k^2 \mathrm{d}k}{\left(\omega_{g\beta} + \omega_q - \omega_i + \dfrac{\hbar k^2}{2\mu}\right)\left(\omega_{g\alpha} - \omega_i + \dfrac{\hbar k^2}{2\mu}\right)}\,, \end{aligned} \tag{5.90}$$

where $\hbar\omega_{g\alpha}$ and $\hbar\omega_{g\beta}$ are the energy gaps for the incident and scattered light. In the case of the zinc blende crystals shown in Fig. 4.17, $\hbar\omega_{g\alpha} = \hbar\omega_{g\beta}$. μ is the reduced mass of the electron–hole pair of the intermediate state and we assume that the masses are the same for the states α and β. Integration of (5.90) is easily performed by using the result shown in the first footnote of in Appendix A.1. When we put the upper bound of the integration as the band edge energies (bandwidths of the joint state of the conduction and valence bands) $\hbar\Delta\omega_\alpha$ and $\hbar\Delta\omega_\beta$, we obtain

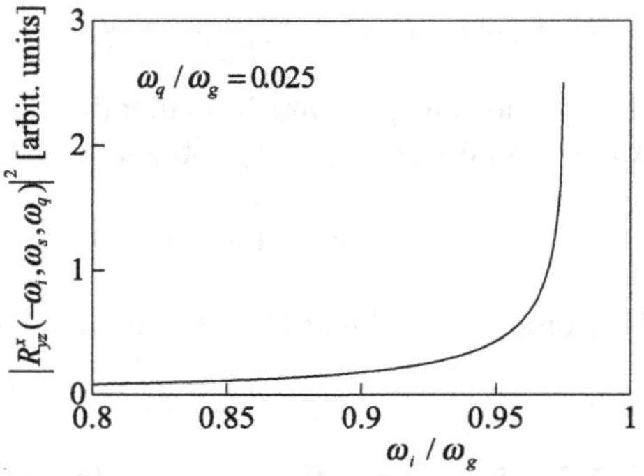

Fig. 5.18 The resonant term of resonant Raman scattering (resonant Brillouin scattering) calculated from (5.92) is plotted as a function of incident photon energy; $\omega_q/\omega_g = 0.025$ assumed. When the incident photon energy approaches the fundamental absorption edge, the scattering efficiency increases resonantly

$$R^x_{yz} = \frac{4}{(2\pi)^2} \cdot \frac{p^z_{0\beta} \Xi^x_{\beta\alpha} p^y_{\alpha 0}}{\omega_{g\beta} - \omega_{g\alpha} + \omega_q} \cdot \left(\frac{2\mu}{\hbar}\right)^{3/2}$$
$$\times \left[\left(\omega_{g\beta} - \omega_s\right)^{1/2} \arctan\left(\frac{\Delta\omega_\beta}{\omega_{g\beta} - \omega_s}\right)^{1/2} - \left(\omega_{g\alpha} - \omega_i\right)^{1/2} \arctan\left(\frac{\Delta\omega_\alpha}{\omega_{g\alpha} - \omega_i}\right)^{1/2} \right] . \tag{5.91}$$

For simplicity we put $\hbar\omega_{g\alpha} = \hbar\omega_{g\beta} = \hbar\omega_g$, and then (5.91) reduces to

$$R^x_{yz}(-\omega_i, \omega_s, \omega_q) = \frac{1}{2\pi}\left(\frac{2\mu}{\hbar}\right)^{3/2} p^z_{0\beta} \Xi^x_{\beta\alpha} p^y_{\alpha 0} \times \frac{1}{\omega_q}\left[\left(\omega_g - \omega_s\right)^{1/2} - \left(\omega_g - \omega_i\right)^{1/2} \right] . \tag{5.92}$$

The scattering efficiency $|R^x_{yz}|^2$ versus photon energy curve calculated from (5.92) is plotted in Fig. 5.18, where we put $\omega_q/\omega_g = 0.025$. It is clear from this figure that the scattering efficiency increases resonantly as the incident phonon energy approaches the band gap. From this reason the phenomenon is called **resonant Raman scattering**. As discussed later, we find a similar result for **resonant Brillouin scattering**, where the difference is the phonon energy ω_q involved.

Next, we will show that Raman scattering is interpreted in terms of the modulation of the dielectric function with respect to the phonon energy. This was first pointed out by Cardona [19, 20]. This idea was introduced from the following investigation [12]. When we take into account the finite lifetime of the intermediate state, the Raman tensor is rewritten by introducing the damping constant or the broadening constant Γ_α as

$$R^x_{yz} = \frac{\hbar}{V}\sum_\alpha \frac{p^z_{0\beta} \Xi^x_{\alpha\alpha} p^y_{\alpha 0}}{(\hbar\omega_s - \mathcal{E}_\alpha + i\Gamma_\alpha)(\hbar\omega_i - \mathcal{E}_\alpha + i\Gamma_\alpha)} . \tag{5.93}$$

We rewrite this equation as

$$R_{yz}^{x} = \frac{p_{0\beta}^{z}\Xi_{\alpha\alpha}^{x}p_{\alpha 0}^{y}}{\omega_{\mathrm{q}}}\sum_{\boldsymbol{k}}\left[\frac{1}{\mathcal{E}_{\mathrm{cv}}(\boldsymbol{k}) - \hbar\omega_{\mathrm{i}} - \mathrm{i}\Gamma_{\alpha}} - \frac{1}{\mathcal{E}_{\mathrm{cv}}(\boldsymbol{k}) - \hbar\omega_{\mathrm{s}} - \mathrm{i}\Gamma_{\alpha}}\right], \tag{5.94}$$

where we put $\hbar\omega_{\alpha} = \mathcal{E}_{\alpha} = \mathcal{E}_{\mathrm{c}}(\boldsymbol{k}) - \mathcal{E}_{\mathrm{v}}(\boldsymbol{k}) \equiv \mathcal{E}_{\mathrm{cv}}(\boldsymbol{k})$. Using the Dirac delta function of Appendix A.1, we find the following relation:

$$\frac{1}{\mathcal{E}_{\mathrm{cv}}(\boldsymbol{k}) - \hbar\omega_{\mathrm{i}} - \mathrm{i}\Gamma_{\alpha}} = \frac{1}{\mathcal{E}_{\mathrm{cv}}(\boldsymbol{k}) - \hbar\omega_{\mathrm{i}}} + \mathrm{i}\pi\delta\left[\mathcal{E}_{\mathrm{cv}}(\boldsymbol{k}) - \hbar\omega_{\mathrm{i}}\right]. \tag{5.95}$$

Comparing (5.95) with (4.127b), we find that the imaginary part of (5.95) corresponds to the imaginary part of the dielectric function (or electric susceptibility). In addition, assuming that the photon energy is close to the fundamental absorption edge ($\mathcal{E}_{\mathrm{cv}}(\boldsymbol{k}) = \mathcal{E}_{\mathrm{g}} + \hbar^2 k^2/2\mu \sim \omega$), we rewrite (4.127b) as

$$\begin{aligned}\sum_{\boldsymbol{k}}\frac{|\boldsymbol{e}\cdot\boldsymbol{p}_{\mathrm{cv}}|^2}{\mathcal{E}_{\mathrm{cv}}(\boldsymbol{k})}\cdot\frac{1}{\mathcal{E}_{\mathrm{cv}}^2 - \hbar^2\omega^2} &= \sum_{\boldsymbol{k}}\frac{|\boldsymbol{e}\cdot\boldsymbol{p}_{\mathrm{cv}}|^2}{\mathcal{E}_{\mathrm{cv}}(\boldsymbol{k})}\cdot\frac{1}{(\mathcal{E}_{\mathrm{cv}} + \hbar\omega)(\mathcal{E}_{\mathrm{cv}} + \hbar\omega)} \\ &\sim \frac{|\boldsymbol{e}\cdot\boldsymbol{p}_{\mathrm{cv}}|^2}{2\hbar^2\omega^2}\sum_{\boldsymbol{k}}\frac{1}{\mathcal{E}_{\mathrm{cv}}(\boldsymbol{k}) - \hbar\omega},\end{aligned} \tag{5.96}$$

and this gives the real part of (5.95). From these considerations we obtain

$$R_{yz}^{x}(-\omega_{\mathrm{i}}, \omega_{\mathrm{s}}, \omega_{\mathrm{q}}) \propto \frac{1}{\hbar\omega_{\mathrm{q}}}\left[\kappa(\omega_{\mathrm{i}}) - \kappa(\omega_{\mathrm{s}})\right]. \tag{5.97}$$

If the phonon energy $\hbar\omega_{\mathrm{q}}$ is much smaller than ω_{i} and ω_{s}, then (5.97) is equivalent to the first derivative of the dielectric function with respect to the energy; in other words, we obtain

$$R_{yz}^{x}(-\omega_{\mathrm{i}}, \omega_{\mathrm{s}}, \omega_{\mathrm{q}}) \propto \frac{\partial\kappa(\omega)}{\partial(\hbar\omega)}. \tag{5.98}$$

Therefore, the scattering efficiency is given by

$$I \propto \left|\frac{\partial\kappa}{\partial\mathcal{E}}\right|^2, \tag{5.99}$$

where we put $\hbar\omega = \mathcal{E}$. This approximation is more accurate because the phonon energy involved is much smaller [20–22]. This approximation is called the **quasi-static approximation**. Experiments on resonant Raman scattering have been carried out by utilizing a laser with a wavelength very close to the band gap. One of the best examples is the experiment on GaP near the $\mathcal{E}_0$ edge, which is shown in Fig. 5.19. In the experiment TO phonon Raman scattering is observed by using several laser lines, which is shown by +. The solid curve is calculated from (5.99) by taking into account the two band edges, the $\mathcal{E}_0$ edge and the spin–orbit split-off band $\mathcal{E}_0 + \Delta_0$, whereas the dashed curve takes into account of the $\mathcal{E}_0$ edge only [23]. As seen in Fig. 5.19 we find good agreement between the experiment and theory.

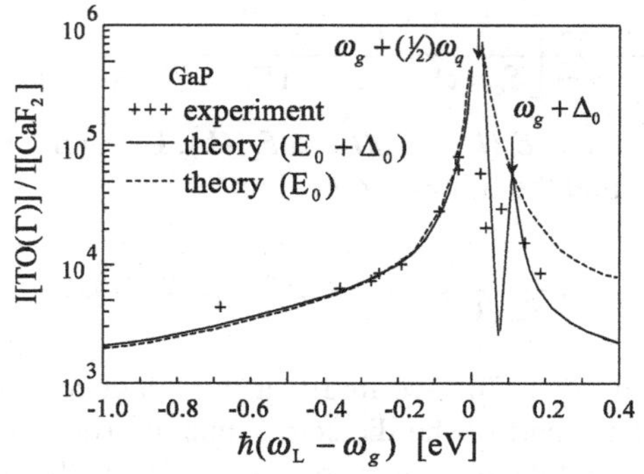

Fig. 5.19 The Raman scattering intensity for TO phonons in GaP is plotted as a function of incident photon energy $\hbar\omega_L$, where resonant enhancement is observed at the $\mathcal{E}_0$ edge and the $\mathcal{E}_0 + \Delta_0$ edge. The *solid curve* is calculated by taking the $\mathcal{E}_0$ and $\mathcal{E}_0 + \Delta_0$ edges into account, whereas the *dashed curve* is calculated by taking account of the $\mathcal{E}_0$ edge only. The theoretical curves are obtained by putting $\hbar(\omega_g + \omega_q)$ for the band gap in the derivative form of the dielectric function (quasi-static approximation)

5.3 Brillouin Scattering

Brillouin scattering involves the interaction of acoustic phonons instead of optical phonons in Raman scattering, and thus its treatment is quite similar to that for Raman scattering. While the optical phonon energy is about several 10 meV, the acoustic phonon energy involved in Brillouin scattering is about 4×10^{-6} eV $= 4\,\mu$eV for 1 GHz phonons and 4 meV for 1 THz phonons. Therefore, it is very difficult to observe Brillouin scattering with the double or triple monochromator used in Raman scattering experiments. Brillouin scattering was predicted theoretically by L. Brillouin in 1922 [24]. Brillouin scattering should be included in the **Pockels effect** reported by F. Pockels in 1889 [25]. The Pockels effect is often referred to as the **photoelastic effect**, where the optical property is changed by a change in the dielectric function induced by elastic deformation of a solid. Elastic deformation includes static stress and thermal disturbance (lattice vibrations or sound propagation); the latter is to be understood as Brillouin scattering. In general, the photoelastic effect is characterized by the photoelastic constant tensor p_{ijkl}. When a strain e_{kl} is applied to a crystal, a change in the inverse dielectric constant $\delta\kappa^{-1}$ is expressed as

$$(\delta\kappa^{-1})_{ij} = p_{ijkl}e_{kl}\,, \tag{5.100}$$

where p_{ijkl} is called the photoelastic constant tensor of Pockels. From this relation we obtain

$$\delta\kappa_{ij}(\boldsymbol{r}, t) = -\kappa_{im}p_{mnkl}e_{kl}\kappa_{nj}\,, \tag{5.101}$$

where we have dropped the summation $\sum$ with respect to the subscripts on the right-hand side. For simplicity we consider the cases for cubic and wurtzite crystals, where their tensors contain diagonal elements only, and then the change in dielectric constant is given by

$$\delta\kappa_{ij}(\boldsymbol{r}, t) = \delta\chi_{ij}(\boldsymbol{r}, t) = -\kappa_{ii}\kappa_{jj}p_{ijkl}e_{kl}\,. \tag{5.102}$$

The scattering efficiency, therefore, is given by an equation similar to (5.66) for Raman scattering and obtained by replacing $\chi_{ij,k}u_k$ with $\delta\kappa_{ij}$ $(= \delta\chi_{ij})$. Theoretical calculations of the Brillouin scattering cross section in cubic crystals have been reported by Benedek and Fritsch [26], in hexagonal crystals by Hamaguchi [27], and in a general form by Nelson, Lazay and Lax [28]. Although $\delta\kappa_{ij}$ is a second-rank tensor and the same as the Raman tensor $\chi_{ij}^k u_k$, the change in the dielectric constant for Brillouin scattering contains the fourth-rank tensor of the photoelastic constant p_{ijkl}, which has three non-zero components p_{11}, p_{12} and p_{44} for cubic crystals and five non-zero components p_{11}, p_{12}, p_{13}, p_{44} and $p_{66} = (1/2)(p_{11} - p_{12})$, where p_{66} is not independent. The Brillouin scattering cross section per unit solid angle $\sigma_{\mathrm{B}}(\omega_{\mathrm{i}})$ (scattering efficiency I : $I \propto \sigma_{\mathrm{B}}(\omega_{\mathrm{i}})$) is given by [28]

$$\sigma_{\mathrm{B}}\Delta\Omega \propto \frac{\omega_{\mathrm{i}}^4 k_{\mathrm{B}}T\Delta\Omega}{8\pi^2c^4\rho v_\mu^2\sin(\theta_{\mathrm{s}}')n^{(\mathrm{s})}n^{(\mathrm{i})}}\left|e_i^{(\mathrm{s})}\chi_{ijkl}e_j^{(\mathrm{i})}b_k a_l\right|^2\,, \tag{5.103}$$

where

$$\chi_{ijkl} = -\frac{1}{2}\kappa_{im}p_{mnkl}\kappa_{nj} \tag{5.104}$$

and ω is the angular frequency of the incident light and the angular frequency of the scattered light is given by $\omega_{\mathrm{s}} = \omega_{\mathrm{i}} \pm \omega_{\mathrm{q}} \simeq \omega_{\mathrm{i}}$. c is the light velocity, ρ the density of the material, v_μ the sound velocity, $\Delta\Omega$ the solid angle of the detector to collect the scattered light, $e_i^{(\mathrm{s})}$ and $e_j^{(\mathrm{i})}$ are the polarization directions of the incident and scattered light, respectively, θ' the scattered angle inside the material, $n^{(\mathrm{s})}$ and $n^{(\mathrm{i})}$ are the refractive indices for the incident and scattered light, respectively, and b_k and a_l are the unit displacement vector and unit wave vector of the elastic waves [28]. It should be noted that the rotational components of the strain tensor are taken into account in [28] and that the authors define a new photoelastic constant $p_{(ij)kl}$ instead of the photoelastic constant p_{ijkl} used in this book. In addition Nelson et al. [28] take into account the indirect effect due to piezoelectricity, the angle between the Poynting vector and the wave vector, and the scattering volume. Therefore, the cross-section of (5.103) is different from used in the original work of Nelson et al. [28].

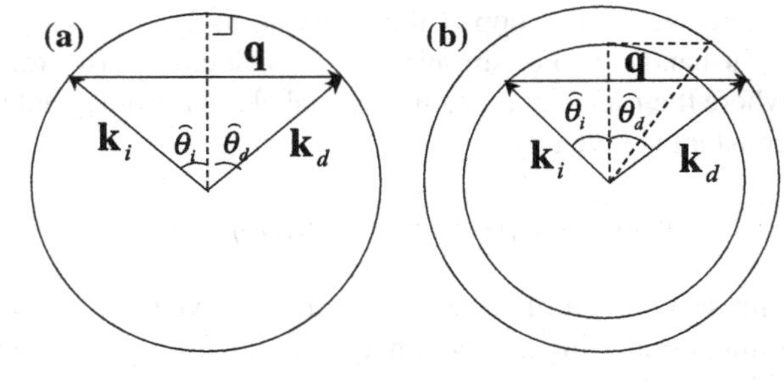

Fig. 5.20 Wave vectors involved in Brillouin scattering for **a** isotropic material and **b** anisotropic material

5.3.1 Scattering Angle

First, we consider the scattering angle in a cubic or isotropic crystal. Let us define the wave vectors of the incident and scattered light inside the crystal as $\boldsymbol{k}_{\mathrm{i}}$ and $\boldsymbol{k}_{\mathrm{d}}$, respectively, and the respective angular frequencies as ω_{i} and ω_{d}. The scattering is assumed to be induced by elastic waves with wave vector $\boldsymbol{q}$ and angular frequency $\omega_\mu = 2\pi f$. The momentum and energy conservation rules are therefore written as

$$\boldsymbol{k}_{\mathrm{i}} \pm \boldsymbol{q} = \boldsymbol{k}_{\mathrm{d}}, \tag{5.105a}$$

$$\hbar\omega_{\mathrm{i}} \pm \hbar\omega_\mu = \hbar\omega_{\mathrm{d}} \,. \tag{5.105b}$$

Using the light velocity c ($\approx 3 \times 10^8$m/s) and the sound velocity v_μ($\approx 3 \times 10^3$m/s), we obtain following relations: $\omega_{\mathrm{i}} = ck_{\mathrm{i}}, \omega_{\mathrm{d}} = ck_{\mathrm{d}}, \omega_\mu = v_\mu q$. Since the sound velocity v_μ is much smaller than the light velocity c ($v_\mu \ll c$), we have $v_\mu q \ll ck_{\mathrm{i}}$ for the elastic waves of which wave vector is comparable to the incident light wave vector, and thus $\omega_\mu \ll \omega_{\mathrm{i}}$, giving rise to

$$\omega_{\mathrm{i}} \cong \omega_{\mathrm{d}}, \quad k_{\mathrm{i}} \cong k_{\mathrm{d}}.$$

Therefore, three vectors k_{i}, k_{d} and q form an isosceles triangle, and the relation is expressed by using a circle as shown in Fig. 5.20a. In this case we find a relation $(\hat{\theta}_{\mathrm{s}}/2 = \hat{\theta}_{\mathrm{i}} = \hat{\theta}_{\mathrm{d}})$, where $\hat{\theta}_{\mathrm{s}}$ is the angle (scattered angle) between the incident and scattered light, and thus we obtain

$$q = 2k_{\mathrm{i}} \sin\left(\frac{\hat{\theta}_{\mathrm{s}}}{2}\right) \tag{5.106}$$

or the frequency of the elastic waves (phonons) f is given by

$$f = \frac{2nv_\mu}{\lambda_0} \sin\left(\frac{\hat{\theta}_{\mathrm{s}}}{2}\right), \tag{5.107}$$

where λ_0 is the wavelength of the light in the air, n is the refractive index, and the relation $k_{\mathrm{i}} = 2\pi n/\lambda_0$ is used. In the case of right-angle (90°) scattering or back scattering shown in Fig. 5.10b or c, the scattering angle inside the crystal is the same as the outside scattering angle, and the above relation is valid. In the case where the incident or scattered light is not perpendicular to the incident plane, however, the angles θ_i and θ_d outside the sample should be calculated by using Snell law, Snell law leads to the relations $\sin\theta_\mathrm{i} = n\sin\hat{\theta}_\mathrm{i}$, $\theta_\mathrm{i} = \theta_\mathrm{d} = \theta_\mathrm{s}/2$ and thus we obtain

$$f = \frac{2v_\mu}{\lambda_0}\sin\frac{\theta_\mathrm{s}}{2}\,. \tag{5.108}$$

The angle θ_s is called the scattering angle outside the crystal.

In a Brillouin scattering event, the scattered light polarization rotates by 90° with respect to the incident light polarization under some scattering configurations. When this rotation occurs in an anisotropic crystal, the anisotropy of the refractive indices induces a pronounced effect on the scattering angles. As an example we consider the case of CdS, and choose the c plane (plane perpendicular to the c axis) as the scattering plane (plane containing the incident and scattered light). Elastic waves are assumed to be transverse and the displacement vector is parallel to the c axis. Since the strain of the transverse elastic waves is e_{zy}, the corresponding photoelastic constant is $p_{44} = p_{zyzy}$, and thus we find that the scattered light polarization is parallel to the c axis for the incident light polarization perpendicular to the c axis. The refractive indices are different for the two different polarizations, which is called **birefringence**. We often use a He–Ne laser for Brillouin scattering and here we discuss the birefringence in CdS at the light wavelength $\lambda_0 = 6328$ nm. The refractive indices are $n_\mathrm{o} = 2.460$ and $n_\mathrm{e} = 2.477$ for the polarization perpendicular and parallel to the c axis, respectively. Although the difference in the refractive indices is quite small, $(n_\mathrm{e} - n_\mathrm{o})/[2(n_\mathrm{e} + n_\mathrm{o})] = 0.7\%$, the birefringence leads to a big effect on the scattering angle.

When we define n_i and n_d for the refractive indices of the incident and diffracted (scattered) light, we find the following relation.

$$k_i = \frac{\omega_\mathrm{i}}{c}n_\mathrm{i} = \frac{2\pi n_\mathrm{i}}{\lambda_0}\,, \tag{5.109a}$$

$$k_d = \frac{\omega_d}{c}n_\mathrm{d} \cong \frac{\omega_\mathrm{i}}{c}n_\mathrm{d} = \frac{2\pi n_\mathrm{d}}{\lambda_0}\,. \tag{5.109b}$$

The momentum conservation rule in the presence of birefringence is shown in Fig. 5.20b, where we find an inequality $\hat{\theta}_\mathrm{i} \neq \hat{\theta}_\mathrm{d}$. From this figure we obtain

$$\frac{n_\mathrm{i}}{\lambda_0}\sin\hat{\theta}_\mathrm{i} + \frac{n_\mathrm{d}}{\lambda_0}\sin\hat{\theta}_\mathrm{d} = \frac{f}{v_\mu}\,, \tag{5.110a}$$

$$\frac{n_\mathrm{i}}{\lambda_0}\cos\hat{\theta}_\mathrm{i} = \frac{n_\mathrm{d}}{\lambda_0}\cos\hat{\theta}_\mathrm{d}\,. \tag{5.110b}$$

This will lead to the following relations [29].

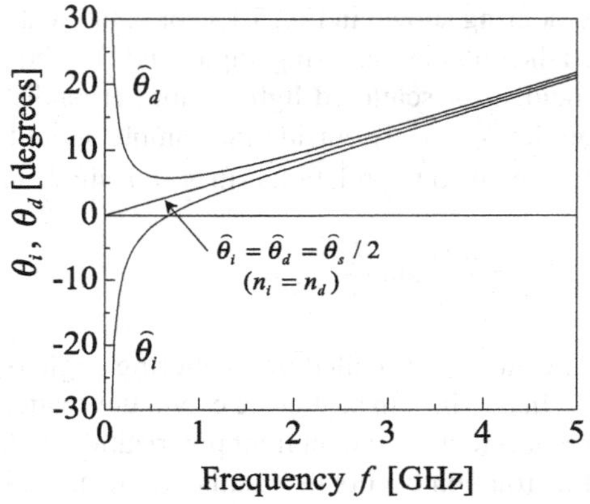

Fig. 5.21 Incident angle $\hat{\theta}_\mathrm{i}$ and diffracted (scattered) angle $\hat{\theta}_\mathrm{d}$ inside an anisotropic crystal (CdS) plotted as a function of elastic wave frequency. The incident and diffracted angles in an isotropic crystal are given by the *straight line*

$$\sin\hat{\theta}_\mathrm{i} = \frac{\lambda_0}{2n_\mathrm{i}v_\mu}\left[f + \frac{v_\mu^2}{f\lambda_0^2}\left(n_\mathrm{i}^2 - n_\mathrm{d}^2\right)\right], \tag{5.111a}$$

$$\cos\hat{\theta}_\mathrm{d} = \frac{\lambda_0}{2n_\mathrm{d}v_\mu}\left[f - \frac{v_\mu^2}{f\lambda_0^2}\left(n_\mathrm{i}^2 - n_\mathrm{d}^2\right)\right]. \tag{5.111b}$$

Therefore, the scattering angle inside the crystal $\hat{\theta}_\mathrm{s} = \hat{\theta}_\mathrm{i} + \hat{\theta}_\mathrm{d}$ is given by

$$\sin\frac{\hat{\theta}_\mathrm{s}}{2} = \frac{f}{2\sqrt{n_\mathrm{i}n_\mathrm{d}}}\left[\left(\frac{f}{v_\mu}\right)^2 - \frac{(n_\mathrm{i} - n_\mathrm{d})^2}{\lambda_0^2}\right]^{1/2}. \tag{5.112}$$

The incident angle $\hat{\theta}_\mathrm{i}$ and diffracted (scattered) angle $\hat{\theta}_\mathrm{d}$ are calculated by (5.111a) and (5.111b) in CdS and plotted as a function of the elastic wave frequency in Fig. 5.21, where the refractive indices used are $n_\mathrm{i} = n_\mathrm{o} = 2.460$ and $n_\mathrm{d} = n_\mathrm{e} = 2.477$. In the figure the straight line is the curve calculated for the incident and diffracted angles $\hat{\theta}_\mathrm{i} = \hat{\theta}_\mathrm{d} = \hat{\theta}_s/2$ for an isotropic crystal assuming that the refractive index $n_\mathrm{i} = n_\mathrm{d} = (n_\mathrm{o} + n_\mathrm{e})/2$. The angles outside the crystal are calculated with the help of Fig. 5.22 and of Snell's law:

$$\sin\theta_\mathrm{i} = n_\mathrm{i}\sin\left\{\sin^{-1}\left(\frac{\lambda_0}{2n_\mathrm{i}v_\mu}\left[f + \frac{v_\mu^2}{f\lambda_0^2}\left(n_\mathrm{i}^2 - n_\mathrm{d}^2\right)\right]\right)\right\}, \tag{5.113}$$

$$\cos\theta_\mathrm{d} = n_\mathrm{d}\sin\left\{\sin^{-1}\left(\frac{\lambda_0}{2n_\mathrm{d}v_\mu}\left[f - \frac{v_\mu^2}{f\lambda_0^2}\left(n_\mathrm{i}^2 - n_\mathrm{d}^2\right)\right]\right)\right\}. \tag{5.114}$$

Note that the angles outside an isotropic crystal are obtained by putting $n_\mathrm{i} = n_\mathrm{d}$. The incident angles θ_i and diffracted angle θ_d outside a CdS crystal are calculated and plotted as a function of elastic wave frequency, which are shown in Fig. 5.23.

The following result is obtained from (5.112). When the frequency of the elastic waves is

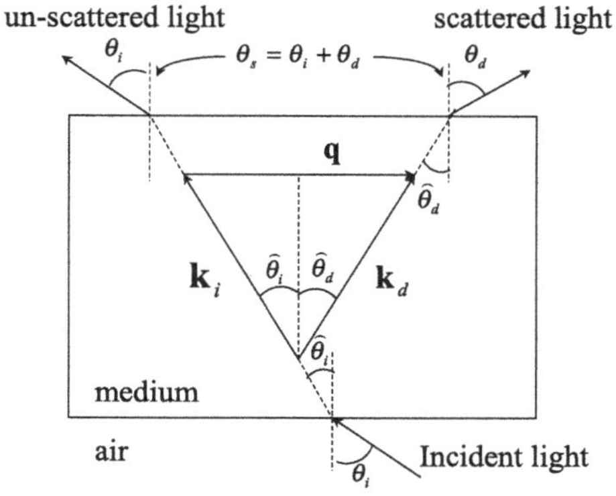

Fig. 5.22 Incident angle $\theta_{\rm i}$, diffracted angle $\theta_{\rm d}$, and scattering angle $\theta_{\rm s} = \theta_{\rm i} + \theta_{\rm d}$ outside a crystal and their relations to the angles inside the crystal $\hat{\theta}_{\rm i}$, $\hat{\theta}_{\rm d}$

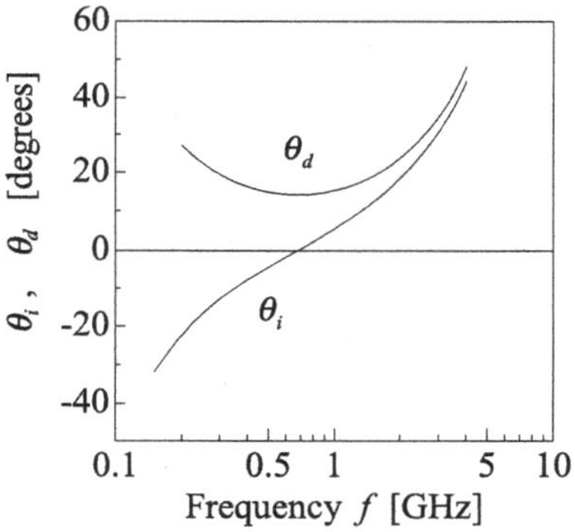

Fig. 5.23 Incident angle θ_i and diffracted (scattered) angle θ_d outside an anisotropic crystal (CdS) as a function of elastic wave frequency, where the refractive indices are $n_{\rm i} = n_{\rm o} = 2.460$ and $n_{\rm d} = n_{\rm e} = 2.477$

$$f_0 = \frac{v_\mu}{\lambda_0}\sqrt{n_{\rm i}^2 - n_{\rm d}^2} = \frac{v_\mu}{\lambda_0}\sqrt{n_{\rm e}^2 - n_{\rm o}^2}\,, \tag{5.115}$$

the incident angle becomes $\hat{\theta}_{\rm i} = 0$. In addition, the minimum observable frequency $f_{\rm min}$ is obtained by putting $\hat{\theta}_{\rm s} = 0$ in (5.112),

$$f_{\rm min} = \frac{v_\mu}{\lambda_0}\,|n_{\rm i} - n_{\rm d}| = \frac{v_\mu}{\lambda_0}\,(n_{\rm e} - n_{\rm o})\ . \tag{5.116}$$

On the other hand, the maximum observable frequency is obtained by putting $\hat{\theta}_{\rm s} = \pi$:

$$f_{\rm max} = \frac{v_\mu}{\lambda_0}\,(n_{\rm i} + n_{\rm d}) = \frac{v_\mu}{\lambda_0}\,(n_{\rm e} - n_{\rm o})\ . \tag{5.117}$$

As typical examples, we present f_0, $f_{\rm min}$, and $f_{\rm max}$ for CdS and ZnO in Table 5.3. From the table we find that $f_0 \approx 1$GHz, $f_{\rm min} \approx 0.5$GHz, and $f_{\rm max} \approx 15$GHz.

Table 5.3 Observable frequencies and other properties of elastic waves from Brillouin scattering experiments in the anisotropic crystals CdS and ZnO

		CdS	ZnO
f_{min}	[GHz]	0.473	0.736
f_0	[GHz]	0.806	1.130
f_{max}	[GHz]	13.7	17.3
$n_{\mathrm{i}} = n_{\mathrm{o}}$		2.460	1.994
$n_{\mathrm{d}} = n_{\mathrm{e}}$		2.477	2.011
v_{μ}	[m/s]	1.76×10^3	2.74×10^3

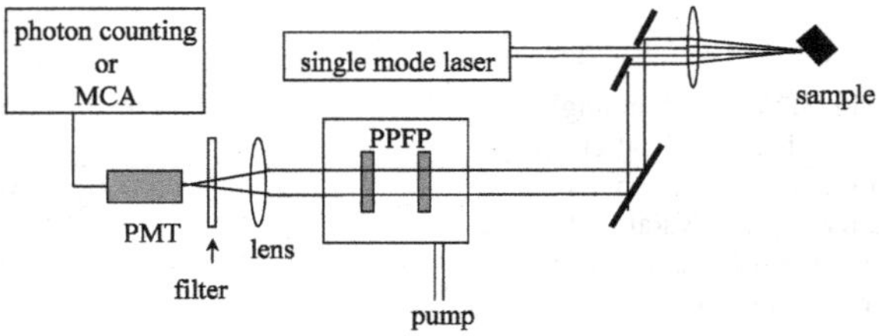

Fig. 5.24 Experimental setup to observe Brillouin scattering. PPFP is a parallel plane Fabry–Perot interferometer. Triple path of the light in the PPFP discriminates undesired components of Rayleigh scattering and improves S/N

5.3.2 *Brillouin Scattering Experiments*

The phonons (elastic waves) involved in Brillouin scattering are for the acoustic modes and their energies are very small compared with the optical phonon energy involved in Raman scattering. Therefore, the frequency shift due to Brillouin scattering is quite small and its detection using a normal monochromator is extremely difficult. Brillouin scattering induced by thermal lattice vibrations (thermal phonons) is usually investigated by using the Fabry–Perot interferometer. The parallel plane Fabry–Perot (PPFP) interferometer is sometimes called the Fabry-Perot etalon, which consists of two parallel plates coated with a high dielectric constant material. When a multiple of the light wavelength is equal to the distance between the plates, interference occurs and the light is transmitted. In order to observe a small change in the wavelength, the thickness of the PPFP is controlled by a piezoelectric element or the PPFP container is evacuated to change the refractive index of the gas (air) inside. Figure 5.24 shows an example of an experimental Fabry-Perot interferometer setup for observing Brillouin scattering. In order to improve the signal-to-noise ration another PPFP is connected in series (tandem type PPFP interferometer) or the light beam is formed into multiple paths by using a prism (multiple path Fabry-Perot interferometer) [30]. Since the scattered signal is extremely weak, the signals are detected by a photon counting system or a multi-channel analyzer (MCA). We present an example of a Brillouin scattering experiment on the Si (100) surface in Fig. 5.25, where the laser line with $\lambda = 488\,\mathrm{nm}$ is used. R and L in Fig. 5.25 are the Rayleigh scattering (no frequency shift) and Brillouin scattering signals due to longitudinal acoustic phonons [30].

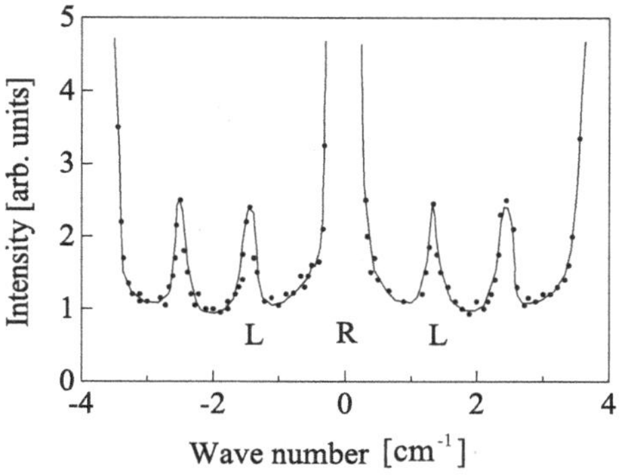

Fig. 5.25 Brillouin scattering spectra of Si (100) surface, where R is Rayleigh scattering and L is Brillouin scattering due to longitudinal acoustic phonons

The study of the elastic properties of solids has a long history and measurements of the velocity of have played a very important role in this aspect. The measurements of the velocity of sound have been performed by using a transducer consisting of a thin piezoelectric plate, which produces a resonant vibration with a frequency determined by the thickness when a.c. field is applied to the transducer. The transducer (glued onto a sample) excites elastic waves by applying an a.c. electric field and the reflected waves are detected by the transducer. The sound velocity is estimated from the delay time of the round trip of the elastic waves and the thickness of the sample. Longitudinal and transverse elastic waves are excited by choosing the crystallographic direction of the transducer, and the anisotropy of the sound velocities are easily measured by choosing the crystal axis of the samples. From these measurements the elastic constants c_{ijkl} are determined. We have to note here that the frequency range produced by a transducer is limited to a low-frequency region from around 1 MHz to several 10 s of MHz. We know that the sound velocity exhibits dispersion and that the sound velocity at high frequency becomes smaller at lower frequencies. Brillouin scattering has been used to investigate the dispersion of the lattice vibrations. Brillouin scattering provides information on the wave vector and the frequency of the lattice vibration and the frequencies are in the region of several GHz. In addition, Brillouin scattering is a kind of non-destructive measurement. From these reasons Brillouin scattering has played an important role in solid state physics.

Another successful application of Brillouin scattering is the investigation of current instabilities caused by acoustoelectric effect, which was one of the exciting topics in physics in 1960–1970. Phonons or acoustic waves (thermal noise) in a crystal are amplified through the acoustoelectric effect and traveling potential waves are excited when the drift velocity of the electron exceeds the sound velocity. Semiconductors such as CdS, ZnO, GaAs and GaSb have no inversion symmetry and thus exhibit piezoelectricity. When an electric field E is applied to such a crystal, electrons move toward the anode (in the anti-parallel direction to the electric field) with drift velocity $v_\mathrm{d} = \mu E$ (μ: drift mobility). When the electron drift velocity v_d exceeds the sound velocity v_μ, $v_\mathrm{d} > v_\mu$, the electrons move by pushing the potential walls which are produced by the piezoelectric effect and move with the velocity of sound, resulting in an energy flow from electrons to acoustic waves and in the amplification of the acoustic waves. The amplification coefficient is proportional to $\gamma = v_d/v_\mu - 1$ and is given by [31]

$$\alpha_e(\omega_\mu, \gamma) = \frac{K^2}{2}\omega_\sigma \left[\frac{\gamma}{\gamma^2 + (\omega_\sigma/\omega_\mu + \omega_\mu/\omega_D)^2}\right], \tag{5.118}$$

where K is the electromechanical coupling coefficient, $\omega_\sigma = \sigma/\kappa\epsilon_0, \omega_D = ev_\mu^2/\mu k_B T$, and $\sigma = ne\mu$ is the conductivity of the semiconductor. The frequency to give the maximum amplification coefficient is given by $f_m = \sqrt{\omega_\sigma \omega_D}/2\pi$, ranging over several GHz. It is well known that the amplification of the acoustic waves excited by a transducer agrees quite well with (5.118). When an electric field is applied to a piezoelectric crystal and the condition $\gamma > 1$ is satisfied, acoustic waves with frequencies around f_m are amplified. Since a crystal that is not uniform gives rise to a higher electric field in a region with higher resistivity, the amplification coefficient becomes higher, resulting in more strongly amplified acoustic waves, in this region of higher electric field. The higher electric field region, high-field domain, moves with the velocity of sound. The phonon intensity in this domain is very high, by factors of 10^5–10^6 higher than the thermal phonon intensity (which exists as thermal noise), and parametric amplification produces phonons with a frequency of half of f_m of intensity about 10^8–10^9 higher than the thermal noise. The parametric mechanism is expressed as the amplification of frequency $f/2$ through the frequency conversion $f = f/2 + f/2$. Such high-intensity phonons are observed with a photomultiplier by adjusting the incident and scattered angles appropriately and a Fabry–Perot interferometer is not required. In addition, the domain passes by in several μs and thus the pulse signal is detected with a high signal-to-noise ratio. An example of experimental data is shown in Fig. 5.26, where we find that phonons of $f_m \simeq 1.4$ GHz are amplified near the cathode and that lower-frequency phonons are more strongly amplified by the parametric mechanism [32]. The loss mechanisms of the amplified phonons were investigated in detail by using this method in addition to spatial and frequency distributions in GaAs [33–35] and in CdS [36].

5.3.3 Resonant Brillouin Scattering

We have discussed Raman scattering and Brillouin Scattering in semiconductors, where all the experiments were carried out by using a laser for the incident light source. On the other hand, Garrod and Bray [37] have shown that Brillouin scattering is possible by using monochromatic light dispersed with a monochromator, if the phonons amplified via the acoustoelectric effect have sufficient intensity. They succeeded in observing the Brillouin scattering intensity by changing the incident light wavelength near the fundamental absorption edge and discovered a very interesting

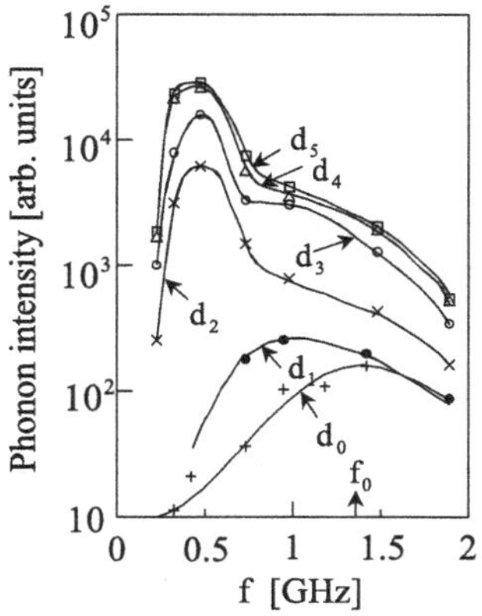

Fig. 5.26 Phonon spectra in acoustoelectric domain in CdS with resistivity 70Ω cm. Transverse acoustic waves (phonons) with displacement vector parallel to the c axis and wave vector perpendicular to the c axis are amplified by applying a high electric field and travel from the cathode to the anode with the velocity of sound. The phonon spectra are obtained at different positions from the cathode, where domains are formed at $d_0 = 0.1$ cm, and $d_1 = 0.132$ cm, $d_2 = 0.20$ cm, $d_3 = 0.26$ cm, $d_4 = 0.322$ cm, and $d_5 = 0.386$ cm (from [32])

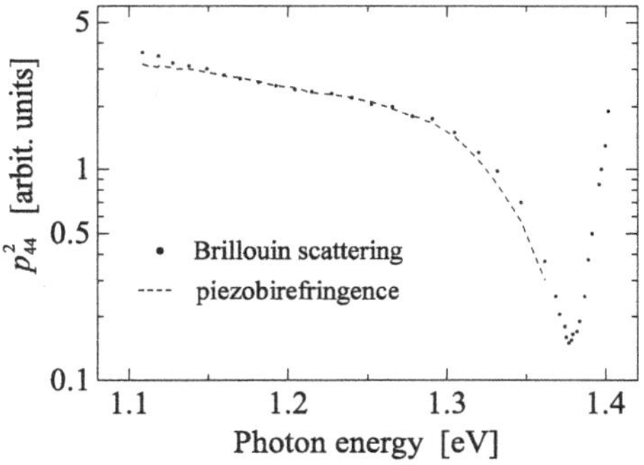

Fig. 5.27 Square of photoelastic constant p_{44} obtained from Brillouin scattering in GaAs as a function of photon energy, where the photoelastic constant is estimated from the Brillouin scattering intensity by assuming that the Brillouin scattering intensity is proportional to p_{44}^2. Phonons, fast transverse phonons of 0.35 GHz, are amplified through the acoustoelectric effect and the Brillouin scattering cross-section is corrected by taking the absorption of the light into account. The *dashed curve* is obtained from piezobirefringence data [37]

feature of the scattering efficiency. Figure 5.27 shows the squared photoelastic constant $|p_{44}|^2$ in GaAs deduced from the Brillouin scattering as a function of incident photon energy, where $|p_{44}|^2$ is proportional to the Brillouin scattering cross-section. A high electric field is applied in the [110] direction to produce an electron drift velocity faster than the sound velocity and the high-intensity phonon domain produced by the acoustoelectric effect provides strong Brillouin scattering signals, enabling us to observe Brillouin scattering near the fundamental absorption edge by using monochromatic light dispersed with a monochromator. The scattering geometry (incident and scattering angles, and polarization) is chosen to observe TA phonons of frequency 0.35 GHz and the incident light pulse is obtained from a mercury lamp dispersed with a monochromator. Since the Brillouin scattering signals are transmitted through the GaAs sample, the absorption has to be corrected near the fundamental absorption edge. The ω_i^4 dependence of the Brillouin scattering cross-section is also taken into account. The squared photoelastic constant $|p_{44}|^2$ is thus obtained as a function of the incident photon energy, which is plotted in Fig. 5.27. The dashed curve in the figure is the photoelastic constant $|p_{44}|^2$ obtained from the piezobirefringence data. We find in Fig. 5.27 that $|p_{44}|^2$ becomes at minimum at around $\hbar\omega_i = 1.38\,\text{eV}$, which produces a sign reversal of p_{44} at the photon energy. Now we will discuss the relation between the photoelastic constant and piezobirefringence.

We have discussed piezobirefringence in Sect. 4.7 and shown that application of stress to a crystal results in a difference in the refractive indices for the directions parallel and perpendicular to the stress. The induced difference of the refractive indices is proportional to the stress when the stress is weak, and the change in the dielectric constant $\Delta\tilde{\kappa}$ corresponding to the stress tensor $\tilde{T}$ is written as

$$\Delta\tilde{\kappa} = \tilde{Q} \cdot \tilde{T}\,, \tag{5.119}$$

where $\tilde{Q}$ is called the piezobirefringence tensor and is related to $\tilde{\alpha}$ defined in Sect. 4.7. When we use the strain tensor defined in Appendix C and use the relation $\tilde{e} = \tilde{s}\tilde{T}$ ($\tilde{s}$: elastic compliance constant), (5.102) reduces to

$$\Delta\tilde{\kappa} = -\tilde{\kappa}\tilde{\kappa}\tilde{p}\tilde{s}\tilde{T} = \tilde{\alpha}\tilde{T}\,, \tag{5.120}$$

and therefore we obtain

$$\tilde{Q} = -\tilde{\kappa}\tilde{\kappa}\tilde{p}\tilde{s} = \tilde{\alpha}\,. \tag{5.121}$$

Next we consider two cases: stress applied to the [100] and [111] directions. For the [100] stress we obtain

$$\alpha(100) = \frac{\Delta\kappa_{\parallel} - \Delta\kappa_{\perp}}{X} = Q_{11} - Q_{12} = -\kappa^2(p_{11} - p_{12})(s_{11} - s_{12})\,, \tag{5.122}$$

and for the [111] stress we obtain

$$\alpha(111) = \frac{\Delta\kappa_{\parallel} - \Delta\kappa_{\perp}}{X} = Q_{44} = -\kappa^2 p_{44}s_{44}\,. \tag{5.123}$$

Now it is apparent from the above results that piezobirefringence experiments will provide the photoelastic constant $\tilde{p}$, which determines the Brillouin scattering-cross section. In Fig. 5.27 the photoelastic constants determined from the piezobirefrin-

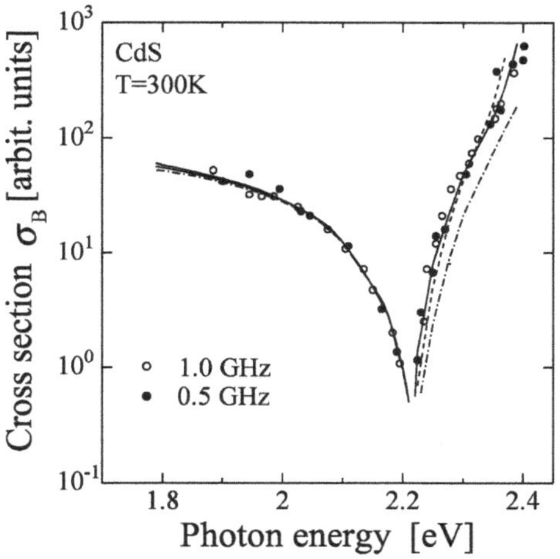

Fig. 5.28 Brillouin scattering cross-sections obtained from Brillouin scattering by acoustoelectric domains in CdS plotted as a function of incident photon energy. Brillouin scattering is observed by using transverse acoustic phonons (elastic waves) of 1 and 0.5 GHz propagating in the c plane of CdS and the cross-sections are estimated by correcting the absorption near the fundamental absorption edge. The *dash-dotted curve* is calculated from (5.91) with $\hbar\omega_{gi} = 2.40(2.494)$ eV and $\hbar\omega_{gs} = 2.38(2.480)$ eV. The *solid curve* is calculated by taking the exciton effect in to account and by putting $\hbar\omega gi = 2.494$ eV and $\hbar\omega_{gs} = 2.480$ eV. Both theoretical curves take into account the resonant cancellation term. (After [39])

gence experiment by Feldman and Horowitz [38] are plotted by the dashed curve, which agrees quite well with the Brillouin scattering data.

In Fig. 5.28 we present experimental result on resonant Brillouin scattering in CdS, where acoustoelectric domains are also used to observe Brillouin scattering cross sections as a function of the incident photon energy. We see very clearly resonant cancellation and resonant enhancement near the fundamental absorption edge in Fig. 5.28 [39]. The theoretical curves in Fig. 5.28 are calculated by taking account of the non-resonant term R_0 in (5.91):

$$\sigma_B \propto |R_{\rm is} - R_0|^2 \, , \tag{5.124}$$

where R_0 is a dispersionless term arising from higher-lying critical points, and resonant cancellation occurs when $R_{\rm is} - R_0 = 0$. The dash-dotted curve in Fig. 5.28 is calculated by putting $\hbar\omega_{\rm gi} = 2.40$ eV and $\hbar\omega_{\rm gs} = 2.38$ eV, and the solid curve is calculated by taking account of the exciton effect and putting $\hbar\omega_{\rm gi} = 2.494$ eV and $\hbar\omega_{\rm gs} = 2.480$ eV. The comparison reveals that neglect of the exciton effect requires energy band gaps lower than the experimentally determined values ($\hbar\omega_{\rm gi} = 2.40$ eV, $\hbar\omega_{\rm gs} = 2.38$ eV) and that the theoretical curve with the exciton effect agrees quite well with the experimental data, where the binding energy of exciton $\mathcal{E}_{\rm ex} = 28$ meV is used. These results indicate the importance of the exciton effect in CdS. In addition to Brillouin scattering by acoustoelectrically amplified phonons, different modes of phonons are excited by bonding a crystal with a different crystallographic axis to the CdS sample and injecting the acoustoelectric domains into the crystal. This phonon injection method provides resonant Brillouin scattering by different phonon modes. The experimental data are interpreted in terms of Loudon's theory and analysis based on the dielectric function [21, 22, 40, 41].

5.4 Polaritons

5.4.1 *Phonon Polaritons*

As discussed in Sect. 6.1.1 of Chap. 6, a crystal with two or more atoms (α atoms) in a unit cell has lattice vibration modes of 3α: $3\alpha - 3$ modes of optical phonons in addition to one longitudinal and two transverse acoustic modes. Here we consider a simple case where a crystal has two atoms in a unit cell with the masses M_1 and M_2. Such semiconductors as Ge and Si with O_h crystal symmetry have the same masses and same electronic charges, and thus the relative displacement of the atoms will not induce electric polarization. Therefore, longitudinal and transverse optical phonons at wave vector $\boldsymbol{q} = 0$ are degenerate and their angular frequency is given by $\omega_0 = \sqrt{2k_0/M_\mathrm{r}}$ ($1/M_\mathrm{r} = 1/M_1 + 1/M_2$, k_0: force constant). In the region $\boldsymbol{q} \neq 0$, the degeneracy of the longitudinal and transverse modes are removed due to elastic anisotropy. On the other hand, GaAs has no inversion symmetry (T_d group), and electronic polarization is induced due to the difference in ionicity of the atoms, resulting in removable of the degeneracy of the longitudinal and transverse optical phonons at $\boldsymbol{q} = 0$.

Let us assume that the lattice consists of atoms with their masses M_+ and M_- and with electronic charge $+e^*$ and $-e^*$. In order to take into account the electric field induced by polarization, we use the local electric field $\boldsymbol{E}_\mathrm{loc}$. The equations of motion for the displacement $\boldsymbol{u}_+$ and $\boldsymbol{u}_-$ are then written as

$$M_+ \frac{\mathrm{d}^2 \boldsymbol{u}_+}{\mathrm{d}t^2} = -2k_0(\boldsymbol{u}_+ - \boldsymbol{u}_-) + e^* \boldsymbol{E}_\mathrm{loc} \tag{5.125a}$$

$$M_- \frac{\mathrm{d}^2 \boldsymbol{u}_-}{\mathrm{d}t^2} = -2k_0(\boldsymbol{u}_- - \boldsymbol{u}_+) - e^* \boldsymbol{E}_\mathrm{loc} , \tag{5.125b}$$

where the local field is defined by $\boldsymbol{E}_\mathrm{loc} = \boldsymbol{E} + \boldsymbol{P}/3\epsilon_0$ by using the applied external electric field $\boldsymbol{E}$. From these equations we obtain

$$M_r \frac{\mathrm{d}^2}{\mathrm{d}t^2}(\boldsymbol{u}_+ - \boldsymbol{u}_-) = -2k_0(\boldsymbol{u}_+ - \boldsymbol{u}_-) + e^* \boldsymbol{E}_\mathrm{loc} . \tag{5.126}$$

Now we assume that the external field is an electromagnetic field with angular frequency ω and that the local field is given by $\boldsymbol{E}_\mathrm{loc} \exp(-\mathrm{i}\omega t)$. The relative displacement is therefore given by

$$u_+ - u_- = \frac{(e^*/M_r)E_{loc}\exp(-i\omega t)}{\omega_0^2 - \omega^2}. \quad (5.127)$$

Since the dipole moment is given by $\mu = e^*(u_+ - u_-)$, the polarization P of a solid with density of unit cells N is

$$P = Ne^*(u_+ - u_-). \quad (5.128)$$

The contribution from the electronic polarization is included by taking account of the Clausius–Mossotti equation. The dielectric constant[2] is then given by

$$\kappa(\omega) = \kappa_\infty + \frac{\kappa_0 - \kappa_\infty}{1 - (\omega/\omega_{TO})^2}, \quad (5.129)$$

where κ_0 is the static dielectric constant and κ_∞ is the high-frequency dielectric constant. The angular frequency of the transverse optical (TO) phonons ω_{TO} is given by

$$\omega_{TO}^2 = \frac{\kappa_\infty + 2}{\kappa_0 + 2}\omega_0^2, \quad (5.130)$$

where we have $\kappa_0 = \kappa_\infty$ without the contribution of ionic polarization and thus $\omega_{TO} = \omega_{LO} = \omega_0$.

Next, we discuss the contribution of ionic polarization to the longitudinal optical phonon frequency ω_{LO}. The longitudinal optical phonon frequency is defined by the frequency required to satisfy the condition $\kappa(\omega) = 0$ in (5.129). Using the angular frequency of the longitudinal optical phonons we find

$$\frac{\omega_{LO}^2}{\omega_{TO}^2} = \frac{\kappa_0}{\kappa_\infty}. \quad (5.131)$$

This relation is called the **Lyddane–Sachs–Teller equation**. With this relation the dielectric function is rewritten as

$$\kappa(\omega) = \kappa_\infty\left(1 + \frac{\omega_{LO}^2 - \omega_{TO}^2}{\omega_{TO}^2 - \omega^2}\right) = \kappa_\infty\frac{\omega_{LO}^2 - \omega^2}{\omega_{TO}^2 - \omega^2}. \quad (5.132)$$

The dielectric constant becomes negative for $\omega_{TO} \le \omega \le \omega_{LO}$, and total reflection occurs in this frequency region.

Now, we consider the interaction between the lattice vibrations and an external electric field

$$E(r, t) = E_0\exp(i[k \cdot r - \omega t]). \quad (5.133)$$

[2]See also Sect. 6.3.6 for the derivation of the following relations.

In the absence of residual charges, the electric displacement $\boldsymbol{D}$ is

$$\mathrm{div}\boldsymbol{D} = 0\,, \tag{5.134}$$

and thus we obtain

$$\kappa(\omega)(\boldsymbol{k} \cdot \boldsymbol{E}_0) = 0\,. \tag{5.135}$$

Therefore, one of the two conditions $\kappa(\omega) = 0$ or $(\boldsymbol{k} \cdot \boldsymbol{E}_0) = 0$ should be satisfied. We will discuss these two cases in the following.

5.4.1.1 (1) Transverse Waves: $(\boldsymbol{k} \cdot \boldsymbol{E}_0) = 0$

Since the wave vector $\boldsymbol{k}$ and electric field vector $\boldsymbol{E}_0$ are perpendicular, the waves are transverse and the external electric field interacts with the transverse waves excited in the crystal. The dielectric function is given by (5.132). The dielectric constant $\kappa(\omega)$ diverges at $\omega = \omega_{\mathrm{TO}}$, and the imaginary part of the dielectric constant behaves like a delta function, giving rise to resonance. The resonance frequency is associated with transverse lattice vibrations and thus called **transverse resonance frequency**.

5.4.1.2 (2) Longitudinal Waves: $\boldsymbol{k} \parallel \boldsymbol{E}_0$ and $\kappa = 0$

When the electric field of the excited waves is longitudinal, we have $\boldsymbol{k} \cdot \boldsymbol{E}_0 \neq 0$ and $\kappa = 0$ is required. The angular frequency ω_{LO} satisfies this condition and is given by (5.131), which tells us that $\omega_{\mathrm{LO}} > \omega_{\mathrm{TO}}$. Since the excited waves are longitudinal, there exists no interaction between the excited waves and the external field. In order to understand these features in more detail we use a simplified model. We define the polarization due to the lattice vibrations by $\boldsymbol{P}_{\mathrm{latt}}$ and the other polarizations by $\boldsymbol{P}_{\mathrm{ele}} = \kappa_\infty \epsilon_0$, which show dispersion in a frequency region higher than the lattice vibrations, where κ_∞ is called the high-frequency dielectric constant or optical dielectric constant. The polarizations $\boldsymbol{P}_{\mathrm{ele}}$ arise from electronic contributions such as the electronic polarization, optical transition, plasma oscillations, and so on. Using these relations we may express the electric displacement by

$$\begin{aligned} \boldsymbol{D} &= \epsilon_0 \boldsymbol{E} + \boldsymbol{P} = \epsilon_0 \boldsymbol{E} + \boldsymbol{P}_{\mathrm{latt}} + \boldsymbol{P}_{\mathrm{ele}} \\ &= \kappa_\infty \epsilon_0 \boldsymbol{E} + \boldsymbol{P}_{\mathrm{latt}} \equiv \kappa(\omega)\epsilon_0 \boldsymbol{E}\,. \end{aligned} \tag{5.136}$$

For $\omega = \omega_{\mathrm{LO}}$, we have $\kappa(\omega_{\mathrm{LO}}) = 0$ and thus $\boldsymbol{D} = 0$. Even if the condition for electric displacement $\boldsymbol{D} = 0$ holds, the lattice vibrations are excited and we have $\boldsymbol{E} \neq 0$. When we define the longitudinal electric field by $\boldsymbol{E}_{\mathrm{latt}}$, we obtain

$$\boldsymbol{E}_{\mathrm{latt}} = -\frac{1}{\kappa_\infty \epsilon_0} \boldsymbol{P}\,. \tag{5.137}$$

For simplicity we neglect the contribution of the polarizations to the local electric field in (5.126) and put $\boldsymbol{E} = \boldsymbol{E}_{\mathrm{loc}}$ and we obtain $\omega_{\mathrm{TO}} = \sqrt{2k_0/M_{\mathrm{r}}}$. Defining the new variable $\tilde{\boldsymbol{u}} = \boldsymbol{u}_+ - \boldsymbol{u}_-$ we get

$$\frac{\mathrm{d}^2}{\mathrm{d}t^2}\tilde{\boldsymbol{u}} = -\omega_{\mathrm{TO}}^2\tilde{\boldsymbol{u}} + e^*\boldsymbol{E}_{\mathrm{latt}}\,. \tag{5.138}$$

Since the polarization of a semiconductor with density of unit cells N is given by $\boldsymbol{P} = Ne^*\tilde{\boldsymbol{u}}$, we put $\boldsymbol{E}_{\mathrm{latt}} = 1/(\kappa_\infty\epsilon_0)Ne^*\tilde{\boldsymbol{u}}$ into (5.138) and obtain for $\omega = \omega_{\mathrm{LO}}$

$$\omega_{\mathrm{LO}}^2 = \omega_{\mathrm{TO}}^2 + \frac{N(e^*)^2}{M_r\kappa_\infty\epsilon_0}\,. \tag{5.139}$$

This result means that the longitudinal optical phonon frequency ω_{LO} is higher than the transverse optical phonon frequency ω_{TO}. This result may be interpreted as showing that the electric field $\boldsymbol{E}_{\mathrm{latt}}$ induced by the polarization $\boldsymbol{P}_{\mathrm{latt}}$ is in the reverse direction to the polarization vector and that the restoring forces of the lattice and the electric field add together, resulting in a LO phonon frequency higher than the TO phonon frequency. Using the relation $\boldsymbol{P} = (\kappa\epsilon_0 - 1)\boldsymbol{E}$ and the static dielectric constant κ_0, we find

$$\kappa_0 - \kappa_\infty = \frac{N(e^*)^2}{M_{\mathrm{r}}\epsilon_0\omega_{\mathrm{TO}}^2} \tag{5.140}$$

and inserting this into (5.139) the Lyddane–Sachs–Teller equation (5.131) is derived.

As stated in Sect. 4.1, electromagnetic waves in a semiconductor are described by Maxwell's equations and for a plane wave we obtain

$$c^2k^2 = \omega^2\kappa(\omega)\,. \tag{5.141}$$

Inserting (5.131) into the above equation, the following relation is obtained:

$$c^2k^2 = \omega^2\left[\kappa_\infty + \frac{\kappa_0 - \kappa_\infty}{1 - (\omega^2/\omega_{\mathrm{TO}}^2)}\right]. \tag{5.142}$$

This relation represents a coupled mode of transverse electromagnetic waves and lattice vibrations and a wave given by this relation is called a **phonon polariton**. To illustrate the dispersion relations of phonon polaritons we plot (5.142) in Fig. 5.29. Figure 5.29 represents the dispersions of phonons, electromagnetic waves and phonon polaritons. The dashed lines (inclined) are the transverse electromagnetic waves not coupled to TO phonons in vacuum and in a medium of dielectric constant κ_∞, respectively, and they are given by $\omega = kc$ and $\omega = kc/\sqrt{\kappa_\infty}$, respectively. The horizontal solid line is the uncoupled LO phonon branch and the horizontal dashed line is the uncoupled TO phonon branch. The solid curves UPL and LPL are the upper and lower polariton branches, respectively, which are both coupled to TO phonons.

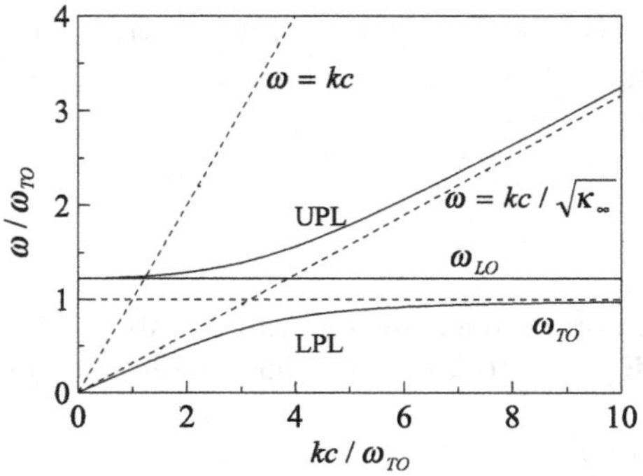

Fig. 5.29 Dispersion curves of phonon polaritons. The *dashed lines* (inclined) are $\omega = kc$ for photons (electromagnetic waves) in vacuum and $\omega = kc/\sqrt{\kappa_\infty}$ for photons in a medium with dielectric constant κ_∞, and they are not coupled with phonons (lattice vibrations). The *horizontal solid line* is for uncoupled LO phonons and the *horizontal dashed line* is for uncoupled TO phonons. The *solid curves* UPL and LPL are upper and lower polariton branches, respectively, which are both coupled modes of photons and phonons. LO phonon branch is drawn with a *solid line* because it is not coupled with photons. The parameters used for the calculation are $\kappa_0 = 15$ and $\kappa_\infty = 10$

Since the LO phonons are not coupled to photons (electromagnetic waves), it is shown by the solid line. The parameters used in the calculation are $\kappa_0 = 15$ and $\kappa_\infty = 10$. It is clear from Fig. 5.29 that solutions for the coupled mode of photons and phonons do not exist in the region $\omega_{\rm TO} \leq \omega \leq \omega_{\rm LO}$. In other words, total reflection will occur in this region. In a real material, however, coupled waves are subject to damping and this effect is incorporated by the replacement of ω^2 in the denominator of (5.142) by $\omega^2 \to \omega^2 + {\rm i}\omega\gamma$ (γ is the damping parameter), giving rise to the penetration of electromagnetic waves into a medium and to reflectivity of less than 100% in this region. Measurements of Raman scattering by phonon polaritons have been done in GaP by Henry and Hopfield [42].

5.4.2 Exciton Polaritons

Let us begin with the simple case where we can neglect the kinetic energy of an exciton, $\hbar^2 K^2/2M$, which is called the approximation without spatial dispersion. The dielectric function of the exciton is given by (4.110) of Sect. 4.5

$$\kappa(\omega) = \kappa_1(\omega) + {\rm i}\kappa_2(\omega) = \frac{C/\pi}{\mathcal{E}_{\rm G} - \mathcal{E}_{\rm ex}/n^2 - (\hbar\omega + {\rm i}\Gamma)} \,. \tag{5.143}$$

Here we will be concerned with the ground state of the exciton and put $n = 1$. We define $\mathcal{E}_{\rm G} - \mathcal{E}_{\rm ex} = \hbar\omega_0$ and the contributions to the dielectric constant except excitons by κ_∞. Therefore, we have

$$\kappa(\omega) = \kappa_\infty \left(1 + \frac{\Delta_{\rm ex}}{\omega_0 - (\omega + {\rm i}\gamma)}\right), \tag{5.144}$$

where we put $\hbar\gamma = \Gamma$. As in the case of phonon polaritons, the electromagnetic wave in a medium in which excitons are excited may be obtained by solving Maxwell's equations. Therefore, we have two solutions, which are the transverse waves of $\boldsymbol{k} \cdot \boldsymbol{E} = 0$ and $\kappa(\omega) \neq 0$, and the longitudinal waves of $\boldsymbol{k} \parallel \boldsymbol{E}$ and $\kappa(\omega) = 0$. For the transverse waves we have

$$c^2k^2 = \omega^2\kappa(\omega), \tag{5.145}$$

and thus we find the dispersion of **exciton polaritons**, which are coupled waves of excitons and photons. For the longitudinal waves we have a solution similar to the case of LO phonons:

$$\kappa(\omega) = 0. \tag{5.146}$$

Therefore, the angular frequency of a longitudinal exciton $\omega_{\rm l}$ is given by

$$\omega_{\rm l} = \omega_0 + \Delta_{\rm ex}. \tag{5.147}$$

The angular frequency of the transverse exciton is given by $\omega_{\rm t} = \omega_0$ and $\Delta_{\rm ex}$ is called the longitudinal–transverse (LT) splitting of the exciton. Assuming that the wave vector $\boldsymbol{k}$ of (5.144) is complex, we put

$$k = k_1 + {\rm i}k_2, \tag{5.148}$$

and then we obtain for the real and imaginary parts of (5.144)

$$\frac{\omega^2\kappa_\infty}{c^2}\left(1 + \frac{\Delta_{\rm ex}}{\omega_0 - \omega}\right) = k_1^2 - k_2^2, \tag{5.149a}$$

$$\pi\delta(\omega - \omega_0)\frac{\omega_0^2\kappa_\infty}{c^2} = 2k_1k_2, \tag{5.149b}$$

where the Dirac delta function in Appendix A is used and κ_∞ is the background dielectric constant (related to electronic transition, other than the exciton). The imaginary part is proportional to $\delta(\omega - \omega_0)$ and exhibits resonance at $\hbar\omega = \hbar\omega_0$ $(= \mathcal{E}_{\rm G} - \mathcal{E}_{\rm ex})$. We may neglect the term k_2 in the region outside the resonance, and so obtain

$$\omega\sqrt{\frac{\omega - \omega_0 - \Delta_{\rm ex}}{\omega - \omega_0}} = \frac{ck_1}{\sqrt{\kappa_\infty}}. \tag{5.150}$$

This equation gives the dispersion of exciton polaritons, which is shown in Fig. 5.30. Figure 5.30 is the dispersion curve of exciton polaritons, where the kinetic energy of the exciton, $\hbar^2K^2/2M$, is neglected (or the momentum of the exciton for center-

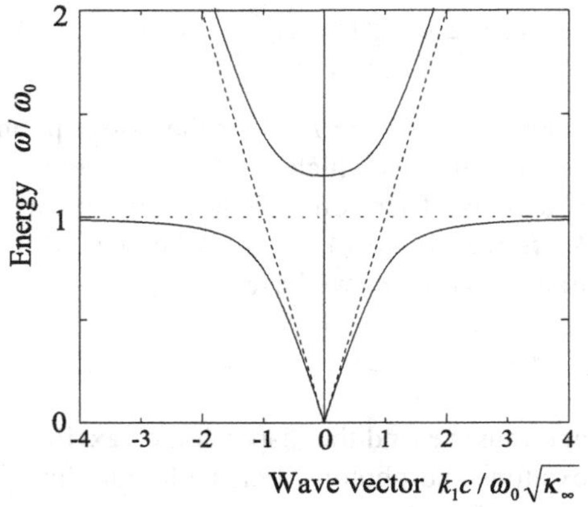

Fig. 5.30 Dispersion curves of exciton polaritons, where spatial dispersion is neglected (momentum for center-of-mass motion is neglected: $\hbar\boldsymbol{K} = 0$). The difference, LT splitting Δ_{ex}, between the angular frequencies of longitudinal excitons ω_l and transverse excitons ω_t is assumed to be $\Delta_{\text{ex}}/\omega_0 = 0.2$, where $\hbar\omega_0 = \mathcal{E}_G - \mathcal{E}_{\text{ex}}$. The *dashed line* shows the photon dispersion in the medium

of-mass motion $\hbar\boldsymbol{K}$ is regarded to be 0). The dispersion curves are calculated by assuming LT splitting to be $\Delta_{\text{ex}}/\omega_0 = 0.2$, where Δ_{ex} is the LT splitting between the angular frequencies of the longitudinal exciton ω_l and the transverse exciton ω_t, and $\Delta_{\text{ex}}/\omega_0 = 0.2$. In the region $\omega \ll \omega_0$ we find that

$$\omega \simeq \frac{ck_1}{\sqrt{\kappa_\infty(1+\Delta_{\text{ex}}/\omega_0)}} \tag{5.151}$$

and that the waves behave like electromagnetic waves (photons) in a material with effective dielectric constant, not equal to $\kappa_\infty(1+\Delta_{\text{ex}}/\omega_0)$, resulting in a slightly slower phase velocity compared to that of light, $c/\sqrt{\kappa_\infty}$. In the region $\omega_0 < \omega < \omega_0 + \Delta_{\text{ex}}$, we have $\kappa(\omega) \le 0$ and thus no solution, resulting in total reflection, just as in the case of phonon polaritons. In the region $\omega \gg \omega_0$ the waves behave like photons (electromagnetic waves) and the phase velocity is $c/\sqrt{\kappa_\infty}$.

Next, we take into account the kinetic energy of the exciton. The dielectric constant is given by

$$\kappa(\omega) = \kappa_\infty\left[1 + \frac{\Delta_{\text{ex}}}{\omega_0 + \hbar K^2/2M - \omega - i\gamma}\right], \tag{5.152}$$

and insertion of this equation into (5.145) leads to

$$\frac{c^2k^2}{\kappa_\infty\omega^2} = 1 + \frac{\Delta_{\text{ex}}}{\omega_0 + \hbar K^2/2M - \omega - i\gamma}. \tag{5.153}$$

It is difficult to obtain an exact solution for this equation, and we approximate such as $\omega_l^2 - \omega_t^2 \simeq 2\omega_t\Delta_{\text{ex}}$ by taking account of $\Delta_{\text{ex}} \ll \omega_0$. Therefore, we obtain the following quadratic equation with respect to ω^2:

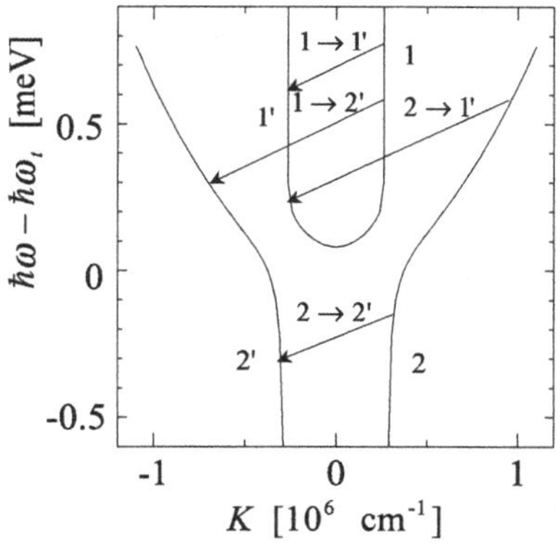

Fig. 5.31 Dispersion curves of exciton polaritons in GaAs, where spatial dispersion is taken into account (momentum for center-of-motion is non-zero: $\hbar \boldsymbol{K} \neq 0$). The LT splitting, the difference between the angular frequencies of longitudinal excitons $\hbar\omega_{\text{l}}$ and transverse excitons $\hbar\omega_{\text{t}} = 1.515\,\text{eV}$, is assumed to be $\hbar\Delta = 0.08\,\text{meV}$ and the exciton mass to be $M = 0.6$. The upper polariton (UPL) and lower polariton (LPL) branches are indicated by 1 and 2, respectively, and the Stokes transitions of Brillouin scattering are indicated by *arrows*. The back-scattering geometry is used and thus the wave vector of scattered polaritons is reversed after scattering

$$\frac{c^2k^2}{\kappa_\infty\omega^2} = 1 + \frac{\omega_{\text{l}}^2 - \omega_{\text{t}}^2}{\omega_{\text{t}}^2 + \omega_{\text{t}}(\hbar K^2/M) - \omega^2} \,. \tag{5.154}$$

The polariton dispersion calculated from (5.154) is shown in Fig. 5.31, where LT splitting between longitudinal excitons $\hbar\omega_{\text{l}}$ and transverse excitons $\hbar\omega_{\text{t}} = 1.515\,\text{eV}$ is assumed to be $\hbar\Delta = 0.08\,\text{meV}$ and the exciton mass to be $M = 0.6$. In Fig. 5.31 the upper polariton (UPL) and lower polariton (LPL) branches are indicated by 1 and 2, respectively, and the Stokes transition of the Brillouin scattering is indicated by arrows, where a back-scattering geometry is assumed and thus the wave vector of scattered polaritons is reversed after scattering. The parameters used here are for GaAs, which are determined from Brillouin scattering by polaritons in GaAs [43].

5.5 Free–Carrier Absorption and Plasmon

Consider a system with electrons of density n, effective mass m^* and electronic charge $-e$. When an electric field $\boldsymbol{E} = \boldsymbol{E}_0 \exp(-\text{i}\omega t)$ is applied to the system, the equation of motion is written as

$$m^* \frac{\text{d}^2}{\text{d}t^2}\boldsymbol{x} + \frac{m^*}{\tau}\frac{\text{d}}{\text{d}t}\boldsymbol{x} = -e\boldsymbol{E}_0 e^{-\text{i}\omega t} \,, \tag{5.155}$$

where τ is the relaxation time of the electron. The displacement vector of the electron is then given by

$$\boldsymbol{x} = \frac{-e\boldsymbol{E}}{m^*(-\omega^2 - \mathrm{i}\omega/\tau)} . \tag{5.156}$$

The polarization $\boldsymbol{P}$ associated with this displacement is defined by

$$\boldsymbol{P} = n(-e)\boldsymbol{x} . \tag{5.157}$$

In addition to the electronic polarization we have to take account of other contributions to the polarization or dielectric constant. This is done by introducing the background polarization $\boldsymbol{P}_\infty$ and background dielectric constant κ_∞, and their relations $\epsilon_0\boldsymbol{E} + \boldsymbol{P}_\infty = \kappa_\infty\epsilon_0\boldsymbol{E}$ and $\epsilon_0\boldsymbol{E} + \boldsymbol{P}_\infty + \boldsymbol{P} = \kappa\epsilon_0\boldsymbol{E}$. Using this definition the dielectric constant reduces to

$$\kappa(\omega) = \kappa_\infty - \frac{ne^2}{\epsilon_0 m^* \omega(\omega + \mathrm{i}/\tau)} . \tag{5.158}$$

When we define the plasma frequency by

$$\omega_\mathrm{p} = \sqrt{\frac{ne^2}{\kappa_\infty \epsilon_0 m^*}} , \tag{5.159}$$

the dielectric constant is rewritten as

$$\kappa(\omega) = \kappa_\infty \left[1 - \frac{\omega_\mathrm{p}^2}{\omega(\omega + \mathrm{i}/\tau)}\right] . \tag{5.160}$$

Let us examine the plasma frequency in semiconductors with electron density $n \simeq 1 \times 10^{18}\,\mathrm{cm}^{-3}$, $m^* = m_0$, $\kappa_\infty = 12$. The plasma frequency is estimated to be $\omega_\mathrm{p} \simeq 1.6 \times 10^{13}$ rad/s and the relaxation time in semiconductors is about $\tau \simeq 10^{-12}$ s, leading to the condition $\omega_\mathrm{p}\tau \gg 1$ in the region of frequency near the plasma frequency (in metals, this condition is satisfied for the frequency of visible light). Therefore, we may safely approximate as

$$\kappa(\omega) = \kappa_\infty \left[1 - \frac{\omega_\mathrm{p}^2}{\omega^2}\right] , \tag{5.161}$$

and the condition $\kappa(\omega_\mathrm{p}) = 0$ is fulfilled at $\omega = \omega_\mathrm{p}$. This result tells us that longitudinal plasma oscillations and thus plasmons are excited in a free electron system as in the cases of phonon polaritons and exciton polaritons. Next we will discuss this longitudinal plasma oscillation in detail. Let us consider the system shown in Fig. 5.32. In the case of a metal, the conduction (free) electrons uniformly occupy the space

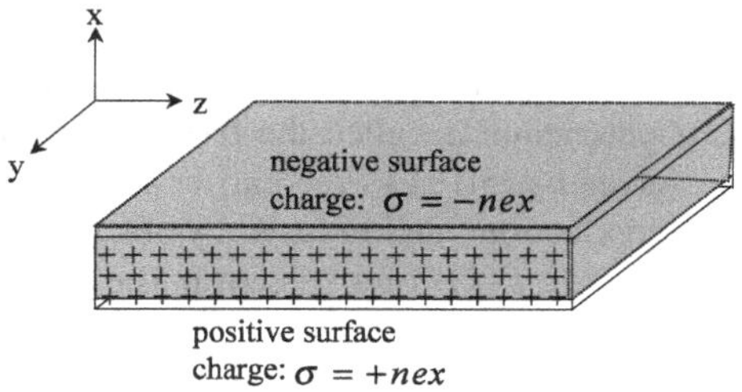

Fig. 5.32 Illustration of plasma oscillations in a metal or in a semiconductor. Metal ions or donor ions in a semiconductor are indicated by the + symbol and the conduction electrons are shown by shading. When electrons in the equilibrium state are displaced together by $\boldsymbol{x}$, a polarization $\boldsymbol{P} = n(-e)\boldsymbol{x}$ results, and a surface charge density $\pm ne|x|$ appears. The electric field induced by the displacement acts so that the displaced electrons are restored, resulting in a collective motion of the electrons (plasma oscillations) with plasma frequency ω_{p}

surrounding the positive metal ions. In the case of a semiconductor, the positive donor ions are surrounded uniformly by the conduction electrons. In Fig. 5.32 the positive ions are indicated by the + symbol and the electrons are shown by the shading. There are no excess charges and thus the medium is neutral. Since the positive ions are not able to move, the displacement of the electrons is taken into account. We assume that an external field is not present and that the electron displacement $\boldsymbol{x}$ occurs in the x direction. This displacement will induce a polarization $\boldsymbol{P} = n(-e)\boldsymbol{x}$. Since the electric displacement is given by $\boldsymbol{D} = \kappa_\infty \epsilon_0 \boldsymbol{E} + \boldsymbol{P} = \kappa(\omega_{\mathrm{p}})\epsilon_0 \boldsymbol{E} = 0$, an electric field $\boldsymbol{E}_1 = -\boldsymbol{P}/\kappa_\infty \epsilon_0 = ne\boldsymbol{x}/\kappa_\infty \epsilon_0$ is induced, and the direction is anti-parallel to the polarization $\boldsymbol{P}$. This electric field acts as a restoring force for electrons, and so the electrons are forced to move toward the equilibrium position. This collective motion of electrons results in a resonant oscillation with plasma frequency ω_{p}. This kind of collective motion in a system with equal positive and negative charges is called a plasma oscillation. Plasma oscillations are known to be excited in gases, where the oscillations are associated with the relative displacement between positive and negative charges.

Next, we will discuss free-carrier absorption. We substitute the complex dielectric constant $\kappa = \kappa_1 + \mathrm{i}\kappa_2$ in (5.158) and obtain for the real and imaginary parts

$$\kappa_1(\omega) = \kappa_\infty - \frac{ne^2\tau^2}{\epsilon_0 m^*(\omega^2\tau^2 + 1)}, \tag{5.162}$$

$$\kappa_2(\omega) = \frac{ne^2\tau}{\epsilon_0 m^* \omega(\omega^2\tau^2 + 1)}. \tag{5.163}$$

As stated in Sect. 4.1, the absorption of electromagnetic waves is determined by the imaginary part of the dielectric constant κ_2 of a medium and the absorption coefficient is given by (4.21)

$$\alpha = \frac{\omega\kappa_2}{n_0 c} = \frac{ne^2\tau}{\epsilon_0 m^* n_0 c(\omega^2\tau^2 + 1)}. \tag{5.164}$$

When an electromagnetic wave of angular frequency ω is applied to a semiconductor with free carriers (electrons or holes), the electromagnetic wave is absorbed by the free carriers. This phenomenon is called the **free-carrier absorption**, and the absorption coefficient is given by (5.164). Free-carrier absorption is caused by electron transitions within a conduction band induced by absorbing photons. Using the initial state $\mathcal{E}(\boldsymbol{k}_\mathrm{i})$ and final state $\mathcal{E}(\boldsymbol{k}_\mathrm{f})$ of an electron, the energy conservation rule $\hbar\omega = \mathcal{E}(\boldsymbol{k}_\mathrm{f}) - \mathcal{E}(\boldsymbol{k}_\mathrm{i})$ and momentum conservation rule $\hbar\boldsymbol{k} = \hbar\boldsymbol{k}_\mathrm{f} - \hbar\boldsymbol{k}_\mathrm{i}$ hold. However, we know that a photon, even with a small wave vector $\boldsymbol{k}$, has a large energy and thus the transition is a direct transition. This means that the energy and momentum conservation rules are not satisfied at the same time. Therefore, free-carrier absorption occurs when impurity scattering or phonon scattering is involved as an intermediate state. Free carrier absorption is very similar to indirect transitions and involves a higher-order perturbation. The treatment of free-carrier absorption has been reported in detail by Fan et al. [44]. When we assume $\omega\tau \gg 1$ in (5.163), κ_2 becomes proportional to ω^{-3} and thus the absorption coefficient is proportional to ω^{-2}. We have to note that the result is obtained by assuming that the relaxation time τ is independent of the angular frequency ω. In experiments the observed absorption coefficient has the form $\alpha(\omega) \propto \omega^{-1.5\sim-3.5}$, and this behavior is believed to be due to the difference of the scattering mechanisms.

Here we will discuss the mobility of electrons or holes in connection with free-carrier absorption. Let us consider the motion of an electron with effective mass m^* in the presence of an electric field $\boldsymbol{E}_0 e^{-\mathrm{i}\omega t}$. The equation of motion is

$$m^* \frac{\mathrm{d}\boldsymbol{v}}{\mathrm{d}t} + \frac{m^* \boldsymbol{v}}{\tau} = -e\boldsymbol{E}_0 e^{-\mathrm{i}\omega t} \tag{5.165}$$

and the electron drift velocity under the a.c. field is

$$\boldsymbol{v} = -\frac{e\boldsymbol{E}}{m^*(-\mathrm{i}\omega + 1/\tau)} . \tag{5.166}$$

The mobility μ defined by $\boldsymbol{v} = -\mu\boldsymbol{E}$ is given by

$$\mu = \frac{e\tau}{m^*(1 - \mathrm{i}\omega\tau)} . \tag{5.167}$$

If the electron density is n, the current density is given by $\boldsymbol{J} = n(-e)\boldsymbol{v} = ne\mu\boldsymbol{E} = \sigma\boldsymbol{E}$, where the conductivity σ is given by

$$\sigma = ne\mu = \frac{ne^2}{m^*} \frac{\tau}{1 - \mathrm{i}\omega\tau} . \tag{5.168}$$

Using the relations (4.24), (4.25a) and (4.25b) described in Sect. 4.1, and taking the background dielectric constant κ_∞ into consideration, (5.168) leads to (5.162) and (5.163).

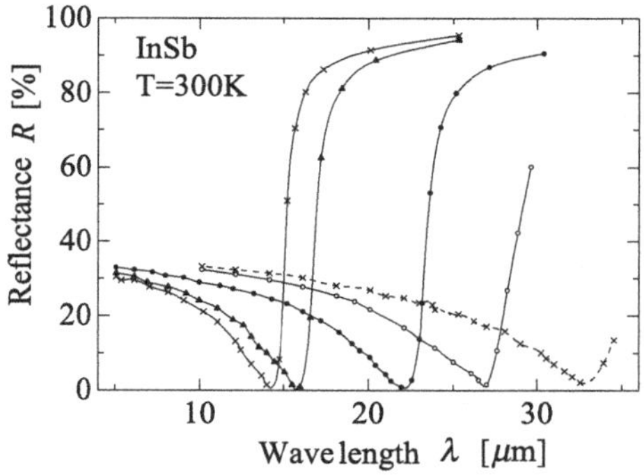

Fig. 5.33 Reflection coefficients of n-InSb with different electron densities near the plasma edge plotted as a function of wavelength. The electron densities are $n = 3.5 \times 10^{17}\text{cm}^{-3}$, $6.2 \times 10^{17}\text{cm}^{-3}$, $1.2 \times 10^{18}\text{cm}^{-3}$, $2.8 \times 10^{18}\text{cm}^{-3}$, and $4.0 \times 10^{18}\text{cm}^{-3}$. (After Spitzer and Fan [45])

Since the dielectric constant becomes zero at the plasma frequency, electromagnetic waves are subject to total reflection. Here we assume that τ is large and the absorption is weak. Then the reflection coefficient given by (4.18) is approximated as

$$R = \frac{(n_0 - 1)^2 + k_0^2}{(n_0 + 1)^2 + k_0^2} \approx \frac{(n_0 - 1)^2}{(n_0 + 1)^2}, \tag{5.169}$$

where we find that the reflection coefficient becomes $R = 1$ at the plasma frequency $\omega = \omega_p$. On the other hand, the reflection coefficient becomes zero at a frequency

$$\omega(R = 0) = \sqrt{\frac{ne^2}{m^*(\kappa_\infty - 1)\epsilon_0}}. \tag{5.170}$$

From these results we may conclude that a reflection edge exists between $\omega(R = 0)$ and ω_p. This feature has been observed in experiments by Spitzer and Fan and their results are shown in Fig. 5.33 [45].

The contribution to the dielectric function from the plasma oscillations is obtained by assuming that there is no dispersion of LO phonons near the region $\omega = \omega_p$, approximating the background dielectric constant by κ_∞. However, the frequency of plasma oscillations is proportional to the square of the carrier density and it is possible to achieve the condition that the plasma frequency is very close to the LO phonon frequency. Under these circumstances transverse optical (TO) phonons are expected not to couple with plasma oscillations. However, longitudinal optical (LO) phonons are expected to couple with plasma oscillations because both are longitudinal. The coupled mode is obtained as follows. The dielectric function without a damping term is obtained from (5.132) and (5.161) as

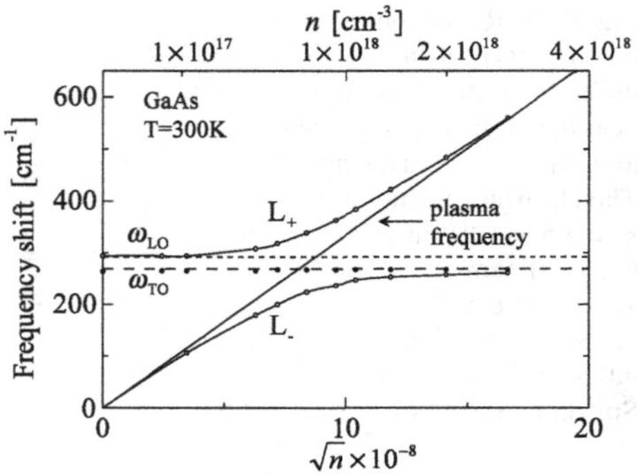

Fig. 5.34 Raman shift of GaAs as a function of the electron density at room temperature. L_+ and L_- are the solutions of plasmons coupled to LO phonons and calculated from (5.172). ∘ and • are experimental data by Mooradian and the *solid curve* is calculated using the parameters $\omega_{LO} = 292\,cm^{-1}$ and $\omega_{TO} = 269\,cm^{-1}$ in (5.172). (After [46])

$$\kappa(\omega) = \kappa_\infty \left[1 - \frac{\omega_p^2}{\omega^2} + \frac{(\kappa_0 - \kappa_\infty)\omega_{TO}}{\omega_{TO}^2 - \omega^2}\right]$$
$$= \kappa_\infty \left[-\frac{\omega_p^2}{\omega^2} + \frac{\omega_{LO}^2 - \omega^2}{\omega_{TO}^2 - \omega^2}\right] . \tag{5.171}$$

The coupled mode is longitudinal and the condition $\kappa(\omega) = 0$ holds, and so we obtain the solution

$$\omega_\pm^2 = \frac{1}{2}\left(\omega_{LO}^2 + \omega_p^2\right) \pm \frac{1}{2}\left[(\omega_{LO}^2 + \omega_p^2) - 4\omega_p^2\omega_{TO}^2\right]^{1/2} , \tag{5.172}$$

where ω_+ and ω_- are the resonance frequencies of the **plasmon–LO phonon coupled mode**, above and below the LO phonon frequency ω_{LO}. Such a coupled mode of plasma oscillations and LO phonons, the plasmon–LO phonon coupled mode, has been predicted by Gurevich et al. [47] and their existence has been proved experimentally by Mooradian and Wright [46, 48]. Figure 5.34 shows the experimental results in GaAs at room temperature by Mooradian [46], where the TO phonons do not couple with the plasmons and the Raman shift is $\omega_{TO} = 269\,cm^{-1}$, and is almost constant, whereas the LO phonons couple with the plasmons and split into two branches. The upper branch L_+ extrapolates to $\omega_{LO} = 292\,cm^{-1}$ and the lower branch L_- approaches zero at lower electron densities. The upper branch L_+ approaches the plasma frequency ω_p and the lower branch L_- approaches the TO phonon frequency ω_{TO} at higher electron densities. The solid curves for the upper L_+ and lower L_- branches are calculated from (5.172) with $\omega_{LO} = 292\,cm^{-1}$ and $\omega_{TO} = 269\,cm^{-1}$, and show good agreement with the experimental data. For a more detailed treatment of plasmon–LO phonon interactions see the paper by Klein [49].

5.6 Problems

(5.1) Derive the Joint density of Two-dimensional band $J_{cv}^{2D}(\hbar\omega - \mathcal{E}_G - \mathcal{E}_x)$ given by (5.15).

(5.2) Calculate Reflectivity near E_0 critical point using (5.27), using the parameters defined for Problems **4.6** of Chap. 4.

(5.3) Compare the calculated curve of Fig. 10.4 of the previous problem with the experimental results shown in Fig. 5.7. Experimental data show a weak peak near the critical point E_0. Explain the difference between Figs. 10.4 and 5.7.

(5.4) Calculate Seraphin coefficients $\alpha(\omega)$ and $\beta(\omega)$ using the model dielectric functions $\kappa_1(\omega)$ and $\kappa_2(\omega)$ for E_0 transition.

(5.5) Electroreflectance experiments provide the critical point energies such as E_0, $E_0 + \Delta_0$, E_1, and $E_1 + \Delta_1$. These critical point energies are determine from the analysis based on the spectra $\Delta R/R$. Calculate the electroreflectance spectra using (5.48) for the cases given below.
(1) Three dimensional case: $m_3 = 5/2$, $C_3 = 0.005$, $\theta_3 = 3.0$, $\mathcal{E}_G = 3.0$ eV, and $\Gamma_3 = 0.050$.
(2) Two dimensional case: $m_2 = 3$, $C_2 = 0.0015$, $\theta_2 = 3.0$, $\mathcal{E}_G = 3.0$ eV, and $\Gamma_2 = 0.050$.

(5.6) Calculate reflectivity due to free carrier plasma using following parameters. $\kappa_\infty = 10$, $\omega_p\tau = 10$. Plot reflectivity as a function of $\omega_p/\omega = \lambda/\lambda_p$, where λ_p is the plasma wavelength.

References

1. M. Abramowitz, I.A. Stegun, *Handbook of Mathematical Functions* (Dover Publications, New York, 1965), p. 446
2. W. Franz, Z. Naturforsch. **13**, 484 (1958)
3. L.V. Keldysh, Zh. Eksp. Teor. Fiz. **34**, 1138 (1958) [English transl.: Sov. Phys.-JETP, **7**, 788 (1958)]
4. B.O. Seraphin, N. Bottka, Phys. Rev. **145**, 628 (1966)
5. D.E. Aspnes, Phys. Rev. **147**, 554 (1966)
6. Y. Hamakawa, P. Handler, F.A. Germano, Phys. Rev. **167**, 709 (1966)
7. H.R. Philipp, H. Ehrenreich, in *Semiconductors and Semimetals*, vol. 3, ed. by R.K. Willardson, A.C. Beer (Academic Press, New York, 1967), pp. 93–124
8. D.E. Aspnes, N. Bottka, Electric-field effects on the dielectric function of semiconductors and insulators, in *Semiconductors and Semimetals*, vol. 9, ed. by R.K. Willardson, A.C. Beer (Academic Press, New York, 1972), pp. 457–543
9. D.E. Aspnes, Phys. Rev. Lett. **28**, 168 (1972)
10. D.E. Aspnes, J.E. Rowe, Phys. Rev. B **5**, 4022 (1972)
11. D.E. Aspnes, Surf. Sci. **37**, 418 (1973)
12. P. Yu, M. Cardona, *Fundamentals of Semiconductors: Physics and Materials Properties* (Springer, Heidelberg, 1996)
13. M. Haraguchi, Y. Nakagawa, M. Fukui, S. Muto, Jpn. J. Appl. Phys. **30**, 1367 (1991)
14. H.M.J. Smith, Phil. Trans. A **241**, 105 (1948)
15. R. Loudon, The Raman effect in crystals. Adv. Phys. **13**, 423–482 (1964); Erratum, Adv. Phys. **14**, 621 (1965)

16. P.A. Temple, C.E. Hathaway, Phys. Rev. B **6**, 3685 (1973)
17. R. Loudon, Proc. Roy. Soc. **A275**, 218 (1963)
18. G.L. Bir, G.E. Pikus, Sov. Phys.-Solid State **2**, 2039 (1961)
19. M. Cardona, Light scattering as a form of modulation spectroscopy. Surf. Sci. **37**, 100–119 (1973)
20. M. Cardona, Surf. Sci. **37**, 100 (1973)
21. S. Adachi, C. Hamaguchi, J. Phys. C, Solid State Phys. **12**, 2917 (1979)
22. S. Adachi, C. Hamaguchi, Phys. Rev. B **19**, 938 (1979)
23. B.A. Weinstein, M. Cardona, Phys. Rev. B **8**, 2795 (1973)
24. L. Brillouin, Ann. Phys. (Paris) **17**, 88 (1922)
25. F. Pockels, Ann. Physik **37**, 144, 372 (1889); **39**, 440 (1890)
26. G.B. Benedek, K. Fritsch, Phys. Rev. **149**, 647 (1966)
27. C. Hamaguchi, J. Phys. Soc. Jpn. **35**, 832 (1973)
28. D.F. Nelson, P.D. Lazay, M. Lax, Phys. Rev. B **6**, 3109 (1972)
29. R.W. Dixon, IEEE J. Quantum Electron. **QE-3**, 85 (1967)
30. J.R. Sandercock, Phys. Rev. Lett. **28**, 237 (1972)
31. D.L. White, J. Appl. Phys. **33**, 2547 (1962)
32. B.W. Hakki, R.W. Dixon, Appl. Phys. Lett. **14**, 185 (1969)
33. D.L. Spears, R. Bray, Phys. Lett. **29A**, 670 (1969)
34. D.L. Spears, Phys. Rev. B **2**, 1931 (1970)
35. E.D. Palik, R. Bray, Phys. Rev. B **3**, 3302 (1971)
36. M. Yamada, C. Hamaguchi, K. Matsumoto, J. Nakai, Phys. Rev. B **7**, 2682 (1973)
37. D.K. Garrod, R. Bray, Phys. Rev. B **6**, 1314 (1972)
38. A. Feldman, D. Horowitz, J. Appl. Phys. **39**, 5597 (1968)
39. K. Ando, C. Hamaguchi, Phys. Rev. B **11**, 3876 (1975)
40. K. Ando, K. Yamabe, S. Hamada, C. Hamaguchi, J. Phys. Soc. Jpn. **41**, 1593 (1976)
41. K. Yamabe, K. Ando, C. Hamaguchi, Jpn. J. Appl. Phys. **16**, 747 (1977)
42. C.H. Henry, J.J. Hopfield, Phys. Rev. Lett. **15**, 964 (1965)
43. R. Ulbrich, C. Weisbuch, Phys. Rev. Lett. **38**, 865 (1977)
44. H.Y. Fan, W. Spitzer, R.J. Collins, Phys. Rev. **101**, 566 (1956)
45. W.G. Spitzer, H.Y. Fan, Phys. Rev. **106**, 882 (1957)
46. A. Mooradian, in *Advances in Solid State Physics*, ed. by O. Madelung (Pergamon, New York, 1969), p. 74
47. V.L. Gurevich, A.I. Larkin, Y.A. Firsov, Fiz. Tverd. Tela **4**, 185 (1962)
48. A. Mooradian, G.B. Wright, Phys. Rev. Lett. **16**, 999 (1966)
49. M.V. Klein, in *Light Scattering in Solids*, ed. by M. Cardona (Springer, New York, 1975), p. 148

Chapter 6
Electron–Phonon Interaction and Electron Transport

Abstract Semiconductor device operation depends on the drift velocity of carriers, electrons and holes, which is determined by the mobility. The mobility μ is expressed by $e\langle\tau\rangle/m^*$, where e, $\langle\tau\rangle$, and m^* are the electronic charge, average of the relaxation time, and the effective mass. The relaxation time or scattering time is limited by various scatterings of carriers. Among them the phonon scattering plays the most important role. In this chapter we begin with the analysis of the lattice vibrations and the derivation of Boltzmann transport equation. Then collision time, relaxation time and mobility are defined. The transition probabilities and transition matrix elements for the scattering due to various modes of phonons, impurity, electron–electron interaction and so on are evaluated using quantum mechanical approach. These results are used to evaluate scattering rates and relaxation times, and finally respective carrier mobility is obtained. In order to get an better insight into the electron transport, scattering rates and mobilities due to the various processes are evaluated numerically and plotted as functions of electron energy, temperature and carrier densities. In addition, electron mobility is evaluated by taking all the relevant scattering processes. Also a theoretical method to evaluate deformation potentials for phonon scattering is given, where the calculated lattice vibrations and the full band structures in the Brillouin zone are properly employed. Many figures obtained by numerical calculations are very informative for an understanding of semiconductor transport.

6.1 Lattice Vibrations

6.1.1 Acoustic Mode and Optical Mode

A crystal consists of a periodic arrangement of atoms and molecules. However, the atoms or molecules of a crystal vibrate around the equilibrium positions at finite temperatures because of their thermal motion. When a atom is displaced from its equilibrium position, the atom is subject to a restoring force depending on the displacement. The restoring force follows Hooke's law and is proportional to the displacement. When each atom vibrates randomly, each atom suffers random forces and

C. Hamaguchi, *Basic Semiconductor Physics*, Graduate Texts in Physics,
DOI 10.1007/978-3-319-66860-4_6

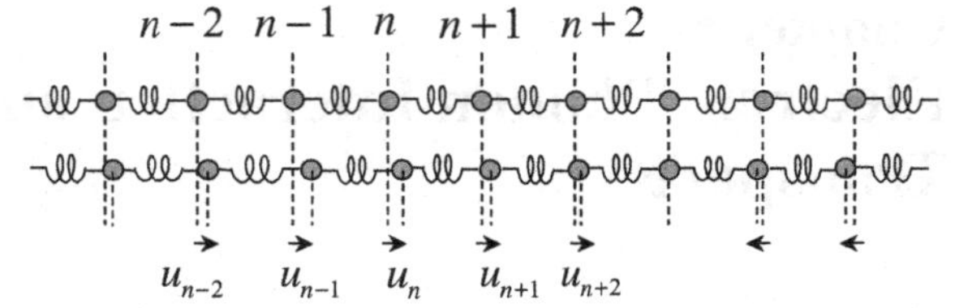

Fig. 6.1 Displacement of atoms from their equilibrium positions (lattice vibration)

its vibration is immediately damped. On the other hand, if each atom vibrates with a small relative displacement from the neighboring atoms, then the vibration will continue with its small vibration energy. For simplicity we consider the one-dimensional lattice shown in Fig. 6.1, where the mass of the atoms is M and the distance between the nearest neighbor atoms (lattice constant) at equilibrium is a. We assume that the inter–atomic forces act on neighboring atoms only and that the atoms are connected to each other by a spring constant (force constant) k_0. Defining the displacement of the atoms from their equilibrium positions by $u_0, u_1, u_2, \ldots, u_n, u_{n+1}, \ldots$, the equation of motion of n-th atom

$$\begin{aligned} M\frac{\mathrm{d}^2}{\mathrm{d}t^2}u_n &= -k_0(u_n - u_{n-1}) - k_0(u_n - u_{n+1}) \\ &= k_0(n_{n-1} + u_{n+1} - 2u_n)\,. \end{aligned} \tag{6.1}$$

As stated above, when the displacement of each atom is independent, an atom will be subject to a strong force from the neighboring atoms and the displacement is damped. Therefore the lowest excitation energy of the lattice vibration corresponds to a wavelike displacement, keeping the displacement of neighboring atoms almost in phase, when we see neighboring atoms. In such a case we may write the solution of (6.1) as

$$u_n = A\exp[\mathrm{i}(qna - \omega t)], \tag{6.2}$$

where $q = \omega/v_\mathrm{s} = 2\pi/\lambda$ is the wave vector, ω is the angular frequency, v_s is the velocity of sound, and λ is wavelength. Inserting (6.2) into (6.1), we obtain

$$M\omega^2 = -k_0\left(\mathrm{e}^{\mathrm{i}qa} + \mathrm{e}^{-\mathrm{i}qa} - 2\right) = 4k_0\sin^2\left(\frac{qa}{2}\right)\,. \tag{6.3}$$

From this we have following relation:

$$\omega = 2\sqrt{\frac{k_0}{M}}\left|\sin\left(\frac{qa}{2}\right)\right|\,. \tag{6.4}$$

When the wavelength is much longer than the lattice constant ($qa \ll 1$), the phase velocity v_s is given by

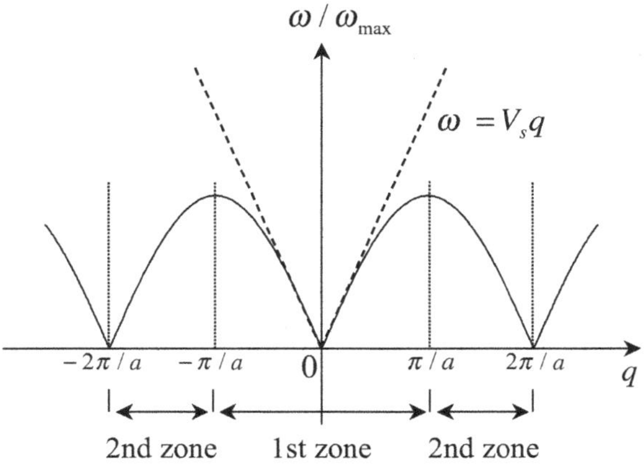

Fig. 6.2 Dispersion curve of one-dimensional lattice vibration

$$v_s = \frac{\omega}{q} = \sqrt{\frac{k_0}{M}}\, a\, \frac{\sin(qa/2)}{(qa/2)} \cong \sqrt{\frac{k_0}{M}}\, a \quad (qa \ll 1)\,. \tag{6.5}$$

This result indicates that a lattice vibration with long wavelength has a constant phase velocity $v_s = \sqrt{k_0/M}a$, which corresponds to the sound velocity. Figure 6.2 shows the calculated curve from (6.4), where we see that the $\omega - q$ curve is a periodic function with a period of $2\pi/a$. The whole dispersion curve is the repetition or displacement of the curve in the period $-\pi/a \leq q \leq \pi/a$, and thus this feature enables us to discuss lattice vibrations in that period. The region of the period is called the **first Brillouin zone**, and the second Brillouin zone is shown in Fig. 6.2. The second, third and other Brillouin zones are equivalent to the first Brillouin zone, which is shown in Sect. 1.4 in detail in connection with the reduced zone scheme. Where there are N atoms, N degrees of freedom of motion exist. Adopting cyclic boundary conditions we can define wave vectors by $q = 2\pi n/(Na)$, where $n = (-N/2 + 1) \sim (+N/2)$, giving rise to N values of the wave vectors. The above defined wave vectors are obtained from the cyclic boundary condition such that the displacement u_n of (6.2) is equivalent for $n = 0$ and $n = N$.

Next, we consider a lattice consisting of two kinds of atoms with masses M_1 and M_2. The atomic distance of the atoms of mass M_1 is a and the same is true for the atoms M_2. When the nearest-neighbor interaction is assumed, the equations of motion are written as in the case of the one-dimensional lattice stated above as

$$\left.\begin{aligned} M_1 \frac{\mathrm{d}^2}{\mathrm{d}t^2} u_{2n+1} &= k_0(u_{2n} + u_{2n+2} - 2u_{2n+1})\,, \\ M_2 \frac{\mathrm{d}^2}{\mathrm{d}t^2} u_{2n} &= k_0(u_{2n-1} + u_{2n+1} - 2u_{2n}) \end{aligned}\right\} \tag{6.6}$$

for the atoms M_1 and M_2. We assume once again wave-type solutions of the same type as before. In addition we may expect another type of solution where atoms of even order and odd order are displaced in the reverse direction to each other, and thus atoms of even order form a wave and odd order atoms form another wave. To satisfy these vibration types we express the displacement as

$$u_{2n+1} = A_1 \mathrm{e}^{\mathrm{i}\{(2n+1)(qa/2)-\omega t\}}, \quad u_{2n} = A_2 \mathrm{e}^{\mathrm{i}\{(2n)(qa/2)-\omega t\}}. \tag{6.7}$$

Inserting these equations into (6.6), we obtain following relations:

$$\left.\begin{aligned} (M_1\omega^2 - 2k_0)A_1 + \{2k_0 \cos(qa/2)\}A_2 = 0, \\ (M_2\omega^2 - 2k_0)A_2 + \{2k_0 \cos(qa/2)\}A_1 = 0. \end{aligned}\right\} \tag{6.8}$$

We easily find that the condition $A_1 = A_2 = 0$ correspond to all atoms at a standstill, and therefore we have to find solutions such that both A_1 and A_2 are not zero simultaneously. This condition is satisfied by requiring the determinant of the simultaneous equations with respect to A_1 and A_2 in (6.8) to be zero, giving rise to

$$\begin{vmatrix} (M_1\omega^2 - 2k_0) & 2k_0 \cos(qa/2) \\ 2k_0 \cos(qa/2) & (M_2\omega^2 - 2k_0) \end{vmatrix} = 0. \tag{6.9}$$

This equation is regarded as a quadratic equation with respect to ω^2, and the solutions ω_-^2 and ω_+^2 are given by

$$\omega_-^2 = k_0 \frac{M_1 + M_2}{M_1 M_2}\left[1 - \sqrt{1 - \frac{4M_1M_2}{(M_1+M_2)^2}\sin^2\left(\frac{qa}{2}\right)}\right], \tag{6.10a}$$

$$\omega_+^2 = k_0 \frac{M_1 + M_2}{M_1 M_2}\left[1 + \sqrt{1 - \frac{4M_1M_2}{(M_1+M_2)}\sin^2\left(\frac{qa}{2}\right)}\right]. \tag{6.10b}$$

It is evident from (6.10a) and (6.10b) that ω_- approaches zero, while ω_+ gives a constant value ($\neq 0$) as $q \to 0$. Taking account of the fact that the angular frequency is positive, ω_- and ω_+ are plotted as a function of the wave vector q for several values of $\alpha = M_1/M_2$ as a parameter in Fig. 6.3, where we define $1/M_\mathrm{r} = 1/M_1 + 1/M_2$. Let us consider the two branches ω_- and ω_+ in the long wavelength limit. For $q = 0$, (6.8), (6.10a) and (6.10b) lead us to the following results:

$$\left.\begin{aligned} \omega_- &= 0 \\ \frac{A_1}{A_2} &= 1 \end{aligned}\right\}, \tag{6.11}$$

$$\left.\begin{aligned} \omega_+ &= \sqrt{2k_0\left(\frac{1}{M_1} + \frac{1}{M_2}\right)} \equiv \sqrt{\frac{2k_0}{M_r}} \\ \frac{A_1}{A_2} &= -\frac{M_2}{M_1} \end{aligned}\right\}. \tag{6.12}$$

When we put $M_1 = M_2$, the ω_- branch gives the same result as a lattice consisting of one kind of atom, corresponding to a sound wave, and thus called the **acoustic branch**, the **acoustic mode** of vibrations or the **acoustic phonon**. Since the acoustic

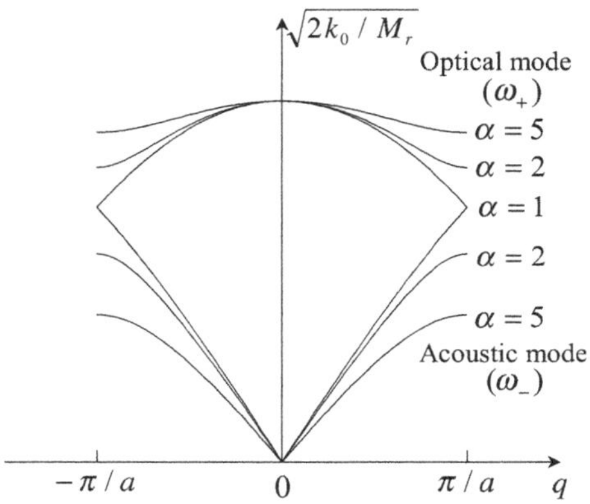

Fig. 6.3 Angular frequency versus wave vector relations for the two vibration modes of a lattice consisting of two kinds of atoms

branch satisfies the condition ($A_1 = A_2$), the neighboring atoms move in the same direction and thus the relative displacement is zero in the long wavelength limit. On the other hand, the ω_+ branch satisfies $A_1/A_2 = -M_2/M_1$ and thus different atoms are displaced in the opposite directions to each other, resulting in zero of the center of mass motion. The ω_+ branch exhibits a relative displacement between the different atoms and induces an electric field when the atoms are ionic. The induced electric field interacts strongly with the external electromagnetic field and absorbs the external waves. This often occurs in the infrared region, leading to it to be called the **optical branch**, **optical mode** of lattice vibrations or the **optical phonon**. Fig. 6.4 shows a schematic illustration of the atomic displacement for (a) the acoustic mode (acoustic phonon) and (b) the optical mode (optical phonon), where we see the difference in the displacement between the two modes of lattice vibrations. The acoustic mode in the limit of $qa \ll 1$ gives the following relation:

$$\omega_- \cong \sqrt{\frac{2k_0}{(M_1 + M_2)}}\, aq \quad (qa \ll 1)\,, \tag{6.13}$$

and thus the sound velocity (phase velocity) is given by

$$v_\mathrm{s} = \frac{\omega_-}{q} = \sqrt{\frac{2k_0}{(M_1 + M_2)}}\, a\,, \tag{6.14}$$

which corresponds to (6.5).

In ionic crystals unusual reflectivity has been observed in the infrared wavelength region of 20–100 μm. For example, the reflectivity of NaCl exhibits maxima around 40 and 60 μm, where the angular frequencies of light for the corresponding wavelengths are $\omega = 2\pi c/\lambda \sim 4 \times 10^{13}\,\mathrm{s}^{-1}$ and their wave vectors are $q = 2\pi/\lambda \sim 10^3\,\mathrm{cm}^{-1}$, which are much smaller than the wave vector at the Brillouin zone edge $\pi/a \simeq 10^8\,\mathrm{cm}^{-1}$. From the energy and momentum conservation

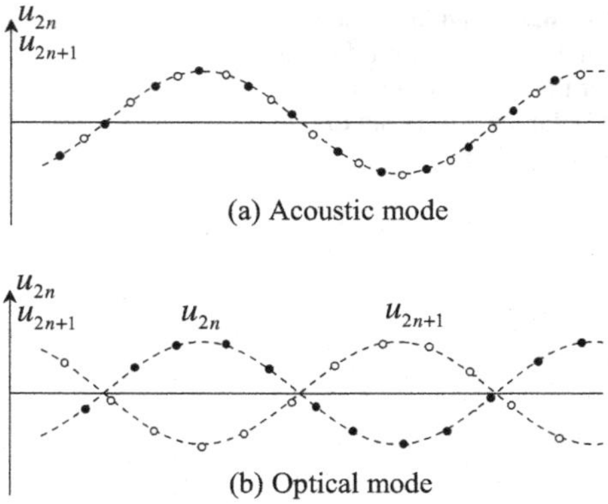

Fig. 6.4 Schematic illustration of two types of lattice vibrations: **a** acoustic mode (acoustic phonon) and **b** optical mode (optical phonon)

rules, such high energy excitation in the small wave vector region is easily found to correspond to the optical phonon branch ω_+.

In general, as stated previously, the optical mode of lattice vibration appears in a crystal with two or more atoms in a unit cell. Since there exist one longitudinal and two transverse acoustic modes, a crystal with s atoms in a unit cell gives rise to $3(s-1)$ optical modes of lattice vibration.

6.1.2 Harmonic Approximation

We consider the motion of the l-th lattice atom of a crystal with mass M_l. The potential energy V of the crystal is a function of the coordinates of the lattice atom $\boldsymbol{x}_l$ and the kinetic energy of the crystal is the sum of the kinetic energy of each atom $\boldsymbol{p}_l^2/2M_l$. Therefore, the total energy of the crystal is given by

$$H = \sum_l \frac{p_l^2}{2M_l} + V(\boldsymbol{x}_1, \boldsymbol{x}_2, \ldots, \boldsymbol{x}_l, \ldots) . \tag{6.15}$$

For simplicity, we consider a crystal consisting of N atoms which are bound together by the nearest-neighbor force only. In equilibrium the position of an atom is expressed in terms of the lattice vector $\boldsymbol{a}$ as

$$\boldsymbol{R}_l = l\boldsymbol{a} . \tag{6.16}$$

For another simplicity we assume a one-dimensional lattice and express the potential energy by

$$V(\boldsymbol{x}_1, \boldsymbol{x}_2, \ldots, \boldsymbol{x}_l, \ldots) = \sum_l f(x_{l+1} - x_l) , \tag{6.17}$$

where $f(x_{l+1} - x_l)$ is a function of the relative distance between the l-th and $(l+1)$-th atoms. We define the displacement u_l of the l-th atom by

$$u_l = x_l - la\,. \tag{6.18}$$

Since we assume the equilibrium condition such that the l-th atom occupies its lattice point la, the following relation holds for every l:

$$\frac{\partial V}{\partial u_l} = 0; \quad u_1 = u_2 = u_3 = \cdots = u_N = 0\,. \tag{6.19}$$

In the case of a small displacement u_l, the potential energy is expanded by the Taylor series

$$V(u_1, u_2, u_3, \ldots, u_N) = V_0 + \frac{1}{2}\sum_{l,l'} u_l u_{l'} \frac{\partial^2 V}{\partial u_l \partial u_{l'}} + \cdots, \tag{6.20}$$

where the first derivative term $\partial V/\partial u_l$ is dropped because of the condition given by (6.19). In addition we may put $V_0 = 0$ by choosing V_0 as the origin of the potential energy. Since the relation

$$V = \sum_l f(x_{l+1} - x_l) = \sum_l f(u_{l+1} - u_l + a) \tag{6.21}$$

holds, we have $\partial f/\partial u_{l+1} = -\partial f/\partial u_l \equiv \partial f/\partial u$, and the summation is rewritten as

$$\begin{aligned}
&\sum_{l,l'} u_l u_{l'} \frac{\partial^2 f}{\partial u_l \partial u_{l'}} \\
&= \sum_l \left(u_{l+1}^2 \frac{\partial^2 f}{\partial u_{l+1}^2} + u_l^2 \frac{\partial^2 f}{\partial u_l^2} + u_{l+1} u_l \frac{\partial^2 f}{\partial u_{l+1} \partial u_l} + u_l u_{l+1} \frac{\partial^2 f}{\partial u_l \partial u_{l+1}} \right) \\
&= \sum_l \left(u_{l+1}^2 + u_l^2 - u_{l+1} u_l - u_l u_{l+1} \right) \frac{\partial^2 f}{\partial u^2} \\
&\equiv g \sum_l (u_{l+1} - u_l)^2\,,
\end{aligned} \tag{6.22}$$

where $g = \partial^2 f/\partial u^2$. Inserting this relation into (6.20) we obtain

$$V(u_1, u_2, u_3, \ldots, u_N) = V_0 + \frac{1}{2} g \sum_l (u_{l+1} - u_l)^2\,. \tag{6.23}$$

From these considerations we find that the Hamiltonian for a crystal of a monoatonic chain can be expressed by the displacement u_l and momentum p_l. These two parameters are independent and satisfy the following commutation relation:

$$[u_l, p_{l'}] = u_l p_{l'} - p_{l'} u_l = \mathrm{i}\hbar\delta_{l,l'}\,, \tag{6.24}$$

where the crystal consists of atoms with the same mass and the momentum operator is defined by $p_{l'} = -\mathrm{i}\hbar\partial/\partial u_{l'}$. The Hamiltonian H for a crystal of a mono-atonic chain is given by

$$H = \frac{1}{2M}\sum_l p_l p_l + \frac{1}{2}g\sum_l (2u_l u_l - u_l u_{l+1} - u_l u_{l-1})\,. \tag{6.25}$$

The Hamiltonian derived above is for a one-dimensional lattice connected by a spring force (or force constant) g, and corresponds to the equation of motion given by (6.1). Therefore, the displacement is given by

$$u_l = A\mathrm{e}^{\mathrm{i}(qla-\omega t)}\,. \tag{6.26}$$

Here we introduce the cyclic boundary condition:

$$u_{l+N} = u_l\,, \tag{6.27}$$

and the wave vector q has to satisfy

$$\exp(\mathrm{i}qaN) = 1 \quad \text{or} \quad q = \frac{2\pi}{aN}l\,, \tag{6.28}$$

where l is an integer. The same relation is obtained by introducing the cyclic boundary condition to (6.25).

Here we show that the Hamiltonian for lattice vibrations given by (6.25) is easily solved by using the Fourier transform. To do this we define

$$Q_q = \frac{1}{\sqrt{N}}\sum_l u_l \mathrm{e}^{-\mathrm{i}qal} \tag{6.29}$$

and then the displacement u_l is obtained by the inverse transform

$$\frac{1}{\sqrt{N}}\sum_q Q_q \mathrm{e}^{\mathrm{i}qal} = \sum_{q,l'} \frac{1}{N} u_{l'} \mathrm{e}^{\mathrm{i}qa(l-l')} = \frac{1}{N}\sum_{l'} u_{l'} N\delta_{ll'} = u_l\,. \tag{6.30}$$

As shown in Appendix A.2, we have the following relation:

$$\sum_q \mathrm{e}^{\mathrm{i}qa(l-l')} = N\delta_{ll'}\,. \tag{6.31}$$

Since this relation is often used in this textbook and is very important for solid state physics, a proof is given here, although a general formula including 3-dimensional case is proved in Appendix A.2. The wave vector is written as

$$q = \frac{2\pi}{aN}n = \frac{2\pi}{L}n,$$

where $L = Na$ is the length of a one-dimensional crystal of N atoms. The wave vector q has N degrees of freedom, which is exactly the same as the number of atoms N. This requires the condition for n such that n ranges from 1 to N, from 0 to $N - 1$, or from $-N/2 + 1$ to $N/2$, resulting in N values for n. This is equivalent to the reduced Brillouin zone scheme and the wave vector q in the first Brillouin zone describes all the features required for lattice vibrations. When we see the wave vectors q corresponding to n in Fig. 6.2, all the features of the relation between ω and q are the repetition of the first Brillouin zone. From this reason the first Brillouin zone is taken in general as

$$\frac{-\pi}{a} < q \leq \frac{\pi}{a}, \quad \left(-\frac{N}{2} < n \leq \frac{N}{2}\right). \tag{6.32}$$

The sign of inequality $<$ (not $\leq$) on the left of the above equation arises from the fact that the difference between $-\pi/a$ and π/a is just the reciprocal lattice vector $2\pi/a$, resulting in the equivalent state. If the sign of inequality $<$ is replaced with $\leq$, the number of degrees of freedom is not N but $N + 1$, which is in disagreement with the lattice degrees of freedom.

For the purpose of calculations we put $m = l - l'$ (m is integer), and we consider the case of $m = 0$ first. Taking account of the result that q has N values, we find

$$\sum_q e^{iqam} = \sum_{n=1}^{N} 1 = N. \qquad (m = 0). \tag{6.33}$$

Next, we consider the case of $m \neq 0$. Since $\exp(iqam) = \exp(i2\pi nm/N) = 1$ for $n = 0$ and $n = N$, we obtain

$$\sum_q e^{iqam} = \sum_{n=0}^{N-1} e^{i2\pi nm/N} = \frac{1 - e^{i2\pi m}}{1 - e^{i2\pi m/N}} = 0. \tag{6.34}$$

Therefore (6.31) is proved.

Let us consider the summation of n for q in the first Brillouin zone $-\pi/a < q \leq \pi/a$ $(N/2 < n \leq N/2)$ and carry out the summation in the range $n = (-N/2+1) \sim N/2$. This is done easily by multiplying by $\exp[i2\pi(N/2-1)m/N]$ and by changing the summation range of n to $n = 0 \sim N - 1$, dividing by the same factor after the summation:

$$\begin{aligned}\sum_q \mathrm{e}^{\mathrm{i}qam} &= \sum_{n=-N/2+1}^{N/2} \mathrm{e}^{\mathrm{i}2\pi nm/N} \\ &= \mathrm{e}^{-\mathrm{i}2\pi(N/2-1)m/N} \sum_{n=-N/2+1}^{N/2} \mathrm{e}^{\mathrm{i}2\pi(n+N/2-1)m/N} \\ &= \mathrm{e}^{-\mathrm{i}2\pi(N/2-1)m/N} \sum_{n'=0}^{N-1} \mathrm{e}^{\mathrm{i}2\pi n'm/N} \\ &= \mathrm{e}^{-\mathrm{i}2\pi(N/2-1)m/N} \frac{1-\mathrm{e}^{\mathrm{i}2\pi m}}{1-\mathrm{e}^{\mathrm{i}2\pi m/N}} = 0\,, \qquad (m \neq 0)\,. \end{aligned} \tag{6.35}$$

From these considerations we find that the following relation holds in general:

$$\sum_q \mathrm{e}^{\mathrm{i}qa(l-l')} = N\delta_{ll'}\,, \tag{6.36}$$

and thus (6.31) is proved. Using this relation we obtain

$$u_l = \frac{1}{\sqrt{N}} \sum_q Q_q \mathrm{e}^{\mathrm{i}qal}, \tag{6.37a}$$

$$Q_q = \frac{1}{\sqrt{N}} \sum_l u_l \mathrm{e}^{-\mathrm{i}qal}\,. \tag{6.37b}$$

Another important method to describe lattice vibrations is the **continuum model**:

$$u(x) = \sum_q Q_q \mathrm{e}^{\mathrm{i}qx}\,, \tag{6.38a}$$

$$Q_q = \frac{1}{L} \int_{-L/2}^{L/2} u(x)\mathrm{e}^{-\mathrm{i}qx}\mathrm{d}x\,, \tag{6.38b}$$

where L is the length of the one-dimensional lattice. These relations for the continuum model are easily proved by referring to the relations given in Appendix A.2 and A.3.

In a similar fashion the momentum operators are defined by

$$p_l = \frac{1}{\sqrt{N}} \sum_q P_q \mathrm{e}^{-\mathrm{i}qal}, \tag{6.39a}$$

$$P_q = \frac{1}{\sqrt{N}} \sum_l p_l \mathrm{e}^{\mathrm{i}qal}\,, \tag{6.39b}$$

and these relations for the continuum model are

$$p(x) = \sum_q P_q \mathrm{e}^{-\mathrm{i}qx}, \tag{6.40a}$$

$$P_q = \frac{1}{L} \int_{-L/2}^{L/2} p(x)\mathrm{e}^{\mathrm{i}qx}\mathrm{d}x\,. \tag{6.40b}$$

These Fourier transform relations are extended to the case of 3-dimensional lattice or d-dimensional lattice in general, where 3-dimensional results are given by (A.41) and (A.42) in Appendix A.3. In the case of d-dimensional lattice we have the Fourier transform

$$f(\boldsymbol{r}) = \sum_{\boldsymbol{q}} F(\boldsymbol{q}) \mathrm{e}^{\mathrm{i}\boldsymbol{q}\cdot\boldsymbol{r}}, \tag{6.41a}$$

$$F(\boldsymbol{q}) = \frac{1}{L^d} \int f(\boldsymbol{r}) \mathrm{e}^{-\mathrm{i}\boldsymbol{q}\cdot\boldsymbol{r}} \mathrm{d}^d\boldsymbol{r}. \tag{6.41b}$$

and the Dirac–δ function and Kronecker δ-function (see also Appendix A.2)

$$\sum_{\boldsymbol{q}} \mathrm{e}^{\mathrm{i}\boldsymbol{q}\cdot(\boldsymbol{r}-\boldsymbol{r}')} = L^d \delta(\boldsymbol{r} - \boldsymbol{r}'), \tag{6.42a}$$

$$\frac{1}{L^d} \int \mathrm{e}^{\mathrm{i}(\boldsymbol{q}-\boldsymbol{q}')\cdot\boldsymbol{r}} \mathrm{d}^d\boldsymbol{r} = \delta_{\boldsymbol{q},\boldsymbol{q}'}. \tag{6.42b}$$

Since the displacement u_l is a Hermite operator, u_l is required to be equivalent to its Hermite conjugate and we have

$$u_l = u_l^\dagger = \frac{1}{\sqrt{N}} \sum_q Q_q \mathrm{e}^{\mathrm{i}qal} = \frac{1}{\sqrt{N}} \sum_q Q_q^\dagger \mathrm{e}^{-\mathrm{i}qal}. \tag{6.43}$$

From this relation we obtain

$$Q_q = Q_{-q}^\dagger \quad \text{or} \quad Q_q^\dagger = Q_{-q}, \tag{6.44}$$

where $Q^\dagger$ is the Hermite conjugate operator of Q. Noting that the summation of q ranges from negative to positive values with the same magnitude, we use the relation $\sum Q_q \exp(\mathrm{i}qal) = \sum Q_{-q} \exp(-\mathrm{i}qal)$ and obtain

$$u_l = \frac{1}{2\sqrt{N}} \sum_q (Q_q \mathrm{e}^{\mathrm{i}qal} + Q_q^\dagger \mathrm{e}^{-\mathrm{i}qal}). \tag{6.45}$$

Using these results we find the following commutation relation:

$$[Q_q, P_{q'}] = \mathrm{i}\hbar\delta_{q,q'}. \tag{6.46}$$

This is easily proved as follows:

$$\begin{aligned} [Q_q, P_{q'}] &= \frac{1}{N} \left[\sum_l u_l \mathrm{e}^{-\mathrm{i}qal}, \sum_{l'} p_{l'} \mathrm{e}^{\mathrm{i}q'al'} \right] \\ &= \frac{1}{N} \sum_{l,l'} [u_l, p_{l'}] \mathrm{e}^{-\mathrm{i}a(ql-q'l')} \end{aligned}$$

$$= \frac{1}{N}\sum_{l,l'} \mathrm{i}\hbar\delta_{ll'}\mathrm{e}^{-\mathrm{i}a(ql-q'l')}$$
$$= \frac{1}{N}\sum_{l} \mathrm{i}\hbar\mathrm{e}^{-\mathrm{i}al(q-q')} = \mathrm{i}\hbar\delta_{q,q'}\,. \tag{6.47}$$

In the case of the continuum model we find

$$[Q_q, P_{q'}] = \frac{1}{L}\int\int [u(x), p(x')]\mathrm{e}^{\mathrm{i}(q'x'-qx)}\mathrm{d}x\mathrm{d}x'$$
$$= \mathrm{i}\hbar\frac{1}{L}\int\int \delta(x-x')\mathrm{e}^{\mathrm{i}(q'x'-qx)}\mathrm{d}x\mathrm{d}x'$$
$$= \mathrm{i}\hbar\frac{1}{L}\int \mathrm{e}^{\mathrm{i}(q'-q)x}\mathrm{d}x = \mathrm{i}\hbar\delta_{q,q'}\,. \tag{6.48}$$

These analyses tell us very important results. In addition to the commutation relation of (6.24) between u_l and $p_{l'}$, the commutation relation of (6.48) holds between Q_q and $P_{q'}$, which are the Fourier transforms of u_l and $p_{l'}$. The Hamiltonian of (6.25) is rewritten as follows by using the operators Q_q and P_q:

$$H = \frac{1}{2M}\sum_{lqq'}\frac{1}{N}P_q P_{q'}\mathrm{e}^{-\mathrm{i}(q+q')al}$$
$$+\frac{1}{2}g\sum_{lqq'}\frac{1}{N}Q_q Q_{q'}\left\{2\mathrm{e}^{\mathrm{i}(q+q')al} - \mathrm{e}^{\mathrm{i}(ql+q'l+q')a} - \mathrm{e}^{\mathrm{i}(ql+q'l-q')a}\right\}$$
$$= \frac{1}{2M}\sum_q P_q P_{-q} + g\sum_q Q_q Q_{-q}\left(1 - \frac{\mathrm{e}^{-\mathrm{i}qa}+\mathrm{e}^{\mathrm{i}qa}}{2}\right)$$
$$= \sum_q\left\{\frac{1}{2M}P_q P_{-q} + gQ_q Q_{-q}[1-\cos(qa)]\right\}\,. \tag{6.49}$$

When we define

$$\omega_q = \sqrt{\frac{2g}{M}[1-\cos(qa)]} = 2\sqrt{\frac{g}{M}}\left|\sin\left(\frac{qa}{2}\right)\right|\,, \tag{6.50}$$

The Hamiltonian is written as

$$H = \sum_q\left(\frac{1}{2M}P_q P_{-q} + \frac{1}{2}M\omega_q^2 Q_q Q_{-q}\right)$$
$$\equiv \sum_q\left(\frac{1}{2M}P_q^2 + \frac{1}{2}M\omega_q^2 Q_q^2\right)\,. \tag{6.51}$$

This results tell us that the Hamiltonian (6.25) for a crystal of N atoms is expressed in terms of the angular frequency ω_q, momentum P_q and displacement Q_q identified by mode q, where P_q and Q_q are the Fourier transforms of p_l and q_l, respectively.

Next, we define the creation operator $a_q^\dagger$ and annihilation operator a_q by

$$a_q^\dagger = \frac{1}{\sqrt{2M\hbar\omega_q}}(M\omega_q Q_{-q} - \mathrm{i}P_q), \tag{6.52a}$$

$$a_q = \frac{1}{\sqrt{2M\hbar\omega_q}}(M\omega_q Q_q + \mathrm{i}P_{-q}). \tag{6.52b}$$

Then the commutation relation between the operators a_q and $a_q^\dagger$ is given by

$$\begin{aligned}[a_q, a_{q'}^\dagger] &= \frac{1}{\sqrt{2M\hbar\omega_q}}\frac{1}{\sqrt{2M\hbar\omega_{q'}}}\{-\mathrm{i}M\omega_q[Q_q, P_{q'}] + \mathrm{i}M\omega_{q'}[P_{-q}, Q_{-q'}]\} \\ &= \delta_{qq'}\,. \end{aligned} \tag{6.53}$$

Therefore, we obtain

$$[a_q, a_q^\dagger] = a_q a_q^\dagger - a_q^\dagger a_q = 1\,. \tag{6.54}$$

Using (6.52a) and (6.52b), the operators P_q and Q_q are expressed by the following relations:

$$P_q = \mathrm{i}\sqrt{\frac{M\hbar\omega_q}{2}}(a_q^\dagger - a_{-q}), \tag{6.55a}$$

$$Q_q = \sqrt{\frac{\hbar}{2M\omega_q}}(a_{-q}^\dagger + a_q)\,. \tag{6.55b}$$

Inserting these relations into (6.51), the Hamiltonian is expressed as

$$H = \frac{1}{2}\sum_q \hbar\omega_q(a_q^\dagger a_q + a_q a_q^\dagger) = \sum_q \hbar\omega_q\left(a_q^\dagger a_q + \frac{1}{2}\right) \equiv \sum_q H_q\,. \tag{6.56}$$

We call

$$\hat{n}_q = a_q^\dagger a_q \tag{6.57}$$

the **number operator** for a boson of mode q. The Hamiltonian H_q for a mode q is written as

$$H_q = \frac{1}{2M}P_q^2 + \frac{1}{2}M\omega_q^2 Q_q^2 = -\frac{\hbar^2}{2M}\frac{\partial^2}{\partial Q_q^2} + \frac{1}{2}M\omega_q^2 Q_q^2\,, \tag{6.58}$$

where we have used

$$P_q = -\mathrm{i}\hbar\frac{\partial}{\partial Q_q}\,. \tag{6.59}$$

Equation (6.58) represents the Hamiltonian of a simple harmonic oscillator with momentum P_q and displacement Q_q. When we define the eigenstate of the Hamiltonian H_q of mode q by $|n_q\rangle$, we obtain

$$H_q|n_q\rangle = \left(n_q + \frac{1}{2}\right)\hbar\omega_q|n_q\rangle \tag{6.60}$$

by analogy to the simple harmonic oscillator solution. The wave function for total Hamiltonian is then given by the product of each eigenfunction such as

$$|n_1, n_2, \ldots, n_q, \ldots\rangle = \Pi|n_q\rangle\,, \tag{6.61}$$

and the total energy is given by

$$\mathcal{E} = \sum_q \left(n_q + \frac{1}{2}\right)\hbar\omega_q\,. \quad (n_q = 0, 1, 2, \ldots)\,. \tag{6.62}$$

We know that the average value of the boson excitation number is given by the Bose–Einstein distribution

$$\overline{n}_q = \frac{1}{\mathrm{e}^{\hbar\omega_q/k_\mathrm{B}T} - 1}\,. \tag{6.63}$$

Here we list some important relations for boson operators, which are easily proved (see Appendix D):

$$a_q^\dagger|n_q\rangle = \sqrt{n_q + 1}|n_q + 1\rangle\,, \tag{6.64a}$$

$$a_q|n_q\rangle = \sqrt{n_q}|n_q - 1\rangle\,, \tag{6.64b}$$

$$a_q^\dagger a_q|n_q\rangle = n_q|n_q\rangle\,, \tag{6.64c}$$

$$a_q a_q^\dagger|n_q\rangle = (n_q + 1)|n_q\rangle\,. \tag{6.64d}$$

Using these boson operators, the displacement of the lattice atoms is written as

$$u_l = \sum_q \sqrt{\frac{\hbar}{2MN\omega_q}}\left(a_q\mathrm{e}^{\mathrm{i}qal} + a_q^\dagger\mathrm{e}^{-\mathrm{i}qal}\right) \tag{6.65}$$

and for the 3-dimensional case we obtain

$$\boldsymbol{u}_l = \sum_{\boldsymbol{q}} \boldsymbol{e}_{\boldsymbol{q}}\sqrt{\frac{\hbar}{2MN\omega_{\boldsymbol{q}}}}\left(a_{\boldsymbol{q}}\mathrm{e}^{\mathrm{i}\boldsymbol{q}\cdot\boldsymbol{R}_l} + a_{\boldsymbol{q}}^\dagger\mathrm{e}^{-\mathrm{i}\boldsymbol{q}\cdot\boldsymbol{R}_l}\right)\,, \tag{6.66}$$

where $\boldsymbol{R}_l$ is the position vector. When we use the continuum model, the displacement is expressed as

$$u(x) = \sum_q \sqrt{\frac{\hbar}{2MN\omega_q}}\left(a_q\mathrm{e}^{\mathrm{i}qx} + a_q^\dagger\mathrm{e}^{-\mathrm{i}qx}\right) \tag{6.67}$$

and for the 3-dimensional case we obtain

$$u(r) = \sum_q e_q \sqrt{\frac{\hbar}{2MN\omega_q}} \left(a_q e^{iq\cdot r} + a_q^\dagger e^{-iq\cdot r}\right) , \tag{6.68}$$

where e_q is the unit vector along the displacement direction.

The above results are for mono-atomic crystals and we have to derive similar relations for a crystal with two atoms of mass M_1 and M_2 in each unit cell. The displacement operator for the acoustic mode is written as

$$u(r) = \sum_q e_q \sqrt{\frac{\hbar}{2(M_1 + M_2)N\omega_q}} \left(a_q e^{iq\cdot r} + a_q^\dagger e^{iq\cdot r}\right) . \tag{6.69}$$

In a similar fashion we may deduce the displacement operator for the optical phonon mode. As stated earlier, optical phonon modes appear in a crystal with two or more atoms in a unit cell. Let us consider the simple case where two atoms exist in a unit cell and define $M = M_1 + M_2$, $M_a = (M_1 + M_2)/2$, or reduced mass M_r $(1/M_r = 1/M_1 + 1/M_2)$. Several papers have been reported to describe the optical phonon modes, but it is not yet clear which mass is correct to express the optical phonon modes: M, M_a or M_r. When the reduced mass is used for this, the relative displacement $u_r(r)$ for optical phonons is written as

$$u_r(r) = \sum_q e_q \sqrt{\frac{\hbar}{2NM_r\omega_q}} \left(a_q e^{iq\cdot r} + a_q^\dagger e^{-iq\cdot r}\right) , \tag{6.70}$$

where $e_q = u_r/u_r$.

6.2 Boltzmann Transport Equation

Let us define the wave vector of an electron by k and discuss the change in its electronic state in an external field. The electron wave vector will be changed under an external force F as

$$\hbar \frac{dk}{dt} = F . \tag{6.71}$$

We define the probability function $f(k, r, t)$ of a particle (electron) which has position vector r and wave vector k at time t, and call it the distribution function of the particle (electron). First, we assume that the electron is not subject to scattering and that the state is changed by an external force (electric field, magnetic field and so on). Then, after a time interval dt, the electron is changed into a new state with position $r + \dot{r}dt$ and wave vector $k + \dot{k}dt$. With the help of Fig. 6.5 we calculate the change in the distribution function (r, k, t) in a time interval dt. A particle that occupied a

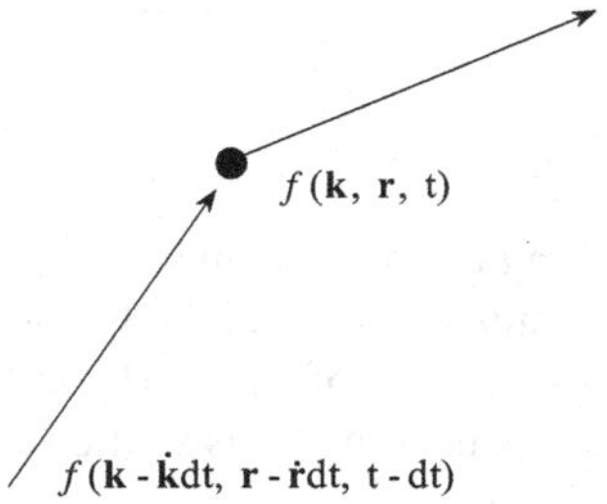

Fig. 6.5 Electron trajectory and change in the distribution function. The distribution function $f(\boldsymbol{k}-\dot{\boldsymbol{k}}\mathrm{d}t, \boldsymbol{r}-\dot{\boldsymbol{r}}\mathrm{d}t, t-\mathrm{d}t)$ at a time $t-\mathrm{d}t$ will become $f(\boldsymbol{k},\boldsymbol{r},t)$ at time t. The distribution function $f(\boldsymbol{k},\boldsymbol{r},t)$ at time t will be changed into another state after a time interval $\mathrm{d}t$. The net change of the distribution function gives rise to drift term (see text for details)

state $\boldsymbol{r}-\dot{\boldsymbol{r}}\mathrm{d}t$ and $\boldsymbol{k}-\dot{\boldsymbol{k}}\mathrm{d}t$ at a time $t-\mathrm{d}t$ will move into a state $f(\boldsymbol{k},\boldsymbol{r},t)$ after a time interval $\mathrm{d}t$ (at time t). Therefore, the rate of change of the distribution function is given by

$$\left(\frac{\mathrm{d}f}{\mathrm{d}t}\right)_{\mathrm{drift}} = [f(\boldsymbol{k}-\dot{\boldsymbol{k}}\mathrm{d}t, \boldsymbol{r}-\dot{\boldsymbol{r}}\mathrm{d}t, t-\mathrm{d}t) - f(\boldsymbol{k},\boldsymbol{r},t)]/\mathrm{d}t\,. \tag{6.72}$$

The right-hand side of above equation is interpreted as follows. The first term of the right-hand side will change into $f(\boldsymbol{k},\boldsymbol{r},t)$ after a time interval $\mathrm{d}t$ (increase in $f(\boldsymbol{k},\boldsymbol{r},t)$) and the second term $f(\boldsymbol{k},\boldsymbol{r},t)$ will change into another state after a time interval $\mathrm{d}t$ (decrease in $f(\boldsymbol{k},\boldsymbol{r},t)$). Since the scattering of a particle is not included in the present analysis, the above rate represents the continuous flow of a particle and thus the term is called a **drift term**.

Now, the Taylor expansion of the first term on the right-hand side in (6.72) gives

$$\begin{aligned} & f(\boldsymbol{k}-\dot{\boldsymbol{k}}\mathrm{d}t, \boldsymbol{r}-\dot{\boldsymbol{r}}\mathrm{d}t, t-\mathrm{d}t) \\ &= f(\boldsymbol{k},\boldsymbol{r},t) - \left[\dot{\boldsymbol{k}}\cdot\frac{\partial f}{\partial \boldsymbol{k}} + \dot{\boldsymbol{r}}\cdot\frac{\partial f}{\partial \boldsymbol{r}} + \frac{\partial f}{\partial t}\right]\mathrm{d}t + \ldots \\ &\equiv f(\boldsymbol{k},\boldsymbol{r},t) - \left[(\dot{\boldsymbol{k}}\cdot\nabla_{\boldsymbol{k}} f) + (\dot{\boldsymbol{r}}\cdot\nabla_{\boldsymbol{r}} f) + \frac{\partial f}{\partial t}\right]\mathrm{d}t + \ldots\,, \end{aligned} \tag{6.73}$$

where

$$\dot{\boldsymbol{k}}\cdot\nabla_{\boldsymbol{k}} f = \dot{\boldsymbol{k}}\cdot\mathrm{grad}_{\boldsymbol{k}} f = \dot{k}_x\frac{\partial f}{\partial k_x} + \dot{k}_y\frac{\partial f}{\partial k_y} + \dot{k}_z\frac{\partial f}{\partial k_z}\,, \tag{6.74}$$

$$\dot{\boldsymbol{r}}\cdot\nabla_{\boldsymbol{r}} f = \dot{\boldsymbol{v}}\cdot\mathrm{grad}_{\boldsymbol{r}} f = \dot{v}_x\frac{\partial f}{\partial x} + \dot{v}_y\frac{\partial f}{\partial y} + \dot{v}_z\frac{\partial f}{\partial z}\,, \tag{6.75}$$

where we have used the definition of $\dot{\boldsymbol{k}}$ as the time derivative of $\boldsymbol{k}$ and where $\boldsymbol{v}=\dot{\boldsymbol{r}}$ is the velocity of the particle. Keeping terms up to the second term in (6.73) and inserting it into (6.72), the drift term is rewritten as

$$\left(\frac{\mathrm{d}f}{\mathrm{d}t}\right)_{\text{drift}} = -\left[\dot{\boldsymbol{k}} \cdot \nabla_{k} f + \boldsymbol{v} \cdot \nabla_{r} f + \frac{\partial f}{\partial t}\right]. \tag{6.76}$$

Since the time variation of the wave vector $\boldsymbol{k}$ under an external force $\boldsymbol{F}$ is given by (6.71), the drift term is given by

$$\left(\frac{\mathrm{d}f}{\mathrm{d}t}\right)_{\text{drift}} = -\left[\frac{1}{\hbar}(\boldsymbol{F} \cdot \nabla_{k} f) + \boldsymbol{v} \cdot \nabla_{r} f + \frac{\partial f}{\partial t}\right]. \tag{6.77}$$

On the other hand, a particle changes its state by scattering (collision) and we define the rate of change in the distribution function due to a collision by $(\mathrm{d}f/\mathrm{d}t)_{\text{coll}}$. Since the distribution function has to satisfy the equilibrium condition (steady state condition) or condition of balance, we have

$$\left(\frac{\mathrm{d}f}{\mathrm{d}t}\right)_{\text{drift}} + \left(\frac{\mathrm{d}f}{\mathrm{d}t}\right)_{\text{coll}} = 0. \tag{6.78}$$

Inserting (6.77) into this we obtain

$$\frac{\partial f}{\partial t} + \frac{1}{\hbar}\boldsymbol{F} \cdot \nabla_{k} f + \boldsymbol{v} \cdot \nabla_{r} f = \left(\frac{\mathrm{d}f}{\mathrm{d}t}\right)_{\text{coll}}. \tag{6.79}$$

This equation is called the **Boltzmann transport equation**.

6.2.1 Collision Term and Relaxation Time

Let us consider a uniform crystal and assume the distribution function to be independent of position $\boldsymbol{r}$. Then the distribution function is written as $f(\boldsymbol{k})$. The rate of change in the distribution function $f(\boldsymbol{k})$ includes two terms. The rate of increase in $f(\boldsymbol{k})$ due to transitions from all possible $\boldsymbol{k}'$ states (excluding $\boldsymbol{k}$) to the $\boldsymbol{k}$ state and the rate of decrease in $f(\boldsymbol{k})$ due to transitions from the $\boldsymbol{k}$ state to other possible $\boldsymbol{k}'$ states. Defining the respective transition rates per unit time by $P(\boldsymbol{k}', \boldsymbol{k})$ and $P(\boldsymbol{k}, \boldsymbol{k}')$, the collision term is written as.

$$\left(\frac{\mathrm{d}f}{\mathrm{d}t}\right)_{\text{coll}} = \sum_{\boldsymbol{k}'} \left\{P(\boldsymbol{k}', \boldsymbol{k}) f(\boldsymbol{k}')[1 - f(\boldsymbol{k})] - P(\boldsymbol{k}, \boldsymbol{k}') f(\boldsymbol{k})[1 - f(\boldsymbol{k}')]\right\}, \tag{6.80}$$

where the prefactor $f(\boldsymbol{k}')[1 - f(\boldsymbol{k})]$ of $P(\boldsymbol{k}', \boldsymbol{k})$ represents the probability of electron occupation in the initial state $\boldsymbol{k}'$ and of electron vacancy in the final state $\boldsymbol{k}$.

For simplicity we consider the case where the Fermi level lies below the bottom of the conduction band in the band gap; then, the conditions $f(\boldsymbol{k}) \ll 1$ and $f(\boldsymbol{k}') \ll 1$

hold. Defining the distribution function for thermal equilibrium by $f_0(\boldsymbol{k})$, we have $(\mathrm{d}f/\mathrm{d}t)_{\mathrm{coll}} = 0$ in (6.80), and thus we obtain the following relation:

$$P(\boldsymbol{k}', \boldsymbol{k}) f_0(\boldsymbol{k}') = P(\boldsymbol{k}, \boldsymbol{k}') f_0(\boldsymbol{k}) . \tag{6.81}$$

This relation is called the **principle of detailed balance**. Using this relation (6.80) is written as

$$\left(\frac{\mathrm{d}f}{\mathrm{d}t}\right)_{\mathrm{coll}} = -\sum_{\boldsymbol{k}'} P(\boldsymbol{k}, \boldsymbol{k}') \left[f(\boldsymbol{k}) - f(\boldsymbol{k}') \frac{f_0(\boldsymbol{k})}{f_0(\boldsymbol{k}')} \right] . \tag{6.82}$$

When we replace the summation of $\boldsymbol{k}'$ by an integral, we have

$$\left(\frac{\mathrm{d}f}{\mathrm{d}t}\right)_{\mathrm{coll}} = -\frac{V}{(2\pi)^3} \int \mathrm{d}^3\boldsymbol{k}' P(\boldsymbol{k}, \boldsymbol{k}') \left[f(\boldsymbol{k}) - f(\boldsymbol{k}') \frac{f_0(\boldsymbol{k})}{f_0(\boldsymbol{k}')} \right] , \tag{6.83}$$

where $V = L^3$ is the crystal volume.[1] When the external force is very weak and the displacement of the distribution function from the thermal equilibrium value is small, we may write (or expand in Taylor series)

$$f(\boldsymbol{k}) = f_0(\boldsymbol{k}) + f_1(\boldsymbol{k}), \quad (f_1(\boldsymbol{k}) \ll f_0(\boldsymbol{k})) . \tag{6.84}$$

Assuming that the energy change due to scattering is small and that the energy $\mathcal{E}(\boldsymbol{k})$ of the initial state $\boldsymbol{k}$ is very close to the energy $\mathcal{E}(\boldsymbol{k}')$ of the final state $\boldsymbol{k}'$ (this condition is called an **elastic collision**), we may put $f_0(\boldsymbol{k}) \cong f_0(\boldsymbol{k}')$. In this case we obtain the following relation from (6.83):

$$\begin{aligned} \left(\frac{\mathrm{d}f}{\mathrm{d}t}\right)_{\mathrm{coll}} &= -f_1(\boldsymbol{k}) \frac{V}{(2\pi)^3} \int \mathrm{d}^3\boldsymbol{k}' P(\boldsymbol{k}, \boldsymbol{k}') \left[1 - \frac{f_1(\boldsymbol{k}')}{f_1(\boldsymbol{k})} \right] , \\ &\equiv -\frac{f_1(\boldsymbol{k})}{\tau(\boldsymbol{k})} \equiv -\frac{f(\boldsymbol{k}) - f_0(\boldsymbol{k})}{\tau(\boldsymbol{k})} , \end{aligned} \tag{6.85}$$

where $\tau(\boldsymbol{k})$ is called the **relaxation time** of the collision and is given by

$$\frac{1}{\tau(\boldsymbol{k})} = \frac{V}{(2\pi)^3} \int \mathrm{d}^3\boldsymbol{k}' P(\boldsymbol{k}, \boldsymbol{k}') \left[1 - \frac{f_1(\boldsymbol{k}')}{f_1(\boldsymbol{k})} \right] , \tag{6.86}$$

which is a function of the electron wave vector $\boldsymbol{k}$ and thus of the energy. This approximation is called the **relaxation approximation**.

After the electron system reaches the steady state under an external field, the external field is removed at time $t = 0$ and we have $\partial f/\partial t = (\partial f/\partial t)_{\mathrm{coll}}$ from the Boltzmann equation (6.79). When we assume the relaxation approximation and we use (6.85), it may be expressed as

[1] We use same notation for the potential and the crystal volume in this book, but it will not introduce any confusion.

$$\frac{\partial f}{\partial t} = \left(\frac{\mathrm{d}f}{\mathrm{d}t}\right)_{\text{coll}} = -\frac{f - f_0}{\tau} . \tag{6.87}$$

For simplicity we assume that the relaxation time τ is constant, and then we obtain

$$f - f_0 = (f - f_0)_{t=0} \exp\left(-\frac{t}{\tau}\right) , \tag{6.88}$$

where $(f - f_0)_{t=0}$ is the shift of the distribution function from thermal equilibrium at time $t = 0$. The above result indicates that the distribution function recovers exponentially with time constant τ toward the thermal equilibrium value after removal of the external field at $t = 0$.

When the relaxation time approximation is valid, the Boltzmann transport equation is written as

$$\frac{\partial f}{\partial t} + \frac{1}{\hbar}\boldsymbol{F} \cdot \nabla_{\boldsymbol{k}} f + \boldsymbol{v} \cdot \nabla_{\boldsymbol{r}} f = -\frac{f - f_0}{\tau} . \tag{6.89}$$

Spatial uniformity results in $\nabla_{\boldsymbol{r}} f = 0$ and the steady state gives rise to $\partial f / \partial t = 0$. Therefore, we obtain

$$\frac{1}{\hbar}\boldsymbol{F} \cdot \nabla_{\boldsymbol{k}} f = -\frac{f_1}{\tau} \tag{6.90}$$

and when the external field is applied in the x direction it reduces to

$$f_1 = -\frac{\tau}{\hbar} F_x \frac{\partial f}{\partial k_x} . \tag{6.91}$$

The energy of the electron is a function of the wave vector and is written as $\mathcal{E} = \hbar^2 k^2 / 2m^*$ for an electron with isotropic effective mass m^*, which leads to the relation $\hbar k_x = m^* v_x$. Therefore, we obtain

$$f_1 = -\frac{\tau}{\hbar} F_x \frac{\partial f}{\partial \mathcal{E}} \frac{\partial \mathcal{E}}{\partial k_x} = -\tau v_x F_x \frac{\partial f}{\partial \mathcal{E}} . \tag{6.92}$$

When we use the relations $f = f_0 + f_1$ and $f_0 \gg f_1$ in the above equation we may approximate it as

$$f_1 = -\tau v_x F_x \frac{\partial f_0}{\partial \mathcal{E}} . \tag{6.93}$$

When we assume elastic scattering and that the relaxation time τ is a function of the energy $\mathcal{E}$, the magnitude of the relaxation time τ is not changed after the scattering event and the relaxation time $\tau(\boldsymbol{k})$ is expressed as follows by inserting (6.93) into (6.86):

$$\frac{1}{\tau(\boldsymbol{k})} = \frac{V}{(2\pi)^3}\int \mathrm{d}^3\boldsymbol{k}' P(\boldsymbol{k},\boldsymbol{k}')\left(1-\frac{k'_x}{k_x}\right)$$
$$= \frac{V}{(2\pi)^3}\int \mathrm{d}^3\boldsymbol{k}' P(\boldsymbol{k},\boldsymbol{k}')(1-\cos\theta)\,, \tag{6.94}$$

where k_x and k'_x are the components of the electron wave vectors $\boldsymbol{k}$ and $\boldsymbol{k}'$ along the direction of the external force (x component) before and after scattering, respectively, and θ is the angle between $\boldsymbol{k}$ and $\boldsymbol{k}'$.

In the case of a degenerate semiconductor, the principle of detailed balance is obtained from (6.80) as

$$P(\boldsymbol{k}',\boldsymbol{k})f_0(\boldsymbol{k}')[1-f_0(\boldsymbol{k})] = P(\boldsymbol{k},\boldsymbol{k}')f_0(\boldsymbol{k})[1-f_0(\boldsymbol{k}')]\,. \tag{6.95}$$

Using this relation, the relaxation time of a degenerate semiconductor is given by

$$\frac{1}{\tau(\boldsymbol{k})} = \frac{1}{1-f_0(\boldsymbol{k})}\sum_{\boldsymbol{k}'} P(\boldsymbol{k},\boldsymbol{k}')[1-f_0(\boldsymbol{k}')]\left(1-\frac{k'_x}{k_x}\right)$$
$$= \frac{1}{1-f_0(\boldsymbol{k})}\frac{V}{(2\pi)^3}\int \mathrm{d}\boldsymbol{k}' P(\boldsymbol{k},\boldsymbol{k}')[1-f_0(\boldsymbol{k}')]\left(1-\frac{k'_x}{k_x}\right)\,. \tag{6.96}$$

The above equation reduces to (6.94) for the non-degenerate case when we let $f_0(\boldsymbol{k}) \ll 1$ and $f_0(\boldsymbol{k}') \ll 1$. It may be pointed out that when the final state is occupied by an electron, $1 - f_0(\boldsymbol{k}') = 0$ and the transition (scattering) is not allowed.

6.2.2 *Mobility and Electrical Conductivity*

When the relaxation time is given by a function of the electron energy or electron wave vector, the electron mobility and electrical conductivity (referred to as mobility and conductivity hereafter, respectively, unless otherwise specified) are calculated as follows. In the presence of an electric field applied in the x direction, we may write $F_x = -eE_x$; (6.84) and (6.93) then give the following relation:

$$f(\boldsymbol{k}) = f_0(\boldsymbol{k}) + eE_x\tau v_x\frac{\partial f_0}{\partial\mathcal{E}}\,. \tag{6.97}$$

Therefore the current density in the x direction is given by

$$J_x = \frac{2}{(2\pi)^3}\int(-e)v_x f(\boldsymbol{k})\mathrm{d}^3\boldsymbol{k}$$
$$= -\frac{e}{4\pi^3}\int v_x f_0(\boldsymbol{k})\mathrm{d}^3\boldsymbol{k} - \frac{e^2E_x}{4\pi^3}\int \tau v_x^2\frac{\partial f_0}{\partial\mathcal{E}}\mathrm{d}^3\boldsymbol{k}\,, \tag{6.98}$$

where the factor 2 arises from the spin degeneracy and $v_x = \hbar k_x/m^*$. The function $f_0(\boldsymbol{k})$ is given by the Fermi (or Fermi–Dirac) distribution function or the Boltzmann distribution function, which is a function of the electron energy $\mathcal{E}(\boldsymbol{k})$. Since $\mathcal{E}$ is an

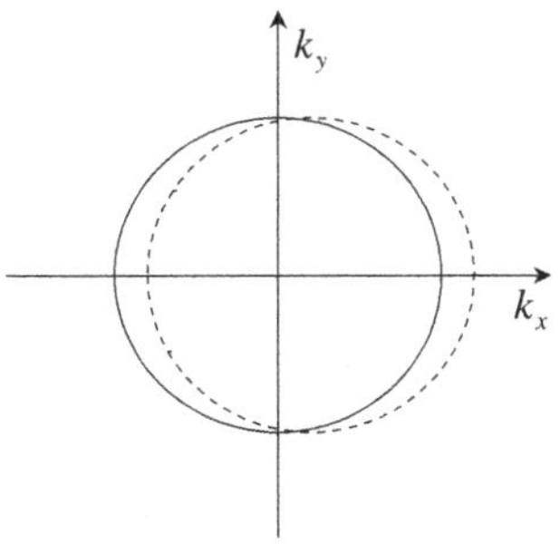

Fig. 6.6 Change in the distribution function under an applied electric field. The distribution function in $\boldsymbol{k}$ space is slightly displaced along the field direction

even function of $\boldsymbol{k}$, $v_x f_0(\boldsymbol{k})$ is an odd function of v_x. Since integration with respect to $\mathrm{d}k_x$ ranges from $-\infty$ to $+\infty$, the first term on the right hand side becomes zero, and only the second term remains, resulting in

$$J_x = -\frac{e^2 E_x}{4\pi^3} \int \tau v_x^2 \frac{\partial f}{\partial \mathcal{E}} \mathrm{d}^3\boldsymbol{k} . \tag{6.99}$$

This result is interpreted with the help of Fig. 6.6. In the absence of an electric field, the distribution function of the thermal equilibrium state is isotropic in $\boldsymbol{k}$ space and thus the current in any direction has the same magnitude, resulting in mutual cancellation and hence in zero current. In the presence of an applied electric field, the distribution function is given by $f = f_0 + f_1$ and is displaced by f_1 in the electric field direction, resulting in a current proportional to the displacement along the electric field direction. Using the electron density n given by

$$n = \frac{2}{(2\pi)^3} \int f_0 \mathrm{d}^3\boldsymbol{k} , \tag{6.100}$$

Equation (6.99) is rewritten as

$$J_x = -e^2 n E_x \frac{\displaystyle\int \tau v_x^2 \frac{\mathrm{d} f_0}{\mathrm{d}\mathcal{E}} \mathrm{d}^3\boldsymbol{k}}{\displaystyle\int f_0 \mathrm{d}^3\boldsymbol{k}} = \frac{e^2 n E_x}{k_\mathrm{B} T} \frac{\displaystyle\int \tau v_x^2 f_0 (1 - f_0) \mathrm{d}^3\boldsymbol{k}}{\displaystyle\int f_0 \mathrm{d}^3\boldsymbol{k}} , \tag{6.101}$$

where the final result is obtained by using the following relations:

$$f_0(\mathcal{E}) = \frac{1}{\mathrm{e}^{(\mathcal{E}-\mathcal{E}_F)/k_B T} + 1} \tag{6.102}$$

$$\frac{\partial f_0}{\partial \mathcal{E}} = -\frac{1}{k_B T} f_0 (1 - f_0) . \tag{6.103}$$

Since the term $\tau f_0(1 - f_0)$ in the integral of (6.101) is a function of the energy $\mathcal{E}$, we express it as $\phi(\mathcal{E})$. Noting $\mathrm{d}^3\boldsymbol{k} = \mathrm{d}k_x \mathrm{d}k_y \mathrm{d}k_z$, we have

$$\int v_x^2 \phi(\mathcal{E}) \mathrm{d}^3 \boldsymbol{k} = \int v_y^2 \phi(\mathcal{E}) \mathrm{d}^3 \boldsymbol{k} = \int v_z^2 \phi(\mathcal{E}) \mathrm{d}^3 \boldsymbol{k} = \frac{1}{3} \int v^2 \phi(\mathcal{E}) \mathrm{d}^3 \boldsymbol{k} . \tag{6.104}$$

The integral with respect to $\mathrm{d}^3\boldsymbol{k}$ is changed into an integral with respect to $\mathcal{E}$ by using the relation $\mathcal{E} = \hbar^2 k^2/2m^*$. We then obtain

$$\mathrm{d}^3 \boldsymbol{k} = 4\pi k^2 \mathrm{d}k = \frac{8\pi m^{*3/2}}{\sqrt{2}\hbar^3} \mathcal{E}^{1/2} \mathrm{d}\mathcal{E} \equiv A \mathcal{E}^{1/2} \mathrm{d}\mathcal{E} . \tag{6.105}$$

In the above equation the effective mass m^* is the density-of-states mass for the isotropic (scalar) effective mass and is often notified by m_{d}^*. In semiconductors such as Ge and Si, we know that the conduction bands consist of a many–valley structure and their constant energy surface is expressed by an ellipsoid with transverse effective mass m_{t} and longitudinal effective mass m_{l}. In such a case the density-of-states mass m_{d}^* is given by $(m_{\mathrm{t}}^2 m_{\mathrm{l}})^{1/3} \equiv m_{\mathrm{d}}^*$. Inserting this relation, (6.104) and (6.105) into (6.101), we obtain

$$J_x = \frac{2e^2 n E_x}{3k_{\mathrm{B}} T m^*} \frac{\displaystyle\int_0^\infty \tau \mathcal{E}^{3/2} f_0 (1 - f_0) \mathrm{d}\mathcal{E}}{\displaystyle\int_0^\infty \mathcal{E}^{1/2} f_0 \mathrm{d}\mathcal{E}} . \tag{6.106}$$

Although the upper limit of the integral should be the upper edge of the band, it is replaced by infinity, taking account of the exponential decay of the distribution function in the higher energy region. As stated above, the density-of-states mass appears in the denominator and numerator and is cancelled, which is evident from (6.101). The effective mass m^* in (6.106) results from the term $v_x = \hbar k_x/m^*$, assuming an isotropic effective mass. The effective mass, usually defined by m_{c}^*, is called the **conductivity effective mass** because it comes from the term for the velocity component. In the case of many–valley semiconductors such as Ge and Si, however, the energy is expressed by an ellipsoid, and the conductivity effective mass is given by $1/m_{\mathrm{c}}^* \equiv (1/3)(2/m_{\mathrm{t}} + 1/m_{\mathrm{l}})$. Next, we carry out the calculation of (6.106) for the following two cases separately.

6.2.2.1 Metal and Degenerate Semiconductor

In this case the Fermi energy $\mathcal{E}_F$ is located in the conduction band, and the term $-\mathrm{d}f_0/\mathrm{d}\mathcal{E} = f_0(1-f_0)/k_{\mathrm{B}}T$ has a large value near the Fermi energy only. Therefore, the term $-\mathrm{d}f_0/\mathrm{d}\mathcal{E}$ may be approximated by the Dirac δ-function. In addition, the following relations,

$$f_0 \cong 1 \quad (\mathcal{E} \leq \mathcal{E}_{\mathrm{F}}) ,$$
$$f_0 \cong 0 \quad (\mathcal{E} > \mathcal{E}_{\mathrm{F}}) ,$$

hold for the Fermi distribution function, and therefore we obtain

$$\frac{1}{k_{\mathrm{B}}T}\int_0^\infty \tau(\mathcal{E})\mathcal{E}^{3/2}f_0(1-f_0)\mathrm{d}\mathcal{E}$$
$$= -\int_0^\infty \tau(\mathcal{E})\mathcal{E}^{3/2}\frac{\mathrm{d}f_0}{\mathrm{d}\mathcal{E}}\mathrm{d}\mathcal{E} \cong \tau(\mathcal{E}_F)\mathcal{E}_{\mathrm{F}}^{3/2} \tag{6.107}$$

$$\int_0^\infty \mathcal{E}^{1/2}f_0\mathrm{d}\mathcal{E} = \int_0^{\mathcal{E}_{\mathrm{F}}} \mathcal{E}^{1/2}\mathrm{d}\mathcal{E} = \frac{2}{3}\mathcal{E}_{\mathrm{F}}^{3/2}. \tag{6.108}$$

Inserting these relations into (6.106), we obtain

$$J_x = \frac{ne^2\tau(\mathcal{E}_{\mathrm{F}})}{m^*}E_x\,, \tag{6.109}$$

where $\tau(\mathcal{E}_{\mathrm{F}})$ is the relaxation time of an electron at $\mathcal{E} = \mathcal{E}_{\mathrm{F}}$ (or at the Fermi surface). The relation between the Fermi energy $\mathcal{E}_{\mathrm{F}}$ and the electron density n is obtained from (6.100) and (6.107), and is given by

$$\mathcal{E}_{\mathrm{F}} = \frac{\hbar^2}{2m^*}(3\pi^2 n)^{2/3}\,. \tag{6.110}$$

6.2.2.2 Non-degenerate Semiconductor

For non-degenerate semiconductors, we have $f_0 \ll 1$ and $1 - f_0 \cong 1$ and we may approximate as

$$f_0 = \exp\left(\frac{\mathcal{E}_{\mathrm{F}}}{k_{\mathrm{B}}T}\right)\exp\left(-\frac{\mathcal{E}}{k_{\mathrm{B}}T}\right) \equiv A_{\mathrm{F}}\exp\left(-\frac{\mathcal{E}}{k_{\mathrm{B}}T}\right), \tag{6.111}$$

and (6.106) reduces to

$$J_x = \frac{2e^2nE_x}{3k_{\mathrm{B}}Tm^*}\frac{\displaystyle\int_0^\infty \tau\mathcal{E}^{3/2}f_0\mathrm{d}\mathcal{E}}{\displaystyle\int_0^\infty \mathcal{E}^{1/2}f_0\mathrm{d}\mathcal{E}}\,. \tag{6.112}$$

Integration by parts gives rise to

$$\int_0^\infty \mathcal{E}^{3/2}f_0\mathrm{d}\mathcal{E} = \frac{3}{2}k_{\mathrm{B}}T\int_0^\infty \mathcal{E}^{1/2}f_0\mathrm{d}\mathcal{E}\,. \tag{6.113}$$

Using this relation we obtain following relation:

$$J_x = \frac{e^2 n E_x}{m^*} \frac{\int_0^\infty \tau \mathcal{E}^{3/2} f_0 \mathrm{d}\mathcal{E}}{\int_0^\infty \mathcal{E}^{3/2} f_0 \mathrm{d}\mathcal{E}} . \tag{6.114}$$

When we rewrite (6.114) as

$$J_x = \frac{ne^2 \langle \tau \rangle}{m^*} E_x , \tag{6.115}$$

the relaxation time $\langle \tau \rangle$ is given by

$$\langle \tau \rangle = \frac{\int_0^\infty \tau \mathcal{E}^{3/2} f_0 \mathrm{d}\mathcal{E}}{\int_0^\infty \mathcal{E}^{3/2} f_0 \mathrm{d}\mathcal{E}} \tag{6.116}$$

and $\langle \tau \rangle$ represents the average of $\tau(\mathcal{E})$.

When we define the average velocity of the electron by $\langle v_x \rangle$ in the presence of an electric field in the x direction, it is also calculated from (6.116). The average velocity is often referred to as the **drift velocity**. This is because the electron contributes to the current via **drift motion** along the electric field direction, suffering from collisions. The average time between the scattering events (collisions) or the scattering rate (collision rate) is given by the average value of the relaxation time $\langle \tau \rangle$, where the scattering rate is not equivalent to the relaxation time in the strict sense. The difference will be discussed later. Using the electron density n, the current density is written as

$$J_x = n(-e)\langle v_x \rangle , \tag{6.117}$$

where the drift velocity $\langle v_x \rangle$ is given by (6.115) as

$$\langle v_x \rangle = -\frac{e\langle \tau \rangle}{m^*} E_x \equiv -\mu E_x . \tag{6.118}$$

Here μ is the average electron velocity (drift velocity) under a unit electric field and gives a measure of the mobility; it is called the **electron mobility**. This mobility is often referred to as the drift mobility to distinguish it from the Hall mobility defined in Chap. 7. Using $\langle \tau \rangle$ the electron mobility is expressed as

$$\mu = \frac{e\langle \tau \rangle}{m^*} , \tag{6.119}$$

and when we define the **conductivity** (electrical conductivity) σ by

$$J_x = \sigma E_x , \tag{6.120}$$

the conductivity is given by

$$\sigma = \frac{ne^2\langle\tau\rangle}{m^*} = ne\mu\,. \tag{6.121}$$

These relations hold for degenerate semiconductors by replacing $\langle\tau(\mathcal{E})\rangle$ with $\tau(\mathcal{E}_{\mathrm{F}})$.

As a simple example we consider the case of the relaxation time τ given by a function of energy $\mathcal{E}$ such as

$$\tau = a\mathcal{E}^{-s}\,. \tag{6.122}$$

Then from (6.116) we obtain

$$\langle\tau\rangle = \frac{a\displaystyle\int_0^\infty \mathcal{E}^{3/2-s}\exp(-\mathcal{E}/k_{\mathrm{B}}T)\mathrm{d}\mathcal{E}}{\displaystyle\int_0^\infty \mathcal{E}^{3/2}\exp(-\mathcal{E}/k_{\mathrm{B}}T)\mathrm{d}\mathcal{E}} = a(k_{\mathrm{B}}T)^{-s}\frac{\Gamma\left(\frac{5}{2}-s\right)}{\Gamma\left(\frac{5}{2}\right)}\,, \tag{6.123}$$

where $\Gamma(s)$ is Γ function and has the following properties as shown in Appendix B

$$\Gamma(s) = \int_0^\infty x^{s-1}\mathrm{e}^{-x}\mathrm{d}x\,, \tag{6.124}$$

$$\Gamma(s+1) = s\Gamma(s)\,, \tag{6.125}$$

$$\Gamma(1) = \Gamma(2) = 1;\ \Gamma(\frac{1}{2}) = \sqrt{\pi},\ \Gamma(\frac{3}{2}) = \frac{\sqrt{\pi}}{2},\ \Gamma(\frac{5}{2}) = \frac{3\sqrt{\pi}}{4}\,. \tag{6.126}$$

6.3 Scattering Probability and Transition Matrix Element

6.3.1 Transition Matrix Element

The transition probability is calculated with the help of quantum mechanics as follows. Let us define the unperturbed Hamiltonian by H_0, the time-dependent perturbing Hamiltonian by H_1 and the total Hamiltonian by H.

$$H = H_0 + H_1\,. \tag{6.127}$$

The eigenstates of the unperturbed Hamiltonian are assumed to be

$$H_0|\boldsymbol{k}\rangle = \mathcal{E}(\boldsymbol{k})|\boldsymbol{k}\rangle\,. \tag{6.128}$$

The scattering rate for electron scattered from an initial state $|\boldsymbol{k}\rangle$ to a final state $|\boldsymbol{k}'\rangle$ is then given by the following relation according to quantum mechanics:

$$P(\boldsymbol{k}, \boldsymbol{k}') = \frac{2\pi}{\hbar}|\langle \boldsymbol{k}'|H_1|\boldsymbol{k}\rangle|^2 \delta\left[\mathcal{E}_{k'} - \mathcal{E}_{k}\right], \tag{6.129}$$

where $\mathcal{E}_k$ and $\mathcal{E}_{k'}$ are the energies of the initial and final states including the perturbation state and ensures the δ-function indicate energy conservation. This relation can also be written as follows. The scattering rate w is calculated by taking all the possible final states into account and is written as

$$\begin{aligned} w &= \frac{2\pi}{\hbar}\sum_{\boldsymbol{k}'}|\langle \boldsymbol{k}'|H_1|\boldsymbol{k}\rangle|^2 \delta\left[\mathcal{E}_{k'} - \mathcal{E}_{k}\right] \\ &= \frac{2\pi}{\hbar}\frac{L^3}{(2\pi)^3}\int \mathrm{d}^3\boldsymbol{k}'|\langle \boldsymbol{k}'|H_1|\boldsymbol{k}\rangle|^2 \delta\left[\mathcal{E}_{k'} - \mathcal{E}_{k}\right] .7 \end{aligned} \tag{6.130}$$

Next we consider electron scattering by phonons. Using (6.58) we obtain

$$H = H_\mathrm{e} + H_\mathrm{l} + H_\mathrm{el}\,, \tag{6.131}$$

$$H_\mathrm{e}|\boldsymbol{k}\rangle = \mathcal{E}(\boldsymbol{k})|\boldsymbol{k}\rangle\,, \tag{6.132}$$

$$H_\mathrm{l}|n_{\boldsymbol{q}}\rangle = \hbar\omega_{\boldsymbol{q}}\left(n_{\boldsymbol{q}} + \frac{1}{2}\right)|n_{\boldsymbol{q}}\rangle\,, \tag{6.133}$$

where H_el is the electron–phonon interaction Hamiltonian. Then the scattering probability for electrons from $|\boldsymbol{k}\rangle$ to $|\boldsymbol{k}'\rangle$ states is written as

$$P(\boldsymbol{k}, \boldsymbol{k}') = \frac{2\pi}{\hbar}|\langle \boldsymbol{k}', \boldsymbol{q}'|H_\mathrm{el}|\boldsymbol{k}, \boldsymbol{q}\rangle|^2 \delta[\mathcal{E}(\boldsymbol{k}') - \mathcal{E}(\boldsymbol{k}) \mp \hbar\omega_{\boldsymbol{q}}]\,, \tag{6.134}$$

where $|n_{\boldsymbol{q}}\rangle$ is written as $|\boldsymbol{q}\rangle$ for simplicity. It should be noted that the phonon energy $\hbar\omega_{\boldsymbol{q}}$ involved with phonon emission or absorption is included in the δ-function.

From the Boltzmann transport equation and the principle of detailed balance, or from (6.94), the relaxation time is given by

$$\begin{aligned} \frac{1}{\tau(\boldsymbol{k})} &= \sum_{\boldsymbol{k}'} P(\boldsymbol{k}, \boldsymbol{k}')\left(1 - \frac{k_x'}{k_x}\right) = \sum_{\boldsymbol{k}'} P(\boldsymbol{k}, \boldsymbol{k}')(1 - \cos\theta) \\ &= \frac{L^3}{(2\pi)^3}\int \mathrm{d}^3\boldsymbol{k}' P(\boldsymbol{k}, \boldsymbol{k}')(1 - \cos\theta)\,, \end{aligned} \tag{6.135}$$

where the volume V is replaced by L^3 in the final relation. As defined previously, $P(\boldsymbol{k}, \boldsymbol{k}')$ is the transition rate of electrons from state $\boldsymbol{k}$ to state $\boldsymbol{k}'$, and θ is the angle between $\boldsymbol{k}$ and $\boldsymbol{k}'$. The transition rate $P(\boldsymbol{k}, \boldsymbol{k}')$ is

$$P(\boldsymbol{k}, \boldsymbol{k}') = \frac{2\pi}{\hbar}|\langle \boldsymbol{k}', \boldsymbol{q}'|V_s|\boldsymbol{k}, \boldsymbol{q}\rangle|^2 \delta[\mathcal{E}_{\boldsymbol{k}',\boldsymbol{q}'} - \mathcal{E}_{\boldsymbol{k},\boldsymbol{q}}]\,, \tag{6.136}$$

where V_s is the scattering potential and is equivalent to the perturbation Hamiltonian H_1 defined above, and $|\boldsymbol{k}, \boldsymbol{q}\rangle$ is given by the product of the electron wave function $|\boldsymbol{k}\rangle$ with wave vector $\boldsymbol{k}$ and the wave function for the scattering center, which is a phonon $|\boldsymbol{q}\rangle$ with wave vector $\boldsymbol{q}$. Therefore, the relaxation time is given by

$$\frac{1}{\tau(\boldsymbol{k})} = \frac{2\pi}{\hbar}\sum_{\boldsymbol{k}'}|\langle\boldsymbol{k}',\boldsymbol{q}'|V|\boldsymbol{k},\boldsymbol{q}\rangle|^2(1-\cos\theta)\delta[\mathcal{E}_{\boldsymbol{k}',\boldsymbol{q}'}-\mathcal{E}_{\boldsymbol{k},\boldsymbol{q}}] \tag{6.137}$$

$$= \frac{L^3}{(2\pi)^2\hbar}\int \mathrm{d}^3\boldsymbol{k}'|\langle\boldsymbol{k}',\boldsymbol{q}'|V|\boldsymbol{k},\boldsymbol{q}\rangle|^2(1-\cos\theta)\delta[\mathcal{E}_{\boldsymbol{k}',\boldsymbol{q}'}-\mathcal{E}_{\boldsymbol{k},\boldsymbol{q}}]\,.$$

In the following we use the relation

$$\sum_{\boldsymbol{k}'} = \frac{L^d}{(2\pi)^d}\int \mathrm{d}^d\boldsymbol{k}'\,. \tag{6.138}$$

Here we will describe a very convenient method to derive the matrix element for scattering, where we use the Fourier transform shown in Appendix A.3. This is done by using cyclic boundary conditions as follows. Let us consider a semiconductor of length L and dimension d. The Fourier transform is written as

$$V_{\mathrm{s}}(\boldsymbol{q}) = \frac{1}{L^d}\int \mathrm{d}^d\boldsymbol{r}\ V_{\mathrm{s}}(\boldsymbol{r})\mathrm{e}^{-\mathrm{i}\boldsymbol{q}\cdot\boldsymbol{r}}\,, \tag{6.139a}$$

$$V_{\mathrm{s}}(\boldsymbol{r}) = \sum_{\boldsymbol{q}} V_{\mathrm{s}}(\boldsymbol{q})\mathrm{e}^{\mathrm{i}\boldsymbol{q}\cdot\boldsymbol{r}}\,. \tag{6.139b}$$

Next we use the relation

$$\frac{1}{L^d}\int \mathrm{d}^d\boldsymbol{r}\ \exp[\mathrm{i}(\boldsymbol{q}-\boldsymbol{q}')\cdot\boldsymbol{r}] = \delta_{\boldsymbol{q},\boldsymbol{q}'}\,, \tag{6.140}$$

of which proof is given in A.2 of Appendix A. Then the matrix element is expressed as

$$\begin{aligned}\langle\boldsymbol{k}',\boldsymbol{q}'|V_{\mathrm{s}}(\boldsymbol{r})|\boldsymbol{k},\boldsymbol{q}\rangle &= \sum_{\boldsymbol{q}''}\langle\boldsymbol{q}'|V_{\mathrm{s}}(\boldsymbol{q}'')|\boldsymbol{q}\rangle\langle\boldsymbol{k}'|\mathrm{e}^{\mathrm{i}\boldsymbol{q}''\cdot\boldsymbol{r}}|\boldsymbol{k}\rangle \\ &= \sum_{\boldsymbol{q}''}\langle\boldsymbol{q}'|V_{\mathrm{s}}(\boldsymbol{q}'')|\boldsymbol{q}\rangle\delta_{\boldsymbol{k}'-\boldsymbol{k},\boldsymbol{q}''} \\ &= \langle\boldsymbol{q}'|V_{\mathrm{s}}(\boldsymbol{k}'-\boldsymbol{k})|\boldsymbol{q}\rangle\,. \end{aligned} \tag{6.141}$$

The above equation is derived by using the electron wave function

$$|\boldsymbol{k}\rangle = \sqrt{\frac{1}{L^3}}\exp(\mathrm{i}\boldsymbol{k}\cdot\boldsymbol{r})\,. \tag{6.142}$$

If we use the Bloch function

$$|\boldsymbol{k}\rangle = \sqrt{\frac{1}{L^3}}u_{\boldsymbol{k}}(\boldsymbol{r})\exp(\mathrm{i}\boldsymbol{k}\cdot\boldsymbol{r}) \tag{6.143}$$

for the electron, then we obtain following result. When we use the cyclic boundary conditions for the Bloch function, the integral is decomposed into the integral in a unit cell and the summation over the unit cells.

$$\begin{aligned}
&\langle \boldsymbol{k}'|\exp(\mathrm{i}\boldsymbol{q}''\cdot\boldsymbol{r})|\boldsymbol{k}\rangle \\
&= \frac{1}{N\Omega}\int \mathrm{d}^3\boldsymbol{r}u^*_{\boldsymbol{k}'}(\boldsymbol{r})u_{\boldsymbol{k}}(\boldsymbol{r})\exp\left[\mathrm{i}(\boldsymbol{k}-\boldsymbol{k}'+\boldsymbol{q}'')\cdot\boldsymbol{r}\right] \\
&= \frac{1}{N}\sum_j \exp\left[\mathrm{i}(\boldsymbol{k}-\boldsymbol{k}'+\boldsymbol{q}'')\cdot\boldsymbol{R}_j\right] \\
&\quad\times\frac{1}{\Omega}\int_\Omega u^*_{\boldsymbol{k}'}(\boldsymbol{r})u_{\boldsymbol{k}}(\boldsymbol{r})\exp\left[\mathrm{i}(\boldsymbol{k}-\boldsymbol{k}'+\boldsymbol{q}'')\cdot\boldsymbol{r}\right]\mathrm{d}^3\boldsymbol{r} \\
&= \delta_{\boldsymbol{k}'-\boldsymbol{k},\boldsymbol{q}''}\frac{1}{\Omega}\int_\Omega u^*_{\boldsymbol{k}'}(\boldsymbol{r})u_{\boldsymbol{k}}(\boldsymbol{r})\mathrm{d}^3\boldsymbol{r} \\
&= I(\boldsymbol{k},\boldsymbol{k}')\delta_{\boldsymbol{k}'-\boldsymbol{k},\boldsymbol{q}''}\,. \qquad (6.144)
\end{aligned}$$

In the above equations the summation with respect to j is non zero only when $\boldsymbol{k}-\boldsymbol{k}'+\boldsymbol{q}''+\boldsymbol{G}=0$ ($\boldsymbol{G}$: reciprocal lattice vector), and thus the Kronecker δ-function should be written as $\delta_{\boldsymbol{k}'-\boldsymbol{k},\boldsymbol{q}''+\boldsymbol{G}}$. Here, however, we assume that the **Umklapp process** of $\boldsymbol{G}\neq 0$ (Umklapp process) does not play in a part in the scattering process and that the **normal process** of $\boldsymbol{G}=0$ takes part in the scattering process. In addition, we have

$$I(\boldsymbol{k},\boldsymbol{k}') = \frac{1}{\Omega}\int_\Omega u^*_{\boldsymbol{k}'}(\boldsymbol{r})u_{\boldsymbol{k}}(\boldsymbol{r})\mathrm{d}^3\boldsymbol{r} \simeq 1\,, \qquad (6.145)$$

where $\boldsymbol{R}_j$ is a lattice vector, N is the number of unit cells and Ω is the volume of a unit cell.

In the following, we will calculate the matrix elements for several scattering processes which play an important role in transport in semiconductors. It should be noted that notations such as $|\langle\boldsymbol{k}'|H_1|\boldsymbol{k}\rangle|$, $|V(\boldsymbol{k}'-\boldsymbol{k})|$ and $|M(\boldsymbol{k},\boldsymbol{k}')|$ are used in this book and that all of them represent the matrix elements for the scattering centers.

6.3.2 Deformation Potential Scattering (Acoustic Phonon Scattering)

The deformation potential theory has been proposed by Bardeen and Shockley, which is based on the energy change of an electron due to lattice deformation. The energy change is related to the volume change of a crystal, $\Delta(\boldsymbol{r})=\delta V/V$ (V: volume), and the electron–phonon interaction Hamiltonian H_{el} is defined by

$$H_{\mathrm{el}} = D_{\mathrm{ac}}\frac{\delta V}{V} = D_{\mathrm{ac}}\,\mathrm{div}\,\boldsymbol{u}(\boldsymbol{r})\,, \qquad (6.146)$$

where D_{ac} is called the **deformation potential** for electron scattering by acoustic phonons. The displacement vector of an acoustic phonon $\boldsymbol{u}(\boldsymbol{r})$ has been derived in Sect. 6.1 and is given by

$$\boldsymbol{u}(\boldsymbol{r}) = \sum_{q} \sqrt{\frac{\hbar}{2MN\omega_q}} \boldsymbol{e}_q \left[a_q \mathrm{e}^{\mathrm{i}\boldsymbol{q}\cdot\boldsymbol{r}} + a_q^\dagger \mathrm{e}^{-\mathrm{i}\boldsymbol{q}\cdot\boldsymbol{r}} \right], \tag{6.147}$$

where $\boldsymbol{e}_q$ is the unit vector along the displacement direction. This result leads to the following relation:

$$H_{\mathrm{el}} = D_{\mathrm{ac}} \sum_{q} \sqrt{\frac{\hbar}{2MN\omega_q}} (\mathrm{i}\boldsymbol{e}_q \cdot \boldsymbol{q}) \left[a_q \mathrm{e}^{\mathrm{i}\boldsymbol{q}\cdot\boldsymbol{r}} - a_q^\dagger \mathrm{e}^{-\mathrm{i}\boldsymbol{q}\cdot\boldsymbol{r}} \right]. \tag{6.148}$$

Next, we calculate the matrix element for this interaction Hamiltonian. Here we replace $|\boldsymbol{q}\rangle$ by $|n_{\boldsymbol{q}}\rangle$ and use the following relations:

$$a_{\boldsymbol{q}} |n_{\boldsymbol{q}}\rangle = \sqrt{n_{\boldsymbol{q}}} |n_{\boldsymbol{q}} - 1\rangle, \tag{6.149a}$$

$$a_{\boldsymbol{q}}^\dagger |n_{\boldsymbol{q}}\rangle = \sqrt{n_{\boldsymbol{q}} + 1} |n_{\boldsymbol{q}} + 1\rangle. \tag{6.149b}$$

Then the matrix element is given by

$$\langle \boldsymbol{k}', n_{\boldsymbol{q}'} | H_{\mathrm{el}} | \boldsymbol{k}, n_{\boldsymbol{q}} \rangle = D_{\mathrm{ac}} \sum_{q} \sqrt{\frac{\hbar}{2MN\omega_q}} (\mathrm{i}\boldsymbol{e}_q \cdot \boldsymbol{q}) \sqrt{n_{\boldsymbol{q}} + \frac{1}{2} \mp \frac{1}{2}} \times \frac{1}{L^3} \int \mathrm{d}^3\boldsymbol{r}\, u_{\boldsymbol{k}'}^*(\boldsymbol{r}) u_{\boldsymbol{k}}(\boldsymbol{r}) \mathrm{e}^{\mathrm{i}(\boldsymbol{k}-\boldsymbol{k}'\pm\boldsymbol{q})\cdot\boldsymbol{r}}. \tag{6.150}$$

Now, we replace the integral over the whole of space with a summation of unit cell integrations, and we obtain

$$\langle \boldsymbol{k}', n_{\boldsymbol{q}'} | H_{\mathrm{el}} | \boldsymbol{k}, n_{\boldsymbol{q}} \rangle = D_{\mathrm{ac}} \sum_{\boldsymbol{q}} \sqrt{\frac{\hbar}{2MN\omega_q}} (\mathrm{i}\boldsymbol{e}_q \cdot \boldsymbol{q}) \sqrt{n_{\boldsymbol{q}} + \frac{1}{2} \pm \frac{1}{2}} I(\boldsymbol{k}, \boldsymbol{k}') \delta_{\boldsymbol{k}',\boldsymbol{k}\pm\boldsymbol{q}}, \tag{6.151}$$

$$I(\boldsymbol{k}, \boldsymbol{k}') = \frac{1}{\Omega} \int_{\Omega} \mathrm{d}^3\boldsymbol{r}\, u_{\boldsymbol{k}'}^*(\boldsymbol{r}) u_{\boldsymbol{k}}(\boldsymbol{r}), \tag{6.152}$$

where the summation with respect to $\boldsymbol{q}$ is simplified by using the relation $\delta_{\boldsymbol{k}',\boldsymbol{k}\pm\boldsymbol{q}}$, where we take into account the normal process only.

Here we will show a more convenient method to derive the above result by using the Fourier transform. When we insert the Fourier transform of H_{el} into (6.141) we obtain the following result directly. Defining

$$H_{\mathrm{el}}(\boldsymbol{r}) = \sum_{\boldsymbol{q}} \left[C(\boldsymbol{q}) a_{\boldsymbol{q}} \mathrm{e}^{\mathrm{i}\boldsymbol{q}\cdot\boldsymbol{r}} + C^\dagger(\boldsymbol{q}) a_{\boldsymbol{q}}^\dagger \mathrm{e}^{-\mathrm{i}\boldsymbol{q}\cdot\boldsymbol{r}} \right], \tag{6.153}$$

$$C(\boldsymbol{q}) = D_{\mathrm{ac}} \sqrt{\frac{\hbar}{2MN\omega_q}} (\mathrm{i}\boldsymbol{e}_q \cdot \boldsymbol{q}), \tag{6.154}$$

we obtain the following relation easily:

$$
\begin{aligned}
H_{\text{el}}(\boldsymbol{q}) &= \frac{1}{L^3}\int \mathrm{d}^3\boldsymbol{r}\; H_{\text{el}}(\boldsymbol{r})\mathrm{e}^{-\mathrm{i}\boldsymbol{q}\cdot\boldsymbol{r}} \\
&= \sum_{\boldsymbol{q}'}\left[C(\boldsymbol{q}')a_{\boldsymbol{q}'}\frac{1}{L^3}\int \mathrm{d}^3\boldsymbol{r}\;\mathrm{e}^{\mathrm{i}(\boldsymbol{q}'-\boldsymbol{q})\cdot\boldsymbol{r}} + C^\dagger(\boldsymbol{q}')a^\dagger_{\boldsymbol{q}'}\frac{1}{L^3}\int \mathrm{d}^3\boldsymbol{r}\;\mathrm{e}^{-\mathrm{i}(\boldsymbol{q}'+\boldsymbol{q})\cdot\boldsymbol{r}}\right] \\
&= \sum_{\boldsymbol{q}'}\left[C(\boldsymbol{q}')a_{\boldsymbol{q}'}\delta_{\boldsymbol{q}',\boldsymbol{q}} + C^\dagger(\boldsymbol{q}')a^\dagger_{\boldsymbol{q}'}\delta_{\boldsymbol{q}',-\boldsymbol{q}}\right] \\
&= C(\boldsymbol{q})a_{\boldsymbol{q}} + C^\dagger(-\boldsymbol{q})a^\dagger_{-\boldsymbol{q}}\,.
\end{aligned}
\tag{6.155}
$$

Inserting this relation into (6.141) we obtain

$$
\begin{aligned}
&\langle n_{\boldsymbol{q}'}|H_{\text{el}}(\boldsymbol{k}'-\boldsymbol{k})|n_{\boldsymbol{q}}\rangle \\
&\quad= \langle n_{\boldsymbol{q}'}|C(\boldsymbol{k}'-\boldsymbol{k})a_{\boldsymbol{k}'-\boldsymbol{k}} + C^\dagger(\boldsymbol{k}-\boldsymbol{k}')a^\dagger_{\boldsymbol{k}-\boldsymbol{k}'}|n_{\boldsymbol{q}}\rangle \\
&\quad= \begin{cases} C(\boldsymbol{q})\sqrt{n_{\boldsymbol{q}}} & (\boldsymbol{k}'=\boldsymbol{k}+\boldsymbol{q};\ \text{absorption}), \\ C^\dagger(\boldsymbol{q})\sqrt{n_{\boldsymbol{q}}+1} & (\boldsymbol{k}'=\boldsymbol{k}-\boldsymbol{q};\ \text{emission})\,. \end{cases}
\end{aligned}
\tag{6.156}
$$

6.3.3 Ionized Impurity Scattering

6.3.3.1 Brooks–Herring Formula

The Coulomb potential due to an i-th point charge ze is expressed in real space as

$$
V_i(r) = \frac{ze^2}{4\pi\epsilon r}\,,
\tag{6.157}
$$

where the dielectric constant ϵ is used to take account of the electronic polarization and ionic polarization due to lattice vibrations. The contribution of valence electrons to the dielectric constant will be discussed later. The Fourier transform of the potential gives $V_i(\boldsymbol{q})$ and the calculation of scattering rate (relaxation time) is straightforward:

$$
V_i(\boldsymbol{q}) = \frac{ze^2}{L^3\epsilon}\frac{1}{q^2}\,.
\tag{6.158}
$$

Although the derivation of the above equation is simple, we consider the screened Coulomb potential for the purpose of calculation and generality, and assume a distribution of ionic charge N_{I} given by

$$
\rho_{\text{I}}(\boldsymbol{r}) = ze\sum_{i=1}^{N_{\text{I}}}\delta(\boldsymbol{r}-\boldsymbol{r}_i)\,,
\tag{6.159}
$$

where $\boldsymbol{r}_i$ is the position vector of an ionized impurity. The Coulomb potential of the ionized impurities is expressed as

$$V(\boldsymbol{r}) = \sum_{i=1}^{N_I} V_i(\boldsymbol{r} - \boldsymbol{r}_i) . \tag{6.160}$$

We will discuss in Sect. 6.3.10 how to take account of the screening, and here we use the result. The screened Coulomb potential $V(\boldsymbol{r} - \boldsymbol{r}_i)$ is given by

$$V(\boldsymbol{r} - \boldsymbol{r}_i) = \frac{ze^2}{4\pi\epsilon} \frac{e^{-q_s|\boldsymbol{r}-\boldsymbol{r}_i|}}{|\boldsymbol{r} - \boldsymbol{r}_i|} , \tag{6.161}$$

where $\boldsymbol{q}_s$ is the inverse of the Debye screening length λ_s and is given by

$$q_s = \frac{1}{\lambda_s} = \left(\frac{nze^2}{\epsilon k_B T}\right)^{1/2} . \tag{6.162}$$

In the above equation n is the density of electrons in the conduction band. The Fourier transform of the Coulomb potential is calculated as follows. For simplicity a single point charge is assumed and we define the potential as

$$V(r) = C \frac{e^{-q_s r}}{r} . \tag{6.163}$$

The Fourier transform is therefore given by

$$\begin{aligned} V(q) &= \frac{1}{L^3} \int V(r) e^{-i\boldsymbol{q}\cdot\boldsymbol{r}} d^3\boldsymbol{r} \\ &= \frac{1}{L^3} \frac{4\pi C}{q^2 + q_s^2} . \end{aligned} \tag{6.164}$$

From this relation we have

$$V(\boldsymbol{q}) = V(q) = \frac{ze^2}{\epsilon L^3} \frac{1}{q^2 + q_s^2} . \tag{6.165}$$

When we put $q_s = 0$ in the above equation, (6.158) is obtained.

From these results the Coulomb potential of ionized impurities N_I is Fourier transformed to give

$$V(\boldsymbol{q}) = \sum_{i=1}^{N_I} \frac{ze^2}{\epsilon L^3} \frac{1}{q^2 + q_s^2} e^{-i\boldsymbol{q}\cdot\boldsymbol{r}_i} . \tag{6.166}$$

Relaxation time is calculated by inserting (6.166) into (6.138) or (6.141) and the squared matrix element of ionized impurity scattering is given by

$$|V(\boldsymbol{k}'-\boldsymbol{k})|^2 = \left(\frac{ze^2}{\epsilon L^3}\right)^2 \frac{1}{(|\boldsymbol{k}'-\boldsymbol{k}|^2+q_s^2)^2}\sum_{i=1}^{N_I}\sum_{j=1}^{N_I}\delta_{ij}$$
$$= N_I\left(\frac{ze^2}{\epsilon L^3}\right)^2 \frac{1}{(|\boldsymbol{k}'-\boldsymbol{k}|^2+q_s^2)^2}\,. \tag{6.167}$$

Therefore, we obtain the relaxation time by inserting this into (6.138), resulting in

$$\frac{1}{\tau_I(\boldsymbol{k})} = \frac{L^3}{(2\pi)^2\hbar}\int d^3\boldsymbol{k}' N_I\left(\frac{ze^2}{\epsilon L^3}\right)^2 \frac{1}{(|\boldsymbol{k}'-\boldsymbol{k}|^2+q_s^2)^2}(1-\cos\theta)$$
$$\times\delta\left[\mathcal{E}_{\boldsymbol{k}'}-\mathcal{E}_{\boldsymbol{k}}\right]\,. \tag{6.168}$$

When we define the impurity density by $n_I = N_I/L^3$, the relaxation time reduces to

$$\frac{1}{\tau_I(\boldsymbol{k})} = \frac{n_I}{(2\pi)^2\hbar}\int d^3\boldsymbol{k}'\left(\frac{ze^2}{\epsilon}\right)^2 \frac{1}{(|\boldsymbol{k}'-\boldsymbol{k}|^2+q_s^2)^2}(1-\cos\theta)\delta\left[\mathcal{E}_{\boldsymbol{k}'}-\mathcal{E}_{\boldsymbol{k}}\right]\,. \tag{6.169}$$

The electron energy is defined by

$$\mathcal{E}_k \equiv \mathcal{E}(\boldsymbol{k}) = \frac{\hbar^2k^2}{2m^*} \tag{6.170}$$

and the integral is carried out in spherical $\boldsymbol{k}$ space:

$$d^3\boldsymbol{k}' = 2\pi k'^2 dk' \sin\theta d\theta\,, \tag{6.171}$$

where θ is the angle between $\boldsymbol{k}$ and $\boldsymbol{k}'$, and $|\boldsymbol{k}'-\boldsymbol{k}| = 2k\sin(\theta/2)$. Using the δ-function we find

$$\int k'^2 dk'\delta\left[\frac{\hbar^2k'^2}{2m^*}-\frac{\hbar^2k^2}{2m^*}\right] = \frac{2m^*}{\hbar^2}\int k'^2dk'\delta[(|\boldsymbol{k}'|-|\boldsymbol{k}|)(|\boldsymbol{k}'|+|\boldsymbol{k}|)]$$
$$= \frac{2m^*}{\hbar^2}\int \frac{k'^2}{2|\boldsymbol{k}|}dk'\delta[(|\boldsymbol{k}'|-|\boldsymbol{k}|)$$
$$= \frac{m^*}{\hbar^2}|\boldsymbol{k}|\,, \tag{6.172}$$

and (6.169) reduces to

$$\frac{1}{\tau_I(\boldsymbol{k})} = \frac{n_I m^* k}{2\pi\hbar^3}\left(\frac{ze^2}{\epsilon}\right)^2 I(\boldsymbol{k})\,, \tag{6.173}$$

where

$$I(\boldsymbol{k}) = \int_0^\pi\left[\frac{1}{\{2k\sin^2(\theta/2)\}^2+q_s^2}\right]^2(1-\cos\theta)\sin\theta d\theta$$
$$= \frac{1}{4k^4}\left\{\log[1+(2k\lambda_s)^2]-\frac{(2k\lambda_s)^2}{1+(2k\lambda_s)^2}\right\}\,. \tag{6.174}$$

Therefore, the relaxation time for ionized impurity scattering is given by the following equation:

$$\frac{1}{\tau_{\mathrm{I}}(\mathcal{E})} = \frac{z^2 e^4 n_{\mathrm{I}}}{16\pi\epsilon^2\sqrt{2m^*}}\mathcal{E}^{-3/2}\left[\log\left(1+\frac{8m^*\lambda_{\mathrm{s}}^2\mathcal{E}}{\hbar^2}\right) - \frac{8m^*\lambda_{\mathrm{s}}^2\mathcal{E}/\hbar^2}{1+(8m^*\lambda_{\mathrm{s}}^2\mathcal{E}/\hbar^2)}\right]. \tag{6.175}$$

The above equation is called the Brooks–Herring formula.

Here we will show that the Fourier transform of a 2-dimensional Coulomb potential is given by

$$V(q) = \frac{e^2}{2\epsilon L^2}\frac{1}{q}. \tag{6.176}$$

The Fourier transform of a 2-dimensional potential $V(r)$ is defined by

$$V(q) = \frac{1}{L^2}\int V(r)\mathrm{e}^{\mathrm{i}\boldsymbol{q}\cdot\boldsymbol{r}}\mathrm{d}^2\boldsymbol{r}. \tag{6.177}$$

When we express the 2-dimensional potential as

$$V(r) = \frac{e^2}{4\pi\epsilon r}, \tag{6.178}$$

then we have

$$V(q) = \frac{e^2}{4\pi\epsilon L^2}\int_0^\infty \mathrm{d}r\int_0^{2\pi}\mathrm{d}\phi\,\mathrm{e}^{\mathrm{i}qr\cos\phi} = \frac{e^2}{2\epsilon L^2 q}\int_0^\infty \mathrm{d}(qr)J_0(qr), \tag{6.179}$$

where $J_0(x)$ is the Bessel function of order zero and

$$\int_0^\infty J_0(x)\mathrm{d}x = 1. \tag{6.180}$$

Therefore, the Fourier transform of the 2-dimensional Coulomb potential is given by

$$V(q) = \frac{e^2}{2\epsilon L^2}\frac{1}{q}. \tag{6.181}$$

6.3.3.2 Conwell–Weisskopf Formula

Conwell and Weisskopf [1] have derived the relaxation time for ionized impurity scattering by adopting Rutherford scattering. Let us define the scattering cross-section of a single impurity by A. In a semiconductor with impurity density n_{I} and with electron velocity $v = (\partial\mathcal{E}/\partial k)/\hbar$, the relaxation time (collision time) of an electron

τ_I is given by $1/\tau_I = n_I v A$. The differential cross-section $\sigma(\theta, \phi)$ is defined by the probability of scattering into a small solid angle $d\Omega = \sin\theta d\theta d\phi$, and in the case of isotropic scattering the cross-section is given by $A = \int \sigma(\theta, \phi) \sin\theta d\theta d\phi$. Assuming elastic scattering and defining the angle θ between the electron wave vectors $\boldsymbol{k}$ and $\boldsymbol{k}'$, we obtain

$$\frac{1}{\tau_{CW}} = n_I v \int_{\phi=0}^{2\pi} \int_{\theta=0}^{\pi} \sigma(\theta, \phi)(1 - \cos\theta) \sin\theta d\theta d\phi$$
$$= 2\pi n_I v \int_0^{\pi} \sigma(\theta)(1 - \cos\theta) \sin\theta d\theta \,. \tag{6.182}$$

Neglecting the screening effect due to conduction electrons, the potential induced by an ionized impurity is given by the Rutherford scattering cross-section:

$$\sigma(\theta) = \frac{1}{4} R^2 \mathrm{cosec}^4\left(\frac{\theta}{2}\right), \quad R = \frac{ze^2}{4\pi\epsilon m^* v^2} \,. \tag{6.183}$$

In a semiconductor with many ionized impurities, the effect of an impurity on a scattering event seems to disappear between the neighboring two impurities. This assumption enables us to cut off the effect of a scattering potential at r_m, where r_m is given by the relation $(2r_m)^{-3} = n_I$. In other words, an electron affected by an impurity will not be scattered by the impurity when the electron is separated from the impurity by a distance r_m. Under this assumption the cut-off angle θ_m, shown in Fig. 6.7, is given by

$$\tan(\theta_m/2) = \frac{R}{r_m} \tag{6.184}$$

and the integral with respect to θ is cut off at the angle θ_m, $(\theta_m < \theta < \pi)$, giving rise to the final result:

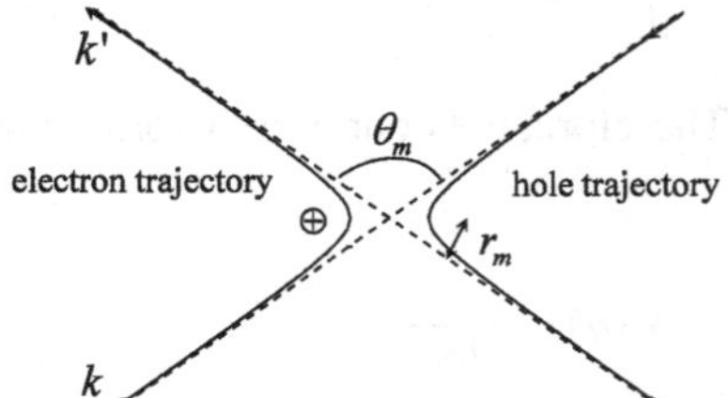

Fig. 6.7 Scattering of electron and hole by an ionized impurity and the cut-off angle θ_m

$$\frac{1}{\tau_{\mathrm{CW}}} = -4\pi n_{\mathrm{I}} v R^2 \ln\left(\sin\frac{\theta_{\mathrm{m}}}{2}\right) = 2\pi n_{\mathrm{I}} v R^2 \ln\left(1 + \frac{r_{\mathrm{m}}}{R^2}\right)$$
$$= \frac{z^2 e^4 n_{\mathrm{I}}}{16\pi\epsilon^2\sqrt{2m^*}} \mathcal{E}^{-3/2} \ln\left[1 + \left(\frac{2\mathcal{E}}{\mathcal{E}_{\mathrm{m}}}\right)^2\right] \tag{6.185}$$

$$\frac{2\mathcal{E}}{\mathcal{E}_{\mathrm{m}}} = \frac{r_{\mathrm{m}}}{R}, \quad \mathcal{E}_{\mathrm{m}} = \frac{ze^2}{4\pi\epsilon r_{\mathrm{m}}}. \tag{6.186}$$

This equation is referred to as the **Conwell–Weisskopf formula** for ionized impurity scattering [1].

6.3.4 *Piezoelectric Potential Scattering*

When a strain S_{kl} is applied to a crystal without inversion symmetry (except the cubic crystal O, class 432), a polarization $\boldsymbol{P}$ is induced. This phenomena is called piezoelectricity. In such a crystal, a potential associated with lattice vibrations is induced and thus the electrons are scattered by this piezoelectric potential. The matrix element for piezoelectric potential scattering is obtained as follows.

The i components of polarization $\boldsymbol{P}$ and electric displacement $\boldsymbol{D}$ in a piezoelectric crystal are written as

$$P_i = e_{ikl} S_{kl} \equiv e_{i\alpha} S_\alpha, \tag{6.187}$$

$$D_i = \epsilon^{\mathrm{s}}_{ij} E_j + P_i = \epsilon^{\mathrm{s}}_{ij} E_j + e_{ijk} S_{kl}$$
$$= \epsilon^{\mathrm{s}}_{ij} E_j + e_{i\alpha} S_\alpha, \tag{6.188}$$

where $\epsilon^{\mathrm{s}}_{ij}$ is the dielectric constant under a constant strain. $e_{ikl} = e_{i\alpha}$ is the piezoelectric constant tensor and it is rewritten as follows by using contraction of subscripts for the coordinate components i, j, k, l:

$$kl = 11\ 22\ 33\ 23, 32\ 13, 31\ 12, 21$$
$$\alpha = 1\ \ 2\ \ 3\ \ \ 4\ \ \ \ 5\ \ \ \ 6$$

On the other hand, the stress tensor $T_{ij} = T_\alpha$ and the strain tensor are related by Hooke's law, and in the absence of piezoelectricity the relation is given by $T_\alpha = c^E_{\alpha\beta} S_\beta$, where $c_{\alpha\beta}$ is elastic constant tensor. In the presence of piezoelectricity, strain and stress are induced by piezoelectricity and thus we obtain

$$T_\alpha = c_{\alpha\beta} S_\beta - e_{i\alpha} E_i\,. \tag{6.189}$$

Equations (6.188) and (6.189) are referred to as the **fundamental equations of piezoelectricity**. Let us define the displacement vector of an elastic wave by $\boldsymbol{u}$, the unit vector of the polarization direction by $\boldsymbol{\pi} = \boldsymbol{u}/|\boldsymbol{u}|$ and the unit vector of the wave vector $\boldsymbol{q}$ by $\boldsymbol{a} = \boldsymbol{q}/|\boldsymbol{q}|$. The elastic wave is then expressed as

$$\boldsymbol{u} = \boldsymbol{\pi} u \mathrm{e}^{\mathrm{i}(q\boldsymbol{a}\cdot\boldsymbol{r} - \omega_q t)} \tag{6.190}$$

and therefore we obtain following relations:

$$S_{kl} = \frac{1}{2}\left(\frac{\partial u_k}{\partial r_l} + \frac{\partial u_l}{\partial r_k}\right)$$
$$= \frac{1}{2}(\mathrm{i}q)(\pi_i a_j + \pi_j a_i)u = \mathrm{i}q\pi_k a_l u \,, \tag{6.191}$$

where the right-hand side is written by omitting the notation of summation for the components. This kind of simplification is often used in tensor equations.

Now we define the piezoelectric potential induced by lattice vibrations as ϕ_{pz}. Then the corresponding electric field $\boldsymbol{E}$ is given by $\boldsymbol{E} = -\mathrm{grad}\ \phi_{\mathrm{pz}}$ or by $E_l = -\mathrm{i}a_j q\phi_{\mathrm{pz}}$. From the relation div $\boldsymbol{D} = 0$ we have

$$\mathrm{div}\boldsymbol{D} = \partial D_i/\partial r_i = (\epsilon^{\mathrm{s}}_{ij} a_i a_j q^2 \phi_{\mathrm{pz}} - q^2 e_{ikl} a_i \pi_k a_l u) = 0 \tag{6.192}$$

and thus

$$\phi_{\mathrm{pz}} = \frac{e_{ikl} a_i \pi_k a_l}{a_i \epsilon^{s}_{ij} a_j} \cdot u \equiv \frac{e^*_{\mathrm{pz}}}{\epsilon^*} u \,, \tag{6.193}$$

where $e^*_{\mathrm{pz}} = e_{ikl} a_i \pi_k a_l$ and $\epsilon^* = a_i \epsilon^*_{ij} a_j$ are the effective piezoelectric constant and effective dielectric constant, respectively. The potential energy associated with the induced piezoelectric potential is given by $V_{\mathrm{pz}} = -e\phi_{\mathrm{pz}}$, which reduces to (dropping the sign for simplicity)

$$V_{\mathrm{pz}} = \frac{e e^*_{\mathrm{pz}}}{\epsilon^*} u \,. \tag{6.194}$$

Here we use the quantized displacement vector given by (6.68) and then the potential energy is rewritten as

$$V_{\mathrm{pz}}(\boldsymbol{r}) = \sum_{\boldsymbol{q}} \frac{e e^*_{\mathrm{pz}}}{\epsilon^*} \sqrt{\frac{\hbar}{2NM\omega_{\boldsymbol{q}}}} \left(a_{\boldsymbol{q}} \mathrm{e}^{\mathrm{i}\boldsymbol{q}\cdot\boldsymbol{r}} + a^\dagger_{\boldsymbol{q}} \mathrm{e}^{-\mathrm{i}\boldsymbol{q}\cdot\boldsymbol{r}}\right) . \tag{6.195}$$

The calculation of the matrix element and relaxation time is straightforward, as in the case of deformation potential scattering. When we rewrite the potential energy as

$$V_{\mathrm{pz}}(\boldsymbol{r}) = \sum_{\boldsymbol{q}} \left[C_{\mathrm{pz}}(\boldsymbol{q}) a_{\boldsymbol{q}} \mathrm{e}^{\mathrm{i}\boldsymbol{q}\cdot\boldsymbol{r}} + C^\dagger_{\mathrm{pz}}(\boldsymbol{q}) \mathrm{e}^{-\mathrm{i}\boldsymbol{q}\cdot\boldsymbol{r}}\right] , \tag{6.196}$$

$$C_{\mathrm{pz}}(\boldsymbol{q}) = \frac{e e^*_{\mathrm{pz}}}{\epsilon^*} \sqrt{\frac{\hbar}{2MN\omega_{\boldsymbol{q}}}} \,, \tag{6.197}$$

the Fourier transform gives rise to

$$\begin{aligned}
V_{\rm pz}(\boldsymbol{q}) &= \frac{1}{L^3}\int \mathrm{d}^3\boldsymbol{r}\, V_{\rm pz}(\boldsymbol{r})\mathrm{e}^{-\mathrm{i}\boldsymbol{q}\cdot\boldsymbol{r}} \\
&= \sum_{\boldsymbol{q}'}\left[C_{\rm pz}(\boldsymbol{q}')a_{\boldsymbol{q}'}\frac{1}{L^3}\int \mathrm{d}^3\boldsymbol{r}\mathrm{e}^{\mathrm{i}(\boldsymbol{q}'-\boldsymbol{q})\cdot\boldsymbol{r}}\right. \\
&\quad \left. + C^{\dagger}_{\rm pz}(\boldsymbol{q}')a^{\dagger}_{\boldsymbol{q}'}\frac{1}{L^3}\int \mathrm{d}^3\boldsymbol{r}\mathrm{e}^{-\mathrm{i}(\boldsymbol{q}'+\boldsymbol{q})\cdot\boldsymbol{r}}\right] \\
&= \sum_{\boldsymbol{q}'} C_{\rm pz}(\boldsymbol{q}')a_{\boldsymbol{q}'}\delta_{\boldsymbol{q}',\boldsymbol{q}} + C^{\dagger}_{\rm pz}(\boldsymbol{q}')a^{\dagger}_{\boldsymbol{q}'}\delta_{\boldsymbol{q}',-\boldsymbol{q}} \\
&= C_{\rm pz}(\boldsymbol{q})a_{\boldsymbol{q}} + C^{\dagger}_{\rm pz}(-\boldsymbol{q})a^{\dagger}_{-\boldsymbol{q}}\,. \qquad (6.198)
\end{aligned}$$

Inserting this equation into (6.141), the matrix element for piezoelectric potential scattering is given by

$$\begin{aligned}
&\langle n_{\boldsymbol{q}'}|V_{\rm pz}(\boldsymbol{k}'-\boldsymbol{k})|n_{\boldsymbol{q}}\rangle \\
&= \langle n_{\boldsymbol{q}'}|C_{\rm pz}(\boldsymbol{k}'-\boldsymbol{k})a_{\boldsymbol{k}'-\boldsymbol{k}} + C^{\dagger}_{\rm pz}(\boldsymbol{k}-\boldsymbol{k}')a^{\dagger}_{\boldsymbol{k}-\boldsymbol{k}'}|n_{\boldsymbol{q}}\rangle \\
&= \begin{cases} C_{\rm pz}(\boldsymbol{q})\sqrt{n_q} & (\boldsymbol{k}' = \boldsymbol{k}+\boldsymbol{q};\ \text{absorption}), \\ C^{\dagger}_{\rm pz}(\boldsymbol{q})\sqrt{n_q+1} & (\boldsymbol{k}' = \boldsymbol{k}-\boldsymbol{q};\ \text{emission}). \end{cases} \qquad (6.199)
\end{aligned}$$

As shown in Sect. 6.4.4 the scattering rates diverge when the electron energy approaches 0. Detailed treatment of the anisotropy of the piezoelectric potential are reported by Zook [2], Hutson [3], and Hutson et al.[4]. Since the scattering rate due to piezoelectric potential is quite small as shown later and its contribution to the electron mobility is negligible, we use the above approximation in this text book.

Screening effect is taken in account by using the similar approach to obtain Brooks–Herring formula for ionized impurity scattering as described in Sect. 6.3.3 [5]. Let's consider the case of a zinc blende crystal, where then non-zero piezoelectric constant is e_{14}, and all other components are zero. Then the two shear acoustic waves give rise piezoelectric potential perpendicular to the propagation direction. We redefine $C^{\rm sc}_{\rm pz}(\boldsymbol{q})$ for the electron–piezoelectric potential interaction by the following relation

$$\begin{aligned}
C^{\rm sc}_{\rm pz}(\boldsymbol{q}) &= \frac{e\cdot e_{14}}{\epsilon}\sqrt{\frac{\hbar}{2MN\omega_q}}\frac{q^2}{q^2+q_{\rm s}^2}\left[2\mathrm{i}(a_x\beta\gamma + a_y\gamma\alpha + \alpha\beta)\right] \\
&\equiv \sqrt{\frac{\hbar}{2MN\omega_q}}\left\langle\frac{e\cdot e^*_{\rm pz}}{\epsilon^*}\right\rangle\frac{2q^2}{q^2+q_{\rm s}^2}, \qquad (6.200)
\end{aligned}$$

where factor 2 arises from two shear acoustic modes, and α, β, and γ are the direction cosines with respect to the crystal axis and of the direction of propagation of the acoustic wave. Therefore we obtain the total scattering rate due to the piezoelectric potential $w_{\rm pz}$

$$w_{\rm pz} = \frac{2e^2\sqrt{m^*}k_{\rm B}T}{4\sqrt{2}\hbar^2}\left\langle\frac{e_{\rm pz}^{*2}}{c^*\epsilon^{*2}}\right\rangle\frac{1}{\sqrt{\mathcal{E}}}\int_0^\infty \frac{q^3}{(q^2+q_{\rm s}^2)^2}{\rm d}q$$
$$= \frac{e^2\sqrt{m^*}k_{\rm B}T}{4\sqrt{2}\hbar^2}\left\langle\frac{e_{\rm pz}^{2}}{c^*\epsilon^{*2}}\right\rangle\frac{1}{\sqrt{\mathcal{E}}}$$
$$\times\left[\log\left(1+\frac{8m^*\mathcal{E}}{\hbar^2 q_{\rm s}^2}\right)-\frac{1}{1+\hbar^2q_{\rm s}^2/8m^*\mathcal{E}}\right], \tag{6.201}$$

where factor 2 is multiplied by taking account of absorption and emission and $q_{\rm s}$ is the screening wave vector given by (6.162)

$$q_{\rm s} = \left(\frac{nze^2}{\epsilon k_{\rm B}T}\right)^{1/2}. \tag{6.202}$$

6.3.5 Non-polar Optical Phonon Scattering

As described in Sect. 6.1, in a crystal with two or more atoms in a unit cell lattice vibrations (optical phonon modes) occurs due to the relative displacement between the atoms in the unit cell and the potential induced by the relative displacement results in electron or hole scattering. The interaction potential is proportional to the relative displacement $\boldsymbol{u}$, where the relative displacement vector of optical phonon modes is derived in Sect. 6.1.2. Also as described in Sect. 5.2, the displacement tensor in diamond type (such as Ge and Si) and zinc blende type crystals (such as GaAs) belongs to the irreducible representation $\Gamma_{25'}$ (Γ_4). In the case of interaction of order 0, therefore, the matrix element disappears for electrons in the s-like conduction bands Γ_1 and $\Gamma_{2'}$ at the Γ point. In other words, interaction between electrons and non-polar optical phonons does not exist in the non-degenerate bands with extrema at the Γ point in the Brillouin zone. It is also known that interaction of order zero disappears in the conduction band minima (many–valley structure such as in Si) in the $\langle 100\rangle$ direction of the Brillouin zone and that the interaction is higher order, resulting in quite a small contribution to electron scattering. However, the interaction is known to be very strong in conduction band minima in the direction $\langle 111\rangle$ such as in Ge and in the valence bands of Si and Ge. The interaction Hamiltonian for non-polar optical phonon scattering is defined as

$$H_{\rm op} = D_{\rm op}\cdot\boldsymbol{u} = \tilde{D}_{\rm op}\cdot G\cdot\boldsymbol{u}, \tag{6.203}$$

where G is the magnitude of the reciprocal lattice vector and $D_{\rm op}$ is the deformation potential for non-polar optical phonon scattering. Since the optical phonon displacement vector $\boldsymbol{u}$ has the dimensions of length, $D_{\rm op}$ is multiplied by G to give the dimensions of energy for $\tilde{D}_{\rm op}$. Here we have to note that the Hamiltonian is also defined in some textbooks by $\tilde{D}_{\rm op}\cdot\boldsymbol{u}/a_0$ with lattice constant a_0. The matrix element for non-polar optical phonon scattering is easily calculated by using the relative displacement vector $\boldsymbol{u}$ of (6.70) derived in Sect. 6.1.2 as the following, where we put $\omega_q \to \omega_0$:

$$|M(\boldsymbol{k}, \boldsymbol{k}')| = \left(\frac{\hbar}{2NM_\mathrm{r}\omega_0}\right)^{1/2} D_\mathrm{op} \times \begin{cases} \sqrt{n_0(\omega_0)}\,, \\ \sqrt{n_0(\omega_0)+1}\,, \end{cases} \tag{6.204}$$

where $n_0(\omega_0)$ is the Bose–Einstein distribution function for optical phonons and $n_0(\omega_0) = 1/[\exp(\hbar\omega_0/k_\mathrm{B}T) - 1]$. In order to correlate the optical phonon deformation potential with acoustic deformation potential we introduce the new deformation potential constant

$$E_\mathrm{lop}^2 \equiv \frac{D_\mathrm{op}^2 v_\mathrm{s}^2}{\omega_0^2} \equiv \frac{\tilde{D}_\mathrm{op}^2 G^2 v_\mathrm{s}^2}{\omega_0^2}\,, \tag{6.205}$$

where v_s is the velocity of sound. In the above expression E_lop has the units of energy similar to the acoustic phonon deformation potential and is referred to as the optical phonon deformation potential. The definition of the new deformation potential has been proposed by Conwell [6] and is very useful to take account of non-polar optical phonon scattering.

6.3.6 Polar Optical Phonon Scattering

As stated in the previous section the lattice vibrations of a crystal with two or more atoms in a unit cell exhibit optical phonon modes due to the relative displacement of the atoms in the unit cell. For simplicity we consider a crystal with two atoms A and B in a unit cell and define the relative displacement vector by $\boldsymbol{u} = \boldsymbol{u}_\mathrm{A} - \boldsymbol{u}_\mathrm{B}$ with respective displacement vectors $\boldsymbol{u}_\mathrm{A}$ and $\boldsymbol{u}_\mathrm{B}$. From (5.126), the equation of motion is written as

$$M_\mathrm{r}\frac{\mathrm{d}^2}{\mathrm{d}t^2}\boldsymbol{u}(\boldsymbol{r}, t) = -M_\mathrm{r}\omega_0^2\boldsymbol{u}(\boldsymbol{r}, t) + e^*\boldsymbol{E}_\mathrm{loc}(\boldsymbol{r}, t) \tag{6.206}$$

for the relative motion of the atoms, where $1/M_\mathrm{r} = 1/M_\mathrm{A} + 1/M_\mathrm{B}$ is the reduced mass and e^* is the effective charge of the pair of atoms and $\boldsymbol{E}_\mathrm{loc}(\boldsymbol{r}, t)$ is the local electric field at position $\boldsymbol{r}$. The polarization associated with the lattice vibrations is given by

$$\boldsymbol{P}(\boldsymbol{r}, t) = Ne^*\boldsymbol{u}(\boldsymbol{r}, t) + N\alpha\boldsymbol{E}_\mathrm{loc}(\boldsymbol{r}, t)\,, \tag{6.207}$$

where the second term on the right-hand side is the contribution from the electronic polarization, N is the number of atomic pairs per unit volume, and α is the electronic polarization constant. In the following analysis we use the Fourier transform method described in Appendix A.3. From the Fourier transform of the electric field $\boldsymbol{E}(\boldsymbol{r},t)$ and polarization $\boldsymbol{P}(\boldsymbol{r},t)$ we obtain the following results:

$$\boldsymbol{P}(\boldsymbol{r},t) = \boldsymbol{P}(\boldsymbol{r})\mathrm{e}^{-\mathrm{i}\omega t}\,, \tag{6.208a}$$

$$\boldsymbol{P}(\boldsymbol{r}) = \frac{L^3}{(2\pi)^3}\int \boldsymbol{P}(\boldsymbol{q})\mathrm{e}^{\mathrm{i}\boldsymbol{q}\cdot\boldsymbol{r}}\mathrm{d}^3\boldsymbol{q}\,. \tag{6.208b}$$

Using the local electric field, $\boldsymbol{E}_{\mathrm{loc}}(\boldsymbol{r},t) = \boldsymbol{E}(\boldsymbol{r},t) + (1/3\epsilon_0)\boldsymbol{P}(\boldsymbol{r},t)$ is Fourier transformed as

$$\boldsymbol{E}_{\mathrm{loc}}(\boldsymbol{q}) = \boldsymbol{E}(\boldsymbol{q}) + \frac{1}{3\epsilon_0}\boldsymbol{P}(\boldsymbol{q})\,. \tag{6.209}$$

Eliminating $\boldsymbol{E}_{\mathrm{loc}}$ and $\boldsymbol{u}$ from (6.209), (6.206) and (6.207), we may obtain the following relation for $\kappa(\omega)$, which has been already shown by (5.131):

$$\boldsymbol{P}(\boldsymbol{q}) = \epsilon_0\chi(\omega)\boldsymbol{E}(\boldsymbol{q}) = \epsilon_0(\kappa(\omega)-1)\boldsymbol{E}(\boldsymbol{q})\,, \tag{6.210}$$

$$\kappa(\omega) = \kappa_\infty\frac{\omega^2-\omega_{\mathrm{LO}}^2}{\omega^2-\omega_{\mathrm{TO}}^2}\,. \tag{6.211}$$

Next we calculate the interaction Hamiltonian for the electron–optical phonon interaction. Defining the potential at $\boldsymbol{r}$ due to the optical phonon by $\phi(\boldsymbol{r})$, the interaction energy is given by $-e\phi(\boldsymbol{r})$ and thus the interaction Hamiltonian for electron–polar optical phonons is written as

$$H_{\mathrm{pop}} = -e\phi(\boldsymbol{r}) = -e\sum_{\boldsymbol{q}}\phi(\boldsymbol{q})\mathrm{e}^{\mathrm{i}\boldsymbol{q}\cdot\boldsymbol{r}}\,, \tag{6.212}$$

where the final relation is obtained by a Fourier transform. Using the relation $-\nabla\phi(\boldsymbol{r}) = \boldsymbol{E}$, we find $\mathrm{i}\boldsymbol{q}\cdot\phi(\boldsymbol{q}) = \boldsymbol{E}(\boldsymbol{q})$. The scalar product of this relation with respect to $\boldsymbol{q}$ gives $\mathrm{i}q^2\phi(\boldsymbol{q}) = (\boldsymbol{q}\cdot\boldsymbol{E}(\boldsymbol{q}))$ and the Hamiltonian for the electron–polar optical phonon interaction is given by

$$H_{\mathrm{pop}} = -e\sum_{\boldsymbol{q}}\left[\frac{\boldsymbol{q}\cdot\boldsymbol{E}(\boldsymbol{q})}{-\mathrm{i}q^2}\right]\mathrm{e}^{\mathrm{i}\boldsymbol{q}\cdot\boldsymbol{r}} = \frac{e}{\mathrm{i}q^2}[\boldsymbol{q}\cdot\boldsymbol{E}(\boldsymbol{r})]\,. \tag{6.213}$$

The electric field $\boldsymbol{E}$ is accompanied by lattice vibrations and only longitudinal optical (LO) phonons with wave vector parallel to the electric field give rise to an electron–polar optical phonon interaction.

In order to estimate the effective charge e^* and to calculate H_{pop}, we use the method of Born and Huang [7]. First we rewrite (6.206) and (6.207) as follows:

$$\ddot{\boldsymbol{w}} = b_{11}\boldsymbol{w} + b_{12}\boldsymbol{E}\,, \tag{6.214}$$

$$\boldsymbol{P} = b_{21}\boldsymbol{w} + b_{22}\boldsymbol{E}\,. \tag{6.215}$$

Here we have introduced new variables defined by

$$w = \sqrt{NM_r}u\,, \tag{6.216a}$$

$$b_{11} = -\omega_0^2 + \frac{Ne^{*2}/3\epsilon_0 M_r}{1 - N\alpha/3\epsilon_0}\,, \tag{6.216b}$$

$$b_{12} = b_{21} = \frac{\sqrt{Ne^{*2}/M_r}}{1 - N\alpha/3\epsilon_0}\,, \tag{6.216c}$$

$$b_{22} = \frac{N\alpha}{1 - N\alpha/3\epsilon_0}\,. \tag{6.216d}$$

The coefficients b_{ij} defined here may be correlated with observable quantities as shown below. Since $\ddot{w} = 0$ for a d.c. field ($\omega = 0$), we obtain $w = -(b_{12}/b_{11})E$ from (6.214). Inserting this relation into (6.215), the polarization P_0 for a d.c. field (due to ionic and electronic polarization) is given by

$$P_0 = \left(-\frac{b_{12}^2}{b_{11}} + b_{22}\right)E\,. \tag{6.217}$$

Therefore, the static dielectric constant κ_0 is written as

$$\kappa_0 = 1 + \chi_0 = 1 + \frac{1}{\epsilon_0}\left(b_{22} - \frac{b_{12}^2}{b_{11}}\right)\,. \tag{6.218}$$

Next, we consider such a high frequency that the atomic displacements cannot follow the electric field. In this case we have $w = 0$, and the corresponding polarization P_∞ is given by

$$P_\infty = b_{22}E\,. \tag{6.219}$$

The corresponding dielectric constant κ_∞ is then written as

$$\kappa_\infty = 1 + \frac{b_{22}}{\epsilon_0}\,. \tag{6.220}$$

The dielectric constant κ_∞ is called the **high-frequency dielectric constant**, and usually corresponds to the dielectric constant in the range of visible light, where only the electronic polarization contributes to the dielectric constant. Using this relation, (6.218) and (6.220) are rewritten as

$$b_{22} = (\kappa_\infty - 1)\epsilon_0\,, \tag{6.221a}$$

$$\frac{b_{12}^2}{b_{11}} = -(\kappa_0 - \kappa_\infty)\epsilon_0\,. \tag{6.221b}$$

On the other hand, the polarization for an electric field with angular frequency ω is obtained by putting $\ddot{w} = -\omega^2 w$ in (6.214):

$$-\omega^2 w = b_{11}w + b_{12}E\,. \tag{6.222}$$

Evaluating $\boldsymbol{w}$ from the above equation and inserting it into (6.215), we find

$$\boldsymbol{P} = \left\{ -\frac{b_{12}b_{21}}{b_{11} + \omega^2} + b_{22} \right\} \boldsymbol{E} \,. \tag{6.223}$$

From these results the dielectric constant $\kappa(\omega)$ at angular frequency ω is given by

$$\kappa(\omega) = 1 + \frac{b_{22}}{\epsilon_0} - \frac{b_{12}b_{21}/\epsilon_0}{b_{11} + \omega^2} = \kappa_\infty + \frac{\kappa_0 - \kappa_\infty}{1 + \omega^2/b_{11}} \,. \tag{6.224}$$

Comparing this result with (5.129) of Sect. 5.4 we find the relation

$$b_{11} = -\omega_{\mathrm{TO}}^2 \,, \tag{6.225}$$

and (5.130) is also derived.

Since $\kappa(\omega_{\mathrm{LO}}) = 0$ for longitudinal optical phonons as stated in Sect. 5.4, we obtain

$$\boldsymbol{P}_{\mathrm{pop}} = -\epsilon_0 \boldsymbol{E} \,. \tag{6.226}$$

Inserting this into (6.215) and eliminating the electric field $\boldsymbol{E}$, we obtain

$$\boldsymbol{P}_{\mathrm{pop}} = b_{21}\boldsymbol{w} - \frac{b_{22}}{\epsilon_0}\boldsymbol{P}_{\mathrm{pop}} \,. \tag{6.227}$$

From (6.221b) and (6.225), the following relation is derived:

$$b_{12}b_{21} = (\kappa_0 - \kappa_\infty)\epsilon_0\omega_{\mathrm{TO}}^2 \,. \tag{6.228}$$

With this relation and (6.221a), (6.227) reduces to

$$\boldsymbol{P}_{\mathrm{pop}} = \epsilon_0^{1/2}(\kappa_0 - \kappa_\infty)^{1/2}\omega_{\mathrm{TO}}\boldsymbol{w} - (\kappa_\infty - 1)\boldsymbol{P}_{\mathrm{pop}} \,. \tag{6.229}$$

This relation may be rewritten as follows by using the Lyddane–Sachs–Teller equation (5.131):

$$\boldsymbol{P}_{\mathrm{pop}} = \epsilon_0^{1/2}\left(\frac{1}{\kappa_\infty} - \frac{1}{\kappa_0}\right)^{1/2} \omega_{\mathrm{LO}} \cdot \boldsymbol{w} \,. \tag{6.230}$$

When $\boldsymbol{w}$ is replaced by the relative displacement $\boldsymbol{u}$, we obtain

$$\boldsymbol{P}_{\mathrm{pop}} = (\epsilon_0 N M_{\mathrm{r}})^{1/2}\left(\frac{1}{\kappa_\infty} - \frac{1}{\kappa_0}\right)^{1/2} \omega_{\mathrm{LO}} \cdot \boldsymbol{u} \equiv N e_{\mathrm{c}}^* \cdot \boldsymbol{u} \,. \tag{6.231}$$

The final result of the above equation is equivalent to the ionic polarization induced by charge e_{c}^* and thus the charge is called the **effective charge**, which is given by

$$e_c^* = \left(\frac{\epsilon_0 M_r}{N}\right)^{1/2} \left(\frac{1}{\kappa_\infty} - \frac{1}{\kappa_0}\right)^{1/2} \omega_{LO} \,. \tag{6.232}$$

Also, we find the relation

$$\omega_{LO}^2 - \omega_{TO}^2 = \frac{\kappa_\infty N}{\epsilon_0 M_r} e_c^{*2} \,. \tag{6.233}$$

The electric field induced by these longitudinal optical phonons may be written as

$$\boldsymbol{E} = -\frac{1}{\epsilon_0}\boldsymbol{P}_{\text{pop}} = -\frac{N e_c^*}{\epsilon_0}\boldsymbol{u} \,. \tag{6.234}$$

Using the relative displacement of (6.70) we obtain

$$\begin{aligned}
\boldsymbol{E} &= -\frac{N e_c^*}{\epsilon_0}\sum_{\boldsymbol{q}}\left(\frac{\hbar}{2L^3 N M_r \omega_{LO}}\right)^{1/2} \boldsymbol{e}_{\boldsymbol{q}}\left(a_{\boldsymbol{q}} e^{i\boldsymbol{q}\cdot\boldsymbol{r}} + a_{\boldsymbol{q}}^\dagger e^{-i\boldsymbol{q}\cdot\boldsymbol{r}}\right) \\
&= -\frac{1}{\epsilon_0}\left(\frac{\epsilon_0 \hbar}{2L^3\omega_{LO}}\right)^{1/2}\left(\frac{1}{\kappa_\infty} - \frac{1}{\kappa_0}\right)^{1/2} \omega_{LO}\sum_{\boldsymbol{q}} \boldsymbol{e}_{\boldsymbol{q}}\left\{a_{\boldsymbol{q}} e^{i\boldsymbol{q}\cdot\boldsymbol{r}} + a_{\boldsymbol{q}}^\dagger e^{-i\boldsymbol{q}\cdot\boldsymbol{r}}\right\} \\
&= -\frac{1}{\epsilon_0}\left(\frac{\hbar}{2L^3\gamma\omega_{LO}}\right)^{1/2}\sum_{\boldsymbol{q}} \boldsymbol{e}_{\boldsymbol{q}}\left(a_{\boldsymbol{q}} e^{i\boldsymbol{q}\cdot\boldsymbol{r}} + a_{\boldsymbol{q}}^\dagger e^{-i\boldsymbol{q}\cdot\boldsymbol{r}}\right) .
\end{aligned} \tag{6.235}$$

Here we have to note that the value of N used to expand the displacement vector in normal modes is the number of atom pairs in the volume $V = L^3$ and that N in this section is the number of atom pairs in a unit volume (N in the normal mode expansion should be read as $L^3 N$ with N as defined in this section). The coefficient γ is defined by

$$\frac{1}{\gamma} = \epsilon_0\left(\frac{1}{\kappa_\infty} - \frac{1}{\kappa_0}\right)\omega_{LO}^2 \,. \tag{6.236}$$

The coefficient γ is often referred to as the **Fröhlich coupling constant**. With this expression the Hamiltonian for the electron–LO–phonon interaction is given by

$$\begin{aligned}
H_{\text{pop}} &= i\frac{1}{\epsilon_0}\left(\frac{e^2\hbar}{2L^3\gamma\omega_{LO}}\right)^{1/2}\sum_{\boldsymbol{q}}\frac{1}{q}\left(a_{\boldsymbol{q}} e^{i\boldsymbol{q}\cdot\boldsymbol{r}} + a_{\boldsymbol{q}}^\dagger e^{-i\boldsymbol{q}\cdot\boldsymbol{r}}\right) \\
&\equiv \sum_{\boldsymbol{q}} C_{\text{pop}}(\boldsymbol{q})\left(a_{\boldsymbol{q}} e^{i\boldsymbol{q}\cdot\boldsymbol{r}} + a_{\boldsymbol{q}}^\dagger e^{-i\boldsymbol{q}\cdot\boldsymbol{r}}\right) ,
\end{aligned} \tag{6.237}$$

where

$$C_{\text{pop}}(\boldsymbol{q}) = i\frac{1}{\epsilon_0}\left(\frac{e^2\hbar}{2L^3\gamma\omega_{LO}}\right)^{1/2}\frac{1}{q} \,. \tag{6.238}$$

Also the strength of the electron–LO–phonon interaction is often expressed by

$$|C_{\text{pop}}(\boldsymbol{q})|^2 = \alpha(\hbar\omega_{\text{LO}})^{3/2}\frac{1}{L^3}\left(\frac{\hbar^2}{2m^*}\right)^{1/2}\frac{1}{q^2}, \tag{6.239}$$

where the dimensionless constant α is given by

$$\alpha = \frac{e^2}{\hbar}\frac{1}{\epsilon_0}\left(\frac{1}{\kappa_\infty} - \frac{1}{\kappa_0}\right)\left(\frac{m^*}{2\hbar\omega_{\text{LO}}}\right)^{1/2}. \tag{6.240}$$

The divergence of H_{pop} at $q = 0$ may be removed by taking account of the screening due to the carriers (electrons or holes) as described in Sect. 6.3.3, where ionized impurity scattering is dealt with. The screening effect (static screening) is taken into account by multiplying by $q^2/(q^2 + q_s^2)$ and thus the coefficient C_{pop} is rewritten as

$$C_{\text{pop}}(\boldsymbol{q}) = \mathrm{i}\frac{1}{\epsilon_0}\left(\frac{e^2\hbar}{2L^3\gamma\omega_{\text{LO}}}\right)^{1/2}\frac{q}{q^2 + q_s^2}. \tag{6.241}$$

The calculation of the matrix elements for polar LO phonon scattering is straightforward and we have

$$\begin{aligned}
|M(\boldsymbol{k}, \boldsymbol{k}')| &= |C_{\text{pop}}|\frac{q}{q^2 + q_s^2} \times \begin{cases}\sqrt{n_q} \\ \sqrt{n_q + 1}\end{cases} \\
&= \left[\frac{e^2\hbar\omega_{\text{LO}}}{2L^3\epsilon_0}\left(\frac{1}{\kappa_\infty} - \frac{1}{\kappa_0}\right)\right]^{1/2}\frac{q}{q^2 + q_s^2} \times \begin{cases}\sqrt{n_q}, \\ \sqrt{n_q + 1},\end{cases}
\end{aligned} \tag{6.242}$$

where n_q is the Bose–Einstein distribution function for LO phonons and is given by

$$n_q = \frac{1}{\exp(\hbar\omega_{\text{LO}}/k_{\text{B}}T) - 1}.$$

6.3.7 *Inter–Valley Phonon Scattering*

We have shown that the wave vectors involved in deformation potential scattering (acoustic phonon), piezoelectric potential scattering, non-polar optical phonon scattering and longitudinal polar optical phonon scattering are small and that the scattering is referred to as long-wavelength phonon scattering. On the other hand, phonons of large wave vector are involved in the scattering in the conduction bands of the many–valley structure such as in Ge and Si, where electrons are scattered between different valleys. The scattering is called **inter-valley phonon scattering**.

As stated previously, electron scattering within a valley by optical phonons is not allowed due to the symmetry properties of the conduction bands and of the optical phonons. The constant energy surfaces of the conduction bands in Si consist of six equivalent ellipsoids with their longitudinal axes along the $\langle 100\rangle$ direction as shown in Fig. 2.9. Referring to Fig. 6.8, we consider the valley located along the k_x and k_y

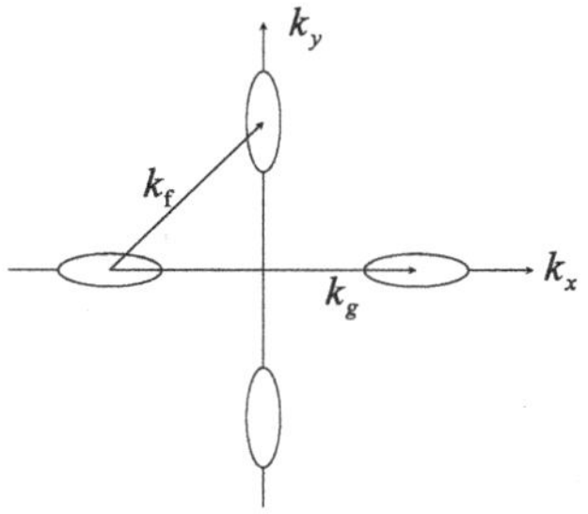

Fig. 6.8 f process and g process of inter-valley phonon scattering in Si

axes and we find that electron scattering between the two different valleys requires a large wave vector change $\boldsymbol{k}_{\rm f}$ or $\boldsymbol{k}_{\rm g}$. The scattering processes are referred to as the f–process and the g–process.[2] In the scattering event of an initial electron from a conduction band valley minimum to the other valley minimum, the electron absorbs or emits a phonon with the wave vector very close to the distance between the two conduction band valley minima. The corresponding phonon energy is given by $\hbar\omega_{\rm f}$ or $\hbar\omega_{\rm g}$. This scattering is very similar to the non-polar optical phonon scattering and we use the results. Defining the deformation potential for scattering between the i and j valleys by D_{ij}, the matrix element is written as below following the result of (6.204):

$$|M(\boldsymbol{k},\boldsymbol{k}')| = \left(\frac{\hbar}{2NM_{\rm r}\omega_{ij}}\right)^{1/2} D_{ij} \times \begin{cases} \sqrt{n(\omega_{ij})}, \\ \sqrt{n(\omega_{ij})+1}\,. \end{cases} \tag{6.243}$$

6.3.8 *Deformation Potential in Degenerate Bands*

As stated in Sects. 2.2 and 2.3, the valence bands in diamond and zinc blende crystals are degenerate at the Γ point ($\boldsymbol{k} = 0$), where the valence bands consist of degenerate heavy and light hole bands and the spin–orbit split-off band due to spin–orbit interaction. In addition we have shown in Sect. 4.7 on piezobirefringence that application of stress results in a change in the valence bands and that the change is well calculated by the effective strain Hamiltonian, (4.171), defined by Picus and Bir. The effective strain Hamiltonian contains three deformation potential constants a, b and d. Under application of uniaxial stress along the directions [100] and [111], the degenerate valence bands $J = 3/2$ are resolved into separate valence bands due to the deformation potential components b and d. On the other hand, the deformation potential a is related to the strain component $e_{xx} + e_{yy} + e_{zz} = \text{div}\,\boldsymbol{u}$ and the divergence of dis-

[2]The names of the f– and g–processes are from the configurations of the valleys for inter-valley scattering, which look like the roman characters f and g, respectively. The reason why the symbol not g but g is used is evident from Fig. 6.8, where we find g is more intuitive.

placement $\boldsymbol{u}$ (div $\boldsymbol{u}$) is equivalent to the volume change $\delta V/V$ of a crystal, where V is the volume. As stated in Sect. 6.3 (Sect. 6.3.2), this term gives rise to an interaction with the longitudinal acoustic phonons and thus the hole mobility depends strongly on the deformation potential a.

We have already shown in Sects. 2.1 and 6.3.7 that the conduction band minima of Ge and Si are not located at Γ ($\boldsymbol{k} = 0$) but lie at the KL point of the [111] direction and at the Δ axes of the [100] direction, respectively, resulting in four equivalent valleys in Ge and six equivalent valleys in Si. Such degenerate conduction bands are called the **many–valley structure**. The stress effect on such a many–valley structure is analyzed as follows. As an example, we consider the case of Si. When a uniaxial stress is applied in the [100] direction, the conduction band valleys in the directions [100] and [$\bar{1}$00] differ in their energy state compared to the other four valleys, and thus the degeneracy is partially removed. On the other hand, the effect of uniaxial stress along [111] on the valleys is equivalent and thus the degeneracy will not be removed. Such behavior of the stress effect in many–valley semiconductors is analyzed by the theory of Herring and Vogt [8]. The stress Hamiltonian of Herring and Vogt is written as

$$H_{\rm HV} = \Xi_{\rm d} \cdot {\rm Trace}(\tilde{e}) + \Xi_{\rm u} \cdot \left[\tilde{m}^{(i)} \cdot \tilde{e} \cdot \tilde{m}^{(i)}\right] , \tag{6.244}$$

where $\tilde{e}$ is the strain tensor given by 3×3 matrix, ${\rm Trace}(\tilde{e}) = e_{xx} + e_{yy} + e_{zz}$, and $\tilde{m}^{(i)}$ is the unit vector of the principal axis of valley (i). Note that the strain tensor S_{ij} is used for the analysis of piezoelectric potential scattering instead of e_{ij}. Here, $\Xi_{\rm d}$ is the hydrostatic deformation potential and $\Xi_{\rm u}$ is the shear deformation potential. Deformation potentials of many valley semiconductors Si and Ge are listed in Table 6.1.

Let us consider a valley in Si whose principal axis is in the (i) direction as shown in Fig. 6.9. For simplicity we consider a valley along the [100] direction, and then (6.244) reduces to

$$H_{\rm HV} = \Xi_{\rm d}(e_{xx} + e_{yy} + e_{zz}) + \Xi_{\rm u} e_{xx} . \tag{6.245}$$

This relation is often referred to as the **Herring–Vogt relation**. In the case of the configuration shown in Fig. 6.9, the deformation potential $D_{\rm ac}$ for the longitudinal acoustic phonon defined in Sect. 6.3.2 is given by

Table 6.1 Deformation potentials $\Xi_{\rm u}$ and $\Xi_{\rm d}$ in [eV] of many valley conduction bands for Si (Δ-valley) and Ge (L–valley) (after Data Book of Landolt–Börnstein [9] and Fischetti [10])

	Si	Ge
$\Xi_{\rm u}$	9.0 ± 2	16.3 (80 K), 19.3 (4 K)
$\Xi_{\rm d}$	$\simeq 5$, 1.1 [10]	-12.3 (4 K), -4.4 [10]

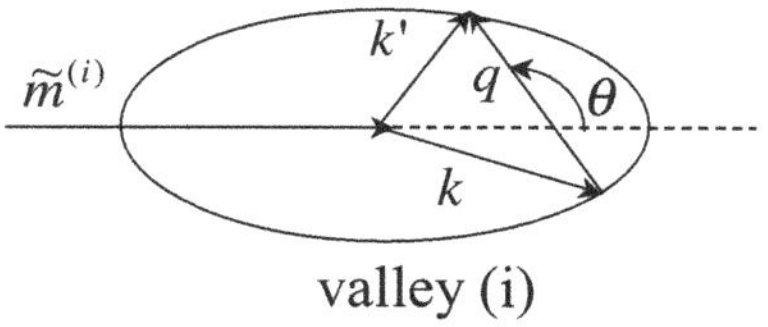

Fig. 6.9 Intra–valley scattering due to the deformation potential. $\boldsymbol{k}$ and $\boldsymbol{k}'$ are the electron wave vectors before and after scattering, respectively, $\boldsymbol{q}$ is the wave vector of the phonon (lattice vibration), and θ is the angle between the phonon wave vector and the principal axis of a valley

$$D_{\rm ac} \to D_{\rm LA} = \Xi_{\rm d} + \Xi_{\rm u} \cos^2 \theta \,. \tag{6.246}$$

We have to note here that scattering by transverse acoustic phonons is allowed. For transverse acoustic phonons with displacement vector $\boldsymbol{u}$ perpendicular to $\boldsymbol{q}$ and parallel to the textbook, we find $e_{xx} = \partial u_x / \partial x \neq 0$. Noting that $u_x = u \sin\theta \exp(\mathrm{i}\boldsymbol{q} \cdot \boldsymbol{r})$, $q_x = q \cos\theta$, we obtain the following relation:

$$D_{\rm ac} \to D_{\rm TA} = \Xi_{\rm u} \sin\theta \cos\theta \,. \tag{6.247}$$

The above treatment is also valid for the conduction band valleys in Ge, which consist of four equivalent valleys located at the L point in the [111] direction of the Brillouin zone, and we may use the relation given by (6.244). Under the application of a [111] uniaxial stress, the valley of [111] has a different energy state compared to the other three valleys for the $[\bar{1}11]$, $[1\bar{1}1]$ and $[11\bar{1}]$ directions.

6.3.9 Theoretical Calculation of Deformation Potentials

The theoretical calculation of the interaction between electrons and phonons (lattice vibrations) is realized by obtaining the effect of the potential energy on an electron induced by the lattice vibrations. The deformation potential due to optical phonons were calculated first by Pötz and Vogl [11], and later the deformation potentials for various phonon modes in III-V semiconductors were calculated by Zollner et al. [12] and the deformation potentials for intra-valley phonons and inter-valley phonons in Si have been reported by Fischetti and Laux [13] and Mizuno et al. [14]. Although various methods to calculate the deformation potentials have been reported, here we will review the **rigid-ionmodel**, where a change in the potential energy due to lattice displacements is the potential energy acting on an electron. It should be noted here that the potential energy acting on an electron is calculated by summing up the contributions from all atoms. Let us define the equilibrium position of an α atom in the l-th unit cell by $\boldsymbol{R}_{l,\alpha}$ and the displacement vector of the atom by $\boldsymbol{u}_{l,\alpha}$. Then the potential energy change due to lattice displacements is given by

$$H_{\text{el-ph}} = \sum_{l,\alpha} \left[V_\alpha(\boldsymbol{r} - \boldsymbol{R}_{l,\alpha} + \boldsymbol{u}_{l,\alpha}) - V_\alpha(\boldsymbol{r} - \boldsymbol{R}_{l,\alpha}) \right]$$
$$= \sum_{l,\alpha} \boldsymbol{u}_{l,\alpha} \cdot \text{grad}\, V_\alpha(\boldsymbol{r} - \boldsymbol{R}_{l,\alpha}) \,. \tag{6.248}$$

The lattice displacement vector $\boldsymbol{u}_{l,\alpha}$ is given by (6.66)[3]

$$\boldsymbol{u}^{(j)}_{l,\alpha} = \sum_{\boldsymbol{q}} \sqrt{\frac{\hbar}{2M_\alpha N \omega^{(j)}_{\boldsymbol{q}}}} \left(a_{\boldsymbol{q}} + a^\dagger_{-\boldsymbol{q}} \right) \boldsymbol{e}^{(j)}_\alpha(\boldsymbol{q}) \mathrm{e}^{\mathrm{i}\boldsymbol{q}\cdot\boldsymbol{R}_{l,\alpha}} \,, \tag{6.249}$$

where $\boldsymbol{e}(\boldsymbol{q})$ is the unit vector of the lattice displacement vector and the superscript (j) represents the mode of lattice vibration. When we put the atomic position vector as $\boldsymbol{R}_{l,\alpha} = \boldsymbol{R}_l + \boldsymbol{\tau}_\alpha$, the Bloch theorem will give rise to the following relation:

$$\langle \boldsymbol{k} + \boldsymbol{q} | \text{grad}\, V_\alpha(\boldsymbol{r} - \boldsymbol{R}_{l,\alpha}) | \boldsymbol{k} \rangle$$
$$= \mathrm{e}^{-\mathrm{i}\boldsymbol{q}\cdot\boldsymbol{R}_l} \langle \boldsymbol{k} + \boldsymbol{q} | \text{grad}\, V_\alpha(\boldsymbol{r} - \boldsymbol{\tau}_\alpha) | \boldsymbol{k} \rangle \,. \tag{6.250}$$

Using the Fermi golden rule, the non-zero matrix elements for this potential energy is given by

$$\langle \boldsymbol{k} \pm \boldsymbol{q}, n_{\boldsymbol{q}} \mp 1 | H_{\text{el-ph}} | \boldsymbol{k}, n_{\boldsymbol{q}} \rangle$$
$$= \sum_\alpha \sqrt{\frac{\hbar}{2NM_\alpha \omega^{(j)}_{\boldsymbol{q}}}} A_\alpha(\boldsymbol{k}, \pm\boldsymbol{q}) \cdot \boldsymbol{e}^{(j)}_\alpha(\pm\boldsymbol{q}) \sqrt{n_{\boldsymbol{q}} + \frac{1}{2} \mp \frac{1}{2}} \,, \tag{6.251}$$

where $A_\alpha(\boldsymbol{k}, \pm\boldsymbol{q})$ is defined by

$$A_\alpha(\boldsymbol{k}, \pm\boldsymbol{q}) = -\langle \boldsymbol{k} \pm \boldsymbol{q} | \text{grad}\, V_\alpha(\boldsymbol{r} - \boldsymbol{\tau}_\alpha) | \boldsymbol{k} \rangle \mathrm{e}^{\mathrm{i}\boldsymbol{q}\cdot\boldsymbol{\tau}_\alpha} \,. \tag{6.252}$$

In order to evaluate the above equations we use the electron wave function $|\boldsymbol{k}\rangle$ calculated by the pseudopotential method discussed in Sect. 1.6 and the atomic potential $V_\alpha(\boldsymbol{r})$ is obtained by using the Fourier transform. These are given below following the definition of Cohen and Bergstresser [15]:

$$|\boldsymbol{k}\rangle = \frac{1}{\sqrt{\Omega}} \sum_{\boldsymbol{G}} C_{\boldsymbol{k}}(\boldsymbol{G}) \mathrm{e}^{\mathrm{i}(\boldsymbol{k}+\boldsymbol{G})\cdot\boldsymbol{r}} \,, \tag{6.253}$$

$$V_\alpha(\boldsymbol{r}) = \frac{1}{2} \sum_{\boldsymbol{G}} V_\alpha(\boldsymbol{G}) \mathrm{e}^{\mathrm{i}\boldsymbol{G}\cdot\boldsymbol{r}} \,, \tag{6.254}$$

where Ω is the unit cell volume and the Fourier coefficient $V_\alpha(\boldsymbol{G})$ is given by

$$V_\alpha(\boldsymbol{G}) = \frac{2}{\Omega} \int_\Omega \mathrm{d}^3\boldsymbol{r}\, V_\alpha(\boldsymbol{r}) \mathrm{e}^{-\mathrm{i}\boldsymbol{G}\cdot\boldsymbol{r}} \,. \tag{6.255}$$

In the above definition, the potential energy is assumed to be periodic with respect to atomic position and thus is expanded in reciprocal lattice vectors. However, it

[3]We have used here the relation: $\sum_{\boldsymbol{q}} a^\dagger_{-\boldsymbol{q}} \exp(\mathrm{i}\boldsymbol{q} \cdot \boldsymbol{R}_{l,\alpha}) = \sum_{\boldsymbol{q}} a^\dagger_{\boldsymbol{q}} \exp(-\mathrm{i}\boldsymbol{q} \cdot \boldsymbol{R}_{l,\alpha})$.

should be pointed out that the potential energy $V_\alpha(\boldsymbol{r})$ includes contributions from the displacement of atomic position due to lattice vibrations, resulting in non-periodic function and that the potential energy $V_\alpha(\boldsymbol{r})$ cannot be expanded in reciprocal lattice vectors. In the following analysis, the Fourier coefficient $V_\alpha(\boldsymbol{G})$ is assumed to be a function of the quasi-continuous vector $\boldsymbol{G}$ such that $V_\alpha(\boldsymbol{G}+\boldsymbol{q})$ is estimated from the curve of $V_\alpha(\boldsymbol{G})$ versus $\boldsymbol{G}$. Then the coefficient $\boldsymbol{A}_\alpha(\boldsymbol{k},\boldsymbol{q})$ of (6.251) in given by

$$\boldsymbol{A}_\alpha(\boldsymbol{k},\boldsymbol{q}) = -\frac{\mathrm{i}}{2}\sum_{\boldsymbol{G},\boldsymbol{G}'} C^*_{\boldsymbol{k}+\boldsymbol{q}}(\boldsymbol{G}')C_{\boldsymbol{k}}(\boldsymbol{G})(\boldsymbol{G}'-\boldsymbol{G}+\boldsymbol{q})V_\alpha(\boldsymbol{G}'-\boldsymbol{G}+\boldsymbol{q}) \times \mathrm{e}^{-\mathrm{i}(\boldsymbol{G}'-\boldsymbol{G})\cdot\boldsymbol{\tau}_\alpha}, \tag{6.256}$$

where it should be noted that $\boldsymbol{A}_\alpha(\boldsymbol{k},\boldsymbol{q})$ is a vector.

Now, we define the deformation potential $D^{(j)}(\boldsymbol{q})$ by the following equation

$$\left|\langle \boldsymbol{k}\pm\boldsymbol{q}, n_{\boldsymbol{q}}\mp 1|H_{\text{el-ph}}|\boldsymbol{k}, n_{\boldsymbol{q}}\rangle\right| = \sqrt{\frac{\hbar}{2NM\omega_{\boldsymbol{q}}^{(j)}}}D^{(j)}(\boldsymbol{k},\pm\boldsymbol{q})\sqrt{n_{\boldsymbol{q}}+\frac{1}{2}\mp\frac{1}{2}}, \tag{6.257}$$

and then we obtain

$$D^{(j)}(\boldsymbol{k},\pm\boldsymbol{q}) = \sqrt{M}\left|\sum_\alpha \frac{1}{\sqrt{M_\alpha}}\boldsymbol{A}(\boldsymbol{k},\pm\boldsymbol{q})\cdot\boldsymbol{e}_\alpha^{(j)}(\pm\boldsymbol{q})\right|, \tag{6.258}$$

where we have used the atomic mass defined by Conwell [6], $M=\sum_\alpha M_\alpha$. Since there exist two atoms in a unit cell of Si, the above equation is written as

$$\begin{aligned} D^{(j)}(\boldsymbol{k},\pm\boldsymbol{q}) &= \sqrt{2}\left|\sum_\alpha \boldsymbol{A}_\alpha(\boldsymbol{k},\pm\boldsymbol{q})\cdot\boldsymbol{e}_\alpha^{(j)}(\pm\boldsymbol{q})\right| \\ &= \frac{1}{\sqrt{2}}\left|\sum_\alpha\sum_{\boldsymbol{G},\boldsymbol{G}'} C^*_{\boldsymbol{k}\pm\boldsymbol{q}}(\boldsymbol{G}')C_{\boldsymbol{k}}(\boldsymbol{G})\mathrm{e}^{-\mathrm{i}(\boldsymbol{G}'-\boldsymbol{G})\cdot\boldsymbol{\tau}_\alpha}\right. \\ &\quad \left.\times\sum_{\boldsymbol{G},\boldsymbol{G}'}\mathrm{e}^{-\mathrm{i}(\boldsymbol{G}'-\boldsymbol{G})}V_\alpha(\boldsymbol{G}'-\boldsymbol{G}\pm\boldsymbol{q})(\boldsymbol{G}'-\boldsymbol{G}\pm\boldsymbol{q})\cdot\boldsymbol{e}_\alpha^{(j)}(\pm\boldsymbol{q})\right|. \end{aligned} \tag{6.259}$$

We find in (6.259) that the deformation potential has the dimensions of energy /length (units of eV/cm) and from comparison with the deformation potential definition stated in Sects. 6.3.2, 6.3.5 and 6.3.7 it corresponds to the deformation potential for optical phonon scattering or inter-valley phonon scattering. On the other hand, it is evident that the deformation potential for acoustic phonon scattering is defined by

$$\Xi = \lim_{q\to 0}\left|\frac{D^{(j)}(\boldsymbol{q})}{q}\right|. \tag{6.260}$$

The procedures for calculating the deformation potential form (6.259) is as follows.

1. First, calculate the lattice vibrations and determine the angular frequency $\omega_{\boldsymbol{q}}^{(j)}$ of mode (i) and the eigenvector of the lattice displacement $\boldsymbol{e}_{\alpha}^{(j)}$.
2. The energy band structure is calculated by using the pseudopotential method, for example, to obtain $C_{\boldsymbol{k}}(\boldsymbol{G})$.
3. The quasi-continuous function $V_{\alpha}(\boldsymbol{q})$ in the region of $\boldsymbol{q} = 0$ to the maximum value of $\boldsymbol{G}$ is estimated from the values of the pseudopotentials $V_{\alpha}(\boldsymbol{G})$ used for the energy band calculation.
4. Then, inserting these values in (6.259), the deformation potential $D^{(j)}(\boldsymbol{q})$ is calculated as a function of the electron wave vector $\boldsymbol{k}$ and phonon wave vector.

Analysis of the lattice vibrations has been made by the shell model, the bond charge model and so on [16–20]. The results shown in this section have been calculated by the bond charge model of Weber et al. [19, 20]. Here we show two different approaches for obtaining the quasi-continuous function $V_{\alpha}(\boldsymbol{q})$ from the pseudopotentials $V_{\alpha}(\boldsymbol{G})$. These two approaches arise from the ambiguity of the value at $q \sim 0$. One is to approximate $V_{\alpha}(0)$ by $-(2/3)\mathcal{E}_{\rm F}$ and extrapolate $V_{\alpha}(\boldsymbol{q})$ by using the pseudopotentials of Cohen and Bergstresser [15], where $\mathcal{E}_{\rm F}$ is the Fermi energy of the valence electrons [21]. The other method is reported by Bednarek and Rössler [22], where $V_{\alpha}(0) = 0$ is assumed at $q = 0$. Quasi-continuous functions of $V(\boldsymbol{q})$ obtained by the two methods are compared in Fig. 6.10. In this book the pseudopotential curve of Bednarek and Rössler [22] is used [23].

Now we will show the results of calculation of the deformation potentials for inter-valley scattering and for acoustic phonon scattering in Si. In general, the deformation potentials depend on the initial and final states of the electron and also on the phonon mode and its wave vector, as seen in the theoretical analysis. In Fig. 6.11a the deformation potentials for inter-valley scattering of f-type phonons are plotted as a function of the angle θ between the principal axis and the electron wave vector of the final state. A similar plot for g–type phonons

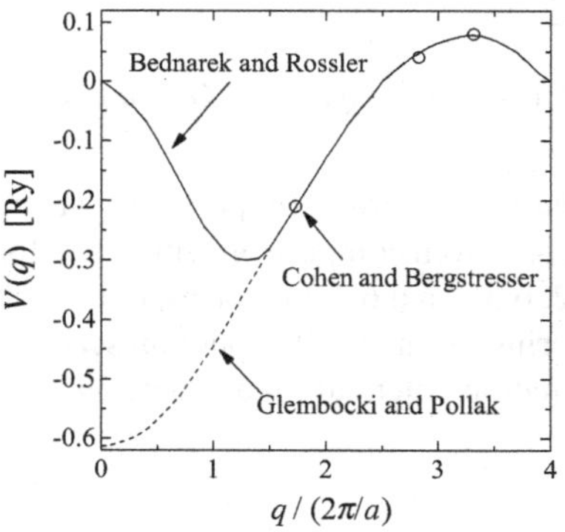

Fig. 6.10 Pseudopotential curve of Si $V(\boldsymbol{q})$ estimated from pseudopotentials of $V(\boldsymbol{G})$ of Cohen and Bergstresser [15]. Two approximation methods are compared: the method of Bednarek and Rössler [22] and the method of Glembocki and Pollak [21]

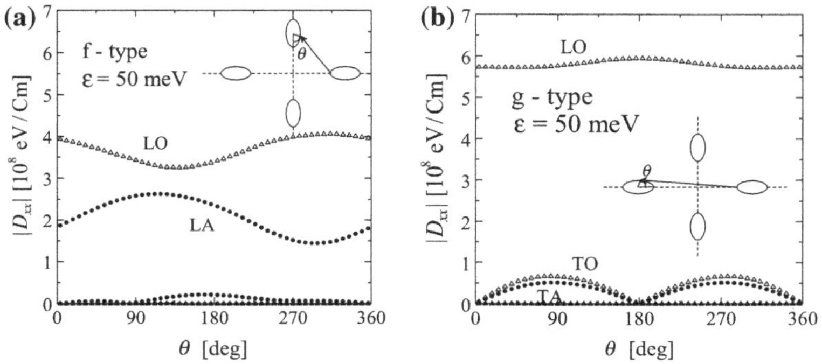

Fig. 6.11 Deformation potentials of Si for **a** f-type and **b** g–type inter-valley phonon scattering as a function of the angle between the principal axis of the valley and the wave vector of the electron. The *inset* shows the constant energy surface of the conduction band valleys in the (100) surface of the Brillouin zone; the electron transition is indicated by an *arrow*

is shown in Fig. 6.11b. The inset of Fig. 6.11 represents the scattering direction of the electron. In the calculations, the initial state of the electron is fixed at $\boldsymbol{k}_\mathrm{i} = (0.85 - 0.098, 0, 0)$ and the final states are taken to be on the constant energy curve of $\mathcal{E} = 50\,\mathrm{meV}$ given by (a) $\boldsymbol{k}_\mathrm{f} = (0.043 \sin\theta, 0.85 - 0.098 \cos\theta, 0)$, and (b) $\boldsymbol{k}_\mathrm{f} = (-0.85 + 0.098 \cos\theta, 0.043 \sin\theta, 0)$.

From the calculations we find that f-type deformation potentials for longitudinal optical (LO) phonons and longitudinal acoustic (LA) phonons play a role in the inter-valley phonon scattering and that the other deformation potentials such as transverse optical (TO) and transverse acoustic (TA) phonons are very weak, enabling us to neglect their contribution to electron scattering. In the case of g–type inter-valley phonon scattering, the deformation potential of LO phonons has a large value and the others are very small. The g–type deformation potentials for TO and TA phonons at the symmetry points at $\theta = 0$ and $180°$ are zero, but they have finite values at other angles, resulting in a weak contribution to the electron scattering.

Next, we will show the calculated results on deformation potentials for intra-valley phonon scattering by the solid curves in Fig. 6.12. As expected and stated before, the valley minima do not exist at the Γ point but along the Δ axis near the X point, and therefore not only LA phonons but also TA phonons take part in the intra-valley scattering. The dotted curve is calculated from the Herring–Vogt relation stated in Sect. 6.3.8, where we put $\Xi_\mathrm{u} = 10.0\,\mathrm{eV}$ and $\Xi_\mathrm{d} = -11.5\,\mathrm{eV}$ in (6.246) and (6.247), and the dotted curve shows good agreement with the deformation potential calculation.

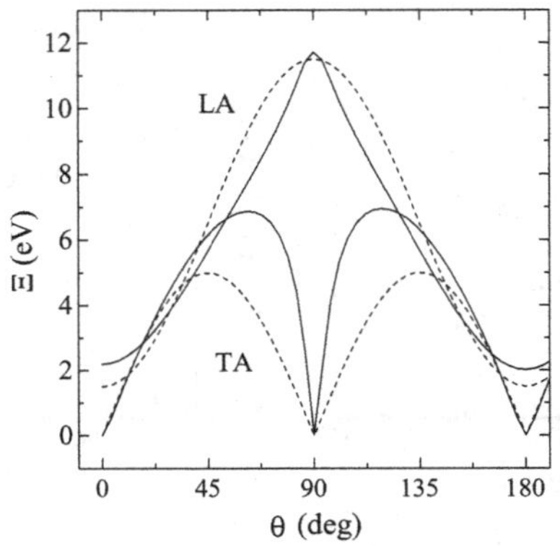

Fig. 6.12 Deformation potentials of Si for intra-valley phonon scattering calculated as a function of the angle θ between the phonon wave vector and the principal axis of the valley. The *dotted curve* is obtained from the theoretical relation of Herring and Vogt [8] by putting $\Xi_u = 10.0\,\text{eV}$ and $\Xi_d = -11.5\,\text{eV}$ in (6.246) and (6.247)

6.3.10 Electron–Electron Interaction and Plasmon Scattering

In this section we will deal with electron–electron interactions such as screening by electrons, electron–electron scattering and plasmon scattering. In the calculations the theory of field quantization is required and the treatment is beyond the purpose of this book. In the following we will show how to derive the dielectric function for many electrons but the details will not be given. However, the outline of the derivation is given in Appendix E. Readers who are interested in a more detailed treatment should refer to the books of Haug and Koch [24], Maham [25], and Ferry and Grondin [26].

In semiconductors with a high electron density, screening by the electron plasma plays an important role in the dielectric constant. The random phase approximation (RPA) leads us to obtain the dielectric function given by the following relation as shown in Appendix E (see also Haug and Koch [24]).

$$\begin{aligned}\kappa(q,\omega) &= 1 - V(\boldsymbol{q})\sum_{\boldsymbol{k}} \frac{f(\boldsymbol{k}-\boldsymbol{q}) - f(\boldsymbol{k})}{\hbar\omega + \mathrm{i}\Gamma + \mathcal{E}(\boldsymbol{k}-\boldsymbol{q}) - \mathcal{E}(\boldsymbol{k})} \\ &= 1 - \frac{e^2}{\epsilon_0 q^2 L^3}\sum_{\boldsymbol{k}} \frac{f(\boldsymbol{k}-\boldsymbol{q}) - f(\boldsymbol{k})}{\hbar\omega + \mathrm{i}\Gamma + \mathcal{E}(\boldsymbol{k}-\boldsymbol{q}) - \mathcal{E}(\boldsymbol{k})} \,. \end{aligned} \tag{6.261}$$

This relation was first derived by Lindhard [27] and is called the **Lindhard formula**. It can easily be shown that this equation in the limit of $q \to 0$ reduces to the dielectric function obtained from classical theory. Let us assume $\Gamma = 0$ for simplicity and expand the second term on the right-hand side of (6.261) in q assuming the limit of $q \to 0$. In the following we assume that the electrons are described by a parabolic band with an isotropic effective mass m. Then we obtain

$$\mathcal{E}(\boldsymbol{k}-\boldsymbol{q})-\mathcal{E}(\boldsymbol{k}) = \frac{\hbar^2}{2m}(k^2-2\boldsymbol{k}\cdot\boldsymbol{q}+q^2)-\frac{\hbar^2k^2}{2m} \simeq -\frac{\hbar^2\boldsymbol{k}\cdot\boldsymbol{q}}{m}, \tag{6.262}$$

$$f(\boldsymbol{k}-\boldsymbol{q})-f(\boldsymbol{k}) = f(\boldsymbol{k})-\boldsymbol{q}\cdot\nabla_{\boldsymbol{k}}f(\boldsymbol{k})+\cdots-f(\boldsymbol{k}) \simeq -\boldsymbol{q}\cdot\nabla_{\boldsymbol{k}}f(\boldsymbol{k}). \tag{6.263}$$

Using these relations in (6.261) we obtain

$$\kappa(0,\omega) = 1-V(q)\sum_{\boldsymbol{k},i}\frac{q_i\,(\partial f/\partial k_i)}{\hbar\omega-\hbar^2\boldsymbol{k}\cdot\boldsymbol{q}/m} \simeq 1-\frac{V(q)}{\hbar\omega}\sum_{\boldsymbol{k},i}q_i\frac{\partial f}{\partial k_i}\left[1+\frac{\hbar\boldsymbol{k}\cdot\boldsymbol{q}}{m\omega}\right], \tag{6.264}$$

where we put $\omega(q\to 0)=\omega_0$. The summation of the first term on the right-hand side of the above equation $\sum\partial f/\partial k_i$ will disappear because of the distribution function f in the limit of $k\to\infty$, and finally we obtain

$$\kappa(0,\omega) = 1+\frac{V(q)}{\hbar\omega}\sum_{\boldsymbol{k},i}q_i\frac{\partial f}{\partial k_i}\frac{\hbar\boldsymbol{k}\cdot\boldsymbol{q}}{m\omega}. \tag{6.265}$$

We change the summation $\sum_{\boldsymbol{k}}$ of the above equation into the integral $2/(2\pi)^3\int dk_x\, dk_y dk_z$ and integrate by parts, and then rewrite it in the form of $\sum_{\boldsymbol{k}}$, resulting in the relation

$$\kappa(0,\omega) = 1-V(q)\frac{q^2}{m\omega^2}\sum_{\boldsymbol{k}}f(\boldsymbol{k}) = 1-V(q)\frac{q^2N_\mathrm{e}}{m\omega^2} = 1-\frac{e^2}{\epsilon_0q^2L^3}\frac{q^2N_\mathrm{e}}{m\omega^2} = 1-\kappa_\infty\frac{\omega_\mathrm{p}^2}{\omega^2}, \tag{6.266}$$

where we have applied (6.158) to the Coulomb potential of electrons and used the relation $V(q)=e^2/\epsilon_0L^3q^2$. We used the relation $\sum f(\boldsymbol{k})=N_\mathrm{e}=L^3n$, where N_e is the total number of electrons and n is the density of electrons. When we define the plasma frequency ω_p by

$$\omega_\mathrm{p}^2 \equiv \frac{e^2n}{\kappa_\infty\epsilon_0m}, \tag{6.267}$$

the angular frequency ω_p agrees with the plasma frequency defined by (5.159). Now, we replace the 1 of the first term on the right-hand side of (6.261) or (6.266) with the background dielectric constant κ_∞, and finally we obtain

$$\kappa(0,\omega) = \kappa_\infty\left[1-\frac{\omega_\mathrm{p}^2}{\omega^2}\right], \tag{6.268}$$

where we find that the result is equivalent to (5.161).

The static dielectric constant including the screening effect is obtained by putting $\hbar\omega + \mathrm{i}\Gamma \to 0$ as

$$\kappa(q,0) = 1 - V(q)\sum_{\boldsymbol{k}} \frac{f(\boldsymbol{k}-\boldsymbol{q}) - f(\boldsymbol{k})}{\mathcal{E}(\boldsymbol{k}-\boldsymbol{q}) - \mathcal{E}(\boldsymbol{k})}. \tag{6.269}$$

This is easily estimated by using the Fermi–Dirac distribution function

$$f(\boldsymbol{k}) = \frac{1}{\mathrm{e}^{(\mathcal{E}(\boldsymbol{k})-\mu)/k_B T} + 1}, \tag{6.270}$$

where μ is the chemical potential, and the chemical potential at $T = 0$ is equal to the Fermi energy $\mathcal{E}_\mathrm{F}$ ($\mu(T=0) = \mathcal{E}_\mathrm{F}$). When we use the relation

$$\sum_i q_i \frac{\partial f(\boldsymbol{k})}{\partial k_i} = -\sum_i q_i \frac{\partial f(\boldsymbol{k})}{\partial \mu}\frac{\partial \mathcal{E}(\boldsymbol{k})}{\partial k_i} = -\sum_i q_i k_i \frac{\hbar^2}{m}\frac{\partial f(\boldsymbol{k})}{\partial \mu}, \tag{6.271}$$

we obtain

$$\begin{aligned}\kappa(q,0) &= \kappa_0 + \frac{e^2}{\epsilon_0 q^2}\frac{\partial}{\partial \mu}\frac{1}{L^3}\sum_{\boldsymbol{k}} f(\boldsymbol{k}) \\ &= \kappa_0 + \frac{e^2}{\epsilon_0 q^2}\frac{\partial n}{\partial \mu} \equiv \kappa_0\left(1 + \frac{q_\mathrm{s}^2}{q^2}\right),\end{aligned} \tag{6.272}$$

where q_s is given by

$$q_\mathrm{s} \equiv \frac{1}{\lambda_\mathrm{s}} = \sqrt{\frac{e^2}{\kappa_0\epsilon_0}\frac{\partial n}{\partial \mu}} \tag{6.273}$$

and is called the inverse of the screening length λ_s. The potential with static screening is written as

$$V_\mathrm{s}(q,0) = \frac{V(q)}{\kappa(q,0)} = \frac{e^2}{\kappa_0\epsilon_0 L^3}\frac{1}{q^2 + q_\mathrm{s}^2} \tag{6.274}$$

and its Fourier transform gives the following relation:

$$V_\mathrm{s}(\boldsymbol{r}) = \sum_{\boldsymbol{q}} \frac{e^2}{\kappa_0\epsilon_0 L^3}\frac{1}{q^2 + q_\mathrm{s}^2}\mathrm{e}^{\mathrm{i}\boldsymbol{q}\cdot\boldsymbol{r}} = \frac{e^2}{4\pi\kappa_0\epsilon_0 r}\mathrm{e}^{-q_\mathrm{s}\cdot r}. \tag{6.275}$$

This result coincides with the expressions (6.161) and (6.165) for electron scattering by screened ionized impurities given in Sect. 6.3.3 except for the screening length. Therefore, we will be concerned with the screening length in the following.

The Fermi energy in a degenerate material is given by (6.110). Putting $\mathcal{E}_F = \mu$, we obtain

$$\frac{\partial n}{\partial \mu} = \frac{3}{2}\frac{n}{\mathcal{E}_F}, \tag{6.276}$$

and thus the inverse of the screening length is given by

$$q_s = \sqrt{\frac{3e^2 n}{2\kappa_0 \epsilon_0 \mathcal{E}_F}} \equiv q_{TF} . \tag{6.277}$$

The quantity q_{TF} is called the **Thomas–Fermi screening wavenumber**. On the other hand, in a non-degenerate semiconductor, the Fermi-Dirac distribution function is approximated by the Boltzmann distribution function, so that the electron density is given by $n = N_c \exp[-(\mathcal{E}_c - \mathcal{E}_F)/k_B T]$, where N_c is the effective density of states of the conduction band. Therefore, we find

$$\frac{\partial \mu}{\partial n} = \frac{k_B T}{n}, \tag{6.278}$$

and the screening wavenumber is given by

$$q_s = \sqrt{\frac{e^2 n}{\kappa_0 \epsilon_0 k_B T}} \equiv q_{DH} , \tag{6.279}$$

where q_{DH} is called the **Debye–Hückel screening wavenumber** or the inverse of the **Debye screening length**, which corresponds to (6.162) and is used to take account of the screening effect on ionized impurity scattering.

Next, we derive the scattering rate of an electron by other electrons or holes, where the screening effect is properly taken into account. Let us define the position vectors of the i-th electron and the j-th electron (or hole) by $\boldsymbol{r}_i$ and $\boldsymbol{r}_j$, respectively. Then the interaction energy of these particles $V_{ee}(\boldsymbol{r}_i - \boldsymbol{r}_j)$ is Fourier transformed to give the following result, where we use (6.166):

$$V_{ee}(\boldsymbol{q}) = \frac{e^2}{\epsilon_0 L^3} \sum_{i \neq j} \sum_{\boldsymbol{q}} \frac{\exp[-i\boldsymbol{q} \cdot (\boldsymbol{r}_i - \boldsymbol{r}_j)]}{q^2} . \tag{6.280}$$

The electron–electron interaction is analyzed by classifying the interaction into two cases by using a cut-off wavenumber q_c. In the case of a short-range interaction, $q > q_c$, the interaction may be treated as two-particle scattering in a screened Coulomb potential. On the other hand, in the case of long-range scattering, $q < q_c$, the interaction is regarded as electron scattering by the collective excitation of electrons (or holes), the plasmon. In the latter case, the plasmon exhibits a coupling to LO phonons and in this case we have to take into account the electron scattering by LO-phonon coupled plasmons. The cut-off wavenumber may be taken to be the screening wavenumber q_s derived above, the Thomas-Fermi screening wavenumber q_{TF} or the Debye Hückel screening wavenumber q_{DH}.

First, we will consider the long-range interaction. Here we will drive the Hamiltonian for the interaction between the electron and the plasmon following the treatment of Kittel [28]. In this analysis we use the approximation of a continuous model, where there are n electrons in a unit volume and the positive charges of the immobile ions are distributed uniformly in the background, preserving charge neutrality ($|e|\rho_0 = n|e|$). For simplicity we neglect the energy change of ions due to the displacement of ions. The Hamiltonian density is therefore given by

$$\mathcal{H} = \frac{\hbar^2}{2nm} p_j p_j + \frac{1}{2} e(\rho - \rho_0) V(\boldsymbol{r}) , \tag{6.281}$$

where nm is the mass density of the gas and the factor $1/2$ in the second term on the right-hand side is used to equate the static energy with the free energy of the electron gas. The static potential is obtained from Poisson's equation as

$$\nabla^2 V = -\frac{1}{\epsilon_\infty} e(\rho - \rho_0) , \tag{6.282}$$

where the background dielectric constant $\kappa_\infty \epsilon_0 = \epsilon_\infty$ is taken into account. The change in the charge density $\delta\rho = (\rho - \rho_0)$ due to the local volume change of electrons is given by

$$\frac{\delta\rho}{\rho} = -\Delta(\boldsymbol{r}) , \tag{6.283}$$

where Δ is the volume change. This leads to the following relation:

$$\delta\rho = -\rho\Delta = -n\frac{\partial R_j}{\partial r_j} , \tag{6.284}$$

where R_j is the j-th component of the displacement. As shown in Sect. 5.5, the plasma oscillation is a longitudinal oscillation, $Q_q \parallel \boldsymbol{k}$, and we obtain

$$R(\boldsymbol{r}) = \sum_q Q_q \mathrm{e}^{\mathrm{i}\boldsymbol{q}\cdot\boldsymbol{r}} , \tag{6.285}$$

$$\delta\rho = -\mathrm{i}n \sum_q q Q_q \mathrm{e}^{\mathrm{i}\boldsymbol{q}\cdot\boldsymbol{r}} . \tag{6.286}$$

The Fourier transform of the potential $V(\boldsymbol{r})$ is easily obtained and we may write

$$V(\boldsymbol{r}) = \sum_q V_q \mathrm{e}^{\mathrm{i}\boldsymbol{q}\cdot\boldsymbol{r}} , \tag{6.287}$$

and so we find that

$$\nabla^2 V(\boldsymbol{r}) = -\sum_q q^2 V_q \mathrm{e}^{\mathrm{i}\boldsymbol{q}\cdot\boldsymbol{r}} . \tag{6.288}$$

Therefore, Poisson's equation may be rewritten as

$$V_q = \mathrm{i}\frac{ne}{\epsilon_\infty}\frac{1}{q}Q_q . \tag{6.289}$$

Using these relations the static potential term of (6.281) is calculated to give the following result:

$$\begin{aligned} \frac{1}{L^3}\int \mathrm{d}^3\boldsymbol{r}\frac{1}{2}e(\rho-\rho_0)V(\boldsymbol{r}) &= \sum_{q,q'}\int \mathrm{d}^3\boldsymbol{r}\frac{n^2e^2}{2\epsilon_\infty}Q_q Q_{q'}\mathrm{e}^{\mathrm{i}(\boldsymbol{q}-\boldsymbol{q}')\cdot\boldsymbol{r}}\frac{k}{k'} \\ &= \frac{n^2e^2}{2\epsilon_\infty}\sum_q Q_q Q_{-q} . \end{aligned} \tag{6.290}$$

Therefore, the Hamiltonian for the plasmon is written as (in the following treatment is exactly the same as in Sect. 6.1.2)

$$\begin{aligned} H &= \sum_q \left[\frac{\hbar^2}{2nm^*}P_q P_{-q} + \frac{n^2e^2}{2\epsilon_\infty}Q_q Q_{-q}\right] \\ &= \sum_q \left[\frac{\hbar^2}{2nm^*}P_q P_{-q} + \frac{nm^*}{2}\omega_p^2 Q_q Q_{-q}\right] . \end{aligned} \tag{6.291}$$

Following the treatment of Sect. 6.1.2, (6.55a), (6.55b) and (6.56) are also derived by putting $M \to nm^*$ and $\omega_q \to \omega_\mathrm{p}$. For example, Q_q is given by

$$Q_q = \sqrt{\frac{\hbar}{2L^3 nm^*\omega_\mathrm{p}}}(a_q + a_{-q}^\dagger) . \tag{6.292}$$

The Hamiltonian for the electron–plasmon interaction is written as

$$\begin{aligned} H_{\text{e-pl}} &= -eV(\boldsymbol{r}) = -e\sum_q V_q \mathrm{e}^{\mathrm{i}\boldsymbol{q}\cdot\boldsymbol{r}} = -\mathrm{i}\sum_q \frac{ne^2}{\epsilon_\infty}\frac{1}{q}Q_q \mathrm{e}^{\mathrm{i}\boldsymbol{q}\cdot\boldsymbol{r}} \\ &= -\mathrm{i}\sum_q \frac{ne^2}{\epsilon_\infty}\frac{1}{q}\sqrt{\frac{\hbar}{2nL^3m^*\omega_\mathrm{p}}}\left(a_q\mathrm{e}^{\mathrm{i}\boldsymbol{q}\cdot\boldsymbol{r}} + a_{-q}^\dagger\mathrm{e}^{-\mathrm{i}\boldsymbol{q}\cdot\boldsymbol{r}}\right) \\ &= -\mathrm{i}\sum_q \frac{1}{\sqrt{L^3}}\frac{e}{\sqrt{2\epsilon_\infty}}\sqrt{\hbar\omega_\mathrm{p}}\frac{1}{q}\left(a_q\mathrm{e}^{\mathrm{i}\boldsymbol{q}\cdot\boldsymbol{r}} + a_{-q}^\dagger\mathrm{e}^{-\mathrm{i}\boldsymbol{q}\cdot\boldsymbol{r}}\right) , \end{aligned} \tag{6.293}$$

where the relation $ne^2/\epsilon_\infty m^* = \omega_\mathrm{p}^2$ is used. Using this result, the matrix element for electron scattering by plasmons is given by

$$\left|M(\boldsymbol{k},\boldsymbol{k}')\right|^2 = \frac{e^2}{2\epsilon_\infty L^3}\hbar\omega_\mathrm{p}\frac{1}{q^2}\left(n_\mathrm{p}+\frac{1}{2}\mp\frac{1}{2}\right)\delta_{\boldsymbol{k}',\boldsymbol{k}\pm\boldsymbol{q}}\,, \tag{6.294}$$

where n_p is the excitation number of a plasmon with energy $\hbar\omega_\mathrm{p}$ and is given by Bose–Einstein statistics as

$$n_\mathrm{p} = \frac{1}{\exp(\hbar\omega_\mathrm{p}/k_\mathrm{B}T)-1}\,.$$

The scattering rate for the electron–plasmon interaction is

$$w_\mathrm{pl} = \frac{2\pi}{\hbar}\sum_{\boldsymbol{k}'}\left|M(\boldsymbol{k},\boldsymbol{k}')\right|^2\delta\left(\mathcal{E}(\boldsymbol{k}')-\mathcal{E}(\boldsymbol{k})\mp\hbar\omega_\mathrm{p}\right)\,. \tag{6.295}$$

Using the relation $\delta_{\boldsymbol{k}',\boldsymbol{k}\mp\boldsymbol{q}}$ and putting $q^2 = k'^2 + k^2 - 2kk'\cos\theta$ (θ is the angle between $\boldsymbol{k}'$ and $\boldsymbol{k}$), and replacing the summation with respect to $\boldsymbol{k}'$ by the integral

$$\sum_{\boldsymbol{k}'} = \frac{L^3}{(2\pi)^3}\int \mathrm{d}^3\boldsymbol{k}' = \frac{L^3}{(2\pi)^3}\int 2\pi k'^2\mathrm{d}k'\int\mathrm{d}(\cos\theta)\,, \tag{6.296}$$

we obtain the following results for the upper limit of the integral q_c. For $\cos(\theta) = \mp 1$ we put

$$q_\mathrm{max}^2 = k'^2 + k^2 + 2k'k = (k'+k)^2 = q_\mathrm{c}^2\,, \tag{6.297}$$

$$q_\mathrm{min}^2 = k'^2 + k^2 - 2k'k = (k'-k)^2\,, \tag{6.298}$$

which give the following result:

$$\int_{-1}^{+1}\frac{1}{q^2}\mathrm{d}(\cos\theta) = \int_{-1}^{1}\frac{1}{k'^2+k^2-2k'k\cos\theta}\mathrm{d}(\cos\theta)$$
$$= -\frac{1}{2k'k}\ln\frac{(k'-k)^2}{(k'+k)^2} = \frac{1}{k'k}\ln\frac{q_\mathrm{max}}{q_\mathrm{min}}\,. \tag{6.299}$$

From these results we obtain the scattering rate for the electron–plasmon interaction as

$$w_\mathrm{pl} = \frac{e^2}{4\pi\hbar\epsilon_\infty}\sqrt{\frac{2m^*}{\hbar^2}}\frac{\hbar\omega_\mathrm{p}}{\sqrt{\mathcal{E}(\boldsymbol{k})}}\left(n_\mathrm{p}+\frac{1}{2}\mp\frac{1}{2}\right)\ln\left(\frac{q_\mathrm{max}}{q_\mathrm{min}}\right)\,, \tag{6.300}$$

where a parabolic energy band structure, $\mathcal{E}(\boldsymbol{k}) = \hbar^2k^2/2m^*$, with scalar effective mass m^* is assumed. The upper limit of the integral q_max is taken to be q_c, which is given by the Thomas–Fermi screening wavenumber q_TF or the Debye–Hückel screening wavenumber q_DH. Using

$$q_{\min} = k\left|1-\sqrt{1\pm\frac{\hbar\omega_{\rm p}}{\mathcal{E}(\boldsymbol{k})}}\right| \equiv k\left|1-\sqrt{\eta_\pm}\right| , \tag{6.301}$$

$$q_{\max} \equiv q_c = q_{\rm TF} \quad (q_{\rm DH}) , \tag{6.302}$$

Equation (6.300) is rewritten as

$$w_{\rm pl} = \frac{e^2}{4\pi\hbar\epsilon_\infty}\sqrt{\frac{2m^*}{\hbar^2}}\frac{\hbar\omega_{\rm p}}{\sqrt{\mathcal{E}(\boldsymbol{k})}} \times \left[n_{\rm p}\ln\left(\frac{q_c}{k\sqrt{\eta_+}-k}\right)\right.$$
$$\left. +(n_{\rm p}+1)\ln\left(\frac{q_{\rm c}}{k-k\sqrt{\eta_-}}\right)u(\eta_-)\right] , \tag{6.303}$$

where $\eta_\pm = 1 \pm \hbar\omega_{\rm p}/\mathcal{E}(\boldsymbol{k})$, $u(\eta_-) = 1$ ($\eta_- \geq 0$), and $u(\eta_-) = 0$ ($\eta_- < 0$).

Next we deal with the short-range interaction $q > q_{\rm c}$. In this case the electron–electron or electron–hole interaction is described by scattering due to a screened Coulomb potential. We assume that the distance between an electron and its counterpart particle, electron or hole, is given by r. As stated in Sect. 6.3.3, the screened Coulomb potential is given by

$$\phi(\boldsymbol{r}) = \frac{e^2}{4\pi\kappa_0\varepsilon_0}\frac{\exp(-q_{\rm s}r)}{r} , \tag{6.304}$$

where $q_{\rm s}$ is the Thomas-Fermi screening wavenumber (inverse of the Thomas–Fermi screening length). Defining the wave vectors of an electron by $\boldsymbol{k}$ and of its counterpart electron or hole by $\boldsymbol{k}_j$, the transition matrix element of the collision is written as (see (6.165) for derivation)

$$M(\boldsymbol{k},\boldsymbol{k}') = \langle \boldsymbol{k}',\boldsymbol{k}'_j|\phi(\boldsymbol{r})|\boldsymbol{k},\boldsymbol{k}_j\rangle$$
$$= \frac{e^2}{\kappa_0\varepsilon_0 L^3}\frac{1}{|\boldsymbol{k}'-\boldsymbol{k}|^2+q_{\rm s}^2}\delta_{\boldsymbol{k}'+\boldsymbol{k}'_j,\boldsymbol{k}+\boldsymbol{k}_j} . \tag{6.305}$$

Finally, we obtain the scattering rate for electron with wave vector $\boldsymbol{k}$ for electron–electron or electron–hole scattering given by the following relation

$$w_{\text{e-j}}(\boldsymbol{k}) = \sum_{\boldsymbol{k}_j}\sum_{\boldsymbol{k}'}\sum_{\boldsymbol{k}'_j}\frac{2\pi}{\hbar}|M|^2 f_j(\boldsymbol{k}_j)$$
$$\times\delta(\mathcal{E}(\boldsymbol{k}')+\mathcal{E}(\boldsymbol{k}'_j)-\mathcal{E}(\boldsymbol{k})-\mathcal{E}(\boldsymbol{k}_j)) , \tag{6.306}$$

where $f_j(\boldsymbol{k}_j)$ is the distribution function of the counterpart carriers.

6.3.11 Alloy Scattering

In a compound semiconductor consisting of three or more elements, each of the three elements is expected not to be periodic in the crystal. As an example let us consider a three-element compound semiconductor such as $A_xB_{1-x}C$; the crystal is usually assumed to consist of $(AC)_x$ and $(BC)_{1-x}$ on average in the ratio $x : (1-x)$. Under this assumption the energy band structure of $A_xB_{1-x}C$ is calculated with the average lattice constant and average pseudopotentials estimated from the ratio $x : (1-x)$ and the result has been shown to agree well with experimental observation. This assumption is called the **virtual-crystal approximation**. This approximation is based on the assumption that the atoms A and B are distributed uniformly in the ratio $x : (1-x)$ around the cation C. In real alloy compounds it is expected that the distribution is not uniform. This non-uniformity results in a local variation of the periodic potential and thus in electron scattering due to the non-uniform potential. This scattering is called **alloy scattering**. (Readers are recommended to refer to as the book of Ridley on alloy scattering [5].)

The alloy potential is Fourier transformed as

$$V_{\text{alloy}}(\boldsymbol{r}) = \sum_{\boldsymbol{q}} V_{\text{alloy}}(\boldsymbol{q}) \exp(\mathrm{i}\boldsymbol{q} \cdot \boldsymbol{r}) . \tag{6.307}$$

The Fourier coefficient $V_{\text{alloy}}(\boldsymbol{q})$ is taken to be the root-mean square of the shift from the average energy and assumed to be independent of $\boldsymbol{q}$. Defining the Fourier coefficients for the A and B atoms by V_a and V_b, the average potential of the Fourier coefficients is given by

$$V_0 = V_a x + V_b(1-x) . \tag{6.308}$$

When the occupation of atom A is changed from x to x', the change in the potential is given by

$$V' - V_0 = (V_a - V_b)(x' - x) . \tag{6.309}$$

Therefore, the root-mean square value of the potential difference is given by

$$|\langle V' - V_0 \rangle| = |V_a - V_b| \left[\frac{x(1-x)}{N_c} \right]^{1/2} , \tag{6.310}$$

where N_c is the number of cations C and corresponds to the unit cell number. The matrix element for the scattering is then written as

$$\langle \boldsymbol{k}'|H'|\boldsymbol{k}\rangle = |V_a - V_b| \left[\frac{x(1-x)}{N_c} \right]^{1/2} \delta_{\boldsymbol{k}\pm\boldsymbol{q},\boldsymbol{k}'} , \tag{6.311}$$

where $\Omega = L^3/N_c$ is the unit cell volume and the scattering rate is given by

$$w_{\text{alloy}}(\boldsymbol{k}) = \frac{2\pi}{\hbar}(V_a - V_b)^2 \Omega x(1-x) \sum_{\boldsymbol{k}} \delta[\mathcal{E}(\boldsymbol{k} \pm \boldsymbol{q}) - \mathcal{E}(\boldsymbol{k})] . \tag{6.312}$$

Since $\sum_{\boldsymbol{k}} \delta[\mathcal{E}(\boldsymbol{k} \pm \boldsymbol{q}) - \mathcal{E}(\boldsymbol{k})]$ is the density of states, we rewrite it as follows

$$\begin{aligned} w_{\text{alloy}}(\boldsymbol{k}) &= \frac{2\pi}{\hbar}(V_{\text{a}} - V_{\text{b}})^2 \Omega x(1-x) \times \frac{2\pi}{(2\pi)^3} \frac{(2m^*)^{3/2}}{\hbar^3} \sqrt{\mathcal{E}(\boldsymbol{k})}, \\ &= (V_{\text{a}} - V_{\text{b}})^2 x(1-x) \Omega \frac{(2m^*)^{2/3}}{2\pi\hbar^4} \sqrt{\mathcal{E}(\boldsymbol{k})} \end{aligned} \tag{6.313}$$

As shown above, alloy scattering depends on $|V_{\text{a}} - V_{\text{b}}|$, but its value is not uniquely determined. The value depends on many factors such as the non-uniformity of the pseudopotentials, the symmetry of the conduction band, the electron affinity and so on, leading us to a difficulty in its determination. Instead, the value $|V_{\text{a}} - V_{\text{b}}|$ is used as a fitting parameter to get a good agreement of the calculated mobility with the experimental mobility.

6.4 Scattering Rate and Relaxation Time

We have to note that the scattering time (the inverse of the scattering rate) differs from the relaxation time for the reasons stated below. The scattering rate $w(\boldsymbol{k})$ is determined by summing all the possible finite states $\boldsymbol{k}'$ of electron scattering from an initial state $\boldsymbol{k}$ to a final state $\boldsymbol{k}'$, and is defined by

$$w(\boldsymbol{k}) = \sum_{\boldsymbol{k}'} P(\boldsymbol{k}, \boldsymbol{k}'), \tag{6.314}$$

where $P(\boldsymbol{k}, \boldsymbol{k}')$ is the transition rate of an electron from the initial state $\boldsymbol{k}$ to final state $\boldsymbol{k}'$ and is given by

$$\begin{aligned} P(\boldsymbol{k}, \boldsymbol{k}') &= \frac{2\pi}{\hbar} \left| \langle \boldsymbol{k}' | H_1 | \boldsymbol{k} \rangle \right|^2 \delta(\mathcal{E}_{\text{f}} - \mathcal{E}_{\text{i}}) \\ &= \frac{2\pi}{\hbar} |M(\boldsymbol{k}, \boldsymbol{k}')|^2 \delta(\mathcal{E}(\boldsymbol{k}') - \mathcal{E}(\boldsymbol{k}) \mp \hbar\omega_q). \end{aligned} \tag{6.315}$$

The scattering rate and relaxation time are often expressed as a function of the electron energy. When a parabolic band with isotropic effective mass or an ellipsoid is assumed, the electron energy $\mathcal{E}$ is expressed as

$$\mathcal{E}(\boldsymbol{k}) = \frac{\hbar^2 k^2}{2m^*}, \tag{6.316}$$

$$\mathcal{E}(\boldsymbol{k}) = \frac{\hbar^2}{2m_{\text{t}}}(k_x^2 + k_y^2) + \frac{\hbar^2}{2m_{\text{l}}} k_z^2. \tag{6.317}$$

The energy band structure exhibits a non-parabolic behavior in some cases and the behavior is well expressed by (2.97):

$$\frac{\hbar^2 k^2}{2m_0^*} \equiv \gamma(\mathcal{E}) = \mathcal{E}\left(1 + \frac{\mathcal{E}}{\mathcal{E}_{\text{G}}}\right). \tag{6.318}$$

In any case, the electron energy $\mathcal{E}$ is expressed as a function of the wave vector $\boldsymbol{k}$. In the following, the scattering rates for various scattering mechanisms are calculated by assuming a parabolic band with an isotropic effective mass.

On the other hand, the relaxation time is defined by (6.94) for the case of elastic scattering and depends on the scattering angle θ, resulting in a difference by a factor $(1 - \cos\theta)$ from the scattering rate. In this section we deal with calculations of the scattering rates and relaxation times for various types of scatterings. First, we discuss the changes in the wave vector and the energy of an electron caused by scattering, assuming the momentum and the energy conservation rules. The collision term $(\mathrm{d}f/\mathrm{d}t)_{\text{coll}}$, which is the change in the distribution function induced by scattering, is given by (6.80) with electron scattering probability $P(\boldsymbol{k}, \boldsymbol{k}')$. By replacing the sum over $\boldsymbol{k}$ by an integral, we obtain

$$\begin{aligned}\left(\frac{\mathrm{d}f}{\mathrm{d}t}\right)_{\text{coll}} &= \sum_{\boldsymbol{k}'} \left\{P(\boldsymbol{k}', \boldsymbol{k}) f(\boldsymbol{k}')[1 - f(\boldsymbol{k})] - P(\boldsymbol{k}, \boldsymbol{k}') f(\boldsymbol{k})[1 - f(\boldsymbol{k}')]\right\} \\ &= \frac{L^3}{(2\pi)^3} \int \mathrm{d}^3\boldsymbol{k}' \left\{P(\boldsymbol{k}', \boldsymbol{k}) f(\boldsymbol{k}')[1 - f(\boldsymbol{k})] \right. \\ &\quad \left. - P(\boldsymbol{k}, \boldsymbol{k}') f(\boldsymbol{k})[1 - f(\boldsymbol{k}')]\right\} . \end{aligned} \tag{6.319}$$

We have shown that the relaxation time τ is defined by (6.85) from the above relation. The collision term $(\mathrm{d}f/\mathrm{d}t)_{\text{coll}}$ may be calculated using the result for matrix elements calculated in the previous section, and thus the relaxation time $\tau(\boldsymbol{k})$ is evaluated. In general, the matrix element for the electron–phonon interaction is written as

$$|M(\boldsymbol{k}, \boldsymbol{k}')|^2 = A(q) \times \begin{cases} n_q & \text{(phonon absorption)}, \\ n_q + 1 & \text{(phonon emission)}. \end{cases} \tag{6.320}$$

It is evident that the matrix element for plasmon scattering is expressed in the same way. In the case of ionized impurity scattering, however, there exists no term related to n_q but instead the elastic scattering condition gives rise to $q = |\boldsymbol{k}' - \boldsymbol{k}|$ and $\mathcal{E}(\boldsymbol{k}') = \mathcal{E}(\boldsymbol{k})$.

In the case of phonon scattering, there are four terms corresponding to transition: (1) from initial state $\boldsymbol{k} + \boldsymbol{q}$ to final state $\boldsymbol{k}$ by emitting a phonon, (2) from $\boldsymbol{k}$ state to $\boldsymbol{k} + \boldsymbol{q}$ state by absorbing a phonon, (3) from $\boldsymbol{k}$ state to $\boldsymbol{k} - \boldsymbol{q}$ state by emitting a phonon, and (4) from $\boldsymbol{k} - \boldsymbol{q}$ state to $\boldsymbol{k}$ state by absorbing a phonon. Therefore, (6.319) should be rewritten as

$$\begin{aligned}\left(\frac{\mathrm{d}f}{\mathrm{d}t}\right)_{\text{coll}} &= \frac{2\pi}{\hbar} \sum_{\boldsymbol{q}} A(q) \Big\{ (n_q + 1) f(\boldsymbol{k} + \boldsymbol{q})[1 - f(\boldsymbol{k})] \\ &\quad - n_q f(\boldsymbol{k})[1 - f(\boldsymbol{k} + \boldsymbol{q})] - (n_q + 1) f(\boldsymbol{k})[1 - f(\boldsymbol{k} - \boldsymbol{q})] \\ &\quad + n_q f(\boldsymbol{k} - \boldsymbol{q})[1 - f(\boldsymbol{k})] \Big\} \delta\left\{\mathcal{E}(\boldsymbol{k}') - \mathcal{E}(\boldsymbol{k}) \mp \hbar\omega_q\right\} , \end{aligned} \tag{6.321}$$

where the summation over $\boldsymbol{k}'$ is replaced by a summation over $\boldsymbol{q}$ using the relation $\boldsymbol{k}' = \boldsymbol{k} \pm \boldsymbol{q}$. When the electron distribution function is approximated by Maxwell–

Boltzmann statistics, we may put $f(\boldsymbol{k} \pm \boldsymbol{q}) \ll 1$ and $f(\boldsymbol{k}) \ll 1$, and then we have

$$\left(\frac{\mathrm{d}f}{\mathrm{d}t}\right)_{\text{coll}} = \frac{2\pi}{\hbar}\sum_{\boldsymbol{q}} A(q)\left\{(n_q + 1)\left[f(\boldsymbol{k}+\boldsymbol{q}) - f(\boldsymbol{k})\right]\right.$$
$$\left. + n_q\left[f(\boldsymbol{k}-\boldsymbol{q}) - f(\boldsymbol{k})\right]\right\}\delta\left\{\mathcal{E}(\boldsymbol{k}\pm\boldsymbol{q}) - \mathcal{E}(\boldsymbol{k}) \mp \hbar\omega_q\right\}. \tag{6.322}$$

Calculation of the above equation is carried out by using the δ-function. The terms $\delta_{\boldsymbol{k}',\boldsymbol{k}\pm\boldsymbol{q}}$ in the matrix element and $\delta[\mathcal{E}(\boldsymbol{k}\pm\boldsymbol{q}) - \mathcal{E}(\boldsymbol{k}) \mp \hbar\omega_q]$ in the transition probability represent the momentum conservation and energy conservation rules, respectively. These relations may be written in the form:

$$\boldsymbol{k}' = \boldsymbol{k} \pm \boldsymbol{q}\,, \tag{6.323}$$

$$\mathcal{E}(\boldsymbol{k}') = \mathcal{E}(\boldsymbol{k}) \pm \hbar\omega_q\,. \tag{6.324}$$

Here we assume a parabolic band with isotropic scalar effective mass m^*, and so these relations lead to

$$\frac{\hbar^2}{2m^*}(\boldsymbol{k} \pm \boldsymbol{q})^2 = \frac{\hbar^2}{2m^*}k^2 \pm \hbar\omega_q\,. \tag{6.325}$$

When we define the angle between $\boldsymbol{k}$ and $\boldsymbol{q}$ by θ as shown in Fig. 6.13, we obtain

$$\pm 2kq\cos\beta + q^2 = \pm\frac{2m^*\omega_q}{\hbar}\,. \tag{6.326}$$

For acoustic phonons the relation $\omega_q = v_s q$ (v_s is the velocity of sound) holds, and we find

$$q = 2k\left(\mp\cos\beta \pm \frac{m^* v_s}{\hbar k}\right). \tag{6.327}$$

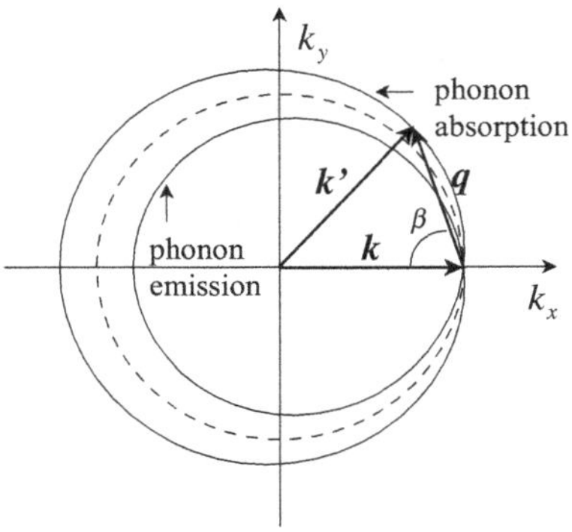

Fig. 6.13 Electron wave vectors before and after scattering with phonon absorption and emission. The *solid curves* are calculated assuming $m^* v_s/\hbar k = v_s/v_{th} = 0.1$

Table 6.2 Allowed range of phonon wave vectors q for electron–phonon scattering

(a) Long wavelength phonon

	Absorption	Emission
$q_{\max}$	$2k\left(1+\dfrac{m^* v_s}{\hbar k}\right)$	$2k\left(1-\dfrac{m^* v_s}{\hbar k}\right)$
$q_{\min}$	0	0

(b) High energy phonon (optical phonon)

	Absorption	Emission
$q_{\max}$	$k\left[\left(1+\dfrac{\hbar\omega_0}{\mathcal{E}}\right)^{1/2}+1\right]$	$k\left[1+\left(1-\dfrac{\hbar\omega_0}{\mathcal{E}}\right)^{1/2}\right]$
$q_{\min}$	$k\left[\left(1+\dfrac{\hbar\omega_0}{\mathcal{E}}\right)^{1/2}-1\right]$	$k\left[1-\left(1-\dfrac{\hbar\omega_0}{\mathcal{E}}\right)^{1/2}\right]$

Let us examine the magnitude of the second term in the brackets on the right-hand side of (6.327). The thermal velocity of the electrons v_{th} is estimated to be $v_{\text{th}} \simeq 2.5 \times 10^7$ cm/s at room temperature. When we assume $\hbar k \sim m^* v_{\text{th}}$, we find $v_s \simeq 5 \times 10^5$ cm/s and thus $m^* v_s/\hbar k \sim 2 \times 10^{-2}$, resulting in a small value at room temperature. The allowed values of q range from 0 to $2k(1 \pm m^* v_s/\hbar k)$ from (6.327), which are summarized in Table 6.2 for the long-wavelength phonon case together with the results for optical phonon scattering (high-energy phonon case).

The solid inner and outer curves of Fig. 6.13 show the scattered electron wave vector $\boldsymbol{k}'$ in the x, y plane for phonon emission and phonon absorption, respectively, where the curves are calculated assuming that $m^* v_s/\hbar k = 0.1$ and that the initial electron wave vector is directed along the x axis. The dotted curve represents $\boldsymbol{k}'$ for the case of $\hbar\omega_q = 0$, which corresponds to elastic scattering. It is seen in Fig. 6.13 that the wave vectors after scattering are distributed along almost circularly shaped curve and when $m^* v_s/\hbar k \ll 1$ is fulfilled, the scattering may be treated as elastic scattering.

On the other hand, in the case of optical phonon scattering or inter–valley phonon scattering, $\hbar\omega_q$ cannot be neglected compared to $\mathcal{E} = \hbar^2 k^2/2m^*$. In the latter case we obtain the following relation from (6.326) (where we put $\omega_q = \omega_0$):

$$(q - k\cos\beta)^2 = k^2\left(\cos^2\beta \pm \frac{\hbar\omega_0}{\mathcal{E}}\right), \tag{6.328}$$

which gives the results for the range of q, $q_{\min}$ and $q_{\max}$, summarized in Table 6.2.

6.4.1 Acoustic Phonon Scattering

In the case of electron–acoustic phonon scattering, the associated acoustic phonon energy is quite small. When the condition $\hbar\omega_q/k_BT \ll 1$ is fulfilled, we may approximate $n_q \simeq n_q + 1 \simeq k_BT/\hbar\omega_q = k_BT/\hbar v_s q$, and then (6.154) and (6.156) lead to

$$A(q)\left(n_q + \frac{1}{2} \mp \frac{1}{2}\right) = D_{ac}^2 q^2 \frac{\hbar}{2L^3\rho\omega_q}\frac{k_BT}{\hbar v_s q} = \frac{D_{ac}^2 k_BT}{2L^3\rho v_s^2}, \tag{6.329}$$

where we put $NM = L^3\rho$ for crystal density ρ. As shown in Fig. 6.14, we define the angle between the electron wave vector $\boldsymbol{k}$ ($\boldsymbol{k}'$) and the electric field $\boldsymbol{E}$ by θ (θ'). The summation or integral with respect to $\boldsymbol{k}'$ is replaced by an integral over q by using the Kronecker δ-function in the matrix element, which may be expressed as

$$\sum_{\boldsymbol{k}'} \to \sum_{\boldsymbol{q}} \to \frac{L^3}{(2\pi)^3}\int_0^{2\pi} d\varphi \int_0^{\pi} \sin\beta d\beta \int_{q_{min}}^{q_{max}} q^2 dq\,. \tag{6.330}$$

Therefore, the scattering rate $w_{ac}(\boldsymbol{k})$ is given by the following relation:

$$\begin{aligned} w_{ac}(\boldsymbol{k}) &= \sum_{\boldsymbol{k}'} P(\boldsymbol{k}, \boldsymbol{k}') \\ &= \frac{L^3}{(2\pi)^3}\frac{2\pi}{\hbar}\int A(q)\left(n_q + \frac{1}{2} \mp \frac{1}{2}\right) q^2 \sin\beta d\beta d\varphi dq \\ &\quad \times\delta\left[\mathcal{E}(\boldsymbol{k}') - \mathcal{E}(\boldsymbol{k}) \mp \hbar\omega_q\right] \\ &= \frac{L^3}{(2\pi)^3}\frac{(2\pi)^2}{\hbar}\int_0^{2k} \frac{D_{ac}^2 k_BT}{2L^3\rho v_s^2} dq \int d(-\cos\beta) \\ &\quad \times\delta\left[\frac{\hbar^2}{2m^*}(2kq\cos\beta + q^2) \pm \hbar\omega_q)\right]. \end{aligned} \tag{6.331}$$

We should note here that the $\pm$ sign in the δ-function represents phonon emission and absorption. Here we use the property of the δ-function given by

$$\delta(Ax) = \frac{1}{A}\delta(x) \quad (A > 0)\,.$$

The scattering rate for acoustic phonon scattering is then given by

$$\begin{aligned} w_{ac} &= \frac{D_{ac}^2 k_BT}{2\pi\hbar\rho v_s^2}\int_0^{2k} q^2 \frac{m^*}{\hbar^2 kq} dq = \frac{D_{ac}^2 k_BT}{2\pi\hbar\rho v_s^2}\frac{m^*}{\hbar^2 k}\int_0^{2k} q dq \\ &= \frac{D_{ac}^2 m^* k_BT}{\pi\hbar^3\rho v_s^2} k = \frac{(2m^*)^{3/2} D_{ac}^2 k_BT}{2\pi\hbar^4\rho v_s^2}\mathcal{E}^{1/2}\,. \end{aligned} \tag{6.332}$$

Next, following the process used to derive (6.94), we calculate the relaxation time τ_{ac} for acoustic phonon scattering from (6.322). Equation (6.322) is rewritten as

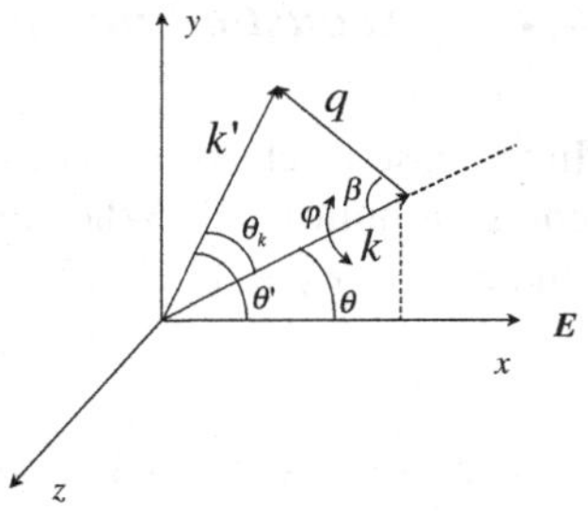

Fig. 6.14 Relation of the angles between initial electron wave vector $\boldsymbol{k}$, phonon wave vector $\boldsymbol{q}$, and scattered electron wave vector $\boldsymbol{k}'$

$$\begin{aligned}\frac{1}{\tau_{\rm ac}} &= \frac{L^3}{(2\pi)^3}\int P(\boldsymbol{k},\boldsymbol{k}')\left(1-\frac{k'\cos\theta'}{k\cos\theta}\right)\mathrm{d}^3\boldsymbol{q}\\ &= \frac{L^3}{(2\pi)^3}\cdot\frac{2\pi}{\hbar}\int A(q)\left\{(n_q+1)\delta\left[\mathcal{E}(\boldsymbol{k}')-\mathcal{E}(\boldsymbol{k})+\hbar\omega_q\right]\right.\\ &\quad \left.+n_q\delta\left[\mathcal{E}(\boldsymbol{k}')-\mathcal{E}(\boldsymbol{k})-\hbar\omega_q\right]\right\}\\ &\quad\times\left(1-\frac{k'\cos\theta'}{k\cos\theta}\right)d\varphi\cdot\sin\beta\mathrm{d}\beta\cdot q^2\mathrm{d}q\,. \end{aligned} \tag{6.333}$$

Referring to the vectors shown in Fig. 6.14, we find the following relation:

$$k'\cos\theta' = k\cos\theta - (q\cos\beta\cos\theta + q\sin\beta\sin\theta\cos\varphi)\,. \tag{6.334}$$

Inserting this into (6.333), integrating over φ, and using the relation $\int_0^{2\pi}\cos\varphi\mathrm{d}\varphi = 0$, we obtain for the relaxation time

$$\begin{aligned}\frac{1}{\tau_{\rm ac}} &= \frac{L^3}{(2\pi)^3}\cdot\frac{2\pi}{\hbar}\cdot 2\pi\int_{\beta=0}^{\pi}\int_{q_{\min}}^{q_{\max}}\frac{q}{k}A(q)q^2\cos\beta\sin\beta\mathrm{d}\beta\mathrm{d}q\\ &\quad\times\left\{(n_q+1)\delta\left[\mathcal{E}(\boldsymbol{k}')-\mathcal{E}(\boldsymbol{k})+\hbar\omega_q\right]\right.\\ &\quad \left.+\,n_q\delta\left[\mathcal{E}(\boldsymbol{k}')-\mathcal{E}(\boldsymbol{k})-\hbar\omega_q\right]\right\}\,. \end{aligned} \tag{6.335}$$

The integral in the above equation may be easily evaluated out by using the property of δ-function as shown below. Here we define a new variable

$$y = \mathcal{E}_{\rm f} - \mathcal{E}_{\rm i} = \frac{\hbar^2}{2m^*}(2kq\cos\beta + q^2) \pm \hbar\omega_q\,, \tag{6.336}$$

which gives

$$\mathrm{d}y = \frac{\hbar^2}{2m^*}2kq\sin\beta\mathrm{d}\beta = \frac{\hbar^2}{m^*}kq\sin\beta\mathrm{d}\beta\,. \tag{6.337}$$

Then the integral with respect to β is written as

$$\int_{\beta=0}^{\pi}\cos\beta\sin\beta\mathrm{d}\beta\left\{(n_q+1)\delta(y_{\rm e}) + n_q\delta(y_{\rm a})\right\}\,, \tag{6.338}$$

where y_e corresponds to phonon emission (+ sign in (6.336)) and y_a to phonon absorption (− sign in (6.336)). From these relations we obtain

$$\int_{\beta=0}^{\pi} \cos\beta \sin\beta d\beta \cdot \delta(y) = \frac{m^*}{\hbar^2 kq} \int \cos\beta dy \cdot \delta(y) = \frac{m^*}{\hbar^2 kq} \cos\beta \,. \tag{6.339}$$

The value of $\cos\beta$ is determined from (6.326) by putting $y = 0$ (arising from the term $dy\delta(y)$). The relaxation time then becomes

$$\frac{1}{\tau_{ac}} = \frac{L^3}{2\pi\hbar} \int_{q_{min}}^{q_{max}} \frac{1}{k} A(q) \left\{ (n_q + 1) \left(\frac{q}{2k} + \frac{m^* v_s}{\hbar k} \right) + n_q \left(\frac{q}{2k} - \frac{m^* v_s}{\hbar k} \right) \right\} \frac{m^* q^2}{\hbar^2 k} dq \,. \tag{6.340}$$

Using (6.154) and (6.156), evaluating $A(q)$ from (6.320), and putting $q/2k \gg m^* v_s/\hbar k$, $n_q + 1 \approx n_q \approx k_B T/\hbar\omega_q$ and $\omega_q = v_s q$, the relaxation time for acoustic phonon scattering is given by (integrate from $q_{min} = 0$ to $q_{max} = 2k$)

$$\frac{1}{\tau_{ac}} = \frac{m^* D_{ac}^2}{4\pi\rho\hbar^2 2k^3} \cdot \frac{2k_B T}{\hbar v_s^2} \cdot \frac{(2k)^4}{4} = \frac{D_{ac}^2 m^* k_B T}{\pi\hbar^3 \rho v_s^2} k = \frac{(2m^*)^{3/2} D_{ac}^2 k_B T}{2\pi\hbar^4 \rho v_s^2} \mathcal{E}^{1/2} \,, \tag{6.341}$$

where $\rho = NM/L^3$ is the crystal density, $\rho v_s^2 = c_{11}$ for pure longitudinal acoustic waves and c_{11} is the elastic constant. The scalar effective mass m^* is to be understood as the density of states mass $m_d^* = (m_l m_t^2)^{1/3}$ for ellipsoidal energy surface like in Si and Ge. When we cannot approximate as $n_q \approx k_B T/\hbar\omega_q$ (equipartition is not valid), the calculation of the relaxation time is complicated but the result has been obtained by Conwell and Brown [29]. Now, comparing the result of the inverse of the relaxation time with the scattering rate of (6.332), we find that both are equivalent, and thus in the case of acoustic phonon scattering (except in the low temperature range) the following relation holds:

$$w_{ac}(\mathcal{E}) = \frac{1}{\tau_{ac}(\mathcal{E})} \,. \tag{6.342}$$

This conclusion may be understood from the fact that the scattering is elastic and isotropic as shown in Fig. 6.13 in the case of $m^* v_s/\hbar k \ll 1$.

The mean-free path l_{ac} of electrons for acoustic phonon scattering is defined by

$$\frac{1}{\tau_{ac}} = \frac{v}{l_{ac}} \,, \tag{6.343}$$

and it is evaluated as

$$l_{ac} = \frac{\pi\hbar^4 \rho v_s^2}{m^{*2} D_{ac}^2 k_B T} \,. \tag{6.344}$$

The above result is obtained for an isotropic scalar effective mass. For an ellipsoidal conduction band the result should be modified by replacing m^* with the effective density of mass $m_\mathrm{d}^* = (m_\mathrm{t}^2 m_\mathrm{l})^{1/3}$. In many–valley semiconductors such as Ge and Si, however, the deformation potentials are expressed by (6.246) and (6.247) of Sect. 6.3.8 and the elastic constants (and thus the velocity of sound) are anisotropic, and therefore the relaxation time becomes very complicated, as shown by Herring and Vogt [8]. In the following we calculate the scattering rates or inverse of relaxation times at room temperature $T = 300$ K by using the following material parameters given in Table 6.3. It should be noted here that various values of material parameters have been reported and thus the calculated results are subject to change with the choice of the parameters (see also Table 6.1). The sound velocity depends on the propagation direction and quasi–longitudinal waves will take part in the electron–acoustic wave interaction. The anisotropy of the quasi–longitudinal acoustic waves is about 10%. Therefore for simplicity, the acoustic weaves are assumed to be pure longitudinal ($\rho v_\mathrm{s}^2 = c_{11}$) and the numerical evaluations of the scattering rate and mobility are carried out by assuming the pure longitudinal acoustic waves.

Table 6.3 Material parameters used to calculate transport properties of Si, Ge, GaAs, and InAs

	Si	Ge	GaAs	InAs
Transverse effective mass: m_t/m	0.19	0.082	–	–
Longitudinal effective mass: m_l/m	0.98	1.58	–	–
Density of states mass: m_d	0.3283	0.2198	–	–
Conductivity effective mass: m_c	0.2598	0.1199	–	–
Effective mass: m^*/m	–	–	0.067	0.022
Static dielectric constant: κ_0	12.0	16.0	12.90	15.15
High frequency dielectric constant: κ_∞	–	–	10.92	12.25
Longitudinal optical phonon energy: $\hbar\omega_\mathrm{LO}(\Gamma_{25'})$ [meV]	64.35	37.30	35.36	29.58
Transverse optical phonon energy: $\hbar\omega_\mathrm{TO}(\Gamma_{25'})$ [meV]	64.35	37.30	33.17	26.94
Mass density: ρ [10^3 kg/m^3]	2.332	5.323	5.36	5.71
Elastic constant: c_{11} [10^{10} N/m^2]	16.58	12.853	11.76	8.329
Elastic constant: c_{12} [10^{10} N/m^2]	6.39	4.826	5.27	4.526
Elastic constant: c_{44} [10^{10} N/m^2]	7.96	6.680	5.96	3.959
Longitudinal sound velocity: v_l (LA) [10^3 m/s]	8.432	4.914	4.684	3.819
Transverse sound velocity: v_t (TA) [10^3 m/s]	5.842	3.543	3.335	2.509
Dilational deformation potential Ξ_d [eV]	$\simeq$5	−12.3	–	–
Uniaxial deformation potential: Ξ_u [eV]	9.2	16.3	–	–
Deformation potential (averaged): Ξ_l † [eV]	(14.5)	(5.91)		
Deformation potential used for mobility analysis: Ξ [eV]	9.0	11.5	14.0	4.9
Piezoelectric constant: e_{14} [C/m^2]	–	–	0.160	0.045

† Ξ_l is the average value for longitudinal acoustic waves calculated by the relation of Herring and Vogt.
(see C. Herring and E. Vogt, Phys. Rev. **101**, (1956) 944)

Scattering rates $1/\tau_{ac}$ for Si and Ge are shown in Figs. 6.15 and 6.16, respectively, and the rates are $1 \sim 2 \times 10^{13}$ [1/s] for electron energy ~0.5 [eV]. In Fig. 6.16 scattering rates for the deformation potential $\Xi = 11.5$ [eV] are also plotted. This deformation potential is estimated to give the electron mobility in n–Ge $\mu_{ac+op} = 3800$ [cm²/Vs] at 300 [K] as shown in Fig. 6.27, in which acoustic phonon scattering and optical phonon scattering are taken into account as described in Sect. 6.5.2 (see Fig. 6.27) and similarly the deformation potential $\Xi = 9.0$ is determined from the comparison of the calculated electron mobility due to acoustic deformation potential and interband phonon scatterings with the experimental results $\mu = 1450$ [cm²/Vs] as discussed later.

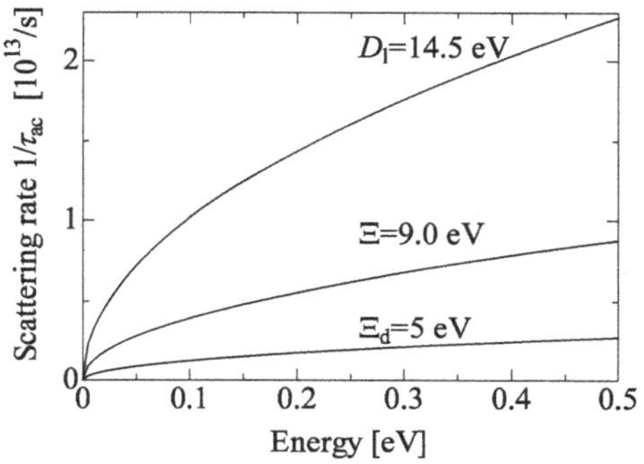

Fig. 6.15 Scattering rates $w = 1/\tau_{ac}$ for deformation potential-type acoustic phonon scattering in Si are plotted as a function of electron energy at $T = 300$ K. The deformation potentials used in the calculation are $\Xi_d = 5$ and $D_{ac} = 14.2$ [eV]. $\Xi = 9.0$ eV is determined from the mobility analysis described in Sect. 6.5

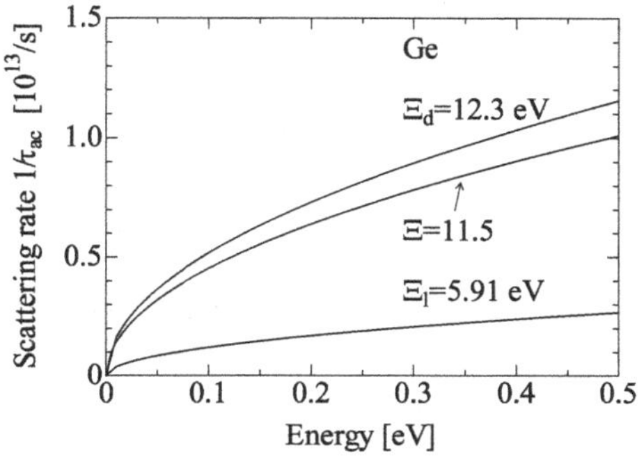

Fig. 6.16 Scattering rates $w = 1/\tau_{ac}$ for deformation potential–type acoustic phonon scattering in Ge are plotted as a function of electron energy at $T = 300$ K. The deformation potentials used are $\Xi_d = -12.3$ and $\Xi_l = 5.91$ [eV]. $\Xi = 11.5$ eV is determined from the mobility analysis described in Sect. 6.5

6.4.2 Non-polar Optical Phonon Scattering

In general the optical phonon energy is $\hbar\omega_0 \approx 50\,\text{meV}$ and is higher than the average energy (thermal energy) of an electron $\mathcal{E} = k_B T \approx 25\,\text{meV}$, at room temperature. Therefore, the electron loses a large amount of energy in scattering by optical phonons, resulting in the invalidity of elastic scattering. This means that the approximation used for evaluating the relaxation time for acoustic phonon scattering cannot be adopted, giving rise to a very difficult task for evaluating the relaxation time. Usually, the θ-dependence is ignored in (6.333), and we put

$$\left(1 - \frac{k' \cos\theta'}{k\cos\theta}\right) = 1\,. \tag{6.345}$$

This approximation is often called the **randomizing collision approximation**. Under this approximation, the inverse of relaxation time is equivalent to the scattering rate. In many books and papers this approximation is adopted and the momentum relaxation time is evaluated from the scattering rate. The scattering rate for non-polar optical phonon scattering is given by

$$\begin{aligned} w_{op} &= \frac{L^3}{(2\pi)^3}\int_{q_{min}}^{q_{max}} A(q)\sin\beta \mathrm{d}\beta q^2 \mathrm{d}q \\ &\times \left\{(n_q+1)\delta\left[\mathcal{E}(\boldsymbol{k}') - \mathcal{E}(\boldsymbol{k}) + \hbar\omega_q\right] + n_q\delta\left[\mathcal{E}(\boldsymbol{k}') - \mathcal{E}(\boldsymbol{k}) - \hbar\omega_q\right]\right\}. \end{aligned} \tag{6.346}$$

Using the following relation

$$\int_{\beta=0}^{\pi} \sin\beta \mathrm{d}\beta \cdot \delta(y) = \frac{m^*}{\hbar^2 kq}\,, \tag{6.347}$$

the scattering rate is evaluated as

$$\begin{aligned} w_{op} &= \frac{L^3}{2\pi\hbar}\frac{m^*}{\hbar^2 k}\left\{\int_{q_{min}}^{q_{max}} A(q)(n_q+1)q\mathrm{d}q + \int_{q_{min}}^{q_{max}} A(q)n_q \cdot qD\right\} \\ &= \frac{(2m^*)^{3/2}}{4\pi\hbar^3\rho}\frac{D_{op}^2}{\omega_0}\left[(n_q+1)\sqrt{\mathcal{E} - \hbar\omega_0} + n_q\sqrt{\mathcal{E} + \hbar\omega_0}\right], \end{aligned} \tag{6.348}$$

where the range of the phonon wave vector q given in Table 6.2 is used. From (6.204) of the previous section the coefficient $A(q)$ is given by $A(q) = D_{op}^2\hbar/(2L^3\rho\omega_0)$. With the deformation potential E_{lop} defined by (6.205), the scattering rate for the electron–non-polar optical phonon interaction is given by

$$w_{op} = \frac{E_{lop}^2}{D_{ac}^2} \cdot \frac{x_0}{2(e^{x_0} - 1)} \cdot \frac{\sqrt{2/m^*}}{l_{ac}}\left[\sqrt{\mathcal{E} + \hbar\omega_0} + e^{x_0}\sqrt{\mathcal{E} - \hbar\omega_0}\right], \tag{6.349}$$

where $x_0 = \hbar\omega_0/k_B T$. For an ellipsoidal conduction band it is evident that the scattering rate is obtained by replacing $m^{*3/2}$ with $(m_t^2 m_l)^{1/2}$.

When the non-polar optical phonon scattering is assumed to be elastic like acoustic phonon scattering and a relation similar to (6.333) is used for the non-polar optical phonon scattering, the momentum relaxation time $1/\tau_{op}$ is written as

$$\frac{1}{\tau_{op}} = \frac{L^3}{2\pi\hbar}\int_{q_{min}}^{q_{max}} \frac{1}{k}A(q)\left\{(n_q+1)\left(\frac{q}{2k}+\frac{m^*\omega_0}{\hbar kq}\right)\right.$$
$$\left. + n_q\left(\frac{q}{2k}-\frac{m^*\omega_0}{\hbar kq}\right)\right\}\frac{m^*q^2}{\hbar^2 k}\mathrm{d}q\,, \tag{6.350}$$

which leads to the following final result for the momentum relaxation time for non-polar optical phonon scattering:

$$\frac{1}{\tau_{op}} = \frac{(2m^*)^{3/2}}{4\pi\hbar^3\rho}\frac{D_{op}^2}{\omega_0}\left[(n_q+1)\sqrt{\mathcal{E}-\hbar\omega_0}+n_q\sqrt{\mathcal{E}+\hbar\omega_0}\right]. \tag{6.351}$$

Again, we find that the inverse of the relaxation time is equal to the scattering rate of (6.351) ($w_{op} = 1/\tau_{op}$). From these results, (6.351) is used for both the scattering time and relaxation time for non-polar optical phonon scattering.

Calculated scattering rates of non-polar optical phonon scattering given by (6.349) are shown in Fig. 6.17, where scattering rates of absorption, emission, and total processes are shown. The used parameters are $E_{lop}^2/D_{ac}^2 = b = 0.4$ and $D_{ac} = 8.55$ [eV], and other parameters are given in Table 6.3.

6.4.3 *Polar Optical Phonon Scattering*

Polar optical phonon scattering is also inelastic and thus we derive the scattering rate w_{pop} according to the procedure we have adopted to obtain the scattering rate for non-polar optical phonon scattering. Equation (6.242) gives rise to the polar optical phonon scattering rate given by

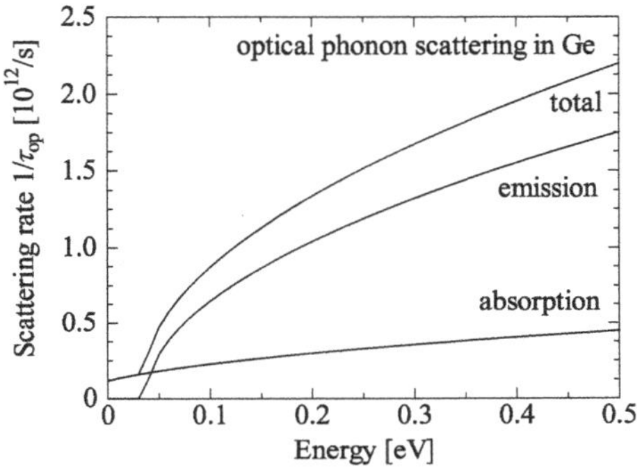

Fig. 6.17 Scattering rates $w = 1/\tau_{op}$ for non-polar optical phonon scattering in Ge are plotted as a function of electron energy at $T = 300$ K, where absorption, emission, and total processes are separately shown for comparison

$$
\begin{aligned}
&w_{\text{pop}}(\mathcal{E}) \\
&= \frac{e^2\omega_{\text{LO}}}{4\pi\epsilon_0}\left(\frac{1}{\kappa_\infty}-\frac{1}{\kappa_0}\right)\frac{m^*}{\hbar^2 k}\left[\int_{q_{\min}}^{q_{\max}}(n_q+1)\frac{\mathrm{d}q}{q}+\int_{q_{\min}}^{q_{\max}} n_q\frac{\mathrm{d}q}{q}\right] \\
&= \frac{e^2\omega_{\text{LO}}}{4\sqrt{2}\pi\epsilon_0\hbar}\left(\frac{1}{\kappa_\infty}-\frac{1}{\kappa_0}\right)\frac{\sqrt{m^*}}{\sqrt{\mathcal{E}}}\left[(n_q+1)\ln\left|\frac{\sqrt{\mathcal{E}}+\sqrt{\mathcal{E}-\hbar\omega_{\text{LO}}}}{\sqrt{\mathcal{E}}-\sqrt{\mathcal{E}-\hbar\omega_{\text{LO}}}}\right|\right. \\
&\quad\left. +n_q\ln\left|\frac{\sqrt{\mathcal{E}+\hbar\omega_{\text{LO}}}+\sqrt{\mathcal{E}}}{\sqrt{\mathcal{E}+\hbar\omega_{\text{LO}}}-\sqrt{\mathcal{E}}}\right|\right] \\
&= \frac{e^2\omega_{\text{LO}}}{2\sqrt{2}\pi\epsilon_0\hbar}\left(\frac{1}{\kappa_\infty}-\frac{1}{\kappa_0}\right)\frac{\sqrt{m^*}}{\sqrt{\mathcal{E}}}\left[(n_q+1)\sinh^{-1}\sqrt{\frac{\mathcal{E}-\hbar\omega_{\text{LO}}}{\hbar\omega_{\text{LO}}}}\right. \\
&\quad\left. +n_q\sinh^{-1}\sqrt{\frac{\mathcal{E}}{\hbar\omega_{\text{LO}}}}\right]. \qquad (6.352)
\end{aligned}
$$

When the relation $\langle\mathcal{E}\rangle \gg \hbar\omega_{\text{LO}}$ holds for the average electron energy $\langle\mathcal{E}\rangle$ and the LO phonon energy ω_{LO}, the scattering is assumed to be elastic and the treatment used for acoustic phonon scattering may be applied. Then the scattering rate for polar optical phonon scattering is given by

$$
\begin{aligned}
w_{\text{pop}}(\mathcal{E}) &= \frac{1}{\tau_{\text{pop}}} \\
&= \frac{e^2\omega_{\text{LO}}}{4\pi\epsilon_0}\left(\frac{1}{\kappa_\infty}-\frac{1}{\kappa_0}\right)\frac{m^*}{\hbar^2 k}\left[\frac{n_q+1}{2k}\int_0^{2k}\mathrm{d}q+\frac{n_q}{2k}\int_0^{2k}\mathrm{d}q\right] \\
&= \frac{e^2\omega_{\text{LO}}}{4\sqrt{2}\pi\epsilon_0\hbar}\left(\frac{1}{\kappa_\infty}-\frac{1}{\kappa_0}\right)\frac{\sqrt{m^*}}{\sqrt{\mathcal{E}}}(2n_q+1)\ (\mathcal{E}\gg\hbar\omega_{\text{LO}}). \qquad (6.353)
\end{aligned}
$$

The same assumption as for non-polar optical phonon scattering gives the momentum relaxation time τ_{pop} shown below from (6.333):

$$
\begin{aligned}
\frac{1}{\tau_{\text{pop}}} &= \frac{L^3}{2\pi\hbar}\int_{q_{\min}}^{q_{\max}}\frac{1}{k}A(q)\left\{(n_q+1)\left(\frac{q}{2k}+\frac{m^*\omega_0}{\hbar k q}\right)\right. \\
&\quad\left.+n_q\left(\frac{q}{2k}-\frac{m^*\omega_0}{\hbar k q}\right)\right\}\frac{m^* q^2}{\hbar^2 k}\mathrm{d}q\,, \qquad (6.354)
\end{aligned}
$$

where $A(q)$ is taken from (6.242) as

$$
A(q) = \frac{e^2\hbar\omega_{\text{LO}}}{2L^3\epsilon_0}\left(\frac{1}{\kappa_\infty}-\frac{1}{\kappa_0}\right)\left(\frac{q}{q^2+q_s^2}\right)^2. \qquad (6.355)
$$

For simplicity, let the screening wavenumber $q_s = 0$. The relaxation time for polar optical phonon scattering is then given by

$$\frac{1}{\tau_{\rm pop}(\mathcal{E})} = \frac{e^2\omega_{\rm LO}}{4\sqrt{2}\pi\epsilon_0\hbar}\left(\frac{1}{\kappa_\infty}-\frac{1}{\kappa_0}\right)\frac{\sqrt{m^*}}{\sqrt{\mathcal{E}}} \times\left[(n_q+1)\left\{\sqrt{1-\frac{\hbar\omega_{\rm LO}}{\mathcal{E}}}+\frac{\hbar\omega_{\rm LO}}{\mathcal{E}}\sinh^{-1}\left(\frac{\mathcal{E}}{\hbar\omega_{\rm LO}}-1\right)^{1/2}\right\} +n_q\left\{\sqrt{1+\frac{\hbar\omega_{\rm LO}}{\mathcal{E}}}-\frac{\hbar\omega_{\rm LO}}{\mathcal{E}}\sinh^{-1}\left(\frac{\mathcal{E}}{\hbar\omega_{\rm LO}}\right)^{1/2}\right\}\right]. \tag{6.356}$$

This equation was first derived by Callen [30].

Polar optical phonon scattering rates (inverse of relaxation time) of polar optical phonon scattering in GaAs at $T = 300$ K are calculated by (6.356) and plotted by the solid curves in Fig. 6.18 as a function of electron energy by solid curves, where the total rate $1/\tau_{\rm pop}$, emission rate and absorption rates are separately shown. In addition we plotted the scattering rate $w_{\rm pop}$ given by (6.352). We find that the scattering rates are below 10^{13} [1/s].

6.4.4 *Piezoelectric Potential Scattering*

Since the phonon energy involved in types of scattering is quite small, we may use a similar approach to that for acoustic phonon scattering. Inserting (6.197) and (6.199) into (6.340), the relaxation time for piezoelectric potential scattering and equivalently the scattering rate is given by

$$\frac{1}{\tau_{\rm pz}} = \frac{e^2 m^* k_{\rm B}T}{2\pi\hbar^3}\left\langle\frac{e_{\rm pz}^{*2}}{c^*\epsilon^{*2}}\right\rangle\frac{1}{k} = \frac{e^2\sqrt{m^*}k_{\rm B}T}{2\sqrt{2}\pi\hbar^2}\left\langle\frac{e_{\rm pz}^{*2}}{c^*\epsilon^{*2}}\right\rangle\frac{1}{\sqrt{\mathcal{E}}}, \tag{6.357}$$

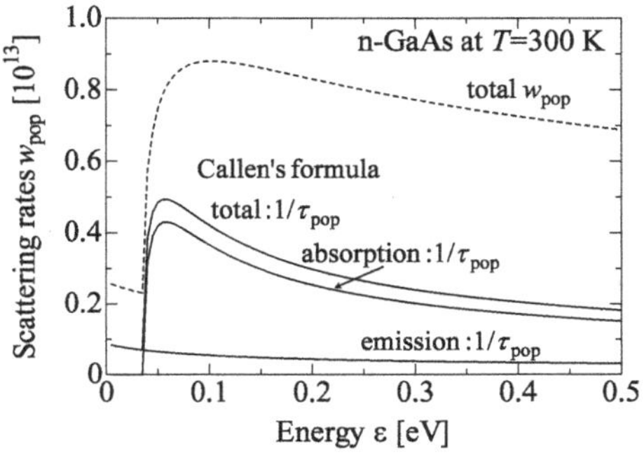

Fig. 6.18 Scattering rates $1/\tau_{\rm pop}$ for polar optical phonon scattering given by (6.356) in GaAs at $T = 300$ K are plotted as a function of electron energy, where absorption, emission, and total processes are separately shown for comparison. In addition scattering rate given by (6.352) are shown by the *dashed curve*

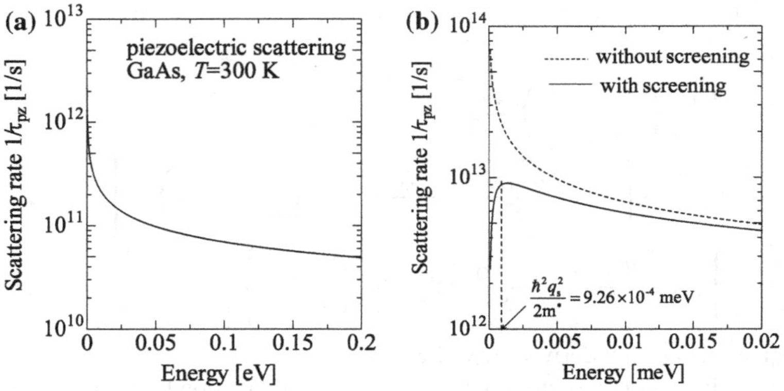

Fig. 6.19 Scattering rates of an electron by piezoelectric potential in GaAs at 300 K, where **a** no noticeable difference between the scattering rates with screening and without screening, except the low electron energy region (see curves in **b**), where the divergence of the scattering rates without the screening is removed in the case with screening effect as shown in **b**. The piezoelectric coupling constant is simplified as $\langle e_{pz}^{*2}/c^*\epsilon^{*2}\rangle = \langle e_{14}^2/c_{44}(\kappa_0\epsilon_0)^2\rangle$

where $\langle\ \rangle$ represents averaged values over the integral of $\boldsymbol{q}$ because $e_{pz}^* = e_{i,kl}a_i\pi_k a_l$, $c^* = c_{ijkl}\pi_i a_j \pi_k a_l$ $(= \rho v_s^2)$ and $\epsilon^* = \epsilon_{ij}^s a_i a_j$ depend on the crystallographic directions. A detailed treatment for the case of an ellipsoidal energy surface is given by Zook [2]. A relation $e_{pz}^{*2}/c^*\epsilon^* = K^{*2}$ is often used in the analysis, where K^* is called the electromechanical coupling coefficient and is known to be an important parameter for describing the strength of piezoelectricity.

The piezoelectric potential scattering rates (inverse of the relaxation time) $1/\tau_{pz}$ in GaAs at 300 K are plotted in Fig. 6.19a, where we take account of the piezoelectric scattering due to transverse acoustic waves propagating in the direction $\langle 110\rangle$ with polarization in the place (001) which gives the strongest interaction with electrons and $e_{pz}^* = e_{14}$ and $c^* = c_{44}$. Here we have to note that the difference in the scattering rates with and without the screening effect in negligible in higher energy region, $\mathcal{E} > \hbar^2 q_s^2/2m^*$, Fig. 6.19b shows the scattering rates with and without the screening effect in the lower energy region, where we find that the screening effect removes the divergence of the scattering rate and the difference in the scattering rates with and without the screening effect is quite small except the energy region $\mathcal{E} \simeq \hbar^2 q_s^2/2m^*$. The scattering rate $1/\tau_{pz}$ with the screening effect in GaAs at $T = 300$ K is plotted in Fig. 6.19a, b for electron density $n = 3.0 \times 10^{13}$ cm^{-3}.

The relaxation time for the piezoelectric scattering with the screening effect (6.201) is derived with the help of the integration for the case of acoustic deformation potential scattering, where the equipartition law holds: $n_{\boldsymbol{q}} = 1/(\exp(\hbar\omega_q/k_B T) - 1)$ $\simeq k_B T/\hbar\omega_q$

$$\frac{1}{\tau_{\mathrm{pz}}} = \frac{e^2\sqrt{m^*}k_{\mathrm{B}}T}{2\sqrt{2}\pi\hbar^2}\left\langle\frac{e_{\mathrm{pz}}^{*2}}{c^*\epsilon^{*2}}\right\rangle\frac{1}{\sqrt{\mathcal{E}}}$$
$$\times\left[1+\frac{1}{1+(8m^*\mathcal{E}/\hbar^2q_{\mathrm{s}}^2)}-\frac{\hbar^2q_{\mathrm{s}}^2}{4m^*\mathcal{E}}\log\left(1+\frac{8m^*\mathcal{E}}{\hbar^2q_{\mathrm{s}}^2}\right)\right] \tag{6.358}$$

When we neglect the screening effect and put $q_{\mathrm{s}} = 0$, (6.358) reduces to (6.357).

6.4.5 Inter–Valley Phonon Scattering

The inter–valley scattering described in Sect. 6.3.7 plays a very important role in electron transport in many–valley semiconductors. It has been also shown that the Gunn effect in GaAs is induced by inter–valley scattering from the Γ valley to the L valleys. In general inter–valley scattering gives rise to the redistribution of electrons in the valleys and the rate of change of the distribution function of the i-th valley is written as

$$\left(\frac{\mathrm{d}f_i}{\mathrm{d}t}\right)_{\mathrm{coll}} = \frac{L^3}{(2\pi)^3}\sum_j^{i\neq j}\int\Big[P(\boldsymbol{k}+\boldsymbol{q},\boldsymbol{k})f_j(\boldsymbol{k}+\boldsymbol{q})\{1-f_i(\boldsymbol{k})\}$$
$$-P(\boldsymbol{k},\boldsymbol{k}+\boldsymbol{q})f_i(\boldsymbol{k})\{1-f_j(\boldsymbol{k}+\boldsymbol{q})\}$$
$$+P(\boldsymbol{k}-\boldsymbol{q})f_j(\boldsymbol{k}-\boldsymbol{q})\{1-f_i(\boldsymbol{k})\}$$
$$-P(\boldsymbol{k},\boldsymbol{k}-\boldsymbol{q})f_i(\boldsymbol{k})\{1-f_j(\boldsymbol{k}-\boldsymbol{q})\}\Big]\mathrm{d}^3\boldsymbol{q}\,. \tag{6.359}$$

Estimated intervalley phonon energies and the deformation potentials are listed in Table 6.4, where the data obtained from magnetophonon resonance (MPR) are listed together with the data from Jacoboni and Reggiani [31]. Analysis of MPR data

Table 6.4 Intervalley phonon energies and their deformation potentials in Si [31]. MPR data are from magnetophonon resonance experiments described in Chap. 7 which are estimated from the fundamental resonance fields $B_f = 21.4$, 70, 84.7, and 106[T]. Intervalley phonon scatterings of g–type $\hbar\omega_{\mathrm{iv,g}} = 22.7$ and/or f-type $\hbar\omega_{\mathrm{iv,f}} = 51.6$ [meV] give rise to the strongest MPR giving rise to $B_f = 84.7$ [T]. Although MPR of $B_f = 21.4$ [T] is weaker than those of 84.7 [T], resonance peaks are clearly assigned as seen in Fig. 7.8

Type	Energy	Deformation potential	Phonon energy [meV] from MPR for fundamental B_f [T]			
	[meV]	[10^8eV/cm]	21.4 [T]	70.0 [T]	84.7 [T]	106 [T]
g-TA	12	0.5	5.7			
g-LA	18.5	0.8		18.8	22.7	28.7
g-LO	61.2	11.0				
f-TA	19.0	0.3	13.0			
f-LA	47.4	2.0		42.3		
f-TO	59.0	2.0			51.6	65.2

are based on the assumption that electrons in the lowest Landau are scattered by the intervalley phonons. Since the magnetic field is applied in the $< 100 >$ direction the lowest Landau level is related to the cyclotron mass $m_{\rm c} = \sqrt{m_{\rm t} m_{\rm l}}$ (= 0.432 m) in the valleys located $< 010 >$ and $< 001 >$. The Si sample used for MPR is very high resistivity (high purity), MPR is caused by the electron transfer form the lowest Landau level to the other valleys of $m_{\rm c} = m_{\rm t}$ (f-type) and $m_{\rm c} = \sqrt{m_{\rm t} m_{\rm l}}$ (g–type). It should be noted here that the phonon modes TA, LA, TO, and LO are not clarified from MPR experiments and thus the intervalley phonon modes for MPR in Table 6.4 are not assigned but tentative. In Table 6.4 the intervalley phonon energies arising from $B_f = 24$ [T] are very low and the assignment is not correct. At low magnetic fields, the Landau level spacing is very small and thus electrons occupy the lower Landau levels arising from the cyclotron $m_{\rm c} = m_{\rm t}$ in addition to $m_{\rm c} = \sqrt{m_{\rm t} m_{\rm l}}$. Therefore the intervalley phonon energy $\hbar\omega_{\rm iv} = 13$ [meV] may be assigned to g–type intervalley phonon energy.

In many–valley semiconductors such as Ge with $\langle 111 \rangle$ valleys and Si with $\langle 100 \rangle$ valleys, the distribution functions of the valleys in a low electric field region are equivalent to each other. The relaxation time for the inter–valley phonon scattering may be treated by assuming constant inter–valley phonon energies $\hbar\omega_{ij}$, enabling us to use the same procedure as for the non-polar optical phonon scattering. Therefore, the scattering rate $w_{\rm int}$ and inverse of relaxation time $1/\tau_{\rm int}$ for inter–valley phonon scattering are given by the following equation:

$$\begin{aligned} w_{\rm int}(\mathcal{E}) &= \frac{1}{\tau_{\rm int}(\mathcal{E})} \qquad (6.360) \\ &= \sum_{j}^{i \neq j} \frac{(2m^*)^{3/2}}{4\pi\hbar^3\rho} \frac{D_{ij}^2}{\omega_{ij}} \left[n_q \sqrt{\mathcal{E} + \hbar\omega_{ij}} + (n_q + 1)\sqrt{\mathcal{E} - \hbar\omega_{ij}} \right], \end{aligned}$$

where m_j^* is the effective density-of-states mass for electrons scattered into the valley j, and $\sum_j^{i \neq j}$ is carried out over the equivalent valleys. In the case of Si, $\sum_j^{i \neq j}$ results in one g–type valley and four f–type $\sum_j^{i \neq j}$. Therefore, the sum is replaced by the valley degeneracy $g_j^{\rm iv}$, and in the case of Si the scattering rate is obtained by using the valley degeneracy $g_{\rm g}^{\rm iv} = 1$ and $g_{\rm f}^{\rm iv} = 4$. There have been reported various data on phonon energies and deformation potentials for inter–valley phonon scattering, but the values have not yet been finalized. In Chap. 7 the magnetophonon resonance effect will be discussed, which is believed to be the best method to determine the types and energies of inter–valley phonons, but the strength of the interaction has not yet been determined [32, 33]. We have shown theoretical calculations of the deformation potentials; the values depend strongly on the parameters used, resulting in ambiguity of the calculated results. Hot carrier transport in Si and device simulation of MOSFETs, where accurate values of deformation potentials and their types are required, have been reported so far by using various parameters. The most reliable values for the deformation potential are believed to be those estimated from Monte Carlo simulation in Si, where the values are determined from a comparison of the calculated drift velocity as a function of electric field with the experimental result

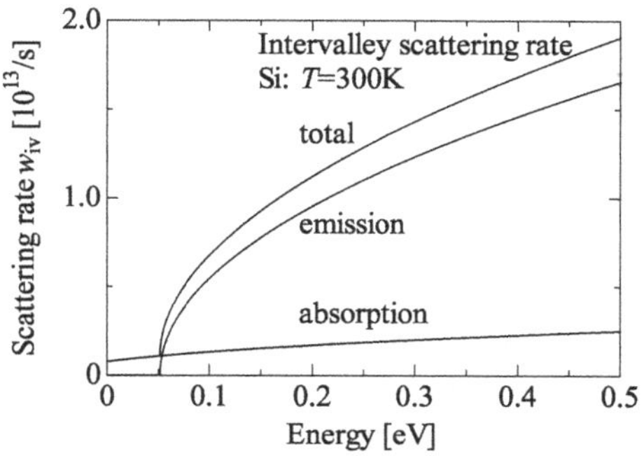

Fig. 6.20 Intervalley phonon scattering rates $w_{\text{iv}} = 1/\tau_{\text{iv}}$ (τ_{iv}: momentum relaxation time) in Si are plotted as a function of electron energy at $T = 300$ [K]. The scattering rates are total, emission and absorption, separately and the used deformation potential for the intervalley phonon $\hbar\omega_{\text{iv}} = 51.6$ [meV] is estimated for $b = 1.73$ to give a good agreement of the observed electron mobility in pure Si as shown in Fig. 6.31

[31]. The result is summarized in Table 6.4. It is evident from Table 6.4 that g–type LO phonons play the most important role in determining the inter–valley phonon scattering in Si. Figure 6.20 shows the calculated scattering rate (or $w_{\text{iv}} = 1/\tau_{\text{iv}}$) as a function of electron energy in Si, intervalley phonon scattering rates of absorption, emission and total rates (sum of absorption and emission rates) are separately plotted. It is seen in Fig. 6.20 that intervalley phonon scattering rates are in the range of 10^{13} [1/s] and much larger than the acoustic deformation potential scattering rate ($<10^{12}$ [1/s] for $\Xi = 9$ [eV]). Therefore the intervalley phonon scattering plays an important role in the lattice scattering at higher temperature in Si. We have to note here again that non-polar optical phonon scattering within the valley is not allowed in Si as stated in Sect. 6.3.7, enabling us to ignore this type of scattering. However, non-polar optical phonon scattering within the valley is much stronger than inter–valley scattering in Ge and the scattering rate is written in a form similar to inter–valley phonon scattering. Therefore, it may be expected that the curves for the scattering rates and mobilities in Ge and Si are quite similar at higher temperatures.

6.4.6 *Ionized Impurity Scattering*

In general, semiconductors contain large amount of donors, acceptors and other impurities, and sometimes they are ionized. Such ionized impurities produce Coulomb potentials which scatter electrons and holes, as described in Sects. 6.3 and 6.3.3. The relaxation time for electron scattering by a screened Coulomb potential (or scattering rate) is given by the Brooks–Herring formula, (6.175):

$$\frac{1}{\tau_{\rm BH}} = \frac{z^2 e^4 n_{\rm I}}{16\pi\epsilon^2\sqrt{2m^*}}\mathcal{E}^{-3/2}\left[\log(1+\xi) - \frac{\xi}{1+\xi}\right], \tag{6.361}$$

where

$$\xi = \frac{8\epsilon m^* k_{\rm B} T}{\hbar^2 e^2 n}\mathcal{E}. \tag{6.362}$$

On the other hand, neglecting the screening effect and using the Rutherford scattering cross-section, Conwell and Weisskopf derived the relaxation time for ionized impurity scattering which is derived from (6.185) as

$$\frac{1}{\tau_{\rm CW}} = \frac{z^2 e^4 n_{\rm I}}{16\pi\epsilon^2\sqrt{2m^*}}\mathcal{E}^{-3/2}\ln\left[1 + \left(\frac{2\mathcal{E}}{\mathcal{E}_{\rm m}}\right)^2\right], \tag{6.363}$$

where

$$\frac{2\mathcal{E}}{\mathcal{E}_{\rm m}} = \frac{r_{\rm m}}{R}\left(\mathcal{E}_{\rm m} = \frac{ze^2}{4\pi\epsilon r_{\rm m}}\right). \tag{6.364}$$

The ionized impurity scattering rate due to Brooks–Herring formula $1/\tau_{\rm BH}$ and Conwell–Weisskopf $1/\tau_{\rm CW}$ in GaAs at $T = 10\,\rm K$ are plotted as a function of the electron energy in Fig. 6.21. In the calculations we assumed the ionize impurity density is $n_{\rm I} = 6.0 \times 10^{13}\,\rm cm^{-3}$ and the half of the impurities give the conduction band electrons. This assumption is based on the fact that some parts of doped donors will fill deep levels, giving rise to ionized impurities and the others are excited in the

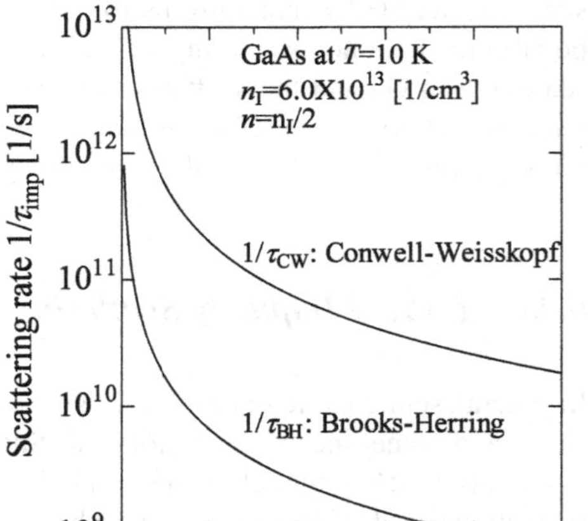

Fig. 6.21 Impurity scattering rates in GaAs given by Brooks–Herring formula and by Conwell–Weisskopf formula as a function of electron energy at lattice temperature $T = 10\,\rm K$, where the ionized impurity density is assumed to be $n_{\rm I} = 6.0 \times 10^{13}\,\rm cm^{-3}$ and a half of the impurities give rise to the conduction band electron density $n = 3.0 \times 10^{13}\,\rm cm^{-3}$

conduction band, giving rise to conduction electrons. Here we find that the screening effect is very important and $1/\tau_{BH}$ is reduced to be about $1/10$ of $1/\tau_{CW}$.

6.4.7 Neutral Impurity Scattering

At low temperatures the donor and acceptor states in semiconductors are occupied by electrons and holes, respectively, and behave as neutral impurities, giving rise to no contribution to Coulomb scattering. The wave function of an electron captured by a shallow donor is spread out over a wide range, as shown in Sects. 3.3 and 3.4 and the effective Bohr radius of the ground state a_I is much larger than the lattice constant. Erginsoy derived the scattering cross-section of an electron by a neutral donor, where the effective Bohr radius and the effective ionization energy are used in the scattering formula for neutral He [34]. Let the electron wave vector be $\boldsymbol{k}$ and $ka_{nI}^* \leq 0.5$. The scattering cross-section may then be approximated as

$$\sigma_{nI} = \frac{20a_{nI}}{k} . \tag{6.365}$$

Let the neutral impurity density be n_{nI}; the scattering probability of an electron traversing a unit distance through unit area is then defined by $n_{nI}\sigma_{nI}$ and thus the mean-free path of the electron is $l_{nI} = \tau_{nI}v = \tau_{nI}\hbar k/m^*$. These relations lead to

$$\frac{1}{l_{nI}} = n_{nI}\sigma_{nI} \tag{6.366}$$

for the mean-free path. Therefore, the relaxation time for neutral impurity scattering (and thus the scattering rate) is given by

$$\frac{1}{\tau_{nI}} = \frac{\hbar k}{m^*} \cdot n_{nI}\sigma_{nI} = \frac{20 n_{nI} a_{nI} \hbar}{m^*} . \tag{6.367}$$

This approximation is based on the assumption that the scattering cross-section is determined by the s orbital of an electron. A better approximation has been given by Sclar [35] and McGill and Baron [36] which is valid for a wider range of electron energy.

6.4.8 Plasmon Scattering

We have derived the matrix elements for plasmon scattering in Sect. 6.3.10 and hence the scattering rate w_{pl}. Since the relaxation time for plasmon scattering is given by $1/\tau_{pl} = w_{pl}$, (6.300) leads to the result

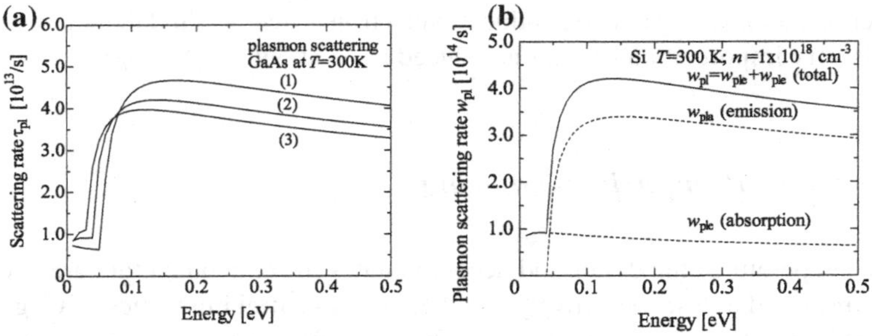

Fig. 6.22 **a** Plasmon scattering rates of GaAs at $T = 300$ [K] are plotted as a function of electron energy up to 0.5 eV for electron density (1) $n = 2 \times 10^{18}$, (2) $n = 1.0 \times 10^{18}$, and (3) $n = 5 \times 10^{17}$ [cm^{-3}]. **b** Plasmon scattering rates of the total w_{pl}, absorption w_{pla}, and emission w_{ple} for GaAs at $T = 300$ [K] are plotted as a function of electron energy up to 0.5 eV for the electron densities $n = 1.0 \times 10^{18}$ [cm^{-3}]. Used parameters are $m^* = 0.067\,\text{m}$ and $\epsilon_\infty = 10.92\epsilon_0$

$$w_{\text{pl}} = \frac{1}{\tau_{\text{pl}}} = \frac{e^2}{4\pi\hbar\epsilon_\infty}\sqrt{\frac{2m^*}{\hbar^2}}\frac{\hbar\omega_{\text{p}}}{\sqrt{\mathcal{E}(\boldsymbol{k})}} \times \left[n_{\text{p}} \ln\left(\frac{q_{\text{c}}}{k\sqrt{\eta_+} - k}\right) \right.$$
$$\left. +(n_{\text{p}}+1)\ln\left(\frac{q_{\text{c}}}{k - k\sqrt{\eta_-}}\right) u(\eta_-) \right], \tag{6.368}$$

where $\omega_{\text{p}} = \sqrt{ne^2/\kappa_\infty\epsilon_0 m^*}$ is the plasma angular frequency, q_{c} is Thomas–Fermi (or Debye–Hückel) screening wavenumber, $\eta_\pm = 1 \pm \hbar\omega_{\text{p}}/\mathcal{E}(\boldsymbol{k})$, $u(\eta_-) = 1$ $(\eta_- \geq 0)$ and $u(\eta_-) = 0$ $(\eta_- < 0)$.

Calculated plasmon scattering rates of GaAs are shown in Fig. 6.22a, where we used scalar effective mass $m^* = 0.0.067m$, dielectric constant $\epsilon_\infty = 10.92\epsilon_0$, and $T = 300$ [K] and the electron densities are 0.5×10^{18}, 1.0×10^{18}, and 2.0×10^{18} [cm^{-3}]. In Fig. 6.22a we find that the scattering rate increases for higher electron densities in higher energy region. In the lower electron energy region, however, the scattering rates are very complicated. In order to get inside of the plasmon scattering, we plot the total scattering rate w_{pl}, absorption rate w_{pla}, and emission absorption rate w_{ple} separately in Fig. 6.22b for GaAs at $T = 300$ [K] as a function of electron energy up to 0.5 eV for electron density $n = 1.0 \times 10^{18}$ [cm^{-3}]. We have to note here that the plasmon emission process is allowed for the case $\hbar\omega_{\text{p}} > \mathcal{E}(\boldsymbol{k})$ and that the plasmon absorption process plays in the lower energy region $\text{e}(\boldsymbol{k}) < \hbar\omega_{\text{p}}$.

6.4.9 Alloy Scattering

Alloy scattering is discussed in Sect. 6.3.11, where the scattering rate w_{alloy} is derived. The relaxation time for alloy scattering (scattering rate) is then written from (6.313) as

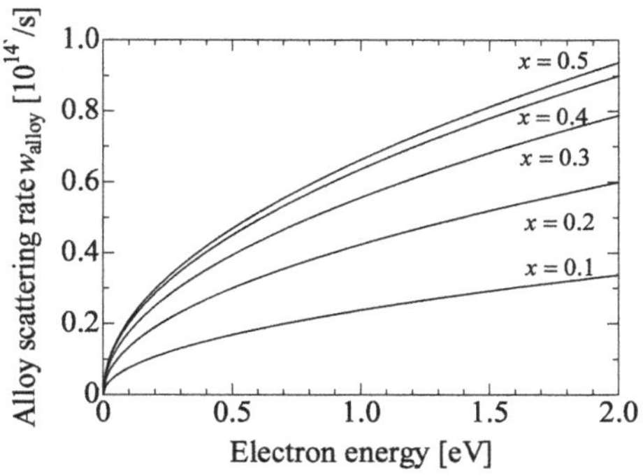

Fig. 6.23 Alloy scattering rates as a function of electron energy for $In_1 - xGa_xAs$ assuming the effective mass is 0.068 m

$$\frac{1}{\tau_{alloy}} = \frac{2\pi}{\hbar}(V_a - V_b)^2 x(1-x)\Omega \frac{(2m^*)^{3/2}\sqrt{\mathcal{E}(\boldsymbol{k})}}{4\pi^2\hbar^3}, \tag{6.369}$$

where Ω is the unit cell volume. As discussed later, estimation $|V_a - V_b|$ is difficult, and we evaluate the electron mobility limited by the alloy scattering by assuming the mobility at 10 K for $x = 0.5$ is 1.0×10^4 [cm^2/Vs], which gives $|V_a - V_b| = 3.185$ [eV] for $m^* = 0.068$ m, $a = 5.66 \times 10^{-10}$ [m]. Using these parameters the scattering rate of the alloy scattering for $x = 0.1 \sim 0.5$ are plotted in Fig. 6.23, where we find that the alloy scattering rate increases with increasing the alloy composition form $x = 0.1$ to 0.5.

6.5 Mobility

In general, the electrons in semiconductors involve many kinds of scattering processes and the total relaxation time τ is given by

$$\frac{1}{\tau} = \sum_j \frac{1}{\tau_j}, \tag{6.370}$$

which assumes that the total scattering probability is defined by the sum of all possible scattering probabilities. Therefore, the electron mobility is obtained from (6.116) (the definition) and (6.119) (average over the distribution function). When the scattering and thus the relaxation time is isotropic, the mobility is calculated from

$$\mu = \frac{e}{m_c^*}\langle\tau\rangle = \frac{e}{m_c^*}\left\langle \frac{1}{\sum_j (1/\tau_j)} \right\rangle, \tag{6.371}$$

where m_c^* is the **conductivity effective mass**. For many–valley semiconductors with ellipsoidal energy surfaces such as in Ge and Si and for an isotropic relaxation time, the conductivity effective mass is given by

$$\frac{1}{m_c^*} = \frac{1}{3}\left(\frac{2}{m_t} + \frac{1}{m_l}\right) . \tag{6.372}$$

It is evident that in semiconductors with isotropic scalar effective mass such as in GaAs, the conductivity effective mass is given by the isotropic effective mass, $m_c^* = m^*$. Here we have to point out that the conductivity effective mass is different in its definition from the effective density-of-states mass. As stated before, the effective mass m^* (referred to as m_d^* hereafter) appearing in the relaxation time $1/\tau$ is the effective density-of-states mass. In Ge and Si with many–valley structures, where the constant energy surface is ellipsoidal, the effective density-of-states mass is given by $m_d^* = (m_t^2 m_l)^{1/3}$. When the effective masses in Si, $m_t = 0.19\,\mathrm{m}$ and $m_l = 0.98\,\mathrm{m}$, are used, we have $m_c = 0.260\,\mathrm{m}$ and $m_d = 0.328\,\mathrm{m}$.

In the following we calculate the electron or hole mobility assuming that the electrons and holes are not degenerate and that their distribution function is given by Maxwell–Boltzmann statistics. In the case of degenerate semiconductors we have to use the Fermi–Dirac distribution function and the results are different from those given here. Sometimes we find a case where two scattering processes play a major part in the scattering or limit the mobility. For example when acoustic phonon scattering and ionized impurity scattering dominate the scattering processes, the relaxation time is written as

$$\frac{1}{\tau} = \frac{1}{\tau_{ac}} + \frac{1}{\tau_I} , \tag{6.373}$$

which may be approximated as follows, except in the region $\langle\tau_{ac}\rangle = \langle\tau_I\rangle$,

$$\frac{1}{\langle\tau\rangle} \simeq \frac{1}{\langle\tau_{ac}\rangle} + \frac{1}{\langle\tau_I\rangle} . \tag{6.374}$$

When this approximation is valid, it is very convenient to estimate the overall behavior of the temperature dependence of the mobility from an averaged value of each relaxation time. For this reason this assumption is often used to calculate the averaged value of the relaxation time and thus the mobility for each scattering process. Although this technique is very convenient, we carried out exact value of the total relaxation time and the average mobility by numerical integration in this textbook. The results for the mobility given here are basically calculated by using the Maxwell–Boltzmann distribution function and (6.116). See the review paper of electron mobility in direct gap polar semiconductors by Rode [37] for more information.

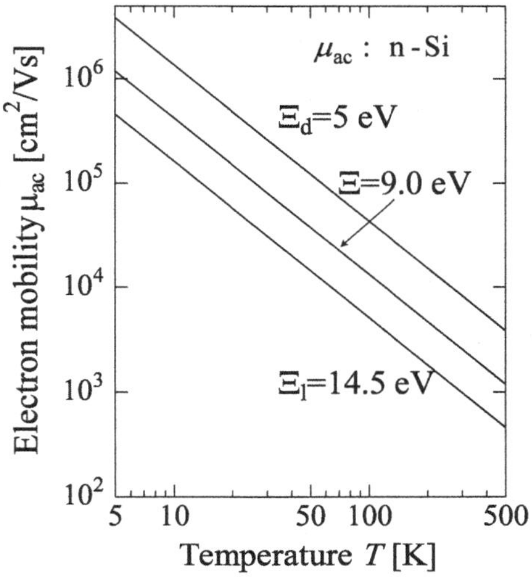

Fig. 6.24 Electron mobility μ_{ac} due to deformation potential-type acoustic phonon scattering in Si are plotted as a function of temperature T [K]. The deformation potentials used in the calculation are dilational $\Xi_d = 5$, average $\Xi_l = 14.5$ [eV], and $\Xi = 9.0$ [eV] is determined from the analysis of temperature dependence of the mobility

6.5.1 Acoustic Phonon Scattering

The relaxation time for acoustic phonon scattering (6.341) gives the mobility

$$\mu_{ac} = \frac{2^{3/2}\sqrt{\pi} e \hbar^4 \rho v_s^2}{3 m_d^{*3/2} m_c^* D_{ac}^2 (k_B T)^{3/2}}, \tag{6.375}$$

where $m_d^{*3/2} m_c^* = m^{*5/2}$ for electrons with isotropic effective mass m^*.

Electron mobilities due to acoustic deformation potential scattering in Si and Ge are plotted in Figs. 6.24 and 6.25, respectively, where the material parameters used in the present calculations are tabulated in Table 6.3. The mobility becomes very high at lower temperatures, and thus we have to take into account of the other scattering processes such as impurity scattering to evaluate the mobility at low temperatures. Also as seen in Figs. 6.24 and 6.25, the mobility due to acoustic phonon scattering depends on the value of Ξ^2 and thus proper value of the deformation potential should be determined by comparing with the experimental results. This may be done by calculating the electron mobility by taking all the scattering processes involved and comparing the calculated curve with the experimental data, which will be shown later.

6.5.2 Non-polar Optical Phonon Scattering

Using (6.351) for the relaxation time of non-polar optical phonon scattering, the electron mobility is given by

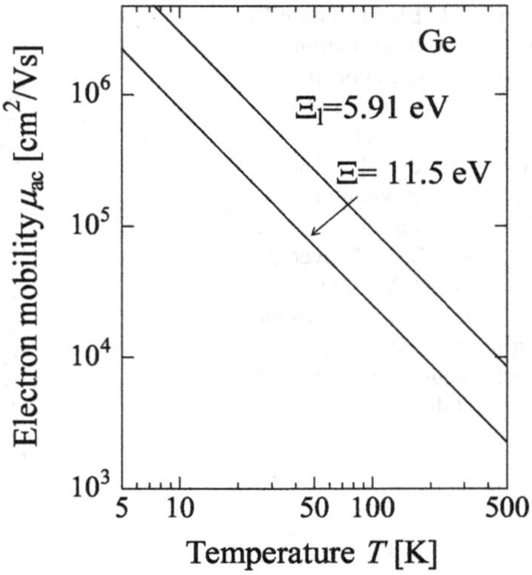

Fig. 6.25 Electron mobility μ_{ac} due to deformation potential-type acoustic phonon scattering in Ge are plotted as a function of temperature T [K]. The deformation potentials used in the calculation are dilational (average) $\Xi_l = 5.91$ [eV] and $\Xi = 11.5$ [eV], where $\Xi = 11.5$ [eV] is determined from the analysis of temperature dependence of the mobility

$$\mu_{op} = \frac{4\sqrt{2\pi}e\hbar^2\rho\sqrt{\hbar\omega_0}}{3m_d^{*3/2}m_c^* D_{op}^2} f(x_0) \,, \tag{6.376}$$

where

$$f(x_0) = x_0^{5/2}\left(e^{x_0} - 1\right) \times \int_0^\infty x e^{-x}\left[\left(1 + \frac{x_0}{x}\right)^{1/2} + e^{x_0}\left(1 - \frac{x_0}{x}\right)^{1/2}\right]^{-1} dx \,, \tag{6.377}$$

$$x_0 = \frac{\hbar\omega_0}{k_B T}, \quad x = \frac{\mathcal{E}}{k_B T} \,. \tag{6.378}$$

When both acoustic phonon and non-polar optical phonon scattering are taken into account, the mobility is evaluated from the following equation:

$$\mu = \mu_{ac} \cdot I(x_0) \,, \tag{6.379}$$

$$I(x_0) = \int_0^\infty x e^{-x}\left[1 + \frac{l_{ac}}{l_{op}}\frac{1}{e^{x_0} - 1}\left\{\left(1 + \frac{x_0}{x}\right)^{1/2} + e^{x_0}\left(1 - \frac{x_0}{x}\right)^{1/2}\right\}\right]^{-1} dx \,, \tag{6.380}$$

$$l_{op} = \frac{2\pi\hbar^3\rho\omega_0}{(D_{op}m_d^*)^2}, \quad \frac{l_{ac}}{l_{op}} = \frac{x_0}{2}\frac{D_{ac}^2}{E_{op}^2} \,. \tag{6.381}$$

Calculated electron mobility μ_{op} due to non-polar optical phonon scattering only is shown in Fig. 6.26, where the deformation potential D_{op} for non-polar optical phonon scattering is determined as follows. This is done by comparing the measured electron mobility with the theoretical calculations by taking account of the acoustic and optical phonon scatterings shown in Fig. 6.27. As given by (6.381) the squared ratio of the

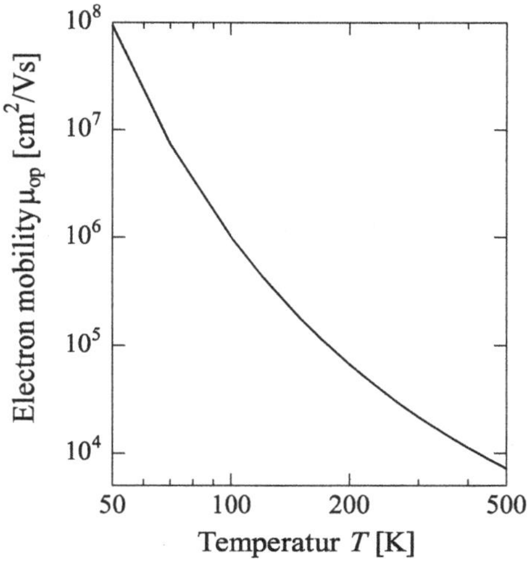

Fig. 6.26 Electron mobility μ_{op} due to non-polar optical phonon scattering in Ge are plotted as a function of temperature T [K]. The deformation potential for non-polar optical phonon scattering is estimated from $D_{ac}^2/E_{op}^2 \equiv b = 0.4$ with $D_{ac} = 11.5$ [eV], which give the mobility $\mu_{ac+op} = 3822$ [cm^2/Vs] due to acoustic phonon scattering and optical phonon scattering at 300 K as shown in Fig. 6.27 (see also text in detail)

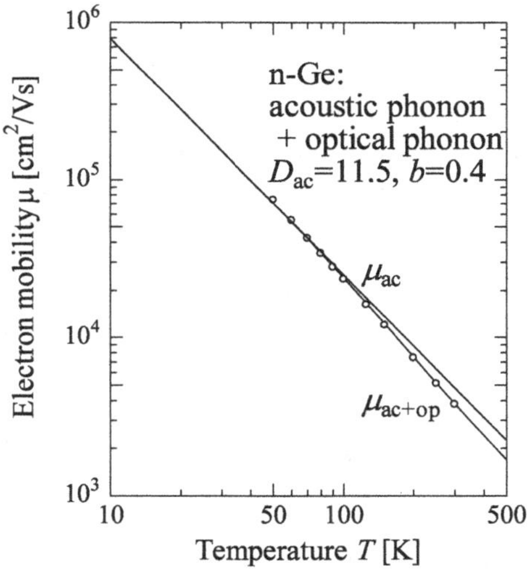

Fig. 6.27 Electron mobility μ_{ac+op} due to deformation potential-type acoustic phonon scattering and non-polar optical phonon scattering in Ge are plotted as a function of temperature T [K]. The used deformation potentials are $D_{ac} = 11.5$ [eV] and $D_{ac}^2/E_{op}^2 \equiv b = 0.4$, which give the electron mobility $\mu_{ac+op} = 3822$ [cm^2/Vs] at 300 [K]. The $\circ$ points represent experimental data, calculated from the empirical formula of Morin [38]

acoustic deformation potential $D_{ac}^2/\Xi_{op}^2 = b$ is determined by comparing the electron mobility with the calculated mobility with acoustic and optical phonon scattering as the following. First we choose a proper value of the acoustic phonon deformation potential $D_{ac}(\equiv \Xi)$ and then determine the ration b by fitting the calculated curve μ_{ac+op} to the experimental results. It is well known that the ratio is $b = 0.4$ for n–Ge and $3.5 \sim 3.8$ for p–Ge as reported by Conwell [6], and Brown and Bray [39]. Using $b = 0.4$ we determined $D_{ac} = \Xi = 11.5$ which gives $\mu_{ac+op} \simeq 3800$ [cm^2/Vs] at $T = 300$ K as shown in Fig. 6.27. The $\circ$ points are obtained from the experimental data of Morin [38], which are expressed by the empirical formula $\mu = 4.90 \times 10^7 T^{-1.66}$ [cm^2/Vs] and agree well with the calculated result.

The electron mobility in n-Ge and n-Si is analyzed by taking the inter–valley phonon scattering into consideration. However, it is known that inter–valley phonon scattering is weak in n-Ge and thus the electron mobility of Ge is analyzed by taking account of acoustic phonon scattering and non-polar optical phonon scattering only. Although the valence band structure in Ge is complicated, the analysis of the hole mobility is simple. Here we will show a comparison of the experimental hole mobility with the theoretical analysis by Brown and Bray [39]. The valence bands of Ge consist of heavy-hole and light-hole bands as discussed in Chaps. 1 and 2, where the constant energy surface of the heavy-hole band is not spherical but warped. In the analysis the hole bands are approximated by a spherical surface with effective mass $m_{\rm h} = 0.35\,{\rm m}$ for heavy hole and $m_{\rm l} = 0.043\,{\rm m}$ for light hole. With this approximation the density ratio of heavy holes to light holes is given by $p_{\rm h}/p_{\rm l} = (m_{\rm h}/m_{\rm l})^{3/2} = 23.2$. Let the heavy–hole mobility be $\mu_{\rm h}$ and light–hole mobility be $\mu_{\rm l}$; the average hole mobility μ is given by

$$\mu = \frac{\mu_{\rm h} p_{\rm h} + \mu_{\rm l} p_{\rm l}}{p_{\rm h} + p_{\rm l}} = \frac{1}{24.2}(\mu_{\rm l} + 23.2\mu_{\rm h}) . \tag{6.382}$$

In order to analyze the hole mobility in p-Ge we have to take account of ionized impurity scattering, acoustic phonon scattering and non-polar optical phonon scattering. At low temperatures the ionized impurity scattering and acoustic phonon scattering dominate the other scattering processes, while at higher temperatures the acoustic phonon scattering and non-polar optical phonon scattering will limit the hole mobility. For these reasons we deal with the mobility analysis in two temperature regions, the lower and higher temperature regions, and two scattering mechanisms in each temperature region are considered. The hole mobility calculated in this approximation is plotted in Fig. 6.28a, b. From the analysis of the hole mobility at lower temperatures the following relations are obtained.

$$\begin{aligned} \mu_{\rm ac} &= 3.37 \times 10^7 T^{-3/2} \quad {\rm cm^2/V \cdot s} , \\ (\mu_h)_{\rm ac} &= 2.60 \times 10^7 T^{-3/2} \quad {\rm cm^2/V \cdot s} , \\ (\mu_l)_{\rm ac} &= 2.12 \times 10^8 T^{-3/2} \quad {\rm cm^2/V \cdot s} . \end{aligned} \tag{6.383}$$

It is very important to point out that intra-band scattering dominates for heavy holes whereas light-hole scattering is dominated by scattering from the light-hole band to the heavy-hole band because of the difference in the effective density-of-states mass. Under this assumption and from (6.375), $\mu_{\rm h} \propto 1/m_{\rm h}^{5/2}$ and $\mu_{\rm l} \propto 1/m_{\rm h}^{3/2} m_{\rm l}$, giving rise to $\mu_{\rm l}/\mu_{\rm h} = m_{\rm h}/m_{\rm l} = 8.1$. This result is supported by the analyzed hole mobility. From the analysis at higher temperatures, they obtained

$$b = \frac{E_{\rm lop}^2}{D_{\rm ac}^2} = \frac{D_{\rm op}^2}{D_{\rm ac}^2} \cdot \frac{v_{\rm s}^2}{\omega_0^2} = 3.8 \tag{6.384}$$

for the squared deformation potential ratio b. Analysis of the electron mobility in n-Ge gives $b = 0.4$. These results tell us that the interaction between holes and non-

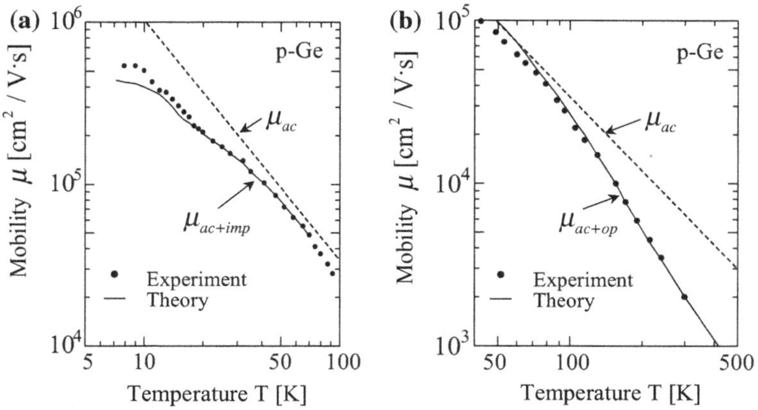

Fig. 6.28 Calculated hole mobility in p-Ge plotted as a function of temperature, along with experimental data. **a** Experimental mobility in the low temperature region is compared with the calculated hole mobility by taking account of ionized impurity scattering and acoustic phonon scattering. **b** Experimental data in the high temperature range are compared with the calculated mobility by taking account of acoustic phonon scattering and non-polar optical phonon scattering. Experimental data are from Brown and Bray [39]

polar optical phonons is much stronger than that between electrons and non-polar optical phonons in n-Ge.

6.5.3 *Polar Optical Phonon Scattering*

Electron mobility due to polar optical phonon scattering is evaluated from the relaxation time given by (6.352) and using the drifted Maxwellian distribution function define by

$$f(\boldsymbol{v}) = \exp\left\{-\frac{m^*(\boldsymbol{v}-\boldsymbol{v}_\mathrm{d})^2}{2k_\mathrm{B}T}\right\}. \tag{6.385}$$

The electron mobility is then given by

$$\mu_\mathrm{pop} = \frac{3(2\pi\hbar\omega_\mathrm{LO})^{1/2}}{4m^{*1/2}E_0 n(\omega_\mathrm{LO})}\frac{1}{x_0^{3/2}\mathrm{e}^{x_0/2}K_1(x_0/2)}, \tag{6.386}$$

where

$$E_0 = \frac{m^* e\hbar\omega_\mathrm{LO}}{4\pi\hbar^2\epsilon_0}\left(\frac{1}{\kappa_\infty}-\frac{1}{\kappa}\right),\ n(\omega_\mathrm{LO}) = \frac{1}{\mathrm{e}^{x_0}-1},\ x_0 = \frac{\hbar\omega_\mathrm{LO}}{k_\mathrm{B}T}, \tag{6.387}$$

$$K_1(t) = t\int_1^\infty \sqrt{z^2-1}\mathrm{e}^{-tz}\mathrm{d}z = \frac{\mathrm{e}^{-t}}{t}\int_0^\infty \sqrt{z(z+2t)}\mathrm{e}^{-z}\mathrm{d}z\,, \tag{6.388}$$

and $K_1(t)$ is the modified Bessel function. In the case of $\mathcal{E} \gg \hbar\omega_{\mathrm{LO}}$, (6.353) gives the electron mobility for polar optical phonon scattering, which is expressed as

$$\mu_{\mathrm{pop}} \cong \frac{8\sqrt{2k_{\mathrm{B}}T}}{3\sqrt{\pi m^*}E_0}\cdot\frac{\mathrm{e}^{x_0}-1}{\mathrm{e}^{x_0}+1}\,. \tag{6.389}$$

The electron mobility due to polar optical phonon scattering is calculated by using the relation (6.356) deduced by Callen [30] and shown in Fig. 6.29. In the calculation we employed numerical integration of the following equation and used the material parameters given in Table 6.3, where the integration range of the electron energy $x = \mathcal{E}/k_{\mathrm{B}}T$ is divided into two regions $[0, \hbar\omega_{\mathrm{pop}}/k_{\mathrm{B}}T]$ and $[\hbar\omega_{\mathrm{pop}}/k_{\mathrm{B}}T], \infty]$ (∞ is replaced by $100\times\hbar\omega_{\mathrm{pop}}/k_{\mathrm{B}}T$].

$$\mu_{\mathrm{pop}} = \frac{e}{m^*} < \tau_{\mathrm{pop}} >\,, \tag{6.390}$$

where τ_{pop} given by (6.356) is used.

In addition we plotted electron mobilities due to acoustic phonon scattering using two different values of deformation potentials $\Xi = 7.0$ [eV] [37] and $\Xi = 14.0$ [eV],

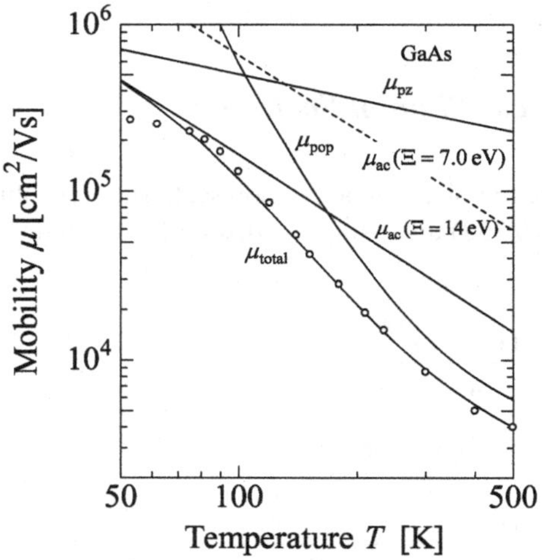

Fig. 6.29 Electron mobility μ_{pop} for polar optical phonon scattering in GaAs is plotted as a function of temperature T, in addition to the mobilities due to acoustic phonon scattering, and piezoelectric potential scattering, where the used parameters are given in Table 6.3. The mobilities due to acoustic phonon scattering for $\Xi = 7.0$ and $\Xi = 14.0$ [eV] are shown to see the difference. The electron mobility μ_{total} due to acoustic ($\Xi = 14.0$ [eV]), piezoelectric, and polar optical phonon scattering is also shown, which gives mobility $\mu = 8,840$ [cm^2/Vs] at $T = 300$ [K]. Data points $\circ$ are obtained by averaging the reported experimental data from Rode [40]

and piezoelectric potential scattering. The mobility due to acoustic phonon scattering deduced from $\Xi = 7.0$ [eV] is too high and the electron mobility due to these three scattering processes does not agree with experimental observation. We calculated the electron mobility due to acoustic, piezoelectric, and polar optical phonon scatterings and $\Xi = 14.0$ [eV] is quite reasonable value of the electron mobility $\mu = 8,840$ [cm^2/Vs] at $T = 300$ K, as shown by $\mu_{\text{total}}=\mu_{\text{ac+pz+pop}}$ in Fig. 6.29. In the calculations we employed numerical integration stated above for the case of combined acoustic deformation potential and polar optical phonon scatterings, where we find a very good agreement between the present calculation and the experimental data. The experimental data are obtained by averaging the reported values of Rode [40]. Here we have to note that the acoustic deformation potential $\Xi = 14.0$ eV gives a very good agreement between the calculations and the experimental data.

6.5.4 Piezoelectric Potential Scattering

As discussed in Sect. 6.4.4, the effect of screening does not play an important role in the piezoelectric potential scattering, we neglect the effect and estimate the electron mobility for piezoelectric potential scattering. Then the mobility is given by the following equation using (6.357).

$$\mu_{\text{pz}} = \frac{16\sqrt{2\pi}\hbar^2}{3m^{*3/2}e\langle e_{\text{pz}}^{*2}/c^*\epsilon^{*2}\rangle}(k_{\text{B}}T)^{-1/2}\,. \tag{6.391}$$

Electron mobility due to piezoelectric potential scattering is shown in Fig. 6.30 as a function of temperature T, where we see the mobility decreased with increasing temperature T. Here we neglected the screening effect which is discussed by Ridley [5]. In the calculation the piezoelectric coupling constant is obtained by using $e_{\text{pz}}^* = e_{14} = 0.16$ and $c^* = c_{44}$, neglecting anisotropy.

6.5.5 Inter–Valley Phonon Scattering

The mobility for inter–valley phonon scattering is evaluated with a relation similar to that for non-polar optical phonon scattering, and is obtained by replacing the optical phonon deformation potential D_{op} by the inter–valley phonon deformation potential D_{ij} and the optical phonon energy $\hbar\omega_{\text{op}}$ by the inter–valley phonon energy $\hbar\omega_{ij}$. In many–valley semiconductors such as Ge and Si, however, there are four and six equivalent valleys, respectively, and the degeneracy factor of the valleys g_{ij}^{iv} for scattered electron must be considered. The electron mobility for inter–valley phonon scattering is given by the following equation using (6.361)

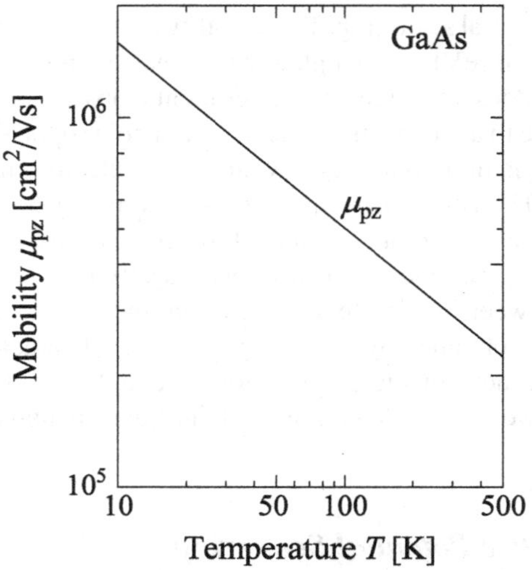

Fig. 6.30 Calculated electron mobility due to piezoelectric potential scattering in GaAs plotted as a function of temperature, where the material parameters are given by Table 6.3, and $e^*_{pz} = e_{14}$ and $c^* = c_{44}$ are used without averaging

$$\mu_{int} = \frac{4g^{iv}_{ij}\sqrt{2\pi}e\hbar^2\rho(\hbar\omega_{ij})^{1/2}}{3m^{*3/2}_{d}m^*_c D^2_{ij}} f(x_{ij}) , \tag{6.392}$$

where $x_{ij} = \hbar\omega_{ij}/k_B T$ and $f(x_{ij})$ is obtained by replacing x_0 in (6.377) by x_{ij}. Let us examine the temperature dependence of the electron mobility in Si. As discussed in Sect. 6.4, the conduction bands of Si consist of equivalent valleys located along the $\langle 100 \rangle$ direction, and g–type and f–type inter–valley scattering will dominate the other scattering processes at higher temperatures. In order to get an insight into the importance of inter–valley scattering, we calculate the electron mobility by taking account of acoustic phonon scattering and f–type inter–valley phonon scattering $\hbar\omega_{iv} = 51.6$ [meV], which is obtained from magnetophonon resonance experiments describe in Sect. 7.4. The assignment of the phonon energy and the types f– and g– are made by assuming that the electrons occupy the lowest Landau level because of the electron density is very low in the used pure samples. The results are summarized in Table 6.4. The deformation potentials for inter–valley phonons listed are from Jacoboni et al. [31] and the intervalley phonon types and energies obtained from magnetophonon experiments are also summarized in Table 6.4. In the present analysis we used the f–type inter–valley phonon energy $\hbar\omega_{iv,f} = 51.6$ [meV] and the intervalley deformation potential, where the degeneracy factor of the valleys are included in the deformation potential. This approximation is adopted to calculate the scattering rate in Si shown in Fig. 6.20. Usually, the calculated electron mobility is adjusted to fit the experimental data by changing the deformation potential parameter $b = \Xi^2_{iv}/D^2_{ac}$. The left hand side of Fig. 6.31 shows the electron mobility limited by

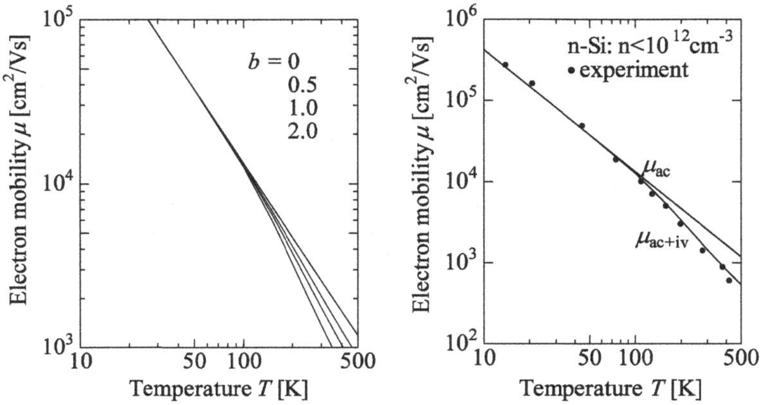

Fig. 6.31 *Left* Calculated temperature dependence of the electron mobility due to acoustic phonon scattering and intervalley phonon scattering in Si for different values of $b = 0, 0.5, 1.0$, and 2.0 from *top to bottom curve*. *Right* Comparison of the temperature dependence of electron mobility in Si between experiments and theoretical calculation with $b = 1.73$. The material parameters are given by Table 6.3, and acoustic deformation potential $\Xi = 9.0$[eV], f–type intervalley phonon energy $\hbar\omega_{\mathrm{iv,f}} = 51.6$ [meV], and the deformation potential ratio $b = 1.73$, which gives electron mobility $\mu = 1452$ [cm^2/Vs] at $T = 300$ K. The experimental data • are from Canali et al. [41] and Jacoboni et al. [42]

lattice scatterings (acoustic and intervalley phonons) as a function of temperature with $b = 0, 0.5, 1.0, 2.0$ as an adjustable parameter. The right hand side of Fig. 6.31 shows the best fitted temperature dependence of the electron mobility in high purity Si with $b = 1.73$, where the experimental data • are from Canali et al. [41] and Jacoboni et al. [42].

The deformation potential parameter b is obtained by replacing the deformation potential for non-polar optical phonon scattering by the deformation potential of inter–valley phonon scattering and is defined as follows.

$$b = \frac{E_{\mathrm{iv}}^2}{D_{\mathrm{ac}}^2}, \qquad E_{\mathrm{iv}}^2 = \frac{D_{\mathrm{iv}}^2 v_{\mathrm{s}}^2}{\omega_{\mathrm{iv}}^2}.$$

Here we have to note that the acoustic phonon deformation potential D_{ac} used here is equivalent to the deformation potential $\Xi = 9, 0$ [eV] determined from the mobility analysis. The parameters used to calculate the scattering rates shown in Fig. 6.20 and the electron mobility in Fig. 6.31 are $\Xi = 9.0$ [eV] and $b = 1.73$. When only acoustic phonon scattering is taken into account ($b = 0$) with acoustic deformation potential $\Xi = 9.0$ [eV], the calculated mobility at $T = 300$ K is $\mu = 2560\,\mathrm{cm^2/Vs}$, which is much larger than the observed mobility $1450\,\mathrm{cm^2/Vs}$, whereas the deformation potential parameter $b = 1.73$ results in $1452\,\mathrm{cm^2/Vs}$, which is very close to the measured value. In the calculation we take into account of only f–type intervalley phonon $\hbar\omega_{\mathrm{iv}} = 51.6$[meV] listed in Table 6.4 which gives rise to the strongest magnetophonon resonance as shown in Sect. 7.4. It is seen in Fig. 6.31 that the electron mobility decreases with increasing b from 0 to 2.0, which means that the electrons are scattered more by inter–valley phonons, resulting in a lower mobility. It may be expected from these results that the calculated electron mobility can be fitted to the experimental results by taking g–type and f–type inter–valley phonon scattering accurately into account and adjusting the deformation potentials. It is also seen in Fig. 6.31 that inter–valley phonon scattering may be disregarded at lower temperatures and that inter–valley phonon scattering affects the electron mobility at temperatures $T > 100\,\mathrm{K}$ [43].

6.5.6 Ionized Impurity Scattering

We have shown that the relaxation time for ionized impurity scattering is given by the Brooks–Herring formula (6.362) and the Conwell–Weisskopf formula (6.363), where these formulae are derived with and without screening by conduction electrons, respectively. The relaxation time becomes a maximum when $\mathcal{E}$ in the logarithmic terms is $3k_{\mathrm{B}}T$ and thus the integral is approximated by replacing $\mathcal{E}$ by $3k_{\mathrm{B}}T$ except for $\mathcal{E}$ in the prefactor. The electron mobility of Brooks–Herring formula is written as

$$\mu_{\mathrm{BH}} = \frac{64\sqrt{\pi}\epsilon^2}{n_{\mathrm{I}} z^2 e^3 m^{*1/2}} (2k_{\mathrm{B}}T)^{3/2} \left[\log(1+\xi_0) - \frac{\xi_0}{1+\xi_0} \right]^{-1}, \tag{6.393}$$

where

$$\xi_0 = \frac{24\epsilon m^* (k_{\mathrm{B}}T)^2}{\hbar^2 e^2 n}. \tag{6.394}$$

The electron mobility of the Conwell–Weisskopf formula is given by

$$\mu_{\mathrm{CW}} = \frac{64\sqrt{\pi}\epsilon^2}{n_{\mathrm{I}} z^2 e^3 m^{*1/2}} (2k_{\mathrm{B}}T)^{3/2} \left[\log\left(1 + \frac{144\pi^2\epsilon^2 k_{\mathrm{B}}^2 T^2}{z^2 e^4 n_{\mathrm{I}}^{2/3}} \right) \right]^{-1}. \tag{6.395}$$

Figure 6.32a is the temperature dependence of calculated mobilities due to ionized impurity scattering by Brooks–Herring formula and by Conwell–Weisskopf formula, evaluated by numerical integration. As expected, however, the logarithmic term is well approximated by replacing $\mathcal{E}$ with $3k_{\mathrm{B}}T$ and the difference between the numerical analysis and the approximation is negligible, such as $\mu_{\mathrm{BH}} = 2.228 \times 10^5$ (numerical) and $\mu_{\mathrm{BH}} = 2.262 \times 10^5$ (approximation) at $T = 10\mathrm{K}$ for $n_{\mathrm{I}} = 6.0 \times 10^{13}\mathrm{cm}^{-3}$ and $n = n_{\mathrm{I}}/2$ in GaAs.

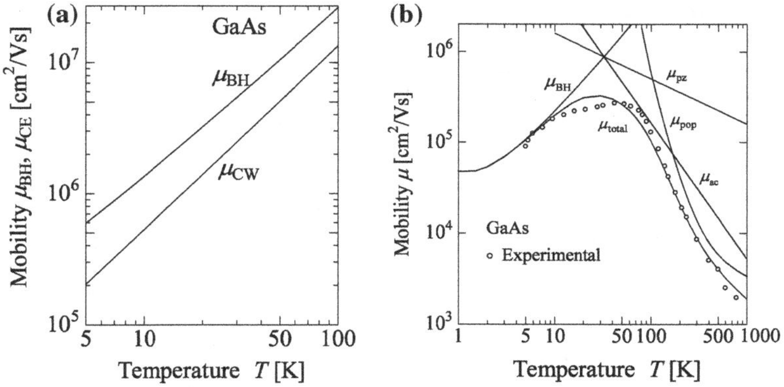

Fig. 6.32 **a** Calculated temperature dependence of the electron mobilities of GaAs due to ionized impurity scattering by Brooks–Herring formula and Conwell–Weisskopf formula, where density of ionized impurity is $n_I = 6.0 \times 10^{13}\text{cm}^{-3}$ and electron density in the conduction band is assumed to be half of the ionized impurity density $n = 3.0 \times 10^{13}\text{cm}^{-3}$. **b** Temperature dependence of electron mobility in GaAs, where ionized impurity scattering, acoustic deformation potential scattering, piezoelectric potential scattering, and polar optical phonon scattering are shown, respectively. The curve μ_{total} is the theoretical curve due to all the scattering, and the parameters are given in Table 6.3. All the mobilities are evaluated by numerical integration. The averaged values of the experimental data are plotted by $\circ$ (see Rode [37, 40])

Figure 6.32b shows temperature dependence of the electron mobility in GaAs, where ionized impurity scattering, acoustic deformation potential scattering, piezoelectric potential scattering, and polar optical phonon scattering are shown, respectively. The curve μ_{total} is the theoretical curve due to over all scatterings, and the used parameters are given in Table 6.3. All of the mobilities are evaluated by numerical integration. The points assigned by $\circ$ is the averaged value of the experimental data of Rode [37, 40].

The total mobility μ_{total} is numerically evaluated by using τ_{total} defined by the following equation

$$\frac{1}{\tau_{\text{total}}} = \frac{1}{\tau_{\text{BH}}} + \frac{1}{\tau_{\text{ac}}} + + \frac{1}{\tau_{\text{pz}}} + \frac{1}{\tau_{\text{pop}}} \tag{6.396}$$

and using (6.119),

$$\mu_{\text{total}} = \frac{e}{m^*} \langle \tau_{\text{total}} \rangle \,. \tag{6.397}$$

6.5.7 Neutral Impurity Scattering

The relaxation time of the Erginsoy formula for neutral impurity scattering leads to

$$\mu_{\text{nI}} = \frac{e}{20 a_{\text{nI}} \hbar} \frac{1}{n_{\text{nI}}} = \frac{e}{20 a_{\text{B}} \hbar} \cdot \frac{m^*/m}{\kappa n_{\text{nI}}} , \tag{6.398}$$

where a_B is Bohr radius.

6.5.8 Plasmon Scattering

The relaxation time for plasmon scattering is given by (6.368) (the inverse of the scattering rate). However, the electron mobility due to the plasmon scattering is not easily evaluated because of the complicated logarithmic terms and thus here we evaluate the mobility by approximating $\mathcal{E}(\boldsymbol{k})$ by $k_{\text{B}}T$ in the logarithmic terms. Then the mobility due to the plasmon scattering is then given by

$$\begin{aligned}
\mu_{\text{pl}} &= \frac{e}{m^*} \langle \tau_{\text{pl}} \rangle \\
&\simeq \frac{32\sqrt{\pi}\hbar^2 \epsilon_\infty}{3\sqrt{2} e (m^*)^{3/2} \hbar\omega_{\text{p}}} (k_{\text{B}}T)^{1/2} \\
&\quad \times \left[n_{\text{p}} \ln \left(\frac{q_{\text{c}}}{\sqrt{\eta_+}\sqrt{2m^* k_{\text{B}} T/\hbar^2} - \sqrt{2m^* k_{\text{B}} T/\hbar^2}} \right) \right. \\
&\quad \left. + (n_{\text{p}} + 1) \ln \left(\frac{q_{\text{c}} u(\eta_-)}{\sqrt{2m^* k_{\text{B}} T/\hbar^2} - \sqrt{\eta_-}\sqrt{2m^* k_{\text{B}} T/\hbar^2}} \right) \right]^{-1} ,
\end{aligned} \tag{6.399}$$

where $\eta_\pm = 1 \pm (\hbar\omega_{\text{p}}/k_{\text{B}}T)$, and other parameters are the same as defined before. Calculated temperature dependence of the mobilities limited by plasmon scattering are shown in Fig. 6.33, where we used scalar effective mass $m^* = 0.067m$ and dielectric constant $\epsilon_\infty = 10.92\epsilon_0$, and the curves are obtained for three different electron densities (1) 0.5×10^{18}, (2) 1.0×10^{18}, and (3) 2.0×10^{18} [cm^{-3}]. These results indicate that the plasmon limited mobility play an important role at higher temperature, which reflects the excited number of the plasmon n_{p}.

In Fig. 6.34 the electron mobility due to plasmon scattering is plotted as a function of electron density. We see a small step around $T = 350$ [K] for the electron density $n \simeq 5.0 \times 10^{17}$ [cm^{-3}], which will be explained in terms that the plasmon emission rate includes the step function $u(\eta_-)$, where η_- becomes 0 at $T = 356.2$ [K] for $\hbar\omega_p = 30.7$ [meV]. The discontinuity at lower electron density is explained in terms that plasmon emission process is forbidden for $k_{\text{B}}T < \hbar\omega_{\text{p}}$. In order to obtain the electron mobility limited by plasmon scattering we approximated by replacing electron energy $\mathcal{E}(\boldsymbol{k})$ by $k_{\text{B}}T$. However, we find that the calculated electron mobility due to plasmon scattering does not show any noticeable change when we change the

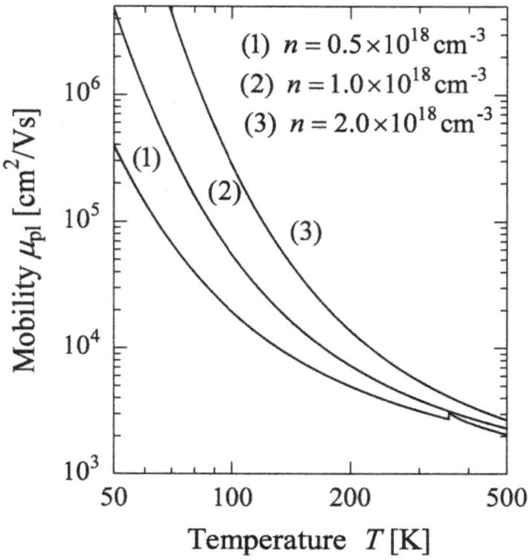

Fig. 6.33 Calculated mobilities limited by plasmon scattering in GaAs are plotted as a function of lattice temperature for electron densities (1) $n = 0.5 \times 10^{18}$, (2) $n = 1.0 \times 10^{18}$, and 2.0×10^{18} [cm^{-3}]. Used parameters are $m^* = 0.067\,\text{m}$ and $\epsilon_\infty = 10.92\epsilon_0$

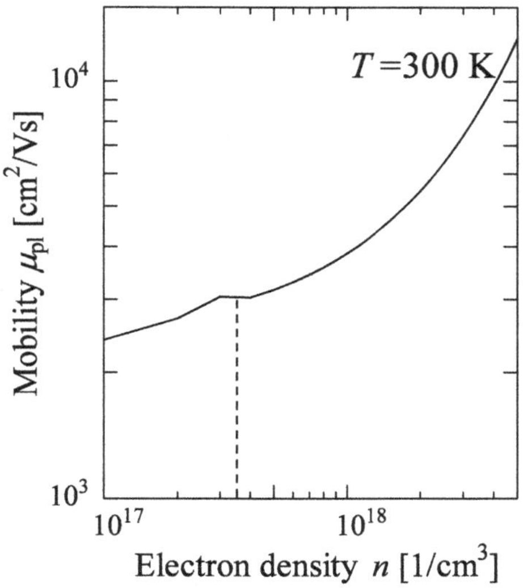

Fig. 6.34 Calculated mobilities limited by plasmon scattering in GaAs are plotted as a function of electron density at $T = 300$. Electron energy in the logarithmic terms are approximated by average electron energy $\langle \mathcal{E}(\boldsymbol{k}) \rangle = k_B T$ and thus the electron density $n = 3.55 \times 10^{23}$ (*dotted bar* in the figure) gives the condition $\hbar\omega_p = k_B T$. In the region of lower electron density, only plasmon absorption is allowed

average electron energy from $k_B T$ to $(3/2)k_B T$, and thus the approximation is valid for the electron mobility due to plasmon scattering.

Now we have to discuss the electron density dependence of the mobility. As shown Figs. 6.33 and 6.34, the electron mobility μ_{pl} increases with increasing the electron density n. The mobility behaves as

$$\mu_{pl} \propto \frac{1}{\hbar\omega_p} \cdot \frac{1}{n_p} = \sqrt{\frac{\epsilon m^*}{ne^2}} \cdot \left[\exp\left(\frac{\hbar\omega_p}{k_B T}\right) - 1\right] \tag{6.400}$$

and thus at higher temperatures or lower electron density such as $\hbar\omega_p/k_B T \ll 1$ the mobility is approximated as $\mu_{pl} \propto \hbar\omega_p/\sqrt{n}$ and thus electron density dependence is very weak as observed in lower electron density n in Fig. 6.34. On the other hands, at lower temperatures or higher electron density such as $\hbar\omega_p/k_B T \gg 1$, the mobility is proportional to $\exp(\hbar\omega_p/k_B T)/\sqrt{n}$, and thus the mobility increase with increasing the electron density n because of the exponential term $\exp(\hbar\omega_p/k_B T)$ as seen the mobility at higher electron density in Fig. 6.34.

6.5.9 Alloy Scattering

The relaxation time for alloy scattering is given by (6.313) or (6.369), and the electron mobility for alloy scattering is evaluated as

$$\mu_{\text{alloy}} = \frac{16e\sqrt{\pi}\hbar^4}{3(2m^*)^{5/2}} \frac{1}{(V_a - V_b)^2 x(1-x)\Omega} (k_B T)^{-1/2}, \tag{6.401}$$

where m^* is the effective mass of the carrier $\Omega = a^3/4$ for a zinc blende crystal with the lattice constant a.

Since we are not able to estimate $|V_a - V_b|$, we evaluate the electron mobility limited by the alloy scattering by assuming the mobility at 10 K for $x = 0.5$ is 1.0×10^4 [cm^2/Vs], which gives $|V_a - V_b| = 3.185$ [eV] for $m^* = 0.067\,\text{m}$, $a = 5.66 \times 10^{-10}$[m]. The calculated electron mobilities due to the alloy scattering are plotted in Fig. 6.35 for $x = 0.1 \sim 0.5$. We find in Fig. 6.35 that the alloy scattering becomes very weak at lower temperatures and depends strongly on the alloy composition. The alloy scattering depends strongly on the potential $|V_a - V_b|^2$ and the composition x. We may expect that the alloy scattering plays an important role at higher temperatures as shown in Fig. 6.35.

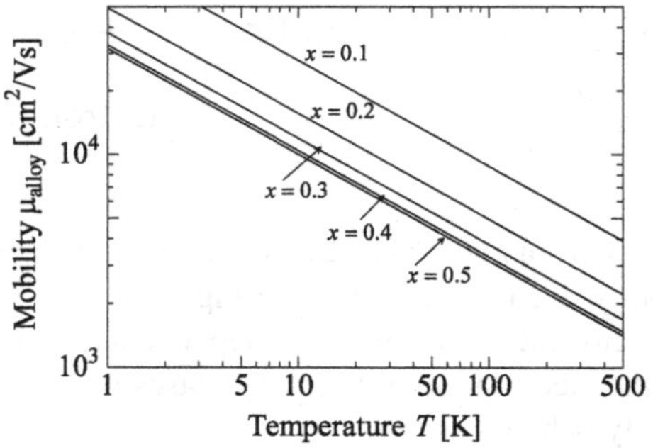

Fig. 6.35 Calculated electron mobility due to alloy scattering as a function of temperature for the alloy composition $x = 0.1 \sim 0.5$. The parameter $|V_a - V_b|$ is determined to give $\mu = 10$ [m^2/Vs] at $T = 10$ [K]. Alloy scattering becomes very weak at lower temperatures and depends strongly on the alloy composition

6.5.10 Electron Mobility in GaN

Devices based on nitrides and their ternary alloys such as GaN and GaInN are very important for light emitting diodes and lasers in the blue ray region. In addition heterobipolar devices based on nitrides are becoming very important for high power and high frequency devices. However, the transport properties are not well known because of their complicated band structures. We deal with the energy band calculations on nitrides and their ternary alloys in Chap. 9. Here electron mobility in GaN is evaluated, assuming that the electron mobility is limited by the scattering processes (1) acoustic phonon deformation potential (ac), (2) piezoelectric potential (pz), (3) ionized impurity (BH: Brooks–Herring formula), (4) polar optical phonon (pop) scatterings. Including these scattering processes, the relaxation time is given by (6.396)

$$\frac{1}{\tau_{\text{total}}} = \frac{1}{\tau_{\text{ac}}} + \frac{1}{\tau_{\text{pz}}} + \frac{1}{\tau_{\text{BH}}} + \frac{1}{\tau_{\text{pop}}}, \tag{6.402}$$

where the relaxation time for the respective scattering process is derived in Chap. 6. The electron mobility is evaluated by (6.397)

$$\mu = \frac{e}{m_{\text{c}}^*} \langle \tau_{\text{total}} \rangle, \tag{6.403}$$

where we assume the effective mass is isotropic and $m_{\text{c}}^* = m_{\text{d}}^* = m^*$. The results are shown in Fig. 6.36. The parameters used in the calculations are listed in Table 6.5.

We find in Fig. 6.36 that the electron mobility of GaN at higher temperatures is limited by the polar optical phonon energy and that the mobility at low temperatures depend on the density of ionized impurities. The electron mobility is expected to depend on the electron effective mass through the density of states mass m_{d}^* and the conductivity effective mass m_{c}^*. Since the electron effective mass is not well determined, it is very interesting to check the effective mass dependence of the

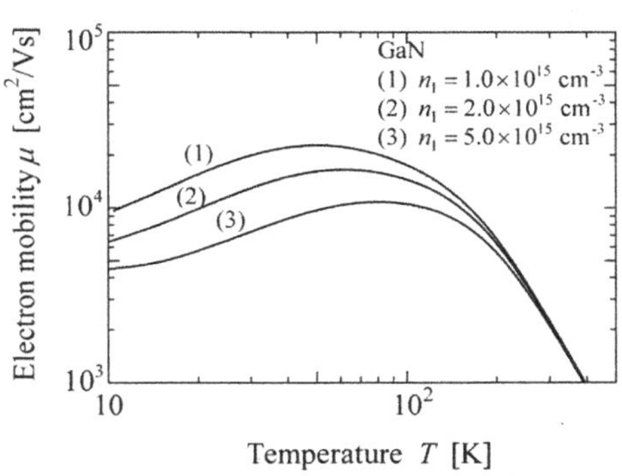

Fig. 6.36 Calculated electron mobility of n-GaN, where the used parameters are given in Table 6.5 and the ionized impurity densities are $n_{\text{I}} = 1.0 \times 10^{15}$, 2.0×10^{15}, and 5.0×10^{15} [cm^{-3}]

Table 6.5 Used parameters to evaluate electron mobility in n-GaN

Material parameters	Symbol [units]	Values
Electron effective mass	m^*/m	0.22
Mass density	ρ [kg/m^3]	6.11×10^3
Elastic constant	c_l [N/m^2]	3.82×10^{11}
Static dielectric constant	κ_0	10.4
High freq. dielectric constant	κ_∞	5.15
Optical phonon equiv temp.	θ_{pop} [K]	1057 (91.1 meV)
Piezoelectric constant	e_{31} [C/m^2]	-0.32
Piezoelectric constant	e_{33} [C/m^2]	0.63
Acoustic deformation potential	E_1 [eV]	13.5
Ionized impurity density	n_{I} [cm^{-3}]	$1.0 \sim 5.0 \times 10^{15}$

Table 6.6 Calculated electron mobility in GaN at $T = 77$ K and $T = 300$ K with the effective mass $m^* = 0.22\,\mathrm{m}$ and $m^* = 0.145\,\mathrm{m}$. The ionized impurity density is assumed to be 1.0×10^{15} cm^{-3}

Electron effective mass m^*	μ (77 K)	μ (300 K)
0.22 m	2.05 m^2/Vs	0.225 m^2/Vs
0.145 m	4.14 m^2/Vs	0.496 m^2/Vs

mobility. We describe the numerical calculations of the electron effective mass and hole mass by the pseudopotential method in Chap. 9, where we obtain $m_\perp/m = 0.145$ and $m_\parallel/m = 0.135$ for perpendicular and parallel to the c–axis of the crystal. On the other hand the reported effective mass is $m^* = 0.22\,\mathrm{m}$ as shown in Table 9.6 (after Vurgaftman and Meyer [59] in Chap. 9). In Table 6.6 the calculated electron mobility at $T = 77$ K and $T = 300$ K with the effective mass $m^* = m^*_{\mathrm{d}} = m^*_{\mathrm{c}} = 0.145\,\mathrm{m}$ and 0.22 m as an example. Here we find that the electron mobility depends on the effective mass and the mobility for $m^* = 0.145$ is approximately twice as the mobility for $m^* = 0.22\,\mathrm{m}$. The effective mass dependence of the mobilities due to the respective scatterings are

$$\mu_{\mathrm{ac}} \propto m^{*-5/2}, \quad \mu_{\mathrm{pop}} \propto m^{*-3/2}, \quad \mu_{\mathrm{pz}} \propto m^{*-3/2}, \quad \mu_{\mathrm{BH}} \propto m^{*-1/2}. \tag{6.404}$$

6.6 Problems

This chapter deals with the scattering processes and evaluate electron mobilities due to various scattering processes in detail. It is very helpful for readers to summarize the processes in a short version. Here the readers are asked to summarize the respective equations for acoustic phonon scattering as an example. This will help the readers to understand other scattering processes.

(6.1) Show the scattering probability $P(\boldsymbol{k}, \boldsymbol{k}')$ for the acoustic phonon scattering, using the interaction Hamiltonian $H_1 = H_{el}$. Describe how to derive the scattering rate w_{ac} for an electron by acoustic phonon deformation potential scattering given by (6.332). Discuss the assumptions used to derive the final result.

(6.2) Show that the relaxation time τ_{ac} for an electron by acoustic phonon deformation potential scattering is given by (6.341) under the $q_{max} = 2k(1 \pm m^* v_s/\hbar k) \simeq 2k$.

(6.3) Calculate the relaxation time without the approximation, and compare the result with the above calculations.

(6.4) Electron mobility of n-GaAs is limited by polar optical phonon scattering. Using the parameters for GaAs listed in Table 6.3, calculate the electron mobility limited by polar optical phonon scattering in GaAs at the range of temperature from 100 to 500 K.

(6.5) Calculate the electron mobility of GaN due to polar optical phonon scattering and show the high temperature mobility is limited by the polar optical phonon scattering. Use the parameters listed in Table 6.5. Also compare the electron mobility for $m^* = 0.22\,\mathrm{m}$ and $m^* = 0.10\,\mathrm{m}$ at $T = 100$ and $T = 300$ K.

References

1. E.M. Conwell, V.F. Weisskopf, Phys. Rev. **77**, 388 (1950)
2. J. Zook, Phys. Rev. **136**, A869 (1964)
3. A.R. Hutson, J. Appl. Phys. **32**, 2287 (1961)
4. A.R. Hutson, D.L. White, J. Appl. Phys. **33**, 40 (1962)
5. B.K. Ridley, *Quantum Processes in Semiconductors* (Oxford University Press, Oxford, 1988)
6. E.M. Conwell, High field transport in semiconductors, *Solid State Physics*, Suppl. 9 (Academic Press, New York, 1967)
7. M. Born, K. Huang, *Dynamical Theory of Crystal Lattice* (Oxford University Press, Oxford, 1988)
8. C. Herring, E. Vogt, Phys. Rev. **101**, 944 (1956)
9. O. Madelung (ed.), Landolt–Börnstein: numerical data and functional relationships in science and technology, *Semiconductors*, vol. 17 (Springer, Berlin, 1982)
10. M.V. Fischetti, J. Appl. Phys. **80**, 2234 (1996)
11. W. Pötz, P. Vogl, Phys. Rev. B **B24**, 2025 (1981)
12. S. Zollner, S. Gopalan, M. Cardona, Appl. Phys. Lett. **54**, 614 (1989); J. Appl. Phys. **68**, 1682 (1990); Phys. Rev. **B44**, 13446 (1991)
13. M. Fischetti, S.E. Laux, Phys. Rev. B **38**, 9721 (1988)
14. H. Mizuno, K. Taniguchi, C. Hamaguchi, Phys. Rev. B **48**, 1512 (1993)
15. M.L. Cohen, T.K. Bergstresser, Phys. Rev. **141**, 789 (1966)
16. K. Kunc, O.H. Nielsen, Comput. Phys. Commun. **16**, 181 (1979)

17. G. Dolling, J.L.T. Waugh, in *Lattice Dynamics*, ed. by R.F. Wallis (Pergamon, Oxford, 1965), pp. 19–32
18. K. Kunc, H. Bilz, Solid State Commun. **19**, 1027 (1976)
19. W. Weber, Phys. Rev. B **15**, 4789 (1977)
20. O.H. Nielsen, W. Weber, Comput. Phys. Commun. **17**, 413 (1979)
21. O.J. Glembocki, F.H. Pollak, Phys. Rev. Lett. **48**, 413 (1982)
22. S. Bednarek, U. Rössler, Phys. Rev. Lett. **48**, 1296 (1982)
23. H. Mizuno, MS Thesis, Osaka University (1995)
24. H. Haug, S.W. Koch, *Quantum Theory of the Optical and Electronic Properties of Semiconductors* (World Scientific, Singapore, 1993). Chaps. 7 and 8
25. G.D. Maham, *Many-Particle Physics* (Plenum Press, New York, 1986). Chap. 5
26. D.K. Ferry, R.O. Grondin, *Physics of Submicron Devices* (Plenum Press, New York, 1991). Chap. 7
27. J. Lindhard, Mat. Fys. Medd. **28**, 8 (1954)
28. C. Kittel, *Quantum Theory of Solids* (Wiley, New York, 1963). Chap. 2
29. E.M. Conwell, A.L. Brown, J. Phys. Chem. Solids **15**, 208 (1960)
30. H. Callen, Phys. Rev. **76**, 1394 (1949)
31. C. Jacoboni, L. Reggiani, Rev. Mod. Phys. **55**, 645 (1983)
32. L. Eaves, R.A. Hoult, R.A. Stradling, R.J. Tidey, J.C. Portal, S. Askenazy, J. Phys. C, Solid State Phys. **8**, 1034 (1975)
33. C. Hamaguchi, Y. Hirose, K. Shimomae, Jpn. J. Appl. Phys. **22**(Suppl. 22–3), 190 (1983)
34. C. Erginsoy, Phys. Rev. **79**, 1013 (1950)
35. N. Sclar, Phys. Rev. **104**, 1548 (1956)
36. T.C. McGill, R. Baron, Phys. Rev. B **11**, 5208 (1975)
37. D.L. Rode, Phys. Rev. B **2**, 1012 (1970)
38. F.J. Morin, Phys. Rev. **93**, 62 (1954)
39. D.M. Brown, R. Bray, Phys. Rev. **127**, 1593 (1962)
40. D.L. Rode, in *Semiconductors and Semimetals*, vol. 10, ed. by R.K. Willardson, A.C. Beer (Academic Press, New York, 1975), pp. 1–89
41. C. Canali, C. Jacoboni, F. Nava, G. Ottavani, A, A1berigi Quaranta, Phys. Rev. **B 12**, 2265 (1975)
42. C. Jacoboni, C. Canali, G. Ottaviani, A. Alberigi Quaranta, Solid-State Electron. **20**, 77 (1977)
43. D. Long, Phys. Rev. **120**, 2024 (1960)

Chapter 7
Magnetotransport Phenomena

Abstract Electron transport in a magnetic field exhibits various interesting characteristics, which is called magnetotransport phenomena. In this chapter we deal with Hall effect, magnetoresistance, and oscillatory magnetoresistance effects in detail. Hall effect is very useful to determine the mobility and density of carriers. Shubnikov–de Haas oscillations are observed in degenerate semiconductors and provide the information of Fermi energy. Magnetophonon resonance arises from inter-Landau level transitions of nondegenerate electron gas and gives the effective mass. Once we know the effective mass we can deduce phonon energy involved with the magnetophonon magnetophonon resonance. These two oscillatory magnetoresistance effects are very widely used to investigate semiconductor parameters. Quantum Hall effect is not dealt with here but discussed in detail in Chap. 8.

7.1 Phenomenological Theory of the Hall Effect

The application of a magnetic field perpendicular to the current through a conducting material results in the generation of an electric field perpendicular to both the current and the magnetic field. This phenomena was discovered in 1879 by E.H. Hall and is called the **Hall effect**. This phenomena has been successfully used to investigate the electronic properties of semiconductors and metals. In this section, first we deal with the Hall effect in a simple way in which all the electrons have the same velocity and later we will discuss a more accurate treatment using the Boltzmann transport equation.

First, let us consider an infinitely long sample, as shown in Fig. 7.1, and assume that a uniform current density $\boldsymbol{J}$ flows along the sample. An electron with drift velocity $\boldsymbol{v}$ experience the Lorentz force $\boldsymbol{F} = -e\boldsymbol{v} \times \boldsymbol{B}$, where $-e$ is the charge of the electron. Let the electron density be n; the current density is then given by $\boldsymbol{J} = n(-e)\boldsymbol{v}$. When the current flows in the $+x$ direction, the velocity component is $v_x < 0$ and thus the electron drifts in the $-x$ direction. In a magnetic field applied in the z direction, the Lorentz force for the electron is given by $F_y = -e(|v_x|B_z) = -e|v_x|B_z < 0$, and thus the electron is forced to move in the $-y$ direction. Noting that the Hall effect is normally measured as the electric field induced in the y direction without any flow

C. Hamaguchi, *Basic Semiconductor Physics*, Graduate Texts in Physics,
DOI 10.1007/978-3-319-66860-4_7

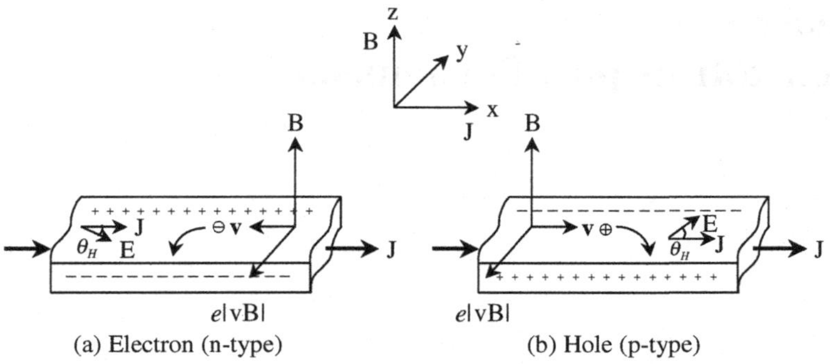

Fig. 7.1 Hall effect in semiconductors. Hall fields for **a** electron (n-type semiconductor) and **b** hole (p-type semiconductor) are reversed

of current in this direction, negative charges due to the electrons will accumulate on the front surface as shown in Fig. 7.1 and immobile positive charges (donors) will appear on the back surface, resulting in an electric field in the $-y$ direction. The induced electric field (Hall field) $E_y < 0$ will balance the Lorentz force and a steady state will be achieved so that $F_y + (-e)E_y = 0$. This electric field E_y is called the **Hall Field**, and we find the following relation:

$$E_y = v_x B_z = -\frac{B_z J_x}{ne} \equiv R_{\rm H} J_x B_z \,, \tag{7.1}$$

where $R_{\rm H}$ is called the **Hall coefficient**. For electrons the Hall coefficient is given by

$$R_{\rm H} = -\frac{1}{ne} \,. \tag{7.2}$$

The angle $\theta_{\rm H}$ between E_x and E_y is called the **Hall angle**, which is given by

$$\tan\theta_{\rm H} = \frac{E_y}{E_x} = -\mu_{\rm e} B_z = \omega_{\rm c}\tau \,. \tag{7.3}$$

In the above equation $J_x = ne\mu_{\rm e}E_x$, $v_x = -\mu E_x$ and

$$\omega_{\rm c} = \frac{eB_z}{m^*} \,, \quad \mu_{\rm e} = \frac{e\tau}{m^*} \tag{7.4}$$

are used and $\omega_{\rm c}$ is the cyclotron angular frequency defined in Chap. 2. The Hall effect is schematically shown in Fig. 7.1a for negative charge carriers (electrons in n-type semiconductors). On the other hand, Fig. 7.1b shows the Hall effect for holes (p-type semiconductors). A hole current in the $+x$ direction is produced by hole drift in the $+y$ direction ($v_x > 0$), and the Lorentz force for a hole in the presence of a magnetic field in the z direction directs it along the $-y$ direction, resulting in positive and negative charge accumulation on the $+y$ surface and $-y$ surface, respectively. This charge accumulation gives rise to a Hall field in the $+y$ direction, which is opposite to the case for electrons. The Hall coefficient for holes with density p is therefore given by

$$R_{\rm H} = \frac{1}{pe} . \tag{7.5}$$

If we let the width in the y direction of a semiconductor sample be w and the thickness in the z direction be t, then we obtain the Hall voltage in the y direction $V_{\rm H} = wE_y$. From current $I_x = J_x wt$ and from (7.1) the Hall voltage is expressed as

$$V_{\rm H} = \frac{R_{\rm H}}{t} I_x B_z . \tag{7.6}$$

Measurement of the Hall effect gives the Hall coefficient $R_{\rm H}$ from (7.6) and the electron density n or hole density p are determined from (7.2) or (7.5), respectively.

It should be noted here that the Hall coefficient given by (7.2) and (7.5) are derived under the assumption that the carrier relaxation time τ is independent of the carrier energy. When the relaxation time τ depends on the carrier energy and the effective mass is not scalar, the Hall coefficient is given by the following relation instead of (7.2):

$$R_{\rm H} = -\frac{r_{\rm H}}{ne} , \tag{7.7}$$

where $r_{\rm H}$ is called the **scattering factor of the Hall coefficient**, and is determined from the scattering mechanisms and distribution function of the carriers. A detailed treatment will be given in next section.

Using (6.121) for the conductivity σ the Hall coefficient (7.2) is rewritten as

$$|R_{\rm H}|\,\sigma = \mu . \tag{7.8}$$

This relation is valid only for $r_{\rm H} = 1$ as seen from (7.7), and in general use the relation

$$|R_{\rm H}|\,\sigma = r_{\rm H}\mu . \tag{7.9}$$

As discussed later, it is very difficult to determine the scattering factor $r_{\rm H}$ in semiconductors. We define the Hall mobility by

$$|R_{\rm H}|\,\sigma = \mu_{\rm H} \tag{7.10}$$

where μ_{H} is called the **Hall mobility**. The Hall mobility defined above has the same dimensions as the drift mobility but differs by a factor r_{H}, or in other words, the following relation holds.

$$\frac{\mu_{\mathrm{H}}}{\mu} = r_{\mathrm{H}} . \tag{7.11}$$

Although the treatment stated above is based on a simplified model, the results give good information to interpret the Hall effect. In the following we will derive relations useful for understanding magnetotransport in semiconductors, which may be extended to the more detailed general case. The equation of motion for an electron is written as

$$m^* \frac{\mathrm{d}\boldsymbol{v}}{\mathrm{d}t} + \frac{m^* \boldsymbol{v}}{\tau} = -e(\boldsymbol{E} + \boldsymbol{v} \times \boldsymbol{B}) . \tag{7.12}$$

When a static electric field is applied to a semiconductor, the steady state condition $\mathrm{d}\boldsymbol{v}/\mathrm{d}t = 0$ gives rise to a steady-state drift velocity $\boldsymbol{v}$ given by

$$\boldsymbol{v} = -\frac{e\tau}{m^*}\boldsymbol{E} \equiv -\mu \boldsymbol{E} . \tag{7.13}$$

Applying the averaging procedure on this relation used in Sect. 6.2.2 gives the same result for the drift mobility as (6.118). Therefore, the velocity is interpreted as the drift velocity. In the presence of a magnetic field $\boldsymbol{B}$, the velocity is given by

$$\boldsymbol{v} = -\frac{e\tau}{m^*}(\boldsymbol{E} + \boldsymbol{v} \times \boldsymbol{B}) . \tag{7.14}$$

If the magnetic field is along the z direction, the above equation leads to

$$v_x = -\frac{e\tau}{m^*}(E_x + v_y B_z) , \tag{7.15a}$$

$$v_y = -\frac{e\tau}{m^*}(E_y - v_x B_z) , \tag{7.15b}$$

$$v_z = -\frac{e\tau}{m^*}E_z . \tag{7.15c}$$

These relations are rewritten as

$$v_x = -\frac{e}{m^*}\left[\frac{\tau}{1+\omega_c^2\tau^2}E_x - \frac{\omega_c\tau^2}{1+\omega_c^2\tau^2}E_y\right] , \tag{7.16a}$$

$$v_y = -\frac{e}{m^*}\left[\frac{\omega_c\tau^2}{1+\omega_c^2\tau^2}E_x + \frac{\tau}{1+\omega_c^2\tau^2}E_y\right] , \tag{7.16b}$$

$$v_z = -\frac{e}{m^*}\tau E_z . \tag{7.16c}$$

Since the current density is given by $\boldsymbol{J} = n(-e)\boldsymbol{v}$, we obtain

$$J_x = \frac{ne^2}{m^*}\left[\frac{\tau}{1+\omega_c^2\tau^2}E_x - \frac{\omega_c\tau^2}{1+\omega_c^2\tau^2}E_y\right], \tag{7.17a}$$

$$J_y = \frac{ne^2}{m^*}\left[\frac{\omega_c\tau^2}{1+\omega_c^2\tau^2}E_x + \frac{\tau}{1+\omega_c^2\tau^2}E_y\right], \tag{7.17b}$$

$$J_z = \frac{ne^2}{m^*}\tau E_z . \tag{7.17c}$$

Therefore, writing the i component of the current as

$$J_i = \sigma_{ij}E_j, \qquad [J] = [\sigma][E] \tag{7.18}$$

we find

$$[\sigma] = \begin{bmatrix} \sigma_{xx} & \sigma_{xy} & 0 \\ \sigma_{yx} & \sigma_{yy} & 0 \\ 0 & 0 & \sigma_{zz} \end{bmatrix}, \tag{7.19}$$

$$\sigma_{xx} = \sigma_{yy} = \frac{ne^2}{m^*}\cdot\frac{\tau}{1+\omega_c^2\tau^2}, \tag{7.20a}$$

$$\sigma_{xy} = -\sigma_{yx} = -\frac{ne^2}{m^*}\cdot\frac{\omega_c\tau^2}{1+\omega_c^2\tau^2}, \tag{7.20b}$$

$$\sigma_{zz} = \frac{ne^2}{m^*}\tau \equiv \sigma_0 . \tag{7.20c}$$

Next, we will discuss the Hall effect, where we assume the current is in the x direction and the magnetic field is applied in the z direction. The current density in the x and y directions are then given by

$$J_x = \sigma_{xx}E_x + \sigma_{xy}E_y , \tag{7.21a}$$

$$J_y = \sigma_{yx}E_x + \sigma_{yy}E_y . \tag{7.21b}$$

Since the Hall effect is measured without current flow in the y direction, i.e. $J_y = 0$, this condition gives rise to the following relation

$$E_y = -\frac{\sigma_{yx}}{\sigma_{yy}}E_x = \frac{\sigma_{xy}}{\sigma_{xx}}E_x = -\omega_c\tau E_x . \tag{7.22}$$

Inserting E_y into J_x of (7.21a), we get

$$J_x = \frac{\sigma_{xx}^2 + \sigma_{xy}^2}{\sigma_{xx}}E_x . \tag{7.23}$$

It is very interesting to point out that (7.20a) and (7.20b) result in

$$J_x = \frac{ne^2}{m^*}\tau E_x = ne\mu E_x = \sigma_0 E_x . \tag{7.24}$$

This result means that the current density J_x is independent of the magnetic field B_z under the assumption we have used above, and that the resistivity is not changed by the application of a magnetic field, resulting in no magnetoresistance effect. We know that this conclusion conflicts with the experimental observations and that all semiconductors exhibit magnetoresistance. This conflicting conclusion arises from the following incorrect assumptions: (1) isotropic effective mass (spherical energy surface), (2) electron relaxation time τ independent of electron energy and all electrons drifting with the same velocity v, (3) shape of the sample not considered (the length is infinite). Although these assumptions will lead to wrong conclusions, the relations derived here give the correct expressions when we introduce the averaging procedures later. For this reason we shall proceed with the theoretical analysis of magnetotransport with the above simplified assumptions.

From (7.21b) and (7.23) we have

$$E_y = \frac{\sigma_{xy}}{\sigma_{xx}} E_x = \frac{\sigma_{xy}}{\sigma_{xx}^2 + \sigma_{xy}^2} J_x \equiv R_\mathrm{H} B_z J_x \,, \tag{7.25}$$

and the Hall coefficient is generally expressed as

$$R_\mathrm{H} = \frac{\sigma_{xy}}{\sigma_{xx}^2 + \sigma_{xy}^2} \frac{1}{B_z} \,. \tag{7.26}$$

Under the above assumption, substituting (7.20a), (7.20b) or (7.24) into (7.25) we obtain

$$E_y = -\omega_\mathrm{c}\tau E_x = -\mu B_z \frac{J_x}{\sigma_0} = -\frac{1}{ne} B_z J_x \tag{7.27}$$

and thus $R_\mathrm{H} = -1/ne$, which is equivalent to (7.2).

Next, we will consider the case where two types of carriers are present. Let us identify two types of carriers by using subscripts 1 and 2. We therefore obtain

$$J_x = \left[\sigma_{xx}^{(1)} + \sigma_{xx}^{(2)}\right] E_x + \left[\sigma_{xy}^{(1)} + \sigma_{xy}^{(2)}\right] E_y \,, \tag{7.28a}$$

$$J_y = \left[\sigma_{xy}^{(1)} + \sigma_{xy}^{(2)}\right] E_x + \left[\sigma_{xx}^{(1)} + \sigma_{xx}^{(2)}\right] E_y \,. \tag{7.28b}$$

Let us define the electrical conductivity for each carrier by σ_1 and σ_2, and the Hall coefficient by R_1 and R_2. Then, from (7.20a) and (7.20b) we have for carrier 1

$$\sigma_{xx}^{(1)} = \frac{\sigma_1}{1 + \sigma_1^2 R_1^2 B_z^2}, \quad \sigma_{xy}^{(1)} = -\frac{\sigma_1^2 R_1 B_z}{1 + \sigma_1^2 R_1^2 B_z^2} \tag{7.29}$$

and similar relations for carrier 2. From the condition for achieving the Hall effect we put $J_y = 0$ and in the same way as we derived (7.26) we obtain

$$R_\mathrm{H} B_z = \frac{\sigma_{xy}^{(1)} + \sigma_{xy}^{(2)}}{\left[\sigma_{xx}^{(1)} + \sigma_{xx}^{(2)}\right]^2 + \left[\sigma_{xy}^{(1)} + \sigma_{xy}^{(2)}\right]^2} \,. \tag{7.30}$$

Inserting (7.29) and the same expression for carrier 2 in this equation we find

$$R_{\mathrm{H}} = \frac{\sigma_1^2 R_1(1+\sigma_2^2 R_2^2 B_z^2) + \sigma_2^2 R_2^2(1+\sigma_1^2 R_1^2 B_z^2)}{(\sigma_1+\sigma_2)^2 + \sigma_1^2\sigma_2^2(R_1+R_2)^2 B_z^2} \,. \tag{7.31}$$

It is very important to point out here that (7.31) is valid even if the relaxation time τ (τ_1 and τ_2) depends on the carrier energy, and thus it gives the general expression for the Hall coefficient with two types of carriers.

Since $\sigma_1|R_1| = \mu_1$ and $\sigma_2|R_2| = \mu_2$, in the weak magnetic field case such as $\mu_1 B_z \ll 1$ and $\mu_2 B_z \ll 1$, (7.31) may be approximated as

$$R_{\mathrm{H}}(0) = \frac{\sigma_1^2 R_1 + \sigma_2^2 R_2}{(\sigma_1+\sigma_2)^2} \,. \tag{7.32}$$

On the other hand, in the high magnetic field case such that $\sigma_1|R_1|B_z = \mu_1 B_z \gg 1$ and $\sigma_2|R_2|B_z = \mu_2 B_z \gg 1$ are satisfied, we obtain

$$R_{\mathrm{H}}(\infty) = \left(\frac{1}{R_1} + \frac{1}{R_2}\right)^{-1} \,. \tag{7.33}$$

As an example, we consider the case of two carriers, electrons and holes, with respective mobilities μ_e and μ_h, densities n and p, and assume $r_H = 1$ for simplicity. Inserting $\sigma_1 = ne\mu_{\mathrm{e}}$, $\sigma_2 = pe\mu_{\mathrm{h}}$, $R_1 = -1/ne$ and $R_2 = 1/pe$ into (7.31) we obtain

$$R_{\mathrm{H}} = \frac{(p - nb^2) + b^2\mu_{\mathrm{h}}^2 B_z^2(p-n)}{(bn+p)^2 + b^2\mu_{\mathrm{h}}^2 B_z^2 (p-n)^2} \cdot \frac{1}{e} \,, \tag{7.34}$$

where $b = \mu_{\mathrm{e}}/\mu_{\mathrm{h}}$ is the mobility ratio. Due to the assumptions described above, the above equation is valid for $r_{\mathrm{H}} = 1$ and thus for the case where the relaxation times are independent of energy and are constant. From this equation we find that the Hall coefficient becomes zero or $R_{\mathrm{H}} = 0$, when the following relation is satisfied:

$$p = \frac{nb^2(1+\mu_{\mathrm{h}}^2 B_z^2)}{1 + b^2\mu_{\mathrm{h}}^2 B_z^2} \,. \tag{7.35}$$

The Hall coefficients at low magnetic fields, $R_{\mathrm{H}}(0)$, and at high magnetic fields, $R_H(\infty)$, are given as follows from (7.32) and (7.33): respectively.

$$R_{\mathrm{H}}(0) = \frac{p\mu_{\mathrm{h}}^2 - n\mu_{\mathrm{e}}^2}{e(p\mu_{\mathrm{h}} + n\mu_{\mathrm{e}})^2} = \frac{p - b^2 n}{e(nb+p)^2} \,, \tag{7.36}$$

$$R_{\mathrm{H}}(\infty) = \frac{1}{e(p-n)} \,. \tag{7.37}$$

7.2 Magnetoresistance Effects

7.2.1 *Theory of Magnetoresistance*

We have discussed the current density in the presence of a magnetic field in Sect. 7.1, where we assumed free electrons with scalar effective mass m^* and constant drift velocity v. Under this assumption we have shown that no magnetoresistance effect appears. Here we discuss a more exact treatment of the magnetoresistance effect with the help of the Boltzmann transport equation.

In the presence of a d.c electric field $\boldsymbol{E}$ and magnetic field $\boldsymbol{B}$, the Lorentz force acting on the electrons given by the right-hand side of (7.12), and insertion of this relation into (6.90) results in

$$-\frac{e}{\hbar}\left(\boldsymbol{E}+\frac{1}{\hbar}\frac{\partial\mathcal{E}}{\partial\boldsymbol{k}}\times\boldsymbol{B}\right)\cdot\frac{\partial f}{\partial\boldsymbol{k}}=-\frac{f-f_0}{\tau}\equiv-\frac{f_1}{\tau}. \tag{7.38}$$

Here, the distribution function f_0 under thermal equilibrium is a function of the electron energy $\mathcal{E}(\boldsymbol{k})$ and we find

$$\frac{\partial f_0}{\partial\boldsymbol{k}}=\frac{\partial f_0}{\partial\mathcal{E}}\cdot\frac{\partial\mathcal{E}}{\partial\boldsymbol{k}}. \tag{7.39}$$

The right-hand side of (7.38), $-f_1/\tau$, is obtained from the first-order approximation by inserting f_0 into f of the right-hand side:

$$-\frac{f_1}{\tau}=-\frac{e}{\hbar^2}\left(\frac{\partial\mathcal{E}}{\partial\boldsymbol{k}}\times\boldsymbol{B}\right)\cdot\frac{\partial\mathcal{E}}{\partial\boldsymbol{k}}\frac{\partial f_0}{\partial\mathcal{E}}. \tag{7.40}$$

Since $[\partial\mathcal{E}(\boldsymbol{k})/\partial\boldsymbol{k}]\times\boldsymbol{B}$ is orthonormal to $\partial\mathcal{E}(\boldsymbol{k})/\partial\boldsymbol{k}$, (7.40) becomes zero. Therefore, only the term f_1 of the distribution function $f = f_0 + f_1$ will contribute to the magnetic effect and (7.38) is approximated as

$$-\frac{e}{\hbar}\boldsymbol{E}\cdot\frac{\partial f_0}{\partial\boldsymbol{k}}-\frac{e}{\hbar^2}\frac{\partial\mathcal{E}}{\partial\boldsymbol{k}}\times\boldsymbol{B}\cdot\frac{f_1}{\partial\boldsymbol{k}}=-\frac{f_1}{\tau}. \tag{7.41}$$

In the following we will solve this equation under the assumption that the distribution function is not significantly changed in a weak electric field. In addition, we will not take account of the quantization of the electronic states.

7.2.2 *General Solutions for a Weak Magnetic Field*

In a weak magnetic field $\boldsymbol{B}$, except for $\boldsymbol{B} = 0$, we may obtain solutions by an iterative method. We start from the solution for zero magnetic field. In zero magnetic field we have from (7.41)

$$f_1 = \frac{e\tau}{\hbar}\boldsymbol{E}\cdot\frac{\partial f_0}{\partial \boldsymbol{k}} \equiv \frac{e\tau}{\hbar}\left(\boldsymbol{E}\cdot\frac{\partial \mathcal{E}}{\partial \boldsymbol{k}}\right)\frac{\partial f_0}{\partial \mathcal{E}} = \frac{e\tau}{\hbar}\left(\sum_j E_j\frac{\partial \mathcal{E}}{\partial k_j}\right)\frac{\partial f_0}{\partial \mathcal{E}}$$
$$\equiv f_1^{(0)}\,, \tag{7.42}$$

where E_j and $\partial\mathcal{E}/\partial k_j$ represent the j component of $\boldsymbol{E}$ and $\partial\mathcal{E}/\partial\boldsymbol{k}$, respectively. In the following we have omitted the symbol of summation. Inserting this f_1 into (7.41), the distribution function in the case of $\boldsymbol{B} \neq 0$ is obtained in the first order of $\boldsymbol{B}$, which is given by

$$f_1 = f_1^{(0)} + \frac{e\tau}{\hbar^2}\left(\frac{\partial \mathcal{E}}{\partial \boldsymbol{k}}\times\boldsymbol{B}\right)\cdot\frac{\partial f_1^{(0)}}{\partial \boldsymbol{k}}$$
$$+\frac{e^2\tau}{\hbar^4}\left(\frac{\partial \mathcal{E}}{\partial \boldsymbol{k}}\times\boldsymbol{B}\right)\cdot\frac{\partial}{\partial \boldsymbol{k}}\left[\tau\left(\frac{\partial \mathcal{E}}{\partial \boldsymbol{k}}\times\boldsymbol{B}\right)\cdot\frac{\partial f_1^{(0)}}{\partial \boldsymbol{k}}\right]$$
$$\equiv f_1^{(2)}\,. \tag{7.43}$$

Inserting this in the equation of current density

$$\boldsymbol{J} = -\frac{e}{4\pi^3}\int \boldsymbol{v} f_1 \mathrm{d}^3\boldsymbol{k} = -\frac{e}{4\pi^3}\int \frac{1}{\hbar}\frac{\partial \mathcal{E}}{\partial \boldsymbol{k}} f_1 \mathrm{d}^3\boldsymbol{k}\,, \tag{7.44}$$

we obtain the current density in the general form in the presence of a weak magnetic field. The equation may be written in the tensor form

$$J_i = \sigma_{ij}E_j + \sigma_{ijl}E_jB_l + \sigma_{ijlm}E_jB_lB_m\,, \tag{7.45}$$

where the first, second and third terms on the right-hand side are respectively obtained from the first, second and third terms on the right-hand side of (7.43). Rewriting (7.44) as

$$J_i = -\frac{e}{4\pi^3\hbar}\int \frac{\partial \mathcal{E}}{\partial k_i} f_1 \mathrm{d}^3\boldsymbol{k}\,, \tag{7.46}$$

we obtain the following relations:

$$\sigma_{ij} = -\frac{e^2}{4\pi^3\hbar^2}\int \mathrm{d}^3\boldsymbol{k}\frac{\partial f_0}{\partial \mathcal{E}}\tau\frac{\partial \mathcal{E}}{\partial k_i}\frac{\partial \mathcal{E}}{\partial k_j}\,, \tag{7.47a}$$

$$\sigma_{ijl} = -\frac{e^3}{4\pi^3\hbar^4}\int \mathrm{d}^3\boldsymbol{k}\frac{\partial f_0}{\partial \mathcal{E}}\tau\frac{\partial \mathcal{E}}{\partial k_i}\left[\frac{\partial \mathcal{E}}{\partial k_r}\frac{\partial}{\partial k_s}\left(\tau\frac{\partial \mathcal{E}}{\partial k_j}\right)\right]\epsilon_{rls}\,, \tag{7.47b}$$

$$\sigma_{ijlm} = -\frac{e^4}{4\pi^3\hbar^6}\int \mathrm{d}^3\boldsymbol{k}\frac{\partial f_0}{\partial \mathcal{E}}\tau\frac{\partial \mathcal{E}}{\partial k_i}$$
$$\times\left\{\frac{\partial \mathcal{E}}{\partial k_r}\frac{\partial}{\partial k_s}\left[\tau\frac{\partial \mathcal{E}}{\partial k_t}\frac{\partial}{\partial k_u}\left(\tau\frac{\partial}{\partial k_j}\right)\right]\right\}\epsilon_{mrs}\epsilon_{ltu}\,, \tag{7.47c}$$

where ϵ_{lrs} is called the permutation tensor and has the following relations. When two of the subscripts lrs are the same, $\epsilon_{lrs} = 0$. When the subscripts are in the order $12312\ldots$ and $21321\ldots$, ϵ_{lrs} are 1 and -1, respectively. The coefficients defined in (7.47a)–(7.47c) are general expressions for the electrical conductivity tensor in the presence of a magnetic field, and its ij component is given by $(\sigma_{ij} + \sigma_{ijl}B_l + \sigma_{ijlm}B_lB_m)$. Here, σ_{ij} is the second-rank tensor of the **electrical conductivity** and the equivalent to the conductivity for $\boldsymbol{B} = 0$. σ_{ijl} is the third-rank **Hall effect tensor** and σ_{ijlm} is the fourth-rank **magnetoconductivity tensor**. The physical meaning of these tensors can be understood from the following examples.

7.2.3 Case of Scalar Effective Mass

The constant energy surface of conduction band is assumed to be spherical and expressed as

$$\mathcal{E}(\boldsymbol{k}) = \frac{\hbar^2}{2m^*}k^2 , \tag{7.48}$$

where m^* is the scalar effective mass. Then we have the following relations:

$$\frac{\partial \mathcal{E}}{\partial k_i} = \frac{\hbar^2}{m^*}k_i, \quad \frac{\partial^2}{\partial k_i \partial k_j} = \frac{\hbar^2}{m^*}\delta_{ij} . \tag{7.49}$$

Inserting these relations into (7.47a)–(7.47c), we carry out the integration using polar coordinates with the polar axis along the electric field $\boldsymbol{E}$ and with the angle θ between $\boldsymbol{k}$ and $\boldsymbol{E}$. In other words, we may put $\mathrm{d}^3\boldsymbol{k} = k^2 \sin\theta \mathrm{d}\theta \mathrm{d}\phi \mathrm{d}k$ and we obtain from (7.47a)

$$\begin{aligned} \sigma_{ij} &= -\frac{e^2\hbar^2}{4\pi^3 m^{*2}} \int_0^\infty \mathrm{d}k \cdot k^2 \int_0^\pi \mathrm{d}\theta \cdot \sin\theta \int_0^{2\pi} \mathrm{d}\phi \cdot k^2 \cdot \cos^2\theta \frac{\partial f_0}{\partial \mathcal{E}} \tau \delta_{ij} \\ &= -\frac{2ne^2}{3m^*} \left[\int_0^\infty \tau \mathcal{E}^{3/2} \frac{\partial f_0}{\partial \mathcal{E}} \mathrm{d}\mathcal{E} \Big/ \int_0^\infty \mathcal{E}^{1/2} f_0 \mathrm{d}\mathcal{E} \right] \delta_{ij} . \end{aligned} \tag{7.50}$$

Using (6.103) in the above equation, we find the same result as (6.106). In the case of non-degenerate semiconductors, the above result leads to (6.114). In a similar fashion, we may calculate (7.47b) and (7.47c), giving rise to the following relations:

$$\sigma_{ij} = \frac{ne^2}{m^*}\langle\tau\rangle\delta_{ij} \equiv \sigma_0\delta_{ij} , \tag{7.51a}$$

$$\sigma_{ijl} = -\frac{ne^3}{m^{*2}}\langle\tau^2\rangle\epsilon_{ijl} \equiv \gamma_0\epsilon_{ijl} , \tag{7.51b}$$

$$\sigma_{ijlm} = \frac{ne^4}{m^{*3}}\langle\tau^3\rangle\epsilon_{mis}\epsilon_{lsj} \equiv \beta_0\epsilon_{mis}\epsilon_{lsj} . \tag{7.51c}$$

Therefore, if we write the current density tensor in a weak magnetic field as

$$[J] = [\sigma(\boldsymbol{B})][E], \tag{7.52}$$

we obtain the electrical conductivity tensor $[\sigma(\boldsymbol{B})]$ given by

$$[\sigma(\boldsymbol{B})] = \begin{bmatrix} \sigma_0 + \beta_0[B_y^2 + B_z^2] & \gamma_0 B_z - \beta_0 B_x B_y & -\gamma_0 B_y - \beta_0 B_x B_z \\ -\gamma_0 B_z - \beta_0 B_x B_y & \sigma_0 + \beta_0(B_x^2 + B_z^2) & \gamma_0 B_x - \beta_0 B_y B_z \\ \gamma_0 B_y - \beta_0 B_x B_z & -\gamma_0 B_x - \beta_0 B_y B_z & \sigma_0 + \beta_0(B_x^2 + B_y^2) \end{bmatrix}. \tag{7.53}$$

Here we will separate the conductivity tensor into terms independent of the magnetic field B, and terms proportional to B and to B^2, and then we find the following relations:

$$\sigma_{ij} = \begin{bmatrix} \sigma_0 & 0 & 0 \\ 0 & \sigma_0 & 0 \\ 0 & & \sigma_0 \end{bmatrix}, \quad \sigma_{ijl} B_l = \begin{bmatrix} 0 & \gamma_0 B_z & -\gamma_0 B_y \\ -\gamma_0 B_z & 0 & \gamma_0 B_x \\ \gamma_0 B_y & -\gamma_0 B_x & 0 \end{bmatrix},$$

$$\sigma_{ijlm} B_l B_m = \begin{bmatrix} \beta_0(B_y^2 + B_z^2) & -\beta_0 B_x B_y & -\beta B_x B_z \\ -\beta_0 B_x B_y & \beta_0(B_x^2 + B_z^2) & -\beta_0 B_y B_z \\ -\beta_0 B_x B_z & -\beta_0 B_y B_z & \beta_0(B_x^2 + B_y^2) \end{bmatrix}. \tag{7.54}$$

Next, we discuss the Hall effect in a weak magnetic field. In order to compare the results of Sect. 7.1, we assume that the magnetic field $\boldsymbol{B}$ is applied in the z direction and $B_x = B_y = 0$. Then we obtain from (7.53) and (7.54)

$$J_x = (\sigma_0 + \beta_0 B_z^2) E_x + \gamma_0 B_z E_y, \tag{7.55a}$$

$$J_y = -\gamma_0 B_z E_x + (\sigma_0 + \beta_0 B_z^2) E_y. \tag{7.55b}$$

Comparing this with (7.20a)–(7.20c), (7.21a) and (7.21b) the following results are derived.

$$\sigma_{xx} = \sigma_{yy} = \sigma_0 + \beta_0 B_z^2 \frac{ne^2}{m^*}\langle\tau\rangle - \frac{ne^2\omega_c^2}{m^*}\langle\tau^3\rangle, \tag{7.56a}$$

$$\sigma_{xy} = -\sigma_{yx} = \gamma_0 B_z = -\frac{ne^2\omega_c}{m^*}\langle\tau^2\rangle. \tag{7.56b}$$

Here the final expressions of the above equations are rewritten by using (7.51a)–(7.51c) and the cyclotron frequency $\omega_c = eB_z/m^*$. When τ is independent of the electron energy, we have $\langle\tau^n\rangle = \tau^n$ and thus (7.20a) and (7.20b) agree with (7.56a) and (7.56b) in the limit of a weak magnetic field $\omega_c\tau \ll 1$. The Hall coefficient is also obtained by introducing the condition $J_y = 0$ as follows. Putting $J_y = 0$ in (7.55b), we calculate E_x and insert it into (7.55a) to obtain

$$E_y = \frac{\sigma_{xy}}{\sigma_{xx}^2 + \sigma_{xy}^2} J_x = \frac{\gamma_0}{(\sigma_0 + \beta_0 B_z^2)^2 + \gamma_0 B_z^2} J_x B_z. \tag{7.57}$$

Therefore, the Hall coefficient R_H is given by

$$R_H = \frac{\sigma_{xy}}{\sigma_{xx}^2 + \sigma_{xy}^2} J_x = \frac{\gamma_0}{(\sigma_0 + \beta_0 B_z^2)^2 + \gamma_0^2 B_z^2} \cong \frac{\gamma_0}{\sigma_0^2}, \tag{7.58}$$

where the final result is obtained by taking account of the fact that $\sigma_0 \gg \beta_0 B_z^2$ and $\sigma_0 \gg \gamma_0 B_z$ in the weak magnetic field limit. Substituting (7.51b) in (7.58), the Hall coefficient in a weak magnetic field is written as

$$R_H = -\frac{r_H}{ne}, \tag{7.59}$$

where r_H is defined by

$$r_H = \frac{\langle \tau^2 \rangle}{\langle \tau \rangle^2}. \tag{7.60}$$

According to the above results we find that the scattering factor of the Hall effect r_H may be calculated when the distribution function and energy dependence of the relaxation time τ are known. As an example, we assume a Maxwellian distribution function and $\tau = a\mathcal{E}^{-s}$, and then from (6.116) the following result is obtained:

$$r_H = \frac{\Gamma(5/2 - 2s)\Gamma(5/2)}{[\Gamma(5/2 - s)]^2}. \tag{7.61}$$

In the case of acoustic phonon scattering as discussed in Sect. 6.4 (see (6.341)), we have $s = 1/2$ and

$$r_H = \frac{3\pi}{8} = 1.18, \tag{7.62}$$

which is very close to 1. In the case of impurity scattering, however, we have $s = -3/2$, as derived in Sect. 6.4, and

$$r_H = \frac{315\pi}{512} = 1.93, \tag{7.63}$$

which means that the Hall mobility is almost twice the drift mobility.

7.2.4 Magnetoresistance

Here we will derive the Hall coefficient and magnetoresistance in a semiconductor with scalar mass in an arbitrary magnetic field. Let us assume the distribution function

$$f_1 = \boldsymbol{c} \cdot \boldsymbol{k}\,, \tag{7.64}$$

and determine the coefficient $\boldsymbol{c}$. Inserting this f_1 into (7.41) we obtain

$$\left[\frac{\hbar e}{m^*}\frac{\partial f_0}{\partial \mathcal{E}}\boldsymbol{E} + \frac{e}{m^*}\boldsymbol{B} \times \frac{\partial}{\partial \boldsymbol{k}}(\boldsymbol{c}\cdot\boldsymbol{k}) - \frac{\boldsymbol{c}}{\tau}\right]\cdot \boldsymbol{k} = 0\,. \tag{7.65}$$

The coefficient $\boldsymbol{c}$ is determined by dividing (7.65) into its components and solving the simultaneous equations of c_x, c_y and c_z. Inserting the components c_x, c_y and c_z into (7.64) and rewriting again in a vector form we obtain

$$\begin{aligned} f_1 &= \frac{\hbar e}{m^*}\frac{\tau}{1+(e\tau/m^*)^2B^2}\frac{\partial f_0}{\partial \mathcal{E}} \\ &\quad \times \left[\boldsymbol{E} + \frac{e\tau}{m^*}\boldsymbol{B}\times\boldsymbol{E} + \left(\frac{e\tau}{m^*}\right)^2(\boldsymbol{B}\cdot\boldsymbol{E})\boldsymbol{B}\right]\cdot\boldsymbol{k}\,. \end{aligned} \tag{7.66}$$

Here we show that the above result coincides with that obtained for the weak magnetic field case. Rewriting the coefficient of the above equation as

$$\frac{\tau}{1+\omega_c^2\tau^2} = \tau - \frac{\omega_c^2\tau^3}{1+\omega_c^2\tau^2}\,,$$

we easily find that it may be approximated $\tau - \omega_c^2\tau^3$ in the case of $\omega_c\tau \ll 1$, leading to conductivity tensor (7.53) for low magnetic field case. From such a comparison the conductivity tensor for an arbitrary magnetic field is obtained from (7.53) by replacing

$$\gamma_0 \to \gamma = -\frac{ne^3}{m^{*2}}\left\langle\frac{\tau^2}{1+\omega_c^2\tau^2}\right\rangle, \tag{7.67a}$$

$$\beta_0 \to \beta = -\frac{ne^4}{m^{*3}}\left\langle\frac{\tau^3}{1+\omega_c^2\tau^2}\right\rangle, \tag{7.67b}$$

where $\langle\cdots\rangle$ is the averaged value over the distribution function as given by (6.116).

In the case of $B_x = B_y = 0$, $B_z \neq 0$, and $E_z = 0$, we obtain

$$\begin{aligned} J_x &= (\sigma_0 + \beta B_z^2)E_x + \gamma B_z E_y \\ &= \frac{ne^2}{m^*}\left[\left\langle\frac{\tau}{1+\omega^2\tau^2}\right\rangle E_x - \left\langle\frac{\omega_c\tau^2}{1+\omega_c^2\tau^2}\right\rangle E_y\right], \end{aligned} \tag{7.68a}$$

$$\begin{aligned} J_y &= -\gamma B_z E_x + (\sigma_0 + \beta B_z^2)E_y \\ &= \frac{ne^2}{m^*}\left[\left\langle\frac{\omega_c\tau^2}{1+\omega_c^2\tau^2}\right\rangle E_x + \left\langle\frac{\tau}{1+\omega_c^2\tau^2}\right\rangle E_y\right]. \end{aligned} \tag{7.68b}$$

From a comparison of these relations with (7.17a) and (7.17b), we find that the difference is the averaging by the distribution function. Therefore, we obtain for the Hall coefficient

$$R_{\mathrm{H}} = -\frac{1}{ne} \frac{\left\langle \dfrac{\tau^2}{1+\omega_c^2\tau^2} \right\rangle}{\left\langle \dfrac{\tau}{1+\omega_c^2\tau^2} \right\rangle^2 + \omega_c^2 \left\langle \dfrac{\tau^2}{1+\omega_c^2\tau^2} \right\rangle^2} \tag{7.69}$$

and thus the scattering factor of the Hall coefficient r_{H} is given by

$$r_{\mathrm{H}} = \frac{\left\langle \dfrac{\tau^2}{1+\omega_c^2\tau^2} \right\rangle}{\left\langle \dfrac{\tau}{1+\omega_c^2\tau^2} \right\rangle^2 + \omega_c^2 \left\langle \dfrac{\tau^2}{1+\omega_c^2\tau^2} \right\rangle^2} . \tag{7.70}$$

Next, deriving E_y from the condition $J_y = 0$ and inserting it into (7.68a), we obtain

$$J_x = \frac{\sigma_{xx}^2 + \sigma_{xy}^2}{\sigma_{xx}} E_x \equiv \sigma(B) E_x , \tag{7.71a}$$

$$\sigma(B) = \frac{ne^2}{m^*} \frac{\left\langle \dfrac{\tau}{1+\omega_c^2\tau^2} \right\rangle^2 + \omega_c^2 \left\langle \dfrac{\tau^2}{1+\omega_c^2\tau^2} \right\rangle^2}{\left\langle \dfrac{\tau}{1+\omega_c^2\tau^2} \right\rangle} . \tag{7.71b}$$

Equation (7.69) leads to the Hall coefficient in two extreme cases. In a weak magnetic field such that $\omega_c\tau \ll 1$, (7.59) and (7.60) are deduced. On the other hand, in a strong magnetic field such that $\omega_c\tau \gg 1$, we obtain

$$R_H \cong -\frac{1}{ne} , \tag{7.72}$$

and the Hall coefficient is independent of the scattering mechanisms and exhibits saturation.

The magnetoresistance is calculated from (7.71b). Since the resistivity ρ is related to the conductivity σ by the relation $\rho = 1/\sigma$, expressing these by

$$\sigma = \sigma_0 \left(1 - \frac{\Delta\sigma}{\sigma_0} \right) , \tag{7.73a}$$

$$\rho = \rho_0 \left(1 + \frac{\Delta\rho}{\rho_0} \right) , \tag{7.73b}$$

the magnetoresistance and magnetoconductance in a weak magnetic field are obtained from (7.71b) and from the assumption of $\Delta\sigma/\sigma_0 \ll 1$ as

$$-\frac{\Delta\sigma}{\sigma_0} = \frac{\Delta\rho}{\rho_0} = \omega_c^2 \frac{\langle\tau^3\rangle\langle\tau\rangle - \langle\tau^2\rangle^2}{\langle\tau\rangle^2} . \tag{7.74}$$

This may be rewritten as the following by defining the Hall coefficient given by (7.59) at low magnetic field as $R_{\mathrm{H}}(0)$ and using (7.60):

$$\frac{\Delta\rho}{\rho_0} = -\frac{\Delta\sigma}{\sigma_0} = \xi R_{\mathrm{H}}(0)^2\sigma_0^2 B_z^2 = \xi(\mu_{\mathrm{H}} B_z)^2 , \tag{7.75a}$$

$$\xi = \frac{\langle\tau^3\rangle\langle\tau\rangle}{\langle\tau^2\rangle^2} - 1 . \tag{7.75b}$$

When we assume a relation $\tau = a\mathcal{E}^{-s}$, ξ is given by

$$\xi = \frac{\Gamma(5/2-3s)\Gamma(5/2-s)}{[\Gamma(5/2-2s)]} - 1 . \tag{7.76}$$

For acoustic phonon scattering, $s = 1/2$, the above expression reduces to

$$\xi = \frac{4}{\pi} - 1 = 0.273 , \tag{7.77}$$

and for ionized impurity scattering, $s = -3/2$, we obtain

$$\xi = \frac{32768}{6615\pi} - 1 = 0.577 . \tag{7.78}$$

In general, $\langle\tau^3\rangle\langle\tau\rangle$ is larger than $\langle\tau^2\rangle^2$, and the relation $\xi \geq 0$ holds. Therefore, in the presence of a magnetic field, the resistivity increases, showing positive magnetoresistance. It is also clear that no magnetoresistance appears in the case of constant τ (τ is independent of electron energy) because we have $\xi = 0$ for this case. Another important conclusion is drawn for a degenerate semiconductor. Since $\partial f_0/\partial\mathcal{E}$ is approximated by the δ-function in degenerate semiconductors, $\langle\tau^n\rangle \cong \tau_F^n$ and $\xi \cong 0$, indicating that the magnetoresistance is extremely small for a degenerate semiconductor with scalar effective mass.

The physical meaning of the magnetoresistance is as follows. In a transverse magnetic field an electron is exposed to a Lorentz force, moving in the direction perpendicular to the magnetic field and the current, and induces the Hall field. When the force due to the Hall field balances the Lorentz force, the electron will move straight through the sample. However, the electrons are distributed over the energy range determined by the distribution function and thus there exist electrons with lower and higher velocities compared to the average velocity which produces the Hall field. The Lorentz force of the electrons with lower and higher velocities will not balance the force due to the Hall field and and thus they spread out, resulting in increasing resistivity (magnetoresistance: $\Delta\rho/\rho_0 \geq 0$).

Finally, we derive the magnetoresitance in a high magnetic field. From (7.71b) the conductivity σ_∞ and resistivity ρ_∞ in the limit of $\omega_c\tau \gg 1$ are given by

$$\frac{\sigma_0}{\sigma_\infty} = \frac{\rho_\infty}{\rho_0} = \langle 1/\tau\rangle\langle\tau\rangle , \tag{7.79}$$

and again assuming the relation $\tau = a\mathcal{E}^{-s}$ we obtain

$$\frac{\rho_\infty}{\rho_0} = \frac{\Gamma(5/2+s)\cdot\Gamma(5/2-s)}{[\Gamma(5/2)]^2} = \begin{cases} \dfrac{32}{9\pi} = 1.13 \quad (s=1/2)\,, \\[2ex] \dfrac{32}{3\pi} = 3.39 \quad (s=-3/2)\,. \end{cases} \tag{7.80}$$

7.3 Shubnikov–de Haas Effect

7.3.1 Theory of Shubnikov–de Haas Effect

In Sect. 2.5 we have discussed electron motion in a magnetic field and explained that the electron cyclotron motion is quantized to form Landau levels. In addition the density of states for the quantized electrons is given by (2.132) and the features are shown in Fig. 2.16. Considering a degenerate semiconductor, and letting the electron density be n and the Fermi energy be $\mathcal{E}_F$, we have in the limit of temperature $T = 0$

$$\mathcal{E}_\mathrm{F} = \frac{\hbar^2}{2m^*}\left(3\pi^2 n\right)^{2/3}\,. \tag{7.81}$$

When a magnetic field applied to the semiconductor is changed, the Fermi energy passes through the bottoms of the Landau levels. In a magnetic field such that the Fermi energy is located at the bottom of one of the Landau levels, where the density of states is a maximum, the scattering rate for electrons is high because there are many final states for scattered electrons. If the Fermi energy is not changed by the presence of magnetic field, the above condition is written as

$$\mathcal{E}_\mathrm{F} = \hbar\omega_\mathrm{c}\left(N+\frac{1}{2}\right) = \frac{eB}{m^*}\left(N+\frac{1}{2}\right)\,, \quad N = 0, 1, 2, \ldots \tag{7.82}$$

and therefore the magnetoresistance oscillates periodically with the inverse magnetic field $1/B$. The period is then given by

$$\Delta\left(\frac{1}{B}\right) = \frac{e\hbar}{m^*\mathcal{E}_\mathrm{F}} = \frac{2e}{\hbar}\left(3\pi^2 n\right)^{-2/3}\,. \tag{7.83}$$

This phenomenon is called the **Shubnikov–de Haas effect**. It is not possible to explain the Shubnikov–de Haas effect by the classical theory of the Boltzmann transport equation. One of the most well-known and understandable methods for this is the density matrix method. The Kubo formula, a generalized linear theory, has been widely used to understand quantum transport including the Shubnikov–de

Haas effect and the detail is found in references [1, 2]. Here we will try to explain it in a simplified manner.

Using the result of the previous section, in a magnetic field applied in the z direction, the transverse magnetoresistance $\rho_\perp$, longitudinal magnetoresistance $\rho_\parallel$ and Hall coefficient R_H are, respectively, given by

$$\rho_\perp = \frac{\sigma_{xx}}{\sigma_{xx}^2 + \sigma_{xy}^2}, \tag{7.84a}$$

$$\rho_\parallel = \frac{1}{\sigma_{zz}}, \tag{7.84b}$$

$$R_H = -\frac{1}{B_z}\frac{\sigma_{yx}}{\sigma_{xx}^2 + \sigma_{yx}^2}. \tag{7.84c}$$

As described later, not only the Shubnikov–de Haas effect but also other quantum effects, namely the **quantum galvanomagnetic effect**, are observed under the condition $\omega_c\tau = \mu B_z \gg 1$. Under this condition $\sigma_{xy} \gg \sigma_{xx}$ is satisfied and thus we obtain the relations

$$\rho_\perp \cong \frac{\sigma_{xx}}{\sigma_{xy}^2}, \tag{7.85a}$$

$$R_H \cong -\frac{1}{B_z\sigma_{yx}}. \tag{7.85b}$$

In the a case where $\omega_c\tau \gg 1$ is satisfied, the magnetoresistance effect cannot be treated by classical theory. This is understood from the discussion stated in Sect. 2.5. When $\omega_c\tau \gg 1$ is satisfied, the electron motion of the Landau orbit results in the formation of Landau levels. When an electric field E_x is applied in the x direction, perpendicular to the magnetic field applied in the z direction, the electron will move with a velocity E_x/B_z in the y direction as shown on the left of Fig. 2.15. In other words, as shown by (2.127), the cyclotron center coordinates (X, Y) are given by

$$\dot{Y} = \frac{E_x}{B_z}, \quad \dot{X} = 0, \tag{7.86}$$

and the electron moves with a constant velocity in the y direction. This gives us the current density $J_y = neE_x/B_y \equiv \sigma_{yx}E_x$, and therefore we obtain the conductivity σ_{yx}

$$\sigma_{yx} = \frac{ne}{B_z}. \tag{7.87}$$

Inserting this conductivity tensor into (7.85b), we find that the result agrees with the Hall coefficient (7.72) for high magnetic field obtained in Sect. 7.2:

$$R_H \cong -\frac{1}{ne}. \tag{7.88}$$

Since no current flows in the x direction, we have $\sigma_{xx} = 0$. In experiments we observe current in the x direction, which may be interpreted as follows. As shown on the right of Fig. 2.15, scattering of an electron results in a change of its cyclotron center, and its component along the x direction gives rise to a current component J_x and thus to $\sigma_{xx} \neq 0$. It is very interesting to point out that the current due to the drift motion is disturbed by scattering according to the classical transport theory, whereas in the presence of a high magnetic field a current component along the electric field direction is induced by electron scattering.

From the example stated above, we find that quantum magnetotransport phenomena are completely different from classical transport phenomena. Also, we have to note again that the electron motion is quantized due to cyclotron motion in the plane perpendicular to the magnetic field, giving rise to one-dimensional motion allowed only along the z direction. This quantization leads to the density of states shown in Fig. 2.16. The one-dimensional density of states results in a divergence of the electron scattering rate at the bottom of the Landau level. To calculate the current density by solving a transport equation, the current density is evaluated with the electron density of states by averaging over the electron energy. Therefore, the density of states appears twice in the calculations, resulting in a logarithmic divergence of the current. This means that the current and thus the conductivity plotted as a function of inverse magnetic field oscillate periodically. Although magnetotransport phenomena are often analyzed by the density matrix method or Kubo formula [2] as stated earlier, here we describe the outline of the method adopted by Roth and Argyres [1].

We define the one-electron Hamiltonian in a magnetic field by H_0, which is written as

$$H_0 = \frac{1}{2m^*}(\boldsymbol{p} + e\boldsymbol{A})^2 + g\mu_B \boldsymbol{s} \cdot \boldsymbol{B}\,, \tag{7.89}$$

where the electron spin is considered, g is the effective g-factor, and $\mu_B = e\hbar/2m$. The eigenvalues for the Hamiltonian (7.89) may be given by the following equation from the analogy of the treatment stated in Sect. 2.5:

$$\mathcal{E}_{nk\pm} = \left(N + \frac{1}{2} \pm \frac{\nu}{2}\right)\hbar\omega_c + \mathcal{E}_z\,; \quad \mathcal{E}_z = \frac{\hbar^2 k_z^2}{2m^*}\,, \tag{7.90}$$

where $\nu = m^* g/2m$. Also, it is evident that the density of states is given by

$$\begin{aligned} g(\mathcal{E}, B_z) &= \frac{1}{V}\sum_{Nk\pm} \delta[\mathcal{E}_{Nk\pm} - \mathcal{E}] \\ &= \left(\frac{2m^*}{\hbar^2}\right)^{1/2} \frac{1}{(2\pi l)^2} \sum_{N,\pm} \left[\mathcal{E} - \left(N + \frac{1}{2} \pm \frac{\nu}{2}\right)\hbar\omega_c\right]^{-1/2}. \end{aligned} \tag{7.91}$$

Here, $V = L^3$ is the crystal volume and $l = (\hbar/eB)^{1/2} = (\hbar/m^*\omega_c)^{1/2}$ is defined by (2.108), which corresponds to the classical cyclotron radius. In general we may express the density of states $g^{(d)}$ for a d-dimensional band as

$$g^{(d)} = \frac{2}{V}\sum_{\boldsymbol{k}} \delta[\mathcal{E}_{\boldsymbol{k}} - \mathcal{E}] = \frac{2}{(2\pi)^d}\int \mathrm{d}^d\boldsymbol{k} \tag{7.92}$$

and for a three-dimensional band we have

$$g^{(3)}(\mathcal{E}) = \frac{2}{(2\pi)^3}\int \mathrm{d}^3\boldsymbol{k} = \frac{2}{(2\pi)^3}\int 4\pi k^2 \mathrm{d}k = \frac{1}{2\pi^2}\left(\frac{2m^*}{\hbar^2}\right)^{3/2}\mathcal{E}^{1/2}. \tag{7.93}$$

The one-dimensional density of states in a magnetic field is evaluated by using the Poisson summation formula:

$$\sum_{m=0}^{\infty}\Phi\left(m+\frac{1}{2}\right) = \sum_{-\infty}^{+\infty}(-1)^r\int_0^{\infty}\Phi(x)\mathrm{e}^{2\pi i r x}\mathrm{d}x \tag{7.94}$$

and the following relation is obtained in the limit of $\hbar\omega_c \gg 1$ using the three-dimensional density of states, $g^{(3)}$:

$$g(\mathcal{E}, B_z) = g^{(3)}\left[1 + \left(\frac{\hbar\omega_c}{2\mathcal{E}}\right)^{1/2}\sum_{r=1}^{\infty}\frac{(-1)^r}{r^{1/2}}\cos\left(\frac{2\pi\mathcal{E}}{\hbar\omega_c}r - \frac{\pi}{4}\right)\cos(\pi\nu r)\right]. \tag{7.95}$$

From these results we find that the density of states oscillates periodically with inverse magnetic field $1/B_z$. This oscillatory feature is reflected in the magnetoresistance oscillations.

The one-electron density matrix, $\rho_{\mathrm{T}}(t)$, satisfies (see Appendix F for the density matrix)

$$i\hbar\frac{\mathrm{d}\rho_{\mathrm{T}}}{\mathrm{d}t} = [H_0 + H' + F, \rho_{\mathrm{T}}], \tag{7.96}$$

where H' is the scattering potential, and F represents the interaction with an electric filed $\boldsymbol{E}$ given by

$$F = e\boldsymbol{E}\cdot\boldsymbol{r}. \tag{7.97}$$

The current density $\boldsymbol{J}$ is evaluated from

$$\boldsymbol{J} = -\frac{e}{V}\mathrm{Tr}(\rho_{\mathrm{T}}\boldsymbol{v}), \tag{7.98}$$

where V is the crystal volume, and $\boldsymbol{v} = (\boldsymbol{p} + e\boldsymbol{A})/m^*$ is the velocity operator stated in Sect. 2.5 and given by

$$\boldsymbol{v} = \frac{1}{m^*}(p_x, \hbar k_y + eBx, \hbar k_z). \tag{7.99}$$

Since the total number of electrons is conserved, we have for the electron density n

$$\frac{1}{V}\mathrm{Tr}(\rho_{\mathrm{T}}) = n\,. \tag{7.100}$$

Rewriting the density matrix as

$$\rho_{\mathrm{T}} = \rho_0 + \rho(t)\,, \tag{7.101}$$

ρ_0 is written using the Fermi–Dirac distribution function as

$$\rho_0 = \frac{1}{\mathrm{e}^{(\mathcal{E}-\mathcal{E}_{\mathrm{F}})/k_{\mathrm{B}}T} + 1}\,. \tag{7.102}$$

Using these results (7.96) is rewritten as

$$\mathrm{i}\hbar\frac{\mathrm{d}\rho(t)}{\mathrm{d}t} = [H_0 + H', \rho(t)] + [F, f(H_0 + H')]\,; \quad \rho(0) = 0\,, \tag{7.103}$$

where f is the Fermi–Dirac distribution function. Solving the above equation for the one-electron density matrix for the electron-phonon interaction, the current density is evaluated from (7.98) and so we derive the magnetoconductivity.

7.3.2 Longitudinal Magnetoresistance Configuration

The case where the magnetic and electric fields are applied in the same direction, the z direction, is called the longitudinal configuration and the magnetoresistance is called the **longitudinal magnetoresistance**. For simplicity we define a new subscript μ instead of the subscripts $N\boldsymbol{k}\pm$. Then the equation of motion for the diagonal element of the density matrix ρ_μ is written as the following [6]:

$$\sum_\nu \left\{w_{\mu\nu}\rho_\mu(1-\rho_\nu) - w_{\nu\mu}\rho_\nu(1-\rho_\mu)\right\} = eE_z v_\mu^z \frac{\mathrm{d}f(\mathcal{E}_\mu)}{\mathrm{d}\mathcal{E}_\mu}\,, \tag{7.104}$$

where v_μ^z is the z component of the velocity operator $\boldsymbol{v}_\mu$ for the Landau state $|\mu\rangle = |N\boldsymbol{k}\pm\rangle$ and $w_{\mu\nu}$ is the scattering rate of an electron between the Landau states $|\mu\rangle$ and $|\nu\rangle$ which is given by

$$w_{\mu\nu} = w_{\nu\mu} = \frac{2\pi}{\hbar}|H'_{\mu\nu}|^2\delta[\mathcal{E}_\mu - \mathcal{E}_\nu]\,. \tag{7.105}$$

Measurements of the Shubnikov–de Haas effect in semiconductors have been carried out at low temperatures to achieve the condition $\omega_c\tau \gg 1$, where impurity scattering and acoustic phonon scattering are dominant. The scattering probabilities are obtained from the method described in Sect. 6.2. It should be noted that electron spin is conserved in the scattering. For impurity scattering we have

$$w_{\mu\nu} = \frac{2\pi}{\hbar}\frac{N_I}{V}\sum_{\boldsymbol{q}} |H'(\boldsymbol{q})|^2 \left|\langle\mu\left|e^{i\boldsymbol{q}\cdot\boldsymbol{r}}\right|\nu\rangle\right|^2 \delta[\mathcal{E}_\mu - \mathcal{E}_\nu]\,, \tag{7.106}$$

and for acoustic phonon scattering we obtain

$$w_{\mu\nu} = \frac{2\pi}{\hbar}\sum_{q} |C(\boldsymbol{q})|^2 \left|\langle\mu\left|e^{i\boldsymbol{q}\cdot\boldsymbol{r}}\right|\nu\rangle\right|^2 \times \left\{(n_q+1)\delta[\mathcal{E}_\mu - \mathcal{E}_\nu - \hbar\omega_q] + n_q\delta[\mathcal{E}_\mu - \mathcal{E}_\nu + \hbar\omega_q]\right\}\,, \tag{7.107}$$

which are shown in Chap. 6. Here, $q = (j, \boldsymbol{q})$ represents mode j and wave vector $\boldsymbol{q}$ for an acoustic phonon with energy $\hbar\omega_q$, and n_q is the Bose–Einstein distribution function given by $n_q = 1/[\exp(\hbar\omega_q/k_BT) - 1]$. The matrix element $\langle\mu|\exp(i\boldsymbol{q}\cdot\boldsymbol{r})|\nu\rangle$ appearing in both impurity and acoustic phonon scattering is evaluated from the eigenfunctions of (2.120). The detail of the evaluation is given by Kubo et al. [2].

We expand the diagonal element of the density matrix as

$$\rho_\mu = f(\mathcal{E}_\mu) + \varphi_\mu \frac{df(\mathcal{E}_\mu)}{d\mathcal{E}_\mu}\,, \tag{7.108}$$

where $f(\mathcal{E}) = 1/[\exp\{(\mathcal{E} - \mathcal{E}_F)/k_BT\} - 1]$ is the Fermi–Dirac distribution function and $df/d\mathcal{E}$ is given by (6.103). Impurity scattering is known to be elastic scattering. Assuming that the acoustic phonon scattering is elastic, we find that $f(\mathcal{E}_\mu)$ before scattering is equal to $f(\mathcal{E}_\nu)$ after scattering. Therefore, the left-hand side of (7.104) is rewritten as $\sum_\nu w_{\mu\nu}(\rho_\mu - \rho_\nu)$, and (7.104) reduces to

$$\sum_\nu w_{\mu\nu}(\varphi_\mu - \varphi_\nu) = eE_z v_\mu^z\,. \tag{7.109}$$

When we adopt the relaxation time approximation as in Sect. 6.2, which is used for solving the classical Boltzmann transport equation, it is possible to express the above relation as

$$\sum_\nu w_{\mu\nu}(\varphi_\mu - \varphi_\nu) = \frac{\varphi_\mu}{\tau_\mu}\,, \tag{7.110}$$

where the following relation holds between the relaxation time τ_μ and scattering probability $w_{\mu\nu}$:

$$\frac{1}{\tau_{N\boldsymbol{k}\pm}} = \sum_{N'\boldsymbol{k}'} w_{N\boldsymbol{k}\pm,N'\boldsymbol{k}'\pm}\left(1 - \frac{k_z'}{k_z}\right)\,. \tag{7.111}$$

As described in Sect. 6.4, the inverse of the relaxation time is equivalent to the scattering rate in the case of isotropic and elastic scattering and therefore we have

$$\varphi_\mu = eE_z\tau_\mu v_\mu^z \,. \tag{7.112}$$

Inserting this into (7.108) and using the result to evaluate (7.98), we obtain the longitudinal magnetoresistance given by

$$\rho_\parallel^{-1} = \sigma_{zz} = -\frac{e^2}{V}\sum_{Nk\pm}\frac{\mathrm{d}f}{\mathrm{d}\mathcal{E}_{Nk\pm}}\tau^\pm(\mathcal{E}_{Nk})\left(v_{Nk}^z\right)^2 \,. \tag{7.113}$$

The Fermi energy of electron in the presence of a magnetic field is determined from (7.91) as

$$n = \int_0^\infty f(\mathcal{E})g(\mathcal{E}, B_z)\mathrm{d}\mathcal{E} \,. \tag{7.114}$$

Assuming that the electron density is conserved, the Fermi energy in a magnetic field $\mathcal{E}_\mathrm{F}$ differs from the Fermi energy for zero magnetic field $\mathcal{E}_\mathrm{F}^0$ and depends on the magnetic field. In general, we may expand the Fermi energy in a magnetic field as

$$\mathcal{E}_\mathrm{F} = \mathcal{E}_\mathrm{F}^{(0)} + \mathcal{E}_\mathrm{F}^{(1)}B_z^2 + \mathcal{E}_\mathrm{F}^{\mathrm{osc}} \,, \tag{7.115}$$

where $\mathcal{E}_\mathrm{F}^{\mathrm{osc}}$ is is a sum of quasiperiodic functions in $1/B_z$, whose amplitude is proportional to $(\hbar\omega_\mathrm{c}/\mathcal{E}_\mathrm{F}^{(0)})^{3/2}\mathcal{E}_\mathrm{F}^{(0)}$ and is small compared to $\mathcal{E}_\mathrm{F}^{(0)}$. In general, neglecting the terms beyond the first term on the right-hand side of (7.115), the Fermi energy is often approximated as $\mathcal{E}_\mathrm{F} = \mathcal{E}_\mathrm{F}^{(0)}$. As discussed in Sect. 6.2.2, $\mathrm{d}f/\mathrm{d}\mathcal{E}$ in (7.113) is given by $-(1/k_\mathrm{B}T)f(1-f)$, which behaves like a δ-function, and thus in the limit of $k_\mathrm{B}T \ll \hbar\omega_\mathrm{c}$ we obtain

$$\rho_\parallel^{-1} = \rho_0^{-1}\frac{4\sqrt{\mathcal{E}_\mathrm{F}^{(0)}}}{n\hbar\omega_\mathrm{c}}\sum_\pm\frac{n_\pm}{\sum_N\left[\mathcal{E}_\mathrm{F} - (N+\frac{1}{2}\pm\frac{1}{2}\nu)\hbar\omega_\mathrm{c}\right]^{-1/2}}, \tag{7.116}$$

where $\rho_0 = m^*/ne^2\tau_0(\mathcal{E}_\mathrm{F}^{(0)})$ is the resistivity for $B_z = 0$, $\tau_0(\mathcal{E}_\mathrm{F}^{(0)})$ is the relaxation time for electrons with Fermi energy at $B_z = 0$, $n_\pm$ are the electron densities for electron spins $\pm$, and $n = n_+ + n_-$. In the case of $\mathcal{E}_\mathrm{F} \gg \hbar\omega_\mathrm{c}$, the longitudinal magnetoresistance for an arbitrary value of $k_\mathrm{B}T/\hbar\omega_\mathrm{c}$ is written as

$$\rho_\parallel \cong \rho_0\left[1 + \sum_{r=1}^\infty b_r\cos\left(\frac{2\pi\mathcal{E}_\mathrm{F}}{\hbar\omega_\mathrm{c}}r - \frac{\pi}{4}\right)\right], \tag{7.117}$$

where

$$b_r = \frac{(-1)^r}{r^{1/2}} \left(\frac{\hbar\omega_c}{2\mathcal{E}_F} \right)^{1/2} \frac{2\pi^2 r k_B T/\hbar\omega_c}{\sinh(2\pi^2 r k_B T/\hbar\omega_c)} \cos(\pi \nu r) e^{-2\pi r \Gamma/\hbar\omega_c}, \quad (7.118)$$

and the broadening of the density of states for electrons (broadening energy Γ) is taken into account. We find from this equation that the longitudinal magnetoresistance oscillates periodically when plotted versus $1/B_z$, as stated earlier.

7.3.3 Transverse Magnetoresistance Configuration

In the transverse magnetoresistance configuration, the electric field is applied perpendicular to the magnetic field and the observed magnetoresistance is called the **transverse magnetoresistance**. Calculations of the transverse magnetoresistance is complicated compared to the longitudinal magnetoresistance. This is understood from the following. In a magnetic field in the z direction, the application of an electric field in the x direction, for example, causes an electron quantized in the x, y plane to move in the y direction with a constant velocity E_x/B_z in the absence of electron scattering and thus to no current flow in the x direction. The electron current is induced by scattering, which gives rise to a change of the cyclotron center in the x direction, and results in the current in the x direction and in $\sigma_{xx} \neq 0$.

In the transverse magnetoresistance configuration, Kubo et al. derived the solution of (7.96) for the elastic scattering case [2, 7]:

$$\sigma_{xx} = \frac{\pi e^2}{\hbar V} \int_{-\infty}^{\infty} d\mathcal{E} \frac{df}{d\mathcal{E}} \mathrm{Tr} \left\{ \delta(\mathcal{E} - H)[H', X] \delta(\mathcal{E} - H)[H', X] \right\}, \quad (7.119)$$

where $H = H_0 + H'$ and X is the operator for the x component of the cyclotron center coordinate. From the above equation the following relation is obtained by taking in account the broadening effect due to collisions [1]:

$$\rho_\perp = \rho_0 \left[1 + \frac{5}{2} \sum_{r=1}^{\infty} b_r \cos\left(\frac{2\pi \mathcal{E}_F}{\hbar\omega_c} r - \frac{\pi}{4} \right) + R \right], \quad (7.120a)$$

$$R = \frac{3}{4} \frac{\hbar\omega_c}{2\mathcal{E}_F} \left\{ \sum_{r=1}^{\infty} b_r \left[\alpha_r \cos\left(\frac{2\pi \mathcal{E}_F}{\hbar\omega_c} r \right) + \beta_r \sin\left(\frac{2\pi \mathcal{E}_F}{\hbar\omega_c} r \right) \right] \right.$$
$$\left. - \log\left(1 - e^{4\pi \Gamma/\hbar\omega_c} \right) \right\}. \quad (7.120b)$$

Here, b_r is given by (7.118), and coefficients α_r and β_r are given by

$$\alpha_r = 2r^{1/2}\sum_{s=1}^{\infty}\frac{1}{[s(r+s)]^{1/2}}\mathrm{e}^{-4\pi s\Gamma/\hbar\omega_c}\,, \tag{7.121a}$$

$$\beta_r = r^{1/2}\sum_{s=1}^{r-1}\frac{1}{[s(r-s)]^{1/2}}\,. \tag{7.121b}$$

These results mean that the transverse magnetoresistance oscillates periodically in the plot versus $1/B_z$ as in the case of longitudinal magnetoresistance. The second term on the right-hand side in (7.120a) is the oscillatory component due to the transition induced by a change in the quantum number N, whereas R is due to the transition without changing the quantum number N and diverges when Γ is zero. This divergence arises from the fact that the integral with respect to the one-dimensional density of states appears twice in the integral containing the scattering rate and distribution function, resulting in a logarithmic divergence for $N = N'$. The contribution from each component depends on the magnitude of $\Gamma/\mathcal{E}_F$ and the contribution from R becomes smaller for $r > 1$ (higher harmonic oscillatory component). Usually, therefore, the analysis has been carried out by neglecting the contribution from R.

The amplitude of the oscillatory component of both the longitudinal and transverse magnetoresistance depends on the magnetic field and the temperature through the term

$$\frac{T}{B_z^{1/2}}\exp\{-2\pi^2 k_B(T+T_D)/\hbar\omega_c\}\,,$$

where

$$T_D = \frac{\Gamma}{\pi k_B} \tag{7.122}$$

is called the **Dingle temperature**. On the other hand, the phase of the oscillatory component is $-\pi/4$ for both longitudinal and transverse configurations, but we have to note that the spin g factor is not included in the above calculations and that a comparison with experiment may be difficult.

Next, a comparison of the theory with experiment is described. In experiments we have to apply electrodes to measure the resistivity in addition to the electrodes for current supply. Measurements of the Shubnikov–de Haas effect are strongly affected by the sample uniformity and thus require many electrode contacts for the resistivity measurement. Applying a constant current, the voltage drops between the pairs of contacts are measured and we choose the pair which shows the largest oscillation. The Shubnikov–de Haas effect is normally observed at low temperatures and measurements of the temperature dependence of the oscillations are required to estimate the effective mass. It is expected from the theory given above that the oscillation amplitude decreases exponentially with temperature, and thus the magnetic field modulation technique is often adopted in order to observe such a weak oscillatory behavior, where the second derivative magnetoresitance is detected. Recently it has been shown that the second-derivative signals directly obtained by computer analysis provides a good signal-to-noise ratio and a good quality of oscillatory structure. As an example, we show in Fig. 7.2 the experimental results on the Shubnikov–de Haas effect in n-GaSb [8], where the longitudinal and transverse magnetoresistance is plotted as a function of the inverse magnetic field and periodic oscillations in the inverse magnetic field $1/B$ are clearly observed, showing good agreement with the theoretical prediction stated above. From the period $2.75\,\mathrm{T}^{-1}$ and with the help of (7.83), the electron density $n = 1.3 \times 10^{18}\,\mathrm{cm}^{-3}$ is obtained.

Here we will discuss how to deduce the effective mass from the analysis of the Shubnikov–de Haas effect. When we put

$$\chi = \frac{2\pi^2 k_\mathrm{B} T}{\hbar\omega_\mathrm{c}},$$

the temperature dependence of the oscillation amplitude is written as

$$\chi \sinh\chi = \frac{2\pi^2 k_\mathrm{B} T/\hbar\omega_\mathrm{c}}{\sinh(2\pi^2 k_\mathrm{B} T/\hbar\omega_\mathrm{c})}. \tag{7.123}$$

Inserting the relation $\omega_\mathrm{c} = eB/m^*$ into the above equation, it is easily seen that the effective mass m^* is estimated from the temperature dependence of the amplitude. The electron effective obtained by this method in n-GaSb is $m^* = 0.05\,\mathrm{m}$ and the effective mass is found to depend on the electron density, probably due to the non-parabolicity of the conduction band. In general, the amplitude of the oscillatory magnetoresistance decays exponentially as the inverse magnetic field $1/B$. This decay is due to the collision broadening of the Landau levels. The amplitude of the oscillatory components of the magnetoresistance due to the Shubnikov–de Haas effect depends on the magnetic field as

$$\left(\frac{\hbar\omega_\mathrm{c}}{\mathcal{E}_\mathrm{F}}\right)^{1/2} \frac{\chi}{\sinh\chi} \mathrm{e}^{-2\pi\Gamma/\hbar\omega_\mathrm{c}} = \left(\frac{\hbar\omega_\mathrm{c}}{\mathcal{E}_\mathrm{F}}\right)^{1/2} \frac{2\pi^2 k_\mathrm{B} T/\hbar\omega_\mathrm{c}}{\sinh(2\pi^2 k_\mathrm{B} T/\hbar\omega_\mathrm{c})} \times \mathrm{e}^{-2\pi^2 k_\mathrm{B} T_\mathrm{D}/\hbar\omega_\mathrm{c}}, \tag{7.124}$$

where $T_\mathrm{D} = \Gamma/\pi k_\mathrm{B}$ is the Dingle temperature defined before. Therefore, when the amplitude of the oscillatory components divided by $[(\hbar\omega_\mathrm{c}/\mathcal{E}_\mathrm{F})^{1/2}\,(\chi/\sinh\chi)]$ is plotted against $1/B$, it will result in a straight line. From the slope of the plot we may estimate the Dingle temperature and thus the strength of the broadening. In Fig. 7.3 such a plot is shown for different electron densities in n-GaSb estimated from the Hall coefficient R_H for n-GaSb at $T = 4.2\,\mathrm{K}$ [8], where we find good agreement with the prediction of the theory. From the analysis the Dingle temperature $T_\mathrm{D} = 6.6\,\mathrm{K}$ is obtained for an electron density $n = 1.3 \times 10^{18}\,\mathrm{cm}^{-3}$. We have define the Dingle temperature to express the Landau broadening and shown how to estimate it from experimental data. Another parameter to give the collision broadening may

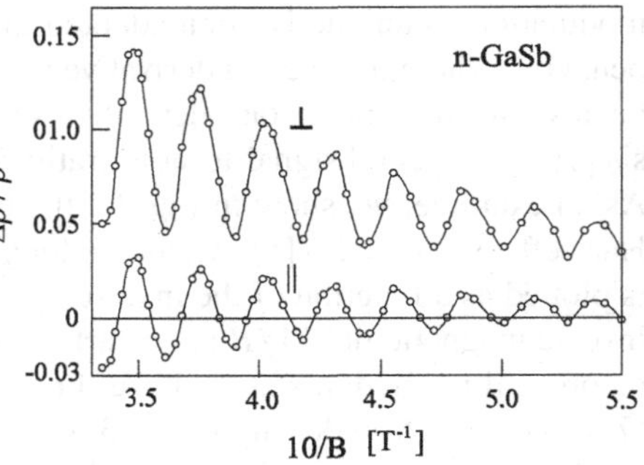

Fig. 7.2 Shubnikov–de Haas effect in n-GaSb ($n = 1.3 \times 10^{18}\,\mathrm{cm}^{-3}$), where the longitudinal and transverse magnetoresistance is plotted as a function of inverse magnetic field $1/B$ (After Becker and Fan [8])

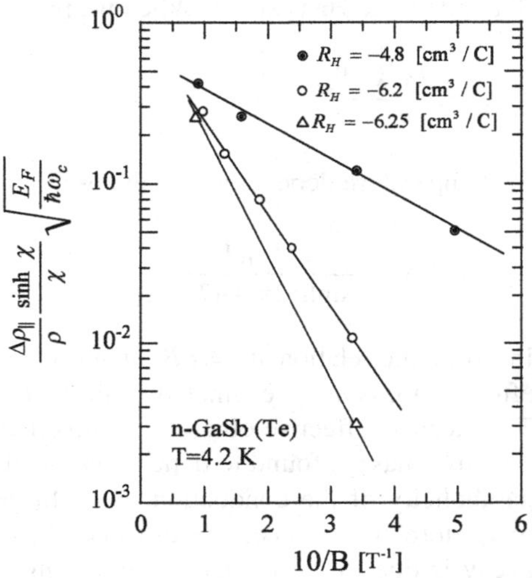

Fig. 7.3 The oscillation amplitude of the longitudinal magnetoresistance in n-GaSb plotted as a function of the inverse magnetic field $1/B$, where we find the exponential decay (Dingle temperature) depends on the electron density estimated from the Hall coefficient R_{H} (After Becker and Fan [8])

be estimated from the electron mobility; it is called the mobility temperature. The mobility temperature is defined by the relation $T_{\mathrm{m}} = \hbar/2\pi k_{\mathrm{B}}\tau_{\mathrm{m}}$, where the relaxation time τ_{m} is estimated from the electron mobility. Becker and Fan have deduced the mobility temperature $T_{\mathrm{m}} = 4.9\,\mathrm{K}$ for n-GaSb with $n = 1.3 \times 10^{18}\,\mathrm{cm}^{-3}$.

7.4 Magnetophonon Resonance

7.4.1 Experiments and Theory of Magnetophonon Resonance

In the previous section we discussed the Shubnikov–de Haas effect, where electrons quantized into Landau levels in the presence of a magnetic field are scattered elastically in the Landau levels and result in a periodic oscillation of the magnetoresistance with the inverse magnetic field $1/B$. On the other hand, we will now consider the oscillatory magnetoresistance induced by inelastic electron scattering, which is known as **magnetophonon resonance**. The magnetophonon resonance is caused by the scattering of electrons between the Landau levels induced by optical phonon absorption and emission, and therefore magnetoresistance maxima appear when the multiple of the Landau level spacing is equal to the optical phonon energy. Strong resonance has been observed in III–V compound semiconductors, where the electron-LO phonon interaction is dominant. The resonance condition is then given by the following relation:

$$N\hbar\omega_c = \hbar\omega_{LO}\,, \quad N = 1, 2, 3, \ldots \tag{7.125}$$

where $\hbar_{LO}$ is the LO phonon energy and $\omega_c = eB/m^*$ is the cyclotron frequency of an electron with effective mass m^*. Observation of the magnetophonon resonance requires excitation of a sufficient number of LO phonons involved with the scattering. Since the excitation number of phonons is given by the Bose–Einstein distribution function

$$n_{LO} = \frac{1}{e^{\hbar\omega_{LO}/k_B T} - 1}\,, \tag{7.126}$$

the number of phonons is very small at such low temperatures that $\hbar\omega_{LO} > k_B T$. At such low temperatures the broadening of the Landau levels is quite small and thus the quantum effect is clearly observed. But the number of excited phonons is limited, resulting in difficulty in observing the magnetophonon resonance at such low temperatures. On the other hand, at higher temperatures, there exist enough phonons to interact with electrons, but the Landau level broadening is quite large, and thus the quantization effect (magnetophonon resonance) is smeared out. Up to now magnetophonon resonance has been observed in most III–V semiconductors in which the electron-LO phonon interaction is strong, and the amplitude of the magnetophonon oscillations exhibits a maximum at the temperature $T \cong 150\,K$ for the reason stated above. It is well known that the observation of magnetophonon resonance is very difficult at temperatures below 40 K and above 350 K. However, at temperatures $T \leq 40\,K$, hot electrons produced by a high applied electric field application make transitions between the Landau levels by emitting LO phonons and exhibit magnetophonon resonance, which is called hot electron magnetophonon resonance.

Magnetophonon resonance was first predicted theoretically in 1961 by Gurevich and Firsov [9], and then in 1963 Puri and Geballe [10] succeeded in observing the oscillatory behavior of the thermoelectric power in InSb in a magnetic field, which is caused by the magnetophonon effect. Magnetophonon resonance in the magnetoresistance in n-InSb was subsequently observed in 1964 by Firsov et al. [11]. When the LO phonon energy involved with the resonance is known, the effective mass of the electron is determined from (7.125) by analyzing the period of oscillation in the plot versus the inverse magnetic field. In addition magnetophonon resonance is observed at temperatures around $T = 150\,\text{K}$; it provides information about the temperature dependence of the effective mass and about various electron-phonon interactions, as will be mentioned later. For this reason the study of magnetophonon resonance provided very important information in the period 1960–1970 and established its position for semiconductor research, especially in the field of quantum transport. The most important contribution to the study of magnetophonon resonance is the discovery of the empirical formula by Stradling and Wood [3], who obtained the following relation to express the oscillatory component from a series of experiments in III–V compound semiconductors:

$$\rho_{xx}^{\text{osc}} \propto \exp\left(-\bar{\gamma}\frac{\omega_{\text{LO}}}{\omega_{\text{c}}}\right)\cos\left(2\pi\frac{\omega_{\text{LO}}}{\omega_{\text{c}}}\right). \tag{7.127}$$

This relation is called the empirical formula of Stradling and Wood. In (7.127) $\bar{\gamma}$ represents the exponential decay of the oscillatory component in the plot versus the inverse magnetic field which arises from the Landau level broadening and corresponds to the Dingle temperature for the Shubnikov–de Haas effect.

Magnetoresistance in the presence of magnetophonon resonance consists of monotonic increase with the magnetic field and an oscillatory component which is periodic with the inverse magnetic field. Clear signals are obtained by removing the monotonic component, and several methods have been proposed to observe the oscillatory components accurately. Stradling and Wood adopted the following method to detect the oscillatory component of the magnetoresistance by removing the monotonic component. The Hall voltage, proportional to σ_{xy}^{-1}, was subtracted from the voltage drop between the sample electrodes. Here the voltage drop is proportional to the magnetoresistance ρ_{xx} and the Hall voltage is proportional to the magnetic field which contains no oscillatory component in it, Later, magnetic field modulation and second-harmonic detection by a lock-in amplifier were successfully used for magnetophonon resonance measurements, where the second-harmonic component is proportional to the second derivative of the magnetoresistance with respect to the magnetic field. Recently, however, the numerical derivative of the magnetoresistance calculated by using a personal computer has provided an excellent method for such purposes.

The magnetic field modulation technique is based on a superposition of a weak ac magnetic field $B_1\cos(\omega t)$ with modulation frequency ω upon a dc magnetic field B and on the detection of the second-harmonic signals (2ω). In general, the magnetoresistance under the application of such a magnetic field is written as

$$\begin{aligned}&\rho[B + B_1\cos(\omega t)]\\ &= \rho(B) + \frac{\mathrm{d}\rho}{\mathrm{d}B}B_1\cos(\omega t) + \frac{1}{2}\frac{\mathrm{d}^2\rho}{\mathrm{d}B^2}[B_1\cos(\omega t)]^2 + \cdots,\end{aligned} \tag{7.128}$$

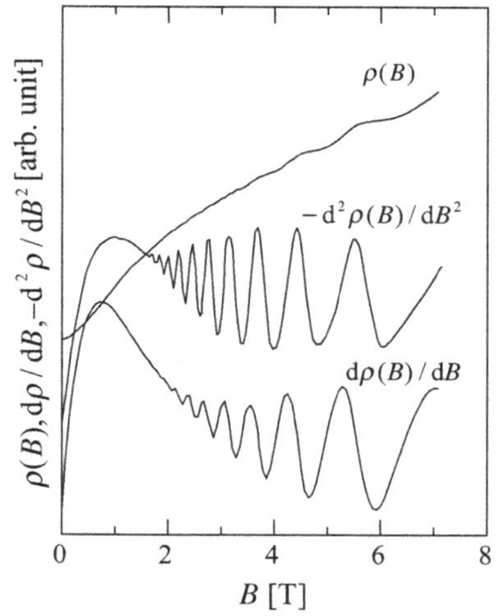

Fig. 7.4 Magnetophonon resonance observed in n-GaAs at $T = 77\,\text{K}$. $\rho(B)$ is the transverse magnetoresistance as a function of magnetic field, $\text{d}\rho/\text{d}B$ is the first derivative of the magnetoresistance measured by magnetic field modulation technique and $-\text{d}^2\rho/\text{d}B^2$ is the second-derivative of the magnetoresistance. The second derivative signals reveal magnetophonon resonance very clearly (After Hazama et al. [12])

where the third term of the right-hand side contains the component oscillating as $\cos(2\omega t)$. Therefore, when the second-harmonic component is detected by a lock-in amplifier, the signal is proportional to the second derivative of the magnetoresistance. This method discriminate the monotonic component from the magnetoresistance and gives $-\text{d}^2\rho/\text{d}B^2$, or the signals very close to the oscillatory component given by (7.127). We have to note here that the second derivative of (7.127) has a contribution from the exponential term and thus the experimentally observed second derivative will not give (7.127). However, a correction is possible to estimate all the parameters required for the analysis of magnetophonon resonance such as ω_{LO} and $\bar{\gamma}$.

As an example, Fig. 7.4 shows the experimental results for magnetophonon resonance in n-GaAs at $T = 160\,\text{K}$ [12]. In Fig. 7.4 three curves are compared: the transverse magnetoresistance and it first derivative and second derivative signals. $\rho(B)$ is the measured transverse magnetoresistance in which slight oscillations are observed as the magnetic field is increased. $\text{d}\rho/\text{d}B$ is the first derivative of the magnetoresistance observed by detecting the fundamental frequency component by a lock-in amplifier and shows clear oscillations, but the signals are 90° out of phase compared to the oscillations of the magnetoresistance. On the other hand, the second-derivative signals $-\text{d}^2\rho/\text{d}B^2$ exhibit very clear oscillations that are in phase with the original transverse magnetoresistance. Using the magnetic field modulation technique, magnetophonon resonance has been observed for up to $N = 18 \sim 20$ peaks of (7.125), and thus it has provided a very accurate determination of the effective mass. Although cyclotron resonance is well known as the most suitable method to determine the effective mass, only one peak is observed. On the other hand, magnetophonon resonance measurements will provide many peaks and thus it is possible to discuss the non-parabolicity of the conduction band in addition to the accurate determination of the effective mass. When the inverse magnetic field of the resonance peak is plotted as a function of N (Landau plot), the electron has a scalar effective mass and the plot exhibit a straight line. However, when the conduction band is nonparabolic, the Landau plot deviates from a straight line and the analysis provides information about the non-parabolicity of the conduction band.

Since magnetophonon resonance is observed under the condition $\omega_c\tau > 1$, the relation $|\sigma_{xy}| > \sigma_{xx}$ is satisfied and we may approximate as

$$\rho_{xx} \cong \frac{\sigma_{xx}}{\sigma_{xy}^2} . \tag{7.129}$$

In the explanation of the Shubnikov–de Haas effect we have shown that (7.87), $\sigma_{yx} \cong ne/B$, holds in a high magnetic field. This relation is valid in the second order of scattering potential V [13]. Therefore we may conclude that the oscillatory component of magnetophonon resonance arises from the oscillatory component of σ_{xx}. As mentioned earlier, in the absence of scattering, an electron moves in the y direction with a constant velocity E/B, and thus we have no current flow in the x direction, giving rise to $\sigma_{xx} = 0$. The current component along the x direction is caused by a change in the cyclotron center induced by electron scattering, resulting in $\sigma_{xx} \neq 0$. This fact means that the conductivity σ_{xx} is proportional to the electron scattering rate. A detailed treatment of the magnetoconductivity will be given later, the above considerations yield the following qualitative conclusions. The electron-LO phonon scattering rate is proportional to the strength of the electron-LO phonon interaction given by (6.239) and to the excitation number of phonons, $n_{LO} + 1$ for phonon emission and n_{LO} for phonon absorption. The scattering rate for the electron-LO phonon interaction is given by the sum of the two terms, absorption and emission. For simplicity, we assume that the scattering rate is proportional to n_{LO}. Expressing the Landau level broadening by Γ and following the treatment of the Shubnikov–de Haas effect, the oscillatory component of the magnetophonon resonance is written as (usually we may put $r = 1$ in the following equations)

$$\begin{aligned}\sigma_{xx} &\sim n_{LO} \sum_{r=1}^{\infty} \exp\left(-2\pi r \frac{\Gamma}{\hbar\omega_c}\right) \cos\left(2\pi r \frac{\omega_{LO}}{\omega_c}\right) \\ &= \frac{1}{\exp(\hbar\omega_{LO}/k_B T) - 1} \sum_{r=1}^{\infty} \exp\left(-\bar{\gamma} r \frac{\omega_{LO}}{\omega_c}\right) \cos\left(2\pi r \frac{\omega_{LO}}{\omega_c}\right),\end{aligned} \tag{7.130}$$

where $\bar{\gamma} = 2\pi\Gamma/\hbar\omega_{LO}$. This simplified estimate of the magnetophonon resonance amplitude is found to agree with the empirical formula of Stradling and Wood for $r = 1$, and the exponential decay of the oscillatory component is well explained in terms of the Landau level broadening (Γ) at low magnetic fields. In addition, the relation explains the temperature dependence of the magnetophonon resonance amplitude. At low temperatures the broadening is small but the number of phonon excitations is low, and thus the oscillations are very weak. At high temperatures, the number of phonon excitations is high but the increase in the broadening constant Γ due to various scattering mechanisms results in an exponential decay of the magnetophonon oscillations. It may be explained also that the oscillation amplitude of magnetophonon resonance exhibits a maximum at around $T = 150\,\text{K}$, decreasing

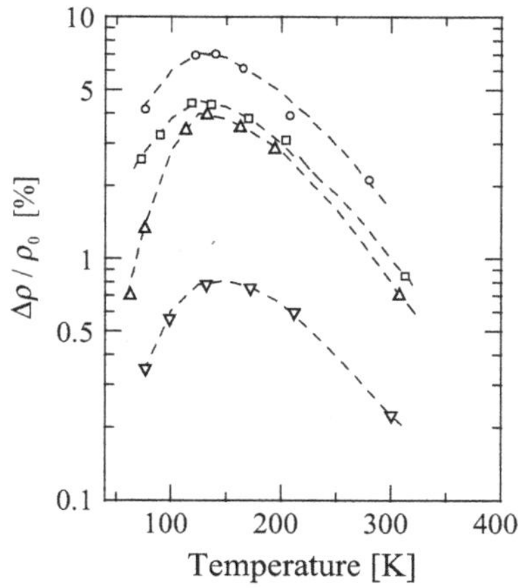

Fig. 7.5 Temperature dependence of the $N = 3$ magnetophonon resonance peak for several n-GaAs samples (After Stradling and Wood [3])

below and above this temperature. The feature is shown in Fig. 7.5, which is obtained from magnetophonon experiments in n-GaAs.

Now we discuss the theory of magnetophonon resonance. According to the Kubo formula [2, 4], the conductance σ_{xx} is given by

$$\begin{aligned}\sigma_{xx} &\\ &= \frac{e^2}{k_\mathrm{B}T}\int_{-\infty}^{+\infty} \mathrm{d}\mathcal{E} \sum_{\boldsymbol{q}} \frac{2\pi}{\hbar}\left(l^2 q_y\right)^2 n_\mathrm{LO}\left(1+n_\mathrm{LO}\right)\left[f(\mathcal{E}) - f(\mathcal{E}+\hbar\omega_\mathrm{LO})\right] \\ &\quad \times \delta(\mathcal{E}+\hbar\omega_\mathrm{LO}-\mathcal{E}_\nu)\delta(\mathcal{E}-\mathcal{E}_{\nu'})\left|\langle \nu'|C(q)\mathrm{e}^{-\mathrm{i}\boldsymbol{q}\cdot\boldsymbol{r}}|\nu\rangle\right|^2 , \end{aligned} \tag{7.131}$$

where $l = (\hbar/eB)^{1/2}$ and $C(q)$ represents the strength of the electron-LO phonon interaction defined by (6.239). Converting the summation over $\boldsymbol{q}$ into an integral and using (2.120) for the electron wave functions in a magnetic field, the above equation reduces to

$$\begin{aligned}\sigma_{xx} = \frac{e^2}{k_\mathrm{B}T}\frac{1}{(2\pi)^3}\int \mathrm{d}q_x\mathrm{d}q_y\mathrm{d}q_z \int_{-\infty}^{+\infty} \mathrm{d}\mathcal{E}\; n_\mathrm{LO}(n_\mathrm{LO}+1)(l^2 q_y)^2 \frac{2\pi}{\hbar}|C(q)|^2 \\ \times \sum_{N,M}\sum_{X,p_z}\left[f(\mathcal{E}-\hbar\omega_\mathrm{LO}) - f(\mathcal{E})\right]\delta(\mathcal{E}-\hbar\omega_\mathrm{LO}-\mathcal{E}_{N,p_z}) \\ \times \delta(\mathcal{E}-\mathcal{E}_{M,p_z-\hbar q_z})\left|J_{NM}(X,q_x,X-l^2 q_y)\right|^2 , \end{aligned} \tag{7.132}$$

where $J_{NM}(X, q_x, X - l^2 q_y)$ is defined by the following equation by using (2.120):

$$\begin{aligned}\langle N, X, p_z|\mathrm{e}^{\mathrm{i}\boldsymbol{q}\cdot\boldsymbol{r}}|M, X', p_z'\rangle = J_{NM}(X, q_x, X - l^2 q_y) \\ \times \delta_{p_z-\hbar q_z, p_z'}\delta_{X-l^2 q_y, X'} . \end{aligned} \tag{7.133}$$

In the case of a scalar electron effective mass, we find

$$\left|J_{NM}(X, q_x, X - l^2 q_y)\right|^2 = \frac{N!}{M!}\zeta^{M-N}\mathrm{e}^{-\zeta}\left[L_N^{M-N}(\zeta)\right]^2 , \tag{7.134}$$

where $L_N^M(\zeta)$ is the Laguerre polynomial and $\zeta = l^2(q_x^2 + q_y^2)/2$ [4]. The integral in (7.132) is evaluated to give [4]

$$\sigma_{xx} = \sigma_1 \sum_{N,M} \exp\left\{-2\bar{\alpha}\left(N + \frac{1}{2}\right)\right\} \exp\left\{-\bar{\alpha}(M - N - P - \delta^P)\right\}$$
$$\times K_0(\bar{\alpha}|N + P - M + \delta^P|) , \tag{7.135a}$$
$$\sigma_1 = \frac{ne^2}{m^*}\frac{\alpha}{\sqrt{\pi}\omega_c} n_{\mathrm{LO}} \left(\frac{\hbar\omega_{\mathrm{LO}}}{k_B T}\right)^{3/2} \sinh\bar{\alpha} , \tag{7.135b}$$

where $\bar{\alpha} = \hbar\omega_c/2k_B T$ and α is defined by (6.240) which is a dimensionless quantity to give the strength of the electron-LO phonon interaction. $\delta^P = \omega_{\mathrm{LO}}/\omega_c - P$ and P is the largest integer contained in $\omega_{\mathrm{LO}}/\omega_c$. The function $K_0(x)$ is the modified Bessel function of order of zero. The function $K_0(x)$ in (7.135b) exhibits a logarithmic divergence in the limit $x \to 0$. Therefore, (7.135b) diverges when $P = N - M$ and $\delta^P = 0$, or when $P = M - N - 1$ and $\delta^P = 1$. This divergence has been already mentioned for the case of the Shubnikov–de Haas effect and arises from the singularity of the density of states. In order to avoid this divergence for evaluation of the equation, the broadening effect of the Landau levels is taken into account and Barker has derived the oscillatory component of the magnetophonon resonance [4] written as

$$\sigma_{\mathrm{osc}} \sim \sum_{r=1}^{\infty} \frac{1}{r} \exp(-2\pi r\Gamma/\hbar\omega_c) \cos\left(2\pi r\frac{\omega_{\mathrm{LO}}}{\omega_c}\right) , \tag{7.136}$$

which is called Barker's formula. When we keep the term for $r = 1$ only in (7.136), it gives the empirical formula of Stradling and Wood. Barker has calculated the broadening constant for various scattering mechanisms. Here we note that we have the following relation between γ, Γ and $\bar{\gamma}$:

$$2\pi\gamma = 2\pi\frac{\Gamma}{\hbar\omega_c} = \bar{\gamma}\frac{\omega_{\mathrm{LO}}}{\omega_c} . \tag{7.137}$$

The values of $\bar{\gamma}$ for various types of scattering are summarized below. For longitudinal optical phonon scattering

$$\bar{\gamma} = 2\pi\left[\frac{\alpha}{2}(1 + n_{\mathrm{LO}})\right]^{2/3}\left(\frac{\omega_c}{\omega_{\mathrm{LO}}}\right)^{2/3} , \tag{7.138}$$

for acoustic phonon scattering

$$\bar{\gamma} = 2\pi \left[\frac{D_{ac}^2}{8\pi \rho v_s^2} \left(\frac{2m^* k_B T}{\hbar^2} \right)^{3/2} (\hbar \omega_c k_B T)^{-1/2} \right]^{2/3} \left(\frac{\omega_c}{\omega_{LO}} \right)^{2/3}, \tag{7.139}$$

and for a single impurity, where an electron does not feel two or more impurities at the same time,

$$\bar{\gamma} = \frac{2\pi}{\hbar \omega_{LO}} \left[\frac{e^4}{4\pi (\kappa \epsilon_0)^2} \left(\frac{\hbar^2}{2m^*} \right)^{1/2} \right]^{2/5} N_I^{2/5}, \tag{7.140}$$

where N_I is the density of impurities. In a semiconductor with a higher density of impurities, an electron is scattered at the same time by two or more impurities and this phenomenon is called the band-tailing effect. For the band-tailing effect

$$\bar{\gamma} = \frac{\sqrt{\pi} e^{3/2}}{\hbar \omega_{LO}} (\kappa \epsilon_0)^{3/4} \left(\frac{k_B T}{4} \right)^{1/4} \left(\frac{N_I^2}{n} \right)^{1/4}, \tag{7.141}$$

where n is the electron density.

At the beginning of this section we mentioned that the electron effective mass is evaluated from (7.125) when we know the LO phonon energy. Since the LO phonon energy has been accurately determined from Raman scattering, magnetophonon resonance experiments provide information about effective mass, non-parabolicity, energy band parameters and so on [3, 5]. In many compound semiconductors, their conduction band minima are located at $\boldsymbol{k} = 0$ (at the Γ point) and the effective mass is given by (2.82) as discussed in Sect. 2.4. When we use this relation, we may estimate the momentum matrix element P. A more accurate expression for the conduction band non-parabolicity is derived by Herman and Weisbuch by the following relation by taking account of contributions from higher-lying conduction bands [14], which is given by (2.238a) and (2.238b)

$$\frac{m}{m^*} - 1 = \frac{P_0^2}{3} \left(\frac{2}{\mathcal{E}_0} + \frac{1}{\mathcal{E}_0 + \Delta_0} \right) + \frac{P_1^2}{3} \left(\frac{2}{\mathcal{E}(\Gamma_8^c) - \mathcal{E}_0} + \frac{1}{\mathcal{E}(\Gamma_7^c) - \mathcal{E}_0} \right) + C, \tag{7.142}$$

where $\mathcal{E}_0$ is the energy gap between the Γ_6^c conduction band and the Γ_8^v valence band, Δ_0 is the spin-orbit splitting energy of the valence bands, $\mathcal{E}(\Gamma_8^c)$ and (Γ_7^c) are the energies of the conduction bands located above the Γ_6^c conduction band, and P_0, (P_1) is the momentum matrix element between the conduction band (higher-lying conduction band) and the valence bands. The constant C represents a contribution from the other higher-lying conduction bands and is determined to be $C = 2$. We have to note here that (7.142) is valid for diamond type crystals and not for III–V compound semiconductors which have no inversion symmetry, and that another consideration is required to validate the similar equation for III–V semiconductors. From the analysis of magnetophonon resonance experiments, Shantharama et al. [15]

has reported that P_1^2 and C are much smaller that the values predicted by Herman, Weisbuch [14]. Hazama et al. [16] have derived a similar expression for zinc blende crystals and evaluated P_0 and P_1 of (7.142) from the $\boldsymbol{k} \cdot \boldsymbol{p}$ theory for GaAs, InSb and InP, where they found that $P_0^2 = 23.55\,\text{eV}$ and $P_1^2 = 0.13\,\text{eV}$ for GaAs and C is -0.8, whereas for InSb and InP they obtained $C = 0$. These values show a good agreement with the values experimentally determined by Shantharama et al.

7.4.2 *Various Types of Magnetophonon Resonance*

The magnetophonon resonance effect discussed above has been theoretically predicted and experimentally observed in various semiconductors, mainly at temperatures $T > 77\,\text{K}$. In addition to this magnetophonon resonance (usually referred to as ordinal magnetophonon resonance), various types of magnetophonon resonance have been observed as discussed below.

7.4.2.1 (a) Impurity Series

At low temperatures $T < 40\,\text{K}$ a high electric field applied to a semiconductor produces **hot electrons** whose average energy is higher than the lattice temperature. Such a hot electron will make a transition from a Landau level to a donor state of an impurity by emitting an LO phonon, resulting in oscillatory behavior of the magnetoresistance. This oscillatory magnetoresistance is called the **impurity series** of magnetophonon resonance [17–19]. The resonance condition is given by

$$N'\hbar\omega_c + E_I(B) = \hbar\omega_{LO}\,, \quad N' = 1, 2, 3, \ldots\,, \tag{7.143}$$

where $E_I(B)$ is the ionization energy of a donor in a magnetic field B. From the analysis of the impurities series, information about impurities, especially the magnetic field dependence of impurity states, has been investigated. An example of the impurity series in the longitudinal magnetoresistance configuration is shown in Fig. 7.6 [19].

7.4.2.2 (b) Two-TA-Phonon Series

In experiments on impurity series other magnetophonon resonance peaks have been often observed by changing the magnitude of the applied electric field, this series is well understood in term of the simultaneous emission of transverse acoustic (TA) phonons. This kind of magnetophonon resonance was first observed by Stradling and Wood [18] and they named it 2 TA magnetophonon series. Such TA phonons have a high density of states, sometimes at the edge of the Brillouin zone, and the transition probability of the simultaneous emission of the TA phonons is high, enabling us

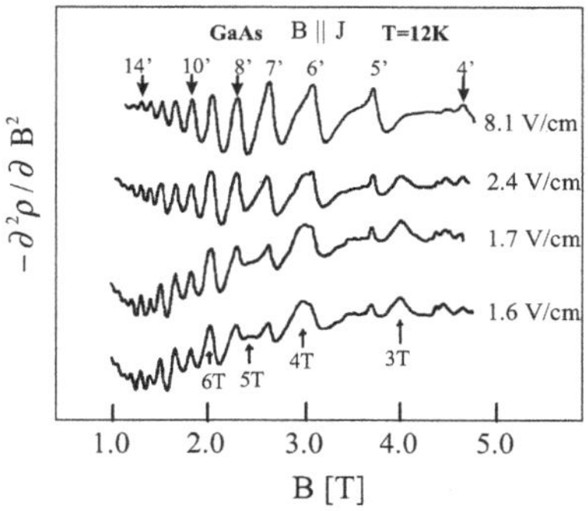

Fig. 7.6 Magnetophonon resonance in the longitudinal magnetoresistance configuration in n-GaAs at $T = 12\,\text{K}$, where the second derivative of the magnetoresistance is plotted as a function of magnetic field. At low electric fields, the 2TA phonon series, indicated by 3T, 4T, . . ., are observed; they are caused by a hot electron transition between Landau levels accompanied by simultaneous two transverse acoustic phonon emission. At higher electric fields, the 2TA series disappears and instead the impurity series, indicated by N', is observed which is caused by an electron transition from a Landau level to the impurity state by emitting an LO phonon (After [19])

to observe the 2 TA series of magnetophonon. The resonance condition of the 2 TA series is given by (see [18])

$$M\hbar\omega_c = 2\hbar\omega_{TA}, \quad M = 1, 2, 3, \ldots, \tag{7.144}$$

here $\hbar\omega_{TA}$ is the TA phonon energy.

The 2 TA series is well observed in Fig. 7.6, where the second derivative of the longitudinal magnetoresistance is obtained by means of magnetic field modulation and the results for n-GaAs at $T = 12\,\text{K}$ are plotted as a function of the magnetic field. Hot electrons are produced by applying an electric field from 1.6 to 8.1 V/cm. At lower electric fields the 2 TA series indicated by 3 T, 4 T, . . ., is clearly seen. At higher electric fields, the 2 TA series disappears and the impurity series is observed. From the 2 TA series and from (7.144), the phonon energy $\hbar\omega_{TA(X)} = 9.8\,\text{meV}$ is determined and this value agrees well with the neutron scattering data, $\hbar\omega_{TA(X)} = 9.7\,\text{meV}$ [20].

7.4.2.3 (c) Inter-valley Phonon Series

We have already shown from the energy band calculations in Chap. 1 and from the analysis of cyclotron resonance in Chap. 2 that the conduction band minima in Si and Ge are not located at the Γ point but on the Δ axes of the $\langle 100 \rangle$ direction and the L point in the $\langle 111 \rangle$ direction of Brillouin zone, respectively. Therefore, there are six equivalent valleys in Si and four equivalent valleys in Ge. As stated previously, such conduction bands are called many-valley structures. In a semiconductor with such a many-valley conduction band structure, the electrons are scattered between

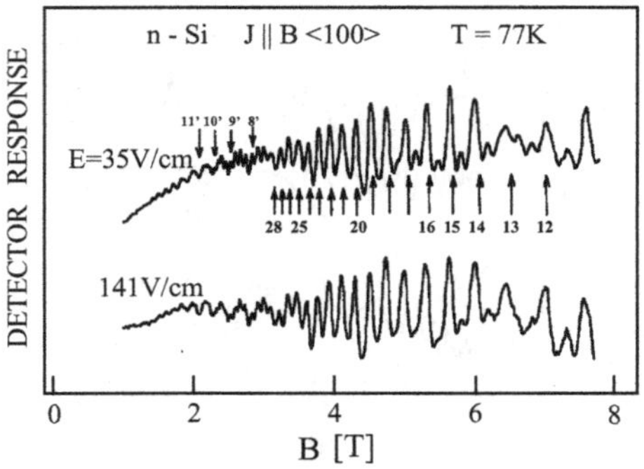

Fig. 7.7 Magnetophonon resonance due to inter-valley phonon scattering (inter-valley phonon series) in n-Si at 77 K, where the second-derivative signals of longitudinal magnetoresistance, $\boldsymbol{J} \parallel \boldsymbol{B}\langle 100\rangle$, are obtained by the magnetic field modulation technique. Several difference series are recognized in the oscillations. The series due to the fundamental magnetic field 84.7 T obtained from the Fourier transform analysis shown in Fig. 7.8 are indicated by 12–28 and the fundamental magnetic field 21.4 T by 8'–11'. Experimental data are obtained at electric fields 35 V/cm and 141 V/cm

the equivalent valleys by absorbing or emitting phonons, which is called inter-valley phonon scattering. There are several types of phonons with specific energies involved in the inter-valley scattering, resulting in the magnetophonon resonance. We have shown in Chap. 6 that the electron mobility in Si is determined by acoustic phonon scattering at low temperatures and by inter-valley phonon scattering at higher temperatures. Magnetophonon resonance due to inter-valley phonon scattering has been reported by Portal et al. [21], Eaves et al. [22] and Hamaguchi et al. [23]. From the analysis of electron mobility in Ge, the inter-valley scattering is found to be very weak, but the magnetophonon resonance in Ge reveals very clearly the inter-valley phonon series [22, 24]. Figure 7.7 shows a typical example of magnetophonon resonance in Si at 77 K due to inter-valley phonon scattering, where the second-derivative signals obtained by the magnetic field modulation technique are plotted against the magnetic field and a clear resonance is seen.

The resonance condition for magnetophonon resonance due to inter-valley phonon scattering is given by the following relation, taking into account the anisotropy of the effective mass:

$$\left(M + \frac{1}{2}\right)\hbar\omega_{c1} - \left(N + \frac{1}{2}\right)\hbar\omega_{c2} = \pm\hbar\omega_{\text{int}}, \quad M, N = 0, 1, 2, \ldots, \tag{7.145}$$

where $\omega_{ci} = \hbar eB/m_{ci}$ is cyclotron frequency for the effective mass m_{ci} (in the plane perpendicular to the magnetic field) of the ith valley, and $\hbar\omega_{\text{int}}$ is the inter-valley phonon energy. From (7.145) two different series of magnetophonon resonance are expected, which are obtained by keeping M or N constant as shown below. The resonance condition for the electron transition from a fixed Landau number N of the valley (2) to Landau levels M of the valley (1) gives the resonance magnetic field B_M

$$\frac{1}{B_M} = \frac{1}{B_{fM}}(M + \gamma_N), \tag{7.146}$$

where

$$\frac{1}{B_{fM}} = \Delta\left(\frac{1}{B}\right)_1 = \frac{e}{m_{c1}\omega_{\rm int}}, \tag{7.147a}$$

$$\gamma_N = \frac{1}{2} - \left(N + \frac{1}{2}\right)\frac{m_{c1}}{m_{c2}}. \tag{7.147b}$$

On the other hand, for the case of a fixed Landau level M we have

$$\frac{1}{B_N} = \frac{1}{B_{fN}}(N + \gamma_M), \tag{7.148}$$

where

$$\frac{1}{B_{fN}} = \Delta\left(\frac{1}{B}\right)_2 = \frac{e}{m_{c2}\omega_{\rm int}}, \tag{7.149a}$$

$$\gamma_M = \frac{1}{2} - \left(M + \frac{1}{2}\right)\frac{m_{c2}}{m_{c1}}. \tag{7.149b}$$

As stated above, inter-valley magnetophonon resonance involves two different valleys with the Landau level indices M and N, and therefore the magnetophonon resonance series are expected to arise from the inter-valley scattering in which the Landau index M or N is changed. In addition there are various kinds of inter-valley phonons; in Si, for example, f- and g-types of acoustic phonons and optical phonons are involved, resulting in many series of magnetophonon resonance and in a very complicated structure of the magnetophonon oscillations. Therefore, it is very difficult to identify the phonon types and the Landau indices from the experimental data. One of the most popular methods to analyze such complicated data is to produce about 200–400 magnetoresistance data points equally spaced with respect to the inverse magnetic field and to obtain the fundamental frequencies (fundamental magnetic fields B_{fM} and B_{fN}) from the Fourier transform. As an example, the magnetophonon resonance data for n-Si in Fig. 7.7 are Fourier transformed and the Fourier spectra are shown in Fig. 7.8. From the Fourier analysis, we obtain the fundamental magnetic field $B_f = 21.4$ and 84.7 T, and at lower electric fields peaks at 70 T and 106 T. The peak at 170 T is due to the second harmonic of the fundamental magnetic field 84.7 T. These fundamental fields are used to assign the magnetophonon resonance series as shown in Fig. 7.7. Using $B_f = 21.4$ [T] we obtain $\hbar\omega_{\rm iv} = 13.04$ [meV] for $m_{c1} = m_{\rm t}$ and 2.53 [meV] for $m_{c1} = m_{\rm l}$ (too small and thus no involved phonons for this), and the former energy gives g–TA phonon scattering. For $B_f = 84.7$ [T], we find f–LA (or f–TO) phonon energy $\hbar\omega_{iv} = 51.6$ [meV] for $m_{c1} = m_{\rm t}$ and g–TA phonon

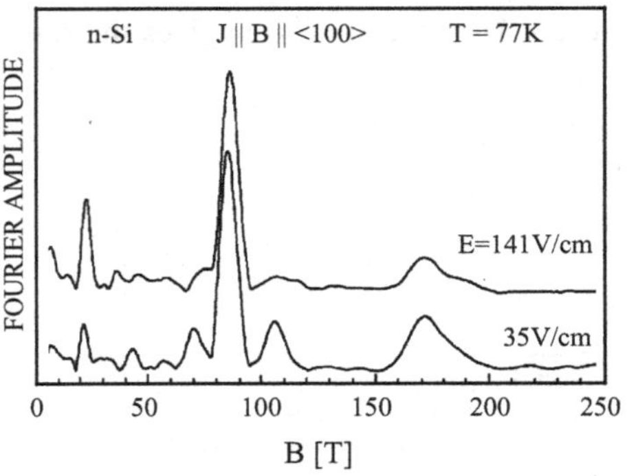

Fig. 7.8 Fourier transform of the data shown in Fig. 7.7. The peaks give the fundamental magnetic fields for the magnetophonon resonance and their higher harmonic components. The types and energies of the inter-valley phonons are elucidated from the fundamental field and the series scattering. The fundamental magnetic fields 84.7 and 21.4 T give rise to the magnetophonon series 12–28 and 8'–11', respectively, in Fig. 7.8

energy 10.0 [meV] for $m_{c1} = m_l$, respectively. These value are listed in Table 6.4 of Chap. 6.

Let us discuss the inter-valley magnetophonon resonance in n-Si in more detail. As stated in Sect. 6.3.7, g- and f-type inter-valley phonons are involved in the magnetophonon processes, where g-type phonons result in the inter-valley scattering between the valleys along the same principal axis, and f-type phonons are associated with the inter-valley scattering between the valleys whose principal axes are perpendicular. When a magnetic field is applied in the $\langle 100 \rangle$ direction, two different series of g-type magnetophonon resonance are expected. One is the inter-valley scattering between the valleys along the $\langle 100 \rangle$ valleys and the other is inter-valley scattering between the valleys in the $\langle 010 \rangle$ (and $\langle 001 \rangle$) axis whose cyclotron frequency is different from that of the $\langle 100 \rangle$ valleys. In g-type magnetophonon resonance, the cyclotron resonance condition reduces to (7.125) derived for magnetophonon resonance due to LO phonon scattering in GaAs. On the other hand, f-type inter-valley phonon scattering involves initial and final Landau levels whose cyclotron masses are different in the case of the transition from the $\langle 100 \rangle$ valley to the other valleys along the different axis. However, for a transition between the $\langle 010 \rangle$ and $\langle 001 \rangle$ valleys, the cyclotron masses of the initial and final states are the same. Therefore, f-type inter-valley magnetophonon gives rise to different series. Since the conduction band minima in Si are located at $0.85 \times k_{100}$ from the Brillouin zone edge k_{100} on the Δ axis, the wave vector of the g-type inter-valley phonon is given by $q_g = 0.3 \times k_{100}$ in the $\langle 100 \rangle$ direction. On the other hand, defining the band-edge wave vector along Σ by k_{110} ($= 3/2\sqrt{2} \times 2\pi/a$), the f-type inter-valley phonon wave vector is given by $q_f = 0.85\sqrt{2} \times 2\pi/a$ and thus we have $q_f = 0.92 \times k_{110}$. It may be possible to estimate the inter-valley phonon energy from the dispersion curves determined from neutron scattering experiments. Since the effective masses of Si are determined very accurately from the cyclotron experiments, the inter-valley phonon energies are estimated from magnetophonon resonance experiments. The analysis of magnetophonon resonance provides various types of phonon modes and their energies, and the assignment is made with the help of neutron scattering data [23]. However, the assignment is ambiguous and will not produce a definite conclusion. We have to note here that the magnetophonon resonance in n-Si has been obtained in very pure single crystals and thus the electrons are not degenerate. In such a case we may expect that the magnetophonon resonance arises from electrons in initial states with a low Landau index (0 or 1).

7.4.3 Magnetophonon Resonance Under High Electric and High Magnetic Fields

The magnetophonon resonance described above is observed under low electric fields and high magnetic fields, and the magnetophonon resonance due to hot electrons is observed at low temperatures in electric fields of up to several 100 V/cm. Several attempts have been made to investigate the high electric field effect in bulk semiconductors where it is not easy to apply a high electric field. In order to overcome this difficulty Eaves et al. [25] proposed to use an $n^{+}nn^{+}$ structure of epitaxially grown GaAs. They succeeded in observing magnetophonon resonance at high electric and high magnetic fields by applying the current in the direction perpendicular to the layers (x direction) and the magnetic field perpendicular to the current (z direction). Figure 7.9 shows the experimental results of Eaves et al., where at low electric fields (low current flow) the ordinal magnetophonon resonance due to LO phonon scattering is observed, whereas at high electric fields (high current flow) the ordinal magnetophonon resonance peaks seem to be depressed and to decay. Eaves et al. interpreted this phenomenon as showing that elastic or quasi-elastic scattering such as impurity scattering and acoustic phonon scattering dominate the transitions between the Landau levels. This process is more clearly explained with the help of Fig. 7.10 as follows. The spatial superposition of the wave functions between the adjacent Landau levels, which is small at low electric fields, increases at higher electric fields, and quasi-inelastic scattering therefore dominates the other scattering mechanisms. As a result, the ordinal magnetophonon resonance due to LO phonon scattering is weakened, causing depression of the magnetophonon peaks. This process is called QUILLS (QUasi-elastic Inter-Landau-Level Scattering). In contrast to this model, Mori et al. [26, 27] proposed the different model named IILLS (Inelastic Inter-Landau-Level Scattering) shown in Fig. 7.11. In this model, two facts are taken into account: (1) Landau levels incline in the presence of a magnetic field, and (2) current (conductivity σ_{xx}) is induced by a change in the electron cyclotron center along the electric field by electron scattering. These two facts are shown schematically in Fig. 7.11, where two different resonances due to LO phonon emission or absorption appear at lower and higher magnetic fields compared to the ordinal magnetophonon resonance. The

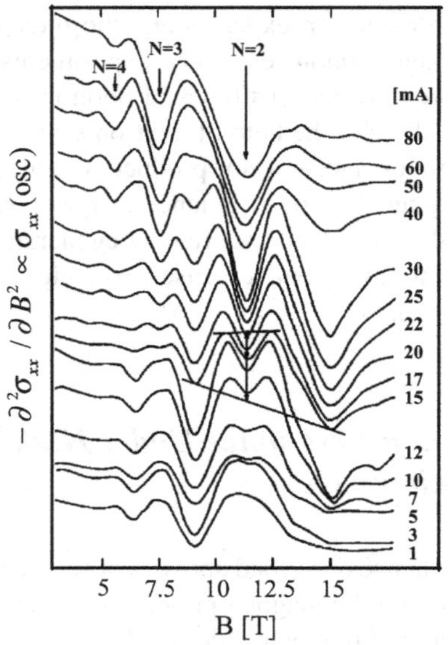

Fig. 7.9 Magnetophonon resonance in an epitaxially grown n $^+$nn$^+$ GaAs structure for a transverse magnetoresistance configuration ($\boldsymbol{J} \perp \boldsymbol{B}$), and high electric and magnetic fields ($T = 300$ K). The high applied electric field is achieved by a high current flow perpendicular to the epitaxial layers. At low electric fields (low current) the ordinal magnetophonon resonance is observed, whereas at higher electric fields, increasing the current, the peaks are depressed (see [25]). The effect was first explained by Eaves et al. in terms of quasi-elastic inter-Landau-level scattering, and later Mori et al. [26] explained it in terms of inelastic inter-Landau-level scattering, resulting in a double-peak structure due to inelastic LO phonon scattering

IILLS model is based on the fact that σ_{xx} is proportional to $(\Delta X)^2 = (l^2 q_y)^2$ according to (7.131). Therefore a vertical transition gives $\Delta X = 0$ and will not contribute to σ_{xx}.

We have described theory of magnetophonon resonance due to LO phonon scattering and the theory of Barker given by (7.131), which is derived from the Kubo formula. Here we will consider the IILLS theory based on the magnetophonon theory. As is well known, the Kubo formula is a kind of linear response theory and the following calculations of σ_{xx} are assumed to be approximated by a linear theory although the Landau levels are changed by the application of an electric field. Equation (7.131) is rewritten as

$$\sigma_{xx} = \frac{e^2\beta}{2}\sum_i\sum_f\sum_{\boldsymbol{q}} (l^2 q_y)^2 (P^{\mathrm{em}}_{i\to f} + P^{\mathrm{ab}}_{i\to f}) \,, \tag{7.150}$$

where $\beta = 1/k_{\mathrm{B}}T$, the second terms on the right-hand side are respectively the probabilities of LO phonon emission and absorption, and i and f represent the initial and final electronic states, respectively. The rates of LO phonon emission and

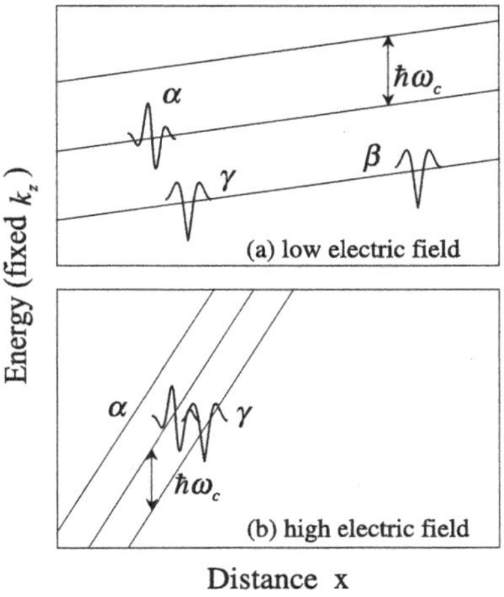

Fig. 7.10 Landau levels (adjacent 3 Landau levels for fixed k_z) **a** at a low electric field and **b** at a high electric field. At a high electric field the wave functions of the two adjacent levels overlap, resulting in the increase in transition probability between the $(n+1)$th and nth Landau levels at the same energy (QUILLS)

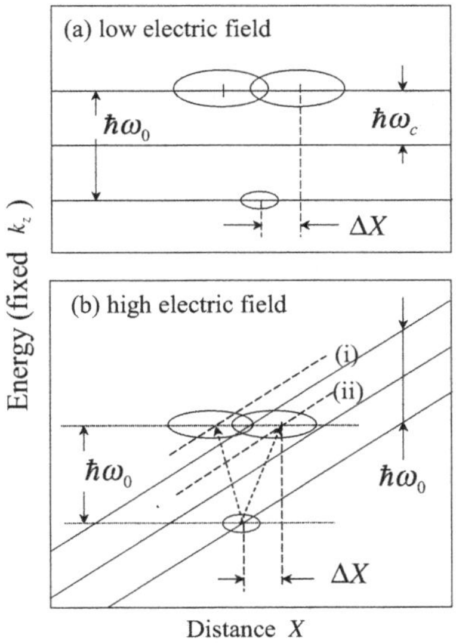

Fig. 7.11 In-elastic inter-Landau-level scattering (IILLS), where three adjacent Landau levels for fixed k_z and classical Landau orbits ($N = 0, 1, 2$) are shown. **a** at low electric fields, a vertical transition will not contribute to σ_{xx}. **b** at a high electric field, a non-vertical transition (i) the high magnetic field side and (ii) the low magnetic field side satisfies the magnetophonon resonance condition, resulting in a double-peak structure. In the figure the separation between the Landau levels of $N = 0$ and $N = 2$ is equal to the LO phonon energy $\hbar\omega_0$

absorption are written as

$$P^{\mathrm{em}}_{i\to f} = \frac{2\pi}{\hbar}(n_{\mathrm{LO}}+1)\left|\langle f|C^*(q)\mathrm{e}^{-\mathrm{i}\boldsymbol{q}\cdot\boldsymbol{r}}|i\rangle\right|^2 \times f(\mathcal{E}_i)\left[1-f(\mathcal{E}_i-\hbar\omega_{\mathrm{LO}})\right]\delta[\mathcal{E}_f-(\mathcal{E}_i-\hbar\omega_{\mathrm{LO}}], \tag{7.151}$$

$$P^{\mathrm{ab}}_{i\to f} = \frac{2\pi}{\hbar}n_{\mathrm{LO}}\left|\langle f|C(q)\mathrm{e}^{\mathrm{i}\boldsymbol{q}\cdot\boldsymbol{r}}|i\rangle\right|^2 \times f(\mathcal{E}_i)\left[1-f(\mathcal{E}_i+\hbar\omega_{\mathrm{LO}})\right]\delta[\mathcal{E}_f-(\mathcal{E}_i+\hbar\omega_{\mathrm{LO}})], \tag{7.152}$$

where $n_{\rm LO}$ is the Bose–Einstein distribution function of the LO phonons and $f(\mathcal{E})$ is the Fermi–Dirac distribution function of the electrons, which is approximated by the Maxwell–Boltzmann distribution function in the types of semiconductor we are concerned with. $\mathcal{E}_i$ and $\mathcal{E}_f$ are the electron energies of the initial and final states. The two terms of the above equation contribute to σ_{xx} in a similar way and we take into account the absorption term only for simplicity.

In the presence of an electric field ($\boldsymbol{E} \parallel x$) and magnetic field ($\boldsymbol{B} \parallel z$), and defining the vector potential by $\boldsymbol{A} = B(0, x, 0)$, the Hamiltonian is written as

$$H = \frac{1}{2m^*}(\boldsymbol{p} + e\boldsymbol{A})^2 + exE\,. \tag{7.153}$$

The eigenfunctions and eigenvalues for the Hamiltonian are given by

$$\psi_n = \exp(\mathrm{i}k_y y)\exp(\mathrm{i}k_z z)\phi_n(x - X)\,, \tag{7.154}$$

$$E_\nu = \left(n + \frac{1}{2}\right)\hbar\omega_0 + \frac{\hbar^2 k_z^2}{2m^*} + eEX_\nu + \frac{1}{2}m^*\left(\frac{E}{B}\right)^2, \tag{7.155}$$

where $\phi_n(x)$ is the solutions for a simple harmonic oscillator and $n = 0, 1, 2, \ldots$. In the y direction the cyclic boundary condition is satisfied and $k_y = 0, \pm 2\pi/L, \pm 4\pi/L, \ldots$, and k_z is the electron wave vector in the direction parallel to the electric field. The center coordinate of the cyclotron motion X_ν is given by

$$X = -\left(l_B^2 k_y + \frac{E}{\omega_{\rm c} B}\right), \tag{7.156}$$

and L is the sample length in the y direction. Considering these results and following the treatment of Barker [4] to derive magnetophonon resonance theory, we obtain the following relations (see N. Mori et al. [26]):

$$\begin{aligned}\sigma_{\rm osc} \sim \sum_{r=1}^{\infty} \frac{1}{r}\exp(-2\pi r\gamma)\Bigg[&\cos 2\pi r\left(\bar{\omega}_0 + \frac{\sqrt{3}}{2}\bar{e}\sqrt{\bar{\omega}_0 + 1}\right)\\ &+ \cos 2\pi r\left(\bar{\omega}_0 - \frac{\sqrt{3}}{2}\bar{e}\sqrt{\bar{\omega}_0 + 1}\right)\Bigg],\end{aligned} \tag{7.157}$$

where

$$\begin{aligned}\bar{e} &= \frac{\sqrt{2}eEl_B}{\hbar\omega_{\rm c}}\,,\\ \bar{\omega}_0 &= \frac{\omega_0}{\omega_{\rm c}}\end{aligned} \tag{7.158}$$

and in the limit of low electric field $\bar{e} = 0$, which gives rise to Barker's formula (7.136). In addition, from (7.157) the resonance condition is given by

$$\bar{\omega}_0 \pm \frac{\sqrt{3}}{2}\bar{e}\sqrt{\bar{\omega}_0 + 1} = N\,, \tag{7.159}$$

and thus the resonance magnetic field is given by

$$B_N^{\pm} \sim \frac{B_0}{N} \pm \sqrt{\frac{3}{2}\frac{m^*}{el_0 B_0}} E \,. \tag{7.160}$$

Here, $B_0/N = B_N$ is the resonance magnetic field for ordinal magnetophonon resonance at low electric fields. At high fields, it is seen from (7.157) and (7.160) that the electric field-induced magnetophonon resonance due to inelastic LO phonon scattering appears at both sides of the ordinal low electric field magnetophonon resonance peak. The oscillatory magnetoconductance σ_{osc} in GaAs in the IILLS model is calculated from (7.157) for electric fields of 0–3.5 kV/cm, and the second derivative of the magnetoconductance with respect to the magnetic field B is plotted in Fig. 7.12 in the magnetic field region near the $N = 2$ ordinal magnetophonon resonance. We find in Fig. 7.12 that the low electric magnetophonon peak at 11 T splits into two peaks on the lower and higher magnetic field sides of the ordinal magnetophonon peak as shown by the arrows. Figure 7.13 shows a plot of the peak positions as a function of the magnetic field, where the experimental data points □ are compared with the theoretical calculation (solid curves) and a good agreement is seen. We have to note here that the QUILLS model is disproved by these considerations and that the IILLS model is based on the linear theory and its validity has not yet been confirmed. In addition the IILLS model is derived by taking only LO phonon absorption into account and the contribution of LO phonon emission is not considered. Taking account of these factors, Wakahara and Ando [28] adopted a more elaborate theory to understand the magnetophonon resonance at high electric and high magnetic fields, and found that the most important mechanism for the peak splitting is due to the IILLS process.

7.4.4 Polaron Effect

In magnetophonon resonance the electrons and phonons are strongly coupled, the electrons being coupled strongly with the phonons by emitting or absorbing LO phonons. Under such conditions an electron is thought of as moving with phonon cloud around it and such a state is called a **polaron**. Once a polaron is formed, the electron moves with the phonons and thus the band-edge effective mass is increased. This feature is explained with the help of the diagrams shown in Fig. 7.14. In general, electron-phonon interaction is represented by two processes shown in Fig. 7.14, where (a) is the process where an electron of state $|\boldsymbol{k}\rangle$ emits a phonon of wave vector $\boldsymbol{q}$ and is transferred to a state $|\boldsymbol{k} - \boldsymbol{q}\rangle$ and (b) is the process where an electron of $|\boldsymbol{k} - \boldsymbol{q}\rangle$ absorbs a phonon of wave vector $\boldsymbol{q}$ and is transferred to a state $|\boldsymbol{k}\rangle$. These two processes are independent. On the other hand, in a system where electrons are strongly coupled with phonons, an electron of $|\boldsymbol{k}\rangle$ emits a phonon of wave vector $\boldsymbol{q}$ and subsequently absorbs the phonon and returns to the initial state $|\boldsymbol{k}\rangle$. This is a

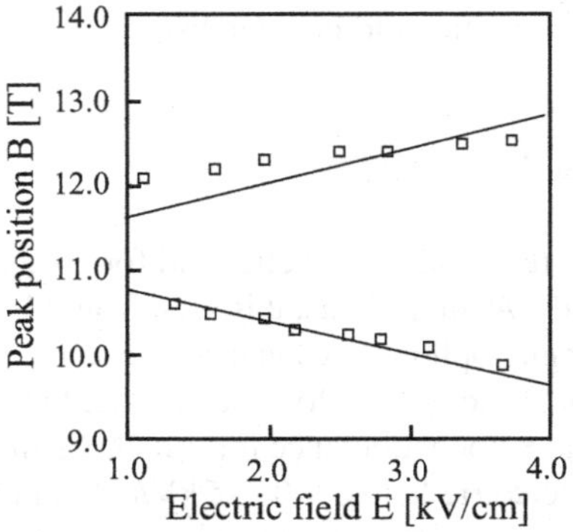

Fig. 7.12 Electric field-dependence of the resonance peaks for electric field-induced magnetophonon resonance due to inelastic inter-Landau-level scattering (IILLS), where the experimental results of Eaves et al. [25] are compared with the theoretical calculations (solid curves). The peak splitting is near the $N = 2$ ordinal magnetophonon resonance

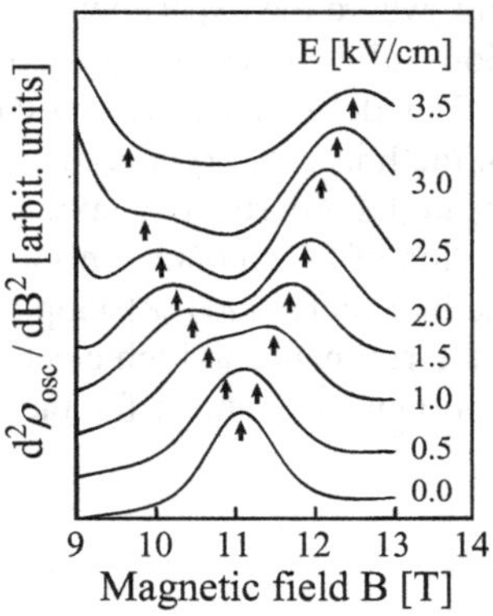

Fig. 7.13 Calculated result of electric-field induced magnetophonon resonance due to inelastic inter-Landau-level scattering (IILLS), where the second derivative of the oscillatory magnetoconductivity $\partial^2\sigma_{osc}/\partial B^2$ in GaAs is plotted as a function of magnetic field B near the $N = 2$ resonance. The peak of 11 T due to ordinal magnetophonon resonance due to LO phonon scattering at zero electric field splits into two peaks with increasing electric field

kind of composite state, the polaron state, as shown in Fig. 7.14c. It is evident from Fig. 7.14c that the polaron state is calculated by second-order perturbation theory.

In the following analysis we assume the very simplified case of a conduction band represented by a parabolic approximation with scalar effective mass; we also assume $T = 0$ and $B = 0$. The Hamiltonian of electron-phonon system under interaction is written as

$$H = H_0 + H' , \tag{7.161}$$

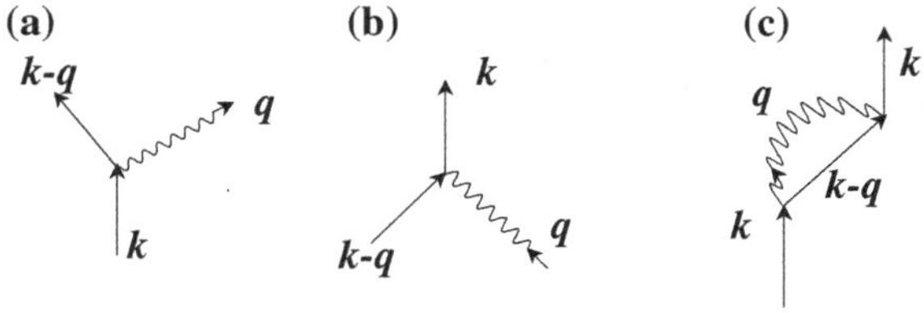

Fig. 7.14 Diagrams for the electron-LO phonon interaction. **a** An electron $|\boldsymbol{k}\rangle$ emits a phonon of wave vector $\boldsymbol{q}$ and is scattered to a state $|\boldsymbol{k}-\boldsymbol{q}\rangle$. **b** an electron $|\boldsymbol{k}-\boldsymbol{q}\rangle$ absorbs a phonon of wave vector $\boldsymbol{q}$ and is scattered to a state $|\boldsymbol{k}\rangle$. **c** When the electron-phonon interaction is strong, an electron $|\boldsymbol{k}\rangle$ subsequently emits and absorbs a phonon of wave vector $\boldsymbol{q}$, and returns to the initial state $|\boldsymbol{k}\rangle$. This is a kind of composite state and the electron moves accompanied by the phonon cloud, and a polaron state is formed

where H_0 is the Hamiltonian for a non-interacting electron and phonon and is given by

$$H_0 = \frac{p^2}{2m^*} + \sum_{\boldsymbol{q}} \hbar\omega_{\rm LO}\left(a_{\boldsymbol{q}}^\dagger a_{\boldsymbol{q}} + \frac{1}{2}\right) . \tag{7.162}$$

The term H' is the Hamiltonian for the electron-phonon interaction and for the Fröhlich-type interaction it may be written as

$$H' = \sum_{\boldsymbol{q}} V_{\boldsymbol{q}}\left(a_{\boldsymbol{q}} e^{i\boldsymbol{q}\cdot\boldsymbol{r}} + a_{\boldsymbol{q}}^\dagger e^{-i\boldsymbol{q}\cdot\boldsymbol{r}}\right) , \tag{7.163}$$

where the term $V_{\boldsymbol{q}}$ is expressed by using (6.239) and (6.240) as

$$V_{\boldsymbol{q}} = \frac{i\hbar\omega_{\rm LO}}{q}\left(\frac{\hbar}{2m^*\omega_{\rm LO}}\right)^{1/4}(4\pi\alpha)^{1/2} , \tag{7.164}$$

$$\alpha = \frac{e^2}{\hbar\epsilon_0}\left(\frac{1}{\kappa_\infty} - \frac{1}{\kappa_0}\right)\left(\frac{m^*}{2\hbar\omega_{\rm LO}}\right)^{1/2} . \tag{7.165}$$

The eigenstate for H_0 is specified by the electron wave vector $\boldsymbol{k}$ and the phonon number n_q of phonon wave vector $\boldsymbol{q}$ and is expressed as $|\boldsymbol{k}, n_q\rangle$, and its eigenenergy is given by

$$\mathcal{E}(\boldsymbol{k}, n_q) = \frac{\hbar^2 k^2}{2m^*} + \sum_{\boldsymbol{q}} \hbar\omega_{\rm LO}\left(n_q + \frac{1}{2}\right) . \tag{7.166}$$

Since we are concerned with a state at $T = 0$, the phonon system is in the vacuum state and the eigenstate for the unperturbed system is given by $|\boldsymbol{k}, 0\rangle$. Then the perturbing energy is obtained as

$$\Delta\mathcal{E} = \langle \boldsymbol{k}, 0|H'|\boldsymbol{k}, 0\rangle + \sum_{n_q \neq 0, \boldsymbol{k}'} \frac{\langle \boldsymbol{k}, 0|H'|\boldsymbol{k}', n_q\rangle\langle \boldsymbol{k}', n_q|H'|\boldsymbol{k}, 0\rangle}{\mathcal{E}(\boldsymbol{k}, 0) - \mathcal{E}(\boldsymbol{k}', n_q)} . \tag{7.167}$$

It is apparent that the second term on the right-hand side of the above equation corresponds to the diagram shown in Fig. 7.14c. The first-order perturbation given by the first term on the right-hand side becomes zero because H' contains the creation and annihilation operators of the phonon. The second term on the right-hand side of (7.167) vanishes except for the term for $n_q = 1$, and the matrix element of the numerator reduces to

$$\langle \boldsymbol{k}', 1_q|H'|\boldsymbol{k}, 0\rangle = V_q \delta(\boldsymbol{k} - \boldsymbol{k}' - \boldsymbol{q}) , \tag{7.168}$$

$$\langle \boldsymbol{k}, 0|H'|\boldsymbol{k}', 1_q\rangle = V_q \delta(-\boldsymbol{k} + \boldsymbol{k}' + \boldsymbol{q}) , \tag{7.169}$$

where (7.168) and (7.169) represent phonon emission and absorption, respectively, and correspond to the diagram of Fig. 7.14c. From the δ-function we obtain the electron wave vector of the virtual state, $\boldsymbol{k}' = \boldsymbol{k} - \boldsymbol{q}$.

The energies of the initial and final states are

$$\mathcal{E}(\boldsymbol{k}, 0) = \mathcal{E}(\boldsymbol{k}) = \frac{\hbar^2 k^2}{2m^*} , \tag{7.170}$$

and the energy of the intermediate state $|\boldsymbol{k} - \boldsymbol{q}\rangle$, where a single phonon of $\boldsymbol{q}$ is excited, is given by

$$\mathcal{E}(\boldsymbol{k}', n_q) = \mathcal{E}(\boldsymbol{k} - \boldsymbol{q}, 1_q) = \frac{\hbar^2}{2m^*}(\boldsymbol{k} - \boldsymbol{q})^2 + \hbar\omega_{\mathrm{LO}} . \tag{7.171}$$

Inserting these results into (7.167) we obtain the following relation:

$$\Delta\mathcal{E}(\boldsymbol{k}) = \sum_{\boldsymbol{q}} \frac{|V_q|^2}{\mathcal{E}(\boldsymbol{k}) - \mathcal{E}(\boldsymbol{k} - \boldsymbol{q}) - \hbar\omega_{\mathrm{LO}}} . \tag{7.172}$$

Converting the summation over $\boldsymbol{k}'$ into a summation over $\boldsymbol{q}$ with the use of the momentum conservation rule and the summation into an integral by using the relation $\sum_{\boldsymbol{q}} = (L^3/8\pi^3) \int \mathrm{d}^3\boldsymbol{q}$, the following result is obtained:

$$\begin{aligned} \Delta\mathcal{E}(\boldsymbol{k}) &= \frac{\alpha\hbar\omega_{\mathrm{LO}} u}{2\pi^2} \int \frac{1}{k^2 - (\boldsymbol{k} - \boldsymbol{q})^2 - u^2 q^2} \cdot \frac{\mathrm{d}^3\boldsymbol{q}}{q^2} \\ &= \frac{\alpha\hbar\omega_{\mathrm{LO}} u}{2\pi k} \int_0^\infty \frac{1}{q} \log\left[\frac{2kq - q^2 - u^2}{2kq + q^2 + u^2}\right] \mathrm{d}q . \end{aligned} \tag{7.173}$$

The derivation of this equation may be done by following the treatment given by (6.296)–(6.299) shown in Sect. 6.3.10. Here, $u = \sqrt{2m^*\omega_{\mathrm{LO}}/\hbar}$ and the upper limit of the integral is set to $q_{\max} \to \infty$. When the integration is carried out, we obtain[1]

$$\Delta\mathcal{E}(\boldsymbol{k}) = -\alpha\hbar\omega_{\mathrm{LO}}\frac{\sin^{-1}(k/u)}{k/u}\,. \tag{7.174}$$

Putting $u = k$, we have $\hbar^2 k^2/2m^* = \hbar\omega_{\mathrm{LO}}$. In the region where the electron energy is less than the LO phonon energy, $k < u$, we may use the asymptotic form $\sin^{-1} x = x + \frac{1}{6}x^3 + \cdots$, and so the electron energy is approximated as

$$\mathcal{E} = \frac{\hbar^2 k^2}{2m^*} + \Delta\mathcal{E} = -\alpha\hbar\omega_{\mathrm{LO}} + \frac{\hbar^2 k^2}{2m^*}\left(1 - \frac{\alpha}{6}\right) + \mathrm{O}(k^4)\,. \tag{7.175}$$

Such an energy in which the perturbation energy is included in the electron energy is called the self-energy of the polaron. The effective mass of the polaron, m^*_{pol} is then given by

$$m^*_{\mathrm{pol}} = \frac{m^*}{1 - \alpha/6}\,, \tag{7.176}$$

or it may be expressed as

$$m^*_{\mathrm{pol}} = \left(1 + \frac{\alpha}{6}\right) m^*\,. \tag{7.177}$$

The effective mass observed by magnetophonon resonance is affected by the polaron effect and thus the band edge effective mass should be corrected by (7.177). In addition (7.175) shows that the polaron state is lower than the bottom of the conduction band by $\alpha\hbar\omega_{\mathrm{LO}}$ and that the polaron can be excited by incident light with photon energy lower than the band gap.

7.5 Problems

(7.1) In order to understand Hall effect, analyze Hall measurement carried out using a device shown in Fig. 7.15. The dimensions of the sample are thickness $t = 1$ mm, width $w = 4$ mm, and length $l = 10$ mm. When magnetic field $B_z = 0.4$ T and current $I_x = 1.0$ mA are applied, Hall voltage $V_{\mathrm{H}} = 1.0$ mV. Hall field is $E_y < 0$. Observed voltage difference V_x between the voltage probes (distance l_x) is 25 mV without the magnetic field. Answer the following questions:
(1) Which is the type of carriers, electrons (n-type) or holes (p-type)?

---

[1] In order to carry out the integration, we assume that the term to be integrated is a function of $\boldsymbol{k}$ and $\boldsymbol{q}$. We then differentiate it with respect to $\boldsymbol{k}$ and integrate the result with respect to $\boldsymbol{q}$, and then integrate it with respect to $\boldsymbol{k}$.

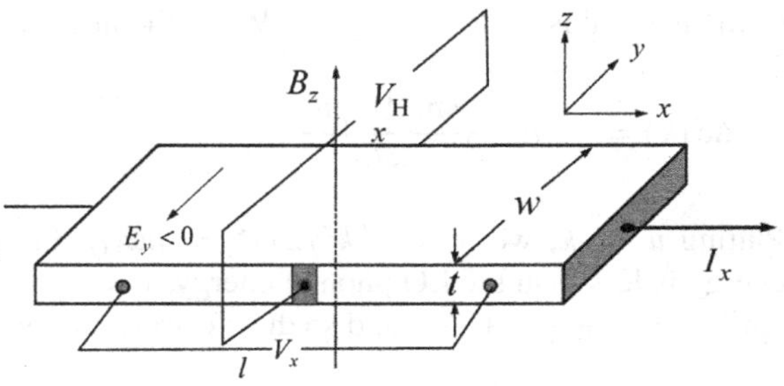

Fig. 7.15 Schematic illustration of Hall effect measurement for a typical sample. Application of magnetic field B_z and current flow I_x, produces Hall voltage V_H

(2) Calculate the density of the carriers.
(3) Calculate the conductivity σ of this sample.
(4) Calculate Hall mobility μ_H of the carriers.

(7.2) Derive (7.51b) for a parabolic band with isotropic effective mass.

(7.3) Derive (7.51c) for a parabolic band with isotropic effective mass..

(7.4) Shubnikov–de Haas oscillations appear periodically with respect to the applied magnetic field B. Consider a material with electron density $n = 2 \times 10^{18}$ [cm^{-3}] and assume the electrons with effective mass $m^* = 0.0135\,\mathrm{m}$ (for InSb) are degenerate.
(1) Calculate Fermi energy $\mathcal{E}_F$ of this material.
(2) Calculate the period of the Shubnikov–de Haas oscillations $(1/B)$.
(3) Calculate $\mathcal{E}_F$ and $(1/B)$ for GaSb shown in Fig. 7.2, where the electron effective mass at the Γ point is known $m^* = 0.039$

(7.5) Calculate oscillatory term for $r = 1$ of magnetophonon resonance given by (7.136) and plot it as a function of magnetic field. Use the parameters $2\pi r\Gamma/\omega_c = 0.5\omega_{LO}/\omega_c$, $m^* = 0.067\,\mathrm{m}$, $\hbar\omega_{LO} = 35.4$ [meV] (for GaAs).

(7.6) Estimate polaron masses in GaAs and InAs using material parameters in Table 6.3 and describe the effective mass correction.

References

1. L.M. Roth, P.N. Argyres, in *Semiconductors and Semimetals*, vol. 1, ed. by R.K. Willardson, A.C. Beer (Academic Press, New York, 1966), pp. 159–202
2. R. Kubo, S.J. Miyake, N. Hashitsume, in *Solid State Physics*, vol. 17, ed. by F. Seitz, D. Turnbull (Academic Press, New York, 1965) p. 279
3. R.A. Stradling, R.A. Wood, J. Phys. C **1**, 1711 (1968)
4. J.R. Barker, J. Phys. C, Solid State Phys. **5**, 1657–1674 (1972)
5. P.G. Harper, J.W. Hodby, R.A. Stradling, Rep. Prog. Phys. **36**, 1 (1973)
6. P.N. Argyres, Phys. Rev. **117**, 315 (1960)
7. R. Kubo, H. Hasegawa, N. Hashitsume, J. Phys. Soc. Jpn. **14**, 56 (1959)
8. W.M. Becker, H.Y. Fan, in *Proceedings of the 7th International Conference on the Physics of Semiconductors, Paris, 1964* (Dumond, Paris and Academic Press, New York, 1964), p. 663
9. V.L. Gurevich, Y.A. Firsov, Zh. Eksp. Teor. Fiz. **40**, 198–213 (1961) [Sov. Phys.-JETP **13**, 137–146 (1961)]

10. S.M. Puri, T.H. Geballe, Bull. Am. Phys. Soc. **8**, 309 (1963)
11. Y.A. Firsov, V.L. Gurevich, R.V. Pareen'ef, S.S. Shalyt, Phys. Rev. Lett. **12**, 660 (1964)
12. H. Hazama, T. Sugimasa, T. Imachi, C. Hamaguchi, J. Phys. Soc. Jpn. **54**, 3488 (1985)
13. P.N. Argyres, L.M. Roth, J. Phys. Chem. Solids **12**, 89 (1959)
14. C. Herman, C. Weisbuch, Phys. Rev. B **15**, 823 (1977)
15. L.G. Shantharama, A.R. Adams, C.N. Ahmad, R.J. Nicholas, J. Phys. C **17**, 4429 (1984)
16. H. Hazama, T. Sugimasa, T. Imachi, C. Hamaguchi, J. Phys. Soc. Jpn. **55**, 1282 (1986)
17. R.A. Stradling, R.A. Wood, Solid State Commun. **6**, 701 (1968)
18. R.A. Stradling, R.A. Wood, J. Phys. C **3**, 2425 (1970)
19. C. Hamaguchi, K. Shimomae, Y. Hirose, Jpn. J. Appl. Phys. **21**(Suppl. 21–3), 92 (1982)
20. J.L.T. Waugh, G. Dolling, Phys. Rev. **132**, 2410 (1963)
21. J.C. Portal, L. Eaves, S. Askenazy, R.A. Stradling, Solid State Commun. **14**, 1241 (1974)
22. L. Eaves, R.A. Hoult, R.A. Stradling, R.J. Tidey, J.C. Portal, S. Skenazy, J. Phys. C **8**, 1034 (1975)
23. C. Hamaguchi, Y. Hirose, K. Shimomae, Jpn. J. Appl. Phys. **22**(Suppl. 22–3), 190 (1983)
24. Y. Hirose, T. Tsukahara, C. Hamaguchi, J. Phys. Soc. Jpn. **52**, 4291 (1983)
25. L. Eaves, P.P. Guimaraes, F.W. Sheard, J.C. Portal, G. Hill, J. Phys. C **17**, 6177 (1984)
26. N. Mori, N. Nakamura, K. Taniguchi, C. Hamaguchi, Semicond. Sci. Technol. **2**, 542 (1987)
27. N. Mori, N. Nakamura, K. Taniguchi, C. Hamaguchi, J. Phys. Soc. Jpn. **57**, 205 (1988)
28. S. Wakahara, T. Ando, J. Phys. Soc. Jpn. **61**, 1257 (1992)

Chapter 8
Quantum Structures

Abstract In this chapter we deal with semiconductor quantum structures. First, two–dimensional electron gas (2DEG) in MOSFET and single heterostructure (High Electron Mobility Transistor: HEMT) are discussed by solving Schrödinder equation, and then transport properties are described by evaluating scattering rate and mobility. Basic idea of superlattices is given by showing Kronig-Penney model and Brillouin zone folding effect. In order to explain the optical properties of superlattices, energy band structures are calculated by using the tight–binding theory, and compared with the photoreflectance experiments. Mesoscopic phenomena observed in a semiconductor structure of the scale in between micro– and macro–structure is discussed by Landauer and Landauer B"uttiker formulas. Aharonov–Bohm effect and ballistic transport are discussed. Quantum Hall effect has attracted many scientists from the reasons due to the development in new physics of semiconductors in addition to the resistance standard, von Klitzing constant. Integral and fractional quantum Hall effects are discussed in different models. In the last part of this chapter, quantum dot (artificial atom) with several electrons is analyzed by employing Slater determinant where Coulomb interaction is exactly taken account, and the analysis of addition energy reveals very interesting features of the shell model and Hund's rule in quantum dots.

8.1 Historical Background

The words "quantum structures" or "quantum effect devices" have been cited very often since the 1980s [1–7]. Early work on the quantization of electrons was initiated in the 1960s, when electrons in the inversion layer of the MOSFET (Metal-Oxide-Semiconductor Field Effect Transistor) [8] were found to be quantized at the interface and to exhibit two-dimensional properties. Then the Quantum Hall effect was discovered in 1980 by van Klitzing et al. [14] by using a Si-MOSFET and thereafter the transport properties of a two-dimensional electron gas has received a great deal of attention.

Another attempt to fabricate quantum structures, which has led to the most important applications in semiconductor physics and devices, was initiated in 1960 by Esaki

C. Hamaguchi, *Basic Semiconductor Physics*, Graduate Texts in Physics,
DOI 10.1007/978-3-319-66860-4_8

and Tsu [15], who introduced the molecular beam epitaxial (MBE) technology to grow layered structures, named superlattices, and succeeded in observing Bloch oscillations. Since then MBE growth techniques have been used by many people to tailor various types of quantum structure and the electron transport and optical properties arising from the quantization effect have been observed. For example, the two-dimensional transport, the quasi-one-dimensional structure (quantum wire), the zero-dimensional structure (quantum dot), and superlattices have been extensively investigated. This research is still in progress and it is very difficult for the author to review them as a chapter of the present textbook. Therefore, in this chapter we will deal with several important phenomena of quantum structures and introduce the basic physics which is required for readers who want to investigate in more detail.

It is best to understand the physics of the two-dimensional electron gas in the MOSFET first and then to investigate the physics of the quantum effects of heterostructures. Therefore, we will study the two-dimensional electron gas in the MOSFET in detail and then look at several important quantum effects in heterostructures. Readers who are interested in more detail are recommended to read the references.

8.2 Two-Dimensional Electron Gas Systems

8.2.1 Two-Dimensional Electron Gas in MOS Inversion Layer

The two-dimensional characteristics of electrons were first observed in MOSFETs and various properties of quantum effects have been investigated in MOSFETs. It is best to study the two-dimensional properties of electrons in the MOSFET for a detailed understanding of the quantum effects in quantum structures fabricated from heterostructures. Therefore in this section we will deal with the two-dimensional electron gas in the MOFSFET.

The structure of the MOSFET is shown in Fig. 8.1, and consists of an insulating SiO_2 film, a metal gate, a source and a drain formed on a p-Si substrate. The gate electrode controls the electron density in the inversion layer (channel) induced below the insulating SiO_2 film by applying a bias, and it thus controls the current flow through the channel. The FET characteristics are similar to those of a triode vacuum tube, which are well described in the textbook of Sze [8]. The energy diagram for the device is shown in Fig. 8.2 for a positive bias applied to the gate electrode. The application of the gate bias lowers the energy of the gate electrode and bends the energy of the semiconductor surface at the interface between the semiconductor and the insulating layer. As a result electrons are induced at the interface of the p-Si, and the layer (or channel) of the electrons is called the inversion layer. The valence band is also bent so that the valence band is filled by electrons, depleting holes in this region, and negatively charged acceptors are left. This layer is therefore called the **depletion layer**, which is indicated by the depth z_d. Beyond the depletion layer

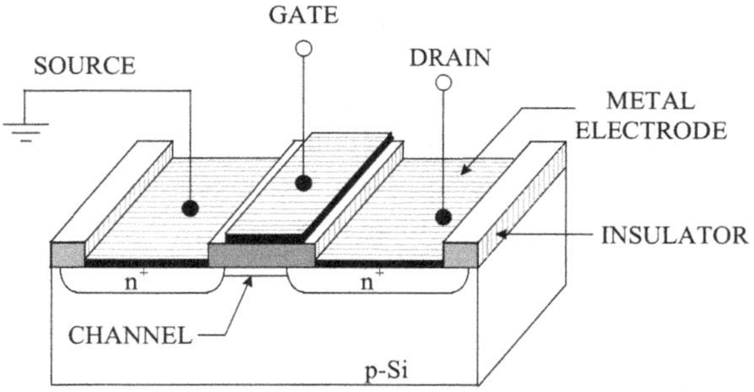

Fig. 8.1 Structure of MOS field effect transistor (MOSFET). The MOSFET is fabricated on a p-Si substrate, with a SiO_2 thin film (insulating film) on the p-Si and a metal gate electrode on the SiO_2. The source and drain electrodes are formed by diffusing p-Si n^+ impurities into the p-Si substrate and then putting metal contacts on the diffused surfaces. Application of a positive voltage bias to the gate electrode induces an n-channel of electrons under the gate electrode, resulting in electron current flow between the source and the drain

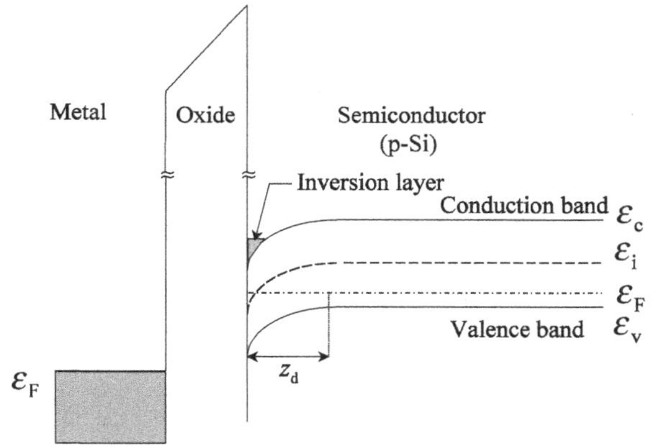

Fig. 8.2 Energy diagram of the MOSFET of Fig. 8.1. It is shown for a positive gate bias, where an inversion layer of electrons is formed near the surface of the p-Si at the interface between the semiconductor (p-Si) and the insulator (SiO_2). The induced electrons are confined in a narrow region at the interface and move only parallel to the interface, exhibiting two-dimensional characteristics. In the region near the interface of the p-Si substrate, the bands bend and holes are depleted leaving ionized acceptors in the region from the surface to z_d. Charge neutrality is achieved by holes and acceptors in the region beyond the depletion layer

($z_d \le z \le \infty$) charge neutrality is preserved by the holes and ionized acceptors, where the Fermi energy is given by the bulk Fermi energy $\mathcal{E}_F$. The electrons in the inversion layer confined at the interface are quantized in the z direction and can move only in the direction x and y (parallel to the interface). From this the electrons in the inversion layer form a two-dimensional state and are called a **two-dimensional electron gas**.

8.2.1.1 (a) Triangular Potential Approximation

It is seen in Fig. 8.2 that electrons in the inversion layer are confined within a narrow region of the interface and that the confining potential looks like a triangular potential. Let the electric field at the interface be E_s and the direction of the confinement be z. The confining potential is then approximated as

$$\phi(z) = -E_s z . \tag{8.1}$$

Assuming that the ionized acceptors N_A and donors N_D ($N_A \gg N_D$) are distributed uniformly along the direction of z from the interface to the position z_d, the potential distribution is obtained by solving Poisson's equation:

$$\phi_d(z) = \frac{e(N_A - N_D)}{2\kappa_s \epsilon_0} z^2 , \tag{8.2}$$

where κ_s is the dielectric constant of Si. Analysis of the potential distribution in this section is carried out by neglecting the contribution of the charge in the depletion region and by taking the contribution of the electrons themselves. The electrons are confined in the z direction and are mobile only in the x, y plane. Letting the area of the x, y plane be A and writing the wave function of an electron in the inversion layer as

$$\psi(x, y, z) = A^{-1/2} \exp(\mathrm{i}k_x x + \mathrm{i}k_y y)\zeta_i(z) , \tag{8.3}$$

The Schrödinger equation in the z direction reduces to

$$-\frac{\hbar^2}{2m_3}\frac{\mathrm{d}^2}{\mathrm{d}z^2}\zeta_i(z) - e\phi(z)\zeta_i(z) = \mathcal{E}_i\zeta_i(z) , \tag{8.4}$$

where m_3 is the electron effective mass in the z direction. Inserting the triangular potential given by (8.1) into the potential energy of (8.4), we obtain the following relation:

$$\frac{\mathrm{d}^2}{\mathrm{d}z^2}\zeta_i(z) + \frac{2m_3 e E_s}{\hbar^2}\left[\frac{\mathcal{E}_i}{eE_s} - z\right]\zeta_i(z) = 0 . \tag{8.5}$$

We have to note here that (8.5) is exactly the same as (5.3a) used to describe the Franz–Keldysh effect in Sect. 5.1.2. Defining

$$z' = \left(\frac{2m_3 e E_s}{\hbar^2}\right)^{1/3}\left[z - \frac{\mathcal{E}_i}{eE_s}\right] , \tag{8.6}$$

(8.5) reduces to

$$\frac{\mathrm{d}^2\zeta_i(z')}{\mathrm{d}z'^2} - z'\zeta_i(z') = 0 . \tag{8.7}$$

The solution of this equation is given by the Airy function as described in Sect. 5.1.2, and is written as

$$\zeta_i(z') = C_i \cdot \mathrm{Ai}(z') . \tag{8.8}$$

This is rewritten as

$$\zeta_i(z) = C_i \cdot \mathrm{Ai}\left([2m_3 eE_s/\hbar^2]^{1/3}[z - \mathcal{E}_i/eE_s]\right) . \tag{8.9}$$

The constant C_i is determined by the normalization condition of (8.9) and given by [1]

$$C_i = \frac{1}{\sqrt{\lambda}\mathrm{Ai}'(x_i)} = \left[\frac{2m_3 eE_s}{\hbar^2}\right]^{1/6} \frac{1}{\mathrm{Ai}'(x_i)} \tag{8.10}$$

The potential barrier at the interface of the insulating SiO_2 film and the p-Si surface is very high and may be approximated an infinite barrier height, enabling us to put $\zeta(z=0)=0$. This assumption gives rise to the solution (see [16])

$$\mathcal{E}_i \approx \left(\frac{\hbar^2}{2m_3}\right)^{1/3} \left[\frac{3}{2}\pi eE_s\left(i + \frac{3}{4}\right)\right]^{2/3} . \tag{8.11}$$

This solution gives a good approximation for a large value of i, and also a reasonably good result for small i. For $i = 0$, for example, the solution is obtained by putting $i + 3/4 = 0.75$, whereas the exact solution is given by inserting 0.7587. The other approximate solutions are closer to the exact solutions. It is concluded from the above results that the energy levels of a confined electron in the z direction are given by

---

[1] The normalization constant (5.8) is evaluated by integrating the squared wave function in the whole region, $-\infty \sim +\infty$, whereas the normalization in MOSFET should be carried out in the region $0 < z < \infty$. Introducing variables, $\lambda = [2m_3 eE_s/\hbar^2]^{-1/3}$ and $x_i = -\mathcal{E}/(\lambda eE_s)$, where x_i is i–th solution of $\mathrm{Ai}(x_i) = 0$, $(i = 0, 1, 2, \ldots)$. Then (8.9) is rewritten as

$$\zeta_i(z) = C_i \cdot \mathrm{Ai}\left(\frac{z}{\lambda} + x_i\right) .$$

Normalization of the wave function (8.9) in the region $0 < z < \infty$ is given by

$$1 = \int_0^\infty |\zeta_i(z)|^2 \mathrm{d}z = C_i^2 \int_0^\infty \mathrm{Ai}^2\left(\frac{z}{\lambda} + z_i\right) \mathrm{d}z$$

Here we change the variable as $x = z/\lambda$, and then

$$\begin{aligned} 1 &= C_i^2 \lambda \int_0^\infty \mathrm{Ai}^2(x + x_i)\mathrm{d}x = C_i^2 \lambda \int_{x_i}^\infty \mathrm{Ai}^2(x)\mathrm{d}x \\ &= C_i^2 \lambda \left[-\{\mathrm{Ai}'(x)\}^2 + x\mathrm{Ai}^2(x)\right]_{x_i}^\infty = C_i^2 \lambda \{\mathrm{Ai}'(x_i)\}^2 . \end{aligned}$$

From this we find the normalization constant C_i given by (8.10).

discrete values due to the quantization. Adding the electron energies of the x and y directions, the total energy of the electron, in the two-dimensional electron gas, is given by

$$\mathcal{E} = \mathcal{E}_i + \frac{\hbar^2 k_x^2}{2m_1} + \frac{\hbar^2 k_y^2}{2m_2} . \tag{8.12}$$

Next we will discuss the density of states for a two-dimensional electron gas and the potential acting on the electrons. The density of states for a two-dimensional electron gas is derived as follows. When the cyclic boundary condition is introduced to a semiconductor of the volume L^3, the number of states in the region between (k_x, k_y) and $(k_x + \mathrm{d}k_x, k_y + \mathrm{d}k_y)$ is given by

$$\frac{2L^2}{(2\pi)^2}\mathrm{d}k_x \mathrm{d}k_y .$$

Therefore, the density of states per unit area is given by

$$\frac{2}{(2\pi)^2}\mathrm{d}k_x \mathrm{d}k_y , \tag{8.13}$$

where the factor 2 in the above equation arises from the spin degeneracy. When we convert the variables by $k_x/\sqrt{m_1/m} = k'_x$ and $k_y/\sqrt{m_2/m} = k'_y$, (8.12) is reduced to

$$\mathcal{E} = \mathcal{E}_i + \frac{\hbar^2}{2m}(k'^2_x + k'^2_y) = \mathcal{E}_i + \frac{\hbar^2}{2m}k'^2 .$$

Taking account of the relations, $\mathrm{d}k_x \mathrm{d}k_y = (\sqrt{m_1 m_2}/m)\mathrm{d}k'_x \mathrm{d}k'_y$ and $2\pi k' \mathrm{d}k' = \pi \mathrm{d}(k'^2) = \pi(2m/\hbar^2)\mathrm{d}(\mathcal{E} - \mathcal{E}_i)$, the following relation is obtained:

$$\frac{2}{(2\pi)^2}\mathrm{d}k_x \mathrm{d}k_y = \frac{2}{(2\pi)^2}\frac{m_\mathrm{d}}{m}2\pi k' \mathrm{d}k' = \frac{m_\mathrm{d}}{\pi\hbar^2}\mathrm{d}(\mathcal{E} - \mathcal{E}_i) ,$$

where the density-of-states mass defined by $m_\mathrm{d} = \sqrt{m_1 m_2}$ is used. This relation gives the density of states for a two-dimensional electron gas, which is shown schematically in Fig. 8.3. As shown in Fig. 8.3, the three-dimensional density of states is expressed by a parabolic curve (dashed curve) but the two-dimensional density-of-states curve (solid curve) is given by step functions. Each step function is called a **subband** and $\mathcal{E}_i$ is the energy of the bottom of the subband and is called the **subband energy**.

As shown in Fig. 2.9, the conduction bands of Si consist of six equivalent valleys, and thus the density of states should be calculated by taking account of the valley degeneracy n_v. For example, in the Si-MOSFET, where the insulating SiO_2 is grown on the (001) surface of Si, we obtain two different series of subbands which are the

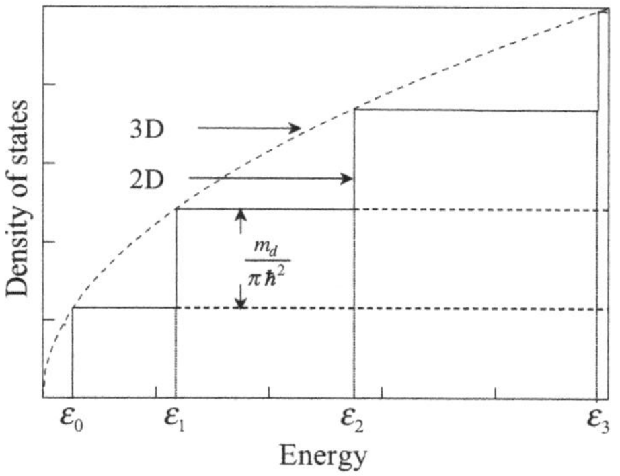

Fig. 8.3 Density of states (solid curve) for a two-dimensional electron gas is shown by the solid curve, a combination of step functions. The step function is called a subband. The energy of the bottom of the subband is given by $\mathcal{E}_i$ and each subband has the step $m_\mathrm{d}/\pi\hbar^2$. The density of states for three-dimensional electrons (dashed curve) is proportional to $\sqrt{\mathcal{E}}$ and thus given by a parabolic curve. In the limit of narrow subband spacing, the two-dimensional density of states approaches the three-dimensional density of states

two equivalent valleys with their principal axis in the direction perpendicular to the interface and four equivalent valleys with the valley axes parallel to the interface; their effective masses in the quantization directions are, respectively, $m_3 = m_\mathrm{l}$ ($n_\mathrm{v} = 2$) and $m_3 = m_\mathrm{t}$ ($n_\mathrm{v} = 4$). Letting the valley degeneracy be $n_{\mathrm{v}i}$ for the subband series $\mathcal{E}_i$, the density of states for a two-dimensional electron gas is written as

$$J_{2\mathrm{D}}^i(\mathcal{E})\mathrm{d}(\mathcal{E} - \mathcal{E}_i) = \frac{n_{\mathrm{v}i} m_{\mathrm{d}i}}{\pi\hbar^2}\mathrm{d}(\mathcal{E} - \mathcal{E}_i) . \tag{8.14}$$

Since electrons obey Fermi–Dirac statistics, the sheet density of the two-dimensional electron gas N_i of the subband E_i is calculated as

$$\begin{aligned} N_i &= \int_{E_i}^{\infty} J_{2\mathrm{D}}^i(\mathcal{E}) \frac{1}{\exp[(\mathcal{E} - \mathcal{E}_\mathrm{F})/k_\mathrm{B}T] + 1} \mathrm{d}\mathcal{E} , \\ &= \frac{n_{\mathrm{v}i} m_{\mathrm{d}i} k_\mathrm{B} T}{\pi\hbar^2} F_0 [(\mathcal{E}_\mathrm{F} - \mathcal{E}_i)/k_\mathrm{B}T] , \qquad (8.15) \\ & F_0(x) = \log(1 + \mathrm{e}^x) , \qquad (8.16) \end{aligned}$$

where $\mathcal{E}_\mathrm{F}$ is the Fermi energy. Then the sheet density of electrons in the inversion layer N_inv is given by

$$N_\mathrm{inv} = \sum_i N_i . \tag{8.17}$$

Since the charge density in the depletion layer is given by

$$\rho_\mathrm{depl}(z) = \begin{cases} -e(N_\mathrm{A} - N_\mathrm{D}), & 0 < z \leq z_\mathrm{d} \\ 0, & z > z_\mathrm{d} \end{cases} \tag{8.18}$$

the corresponding sheet density of the ionized impurities in the depletion layer N_{depl} is defined by

$$N_{\mathrm{depl}} = (N_{\mathrm{A}} - N_{\mathrm{D}})z_{\mathrm{d}}\,, \tag{8.19}$$

and the depth of the depletion layer z_{d} is estimated from the curvature of the potential ϕ_{d} in (8.2) [17]. As seen in Fig. 8.2, the potential curvature is given by the following relation assuming the energy at $z \to \infty$ (the bulk region of the semiconductor) as $(\mathcal{E}_{\mathrm{c}} - \mathcal{E}_{\mathrm{F}})_{\mathrm{bulk}}$ and neglecting the contribution from the inversion electrons. At $T = 0$ and at zero bias between the source and drain, the potential curvature is determined as

$$\phi_{\mathrm{d}} = (\mathcal{E}_{\mathrm{c}} - \mathcal{E}_{\mathrm{F}})_{\mathrm{bulk}} + \mathcal{E}_{\mathrm{F}} - \frac{eN_{\mathrm{inv}}z_{\mathrm{av}}}{\kappa_{\mathrm{s}}\epsilon_0}\,, \tag{8.20}$$

where $\mathcal{E}_{\mathrm{F}}$ is the Fermi energy measured from the bottom of the conduction band at the Si–SiO$_2$ interface, and the following parameters are used

$$z_{\mathrm{av}} = \frac{\sum_i N_i z_i}{N_{\mathrm{inv}}}\,, \tag{8.21}$$

$$z_i = \int z|\zeta_i(z)|^2 \mathrm{d}z \bigg/ \int |\zeta_i(z)|^2 \mathrm{d}z\,. \tag{8.22}$$

The electric field at the Si surface is determined by the total sheet density of the charges. When we define it by $-E_{\mathrm{s}}$, it is given by

$$E_{\mathrm{s}} = \frac{e(N_{\mathrm{inv}} + N_{\mathrm{depl}})}{\kappa_{\mathrm{s}}\epsilon_0}\,, \tag{8.23}$$

where $N_{\mathrm{depl}} = z_{\mathrm{d}}(N_{\mathrm{A}} - N_{\mathrm{D}})$. The potential distribution $\phi(z)$ is determined by Poisson's equation

$$\frac{\mathrm{d}^2\phi(z)}{\mathrm{d}z^2} = -\frac{1}{\kappa_{\mathrm{s}}\epsilon_0}\left[\rho_{\mathrm{depl}}(z) - e\sum_i N_i\zeta_i^2(z)\right]. \tag{8.24}$$

8.2.1.2 (b) Solutions by the Variational Method

More accurate solutions may be obtained by using the variational principle. The variational principle is based on the assumption of a trial wave function and on determination of the wave function to minimize the energy. The most popular trial function is the Fang–Howard function [18] $\zeta_0(z)$ for the ground state given by the following equation, where $\zeta(z = 0) = 0$ at $z = 0$ and $\zeta(z = \infty) = 0$ at $z \to \infty$:

$$\zeta_0(z) = \sqrt{\frac{1}{2}b^3}\, z\mathrm{e}^{-bz/2}\,. \tag{8.25}$$

This assumption gives a good result when the electron density in the inversion layer is low and the electrons occupy the lowest subband. The variational parameter b of the trial function is determined to minimize the subband energy. Inserting (8.25) into the one-dimensional Schrödinger Equation (8.4), the expectation value of the energy is evaluated as

$$\mathcal{E}_0 = \frac{\hbar^2 b^2}{8m_3} + \frac{3e^2}{\kappa_s \epsilon_0 b}\left[N_{\text{depl}} + \frac{11}{16}N_{\text{inv}} - \frac{2}{b}(N_A - N_D)\right]. \tag{8.26}$$

Various parameters are determined to minimize the energy $\mathcal{E}_0$ and are given below:

$$b_0 = \left(\frac{12m_3 e^2 N^*}{\kappa_s \epsilon_0 \hbar^2}\right)^{1/3}, \tag{8.27a}$$

$$z_0 = \left(\frac{9\kappa_s \epsilon_0 \hbar^2}{4m_3 e^2 N^*}\right)^{1/3}, \tag{8.27b}$$

$$\mathcal{E}_0 = \left(\frac{3}{2}\right)^{5/3}\left(\frac{e^2\hbar}{m_3^{1/2}\kappa_s\epsilon_0}\right)^{2/3}\left(N_{\text{depl}} + \frac{55}{96}N_{\text{inv}}\right)\frac{1}{N^{*1/3}}, \tag{8.27c}$$

where $N^* = N_{\text{depl}} + \frac{11}{32}N_{\text{inv}}$. It has been shown by Stern [17] that the subband energy $\mathcal{E}_0$ in Si calculated by this method shows good agreement with the result of self-consistent calculations.

8.2.1.3 (c) Solutions by the Self-Consistent Method

The Schrödinger equation, Poisson's equation and the electron sheet density are written as

$$\left[-\frac{\hbar^2}{2m_3}\frac{d^2}{dz^2} + [\mathcal{E}_i - e\phi(z)]\right]\zeta_i(z) = 0, \tag{8.28}$$

$$\frac{d^2\phi(z)}{dz^2} = -\frac{1}{\kappa_s\epsilon_0}\left[\rho_{\text{depl}} - e\sum_i N_i|\zeta_i(z)|^2\right], \tag{8.29}$$

$$N_i = \frac{n_v m_d k_B T}{\pi\hbar^2}\log\left(1 + \exp\left[\frac{E_F - \mathcal{E}_i}{k_B T}\right]\right). \tag{8.30}$$

The simultaneous equations may be solved self-consistently by using numerical analysis as follows. Let the Si/SiO_2 interface be at the origin $z = 0$, and choose a distance $z = L$ where the wave function converges (for example, 20 nm). The region $[0, L]$ is discretized into 200 meshes and we define $h = L/200$. The following calculations are usually solved by introducing dimensionless values for energy and distance. Here we show how to solve by the difference method. Differentiation of a function $f(z)$ is expressed as

$$\frac{\mathrm{d}f}{\mathrm{d}z} \to \frac{f_{(j+1)} - f_{(j)}}{h}, \tag{8.31}$$

$$\frac{\mathrm{d}^2 f}{\mathrm{d}z^2} \to \frac{f_{(j+1)} - 2f_{(j)} + f_{(j-1)}}{h^2}. \tag{8.32}$$

The Schrödinger equation (8.28) and Poisson's equation (8.29) are then expressed as 200×200 matrix equations. Since both equations are second-order differential equations, the corresponding matrices have diagonal elements and adjacent components, or in other words the matrices are tridiagonal, and thus the solutions are easily obtained by diagonalization. From those procedures we obtain 200 eigenvalues and 200 eigenfunctions from the Schrödinger equation because we have discretized the equation into 200 meshes. Therefore, this difference method gives the solutions of the Schrödinger equation with the accuracy of the 200 mesh discretization. This procedure provides smooth solutions for lower eigenstates.[2] On the other hand, Poisson's equation is solved easily by the iteration method when the initial value E_s is known. Another method to solve Poisson's equation is to diagonalize the tridiagonal matrix using the values at $z = 0$ and $z = L$.

The simultaneous equations, (8.28) and (8.29), contain the wave function $\zeta_i(z)$ and potential $\phi(z)$ and they are not independent. Here we will show a method to obtain self-consistent solutions for the simultaneous equations.

1. Determine the surface of Si. In the case of the (001) surface of Si, we have to solve two Schrödinger equations for $m_3 = m_\mathrm{l}$, $n_\mathrm{v} = 2$ and $m_3 = m_\mathrm{t}$, $n_\mathrm{v} = 4$.
2. Determine the initial values of N_inv and $N_\mathrm{A} - N_\mathrm{D}$.
3. The total sheet electron density N_inv is assumed to occupy the ground state subband $\mathcal{E}_0$, and parameters such as z_av, z_d, ϕ_d, N_depl, and $\mathcal{E}_\mathrm{s}$ are determined using the Fang–Howard trial function. In the next iteration these parameters are replaced by new values.
4. Using these initial parameters, solve Poisson equation, (8.29), and obtain an approximate solution of the potential $\phi(z)$.[3]
5. Inserting the potential into the Schrödinger equation, (8.28), obtain the wave function $\zeta_i(z)$ and eigenvalue $\mathcal{E}_i$.
6. Calculate the electron sheet density of each subband from (8.30).
7. Replace the data of (3) by new data, and repeat the processes (3–6). This iteration is continued till the energy $\mathcal{E}_i$ converges.

Self-consistent solutions for the electronic states in Si-MOSFET fabricated on the (100) surface are shown in Fig. 8.4, where the electron sheet density $N_\mathrm{inv} = 1.0 \times 10^{13}\,\mathrm{cm}^{-2}$, $N_\mathrm{A} = 5.0 \times 10^{17}\,\mathrm{cm}^{-3}$ and the parameters used are $m_\mathrm{l} = 0.916\,m$,

---

[2] Although the Schrödinger equation is a continuous function, the method discretizes the wave function into 200 meshes, giving rise to a 200×200 matrix equation. The eigenvalues and eigenfunctions are obtained by diagonalizing the matrix. In general we are interested in the lower energy states and some of the lowest values of the energies and wave functions. Details of the calculations are found in textbooks of numerical analysis.

[3] Neglecting the contribution from N_depl and assuming $-E_\mathrm{s} = eN_\mathrm{inv}/\kappa_\mathrm{s}\epsilon_0$ and a triangular potential, the Schrödinger equation may be solved by the iteration method to give converged solutions.

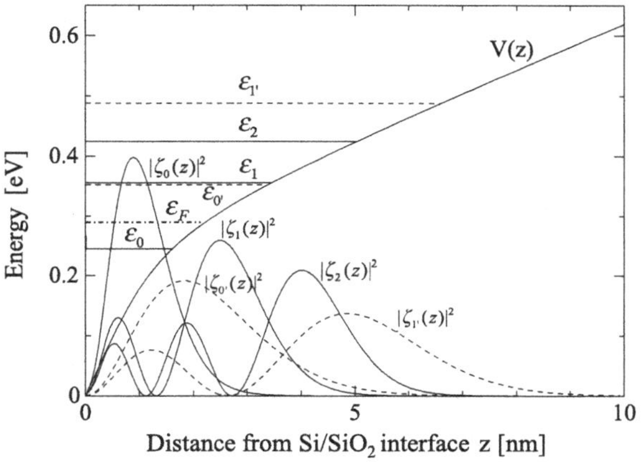

Fig. 8.4 Self-consistent solutions for inversion electrons in the Si (001) surface. The potential energy $V(z)$, wave function $|\zeta_i(z)|^2$, subband energy $\mathcal{E}_i$ and Fermi energy (horizontal dot-dashed line) are shown. The subband energies of the $\mathcal{E}_i$ series are indicated by the horizontal solid lines and the $\mathcal{E}_{i'}$ series by the horizontal dashed lines, and the corresponding wave functions are shown by the solid and dashed curves. $N_{\text{inv}} = 1.0 \times 10^{13}\,\text{cm}^{-2}$, $N_{\text{A}} = 5.0 \times 10^{17}\,\text{cm}^{-3}$, and $T = 300\,\text{K}$. The Fermi energy is determined to be $\mathcal{E}_{\text{F}} = 292\,\text{meV}$

Table 8.1 Self-consistent solutions for the electronic states in the inversion layer of the (001) surface of Si-MOSFET. $N_{\text{inv}} = 1.0 \times 10^{13}\,\text{cm}^{-2}$, $N_{\text{A}} = 5.0 \times 10^{17}\,\text{cm}^{-3}$, and $T = 300\,\text{K}$

Notation	Subband energy (meV)	Subband occupation (%)	Average distance z_{av} (nm)
$\mathcal{E}_0$	245.3	79.9	1.11
$\mathcal{E}_1$	355.4	3.4	2.39
$\mathcal{E}_2$	424.5	0.2	3.44
$\mathcal{E}_{0'}$	352.5	16.4	2.28
$\mathcal{E}_{1'}$	488.3	0.1	4.46
$\mathcal{E}_{2'}$	590.2	0.0	6.19
$\mathcal{E}_{3'}$	676.7	0.0	7.75

$m_{\text{t}} = 0.190\,m$ and $\kappa_{\text{s}} = 11.7$. In the case of the Si (001) surface, we have a subband series of $\mathcal{E}_0, \mathcal{E}_1, \mathcal{E}_2, \ldots$ for $m_3 = m_{\text{l}}$ and $n_{\text{v}} = 2$, and series of $\mathcal{E}_{0'}, \mathcal{E}_{1'}, \mathcal{E}_{2'}, \ldots$ for $m_3 = m_{\text{t}}$ and $n_{\text{v}} = 4$. The solid curve in Fig. 8.4 is the conduction band profile $V(z)$, where the conduction band energy at the Si/SiO$_2$ interface is set to be zero. The horizontal solid lines are $\mathcal{E}_i$ ($i = 0, 1, 2$) and the horizontal dashed lines are $\mathcal{E}_{i'}$ ($i' = 0, 1$), and the Fermi energy is shown by the dot-dashed line. The squared wave functions for the $\mathcal{E}_i$ series are plotted by solid curves and those for the $\mathcal{E}_{i'}$ series by dashed curves. It is very interesting to point out that the subband energies $\mathcal{E}_1$ and $\mathcal{E}_{0'}$ are very close. In Table 8.1 the self-consistent solutions for the subband energies, electron occupation and the average extension (distance) of the wave function z_{av} are summarized. The Fermi energy is $\mathcal{E}_{\text{F}} = 292\,\text{meV}$.

From these results it is clear that electrons in the inversion layer of Si-MOSFET are quantized to form a two-dimensional electron gas. Experiments on the Shubnikov–de Haas effect and so on have been reported so far. The most prominent work has been the discovery of the quantum Hall effect in 1980, which will be discussed in detail later.

8.2.2 Quantum Wells and HEMT

The semiconductors which exist in nature are elements and compound materials with periodic structures such as Si, Ge, and GaAs. In the late 1960 s to early 1970s, however, Esaki and Tsu proposed a completely new idea for growing new materials called **superlattices**, which do not exist in nature, which is based on the crystal growth technique, **molecular beam epitaxy**. They fabricated a new structure, called the **superlattice**, where n layers of GaAs and m layers of AlAs were grown periodically in a very high vacuum. Since then various types of quantum structures have been proposed and investigated experimentally. The purpose of Esaki and Tsu was to observe Bloch oscillations in superlattices. Bloch oscillations are achieved when electrons are accelerated from the bottom of a conduction band to the Brillouin zone edge without any scattering, and thus a very high field is required. In such a high electric field electrical breakdown readily occurs in a bulk semiconductor. On the other hand, in a superlattice, where the Brillouin zone is folded as stated later and mini-bands are formed, it may become possible to observe Bloch oscillations.

Here we have to note that there exist several restrictions in order to grow a different semiconductor on a substrate. When different materials A and B are grown on a same substrate, the structure is called a **heterostructure**, and the characteristics of the heterostructure depend strongly on the physical and chemical properties of both materials. In order to get a good structure it is essential that the lattice constants of both materials are equal or, in other words, the two materials are **lattice matched**. Under the lattice-matched condition, controlled crystal growth provides lattice-matched crystal growth, **pseudomorphic growth**. Lattice mismatch of crystal A with lattice constant $a_{\rm A}$ and B with lattice constant $a_{\rm B}$ is defined by $\eta = 2|a_{\rm A} - a_{\rm B}|/(a_{\rm A} + a_{\rm B})$. For a heterostructure with a small lattice mismatch, pseudomorphic growth is achieved, while for a heterostructure with a large lattice mismatch lattice defects (misfit dislocation) are introduced in order to relax the strain due to the lattice mismatch. In order to avoid the defects, thin layer superlattices are grown alternately by limiting the number of the layers to less than the critical layer thickness. Such a superlattice grown without strain relaxation is called a **strained-layer superlattice**. From these considerations we find that the growth of superlattices and quantum structures without strain effects is achieved by choosing materials with small lattice mismatch, or lattice-matched materials. In Fig. 8.5 the energy gaps versus the lattice constants are plotted for typical III–V compound semiconductors (cubic zinc blende crystals), where the solid and open circles represent direct and indirect band gaps, respectively. For example, the lattice constants of GaAs and AlAs are 5.653 Å and 5.660 Å,

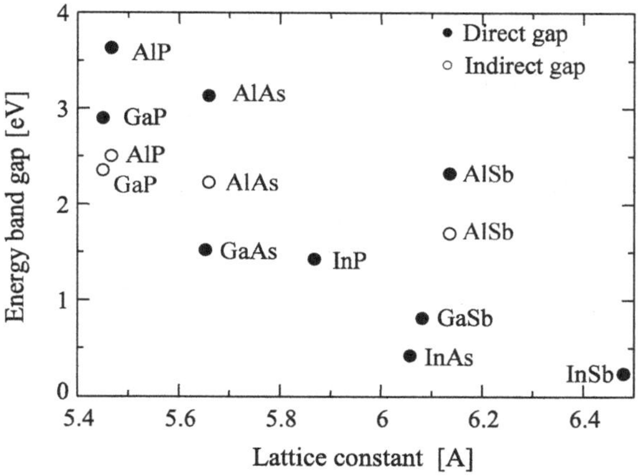

Fig. 8.5 Energy band gaps plotted as a function of the lattice constant for typical III–V compound semiconductors, where the solid and open circles represent direct and indirect gaps, respectively

respectively, and the lattice mismatch is $\eta = 0.12\%$. In the case of such a small lattice mismatch, pseudomorphic crystal growth is possible. From this reason early studies of quantum structures have been carried out intensively in heterostructures with GaAs and $Al_xGa_{1-x}As$ ($x = 0 \sim 1$). In materials of large lattice mismatch, it is possible to achieve lattice-matched crystal growth by using ternary alloys, such as a combination of GaInAs and InP. Compounds of four elements will provide a wide variety of pseudomorphic crystal growth.

The energy band diagrams of heterostructures with lattice match are classified into three cases, as shown in Fig. 8.6, which are type I, type II and type II after the notation

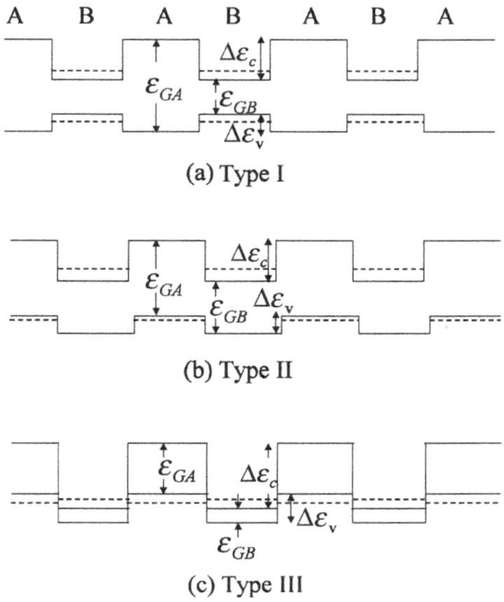

Fig. 8.6 Three types of heterostructure. **a** type I: electrons and holes are confined in the same space. **b** type II (staggered): Confinement of electrons and holes are spatially different, resulting in spatially indirect transition. **c** type III (misaligned): zero-gap

of Esaki et al. In type I the conduction band of semiconductor B with the smaller band gap becomes a minimum and the valence band a maximum, and therefore electrons and holes are excited in this region. In type II (staggered) the minimum of the conduction band of semiconductor A and the maximum of the valence band of semiconductor B are spatially different, resulting in a spatially indirect band gap. In type III (misaligned) the heterostructures behave as a zero-gap material (semimetal) or as a narrow gap material. Of these three types the type I heterostructures have been most intensively investigated.

Here we will consider the **quantum wells** of type I heterostructures. The regions of the conduction band minimum and maximum are called the **well layer** and **barrier layer**, respectively. When the barrier layer is thick, electrons are not able to penetrate and are confined in the quantum well region, resulting in quantization of the electron motion perpendicular to the heterointerface, forming a subband structure. Since electrons are able to move in the plane parallel to the heterointerface, the electrons exhibit two-dimensional characteristics. These features are shown in Fig. 8.7. Figure 8.7a is the case for a non-doped heterostructure and both the well and barrier regions are electronically neutral, giving rise to zero electric field. Figure 8.7b is the uniformly doped case in which donors and electrons excited from the donors are confined in the well region, giving rise to two-dimensional electron gas formation in the

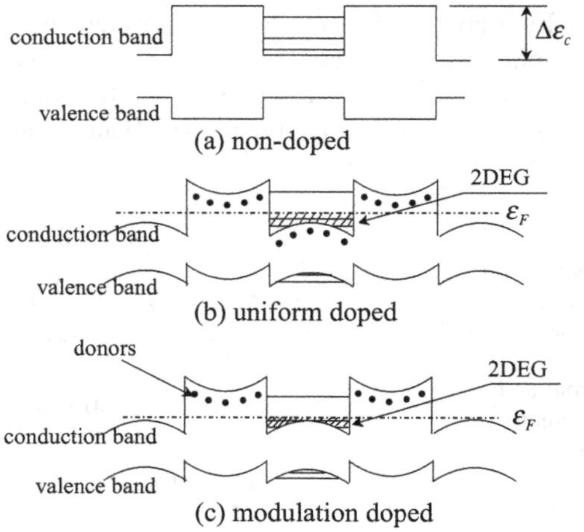

Fig. 8.7 Modulation doping in heterostructures. **a** non-doped case, where no electric field exists. When low densities of electrons and holes are excited by photon absorption, they occupy the subbands in the conduction band and the valence band, respectively, and they recombine to end up at the initial states. **b** is the case where impurity donors are uniformly doped in both the well and barrier regions. The electrons confined in the well regions form two-dimensional states and are subject to impurity scattering in the well region, resulting in reduction of the electron mobility at low temperatures. **c** is the case called modulation doped, where donor impurities are doped only in the barrier regions, and electrons confined in the well regions are separated from the ionized impurities. Therefore, the high mobility of two-dimensional electron gas is achieved

well region. The electron mobility in such a structure is affected by ionized impurity scattering due to the donors in the well region and is reduced at low temperatures. To avoid this decrease in mobility a new idea was proposed, which is shown in Fig. 8.7c, where electrons in the well regions are supplied from donors in the barrier regions and confined in the well regions, resulting in a reduction of the impurity scattering and an increase of the mobility at low temperatures. Note that impurity potential is a long-range force and thus electrons are affected by the ionized donors in the barrier region. A more dramatic reduction of impurity scattering is achieved by putting a non-doped spacer region near the well region. This method is called **modulation doping** [19].

In such structures, electrons are confined in the well region by the potential walls of the conduction band discontinuity $\Delta\mathcal{E}_c$ at the interface and are quantized in the direction perpendicular to the barriers, whereas the electrons can move in the plane parallel to the interface. Therefore, a two-dimensional electron gas is formed in the well regions. For simplicity, we assume that the potential wall is infinity and that the well region is from $z = 0$ to $z = L$ (well width L). Then the electron wave functions (called envelope functions) $\zeta_n(z)$ are obtained by solving the one-dimensional Schrödinger equation

$$-\frac{\hbar^2}{2m^*}\frac{\mathrm{d}^2}{\mathrm{d}z^2}\zeta_n(z) = \mathcal{E}_n\zeta_n(z)\,,$$

which will give solutions

$$\zeta_n(z) = A\sin\left(\frac{\pi}{L}nz\right)\,, \tag{8.33}$$

where A is the normalization constant, $A = \sqrt{2/L}$. The corresponding eigenvalues are

$$\mathcal{E}_n = \frac{\hbar^2}{2m^*}\left(\frac{\pi}{L}n\right)^2 \tag{8.34}$$

and the total electron energy is given by

$$\mathcal{E} = \frac{\hbar^2}{2m^*}\left(k_x^2 + k_y^2\right) + \frac{\hbar^2}{2m^*}\left(\frac{\pi}{L}n\right)^2 \equiv \frac{\hbar^2}{2m^*}\left(k_x^2 + k_y^2\right) + \mathcal{E}_n\,. \tag{8.35}$$

The eigenfunctions $\zeta_n(z)$ are orthogonal for different n. The ground state is given by $n = 1$ and the number of nodes of the function is $n - 1$. In other words, the greater is the number of nodes $(n - 1)$, the higher is the energy. This relation holds for the self-consistent solutions described next.

Now, we will deal with self-consistent calculations of the two-dimensional electron gas in heterostructures. We adopt the effective mass approximation to obtain the states of the two-dimensional electron gas. In general the electron effective mass and dielectric constant depend on the materials. Assuming that the particle velocity and electric field are along the direction perpendicular to the heterointerface, the

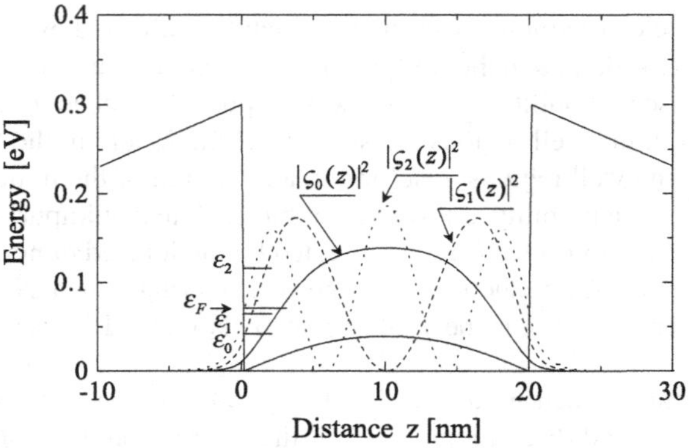

Fig. 8.8 Self-consistent solutions for a two-dimensional electron gas in an $Al_{0.3}Ga_{0.7}As/GaAs$ modulation-doped quantum well with well width 20 nm. The results are at $T = 4.2\,\text{K}$ for electron sheet density $N_s = 10^{12}\,\text{cm}^{-2}$ and spacer layer of 10 nm in the AlGaAs barriers. The subband energies and Fermi energy are indicated by the horizontal lines and the square of the wave function $|\zeta_n(z)|^2$ is plotted

simultaneous equations to be solved are as the following. The Schrödinger equation is given by

$$-\frac{\hbar^2}{2}\frac{\mathrm{d}}{\mathrm{d}z}\frac{1}{m^*(z)}\frac{\mathrm{d}}{\mathrm{d}z}\zeta_n(z) + V(z)\zeta_n(z) = \mathcal{E}_n\zeta_n(z)\,, \tag{8.36}$$

and Poisson's equation for the potential $\phi(z) = -(1/e)V(z)$ is written as

$$\frac{\mathrm{d}}{\mathrm{d}z}\kappa(z)\frac{\mathrm{d}}{\mathrm{d}z}\phi(z) = -\frac{1}{\epsilon_0}\left(\rho(z) + e\sum_n N_n|\zeta_n(z)|^2\right), \tag{8.37}$$

where the spatial dependence of the effective mass $m^*(z)$ and the dielectric constant $\epsilon(z)$ are taken into account. The simultaneous equations are solved in a similar fashion to the self-consistent calculations for the MOSFET. Here we will show the results for quantum wells of well width $L = 20$ nm and 50 nm in Figs. 8.8 and 8.9, respectively. In the calculations, the effective mass $m^* = 0.068\,m$ and the dielectric constant $\kappa = 12.9$ are assumed to be spatially uniform, for simplicity. The barrier height and electron sheet density are set to be $\Delta\mathcal{E}_c = 0.3\,\text{eV}$ and $N_s = 10^{12}\,\text{cm}^{-2}$, respectively and the solutions are obtained for $T = 4.2\,\text{K}$. The subband energy $\mathcal{E}_n$, Fermi energy $\mathcal{E}_F$ and electron sheet density N_n in each subband obtained from the self-consistent calculations are summarized in Table 8.2.

We have already mentioned that the modulation doping will improve the electron mobility. Mimura has been involved with the development of the high frequency GaAs MESFET (**Me**tal **S**emiconductor **F**ield **E**ffect **T**ransistor) and arrived at the conclusion that the surface states of GaAs play the most important role in its performance. During the process to find a method for improvement, he got the idea

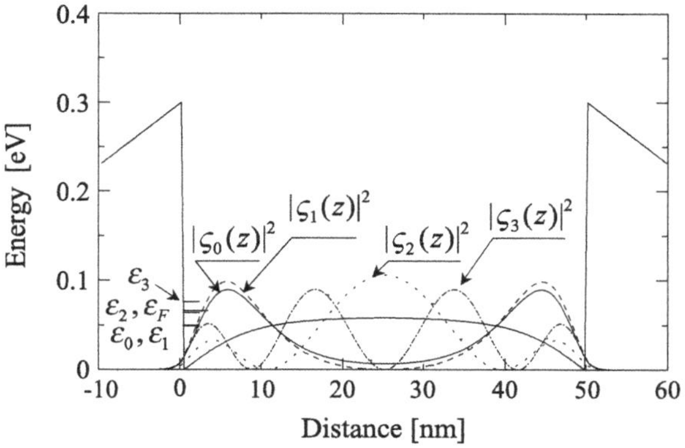

Fig. 8.9 Self-consistent solutions for a two-dimensional electron gas in an $Al_{0.3}Ga_{0.7}As$/GaAs modulation-doped quantum well with well width 50 nm. The results are at $T = 4.2$ K for electron sheet density $N_s = 10^{12}\,cm^{-2}$ and spacer layer of 10 nm in the AlGaAs barriers. The subband energies and Fermi energy are indicated by the horizontal lines and the square of the wave function $|\zeta_n(z)|^2$ is plotted

Table 8.2 Fermi energy, subband energies and electron sheet density in each subband obtained from self-consistent calculations for $Al_{0.3}Ga_{0.7}As$/GaAs quantum wells with well width 20 nm and 50 nm. The parameters used are $m^* = 0.068\,m$, $\epsilon = 12.9$, $N_s = 10^{12}\,cm^{-2}$, and $T = 4.2$ K

	$L = 20$ nm		$L = 50$ nm	
n	$\mathcal{E}_n$ (eV)	N_n ($10^{11}cm^{-2}$)	$\mathcal{E}_n$ (eV)	N_n ($10^{11}cm^{-2}$)
0	0.042	8.3	0.049	4.8
1	0.065	1.7	0.050	4.7
2	0.116	0.0	0.064	0.55
$\mathcal{E}_F$	0.071		0.066	

of using an AlGaAs/GaAs heterostructure. With the help of Hiyamizu, they grew a heterostructure FET and confirmed its FET characteristics in 1980 [20]. They named the device the **High Electron Mobility Transistor** (HEMT). The structure of the HEMT is shown in Fig. 8.10a. High purity GaAs is grown on a GaAs substrate, and an $Al_xGa_{1-x}As$ ($x \cong 0.3$) layer of 60–100 Å thickness is then grown, followed by the growth of an $Al_xGa_{1-x}As$ layer with Si donors. Usually a cap layer of the GaAs is grown on the surface of AlGaAs in order to avoid oxidation of the AlGaAs. Metals such as AuGeNi are deposited on the GaAs surface and the ohmic contacts of the source and drain are formed by thermal diffusion. The gate electrode is fabricated by the deposition of Al metal between the source and drain contacts, which controls the sheet density of the two-dimensional electron gas by the gate voltage. The energy band diagram near the region of the AlGaAs/GaAs interface is shown in Fig. 8.10b, where the ionized donors in the AlGaAs layer and the two-dimensional electron gas in the GaAs layer are separated. The electron mobility and sheet electron density of such a HEMT structure are plotted as a function of temperature in Fig. 8.11. The elec-

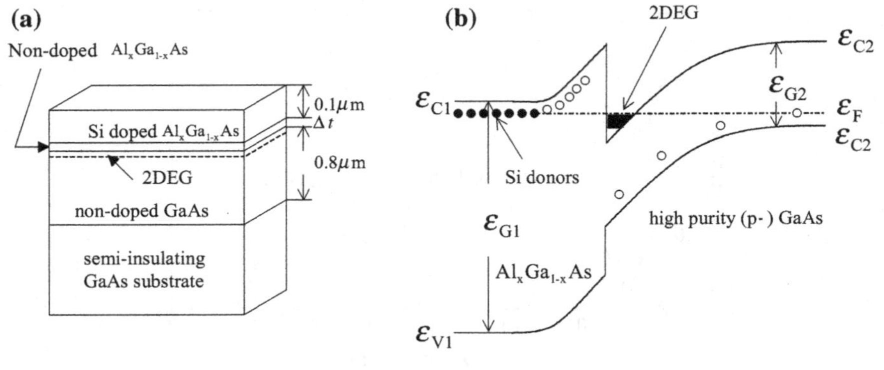

Fig. 8.10 **a** Structure of HEMT, and **b** its energy band diagram (see [21])

Fig. 8.11 Temperature dependence of the electron sheet density and electron mobility in a HEMT, where △ and ○ are the electron sheet density and electron mobility under illumination, respectively. (After [21])

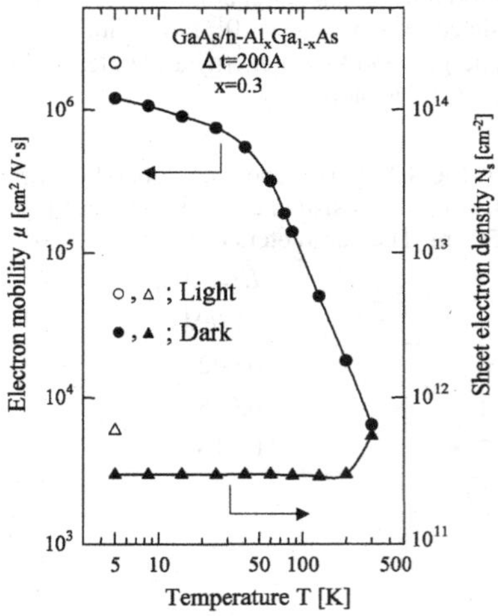

tron sheet density is almost constant in the temperature region shown, whereas the electron mobility becomes very high at low temperature because of the reduction of impurity scattering [21]. Later, improvements in the crystal growth and the structure led to a high electron mobility greater than $1 \times 10^7\,\text{cm}^{-2}/\text{Vs}$. In Fig. 8.11, the effect of light illumination on the electronic characteristics is shown, causing an increase in the electron sheet density and electron mobility as indicated by the symbols △ and ○, respectively [21].

An example of a self-consistent calculation of the two-dimensional electron gas in a HEMT at $T = 4.2\,\text{K}$ is shown in Fig. 8.12, where we have assumed $N_s = 6 \times 10^{11}\,\text{cm}^{-2}$ and $\Delta t = 10\,\text{nm}$. The Fermi energy is $\mathcal{E}_F = 0.085\,\text{eV}$ and all the electrons occupy only the lowest subband below the Fermi energy, resulting in a degenerate condition. These electrons are confined in a narrow region about 15 nm thickness at the interface and thus the reason for the two-dimensionality of the electrons is clear. In addition, the electrons are well separated from the ionized donors in the

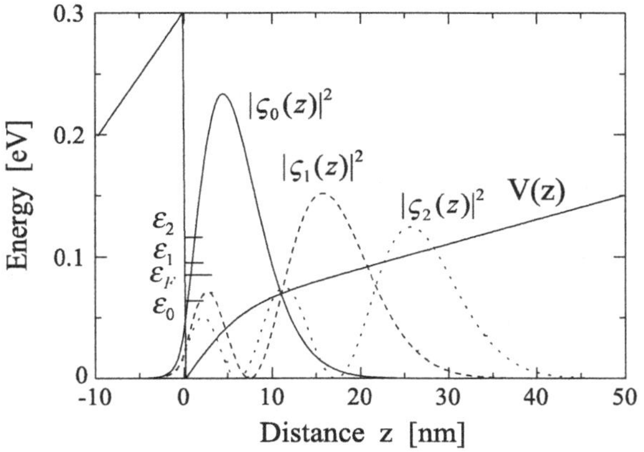

Fig. 8.12 Self-consistent solutions for a two-dimensional electron gas in a HEMT fabricated from $Al_xGa_{1-x}As/GaAs$ ($x = 0.3$), where $N_s = 6 \times 10^{11}\,\mathrm{cm}^{-2}$, $T = 4.2\,\mathrm{K}$, $\Delta\mathcal{E}_c = 0.3\,\mathrm{eV}$, spacer layer thickness $\Delta t = 10\,\mathrm{nm}$, and $m^* = 0.068\,m$. The bottom of the conduction band is plotted by the solid curve and the squared wave functions for each subband are shown. The values obtained are $\mathcal{E}_F = 0.085$, $\mathcal{E}_0 = 0.064$, $\mathcal{E}_1 = 0.095$, $\mathcal{E}_2 = 0.115\,\mathrm{eV}$, $N_0 = 6 \times 10^{11}\,\mathrm{cm}^{-2}$, and $N_1, N_2 \approx 0$

AlGaAs layer and the impurity scattering is reduced. We have to note that impurity scattering is a long-range force interaction and thus remote impurity scattering plays an important role in the mobility at low temperatures.

8.3 Transport Phenomena of Two-Dimensional Electron Gas

8.3.1 Fundamental Equations

In this section we will consider transport phenomena in a two-dimensional electron gas; the theoretical calculations are based on references [2–5]. After these original works many papers have been published, and more detailed calculations are found in [22, 23] (see also references therein). The electrons in the conduction bands of GaAs and AlAs are treated by the effective mass approximation and the wave function of the bulk state is expressed as

$$\Psi = V^{-1/2} \sum_{\boldsymbol{k}} C_{\boldsymbol{k}} \exp(\mathrm{i}\boldsymbol{k} \cdot \boldsymbol{r}) , \tag{8.38}$$

where Ψ is normalized by the volume V and $\sum |C_{\boldsymbol{k}}|^2 = 1$. The band-edge energy is given by

$$\mathcal{E}(\boldsymbol{k}) = \frac{\hbar^2 k^2}{2m^*} . \tag{8.39}$$

In the following we assume that the effective mass approximation is valid for heterostructures. When electrons are confined in the z direction, the wave function of the two-dimensional electron gas is given by (8.3) and written as

$$\Psi = \zeta(z) A^{-1/2} \exp(\mathrm{i}\boldsymbol{k}_\parallel \cdot \boldsymbol{r}_\parallel) , \tag{8.40}$$

where A is the area of the x, y plane, L is the width of the z direction, and the volume is given by $V = AL$. The normalization condition is written as

$$\int |\zeta(z)|^2 \mathrm{d}z = 1 . \tag{8.41}$$

Writing the subband energy as $\mathcal{E}_n$, we have shown that the electron energy in the quantum well of the heterostructure is defined by

$$\mathcal{E}(\boldsymbol{k}) = \mathcal{E}_n + \mathcal{E}(\boldsymbol{k}_\parallel) , \tag{8.42}$$

where $\boldsymbol{k}_\parallel$ is the wave vector parallel to the heterointerface given by

$$k_\parallel^2 = k_x^2 + k_y^2 \tag{8.43}$$

and the energy in the plane parallel to the heterointerface is given by

$$\mathcal{E}(k_\parallel) = \frac{\hbar^2}{2m^*}(k_x^2 + k_y^2) . \tag{8.44}$$

Next, we assume infinite barriers and a well width of W for simplicity. Under this assumption, the wave function and energy are given by the following relations as shown in Sect. 8.2.2:

$$\zeta_n(z) = \sqrt{\frac{2}{W}} \sin\left(\frac{n\pi}{W} z\right) , \tag{8.45}$$

$$\mathcal{E}_n = \frac{\hbar^2}{2m^*}\left(\frac{\pi}{W}\right)^2 n^2 . \tag{8.46}$$

In the following we show the calculations based on (8.45). It is very easy to perform numerical calculations with the use of self-consistent solutions. First, we present the fundamental equations derived by Price [2].

Phonons are assumed to be three dimensional and their wave vector is defined by $\boldsymbol{q}^2 = \boldsymbol{q}_\parallel^2 + q_z^2$ for the purpose of the calculations. The wave vector of an electron is changed by $\boldsymbol{q}_\parallel$ after phonon emission or absorption, but such a conservation rule does not hold in the z direction. Instead, the following integral results from the matrix element:

$$I_{mn}(q_z) = \int_0^W \zeta_m(z)^* \zeta_n(z) \exp(\mathrm{i}q_z z)\mathrm{d}z\,, \tag{8.47}$$

where $m = n$ corresponds to an intra-subband transition and $m \neq n$ corresponds to an inter-subband transition. From the conditions of normalization and orthogonality, the integral has the property of a δ-function, i.e.

$$I_{mn}(0) = \begin{cases} 1 & (\text{for } m = n) \\ 0 & (\text{for } m \neq n)\,. \end{cases} \tag{8.48}$$

Next, we will present some important relations for the calculations of the scattering probability and the numerical results.

$$\int_{-\infty}^{\infty} |I_{mn}(q_z)|^2 \,\mathrm{d}q_z = 2\pi \int_0^W \Phi_{mn}^2 \mathrm{d}z\,, \tag{8.49}$$

where

$$\Phi_{mn}(z) = \zeta_m^*(z)\zeta_n(z)\,. \tag{8.50}$$

It is very convenient to introduce the following coefficient:

$$\frac{1}{b_{mn}} = 2\int_0^W \Phi_{mn}^2 \mathrm{d}z\,. \tag{8.51}$$

Using this relation we may obtain the following relation

$$\int_{-\infty}^{\infty} |I_{nn}|^2 \,\mathrm{d}q_z = \frac{\pi}{b_{nn}}\,. \tag{8.52}$$

In a similar fashion

$$\int_{-\infty}^{\infty} |I_{mn}|^2 \,\mathrm{d}q_z = \frac{\pi}{b_{mn}}\,. \tag{8.53}$$

Using (8.45) for the wave function $\zeta_n(z)$, we find

$$b_{nn} = \frac{W}{3}\,, \tag{8.54}$$

$$b_{mn} = \frac{W}{2}\,, \quad (n \neq m) \tag{8.55}$$

which are independent of the subband indices m and n. The above results indicate that the coefficient b_{mn} is different for intra-subband transitions ($m = n$) and inter-subband transitions ($m \neq n$). It may be understood that the value $1/b$ gives a measure of the decay of $|I|^2$ with q. When we use the wave function given by (8.45), $I_{mn}(q)$ is written as

$$I_{nn}(q) = \frac{\sin(\frac{1}{2}Wq)}{\frac{1}{2}Wq} \frac{n^2}{n^2 - (Wq/2\pi)^2} P\,, \tag{8.56}$$

$$I_{mn}(q) = \frac{{}^{\sin}_{\cos}(\frac{1}{2}Wq)}{\frac{1}{2}Wq} \frac{4mn(Wq/\pi)^2}{4m^2n^2 - \left[m^2 + n^2 - (Wq/\pi)^2\right]^2} P\,, \tag{8.57}$$

where sin() in the symbol ${}^{\sin}_{\cos}()$ appearing on the right-hand side of $I_{mn}(q)$ is for the case that both m and n are even or odd, and cos() is for the case that one of the m and n is even and the other is odd, and P is the phase factor such that $|P| = 1$ with phase angle $\pm\frac{1}{2}Wq$.

8.3.2 Scattering Rate

8.3.2.1 (a) Acoustic Phonon Scattering and Non-polar Optical Phonon Scattering

Detailed discussion has been given of the matrix elements of electron scattering in the case of three-dimensional (bulk) semiconductors in Sect. 6.3. Here, we will calculate the scattering rates for a two-dimensional electron gas using the result for the three-dimensional case. Therefore, we first summarize the results for the three-dimensional case. Letting the electron–phonon interaction Hamiltonian be H_1 and using (6.156), the scattering matrix element is written as

$$\langle \boldsymbol{k}'|H_1^{\pm}|\boldsymbol{k}\rangle = \left(n(\boldsymbol{q}) + \frac{1}{2} \pm \frac{1}{2}\right)^{1/2} \delta^{\pm}_{\boldsymbol{k},\boldsymbol{k}'\pm\boldsymbol{q}} C_{\boldsymbol{q}}\,, \tag{8.58}$$

where the upper (lower) symbol of $\pm$ corresponds to phonon emission (absorption) of a phonon with wave vector $\boldsymbol{q}$ and energy $\hbar\omega$, $\boldsymbol{k}$ and $\boldsymbol{k}'$ represent the electron initial and final states, respectively, and $n(\boldsymbol{q})$ is the excitation number of phonons, which is given by the Bose–Einstein distribution function. The wave vector $\boldsymbol{q}$ is expressed as

$$\pm\boldsymbol{q} = \boldsymbol{k} - \boldsymbol{k}'\,. \tag{8.59}$$

The scattering rate is derived as follows from the results stated in Sect. 6.4:

$$\begin{aligned} w_{\mathrm{III}}(\boldsymbol{k}, \boldsymbol{k}') &= w_{\mathrm{III}}^{+} + w_{\mathrm{III}}^{-} \\ &= \frac{2\pi}{\hbar} \sum_{\boldsymbol{k}'} \left|C_{\boldsymbol{q}}\right|^2 \left\{ n(\boldsymbol{q}) \delta\left[\mathcal{E}(\boldsymbol{k}) - \mathcal{E}(\boldsymbol{k}') + \hbar\omega\right] \right. \\ &\quad \left. + \left[(n(\boldsymbol{q}) + 1)\right] \delta\left[\mathcal{E}(\boldsymbol{k}) - \mathcal{E}(\boldsymbol{k}') - \hbar\omega\right] \right\}\,. \end{aligned} \tag{8.60}$$

Noting that electron spin is conserved in scattering, and using (8.59) and replacing the summation of the final states with an integral, we find

$$\sum_{\boldsymbol{k}'} = \frac{1}{(2\pi)^3}\int \mathrm{d}^3\boldsymbol{k}' = \frac{1}{(2\pi)^3}\int \mathrm{d}^3\boldsymbol{q}\,. \tag{8.61}$$

When the excitation number of phonons is approximated as $n(\boldsymbol{q}) \sim n(\boldsymbol{q}) + 1 \sim k_B T/\hbar\omega$, the scattering rate reduces to

$$w_{\mathrm{III}} = \frac{2\pi}{\hbar}\frac{2}{(2\pi)^3} S_{\mathrm{III}} \int \mathrm{d}^3\boldsymbol{k}'\delta\left(\mathcal{E}(\boldsymbol{k}) - \mathcal{E}(\boldsymbol{k}')\right)\,, \tag{8.62}$$

where

$$S_{\mathrm{III}} = \frac{k_{\mathrm{B}}T}{\hbar\omega}|C|^2 = \frac{k_{\mathrm{B}}T D_{\mathrm{ac}}^2}{2\rho v_{\mathrm{s}}^2} \tag{8.63}$$

and the integral, which includes the δ-function, gives rise to the three-dimensional density of states

$$\begin{aligned} g_{\mathrm{III}} &= \frac{1}{(2\pi)^3}\int \delta\left(\mathcal{E}(\boldsymbol{k}) - \mathcal{E}(\boldsymbol{k}')\right) d^3\boldsymbol{k}' = \frac{1}{(2\pi)^3} 4\pi k^2 \frac{\mathrm{d}k}{\mathrm{d}\mathcal{E}} \\ &= \frac{1}{(2\pi)^3}\left[\frac{2\pi(2m^*)^{3/2}}{\hbar^3}\right]\mathcal{E}^{1/2}\,. \end{aligned} \tag{8.64}$$

The result means that the number of states in the energy range $\mathcal{E}$ and $\mathcal{E} + d\mathcal{E}$ per unit volume is given by $g_{\mathrm{III}} d\mathcal{E}$, where the spin flip transition is disregarded and the spin degeneracy is set to be 1.

In scattering in a two-dimensional electron gas, the spatial integration of the Bloch function (8.40) is divided into integrals for the perpendicular direction $\mathrm{d}z$ to the heterointerface and for the parallel direction $\mathrm{d}\boldsymbol{r}_{\parallel}$. Therefore, two terms, $I_{mn}(q_z)$ defined in Sect. 8.3.1 and $\sum_{\boldsymbol{k}'} \delta[\mathcal{E}(\boldsymbol{k}_{\parallel}) + \mathcal{E}_m - \mathcal{E}(\boldsymbol{k}'_{\parallel}) - \mathcal{E}_n \pm \hbar\omega]$, appear in the matrix element, where $\boldsymbol{k}_{\parallel}$ ($\boldsymbol{q}_{\parallel}$) and $\boldsymbol{k}'_{\parallel}$ ($\boldsymbol{q}'_{\parallel}$) represent the wave vectors parallel to the heterointerface. Replacing the summation of the final states by $(1/(2\pi)^3)\mathrm{d}^2\boldsymbol{q}_{\parallel}\mathrm{d}q_z$ and integral with respect to $\boldsymbol{q}_{\parallel}$ by $\mathrm{d}^2\boldsymbol{k}_{\parallel}$ using the property of the δ-function, we obtain the following results:

$$\begin{aligned} w_{\mathrm{II}} &= \frac{2\pi}{\hbar}\frac{1}{(2\pi)^3}\int |I_{mn}(q_z)|^2 \mathrm{d}q_z \\ &\times \int |C|^2 \left\{n(\boldsymbol{q})\delta\left[\mathcal{E}(\boldsymbol{k}_{\parallel}) - \mathcal{E}(\boldsymbol{k}'_{\parallel}) + \Delta\mathcal{E}_{mn} + \hbar\omega\right]\right. \\ &\left. + \left[(n(\boldsymbol{q}) + 1)\right]\delta\left[\mathcal{E}(\boldsymbol{k}_{\parallel}) - \mathcal{E}(\boldsymbol{k}'_{\parallel}) + \Delta\mathcal{E}_{mn} - \hbar\omega\right]\right\} \mathrm{d}^2\boldsymbol{k}'_{\parallel}\,, \end{aligned} \tag{8.65}$$

where

$$\Delta\mathcal{E}_{mn} = \mathcal{E}_m - \mathcal{E}_n = -\Delta\mathcal{E}_{nm}\,. \tag{8.66}$$

Adopting the same assumption as for the three-dimensional case, it may be evident that we have the result

$$S_{\mathrm{II}} = \int S_{\mathrm{III}}\, |I_{mn}(q_z)|^2 \, \mathrm{d}q_z \, . \tag{8.67}$$

Since the two-dimensional density of states is given by

$$g_{\mathrm{II}}\mathrm{d}\mathcal{E} = \int \delta\left[\mathcal{E}(\boldsymbol{k}_{\parallel}) - \mathcal{E}(\boldsymbol{k}'_{\parallel})\right] \mathrm{d}^2\boldsymbol{k}'_{\parallel} = \frac{2}{(2\pi)^2} 2\pi k_{\parallel} \mathrm{d}k_{\parallel} = \frac{m^*}{\pi\hbar^2}\mathrm{d}\mathcal{E} \, , \tag{8.68}$$

we have

$$w_{\mathrm{II}} = \frac{1}{b}\frac{m^*}{\hbar^3}\frac{k_{\mathrm{B}}T D_{\mathrm{ac}}^2}{2\rho v_{\mathrm{s}}^2}\, u(\mathcal{E}(\boldsymbol{k}_{\parallel}) - \Delta\mathcal{E}_{nm}) \, , \tag{8.69}$$

where $u(x)$ is the step function which satisfies $u(x \geq 0) = 1$ and $u(x < 0) = 0$. Here we discuss b in detail. In the case of intra-subband scattering within the nth subband, $1/b$ may be written as $1/b = 3/W$, and thus this relation is used for scattering within the ground-state subband. In the case of inter-subband scattering, on the other hand, we have $1/b = 2/W$, and thus the effective value of b is given by the following relation when the electron with energy $\mathcal{E}$ in the range $\mathcal{E}_n < \mathcal{E}_{n+1}$ is scattered into the mth subband by acoustic phonons:

$$g_{\mathrm{II}} \cdot \frac{\pi}{b} = g_{\mathrm{II}}\left[\frac{\pi}{b_{mm}} + \sum_j \frac{\pi}{b_{mj}}\right] = g_{\mathrm{II}}\left[\frac{3\pi}{W} + (n-1)\frac{2\pi}{W}\right] , \tag{8.70}$$

where $m \leq n$. For a large value of n, the prefactor of g_{II} in the above equation reduces to $\cong (2m^*\mathcal{E})^{1/2} 2/\hbar = g_{\mathrm{III}}/g_{\mathrm{II}}$ and (8.70) is approximated as $\cong g_{\mathrm{III}}$, giving rise to three-dimensional scattering.

A similar treatment is possible for non-polar optical phonon scattering and the scattering rate for a two-dimensional electron gas is written as

$$w_{\mathrm{op}} = \frac{1}{b}\frac{D_{\mathrm{op}}^2 m^*}{4\rho\hbar^2\omega_0}\left[n(\omega_0) + \frac{1}{2} \pm \frac{1}{2}\right] u(\mathcal{E}(\boldsymbol{k}_{\parallel}) \mp \hbar\omega_0 - \Delta\mathcal{E}_{nm}) \, . \tag{8.71}$$

Here we introduce new parameter

$$x_0 = \frac{\hbar\omega_0}{k_{\mathrm{B}}T} , \quad n(x_0) = \frac{1}{\mathrm{e}^{x_0} - 1} , \tag{8.72}$$

and we find that

$$N_{\rm opnm} = \left[n(x_0) + \frac{1}{2} \pm \frac{1}{2}\right] u(\mathcal{E} \mp \hbar\omega_0 - \Delta\mathcal{E}_{nm}) \\ = \frac{1}{{\rm e}^{x_0} - 1}\left[1 + {\rm e}^{x_0} u(x - x_0 - x_{nm})\right] , \tag{8.73}$$

and for intra–subband transitions, putting $\Delta\mathcal{E}_{nm} = 0$ we obtain

$$N_{\rm op} = \frac{1}{{\rm e}^{x_0} - 1}\left[1 + {\rm e}^{x_0} u(x - x_0)\right] , \tag{8.74}$$

where we used $x_{mn} = \Delta\mathcal{E}_{nm}/\hbar\omega_0$. Therefore (8.71) is rewritten as

$$w_{\rm op} = \frac{1}{b}\frac{D_{\rm op}^2 m^*}{4\rho\hbar^2\omega_0} N_{\rm opmn} , \quad w_{\rm op} = \frac{1}{b}\frac{D_{\rm op}^2 m^*}{4\rho\hbar^2\omega_0} N_{\rm op} \ ({\rm for}\, x_{nm} = 0) . \tag{8.75}$$

See evaluation of the electron mobility for Problems and Answers of this Chapter.

8.3.2.2 (b) Inter-Valley Phonon Scattering

In many-valley conduction bands such as in Si, electrons are scattered between the valleys, absorbing or emitting a large wave vector phonon. This inter-valley phonon scattering is described in detail in Sect. 6.3.7 for the case of bulk semiconductors and is well interpreted in terms of deformation potential theory. Defining the involved phonon angular frequency as ω_{ij} and the deformation potential as D_{ij}, the scattering rate $w_{\rm int} = 1/\tau_{\rm int}$ from the initial state of an electron in valley i, subband m and wave vector $\boldsymbol{k}_\parallel$ to the final sate in valley j, subband n and wave vector $\boldsymbol{k}'_\parallel$ is calculated as follows.

The strength of the interaction for inter-valley phonon scattering may be expressed as

$$C_q = \sqrt{\frac{\hbar}{2\rho\omega_{ij}}} D_{ij} . \tag{8.76}$$

The energies of the initial and final states of the electron are respectively written as

$$\mathcal{E}_{im}(\boldsymbol{k}_\parallel) = \frac{\hbar^2 \boldsymbol{k}_\parallel^2}{2m_i^*} + \mathcal{E}_{im} , \tag{8.77}$$

$$\mathcal{E}_{jn}(\boldsymbol{k}_\parallel) = \frac{\hbar^2 \boldsymbol{k}_\parallel'^2}{2m_j^*} + \mathcal{E}_{jn} , \tag{8.78}$$

where m_i^* and m_j^* are the isotropic effective masses of the valley i and j, and in the presence of anisotropy the final results may be obtained by replacing the effective masses with the density of states masses. Note that $\mathcal{E}_{im}$ and $\mathcal{E}_{jn}$ are the energies of the bottom of the subbands m and n. Using these results the scattering rate is calculated as

$$w_{\rm int} = \frac{2\pi}{\hbar}\frac{\hbar D_{ij}^2}{2\rho\omega_{ij}}\frac{\nu_j}{(2\pi)^2}\int_{\boldsymbol{k}'_\parallel}\frac{1}{2\pi}\int_{q_z}|I_{mn}(q_z)|^2 \times\delta\left[\mathcal{E}_{jn}(\boldsymbol{k}'_\parallel) - \mathcal{E}_{im}(\boldsymbol{k}_\parallel) \pm \hbar\omega_{ij}\right]\left(n_q + \frac{1}{2} \mp \frac{1}{2}\right)\mathrm{d}q_z\mathrm{d}^2\boldsymbol{k}'_\parallel\,, \tag{8.79}$$

where ν_j is the valley degeneracy of the final state. When we put

$$D = \frac{m_j^*}{m_i^*}\boldsymbol{k}_\parallel^2 + \frac{2m_j^*}{\hbar^2}\left(\mathcal{E}_{im} - \mathcal{E}_{jn} \mp \hbar\omega_{ij}\right)\,, \tag{8.80}$$

the integral of (8.79) with the δ-function reduces to

$$\int_{\boldsymbol{k}'_\parallel}\delta\left[\frac{\hbar^2}{2m_j^*}(k_\parallel'^2 - D)\right]\mathrm{d}^2\boldsymbol{k}'_\parallel = \int_0^\infty \delta\left[\frac{\hbar^2}{2m_j^*}(k_\parallel'^2 - D)\right]2\pi k'_\parallel \mathrm{d}k'_\parallel = \frac{2\pi m_j^*}{\hbar^2}u(D)\,. \tag{8.81}$$

Since the integral with respect to q_z is the same as for the case of acoustic phonon scattering, the inter-valley phonon scattering rate is given by

$$w_{\rm int} = \frac{\nu_j D_{ij}^2 m_j^*}{2\rho\hbar^2\omega_{ij}}\cdot\frac{1}{2b_{mn}}\left(n_q + \frac{1}{2} \mp \frac{1}{2}\right)\cdot u(\mathcal{E}(\boldsymbol{k}_\parallel) \mp \hbar\omega_{ij} - \Delta\mathcal{E}_{nm}) = \frac{\nu_j D_{ij}^2 m_j^*}{2\rho\hbar^2\omega_{ij}}\cdot\frac{1}{2b_{mn}}N_{ij}\,. \tag{8.82}$$

where N_{ij} is the excited number of the intervalley phonons and given by (8.73) replacing $\omega_{\rm op}$ with ω_{ij}.

8.3.2.3 (c) Polar Optical Phonon Scattering

Next, we will discuss polar optical phonon scattering. Since the matrix element for polar optical phonon scattering is proportional to the inverse of the phonon wave vector, q^{-1}, the previous treatments cannot be applied. From (6.239) derived in Sect. 6.3.6, we obtain the electron–polar optical phonon interaction

$$|C_{\rm pop}(\boldsymbol{q})|^2 = \alpha(\hbar\omega_{\rm LO})^{3/2}\frac{1}{L^3}\left(\frac{\hbar^2}{2m^*}\right)^{1/2}\frac{1}{q^2}\,, \tag{8.83}$$

where α is dimensionless quantity and is given by (6.240) and $L^3 = V$ is the volume of the three-dimensional crystal. In the present case we are using the normalized wave function given by (8.40), and thus we can disregard the factor L^3. Therefore we rewrite it as

$$|C_{\rm pop}(\boldsymbol{q})|^2 = \alpha(\hbar\omega_{\rm LO})^{3/2}\left(\frac{\hbar^2}{2m^*}\right)^{1/2}\frac{1}{q^2} = \frac{e^2\hbar\omega_0}{2\epsilon_0}\left(\frac{1}{\kappa_\infty} - \frac{1}{\kappa_0}\right)\frac{1}{q^2}, \tag{8.84}$$

and calculate the scattering rate as follows. Separating the phonon wave vector $\boldsymbol{q}$ into the component perpendicular to the heterointerface, q_z, and the component parallel to the interface, $\boldsymbol{Q} = \pm(\boldsymbol{k}_\parallel - \boldsymbol{k}'_\parallel)$, we obtain

$$J_{mn}(Q) \equiv \int_{-\infty}^{+\infty}\frac{|I_{mn}(q_z)|^2}{q_z^2 + Q^2}\mathrm{d}q_z\,. \tag{8.85}$$

Then the scattering rate for a two-dimensional electron gas for polar optical phonon scattering is given by

$$w_{\rm pop} = \int\frac{e^2\omega_0}{2(2\pi)^2\epsilon_0}\left(\frac{1}{\kappa_\infty} - \frac{1}{\kappa_0}\right)J_{mn}(Q)\mathrm{d}^2\boldsymbol{k}'_\parallel \times\left\{\left(n(\omega_0) + \frac{1}{2} \pm \frac{1}{2}\right)\delta\left[\mathcal{E}(\boldsymbol{k}_\parallel) - \mathcal{E}(\boldsymbol{k}'_\parallel) \mp \hbar\omega_0 - \Delta\mathcal{E}_{nm}\right]\right\}. \tag{8.86}$$

Equation (8.85) may be rewritten as the following by using (8.50):

$$J_{mn}(Q) = \frac{\pi}{Q}\int\int\mathrm{d}z_1\mathrm{d}z_2\Phi_{mn}(z_1)\Phi_{mn}(z_2)\exp(-Q|z_1 - z_2|)\,. \tag{8.87}$$

Following the treatment of Price [2], we discuss the following two extreme cases. As stated before, $|I_{mn}|^2$ decreases rapidly with increasing q. When the parameter Q is smaller than the critical value of q, $|I_{nn}(q_z)|^2$ may be replaced with the value at $q_z = 0$ and put outside the integral, giving rise to the approximate result

$$J_{nn}(0) \simeq \frac{\pi}{Q} \quad \text{(for small } Q)\,. \tag{8.88}$$

The value of $J_{mn}(0)$ may be approximated by expanding the exponential term as $1 - Q|z_1 - z_2| + \cdots$ to give

$$J_{mn}(0) \simeq \frac{2W}{\pi}\frac{m^2 + n^2}{(m^2 - n^2)^2} \quad \text{(for small } Q)\,, \tag{8.89}$$

where the wave function derived for a quantum well with infinite barrier height, (8.45), is used.

On the other hand, for large values of Q, $\sin(Wq_z/2)/(Wq_z/2)$ in $I_{mn}(q_z)$ behaves like a δ-function and it may be approximated as

$$J_{mn} \simeq \frac{\pi}{b_{mn}Q^2} \quad \text{(for large } Q\text{)} . \tag{8.90}$$

Using these results, $|I_{nn}(q_z)|^2$ and $J_{nn}(Q)$ which determine the strength of intra-subband scattering and $|I_{mn}(q_z)|^2$ and $J_{mn}(Q)$ $(m \neq n)$ which determine the strength of inter-subband scattering are calculated and plotted in Figs. 8.13 and 8.14 for several parameters, respectively. For the intra-subband scattering, $J_{11}(Q)$ and $J_{22}(Q)$ are almost the same in magnitude and therefore the former values are plotted in Fig. 8.13. The curves of $|I_{mn}(q_z)|^2$ represent intra-subband $(m = n)$ and inter-subband scattering $(m \neq n)$ for acoustic phonon scattering and non-polar optical phonon scattering. The curves of $J_{mn}(Q)$ represent the strength of the polar optical phonon scattering.

Here we will estimate the scattering rate for polar optical phonon following the treatment of Price [2]. It is evident from (8.86) that the scattering rate depends on $Q = |\boldsymbol{k}_\parallel - \boldsymbol{k}'_\parallel|$. We will consider the case where the inter-subband transition plays an important role and we will consider the temperature range

$$T < \frac{\hbar\omega_0}{k_B} \equiv T_0 . \tag{8.91}$$

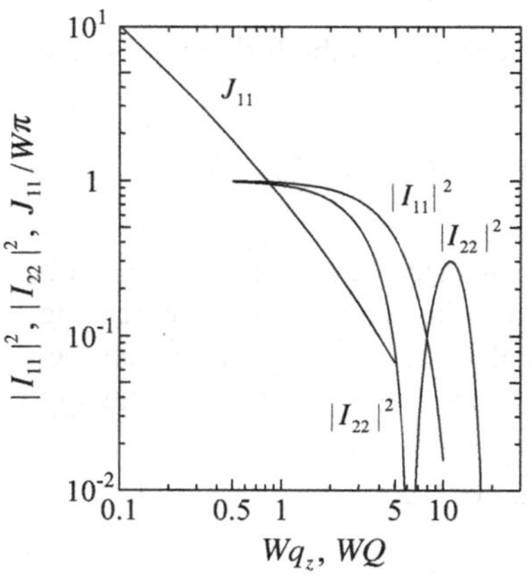

Fig. 8.13 Scattering strength plotted as a function of q_z and Q. The scattering strength for acoustic phonon scattering and non-polar optical phonon scattering is shown for the intra-subband in the ground subband $|I_{11}(q_z)|^2$ and in the second subband $|I_{22}(q_z)|^2$ as a function of Wq_z. For polar optical phonon scattering the scattering strength for the intra-subband scattering in the ground subband $J_{11}(Q)/WQ$ is shown as a function of WQ. The intra-subband scattering strength in the second subband $J_{22}(Q)/WQ$ is almost the same as $J_{11}(Q)/WQ$ and thus omitted in the figure

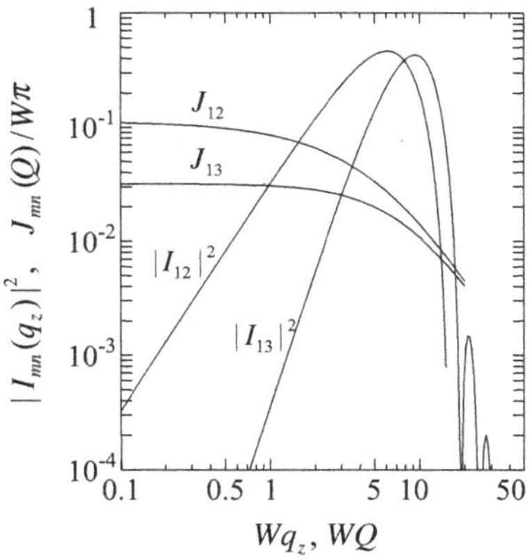

Fig. 8.14 Scattering strength for the inter-subband transition plotted as a function of q_z and Q. The scattering strength for the inter-subband transition between the ground subband and the second or the third subband for acoustic phonon scattering and non-polar optical phonon scattering, $|I_{12}(q_z)|^2$ and $|I_{13}(q_z)|^2$, are plotted as a function of Wq_z. The scattering strength for polar optical phonons is shown for the inter-subband transitions between the ground subband and the second or the third subband. See the plots of $J_{12}(Q)/WQ$ and $J_{13}(Q)/WQ$ versus WQ

Then the value of Q for the inter-subband transition is expected to be very close to the wave vector to satisfy the relation $\mathcal{E}(k_0) = \hbar\omega_0$ and thus to the value

$$k_0 = \left(\frac{2m^*\omega_0}{\hbar}\right)^{1/2} . \tag{8.92}$$

When $\Delta\mathcal{E}_{12}$ is larger than $\hbar\omega_0$ by a factor of several times $k_B T$, the inter-subband transitions can be ignored. Under such conditions, there are not enough electrons to emit phonons and thus phonon absorption dominates. Therefore, it may be possible to keep the term proportional to $n(\omega_0)$ in (8.86) and to replace $J_{11}(Q)$ with $J_{11}(k_0)$, giving rise to the scattering rate

$$w_{\text{pop}} \simeq J_{11}(k_0)k_0 \frac{e^2}{8\pi\hbar\epsilon_0}\left(\frac{1}{\kappa_\infty} - \frac{1}{\kappa_0}\right) \times \left(n(\omega_0) + \frac{1}{2} \mp \frac{1}{2}\right) u(\mathcal{E}(\boldsymbol{k}_\parallel) \mp \hbar\omega_0 - \Delta\mathcal{E}_{nm}) \tag{8.93}$$

$$= J_{11}(k_0)k_0 \frac{e^2}{8\pi\hbar\epsilon_0}\left(\frac{1}{\kappa_\infty} - \frac{1}{\kappa_0}\right) N_{\text{popnm}} , \tag{8.94}$$

where N_{popnm} for inter–subband scattering and N_{pop} for intra–subband scattering are given by (8.73) and (8.74) by replacing the phonon energy with polar optical phonon energy, respectively. From (8.46) and (8.92) we find the following relations:

$$\left(\frac{1}{3}Wk_0\right)^2 = \frac{\pi^2}{3}\frac{\hbar\omega_0}{\Delta\mathcal{E}_{12}}. \tag{8.95}$$

Since $b_{11} = (1/3)\,W$ as shown before, k_0 is replaced by Q and the left-hand side of (8.95) may be replaced by $(bQ)^2$. When the relation $\hbar\omega_0 < \Delta\mathcal{E}_{12}$ is satisfied, (8.95) leads to the result that bQ can take the values in the range from 0 to $\pi/\sqrt{3} = 1.814$ and the scattering rate given by (8.94) depends on Wk_0 through $J_{11}(k_0)$. Since $k_0 = 2.50\times 10^6\,\mathrm{cm}^{-1}$ in the case of GaAs, the scattering rate strongly depends on which parameter is larger, the well width of the two-dimensional electron gas W or $1/k_0$. Since $1/k_0 \simeq 4\,\mathrm{nm}$ and the well width of a typical two-dimensional electron device is 10–100 nm or more, we may expect the condition $Wk_0 > 1$. Then using (8.90), (8.94) reduces to

$$\begin{aligned} w_{\mathrm{pop}} &\simeq \frac{\pi}{b_{11}}\frac{e^2}{8\pi\hbar\epsilon_0}\left(\frac{1}{\kappa_\infty} - \frac{1}{\kappa_0}\right)N_{\mathrm{pop}} = \frac{3}{W}\frac{e^2}{8\hbar\epsilon_0}\left(\frac{1}{\kappa_\infty} - \frac{1}{\kappa_0}\right)N_{\mathrm{pop}} \\ &\sim 1.45\times 10^5 \times \frac{1}{W}N_{\mathrm{pop}}\left[\mathrm{s}^{-1}\right]. \end{aligned} \tag{8.96}$$

For example, in a quantum well of 10 nm well width, the phonon excitation number at $T = 300\,\mathrm{K}$ is $n(\omega_0) = 0.342$, which gives $w_{\mathrm{pop}} = 4.95\times 10^{12}\,\mathrm{s}^{-1}$, and $n(\omega_0) = 0.0168$ at $T = 100\,\mathrm{K}$, which gives $w_{\mathrm{pop}} = 2.43\times 10^{11}\,\mathrm{s}^{-1}$.

In a narrow quantum well, on the other hand, using (8.88), (8.94) reduces to

$$w_{\mathrm{pop}} = k_0\frac{e^2}{8\hbar\epsilon_0}\left(\frac{1}{\kappa_\infty} - \frac{1}{\kappa_0}\right)N_{\mathrm{pop}} \sim 1.20\times 10^{13}\times N_{\mathrm{pop}}\left[\mathrm{s}^{-1}\right]. \tag{8.97}$$

In a similar fashion to the wide well case, w_{pop} may be estimated as $4.12\times 10^{12}\,\mathrm{s}^{-1}$ at $T = 300\,\mathrm{K}$ and $1.85\times 10^{11}\,\mathrm{s}^{-1}$ at $T = 100\,\mathrm{K}$. From these results we may conclude that both approximations give almost the same magnitude for the scattering rate.

For deformation-potential type acoustic phonon scattering, the scattering rate in a quantum well with well width $W = 10\,\mathrm{nm}$ is estimated from (8.69) to be $1.16\times 10^{12}\,\mathrm{s}^{-1}$ at $T = 300\,\mathrm{K}$ and $0.202\times 10^{12}\,\mathrm{s}^{-1}$ at $T = 100\,\mathrm{K}$. Optical phonon scattering, polar and non-polar, is reduced very rapidly at lower temperatures because of the low excitation number of phonons involved. Therefore, in the lower temperature range ($T < 100\,\mathrm{K}$), acoustic phonon scattering dominates, but at higher temperatures optical phonon scattering plays an important role. See Problems and Answers of this Chapter for evaluation of electron mobility of the respective model.

8.3.2.4 (d) Piezoelectric Potential Scattering

Many III–V compound semiconductors belong to the zinc blende crystal structure, and piezoelectric field waves are excited along with the acoustic phonon propagation. These piezoelectric potential waves induce electron scattering and the scattering processes in the three-dimensional case is described in detail in Sect. 6.3.4. Here the lattice vibrations are dealt with as three-dimensional phonons to derive the scattering rate in a two-dimensional electron gas for piezoelectric potential scattering. Using

(6.197) we may write

$$|C_{\rm pz}(\boldsymbol{q})|^2 = \left(\frac{e \cdot e^*_{\rm pz}}{\epsilon^*}\right)^2 \frac{\hbar}{2MN\omega_q}, \tag{8.98}$$

where $MN = \rho$ is the crystal density, $e^*_{\rm pz}$ is the effective piezoelectric constant, and $\epsilon = \kappa\epsilon_0$. Using the relation $\omega_q = \boldsymbol{v}_{\rm s} \cdot \boldsymbol{q}$ and approximating as $n(\omega_q) = k_{\rm B}T/\hbar\omega_q$ (assumption of equipartition rule), and from the comparison with the deformation potential type acoustic phonon scattering, we obtain the relation

$$D^2_{\rm ac} \to \left(\frac{e \cdot e_{14}}{\epsilon}\right)^2 \frac{A}{Q^2 + q_z^2}, \tag{8.99}$$

where $A(\boldsymbol{Q}, q_z)$ is a dimensionless parameter of anisotropy which depends on the phonon propagation direction. In the case of acoustic phonon scattering for intra-subband transitions, the associated phonon wave vector Q is small and all the terms except $|I_{nn}|^2$ in the integral decrease rapidly, enabling us to approximate $|I_{nn}|^2 = 1$. Then we have

$$D^2_{\rm ac}\frac{\pi}{b} \to \left(\frac{e \cdot e_{14}}{\epsilon}\right)^2 \frac{\pi}{W} B, \tag{8.100}$$

where

$$B = \frac{Q}{\pi}\int_{\infty}^{\infty} \frac{A}{Q^2 + q_z^2} \mathrm{d}q_z. \tag{8.101}$$

Since transverse and longitudinal acoustic waves contribute to the piezoelectric scattering, defining the respective term as $B_{\rm t}$ and $B_{\rm l}$, we may rewrite

$$S_{\rm II} = \frac{k_{\rm B}T}{\epsilon^2 Q}\frac{\pi}{2}\left[\frac{(e \cdot e_{\rm l})^2 B_{\rm l}}{\rho v_{\rm l}^2} + 2\frac{(e \cdot e_{\rm t})^2 B_{\rm t}}{\rho v_{\rm t}^2}\right], \tag{8.102}$$

where $e_{\rm l}$ and $e_{\rm t}$ are the piezoelectric constants for longitudinal and transverse acoustic waves, respectively, and in the case of GaAs, for example, $e_{\rm t} = e_{14}$ and $e_{\rm l} = 0$. The piezoelectric constant e_{14} of GaAs is known to be $-0.160\,{\rm C/m^2}$. According to the calculations of Price [2] the coefficients $B_{\rm l}$ and $B_{\rm t}$ are

$$B_{\rm l} = \frac{9}{32}, \quad 2B_{\rm t} = \frac{13}{32}. \tag{8.103}$$

Subband transitions due to the piezoelectric potential may be treated as elastic scattering and we have the following relation between Q and $\boldsymbol{k}_{\parallel}$:

$$Q^2 = 2(1 - \cos\theta)k_{\parallel}^2, \tag{8.104}$$

where θ is the angle between the electron wave vectors $\boldsymbol{k}_\parallel$ of the initial and final states. The scattering rate in a two-dimensional electron gas due to the piezoelectric potential is given by

$$w_{\text{piez}} = \frac{m^*}{\pi\hbar^3}\langle S_{\text{II}}\rangle\,, \tag{8.105}$$

where $\langle\,\rangle$ is the average over the angle θ. The relaxation time which determines the electron mobility is obtained from the scattering rate multiplied by $(1 - \cos\theta)$ and is averaged as shown in Sect. 6.2. Therefore, the relaxation time is given by

$$\frac{1}{\tau} = \frac{m^*}{\pi\hbar^3}\langle(1-\cos\theta)S_{\text{II}}\rangle\,. \tag{8.106}$$

As seen in (8.102), the scattering factor for the piezoelectric potential S_{II} is proportional to $1/Q$. Using the relation of (8.104), the average of (8.106) results in

$$\left\langle \frac{k_\parallel}{Q}(1-\cos\theta)\right\rangle = \frac{2}{\pi}\,. \tag{8.107}$$

Therefore, the relaxation time of the electron in the subband $\mathcal{E}_m$ is given by the following relation:

$$\begin{aligned}\frac{1}{\tau_{\text{piez}}} &= \frac{m^*}{\pi\hbar^3}\frac{2}{\pi}(QS_{\text{II}})\frac{1}{k_\parallel}\\ &= \frac{\sqrt{m^*}}{\sqrt{2}\pi\hbar^2\epsilon^2}\left[\frac{(e\cdot e_l)^2B_{\text{l}}}{\rho v_{\text{l}}^2} + \frac{2(e\cdot e_{\text{t}})^2B_{\text{t}}}{\rho v_{\text{t}}^2}\right]\frac{k_{\text{B}}T}{\sqrt{\mathcal{E}-\mathcal{E}_m}}\,.\end{aligned} \tag{8.108}$$

8.3.2.5 (e) Ionized Impurity Scattering

Electrons are scattered by the Coulomb potential induced by ionized impurities [3, 24, 25]. Assuming that a point charge $+e$ is located at the position $(\boldsymbol{r}_{0\parallel}, z_0)$, the static potential induced by the charge $\phi(\boldsymbol{r}_\parallel, z)$ is written as

$$\phi(\boldsymbol{r}_\parallel, z) = \frac{e}{4\pi\kappa\epsilon_0}\frac{1}{\sqrt{(\boldsymbol{r}_\parallel - \boldsymbol{r}_{\parallel 0})^2 + (z-z_0)^2}}\,, \tag{8.109}$$

where $\kappa\epsilon_0$ is the dielectric constant of the semiconductor we are interested in. The scattering Hamiltonian for this ionized impurity is written as

$$H_{\text{ion}} = -e\phi\,, \tag{8.110}$$

and the scattering matrix element of an electron from a state in subband m and wave vector $\boldsymbol{k}_\parallel$ into a state in subband n and wave vector $\boldsymbol{k}'_\parallel$ is given by

$$\langle n, \boldsymbol{k}'_{\parallel} | H_{\rm ion} | m, \boldsymbol{k}_{\parallel} \rangle = -\frac{e^2}{4\pi\kappa\epsilon_0 A} \int \zeta_n^*(z)\zeta_m(z) {\rm d}z \times \int \frac{\exp[i(\boldsymbol{k}_{\parallel} - \boldsymbol{k}'_{\parallel}) \cdot \boldsymbol{r}_{\parallel}]}{\sqrt{(\boldsymbol{r}_{\parallel} - \boldsymbol{r}_{\parallel 0})^2 + (z - z_0)^2}} {\rm d}^2\boldsymbol{k}_{\parallel} , \tag{8.111}$$

where A is the area of normalization. Putting $\boldsymbol{Q} = \boldsymbol{k}_{\parallel} - \boldsymbol{k}'_{\parallel}$, the squared matrix element is

$$\left| \langle n, \boldsymbol{k}'_{\parallel} | H_{\rm ion} | m, \boldsymbol{k}_{\parallel} \rangle \right|^2 = \left(\frac{e^2}{2\pi\kappa\epsilon_0} \right)^2 \frac{1}{A^2 Q^2} \left| I_{mn}(Q, z_0) \right|^2 , \tag{8.112}$$

where

$$I_{mn}(Q, z_0) = \int \zeta_m^*(z)\zeta_n(z) \exp(-Q|z - z_0|) {\rm d}z . \tag{8.113}$$

The scattering rate for a two-dimensional electron from a state $|m, \boldsymbol{k}_{\parallel}\rangle$ to a state $|n, \boldsymbol{k}'_{\parallel}\rangle$ scattered by a point charge $+e$ located at $(\boldsymbol{r}_{\parallel 0}, z_0)$ is then given by

$$P(m, \boldsymbol{k}_{\parallel} \to n, \boldsymbol{k}'_{\parallel}; z_0) = \frac{2\pi}{\hbar} \left| \langle n, \boldsymbol{k}'_{\parallel} | H_{\rm ion} | m, \boldsymbol{k}_{\parallel} \rangle \right|^2 \delta \left[\mathcal{E}(\boldsymbol{k}'_{\parallel}) - \mathcal{E}(\boldsymbol{k}_{\parallel}) \right] , \tag{8.114}$$

where

$$\mathcal{E}(\boldsymbol{k}_{\parallel}) = \frac{\hbar^2 k_{\parallel}^2}{2m^*} + \mathcal{E}_m , \tag{8.115}$$

$$\mathcal{E}(\boldsymbol{k}'_{\parallel}) = \frac{\hbar^2 k_{\parallel}'^2}{2m^*} + \mathcal{E}_n . \tag{8.116}$$

When the ionized impurities are distributed uniformly in the x, y plane and the volume density in the z direction is given by $g_{\rm ion}(z)$, the scattering probability of (8.114) reduces to

$$P(m, \boldsymbol{k}_{\parallel} \to n, \boldsymbol{k}'_{\parallel}) = A \int P(m, \boldsymbol{k}_{\parallel} \to n, \boldsymbol{k}'_{\parallel}; z_0) g_{\rm ion}(z_0) {\rm d}z_0 = \frac{2\pi}{\hbar} \left(\frac{e^2}{2\pi\kappa\epsilon_0} \right)^2 \frac{1}{Q^2 A} J_{mn}^{\rm ion}(Q) \delta \left[\mathcal{E}(\boldsymbol{k}'_{\parallel}) - \mathcal{E}(\boldsymbol{k}_{\parallel}) \right] , \tag{8.117}$$

where $g_{\rm ion}(z_0)$ is the density of ions per unit volume and $N_{\rm ion} = \int g_{\rm ion}(z_0) {\rm d}z_0$ is the ion density per unit area. $J_{mn}^{\rm ion}(Q)$ is given by

$$J_{mn}^{\rm ion}(Q) = \int_0^W \left| I_{mn}(Q, z_0) \right|^2 g_{\rm ion}(z_0) {\rm d}z_0 . \tag{8.118}$$

When impurity ions are uniformly distributed and $g_{\text{ion}}(z_0) = n_{\text{ion}}$, the above integration leads to (assuming $Q = 0$, $J_{nn} = 1$)

$$J_{nn}^{\text{ion}}(Q) \simeq= A n_{\text{ion}} W \,. \tag{8.119}$$

Since the impurity scattering in a two-dimensional electron gas can be treated as elastic scattering, the relaxation time is given by

$$\frac{1}{\tau_{\text{ion}}} = \sum_{n} \sum_{\boldsymbol{k}'_{\parallel}} P(m, \boldsymbol{k}_{\parallel} \to n, \boldsymbol{k}'_{\parallel}) \left(1 - \frac{k'_x}{k_x}\right) \,. \tag{8.120}$$

Replacing the summation with respect to $\boldsymbol{k}'_{\parallel}$ by an integral and defining the scattering angle by $k'_x/k_x = \cos\theta$, we obtain

$$\frac{1}{\tau_{\text{ion}}} = \sum_{n} \int \frac{2\pi}{\hbar} \left(\frac{e^2}{2\pi\kappa\epsilon_0}\right)^2 \frac{1}{AQ^2} J_{mn}^{\text{ion}}(Q) \delta\left[\mathcal{E}_n - \mathcal{E}_m + \frac{\hbar^2 k'^2_{\parallel}}{2m^*} - \frac{\hbar^2 k^2_{\parallel}}{2m^*}\right]$$
$$\times (1 - \cos\theta) \mathrm{d}^2 \boldsymbol{k}'_{\parallel} \,. \tag{8.121}$$

When we define

$$D = k^2 - \frac{2m^*}{\hbar^2} (\mathcal{E}_n - \mathcal{E}_m) \,, \tag{8.122}$$

$$Q = \sqrt{k^2 + D - 2k\sqrt{D}\cos\theta} \,, \tag{8.123}$$

the property of the δ-function gives rise to the following result:

$$\frac{1}{\tau_{\text{ion}}} = \sum_{n} \left(\frac{e^4}{2\pi\hbar(\kappa\epsilon_0)^2}\right) \int \frac{J_{mn}^{\text{ion}}(Q)}{A[Q(\theta)]^2} (1 - \cos\theta) \mathrm{d}\theta \cdot u(D) \,. \tag{8.124}$$

As will be discussed later, the screening by a two-dimensional electron gas of sheet density N_s modifies the relaxation time as

$$\frac{1}{\tau_{\text{ion}}} = \sum_{n} \left(\frac{e^4}{2\pi\hbar(\kappa\epsilon_0)^2}\right) \int \frac{J_{mn}^{\text{ion}}(Q(\theta))\,(1 - \cos\theta)}{A\left[Q(\theta) + P H_{mn}(Q(\theta))\right]^2} \mathrm{d}\theta \cdot u(D) \,, \tag{8.125}$$

where the screening parameters P and H_{mn} are given by

$$P = \frac{e^2 N_s}{2\kappa\epsilon_0 k_B T} \,, \tag{8.126}$$

$$H_{mn}(Q(\theta)) = \int\int \zeta_m(z_1)\zeta_m(z_2)\zeta_n(z_1)\zeta_n(z_2)$$
$$\times \exp\left[-Q(\theta)|z_1 - z_2|\right] \mathrm{d}z_1 \mathrm{d}z_2 \,. \tag{8.127}$$

8.3.2.6 (f) Surface Roughness Scattering

A two-dimensional electron gas is confined by potential barriers, which are the interface of SiO_2 and Si in the case of a Si-MOSFET and are the GaAs/AlGaAs interfaces in the case of heterostructures. It is almost impossible to grow such surfaces with flatness of atomic order and thus the electrons are subject to scattering due to the surface roughness. The surface roughness scattering has been investigated theoretically by Ando and Matsumoto [1, 26, 27].

Let the plane parallel to the interface be the x, y plane and the perpendicular direction be the z axis. When the interface has an irregular fluctuation $\Delta(x, y)$ in the z direction, the fluctuation results in a fluctuation in the potential $V(z)$, which induces electron scattering. Assuming that the fluctuation $\Delta(x, y)$ is small and varies very slowly, then the expansion of $\Delta(x, y)$ may be approximated to the first order, which gives the perturbation Hamiltonian for surface roughness H_{sr} as

$$H_{\mathrm{sr}} = -\frac{\mathrm{d}V}{\mathrm{d}z}\Delta(x, y)\,. \tag{8.128}$$

In the following we will consider intra-subband transitions for simplicity. The matrix element between the initial state of wave vector $\boldsymbol{k}_\parallel$ and the final state $\boldsymbol{k}'_\parallel$ is written as

$$\begin{aligned}\langle \boldsymbol{k}'_\parallel | H_{\mathrm{sr}} | \boldsymbol{k}_\parallel \rangle &= -\int_{-\infty}^{\infty} \zeta^*(z)\frac{\mathrm{d}V}{\mathrm{d}z}\zeta(z)\mathrm{d}z \frac{1}{A}\int_A \Delta(x, y)\mathrm{e}^{\mathrm{i}(\boldsymbol{k}_\parallel - \boldsymbol{k}'_\parallel)\cdot \boldsymbol{r}_\parallel}\mathrm{d}^2\boldsymbol{r}_\parallel \\ &= -F_{\mathrm{eff}}\Delta(\boldsymbol{k}_\parallel - \boldsymbol{k}'_\parallel)\,,\end{aligned} \tag{8.129}$$

where

$$F_{\mathrm{eff}} = \int_{-\infty}^{\infty} \zeta^*(z)\frac{\mathrm{d}V}{\mathrm{d}z}\zeta(z)\mathrm{d}z\,, \tag{8.130}$$

$$\Delta(\boldsymbol{Q}) = \frac{1}{A}\int_A \Delta(\boldsymbol{r}_\parallel)\mathrm{e}^{\mathrm{i}\boldsymbol{Q}\cdot\boldsymbol{r}_\parallel}\mathrm{d}^2\boldsymbol{r}_\parallel\,. \tag{8.131}$$

Here, $\boldsymbol{Q} = \boldsymbol{k}_\parallel - \boldsymbol{k}'_\parallel$, $\boldsymbol{r}_\parallel$ is the position vector in the x, y plane, A is the area of normalization and $\zeta(z)$ is the envelope function obtained by solving the one-dimensional Schrödinger equation in the effective mass approximation. F_{eff}/e is expressed by the effective electric field at the interface E_{eff} as follows [26]:

$$F_{\mathrm{eff}} = \frac{e^2}{\kappa\epsilon_0}\left(\frac{1}{2}N_{\mathrm{s}} + N_{\mathrm{depl}}\right) = eE_{\mathrm{eff}}\,, \tag{8.132}$$

where N_{s} is the sheet electron density and N_{depl} is the space charge density in the depletion layer. Note that this relation is derived for electrons in the inversion layer of a Si-MOSFET. Usually the correlation of the surface roughness is expressed by using a Gaussian function, which leads to the result

$$\langle \Delta(\boldsymbol{r})\Delta(\boldsymbol{r}')\rangle = \Delta^2 \exp\left[\frac{(\boldsymbol{r}-\boldsymbol{r}')^2}{\Lambda^2}\right], \tag{8.133}$$

where Δ is the average height of the roughness in the z direction and Λ is the spatial extent of the roughness in the direction parallel to the interface. The Fourier transform of (8.133) reduces to

$$\langle |\Delta(\boldsymbol{Q})|^2\rangle = \frac{\pi\Delta^2\Lambda^2}{A}\exp\left[-\frac{Q^2\Lambda^2}{4}\right]. \tag{8.134}$$

With this approximation the scattering rate or the relaxation time τ_{sr} due to surface roughness scattering is derived as

$$\frac{1}{\tau_{\mathrm{sr}}} = \frac{2\pi}{\hbar}\sum_{\boldsymbol{k}'_{\parallel}} \left|\langle \boldsymbol{k}'_{\parallel}|H_{\mathrm{sr}}|\boldsymbol{k}_{\parallel}\rangle\right|^2 \delta\left[\mathcal{E}(\boldsymbol{k}'_{\parallel}) - \mathcal{E}(\boldsymbol{k}_{\parallel})\right]\left(1-\frac{k'_x}{k_x}\right). \tag{8.135}$$

Replacing $(1-k'_x/k_x)$ in (8.135) with $(1-\cos\theta)$, taking account of the screening effect, and replacing the summation with respect to $\boldsymbol{k}'_{\parallel}$ by an integral with respect to $\boldsymbol{Q}$, we obtain

$$\frac{1}{\tau_{\mathrm{sr}}} = \frac{2\pi}{\hbar}\sum_{\boldsymbol{Q}} \frac{\pi}{A}\left[\frac{\Delta\Lambda e^2 N^*}{\epsilon(Q)}\right]^2 \exp\left[-\frac{Q^2\Lambda^2}{4}\right] \times\delta\left[\mathcal{E}(\boldsymbol{k}_{\parallel}-\boldsymbol{Q}) - \mathcal{E}(\boldsymbol{k}_{\parallel})\right](1-\cos\theta), \tag{8.136}$$

where $N^* = (N_{\mathrm{s}}/2 + N_{\mathrm{depl}})$ and $\epsilon(Q)$ is the dielectric constant with screening effect given by

$$\epsilon(Q) = \kappa\epsilon_0\left[1 + \frac{1}{\kappa\epsilon_0}\frac{1}{Q}\frac{e^2 m^*}{2\pi\hbar^2}F(Q)\right], \tag{8.137}$$

and $\kappa\epsilon_0$ is the dielectric constant of the semiconductor without the screening effect.[4] $F(Q)$ is defined by

$$F(Q) = \int \mathrm{d}z_1 \int \mathrm{d}z_2\, |\zeta(z_1)|^2\, |\zeta(z_2)|^2 \exp(-Q|z_1 - z_2|). \tag{8.138}$$

Approximation of (8.136) leads to the following relation:

$$\frac{1}{\tau_{\mathrm{sr}}(Q)} \simeq \frac{\pi m^*\Delta^2\Lambda^2 e^4 N^{*2}}{\hbar^3\,[\epsilon(Q)]^2}\exp\left(-\frac{Q^2\Lambda^2}{4}\right). \tag{8.139}$$

[4] In the case of a Si-MOSFET the dielectric constant should be estimated by taking contributions from the semiconductor and the insulator (SiO_2). See Reference [1].

From these results we find that the surface roughness scattering results in an electron mobility proportional to $1/N^{*2}$ and therefore the electron mobility is expected to decrease with the square of the electron density. This feature has been experimentally confirmed in the Si-MOSFET. For example, Hartstein et al. [28] have reported that the decrease in electron mobility is explained in terms of $(N_s + N_{\rm depl})^2$ instead of $(N_s + 2N_{\rm depl})^2$ [1]. It has been pointed out that the surface roughness approximated by an exponential function explains the experimental data better than the Gaussian function approximation [29, 30].

8.3.2.7 (g) Screening Effect

Here we will discuss the effect of screening on the impurity potential. We assume that a change in the potential $\delta\phi(\boldsymbol{r})$ is induced by an ionized impurity and that the occupation of electrons is modified by the potential change. Then the subband energy $\mathcal{E}_i$ of the subband i is changed by the amount

$$\delta\mathcal{E}_i(\boldsymbol{r}_\parallel) = -e\bar{\phi}(\boldsymbol{r}_\parallel) = -e\int_{-\infty}^{\infty} \delta\phi(\boldsymbol{r})|\zeta_i(z)|^2 \mathrm{d}z\,. \tag{8.140}$$

Therefore, the charge density induced by the ionized impurity is given by

$$\begin{aligned} \rho_{\rm ind}(\boldsymbol{r}) &= -e\sum_i \delta N_i|\zeta(z)|^2 = -e\sum_i \frac{\partial N_i}{\partial \mathcal{E}_i}\delta\mathcal{E}_i|\zeta(z)|^2 \\ &= e^2\sum_i \frac{\partial N_i}{\partial \mathcal{E}_i}\bar{\phi}_i(\boldsymbol{r}_\parallel)|\zeta(z)|^2 = -e^2\sum_i \frac{\partial N_i}{\partial \mathcal{E}_{\rm F}}\bar{\phi}_i(\boldsymbol{r}_\parallel)|\zeta(z)|^2\,, \end{aligned} \tag{8.141}$$

where the sheet electron density N_i of the subband i is given by (8.15) and thus we have

$$\frac{\partial N_i}{\partial \mathcal{E}_{\rm F}} = \frac{m_{{\rm d}i}n_{{\rm v}i}}{\pi\hbar^2}\frac{1}{1+\exp\left[(\mathcal{E}_i - \mathcal{E}_{\rm F})/k_{\rm B}T\right]}\,. \tag{8.142}$$

Inserting this relation into (8.142), the induced charge density is reduced to

$$\rho_{\rm ind}(\boldsymbol{r}) = -2\kappa\epsilon_0\sum_i P_i\bar{\phi}(\boldsymbol{r}_\parallel)|\zeta(z)|^2\,, \tag{8.143}$$

where P_i is defined by the following equation:

$$P_i = \frac{e^2}{2\kappa\epsilon_0}\frac{m_{{\rm d}i}n_{{\rm v}i}}{\pi\hbar^2}\frac{1}{1+\exp\left[(\mathcal{E}_i - \mathcal{E}_{\rm F})/k_{\rm B}T\right]}\,. \tag{8.144}$$

Using these results, Poisson's equation under the influence of the screening effect is obtained as

$$\nabla^2 \delta\phi(\boldsymbol{r}) = -\frac{1}{\kappa\epsilon_0}\left[\rho_{\text{ext}}(\boldsymbol{r}) + \rho_{\text{ind}}(\boldsymbol{r})\right] . \tag{8.145}$$

Replacing the potential due to ionized impurity $\delta\phi(\boldsymbol{r})$ with $\tilde{\phi}(\boldsymbol{r})$ and inserting (8.143) into (8.145), we obtain

$$\nabla^2 \tilde{\phi}(\boldsymbol{r}) - 2\sum_i P_i|\zeta(z)|^2 \int_{-\infty}^{\infty} \tilde{\phi}(\boldsymbol{r})|\zeta(z)|^2 \mathrm{d}z = -\frac{1}{\kappa\epsilon_0}\rho_{\text{ext}}(\boldsymbol{r}) . \tag{8.146}$$

The positive charge at $z = z_0$ is defined by

$$\rho_{\text{ext}}(\boldsymbol{r}) = e \cdot \delta(\boldsymbol{r}_\parallel)\delta(z - z_0) . \tag{8.147}$$

Inserting this into (8.146) we find

$$\nabla^2 \tilde{\phi}(\boldsymbol{r}) - 2\sum_i P_i|\zeta_i(z)|^2 \int_{-\infty}^{\infty} \tilde{\phi}(\boldsymbol{r})|\zeta_i(z)|^2 \mathrm{d}z$$
$$= -\frac{e}{\kappa\epsilon_0}\delta(\boldsymbol{r}_\parallel)\delta(z - z_0) . \tag{8.148}$$

We apply the Fourier transform

$$\tilde{\Phi}(\boldsymbol{Q}, z) = \int \tilde{\phi}(\boldsymbol{r}_\parallel, z) \mathrm{e}^{\mathrm{i}\boldsymbol{Q}\cdot\boldsymbol{r}_\parallel} \mathrm{d}^2\boldsymbol{r}_\parallel \tag{8.149}$$

to both sides of (8.148), and the result is

$$\left(\frac{\partial^2}{\partial z^2} - Q^2\right)\tilde{\Phi}(\boldsymbol{Q}, z) - 2\sum_i P_i|\zeta_i(z)|^2 \bar{\Phi}(\boldsymbol{Q}) = -\frac{e}{\kappa\epsilon_0}\delta(z - z_0) , \tag{8.150}$$

where

$$\bar{\Phi}_i(\boldsymbol{Q}) = \int_{-\infty}^{\infty} \tilde{\Phi}(\boldsymbol{Q}, z)|\zeta_i(z)|^2 \mathrm{d}z . \tag{8.151}$$

Defining

$$f(z) = -\frac{e}{\kappa\epsilon_0}\delta(z - z_0) + 2\sum_i P_i|\zeta_i(z)|^2 \bar{\Phi}_i(\boldsymbol{Q}) , \tag{8.152}$$

(8.150) reduces to

$$\left(\frac{\partial^2}{\partial z^2} - Q^2\right)\tilde{\Phi}(\boldsymbol{Q}, z) = f(z) . \tag{8.153}$$

Using the Fourier transform of the above equation

$$\hat{\psi}(\boldsymbol{Q}, q_z) = \int_{-\infty}^{\infty} \tilde{\Phi}(\boldsymbol{Q}, z)\mathrm{e}^{\mathrm{i}q_z z}\mathrm{d}z\,, \tag{8.154}$$

$$\hat{f}(q_z) = \int_{-\infty}^{\infty} \tilde{\Phi}(\boldsymbol{Q}, z)\mathrm{e}^{\mathrm{i}q_z z}\mathrm{d}z\,, \tag{8.155}$$

(8.153) is rewritten as

$$\hat{\psi}(\boldsymbol{Q}, q_z) = -\frac{\hat{f}(q_z)}{q_z^2 + Q^2}\,. \tag{8.156}$$

The inverse Fourier transform of the above equation leads to the following relation:

$$\begin{aligned}
\tilde{\Phi}(\boldsymbol{Q}, z) &= \frac{1}{2\pi}\int_{-\infty}^{\infty} \hat{\psi}(\boldsymbol{Q}, q_z)\mathrm{e}^{-\mathrm{i}q_z z}\mathrm{d}q_z = -\frac{1}{2\pi}\int_{-\infty}^{\infty} \frac{\hat{f}(q_z)\mathrm{e}^{-\mathrm{i}q_z z}}{q_z^2 + Q^2}\mathrm{d}q_z \\
&= -\frac{1}{2\pi}\int_{-\infty}^{\infty} f(z')\mathrm{d}z' \int_{-\infty}^{\infty} \frac{\mathrm{e}^{-\mathrm{i}q_z z}}{q_z^2 + Q^2}\mathrm{d}q_z = -\int_{-\infty}^{\infty} \frac{\mathrm{e}^{-Q|z-z'|}}{2Q} f(z')\mathrm{d}z' \\
&= -\int_{-\infty}^{\infty} \frac{\mathrm{e}^{-Q|z-z'|}}{2Q}\left[-\frac{e}{\kappa\epsilon_0}\delta(z'-z_0) + 2\sum_i P_i|\zeta_i(z')|^2\bar{\Phi}_i(\boldsymbol{Q})\right]\mathrm{d}z' \\
&= \frac{e}{\kappa\epsilon_0}\frac{\mathrm{e}^{-Q|z-z_0|}}{2Q} - \sum_i P_i\bar{\Phi}_i(\boldsymbol{Q})\int_{-\infty}^{\infty} \frac{\mathrm{e}^{-Q|z-z'|}}{Q}|\zeta_i(z')|^2\mathrm{d}z'\,.
\end{aligned} \tag{8.157}$$

Let us consider the scattering rate due to ionized impurities with the screening effect. The matrix element $\tilde{M}$ for impurity scattering from an initial state in subband m and with wave vector $\boldsymbol{k}_\parallel$ to a final state in subband n and with wave vector $\boldsymbol{k}'_\parallel$ is given by

$$\begin{aligned}
\tilde{M} &= \langle n, \boldsymbol{k}'_\parallel|e\tilde{\phi}|m, \boldsymbol{k}_\parallel\rangle = \int_{-\infty}^{\infty} \zeta_n^*(z)\zeta_m(z)\cdot\frac{e}{A}\int_A \tilde{\phi}(\boldsymbol{r})\mathrm{e}^{\mathrm{i}(\boldsymbol{k}_\parallel - \boldsymbol{k}'_\parallel)}\cdot\boldsymbol{r}_\parallel\mathrm{d}^2\boldsymbol{r}_\parallel \\
&= \frac{e}{A}\int_{-\infty}^{\infty} \zeta_n^*(z)\zeta_m(z)\tilde{\Phi}(\boldsymbol{Q}, z)\mathrm{d}z \\
&= \frac{e^2}{2\kappa\epsilon_0 QA}\int_{-\infty}^{\infty} \zeta_n^*(z)\zeta_m(z)\mathrm{e}^{-Q|z-z_0|}\mathrm{d}z \\
&\quad -\sum_i \frac{e}{A}P_i\bar{\Phi}(\boldsymbol{Q})\frac{1}{Q}\int_{-\infty}^{\infty}\mathrm{d}z\int_{-\infty}^{\infty}\mathrm{d}z'\zeta_n^*(z)\zeta_m(z)|\zeta_i(z')|^2\mathrm{e}^{-Q|z-z'|} \\
&= \langle n, \boldsymbol{k}'_\parallel|e\phi|m, \boldsymbol{k}_\parallel\rangle - \sum_i P_i\langle i, \boldsymbol{k}'_\parallel|e\tilde{\phi}|i, \boldsymbol{k}_\parallel\rangle\left[\frac{H_{mn}^i(Q)}{Q}\right]\,.
\end{aligned} \tag{8.158}$$

Now we define following relations

$$M = \langle n, \boldsymbol{k}'_{\parallel} | e\phi | m, \boldsymbol{k}_{\parallel} \rangle \,, \tag{8.159}$$

$$\tilde{M} = \langle n, \boldsymbol{k}'_{\parallel} | e\tilde{\phi} | m, \boldsymbol{k}_{\parallel} \rangle \,, \tag{8.160}$$

$$H^i_{mn}(Q) = \int_{-\infty}^{\infty} \mathrm{d}z \int_{-\infty}^{\infty} \mathrm{d}z' \zeta_n^*(z)\zeta_m(z)|\zeta_i(z')|^2 \mathrm{e}^{-Q|z-z'|} \,, \tag{8.161}$$

where ϕ and M are the potential and matrix element without screening by the two-dimensional electron gas. Therefore, the matrix element is rewritten as follows by using the above relations and (8.158).

$$\tilde{M} = M - P_i \tilde{M} \frac{H^i_{mn}}{Q} \,, \tag{8.162}$$

$$\tilde{M} = \frac{Q}{Q + P_i H^i_{mn}} M \,. \tag{8.163}$$

8.3.2.8 (h) Remote Ionized Impurity Scattering

The discovery of modulation doping by Dingle et al. [19] in 1978 has enabled to improve the low-temperature mobility in AlGaAs/GaAs heterostructures. The most important factor to achieve high electron mobility is to separate 2DEG from ionized impurities in modulation-doped AlGaAs/GaAs heterostructures, where the 2DEG in GaAs layer at the interface is supplied from the donors doped in the barrier layer AlGaAs. This structure will reduce ionized impurity scattering because the GaAs layer is not doped intentionally and thus very low density of acceptors is introduced unintentionally. At low temperatures 2DEG is suffered from acoustic phonon scattering and ionized impurity scattering. Therefore the ionized impurities introduced in AlGaAs layer are the most important source of electron scattering. Interaction potential of 2DEG with such remote ionized impurities in AlGaAs is long range and expected to be the source of scattering potential for 2DEG in GaAs layer. It is well known that the introduction of a spacer layer, non-doped layer in the barrier AlGaAs at the interface, will increase the electron mobility quite a bit [21]. The first observation of electron mobility exceeding $10^6\,\mathrm{cm^2/Vs}$ was made by Hiyamizu et al. [21]. In addition they reported an increase in the electron mobility after light exposure. Improvement of molecular beam epitaxy has let to produce heterostructures with low temperature mobility values exceeding $1 \times 10^7\,\mathrm{cm^2/Vs}$. Pfeiffer et al. [31] observed Hall mobility of $1.17 \times 10^7\,\mathrm{cm^2/Vs}$ at carrier density of 2.4×10^{11} electrons/cm^2 with a spacer layer of 700 Å after exposure to light at 0.35 K. Saku et al. [32] reported electron mobility $1.05 \times 10^7\,\mathrm{cm^2/Vs}$ in modulation-doped AlGaAs/GaAs with a spacer layer thickness 750 Å and with an electron sheet density of $3 \times 10^{11}\,\mathrm{cm^{-2}}$. Umansky et al. [33] reported a maximum electron mobility of $1.44 \times 10^7\,\mathrm{cm^2/Vs}$ at 0.1 K with a spacer thickness of 680 Å and a sheet electron density of $2.4 \times 10^{11}\,\mathrm{cm^{-2}}$. All the data reported so far have revealed the importance of the doping profile in the barrier layer AlGaAs and thus the electron mobility is affected strongly by the remote ionized impurities.

Hess [24] was the first to calculate the remote impurity scattering, but the results did not agreement with experimental observation. Ando et al. [1] reported a detailed model how to deal with the remote ionized impurity scattering, but they did not show any calculated results. Using the model Ando [34] reported an elaborate analysis of Coulomb scattering in addition to interface-roughness scattering and alloy-disorder scattering, and concluded that the Coulomb scattering due to remote impurities plays an important role in determining the low-temperature mobility. In the calculations two subbands are taken into account. Lee et al. [35] reported detailed analysis of low field mobility of 2DEG in modulation doped AlGaAs/GaAs layer, where they took into account of remote impurity scattering, scattering due to background impurities, acoustic deformation potential scattering, polar optical phonon scattering and piezoelectric scattering. They pointed out that the low temperature mobility is well explained in terms of scattering by remote and background impurities. In addition they point out the importance of acoustic phonon scattering. Gold [36] also has reported theoretical analysis of the electron mobility of 2DEG in AlGaAs/GaAs heterostructures and interpreted experimental data in terms of multiple scattering mechanisms including remote impurity scattering. These reported analyses [31, 33, 35, 36] reveal that the scattering by background impurities plays an important role in the low temperature mobility.

Here the scattering rate of 2DEG by ionized impurities is derived along the theoretical treatment made by Ando et al. [1]. For simplicity we assume that there exists only one type of impurities with the electronic charge $+Ze$. The dimensions of the sample are taken to be L for the length, $A = L^2$ for the area and $V = L^3$ for the volume. An impurity located at $(\boldsymbol{r}_i, z_i)$ will produce a potential energy given by the following relation for $z_i < 0$

$$V(\boldsymbol{r}, z) = \sum_i V_i(\boldsymbol{r} - \boldsymbol{r}_i, z - z_i), \tag{8.164}$$

$$V_i(\boldsymbol{r}, z) = -\frac{Ze^2}{4\pi\epsilon\sqrt{(\boldsymbol{r} - \boldsymbol{r}_i)^2 + (z - z_i)^2}}, \tag{8.165}$$

where $\epsilon = \kappa\epsilon_0$ is the average dielectric constant of AlGaAs and GaAs layers and the dielectric constants of the two materials are assumed to be the same for simplicity in the following numerical calculations. Since we are interested in the electron mobility at low temperatures, the electrons are assumed to occupy the lowest subband (the electric quantum limit) with the wave function $\zeta_0(z)$ and the subband energy $\mathcal{E}_0$, and thus the effective potential for the 2DEG is given by

$$v_i(\boldsymbol{r}) = \int_0^\infty |\zeta_0(z)|^2 V_i(\boldsymbol{r}, z)\, \mathrm{d}z\,. \tag{8.166}$$

Fourier transform of the above equation with respect to two-dimensional directions $\boldsymbol{r} = (x, y)$ will give the following relation

$$v_i(\boldsymbol{r}) = \sum_{\boldsymbol{Q}} v_{\boldsymbol{Q}}(z_i) \mathrm{e}^{\mathrm{i}\boldsymbol{Q}\cdot(\boldsymbol{r}-\boldsymbol{r}_i)} . \tag{8.167}$$

The inverse transform of the above equation results in

$$v_{\boldsymbol{Q}}(z_i) = -\frac{1}{L^2}\frac{Ze^2}{2\epsilon Q} F(Q, z_i), \tag{8.168}$$

$$F(Q, z_i) = \mathrm{e}^{-Q|z_i|} \int_0^{\infty} \mathrm{e}^{-Q|z|} |\zeta_0(z)|^2 \mathrm{d}z . \tag{8.169}$$

When the screening of the potential energy by the 2DEG is taken into account, Q in the denominator of (8.168) should be replaced as follows [4] (and see Sect. 8.3.2.7(g))

$$Q \to Q + PH_{mn} \equiv Q + Q_{\mathrm{s}} , \tag{8.170}$$

where

$$P = \frac{e^2}{2\epsilon}\frac{\partial N_{\mathrm{s}}}{\partial \mathcal{E}_{\mathrm{F}}} = \frac{e^2}{2\epsilon}\frac{m^*}{\pi\hbar^2}\frac{1}{1+\exp\left[(\mathcal{E}_0 - \mathcal{E}_{\mathrm{F}})/k_{\mathrm{B}}T\right]} , \tag{8.171}$$

$$\begin{aligned} H_{mn} &= H_{mn}\left(Q(\theta)\right) \\ &= \iint \zeta_m(z_1)\zeta_m(z_2)\zeta_n(z_1)\zeta_n(z_2)\exp(-Q|z_1 - z_2|)\mathrm{d}z_1\mathrm{d}z_2 . \end{aligned} \tag{8.172}$$

Here we evaluate $H_{mn}(m = n = 0)$ given by (8.172) using the Fang-Howard formula (8.25) [18] for the ground state of the 2DEG and obtain the following relation:

$$H_{00} = \frac{1 + (9/8)(Q/b) + (3/8)(Q/b)^2}{(1 + Q/b)^3} . \tag{8.173}$$

The result is shown in Fig. 8.15, where $H_{00}(Q)$ is plotted as a function of Q/b.

Now we evaluate the relaxation time due to remote ionized impurity scattering. A schematic drawing of the modulation-doped heterostructure used in the present calculations is shown in Fig. 8.10a, b, where we assume AlGaAs layer is uniformly doped by Si donors (N_{i} [cm^{-3}]) from $z = -d_{\mathrm{s}}$ to $z = -(d_{\mathrm{N}} + d_{\mathrm{s}})$. The layer with thickness d_{s} is called the spacer layer. The relaxation time due to the remote ionized impurity scattering is then written as

$$\begin{aligned} \frac{1}{\tau_{\mathrm{rmt}}} &= \frac{2\pi}{\hbar}\sum_{\boldsymbol{k}'} \int_{-(d_{\mathrm{s}}+d_{\mathrm{N}})}^{-d_{\mathrm{s}}} L^2 N_{\mathrm{i}}(z_i)\mathrm{d}z_i \\ &\quad \times |v_{\boldsymbol{k}-\boldsymbol{k}'}(z_i)|^2 (1 - \cos\theta_{\boldsymbol{k}\boldsymbol{k}'})\delta\left[\mathcal{E}(\boldsymbol{k}) - \mathcal{E}(\boldsymbol{k}')\right] , \end{aligned} \tag{8.174}$$

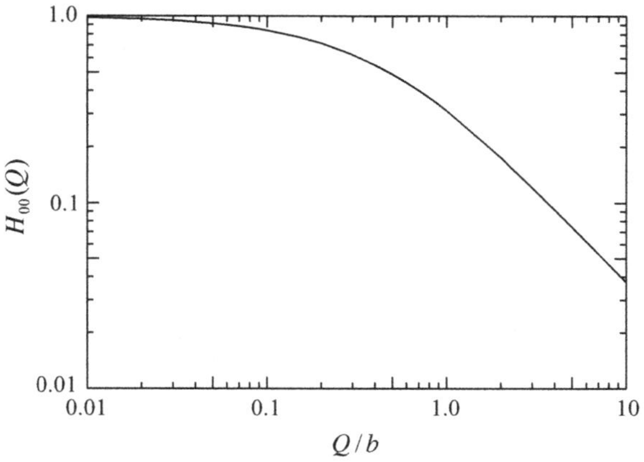

Fig. 8.15 The screening factor $H_{00}(Q)$ given by (8.173) is plotted as a function of Q/b

where $\theta_{\boldsymbol{k}\boldsymbol{k}'}$ is the angle between the wave vectors of 2DEG $\boldsymbol{k}$ and $\boldsymbol{k}'$, and the energy of the 2DEG is given by $\mathcal{E}(\boldsymbol{k}) = \hbar^2 k^2/2m^*$. Summation with respect to $\boldsymbol{k}'$ in (8.174) is expressed in an integral form shown below.

$$\sum_{\boldsymbol{k}'} = \frac{L^2}{(2\pi)^2} \int \mathrm{d}^2\boldsymbol{k}' \, . \tag{8.175}$$

When the distribution of the ionized impurities is uniform with the density N_i, (8.174) is reduced to

$$\frac{1}{\tau_\mathrm{rmt}} = \frac{2\pi}{\hbar} \frac{L^2}{(2\pi)^2} \int \mathrm{d}^2\boldsymbol{k}' L^2 N_\mathrm{i} \int_{-(d_\mathrm{s}+d_\mathrm{N})}^{-d_\mathrm{s}} \mathrm{d}z_i \\ \times |v_{\boldsymbol{k}-\boldsymbol{k}'}(z_i)|^2 (1 - \cos\theta_{\boldsymbol{k}\boldsymbol{k}'}) \delta\left[\mathcal{E}(\boldsymbol{k}') - \mathcal{E}(\boldsymbol{k})\right] \, , \tag{8.176}$$

where we used the conservation laws for the wave vectors of 2DEG $\boldsymbol{k}' = \boldsymbol{k} \pm \boldsymbol{Q}$ and the energy $\mathcal{E}(\boldsymbol{k}') = \mathcal{E}(\boldsymbol{k})$. Using Fang-Howard formula (8.25) for the ground state of 2DEG, (8.169) is written as

$$F(Q, z_i) = \mathrm{e}^{-Q|z_i|} \frac{b^3}{(Q+b)^3} \, . \tag{8.177}$$

The average wave vector of 2DEG is given by $k = \sqrt{2m^* k_\mathrm{B} T/\hbar^2}$ ($k < 10^8$ [1/m], $T < 50$) for non-degenerate case (Boltzmann statistics) and given by the Fermi wave vector $k_\mathrm{F} = \sqrt{2m^* \mathcal{E}_\mathrm{F}/\hbar} = \sqrt{2\pi N_\mathrm{s}}$ (N_s is the sheet density of 2DEG) for degenerate case. Since $b = 5.3 \times 10^8$[1/m] and $k_\mathrm{F} \sim 2.5 \times 10^8$ [1/m] for $N_\mathrm{s} = 10^{16}$ [1/m^2], the relation $b > Q = 2k\sin(\theta/2)$ holds at low temperatures or for the sheet electron density $N_\mathrm{s} < 2 \times 10^{12}$[1/cm^2], and thus we may approximate $b^3/(b+Q)^3 \simeq 1$

in (8.177) and $H_{00}(Q) \simeq 1$. Then, (8.176) reduces to the following relation with $Q = 2k\cos\theta_{k'k}$, where we used $\theta_{k'k} = \theta$ for simplicity.

$$\frac{1}{\tau_{\rm rmt}} = \frac{1}{2\pi\hbar}\left(\frac{Ze^2}{2\epsilon}\right)^2 \int k'{\rm d}k' \int_0^{2\pi} {\rm d}\theta\, N_{\rm i} \int {\rm d}z_i\, {\rm e}^{-4k\sin(\theta/2)|z_i|} \times \frac{(1-\cos\theta)}{(2k\sin(\theta/2)+Q_{\rm s})^2}\,\delta\left[\frac{\hbar^2}{2m^*}\left(|k'|-|k|)(|k'|+|k|\right)\right]. \tag{8.178}$$

Using the property of the δ-function and integrating over z_i, the above equation reduces to

$$\begin{aligned}\frac{1}{\tau_{\rm rmt}} &= \frac{N_{\rm i}m^*}{8\pi\hbar^3}\left(\frac{Ze^2}{\epsilon}\right)^2 \int_0^{2\pi} {\rm d}\theta \frac{2\sin^2(\theta/2)}{[2k\sin(\theta/2)+Q_{\rm s}]^2} \\ &\quad \times \left[\frac{\exp[-4d_{\rm s}k\sin(\theta/2)] - \exp[-4(d_{\rm s}+d_{\rm N})k\sin(\theta/2)]}{4k\sin(\theta/2)}\right] \\ &= \frac{N_{\rm i}m^*}{64\pi\hbar^3}\left(\frac{Ze^2}{\epsilon}\right)^2 \frac{2}{k^3}\int_0^{2\pi} {\rm d}(\theta/2)\frac{\sin(\theta/2)}{[\sin(\theta/2)+Q_{\rm s}/2k]^2} \\ &\quad \times \{\exp[-4d_{\rm s}k\sin(\theta/2)] - \exp[-4(d_{\rm s}+d_{\rm N})k\sin(\theta/2)]\}\,.\end{aligned} \tag{8.179}$$

Replacing $\theta/2$ by θ and taking into account of the symmetry of the function and changing the range of integration into $[0, \pi/2]$, we obtain the following relation.

$$\frac{1}{\tau_{\rm rmt}} = \frac{N_{\rm i}m^*}{16\pi\hbar^3}\left(\frac{Ze^2}{\epsilon}\right)^2 \frac{1}{k^3}\int_0^{\pi/2} {\rm d}\theta\frac{\sin\theta}{[\sin\theta+Q_{\rm s}/2k]^2} \times \{\exp[-4d_{\rm s}k\sin\theta] - \exp[-4(d_{\rm s}+d_{\rm N})k\sin\theta]\}\,. \tag{8.180}$$

Equation (8.180) is shown to be equivalent to (6) of Hess [24] and to (6) of Lee et al. [35]. Integral with respect to θ of the above equation will not give an analytical solution and thus we carry out the integral numerically to evaluate the electron mobility in two cases, degenerate and non-degenerate 2DEG.

(1) Degenerate Case:

When the 2DEG is degenerate ($\mathcal{E}_{\rm F} - \mathcal{E}_0 > 0$) and occupies only the ground subband, we obtain the following relations

$$N_{\rm s} = \frac{m^*}{\pi\hbar^2}(\mathcal{E}_{\rm F} - \mathcal{E}_0)\,, \tag{8.181}$$

$$P = \frac{e^2}{2\epsilon}\frac{\partial N_{\rm s}}{\partial\mathcal{E}_{\rm F}} = \frac{e^2}{2\epsilon}\frac{m^*}{\pi\hbar^2} \equiv Q_{\rm s}\,, \tag{8.182}$$

where $m^*/\pi\hbar^2$ is the density of states of 2DEG. The electron mobility of degenerate 2DEG is calculated by replacing k of $\tau(k)$ by Fermi wave vector $k_{\rm F}$ as shown later in (8.211). Since Fermi wave vector is defined by $k_{\rm F} = [2m^*(\mathcal{E}_{\rm F} - \mathcal{E}_0)/\hbar^2]^{1/2}$, we obtain $\langle k \rangle = k_F = \sqrt{2\pi N_s}$ from (8.181). The mobility is, then, evaluated by numerical integration of (8.180). We introduce $I_1(Q_s/2k_F, d)$ defined by

$$I_1(Q_s/2k_F, d) = \int_0^{\pi/2} d\theta \frac{\sin\theta}{[\sin\theta + Q_s/2k_F]^2} \exp[-4d\, k_F \sin\theta]\,, \tag{8.183}$$

where d is d_s or $d_s + d_N$. Inserting (8.180) and (8.183) into (8.211), the mobility of degenerate 2DEG is given by

$$\begin{aligned} \mu_{rmt} &= \frac{e\tau(k_F)}{m^*} \\ &= \frac{16\pi\hbar^3 e}{N_{is}m^{*2}} \left(\frac{\epsilon}{Ze^2}\right)^2 \frac{k_F^3}{I_1(Q_s/2k_F, d_s) - I_1(Q_s/2k_F, d_s + d_N)}\,. \end{aligned} \tag{8.184}$$

As stated earlier higher electron mobility is achieved by introducing different doping profile. Among them Pfeiffer et al. [31] reported that δ-doping in AlGaAs layer results in a high electron mobility of $11.7 \times 10^6\,\mathrm{cm^2/Vs}$ at a carrier density of 2.4×10^{11} electrons/cm^2 with a spacer layer of 700 Å after exposure to light at 0.35 K. Electron mobility limited by δ-doped remote impurity scattering is easily calculated. The integral with respect to z_i in (8.178) is evaluated as

$$\int dz_i\, N_i\, e^{-4k\sin(\theta/2)z_i} = N_{is}e^{-4k\sin(\theta/2)d_s}\,, \tag{8.185}$$

where N_{is} is the sheet density of δ-doped impurities. Using this relation, (8.179) is written as

$$\begin{aligned} \frac{1}{\tau_{rmt}} = \frac{N_{is}m^*}{8\pi\hbar^3}\left(\frac{Ze^2}{\epsilon}\right)^2 \frac{1}{k^2}\int_0^{2\pi} d(\theta/2)\frac{\sin^2(\theta/2)}{[\sin(\theta/2) + Q_s/2k]^2} \\ \times \exp[-4d_s k\sin(\theta/2)]\,. \end{aligned} \tag{8.186}$$

The integral with respect to θ of (8.186) does not give any analytical expression and thus we carry out integration for typical values of Q_s, $k = k_F$ and d_s, defining $I(Q_s/2k_F)$ by

$$I_2(Q_s/2k_F) = \int_0^{\pi/2} \frac{\sin^2\theta}{(\sin\theta + Q_s/2k_F)^2} \exp(-4d_s k_F \sin\theta) d\theta\,. \tag{8.187}$$

Using (8.211) electron mobility due to a δ-doped heterostructure at low temperatures is therefore given by

$$\mu_{rmt} = \frac{e\tau(k_F)}{m^*} = \frac{4\pi\hbar^3 e}{N_{is}m^{*2}}\left(\frac{\epsilon}{Ze^2}\right)^2 k_F^2 \frac{1}{I_2(Q_s/2k_F)}\,. \tag{8.188}$$

(2) Non-degenerate Case:

In this case the Fermi energy is lower than the ground subband energy, ($\mathcal{E}_F < \mathcal{E}_0$), and the Fermi–Dirac distribution function may be approximated as $1/[1+\exp\{(\mathcal{E} - \mathcal{E}_F)/k_B T\}] \simeq \exp\{-(\mathcal{E} - \mathcal{E}_F)/k_B T\}$. Since $H_{mn} \simeq 1$ ($m = n = 0$) for $b \gg Q$, we obtain $PH_{mn} \sim P \equiv Q_s$ and the following relations

$$N_s = \frac{m^*}{\pi\hbar^2}\int_{\mathcal{E}_0}^{\infty} \exp\left[-\frac{\mathcal{E} - \mathcal{E}_F}{k_B T}\right] d\mathcal{E} = \frac{m^*}{\pi\hbar^2} k_B T \exp\left[\frac{-(\mathcal{E}_0 - \mathcal{E}_F)}{k_B T}\right], \tag{8.189}$$

$$P = \frac{e^2}{2\epsilon}\frac{\partial N_s}{\partial \mathcal{E}_F} = \frac{e^2 N_s}{2\epsilon k_B T} \equiv Q_s, \tag{8.190}$$

where $m^*/\pi\hbar^2$ is the density of states for 2DEG as shown before for degenerate 2DEG. Here only the ground subband with its energy $\mathcal{E}_0$ is considered. Inserting (8.180) into (8.208) and (8.209) we may evaluate the mobility. However, the integral of (8.179) does not give an analytical solution we assume that k in the integrand is given by using the average k-vector $\langle k\rangle = [2m * k_B T/\hbar^2]^{1/2}$. Replacing k_F by $\langle k\rangle$ in (8.183), the electron mobility is evaluated from (8.208) and (8.209) as

$$\mu_{\mathrm{rmt}} = \frac{60\sqrt{2}\pi^{3/2}\epsilon^2(k_B T)^{3/2}}{N_i\sqrt{m^*}e^3} \times \frac{1}{I_1(Q_s/2\langle k\rangle, d_s) - I_1((Q_s/2\langle k\rangle, d_s + d_N)}. \tag{8.191}$$

Now we evaluate the mobility numerically and compare with experimental results shown in Fig. 8.11. For the purpose of comparison with the experimental data of Hiyamizu et al. [21] as an example, we assume $m^* = 0.067m$, $\epsilon = \kappa\epsilon$ with $\kappa = 13$, and the spacer layer thickness 200 Å. When a perfect ionization of donors is assumed, we obtain the thickness d_N from the relation $N_s = d_N N_i$, which gives $d_N = 0.5\times 10^{-6}$ cm =50 Å for $N_i = 10^{18}$ cm^{-3} and $N_s = 5 \times 10^{11}$ cm^2 and $d_N = 30$ Å for $N_s = 3\times 10^{11}$ cm^{-2}. The former and latter cases of N_s correspond to the parameters of Hiyamizu et al. [21] with and without exposure to light, respectively. Fermi energy at low temperatures is $\mathcal{E}_F - \mathcal{E}_0 = 3.55$ meV for the 2DEG density $N_s = 10^{11}$ cm^{-2} and 35.5 meV for $N_s = 10^{12}$ cm^{-2} from (8.181). In such cases 2DEG is treated as degenerate.

The mobility of 2DEG limited by the remote ionized impurity scattering calculated from (8.184) is shown in Fig. 8.16, where the mobility is plotted as a function of 2DEG sheet density N_s for three different spacer layer thickness $d_s = 200, 250, 300$ Å and for the impurity density $N_i = 10^{18}$ cm^{18}. For comparison calculated mobility for $N_i = 1.5^{17}$ cm^{-3} and $d_s = 200$ Å is plotted by the dashed curve. The calculated mobility increases with increasing the 2DEG sheet density, which is interpreted in terms of screening effect. Higher the sheet density is, the screening becomes more effective, resulting in higher electron mobility.

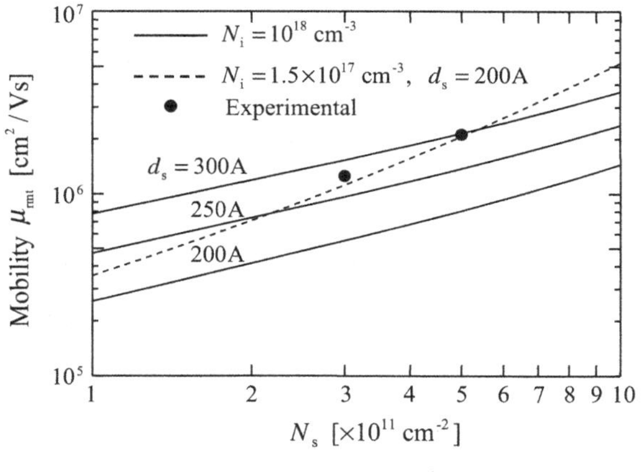

Fig. 8.16 Calculated mobility μ_{rmt} of 2DEG limited by the remote ionized impurity scattering in modulation-doped AlGaAs/GaAs heterostructures (HEMT) is plotted as a function of the electron sheet density N_S. Three curves are calculated for different spacer layer thickness $d_S = 200, 250, 300$ Å. The donor density in AlGaAs layer is $N_i = 10^{18}$ cm^{-3} and the mobility is calculated from (8.184) assuming that 2DEG is degenerate. The donors are assumed to be completely ionized and the doped layer thickness is estimated by using the relation $d_N = N_s/N_i$. The mobility increases with increasing the spacer layer thickness d_S. Experimental data of Hiyamizu et al. [21] are plotted by black circles. The calculated mobility for $N_i = 1.5 \times 10^{18}$ cm^{-3} and $d_S = 200$ Å is plotted by the dashed curve for comparison

Here we compare with the experimental data of Hiyamizu et al. [21]. Their data are plotted by the solid circles in Fig. 8.16, where $\mu = 1.25\times10^6$ (2.12×10^6) cm^2/Vs at $T \sim 5$ K for a sample with the impurity density $N_i = 1.0 \times 10^{18}$ cm^{-3}, $N_s = 3\times10^{11}$ cm^{-2} (5.04×10^{11} cm^{-2} after light exposure) and the spacer layer thickness $d_s = 200$ Å. The calculated mobility is $\mu_{rmt} = 0.55 \times 10^6$ cm^2/Vs for $N_s = 3 \times 10^{11}$ cm^{-2} and 0.81×10^6 cm^2/Vs for $N_s = 5.04 \times 10^{11}$ cm^{-2} when the ionized impurity density is assumed to be the doped donor density $N_i = 10^{18}$ cm^{-3}. The calculated mobility is very small compared with the experimental data. This discrepancy may be understood when we take into account of the ionization rate of the donors. As shown later the ionization of the donors is not perfect and expected to be 14 to 25 % in the case of δ-doped AlGaAs/GaAs. The dashed curve in Fig. 8.16 is calculated assuming the ionized donor density of $N_i = 1.5 \times 10^{17}$ cm^{-3} in the sample of Hiyamizu et al. [21]. The calculated electron mobility is $\mu_{rmt} = 1.1 \times 10^6$ cm^2/Vs for $N_s = 3 \times 10^{11}$ cm^{-2} and 2.09×10^6 cm^2/Vs for $N_s = 5.04 \times 10^{11}$ cm^{-2} after exposure to light. These values are in good agreement with the experimental observation of Hiyamizu et al. [21]. The present results show an importance of the spacer layer to achieve high electron mobility in modulation-doped heterostructures. We have to note here that the ionized donor density after exposure to light is higher than 1.5×10^{17} cm^{-3}, which will reduces the mobility a little bit.

As stated earlier Ando [34] reported the calculations of the mobility in AlGaAs/GaAs heterostructure using trial functions of the electron ground state given by

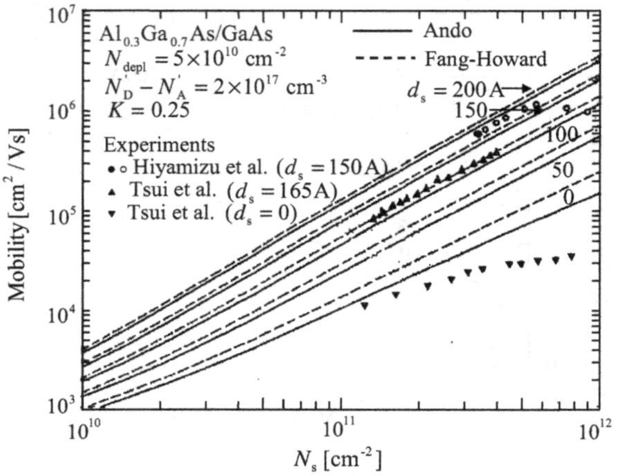

Fig. 8.17 Mobility calculated by Ando [34] including Coulomb scattering by impurities in GaAs and AlGaAs layers. Wave function "Ando" takes account of penetration into AlGaAs layer (solid curves) and "Fang-Howard" is given by (8.25) (dashed curves). In the calculations ionized impurities in GaAs, depletion charge density $N_{\rm depl} = 5 \times 10^{10}\,{\rm cm}^{-2}$ and compensated donors in AlGaAs, $N'_{\rm D} - N'_{\rm A} = 2 \times 10^{17}\,{\rm cm}^{-3}$ with the compensation rate $K = 0.25$ are assumed. Calculated mobilities are plotted as a function of the electron sheet density $N_{\rm s}$ for different spacer layer thickness $d_{\rm s} = 200, 150, 100, 50,$ and 0 A. Experimental data of Hiyamizu et al. [21] and Tsui et al. [37] are also plotted for comparison. After Ando [34]

Fang and Howard (8.25) and by the wave function with a nonvanishing amplitude in AlGaAs, here referred to as "Ando". The results are shown in Fig. 8.17. We find there exists only a slight difference between the two approximations of the wave functions "Ando" and "Fang-Howard". The most important difference in the assumption between the analysis shown in Figs. 8.16 and 8.17 is the acceptors impurities in GaAs and AlGaAs layers. Ando takes account of the ionized acceptors of the depletion region in the GaAs layer and the compensation in the AlGaAs layer. Therefore the calculated mobility Fig. 8.17 is lower than the result shown in Fig. 8.16. However, the overall features of the results are very similar and reveal the importance of the screening effect in the Coulomb scattering. The solid and dashed curves in Fig. 8.17 are calculated for the wave functions of "Ando" and "Fang-Howard", respectively. Experimental data of Hiyamizu et al. [21] and Tsui et al. [37] are plotted along with the calculated curves, where the black dot corresponds to the electron sheet density under the equilibrium condition (in the dark) and the open circles are obtained under exposure to light. The calculated curves by Ando [34] shows a reasonable agreement with the experimental data, except the decrease in mobility at high electron density observed by Hiyamizu et al.

Next we calculate electron mobility in δ-doped AlGaAs/GaAs. In the calculation we have to estimate the ionized donor density in the δ-doped AlGaAs. Since the 2DEG in GaAs is supplied by the ionized donors, it is very reasonable to assume that $N_{\rm is} = N_{\rm s}$. The calculated result is shown by the solid curve in Fig. 8.18 as a function

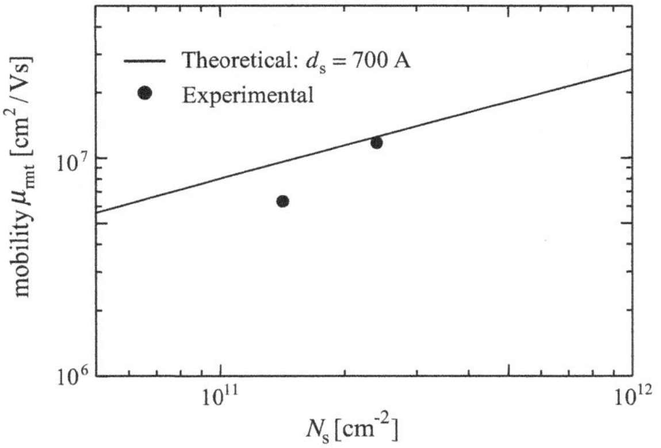

Fig. 8.18 Calculated mobility μ_{rmt} of 2DEG limited by remote ionized impurity scattering in the δ-doped AlGaAs/GaAs heterostructures is plotted as a function of the electron sheet density N_s. The solid curve is calculated for spacer layer thickness $d_s = 700\,\text{Å}$. For comparison the experimental data of Pfeiffer et al. [31] are plotted by the solid circles we assumed that the ionized donor density is $N_{is} = N_s$ (the δ-doped donors of $1.0 \times 10^{12}\,\text{cm}^{-2}$ are partially ionized). The calculated mobility increases with increasing 2DEG density due to the screening effect

of the sheet density of 2DEG N_s, and the experimental date of Pfeiffer et al. [31] are plotted by the solid circles, one in the dark and the other in the light exposure. We find in Fig. 8.18 that the calculated mobility increases with increasing the sheet density of 2DEG and that the calculated result agrees well with the experimental values. The calculated mobility is $\mu_{rmt} = 12.5 \times 10^6\,\text{cm}^2/\text{Vs}$ which is in good agreement with the observed mobility $11.7 \times 10^6\,\text{cm}^2/\text{Vs}$ of Pfeiffer [31]. The mobility of the same sample in the dark is estimated to be $9.6 \times 10^6\,\text{cm}^2/\text{Vs}$ at 0.35 K which is in reasonable agreement with the experimental mobility $6.3 \times 10^6\,\text{cm}^2/\text{Vs}$. We can conclude here again that the spacer layer plays an important role in achieving high electron mobility in modulation-doped heterostructures.

2DEG of AlGaAs/GaAs heterostructures is usually degenerate as shown above. Here we will show the mobility of non-degenerate 2DEG limited by the remote impurity scattering for comparison. The calculated mobility of non-degenerate 2DEG limited by remote impurity scattering using (8.191) is shown in Fig. 8.19, where the doping density in AlGaAs layer is $N_i = 1.0 \times 10^{17}\,\text{cm}^{-3}$, spacer layer thickness $d_s = 200\,\text{Å}$ and sheet density of 2DEG $N_s = 10^{10}$, 5×10^{10} and $10^{11}\,\text{cm}^{-2}$. We find in Fig. 8.19 that the mobility increases with increasing the 2DEG density N_s at low temperatures due to screening effect. At high temperatures the screening effect is weakened for lower 2DEG density and thus the mobility behaves like $\mu_{rmt} \propto T^{3/2}$ as in the case of 3D electrons.

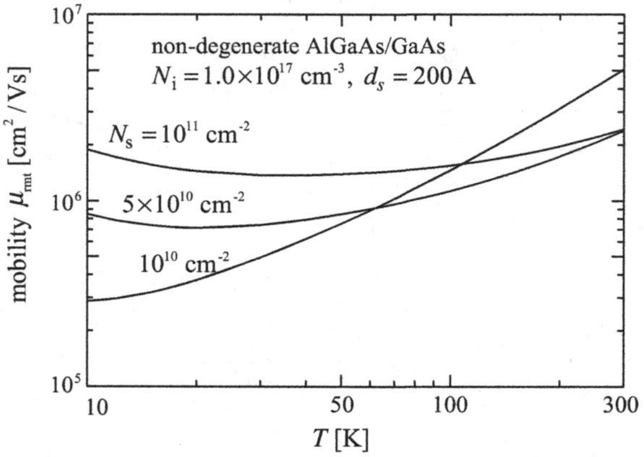

Fig. 8.19 Mobility μ_{rmt} of non-degenerate 2DEG limited by remote impurity scattering in modulation-doped AlGaAs/GaAs heterostructure (HEMT), where the mobility is evaluated from (8.191) for $N_{\text{i}} = 10^{17}\ \text{cm}^{-3}$, $d_{\text{s}} = 200\,\text{Å}$ and electron sheet density $N_{\text{s}} = 10^{10}$, 5×10^{10} and $10^{11}\ \text{cm}^{-2}$

8.3.3 *Mobility of a Two-Dimensional Electron Gas*

In general the electron mobility can be evaluated by solving the Boltzmann transport equation. We have already presented the treatment of three-dimensional and bulk semiconductors in Chap. 6. A similar treatment is possible for the case of a two-dimensional electron gas and the difference arises from the relaxation time τ and the density of states of the two-dimensional electron gas. According to the treatment of Sect. 6.2, we assume that the external electric field is weak and that the change in the distribution function due to the external field is small compared to the thermal equilibrium value f_0. The distribution function is written as

$$f = f_0 + f_1\,, \quad (f_1 << f_0) \tag{8.192}$$

and the relaxation time τ is approximated in a similar fashion to (6.93) as

$$f_1 = -\tau v_x F_x \frac{\partial f_0}{\partial \mathcal{E}}\,. \tag{8.193}$$

In the presence of an electric field E_x in the x direction, $F_x = -eE_x$ and we obtain

$$f(\boldsymbol{k}_\parallel) = f_0(\boldsymbol{k}_\parallel) + eE_x \tau v_x \frac{\partial f_0}{\partial \mathcal{E}}\,. \tag{8.194}$$

Therefore, the current density in the x direction is given by

$$\begin{aligned} J_x &= \frac{2}{(2\pi)^2}\int (-e) v_x f(\boldsymbol{k}_\parallel) \mathrm{d}^2\boldsymbol{k}_\parallel \\ &= -\frac{e}{2\pi^2}\int v_x f_0(\boldsymbol{k}_\parallel)\mathrm{d}^2\boldsymbol{k}_\parallel - \frac{e^2 E_x}{2\pi^2}\int \tau v_x^2 \frac{\partial f_0}{\partial \mathcal{E}} \mathrm{d}^2\boldsymbol{k}_\parallel\,, \end{aligned} \tag{8.195}$$

where the factor 2 represents the spin degeneracy. Since $v_x f_0$ in the first term on the right-hand side of the second equation is an odd function with respect to k_x, integration from $-\infty$ to $+\infty$ results in zero. Therefore, we obtain

$$J_x = -\frac{e^2 E_x}{2\pi^2} \int \tau v_x^2 \frac{\partial f_0}{\partial \mathcal{E}} \mathrm{d}^2 \boldsymbol{k}_\parallel \,. \tag{8.196}$$

The sheet density of the two-dimensional electron gas, N, where only the ground subband is taken into consideration, is given by

$$N = \frac{2}{(2\pi)^2} \int f_0 \mathrm{d}^2 \boldsymbol{k}_\parallel \,. \tag{8.197}$$

Therefore, (8.196) for the current density reduces to

$$J_x = -e^2 N E_x \frac{\displaystyle\int_0^\infty \tau v_x^2 \frac{\partial f_0}{\partial \mathcal{E}} \mathrm{d}^2 \boldsymbol{k}_\parallel}{\displaystyle\int_0^\infty f_0 \mathrm{d}^2 \boldsymbol{k}_\parallel} \,. \tag{8.198}$$

The total energy of the two-dimensional electron gas in the ith subband is written as follows when the electrons have isotropic effective mass m^* in the (k_x, k_y) plane:

$$\mathcal{E} = \mathcal{E}_i + \frac{\hbar^2}{2m^*} \boldsymbol{k}_\parallel^2 = \mathcal{E}_i + \frac{1}{2} m^* v_\parallel^2 \,, \tag{8.199}$$

where $\hbar \boldsymbol{k}_\parallel = m^* \boldsymbol{v}_\parallel$. When we write the squared average of v_x^2 as $\langle v_x^2 \rangle$, we have $\langle v_x^2 \rangle = \langle v_y^2 \rangle$ and

$$\langle v_x^2 \rangle = \frac{1}{2} \langle v_x^2 + v_y^2 \rangle = \frac{1}{2} \langle v_\parallel^2 \rangle = \frac{1}{m^*} \langle \mathcal{E} - \mathcal{E}_i \rangle \,. \tag{8.200}$$

Inserting this relation into (8.198) and replacing the summation with respect to $\boldsymbol{k}_\parallel$ by an integral with respect to energy $\mathcal{E}$, the following result is obtained:

$$J_x = -\frac{e^2 N E_x}{m^*} \frac{\displaystyle\int_{\mathcal{E}_i}^\infty \tau (\mathcal{E} - \mathcal{E}_i) \frac{\mathrm{d} f_0(\mathcal{E})}{\mathrm{d}\mathcal{E}} \mathrm{d}\mathcal{E}}{\displaystyle\int_{\mathcal{E}_i}^\infty f_0(\mathcal{E}) \mathrm{d}\mathcal{E}} \,. \tag{8.201}$$

Putting $\mathcal{E} - \mathcal{E}_i = \mathcal{E}'$ and once again replacing $\mathcal{E}'$ by $\mathcal{E}$, the above equation results in

$$J_x = -\frac{e^2 N E_x}{m^*} \frac{\displaystyle\int_0^\infty \tau \mathcal{E} \frac{\mathrm{d} f_0(\mathcal{E} + \mathcal{E}_i)}{\mathrm{d}\mathcal{E}} \mathrm{d}\mathcal{E}}{\displaystyle\int_0^\infty f_0(\mathcal{E} + \mathcal{E}_i) \mathrm{d}\mathcal{E}} \,. \tag{8.202}$$

Now, we define the current density by

$$J_x = \frac{Ne^2 E_x}{m^*} \langle \tau \rangle \,, \tag{8.203}$$

where

$$\langle \tau \rangle = -\frac{\displaystyle\int_0^\infty \tau \mathcal{E} \frac{\mathrm{d} f_0(\mathcal{E} + \mathcal{E}_i)}{\mathrm{d}\mathcal{E}} \mathrm{d}\mathcal{E}}{\displaystyle\int_0^\infty f_0(\mathcal{E} + \mathcal{E}_i) \mathrm{d}\mathcal{E}} = \frac{\displaystyle\int_0^\infty \tau \mathcal{E} \frac{\mathrm{d} f_0(\mathcal{E} + \mathcal{E}_i)}{\mathrm{d}\mathcal{E}} \mathrm{d}\mathcal{E}}{\displaystyle\int_0^\infty \mathcal{E} \frac{\mathrm{d} f_0(\mathcal{E} + \mathcal{E}_i)}{\mathrm{d}\mathcal{E}} \mathrm{d}\mathcal{E}} \,. \tag{8.204}$$

It is evident that $\langle \tau \rangle$ represents the average of the relaxation time over the distribution function.

When the two-dimensional electron gas is degenerate, the Fermi–Dirac distribution function should be used for f_0:

$$f_0(\mathcal{E}) = \frac{1}{1 + \exp\left(\dfrac{\mathcal{E} - \mathcal{E}_\mathrm{F}}{k_\mathrm{B} T}\right)} \,, \tag{8.205}$$

and (8.204) is reduced to

$$\langle \tau \rangle = \frac{\displaystyle\int_0^\infty \mathcal{E} \tau f_0(\mathcal{E} + \mathcal{E}_i) \left[1 - f_0(\mathcal{E} + \mathcal{E}_i)\right] \mathrm{d}\mathcal{E}}{\displaystyle\int_0^\infty \mathcal{E} f_0(\mathcal{E} + \mathcal{E}_i) \left[1 - f_0(\mathcal{E} + \mathcal{E}_i)\right] \mathrm{d}\mathcal{E}} \,. \tag{8.206}$$

On the other hand, in the case of a non-degenerate electron gas, the Maxwell–Boltzmann distribution function is used for f_0:

$$f_0(\mathcal{E}) = A \exp\left(-\frac{\mathcal{E}}{k_\mathrm{B} T}\right) \,, \tag{8.207}$$

and then the average value of the relaxation time is given by

$$\langle \tau \rangle = \frac{\displaystyle\int_0^\infty \mathcal{E} \tau f_0 \mathrm{d}\mathcal{E}}{\displaystyle\int_0^\infty \mathcal{E} f_0 \mathrm{d}\mathcal{E}} \,. \tag{8.208}$$

The mobility μ of a two-dimensional electron gas is evaluated from the averaged relaxation time $\langle \tau \rangle$ by

$$\mu = \frac{e \langle \tau \rangle}{m^*} \,, \tag{8.209}$$

where the relaxation time τ is determined by the various scattering processes and the scattering rates are as discussed in the previous section. Expressing the relaxation time as τ_i, the relaxation time due to all the scattering processes is given by the following relation in the same way as in (6.370):

$$\frac{1}{\tau} = \sum_i \frac{1}{\tau_i} . \tag{8.210}$$

Inserting this relaxation time τ into (8.206) or (8.208), we may evaluate the average value of the relaxation time and thus the electron mobility.

At low temperatures the average relaxation time is evaluated by replacing $\mathcal{E}$ with $\mathcal{E}_{\rm F}$, $\langle\tau\rangle = \tau(\mathcal{E}_{\rm F})$, and thus the mobility is given by

$$\mu = \frac{e\,\tau(\mathcal{E}_{\rm F})}{m^*} = \frac{e\,\tau(k_{\rm F})}{m^*} . \tag{8.211}$$

We will now compare the mobility of a two-dimensional electron gas deduced from experiments and from theoretical calculations. As a typical example, we will consider the electron mobility in the inversion layer of a Si-MOSFET. We have already shown that positive gate voltage applied to an n-channel MOSFET formed on a p-type substrate induces electrons at the Si/SiO_2 interface, resulting in the formation of an inversion layer, as discussed in Sect. 8.2.1. The MOSFET used for the experiment has an acceptor density $N_{\rm A} = 8 \times 10^{16}\,{\rm cm}^{-3}$. The sheet electron density is evaluated from the relation

$$N_{\rm s} = \frac{C_{\rm ox}}{e}(V_{\rm g} - V_{\rm th}) , \tag{8.212}$$

where $C_{\rm ox}$ is the capacity of the gate oxide, $V_{\rm g}$ is the applied gate voltage and $V_{\rm th}$ is the threshold voltage. The effective electric field $E_{\rm eff}$ induced at the interface is evaluated from (8.132). The surface concentration of the depletion charge $N_{\rm depl}$ is determined from the following relations.

$$N_{\rm depl} = \left(\frac{4\kappa\varepsilon_0\phi_{\rm B}N_{\rm sub}}{e}\right)^{1/2} , \tag{8.213a}$$

$$\phi_{\rm B} = \frac{k_{\rm B}T}{e}\log\left(\frac{N_{\rm sub}}{n_{\rm i}}\right) , \tag{8.213b}$$

where $\phi_{\rm B}$ is the bulk Fermi energy, $N_{\rm sub}$ is the substrate impurity concentration ($N_{\rm sub} = N_{\rm A}$: acceptor density), $n_{\rm i}$ is the intrinsic carrier concentration.

Figure 8.20 shows the measured electron mobility $\mu_{\rm exp}$ as a function of the effective electric field at $T = 77$ K, which is compared with the theoretical calculations [38]. In Fig. 8.20 are plotted the respective mobilities calculated as a function of the effective electric field for ionized impurity scattering $\mu_{\rm ion}$ and the sum of three scattering mechanisms: acoustic phonon scattering, inter-valley phonon scattering and surface

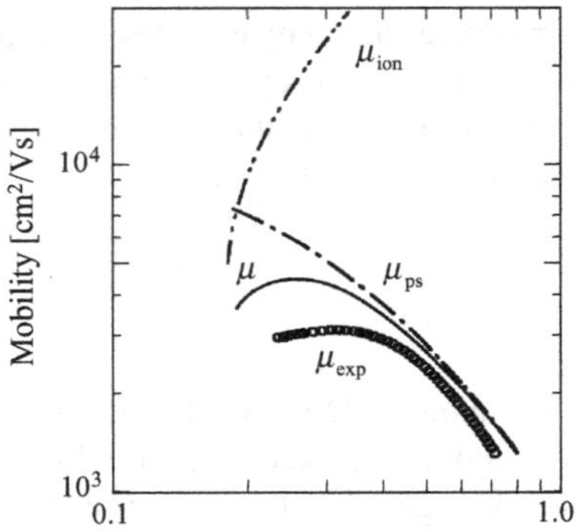

Fig. 8.20 Electron mobility in Si-MOSFET plotted as a function of the effective gate electric field at $T = 77\,K$. Experimental data for the effective mobility μ_{exp} determined from conductance measurements are shown by ○ and the solid curves of μ are theoretically calculated for proper scattering processes and for a surface roughness parameter of $\Delta\Lambda = 25 \times 10^{-20}\,m^2$, where μ_{ion} is the mobility due to ionized impurity scattering, μ_{ps} is the mobility due to the sum of acoustic phonon, inter-valley phonon and surface roughness scattering

roughness scattering, μ_{ps}, where the parameter of the surface roughness is taken to be $\Delta\Lambda = 25 \times 10^{-20}\,m^2$. The mobility μ calculated for all the scattering processes is shown by the solid curve. It is very interesting to point out that theory and experiment exhibit a E_{eff}^{-2} dependence at higher effective electric fields E_{eff}.

Figure 8.21 shows the temperature dependence of the electron mobility for sheet electron density $N_s = 3 \times 10^{12}\,cm^{-2}$, where the experimental results and the theoretical calculations are compared. The experimental data are shown by μ_{exp}. The electron mobilities for the respective scattering processes are shown by μ_{ion} for ionized impurity scattering, μ_{int} for inter-valley phonon scattering, μ_{ac} for acoustic phonon scattering and μ_{sr} for surface roughness scattering. In addition, the total mobility calculated by taking all the scattering processes into account is shown by the curve μ_{total} [39], where we find good agreement between the experimental data and the theoretical calculations. Also, we find that the surface roughness scattering dominates over the other scattering mechanisms at high effective electric fields, resulting in a E_{eff}^{-2} dependence. Similar results have been reported by Takagi et al. [40, 41].

When the electron or hole mobility in inversion layer of MOSFET is measured as a function of the electron sheet density, the mobility approaches to a universal curve at higher electron density, independent of the doping density in the substrate. This behavior is called the universality of the mobility . The most typical example has been reported by Takagi et al. [40] as shown in Fig. 8.22, where the electron mobility of n-channel MOSFET formed in the (100) plane of Si substrate is plotted as a function of the effective electric field E_{eff} (equivalent to the electron sheet density N_s). Here, the effective electric field E_{eff} is related to the inversion electron density N_s and to the acceptor density in the depletion layer N_{depl} according to (8.132). In Fig. 8.22

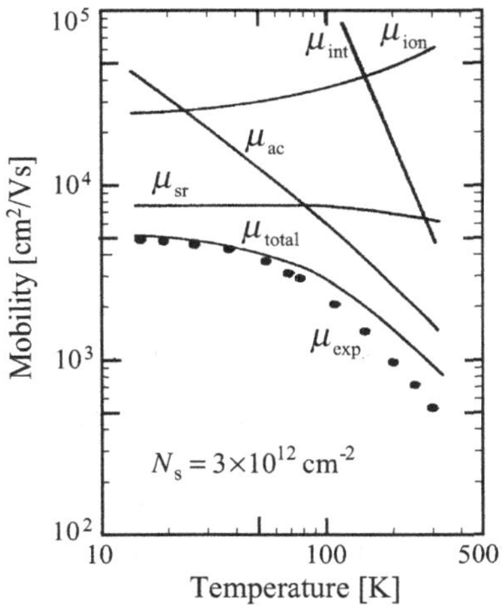

Fig. 8.21 Temperature dependence of electron mobility in Si-MOSFET determined from conductance measurements compared with the theoretical calculations for the sheet electron density $N_s = 3 \times 10^{12}\,\mathrm{cm}^{-2}$, where μ_{exp} is the experimental result. The calculated results are shown by μ_{ion} for ionized impurity scattering, μ_{int} for inter-valley phonon scattering, μ_{ac} for acoustic phonon scattering, μ_{sr} for surface roughness scattering and μ_{total} is calculated by taking all the scattering processes into account

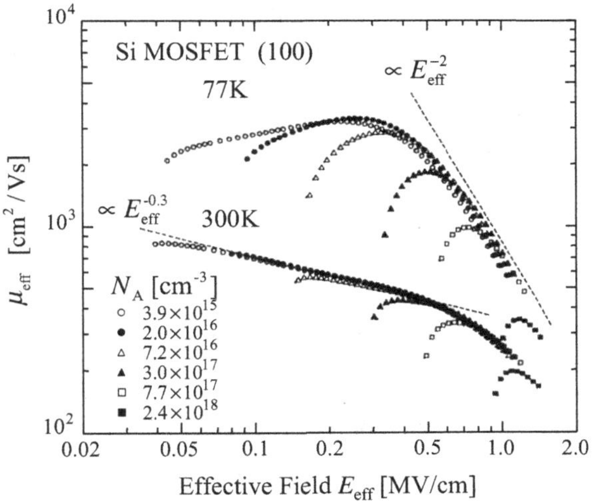

Fig. 8.22 Universality of the electron mobility in MOSFET. Electron mobility of n-channel MOSFET formed in the (100) surface of Si substrate is plotted as a function of the effective electric field E_{eff} measure by Takagi et al. [40] at 300 K and 77 K. The effective field is defined by (8.132) as $E_{eff} = e\cdot(N_{depl}+N_s/2)/\kappa\epsilon_0$. The parameter of the experimental data is the acceptor density in the substrate. After Takagi et al. [40]

we find that the electron mobility at 77 K approaches the curve $\mu \propto E_{eff}^{-2}$ at higher electric filed and thus at higher electron density, and that the behavior is the similar for different acceptor density. This behavior may be explained in terms that the surface roughness scattering dominates the scattering, resulting in $1/\tau_{sr} \propto E_{eff}^2$ from (8.139) and giving rise to $\mu_{sr} \propto E_{eff}^{-2}$. On the other hand, the mobility at 300 K is found to be

$\mu \propto E_{\text{eff}}^{-0.3}$ independent of the acceptor density and the mobility at higher effective field approaches the mobility limited by surface roughness scattering $\mu_{\text{sr}} \propto E_{\text{eff}}^{-2}$. The universality given by the curve $\mu \propto E_{\text{eff}}^{-0.3}$ in the region $E_{\text{eff}} < 0.5$ MV/cm is explained by Takagi et al. in terms of the two dominant scatterings, acoustic phonon scattering and surface roughness scattering. The relation, however, is derived from the assumption that the mobility limited by the surface roughness scattering is given by $\mu_{\text{rs}} \propto E_{\text{eff}}^{-2.6}$.

It is well known that in the development of extremely small MOSFETs the device simulation is very important. Such a device simulator requires accurate determination of the electron mobility as a function of the electric field. Although several kinds of models have been proposed for the field dependence of the mobility, here experimentally determined relations will be presented. The universality of the mobility observed by Takagi et al. [40] is not the drift mobility as a function of electric field along the current channel. An accurate determination of the drift mobility in a MOS structure was made by Cooper and Nelson [42]. They used a MOS structure with doped poly silicon gate so that uniform electric field between the source and drain is achieved along the channel. Drift mobility was measured by the time-of-flight of the electrons excited by a short laser pulse near the surface of Si at the Si/SiO_2 interface. For this purpose two apertures of $10\,\mu$m with a center-to-center spacing of $60\,\mu$m were prepared between the source and drain so that the discrete charge packets introduced by a laser pulse drift in a region of uniform electric field along the channel. The current through the drain contact was measured and the time difference between the two current peaks give the transit time of electrons between the two apertures. The measured low filed mobility μ_{eff} is plotted as a function of the effective normal filed E_{eff} in Fig. 8.23 for $N_{\text{s}} = 3.5 \times 10^{10}\,\text{cm}^{-2}$. The drift velocity v_{d} was measured as a function of the electric field along the channel E_{ch}, which is plotted in Fig. 8.24 for the effective filed $E_{\text{eff}} = 4,\ 9,\ 20,\ 30,\ 50\ \ [\times 10^4\ \text{V/cm}]$. As shown in Fig. 8.23 the electron mobility is well expressed by the solid curve given by

$$\mu_{\text{eff}} = \frac{\mu_0}{1 + (E_{\text{eff}}/E_{\text{c}})^{0.657}}, \tag{8.214}$$

where $\mu_0 = 1105$ cm^2/Vs is the mobility at low effective field E_{eff} and $E_{\text{c}} = 30.5 \times 10^4$ V/cm. This behavior of the mobility may be interpreted in terms of acoustic phonon scattering and surface roughness scattering. On the other hand, the empirical formula for the drift velocity is given by

$$v_{\text{d}} = \frac{\mu_{\text{eff}} E_{\text{ch}}}{[1 + (\mu_{\text{eff}} E_{\text{ch}}/v_{\text{s}})^{\alpha}]^{1/\alpha}}, \tag{8.215}$$

where $v_{\text{s}} = 9.23 \times 10^6$ cm/s is the saturation velocity of electrons and $\alpha = 1.92$. The empirical formula is called Cooper–Nelson formula. The solid curves in Fig. 8.24 are calculated from the empirical formula, where the saturation velocity is found to be

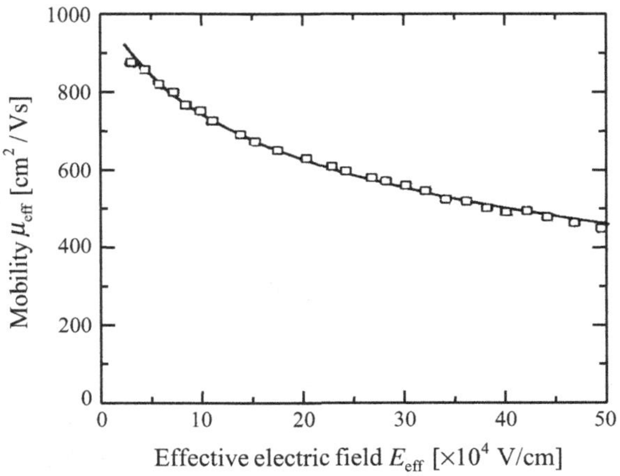

Fig. 8.23 The electron drift mobility is plotted as a function of effective normal field. The data are measured in a Si-MOS structure by the time-of-flight method and the solid curve is calculated from Cooper–Nelson formula (8.214). $T = 300\,{\rm K}$. After Cooper and Nelson [42]

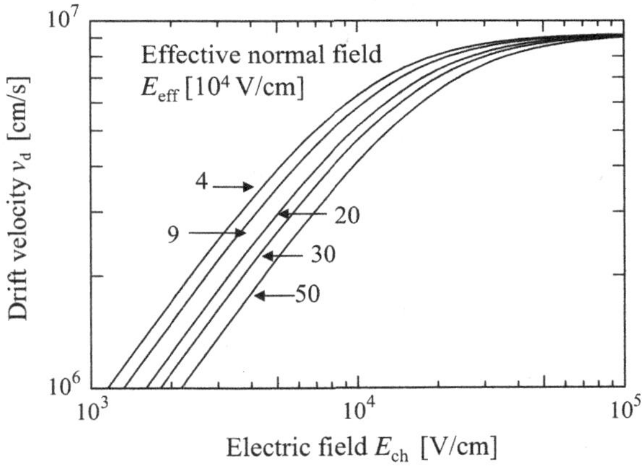

Fig. 8.24 Electron drift velocity is plotted as a function of electric filed along the channel for several effective normal fields where experiments were made by the time-of-flight method in a MOS structure and the fitted curves by Cooper–Nelson formula (8.215) are shown. $T = 300\,{\rm K}$. After Cooper and Nelson [42]

9.2×10^6 cm/s independent of the effective normal field. The effective normal field dependence of the drift velocity may be explained in terms of the surface roughness scattering.

8.4 Superlattices

8.4.1 Kronig–Penney Model

Many review articles on superlattices have been published. Among them Chap. 9 of the textbook by Y. Yu and M. Cardona [6] gives a very detailed discussion of quantum confinement, energy band structures, lattice vibrations and so on. In this chapter we

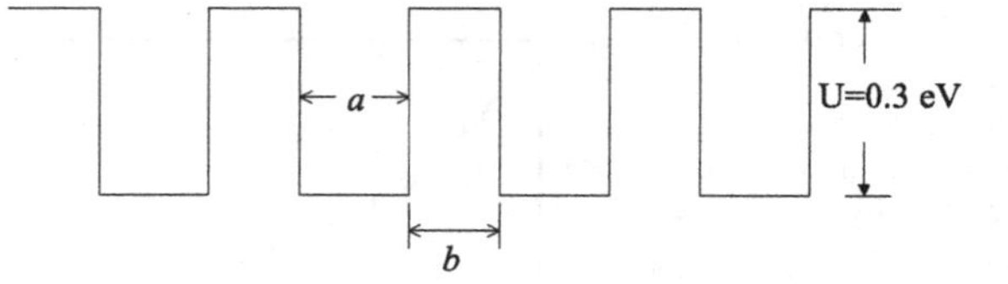

Fig. 8.25 Superlattice model with period $a+b$ for the Kronig–Penney model. Well width a, barrier width b and conduction band discontinuity 0.3 eV are assumed. The calculation presented in the text is obtained by assuming $a = b$ and the same effective mass $m^* = 0.067\,m$ for the conduction band edges in the well region and the barrier regions

will discuss the energy band structures of superlattices in detail, which are very important for understanding the electronic and optical properties of superlattices. It should be noted that there have been reported various types of theories to deal with the energy band structures of superlattices and that to review all of them is beyond the ability of the author. Instead, we will discuss here the tight binding approximation, which we believe to be easy to understand and to provide a good interpretation of the physical properties of superlattices. Then we will try to compare the theoretical results with experiment.

In order to get an insight into the energy band structures of superlattices, we discuss first the most common method, **Kronig–Penney model**. This is well known as one of the methods to understand the quasi-one-dimensional band structure. Let us consider a superlattice grown in the z direction in a semiconductor system with very small lattice mismatch such as GaAs/$Al_xGa_{1-x}As$. When we disregard electron motion in the x, y plane ($k_x = k_y = 0$), the electronic properties are obtained from the energy states in the z direction only. Consider a superlattice with a well width a, barrier width b and superlattice period $d = a+b$ as shown in Fig. 8.25. For simplicity, we assume that the electronic properties of both semiconductors are almost the same but that the difference in electron affinities results in a conduction band discontinuity, $U = 0.3$ eV. Expressing the effective masses of the conduction band edges in the semiconductors as m^*_{A} and m^*_{B}, the Bloch wave vector as $k_z = k$, and the energy as $\mathcal{E}$, the Kronig–Penney model leads to the following relations:

$$\cos(kd) = \cos(k_1 a)\cos(k_2 b) - \frac{k_1^2 + k_2^2}{2k_1 k_2}\sin(k_1 a)\sin(k_2 b) \quad \text{for } \mathcal{E} > U\,, \tag{8.216}$$

and

$$\cos(kd) = \cos(k_1 a)\cosh(\kappa b) - \frac{k_1^2 - \kappa^2}{2k_1 \kappa}\sin(k_1 a)\sinh(\kappa b) \quad \text{for } \mathcal{E} < U\,, \tag{8.217}$$

where k_1, k_2 and κ are defined by the following relations:

$$\mathcal{E} = \frac{\hbar^2 k_1^2}{2m_A^*}, \tag{8.218a}$$

$$\mathcal{E} - U = \frac{\hbar^2 k_2^2}{2m_B^*} \quad \text{for } \mathcal{E} > U, \tag{8.218b}$$

$$U - \mathcal{E} = \frac{\hbar^2 \kappa^2}{2m_B^*} \quad \text{for } \mathcal{E} < U. \tag{8.218c}$$

Equations (8.216) and (8.217) are easily solved numerically. For simplicity we assume $a = b$, $d = 2a$ and $m_A^* = m_B^* = m^* = 0.067\,m$ in Fig. 8.25 and the calculated result is shown in Fig. 8.26, where the horizontal axis is the well width or the barrier width a and the shaded areas represent the regions where energy eigenvalues exist. When the barrier width a is large, the probability for electrons to penetrate through the barrier decreases and electrons are confined in the well, giving rise to discrete energy levels. In this case, therefore, electrons are confined in the well regions and the electronic states are the same as the two-dimensional electron gas we have already discussed. Even in the case of large a, electrons with high energy

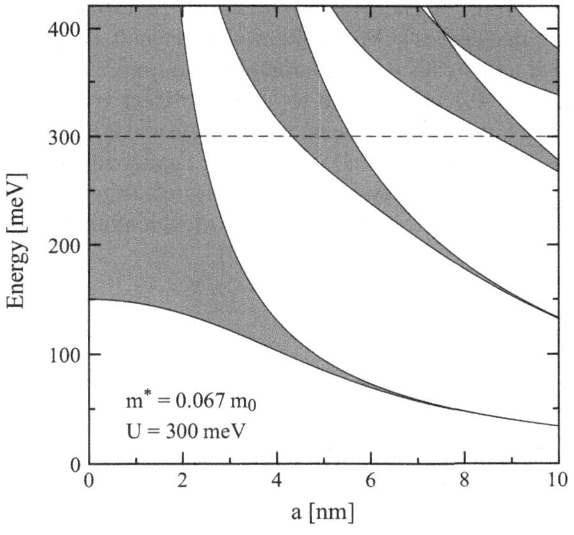

Fig. 8.26 Electronic states of superlattices calculated from the Kronig–Penney model plotted as a function of the barrier width (well width), with the assumption of the same width for the well a and the barrier b ($a = b$), a superlattice period $d = a + a$, the same effective mass of the conduction band edges, $m^* = 0.067\,m$, and a conduction band discontinuity $U = 0.3\,\text{eV}$. For large a, the electrons are confined within the well layers, resulting in discrete energy levels, whereas for higher energies or smaller a, the electrons penetrate into the barrier layers and form mini-bands

feel effectively lower barrier height and such electrons can penetrate into the barrier, resulting in the formation of energy bands (mini-bands). On the other hand, when the barrier width a is thin, electrons can penetrate into the barrier and form energy bands (mini-bands). Such bands are called **mini-bands**.

8.4.2 *Effect of Brillouin Zone Folding*

As seen in the discussion of the previous section, a superlattice consisting of materials with lattice constants a and b has a period $a + b$. Therefore, the first Brillouin zone edge of such a superlattice is given by $2\pi/(a + b)$, indicating that a larger unit cell results in a smaller Brillouin zone. In other words, Brillouin zones of bulk materials, $2\pi/a$ and $2\pi/b$, are folded into a smaller zone with $2\pi/(a+b)$ in a superlattice. As a result, many mini-bands are formed in a superlattice. This is a typical feature for superlattices and is called the **zone-folding effect**. It is very interesting to investigate the zone-folding effect of direct and indirect transition semiconductors. Figure 8.27a,b shows the energy band structures of direct and indirect bad gap semiconductors, respectively. Let us assume that we have two semiconductors with similar electronic properties and that a superlattice with mono-atomic layers is grown layer by layer. Therefore, the energy band structure is zone-folded at the point $2\pi/2a$, as shown by the dotted curves in Fig. 8.27a or b. As a result, the zone-folding of indirect gap semiconductors causes the conduction band minimum of the superlattice to be located at $\boldsymbol{k} = 0$ and therefore to a **quasi-direct transition**. For example, GaAs and AlAs have almost the same lattice constant and it is well known that GaAs and AlAs are respectively direct and indirect gap semiconductors. From the above considerations, therefore, we may expect a quasi-direct transition in the superlattice $(GaAs)_1/(AlAs)_1$ due to the zone-folding effect in which the conduction band edge

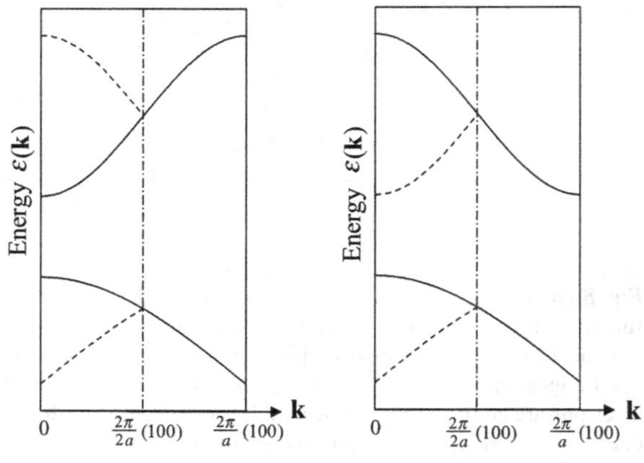

Fig. 8.27 Brillouin zone-folding effect in superlattices, where the band structures are illustrated for the change in the lattice constant from a to $2a$, for **a** a direct band gap semiconductor and **b** an indirect band gap semiconductor, where the dotted curves show zone-folded bands schematically

at the X point of AlAs is folded to the Γ point. It has been pointed out that the use of superlattices makes it possible to control energy band structures and to produce materials with new properties. For this reason, this kind of technology is sometimes called **band gap engineering**.

The crystal structure of GaAs and AlAs is the zinc blende type as shown in Fig. 1.4, where face-centered cubic lattices with different atoms are displaced by a distance of $a/4$ from each other in the diagonal direction. The bulk crystals belong to the point group $T_d(\bar{4}3\,m)$ but the superlattices of the layer structures of these crystals have lower symmetry. For example, the Bravais lattice of a superlattice $(GaAs)_n/(AlAs)_m$ with n layers of GaAs and m layers of AlAs depends on the total number of layers in a period, $n + m$. If it even, the lattice is simple tetragonal, belonging to the space group D_{2d}^5 $(P\bar{4}m2)$; if it is odd, the lattice is body-centered tetragonal, belonging to the space group D_{2d}^9 $(I\bar{4}m2)$. In the following we assume that GaAs and AlAs are lattice matched with the lattice constant a_0. Figure 8.28 shows the crystal structure of the $(GaAs)_1/(AlAs)_1$ superlattice. We may show that the lattice constants of the superlattice $(GaAs)_n/(AlAs)_m$ are given by $a = a_0/\sqrt{2}$ and $c = (n + m)a_0/2$. Therefore, the unit cell volume of the superlattice $(GaAs)_n/(AlAs)_m$ is evaluated to be $a_0^3/4$ multiplied by $n+m$. The reciprocal lattice vectors are easily calculated from this result.

Using the reciprocal lattice vectors, the first Brillouin zone of the $(GaAs)_n/(AlAs)_m$ superlattice is calculated which is presented in Figs. 8.29 and 8.30. Figure 8.29 shows the first Brillouin zone for $n + m = 2$ $(n = m = 1)$ and Fig. 8.30 is for $n + m = 3$, where the points at the zone boundaries of the superlattice are marked by labels such as $\underline{Z}$ and $\underline{X}$ in order to distinguish the critical points of the superlattice from those of zinc-blende crystal. In the text, the notation without an underline is used for the energy band structures shown later. The first Brillouin zone for other combinations, $n + m$, of the $(GaAs)_n/(AlAs)_m$ superlattice may be calculated in a similar fashion, giving rise to a simple tetragonal the Bravais lattice for an even value of $n+m$ and to a body-centered tetragonal for an odd value of $n + m$. With these results the energy band structures of superlattices are calculated.

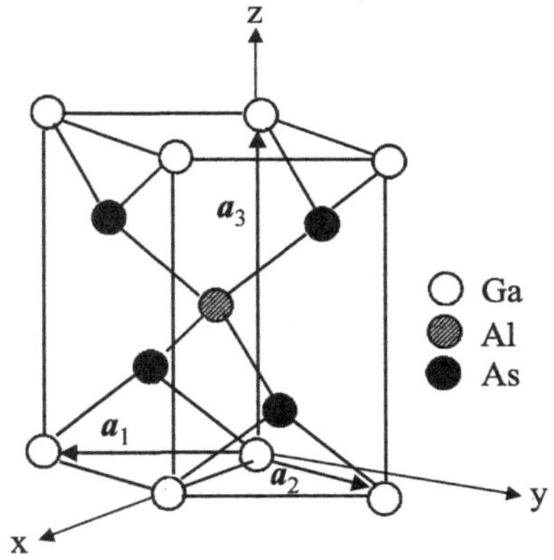

Fig. 8.28 Crystal structure of $(GaAs)_1/(AlAs)_1$ superlattice. Two zinc blende crystals are layered along the growth direction

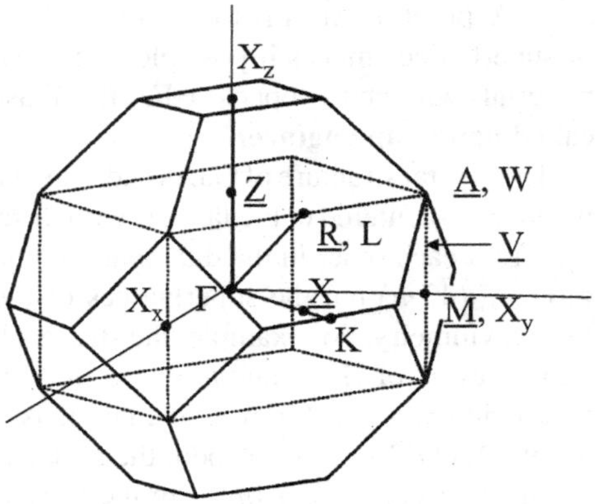

Fig. 8.29 The first Brillouin zone of the $(GaAs)_n/(AlAs)_m$ superlattice for $n + m = 2$ along with the first Brillouin zone of the zinc blende crystal. In general, the first Brillouin zone of a superlattice with an even value of $n + m$ differs only in the folding in the direction X_z, resulting in a simple tetragonal lattice

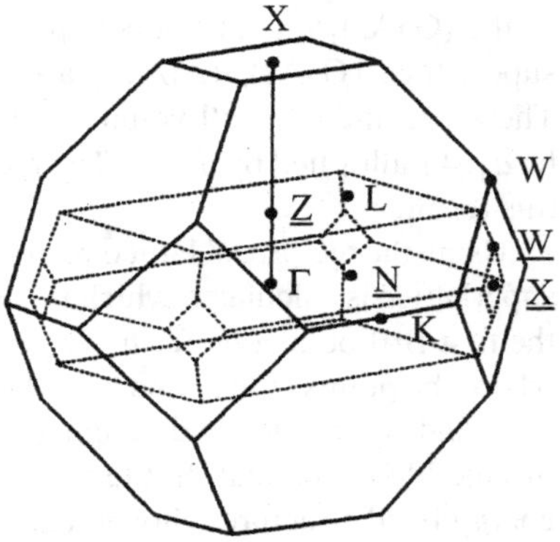

Fig. 8.30 The first Brillouin zone of the $(GaAs)_n/(AlAs)_m$ superlattice for $n + m = 3$ along with the first Brillouin zone of the zinc blende crystal. In general, the first Brillouin zone of a superlattice with an odd value of $n + m$ differs only in the folding in the direction X_z, resulting in a body-centered tetragonal lattice

Here and in the following sections we will compare the theoretical calculations with the experimental results for $(GaAs)_n/(AlAs)_n$ superlattices as an example. There have been published many papers on the theoretical calculations of superlattice energy band structures and experiments so far, and it is not the aim of this textbook to cover all of them. Therefore, we will present the theory of the tight binding approximation and compare the calculations with the experimental results for the photoreflectance and photoluminescence [43–45]. The bottom of the conduction band of GaAs is located at the Γ point and that of AlAs is at the X point. Therefore, if we change the layer number n of the $(GaAs)_n/(AlAs)_n$ superlattice, we can expect a crossing of the two conduction band edges, resulting in a crossing between type I and type II superlattices. A typical experiment to show the crossing is the photoluminescence method [46–57]. On the other hand, energy band calculations for $(GaAs)_n/(AlAs)_n$ superlattices have been carried out by the envelope function model [58], tight binding approximation [59–66], empirical and self-consistent pseudopotential methods [67–70], local density functional method [71–76] and augmented-spherical-wave method [77].

8.4.3 Tight Binding Approximation

The tight binding approximation is based on the linear combination of atomic orbitals, where the chemical bonds of a few atomic orbitals are considered, and has been successfully applied to calculation of the energy band structures of solids [78]. Although an increase in the number of atomic orbitals gives better accuracy for the calculated energy band structure, the time required for computation increases. The most popular tight binding method to calculate the energy band structure of semiconductors takes account of four atomic orbitals of s, p_x, p_y and p_z. In general a set of orthogonal and normalized atomic orbitals is obtained by the method of Löwin [79].

The wave functions $|\psi\rangle$ of a system satisfy the following eigenvalue equation:

$$H|\psi\rangle = \mathcal{E}|\psi\rangle \,, \tag{8.219}$$

where $\mathcal{E}$ is the energy eigenvalue. When the wave function $|\psi\rangle$ is expanded in atomic orbital functions $|\phi_j(\boldsymbol{r})\rangle$, we obtain

$$|\psi\rangle = \sum_j C_j|\phi_j\rangle \,. \tag{8.220}$$

According to the variational principle, the energy

$$\mathcal{E} = \frac{\langle\psi|H|\psi\rangle}{\langle\psi|\psi\rangle} \,, \tag{8.221}$$

is minimized by changing the expansion coefficient C_j. From this we find

$$\left(H_{ij} - \mathcal{E}\delta_{ij}\right) C_j = 0 \,. \tag{8.222}$$

The condition for the above equation to have a reasonable solution is

$$\det\left|H_{ij} - \mathcal{E}\delta_{ij}\right| = 0 \,, \tag{8.223}$$

where

$$H_{ij} = \langle\phi_i|H|\phi_j\rangle \,. \tag{8.224}$$

When the basis functions $|\phi_j\rangle$ consist of a limited number of atomic orbitals, the functions do not satisfy the condition of normalization, orthogonality and completeness. Therefore, a set of basis functions which satisfy normalization and orthogonality should be chosen. Usually the energy band calculations are carried out for Löwdin orbitals composed of atomic orbitals. The Löwdin orbital is defined by the following relation:

$$|\phi_i^{\mathrm{L}}\rangle = \sum_{ij} S_{ij}^{-1} |\phi_j\rangle\,, \tag{8.225}$$

$$S_{ij} = \langle \phi_i | \phi_j \rangle\,. \tag{8.226}$$

Using the Löwdin orbitals $|\phi_i^{\mathrm{L}}\rangle$ for the basis functions, (8.223) reduces to

$$\det \left| H_{ij}^{\mathrm{L}} - \mathcal{E}\delta_{i,j} \right| = 0\,, \tag{8.227}$$

where

$$H_{ij}^{\mathrm{L}} = \langle \phi_i^{\mathrm{L}} | H | \phi_j^{\mathrm{L}} \rangle\,. \tag{8.228}$$

In the following tight binding approximation, the Löwdin orbitals are used as the atomic orbitals.

Energy band calculations have been carried out utilizing the cyclic boundary condition. Here we assume zinc blende crystals and define the Bloch sum (quasi-atomic functions) defined by the following relation, which are composed of atomic orbitals and satisfy the cyclic boundary condition:

$$|nb\boldsymbol{k}) = \frac{1}{\sqrt{N}} \sum_{\boldsymbol{R}_i} \exp\left[\mathrm{i}\boldsymbol{k}\cdot(\boldsymbol{R}_i + \boldsymbol{v}_b)\right] |nb\boldsymbol{R}_i)\,, \tag{8.229}$$

where N is the number of unit cells, b represents a (anion) or c (cation), $\boldsymbol{R}_i$ is the position of the anion, $\boldsymbol{v}_b$ is $\delta_{b,c}(a_0/4)(1, 1, 1)$, and n is the orbital function s, p_x, p_y or p_z. The Schrödinger equation for the Bloch function $|\boldsymbol{k}\lambda\rangle$ is written as

$$[H - \mathcal{E}(\boldsymbol{k}, \lambda)]\, |\boldsymbol{k}\lambda\rangle = 0\,, \tag{8.230}$$

where λ is the band index. The above equation may be expressed by matrix representation as

$$\sum_{n,b'} \left[(nb\boldsymbol{k}|H|n'b'\boldsymbol{k}) - \mathcal{E}(\boldsymbol{k}, \lambda)\delta_{n,n'}, \delta_{b,b'}\right] (n'b'\boldsymbol{k}|\boldsymbol{k}\lambda\rangle = 0\,. \tag{8.231}$$

The Bloch function $|\boldsymbol{k}\lambda\rangle$ is written as the following with the use of the Bloch sum $|nb\boldsymbol{k})$:

$$|\lambda\boldsymbol{k}\rangle = \sum_{n,b} |nb\boldsymbol{k})(nb\boldsymbol{k}|\boldsymbol{k}\lambda\rangle\,. \tag{8.232}$$

The matrix element $(nb\boldsymbol{k}|H|n'b'\boldsymbol{k})$ is written as

$$(nb\boldsymbol{k}|H|n'b'\boldsymbol{k}) = \frac{1}{N} \sum_{\boldsymbol{R}_i,\boldsymbol{R}_j} \mathrm{e}^{\mathrm{i}\boldsymbol{k}\cdot(\boldsymbol{R}_i - \boldsymbol{R}_j + \boldsymbol{v}_b - \boldsymbol{v}_{b'})} (nb\boldsymbol{R}_j|H|n'b'\boldsymbol{R}_i)\,. \tag{8.233}$$

Table 8.3 Inter-atomic matrix elements $(nb0|H|n'b'\boldsymbol{R}_m)$. l, m and n are the direction cosines of $\boldsymbol{R}_m$ and the notation in the right-hand column is the representation of Slater-Koster. (After [78])

$\mathcal{E}_{ss} \equiv (sb0|H|sb'\boldsymbol{R}_m)$	$(ss\sigma)$
$\mathcal{E}_{sx} \equiv (sb0|H|p_x b'\boldsymbol{R}_m)$	$l(sp\sigma)$
$\mathcal{E}_{xx} \equiv (p_x b0|H|p_x b'\boldsymbol{R}_m)$	$l^2(pp\sigma) + (1-l^2)(pp\pi)$
$\mathcal{E}_{xy} \equiv (p_x b0|H|p_y b'\boldsymbol{R}_m)$	$lm(pp\sigma) - lm(pp\pi)$
$\mathcal{E}_{xz} \equiv (p_x b0|H|p_z b'\boldsymbol{R}_m)$	$ln(pp\sigma) - ln(pp\pi)$

Here, we assume that $(nb\boldsymbol{R}_j|H|n'b'\boldsymbol{R}_i)$ depends only on $\boldsymbol{R}_m = \boldsymbol{R}_i - \boldsymbol{R}_j + \boldsymbol{v}_b - \boldsymbol{v}_{b'}$. This assumption is called the **two-center approximation** [78]. Using this assumption (8.233) reduces to

$$(nb\boldsymbol{k}|H|n'b'\boldsymbol{k}) = \sum_{\boldsymbol{R}_m} \exp(i\boldsymbol{k} \cdot \boldsymbol{R}_m)(nb0|H|n'b'\boldsymbol{R}_m), \tag{8.234}$$

where $(nb0|H|n'b'\boldsymbol{R}_m)$ is called the inter-atomic matrix element and is given by Table 8.3. $(nn'\sigma)$ and $(nn'\pi)$ are, respectively, the inter-atomic matrix elements for the σ bonding and π bonding between the atomic orbitals n and n'. These inter-atomic matrix elements are determined empirically or semi-empirically and the accuracy of the tight binding approximation is determined by these values. It should be pointed out here that the inter-atomic matrix elements are proportional to the square of the bond length d (d^{-2}) [9]. Chadi and Cohen [80] have calculated the energy band structures of semiconductors by taking account of the atomic orbitals s, p_x, p_y and p_z and the nearest-neighbor interactions, but they failed to obtain the band structures of indirect band gap semiconductors. Two different methods have been reported in order to solve this inconsistency. One is to take account of the second nearest-neighbor interactions of the atomic orbitals in addition to the nearest neighbor interaction [43]. The other is to use the excited state s^* in addition to the sp^3 orbitals [10]. In the following we will discuss the sp^2s^* tight binding approximation first and then the second nearest-neighbor approximation in connection with the energy band calculation of superlattices.

8.4.4 sp^3s^* Tight Binding Approximation

It has been pointed out that the tight binding approximation interprets the valence band structure but fails in describing conduction bands with high energy [9]. The reason for this disagreement is understood to be that the wave function of an electron in a conduction band, not localized in the crystal, cannot be described by the wave functions localized around the atoms. This is also the reason why the sp^3 tight binding approximation fails in describing the conduction band in indirect gap semi-

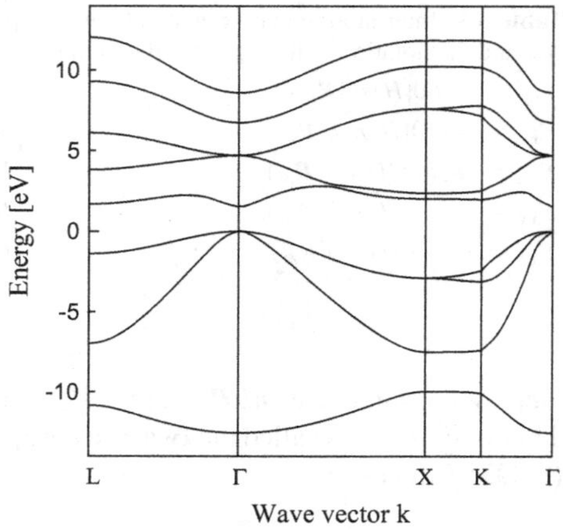

Fig. 8.31 Energy band structure of GaAs calculated by the sp^3s^* tight binding approximation, which gives the direct band gap at the Γ point

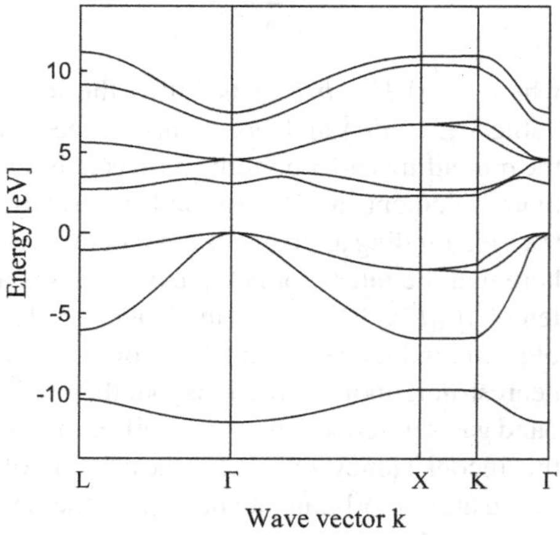

Fig. 8.32 Energy band structure of AlAs calculated by the sp^3s^* tight binding approximation, which gives the indirect band gap in contrast to the sp^3 tight binding approximation

conductors. Vogl et al. [10] have proposed a new method to describe the conduction band by introducing an excited s state, the s^* orbital, in addition to the s and p orbitals for the tight binding basis functions. This method is called the sp^3s^* tight binding approximation. First we will show the method proposed by Vogl et al. and the energy band structures of III–V compound semiconductors.

In the calculation of the sp^3s^* tight binding approximation, the atomic orbitals n of (8.232) are five orbitals for the anion and 5 orbitals for cations, a total of ten orbitals. Therefore, Hamiltonian matrix for the basis functions $|n\boldsymbol{b}\boldsymbol{k}\rangle$ results in a 10×10 matrix. These elements are given by Table (A) of reference [10], and are also shown in Sect. 8.4.5 of this textbook. From the solutions of the determinant the band structures of III–V semiconductors are obtained. As an example, the band structures of GaAs and AlAs are calculated by using the parameters given by Vogl et al. [10], which are shown in Figs. 8.31 and 8.32, respectively. It is found in Fig. 8.32 that the

sp^3 tight binding approximation gives the indirect band gap band structure for AlAs, which explains the experimental results quite well, whereas the sp^3 tight binding approximation failed in explaining it. The conduction band minima are located near the X point.

8.4.5 Energy Band Calculations for Superlattices

Let us consider a superlattice $(\mathrm{ca})_n/(\mathrm{CA})_m$ consisting of a layer structure of two different zinc blende-type materials such as n layers of ca and m layers of CA, periodically grown on the (001) surface, where c and C are cations, and a and A are anions. The superlattice $(\mathrm{ca})_n/(\mathrm{CA})_m$ has $2(n+m)$ atoms in the unit cell (the position vector is $\boldsymbol{R}_i$ and i is the index for each unit cell) and the respective atom has five orbitals of s, p_x, p_y, p_z and s^*. Then the Bloch sum is written as the following:

$$|\chi_j^{\alpha}(\boldsymbol{k})\rangle = \frac{1}{\sqrt{N}} \sum_{\boldsymbol{R}_i} \exp\left[i\boldsymbol{k} \cdot (\boldsymbol{R}_i + \boldsymbol{\tau}_j) \right] |\alpha j\rangle , \tag{8.235}$$

where $\alpha = s, p_x, p_y, p_z, s^*$ and the subscript j indicates the atom in the unit cell with the position vector $\boldsymbol{\tau}_j$. The Hamiltonian matrix of the superlattice for the basis functions is given by

$$\hat{H} = \begin{array}{c} \\ 1 \\ 2 \\ 3 \\ \vdots \\ \vdots \\ 2n \\ 1 \\ 2 \\ 3 \\ \vdots \\ \vdots \\ 2m \end{array} \begin{array}{c} \begin{array}{cccccccccccc} 1 & 2 & 3 & \cdots & \cdots & 2n & 1 & 2 & 3 & \cdots & \cdots & 2m \end{array} \\ \left(\begin{array}{cccccccccccc} \hat{a} & \widehat{ac} & & & & & & & & & & \widehat{Ca}^{\dagger} \\ & \hat{c} & \widehat{ca} & & & & & & & & & \\ & & \hat{a} & \ddots & & & & & 0 & & & \\ & & & \ddots & \ddots & & & & & & & \\ & & & & \ddots & \widehat{ac} & & & & & & \\ & & & & & \hat{c} & \widehat{cA} & & & & & \\ & & & & & & \hat{A} & \widehat{AC} & & & & \\ & & & & & & & \hat{C} & \widehat{CA} & & & \\ & & & & & & & & \hat{A} & \ddots & & \\ & & & & & & & & & \ddots & \ddots & \\ & h.c. & & & & & & & & & \hat{A} & \widehat{AC} \\ & & & & & & & & & & & \hat{C} \end{array} \right) \end{array} \tag{8.236}$$

where $h.c.$ means the Hermite conjugate and the matrix elements of the above equation are given by the following matrix:

$$\widehat{a} = \begin{array}{c} s \\ p_x \\ p_y \\ p_z \\ s^* \end{array} \overset{\begin{array}{ccccc} s & p_x & p_y & p_z & s^* \end{array}}{\begin{pmatrix} \mathcal{E}_{sa} & 0 & 0 & 0 & 0 \\ 0 & \mathcal{E}_{pa} & 0 & 0 & 0 \\ 0 & 0 & \mathcal{E}_{pa} & 0 & 0 \\ 0 & 0 & 0 & \mathcal{E}_{pa} & 0 \\ 0 & 0 & 0 & 0 & \mathcal{E}_{s^*a} \end{pmatrix}} \tag{8.237a}$$

$$\widehat{c} = \begin{array}{c} s \\ p_x \\ p_y \\ p_z \\ s^* \end{array} \overset{\begin{array}{ccccc} s & p_x & p_y & p_z & s^* \end{array}}{\begin{pmatrix} \mathcal{E}_{sc} & 0 & 0 & 0 & 0 \\ 0 & \mathcal{E}_{pc} & 0 & 0 & 0 \\ 0 & 0 & \mathcal{E}_{pc} & 0 & 0 \\ 0 & 0 & 0 & \mathcal{E}_{pc} & 0 \\ 0 & 0 & 0 & 0 & \mathcal{E}_{s^*c} \end{pmatrix}} \tag{8.237b}$$

$$\widehat{ac} = \begin{array}{c} s \\ p_x \\ p_y \\ p_z \\ s^* \end{array} \overset{\begin{array}{ccccc} s & p_x & p_y & p_z & s^* \end{array}}{\begin{pmatrix} V_{ss}g_0 & V_{sapc}g_1 & V_{sapc}g_1 & V_{sapc}g_0 & 0 \\ -V_{scpa}g_1 & V_{xx}g_0 & V_{xy}g_0 & V_{xy}g_1 & -V_{s^*cpa}g_1 \\ -V_{scpa}g_1 & V_{xy}g_0 & V_{xx}g_0 & V_{xy}g_1 & -V_{s^*cpa}g_1 \\ -V_{scpa}g_0 & V_{xy}g_1 & V_{xy}g_1 & V_{xx}g_0 & -V_{s^*cpa}g_0 \\ 0 & V_{s^*apc}g_1 & V_{s^*apc}g_1 & V_{s^*apc}g_0 & 0 \end{pmatrix}}$$

$$\widehat{ca} = \begin{array}{c} s \\ p_x \\ p_y \\ p_z \\ s^* \end{array} \overset{\begin{array}{ccccc} s & p_x & p_y & p_z & s^* \end{array}}{\begin{pmatrix} V_{ss}g_2 & -V_{scpa}g_3 & V_{scpa}g_3 & V_{scpa}g_2 & 0 \\ V_{sapc}g_3 & V_{xx}g_2 & -V_{xy}g_2 & -V_{xy}g_3 & V_{s^*apc}g_3 \\ -V_{sapc}g_3 & -V_{xy}g_2 & V_{xx}g_2 & V_{xy}g_3 & -V_{s^*apc}g_3 \\ -V_{sapc}g_2 & -V_{xy}g_3 & V_{xy}g_3 & V_{xx}g_2 & -V_{s^*apc}g_2 \\ 0 & -V_{s^*cpa}g_3 & V_{s^*cpa}g_3 & V_{s^*cpa}g_2 & 0 \end{pmatrix}}$$

The phase factors g_i $(i = 0 \ldots 4)$ are given by the following equations with $\boldsymbol{k} = (\xi, \eta, \zeta)2\pi/a_0$: (see footnote[5] for the derivation)

[5]The phase factors are obtained as follows. For example, in a bulk material, let the position vectors of the four atoms next to a central anion $\boldsymbol{d}_1 = (111)a/4$, $\boldsymbol{d}_2 = (\bar{1}1\bar{1})a/4$, $\boldsymbol{d}_3 = (\bar{1}\bar{1}1)a/4$, $\boldsymbol{d}_4 = (1\bar{1}\bar{1})a/4$ and we have

$$g_0 = e^{i\boldsymbol{k}\cdot\boldsymbol{d}_1} + e^{i\boldsymbol{k}\cdot\boldsymbol{d}_2} + e^{i\boldsymbol{k}\cdot\boldsymbol{d}_3} + e^{i\boldsymbol{k}\cdot\boldsymbol{d}_4}$$
$$g_1 = e^{i\boldsymbol{k}\cdot\boldsymbol{d}_1} + e^{i\boldsymbol{k}\cdot\boldsymbol{d}_2} - e^{i\boldsymbol{k}\cdot\boldsymbol{d}_3} - e^{i\boldsymbol{k}\cdot\boldsymbol{d}_4}$$
$$g_2 = e^{i\boldsymbol{k}\cdot\boldsymbol{d}_1} - e^{i\boldsymbol{k}\cdot\boldsymbol{d}_2} + e^{i\boldsymbol{k}\cdot\boldsymbol{d}_3} - e^{i\boldsymbol{k}\cdot\boldsymbol{d}_4}$$
$$g_3 = e^{i\boldsymbol{k}\cdot\boldsymbol{d}_1} - e^{i\boldsymbol{k}\cdot\boldsymbol{d}_2} - e^{i\boldsymbol{k}\cdot\boldsymbol{d}_3} + e^{i\boldsymbol{k}\cdot\boldsymbol{d}_4}$$

$$g_0 = \frac{1}{2}\exp\left(\mathrm{i}\frac{\zeta}{2}\right)\cos\left(\frac{\xi+\eta}{2}\right), \tag{8.238a}$$

$$g_1 = \frac{\mathrm{i}}{2}\exp\left(\mathrm{i}\frac{\zeta}{2}\right)\sin\left(\frac{\xi+\eta}{2}\right), \tag{8.238b}$$

$$g_2 = \frac{1}{2}\exp\left(\mathrm{i}\frac{\zeta}{2}\right)\cos\left(\frac{\xi-\eta}{2}\right), \tag{8.238c}$$

$$g_3 = \frac{\mathrm{i}}{2}\exp\left(\mathrm{i}\frac{\zeta}{2}\right)\sin\left(\frac{\xi-\eta}{2}\right). \tag{8.238d}$$

The position vectors of a superlattice are also obtained in a similar fashion. In addition the position vectors of the next nearest neighbors $g_4 \ldots g_{11}$ are also derived in a similar way. The matrix elements $\widehat{ac}$, $\widehat{ca}$ and so on represent the nearest neighbor interaction in a bulk material, and $\widehat{cA}$ and $\widehat{Ca}$ indicate the interaction at the heterointerface.

We have derived the Hamiltonian matrix to calculate the energy band structure of an arbitrary $n+m$ superlattice, which will give us eigenvalues and eigenstates at an arbitrary point $\boldsymbol{k}$ in the Brillouin zone. However, we have to note that the energy band calculations require several parameters to be determined empirically. Before calculating the energy band structures of superlattices, we shall derive the Hamiltonian matrix of a bulk semiconductor such as GaAs and AlAs. The bulk matrix elements are obtained by putting $n=1$ and $m=0$.

$$\widehat{H}(\text{bulk}) = \widehat{H}(n=1, m=0) = \begin{vmatrix} \widehat{a} + \widehat{aa} + \widehat{aa}^{\dagger} & \widehat{ac} + \widehat{ca}^{\dagger} \\ \widehat{ac}^{\dagger} + \widehat{ca} & \widehat{c} + \widehat{cc} + \widehat{cc}^{\dagger} \end{vmatrix} \tag{8.239}$$

represents the bulk Hamiltonian matrix elements of the tight binding approximation for a bulk semiconductor such as GaAs, and reduces to the matrix of Table (A) of Vogl et al. [10]. The parameters for the tight binding approximation are given by the following, which are determined empirically:

$$\mathcal{E}_{s\pm} = \left[\mathcal{E}(\Gamma_1^{\mathrm{c}}) + \mathcal{E}(\Gamma_1^{\mathrm{v}}) \pm \Delta\mathcal{E}_s\right], \tag{8.240a}$$

$$\mathcal{E}_{p\pm} = \mathcal{E}_{xx\pm} = \left[\mathcal{E}(\Gamma_{15}^{\mathrm{c}}) + \mathcal{E}(\Gamma_{15}^{\mathrm{c}})_{15}^{\mathrm{v}} \pm \Delta\mathcal{E}_p\right], \tag{8.240b}$$

$$V_{ss} = 4\mathcal{E}_{ss}\left(\frac{1}{2}\frac{1}{2}\frac{1}{2}\right) = -\sqrt{\mathcal{E}_{s+}\mathcal{E}_{s-} - \mathcal{E}(\Gamma_1^{\mathrm{c}})\mathcal{E}(\Gamma_1^{\mathrm{v}})}, \tag{8.240c}$$

$$V_{xx} = 4\mathcal{E}_{xx}\left(\frac{1}{2}\frac{1}{2}\frac{1}{2}\right) = \sqrt{\mathcal{E}_{p+}\mathcal{E}_{p-} - \mathcal{E}(\Gamma_{15}^{\mathrm{c}})\mathcal{E}(\Gamma_{15}^{\mathrm{v}})}, \tag{8.240d}$$

$$V_{xy} = 4\mathcal{E}_{xy}\left(\frac{1}{2}\frac{1}{2}\frac{1}{2}\right) = \sqrt{\mathcal{E}_{p+}\mathcal{E}_{p-} - \mathcal{E}(X_5^{\mathrm{c}})\mathcal{E}(X_5^{\mathrm{v}})}, \tag{8.240e}$$

$$V_{s\pm p\mp} = 4\mathcal{E}_{s\pm p\mp}\left(\frac{1}{2}\frac{1}{2}\frac{1}{2}\right) = \sqrt{[\mathcal{E}_{s\pm}C_{\pm} - D_{\pm}]/[\mathcal{E}_{s\pm} - \mathcal{E}_{s^*\pm}]}, \tag{8.240f}$$

$$V_{s^*\pm p\mp} = 4\mathcal{E}_{s^*\pm p\mp}\left(\frac{1}{2}\frac{1}{2}\frac{1}{2}\right) = \sqrt{[\mathcal{E}_{s^*\pm}C_{\pm} - D_{\pm}]/[\mathcal{E}_{s^*\pm} - \mathcal{E}_{s\pm}]}. \tag{8.240g}$$

where

$$
\begin{aligned}
C_\pm &= \mathcal{E}_{s\pm}\mathcal{E}_{p\mp} - \mathcal{E}(X_\pm^{\rm v})\mathcal{E}(X_\pm^{\rm c}) \\
&\quad +[\mathcal{E}_{s\pm} + \mathcal{E}_{p\mp} - \mathcal{E}(X_\pm^{\rm v}) - \mathcal{E}(X_\pm^{\rm c})][\mathcal{E}_{s^*\pm} - \mathcal{E}(X_\pm^{\rm v}) - \mathcal{E}(X_\pm^{\rm c})] \,,
\end{aligned}
\tag{8.241a}
$$

$$
D_\pm = \det \begin{vmatrix} \mathcal{E}_{s\pm} & \mathcal{E}(X_\pm^{\rm c}) & \mathcal{E}(X_\pm^{\rm v}) \\ \mathcal{E}(X_\pm^{\rm c}) & \mathcal{E}_{p\mp} & \mathcal{E}(X_\pm^{\rm v}) \\ \mathcal{E}(X_\pm^{\rm c}) & \mathcal{E}(X_\pm^{\rm c}) & \mathcal{E}_{s^*\pm} \end{vmatrix} \,. \tag{8.241b}
$$

Here $\mathcal{E}(X_+) = \mathcal{E}(X_3)$, $\mathcal{E}(X_-) = \mathcal{E}(X_1)$, where the subscripts + and − represent the anion and the cation, respectively, and $\Delta\mathcal{E}(\alpha) = \mathcal{E}_{\alpha c} - \mathcal{E}_{\alpha a}$. In the above equation the representation of Slater–Koster is used (see [78]). A convenient method to deduce the parameters of the tight binding approximation has been reported by Yamaguchi [64].

A comparison between the energy band structures of superlattices calculated from the sp^3s^* tight binding approximation and deduced from experiment has been reported by Fujimoto et al. [44], where the transition energies determined from photoreflectance and photoluminescence experiments on $(\mathrm{GaAs})_n/(\mathrm{AlAs})_n$ ($n = 1 \ldots 15$) are compared with the theoretical calculations. Later Matsuoka et al. [45] pointed out that the assumptions used for the calculations are not correct. Photoreflectance signals are produced by the following mechanisms. Electron–hole pairs are excited by illumination of a weak intensity laser with photon energy larger than the energy band gap. The excited electrons (or holes) recombine at the sample surface and the other carriers, i.e. holes (electrons) induce an electric field at the surface, and thus the surface electric field is modulated without electrode contacts. Photoreflectance is therefore one of the techniques of modulation spectroscopy.

The experimental results at $T = 200\,\mathrm{K}$ for photoreflectance (PR) and photoluminescence (PL) on $(\mathrm{GaAs})_n/(\mathrm{AlAs})_n$ superlattices with $n = 8$ and $n = 12$ are shown in Figs. 8.33 and 8.34, respectively, where the best fit theoretical curves based on the Aspnes theory described in Sect. 5.1.3 are also plotted and the transition energies obtained from the best fit curves are indicated by the arrows. The arrows in Fig. 8.33 for the $(\mathrm{GaAs})_8/(\mathrm{AlAs})_8$ superlattice indicate the transition energies, 1.797 eV, 1.897 eV, 1.915 eV, and 1.951 eV, which are determined from the best fit of the theoretical curve to the experimental data. In this superlattice we find a weak structure around 1.797 eV below the main peak at about 1.9 eV. Since the direct transition results in a strong peak in the PR spectra, the weak structure seems to arise from a direct transition with a very weak transition probability. In other words, the weak transition is ascribed to the quasi-direct transition between the zone-folded X_z conduction band and the heavy-hole band (X_z–$\Gamma_{\rm h}$ transition). The main peak is ascribed to the direct transition between the Γ conduction band and the heavy-hole band (Γ–$\Gamma_{\rm h}$ transition), and the peak at about 1.95 eV is due to the direct transition between the Γ conduction band and the light-hole band (zone-folded valence band) (Γ–$\Gamma_{\rm l}$ transition). We find in Fig. 8.33 that PL peaks appear at 1.8515 eV and 1.902 eV and these peaks exhibit good agreement with the weak structure at 1.797 eV and the main peak at 1.9 eV observed in the PR spectra.

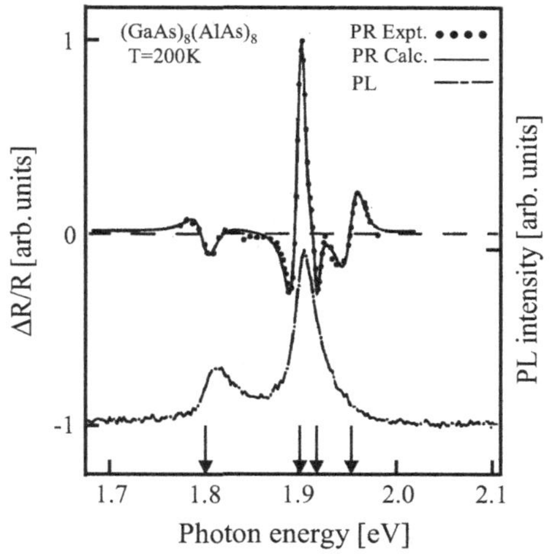

Fig. 8.33 Photoreflectance (PR) and photoluminescence (PL) spectra of the $(GaAs)_8/(AlAs)_8$ superlattice, where the solid curves are the theoretical results based on the Aspnes theory and the arrows indicate the transition energies determined from the analysis. The dot-dashed curves are photoluminescence data. ($T = 200\,K$)

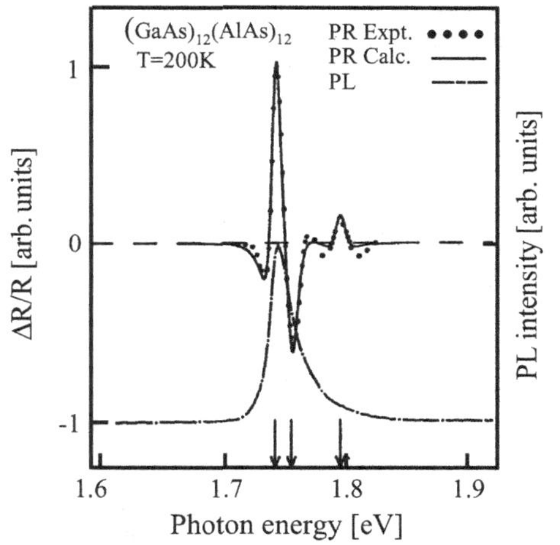

Fig. 8.34 Photoreflectance (PR) and photoluminescence (PL) spectra of the $(GaAs)_{12}/(AlAs)_{12}$ superlattice, where the solid curves are the theoretical results based on the Aspnes theory and the arrows indicate the transition energies determined from the analysis. The dot-dashed curves are photoluminescence data. ($T = 200\,K$)

Figure 8.34 shows the PR and PL spectra at 200 K for $(GaAs)_{12}/(AlAs)_{12}$ superlattice along with the best fit PR curve from the Aspnes theory. From the analysis of the PR spectra three transitions at 1.740 eV, 1.754 eV and 1.793 eV are obtained in the narrow region of photon energy from 1.7 eV to 1.85 eV. However, it is very interesting to point out that any weak structure such as observed in the PR spectra of the $(GaAs)_8/(AlAs)_8$ superlattice of Fig. 8.33 has not been resolved in the $(GaAs)_{12}/(AlAs)_{12}$ superlattice. This may be interpreted as showing that the X_z conduction band is located on the higher energy side than the Γ conduction band or that these two bands are mixed. The PL spectra show a peak at about 1.75 eV which is in good agreement with the peaks at 1.74 eV and 1.754 eV of the PR spectra. In the

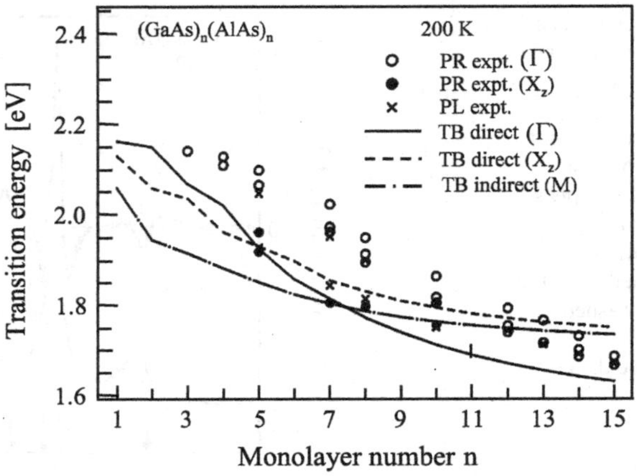

Fig. 8.35 Transition energies of $(GaAs)_n/(AlAs)_n$ superlattices calculated by the sp^3s^* tight binding approximation are shown as a function of the layer number n, where the solid, dotted and dot-dashed curves are for direct allowed, quasi-direct, and indirect transition energies. In the figure the experimental data obtained by Matsuoka et al. [45] are also shown. The calculations based on the sp^3s^* tight binding approximation show that a crossing from direct to indirect transitions occurs at $n = 8$ and the quasi-direct transition appears on the lower-energy side for $n < 5$

following a comparison will be described between these experimental results and calculations based on the tight binding approximation.

A comparison between experiments and calculations based on the sp^3s^* tight binding approximation is shown in Fig. 8.35, where a similar comparison has been reported by Fujimoto et al. [44]. In the calculations the valence band discontinuity is assumed to be 0.54 eV. It is found in Fig. 8.35 that the theoretical calculations reveal a crossing from direct to indirect transitions occurs at $n = 8$ and to a quasi-direct transition for $n < 5$. However, the experimental results exhibit the crossing between the direct and indirect or quasi-direct (X_z) transitions around $n = 12$ and that there exists a considerable disagreement between the experimental data and the calculations in the region of small n. The calculations show that the indirect transition energy is the lowest for $n < 8$, which corresponds to the R point for $n = 1$ and the M point otherwise. The reason for this disagreement may be ascribed to the fact that the sp^3s^* tight binding approximation does not take account of the correct value of the effective mass of the X valleys of AlAs [45, 81].

8.4.6 Second Nearest-Neighbor sp^3 Tight Binding Approximation

We have already pointed out that Brillouin zone folding plays an important role in the electronic and optical properties of superlattices. In particular, the states X_z and

X_{xy} strongly reflect the magnitude and anisotropy of the effective mass of the X valleys. Therefore, the energy band calculations require the exact effective mass of the X valleys. It has been shown that the sp^3s^* tight binding approximation explains the indirect transition in some III–V semiconductors. However, the method will not guarantee the anisotropy of the effective mass at the bottom of the conduction band valleys. We may easily understand that more parameters are required to calculate the effective mass anisotropy of the X valleys from the tight binding theory. The states X_z and X_{xy} in superlattices may be calculated accurately by taking account of the precise anisotropy of the effective mass at the X point. Focusing on this point, Lu and Sham [43] have calculated the energy band structure by using the second nearest-neighbor sp^3 tight binding approximation and succeeded in obtaining good agreement with the experimental results. Also Matsuoka et al. [45] have performed energy band calculations for $(GaAs)_n/(AlAs)_n$ superlattices by the second nearest-neighbor sp^3 tight binding approximation and explained their experimental results.

The $(ca)_n(CA)_m$ superlattice contains $2(n+m)$ atoms in a unit cell $\boldsymbol{R}_i$ (i indicates the ith unit cell), and in the second nearest-neighbor approximation four orbitals s, p_x, p_y and p_z are taken into account. The Hamiltonian matrix of a superlattice for these orbitals is similar to that for the sp^3s^* tight binding approximation and reduces to a similar expression to (8.236). However, the elements are different and given by

$$\widehat{b} = \begin{array}{c} \\ s \\ p_x \\ p_y \\ p_z \end{array} \begin{array}{c} \begin{array}{cccc} s & p_x & p_y & p_z \end{array} \\ \begin{pmatrix} \mathcal{E}_{sb} + \mathcal{E}_{sbsb} g_7 & \mathcal{E}_{sbxb} g_{10} & \mathcal{E}_{sbxb} g_{11} & 0 \\ -\mathcal{E}_{sbxb} g_{10} & \mathcal{E}_{pb} + \mathcal{E}_{xbxb} g_7 & -\mathrm{i}\lambda_b + \mathcal{E}_{xbyb} g_4 & \lambda_b \\ -\mathcal{E}_{sbxb} g_{11} & \mathrm{i}\lambda_b + \mathcal{E}_{xbyb} g_4 & \mathcal{E}_{pb} + \mathcal{E}_{xbxb} g_7 & -\mathrm{i}\lambda_b \\ 0 & \lambda_b & \mathrm{i}\lambda_b & \mathcal{E}_{pb} + \mathcal{E}_{zbzb} g_7 \end{pmatrix} \end{array}$$

$$\widehat{ac} = \begin{array}{c} \\ s \\ p_x \\ p_y \\ p_z \end{array} \begin{array}{c} \begin{array}{cccc} s & p_x & p_y & p_z \end{array} \\ \begin{pmatrix} V_{ss} g_0 & V_{sapc} g_1 & V_{sapc} g_1 & V_{sapc} g_0 \\ -V_{scpa} g_1 & V_{xx} g_0 & V_{xy} g_0 & V_{xy} g_1 \\ -V_{scpa} g_1 & V_{xy} g_0 & V_{xx} g_0 & V_{xy} g_1 \\ -V_{scpa} g_0 & V_{xy} g_1 & V_{xy} g_1 & V_{xx} g_0 \end{pmatrix} \end{array}$$

$$\widehat{ca} = \begin{array}{c} \\ s \\ p_x \\ p_y \\ p_z \end{array} \begin{array}{c} \begin{array}{cccc} s & p_x & p_y & p_z \end{array} \\ \begin{pmatrix} V_{ss} g_2 & -V_{scpa} g_3 & V_{scpa} g_3 & V_{scpa} g_2 \\ V_{sapc} g_3 & V_{xx} g_2 & -V_{xy} g_2 & -V_{xy} g_3 \\ -V_{sapc} g_3 & -V_{xy} g_2 & V_{xx} g_2 & V_{xy} g_3 \\ -V_{sapc} g_2 & -V_{xy} g_3 & V_{xy} g_3 & V_{xx} g_2 \end{pmatrix} \end{array}$$

$$\widehat{bb} = \begin{array}{c} \\ s \\ p_x \\ p_y \\ p_z \end{array} \begin{array}{c} \begin{array}{cccc} s & p_x & p_y & p_z \end{array} \\ \begin{pmatrix} \mathcal{E}_{sbsb}(g_8 + g_9) & -\mathcal{E}_{sbxb} g_5 & -\mathcal{E}_{sbxb} g_6 & \mathcal{E}_{sbxb}(g_8 + g_9) \\ \mathcal{E}_{sbxb} g_5 & \mathcal{E}_{xbxb} g_8 + \mathcal{E}_{zbzb} g_9 & 0 & -\mathcal{E}_{xbyb} g_5 \\ \mathcal{E}_{sbxb} g_6 & 0 & \mathcal{E}_{xbxb} g_9 + \mathcal{E}_{zbzb} g_8 & -\mathcal{E}_{xbyb} g_6 \\ -\mathcal{E}_{sbxb}(g_8 + g_9) & -\mathcal{E}_{xbyb} g_5 & -\mathcal{E}_{xbyb} g_6 & \mathcal{E}_{xbxb}(g_8 + g_9) \end{pmatrix} \end{array}$$

where

$$\mathcal{E}_{\alpha b} = \mathcal{E}_{\alpha}(000)_b, \quad V_{ss} = 4\mathcal{E}_{ss}(\tfrac{1}{2}\tfrac{1}{2}\tfrac{1}{2}), \quad V_{xx} = 4\mathcal{E}_{xx}(\tfrac{1}{2}\tfrac{1}{2}\tfrac{1}{2}),$$
$$V_{xy} = 4\mathcal{E}_{xy}(\tfrac{1}{2}\tfrac{1}{2}\tfrac{1}{2}), \quad V_{sapc} = 4\mathcal{E}_{sx}(\tfrac{1}{2}\tfrac{1}{2}\tfrac{1}{2})_{ac}, \quad V_{scpa} = 4\mathcal{E}_{sx}(\tfrac{1}{2}\tfrac{1}{2}\tfrac{1}{2})_{ca},$$
$$\mathcal{E}_{sbsb} = 4\mathcal{E}_{ss}(110)_b, \quad \mathcal{E}_{sbxb} = 4\mathcal{E}_{sx}(110)_b, \quad \mathcal{E}_{xbxb} = 4\mathcal{E}_{xx}(110)_b,$$
$$\mathcal{E}_{xbyb} = 4\mathcal{E}_{xy}(110)_b, \quad \mathcal{E}_{zbzb} = 4\mathcal{E}_{xx}(011)_b,$$

and λ_b is the spin–orbit interaction of the p orbitals. The renormalized spin–orbit splitting of the anion and cation, Δ_a and Δ_c, is defined by

$$\lambda_b = \frac{1}{3}\Delta_b, \quad (b = a, c) \tag{8.242}$$

and the following relations hold for $|p_x b\alpha)$, $|p_y b\alpha)$ and $|p_z b\alpha)$:

$$(p_x b\alpha|H_{so}|p_y b\alpha) = -\mathrm{i}\lambda_b\,, \tag{8.243a}$$
$$(p_x b\alpha|H_{so}|p_z b\alpha) = \lambda_b\,, \tag{8.243b}$$
$$(p_y b\alpha|H_{so}|p_x b\alpha) = \mathrm{i}\lambda_b\,, \tag{8.243c}$$
$$(p_y b\alpha|H_{so}|p_z b\alpha) = -\mathrm{i}\lambda_b\,, \tag{8.243d}$$
$$(p_z b\alpha|H_{so}|p_x b\alpha) = \lambda_b\,, \tag{8.243e}$$
$$(p_z b\alpha|H_{so}|p_y b\alpha) = \mathrm{i}\lambda_b\,. \tag{8.243f}$$

The renormalized spin–orbit splitting energies for the p orbitals are tabulated in Table 8.4. The phase factors g_i $(i = 0\ldots 11)$ are given by (8.238a–8.238d) for $g_0 \ldots g_3$, and the other factors are as follows:

$$g_4 = \sin(\xi)\sin(\eta)\,, \tag{8.244a}$$
$$g_5 = -\frac{\mathrm{i}}{2}\exp(\mathrm{i}\zeta)\sin(\xi)\,, \tag{8.244b}$$
$$g_6 = -\frac{\mathrm{i}}{2}\exp(\mathrm{i}\zeta)\sin(\eta)\,, \tag{8.244c}$$
$$g_7 = \cos\xi\cos\eta\,, \tag{8.244d}$$
$$g_8 = \frac{1}{2}\exp(\mathrm{i}\zeta)\cos\xi\,, \tag{8.244e}$$

Table 8.4 The renormalized spin–orbit splitting energies for the p orbitals [eV]

Al	Si	P	S
0.024	0.044	0.067	0.074

Table 8.5 Parameters for the tight binding approximation shown with the notation of Slater-Koster (see [78])

SK notation	GaAs	AlAs
$\mathcal{E}_{ss}(000)_a$	7.0012	7.3378
$\mathcal{E}_{ss}(000)_c$	7.2004	6.1030
$\mathcal{E}_{xx}(000)_a$	−0.6498	0.4592
$\mathcal{E}_{xx}(000)_c$	5.7192	6.0433
$\mathcal{E}_{ss}(\frac{1}{2}\frac{1}{2}\frac{1}{2})$	0.6084	0.4657
$\mathcal{E}_{xx}(\frac{1}{2}\frac{1}{2}\frac{1}{2})$	−0.5586	−0.5401
$\mathcal{E}_{xy}(\frac{1}{2}\frac{1}{2}\frac{1}{2})$	−1.2224	−1.4245
$\mathcal{E}_{sx}(\frac{1}{2}\frac{1}{2}\frac{1}{2})_{ac}$	−0.6375	−0.4981
$\mathcal{E}_{sx}(\frac{1}{2}\frac{1}{2}\frac{1}{2})_{ca}$	−1.8169	−1.8926
$\mathcal{E}_{ss}(110)_a$	−0.3699	−0.2534
$\mathcal{E}_{sx}(110)_a$	−0.5760	−0.8941
$\mathcal{E}_{xx}(110)_a$	0.2813	0.1453
$\mathcal{E}_{xx}(011)_c$	−0.6500	−0.7912

$$g_9 = \frac{1}{2}\exp(\mathrm{i}\zeta)\cos\eta\,, \tag{8.244f}$$

$$g_{10} = \mathrm{i}\sin\xi\cos\eta\,, \tag{8.244g}$$

$$g_{11} = \mathrm{i}\cos\xi\sin\eta\,. \tag{8.244h}$$

The parameters used in the calculations are those determined by Lu and Sham, which are shown in Table 8.5. The other parameters not shown in the table are assumed to be 0. Also, the spin–orbit interaction is neglected for simplicity.

Here we have to point out that some additional parameters are required to calculate the energy band structure of a superlattice such as GaAs/AlAs. The additional parameters to be determined are those of the atom at the interface (As atom in the case of GaAs/AlAs superlattice), which are usually approximated by the following method.

1. The parameters $\mathcal{E}_{sb}$, $\mathcal{E}_{pb}$ and λ_b of the interface atom As are estimated from the average values of GaAs and AlAs.
2. The nearest-neighbor interaction between the interface atom As and the Ga atom or Al atom is approximated by the interaction in GaAs or AlAs, respectively.
3. The second nearest-neighbor interaction between the Ga atom and the Al atom is approximated by the average value of the Ga–Ga interaction in GaAs and the Al–Al interaction in AlAs.
4. The second nearest-neighbor interaction between the As atoms at the interface is approximated by the average value of the second nearest-neighbor interaction in GaAs and AlAs. The second nearest-neighbor interaction between the interface atom As and the As atom in GaAs or AlAs is approximated by the interaction in GaAs and AlAs.

These approximations will give a good result when the parameters associated with the As atoms of GaAs and AlAs are almost the same. Lu and Sham [43] have reported that

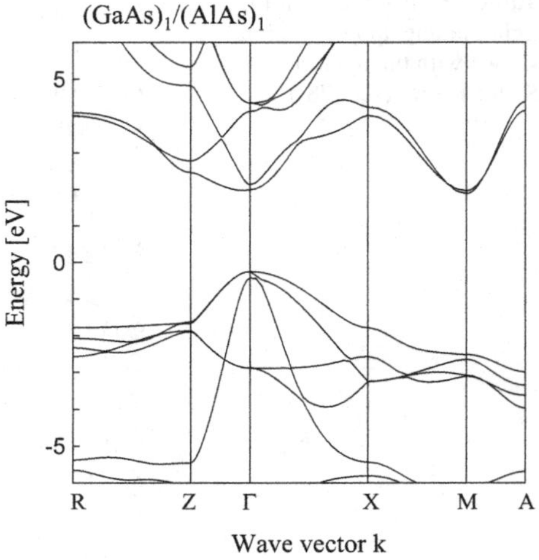

Fig. 8.36 Energy band structure of the $(GaAs)_1/(AlAs)_1$ superlattice calculated by the second nearest-neighbor sp^3 tight binding approximation

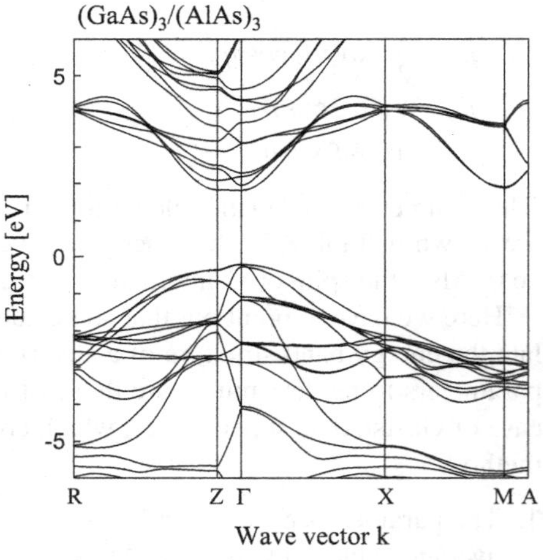

Fig. 8.37 Energy band structure of the $(GaAs)_3/(AlAs)_3$ superlattice calculated by the second nearest-neighbor sp^3 tight binding approximation

such approximations will not introduce a considerable effect in the case of $n > 3$. In addition, the approximations are essential for the wave functions which are required to satisfy their symmetry [81].

Using the parameters estimated from these assumptions and those reported by Lu and Sham [43], and assuming the valence band discontinuity $\Delta\mathcal{E}_v = 0.55\,\text{eV}$, the energy band structures of $(GaAs)_n/(AlAs)_n$ are calculated. Typical examples of the energy band structures are shown in Figs. 8.36, 8.37, 8.38 and 8.39 for superlattices of $n = 1$, 3, 8 and 12. In Fig. 8.40 the transition energies of $(GaAs)_n/(AlAs)_n$ superlattices as a function of the atomic layer number n calculated from the second nearest-neighbor sp^3 tight binding approximation, along with the experimental data, where the experimental data are the same as those shown in Fig. 8.35, which are

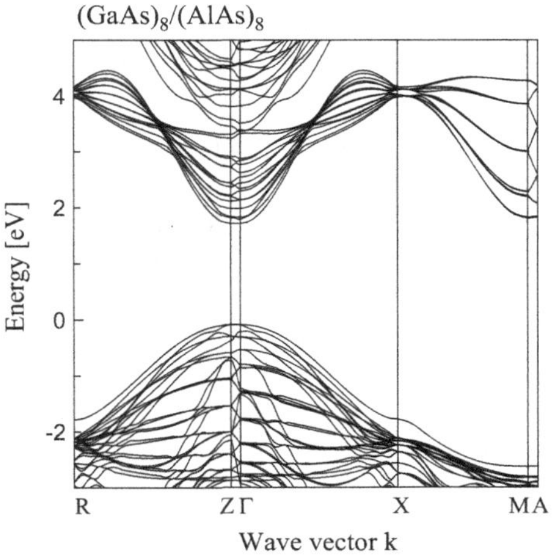

Fig. 8.38 Energy band structure of the $(GaAs)_8/(AlAs)_8$ superlattice calculated by the second nearest-neighbor sp^3 tight binding approximation

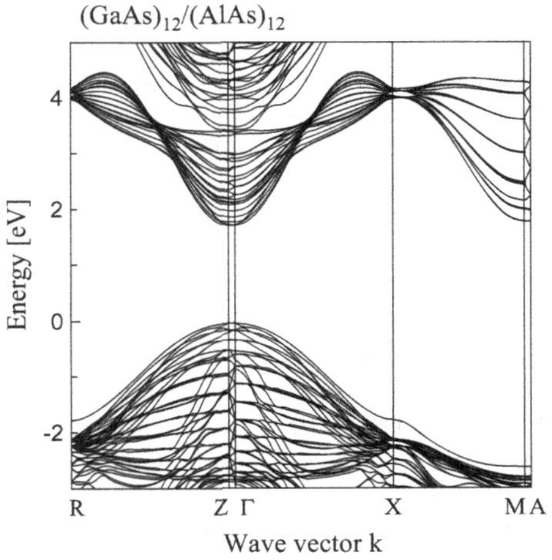

Fig. 8.39 Energy band structure of the $(GaAs)_{12}/(AlAs)_{12}$ superlattice calculated by the second nearest-neighbor sp^3 tight binding approximation

obtained from photoreflectance (PR) and photoluminescence (PL) experiments. It is found from the calculated results that the lowest transition energy is associated with a quasi-direct transition involving the X_z conduction band for $n < 12$ except $n = 1$. In addition, we find that the conduction band at the Z point (indirect band) lies almost at the same level as the X_z conduction band for $n = 3 \ldots 12$ and that the X_z conduction band is lower than the X_{xy} conduction band (M point) except for $n = 1$. These calculated results show good agreement with the calculations based on the effective mass approximation [82, 83], and indicate the importance of the effective mass anisotropy of the X valleys in AlAs. It should be noted that the lowest conduction band for $n = 1$ is the X_{xy} conduction band located at the M point, but the results depend strongly on the parameters used. In addition, it seems to be

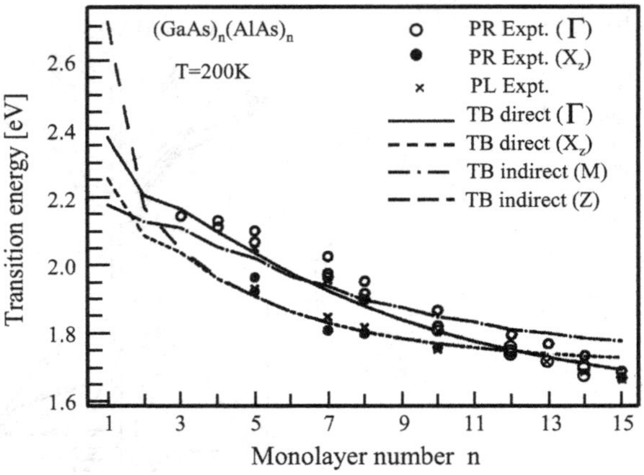

Fig. 8.40 Transition energies as a function of layer number n in the $(GaAs)_n/(AlAs)_n$ superlattice calculated by the second nearest-neighbor sp^3 tight binding approximation. The solid and dashed curves are respectively the direct (Γ) and quasi-direct (X_z) transitions calculated by the second nearest-neighbor sp^3 tight binding approximation and the dot-dashed curve is the indirect transition energy associated with the M point. Experimental results for photoreflectance (PR) and photoluminescence (PL) are also shown for comparison. The theory indicates that the crossing between the direct allowed and quasi-direct transitions occurs at $n = 12$ and that the bottom of the X_{xy} conduction band (M point) is higher that the bottom of the X_z conduction band except for $n = 1$. The conduction band at the Z point (indirect band) lies at almost the same level as the X_z conduction band for $n = 3$–12 (see [45])

almost impossible to grow a perfect superlattice of $(GaAs)_1/(AlAs)_1$ and thus we are not able to determine conclusively the band structure of the superlattice for $n = 1$. Ge et al. [84] have reported experiments on PL, PLE (photoluminescence excitation spectra) and PL under the application of pressure and explained the results as showing that the lowest conduction band is X_{xy} for $n < 3$ and thus that the superlattices are the indirect transition type. In order to draw this conclusion we need to do accurate band structure calculations by taking into account the conduction band at the L point and by making a careful comparison between experiment and theory.

Finally, we will discuss the contribution of the atomic orbitals to the band edges of superlattices. The eigenstates of the eigenvalues consist of the mixture of the atomic orbitals which give the measure of the contribution by the eigenstates. Therefore, the electron distribution in the unit cell is given by

$$|\langle\lambda\boldsymbol{k}|\lambda\boldsymbol{k}\rangle|^2 = \sum_n |(nbj\boldsymbol{k}|\lambda\boldsymbol{k}\rangle|^2 , \qquad (8.245)$$

where $n = s,\ p_x,\ p_y$ and p_z, $b = a$ or c, and j indicates the atom position in the unit cell with $j = 1 - 2(n + m)$. Figure 8.41a–c shows respectively the density distribution of the atomic orbitals at the top of the valence band (Γ_v), at the bottom of the X_z conduction band and at the bottom of the Γ conduction band (Γ_c) in the

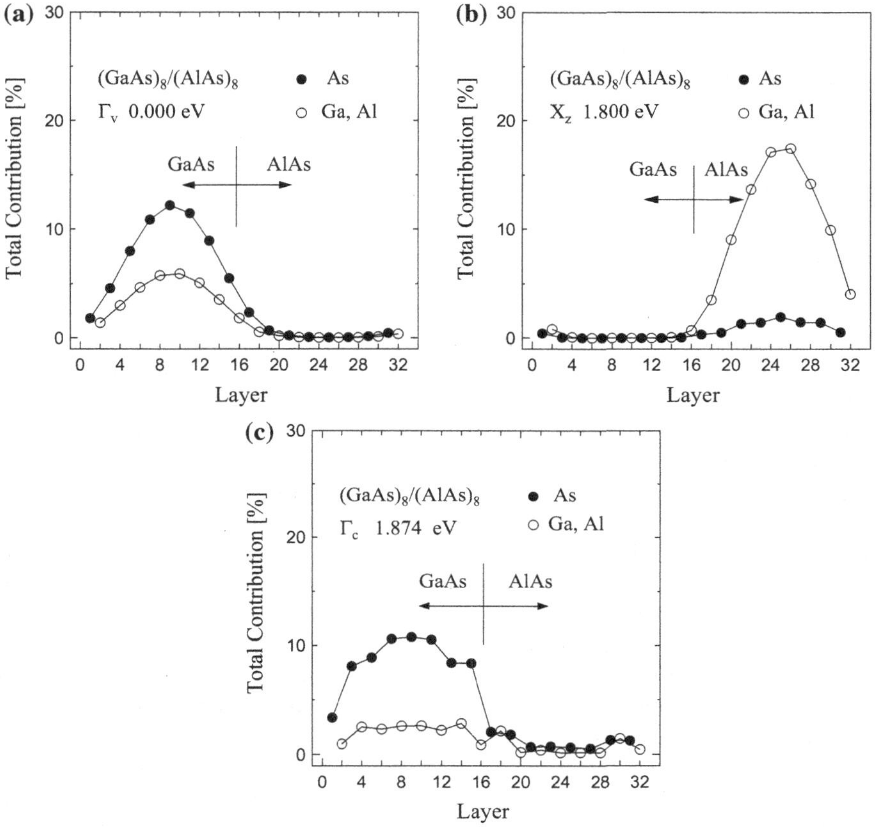

Fig. 8.41 Contribution of electrons from each atom in the superlattice $(GaAs)_8/(AlAs)_8$. The electron contribution is shown **a** the valence band (Γ_v), where the valence band Γ_v is found to consist mostly of electrons in the Ga and As atoms, **b** the X_z conduction band (the X is zone folded at the Γ point), where the electrons of the Al atoms in AlAs mostly contribute to the zone-folded X_z conduction band, and **c** the Γ_c conduction band, where the contribution of electrons from the As atoms in GaAs dominates

$(GaAs)_8/(AlAs)_8$ superlattice. From the figure we find the following result. The top of the valence band (Γ_v) dominated by the atomic orbitals in GaAs, the X_z conduction band edge (X_z is zone folded at the Γ point) consists of the atomic orbitals in the AlAs layer, especially of the orbitals of the As atoms, and the Γ_c conduction band edge consists of the atomic orbitals in GaAs, mostly of As atom orbitals. From these considerations we find the contribution of the bands from the atomic orbitals, the symmetry of the valence and conduction bands, and the contribution of the energy bands from the X point in AlAs and the Γ point of GaAs. In addition, we find that the electron distributions satisfy the symmetry operation IC_4, indicating the correctness of the assumptions.

8.5 Mesoscopic Phenomena

8.5.1 Mesoscopic Region

Many aspects of this section are based on the textbooks edited by Namba [11] and by Ando et al. [12]. The technical word **macroscopic** is often cited against the word **microscopic**. However, the definition of "microscopic" has changed along with the development of LSI (large scale integration). For example, vacuum tubes have the size which we can handle by our hands and the electric circuits for vacuum tubes are assembled by soldering, although they were improved and miniaturized from year to year after their invention in 1906. Therefore, vacuum tubes are macroscopic devices. On the other hand, transistors are very small in size and are classified as microscopic devices. The operating principle of transistors is based on the mechanisms that minority carriers (holes in n-type semiconductors or electrons in p-type semiconductors) emitted from the emitter to the base region diffuse through the base region and arrive at the collector region, giving rise to the collector current signal induced by the emitter current signal. This base region is of the size of microns (μm) and therefore a transistor may be classified as a microscopic device. Several tens of millions of such transistors or MOSFETs are integrated in a Si substrate of about 1 cm^2 area, which is called ULSI (ultra-large scale integration). Compared to ULSI semiconductor devices, the old transistor is no longer a microscopic device. The trend of ULSI size is shown in Fig. 8.42, where we find that the capacity of memories is increasing and the corresponding size of the devices is decreasing from

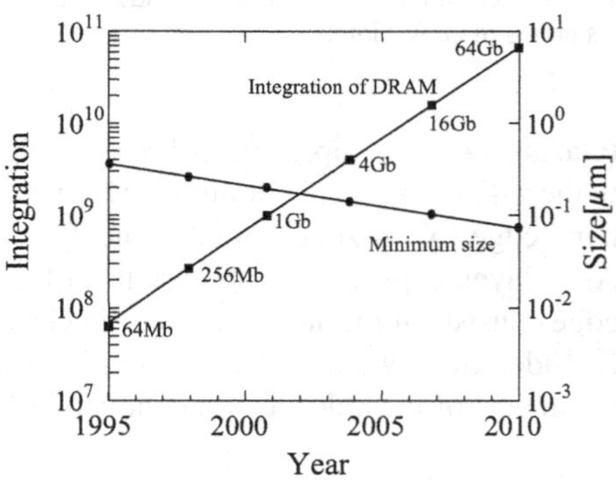

Fig. 8.42 Trend of integration of semiconductor memories. Integration of DRAM and its size are plotted as a function of year. The size of submicron region is expected to be achieved in the next generation

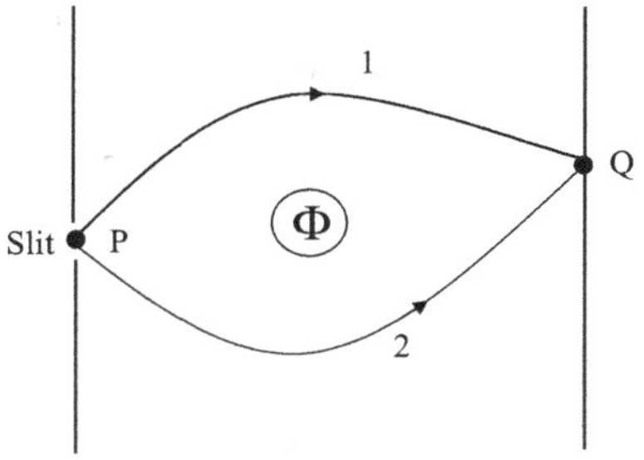

Fig. 8.43 Interference of electron waves. Electron waves transmitted from a slit transit through the region of magnetic flux Φ of a solenoid and the interference appears on a screen Q

year to year. If this trend continues, DRAMs of 1 Gb will be fabricated with a device size of 0.1 μm in the early part of the 21st century.

In the 1980 s metal rings, semiconductor wires, point contact devices and so on were fabricated by using semiconductor LSI technology and the physical properties of these devices were reported. Such work was triggered by an idea to prove the paper published in 1959 by Aharonov and Bohm [85]. In the paper they showed that electrons in a solid have a phase factor which is affected by an external magnetic field. This phenomenon is called the **Aharonov–Bohm effect**. This effect is not observed in a system with inelastic scattering because the electrons lose the phase information after the inelastic scattering. As an example, let us consider electron waves with a finite energy that travel through a region of magnetic flux Φ produced by a small solenoid in vacuum as shown in Fig. 8.43. In such a system, interference of the electron waves occurs and conductance or resistance oscillations with the period of the flux quanta $\Phi_0 = h/e$ or $\Phi_0/2$ are observed. In experiments, metals rings were used and the uniform magnetic field and the resistance exhibited oscillations with a period of h/e [86].

This interference effect may be explained as follows. The electron wave $\psi(\boldsymbol{r})$ is described by the Schrödinger equation

$$\left[\frac{1}{2m}(\boldsymbol{p}+e\boldsymbol{A})^2 + V(\boldsymbol{r})\right]\psi(\boldsymbol{r}) = \mathcal{E}\psi(\boldsymbol{r}), \tag{8.246}$$

where m is the electron effective mass, $-e$ is the electron charge, $\boldsymbol{A}$ is the vector potential and $\mathcal{E}$ is the energy eigenvalue. For simplicity we assume $V(\boldsymbol{r}) = 0$ and the electron wave is expressed as $\psi(\boldsymbol{r}) \propto \mathrm{e}^{-\mathrm{i}\boldsymbol{k}\cdot\boldsymbol{r}}$ ($k = \sqrt{2m\mathcal{E}/\hbar^2}$). Let the electron wave functions on the screen Q pass through the different channels $\psi_i(\mathrm{Q})$ ($i = 1, 2$). The interference intensity on the screen Q is proportional to $\Re[\psi_1^*(\mathrm{Q})\psi_2(\mathrm{Q})] \propto \cos\theta_{12}(\mathrm{Q})$, where $\theta_{12}(\mathrm{Q})$ is the phase difference between the electrons in the difference paths. Let us express the electron wave in the presence of a magnetic flux by

$$\psi_i(\boldsymbol{r}) = \exp[-\mathrm{i}\theta_i(\boldsymbol{r})]\psi_i(\boldsymbol{r}), \tag{8.247}$$

$$\theta_i(\boldsymbol{r}) = \frac{2\pi}{\Phi_0}\int_{\mathrm{P}}^{\mathrm{Q}} \boldsymbol{A}\cdot \mathrm{d}\boldsymbol{s}_i, \tag{8.248}$$

where $\Phi_0 = h/e$ is the **flux quanta** and $\int \mathrm{d}s_i$ is the line integral along the path i. The interference intensity of the electron waves on the screen Q is given by

$$|\psi_1(\mathrm{Q}) + \psi_2(\mathrm{Q})|^2 = |\psi_1(\mathrm{Q})|2 + |\psi_2(\mathrm{Q})|2 + 2\Re[\psi_1^*(\mathrm{Q})\psi_2(\mathrm{Q})]$$
$$\simeq 2|\psi_1^0(\mathrm{Q})|^2 \{1 + \cos[\xi_E(\mathrm{Q}) + (\theta_1 - \theta_2]\} , \quad (8.249)$$

$$\theta_1 - \theta_2 = \frac{2\pi}{\Phi_0}\left[\int_{\mathrm{P}}^{\mathrm{Q}} \boldsymbol{A}(\boldsymbol{s}) \cdot \mathrm{d}\boldsymbol{s}_1 - \int_{\mathrm{P}}^{\mathrm{Q}} \boldsymbol{A}(\boldsymbol{s}) \cdot \mathrm{d}\boldsymbol{s}_2\right]$$
$$= \frac{2\pi}{\Phi_0}\oint \boldsymbol{A}(\boldsymbol{s}) \cdot \mathrm{d}\boldsymbol{s} = 2\pi\frac{\Phi}{\Phi_0} , \quad (8.250)$$

where we have assumed $|\psi_1(\mathrm{Q})|^2 \sim |\psi_2(\mathrm{Q})|^2 \sim |\psi_1^0(\mathrm{Q})|^2$ and $\xi_E(\mathrm{Q})$ is defined by $\psi_1^{0*}(\mathrm{Q})\psi_2^0(\mathrm{Q}) = |\psi_1^0|^2 \exp[\mathrm{i}\xi_E(\mathrm{Q})]$. It is evident from these equations that the interference intensity of electron waves on the screen Q is given by a function of Φ/Φ_0 and thus the interference results in periodic oscillations. This phenomenon is called the **Aharonov–Bohm effect** or **AB effect** [85].

The sample used for observation of the AB effect is a gold ring with diameter 825 nm ($\sim$ 0.8 μm and line width 49 nm), which is not an extremely small device [86]. The experimental data will be shown in Sect. 8.5.4. Later Ishibashi et al. [87] observed resistance oscillations of period e/h in a small ring of semiconductor, where the sample diameter of the sample used in the experiment was 1 μm. It is very interesting to point out that the samples used in the experiments were not extremely small and that quantum effects are very obviously observed in samples with a size of below and beyond μm or of several μm. The size ranges between the microscopic and macroscopic region and thus the system is called the **mesoscopic system** [11]. These investigations clarified the difference between microscopic and macroscopic systems and the term "microscopic" was then used to describe the atomic size. On the other hand, the measure of a mesoscopic system depends the state of the electrons in the system and the size of the mesoscopic structure is determined by the electron mean free path and diffusion length. Roughly speaking, the size of a mesoscopic system is in the range around μm and is defined as a system in which obvious quantum effects are observed. In such a mesoscopic system, between microscopic and macroscopic systems, various new phenomena have been observed experimentally and interpreted theoretically in the 1990 s and this trend is still continuing. In addition, various phenomena observed and explained previously are again interpreted in terms of the new theory for the mesoscopic region.

8.5.2 *Definition of Mesoscopic Region*

Here we will define physical parameters that are useful for understanding mesoscopic phenomena [11]. Electrical conduction in semiconductors (solids) is described by the mean free path or Fermi wavelength. The mean free path Λ is the average distance of electron transit between collisions and is defined by using the average scattering time or relaxation time τ and Fermi velocity $v_{\rm F}$:

$$\Lambda = v_{\rm F}\tau\,. \tag{8.251}$$

The electrical conductivity σ and the electron mobility μ

$$\sigma = \frac{ne^2\tau}{m} = ne\mu\,, \tag{8.252}$$

$$\mu = \frac{e\tau}{m}\,, \tag{8.253}$$

where n is the electron density, m is the electron effective mass and $-e$ is the electron charge. The Fermi wavelength $\lambda_{\rm F}$ is the wave length of the electron with Fermi energy $\mathcal{E}_{\rm F}$, and is derived as

$$\frac{\hbar^2 k_{\rm F}^2}{2m} = \mathcal{E}_{\rm F}\,, \tag{8.254}$$

$$\lambda_{\rm F} = \frac{2\pi}{k_{\rm F}}\,. \tag{8.255}$$

The electron density in a metal is very high and the Fermi energy is large, giving rise to the Fermi wavelength $\lambda_{\rm F} \sim 1\,\text{Å}$. On the other hand, the electron density in a semiconductor is low and the Fermi wavelength is very large. For example, for the Fermi wavelength of two-dimensional electron gas in a GaAs/AlGaAs heterostructure we have

$$\lambda_{\rm F} \sim 400\text{Å}\ (n \sim 3 \times 10^{11}\ {\rm cm}^{-2})\,, \tag{8.256}$$

$$\Lambda = 1 \sim 100\,\mu{\rm m}\,. \tag{8.257}$$

This size is larger than the size of the LSI microstructures and rings we have discussed, and we may expect new features of electrical conduction which have not yet been observed in normal size devices.

In addition to these physical parameters, the diffusion coefficient D and diffusion length DL_T also give a measure to understand the phenomena in mesoscopic region. These parameters are given by

$$D \sim v_{\rm F}^2\tau = \frac{\Lambda^2}{\tau} = \Lambda v_{\rm F}\,, \tag{8.258}$$

$$L_T = \sqrt{\frac{D\hbar}{k_{\rm B}T}}\,. \tag{8.259}$$

In the analysis of the electron interference effect, the phase coherence length is very important and is defined by

$$L_\phi = \sqrt{D\tau_\phi} = \Lambda\sqrt{\frac{\tau_\phi}{\tau}}\,, \tag{8.260}$$

where τ_ϕ is the phase relaxation time governed by inelastic scattering.

When the size of a sample is large enough for electrons to experience repeated scattering and to drift between the electrode contacts and the electron motion is governed by the classical Boltzmann transport equation, the electrons are in the **diffusion region**. On the other hand, when electrons transit between the electrode contacts without suffering any scattering, the electrons are in the **ballistic region**. The **mesoscopic system** which is between the microscopic and macroscopic regions in size is characterized by the phase coherence length L_ϕ. When the sample length becomes shorter than the phase coherence length, a quantum mechanical effect, that is characteristic of the system structure results. Therefore, the mesoscopic region includes a part of the diffusion region and the ballistic region and is described as follows by using the physical parameters defined above [11]:

$$\textbf{mesoscopic region} = \begin{cases} \text{diffusion region } (L \gg \Lambda) \\ \text{ballistic region } \ (L \ll \Lambda) \end{cases}, \tag{8.261}$$

where L is the system size and the electrical conduction is independent of L in the diffusion region but depends on L in the ballistic region. The electron mean free path in a high-mobility semiconductor is several tens μm and the ballistic region is easily achieved in such a semiconductor. In addition, the electron Fermi wavelength in a high-mobility semiconductor is several 100 μm and thus quantum effects are easily observed in confined electron system of heterostructures.

8.5.3 Landauer Formula and Büttiker–Landauer Formula

Let us explain the Landauer formula with the help of the system shown in Fig. 8.44. Ideal conductors are connected to a conductor specimen and the ideal conductors are connected to ideal electrodes called reservoirs. Let the chemical potentials of the reservoirs on the left- and right-hand sides be μ_1 and μ_2, respectively, and the chemical potentials of the ideal conductors be μ_A and μ_B, respectively. Assuming that the electron channel is one-dimensional and its energy is given by $\mathcal{E} = \hbar^2 k_x^2/2m^*$, then the density of electrons moving in one direction (positive direction) is given by $\partial n_+/\partial\mathcal{E} = 1/\pi\hbar v_x$, where the electron spins are taken into account. The electron velocity v_x is given by $m^* v_x = \hbar k_x$. Letting the electron transmission coefficient through this channel be T and the reflectivity R, we have the relation $T + R = 1$. The electron current through this system is then defined as

$$I = (-e)v_x \frac{\partial n_+}{\partial \mathcal{E}} T(\mu_1 - \mu_2) = -\frac{e}{\pi\hbar} T(\mu_1 - \mu_2) . \tag{8.262}$$

Since the voltage difference between the electrodes is given by $-eV_{21} = \mu_1 - \mu_2$, the conductance between the two terminals is obtained as

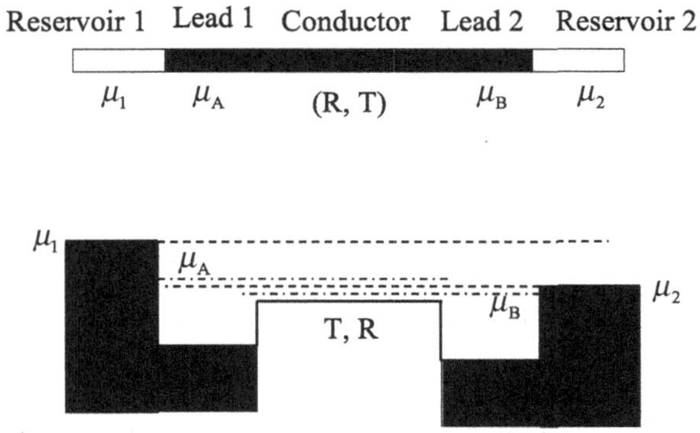

Fig. 8.44 Schematic illustration of a system used to derive the Landauer formula and its energy diagram. Ideal conductors are connected to both sides of a system (conductor) and the ideal conductors are contacted to ideal electrodes called reservoirs. The current through the system flows due to the chemical potential difference between the reservoirs

$$G = \frac{I}{V_{21}} = \frac{e^2}{\pi\hbar}T = \frac{2e^2}{h}T\,. \tag{8.263}$$

The conductance measured in the system is defined by the voltage applied to the system and the current through the system and not by the voltage drop between the electrodes and the current through the electrodes. When the reflectivity R is unity, the left-hand and right-hand conductors are equilibrated with electrodes 1 and 2, respectively, resulting in $-eV = \mu_1 - \mu_2$. On the other hand, when the transmission coefficient T is unity, we have the relation $-eV = 0$. In general, there holds the following relation for arbitrary transmission coefficient T and reflection coefficient R (see Ando [11], Chap. 2):

$$-eV = R(\mu_1 - \mu_2)\,. \tag{8.264}$$

Therefore the conductance of the system is written as

$$G = \frac{e^2}{\pi\hbar}\frac{T}{R} = \frac{e^2}{\pi\hbar}\frac{T}{1-T}\,. \tag{8.265}$$

This relation is called the **Landauer formula** [88].

Generalized expressions for the conductance of a system with two or more terminals have been derived by Landauer and Büttiker, where the Landauer formula is extended to a system with multi-terminals and called the **Büttiker-Landauer formula** [89]. In this textbook we follow the treatment of Büttiker [90]. Figure 8.45 shows a conductor with four terminals connected via perfect leads (unshaded) to four reservoirs at chemical potentials μ_1, μ_2, μ_3 and μ_4. The shaded region is a disordered conductor. The reservoirs serve both as a source and as a sink of carriers and of

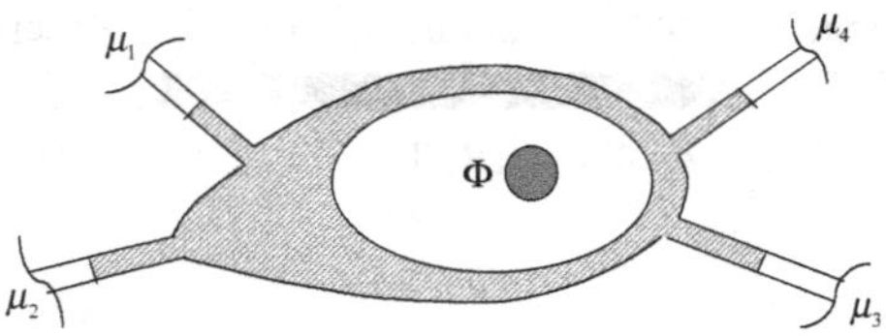

Fig. 8.45 Schematic illustration of a device which consists of a disordered normal conductor with four terminals via perfect leads (unshaded) to four reservoirs at chemical potentials μ_1, μ_2, μ_3 and μ_4. An Aharonov–Bohm flux Φ is applied through the hole in the sample

energy. At $T = 0\,\text{K}$ they can supply carriers with energy up to μ_i to the leads. A carrier supplied through the lead to the reservoir is absorbed by the reservoir depending on the phase and energy of the incident carrier. The unshaded regions of the leads are perfect conductors, free of elastic scattering, between the disordered terminals and the reservoirs.

First, we assume that the perfect leads are one-dimensional quantum channels. Then there exist two types of carriers at the Fermi level, one with positive velocity leaving the reservoir and one with negative velocity. Scattering in the sample is assumed to be elastic; inelastic scattering will occur only in the reservoirs. Therefore, the elastic scattering in the sample is described by an S-matrix. Designating the amplitudes of the incident currents as α_i and the amplitudes of the outgoing currents as α'_i $(i = 1, \cdots, 4)$, we have the following relation between these amplitudes:

$$\alpha'_i = \sum_{i=1}^{i=4} s_{ij}\alpha_j . \tag{8.266}$$

Since the current is conserved, the S-matrix is unitary, $S^\dagger = S^{-1}$, where $S^\dagger$ is the Hermitian conjugate of S. Time reversal requires the relation $S^*(-\Phi) = S^{-1}(\Phi)$ ($*$ means complex conjugate). From these relations the S-matrix obeys the reciprocity relations $s_{ij}(\Phi) = s_{ji}(-\Phi)$. The coefficient $s_{ij}(\Phi)$ is the transmission amplitude for a carrier leaving contact j to reach contact i in the presence of a flux Φ and is the same as that of a carrier leaving contact i to reach contact j if the flux is reversed. We define the transmission probabilities for carriers leaving lead j to reach lead i by $T_{ij} = |s_{ij}|^2$, $i \neq j$, and the reflection probabilities for carriers leaving lead i to be reflected to lead i by $R_{ii} = |s_{ii}|^2$. Then the reciprocity symmetry of the S-matrix implies that

$$R_{ii}(\Phi) = R_{ii}(-\Phi), \quad T_{ij}(\Phi) = T_{ji}(-\Phi) . \tag{8.267}$$

Let us calculate the current flowing through the leads. We assume that the potential differences between the leads are small and that the transmission and reflection probabilities are independent of the carrier energy. First we introduce a chemical potential that is small compared to the chemical potentials μ_i of the four conductors. When the carrier energy is below the chemical potential μ_0, the states of positive and negative velocities are both occupied and thus the net current is zero. Therefore, we can analyze the channel transport by taking account of the states above μ_0, $\Delta\mu_i = \mu_i - \mu_0$. Referring to the derivation of (8.262) for the current in one-dimensional channel, the current injected from reservoir i into the lead i is given by $ev_i(\mathrm{d}n/\mathrm{d}\mathcal{E})\Delta\mu_i$, where v_i is the velocity at the Fermi energy in lead i and $\mathrm{d}n/\mathrm{d}\mathcal{E} = 1/2\pi\hbar v_i$ is the density of states (each state of electron spin) for carriers with negative or with positive velocity v_i at the Fermi energy. Therefore, the current injected by reservoir i is given by $(e/h)\Delta\mu_i$. When we consider the current in lead 1, a current $(e/h)(1 - R_{11})\Delta\mu_1$ is reflected back to lead 1. Carriers which are injected by reservoir 2 into lead 1 reduce the current in lead 1 by $-(e/h)T_{12}\Delta\mu_2$. Similarly, from the current fed into leads 3 and 4 we obtain a current in lead 1 of $-(e/h)(T_{13}\Delta\mu_3 + T_{14}\Delta\mu_4)$. Summing these results and applying similar considerations to determine the currents in the other leads results in

$$I_i = \frac{e}{h}\left[(1 - R_{ii})\mu_i - \sum_{i \neq j} T_{ij}\mu_j\right], \tag{8.268}$$

where the currents are independent of the reference potential μ_0, since the coefficients multiplying the potentials add to zero. This formula is called the **Büttiker–Landauer formula**. We have to note that the following relation holds for all the leads i:

$$R_{ii} + \sum_{j \neq i} T_{ij} = 1\,. \tag{8.269}$$

The above results are derived for a current in single channel. In general, electrons are confined in the directions perpendicular to the current, giving rise to discrete quantum levels $\mathcal{E}_n, n = 1, 2, \cdots$. Therefore, the number of channels will be changed by the relation between the quantum level $\mathcal{E}_n$ and the Fermi energy $\mathcal{E}_\mathrm{F}$. For a number of quantum channels N_i the scattering matrix contains the elements $(\sum N_i)^2$. Here we define the element by $s_{ij,mn}$ which gives the transmission amplitude for a carrier incident in channel n in lead j to reach channel m in lead i. The probability for a carrier incident in channel n in lead i to be reflected into the same lead into channel m is denoted by $R_{ii,mn} = |s_{ii,mn}|^2$, and the probability for a carrier incident in lead j in channel n to be transmitted into lead i in channel m is $T_{ij,mn} = |s_{ij,mn}|^2$. The current in lead i due to carriers injected in lead j is

$$I_{ij} = -\frac{e}{h}\sum_{mn} T_{ij,mn}\Delta\mu_j\,. \tag{8.270}$$

Therefore, if we introduce

$$R_{ii} = \sum_{mn} R_{ii,mn}\,, \quad T_{ij} = \sum_{mn} T_{ij,mn}\,, \tag{8.271}$$

the currents flowing from the reservoir toward the conductor is

$$I_i = \frac{e}{h}\left[(N_i - R_{ii})\mu_i - \sum_{i \neq j} T_{ij}\mu_j\right]. \tag{8.272}$$

where N_i is the number of channels in lead i.

The result given by (8.272) indicates that the conductance is evaluated from (e/h) multiplied by the term [ ] when the currents and chemical potentials are measured simultaneously at all the probes. Note that experiments are carried out by choosing appropriate current leads and potential probes. As an example, consider the four-probe device shown in Fig. 8.45, where a current I_1 is fed into lead 1 and is taken out in lead 3, and a current I_2 is fed into lead 2 and leaves the sample through lead 4. Then we have to solve (8.272) under the condition that $I_1 = -I_3$ and $I_2 = -I_4$. This will give the two currents as a function of potential difference $V_i = \mu_i/e$,

$$I_1 = \alpha_{11}(V_1 - V_3) - \alpha_{12}(V_2 - V_4)\,, \tag{8.273}$$

$$I_2 = = -\alpha_{21}(V_1 - V_3) + \alpha_{22}(V_2 - V_4)\,, \tag{8.274}$$

where the conductance matrix α_{ij} is given by

$$\alpha_{11} = (e^2/h)[(1 - R_{11})S - (T_{14} + T_{12})(T_{41} + T_{21})]/S\,, \tag{8.275a}$$

$$\alpha_{12} = (e^2/h)(T_{12}T_{34} - T_{14}T_{32})/S\,, \tag{8.275b}$$

$$\alpha_{21} = (e^2/h)(T_{21}T_{43} - T_{23}T_{41})/S\,, \tag{8.275c}$$

$$\alpha_{22} = (e^2/h)[(1 - R_{22})S - (T_{21} + T_{23})(T_{32} + T_{12})]/S\,, \tag{8.275d}$$

$$S = T_{12} + T_{14} + T_{32} + T_{34} = T_{21} + T_{41} + T_{23} + T_{43}\,. \tag{8.275e}$$

From these results we find that the diagonal elements are symmetric in the magnetic flux $\alpha_{11}(\Phi) = \alpha_{22}(-\Phi)$, $\alpha_{22}(\Phi) = \alpha_{22}(-\Phi)$, and the off-diagonal elements satisfy $\alpha_{12}(\Phi) = \alpha_{21}(-\Phi)$.

Next we will discuss how to derive the resistance of the system from (8.273) and (8.274). In the four-probe system shown in Fig. 8.45, a current is fed between the two leads and the chemical potentials of the two leads is measured. For example, when a current is fed through lead 1 and taken out through lead 3, the current between leads 2 and 4 is zero, and the chemical potentials are $\mu_2 = eV_2$ and $\mu_4 = eV_4$, we obtain from (8.274) $V_2 - V_4 = (\alpha_{21}/\alpha_{22})(V_1 - V_3)$ for $I_2 = 0$. Inserting this into (8.273), the current I_1 is given by a function of $(V_2 - V_4)$. Therefore the resistance is expressed by the following relation when a current flows between lead 1 and 3 and the potentials are measured at leads 2 and 4:

$$\mathcal{R}_{13,24} = \frac{V_2 - V_4}{I_1} = \frac{\alpha_{21}}{(\alpha_{11}\alpha_{22} - \alpha_{12}\alpha_{21})} . \tag{8.276}$$

Since α_{21} is in general not symmetric, the resistance $\mathcal{R}_{13,24}$ is also not symmetric. Now we exchange the current and the voltage probes but keep the flux fixed. The resistance is then given by

$$\mathcal{R}_{24,13} = \frac{\alpha_{12}}{(\alpha_{11}\alpha_{22} - \alpha_{12}\alpha_{21})} . \tag{8.277}$$

The sum of these resistances, $\mathcal{S}_\alpha = (\mathcal{R}_{13,24} + \mathcal{R}_{24,13})/2$, is symmetric.

In general under a flux Φ, if a current is fed into lead m and taken out from lead n, and if the potential difference between leads k and l is measured, the resistance is defined by the following relation [90]:

$$\mathcal{R}_{mn,kl} = \frac{h}{e^2}\frac{(T_{km}T_{ln} - T_{kn}T_{lm})}{D} , \tag{8.278}$$

where $D = (h/e^2)^2(\alpha_{11}\alpha_{22} - \alpha_{12}\alpha_{21})/S$. Since D is independent of the exchange of mn and kl, the relation $\mathcal{R}_{mn,kl} = -\mathcal{R}_{mn,lk} = -\mathcal{R}_{nm,kl}$ holds.

Here we will discuss the relation between the Büttiker–Landauer formula and Landauer formula. Consider a device such as that shown in Fig. 8.46, where a current is fed through lead 1 and taken out from lead 2, and the potential difference is measured between leads 3 and 4, which are weakly connected to the conductor through tunnel barriers. In this case the following relation is derived [89]:

$$\mu_3 - \mu_4 = \frac{T_{31}T_{42}}{(T_{31} + T_{32})(T_{41} + T_{42})}(\mu_1 - \mu_2) . \tag{8.279}$$

Since the voltage probes are connected to the conductor through the perfect leads and only elastic scattering occurs in the conductor between the voltage probes, the system is characterized by a transmission probability T and a reflection probability R. Then we have the relations, $T_{12} = T_{21} = T$, $T_{31} = T_{13} = T_{42} = T_{24} = 1 + R$, $T_{32} = T_{23} = T_{14} = T_{41} = T$, and $T_{43} = T_{34} = T$. Inserting these relations into (8.279) we obtain

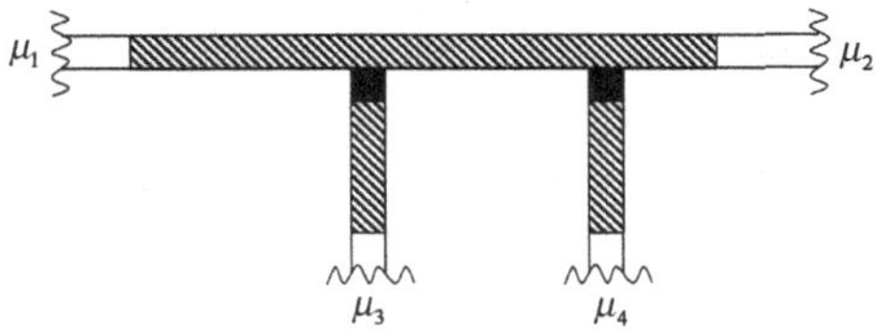

Fig. 8.46 Four-terminal conductor. The two current probes are 1 and 2, while the two potential probes are connected through the tunneling barrier junctions (dark areas)

$$\mu_3 - \mu_4 = \frac{1}{4}[(1+R)^2 - (1-R)^2](\mu_1 - \mu_2) = R(\mu_1 - \mu_2)\,. \tag{8.280}$$

Since the current is given by $I = (e/h)T(\mu_1 - \mu_2)$, the resistance $\mathcal{R}$ (conductance $: G = 1/\mathcal{R}$) is expressed as

$$\mathcal{R} = \frac{h}{e^2}\frac{R}{T}\,. \tag{8.281}$$

When we take into account the spin degeneracy factor 2, the above equation is equivalent to the Landauer formula, (8.265).

8.5.4 Research in the Mesoscopic Region

Recently, many scientists have been attracted to do research on mesoscopic structures and a variety of phenomena have been observed so far. The definition of mesoscopic structures has become clear from these investigations. In the following we discuss some important and advanced work related to mesoscopic phenomena. Some well-known phenomena have been reinterpreted in terms of mesoscopic phenomena. The quantum Hall effect, for example, is also interpreted in terms of the Büttiker–Landauer formula for edge channels, which is one of the most important theories of mesoscopic phenomena. As stated previously, research on mesoscopic structures is being intensively carried out in the fields of semiconductor physics, new functional devices fabricated in semiconductors and advanced semiconductor devices, and many other interesting work has been done beside the work discussed here.

8.5.5 Aharonov–Bohm Effect (AB Effect)

The Aharonov–Bohm effect is based on the prediction reported by Aharonov and Bohm in 1959 for a quantum mechanical investigation, as discussed in Sect. 8.5.1. When electron waves with a finite energy transit through a small solenoid and the magnetic field (vector potential) is changed, the conductance of a metal ring oscillates periodically as the magnetic flux becomes a multiple of the flux quantum $\Phi_0 = h/e$. This phenomenon was first proved by Webb et al. [86, 91, 92] and an example is shown in Fig. 8.47, where the inset shows the metal ring structure used in the experiment and the line width is 40 nm and ring diameter is 0.8 μm. The ring has Au leads attached to symmetric positions on the ring and the magnetoresistance can be measured through the outer circuits connected to the Au leads. Electron waves entering from one of the leads are separated into two parts and are subjected to the magnetic field, resulting in interference at the other lead. The phase of the interference changes by 2π as the flux is increased by a multiple of h/e and the

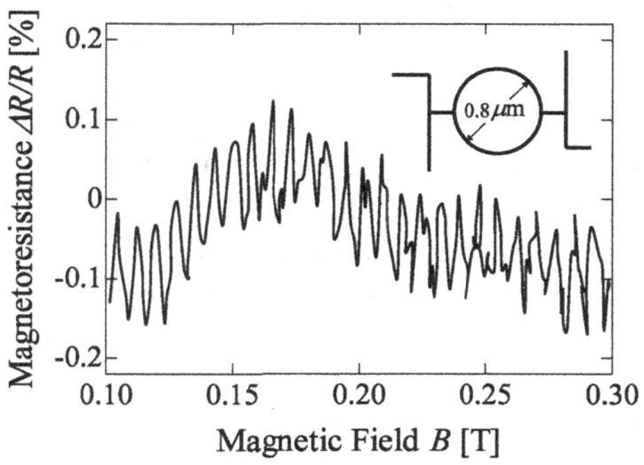

Fig. 8.47 Periodic oscillations of magnetoresistance in a Au ring of diameter 0.8 μm and line width 40 nm shown in the inset. A period of h/e and its second-harmonic component $2\,h/e$ are observed in a magnetic field B=0–8 T and at temperature $T = 50\,\text{mK}$

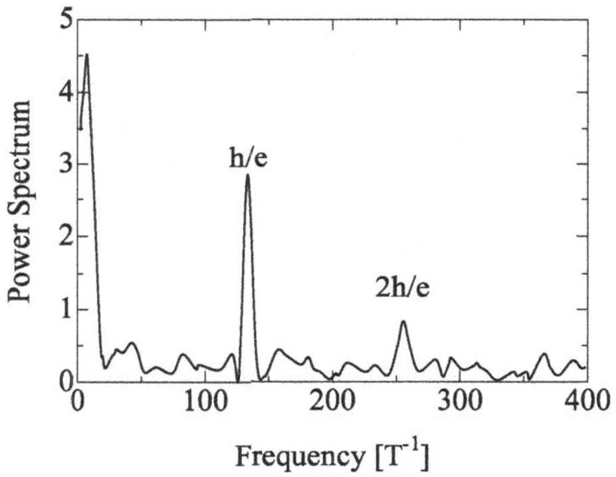

Fig. 8.48 Fourier transform of Fig. 8.47. The period h/e ($1/0.0076\,\text{T}^{-1}$) corresponds to the condition where the flux quanta pass through the metal ring. The weak second-harmonic component $2\,h/e$ ($1/0.0038\,\text{T}^{-1}$) is also observed

magnetoresistance oscillates with period h/e. The experimental results are shown in Fig. 8.47, where the oscillatory magnetoresistance is plotted as a function of magnetic field over the range 0–8 T and the period is found to be 0.0076 T ($1/0.0076\,\text{T}^{-1}$). This period corresponds to the flux quanta h/e through the Au ring. It is very interesting to point out that more than 1000 Aharonov–Bohm oscillations are observed in the magnetic field region 0–8 T. The Fourier transform of the data is shown in Fig. 8.48, where two peaks corresponding to the period h/e and its weak second-harmonic component $2\,h/e$ ($1/0.0038\,\text{T}^{-1}$) are identified. These two periods are observed in the whole experimental range of magnetic field up to 8 T. We have to note here that the AB effect is shown theoretically to be related to fluctuations. The interference is caused by electrons from many channels and thus random motion of electrons results in fluctuations of the conductance. Therefore, the AB effect is not observed in an electron system with many channels or, in other words, in a macroscopic system. In such a macroscopic system the AAS effect (Altshuler–Aharonov–Spivak effect) with the period of flux $\Phi_0/2$ is observed instead of the AB effect [93].

8.5.6 Ballistic Electron Transport

ULSI (Ultra-Large Scale Integration) technology and microfabrication of semiconductor devices enable us to produce an extremely short channel device in which electrons can transit ballistically between the electrode contacts without suffering any scattering when the electrode distance is shorter than the electron mean free path. This phenomenon is interpreted as an analogy to electron transport in a vacuum tube and has attracted semiconductor engineers from an aspect of developing new high-frequency devices. [94]. However, this interpretation does not take account of the effects of electrical leads and reservoirs and thus it is not correct. It was later pointed out that such a device should be interpreted with the help of the Landauer formula. The most striking finding is the quantization of conductance in a one-dimensional channel, which is well explained by the Landauer formula and accepted as a typical example of a mesoscopic phenomenon. The device structure is shown in Fig. 8.49, where the split-gate structure is formed on AlGaAs/GaAs and the point contacts constrict the channel of the two-dimensional electron gas. The Electron current through the point contact structure shows quantized conductance in multiples of $2\,e^2/h$. The phenomenon has been discovered independently by van Wees et al. [95] and Wharam et al. [96]. An example of the experimental data is shown in Fig. 8.50.

When a voltage is applied to the split-gate, the electron channel is constricted due to the expansion of the depletion region and electrons in a limited number of channel mode can pass through the constriction. The experimental result is explained in terms of the Landauer formula as follows. In the case where the effect of reflection can be neglected and the transmission coefficient T_n depends on the channel mode, the conductance G is given by

$$G = \frac{2e^2}{h}\sum_{n=1}^{N} T_n\,. \tag{8.282}$$

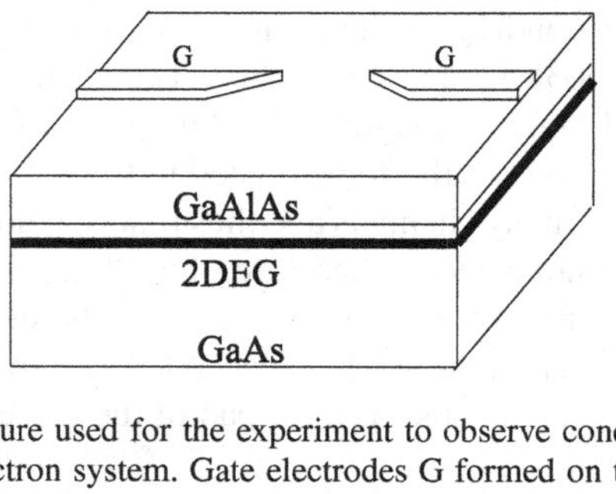

Fig. 8.49 Split-gate structure used for the experiment to observe conductance quantization in a quasi-one dimensional electron system. Gate electrodes G formed on the AlGaAs surface above the two-dimensional electron gas system constrict the electron channels between the point contacts and the conductance due to electron flow through the constricted region is measured

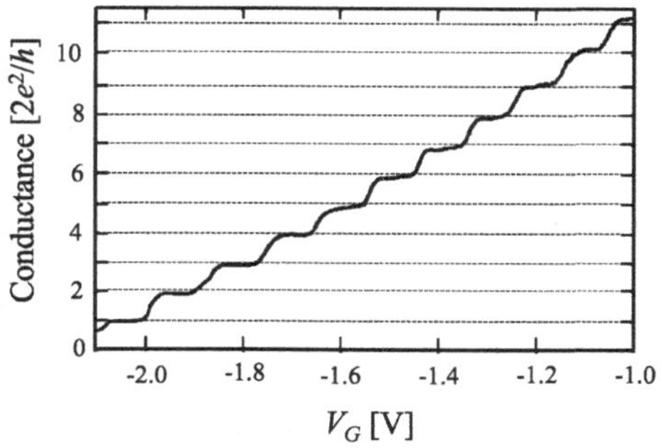

Fig. 8.50 An example of experimental data for conductance quantization in the split-gate structure. The horizontal axis is the voltage applied to the gate electrodes G and the vertical axis is the conductance in units of $2\,e^2/h$. The conductance quantized in units of $2\,e^2/h$ is well resolved in the experimental range as staircases. (After van Wees et al. [95])

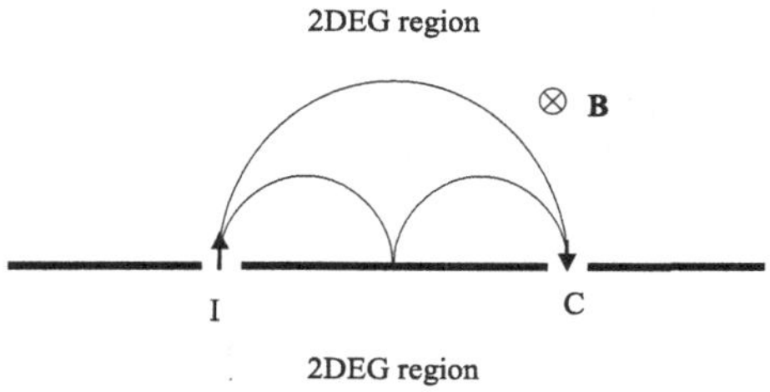

Fig. 8.51 A schematic illustration of a sample used for observation of the magnetic focusing effect. In the presence of a magnetic field an electron beam is injected through the injection channel I and the voltage variation is detected through the detection channel C. The detected voltage oscillates periodically when the magnetic field is swept

Assuming that electrons of modes $n = 1, 2, \cdots, N$ can pass through the point contacts channel and approximating the transmission coefficient to be $T_n \approx 1$ for the channels, then we obtain

$$G \approx \frac{2e^2}{h} N \,. \tag{8.283}$$

A change in the applied voltage on the gates of the sample shown in Fig. 8.49 results in a change in the channels of electrons (modes), and the conductance is quantized as multiples of $2\,e^2/h$. This feature is clearly seen in Fig. 8.50.

Another well-known phenomenon of electron ballistic transport is the magnetic focusing effect. This effect is observed in the sample shown in Fig. 8.51, where two point contacts are separated by distance L. When a multiple of the cyclotron radius $2\,r_c$ becomes equal to the distance L in a magnetic field perpendicular to the surface, ($2Nr_c = L$, $N = 1, 2, 3, \cdots$), electrons emitted from the injection channel enter the detection channel and cause a change in the electric current or voltage [97, 98].

8.6 Quantum Hall Effect

The quantum Hall effect discovered in 1980 by Klaus von Klitzing et al. [14] has made a great impact upon semiconductor physics. The importance of the quantum Hall effect can be understood from the fact that the Hall resistance has been adopted as the international standard of resistance. Figure 8.52 shows the first observation of the quantum Hall effect in a Si-MOSFET, where a two-dimensional electron gas is induced in the inversion layer by applying a gate voltage and the Hall effect due to the two-dimensional electron gas is measured. In the presence of a magnetic field perpendicular to the interface and of a constant current between the source and the drain, the Hall conductivity σ_{xy} in the classical theory is expected to be proportional to the inversion electron density N_s which is changed by the gate voltage. The Hall voltage is proportional to the Hall resistance and thus to the inversion electron density. However, the experiments on Si MOSFETs by von Klitzing et al. revealed that the Hall voltage is not proportional to the inversion electron density but that instead the Hall voltage exhibits plateaus in some regions of the electron sheet density as shown in Fig. 8.52. The experiment was carried out with a source-drain current $I = 1\,\mu$A, magnetic field $B = 18$ T and at temperature $T = 1.5$ K. The Hall voltage shows a

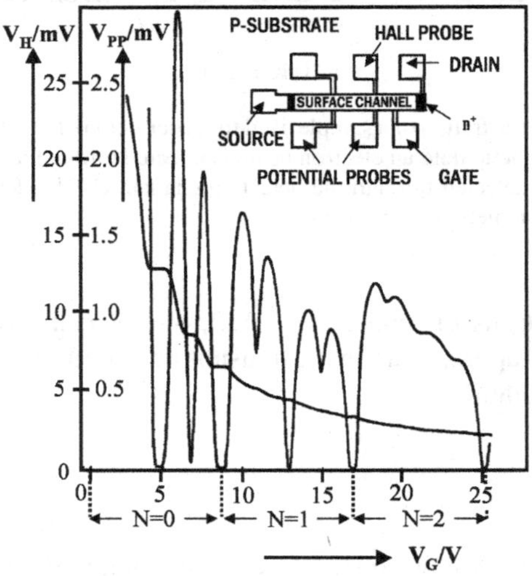

Fig. 8.52 Gate voltage V_G dependence of Hall voltage V_H and potential difference V_{pp} between two potential probes in the two-dimensional electron gas of a Si-MOSFET, for temperature $T = 1.5$ K, magnetic field $B = 18$ T, and source–drain current $I = 1$ mA. The inset shows the device structure with length $L = 400\,\mu$m, width $W = 50\,\mu$m, and the distance between the potential probes $L_{pp} = 130\,\mu$m. The voltage drop V_{pp} shows Shubnikov–de Haas oscillations, which are proportional to σ_{xx}. When V_{pp} becomes a minimum, the Hall voltage V_H (proportional to σ_{xy}) exhibits a plateau and the Hall resistance $\mathcal{R}_K = V_H/I$ is quantized. In the experiment the filling factor of the Landau levels of the two-dimensional electron gas is changed by the gate voltage. N is the Landau index. (After von Klitzing et al. [14])

plateau in the region where the voltage V_{pp} (proportional to σ_{xx}) becomes a minimum. Under these conditions the Hall resistance defined by the Hall voltage divided by the source-drain current is found to be

$$\mathcal{R}_H = \frac{h}{e^2} \cdot \frac{1}{i} = \frac{25813}{i} [\Omega] , \tag{8.284}$$

where $i = 1, 2, 3, \cdots$ and thus the Hall resistance $\mathcal{R}_H$ is quantized. Here the Hall resistance is defined by $\mathcal{R}_H$ in order to distinguish it from the Hall coefficient R_H of Sect. 7.1. Later the quantized Hall resistance has been measured at various institutions in the world and found to agree with 10 digits in the accuracy of and the quantized Hall resistance is approved as the international standard of resistance:

$$R_K = 25812.8074555(59)\,\Omega . \tag{8.285}$$

The constant is therefore called the **von Klitzing constant**.[6] Later (in 1982) the Hall resistance and magnetoresistance were found to exhibit an anomaly for fractional values of i. Since then the former phenomenon (i is integer) is called the **integer quantum Hall effect** (IQHE) and the latter phenomenon (i is fractional) is called the **fractional quantum Hall effect** (FQHE). The discovery of the quantum Hall effect was stimulated by the early work on transport in a two-dimensional electron gas by T. Ando, Y. Uemura and S. Kawaji in Japan.

Figure 8.53 shows the quantum Hall effect in the AlGaAs/GaAs HEMT structure, where the high mobility of the two-dimensional electron gas results in a clear quantum Hall effect [99]. Kawaji et al. reported a detailed investigation of the accuracy of the quantized Hall resistance in order to adopt the value for the international standard [13]. Komiyama, Kawaji et al. have carried out detailed experiments to show that the plateaus of the quantum Hall resistance disappear under the application of a high electric current [11, 13].

We start from (7.20a) and (7.20b) derived for three dimensional case of Sect. 7.1. Denoting the sheet density of the two-dimensional electron gas by N_s [1/m^2] and the cyclotron angular frequency by $\omega_c = eB/m^*$, then the conductivity in a magnetic field is given by expanding (7.20b) in terms of $1/(\omega_c\tau)^2$ for $\omega_c\tau \geq 1$

$$\sigma_{xy} = -\frac{N_s e}{B} + \frac{\sigma_{xx}}{\omega_c \tau} \equiv -\frac{N_s e}{B} + \Delta\sigma_{xy} . \tag{8.286}$$

When $\omega_c\tau \gg 1$ is satisfied, the following approximation is valid:

$$\sigma_{xy} = -\frac{N_s e}{B} . \tag{8.287}$$

---

[6]From 2014 CODADA recommended value and the number in parentheses is standard uncertainty in the last two digits of the given value (http://physics.nist.gov/cgi-bin/cuu/Value?rk). See also the fundamental constants of Physics and Chemistry reported by P. J. Mohr, B. N. Taylor, and D. B. Newell; *Rev. Mod. Phys.*, **84**, (2012) 1527.

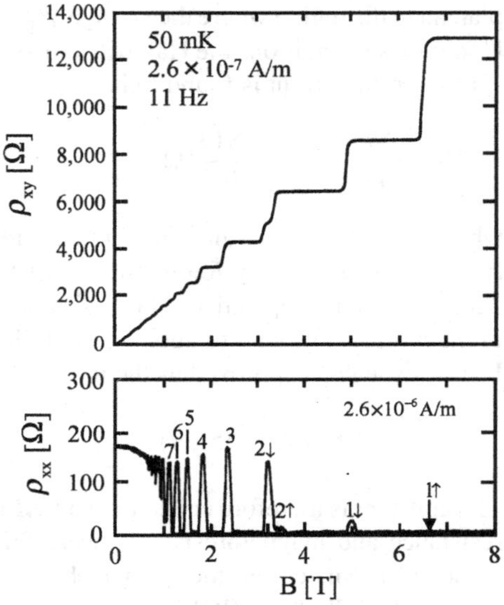

Fig. 8.53 Quantum Hall effect in AlGaAs/GaAs. The high mobility of the two-dimensional electron gas in AlGaAs/GaAs results in clear plateaus in the Hall conductance σ_{xy}. The lower part of this figure shows the magnetoconductance σ_{xx} as a function of the magnetic field, where the indices are the Landau quantum number and spin polarization. The measurements are for $T = 50\,\text{mK}$ and $I = 2.6\,\mu\text{A/m}$

We know that the density of states for a two-dimensional system is given by a constant value of $m^*/2\pi\hbar^2$. When the cyclotron radius of the ground Landau state is given by l, the density of states for each Landau level in a two-dimensional system in a magnetic field is defined by (2.108) of Sect. 2.5

$$\frac{1}{2\pi l^2} = \frac{m^*}{2\pi\hbar^2} \cdot \hbar\omega_c = \frac{eB}{2\pi\hbar} = \frac{eB}{h}\,. \tag{8.288}$$

When the Landau levels are degenerate for spins, the density of states defined above is multiplied by a factor 2. For simplicity, we neglect the spin degeneracy and take into account one of the spins. In addition we neglect the broadening of the density of states and assume that it is given by a delta function. Under these simplified assumptions we examine how the Fermi energy behaves in the presence of magnetic field. The Landau levels are given by

$$\mathcal{E}_N = \left(N + \frac{1}{2}\right)\hbar\omega_c, \quad N = 0, 1, 2, 3, \ldots,$$

which are plotted in Fig. 8.54 for $m^* = 0.067\,\text{m}$. When we assume a constant electron sheet density $N_s = 4.0 \times 10^{11}\,\text{cm}^{-2}$, a higher magnetic field in which the electrons occupy the lowest Landau level $N = 0$ only is given by

$$\frac{eB_1}{h} = N_s, \quad B_1 = 16.6\,[\text{T}]\,.$$

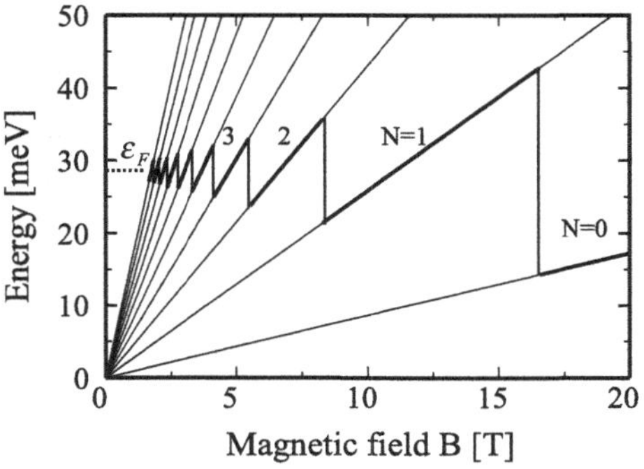

Fig. 8.54 Landau levels and Fermi energy are plotted as a function of magnetic field for two-dimensional electron gas with the effective mass $m^* = 0.67\,\text{m}$ and sheet electron density $N_s = 4.0 \times 10^{-11}\,\text{cm}^{-2}$. The density of states is assumed to be given by a delta function of aerial density $2\,eB/h$ and the Fermi energy oscillates with magnetic field. The oscillatory behavior of the Fermi energy result in Shubnikov-de Haas oscillations in σ_{xx}. The electron spins are not taken into account

In a magnetic field lower than B_1, the Landau level $N = 0$ cannot contain all the electrons and some of the electrons occupy the second Landau level $N = 1$. The magnetic field is lowered below $B_2 = B_1/2$, and electrons occupy the third Landau level $N = 2$. The Fermi level at $T = 0\,\text{K}$ thus obtained is plotted by the solid lines in Fig. 8.54.

When electrons occupy up to the ith Landau levels ($i = N + 1$), we find

$$N_s = i \cdot \frac{eB}{h}, \qquad i = 1, 2, 3, \ldots, \tag{8.289}$$

where i is called as "**filling factor**" and thus the Hall conductivity is obtained as

$$\sigma_{xy} = -i \cdot \frac{e^2}{h}, \qquad i = 1, 2, 3, \ldots. \tag{8.290}$$

Here we consider dimension of two–dimensional electron gas in a magnetic field applied in the perpendicular to the electron sheet (z direction). N_s is defined in units of $[1/\text{m}^2]$ (n: $[1/\text{m}^3]$ for three dimensional electrons) as described before. Assuming the width of the current channel of two dimensional electrons as W [m], the parameters in (7.20a) and (7.20b) are redefined as $I_x = J_x \cdot W$, $V_H = E_y \cdot W$ and thus we have the following relations

$$J_i\ [\text{A/m}],\quad E_j\ [\text{V/m}],\quad \sigma_{ij}\ [\text{A/V}] = [1/\Omega]\,.$$

Therefore the Hall resistance is given by $\mathcal{R}_H = V_H/I_x = E_y/J_x = 1/\sigma_{xy}$ and has dimension of $[\text{V/A}] = [\Omega]$. Therefore the Hall resistance is given by for filling factor $i = 1$ (Landau levels are filled up to i–th level),

$$\mathcal{R}_\mathrm{H} = \frac{1}{i} \cdot \frac{h}{e^2} = \frac{25813}{i} \, [\Omega] \tag{8.291}$$

and the relation of (8.284) is derived. It is very important to note the following remarks. The above treatment shows only that the value of the Hall resistance $\mathcal{R}_\mathrm{H}$ is quantized at a point where the Landau level occupation number i changes. In other words, the treatment does not explain the plateau of the Hall resistance in certain magnetic field regions.

In classical theory, the Hall voltage is proportional to the Hall resistance

$$\mathcal{R}_\mathrm{H} = \frac{B}{N_\mathrm{s} e} \, . \tag{8.292}$$

Therefore, the Hall resistance is proportional to the magnetic field and inversely proportional to the electron sheet density. Such a state that the electrons occupy up to the ith Landau levels is given by (8.289), and this condition leads to the relation $N_\mathrm{s} = ieB/h$ between the electron sheet density N_s and the magnetic field B for the quantized Hall state. The density of states $(eB/h)\delta(\mathcal{E} - \mathcal{E}_N)$ is obtained without broadening of the Landau level, but electronic states in a real semiconductors is broadened by scattering. In such a case increasing N_s beyond filling factor $i = 1$, and some electrons occupy the non-conductive broadened states and σ_{xx} in the second term of (8.286) is zero, resulting in constant σ_{xy} until the electrons occupy the mobile states of the next Landau level (N=1; i=2). This feature may explain the plateau of the Quantum Hall effect as follows.

Figure 8.55 illustrate the density of states of a Landau level ($N = 0, \ i = 1$) with the localized states shown by hatched area, conductivity σ_{xx}, and Hall conductivity σ_{xy}. When the sheet electron density is increased, Fermi level moves as $N_\mathrm{s} 2\pi l^2$ up to 1 (the Landau level filled). When Fermi level lies in the localized region, $\sigma_{xy} = 0$ because of non–conducting electrons ($N_\mathrm{s}\tau = 0$; by putting $n \to N_\mathrm{s}$ in (7.20b)), and plateau appears when Fermi level lies in the higher hatched region as shown in Fig. 8.55.

As discussed above, when the Landau levels have a finite width, the electronic states at the edge of the Landau levels are localized and the electrons in these states will not contribute to the current. This feature is illustrated in Fig. 8.56 for more general discussion. When the Fermi level is located in the hatched region (localized state), the electrons do not contribute to the current, and $\sigma_{xx} = 0$ and $\sigma_{xy} = -N_\mathrm{s} e/B = -ie^2/h$, resulting in a plateau of σ_{xy}. On the other hand, when the Fermi level is in a non-localized state, the Hall conductance is given by $\sigma_{xy} \simeq -N_\mathrm{s} e/B + \sigma_{xx}/\omega_\mathrm{c}\tau$ and expected to behave like a curve as shown in Fig. 8.55. With increasing the electron density further until the next Landau level is occupied by electrons, Hall conductance behaves very similarly and gives rise to the quantized Hall conductance $\sigma_{xy} = -(i+1)e^2/h$. These features are depicted in Fig. 8.56.

It is even more interesting to point out that the condition leads to a singularity in the magnetic flux quantization. Let us consider the case where electrons occupy the lowest Landau level ($i = 1$; $N = 0$). As shown before, the flux quantum is given by

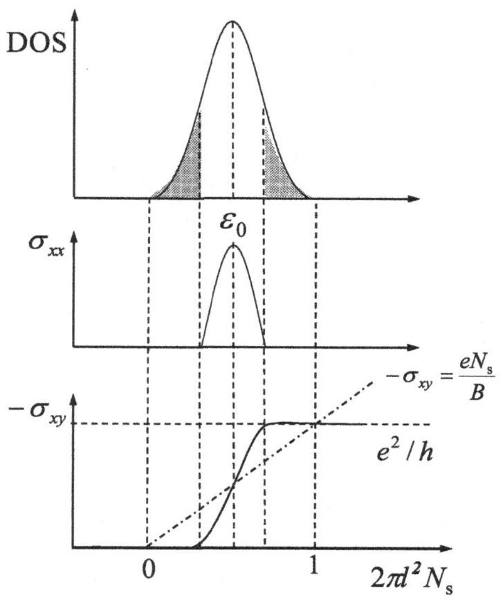

Fig. 8.55 Landau levels and conductivity. **a** Landau level with localized states (hatched region), **b** conductivity σ_{xx} contributed from the electrons in the non-localized region, and Hall conductivity σ_{xy} as a function electron sheet density N_s below Fermi level. (After Aoki and Ando [100])

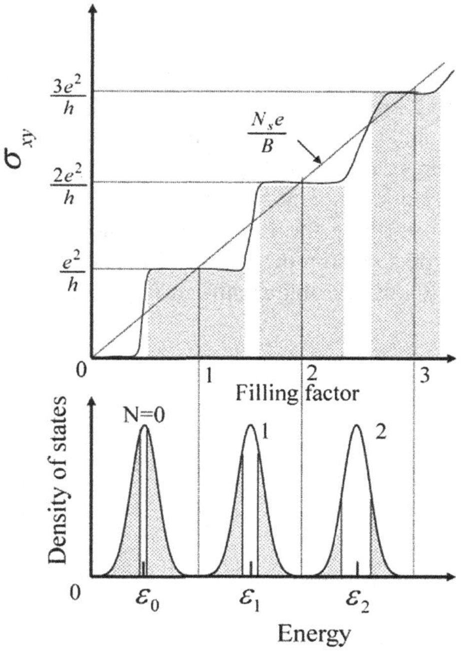

Fig. 8.56 Origin of the plateau in the quantum Hall effect. In this figure electron filling is illustrated as the sheet electron density is changed under a constant applied magnetic field. In the presence of localized states of the Landau levels the electrons occupying the localized states cannot contribute to the electron current and the plateaus in the quantized Hall states result, depending on the filling factor of the electrons. (After Aoki and Ando [100])

h/e, and the area occupied by an electron is $2\pi l^2 = h/(eB)$. When we put $i = 1$ in (8.289), we find $B/N_s = h/e$ and a single magnetic flux goes through a Landau orbit for an electron. In general, we find that i flux quanta go through the Landau orbit for the quantized Hall state with filling factor i. Therefore, we may conclude that the quantized Hall state is accompanied by magnetic flux quantization.

The first theoretical interpretation of the quantum Hall effect was given by Aoki and Ando [100] by using the linear response theory. Later, a theory based on the gauge transform was reported by Laughlin [101]. In addition, the importance of the edge current in the quantum Hall regime was pointed out by Halperin [102], and Büttiker [103] succeeded in explaining the integer quantum Hall effect in terms of the edge current by the Büttiker–Landauer formula, which is an extension of the Landauer formula. Reports on the quantum Hall effects are found in review papers such as [104–106].

This textbook does not intend to review the theories of the quantum Hall effects, but to give an introduction for experimentalists. First, an outline of the linear response theory of Aoki and Ando was given to understand the Quantum Hall Effect. When we neglect the broadening of the Landau levels and the localization of the electrons, the electrons are quantized into Landau orbits, and under the condition of the integer quantum Hall regime and for $i = 1$, in the classical picture the Landau orbits of the N_i electrons occupy the whole area of the sample. Under this condition we obtain $2\pi l^2 \times N_i = 1$ from (8.288) and (8.289), which corresponds to the classical model. This model will not give the electron drift along the electric field direction and thus no current in the longitudinal direction, which gives rise to $\sigma_{xx} = 0$.

In experiments plateaus are observed in the region near this magnetic field, which may be interpreted in terms of the localization of electronic states as described before. When the potential fluctuates slowly compared to the magnetic length l, electrons move along the equipotential lines (classical orbits) near the hills and valleys as expected from the classical theory as shown in Fig. 8.57. Such localized electrons will not contribute to the current. With increasing sheet electron density, the electron

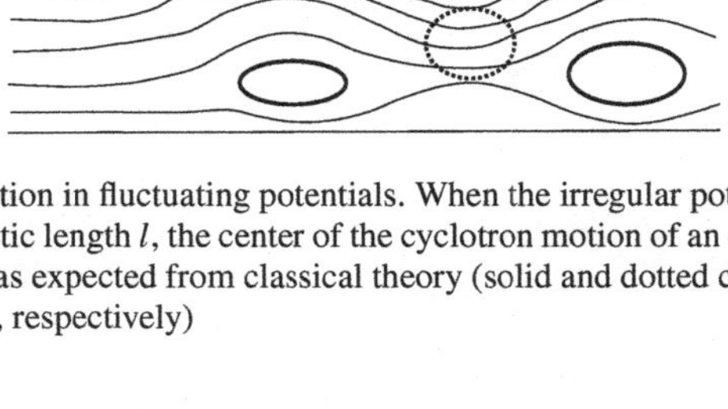

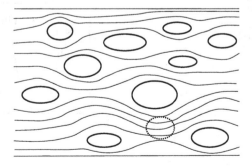

Fig. 8.57 Electron motion in fluctuating potentials. When the irregular potential fluctuates slowly compared to the magnetic length l, the center of the cyclotron motion of an electron traverses along the equipotential lines as expected from classical theory (solid and dotted curves near the hills and valleys of the potential, respectively)

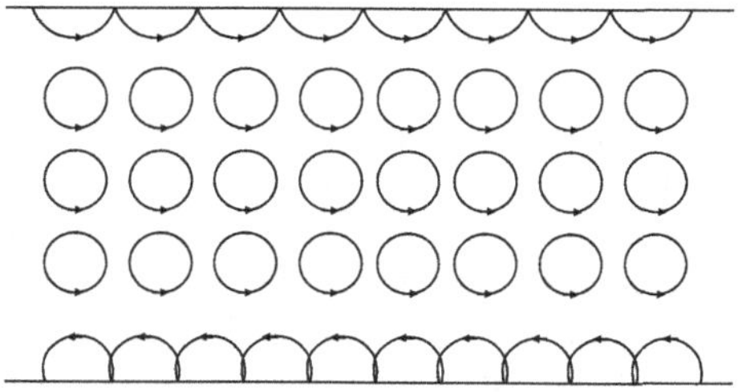

Fig. 8.58 Motion of a two-dimensional electron gas in a high magnetic field is illustrated schematically. In the middle region of the sample the electrons complete cyclotron orbits, whereas electrons near the sample edge cannot make cyclotron motion but a skipping motion instead because of the existence of the potential wall at the edges. In general, the skipping motions of the upper and lower edges are different in their distance between the cyclotron centers and the edge. This skipping motion gives rise to the edge channel in the quantum Hall effect

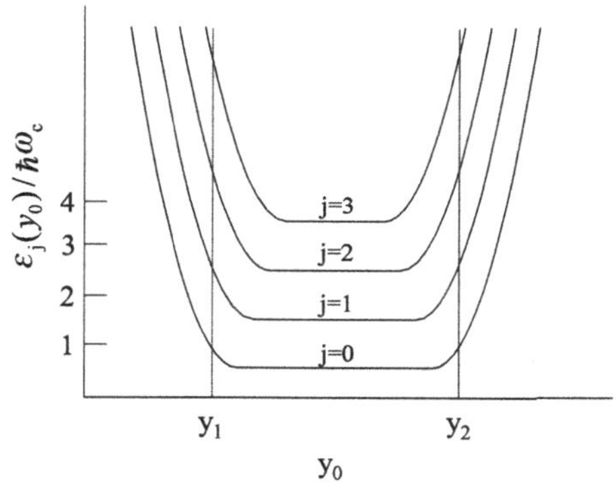

Fig. 8.59 Energy spectrum of electrons in a high magnetic field in a semiconductor with a rectangular confining potential (walls at y_1 and y_2) at the semiconductor surfaces. The Landau levels near the center are $\mathcal{E}_j = \hbar\omega_c(j + 1/2)$ (flat) but are strongly bent upward near the edges. y_0 is the center of the harmonic oscillator wave functions. (After [102])

orbits are not closed and electron current is induced along the electric field direction. Next, we consider the edge channel. As shown schematically in Fig. 8.58, electrons near the center can complete cyclotron orbits and are Landau quantized. However, electrons near the edges are affected by the confining potential and perform a skipping motion. In Fig. 8.58 the magnetic field is upward normal to the page and the direction of the electron motion is shown by the arrows. Under these boundary conditions the electronic states in a magnetic field are schematically shown by Fig. 8.59, where the electron energy in the y direction is plotted for the case of the current in the x direction and the magnetic field in the z direction, and y_1 and y_2 are the sample boundaries [102]. In the central part of the y direction, an electron performs cyclotron motion and its energy is flat, whereas electrons near the edges at y_1 and y_2 are affected by the confining potential, resulting in a skipping motion flowing in the x direction.

The associated current is called the **edge current** and the channel is called the **edge channel**.

Electrons in the edge channels are expected to be ballistic and thus their transport properties are analyzed by the Landauer formula or the Büttiker–Landauer formula [103]. Here we will discuss the quantum Hall effect by the method introduced by Büttiker. It should be noted here that the samples used in the measurements of the quantum Hall effect have electrode contacts for the Hall voltage in addition to the current probes. Therefore, the devices are multi-terminals. First we discuss the edge states by taking account of the boundary condition (confining potential). Defining the vector potential $\boldsymbol{A} = (-By, 0, 0)$, the Hamiltonian for an electron with effective mass m^* and electronic charge $-e$ is given by the following equation according to the results presented in Sect. 2.5.

$$H = \frac{1}{2m^*}\left[(p_x + eBy) + p_y^2\right] + V(y)\,. \tag{8.293}$$

In the presence of a magnetic field, an electron performs cyclotron motion in the xy plane, and the wave function of the electron is written as $\psi_{j,k} = \exp(ikx)F_j(y)$ and the function F is given by an eigenfunction of the following equation:

$$\left[-\frac{\hbar^2}{2m^*}\frac{\partial^2}{\partial y^2} + \frac{1}{2}m^*\omega_c^2(y - y_0)^2 + V(y)\right]F_j(y) = \mathcal{E}_j F_j(y)\,, \tag{8.294}$$

where $\omega_c = eB/m^*$ is the cyclotron angular frequency. The eigenvalues of (8.294) depend on

$$y_0 = -\frac{\hbar k}{m^*}\frac{1}{\omega_c} = -kl^2 \tag{8.295}$$

and $l = \sqrt{\hbar/eB}$. In Fig. 8.59, letting the potential $V(y) \equiv 0$ in the flat region, the solution of (8.294) is given by

$$\mathcal{E}_{jk} = \hbar\omega_c\left(j + \frac{1}{2}\right)\,, \tag{8.296}$$

with $j = 0, 1, 2, \ldots$ as shown in Sect. 2.5. The solution given by (8.296) is independent of the parameter y_0 and thus of k. However, in the region near the edges y_1 and y_2, as shown in Fig. 8.59, the electrons cannot complete cyclotron orbits and instead perform a skipping motion. As a result, the energy eigenvalue of an electron is a function of y_0 and deviates from the value given by (8.296), giving rise to bending upward near the edge. In the edge region the electron energy depends on the distance $|y_1 - y_0|$ or $|y_2 - y_0|$. Figure 8.59 illustrates the electron energies by taking account of the confining potential. The electron energy is then expressed as

$$\mathcal{E}_{jk} = \mathcal{E}[j, \omega_c, y_0(k)]\,. \tag{8.297}$$

The velocity along the sample edge of a carrier in such an edge state is defined as

$$v_{jk} = \frac{1}{\hbar}\frac{\mathrm{d}\mathcal{E}_{jk}}{\mathrm{d}k} = \frac{1}{\hbar}\frac{\mathrm{d}\mathcal{E}_{jk}}{\mathrm{d}y_0}\frac{\mathrm{d}y_0}{\mathrm{d}k}. \tag{8.298}$$

Here we find that the velocity along the sample edge is proportional to the slope of the Landau level, $\mathrm{d}\mathcal{E}/\mathrm{d}y_0$. $\mathrm{d}\mathcal{E}/\mathrm{d}y_0$ is negative at the upper edge y_2 and is positive at the lower edge y_1 (see Fig. 8.59). When a magnetic field is applied in the upward direction with respect to the page in Fig. 8.58, $\mathrm{d}y_0/\mathrm{d}k$ is negative and therefore the velocity is positive at the upper edge y_2 and negative at the lower edge y_1. In the region of the bulk Landau levels or the flat region of the Landau levels shown in Fig. 8.59, $\mathcal{E}$ is independent of y_0 and thus the carrier velocity in this region is zero. This is again understood from the fact that the carriers complete cyclotron orbits, the motion is quantized in the plane perpendicular to the magnetic field and the center of the cyclotron orbit will not move. The density of states in this region is discussed in Sect. 2.5. On the other hand, the density of states of a carrier along the edge state of Landau level $\mathcal{E}_j$ is one-dimensional and given by $\mathrm{d}n/\mathrm{d}k = 1/2\pi$ or $\mathrm{d}n/\mathrm{d}k = \mathrm{d}n/\mathrm{d}y_0|\mathrm{d}y_0/\mathrm{d}k|$. Since we obtain $\mathrm{d}n/\mathrm{d}y_0 = 1/(2\pi l^2)$ from (8.295), the density of states in energy space is related to the velocity by the following relation:

$$\left[\frac{\mathrm{d}n}{\mathrm{d}\mathcal{E}}\right]_j = \frac{\mathrm{d}n}{\mathrm{d}k}\left[\frac{\mathrm{d}k}{\mathrm{d}\mathcal{E}}\right]_j = \frac{1}{2\pi\hbar v_{jk}}. \tag{8.299}$$

The density of states at the Fermi energy is evaluated from (8.297) by replacing $\mathcal{E}_{jk}$ with $\mathcal{E}_\mathrm{F}$. From this relation we may obtain k at the Fermi energy, and the states consist of discrete values of $n = 1, 2, \ldots, N$. Here we have to note that N should be taken for positive and negative values of k. When the Fermi energy passes through a Landau level, the number of edge states changes from N to $N-1$.

First, we consider the case of a two-terminal circuit and calculate the current injected into the edge by using the Landauer formula. The current through the edge state is written as

$$I = ev_j\left[\frac{\mathrm{d}n}{\mathrm{d}\mathcal{E}}\right]_j(\mu_1-\mu_2) = \frac{e}{h}\Delta\mu. \tag{8.300}$$

Since the current injected into the edge states is equivalent to the current injected into quantum channels, the resistance of this two-terminal circuit is given by

$$\mathcal{R} = \frac{h}{e^2}\frac{1}{N}, \tag{8.301}$$

where N is the number of edge states (the number of one-dimensional channels for positive velocity). We have to note that the resistance given by (8.301) is a two-terminal resistance and not the Hall resistance. The result is just a repeat of the calculation described in Sect. 8.5.3.

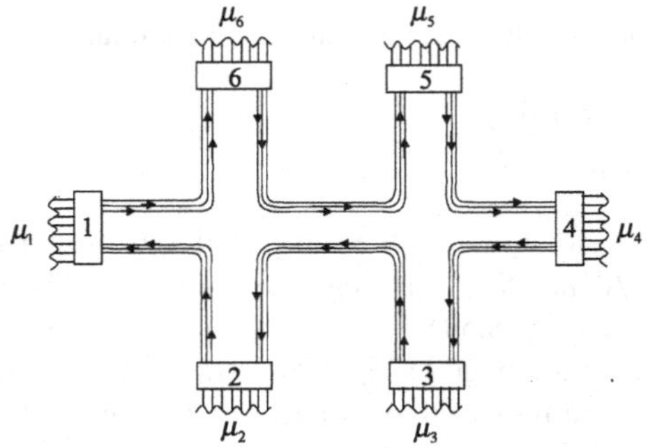

Fig. 8.60 Electron flow of the edge states in a sample with Hall bar geometry in a magnetic field applied perpendicular and upward with respect to the page. In the figure two edge channels are assumed to be active in. Coherent electron motion from one contact to the other is prevented by phase randomizing reservoirs. The electrode distance is assumed to be longer than the inelastic scattering length. Under these conditions The Hall plateau is explained as described in the text. In the figure the direction of electron flow is indicated by the arrow (see [103])

Next, we discuss the quantization of Hall resistance on the basis of the edge current in a sample with Hall bar geometry. Consider the sample with six electrode contacts shown in Fig. 8.60, where we identify the electrodes by labels 1, 2, 3, ..., 6. Current is injected from electrode 1 (source) and taken out from electrode 4 (drain). The current direction is given by the electron flow multiplied by $-e$ and the direction is reversed with respect to the electron flow direction. The arrows in Fig. 8.60 show the direction of electron flow. Although the Hall contacts for normal measurements are taken to be the pair of 2, 6 or 3, 5, we find in general that the pair 2, 5 or 3, 6 may give the same result. The electrons in the edge channels are one-dimensional and the electrons may bend around a corner because of the Lorentz force. The current injected from contact 1 enters into the voltage probe 6. Since the current is not taken out of probe 6, the same amount of current is fed from the other side of the contact. If the contacts 5, 3, and 2 are voltage probes, the same condition should be fulfilled. Since electrical current flows in or out electrode contacts 1 and 4, the difference between the current flowing from the sample toward the contact and the current flowing from the other contact into the sample is the net current in the device. For simplicity, the edge is assumed to contain $N_i \equiv N$ edge channels (the Landau levels up to N_i are filled by electrons). The current in the edge channels is evaluated from the Büttiker–Landauer formula of (8.272). We assume perfect contact between the electrodes and the two-dimensional electron gas system, and thus $R_{ii} = 0$. The current at the source contact is $I_1 = -I$ and the current at the drain is $I_4 = +I$. The current at the other potential probes is 0. Since $N_i = N$ at all the edges, we find $T_{61} = N$, $T_{56} = N$, $T_{45} = N$, $T_{34} = N$, $T_{23} = N$ and $T_{12} = N$, and all the other transmission probabilities are zero. From these results the current flowing from each reservoir into the semiconductor is given by

$$\begin{aligned} I_1 &= \tfrac{Ne}{h}(\mu_1 - \mu_2) = -I\,, & I_4 &= \tfrac{Ne}{h}(\mu_4 - \mu_5) = I\,, \\ I_2 &= \tfrac{Ne}{h}(\mu_2 - \mu_3) = 0\,, & I_3 &= \tfrac{Ne}{h}(\mu_3 - \mu_4) = 0\,, \\ I_5 &= \tfrac{Ne}{h}(\mu_5 - \mu_6) = 0\,, & I_6 &= \tfrac{Ne}{h}(\mu_6 - \mu_1) = 0\,. \end{aligned} \tag{8.302}$$

Therefore, we obtain

$$\mathcal{R}_{14,62} = \frac{\mu_6 - \mu_2}{-eI} = \frac{h}{e^2} \cdot \frac{1}{N}\,, \tag{8.303}$$

where we find that the result agrees with (8.284) ($N \to i$), and thus the quantum Hall effect is explained. It is obvious that $\mathcal{R}_{14,53} = \mathcal{R}_{14,63} \equiv \mathcal{R}_{14,62}$. In addition, we find

$$\mathcal{R}_{xx} = \mathcal{R}_{14,23} = \mathcal{R}_{14,56} = 0\,, \tag{8.304}$$

and the longitudinal resistance $\mathcal{R}_{xx} = 0$, which is in agreement with the experimental result of $\sigma_{xx} = 0$. Another theory of the quantum Hall besides those mentioned here is the theory based on the gauge transform by Laughlin [101].

The discovery of the quantum Hall effect has made a great contribution to semiconductor physics. Later, Tsui, Stormer and Gossard reported Hall resistance plateaus and of vanishing of ρ_{xx} for a filling factor ν at $\nu = 1/3$ [107], and this is called the **fractional quantum Hall effect** (FQHE). More detailed investigations revealed obvious fractional quantum Hall effects for the filling factors $\nu = p/q$ (q is always odd) such as $\nu = 5/3, 4/3, 2/3, 3/5, 4/7, 4/9, 3/7, 2/5, \ldots$. The research in the field of quantum Hall effect is still making great progress and expanding widely. Another exciting subject is the existence of the Wigner crystal predicted by Tsui et al. [107] and this is still an attractive field of research for both theory and experiment. The theory of the fractional quantum Hall effect has been developed by Laughlin, who took account of many-body effects and succeeded in explaining various experimental observations [108], but it is believed to be incomplete. In addition, recently various models such as the composite boson model, the composite fermion model, and so on have been proposed. A detailed review is given by Aoki, and readers who are interested in this field are recommended to read the paper by Aoki [13].

Figure 8.61 shows an example of experimental data on the fractional quantum Hall effect, where the diagonal component of magnetoresistance ρ_{xx} is plotted as a function of magnetic field in the region near the filling factor $\nu = 1/2$ and 1/4 at a temperature of $T = 40$ mK [109]. The physical explanation of the fractional quantum Hall effect is as follows. At extremely low temperatures, the Coulomb interaction dominates in two-dimensional electron gas systems, and the electrons are condensed into quantum liquids at a filling factor $\nu = p/q$ (q is an odd number). The filling factor is different from the case of the integer quantum Hall effect and the Landau levels are partially filled by electrons. The most remarkable condensation occurs at the filling factor $\nu = 1/q$, which is interpreted in terms of the many-body wave functions derived by Laughlin [108]. A gap appears above this ground state and the

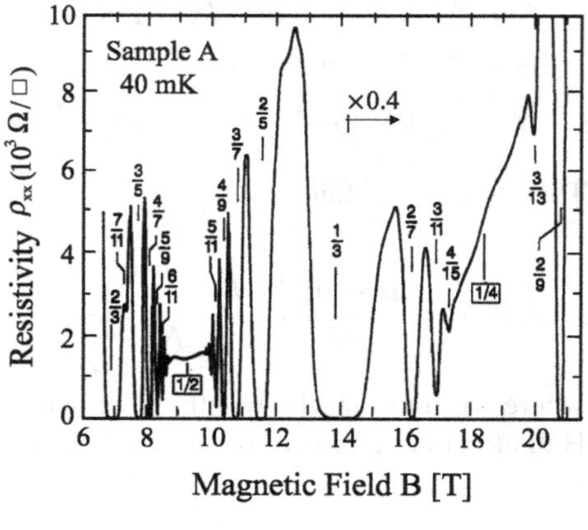

Fig. 8.61 Diagonal component of magnetoresistance ρ_{xx} plotted as a function of magnetic field in the region near the filling factor $\nu = 1/2$ and 1/3 at $T = 40\,\text{mK}$, where the data for magnetic field higher than 14 T are divided by a factor 2.5. The series of $p/(2\,p \pm 1)$ is clearly observed near the filling factor $\nu = 1/2$

conductance σ_{xx} disappears around the filling factor. Therefore, the Hall conductance $\sigma_{xy} = hq/e^2$ is also quantized at the filling factor. Excitation beyond this gap results in a creation of a quasi-particle with fractional charge e/q. Such a condensation is known to occur for a different electron density near the filling factor $\nu = 1/q$, and in general the fractional quantum Hall effect at $\nu = p/q$ is observed. However, the experimental data reveals fractional quantum Hall effects besides the condition $\nu = p/q$, and the theories reported so far have not been brought together under a common interpretation. Of the various reported experiments the work of Jiang et al. has attracted many researchers [110]. They observed a very deep minimum in σ_{xx} at the filling factor $\nu = 1/2$, and found that the temperature dependence of the minimum is quite different from other fractional quantum Hall effects. As shown clearly in Fig. 8.61, a series of $p/(2\,p \pm 1)$ is observed for up to $\nu = 9/19$ and $9/17$ for the filling factor $\nu = 1/2$, and for the fractional quantum Hall states of $\nu = 1/4$ the series of $p/(4\,p \pm 1)$ such as $\nu = 1/3, 2/5, 3/7, 4/9$ and $5/11$ are observed. The most probable theory to explain the fractional quantum Hall effect is known to be the composite fermion model of Jain [111], but all the observation has not explained yet.

8.7 Coulomb Blockade and Single Electron Transistor

The semiconductor devices in common use utilize many electrons in each device. For example, 10^{11} to 10^{12} electrons exist in the $1\,\text{cm}^2$ area of a typical MOSFET device, and thus in a MOSFET with an area as small as $1\,\mu\text{m} \times 1\,\mu\text{m}$ 10^3–10^4 electrons are involved in the device operation. If device fabrication technology continues to advance and a device size below $0.1\,\mu\text{m}$ becomes possible, then the number of electrons involved in the operation decreases and in the limit a device operated by a single electron will be achieved in the future. However, it is expected that new devices based on operating principles different from those in present devices may

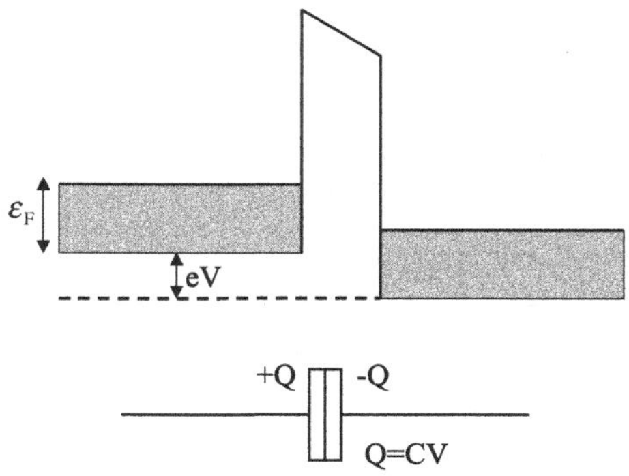

Fig. 8.62 Schematic illustration of a tunnel junction biased by a voltage V. The lower part shows its equivalent circuit

be developed, where single or several electrons take part in the device operation. Research projects on new functional devices with few electrons are in progress. The most important principle for such devices is believed to be the Coulomb blockade [112, 113]. In this section the basic principle of the Coulomb blockade will be described [114–116].

First, consider the tunnel junction shown in Fig. 8.62. The energy stored in the junction is given by

$$U = \frac{Q^2}{2C} . \tag{8.305}$$

In the presence of a bias voltage V, an electron at the source electrode with kinetic energy $\mathcal{E}_s(k)$ will tunnel into the state at the drain electrode with kinetic energy $\mathcal{E}_d(k')$, and thus we find the following relation:

$$\mathcal{E}_s(k) + \frac{1}{2}CV^2 = \mathcal{E}_d(k') + \frac{(CV - e)^2}{2C} . \tag{8.306}$$

Since the electron tunneling through the junction has to satisfy the Pauli exclusion principle the following inequalities are required.

$$\mathcal{E}_s(k) < \mathcal{E}_F - k_B T , \quad \mathcal{E}_d(k') > \mathcal{E}_F + k_B T . \tag{8.307}$$

This will lead to

$$\mathcal{E}_d(k') - \mathcal{E}_s(k) > 2k_B T . \tag{8.308}$$

Therefore, the tunneling condition is given by

$$eV \geq \frac{e^2}{2C} - 2k_B T , \tag{8.309}$$

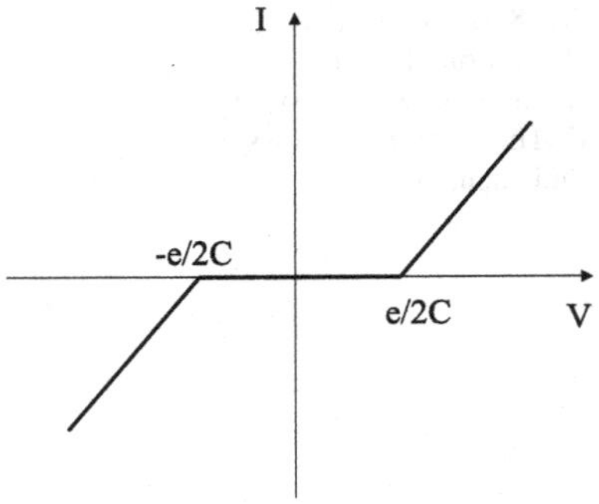

Fig. 8.63 A finite energy is required for an electron to tunnel through the junction and in the limit of temperature $T = 0$ no electron current flows at the bias voltage below $V = e/2C$. This bias region is called the Coulomb gap

and no current flows at a bias below a threshold voltage which depends on the temperature. At $T = 0$, no current results below $V = e/2C$. This bias region is called the **Coulomb gap** and the tunneling characteristics of the Coulomb gap are schematically illustrated in Fig. 8.63. The prohibition of tunneling is called the **Coulomb blockade**.

We discuss the Coulomb blockade phenomenon in a little more detail by investigating the change in the electrostatic energy at the tunnel junction shown in Fig. 8.62. Letting the electrostatic capacitance of a small junction be C, we calculate the change in the electrostatic energy due to the tunneling of an electron. In this system the change in electrostatic energy before and after the tunneling is estimated to be of the order of the Coulomb energy of an electron given by

$$\mathcal{E}_C = \frac{e^2}{2C}.$$

This energy is quite small in a junction of macroscopic size and thus the energy is washed by the thermal noise, making detection impossible. However, in a tunneling junction with an area of 0.01 μm^2 with an insulating film of 1 nm thickness, the energy becomes equivalent to temperature of about 1 K. Therefore, if the temperature of the tunnel junction is kept below 1 K, it is expected that the tunneling probability will be controlled by the Coulomb energy $\mathcal{E}_C$.

When charges of $\pm Q$ are stored at the junction surfaces, the electrostatic energy of this system is given by $Q^2/2C$. Let us consider the case where an electron tunnels from the negative electrode to the positive electrode in this system. The charges at the positive and negative electrodes will therefore be changed to $\pm(Q - e)$. The change in the electrostatic energy before and after the tunneling is then given by

$$\Delta\mathcal{E} = \frac{(Q-e)^2}{2C} - \frac{Q^2}{2C} = \frac{e}{C}\left(\frac{e}{2} - Q\right) = \mathcal{E}_C - eV, \tag{8.310}$$

where the final relation is rewritten by using the voltage applied to the junction $V = Q/C$. This result tells us the following fact. When an electron tunnels in the system, the system loses its own Coulomb energy $\mathcal{E}_C$ and receives energy eV from the voltage source. Therefore, the tunneling condition for an electron in this system is such that the voltage applied to the junction has to satisfy $V > \mathcal{E}_C/e$. In other words, under the condition

$$V < \frac{\mathcal{E}_C}{e}, \tag{8.311}$$

and at low temperatures such that $k_B T \ll \mathcal{E}_C$, the electron is not allowed to tunnel through the junction.

In a similar fashion, the change in the charges due to the tunneling of an electron from the positive electrode to the negative electrode is given by $\pm(Q + e)$ and thus the change in the electrostatic energy is shown to be

$$\Delta\mathcal{E} = \frac{(Q+e)^2}{2C} - \frac{Q^2}{2C} = eV + \mathcal{E}_C\,. \tag{8.312}$$

Therefore, for an applied bias voltage such that

$$-\frac{\mathcal{E}_C}{e} < V, \tag{8.313}$$

the electron is not allowed to tunnel through the junction. From these results we find that electron tunneling is forbidden for $|V| < 2/2\,C$ or $|Q| < e/2$, and the Coulomb gap appears.

Next, let us discuss the characteristics of a single electron transistor (SET). An example of a SET circuit is shown in Fig. 8.64. The characteristics of the SET are interpreted in terms of the Coulomb blockade phenomenon. In the device two tunnel junctions are connected in series and the isolated area in between is called a **Coulomb island**. The electron number in the Coulomb island is controlled externally through the junction capacitance C_G, and the capacitance is called the gate electrode or gate capacitance. In such a circuit tunneling through the gate capacitance can be neglected.

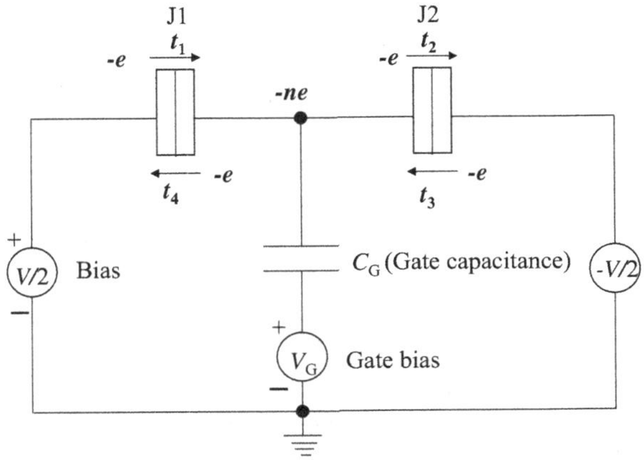

Fig. 8.64 Single electron transistor circuit consisting of two tunneling junctions and a condenser to control electron tunneling. The arrows indicate four possible tunneling processes of an electron when n electrons exist in the Coulomb island

In order to compare it with a transistor, the contact leading to the left tunnel junction is called the source electrode and the contact outside the right tunnel junction is called the drain electrode. We consider the case where voltages $+V/2$ and $-V/2$ are applied to the source and drain electrodes, respectively, and the gate electrode is biased independently by V_G as shown in Fig. 8.64. From (8.310) it may be understood that electron tunneling through the junction is possible in the case where the change in the electrostatic energy $\Delta\mathcal{E}$ is zero or negative. This is because a more stable state of lower energy exists after the tunneling. On the contrary, in the case of $\Delta\mathcal{E} > 0$, electron tunneling is not allowed.

Assume that there are n excess electrons and thus $-ne$ charges in the Coulomb island. Four tunneling processes, t_1, t_2, t_3 and t_4, are possible, as shown in Fig. 8.64, where an electron is added to or removed from the n electrons due to the tunneling. We define the change in the electrostatic energy due to the four tunneling processes by $\Delta\mathcal{E}_1, \ldots, \Delta\mathcal{E}_4$. When all of them are positive, tunneling is forbidden and the number of electrons in the Coulomb island is unchanged, giving rise to zero electron current. For example, the change in electrostatic energy $\Delta\mathcal{E}_1$ due to the tunneling process t_1 is given by

$$\Delta\mathcal{E}_1 = \frac{e}{C_\Sigma}\left[\frac{C_\Sigma}{2}V - C_G V_G - e\left(n + \frac{1}{2}\right)\right], \tag{8.314}$$

$$C_\Sigma = 2C + C_G, \tag{8.315}$$

where C_Σ is the total electrostatic capacitance seen from the Coulomb island. In a similar fashion, $\Delta\mathcal{E}_2$, $\Delta\mathcal{E}_3$ and $\Delta\mathcal{E}_4$ are also calculated, and the results are given by (8.314) by replacing the last term on the right–hand side with one of the terms $\pm(n \pm 1/2)$. Equation (8.314) is plotted in Fig. 8.65a with horizontal axis $C_G V_G$ and vertical axis $C_\Sigma V$, where the hatched region satisfies $\Delta\mathcal{E}_1 > 0$ and tunneling is forbidden. The region where all the four tunneling processes are forbidden is shown by the hatched diamonds in Fig. 8.65b. This region is called the **Coulomb diamond**. When the number of electrons n is not fixed, we find the region for forbidden tunneling for different n as shown in Fig. 8.65c.

Next, we discuss the tunneling current in the SET device. Keeping the voltage applied to the source–drain electrodes constant and changing the gate voltage V_G, the SET characteristics pass through along the dotted line in Fig. 8.66 and thus pass

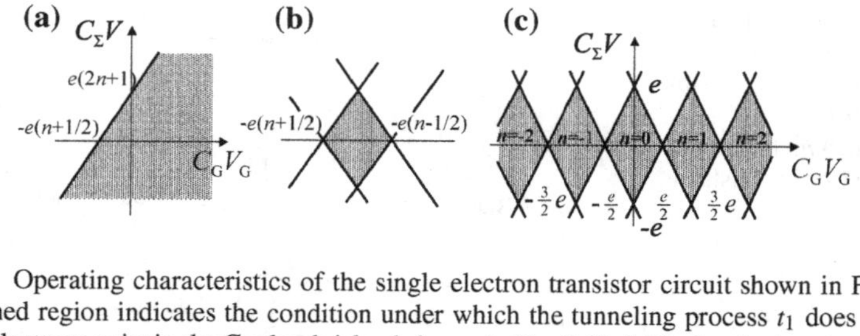

Fig. 8.65 Operating characteristics of the single electron transistor circuit shown in Fig. 8.64. **a** The hatched region indicates the condition under which the tunneling process t_1 does not occur, where n electrons exist in the Coulomb island shown in Fig. 8.64. **b** The hatched region shows the condition under which any of the tunneling processes t_1, t_2, t_3 and t_4 will not occur. **c** Condition for no tunneling is shown by the hatched region for different values of the number n

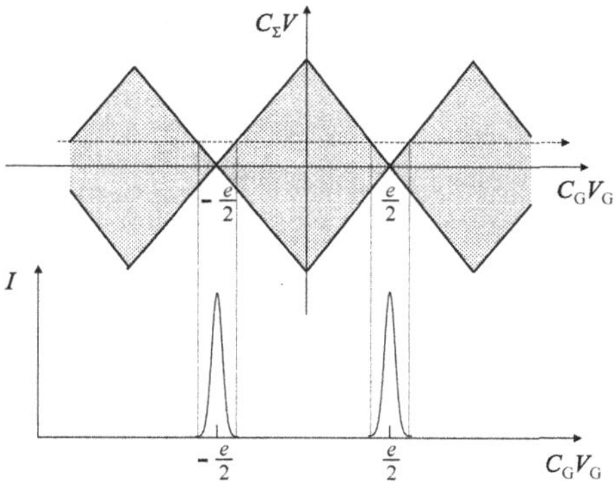

Fig. 8.66 SET characteristics when the source and drain currents are controlled by the gate voltage V_G. Keeping source–drain voltage constant and changing the gate voltage V_G, the characteristics pass through the forbidden and allowed tunneling regions alternately. When the gate voltage V_G passes through the region without hatching, the tunneling current and source–drain current flow as shown in the lower part of the figure

through the diamond of forbidden tunneling and outside the diamond, allowing tunneling, alternately. Therefore, the source–drain current flows periodically as shown in the lower part of Fig. 8.66. It is therefore evident from Fig. 8.66 that the change in the electronic charge at the gate electrode is less than e, the charge of one electron. It means that the device is operated by a charge of less than a single electron. The term "**single electron transistor**" (SET) was named after the fact that such a circuit consisting of tunneling junctions is operated by controlling the source–drain current by the gate voltage, and the device characteristics are comparable to those of the MOSFET and MESFET.

In order to apply the Coulomb blockade effect to a real device operated at a temperature near 300 K, $e^2/2C$ is required to be larger than $k_B T$ and thus we have to achieve a device with very small capacitance. At $T = 300$ K, the capacitance should be $C \leq 3.1 \times 10^{-18}$ F $= 3.1$ aF. It may be possible to fabricate a device of this size by using microfabrication technology, but its large scale integration seems to be impossible. This is because in such a small device the noise, radiation hardness, reproducibility and uniformity of large scale integration pose very serious problems for fabrication process. There have been reported so far various types of SET circuits and related functional devices, but most of them are concerned with the investigation of the physical properties of the devices. Recently several important results for SET devices have been reported. Transistors operated by a single electron have been fabricated and the device operation has been confirmed by observing the Coulomb blockade phenomenon at room temperature [117–121]. In addition Yano et al. have succeeded in fabricating very small MOSFETs by using poly Si and in observing SET characteristics by charging and discharging one electron at a trap near the channel, resulting in a change in the threshold voltage. They have integrated SETs and fabricated SET memories [122].

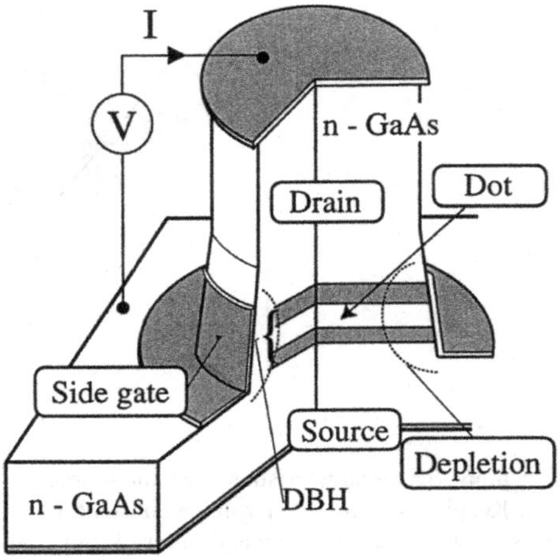

Fig. 8.67 Quantum dot structure used by Tarucha et al. [125] for the investigation of Coulomb oscillations and addition energy *vs* electron number. Electrons are confined in the z direction by the double barrier heterostructure AlGaAs/$In_{0.05}Ga_{0.95}As$/AlGaAs and then the side gate confines electrons in the (x, y) plane, forming quantum dot disk

8.8 Quantum Dots

In this section we present a good example of an artificial atom and discuss Coulomb interaction of electrons confined in a disk composed heterostructure GaAs/AlGaAs/InGaAs.[7] It was pointed out that electrons in a quantum dot with a good symmetry exhibit the shell structure of atoms and thus the electronic states reflect Hund's rule [123–125]. From this reason such a quantum dot is called an artificial atom, in which electronic states are controlled by the unit of an electron. Here we summarize the experimental date of Tarucha et al. [125] and theoretical explanation of the observed results. Figure 8.67 shows the quantum dot structure used by Tarucha et al. [125], where double barrier tunnel structure is cut into a disk shape and the side gate confines electrons with its parabolic potential. First we disregard Coulomb interaction and show that the assumption fails in explaining typical features. Then we deal with many electron system by the method of diagonalization of the coupled N–electron Hamiltonian with Coulomb interaction.

8.8.1 Addition Energy

Tarucha et al. used devices such as shown in Fig. 8.67, where the double barrier heterostructure (DBH) consists of an undoped 12.0 nm $In_{0.05}Ga_{0.95}As$ and undoped $Al_{0.22}Ga_{0.78}As$ barriers of thickness 9.0 (upper barrier) and 7.5 nm (lower barrier). Figure 8.68 shows the current flowing through the device as a function of the gate voltage under the drain bias of $V = 150\,\mu V$, where we see clear Coulomb oscillations

---

[7]This section is based on the Ph.D Thesis (in Japanese) by Tatsuya Ezaki, submitted to Osaka University (1997).

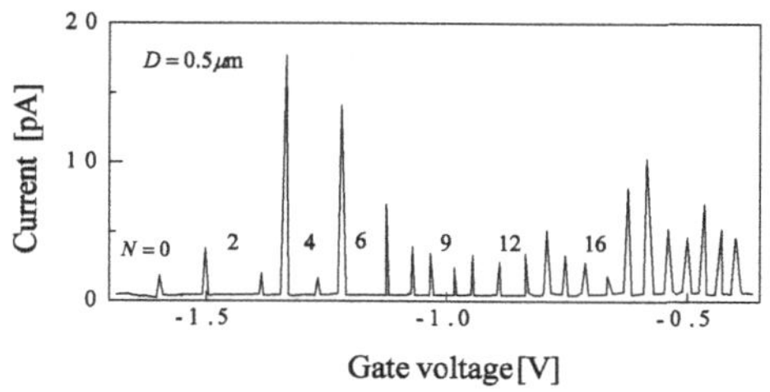

Fig. 8.68 Coulomb oscillations observed in the quantum dot of $D = 0.5\,\mu$m shown in Fig. 8.67, where the current through the dot is plotted as a function of the gate voltage under the drain voltage of $V_d = 150\,\mu$V [125]

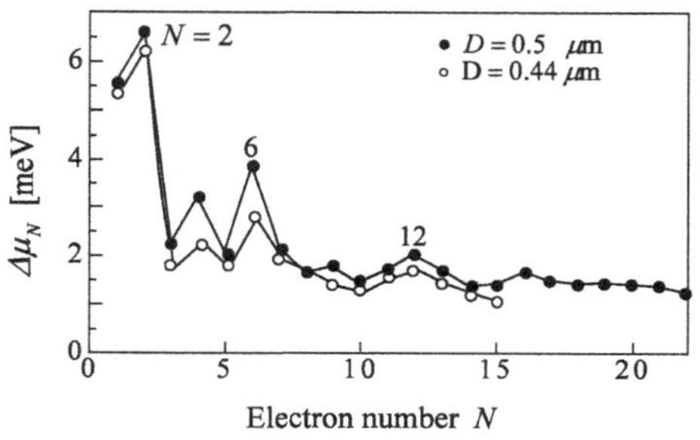

Fig. 8.69 Chemical potential difference *vs* electron number for two different dots with $D = 0.5$ and $0.44\,\mu$m. See text for the definition of the addition energy and the chemical potential difference [125]

for $V_g > -1.6$ V with each peak corresponding to a change of exactly one electron in the dot. The spacing between the current peaks in Fig. 8.68 reflects the energy required to add one more to a dot containing N electrons. The spacing at N=2 and 3, N=4 and 5, and N=6 and 7 are larger than the other spacing, which means the difference in the addition energy defined by $\Delta\mu_N$ below is higher at $N = 2, 4, 6, \ldots$ than the other spacing. Tarucha et al. estimated the chemical potential difference (difference in addition energy) from their experiments and plotted as a function of electron number in Fig. 8.69 for two different devices.

Addition energy μ_N and the chemical potential difference $\Delta\mu_N$ are defined below, which are illustrated in Fig. 8.70. When we neglect Coulomb interaction, the energy $\mathcal{E}_N$ of N electrons is given by summing up the energy levels ε_j of the one–electron states,

$$\mathcal{E}_N = \sum_{j=1}^{N} \varepsilon_j \,. \tag{8.316}$$

Then the addition energy (chemical potential) required for adding the N–th electron to the electronic state with N-1 electrons is define by

$$\mu_N = \mathcal{E}_N - \mathcal{E}_{N-1} = \varepsilon_N \,. \tag{8.317}$$

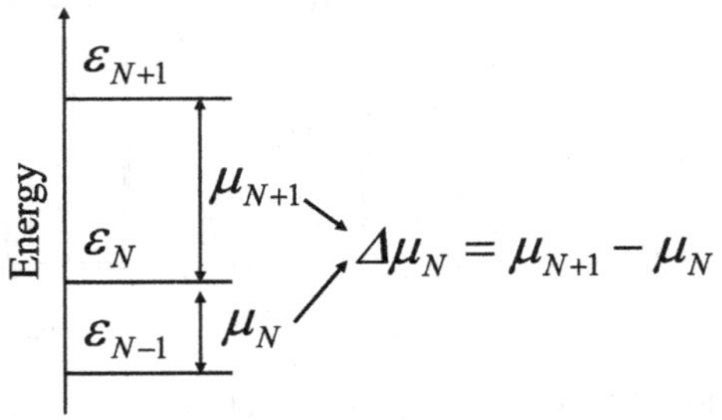

Fig. 8.70 Relation between electronic state $\mathcal{E}_N$ of N electrons, the chemical potential (addition energy) μ_N and the chemical potential difference $\Delta\mu_N$ are illustrated

In the presence of Coulomb interaction we have to calculate the total energy $\mathcal{E}_N$ by taking account of many body effects, which will be discussed later. Figure 8.70 illustrates schematically, the electronic energies for N-1, N and N+1 electrons, the chemical potentials and the difference in chemical potential. The difference in the chemical potential $\Delta\mu_N$ is given by

$$\Delta\mu_N = \mu_{N+1} - \mu_N \,. \tag{8.318}$$

In Fig. 8.69 we find that the chemical potential difference exhibits clear peaks at $N = 2,\ 4,\ 6,$ and, 12. These features may be understood by investigating the filling of the shell structure. Let's consider electronic states without Coulomb interaction, and then one electron Hamiltonian $h_0(\boldsymbol{r}, \boldsymbol{s})$ in the presence of a magnetic field $\boldsymbol{B}$ is written as

$$h_0(\boldsymbol{r}, \boldsymbol{s}) = \frac{1}{2m^*}(\boldsymbol{p} + e\boldsymbol{A})^2 + V(x, y) + V(z) \,, \tag{8.319}$$

where $\boldsymbol{r}$ is the three dimensional coordinates and $\boldsymbol{s}$ is the spin coordinate. In the z direction the electrons in the disk are confined by the double barriers of the heterostructure as shown earlier. Here we are interested in the system of few electrons, and then the lowest subband is taken into account. In the (x, y) plane, the potential $V(x, y)$ is produced by the depletion layer controlled by the gate voltage, and the shape is parabolic. For the purpose of the later discussion we express the potential defined by Ezaki et al. [126] and Ezaki [127],

$$V(x, y) = \frac{1}{2}m^* \left(\omega_x^2 x^2 + \omega_y^2 y^2\right) \left[1 + \alpha\frac{2}{7}\cos(3\phi)\right] , \tag{8.320}$$

where $\hbar\omega_x$ and $\hbar\omega_y$ are the confinement potentials in the x and y directions, respectively, and α is introduced to modify the potential shapes as follows. When we put $\alpha = 0$, we obtain a circular dot for $\omega_x = \omega_y$ and an ellipsoidal disk for $\omega_x \neq \omega_y$, and a triangular dot for $\alpha = 1$ and $\omega_x = \omega_y$. First we consider the simplest case of a circular dot, and put $\omega_x = \omega_y = \omega_0$. Also note that ω_{L} is the Lamor frequency and related to the cyclotron frequency by $\omega_{\mathrm{L}} = \omega_{\mathrm{c}}/2$. In the z direction we assume infinite barrier potentials with the well width W, and thus

$$\zeta(z) = \sqrt{\frac{2}{W}} \sin\left(\frac{\pi}{W} z\right) . \tag{8.321}$$

Since the system is symmetrical in the (x, y) plane, we use the polar coordinate (ρ, ϕ), and then we may write the wave function of the electron in the disk as

$$\psi_{mn}(\rho, \phi, z) = \frac{1}{l_0}\sqrt{\frac{2!}{(|m|+n)!}} \exp\left[-\frac{1}{2}\left(\frac{\rho}{l_0}\right)^2\right] \times \left(\frac{\rho}{l_0}\right)^{|m|} L_n^{|m|}\left(\frac{\rho}{l_0}\right) \varphi_m(\phi)\zeta(z) , \tag{8.322a}$$

$$\varphi_m(\phi) = \frac{1}{\sqrt{2\pi}} \mathrm{e}^{-\mathrm{i}m\phi} , \tag{8.322b}$$

$$l_0 = \sqrt{\frac{\hbar}{m^* \Omega}} , \tag{8.322c}$$

$$\Omega = \sqrt{\omega_0^2 + \omega_\mathrm{L}^2} , \tag{8.322d}$$

$$\omega_\mathrm{L} = \frac{eB}{2m^*} , \tag{8.322e}$$

where $L_n^{|m|}$ is the generalized Laguerre polynomial. The energy of one–electron state is

$$\varepsilon_{mn} = \hbar\Omega(2n + |m| + 1) - m\hbar\omega_\mathrm{L} + \varepsilon_z , \tag{8.323}$$

where ε_z is the ground subband energy of electrons confined in the z direction. The potentials for elliptic and triangular shapes are also expressed as follows for the purpose to expand the electronic states by Slater determinant. In the case of elliptic potential

$$V(\rho, \phi) = \frac{1}{2} m^* \omega \rho^2 + \frac{1}{4} m^* \left(\omega_x^2 - \omega_y^2\right) \rho^2 \cos(2\phi) , \tag{8.324a}$$

$$\omega = \sqrt{\frac{\omega_x^2 + \omega_y^2}{2}} , \tag{8.324b}$$

and for the triangular potential

$$V(\rho, \phi) = \frac{1}{2} m^* \omega_0^2 \rho^2 + \frac{1}{7} m^* \omega_0^2 \rho^2 \cos(3\phi) . \tag{8.324c}$$

In order to understand the peaks of the chemical potential difference in Fig. 8.69, we first deal with the case of no Coulomb interaction in a circular disk. Then the one–electron energy is given by

$$\varepsilon_{mn} = \hbar\omega_0(2n + |m| + 1) \quad (+\varepsilon_z) . \tag{8.325}$$

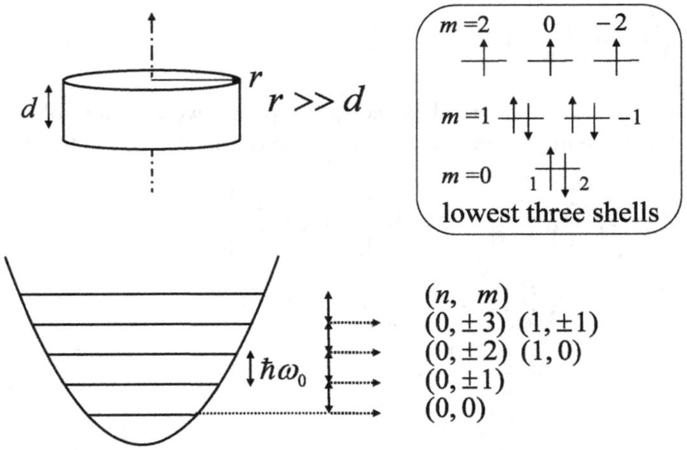

Fig. 8.71 Electronic states in a circular disk with parabolic confinement potential. The lower left figure shows the quantized energy levels with equal spacing $\hbar\omega_0$ and several energy states are shown by the quantum number (n, m) in the lower right figure. The upper right figure shows the filling of 9 electrons with spin up and down, where the Coulomb interaction is neglected

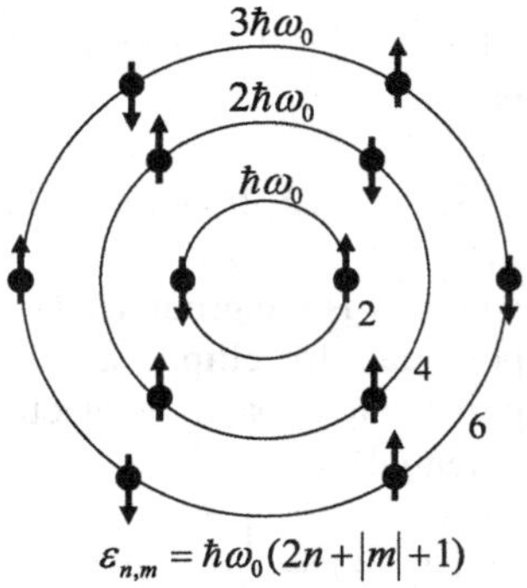

Fig. 8.72 Shell structure is shown, where the energy is given by $\varepsilon_{nm} = \hbar\omega_0(2n + |m| + 1)$. The filled shells are illustrated with the spin orientations. From this figure the chemical potential difference without Coulomb interaction in a circular disk is easily deduced, which is shown in Fig. 8.73

Therefore the electronic states are well expressed by the quantum number (n, m). The lower states are then expressed by $(n, m) = (0, 0), (0, \pm1), (0, \pm2), (1, 0), (0, \pm3), (1, \pm1), \ldots$, which are illustrated in Fig. 8.71.

In Fig. 8.72 such a shell filling is represented by using an atomic orbital model. We may easily evaluate the difference in addition energies $\Delta\mu_N$ as shown in Fig. 8.73, where we find the peaks with the height of $\hbar\omega_0$ at N=2, 4, and 6. Therefore the peaks found by Tarucha et al. shown in Fig. 8.69 are qualitatively explained. In order to explain the detailed features of Fig. 8.69, however, we have to take into account the Coulomb interaction of N electrons in a disk.

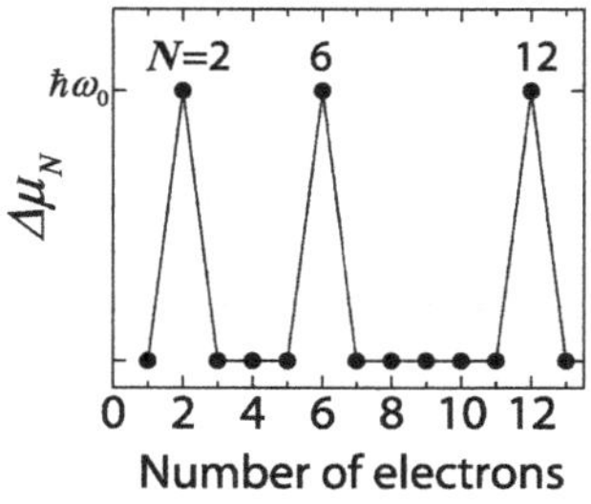

Fig. 8.73 The shell structure of the electronic states in a circular disk and the chemical potential difference is plotted as a function of electron number, where the Coulomb interaction is neglected

8.8.2 Exact Diagonalization Method

It is well known that the electron–electron interaction plays an important role in many electron system and that the energy eigenstates derived for a single electron Hamiltonian fail in explaining various aspect of the observed features. In Sect. 8.8.1 such a single electron approximation fails in quantitative explanation of the experimental data on the chemical potential change in a quantum dot. In order to analyze a many electron system with including Coulomb interaction, we have to use many electron wave functions. We know that Slater determinant describes the properties of fermion system, which is written for N electrons as

$$|1,2,\cdots,N\rangle = \frac{1}{\sqrt{N!}}\begin{vmatrix} \chi_1(\boldsymbol{x}_1) & \chi_2(\boldsymbol{x}_1) & \cdots & \chi_N(\boldsymbol{x}_1) \\ \chi_1(\boldsymbol{x}_2) & \chi_2(\boldsymbol{x}_2) & \cdots & \chi_N(\boldsymbol{x}_2) \\ \vdots & \vdots & \ddots & \vdots \\ \chi_1(\boldsymbol{x}_N) & \chi_2(\boldsymbol{x}_N) & \cdots & \chi_N(\boldsymbol{x}_N) \end{vmatrix}, \tag{8.326}$$

where $\boldsymbol{x}_i$ denotes the Cartesian coordinates $\boldsymbol{r}_i$ plus spin coordinates for electron i and is equivalently expressed as $\chi_i(\boldsymbol{x}_i) = \psi_i(\boldsymbol{r}_i)|\alpha\rangle$ or $\chi_i(\boldsymbol{x}_i) = \psi_i(\boldsymbol{r}_i)|\beta\rangle$ with one–electron eigenfunction $\psi_i(\boldsymbol{r}_i)$ and spin functions $|\alpha\rangle$ and $|\beta\rangle$. One–electron eigenfunction $\psi_i(\boldsymbol{r})$ is given by

$$h_0(\boldsymbol{r})\psi_i(\boldsymbol{r}) = \varepsilon_i\psi_i(\boldsymbol{r}), \tag{8.327}$$

where $h_0(\boldsymbol{r})$ is one–electron Hamiltonian and ε_i is the eigenenergy. Using the solutions of one–electron Hamiltonian $h_0(\boldsymbol{x})$, we rewrite the eigenfunction as $\psi_{i\sigma}(\boldsymbol{x})$ and the eigenenergy as $\varepsilon_{i\sigma}$, where we introduced spin quantum number σ. Then the second quantized Hamiltonian is written as

$$\mathcal{H}_0 = \sum_{i\sigma}\varepsilon_{i\sigma}C^\dagger_{i\sigma}C_{i\sigma}, \tag{8.328}$$

where i is the quantum number for the spatial part of wave function and σ is the spin quantum number. $C^\dagger_{i\sigma}$ and $C_{i\sigma}$ represent the creation and annihilation operators of the electron state $i\sigma$, respectively. When 2 or more electrons are contained in a quantum dot, we have to take into account of the electron–electron interactions, among which Coulomb interaction plays the most important role. The Hamiltonian with Coulomb interaction is written as

$$\mathcal{H}_{\mathrm{C}} = \frac{1}{2}\sum_{i'j'ji}\sum_{\sigma\sigma'} u_{i'j'ji} C^{\dagger}_{i'\sigma} C^{\dagger}_{j'\sigma'} C_{j\sigma'} C_{i\sigma} \,, \tag{8.329a}$$

$$u_{i'j'ji} = \iint \psi^{*}_{i'}(\boldsymbol{r}_1)\psi^{*}_{j'}(\boldsymbol{r}_2) u(\boldsymbol{r}_1 - \boldsymbol{r}_2)\psi_j(\boldsymbol{r}_2)\psi_i(\boldsymbol{r}_1)\, d\boldsymbol{r}_1\, d\boldsymbol{r}_2 \,, \tag{8.329b}$$

$$u(\boldsymbol{r}_1 - \boldsymbol{r}_2) = \frac{e^2}{4\pi\epsilon|\boldsymbol{r}_1 - \boldsymbol{r}_2|} \,. \tag{8.329c}$$

Then the total Hamiltonian is given by

$$\mathcal{H} = \sum_{i\sigma} \varepsilon_{i\sigma} C^{\dagger}_{i\sigma} C_{i\sigma} + \frac{1}{2}\sum_{i'j'ji}\sum_{\sigma\sigma'} u_{i'j'ji} C^{\dagger}_{i'\sigma} C^{\dagger}_{j'\sigma'} C_{j\sigma'} C_{i\sigma} \,. \tag{8.330}$$

The wave function $|\Psi\rangle$ of N electrons in a quantum dot is expressed by a linear combination of the orthonormal system $|I\rangle$ (a linear combination of Slater determinants)

$$|\Psi\rangle = \sum_{I} d_I |I\rangle \,. \tag{8.331}$$

In this book the basis function of N electrons is given by

$$|I\rangle = |n_{\lambda 1}, n_{\lambda 2}, \cdots, n_{\lambda k}, \cdots\rangle \quad \begin{pmatrix} n_{\lambda_i} = 1, \ (i = 1, 2, \cdots, N) \\ n_{\lambda j} = 0, \ (\lambda j \neq \lambda_i) \end{pmatrix} \tag{8.332a}$$

$$= C^{\dagger}_{\lambda_N} C^{\dagger}_{\lambda_{N-1}} C^{\dagger}_{\lambda_1} |0\rangle \,, \tag{8.332b}$$

where $\lambda_1 < \lambda_2 < \cdots < \lambda_N$ and $|0\rangle$ is the vacuum state. Using the orthonormal system of N particles $|I\rangle$ we may solve the Hamiltonian matrix $\mathcal{H}_{IJ} = \langle I|\mathcal{H}|J\rangle = \langle I|\mathcal{H}_0 + \mathcal{H}_{\mathrm{C}}|J\rangle$ and obtain the many particle states exactly.

8.8.3 Hamiltonian for Electrons in a Quantum Dot

As described in Sect. 8.8.2, Hamiltonian for interacting electrons is given by

$$\mathcal{H} = \sum_{i\sigma} \varepsilon_{i\sigma} C^{\dagger}_{i\sigma} C_{i\sigma} + \frac{1}{2}\sum_{i'j'ji}\sum_{\sigma\sigma'} u_{i'j'ji} C^{\dagger}_{i'\sigma} C^{\dagger}_{j'\sigma'} C_{j\sigma'} C_{i\sigma} \,, \tag{8.333}$$

where the second term on the right hand side is the energy of Coulomb interaction. Here we will expand the states of N electrons in terms of orthonormal sets, which is done by using Slater determinant of N electrons. We compose Slater determinant of N electrons by the wave functions of (8.322e) obtained for one–electron solutions confined in the two dimensional simple harmonic type potential ($\omega_x = \omega_y = \omega_0$) in the plane parallel to the hetero–interfaces and in the infinite potentials perpendicular to the hetero–interfaces (z direction).

In the later section we mention the results on the disks with elliptic and triangular potentials. For this purpose we divide the potentials into two terms, circularly symmetric and asymmetric parts, which are given for the elliptic potential

$$V(\rho, \phi) = \frac{1}{2}m^*\omega^2\rho^2 + \frac{1}{4}m^*(\omega_x^2 - \omega_y^2)\rho^2 \cos 2\phi\,, \tag{8.334a}$$

$$\omega = \sqrt{\frac{\omega_x^2 + \omega_y^2}{2}}\,, \tag{8.334b}$$

and for the triangular potential

$$V(\rho, \phi) = \frac{1}{2}m^*\omega_0^2\rho^2 + \frac{1}{7}m^*\omega_0^2\rho^2 \cos 3\phi\,. \tag{8.335}$$

Expressing the asymmetric part of the potential by $V'(\boldsymbol{r})$, Hamiltonian of N electrons is given by

$$\begin{aligned}\mathcal{H} &= \sum_{i\sigma} \varepsilon_{i\sigma} C_{i\sigma}^{\dagger} C_{i\sigma} + \sum_{i'i}\sum_{\sigma} v'_{i'i} C_{i'\sigma}^{\dagger} C_{i\sigma} \\ &\quad + \frac{1}{2}\sum_{i'j'ji}\sum_{\sigma\sigma'} u_{i'j'ji} C_{i'\sigma}^{\dagger} C_{j'\sigma'}^{\dagger} C_{j\sigma'} C_{i\sigma}\,,\end{aligned} \tag{8.336a}$$

$$u_{i'j'ji} = \iint \psi_{i'}^*(\boldsymbol{r}_1)\psi_{j'}^*(\boldsymbol{r}_2) u(\boldsymbol{r}_1 - \boldsymbol{r}_2)\psi_j(\boldsymbol{r}_2)\psi_i(\boldsymbol{r}_1)\, \mathrm{d}^3\boldsymbol{r}_1\, \mathrm{d}^3\boldsymbol{r}_2, \tag{8.336b}$$

$$v'_{i'i} = \int \psi_{i'}^*(\boldsymbol{r}) V'(\boldsymbol{r}) \psi_i(\boldsymbol{r})\, \mathrm{d}^3\boldsymbol{r}\,, \tag{8.336c}$$

where Fourier expansion of the Coulomb interaction is written as

$$u(\boldsymbol{r}_1 - \boldsymbol{r}_2) = \sum_{Q} \frac{e^2}{2\epsilon Q} \mathrm{e}^{i\,Q\cdot(\text{æ}_1 - \text{æ}_2)} \mathrm{e}^{-Q|z_1 - z_2|}\,. \tag{8.337}$$

Inserting the one–electron wave functions of (8.322e) into (8.336b), we obtain

$$u_{i'j'ji} = \frac{e^2}{4\pi\epsilon}\int_0^{\infty} F(Q) g_{i'i}(Q) g_{jj'}^*(Q)\, \mathrm{d}Q \times \delta_{m_i + m_j, m_{i'} + m_{j'}}\,, \tag{8.338a}$$

$$g_{i'i}(Q) = i^{m_i - m_{i'}} \int_0^\infty \psi_{i'}^*(\rho)\psi_i(\rho) J_{m_i - m_{i'}}(Q\rho)\rho\,\mathrm{d}\rho\,, \tag{8.338b}$$

$$\begin{aligned} F(Q) &= \iint |\xi(z_1)|^2 |\xi(z_2)|^2 \mathrm{e}^{-Q|z_1 - z_2|}\,\mathrm{d}z_1\,\mathrm{d}z_2 \\ &= \frac{2}{x^2 + (2\pi)^2}\left[\frac{3}{2}x + \frac{(2\pi)^2}{x} - \frac{(2\pi)^4}{x^2(x^2 + (2\pi)^2)}(1 - \mathrm{e}^{-x})\right] \\ x &= QW\,, \end{aligned} \tag{8.338c}$$

where $J_n(x)$ is Bessel function and $\delta_{m_i+m_j,m_{i'}+m_{j'}}$ of (8.338a) represents the momentum conservation rule.

The asymmetric part of the elliptic confinement potential $V'(\boldsymbol{r})$ is written by using (8.334a) and (8.335) as

$$V'(\boldsymbol{r}) = \frac{1}{4}m^*(\omega_x^2 - \omega_y^2)\rho^2\cos 2\phi\,, \tag{8.339}$$

and inserting it into (8.336c) we obtain the matrix element of the asymmetric potential

$$v'_{i'i} = \frac{1}{8}m^*(\omega_x^2 - \omega_y^2)\int \psi_{i'}^*(\rho)\psi_i(\rho)\rho^3\,d\rho \times (\delta_{m',m+2} + \delta_{m',m-2})\,. \tag{8.340}$$

In the case of the triangular confinement potential, the asymmetric part is

$$V'(\boldsymbol{r}) = \frac{1}{7}m^*\omega_0^2\rho^2\cos 3\phi\,, \tag{8.341}$$

and inserting it into (8.336c) we obtain the matrix element of the asymmetric part of potential $v'_{i'i}$

$$v'_{i'i} = \frac{1}{14}m^*\omega_0^2\int \psi_{i'}^*(\rho)\psi_i(\rho)\rho^3\,d\rho \times (\delta_{m',m+3} + \delta_{m',m-3})\,. \tag{8.342}$$

In the following analysis we use a quantum dot model shown in Fig. 8.74, where the dot is formed by AlGaAs/$In_{0.05}Ga_{0.95}As$/AlGaAs double heterostructure and constriction in the (x, y) plane. The double heterostructure confines electrons in the z direction with the well width W. The bird's–eye–view of the model potential in the (x, y) plane is calculated from (8.320) and plotted in Fig. 8.75 for the ellipsoidal potential and in Fig. 8.76 for the triangular potential, where the equi–energy lines are shown. The elliptic potential is obtained by putting $\alpha = 0$ and $\omega_x \neq \omega_y$ in (8.320) and the circular potential is obtained by putting $\omega_x = \omega_y$. The triangular potential is shown in Fig. 8.76, where we put $\alpha = 1$ and $\omega_x = \omega_y$ in (8.320).

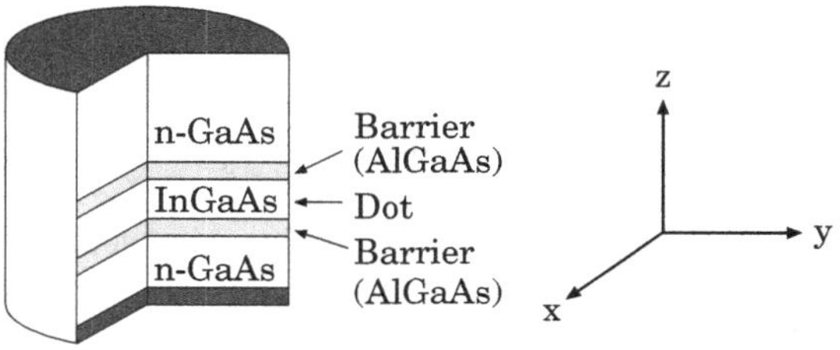

Fig. 8.74 A model of a quantum dot used for the analysis. The dot is formed by AlGaAs/InGaAs/AlGaAs double heterostructure and constricted in the (x, y) plane by etching and gate voltage [126]

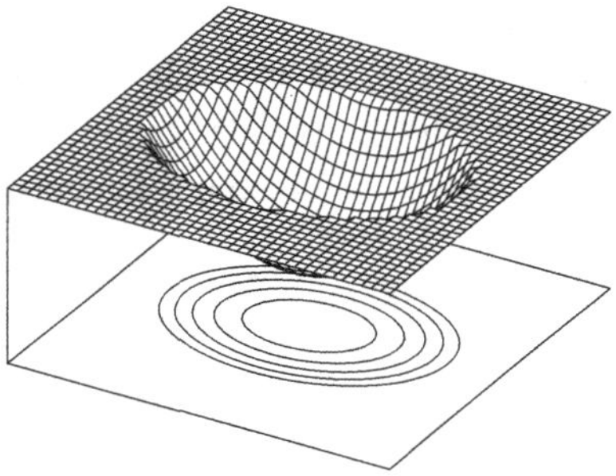

Fig. 8.75 Elliptic confining potential in the (x, y) plane sandwiched by the double heterostructure. The potential is calculated by putting $\alpha = 0$ and $\omega_x \neq \omega_y$ in (8.320). When we put $\omega_x = \omega_y$, we obtain the circular confining potential [126]

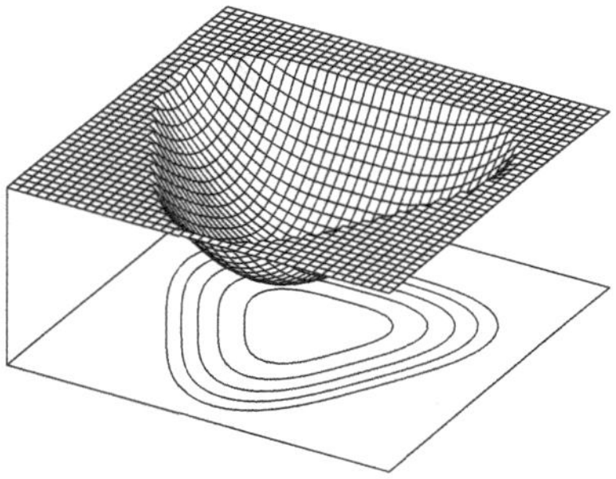

Fig. 8.76 Triangular confining potential in the (x, y) plane sandwiched by the double heterostructure. The potential is calculated by putting $\alpha = 1$ and $\omega_x = \omega_y$ in (8.320) [126]

8.8.4 Diagonalization of N Electrons Hamiltonian Matrix

The wave function $|\Psi\rangle$ of N electrons is expressed by a linear combination of the orthonormal system, in other words N particle Slater determinant $|I\rangle$ given by (8.331),

$$|\Psi\rangle = \sum_I d_I |I\rangle \,. \tag{8.343}$$

Here we define the difference $|I - J|$ between Slater determinants of N electrons $|I\rangle$ and $|J\rangle$ by the number of the different pairs of one–electron states.

$$|I\rangle = |m_1, m_2, \cdots, m_k, \cdots\rangle \,, \tag{8.344}$$

$$|J\rangle = |n_1, n_2, \cdots, n_k, \cdots\rangle \,. \tag{8.345}$$

Using the above expressions we find the following results. When $m_i = n_i$ ($i = 1, 2, \cdots$) and thus $|I\rangle = |J\rangle$, we obtain $|I - J| = 0$. When n_p–th and n_q–th one electron states of $|I\rangle$ and $|J\rangle$ are interchanged, for example,

$$|I\rangle = |\cdots, \overset{n_p}{1_p}, \cdots, \overset{n_q}{0_q}, \cdots\rangle, \qquad |J\rangle = |\cdots, \overset{n_p}{0_p}, \cdots, \overset{n_q}{1_q}, \cdots\rangle \,, \tag{8.346}$$

we obtain $|I - J| = 1$. For the purpose of further calculations we classify the matrix elements of N electrons Hamiltonian $H_{IJ} = \langle I|\mathcal{H}|J\rangle = \langle I|\mathcal{H}_0 + V' + \mathcal{H}_{\mathrm{C}}|J\rangle$ into the following four cases by using the difference between Slater determinants $|I - J|$.

1. $|I - J| = 0$ (diagonal elements)

$$\langle I|\mathcal{H}_0|J\rangle = \sum_{i=1}^{N} \varepsilon_{\lambda_i} \,, \tag{8.347a}$$

$$\langle I|V'|J\rangle = \sum_{i=1}^{N} v'_{\lambda_i\lambda_i} \,, \tag{8.347b}$$

$$\langle I|\mathcal{H}_{\mathrm{C}}|J\rangle = \sum_{i=1}^{N-1} \sum_{j=i+1}^{N} \left(u_{\lambda_i\lambda_j\lambda_j\lambda_i} \,, -u_{\lambda_i\lambda_j\lambda_i\lambda_j} \delta_{\sigma_{\lambda_i}\sigma_{\lambda_j}}\right) \,. \tag{8.347c}$$

2. $|I - J| = 1$

$$\begin{aligned} |I\rangle &= |\cdots, \overset{n_p}{1_p}, \cdots, \overset{n_q}{0_q}, \cdots\rangle \,, \\ |J\rangle &= |\cdots, \overset{n_p}{0_p}, \cdots, \overset{n_q}{1_q}, \cdots\rangle \,, \end{aligned} \tag{8.348a}$$

$$\langle I|\mathcal{H}_0|J\rangle = 0 \,, \tag{8.348b}$$

$$\langle I|V'|J\rangle = (-1)^{n_p+n_q} v'_{qp} \delta_{\sigma_p\sigma_q} \,, \tag{8.348c}$$

$$\langle I|\mathcal{H}_{\mathrm{C}}|J\rangle = (-1)^{n_p+n_q} \sum_{i=1}^{N} \left(u_{q\lambda_i\lambda_i p}\delta_{\sigma_p\sigma_q} - u_{\lambda_i q\lambda_i p}\delta_{\sigma_q\sigma_{\lambda_i}}\right) . \tag{8.348d}$$

3. $|I - J| = 2$

$$\begin{aligned} |I\rangle &= |\cdots, \overset{n_p}{1_p}, \cdots, \overset{n_r}{0_r}, \cdots, \overset{n_q}{1_q}, \cdots, \overset{n_s}{0_s}, \cdots\rangle \,, \\ |J\rangle &= |\cdots, \overset{n_p}{0_p}, \cdots, \overset{n_r}{1_r}, \cdots, \overset{n_q}{0_q}, \cdots, \overset{n_s}{1_s}, \cdots\rangle \,, \end{aligned} \tag{8.349a}$$

$$\langle I|\mathcal{H}_0|J\rangle = \langle I|V'|J\rangle = 0 \,, \tag{8.349b}$$

$$\begin{aligned} \langle I|\mathcal{H}_{\mathrm{C}}|J\rangle &= (-1)^{n_p+n_q+n_r+n_s} \\ &\quad \times \left(u_{rsqp}\delta_{\sigma_p\sigma_r}\delta_{\sigma_q\sigma_s} - u_{srqp}\delta_{\sigma_p\sigma_s}\delta_{\sigma_q\sigma_s}\right) . \end{aligned} \tag{8.349c}$$

4. $|I - J| \geq 3$

$$\langle I|\mathcal{H}_0|J\rangle = \langle I|V'|J\rangle = \langle I|\mathcal{H}_C|J\rangle = 0\,. \tag{8.350}$$

8.8.5 Electronic States in Quantum Dots

Here we will show the results calculated by means of the exact diagonalization. The parameters of $In_{0.05}Ga_{0.95}As$ such as the effective mass and dielectric constants are estimated by extrapolating the parameters of InAs and GaAs, and we use the effective mass $m^* = 0.065m_0$ and the dielectric constant $\epsilon = 12.9\epsilon_0$. The confinement potential is determined so that the average value of $\hbar\omega_x$ and $\hbar\omega_y$ remains constant and we set $(\hbar\omega_x + \hbar\omega_y)/2 = \hbar\omega_0 = 3$ [meV]. The well width in the z direction is assumed to be $W = 12$ [nm]. Slater determinant is composed by 20 one–electron states from the lowest ground state to upper 20–th states. The separation of the energy levels between the ground state and the first excited state levels due to the confinement by the heterobarriers in the z direction estimated from (8.321) is 121 [meV], while the energy difference between the ground state and the 20–th state is 9 [meV]. Therefore we consider only the ground state in the z direction confinement.

First we present a comparison between the calculated results by the exact diagonalization method and the experimental data of Tarucha et al. for a circular quantum dot. Figure 8.77 shows the chemical potential difference $\Delta\mu_N = \mu_{N+1} - \mu_N$ as a function of electron number N, where the calculated results are shown by ∘ and the experimental data of Tarucha et al. [125] by •. Here we note again that the chemical potential μ_N of a quantum dot with N electrons is defined by $\mu_N \equiv \mathcal{E}_{N+1} - \mathcal{E}_N$ where the ground state energy of the quantum dot with N electrons is given by $\mathcal{E}_N$. Therefore the chemical potential difference $\Delta\mu_N$ is the energy required to add another

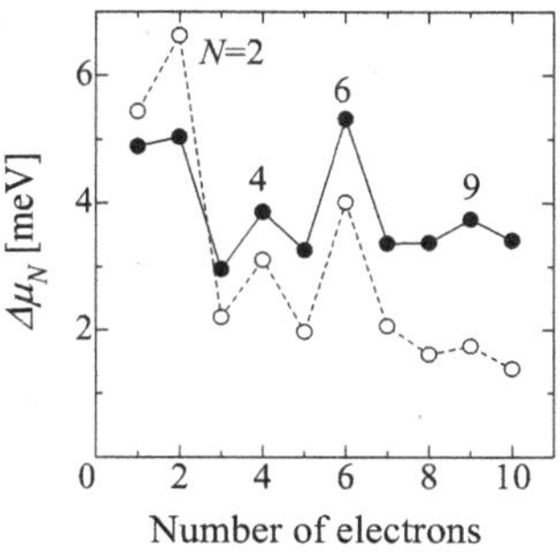

Fig. 8.77 The chemical potential difference $\Delta\mu_N = \mu_{N+1} - \mu_N$ calculated by the exact diagonalization method (∘) *vs.* electron number N in a circular dot is compared with the experimental data (•) of Tarucha et al. [125]. In the calculation the confinement potential by the heterointerfaces is assumed to be $\hbar\omega_0 = 3$ meV

1 electron to the dot and thus called charging energy. A state of a large chemical potential difference $\Delta\mu_N$ corresponds to a state with more stable energy. In Fig. 8.77 we find that the chemical potential difference $\Delta\mu_N$ exhibits a large value for electron number $N = 2$ and 6, and thus the states are more stable. This feature may be easily understood from the results shown in Figs. 8.72 and 8.73, where the shell structure is calculated in the case of no–Coulomb interaction or one–electron approximation. As given by (8.325) the one–electron energy states are expressed by the quantum number (m, n) or $\varepsilon_{mn} = \hbar\omega_0(2n + |m| + 1)$, and thus the energy is degenerate by the factor $2(2n + |m| + 1)$ with the spin degeneracy 2. A circular quantum dot with electrons $N = 2, 6, 12, \cdots$ exhibits closed shell structures, and the energy states become stable, resulting in a large chemical potential difference $\Delta\mu_N$.

In addition to $N = 2$ and $N = 6$, we find weak peaks at $N = 4$ and $N = 9$. These two peaks are stable a little and interpreted in the following way. Let's consider the case of $N = 4$ electrons. Two of them will occupy the lowest state and the other 2 electrons will occupy a half of the states of $2n + |m| = 1$, resulting in a half–filled shell structure. The exact diagonalization shows that the spin triplet state of the total spins $S = \hbar$ $(S_z = 0, \pm\hbar)$ is the ground state. The outer shell in the quantum dot with $N = 4$ electrons contains 2 electrons with the same spin orientation. This may be explained in terms of Pauli's exclusion principle as follows. Since the 2 electrons with the parallel spins of the outer–shell keep away due to Pauli's exclusion principle, and thus the Coulomb energy becomes smaller, resulting in a stable state of the spin–triplet state $S = \hbar$ $(S_z = 0, \pm\hbar)$. When the quantum dot contains 9 electrons, the ground state is formed by the spin quadruplet state with the total spins $S = 3\hbar/2$ $(S_z = \pm\hbar/2, \pm 3\hbar/2)$ and a more stable state is achieved. We know that the atomic states obey Hund's rule. From the exact diagonalization method, a ground state of an incomplete shell in a quantum dot is formed so that the total spin becomes a maximum and thus quantum dot states also obey Hund's rule [126, 127].

8.8.6 Quantum Dot States in Magnetic Field

When a magnetic field is applied in the z direction, the one–electron energy levels of a quantum dot are given by (8.323), which are shown in Fig. 8.78(a). Since each state is doubly degenerate with respect to spin, total number of the electrons filling the levels at $B = 0$ is $N = (0), 2, 6, 12, 20, \cdots$ as discussed before. The degenerate energy states with the same value n at $B = 0$ split into several levels at $B \neq 0$ and the energy level with the angular momentum parallel to the magnetic field ($m > 0$ becomes the lowest). Such a simple feature is dramatically changed when we take Coulomb interaction into account. Here we present the results calculated by the exact diagonalization method for electrons $N = 4$ in Fig. 8.78(b), where the quantum dot model is the same as the circular dot used for the results of $B = 0$ and the calculated energy levels are plotted as a function of magnetic Field B. In Fig. 8.78(b), •, □ and ○ are the states with the total angular momentum 0, $\pm\hbar$ and $\pm 2\hbar$, respectively. Δ is the energy difference between the singlet and triplet states due to the exchange

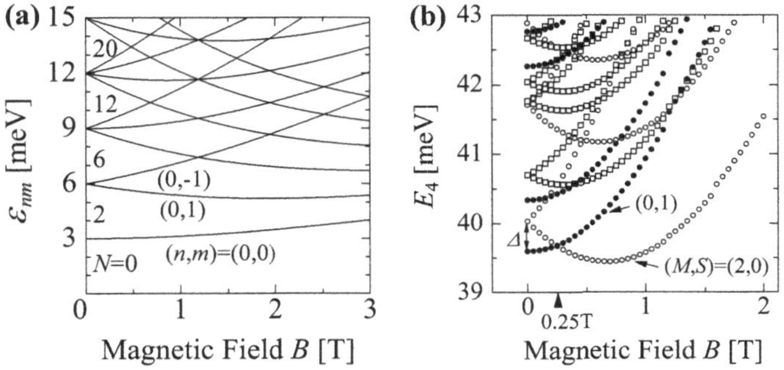

Fig. 8.78 (a)One–electron states of a quantum dot in magnetic field, where the states are doubly degenerate with respect to spins. The state with the angular momentum parallel to the magnetic field (m is positive) is lower than the other states with the same n. (b)Energy states in a circular dot with 4 electrons calculated by exact diagonalization method, where •, □ and ∘ are the states with the total angular momentum 0, $\pm\hbar$ and $\pm 2\hbar$, respectively. Δ is the energy difference between the singlet and triplet states due to the exchange interaction. The ground state energy of the spin triplet $S = 1$ with the total angular momentum $M = 0$ increases with the magnetic field, while the singlet state $S = 0$ with total angular momentum $M = 2$ decreases, and the transition of the ground state occurs at $B \approx 0.25$ [T]

interaction, which is estimated to be $\Delta = 0.44$ [meV]. The ground state energy of the spin triplet $S = 1$ with the total angular momentum $M = 0$ increases with the magnetic field, while the singlet state $S = 0$ with total angular momentum $M = 2$ decreases, and the transition of the ground state occurs at $B \approx 0.25$ [T]. It is also shown by the exact diagonalization that transition of the ground state for electrons $N = 5$ occurs from the state of the total angular momentum $M = 1$ to the state $M = 4$ at $B \approx 0.6$ [T].

8.8.7 Electronic States in Elliptic and Triangular Quantum Dots

As described before we may expect the chemical potential difference depends on the structure of quantum dots, and therefore it is very interesting to show the calculated results on elliptic and triangular quantum dots. In Fig. 8.79 the chemical potential difference is shown as a function of electron number N in the two different elliptic dots with $(\hbar\omega_x, \hbar\omega_y) = (2.5\text{meV}, 3.5\text{meV})$ and $(\hbar\omega_x, \hbar\omega_y) = (2\text{meV}, 4\text{meV})$ by □ and Δ, respectively, where the data of • are the chemical potential difference in a circular quantum dot with $\hbar\omega_x = \hbar\omega_y = 3\text{meV}$. The well width in the z direction is the same for the above three quantum dots and $W = 12\text{nm}$. We find in Fig. 8.79 that peaks except $N = 6$ are not well resolved compared to the case of the circular dot. This feature is explained in terms that the degeneracy of one–electron states is

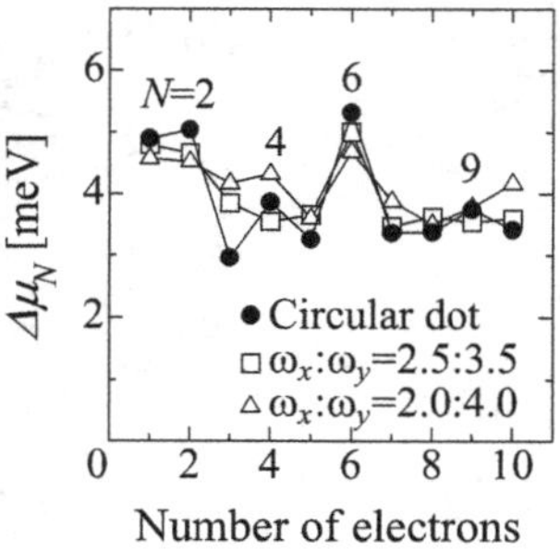

Fig. 8.79 Calculated chemical potential difference $\Delta\mu_N$ in two kinds of elliptic dots compared with the results of a circular dot. □ and ○ are the results for the elliptic dots with the confinement potentials in x and y directions $(\hbar\omega_x, \hbar\omega_y) = (2.5, 3.5\,\text{meV})$ and $(2, 4\,\text{meV})$, respectively. The results for the circular dot of $\hbar\omega_0 = 3\text{meV}$ are shown by • for comparison [126]

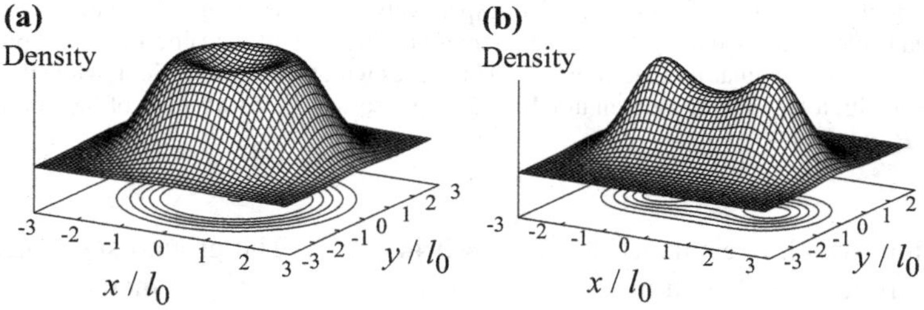

Fig. 8.80 Calculated electron distributions for $N = 4$ in (a) the circular dot ($\hbar\omega_0 = 3\text{meV}$) and (b) the elliptic dot $(\hbar\omega_x, \hbar\omega_y) = (2\text{meV}, = 4\text{meV})$, where $\ell_0 = 20\text{nm}$

removed due to the weakened symmetry in elliptic quantum dots, resulting in orbits of mixed angular momentum.

The electron distributions of (a) the circular and (b) elliptic dots in the (x, y) plane calculated by the exact diagonalization method are shown in Fig. 8.80, where 4 electrons are contained in the circular dot (a) of $\hbar\omega_0 = 3\,\text{meV}$ and in the elliptic dot (b) of $(\hbar\omega_x, \hbar\omega_y) = (2\text{meV}, 4\text{meV})$. As seen in Fig. 8.80(a) the electron distribution in the circular dot is symmetric in the plane, while (b) in the elliptic dot the electron distribution has double peaks along the long axis.

For comparison, calculated results of the chemical potential difference for a triangular dot (•) with $\hbar\omega_0 = 3\text{meV}$ are shown together with the results for a circular dot (△) in Fig. 8.81. We find no significant difference between the two dots in Fig. 8.81. Here we have to note that the corners of the potential of the triangular dot used in the calculations is smoothed, resulting in a small change in the chemical potential difference compared to the circular dot. However, we find that the states for electrons $N = 3$, 6 and 9 are stable a little compared to the circular dot. This feature may be explained in terms of the electron distribution shown in Fig. 8.82, where the results for $N = 3$ are shown in Fig. 8.82(a) for the circular dot and in Fig. 8.82(b) for the

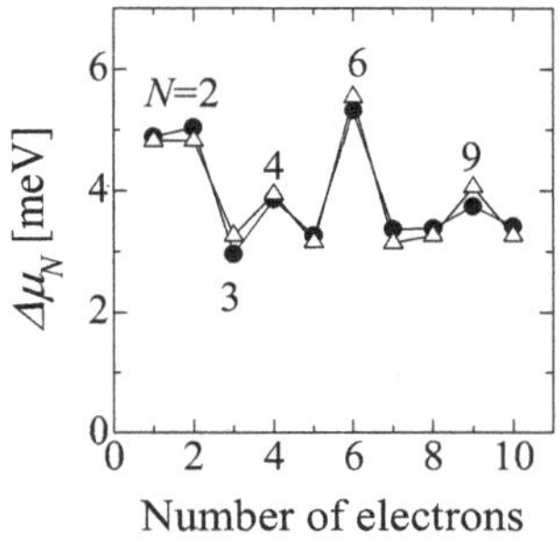

Fig. 8.81 Calculated chemical potential difference $\Delta\mu_N$ in a triangular dot is plotted as a function of electron number N, where the confinement potential is assumed to be $\hbar\omega_0 = 3$meV. Calculated results for the circular dot (Fig. 8.77) are shown for comparison. △ and • are the results for the triangular dot and the circular dot

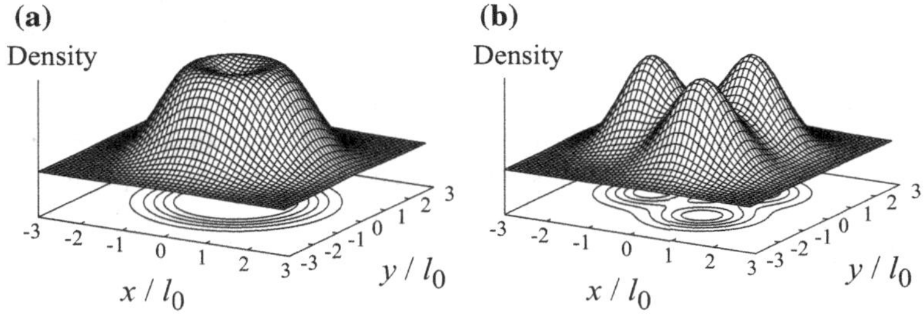

Fig. 8.82 A comparison of the electron wave function distribution for electrons $N = 3$ in a circular dot (left curve) and in a triangular dot (right curve), where,$\ell_0 = 20$nm

triangular dot. The electron distribution of the circular dot in Fig. 8.82(a) is symmetric but the central part of the distribution is low due to the Coulomb repulsion. In the triangular dot, on the contrarily, each electron occupies the corner as shown in Fig. 8.82(b), resulting in a little bit stable states for electrons 3, 6 and 9.

8.9 Problems

(8.1) Show that two–dimensional electron gas density of the subband $\mathcal{E}_i$ is given by (8.15).

(8.2) Show the scattering rate and the mobility due to acoustic deformation potential scattering. Calculate temperature dependence of two–dimensional electron gas due to the scattering.

(8.3) Show the scattering rate and the mobility due to non–polar optical phonons. Calculate temperature dependence of two–dimensional electron gas due to the scattering.

(8.4) Show the scattering rate and the mobility due to polar optical phonons. Calculate temperature dependence of two–dimensional electron gas due to the scattering.

(8.5) Show the scattering rate and the mobility due to piezoelectric potential. Calculate temperature dependence of two–dimensional electron gas due to the scattering. Consider transverse acoustic waves with v_t and $e_{14} = 0.160$ [C/m^2].

(8.6) Show the scattering rate and the mobility due to ionized ion scattering. Calculate temperature dependence of two–dimensional electron gas due to the scattering.

(8.7) Draw the mobilities of two–dimensional electron gas due to all the scattering processes shown above as a function of temperature using the parameters for GaAs given in Table 6.3 and assume quantum well width $W = 10$ nm and optical phonon deformation potential $E_{\rm op} = D_{\rm ac}/0.4$ (relation used for electrons of Ge) and $D_{\rm op} = E_{\rm op}\omega_0/v_s$.

(8.8) Compare the calculated mobilities of two–dimensional electron gas due to polar optical phonon scattering between the two models given by (8.97) (8.96) and (8.97), changing the quantum well width $W = 5, 10, 20, 50$ [nm].

(8.9) Calculate the mobility of two dimensional electron gas due to the scattering processes (1) acoustic deformation potential, (2) piezoelectric potential scattering, (3) ionized impurity scattering, and (4) polar optical phonon scattering. For this purpose a simplified approximate formula given by the following relation;

$$\frac{1}{\mu_{2\rm D}} = \frac{1}{\mu_{\rm ac}} + \frac{1}{\mu_{\rm piez}} + \frac{1}{\mu_{\rm ion}} + \frac{1}{\mu_{\rm pop}(k)}.$$

(8.10) In order to understand the filling factor of integral Quantum Hall effect, consider the following cases. The density of Landau level is given by δ function $(eB/h)\delta(\mathcal{E} - \mathcal{E}_N)$. For simplicity we assume each Landau level state is given straight narrow bar with its areal density of (eB/h). Using this model show the filling states for $i = 1$ and $i = 4$.

(8.11) In this chapter the electron–phonon interactions are treated as the confined electrons (2DEG) scattered by the bulk phonons. However, phonon modes (lattice vibrations) are also affected by the confinement. Comment on this interactions.

References

1. T. Ando, A.B. Fowler, F. Stern, Electronic properties of two-dimensional systems. Rev. Mod. Phys. **54**, 437–672 (1982)
2. P.J. Price, Two-dimensional electron transport in semiconductor layers: i. phonon scattering. Ann. Phys. **133**, 217–239 (1981)
3. P.J. Price, Two-dimensional electron transport in semiconductor layers II. screening. J. Vac. Sci. Technol. **19**, 599–603 (1981)

4. P.J. Price, Electron transport in polar heterolayers. Surf. Sci. **113**, 199–210 (1982)
5. F.A. Riddoch, B.K. Ridley, On the scattering of electrons by polar phonons in quasi-2D quantum wells. J. Phys. C Solid Sate Phys. **16**, 6971–6982 (1983)
6. P.Y. Yu, M. Cardona, *Fundamentals of Semiconductors* (Springer, Heidelberg, 1996) (Chap. 9)
7. C. Weisbuch, B. Vinter, *Quantum Semiconductor Structures: Fundamentals and Applications* (Academic Press, New York, 1991)
8. S.M. Sze, *Physics of Semiconductor Devices* (Wiley-Interscience, New York, 1969) (Chaps. 9–10, pp. 425–566)
9. W. Harrison, *Electronic Structure and the Properties of Solids* (Freeman, San Francisco, 1980)
10. P. Vogl, H.P. Hjalmarson, J.D. Dow, A semi-empirical tight-binding theory of the electronic structure of semiconductors. J. Phys. Chem. Solids **44**, 365 (1983)
11. S. Namba (ed.), *Fundamentals of Mesoscopic Phenomena* (Ohm-sha, Tokyo, 1994) (In Japanese. This book contains a review of mesoscopic phenomena)
12. T. Ando, Y. Arakawa, S. Komiyamiyama, H. Nakashima: *Mesoscopic Physics and Electronics* (Springer, New York, 1998) (In reference [16] most of the contents of reference [16] are revised and translated into English)
13. T. Ando (ed.), *Quantum Effect and Magnetic Field* (Maruzen, Tokyo, 1995) (in Japanese)
14. K. von Klitzing, G. Dorda, M. Pepper, Phys. Rev. Lett. **45**, 494 (1980)
15. L. Esaki, R. Tsu, IBM J. Res. Dev. **14**, 61 (1970)
16. M. Abramowitz, I.A. Stegun, *Handbook of Mathematical Functions* (Dover Publications, New York, 1965), p. 446
17. F. Stern, Phys. Rev. B **5**, 4891 (1972)
18. F.F. Fang, W.E. Howard, Phys. Rev. Lett. **16**, 797 (1966)
19. R. Dingle, H.L. Störmer, A.C. Gossard, W. Wiegmann, Appl. Phys. Lett. **33**, 665 (1978)
20. T. Mimura, S. Hiyamizu, T. Fujii, K. Nambu, Jpn. J. Appl. Phys. **19**, L255 (1980)
21. S. Hiyamizu, J. Saito, K. Nambu, T. Ishikawa, Jpn. J. Appl. Phys. **22**, L609 (1983)
22. K. Hirakawa, H. Sakaki, Phys. Rev. B **33**, 8291 (1986)
23. T. Kawamura, S. Das, Sarma. Phys. Rev. B **45**, 3612 (1992)
24. K. Hess, Appl. Phys. Lett. **35**, 484 (1979)
25. S. Mori, T. Ando, J. Phys. Soc. Jpn. **48**, 865 (1980)
26. Y. Matsumoto, Y. Uemura, Proceedings of the 6th International Vacuum Congress and 2nd International Conference on Solid Surfaces, Kyoto, (1974); Jpn. J. Appl. Phys. **13**, (1974) Suppl. 2, p. 367
27. T. Ando, J. Phys. Soc. Jpn. **43**, 1616 (1977)
28. A. Hartstein, A.B. Fowler, in *Proceedings of the 13th International Conference on the Physics of Semiconductors*, Rome, ed. by F.G. Fumi (North-Holland, Amsterdam, 1976), pp. 741–745
29. S.M. Goodnick, D.K. Ferry, C.W. Wilmsen, Z. Liliental, D. Fathy, O.L. Klivanek, Phys. Rev. B **32**, 8171 (1985)
30. S. Yamakawa, H. Ueno, K. Taniguci, C. Hamaguchi, J. Appl. Phys. **79**, (1996) (911. See also reference [34])
31. L. Pfeiffer, K.W. West, H.L. Stormer, K.W. Baldwin, Appl. Phys. Lett. **55**, 1888 (1989)
32. T. Saku, Y. Horikoshi, Y. Tokura, Jpn. J. Appl. Phys. **35**, 34 (1996)
33. V. Umansky, R. de Picciotto, M. Heiblum, Appl. Phys. Lett. **71**(1997), 683 (1997)
34. T. Ando, J. Phys. Soc. Jpn. **51**, 3900 (1982)
35. K. Lee, M.S. Shur, T.J. Drummond, H. Morkoç, J. Appl. Phys. **54**, 6432 (1983)
36. A. Gold, Appl. Phys. Lett. **54**, 2100 (1989)
37. D.C. Tsui, A.C. Gossard, G. Kaminski, W. Wiegmann, Appl. Phys. Lett. **39**, 712 (1981)
38. K. Masaki, Ph.D thesis, Osaka University (1992)
39. K. Masaki, K. Taniguchi, C. Hamaguchi, M. Iwase, Jpn. J. Appl. Phys. **30**, 2734 (1991)
40. S. Takagi, A. Toriumi, M. Iwase, H. Tango, I.E.E.E. Trans, Electron Devices **41**, 2357–2363 (1994)
41. S. Takagi, J.L. Hoyt, J.J. Welser, J.F. Gibons, J. Appl. Phys. **80**, 1567 (1996)
42. J.A. Cooper Jr., D.F. Nelson, J. Appl. Phys. **54**, 1445 (1983)
43. Y.-T. Lu, L.J. Sham, Phys. Rev. B **40**, 5567 (1989)

44. H. Fujimoto, C. Hamaguchi, T. Nakazawa, K. Taniguchi, K. Imanishi, Phys. Rev. B **41**, 7593 (1990)
45. T. Matsuoka, T. Nakazawa, T. Ohya, K. Taniguchi, C. Hamaguchi, Phys. Rev. B **43**, 11798 (1991)
46. A. Ishibashi, Y. Mori, M. Itahashi, N. Watanabe, J. Appl. Phys. **58**, 2691 (1985)
47. E. Finkman, M.D. Sturge, M.C. Tamargo, Appl. Phys. Lett. **49**, 1299 (1986)
48. E. Finkman, M.D. Sturge, M.-H. Meynadier, R.E. Nahory, M.C. Tamargo, D.M. Hwang, C.C. Chang, J. Lumin. **39**, 57 (1987)
49. G. Danan, B. Etienne, F. Mollot, R. Planel, A.M. Jeaqn-Louis, F. Alexandre, B. Jusserand, G. Le Roux, J.Y. Marzin, H. Savary, B. Sermage, Phys. Rev. B **35**, 6207 (1987)
50. J. Nagle, M. Garriga, W. Stolz, T. Isu, K. Ploog, J. Phys. (Paris) Colloq. **48**, C5–495 (1987)
51. F. Minami, K. Hirata, K. Era, T. Yao, Y. Matsumoto, Phys. Rev. B **36**, 2875 (1987)
52. D.S. Jiang, K. Kelting, H.J. Queisser, K. Ploog, J. Appl. Phys. **63**, 845 (1988)
53. K. Takahashi, T. Hayakawa, T. Suyama, M. Kondo, S. Yamamoto, T. Hijikata, J. Appl. Phys. **63**, 1729 (1988)
54. M.-H. Meynadier, R.E. Nahory, J.M. Workock, M.C. Tamargo, J.L. de Miguel, M.D. Sturge, Phys. Rev. Lett. **60**, 1338 (1988)
55. J.E. Golub, P.F. Liao, D.J. Eilenberger, J.P. Haribison, L.T. Florez, Y. Prior, Appl. Phys. Lett. **53**, 2584 (1988)
56. K.J. Moore, P. Dawson, C.T. Foxon, Phys. Rev. B **38**, 3368 (1988); K.J. Moore, G. Duggan, P. Dawson, C.T. Foxon. Phys. Rev. B **38**, 5535 (1988)
57. H. Kato, Y. Okada, M. Nakayama, Y. Watanabe, Solid State Commun. **70**, 535 (1989)
58. G. Bastard, Phys. Rev. B **25**, 7584 (1982)
59. J.N. Schulman, T.C. McGill, Phys. Rev. Lett. **39**, 1681 (1977); Phys. Rev. B **19**, 6341 (1979); J. Vac. Sci. Technol. **15**, 1456 (1978); Phys. Rev. B **23**, 4149 (1981)
60. H. Rücker, M. Hanke, F. Bechstedt, R. Enderlein, Superlattices Microstruct. **2**, 477 (1986)
61. J. Ihm, Appl. Phys. Lett. **50**, 1068 (1987)
62. L. Brey, C. Tejedor, Phys. Rev. B **35**, 9112 (1987)
63. S. Nara, Jpn. J. Appl. Phys. **26**(690), 1713 (1987)
64. E. Yamaguchi, J. Phys. Soc. Jpn. **56**, 2835 (1987)
65. K.K. Mon, Solid State Commun. **41**, 699 (1982)
66. W.A. Harrison, Phys. Rev. B **23**, 5230 (1981)
67. E. Caruthers, P.J. Lin-Chung, Phys. Rev. B **17**, 2705 (1978)
68. W. Andreoni, R. Car, Phys. Rev. B **21**, 3334 (1980)
69. M.A. Gell, D. Ninno, M. Jaros, D.C. Herbert, Phys. Rev. B **34**, 2416 (1986); M.A. Gell, D.C. Herbert, *ibid.* B **35**, 9591 (1987); Jian-Bai Xia: Phys. Rev. B **38**, 8358 (1988)
70. J.S. Nelson, C.Y. Fong, Inder P. Batta, W.E. Pickett, B.K. Klein, Phys. Rev. B **37**, 10203 (1988)
71. T. Nakayama, H. Kamimura, J. Phys. Soc. Jpn. **54**, 4726 (1985)
72. I.P. Batra, S. Ciraci, J.S. Nelson, J. Vac. Sci. Technol. B **5**, 1300 (1987)
73. D.M. Bylander, L. Kleiman, Phys. Rev. B **36**, 3229 (1987)
74. S.-H. Wei, A. Zunger, J. Appl. Phys. **63**, 5795 (1988)
75. Y. Hatsugai, T. Fujiwara, Phys. Rev. B **37**, 1280 (1988)
76. S. Massidda, B.I. Min, A.J. Freeman, Phys. Rev. B **38**, 1970 (1988)
77. R. Eppenga, M.F.H. Schuurmans, Phys. Rev. B **38**, 3541 (1988)
78. J.C. Slater, G.F. Koster, Phys. Rev. **94**, 1498 (1954)
79. P.O. Löwdin, J. Chem Phys. **18**, 365 (1950)
80. D.J. Chadi, M.L. Cohen, Phys. Status Solidi **68**, 405 (1975)
81. T. Matsuoka: Ph.D. thesis, Osaka University (1991)
82. K.J. Moore, G. Duggan, P. Dawson, C.T. Foxon, Phys. Rev. B **38**, 1204 (1988)
83. M. Nakayama, I. Tanaka, I. Kimura, H. Nishimura, Jpn. J. Appl. Phys. **29**, 41 (1990)
84. W. Ge, M.D. Sturge, W.D. Schmidt, L.N. Pfeiffer, K.W. West, Appl. Phys. Lett. **57**, 55 (1990)
85. Y. Aharonov, D. Bohm, Phys. Rev. **115**, 485 (1959)
86. R.A. Webb, S. Washburn, C.P. Umbach, R.B. Laibowitz, Phys. Rev. Lett. **54**, 2696 (1985)

87. K. Ishibashi, Y. Takagaki, K. Gamo, S. Namba, K. Murase, Y. Aoyagi, M. Kawabe, Solid State Commun. **64**, 573 (1987)
88. R. Landauer: IBM J. Res. Dev. **1**, 223 (1957); Philos. Mag. **21**, 863 (1970)
89. M. Büttiker, Phys. Rev. B **35**, 4123 (1987)
90. M. Büttiker, IBM J. Res. Dev. **32**, 317 (1988)
91. B. Schwarzschild, Phys. Today **39**, 17 (1986)
92. R.A. Webb, S. Washburn, A.D. Scott, C.P. Umbach, R.B. Laibowitz, Jpn. J. Appl. Phys. **26**(Suppl), 26 (1987)
93. L. Altshuler, A.G. Aronov, B.Z. Spivak, JETP Lett. **33**, 101 (1981)
94. M.S. Shur, L.F. Eastman, IEEE Trans. Electron Devices ED-**26**, 1677 (1979)
95. B.J. van Wees, H. van Houten, C.W.J. Beenakker, J.G. Williamson, L.P. Kouwenhoven, D. van der Marel, C.T. Foxon, Phys. Rev. Lett. **60**, 848 (1988)
96. D.A. Wharam, T.J. Thornton, R. Newbury, M. Pepper, H. Ahmed, J.E.F. Frost, D.G. Hasko, D.C. Peakock, D.A. Ritchie, G.A.C. Jones, J. Phys. C **21**, L209 (1988)
97. H. van Houten, B.J. van Wees, J.E. Mooij, C.W.J. Beenakker, J.G. Williamson, C.T. Foxon, Europhys. Lett. **5**, 721 (1988)
98. B.J. van Wees, H. van Houten, C.W.J. Beenakker, L.P. Kouwenhoven, J.G. Williamson, J.E. Mooij, C.T. Foxon, J.J. Harris, in *Proceedings of the 19th International Conference on Physics of Semiconductors*, Warsaw, 1988, ed. by W. Zawadzki (1988), p. 39
99. M.A. Paalanen, D.C. Tsui, A.C. Gossard, Phys. Rev. B **25**, 5566 (1982)
100. H. Aoki, T. Ando, Solid State Commun. **38**, 1079 (1981)
101. R.B. Laughlin, Phys. Rev. B **23**, 5632 (1981)
102. B.I. Halperin, Phys. Rev. B **25**, 2185 (1982)
103. M. Büttiker, Phys. Rev. B **38**, 9375 (1988)
104. Halperin, Helv. Phys. Acta **56**, 75 (1983)
105. D.R. Yennie, Rev. Mod. Phys. **59**, 781 (1987)
106. H. Aoki, Rep. Prog. Phys. **50**, 655 (1987)
107. D.C. Tsui, H.L. Stormer, A.C. Gossard, Phys. Rev. Lett. **48**, 1559 (1982)
108. R.B. Laughlin, Phys. Rev. Lett. **50**, 1395 (1983)
109. H.L. Stormer, R.R. Du, W. Kang, D.C. Tsui, L.N. Pfeiffer, K.W. Baldwin, K.W. West, Semicond. Sci. Technol. **9**, 1853 (1994)
110. H.W. Jiang, H.L. Stormer, D.C. Tsui, L.N. Pfeiffer, K.W. West, Phys. Rev. B **46**, 12013 (1989); Phys. Rev. B **46**, 10468 (1992); Phys. Rev. B **46**, 10468 (1992)
111. J.K. Jain, Phys. Rev. B **40**, 8079 (1989); Phys. Rev. B **41**, 7653 (1990)
112. D.V. Averin, K.K. Likarev, in *Mesoscopic Phenomena in Solids*, ed. by B.L. Al'tshuler, P.A. Lee, R.A. Webb (Elsevier, Amsterdam, 1991)
113. H. Grabert, M. Devoret (eds.), *Single Charge Tunneling* (Plenum, New York, 1992)
114. M. Stopa, see reference [16], pp. 56–66
115. M. Ueda, *Oyo Buturi.* (Jpn. Soc. Appl. Phys.) **62**, 889–897 (1993) (in Japanese)
116. S. Kastumoto, *Kagaku (Science)* **63**, 530–537 (1993) (in Japanese)
117. T.A. Fulton, G.J. Dolan, Phys. Rev. Lett. **95**, 109 (1987)
118. D.C. Ralph, C.T. Black, M. Tinkham, Phys. Rev. Lett. **74**, 3241 (1995)
119. W. Chen, H. Ahmed, K. Nakazato, Appl. Phys. Lett. **66**, 3383 (1995)
120. D.L. Klein, P.L. McEuen, J.E. Bowen Katari, R. Roth, A.P. Alivisatos, Appl. Phys. Lett. **68**, 2574 (1996)
121. K. Matsumoto, M. Ishii, K. Segawa, Y. Oka, B.J. Vartanian, J.S. Harris, Appl. Phys. Lett. **68**, 34 (1996)
122. K. Yano, T. Ishii, T. Hashimoto, T. Kobayashi, F. Murai, K. Seki, IEEE Trans. Electron Devices, **41**, 1628 (1994); Appl. Phys. Lett. **67**, 828 (1995)
123. M.A. Kastner, Phys. Today **46**, 24 (1993)
124. N. Johnson, Condens. Matter **7**, 965 (1995)
125. S. Tarucha, D.G. Austing, T. Honda, R.J. van der Hage, L.P. Kouwenhoven, Phys. Rev. Lett. **77**, 3613 (1996)
126. T. Ezaki, N. Mori, C. Hamaguchi, Phys. Rev. B **56**, 6428 (1997)
127. T. Ezaki, *Study on Electronic States and Electrical Conduction in Semiconductor Quantum Devices*, Ph.D. thesis (Osaka University, (1997)), pp. 44–59

Chapter 9
Light Emission and Laser

Abstract In this chapter, physics of luminescence and laser oscillations are treated in detail. First the definition of Einstein coefficients is introduced to connect absorption with spontaneous emission and stimulated emission. Then the spontaneous and stimulated emission rates are derived from the perturbation theory. Absorption and emission rates are strongly affected by the density of states, The density of states in the presence of impurities are shown to result in the band tail effect, where Kane's model is used. Using these results the gain of laser oscillations are discussed. The results are also used to explain various types of luminescence. Since semiconductor lasers are fabricated in heterostructures and light emission is confined by the heterointerfaces, we discuss optical wave guide analysis to reveal the importance of double heterostructure. Mode analysis of the waveguide is given in detail to explain TE and TM modes and confinement factor, in addition to Fabry–Perot analysis. Since most of the laser diodes (LDs) are fabricated in quantum well structures, confinements of electrons in the conduction band and holes in the valence bands play an important role in the laser mode and the gain. In order to obtain the quantized states in valence bands, we show how to solve 6×6 Luttinger Hamiltonian including strain effect. Final part of this chapter is devoted to the physics of GaN based lasers, where we show the strain effect plays an important role in the laser oscillations. Here we present how to deal with the energy band structure and Luttinger Hamiltonian in wurtzite crystals.

In Chaps. 4 and 5 we dealt with optical properties of semiconductors, where electron transition induced by absorbing photon plays the most important role. It is also possible for an electron to make transition from a higher energy state to a lower energy state by emitting a photon. In other words electrons recombine with holes by emitting light. The latter process gives rise to a light emission (luminescence) and laser action, and provides well known devices such as **light emitting diode (LED)** and **laser diode (LD)**. In this chapter we deal with light emission from semiconductors and laser diodes. First we deal with spontaneous emission, stimulated emission and absorption, where spontaneous emission is related to luminescence and stimulated emission plays the most important role in the laser action.

C. Hamaguchi, *Basic Semiconductor Physics*, Graduate Texts in Physics,
DOI 10.1007/978-3-319-66860-4_9

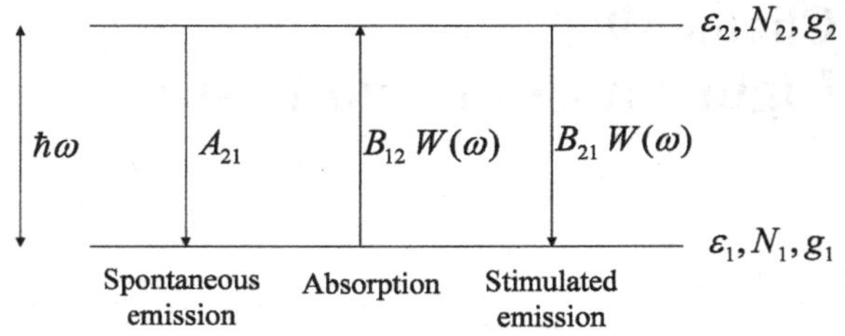

Fig. 9.1 Three optical processes of radiation proposed by Einstein: spontaneous emission, absorption and stimulated emission

9.1 Einstein Coefficients *A* and *B*

Einstein proposed the optical processes of absorption, spontaneous emission and stimulated emission in 1917.[1] In the paper Einstein dealt with emission and absorption of light and pointed out for the first time the concept of stimulated emission. We know that the basic mechanism of the laser action comes from the stimulated emission, but the discovery of lasers was accomplished after his death. The concept proposed by Einstein explains not only the principle on laser action but also the optical processes of absorption and emission. Here we will describe Einstein's theory first and then derive so called Einstein coefficients A and B. We consider an interaction between radiation field (light or photon) and N atoms or molecules in a cavity, where the atoms or molecules have energy levels $\mathcal{E}_1$ and $\mathcal{E}_2$, and we assume that the atoms or molecules are allowed to make transition between the two levels by absorbing or emitting the photon energy given by

$$\hbar\omega = \mathcal{E}_2 - \mathcal{E}_1 , \tag{9.1}$$

where the atomic energies are degenerate with the degeneracy g_1 and g_2, and the number of atoms or molecules are N_1 and N_2, respectively. The system is shown schematically in Fig. 9.1.

The average density of radiation field $W(\omega)$ depends on the thermal equilibrium value $W_{\mathrm{T}}(\omega)$ and the external excitation $W_{\mathrm{E}}(\omega)$, and is given by the following relation

$$W(\omega) = W_{\mathrm{T}}(\omega) + W_{\mathrm{E}}(\omega) . \tag{9.2}$$

Here we have to note that $W_{\mathrm{E}}(\omega)$ is not uniform in the cavity in general case.

Emission and absorption of photon are defined in the following way. We define transition rate of an atom from the energy level 2 to the energy level 1 by A_{21}. An atom of the energy level 1 is not possible to make a transition to the energy level 2 without radiation field, because the energy conservation is not fulfilled in the transition. However, the transition $1 \to 2$ is possible by absorbing a photon $\hbar\omega$. We define

[1]A. Einstein: "Zur Quantentheorie der Strahlung," Phys. Z. 18, 121 (1917), The title in English is "On the Quantum Theory of Radiation".

the transition rate $1 \to 2$ by B_{12} in the presence of photon field with the radiation density $W(\omega)$. These two processes are easily understood, but careful consideration is required for the transition from the upper level 2 to the lower level 1 in the presence of the radiation $W(\omega)$. It is easily expected that the transition $2 \to 1$ is enhanced in the presence of the radiation field. This transition enhanced by the radiation is called **stimulated emission** and the rate is defined by B_{21}. These three transition processes are schematically shown in Fig. 9.1. These optical processes were defined by Einstein in 1917, and the coefficients A_{21}, B_{12} and B_{21} are called **Einstein coefficients**. We have to note here that the Einstein coefficients A_{21}, B_{12} and B_{21} are independent of the radiation density $W(\omega)$. In the above case absorption and stimulated emission are assumed to be proportional to the radiation density $W(\omega)$. This assumption is valid when the radiation density is a slowly varying function of transition frequency ω. The transition A_{21} is called **spontaneous emission** and independent of the radiation density.

The occupation densities N_1 of the level 1 and N_2 of the levels 2 are governed by the following rate equation,

$$\frac{\mathrm{d}N_1}{\mathrm{d}t} = -\frac{\mathrm{d}N_2}{\mathrm{d}t} = N_2 A_{21} - N_1 B_{12} W(\omega) + N_2 B_{21} W(\omega) \,. \tag{9.3}$$

In order to investigate the relations between the Einstein coefficients, first we consider the case of thermal equilibrium. In the thermal equilibrium the distribution densities of the two levels are constant and then (9.3) leads to the following relation,

$$N_2 A_{21} - N_1 B_{12} W_T(\omega) + N_2 B_{21} W_T(\omega) = 0 \,. \tag{9.4}$$

This results in

$$W_T(\omega) = \frac{A_{21}}{(N_1/N_2) B_{12} - B_{21}} \,. \tag{9.5}$$

In the thermal equilibrium, the occupation densities of N_1 and N_2 are given by Boltzmann statistics,

$$\frac{N_1}{N_2} = \frac{g_1 \exp(-\beta \mathcal{E}_1)}{g_2 \exp(-\beta \mathcal{E}_2)} = \frac{g_1}{g_2} \exp(\beta \hbar \omega) \,, \tag{9.6}$$

where $\beta = 1/k_{\mathrm{B}} T$. Using this relation (9.5) is rewritten as

$$W_T(\omega) = \frac{A_{21}}{(g_1/g_2) \exp(\beta \hbar \omega) B_{12} - B_{21}} \,, \tag{9.7}$$

where g_1 and g_2 represent the degeneracy of the states. We have to note here that the above equation is derived by using Boltzmann statistics and that the result of Einstein theory is easily proved to be consistent with Planck's radiation theory. From the theory of Planck, the radiation density is given by

$$W_T(\omega)\mathrm{d}\omega = \bar{n}\hbar\omega\rho_\omega \mathrm{d}\omega = \frac{\bar{n}\hbar\omega^3}{\pi^2 c^3}\mathrm{d}\omega = \frac{\hbar\omega^3}{\pi^2 c^3}\frac{\mathrm{d}\omega}{\exp(\beta\hbar\omega) - 1}, \tag{9.8}$$

where $\bar{n}$ is the occupation density of photons. Equation (9.7) should coincide with (9.8), which is validated by the following equations,

$$\frac{g_1}{g_2}B_{12} = B_{21}, \tag{9.9}$$

$$\frac{\hbar\omega^3}{\pi^2 c^3}B_{21} = A_{21}, \tag{9.10}$$

where the above relations should hold at all temperature range. From these considerations we find that the three Einstein coefficients are related by the above two equations. In addition it is evident from (9.7) that Einstein theory with stimulated emission term is equivalent with Planck's radiation theory. When $g_1 = g_2$, we obtain the following relation

$$B_{12} = B_{21}. \tag{9.11}$$

9.2 Spontaneous Emission and Stimulated Emission

As shown later, laser oscillations in semiconductors are induced by a strong emission by recombination. Recombination emissions consists of two terms, **spontaneous emission** and **stimulated emission**. Among them, stimulated emission plays the most important role in the laser oscillations in semiconductors. Here we deal with the recombination emissions in semiconductors in detail.

First, we consider the case shown in Fig. 9.2, where excited electrons in the conduction band recombine with the holes in the valence band by emitting light. Figure 9.2 shows a process that an electron in a valence band makes a transition by absorb a light, and the reverse process that the excited electron recombine with the hole by emitting a light. Here the reverse process has two distinct emission processes, spontaneous and stimulated emissions. Before dealing with these

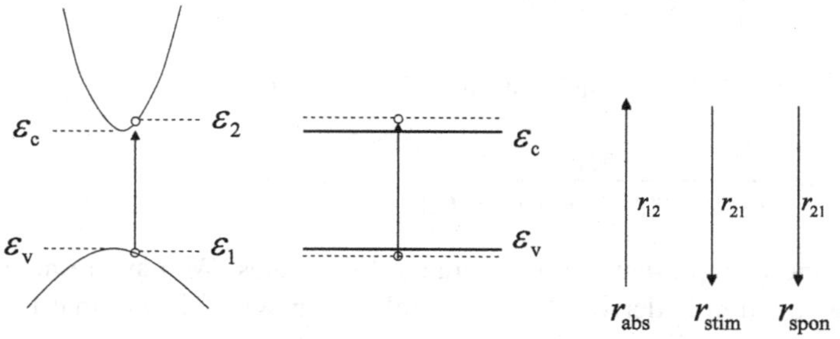

Fig. 9.2 Optical absorption and recombination emission

processes by quantum mechanics, we will briefly mention how to excite electrons from the valence bands to the conduction band. When we create electron–hole pairs in a semiconductor by light incidence, the resulting recombination emission is called **photoluminescence**. In the case where an electric field excites electron–hole pairs or the electric field injects electrons and holes in p–n junction, the resulting light emission is called **electro-luminescence**. Sometimes electron beams are used to excite electron–hole pairs, and the light emission is called **cathode luminescence**. Light emitting diode (LED) and laser diode (LD) are excited by electron–hole injection. Also we have to note that excited electrons or holes recombine by absorbing or emitting phonons (lattice vibrations), and thus no photons are emitted. This process is called **non-radiative recombination**.

Let's discuss the excitation of electrons of $\mathcal{E}_1$ in the valence band into the conduction band $\mathcal{E}_2$ by incident photon and light emission by recombination as shown in Fig. 9.2. The rate of photon emission in the energy space $\mathrm{d}\mathcal{E}$ per unit time and unit solid angle Ω is given by the following relation according to Lasher and Stern [1]

$$r(\mathcal{E})\mathrm{d}\mathcal{E}(\mathrm{d}\Omega/4\pi) = \left[r_{\mathrm{spon}}(\mathcal{E}) + n_{\mathrm{photon}} r_{\mathrm{stim}}(\mathcal{E})\right] \mathrm{d}\mathcal{E}(\mathrm{d}\Omega/4\pi)\,, \tag{9.12}$$

where n_{photon} is the photon numbers per mode given for thermal equilibrium by

$$n_0 = \frac{1}{\exp(\mathcal{E}/k_{\mathrm{B}}T) - 1}\,, \tag{9.13}$$

k_{B} is Boltzmann's constant and T is the absolute temperature. In (9.12) the term $r_{\mathrm{spon}}(\mathcal{E})$ represents the spontaneous transition of electrons from the upper level to the lower level and, $n_{\mathrm{photon}} r_{\mathrm{stim}}$ is the difference between the upward transition (absorption) and stimulated downward transition. For band–to–band transition the spontaneous and stimulated emission rates are given by [6, 9] (see also Appendix G for the derivation)

$$r_{\mathrm{spon}}(\mathcal{E})\mathrm{d}\mathcal{E} = \sum \frac{n_{\mathrm{r}} e^2 \mathcal{E}}{\pi \epsilon_0 m^2 \hbar^2 c^3} |M|^2 f_2(1 - f_1)\mathrm{d}\mathcal{E}\,, \tag{9.14}$$

$$r_{\mathrm{stim}}(\mathcal{E})\mathrm{d}\mathcal{E} = \sum \frac{n_{\mathrm{r}} e^2 \mathcal{E}}{\pi \epsilon_0 m^2 \hbar^2 c^3} |M|^2 (f_2 - f_1)\mathrm{d}\mathcal{E}\,, \tag{9.15}$$

where the factor $(f_2 - f_1)$ of (9.15) is derived from the net rate of stimulated transition between the downward factor $f_2(1-f_1)$ and the upward factor $f_1(1-f_2)$; $f_2(1-f_1) - f_1(1-f_2) = (f_2 - f_1)$. Here f_2 and f_1 are the electron occupation probabilities of the upper and lower levels, respectively, and n_{r} is the refractive index. The summation $\sum$ is taken over all pairs of states in the conduction and valence bands per unit volume whose energy difference is $\mathrm{d}\mathcal{E}$ and $\mathcal{E} + \mathrm{d}\mathcal{E}$. The squared momentum matrix element $|M|^2$ is defined in Chap. 4 and for the purpose of simplicity we assume the value averaged over the all polarizations of the incident light as

$$|M|^2 = \frac{1}{3}\left(|M_x|^2 + |M_y|^2 + |M_z|^2\right), \tag{9.16a}$$

$$M_x = -\mathrm{i}\hbar\langle\psi_2|\exp(\mathrm{i}\boldsymbol{k}\cdot\boldsymbol{r})\frac{\partial}{\partial x}|\psi_1\rangle, \tag{9.16b}$$

where $|\psi_1\rangle$ and $|\psi_2\rangle$ are the lower and higher bands wave functions, respectively. As stated in Chap. 4, for band–to–band transition the electron wave vectors $\boldsymbol{k}_2$ and $\boldsymbol{k}_1$, and photon wave vector $\boldsymbol{k}$ are governed by the conservation rule $\delta(\boldsymbol{k}_2 - \boldsymbol{k}_1 \pm \boldsymbol{k})$. Since photon wave vector $\boldsymbol{k}$ is negligibly small, we can use the relation $\boldsymbol{k}_2 = \boldsymbol{k}_1$.[2] When the upper and lower energy bands $\mathcal{E}_2$ and $\mathcal{E}_1$ are expressed by the functions of $k = |\boldsymbol{k}_2| = |\boldsymbol{k}_1|$, the density of states for the band–to–band transition (reduced density of states) ρ_{red} is easily evaluated. Then the transition rates for photon energy $\mathcal{E}$ are given by

$$r_{\mathrm{spon}}(\mathcal{E}) = \frac{n_{\mathrm{r}}e^2\mathcal{E}}{\pi\epsilon_0 m^2\hbar^2 c^3}|M|^2\rho_{\mathrm{red}}(\mathcal{E})f_2(1-f_1), \tag{9.17}$$

$$r_{\mathrm{stim}}(\mathcal{E}) = \frac{n_{\mathrm{r}}e^2\mathcal{E}}{\pi\epsilon_0 m^2\hbar^2 c^3}|M|^2\rho_{\mathrm{red}}(\mathcal{E})(f_2-f_1), \tag{9.18}$$

where the reduced density of states for the band–to–band transition is given by for one direction of spin

$$\rho_{\mathrm{red}}(\mathcal{E}) = \frac{1}{(2\pi^2)}k^2\frac{\mathrm{d}(\mathcal{E}_2-\mathcal{E}_1)}{\mathrm{d}k}. \tag{9.19}$$

From the energy conservation we have $\mathcal{E}_2 - \mathcal{E}_1 = \mathcal{E}$, and thus we can define the following relation,

$$\mathcal{R}_{\mathrm{spon}} = \int r_{\mathrm{spon}}(\mathcal{E})\mathrm{d}\mathcal{E}. \tag{9.20}$$

The above equation is evaluated in the three extreme cases by Lasher and Stern [1].

(1) In the case of degenerate valence band, we can put $f_1 = 0$ for all transitions.

$$\mathcal{R}_{\mathrm{spon}} = \frac{2n_{\mathrm{r}}e^2\mathcal{E}}{\pi\epsilon_0 m^2\hbar^2 c^3}\langle|M_{\mathrm{b}}|^2\rangle_{\mathrm{av}}\left[\left(1+\frac{m_{\mathrm{e}}}{m_{\mathrm{hh}}}\right)^{-3/2} + \left(1+\frac{m_{\mathrm{e}}}{m_{\mathrm{lh}}}\right)^{-3/2}\right]n, \tag{9.21}$$

where n is the electron density, and $\langle|M_{\mathrm{b}}|^2\rangle_{\mathrm{av}}$ is the average matrix element connecting states near the band edges.

(2) In the case of non-degenerate conduction and valence bands

$$\mathcal{R}_{\mathrm{spon}} = \frac{n_{\mathrm{r}}e^2\mathcal{E}}{\pi\epsilon_0 m^2\hbar^2 c^3}\left(\frac{2\pi\hbar^2}{mk_{\mathrm{B}}T}\right)^{3/2}\langle|M_{\mathrm{b}}|^2\rangle_{\mathrm{av}} \times \frac{[m_{\mathrm{lh}}/(m_{\mathrm{e}}+m_{\mathrm{lh}})]^{3/2} + [m_{\mathrm{hh}}/(m_{\mathrm{e}}+m_{\mathrm{hh}})]^{3/2}}{(m_{\mathrm{lh}}/m)^{3/2} + (m_{\mathrm{hh}}/m)^{3/2}} \tag{9.22}$$

[2] As described later this assumption is not valid for the transition including the band tail sates.

(**3**) In the case where the matrix elements are all the same for all the transitions (no–selection rule case), the following general equation is derived from (9.14) and (9.15).

$$r_{\text{spon}}(\mathcal{E}) = B\int \rho_{\text{c}}(\mathcal{E}')\rho_{\text{v}}(\mathcal{E}'-\mathcal{E})f_c(\mathcal{E}')\left[1-f_{\text{v}}(\mathcal{E}'-\mathcal{E})\right]\mathrm{d}\mathcal{E}', \tag{9.23}$$

$$r_{\text{stim}}(\mathcal{E}) = B\int \rho_{\text{c}}(\mathcal{E}')\rho_{\text{v}}(\mathcal{E}'-\mathcal{E})\left[f_c(\mathcal{E}')-f_{\text{v}}(\mathcal{E}'-\mathcal{E})\right]\mathrm{d}\mathcal{E}', \tag{9.24}$$

$$\mathcal{E} = \hbar\omega, \qquad \mathcal{E}' = \mathcal{E}_2 = \mathcal{E}_1 + \mathcal{E} = \mathcal{E}_1 + \hbar\omega, \tag{9.25}$$

where ρ_{c} and ρ_{v} are the density of states of the conduction and valence bands with two spin directions, respectively. For the valence bands the heavy hole and light hole bands should be taken into account. These relations are valid for large amounts of electron and hole injection and the conservation of energies is assumed but the conservation of wave vectors is not included. As mentioned later, the conservation of wave vectors is not required for the transitions including the band tail states, and thus the above equations may be used. For the transitions with the wave vector conservation we have to use the following relations.

$$\mathcal{E}_1 = \mathcal{E}_{\text{v}} - \frac{m_\mu}{m_{\text{h}}}(\hbar\omega - \mathcal{E}_{\text{G}}), \tag{9.26}$$

$$\mathcal{E}_2 = \mathcal{E}_{\text{c}} + \frac{m_\mu}{m_{\text{e}}}(\hbar\omega - \mathcal{E}_{\text{G}}), \tag{9.27}$$

$$\frac{1}{m_{\text{red}}} = \frac{1}{m_{\text{e}}} + \frac{1}{m_{\text{h}}}. \tag{9.28}$$

The coefficient B is given by

$$B = \frac{n_{\text{r}}e^2\mathcal{E}}{\pi\epsilon_0 m^2\hbar^2c^3}\langle|M|^2\rangle_{\text{av}}V, \tag{9.29}$$

where $\langle|M|^2\rangle_{\text{av}}$ is the average value of the squared matrix element over spins in the upper and lower bands, and V is the crystal volume.

Total rate of spontaneous emission is easily evaluated if the selection rule is disregarded and then we obtain the following relation by carrying integration over the conduction and valence bands..

$$\mathcal{R}_{\text{spon}} = Bnp, \tag{9.30}$$

where n and p are the injected electron and hole densities. The total rate $\mathcal{R}_{\text{spon}}$ is the spontaneous emission rate per unit volume and the dimension is m^3/sec.

Distribution functions of the upper $\mathcal{E}_2$ (conduction band) and the lower $\mathcal{E}_1$ (valence band) states are given by the following relations by using quasi–Fermi level $\mathcal{E}_{\text{Fn}}$ in the conduction band and quasi–Fermi level $\mathcal{E}_{\text{Fp}}$ in the valence band,

$$f_2 = \frac{1}{1 + \exp[(\mathcal{E}_2 - \mathcal{E}_{\mathrm{Fn}})/k_{\mathrm{B}}T]} , \tag{9.31}$$

$$f_1 = \frac{1}{1 + \exp[(\mathcal{E}_1 - \mathcal{E}_{\mathrm{Fp}})/k_{\mathrm{B}}T]} . \tag{9.32}$$

Using these equations in (9.14) and (9.15), we obtain the following relation,

$$r_{\mathrm{stim}}(\mathcal{E}) = r_{\mathrm{spon}}(\mathcal{E}) \{1 - \exp[(\mathcal{E} - \Delta\mathcal{E}_{\mathrm{F}})/k_{\mathrm{B}}T]\} , \tag{9.33}$$

where $\Delta\mathcal{E}_{\mathrm{F}} = \mathcal{E}_{\mathrm{Fn}} - \mathcal{E}_{\mathrm{Fp}}$ is the difference of the quasi–Fermi energies and becomes 0 in the thermal equilibrium.

Stimulated emission r_{stim} is related to the absorption coefficient $\alpha(\mathcal{E})$ defined in Chap. 4 and given by

$$\alpha(\mathcal{E}) = -\frac{\pi^2 c^2 \hbar^3}{n_{\mathrm{r}}^2 \mathcal{E}^2} r_{\mathrm{stim}}(\mathcal{E}) , \tag{9.34}$$

where $-$ sign is used to express that the stimulated emission $r_{\mathrm{stim}}(\mathcal{E})$ has an opposite sign compared to the absorption coefficient $\alpha(\mathcal{E})$. Equation (9.34) is easily derived from (G.14) and (9.18). When we put $g(\hbar\omega) = -\alpha(\hbar\omega)$ for the semiconductor laser gain,

$$\begin{aligned} g(\hbar\omega) = -\alpha(\hbar\omega) &= \frac{\pi^2 c^2 \hbar^3}{n_{\mathrm{r}}^2 (\hbar\omega)^2} r_{\mathrm{stim}}(\hbar\omega) \\ &= \frac{\pi e^2}{n_{\mathrm{r}} c \epsilon_0 m^2 \omega} |M|^2 (f_2 - f_1) \rho_{\mathrm{red}}(\hbar\omega) . \end{aligned} \tag{9.35}$$

Here we find that the gain is positive ($g > 0$) when $f_2 - f_1 > 0$, which means that the population in the upper level f_2 is higher than the lower level f_1. This condition is called **population inversion**.

We show in Figs. 9.3 and 9.4 the calculated results of inversion factor $f_2 - f_1$ and gain factor $(f_2 - f_1)\rho_{\mathrm{red}}$, respectively, as a function of the emitted photon energy $\hbar\omega$, where we used (9.26) $\sim$ (9.28), (9.31), and (9.32), with $m_{\mathrm{e}} = 0.068\,\mathrm{m}$, $m_{\mathrm{h}} = 0.59\,\mathrm{m}$, $\mathcal{E}_{\mathrm{G}} = 1.43\,\mathrm{eV}$, and $T = 300\,\mathrm{K}$. The injected electrons and holes are estimated by choosing proper values of the quasi–Fermi levels. The curves of the inversion factors in Fig. 9.3 correspond to the quasi–Fermi levels, $\mathcal{E}_{\mathrm{Fn}} - \mathcal{E}_{\mathrm{c}} = \mathcal{E}_{\mathrm{v}} - \mathcal{E}_{\mathrm{Fp}} = -20, -10, 0, +5, +10, +15, +20$ meV from the bottom to the upper curves. The gain factors $(f_2 - f_1)\rho_{\mathrm{red}}$ in Fig. 9.4 are estimated by assuming $\mathcal{E}_{\mathrm{Fn}} - \mathcal{E}_{\mathrm{c}} = \mathcal{E}_{\mathrm{v}} - \mathcal{E}_{\mathrm{Fp}} = -20, -15, -10, -5, 0, 5, 10, 15, 20$ meV from the bottom to the upper curves. We see in Figs. 9.3 and 9.4 that the inversion factor and the gain factor become positive near the band edge for higher injection levels.

From the results we may write

$$r_{\mathrm{spon}}(\mathcal{E}) = \frac{4\epsilon_0 n_{\mathrm{r}}^2 \mathcal{E}^2}{\pi c^2 \hbar^3} \alpha(\mathcal{E}) \frac{1}{\exp[(\mathcal{E} - \Delta\mathcal{E}_{\mathrm{F}})/k_{\mathrm{B}}T] - 1} . \tag{9.36}$$

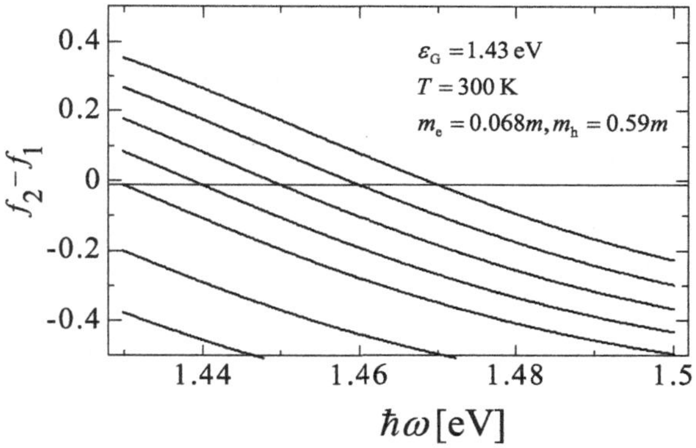

Fig. 9.3 Inversion factor $f_2 - f_1$ of pn junction is plotted as a function of emitted photon energy, where the curves are obtained by putting $\mathcal{E}_{Fn} - \mathcal{E}_c = \mathcal{E}_v - \mathcal{E}_{Fp} = -20, -10, 0, +5, +10, +15, +20$ meV from the bottom to the upper curves. $m_e = 0.068$ m, $m_h = 0.59$ m, $\mathcal{E}_G = 1.43$ eV, and $T = 300$ K

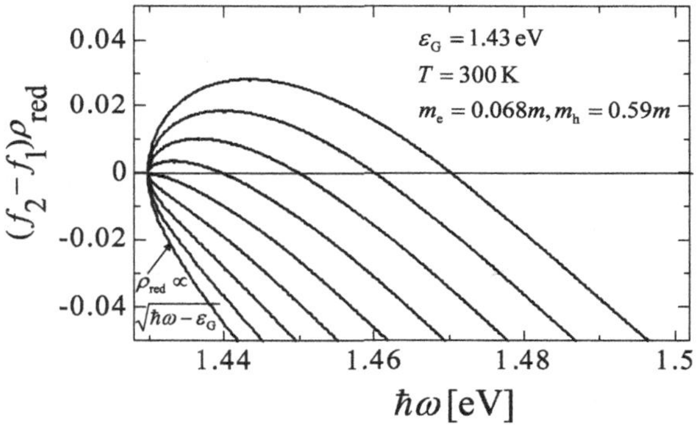

Fig. 9.4 The gain factor $(f_2 - f_1)\rho_{red}$ is plotted as a function of emitted photon energy, where the curves are obtained by putting $\mathcal{E}_{Fn} - \mathcal{E}_c = \mathcal{E}_v - \mathcal{E}_{Fp} = -20, -15, -10, -5, 0, 5, 10, 15, 20$ meV. $m_e = 0.068$ m, $m_h = 0.59$ m, $\mathcal{E}_G = 1.43$ eV, and $T = 300$ K

Summarizing these results the laser gain due to the stimulated emission is equivalent to the negative absorption coefficient and given by

$$g(\omega) = \frac{\pi c^2 \hbar^3}{4\epsilon_0 n_r^2 (\hbar\omega)^2} r_{stim}(\omega)$$
$$= \frac{\pi c^2 \hbar^3}{4\epsilon_0 n_r^2 (\hbar\omega)^2} r_{spon}(\omega) \left\{1 - \exp\left[\frac{\hbar\omega - \Delta\mathcal{E}_F}{k_B T}\right]\right\}. \quad (9.37)$$

The above equation expresses the laser action. The gain g is positive ($g > 0$) for $\hbar\omega > \Delta\mathcal{E}_F = (\mathcal{E}_{Fn} - \mathcal{E}_{Fp})$. On the other hand, for $\hbar\omega < \Delta\mathcal{E}_F$ the gain is negative and light absorption occurs.

Spontaneous and stimulated emission rates of GaAs calculated for no–selection rule case from (9.23) and (9.24) are shown in Fig. 9.5, where the parabolic energy bands are assumed and temperature is $T = 80$ K. In the calculations quasi–Fermi levels are assumed to be 11.8 meV for holes, and 5 meV and 15 meV for electrons. The coefficient C for GaAs is estimated from (9.38) and (9.40) [1], which is given by

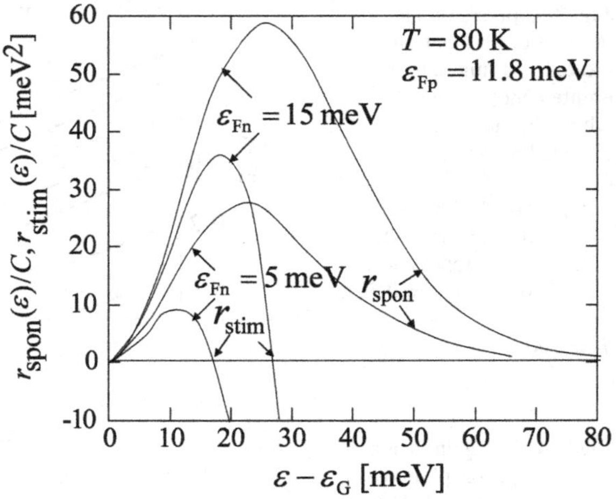

Fig. 9.5 Spontaneous emission rate $r_{\rm spon}$ and stimulated emission rate $r_{\rm stim}$ are calculated from (9.23) and (9.24) for GaAs assuming parabolic energy bands at 80 K, where the quasi–Fermi levels are 11.8 meV for holes, and 5 and 15 meV for electrons. Coefficient C is for GaAs and defined by (9.38) and (9.40) [1]

$$C = \frac{128}{3\pi^3\hbar^2 m e^4 c^3} \cdot \frac{m_{\rm c}^{1/2} m_{\rm v}^{3/2} m}{m^{*3}} \cdot \frac{\mathcal{E}_{\rm G} + \Delta}{\mathcal{E}_{\rm G} + (2/3)\Delta} n_{\rm r} \mathcal{E} \mathcal{E}_{\rm G} \kappa^3 . \tag{9.38}$$

Using the values $\mathcal{E}_{\rm G} = 1.521$ eV, $m_{\rm c} = 0.072$m, $\mathcal{E} = 1.47$ eV, $\Delta = 0.33$ eV, $n_{\rm r} = 3.6$, and $\kappa = 12.5$, we obtain

$$B = 0.75 \times 10^{-9} {\rm cm}^2/{\rm sec}, \tag{9.39}$$

$$C = 2.6 \times 10^{23} {\rm cm}^{-3}\ {\rm sec}^{-1} {\rm meV}^{-3} . \tag{9.40}$$

9.3 Band Tail Effect

In the previous section we are concerned with the emission rates in semiconductors with parabolic bands, and assumed that no electronic states in the band gap. In such a case distributions of electrons and holes are schematically shown as in Fig. 9.6.

However, Kane [2] pointed out that the electronic states are deformed by the potential due to impurities (cluster of impurities). Taking account of the Thomas–Fermi screening (see Sect. 6.3.10) by electrons and assuming Gaussian distribution for the potential fluctuations, the density of states is approximated as

$$\rho_{\rm c}(\mathcal{E}') = \frac{m_{\rm c}^{3/2}}{\pi^2\hbar^3} (2\eta_{\rm c})^{1/2} y \left(\frac{\mathcal{E}' - \mathcal{E}_{\rm c}}{\eta_{\rm c}} \right) , \tag{9.41}$$

where $m_{\rm c}$ is the electron effective mass and $\mathcal{E}_{\rm c}$ is the conduction band bottom energy obtained without the impurity potentials. Putting $\mathcal{E}' - \mathcal{E}_{\rm c} = \mathcal{E}$ and $x = \mathcal{E}/\eta_{\rm c}$ the function $y(x)$ is written as

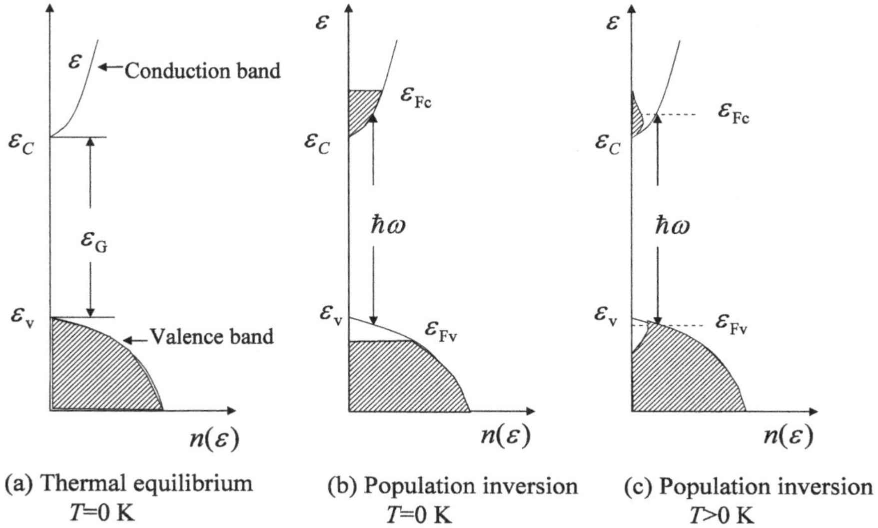

Fig. 9.6 Electron and hole distribution in parabolic bands, **a** thermal equilibrium, **b** population inversion at $T = 0$ K, and **c** population inversion at $T > 0$ K

$$y(x) = \frac{1}{\pi^{1/2}} \int_{-\infty}^{x} (x - z)^{1/2} \exp(-z^2) \mathrm{d}z \,. \tag{9.42}$$

Here η_c represents the extension of the band tail, and

$$\eta_c = \frac{e^2}{4\pi\kappa\epsilon} (4\pi N_D L_{scr})^{1/2} \,, \tag{9.43}$$

where $\kappa\epsilon_0$ is the dielectric constant, N_D is the donor density and L_{scr} is the screening length, defined by $1/q_s$ in Sect. 6.3.10 (see (6.277) or (6.279)). The fluctuation of the potentials results in a Gaussian type modification of the band edge and gives rise to extended electronic states in the band gap. This band states are called **band tail states** and the effect is referred to as **band tail effect**. The same effect is expected in the valence band. When both of the conduction and valence bands are affected by such potential fluctuations, the donor density N_D^+ in (9.43) is replaced as $N_D \to N_D^+ + N_A^-$, where N_D^+ is the ionized donor density and N_A^+ is the ionized acceptor density. Figure 9.7 shows a comparison between the density of states with the band tail effect, (9.42), and the density of states for the parabolic band $y(x) = \sqrt{x}$ without the band tail effect. It is clear from (9.42) that $\rho \propto \mathcal{E}^{1/2}$ at higher energies and $\rho \propto \exp(-\mathcal{E}^2/\eta_c^2)$ at lower energies.

Here we will summarize the theory of band tail effect proposed by Kane [2]. Hamiltonian of an electron is written as

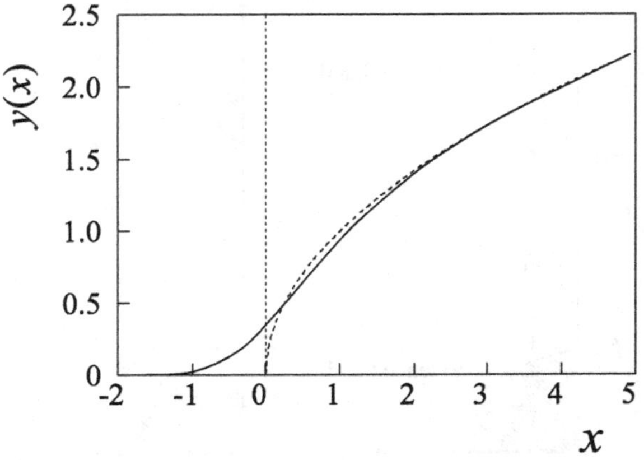

Fig. 9.7 Band edge states with the band tail effect $y(x)$ of (9.42) is compared with the parabolic band states given by $y(x) = \sqrt{x}$

$$H = H_0 + H_{\mathrm{I}}, \tag{9.44}$$

$$H_0 = -\frac{\hbar^2}{2m^*}\nabla^2, \tag{9.45}$$

$$H_{\mathrm{I}} = \sum_i v(\boldsymbol{r} - \boldsymbol{r}_i), \tag{9.46}$$

$$v(r) = -\frac{e^2}{4\pi\kappa\epsilon_0 r}\exp\left(-\frac{r}{L_{\mathrm{scr}}}\right), \tag{9.47}$$

where H_0 is the unperturbed Hamiltonian of a perfect single crystal with the effective mass m^* and the dielectric constant $\kappa\epsilon_0$, and perturbation H_{I} arises from randomly distributed impurities over lattice sites $\boldsymbol{r}_i$. The potential energy of an impurity $v(r)$ is screened by electrons with the density n, and the screening length $L_{\mathrm{scr}} = (1/q_{\mathrm{s}})$ is given by (6.273) as shown in Sect. 6.3.10,

$$\frac{1}{L_{\mathrm{scr}}} = \sqrt{\frac{e^2}{\kappa\epsilon_0}\frac{\partial n}{\partial \mathcal{E}_{\mathrm{F}}}}. \tag{9.48}$$

As described in Sect. 6.3.10, the screening length in a degenerate semiconductors is given by Thomas–Fermi screening length, (6.277),

$$\frac{1}{L_{\mathrm{scr}}} = \sqrt{\frac{3e^2 n}{2\kappa\epsilon_0\mathcal{E}_{\mathrm{F}}}} \equiv q_{\mathrm{TF}}, \tag{9.49}$$

or in a non-degenerate semiconductor the screening length is given by Debye–Huckel screening length, (6.279),

$$\frac{1}{L_{\mathrm{scr}}} = \sqrt{\frac{e^2 n}{\kappa\epsilon_0 k_{\mathrm{B}} T}} \equiv q_{\mathrm{DH}}. \tag{9.50}$$

We define the potential distribution function, $F(\mathcal{V})$, by

$$\Delta p = F(\mathcal{V})\Delta\mathcal{V}, \tag{9.51}$$

where Δp is the probability of finding the potential between $\mathcal{V}$ and $\mathcal{V} + \Delta\mathcal{V}$. The total density of states is given by

$$\rho(\mathcal{E}) = \frac{\sqrt{2}m^{*3/2}V}{\pi^2\hbar^3}\int_{-\infty}^{\mathcal{E}}(\mathcal{E} - \mathcal{V})^{1/2}F(\mathcal{V})\mathrm{d}\mathcal{V}. \tag{9.52}$$

Assuming Gaussian distribution of the impurities, Kane [2] derived

$$F(\mathcal{V}) = \frac{1}{\sqrt{\pi}\eta}\exp\left(-\frac{\mathcal{V}^2}{\eta^2}\right), \tag{9.53a}$$

$$\eta = \frac{e^2}{4\pi\kappa\epsilon_0}(4\pi n L_{\mathrm{scr}})^{1/2}, \tag{9.53b}$$

where η has been defined by (9.43), and

$$\int v^2(r)\mathrm{d}^3\boldsymbol{r} = \frac{2\pi e^4}{(4\pi\kappa\epsilon_0)^2}L_{\mathrm{scr}} \tag{9.54}$$

for the screened Coulomb potential. Substituting (9.53a) into (9.52), Kane derived (9.41) for the density of states with the band tail effect.

Distributions of the electron and hole densities with the band tail effect are shown in Fig. 9.8. Spontaneous emission spectra r_{spon} calculated by Stern [3] with the band tail effect are shown as a function of photon energy in Fig. 9.9, where the arrows are the peak positions of $-\alpha(\mathcal{E})$. The spectra are calculated for the gain $g = 100\,\mathrm{cm}^{-1}$.

The gain g *vs.* current density in GaAs is shown in Fig. 9.10, where curves are plotted for several temperatures, and the donor and acceptor densities are $N_{\mathrm{D}} = 1 \times 10^{18}$ and $N_{\mathrm{A}} = 4 \times 10^{18}\,\mathrm{cm}^{-3}$, respectively. The solid and dashed curves are the

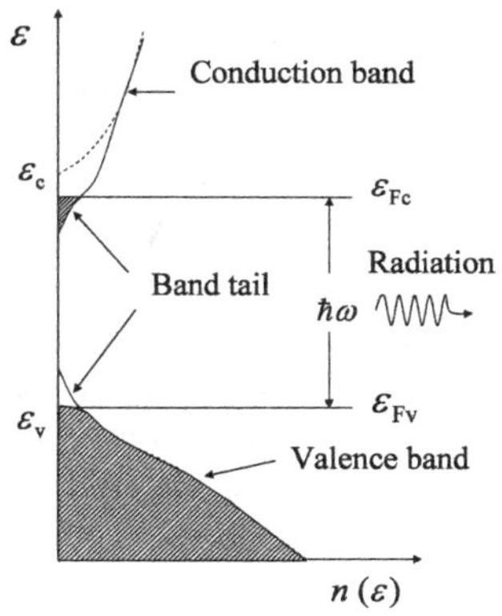

Fig. 9.8 Electron and hole distribution in the conduction band and valence band, where the band tail effect is included

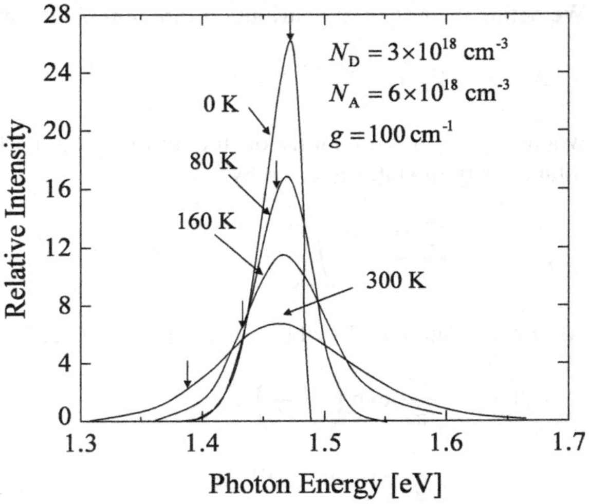

Fig. 9.9 The calculated spontaneous emission spectra in GaAs with $N_D = 3 \times 10^{18}\text{cm}^{-3}$ and $N_A = 6 \times 10^{18}\text{cm}^{-3}$ at several temperatures. The *arrows* show the peak positions of $-\alpha(\mathcal{E})$. (from Sterm [3])

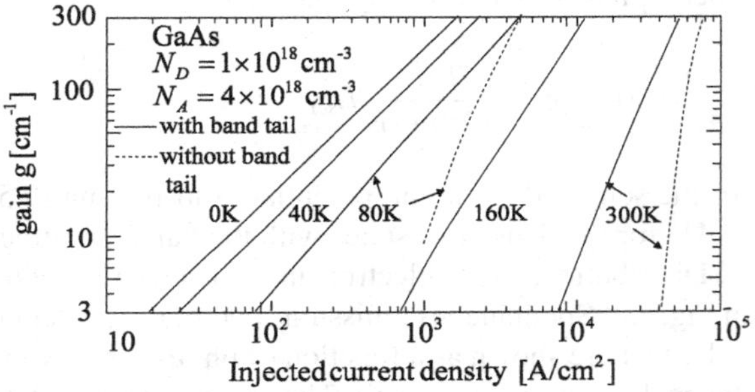

Fig. 9.10 The gain g is plotted as a function of injected current density in GaAs with donor density $N_D = 1 \times 10^{18}$ and acceptor density $N_A = 4 \times 10^{18}$ cm^{-3}. The solid and dashed curves are calculated with and without the band tail effect, respectively. (from Stern [3])

gains with and without the band tail effect, respectively. In Fig. 9.11, current density required to get the gain $g = 100\,\text{cm}^{-1}$ is plotted as a function of temperature for different impurity densities, where $N_A - N_D = 3 \times 10^{18}$ cm^{-3} and donor densities $N_D = 3 \times 10^{17}$, 3×10^{18}, 1.0×10^{19}, 3×10^{19} cm^{-3}.

Semiconductor laser diodes consist of highly doped pn junctions and thus the band tail effect plays an important role in the laser action. In such a case the momentum (wave vector) conservation rule is not fulfilled, and thus the laser gain is calculated by taking account of the density of states with the band tail effect and the energy conservation rule. We will discuss laser oscillations in Sect. 9.4 in connection with luminescence and detailed mechanisms in Sect. 9.5. The difference between LDs (Laser Diodes) and LEDs (Light Emitting Diodes) is the line width of the emission spectra. The spectral line shapes of LDs are very narrow with high output but LEDs exhibit wide line width. In the next section we deal with the emission of LDs and LEDs in various materials and discuss the mechanisms of the luminescence.

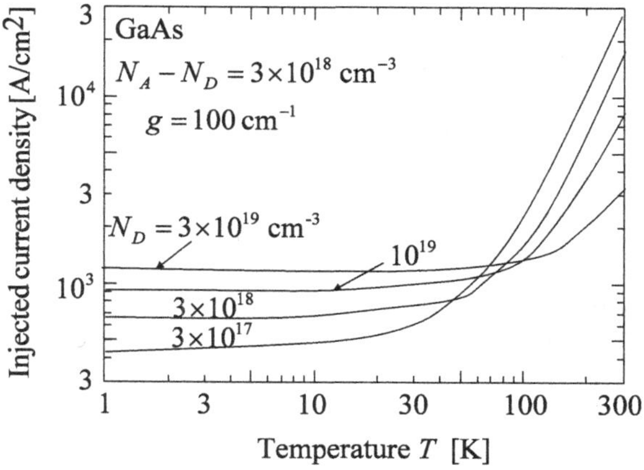

Fig. 9.11 Injected current density required to get gain $g = 100\,\text{cm}^{-1}$ is plotted as a function of temperature for GaAs with the impurity density $N_A - N_D = 3 \times 10^{18}$ cm^{-3}. (from Stern [3])

9.4 Luminescence

Excited electrons and holes in semiconductors recombine to emit photons. This process is called **luminescence**. The excitations (pumping) by light (photons), cathode ray (electron beams) and electron–hole injection by forward current in pn junction are called, respectively, photoluminescence, cathode luminescence, and electro–luminescence. Luminescence in semiconductors is based on the light emission due to the recombination of excited electrons and holes. Various mechanisms are involved in the recombination processes, in which the direct recombination of electrons and holes in direct gap semiconductors play the most important role. In indirect semiconductors the recombination involves phonon emission or absorption and thus the intensity is very weak. Luminescence is also induced by other mechanisms such as (1) recombination of excited electrons in the conduction band with the holes trapped in acceptors, (2) recombination of trapped electrons in donors with holes in the valence band, (3) recombination of the pair states of donor and acceptor, and so on. At low temperatures an electron bound to a donor attracts an hole and forms an electron–hole pair (exciton) bound at a donor (DX pair) which dissociates to emit light. In addition isoelectronic traps in GaP such as GaP:N results in strong light emission in indirect semiconductors. In this section we will concern with the mechanisms of luminescence in semiconductors. See the review articles of Bebb and Williams [4], Williams and Bebb [5], and Holonyak and Lee [6] for the details of the theories and experiments on luminescence.

9.4.1 Luminescence Due to Band to Band Transition

In semiconductors without the band tail effect luminescence is governed by band to band transition, and the luminescence intensity I is proportional to the spontaneous

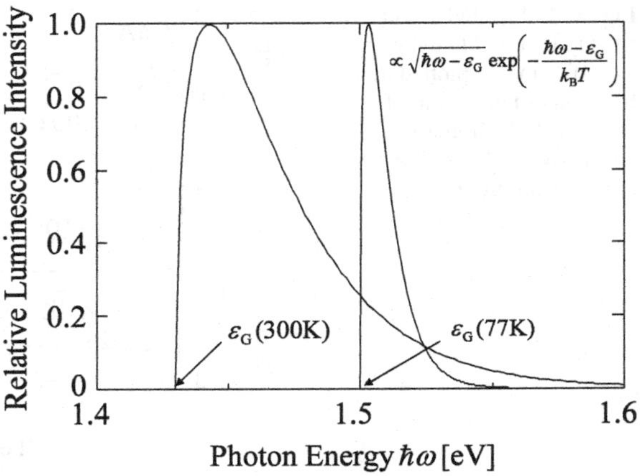

Fig. 9.12 Calculated luminescence spectra for the band to band transition in GaAs in GaAs, where the distribution function of the excited electrons is assumed to be given by Boltzmann distribution function and the temperature dependence of the band gap is taken account. $\mathcal{E}_G(300\text{K}) = 1.43\,\text{eV}$ at $T = 300\text{K}$ and $\mathcal{E}_G(77\text{K}) = 1.50\,\text{eV}$ at $T = 77\text{K}$

emission rate r_{spon}, which is given by the following relation with the emitted photon energy $\hbar\omega$ from (9.17),

$$I(\hbar\omega) \propto r_{\text{spon}} \propto \sqrt{\hbar\omega - \mathcal{E}_G} \exp\left(-\frac{\hbar\omega - \mathcal{E}_G}{k_B T}\right), \tag{9.55}$$

where the excited electron density is approximated by Boltzmann distribution $f_2 \cong \exp(-\hbar\omega/k_B T)$ and the hole distribution function is assumed to be $f_1 \cong 1$. Using this relation calculated luminescence spectra are shown in Fig. 9.12, where the normalized intensities at $T = 300\,\text{K}$ and $T = 77\,\text{K}$ are compared in GaAs. In Fig. 9.12 the temperature dependence of the band gap is taken into account; $\mathcal{E}_G(300\,\text{K}) = 1.43\,\text{eV}$ at $T = 300\,\text{K}$ and $\mathcal{E}_G(77\,\text{K}) = 1.50\,\text{eV}$ at $T = 77\,\text{K}$. We see in the figure that the band width of the luminescence becomes narrower at lower temperatures.

9.4.2 Luminescence Due to Excitons

We discussed in Sect. 4.5 excitons in an direct band gap semiconductor and in indirect band gap semiconductor. Luminescence is the reverse process of the absorption.

9.4.2.1 (1) Luminescence Due to Free Excitons

In Sect. 4.5 we discussed Coulomb interaction of an electron–hole pair (exciton) by solving Hamiltonian, and the exciton state is described by the bound state with the dissociation energy $\mathcal{E}_x/n^2$, $n = 1, 2, 3, \ldots$) and the continuum state with the center of mass motion. These excitons are possible to move freely in the crystal (called free excitons) and are distinguished from the excitons bound to donors or acceptors, **bound excitons**, which will be discussed later. From the analogy to hydrogen atom model, using the Bohr radius of an exciton $a_{\mathrm{x}} = (m/\mu)(\epsilon/\epsilon_0)a_{\mathrm{B}}$ and the exciton reduced mass $1/\mu = 1/m_{\mathrm{e}} + 1/m_{\mathrm{h}}$, we obtain $\mathcal{E}_{\mathrm{x}} = (\mu/m)(\epsilon_0/\epsilon)^2\mathcal{E}_{\mathrm{H}} = (\mu e^4)/[2(4\pi\epsilon)^2\hbar^2]$, and the absorption coefficient of the exciton is given by (see Sect. 4.5).

$$\alpha(\hbar\omega) \propto \frac{1}{(\hbar\omega - \mathcal{E}_{\mathrm{G}} + \mathcal{E}_{\mathrm{x}})^2 + \Gamma^2}, \tag{9.56}$$

where the dielectric function of the exciton is approximated by Lorentzian. Therefore the luminescence intensity has a peak around the energy $\mathcal{E}_{\mathrm{x}}$ below the band edge. The luminescence near the band edge is modified by the factor $\exp(-\hbar\omega/k_{\mathrm{B}}T)$ and the peak position is shifted to the lower energy side by $k_{\mathrm{B}}T$. Therefore it is very difficult to determine the energy gap exactly from the luminescence spectrum. The luminescence spectrum of the free exciton is given by

$$I(\hbar\omega) \propto \frac{1}{(\hbar\omega - \mathcal{E}_{\mathrm{G}} + \mathcal{E}_{\mathrm{x}})^2 + \Gamma^2} \exp\left[-(\hbar\omega - \mathcal{E}_{\mathrm{G}} + \mathcal{E}_{\mathrm{x}})/k_{\mathrm{B}}T\right], \tag{9.57}$$

where Γ is the broadening energy.

9.4.2.2 (2) Luminescence Due to Indirect Excitons

The locations of the conduction band bottom and valence band top in a indirect gap semiconductor are different in the $\boldsymbol{k}$ space, such as in GaP. As described in Sec. 4.5 light absorption occurs accompanying absorption or emission of a phonon. Since various types of phonons are involved in the absorption, complicated structures of the absorption spectra are observed. The absorption coefficient due to indirect excitons is expressed as follows (see Sec. 4.5.2)

$$\alpha(\hbar\omega) \propto \left[\hbar\omega - \mathcal{E}_{\mathrm{G}} \mp \hbar\omega_{\boldsymbol{q}}^{\alpha} + \mathcal{E}_{\mathrm{x}}/n^2\right]^{1/2}, \tag{9.58}$$

where $\hbar\omega_{\boldsymbol{q}}^{\alpha}$ is the phonon energy involved in the absorption and $\mp$ corresponds to absorption and emission of a phonon. Therefore luminescence involving phonon emission appears below the band gap energy and is well resolved. Thus the luminescence spectrum is given by

$$I(\hbar\omega) \propto \left[\hbar\omega - \mathcal{E}_{\mathrm{G}} + \hbar\omega_{\boldsymbol{q}}^{\alpha} + \mathcal{E}_{\mathrm{x}}/n^2\right]^{1/2} \times \exp\left[-(\hbar\omega - \mathcal{E}_{\mathrm{G}} + \hbar\omega_{\boldsymbol{q}}^{\alpha} + \mathcal{E}_{\mathrm{x}}/n^2)/k_{\mathrm{B}}T\right]. \tag{9.59}$$

Comparing with (9.59) with (9.55) we find the line shape of the photoluminescence due to the indirect exciton is similar to the line shape of the direct band to band transition. From these considerations the luminescence spectrum due to indirect exciton accompanies well resolved phonon peaks.

9.4.2.3 (3) Luminescence Due to Excitons Bound to Impurities

The state in which an electron and a hole are bound to an impurity is called **bound exciton** and the energy state depends on the mass ratio of an electron and a hole, $\sigma = m_e/m_h$. See the references for the detailed analysis of the bound excitons. Here we summarize the luminescence due to the bound excitons.

The bound state of the exciton–ionized donor (an exciton bound to an ionized donor) is expressed as

$$(\mathrm{D}^+, \mathrm{x}), \quad \oplus - +, \quad \text{or} \quad \mathrm{D}^+\mathrm{e\,h}\,. \tag{9.60}$$

The exciton bound to ionized acceptor is expressed by

$$(\mathrm{A}^-, \mathrm{x}), \quad \ominus + -, \quad \text{or} \quad \mathrm{A}^-\mathrm{e\,h}\,, \tag{9.61}$$

where other symbols are also used in papers reported so far. The exciton bound to a neutral donor consists of a neutral donor D^0, two electrons $--$ and a hole $+$, and expressed as

$$(\mathrm{D}^0, \mathrm{x}), \quad \oplus - - + \quad \text{or} \quad \mathrm{D}^+\mathrm{e\,e\,h}\,. \tag{9.62}$$

The exciton bound to a neutral acceptor is shown by

$$(\mathrm{A}^0, \mathrm{x}), \quad \ominus + + - \quad \text{or} \quad \mathrm{A}^-\mathrm{h\,h\,e}\,. \tag{9.63}$$

The emitted photon energy due to the dissociation of the exciton complex in GaAs is written as [7]

$$\hbar\omega(\mathrm{D}^0, \mathrm{x}) = \mathcal{E}_\mathrm{G} - \mathcal{E}_\mathrm{x} - 0.13\mathcal{E}_\mathrm{D}\,, \tag{9.64a}$$

$$\hbar\omega(\mathrm{D}^+, \mathrm{x}) = \mathcal{E}_\mathrm{G} - \mathcal{E}_\mathrm{D} - 0.06\mathcal{E}_\mathrm{D}\,, \tag{9.64b}$$

$$\hbar\omega(\mathrm{A}^0, \mathrm{x}) = \mathcal{E}_\mathrm{G} - \mathcal{E}_\mathrm{x} - 0.0.07\mathcal{E}_\mathrm{A}\,, \tag{9.64c}$$

$$\hbar\omega(\mathrm{A}^-, \mathrm{x}) = \mathcal{E}_\mathrm{G} - \mathcal{E}_\mathrm{A} - 0.4\mathcal{E}_\mathrm{A}\,. \tag{9.64d}$$

9.4.2.4 (4) Luminescence Due to Donor–Acceptor Pairs

A donor and an acceptor occupy different sites of the lattices. In the case where the wave functions of the donor and the acceptor overlap, the donor ion D^+, the acceptor ion A^-, an electron $-$ and a hole $+$ compose a complex state, and luminescence due to the dissociation of the complex is observed. Such a complex is written as

$$(\mathrm{D}^0, \mathrm{A}^0), \quad (\mathrm{D}^+\mathrm{A}^-, \mathrm{x}), \quad \oplus \ominus - + \quad \mathrm{D}^+\mathrm{A}^-\mathrm{e\,h}\,. \tag{9.65}$$

This type of a complex is called an exciton bound to donor–acceptor pair D^+A^- and the bound energy depends on the distance between the donor D^+ and the acceptor A^-.

9.4.2.5 (5) Exciton Molecules

Another important example of complexes is exciton molecules (exciton–exciton complex) which is expressed as

$$(\mathrm{x\,x}), \quad ++--, \quad \text{or} \quad \mathrm{h\,h\,e\,e}. \tag{9.66}$$

The excitation energy of an exciton is given by

$$\mathcal{E}(\mathrm{x}) = \mathcal{E}_\mathrm{G} - \mathcal{E}_\mathrm{x}. \tag{9.67}$$

Letting the interaction energy of two excitons D_0, the excitation energy of two excitons is approximated as $2\mathcal{E}(\mathrm{x}) - D_0$. Therefore we obtain the following relation for the excitation energy of an exciton molecule,

$$\mathcal{E}(\mathrm{x}_1\,\mathrm{x}_2) = \mathcal{E}(\mathrm{x}_1) + \mathcal{E}(\mathrm{x}_2) - D_0. \tag{9.68}$$

When an exciton molecule dissociates, one of the exciton of the molecule dissociates and leaves a free exciton of the energy $\mathcal{E}(\mathrm{x}_2)$ and a photon $\hbar\omega$. Since the final state of the energy $\hbar\omega + \mathcal{E}(\mathrm{x}_2)$ is equivalent to the initial state $\mathcal{E}(\mathrm{x\,x})$, the following relation is obtained,

$$\hbar\omega = \mathcal{E}(\mathrm{x}_1) - D_0 \equiv \mathcal{E}_\mathrm{G} - \mathcal{E}_\mathrm{x} - D_0, \tag{9.69}$$

where the crystal still has the energy $\mathcal{E}(\mathrm{x}_2)$ and thus emits the second photon of energy $\hbar\omega' = \mathcal{E}_\mathrm{G} - \mathcal{E}_\mathrm{x}$.

9.4.3 Luminescence via Impurities

The luminescence via impurities plays an important role in the recombination in semiconductors. Here we will deal with the basic mechanisms of luminescence associated with impurities. See the papers by Dumke [8] and Eagles [9] for the theoretical treatments. These two papers are based on different expressions of the wave functions of the impurity states, but the obtained results are the same. We will describe the method of Dumke in detail and briefly refer the result of Eagles.

For simplicity, we calculate the recombination rate between a hole bound to the ground state of an acceptor and an electron in the conduction band. It is apparent that the recombination rate between an electron bound to a donor and a hole in the valence band as seen from the following treatment. The wave function (envelop function) $F(\boldsymbol{r})$ of a particle bound to an impurity ion is given by the effective mass approximation,

$$\left[-\frac{\hbar^2}{2m^*}\nabla^2 - \frac{e^2}{4\pi\kappa\epsilon_0 r} - \mathcal{E}\right] F(\boldsymbol{r}) = 0\,. \tag{9.70}$$

Since a hole near the valence band edge $\boldsymbol{k} \cong 0$ is captured by an acceptor, the wave function may be expressed by the envelop function given by the above equation. Therefore the wave function of a hole captured by an acceptor is given by the product of the Bloch function $\Psi_{v,\boldsymbol{k}}(\boldsymbol{r}) = \mathrm{e}^{\mathrm{i}\boldsymbol{k}\cdot\boldsymbol{r}} u_{v,\boldsymbol{k}}(\boldsymbol{r})$ and the envelop function $F(\boldsymbol{r})$. In other words,

$$\Psi_{\mathrm{A}}(\boldsymbol{r}) = u_{v,\boldsymbol{k}}(\boldsymbol{r})\mathrm{e}^{\mathrm{i}\boldsymbol{k}\cdot\boldsymbol{r}} F(\boldsymbol{r}) \cong u_{v,0}(\boldsymbol{r})F(\boldsymbol{r}) \tag{9.71}$$

The eigenvalue of the resulting effective mass equation is given by

$$\mathcal{E}_n = -\frac{m^* e^4}{2(4\pi\kappa\epsilon_0)^2\hbar^2} \cdot \frac{1}{n^2}, \qquad n = 1, 2, \ldots\,, \tag{9.72}$$

and the eigenfunction of the ground state ($n = 1$) is written as

$$F_1(\boldsymbol{r}) = \sqrt{\frac{1}{\pi a_{\mathrm{A}}^3}}\mathrm{e}^{-r/a_{\mathrm{A}}} \tag{9.73}$$

$$a_{\mathrm{A}} = \frac{4\pi\hbar^2\kappa\epsilon_0}{m^* e^2}\,. \tag{9.74}$$

Using these results the momentum matrix element for the transition of the electron in the conduction band to the acceptor is obtained as follows.

Here we will present the calculations of the momentum matrix element proposed by Dumke [4, 8]. The momentum matrix element is written as

$$(\boldsymbol{e}\cdot\boldsymbol{p})_{c\mathrm{A}} = \langle c, \boldsymbol{k}|\boldsymbol{e}\cdot\boldsymbol{p}|\mathrm{A}\rangle = \int \mathrm{e}^{-\mathrm{i}\boldsymbol{k}\cdot\boldsymbol{r}} u_{c,\boldsymbol{k}}(\boldsymbol{e}\cdot\boldsymbol{p})\Psi_{\mathrm{A}}(\boldsymbol{r})\mathrm{d}^3\boldsymbol{r}\,, \tag{9.75}$$

where $\boldsymbol{e}$ is the unit vector of the light polarization and $\boldsymbol{p}$ is the momentum operator. The calculation of the above momentum matrix element is carried out by operating $\boldsymbol{p} = -\mathrm{i}\hbar\nabla$ to $\Psi_{\mathrm{A}}(\boldsymbol{r})$ and the following two terms appear.

$$(\boldsymbol{e}\cdot\boldsymbol{p})_{c\mathrm{A}} = \int \mathrm{e}^{-\mathrm{i}\boldsymbol{k}\cdot\boldsymbol{r}} F(\boldsymbol{r})\left[u_{c,\boldsymbol{k}}(\boldsymbol{r})(\boldsymbol{e}\cdot\boldsymbol{p})u_{v,0}(\boldsymbol{r})\right]\mathrm{d}^3\boldsymbol{r} + \int \mathrm{e}^{-\mathrm{i}\boldsymbol{k}\boldsymbol{r}}\left[(\boldsymbol{e}\cdot\boldsymbol{p})F(\boldsymbol{r})\right]u_{c,\boldsymbol{k}}(\boldsymbol{r})u_{v,0}(\boldsymbol{r})\mathrm{d}^3\boldsymbol{r}\,. \tag{9.76}$$

The above integration is not easy, but may be carried out under the following assumption. When the functions $F(\boldsymbol{r})$ and $\mathrm{e}^{-\mathrm{i}\boldsymbol{k}\cdot\boldsymbol{r}}$ are assumed to change slowly compared to the period of the Bloch function $u(\boldsymbol{r} + \boldsymbol{R}_m) = u(\boldsymbol{r})$, the integral is replaced by the summation of the unit cells Ω; $\int \to \sum_{\boldsymbol{R}_m} \int_\omega$. The slowly varying function $F(\boldsymbol{r})$ may be assumed to be constant within a unit cell and approximated by the value at the center of the unit cell $F(\boldsymbol{R}_m)$. Thus we obtain

$$\Omega \sum_{\boldsymbol{R}_m} \mathrm{e}^{-\mathrm{i}\boldsymbol{k}\cdot\boldsymbol{R}_m} F(\boldsymbol{R}_m) \frac{1}{\Omega} \int_\Omega u_{c,\boldsymbol{k}}(\boldsymbol{r})(\boldsymbol{e}\cdot\boldsymbol{p}) u_{v,0}(\boldsymbol{r}) \mathrm{d}^3\boldsymbol{r}$$
$$+\Omega \sum_{\boldsymbol{R}_m} \mathrm{e}^{-\mathrm{i}\boldsymbol{k}\cdot\boldsymbol{R}_m} [(\boldsymbol{e}\cdot\boldsymbol{p})F(\boldsymbol{r})]_{\boldsymbol{R}_m} \frac{1}{\Omega} \int_\Omega u_{c,\boldsymbol{k}}(\boldsymbol{r}) u_{v,0}(\boldsymbol{r}) \mathrm{d}^3\boldsymbol{r}\,. \tag{9.77}$$

The second term of the above equation vanishes due the orthogonality of the Bloch functions. The summation of the first term may be transformed into an integral form by using an approximation

$$\Omega \mathrm{e}^{-\mathrm{i}\boldsymbol{k}\cdot\boldsymbol{R}_m} F(\boldsymbol{R}_m) = \int_\Omega \mathrm{e}^{-\mathrm{i}\boldsymbol{k}\cdot\boldsymbol{R}_m} F(\boldsymbol{R}_m) \mathrm{d}^3\boldsymbol{r} \simeq \int_\Omega \mathrm{e}^{-\mathrm{i}\boldsymbol{k}\cdot\boldsymbol{r}} F(\boldsymbol{r}) \mathrm{d}^3\boldsymbol{r}\,, \tag{9.78}$$

and therefore the summation over the unit cells is rewritten by the integral over the crystal volume V

$$\sum_{\boldsymbol{R}_m} \Omega \mathrm{e}^{-\mathrm{i}\boldsymbol{k}\cdot\boldsymbol{R}_m} F(\boldsymbol{R}_m) \simeq \int_V \mathrm{e}^{-\mathrm{i}\boldsymbol{k}\cdot\boldsymbol{r}} F(\boldsymbol{r}) \mathrm{d}^3\boldsymbol{r}\,. \tag{9.79}$$

The momentum matrix element of (9.76) is thus given by

$$(\boldsymbol{e}\cdot\boldsymbol{p})_{c,\mathrm{A}} = \langle c, \boldsymbol{k}|\boldsymbol{e}\cdot\boldsymbol{p}|\mathrm{A}\rangle = \int \mathrm{e}^{-\mathrm{i}\boldsymbol{k}\cdot\boldsymbol{r}} F(\boldsymbol{r}) \mathrm{d}^3\boldsymbol{r}\cdot p_{cv} \equiv a(\boldsymbol{k})\cdot p_{cv}\,, \tag{9.80a}$$

where

$$p_{cv}(\boldsymbol{k}) = \frac{1}{\Omega} \int_\Omega u_{c,\boldsymbol{k}}(\boldsymbol{r})(\boldsymbol{e}\cdot\boldsymbol{p}) u_{v,0}(\boldsymbol{r}) \mathrm{d}^3\boldsymbol{r}\,. \tag{9.80b}$$

Since the matrix element $p_{cv}(\boldsymbol{k})$ varies very slowly with the $\boldsymbol{k}$ value, we use the value at the band edge $p_{cv}(0)$

$$p_{cv}(0) = \frac{1}{\Omega} \int_\Omega u_{c,0}(\boldsymbol{r})(\boldsymbol{e}\cdot\boldsymbol{p}) v_{v,0}(\boldsymbol{r}) \mathrm{d}^3\boldsymbol{r}\,. \tag{9.80c}$$

The integral of $a(\boldsymbol{k})$ is evaluated to obtain

$$\begin{aligned} a(\boldsymbol{k}) &= \int \mathrm{e}^{-\mathrm{i}\boldsymbol{k}\cdot\boldsymbol{r}} F(\boldsymbol{r}) \mathrm{d}^3\boldsymbol{r} \\ &= \frac{1}{(\pi a_\mathrm{A}^3)^{1/2}} \int \mathrm{e}^{-\mathrm{i}kr\cos\theta} \mathrm{e}^{-r/a_\mathrm{A}} r^2 \sin\theta \mathrm{d}\theta \mathrm{d}\phi \mathrm{d}r \\ &= \frac{8\left(\pi a_\mathrm{A}^3\right)^{1/2}}{\left[1+(ka_\mathrm{A})^2\right]^2}\,. \end{aligned} \tag{9.81}$$

Eagles expressed the wave function of the impurity by a linear combination of Bloch functions [9] and the shallow impurity state by using the valence band wave functions, and obtained the same results of Dumke described above.

Using the momentum matrix element for a bulk crystal $M_{\rm cv}(\equiv p_{\rm cv})$, the squared matrix element of an electron transition between the conduction band and the acceptor level is given by

$$\langle |M|^2 \rangle_{\rm av} = \langle |M_{\rm b}|^2 \rangle_{\rm av} \frac{1}{V} \frac{64\left(\pi a_{\rm A}^3\right)}{\left[1+(ka_{\rm A})^2\right]^4}, \tag{9.82}$$

where V is the volume of the crystal. $\langle |M_{\rm cv}|^2 \rangle_{\rm av}$ is the average matrix element connecting bulk states near the band edges and given by the following relation using the interband matrix element P defined in Chap. 2

$$\langle |M_{\rm cv}|^2 \rangle_{\rm av} = \frac{m^2 P^2}{6\hbar^2} \simeq \frac{m^2}{12 m_{\rm e}^*} \frac{\mathcal{E}_{\rm G}(\mathcal{E}_{\rm G}+\Delta_0)}{\mathcal{E}_{\rm G}+2\Delta_0/3}, \tag{9.83}$$

where $m_{\rm e}^*$ is the band edge effective mass of the conduction band and given by (2.158). The above relation is easily deduced from the wave functions of the heavy and light hole bands given by (2.63a), (2.63b), (2.63e), (2.63f) and $P = (\hbar/m)\langle S|p_z|Z\rangle = (\hbar/m)P_0$, and the conduction band wave function $|S\rangle$.[3]

[3]The term $m^2P^2/6\hbar^2$ of (9.83) should be replaced by $4m^2P^2/3\hbar^2$, when we take the heavy and light hole bands into account and neglect the spin–orbit split–off band. For spin orientation $|\uparrow\rangle$, we obtain

$$\left\langle S \middle| p_x \middle| \frac{3}{2}, \frac{3}{2} \right\rangle^2 = \left\langle S \middle| p_y \middle| \frac{3}{2}, \frac{3}{2} \right\rangle^2 = \frac{1}{2}\frac{m^2}{\hbar^2}P^2,$$
$$\left\langle S \middle| p_x \middle| \frac{3}{2}, -\frac{1}{2} \right\rangle^2 = \left\langle S \middle| p_y \middle| \frac{3}{2}, -\frac{1}{2} \right\rangle^2 = \frac{1}{6}\frac{m^2}{\hbar^2}P^2,$$
$$\left\langle S \middle| p_z \middle| \frac{3}{2}, \frac{1}{2} \right\rangle^2 = \frac{4}{6}\frac{m^2}{\hbar^2}P^2.$$

Then we find the momentum matrix element for spin orientation $|\uparrow\rangle$

$$|M_x|^2 = \left\langle S \middle| p_x \middle| \frac{3}{2}, \frac{3}{2} \right\rangle^2 + \left\langle S \middle| p_x \middle| \frac{3}{2}, -\frac{1}{2} \right\rangle^2 = \left(\frac{1}{2}+\frac{1}{6}\right)\frac{m^2}{\hbar^2}P^2.$$

Summing up all the components and multiplying by 2 of spin degeneracy, we obtain

$$\langle |M_{\rm b}|^2 \rangle_{\rm av} = \frac{2}{3}\left(|M_x|^2 + |M_y|^2 + |M_z|^2\right) = \frac{4m^2P^2}{3\hbar^2},$$

where we used the following definitions,

$$\langle S|p_x|X\rangle^2 = \langle S|p_y|Y\rangle^2 = \langle S|p_z|Z\rangle^2 = \frac{m^2}{\hbar^2}P^2.$$

When the overlapping of the acceptor wave functions is negligible, then the spontaneous emission rate is obtained from (G.28) of Appendix G

$$r_{\rm spon}(\hbar\omega) = \frac{n_{\rm r}e^2\hbar\omega}{\pi\epsilon_0 m^2\hbar^2 c^3}\sum_{\boldsymbol{k}} N_{\rm A}\langle|M_{\rm b}|^2\rangle_{\rm av}\frac{64\left(\pi a_{\rm A}^3\right)}{\left[1+(ka_{\rm A})^2\right]^4} \times\delta\left[\mathcal{E}_{\rm c}(\boldsymbol{k})-(\mathcal{E}_{\rm v,0}+\mathcal{E}_{\rm A})-\hbar\omega\right]f(\mathcal{E}_2)[1-f(\mathcal{E}_1)], \tag{9.84}$$

where $N_{\rm A}$ is the density of the acceptors, and $f(\mathcal{E}_2)$ and $f(\mathcal{E}_1)$ are the occupation functions of the electrons in the conduction band and the acceptor level, respectively. Similar expression is obtained for the transition of electron from the donor state to the valence band.

From the above considerations the luminescence intensity due to the transition of the electrons in the conduction band to the acceptor level is given by the following relation

$$I(\hbar\omega) \propto r_{\rm spon} \propto \sqrt{\hbar\omega-\mathcal{E}_{\rm G}+\mathcal{E}_{\rm A}}\exp\left(-\frac{\hbar\omega-\mathcal{E}_{\rm G}+\mathcal{E}_{\rm A}}{k_{\rm B}T}\right), \tag{9.85}$$

which is quite similar to the band to band transition of (9.55), and the luminescence peak shifts to the lower energy site by $\mathcal{E}_{\rm A}$ compared to the band to band transition. The luminescence line shape is expected to be similar to Fig. 9.12. Photoluminescence spectra of GaAs:Cd (Cd doped GaAs) are compared with (9.85) at $T = 20$ K and 80 K in Fig. 9.13, where we find the luminescence peak appears at $\mathcal{E}_{\rm G} - \mathcal{E}_{\rm A}$ but strong tails are observed at the lower energy side. Since the acceptor density is $N_{\rm A} = 4 \times 10^{16}$ cm^{-3}, the band tail effect of the conduction band is negligible and thus the tails of the luminescence is ascribed to the excitons bound to acceptors (an electron is bound to an acceptor capturing a hole). Assuming the band gap $\mathcal{E}_{\rm G} = 1.521$ eV at $T = 20$ K, the binding energy of the acceptor is $\mathcal{E}_{\rm A} = 34.5$ meV. The band gap at $T = 80$ K is estimated as $\mathcal{E}_{\rm G}(T = 80) = 1.512$ eV.

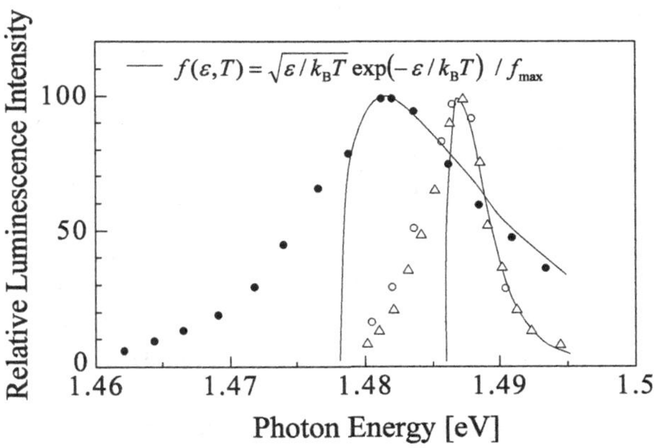

Fig. 9.13 Experimental data of photoluminescence from Cd doped GaAs (GaAs:Cd) at $T = 20$K and 80 K are compared with (9.85). ∘ and • are for the sample 1 at $T = 20$ and $T = 80$K, respectively, and △ are for the sample 2 at $T = 20$K. (after Williams and Webb [10])

9.4.4 Luminescence in GaP and GaAsP via N Traps

It is well known that the bottoms of the conduction bands of GaP and $GaAs_xP_{1-x}$ ($x \geq 0.47$) are located near the X point in the Brillouin zone, and therefore optical transition is indirect with emission or absorption of phonons. However, strong luminescence is observed in nitrogen (N) doped samples. Here we explain the mechanisms of the luminescence due to N levels. For more detailed treatments see the review paper by Holonyak and Lee [6] and references cited in the text. LED (Light–Emitting Diode) based on $GaAs_xP_{1-x}$ crystals are now commercially popular devices. Success of such LEDs are due to the following two factors. One is the unique mechanisms of the luminescence and the other is the crystal growth technologies of $GaAs_xP_{1-x}$ on large GaP or GaAs substrates utilizing open–tube vapor–phase–epitaxy (VPE) by transport of AsH_3-PH_3. The luminescence of $GaAs_xP_{1-x}$ in the direct transition region $x \leq 0.46$ ($T = 77$ K); $x \leq 0.49$ (300K) is quite similar to GaAs and in the region of direct–indirect transition the luminescence is ruled by the emission via the X conduction band with the large effective mass and the donor level associated with the X conduction band. In the indirect transition region of $GaAs_xP_{1-x}$ ($x \geq 0.46$) the luminescence is dominated by the isoelectronic N level.

Nitrogen (N) atoms doped in GaP and $GaAs_xP_{1-x}$ substitute phosphorous (P) atoms. The N atoms are electronically equivalent to P atoms and thus called isoelectronic traps. The electronegativity of N is larger than P. Lattices surrounding doped N atoms relax and the negativity of N atom is reduced but electrons are localized around the N atoms. It is known that the potential on N atom is different from the Coulomb potential of usual donors screened up to several atomic distance and that the potential attracts an electron toward the central part (central cell potential). The dissociation energy of a trapped electron at N atom in GaP is quite small and is estimated to be about 10meV. Since the electron is captured by the N trap, the wave function ψ is expected to be spread in $\boldsymbol{k}$ space as shown in Fig. 9.14 from the uncertainty principle and strong luminescence similar to the direct transition is observed although GaP is an indirect transition semiconductor. Experimental data indicate that the wave function at $\boldsymbol{k} = 0$ of the electron trapped at N is about 100 times of the wave function of the electron bound to a regular donor. In addition the wave function of an electron captured by the N trap overlaps with a hole wave function at $\boldsymbol{k} = 0$ (Γpoint), resulting in a formation of an exciton, and luminescence due to the exciton is observed. When we increase doping of N atoms in GaP, NN pairs are formed. The nearest neighbor pair is expressed by NN_1 (NN_2, NN_3, ... are the second, the third, ... nearest pairs). As shown in Fig. 9.15, luminescence of NN_1 and NN_3 pairs are observed and such a pair luminescence is distinguished from the luminescence due to the isolated N atoms (N_X and A–B line). The series of N pairs NN_1, NN_2, ..., NN_{10} are resolved by the fluorescence (photoluminescence) spectroscopy [12, 13]. In $GaAs_xP_{1-x}$ ($x \cong 0.28$) overlapped luminescence of N_X and of the $\mathcal{E}_\Gamma$ band is observed.

A typical example of the experimental results of luminescence of excitons trapped at N isoelectronic traps is shown in Fig. 9.16. The energy gap of GaP is $\mathcal{E}_G(T =$

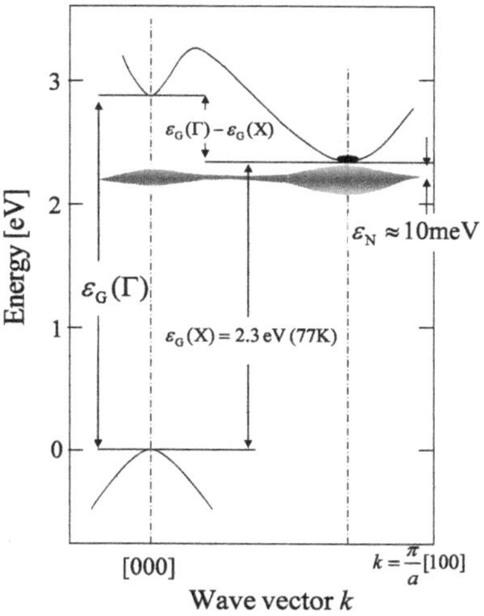

Fig. 9.14 Band structure of GaP:N, where the shaded region represents the amplitude of the electron wave function $\psi_N(\boldsymbol{k})$ bound to the isoelectronic N trap and has large amplitudes near the X and Γ points reflecting the respective conduction band minima. The binding energy of the N trap is about 10 meV and the potential is localized near the N atom (broadened in the $\boldsymbol{k}$ space), the wave function $\psi_N(\boldsymbol{k})$ has a large amplitude near the region $\boldsymbol{k} = 0$, resulting in a strong quasi–direct transition. (Figure after Holonyak and Lee [6])

Fig. 9.15 The energy levels of the Γ conduction band, X conduction band, and N and NN pairs in N–doped $GaAs_xP_{1-x}$ at 77K. N_Γ is the donor level ($x \leq 0.55$) associated with the Γ conduction band. NN_1 and NN_3 pairs are experimentally resolved for $x \geq 0.90$, but in the other region observed as a broad N_X level. Such a broad luminescence in GaP:N is called A–line. (Figure after Holonyak and Lee [6])

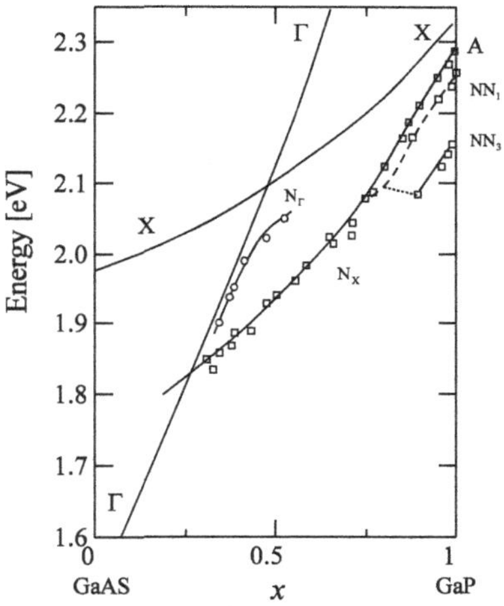

0K) $= \mathcal{E}(X) - \mathcal{E}(\Gamma) = 2.339$ eV. The peaks of A and B are the no–phonon lines of the excitons bound to N atoms (N and NN pairs) and phonon lines are well resolved. Figure 9.17 shows photoluminescence spectra of $GaAs_xP_{1-x}(x = 0.34)$:N^+, $N_D = 1.8 \times 10^{17}\,\text{cm}^{-3}$, where N^+ means heavy doping of N atoms ($\simeq 10^{19}\,\text{cm}^{-3}$). In the region of weak excitation (500 W/cm^2), luminescence from the Γ conduction band

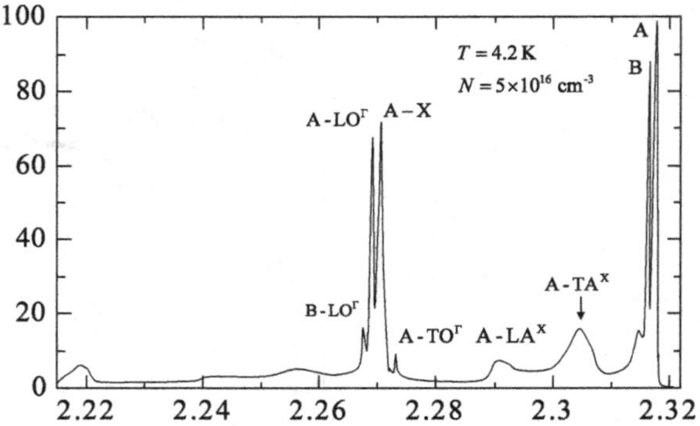

Fig. 9.16 Photoluminescence of excitons trapped at N isoelectronic traps in GaP. The peaks A and B are no–phonon lines. $T = 4.2$K and $N = 5 \times 10^{16}\,\text{cm}^{-3}$. The energy gap of GaP $\mathcal{E}_G(T = 0\text{K}) = \mathcal{E}(X) - \mathcal{E}(\Gamma)$ = 2.339 eV. [Dean [11]]

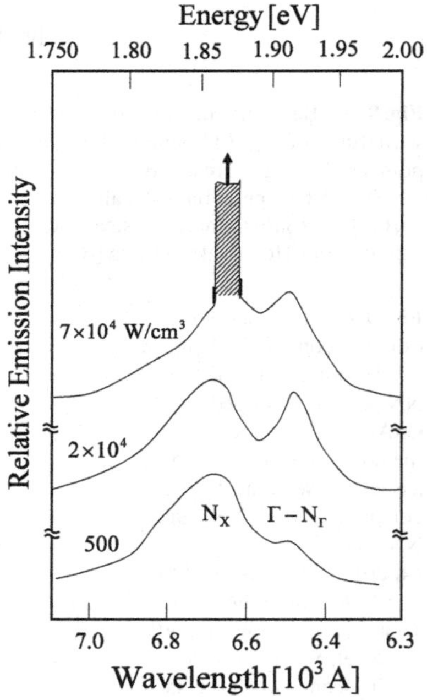

Fig. 9.17 Photoluminescence spectra of $GaAs_xP_{1-x}(x = 0.34)$:N^+, $N_D = 1.8 \times 10^{17}\,\text{cm}^{-3}$. In a low excitation (500 W/cm^2), the band to band transition associated with the Γ conduction band and emission from the N_X level are well resolved. Increasing the excitation power the emission of the band to band dominates the emission of N_X level. For excitation power of 7×10^4 W/cm^2 laser oscillations of N_X level are observed, where the shaded region consists of 10 lines of laser oscillations. (After Holonyak and Lee [6])

and from N_X is well resolved. Increasing the excitation power luminescence from the Γ conduction band dominates the luminescence from N_X level. For excitation power 7×10^4 W/cm^2, laser oscillations from N_X level are observed. The shaded region consists of well resolved 10 lines of laser oscillations.

Luminescence of p–type $GaAs_xP_{1-x}(x = 0.34)$:N^+:Zn is shown in Fig. 9.18, where luminescence from N doped GaAsP substrate is shown by the dashed curve. Luminescence from a p–type substrate with additional doping by Zn exhibit an emission peak from N_X at 29 meV lower than the emission peak of N_X in n–type

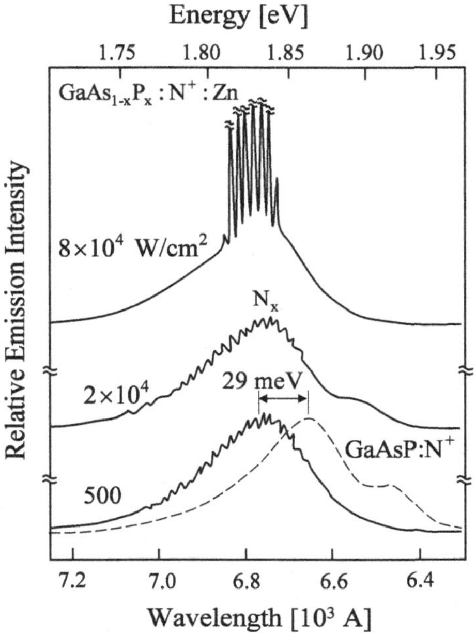

Fig. 9.18 Luminescence spectra of p–type $GaAs_xP_{1-x}(x = 0.34)$:N^+:Zn, $N_D = 1.8 \times 10^{17}\,cm^{-3}$. For low excitation (500 W/cm^2), the band to band transition associated with the Γ conduction band and emission from the N_X level are well resolved. Increasing the excitation power the emission of the band to band dominates the emission of N_X level. For excitation power 8×10^4 W/cm^2 laser oscillations of N_X level are observed, where the dashed curve shows the luminescence from the p–type $GaAs_xP_{1-x}(x = 0.34)$:N^+ in Fig. 9.17 and Zn doped p type $GaAs_xP_{1-x}(x = 0.34)$:N^+:Zn exhibits luminescence peak 29meV lower than the substrate without Zn doping. (After Holonyak and Lee [6])

GaAsP:N^+. This emission peak is ascribed to the recombination of the electrons captured by N_X with the holes captured at Zn acceptors.

In GaAs doped with Zn and O, oxygen O behaves as a donor and Zn as a shallow acceptor. When these two dopants occupy the nearest lattice sites, a pair of Zn·O is formed. Electrons far from the pair see an electronically neutral complex, but for an electron close to the pair the electronegativity of O is larger than Zn. Therefore O atom forms electron trap of binding energy about 0.3 eV. When an electron is captured by the Zn·O complex, the complex is negatively charged and the resulting Coulomb potential attracts a hole, forming an exciton. The photoluminescence due to the recombination of the electron–hole pair exhibits a peak at about 1.8eV, emitting red light.

9.4.5 Luminescence from GaInNAs

In recent years GaInNAs (called gainnas) has attracted considerable attention as a material for near infrared light emitting devices. In direct transition semiconductor $Ga_{1-x}In_xN_yAs_{1-y}$, nitrogen atom N with low composition $y < 0.01$ acts as an impurity. In the range $y \sim 1\%$, however, the distortion of the lattices surrounding the N atom and its electronegativity deform the energy bands of the host crystal and the bottom of the conduction band is lowered considerably. Therefore the wave length of the near infrared light emission can be changed by introducing a small amount of N impurity into GaInAs. From these reasons GaInNAs is believed to be one of the best candidates for tuning the wave length in the near infrared region and interesting results have been reported. [14] ~ [15]

The effect of nitrogen (N) atom is explained with the help of Fig. 9.15. Impurity level of nitrogen (N) of $GaAs_{1-x}P_x$ is located in the Γ conduction band for the composition $x \leq 0.2$ [16]. In such a case a strong interaction between the N level and the electrons in the Γ conduction band results in a deformation of the conduction band. In other words, the interaction between N impurity level and the conduction band gives rise to a band–crossing as shown in Fig. 9.19 and the conduction band bottom shifts to the lower energy side. The band anti-crossing model was proposed by Shan et al. [17] and shown below.

Let the conduction band of GaInNAs $\mathcal{E}_M(\boldsymbol{k})$, and assume the nitrogen level without interaction $\mathcal{E}_N$. These energy levels are measured from the top of the valence band. When an interaction between these two levels, the conduction band is perturbed. Assuming the interaction energy between the two levels as V_{MN}, the energy levels are obtained by solving the following eigenvalue equation.

$$\begin{vmatrix} \mathcal{E}_M - \mathcal{E} & V_{MN} \\ V_{MN} & \mathcal{E}_N - \mathcal{E} \end{vmatrix} = 0 . \tag{9.86}$$

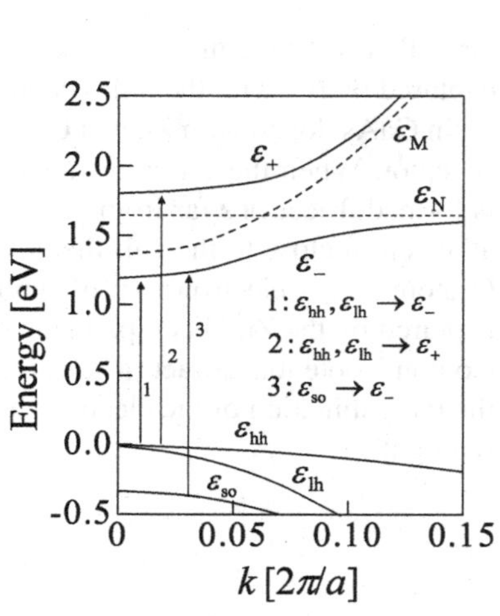

Fig. 9.19 Anti-crossing model for $Ga_{0.96}In_{0.04}As_{0.99}N_{0.01}$ induced by the interaction between the Γ conduction band and the N level located in the conduction band. The *arrows* are the optical transitions observed by experiments. ([17, 18])

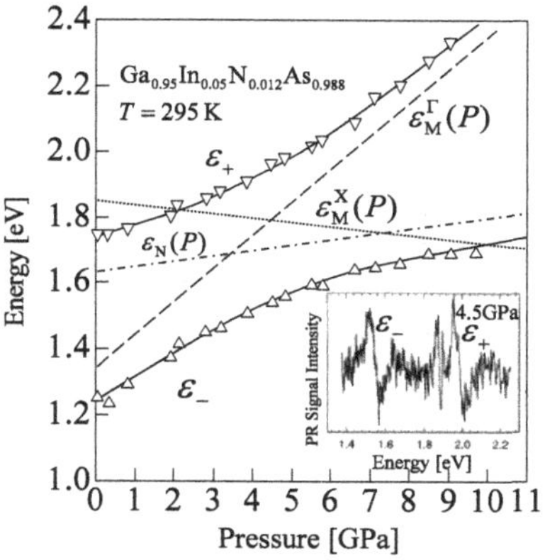

Fig. 9.20 Pressure dependence of the optical transition energies to the $\mathcal{E}_\pm$ levels in $Ga_{0.95}In_{0.05}As_{0.988}N_{0.012}$. The solid curves are calculated by (9.86), $\triangle$ and ∇ are the experimental data obtained by photoreflectance. The calculated curves are obtained by taking account of the pressure dependence of Γ and the X conduction bands, and N_X level shown in the figure. ([17])

The interaction energy V_{MN} gives rise to the mixing of the two levels and anti-crossing of the two levels occurs. The eigenvalues of (9.86) are given by

$$\mathcal{E}_\pm(\boldsymbol{k}) = \frac{1}{2}\left(\mathcal{E}_N + \mathcal{E}_M(\boldsymbol{k}) \pm \sqrt{[\mathcal{E}_N + \mathcal{E}_M(\boldsymbol{k})]^2 + 4V_{MN}^2}\right). \tag{9.87}$$

The calculated results are shown in Fig. 9.19, where we assumed $\mathcal{E}_N$=1.65 eV, V_{MN}=0.27 eV, and $\mathcal{E}_M$=1.3 eV for the conduction band edge without interaction measured from the top of the valence band. This figure well explains the band–crossing effect in GaInNAs and the experimental results reported so far on optical absorption, ellipsometry and modulation spectroscopy are well explained. Figure 9.20 shows a comparison of the calculated results with the pressure dependence of the conduction band edge measured from the top of the valence band obtained by photoreflectance measurements, where we find a good agreement with each other. The calculations based on the band–crossing model are carried out using (9.87) with the measured pressure dependence of the conduction band edge and N level for $Ga_{0.95}In_{0.05}N_{0.012}As_{0.988}$

$$\mathcal{E}_M = 1.35 + 0.1P \text{ [eV]}, \tag{9.88a}$$

$$\mathcal{E}_N = 1.65 + 0.0015P \text{ [eV]}, \tag{9.88b}$$

$$V_{MN} = 0.2\,\text{eV}, \tag{9.88c}$$

where the hydrostatic pressure P is expressed in units of [Gpa]. In Fig. 9.20 the pressure dependences of the Γ conduction band bottom $\mathcal{E}_M^\Gamma$ and the X conduction band bottom $\mathcal{E}_M^X$ and the N level $\mathcal{E}_N$ are also shown. From a comparison between the experimental and calculated results the interaction energy V_{MN} of $Ga_{1-x}In_xN_yAs_{1-y}$ varies from V_{MN}=0.12 eV for $x = 0.009$ to 0.4 eV for $x = 0.023$.

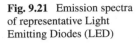

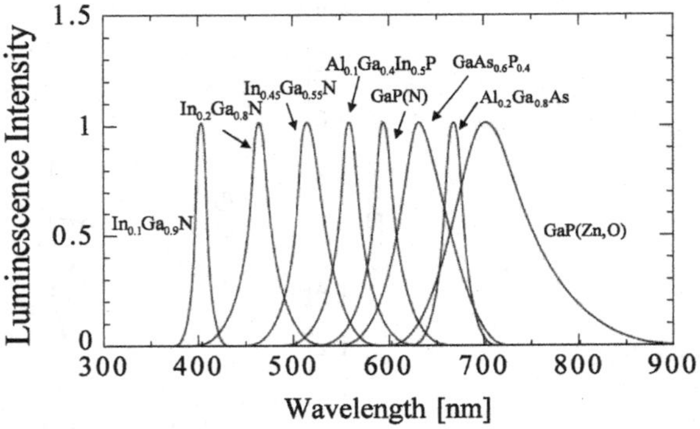

Fig. 9.21 Emission spectra of representative Light Emitting Diodes (LED)

9.4.6 Light Emitting Diodes (LEDs) in Visible Region

Commercially available LEDs cover visible light range using various combinations of compound semiconductors and doping technologies. Also laser diodes (LDs) are produced. The LEDs and LDs consist of pn junctions and the stimulated emission due to the recombination of injected electrons and holes into the junction region play an important role in the laser oscillations. Changing the composition x of $GaAs_{1-x}P_x$ and doping N, Zn, O, and so on, LEDs and LDs of the red light region are fabricated. Here we will show emission spectra of typical light emitting diodes in Fig. 9.21.

9.5 Heterostructure Optical Waveguide

Semiconductor laser oscillations were first observed by Nathan et al. [19], Hall et al. [20], and by Quist et al. [21] in GaAs p–n junctions in 1962. Also visible stimulated emission from $GaAs_{1-x}P_x$ was reported by Holonyak and Bevacqua [22] in 1962. Such laser diodes (Laser Diodes: LDs) were fabricated in the form of pn junctions with bulk materials, where electrons and holes are injected into the pn junction region and electron–hole recombination results in stimulated emission. The spectrum has a very sharp peak much narrower compared with LED and the output power is very high. Later, double heterostructures were proposed by Hayashi et al. [23] using AlGaAs/GaAs/AlGaAs, where the active region of GaAs layer is sandwiched by the barrier layers of AlGaAs. The double heterostructure laser diode is schematically illustrated in Fig. 9.22; (a) the energy band structure under the thermal equilibrium, (b) injection of electrons from n–AlGaAs and holes from p–AlGaAs into the active GaAs region under the forward bias, (c) the distribution of the refractive indices, and (d) the confinement of the emitted light. The double heterostructure confines emitted light in the GaAs layer in addition to the confinement of electrons and holes in the active region. Therefore very high efficiency of laser emission is achieved. Since then double heterostructure laser diodes (DH–LDs) became very popular and

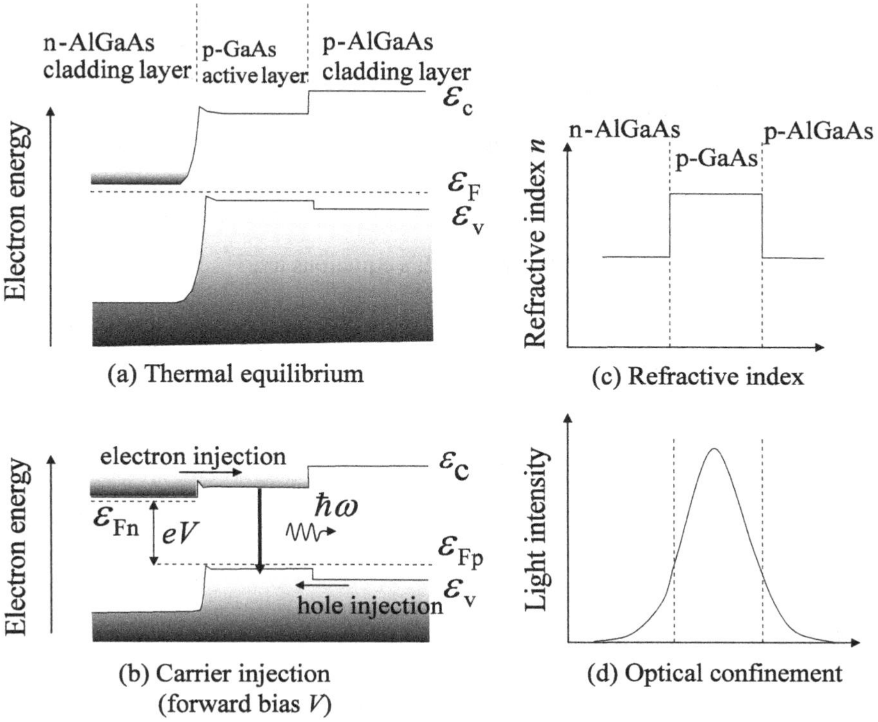

Fig. 9.22 Double heterostructure semiconductor laser diode (DH–LD) of AlGaAs/GaAs/AlGaAs. **a** Thermal equilibrium of double heterostructure, **b** carrier injection under forward bias V, **c** refractive index n in the direction perpendicular to the heterointerface, and **c** confinement of the laser emission

thus all the LDs used in various applications are fabricated in the form of DH–LDs. Another approach was made to decrease the depth of the active region, resulting in a formation of quantum well, which is called quantum well laser diode (QW–LD). Here we deal with optical waveguide, and discuss optical confinement, propagation and gain in double heterostructures.

9.5.1 Wave Equations for Planar Waveguide

In Chap. 4 we derived the optical absorption coefficient by solving Maxwell's equations. In this section we present the analysis of optical wave guide to be applied to laser diodes (see Textbook by Suhara [24] for detailed treatments). First we deal with a sandwiched structure by the planes of different refractive indices, where we neglect the quantum confinement of electrons and holes which will be discussed in the later section. Consider a waveguide shown in Fig. 9.23 with distribution of the refractive indices $n(\boldsymbol{r})$. Dielectric constant is related to refractive index by $\epsilon/\epsilon_0 = n^2$. Let the refractive index of the core (guiding) region $n_{\rm G}$ sandwiched by the upper cladding layer of $n_{\rm UC}$, and the lower cladding layer of $n_{\rm LC}$. The width of the core region is d. In the planar waveguide shown in Fig. 9.23 we may write the dielectric constant as

$$\kappa(z) = [n(z)]^2\,, \tag{9.89}$$

where $n(z)$ is the refractive index. Maxwell's equations are

$$\nabla \times \boldsymbol{H} = \epsilon\frac{\partial \boldsymbol{E}}{\partial t} + \sigma \boldsymbol{E}\,, \tag{9.90a}$$

$$\nabla \times \boldsymbol{E} = -\mu\frac{\partial \boldsymbol{H}}{\partial t}\,, \tag{9.90b}$$

$$\nabla \cdot (\epsilon \boldsymbol{E}) = 0\,, \tag{9.90c}$$

$$\nabla \cdot (\mu \boldsymbol{H}) = 0\,. \tag{9.90d}$$

We assume plane waves for the electric field $\boldsymbol{E}$ and magnetic field $\boldsymbol{H}$ with the time dependence $\exp(-\mathrm{i}\omega t)$, and the wave guide is lossless ($\sigma = 0$). Using the vector formula

$$\nabla \times \nabla \times \boldsymbol{E} = \nabla(\nabla \cdot \boldsymbol{E}) - \nabla^2 \boldsymbol{E}\,, \tag{9.91}$$

the wave equation for the electric field $\boldsymbol{E}$ is written as

$$\nabla^2 \boldsymbol{E} + \omega^2 \epsilon\mu \boldsymbol{E} = \nabla\left(\boldsymbol{E}\frac{\nabla\epsilon}{\epsilon}\right). \tag{9.92}$$

In a similar fashion replacing the electric field $\boldsymbol{E}$ of (9.91) by the magnetic field $\boldsymbol{H}$, and using (9.90a) and (9.90b), the following equation is derived for the wave function of the magnetic field $\boldsymbol{H}$.

$$\nabla^2 \boldsymbol{H} + \omega^2 \epsilon\mu \boldsymbol{H} = \frac{\nabla\epsilon}{\epsilon} \times (\nabla \times \boldsymbol{H})\,. \tag{9.93}$$

For simplicity we assume the refractive index n_i within the core (guiding) and cladding layers is uniform, and then the right hand sides of (9.92) and (9.93) become 0. Therefore we obtain the following wave equations

$$\nabla^2 \boldsymbol{E} + \omega^2 \epsilon\mu \boldsymbol{E} = 0\,, \tag{9.94a}$$

$$\nabla^2 \boldsymbol{H} + \omega^2 \epsilon\mu \boldsymbol{H} = 0\,. \tag{9.94b}$$

Let the propagation direction of the light along the y axis as shown in Fig. 9.23, and electric and magnetic fields propagate along the y axis with the form $\propto \exp[\mathrm{i}(\beta y - \omega t)]$, where β is called as **propagation constant**. Then the wave functions are written as

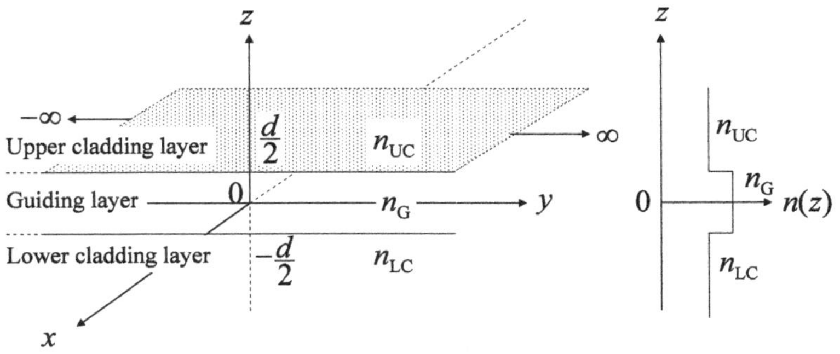

Fig. 9.23 Parallel waveguide structure with step–refractive index used for the analysis of optical waveguide

$$\boldsymbol{E}(x, y, z) = \boldsymbol{E}(x, z) \exp(\mathrm{i}\beta y) \tag{9.95a}$$

$$\boldsymbol{H}(x, y, z) = \boldsymbol{H}(x, z) \exp(\mathrm{i}\beta y) \tag{9.95b}$$

Inserting (9.95a) into (9.94a) and approximating $\nabla\kappa(z) \cong 0$, (in a similar fashion for $\boldsymbol{H}$)

$$\left[\nabla_{\perp}^2 + \left\{k_0^2\kappa(z) - \beta^2\right\}\right] \boldsymbol{E}(x, z) = 0\,. \tag{9.96}$$

Here we defined the wavenumber k_0

$$k_0 = \omega\sqrt{\epsilon_0\mu_0} = \frac{\omega}{c}\,, \tag{9.97}$$

where c is the light velocity in free space.[4] In (9.96) we used $\nabla_{\perp}^2 = \partial^2/\partial x^2 + \partial^2/\partial z^2$. In the present analysis we assume the magnetic permeability is the value of the vacuum, $\mu = \mu_0$. When the wave $\boldsymbol{E}(x, z)$ is finite near the middle of the planes and approaches 0 far from the center ($z \to \infty$ in the case of Fig. 9.23), the waves are confined and thus are called "guided modes". The guided modes consist of finite number of the solutions with discrete propagation constants. On the other hand, when $\boldsymbol{E}(x, z)$ diverges for $z \to \infty$, the waves radiate outward the center of the guide and thus the waves are called "radiation modes". In this textbook we define the propagation constant

$$\beta = \tilde{n}k_0\,, \tag{9.98a}$$

$$n^2 = \epsilon(z)/\epsilon_0 = \kappa(z)\,, \tag{9.98b}$$

where $\tilde{n}$ is called the **effective refractive index** or **mode refractive index**.

Since the electric and magnetic fields in the waveguide shown in Fig. 9.23 are independent in the x direction, we put $\partial/\partial x = 0$, the other components of the

[4] In Chap. 4 we defined the extinction coefficient by k_0 which is the imaginary part of the complex refractive index. Note the difference of k_0 used here from the extinction coefficient.

electric and magnetic waves are obtained by solving Maxwell's equations (9.90a) (putting $\sigma = 0$) and (9.90b) as follows.[5]

$$\frac{\partial H_z}{\partial y} - \frac{\partial H_y}{\partial z} = \mathrm{i}\beta H_z - \frac{\partial H_y}{\partial z} = -\mathrm{i}\omega\epsilon E_x \,, \tag{9.99a}$$

$$\frac{\partial H_x}{\partial z} - \frac{\partial H_z}{\partial x} = \frac{\partial H_x}{\partial z} = -\mathrm{i}\omega\epsilon E_y \,, \tag{9.99b}$$

$$\frac{\partial H_y}{\partial x} - \frac{\partial H_x}{\partial y} = -\frac{\partial H_x}{\partial y} = -\mathrm{i}\beta H_x = -\mathrm{i}\omega\epsilon E_z \,, \tag{9.99c}$$

$$\frac{\partial E_z}{\partial y} - \frac{\partial E_y}{\partial z} = \mathrm{i}\beta E_z - \frac{\partial E_y}{\partial z} = \mathrm{i}\omega\mu_0 H_x \,, \tag{9.99d}$$

$$\frac{\partial E_x}{\partial z} - \frac{\partial E_z}{\partial x} = \frac{\partial E_x}{\partial z} = \mathrm{i}\omega\mu_0 H_y \,, \tag{9.99e}$$

$$\frac{\partial E_y}{\partial x} - \frac{\partial E_x}{\partial y} = -\frac{\partial E_x}{\partial y} = -\mathrm{i}\beta E_x = \mathrm{i}\omega\mu_0 H_z \,, \tag{9.99f}$$

We find from (9.99a) $\sim$ (9.99f) the two solutions as follows.

9.5.1.1 (1) TE Modes

Putting $H_x = 0$ in (9.99b) and (9.99c), we find $E_y = E_z = 0$ and thus H_y and H_z are written as

$$H_z = -\frac{\beta}{\omega\mu_0} E_x \,, \qquad H_y = -\frac{\mathrm{i}}{\omega\mu_0}\frac{\partial E_x}{\partial z} \tag{9.100}$$

Since the electric filed vector has the component E_x only and is perpendicular to the propagation direction y, the electromagnetic waves are called **transverse electric modes** (TE modes). Inserting (9.100) into (9.99a) we obtain the wave equation for the electric field

$$\frac{\partial^2 E_x}{\partial z^2} + [k_0^2 n(z)^2 - \beta^2] E_x = 0 \,. \tag{9.101}$$

TE modes are well expressed by

$$\text{TE modes: } \boldsymbol{E} = (E_x, 0, 0), \ \boldsymbol{H} = (0, H_y, H_z) \,.$$

---

[5]Equation (9.90a) may be rewritten in vector components as

$$\left(\boldsymbol{i}\frac{\partial}{\partial x} + \boldsymbol{j}\frac{\partial}{\partial y} + \boldsymbol{k}\frac{\partial}{\partial z}\right) \times (\boldsymbol{i}H_x + \boldsymbol{j}H_y + \boldsymbol{k}H_z) = \epsilon\frac{\partial}{\partial t}(\boldsymbol{i}E_x + \boldsymbol{j}E_y + \boldsymbol{k}H_z)$$

and similarly for (9.90b). Putting $\partial/\partial x = 0$ and $\partial/\partial y = \mathrm{i}\beta$, and equating the same components of the left and right hands we obtain (9.99a) $\sim$ (9.99f).

9.5.1.2 (2) TM Modes

On the other hand, when we put $E_x = 0$ in (9.99e) and (9.99f), we obtain $H_y = H_z = 0$, giving rise to the following results

$$E_z = \frac{\beta}{\omega\epsilon} H_x\,, \qquad E_y = \frac{\mathrm{i}}{\omega\epsilon}\frac{\partial H_x}{\partial z}\,. \tag{9.102}$$

Since the magnetic field vector has the component H_x only and is perpendicular to the propagation direction y, the electromagnetic waves are called **transverse magnetic modes** (TM modes). The wave equation for the magnetic field H_y is obtained by inserting (9.102) into (9.99d)

$$n(z)^2 \frac{\partial}{\partial z}\left(n(z)^{-2}\frac{\partial H_x}{\partial x}\right) + \left[k_0^2 n(z)^2 - \beta^2\right] H_x = 0\,. \tag{9.103}$$

TM modes are well expressed by

$$\text{TM modes: } \boldsymbol{E} = (0, E_y, E_z),\ \boldsymbol{H} = (H_x, 0, 0)\,.$$

In the following we will discuss wave propagation of TE and TM modes in the waveguide shown in Fig. 9.23. The refractive index distribution $n(z)$ in Fig. 9.23 as

$$n(z) = \begin{cases} n_{\mathrm{UC}} & (z > d/2) \\ n_{\mathrm{G}} & (-d/2 < z < d/2) \qquad n_{\mathrm{G}} > n_{\mathrm{UC}},\, n_{\mathrm{LC}} \\ n_{\mathrm{LC}} & (z < -d/2) \end{cases} \tag{9.104}$$

When the refractive indices are uniform in each region, the wave equations (9.101) for TE modes and (9.103) for TM modes are the same in the form. We define the following coefficients using $\beta = \tilde{n}k_0$,

$$\gamma_{\mathrm{UC}} = k_0(\tilde{n}^2 - n_{\mathrm{UC}}^2)^{1/2} \tag{9.105a}$$
$$\kappa_{\mathrm{G}} = k_0(n_{\mathrm{G}}^2 - \tilde{n}^2)^{1/2} \tag{9.105b}$$
$$\gamma_{\mathrm{LC}} = k_0(\tilde{n}^2 - n_{\mathrm{UL}}^2)^{1/2} \tag{9.105c}$$

where $\tilde{n}$ is **effective refractive index** and determined from the boundary conditions as discussed later.

Since we are interested in the wave functions confined mostly in the guiding layer and decays in the cladding layers at $x \to \pm\infty$, the solutions of the wave equations (9.101) and (9.103) may be $E_x(z),\ H_x(z) \propto \exp(\pm \mathrm{i}\kappa z)$ ($\propto \cos(z)$ or $\propto \sin(z)$) in the guiding layer and $\propto \exp(\pm\gamma z)$ in the cladding layers. For such guiding modes, κ and γ should be real and thus the following relation holds for the refractive indices,

$$n_{\mathrm{UC}},\, n_{\mathrm{LC}} < \tilde{n} < n_{\mathrm{G}}\,. \tag{9.106}$$

9.5.2 Transverse Electric Modes

Using the wave equations derived above, we drive the solutions of the guided TE modes,

$$E_x(z) = \begin{cases} E_{\rm UC} \exp\left[-\gamma_{\rm UC}\left(z - \dfrac{d}{2}\right)\right] & (z > d/2) \\ E_{\rm G} \cos\left[\kappa_{\rm G}\left(z - \dfrac{d}{2}\right) + \Phi_{\rm UC}\right] & (-d/2 < z < d/2) \\ E_{\rm LC} \exp\left[+\gamma_{\rm LC}\left(z + \dfrac{d}{2}\right)\right] & (z < -d/2) \end{cases} \tag{9.107}$$

The other components H_y and H_z are obtained from (9.100). From the boundary conditions and continuities of E_x and $\mathrm{d}E_x/\mathrm{d}z$ we may correlate the coefficients introduced in (9.107). The boundary conditions at $z = \pm d/2$ give

$$\kappa_{\rm G} d - \Phi_{\rm UC} - \Phi_{\rm LC} = m\pi \quad (m = 0,\ 1,\ 2,\ \ldots), \tag{9.108a}$$

$$\Phi_{\rm UC} = \tan^{-1}\left(\frac{\gamma_{\rm UC}}{\kappa_{\rm G}}\right), \quad \Phi_{\rm LC} = \tan^{-1}\left(\frac{\gamma_{\rm LC}}{\kappa_{\rm G}}\right), \tag{9.108b}$$

$$E_{\rm UC} = E_{\rm G} \cos(\Phi_{\rm UC}), \quad E_{\rm LC} = (-1)^m E_{\rm G} \cos(\Phi_{LC}). \tag{9.108c}$$

The power flow in the y direction P_y per unit width along the x direction is calculated from the real part of Poynting vector $\boldsymbol{P} = (1/2)\boldsymbol{E} \times \boldsymbol{H}^*$ as

$$P_y = \frac{\beta}{2\omega\mu_0} \int E_x^2(z)\mathrm{d}z = \frac{\tilde{n}}{4}\left(\frac{\epsilon_0}{\mu_0}\right)^{1/2} E_{\rm G}^2 d_{\rm eff}, \tag{9.109}$$

where

$$d_{\rm eff} = d + \frac{1}{\gamma_{\rm UC}} + \frac{1}{\gamma_{\rm LC}}. \tag{9.110}$$

In the case of symmetric waveguides with $n_{\rm UC} = n_{\rm LC} = n_{\rm C}$ we obtain the following solutions from (9.107)

$$E_x(z) = \begin{cases} E_{\rm G} \cos(\Phi_{\rm C}) \exp\left[-\gamma_{\rm C}\left(z - \dfrac{d}{2}\right)\right] & (z > d/2) \\ E_{\rm G} \cos\left(\kappa_{\rm G} z - \dfrac{m\pi}{2}\right) & (|z| < d/2) \\ (-1)^m E_{\rm G} \cos(\Phi_{\rm C}) \exp\left[+\gamma_{\rm C}\left(z + \dfrac{d}{2}\right)\right] & (z < -d/2) \end{cases} \tag{9.111}$$

where

$$\kappa_G d - 2\Phi_C = m\pi\,, \qquad \Phi_C = \tan^{-1}\left(\frac{\gamma_C}{\kappa_G}\right), \tag{9.112}$$

and note here that $E_x(z)$ is symmetric with respect to z for even m and antisymmetric for odd m. Therefore the wave functions are call **even modes** for even number $m = 0$, 2, 4, ..., and **odd modes** for odd number $m = 1, 3, 5, \ldots$. These waves are named as TE_0, TE_1, TE_2, ...

9.5.3 Transverse Magnetic Modes

In a similar fashion the wave equation for the transverse magnetic modes is solved, and we obtain expressions quite similar to the transverse electric modes.

$$H_x(z) = \begin{cases} H_{UC}\exp\left[-\gamma_{UC}\left(z - \frac{d}{2}\right)\right] & (z > d/2) \\ H_G\cos\left[\kappa_G\left(z - \frac{d}{2}\right) + \Phi_{UC}\right] & (-d/2 < z < d/2) \\ H_{LC}\exp\left[+\gamma_{LC}\left(z + \frac{d}{2}\right)\right] & (z < -d/2) \end{cases} \tag{9.113}$$

The other components E_y and E_z are obtained from (9.102). From the boundary conditions and continuities of $H_x(z)$ and $n^{-2}\mathrm{d}H_x/\mathrm{d}z$ we may correlate the coefficients introduced in (9.113). The boundary conditions at $z = \pm d/2$ give

$$\kappa_G d - \Phi_{UC} - \Phi_{LC} = m\pi \quad (m = 0,\ 1,\ 2,\ \ldots)\,, \tag{9.114}$$

$$\Phi_{UC} = \tan^{-1}\left[\left(\frac{n_G}{n_{UC}}\right)^2\left(\frac{\gamma_{UC}}{\kappa_G}\right)\right], \tag{9.115}$$

$$\Phi_{LC} = \tan^{-1}\left[\left(\frac{n_G}{n_{LC}}\right)^2\left(\frac{\gamma_{LC}}{\kappa_G}\right)\right], \tag{9.116}$$

$$H_{UC} = H_G\cos(\Phi_{UC}), \qquad H_{LC} = (-1)^m H_G\cos(\Phi_{LC})\,. \tag{9.117}$$

The power flow in the y direction P_y per unit width along the x direction is

$$P_y = \frac{2\beta}{\omega\epsilon_0}\int\frac{H_x^2(z)}{n^2}\mathrm{d}z = \frac{\tilde{n}}{4}\left(\frac{\mu_0}{\epsilon_0}\right)^{1/2} H_G^2 d_{\mathrm{eff}}\,, \tag{9.118}$$

where

$$d_{\text{eff}} = d + \frac{1}{\gamma_{\text{UC}} q_{\text{UC}}} + \frac{1}{\gamma_{\text{LC}} q_{\text{LC}}}, \tag{9.119a}$$

$$q_{\text{UC}} = \left(\frac{\tilde{n}}{n_{\text{G}}}\right)^2 + \left(\frac{\tilde{n}}{n_{\text{UC}}}\right)^2 - 1, \tag{9.119b}$$

$$q_{\text{LC}} = \left(\frac{\tilde{n}}{n_{\text{G}}}\right)^2 + \left(\frac{\tilde{n}}{n_{\text{LC}}}\right)^2 - 1. \tag{9.119c}$$

In the case of symmetric waveguides with $n_{\text{UC}} = n_{\text{LC}} = n_{\text{C}}$ we obtain the following solutions from (9.113)

$$H_x(z) = \begin{cases} H_{\text{G}} \cos(\Phi_{\text{C}}) \exp\left[-\gamma_{\text{C}}\left(z - \frac{d}{2}\right)\right] & (z > d/2) \\ H_{\text{G}} \cos\left(\kappa_{\text{G}} z - \frac{m\pi}{2}\right) & (|z| < d/2) \\ (-1)^m H_{\text{G}} \cos(\Phi_{\text{C}}) \exp\left[+\gamma_{\text{C}}\left(z + \frac{d}{2}\right)\right] & (z < -d/2) \end{cases} \tag{9.120}$$

where

$$\kappa_{\text{G}} d - 2\Phi_{\text{C}} = m\pi, \quad \Phi_{\text{C}} = \tan^{-1}\left[\left(\frac{n_{\text{G}}}{n_{\text{C}}}\right)^2 \left(\frac{\gamma_{\text{C}}}{\kappa_{\text{G}}}\right)\right], \tag{9.121}$$

The electric field distributions of TE modes calculated from (9.111) are plotted in Fig. 9.24, where we put the refractive indices as $n_{\text{G}} = 3.5$ and $n_{\text{C}} = 3.3$. Here we have to note that the effective refractive index $\tilde{n}$ depends on the mode number and thus we have to determine it by solving (9.112), and we find $\tilde{n} = 3.47808, 3.41418$, and 3.32125 for the modes $m = 0$, 1, and 2. In the figure we find that the fields

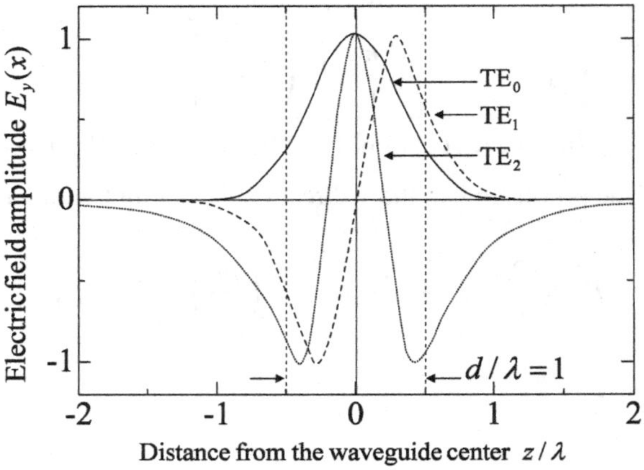

Fig. 9.24 Electric field distribution of the TE mode in a parallel waveguide structure with step–refractive index $n_{\text{G}} = 3.5$ and $n_{\text{C}} = 3.3$. The effective refractive index is evaluated to be $\tilde{n} = 3.47808, 3.41418$, and 3.32125 for TE_0, TE_1, and TE_2. (see text)

are confined in the guiding layer and decay exponentially for $|x| > d/2$ and that m represents the number of nodes.

9.5.4 Effective Refractive Index

Here we will show how to obtain the effective refractive index $\tilde{n}$ numerically. First, we define

$$V_{\mathrm{d}} = k_0 d (n_{\mathrm{G}}^2 - n_{\mathrm{LC}}^2)^{1/2} = \frac{\omega}{c} d (n_{\mathrm{G}}^2 - n_{\mathrm{LC}}^2)^{1/2} \tag{9.122}$$

where V_{d} is called the normalized frequency. The degree of asymmetry of the upper and lower cladding layers is defined by

$$a_{\mathrm{TE}} = \frac{n_{\mathrm{LC}}^2 - n_{\mathrm{UC}}^2}{n_{\mathrm{G}}^2 - n_{\mathrm{LC}}^2} . \tag{9.123}$$

We define another parameter for the effective refractive index

$$b = \frac{\tilde{n}^2 - n_{\mathrm{UC}}^2}{n_{\mathrm{G}}^2 - n_{\mathrm{LC}}^2} . \tag{9.124}$$

Using these parameters we rewrite (9.108a) $\sim$ (9.108c) as

$$V(1-b)^{1/2} - \tan^{-1}\left(\frac{b}{1-b}\right)^{1/2} - \tan^{-1}\left(\frac{a+b}{1-b}\right)^{1/2} = m\pi , \tag{9.125}$$

where V_{d} and a_{TE} are simplified as V and a.

In a symmetric waveguide putting $n_{\mathrm{UC}} = n_{\mathrm{LC}} = n_{\mathrm{C}}$ and thus $a = 0$, we obtain the following relation.

$$V(1-b)^{1/2} - 2\tan^{-1}\left(\frac{b}{1-b}\right)^{1/2} = m\pi , \tag{9.126}$$

or

$$\tan\left(\frac{1}{2}V\sqrt{1-b} - m\frac{\pi}{2}\right) = \frac{\sqrt{b}}{\sqrt{1-b}} . \tag{9.127}$$

Equations (9.126) and (9.127) are transcendental and give no analytical solutions. We present numerical solutions in Fig. 9.25. As shown in Fig. 9.25 the number of guided modes increases with increasing the normalized frequency V or the guiding layer thickness d. In the region of $V < \pi$ only the lowest mode $m = 0$ is excited

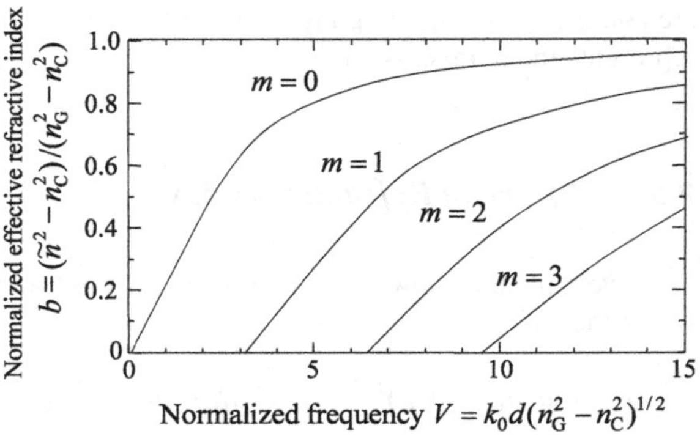

Fig. 9.25 Normalized effective refractive index $b = (\tilde{n}^2 - n_C^2)/(n_G^2 - n_C^2)$ is plotted as a function of normalized frequency $V = k_0 d\sqrt{n_G^2 - n_C^2}$ for $m = 0, 1, 2$, and 3, in a planar waveguide

and the mode is called the fundamental mode. Such a waveguide is referred to as a single–mode waveguide. In the region $V > \pi$ higher modes $m = 1, 2, 3, \ldots$ are excited and such a waveguide is called a multimode waveguide. The effective refractive index $\tilde{n}$ of each mode increases monotonically with increasing V. On the other hand when V is decreased, b approaches 0 and we find $\tilde{n} = n_C$, resulting in no mode confinement. Therefore the electromagnetic field propagate uniformly without decaying. This condition is called the cutoff. The cutoff frequency of the mth mode is given by

$$V_m^{\text{cutoff}} = m\pi \tan^{-1}(\sqrt{a})\,. \tag{9.128}$$

In a symmetrical waveguide we have $a = 0$ and $V_0^{\text{cutoff}} = 0$. From these considerations the single–mode waveguide is achieved under the condition

$$\tan^{-1}\sqrt{a} < V < \pi + \tan^{-1}\sqrt{a}\,, \tag{9.129}$$

or in a symmetrical waveguide $0 < V < \pi$.

In the case of $\tilde{n} < n_C$, we find $\gamma_C = ik_0\sqrt{n_C^2 - \tilde{n}^2}$ from (9.105a), and thus the electric field in the clad layers is given by $\propto \exp[i|\gamma_C|(|x| - d/2)]$. Such electromagnetic waves propagate in the clad layers without decaying, and are called **radiative modes** with the propagation coefficient $|\beta|$ continuous in the range $|\beta| < n_C k_0$.

9.5.5 *Confinement Factor*

The electromagnetic field of the guided mode is distributed not only in the guiding (active) layer but also in the cladding layers and thus the gain of the guided mode strongly depend on the power distribution in the guided layer. Here we introduce the **confinement factor**, which is defined as the ratio of the electromagnetic wave power flowing in the guiding layer to the total power of the guided mode. When the confinement factor is small, enough power of the laser oscillations is not achieved and thus higher threshold current is required for the laser oscillations. The mode confinement factor Γ in a planar wave guide is expressed as

$$\Gamma = \frac{\displaystyle\int_{-d/2}^{d/2} |E_x(z)|^2 \mathrm{d}z}{\displaystyle\int_{-\infty}^{+\infty} |E_x(z)|^2 \mathrm{d}z} = \frac{\displaystyle\int_{0}^{d/2} |E_x(z)|^2 \mathrm{d}z}{\displaystyle\int_{0}^{+\infty} |E_x(z)|^2 \mathrm{d}z} . \tag{9.130}$$

Inserting (9.111) into (9.130) the confinement factor in a planar waveguide is evaluated as

$$\Gamma = \frac{1 + 2\gamma_c d / V_d^2}{1 + 2/\gamma_c d} = \frac{V_d + 2\sqrt{b}}{V_d + 2/\sqrt{b}} . \tag{9.131}$$

Since the confinement factor Γ is expressed by the normalized frequency V and the normalized effective refractive index b as shown in (9.131), V *vs* b curve is uniquely determined. Figure 9.26 shows the confinement factor Γ as a function of the normalized frequency for the fundamental mode TE_0 ($m = 0$) in a planar waveguide, where b is calculated from (9.126) or (9.127) and substituted in (9.131). Once we know materials of a waveguide, then using the refractive indices of the cladding and guiding layers, and the thickness of the guiding layer d, b and V are estimated from Fig. 9.25 and finally we obtain the confinement factor Γ. The calculated confinement factor in double–heterostructures of $\mathrm{Al}_x\mathrm{Ga}_{1-x}\mathrm{As/GaAs/Al}_x\mathrm{Ga}_{1-x}\mathrm{As}$ is shown in Fig. 9.27, where the confinement factor Γ is plotted as a function of the active layer thickness d with the composition x as a parameter [25].

From the above analysis the confinement factor increases with increasing the active layer thickness d and the optical power is effectively confined in the active (guiding) layer, but the threshold current density for the laser oscillation increases linearly with the active layer thickness. Decreasing the active layer thickness, on

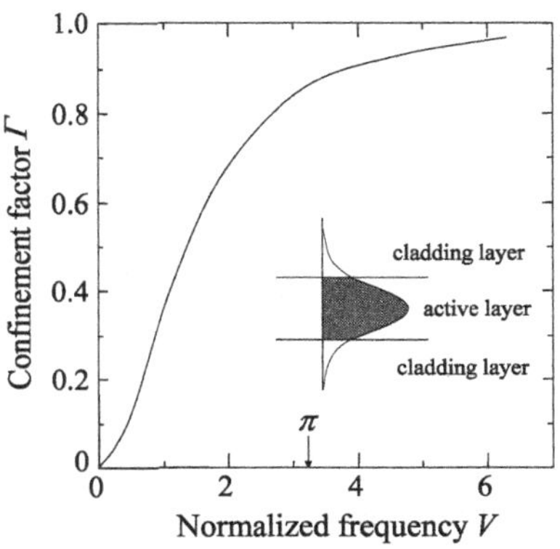

Fig. 9.26 Confine factor Γ as a function of the normalized frequency V in a planar waveguide, where the results for the fundamental mode are shown

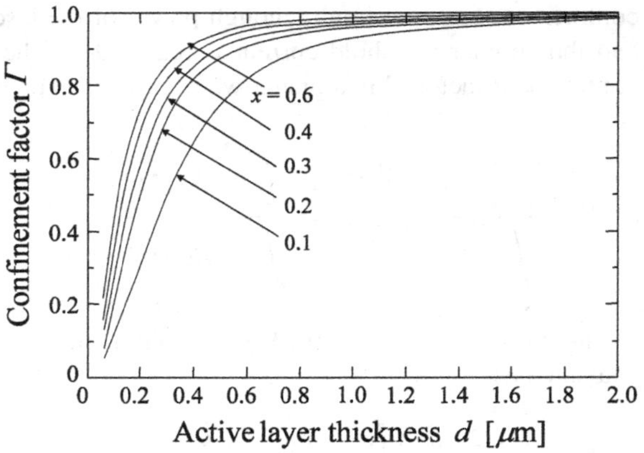

Fig. 9.27 The confinement factor Γ as a function of the active layer thickness d in $Al_xGa_{1-x}As/GaAs/Al_xGa_{1-x}As$ double–heterostructure laser $GaAs/Al_xGa_{1-x}As$. (After Casey [25])

the other hand, optical power is not effectively confined and the threshold current density increases. Therefore the minimum threshold current density is estimated to be $d \cong 0.2 \sim 0.3\,\mu m$.

9.5.6 *Laser Oscillations*

A schematic structure of semiconductor lasers fabricated on semiconductor substrates is shown in Fig. 9.28, where the Fabry–Perot resonator consists of the cleaved parallel surfaces with the resonator length L. For simplicity we assume the reflectivity of the left and right mirrors is R. We define the complex propagation constant $\tilde{\beta}$ by

$$\tilde{\beta} = \beta + i\frac{\alpha}{2}, \quad \beta = \tilde{n}k_0 = \tilde{n}\frac{\omega}{c}, \tag{9.132}$$

where $\alpha\,(> 0)$ is the internal loss due to the free carrier absorption and the internal reflection, and $\alpha\,(< 0)$ is the gain given by $\alpha = -\Gamma g$. The optical wave in the

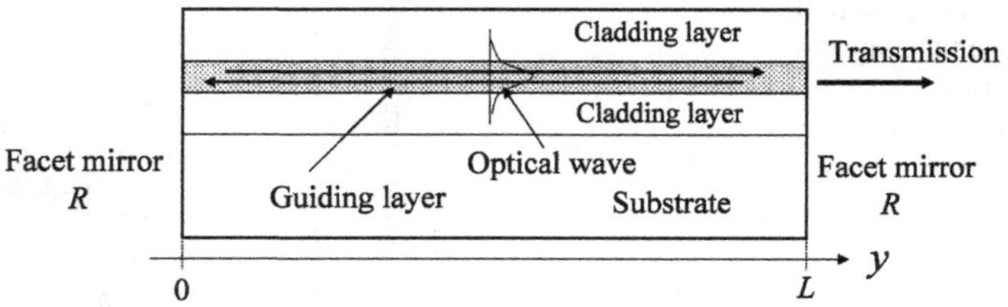

Fig. 9.28 Fabry–Perot resonator of semiconductor laser used for the analysis of laser oscillations. The reflectivity is assumed to be the same for the *right* and *left* mirrors

waveguide resonator may be written as

$$\boldsymbol{E}(x, y, z) = \boldsymbol{E}(x, z)\left[A^{+}\exp(+\mathrm{i}\tilde{\beta}y) + A^{-}\exp(+\mathrm{i}\tilde{\beta}y)\right], \tag{9.133}$$

where the first and second terms in the square brackets represent the forward and backward waves, respectively. Putting R be the power reflectivity of the mirrors, the amplitude A_{in} of the wave incident from the left–hand side of the resonator and A_{tr} of the wave transmitted into the right–hand side are given by the following relations from the boundary conditions for both facets,

$$A^{+} = (1-R)^{1/2}A_{\text{in}} + R^{1/2}A^{-}, \tag{9.134a}$$

$$A^{-}\exp(-\mathrm{i}\tilde{\beta}L) = R^{1/2}A^{+}\exp(+\mathrm{i}\tilde{\beta}L), \tag{9.134b}$$

$$A_{\text{tr}} = (1-R)^{1/2}A^{+}\exp(+\mathrm{i}\tilde{\beta}L). \tag{9.134c}$$

Using these relations the power transmissivity T of the resonator is given by

$$\begin{aligned} T &= \frac{|A_{\text{tr}}|^2}{|A_{\text{in}}|^2} \\ &= \frac{(1-R)^2\exp(-\alpha L)}{1 + R^2\exp(-2\alpha L) - 2R\exp(-\alpha L)\cos(2\beta L)}. \end{aligned} \tag{9.135}$$

When the optical wave starts to propagate at the left–hand mirror $z = 0$, the power transmissivity T reduces to

$$T = \frac{(1-R)\exp(-\alpha L)}{1 + R^2\exp(-2\alpha L) - 2R\exp(-\alpha L)\cos(2\beta L)}. \tag{9.136}$$

The transmissivity exhibits maximum at $2\beta L = 2m\pi$ (m is an integer). The maximum condition is caused by the successive mirror reflections and the round trips in the waveguide are superimposed in phase. Under this condition the amplitudes of the wave increase, resulting in resonance. The calculated transmissivity T as a function of the optical frequency is shown in Fig. 9.29. From (9.132) we find $\delta(2\beta L) = 2L(\partial\beta/\partial\omega)\delta\omega$, and thus the separation of angular frequency $\delta\omega$ between the adjacent peaks is given by

$$\delta\omega = \frac{2\pi}{2L\partial\beta/\partial\omega} = 2\pi\frac{v_{\text{g}}}{2L} = 2\pi\frac{c}{2Ln_{\text{g}}} \tag{9.137a}$$

$$v_{\text{g}} = \frac{1}{\partial\beta/\partial\omega} = \frac{c}{n_{\text{g}}}, \quad n_{\text{g}} = \tilde{n} + \omega\frac{\partial\tilde{n}}{\partial\omega}, \tag{9.137b}$$

where v_{g} is the group velocity, at which the optical power propagates in the waveguide, and n_{g} is called the group index of refraction.

Let's consider the conditions for laser oscillation using Fig. 9.30. In a real semiconductor laser the optical power is not constant, but the optical waves are amplified during the propagation back and forth in the resonator. When the gain factor of the active region is given by g and the confinement factor by Γ, the effective gain factor is given by Γg. In a waveguide we have to take account of the propagation loss α_{int}

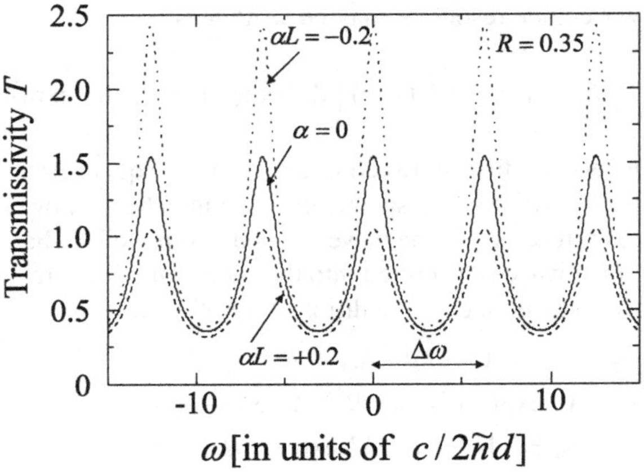

Fig. 9.29 Transmissivity of a Fabry–Perot resonator as a function of optical frequency calculated from (9.136), with the reflectivity $R = 0.35$ and the internal loss of $\alpha L = -0.2,\ 0,\ 0.2$. The negative value of α corresponds to a positive gain (amplification)

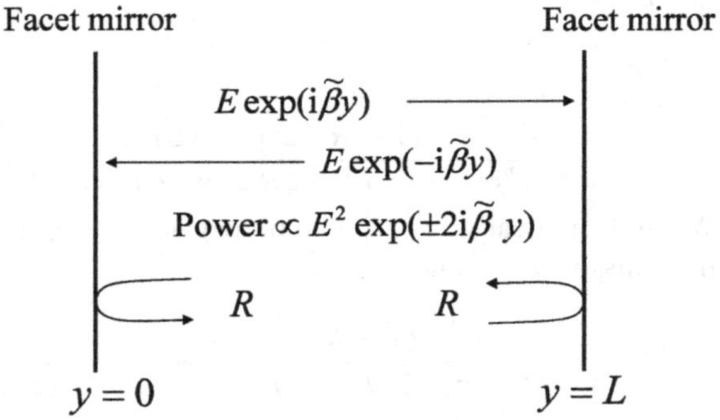

Fig. 9.30 Fabry–Perot cavity to calculate the threshold of the laser oscillations. The cavity length is L and the reflectivity is R. Optical waves after a round trip should be in phase and the power amplitude is equal to or larger than the initial wave

due to absorption and scattering. Then the net gain factor is given by $g_{\mathrm{net}} = \Gamma g - \alpha_{\mathrm{int}}$. Defining $\tilde{\beta} = \beta + \mathrm{i}g_{\mathrm{net}}/2$, the optical amplification occurs when the amplitude after a round trip is equal to or greater than the initial amplitude

$$R\exp[2\mathrm{i}\tilde{\beta}L] \cdot R\exp[2\tilde{\beta}L] \geq 1\,, \tag{9.138a}$$

and thus we find the following relation

$$\exp[2\mathrm{i}\tilde{\beta}L] \geq \frac{1}{R}\,. \tag{9.138b}$$

The above equations may be rewritten as

$$\exp[2\mathrm{i}\tilde{\beta}L] = \exp\left[2\mathrm{i}\beta L - g_{\mathrm{net}}L\right] \geq \frac{1}{R} \tag{9.139}$$

This equation may be rearranged by separating the real and imaginary components as

$$\exp[g_{\text{net}}L] \geq \frac{1}{R}, \qquad \Gamma g \geq \alpha_{\text{int}} + \frac{1}{L}\ln\frac{1}{R}, \tag{9.140a}$$

$$2\beta(\omega)L = 2\pi m \ (m : \text{integer}). \tag{9.140b}$$

From (9.140a) we obtain

$$(\Gamma g)_{\text{th}} = \alpha_{\text{int}} + \frac{1}{L}\ln\left(\frac{1}{R}\right). \tag{9.141}$$

Since the gain factor g is proportional to the injected current density, the relation of (9.141) well explains the experimental results on the threshold current *vs* the inverse cavity length $1/L$ given by

$$\eta J_{\text{th}} \geq \alpha_{\text{int}} + \frac{1}{L}\ln\frac{1}{R}, \tag{9.142}$$

where η is a constant.

Equation (9.140b) gives the same condition we discussed below (9.136). In other words, the optical wave after a round trip propagation (distance is $2L$) is in phase with the original wave. The frequencies ω_m satisfying the condition of (9.140b) are called **longitudinal mode frequencies**. The frequency separation $\delta\omega$ is quite small compared to the frequency of the laser oscillations. For example assuming the laser wave length $\lambda = 750$ [nm], the cavity length =1[mm] and the refractive index $\tilde{n} \cong n_{\text{g}} \cong 3.5$, the separation of the emission peaks is given by

$$\frac{\delta\omega}{\omega} = \frac{2\pi c/2Ln_{\text{g}}}{2\pi c/\lambda} = \frac{\lambda}{2Ln_{\text{g}}} \cong 10^{-4}. \tag{9.143}$$

Thus a high resolution monochromator is required to resolve the multiple peaks.

The threshold of the laser oscillations is well illustrated in Fig. 9.31, where the gain curves are schematically drawn using the curves in Figs. 9.4 and 9.5. When the injection current is increased, the mode gain $\Gamma g(\omega_m)$ at the longitudinal mode

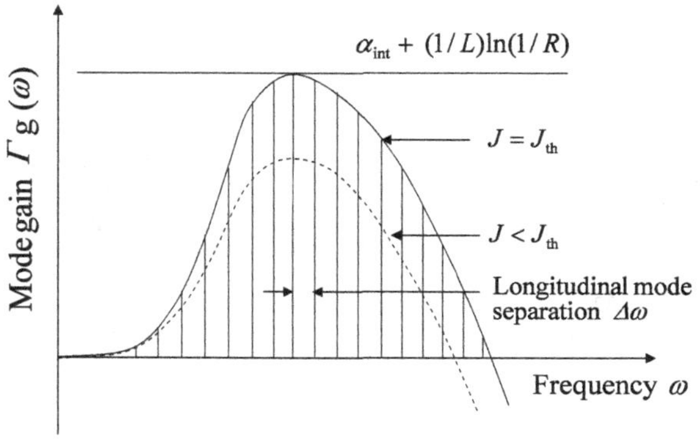

Fig. 9.31 Schematic illustration of the threshold condition for the laser oscillations in a Fabry–Perot semiconductor laser, where the longitudinal mode gain $\Gamma g(\omega)$ is plotted together with the longitudinal mode frequencies ω_m. The horizontal line shows the sum of the internal loss and the reflection loss given by (9.140a)

frequency reach the value of the right–hand side in (9.140a) and the condition of (9.141) and thus (9.142) is fulfilled, resulting in the start of laser oscillations.

9.6 Stimulated Emission in Quantum Well Structures

In the previous Sect. 9.5 (Sect. 9.5.5) we described the confinement of the optical wave in a waveguide by utilizing the difference in the refractive indices. This result reminds us that confinement of the electrons and holes in the same region will produce a higher efficiency of stimulated emission. When the well width of a double heterostructure is reduced, the electrons and holes will be quantized to produce two dimensional carriers as discussed in Chap. 8 (Sect. 8.2). Figure 9.32 shows a schematic structure of a single quantum well laser AlGaAs/GaAs/AlGaAs, where the lowest subband energy of the electrons in the well of conduction band and the highest subband energy of the holes in the well of the valence band are shown by the dashed lines.

Electrons in the subband recombine with the holes in the subband and emit photons of energy $\hbar\omega$. Quantized wave functions of electrons are easily calculated by solving Schrödinger equation, but the wave functions of holes in the valence bands are not obtained by a simple calculation, which will be discussed later. In order to evaluate the subband energies and wave functions (envelop functions) we have to determine the band discontinuities in the conduction and valence bands. The band discontinuities are determined from the difference between the band gaps of the barrier region $\mathcal{E}_{\mathrm{GB}}$ and the well region $\mathcal{E}_{\mathrm{GW}}$, $\delta\mathcal{E}_{\mathrm{G}} = \mathcal{E}_{\mathrm{BG}} - \mathcal{E}_{\mathrm{GW}}$, and the allocation of it on the conduction band and the valence band. The allocation ratio depends on the materials and a variety of the ratios are reported so far. For example, the band offset parameters

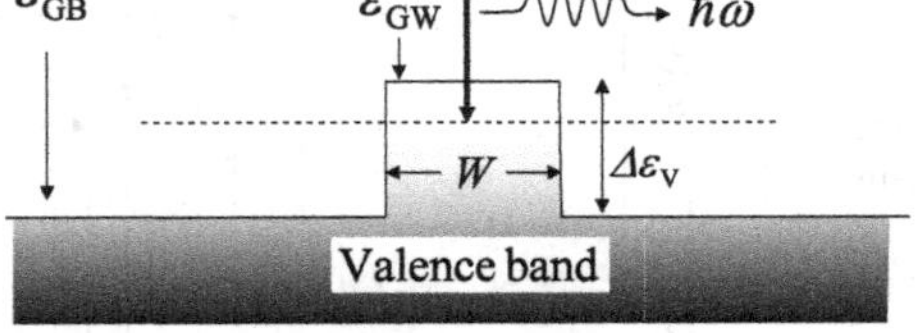

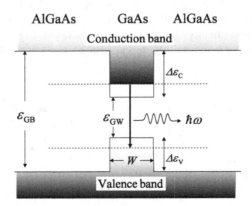

Fig. 9.32 Schematic illustration of AlGaAs/GaAs/AlGaAs single quantum well laser. The lowest subband of the conduction band and the highest subband of the valence band are considered here. The two dimensional electron gas recombines with the two dimensional hole gas to emit photon $\hbar\omega$ with high efficiency. The complicated subband structures of hole states is not shown here, which will be discussed in the text

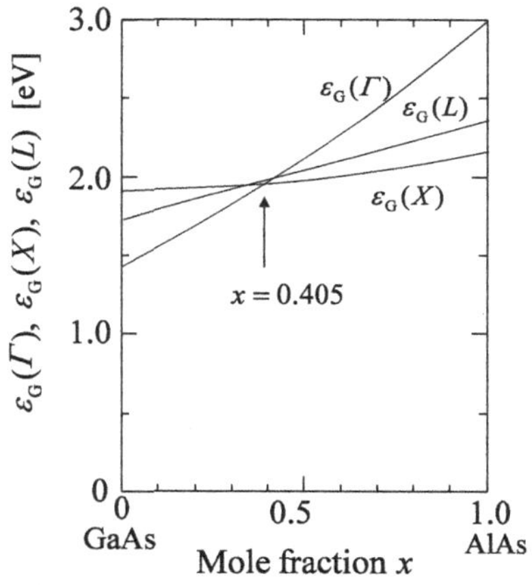

Fig. 9.33 Band gaps at Γ, X, and L point of $Al_xGa_{1-x}As$ at room temperature ($T = 295$K) are plotted as a function of the mole fraction x. Band crossing of the *Gamma* and X points occurs at $x = 0.405$. The data are from Lee et al. [26]

$Q_c = \delta\mathcal{E}_c/\delta\mathcal{E}_G$ are reported $Q_c = 0.661 \pm 0.015$ [27], 0.62 [28], 0.65 [29], and so on. Here we use the following parameters in the present analysis,

$$\Delta\mathcal{E}_c/\Delta\mathcal{E}_G = 0.67 \pm 0.01\,, \tag{9.144a}$$

$$\Delta\mathcal{E}_v/\Delta\mathcal{E}_G = 0.33 \pm 0.01\,, \tag{9.144b}$$

for GaAs/AlGaAs heterostructures [24, 30], and

$$\Delta\mathcal{E}_c/\Delta\mathcal{E}_G = 0.39 \pm 0.01\,, \tag{9.145a}$$

$$\Delta\mathcal{E}_v/\Delta\mathcal{E}_G = 0.61 \pm 0.01\,, \tag{9.145b}$$

for InP/InGaAsP heterostructures [24, 31].

The energies of the band gaps at Γ, X, and L points in $Al_xGa_{1-x}As$ are plotted as a function of the mole fraction in Fig. 9.33. The data are from Lee et al. [26]. The analytical expressions are shown by the following relations.

$$\mathcal{E}_G(\Gamma) = 1.425 + (2.980 - 1.425)x - 0.37x(1 - x)\,, \tag{9.146a}$$

$$\mathcal{E}_G(X) = 1.911 + (2.161 - 1.911)x - 0.245x(1 - x)\,, \tag{9.146b}$$

$$\mathcal{E}_G(L) = 1.734 + (2.363 - 1.734)x - 0.055x(1 - x)\,. \tag{9.146c}$$

Using these data we may obtain $\Delta\mathcal{E}_G$ for $Al_xGa_{1-x}As$ ($x = 0.3$) = 0.3888 eV, which gives us $\Delta\mathcal{E}_c = 0.2605$ and $\Delta\mathcal{E}_v = 0.1283$ eV at $T = 300$ K.

9.6.1 Confinement in Quantum Well

Here we discuss the confinement of electrons and holes in a quantum well. We have already shown how to solve Schrödinger equation for electrons in a quantum well. The conduction band of GaAs is well expressed by a parabolic function with scalar effective mass m_e^* but the valence bands are specified by the heavy hole, light hole

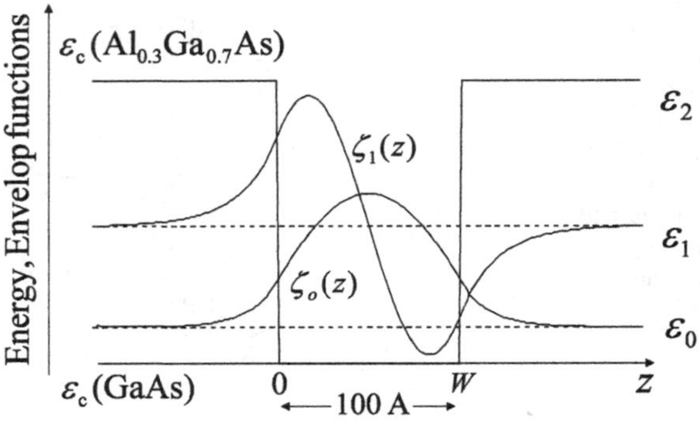

Fig. 9.34 The envelop functions for the electrons obtained by solving Schrödinger equation (9.147) are shown for the lowest two states in AlGaAs/GaAs/AlGaAs quantum well with 100 A well width. Here the conduction band discontinuity is assumed to be 0.26 eV, and the subband energies are shown by the dashed lines. The electron effective mass is assumed to be $m_e^* = 0.068\,\mathrm{m}$

and spin–orbit split–off bands. When we assume that the confinement direction is along the z axis and $k_y = k_z = 0$, then the wave functions of electrons and holes (envelop functions) are obtained by solving the following Schrödinger equation,[6]

$$\left[-\frac{\hbar^2}{2m_{jz}}\frac{\mathrm{d}^2}{\mathrm{d}z^2} + V(z)\right]\psi_{nj}(z) = \mathcal{E}_{nj}\psi_{nj}(z)\,, \qquad j = \mathrm{e,\ hh,\ lh,\ so} \tag{9.147}$$

where j = e, hh, lh, and so correspond to the electron, heavy hole, light hole and spin–orbit split–off hole, respectively. The electron energy is measured from the conduction band edge and the hole energy is measured from the top of the valence band. When we neglect the potential produced by the electrons and holes, the solution of (9.147) is easily obtained by discretizing (9.147) in N sections as described in Chap. 8 (Sect. 8.2) and diagonalizing the $N \times N$ matrix. An example is shown in Fig. 9.34 for electrons in AlGaAs/GaAs/AlGaAs single quantum well of 100 A well width, where we assumed the electron effective mass is 0.068 m and the conduction band discontinuity is 0.26 eV. The envelop functions of the valence bands are shown in Fig. 9.35 in AlGaAs/GaAs/AlGaAs single quantum well of 100 A well width, where we assumed the band offset is $\Delta\mathcal{E}_v = 0.1283$ eV, and the effective masses of heavy hole and light hole $m_{hh}^* = 0.377$ m and $m_{lh}^* = 0.091$ m, respectively. These band parameters will be discussed later in connection with Luttinger parameters.

Quantization of holes in the valence bands is complicated because of the band structures, where the heavy hole band, light hole band and spin–orbit split–off bands arises from the mixing of the wave functions $\left|\frac{3}{2}, \pm\frac{3}{2}\right\rangle$, $\left|\frac{3}{2}, \pm\frac{1}{2}\right\rangle$ and $\left|\frac{1}{2}, \pm\frac{1}{2}\right\rangle$ as shown in Chap. 2. When the spin–orbit interaction is taken into account, the matrix elements of the valence bands are given by (2.64). Let's consider the case of GaAs where the

[6] In this section the quantum well confinement is taken to be in the z direction, although we assumed the waveguide confinement is in the x direction in Sect. 9.5.

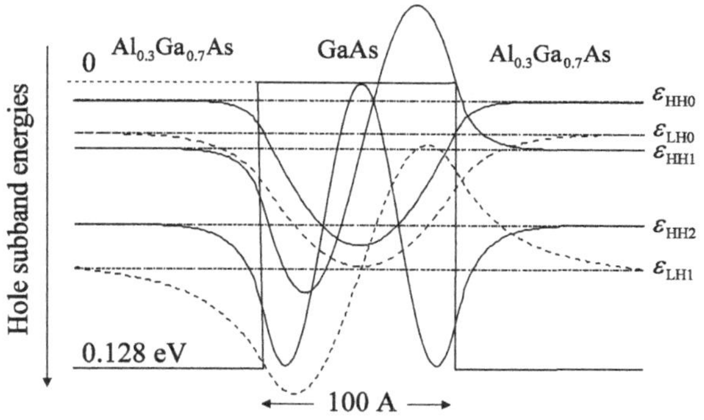

Fig. 9.35 The envelop functions of holes in the valence bands for $k_x = k_y = 0$ obtained by solving Schrödinger equation (9.147) are shown for the lowest three states of the heavy hole band (*solid curves*) and for the lowest two states of the light hole band (*dashed curves*). The subband energies are indicated by the *dot–dashed lines*. Parameters used are $\Delta\mathcal{E}_v = 0.1283\,\text{eV}$, $m^*_{hh} = 0.377\,\text{m}$ and $m^*_{lh} = 0.091\,\text{m}$

spin–orbit splitting is large (0.34 eV) and the spin–orbit split–off band is neglected. Then the Hamiltonian matrix is approximated by 4×4 matrix, which is deduced from (2.64) or given by Luttinger and Kohn [32] and Ando [33]

$$\mathcal{H} = \begin{vmatrix} \frac{1}{2}P & L & M & 0 \\ L^* & \frac{1}{6}P + \frac{2}{3}Q & 0 & M \\ M^* & 0 & \frac{1}{6}P + \frac{2}{3}Q & -L \\ 0 & M^* & -L^* & \frac{1}{2}P \end{vmatrix}, \tag{9.148}$$

where

$$P = \frac{\hbar^2}{2m}\left[(A+B)(k_x^2 + k_y^2) + 2Bk_z^2\right], \tag{9.149a}$$

$$Q = \frac{\hbar^2}{2m}\left[B(k_x^2 + k_y^2) + Ak_z^2\right], \tag{9.149b}$$

$$L = -\frac{1}{\sqrt{3}}\frac{\hbar^2}{2m}C(k_x - \mathrm{i}k_y)k_z, \tag{9.149c}$$

$$M = \frac{1}{\sqrt{12}}\frac{\hbar^2}{2m}\left[(A-B)(k_x^2 - k_y^2) + 2\mathrm{i}Ck_xk_y\right], \tag{9.149d}$$

with m being the free electron mass. The coefficients A, B and C $(\times\hbar^2/2m)$ correspond to L, M and N of (2.46a) $\sim$ (2.46c), respectively, and are related to Luttinger parameters [34] γ_1, γ_2 and γ_3 through,

$$\gamma_1 = -\frac{1}{3}(A + 2B)\,, \tag{9.150a}$$

$$\gamma_2 = -\frac{1}{6}(A - B)\,, \tag{9.150b}$$

$$\gamma_3 = -\frac{1}{6}C\,, \tag{9.150c}$$

or

$$A = -(\gamma_1 + 4\gamma_2)\,, \tag{9.151a}$$

$$B = -(\gamma_1 - 2\gamma_2)\,, \tag{9.151b}$$

$$C = -6\gamma_3\,. \tag{9.151c}$$

Let's consider a simplified case of $k_x = k_y = 0$. Then (9.148) is written by two sets of single component equations as shown by (9.147) with $m_{\text{hhz}}/m = -1/B$ and $m_{\text{lhz}}/m = -3/(2A + B)$. The solutions of quantized hole states are evaluated in the similar fashion of two dimensional electron gas. The results are shown in Fig. 9.35. On the other hand, when the condition of $k_x = k_y = 0$ is not fulfilled, non-diagonal components of matrix equation (9.148) result in mixing of the wave functions $\left|\frac{3}{2}, \pm\frac{3}{2}\right\rangle$ and $\left|\frac{3}{2}, \pm\frac{1}{2}\right\rangle$ and thus the quantized states depend on k_x and k_y. Here we will show how to obtain the quantized states of the valence bands. Ando reported two different methods to solve 4×4 matrix Hamiltonian (see paper by Ando [33] for detail). We deal with the 4×4 Hamiltonian matrix (9.148) and the present method will be easily extended to the case of 6×6 Hamiltonian matrix.[7] We are interested in quantization of holes in the z direction and thus k_x and k_y are treated as constants, while k_z is given by

$$k_z = -\mathrm{i}\frac{\partial}{\partial z}\,. \tag{9.152}$$

In order to avoid the complexity arising from the wave functions, we define the four eigenfunctions as

$$\left|\frac{3}{2}, \frac{3}{2}\right\rangle = F_1(z) \tag{9.153a}$$

$$\left|\frac{3}{2}, \frac{1}{2}\right\rangle = F_2(z) \tag{9.153b}$$

$$\left|\frac{3}{2}, -\frac{1}{2}\right\rangle = F_3(z) \tag{9.153c}$$

$$\left|\frac{3}{2}, -\frac{3}{2}\right\rangle = F_4(z)\,, \tag{9.153d}$$

and discretize the z direction into N segments. The length L is taken to be larger than the well width W so that the wave functions vanish at $z = 0$ and $z = L$. The segment is then defined as $\mathrm{d}z \equiv h = L/N$ and the eigenfunction is written as

[7] The 6×6 Hamiltonian matrix is given by (V.13) of Luttinger and Kohn [35].

$$F_i(z) = F_i(h \cdot j) \rightarrow F_{ij} \quad (9.154)$$
$$i = 1, 2, 3, 4; \; j = 1, 2, 3, \ldots, N-1, N\,.$$

Using the definition of (9.152), the Hamiltonian matrix (9.148) is expressed by the following components

$$\frac{\mathrm{d}}{\mathrm{d}z} F_i(z) = \frac{F_{i,j+1} - Fi,\, j}{h} \quad (9.155)$$

$$\frac{\mathrm{d}^2}{\mathrm{d}z^2} F_i(z) = \frac{F_{i,j+1} - 2F_{i,j} + F_{i,j-1}}{h^2}\,. \quad (9.156)$$

Therefore the 4×4 matrix of (9.148) is rewritten by $4N \times 4N$ matrix. Diagonalizing $4N \times 4N$ matrix, we obtain the eigenvalues and corresponding eigenstates, which are dependent of k_x and k_y values. As a result the energy of a subband is not parabolic with respect to k_x and k_y and the constant energy surface is warped in the (k_x, k_y) plane.

In order to obtain self–consistent solutions for a specific sheet density of holes, the density of states is required which may be evaluated from the calculated eigenstates in the (k_x, k_y) plane. Since both electrons and holes exist in a quantum well laser and the resulting Hartree potentials cancel each other (not completely, but weakened), we neglect the self–consistency for simplicity in the present calculations. A typical example of the calculated subband energies as a function of k_x (solid curves in the [10] direction) and $k_\parallel = \sqrt{2}k_x = \sqrt{2}k_y$ (dashed curves in the [11] direction) are shown in the left figure of Fig. 9.36, where the results are obtained for a $Al_xGa_{1-x}As/GaAs/Al_xGa_{1-x}As$ ($x = 0.3$) quantum well of 80 A well–width and the valence band offset $\Delta\mathcal{E}_v = 0.15$ eV. We used the notation HH0 and HH1 for the heavy hole subbands and LH0 for the light hole subband. The right figure shows the density of states as a function of hole energy. Here we find that the density of states is no more a simple step–like function.

9.6.2 *Optical Transition in Quantum Well Structures*

Here we will concern with the optical transition in a quantum well structure based on the direct transition model. The spontaneous and stimulated transition rates for photon energy $\hbar\omega$ are given by using (9.17) and (9.18),

$$r_{\mathrm{spon}}(\hbar\omega) = \frac{n_{\mathrm{r}} e^2 \omega}{\pi \epsilon_0 m^2 \hbar c^3} |M|^2 \rho_{\mathrm{2Dred}}(\hbar\omega) f_2(1 - f_1)\,, \quad (9.157)$$

$$r_{\mathrm{stim}}(\hbar\omega) = \frac{n_{\mathrm{r}} e^2 \omega}{\pi \epsilon_0 m^2 \hbar c^3} |M|^2 \rho_{\mathrm{2Dred}}(\mathcal{E})(f_2 - f_1)\,, \quad (9.158)$$

where ρ_{2Dred} is the two–dimensional reduced density of states per unit area defined by

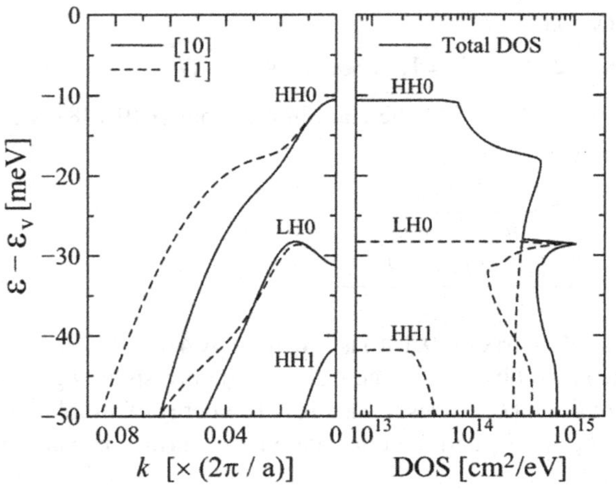

Fig. 9.36 The *left figure* shows the subband energies as a function of wave vectors obtained by solving 4 × 4 matrix Hamiltonian given by (9.148). The *solid* and *dashed curves* are the dispersion of the subband energies in the [10] and [11] directions of wave vectors (in units of $2\pi/a$), where the heavy–hole subbands HH0 and HH1, and the light–hole subband LH0 are plotted. The *right figure* shows the density of states as a function of the hole energy, where the *dashed curves* are the density of states for the subbands HH0, HH1 and LH0, and the solid curve is the total density of states. Parameters used are well–width $W = 80$ A, $\Delta\mathcal{E}_v = 0.150$ eV, $\gamma_1 = -6.85$, $\gamma_2 = -2.10$ and $\gamma_3 = -2.90$

$$\rho_{2\text{Dred}}(k_x k_y)\mathrm{d}k_x\mathrm{d}k_y = \frac{1}{(2\pi)^2}\mathrm{d}k_x\mathrm{d}k_y\,, \tag{9.159a}$$

$$\rho_{2\text{Dred}}(\hbar\omega)\mathrm{d}(\hbar\omega) = \frac{m_{\text{red}}}{2\pi\hbar^2}\mathrm{d}(\mathcal{E}_1 + \hbar\omega - \mathcal{E}_2)\,, \tag{9.159b}$$

where m_{red} is the reduced mass and the spin degeneracy is not included because it is already included in the momentum matrix element. More detailed discussion will be given in Sect. 9.6.3. The spontaneous and stimulated transition rates are defined per unit area or per sheet density of carriers. Then the gain and absorption coefficients given by (9.35) may be rewritten for the subband transition as

$$\begin{aligned} g(\hbar\omega) = -\alpha(\hbar\omega) &= \frac{\pi^2 c^2 \hbar^3}{n_r^2(\hbar\omega)^2} r_{\text{stim}}(\hbar\omega) \\ &= \frac{\pi e^2}{n_r c\epsilon_0 m^2\omega}\,|M|^2\,(f_2 - f_1)\rho_{2\text{Dred}}(\hbar\omega)\,. \end{aligned} \tag{9.160}$$

Equations (9.157)–(9.160) are expressed per unit area and thus the values per unit volume are obtained by dividing these expressions with the quantum well width L_z.

The momentum matrix element $|M|$ used in (9.157), (9.158) and (9.160) are not the same as the bulk matrix elements discussed in Sect. 9.4.3. In order to discuss the gains of TE and TM modes in a quantum well structure we deal with a more general analysis taking account of the three bands, the heavy, light and spin–orbit split–off bands. In a quantum well structure of lattice matched AlGaAs/GaAs/AlGaAs the strain effect is neglected and we may obtain the subband quantization using 4×4 Hamiltonian matrix. In a lattice mismatched quantum well like GaInP–AlGaInP, however, the spin–orbit split–off is quite small, 0.11 eV, and thus the strain effect introduces mixing of the three hole bands. In this text we will not deal with the subband structure of GaInP–AlGaInP laser, but we present how to analyze such a general case. The 6×6 Hamiltonian given by Luttinger and Kohn [35] is easily extended to include the strain effect, using the basis functions of $|\frac{3}{2}, \frac{3}{2}\rangle$, $|\frac{3}{2}, \frac{1}{2}\rangle$, $|\frac{3}{2}, -\frac{1}{2}\rangle$, $|\frac{3}{2}, -\frac{3}{2}\rangle$, $|\frac{1}{2}, \frac{1}{2}\rangle$ and $|\frac{1}{2}, -\frac{1}{2}\rangle$. The terms of the uniaxial and biaxial strain effect in the 6×6 Hamiltonian are given by (4.173) and (4.174a) $\sim$ (4.174c), and thus we obtain the following Luttinger Hamiltonian

$$\begin{vmatrix} \frac{1}{2}P_{\text{hh}} & L & M & 0 & \mathrm{i}L/\sqrt{2} & -\mathrm{i}\sqrt{2}M \\ L^* & \frac{1}{6}P_{\text{lh}}+\frac{2}{3}Q & 0 & M & -\mathrm{i}N & \mathrm{i}\sqrt{3}L/\sqrt{2} \\ M^* & 0 & \frac{1}{6}P_{\text{lh}}+\frac{2}{3}Q & -L & -\mathrm{i}\sqrt{3}L^*/\sqrt{2} & -\mathrm{i}N \\ 0 & M^* & -L^* & \frac{1}{2}P_{\text{hh}} & -\mathrm{i}\sqrt{2}M^* & -\mathrm{i}L^*/\sqrt{2} \\ -\mathrm{i}L^*/\sqrt{2} & \mathrm{i}N & \mathrm{i}\sqrt{3}L/\sqrt{2} & \mathrm{i}\sqrt{2}M & \frac{1}{3}(P+Q)-\Delta_{\text{so}} & 0 \\ \mathrm{i}\sqrt{2}M^* & -\mathrm{i}\sqrt{3}L^*/\sqrt{2} & \mathrm{i}N & \mathrm{i}L/\sqrt{2} & 0 & \frac{1}{3}(P+Q)-\Delta_{\text{so}} \end{vmatrix} + V(z)\,, \tag{9.161}$$

where $V(z)$ is the potential energy at the band edge, Δ_{so} is the spin–orbit split–off energy with strain effect as given below. The following expressions for the matrix elements are newly defined,

$$\frac{1}{2}P_{\text{hh}} = \frac{1}{2}P + \left(\Delta^{\text{hydro}} + \Delta^{\text{shear}}\right), \tag{9.162a}$$

$$\frac{1}{6}P_{\text{lh}} = \frac{1}{6}P + \left(\Delta^{\text{hydro}} - \Delta^{\text{shear}}\right), \tag{9.162b}$$

$$N = \frac{P-2Q}{3\sqrt{2}} + \left(\sqrt{2}\Delta^{\text{shear}}\right), \tag{9.162c}$$

$$\Delta_{\text{so}} = \Delta_0 + \left(\Delta^{\text{hydro}}\right), \tag{9.162d}$$

where the terms in the parentheses are the contribution from the strain. The terms Δ^{hydro} and Δ^{shear} are hydrostatic and shear deformation energies, respectively, and are given by the following relations using the strain Hamiltonian given by (4.171) and (4.172a) under the strain (see also, (4.175a) and (4.175b)),

$$H_{\text{s}} = -a_{\text{v}}(e_{xx}+e_{yy}+e_{zz}) - 3b\left[(L_x^2 - \frac{1}{3}\boldsymbol{L}^2)e_{xx} + \text{c.p.}\right], \tag{9.163a}$$

$$\Delta^{\text{hydro}} = -a_{\text{v}}\left(e_{xx}+e_{yy}+e_{zz}\right), \tag{9.163b}$$

$$\Delta^{\text{shear}} = +\frac{b}{2}\left(e_{xx}+e_{yy}-2e_{zz}\right). \tag{9.163c}$$

The hydrostatic and shear deformation energies under a biaxial stress are rewritten as[8]

$$\Delta^{\text{hydro}} = -2a_{\text{v}}\left(\frac{c_{11}-c_{12}}{c_{11}}\right)e_{\parallel}\,, \tag{9.164a}$$

$$\Delta^{\text{shear}} = -b\left(\frac{c_{11}+2c_{12}}{c_{11}}\right)e_{\parallel}\,. \tag{9.164b}$$

Here we used the notations c_{11} and c_{12} for the elastic stiffness constants, $e_{\parallel}$ is the biaxial strain, and a_{v} is the hydrostatic deformation potential in the valence band and b is the shear deformation potential. The strain in the heterostructure is defined by using the lattice constants a_{sub} for the substrate and a_{epi} for the epitaxial layer as

$$e_{xx} = e_{yy} = \frac{a_{\text{sub}}-a_{\text{epi}}}{a_{\text{epi}}} \equiv e_{\parallel} \tag{9.165a}$$

$$e_{zz} = -2\frac{c_{12}}{c_{11}}e_{xx} \equiv -2\frac{c_{12}}{c_{11}}e_{\parallel}\,. \tag{9.165b}$$

From the definition of (9.165a), we find the following conditions. When the lattice constant of an epitaxial layer is larger than that of a substrate, $a_{\text{epi}} > a_{\text{sub}}$, a compressive strain is induced. In the case of $a_{\text{epi}} < a_{\text{sub}}$, on the other hand, the epitaxial layer is expanded and thus a tensile strain is induced.

The subband wave functions (envelop functions) of the valence band in a heterostructure are obtained by solving the Hamiltonian $H(\boldsymbol{k}_{\parallel}, k_z \to -\mathrm{i}\frac{\partial}{\partial z})$ discretized into $6N \times 6N$ matrix (4×4 matrix neglecting the spin–orbit split–off band). Then the subband wave function is given by a linear combination of the 6 basis functions, with the notation ν for the subband index and $(k_x, k_y) = \boldsymbol{k}_{\parallel}$,

$$\begin{aligned}\psi_{\text{v}}^{\nu}(\boldsymbol{k}_{\parallel}, z) = {}& \phi_{1\text{h}}^{\nu}(\boldsymbol{k}_{\parallel})\left|\frac{3}{2},\frac{3}{2}\right\rangle + \phi_{1\text{l}}^{\nu}(\boldsymbol{k}_{\parallel})\left|\frac{3}{2},\frac{1}{2}\right\rangle + \phi_{2\text{l}}^{\nu}(\boldsymbol{k}_{\parallel})\left|\frac{3}{2},-\frac{1}{2}\right\rangle \\ & + \phi_{2\text{h}}^{\nu}(\boldsymbol{k}_{\parallel})\left|\frac{3}{2},-\frac{3}{2}\right\rangle + \phi_{1\text{s}}^{\nu}(\boldsymbol{k}_{\parallel})\left|\frac{1}{2},\frac{1}{2}\right\rangle + \phi_{2\text{s}}^{\nu}(\boldsymbol{k}_{\parallel})\left|\frac{1}{2},-\frac{1}{2}\right\rangle,\end{aligned} \tag{9.166}$$

where $\phi_{i\text{h}}^{\nu}(\boldsymbol{k}_{\parallel})$, $\phi_{i\text{l}}^{\nu}(\boldsymbol{k}_{\parallel})$, and $\phi_{i\text{s}}^{\nu}(\boldsymbol{k}_{\parallel})$ $(i = 1, 2)$ are envelop functions of the heavy hole ($|3/2, \pm 3/2\rangle$), light hole ($|3/2, \pm 1/2\rangle$), and spin split hole ($|1/2, \pm 1/2\rangle$). Note here that $\phi_{i\text{v}}^{\nu}$ (v= h, l, and so) include z direction dependence which is obtained from

---

[8] Those relations are easily deduced by using the stress tensors, $T_{xx} = T_1 = X$, $T_{yy} = T_2 = X$, and $T_{zz} = T_3 = 0$. The strain tensor is given by $e_{ij} \equiv e_{\alpha} = s_{\alpha\beta}T_{\beta}$ $(\alpha, \beta = 1, 2, 3, \ldots, 6)$. Therefore we obtain

$$\begin{aligned} e_{xx} &= e_{yy} = s_{11}T_1 + s_{12}T_2 = (s_{11}+s_{12})X \\ e_{zz} &= s_{12}T_1 + s_{12}T_2 = 2s_{12}X\,, \end{aligned}$$

where s_{11} and s_{12} are the elastic compliance tensors and related to $c_{\alpha\beta}$ by

$$s_{11} = \frac{c_{11}+c_{12}}{(c_{11}-c_{12})(c_{11}+2c_{12})}\,, \qquad s_{12} = -\frac{c_{12}}{(c_{11}-c_{12})(c_{11}+2c_{12})}\,.$$

the diagonalization of the Hamiltonian matrix. Writing the envelop functions of the electron in the conduction band and holes in the valence bands as

$$\psi_{\mathrm{c}}^{\mu}(x, y, z) = \phi_{\mathrm{c}}^{\mu}(z)|S\rangle\,, \tag{9.167}$$

$$\psi_{\mathrm{v}}^{\nu}(x, y, z) = \sum_{\mathrm{v}} \phi_{\mathrm{v}}^{\nu}(\boldsymbol{k}_{\|}, z)|\Gamma_{25'}\rangle\,, \tag{9.168}$$

we may evaluate the momentum matrix element for the transition between the conduction and valence bands,

$$|M_{\mathrm{cv}}^{\mu\nu}(\boldsymbol{k}_{\|})|^2 = \left|\langle\psi_{\mathrm{c}}^{\mu}|\boldsymbol{p}|\psi_{\mathrm{v}}^{\nu}\rangle\right|^2 . \tag{9.169}$$

Since we are interested in a quantum well structure with the confinement direction parallel to the z axis, we assume the propagation direction of the light along the y axis. Then the TE mode gives rise to the polarization of the $\boldsymbol{E} = (E_x, 0, 0)$ and $\boldsymbol{H} = (0, H_y, H_z)$. On the other hand TM mode is shown by $\boldsymbol{H} = (H_x, 0, 0)$ and $\boldsymbol{E} = (0, E_y, E_z)$. Therefore the momentum matrix element for TE mode is given by

$$\begin{aligned} \left|M_{\mathrm{cv}}^{\mu\nu}(\boldsymbol{k}_{\|})\right|^2 &= 2\langle S|p_x|X\rangle^2 \\ &\times \left[\frac{1}{2}\langle\phi_{\mathrm{c}}^{\mu}|\phi_{1\mathrm{h}}^{\nu}(\boldsymbol{k}_{\|})\rangle^2 + \frac{1}{6}\langle\phi_{\mathrm{c}}^{\mu}|\phi_{2\mathrm{l}}^{\nu}(\boldsymbol{k}_{\|})\rangle^2 + \frac{1}{3}\langle\phi_{\mathrm{c}}^{\mu}|\phi_{2\mathrm{s}}^{\nu}(\boldsymbol{k}_{\|})\rangle^2\right], \end{aligned} \tag{9.170}$$

and for TM mode

$$\left|M_{\mathrm{cv}}^{\mu\nu}(\boldsymbol{k}_{\|})\right|^2 = 2\langle S|p_z|Z\rangle^2 \left[\frac{2}{3}\langle\phi_{\mathrm{c}}^{\mu}|\phi_{1\mathrm{l}}^{\nu}(\boldsymbol{k}_{\|})\rangle^2 + \frac{1}{3}\langle\phi_{\mathrm{c}}^{\mu}|\phi_{1\mathrm{s}}^{\nu}(\boldsymbol{k}_{\|})\rangle^2\right] . \tag{9.171}$$

The factor 2 in (9.170) and (9.171) is introduced to include the spin degeneracy, and $\langle\phi_{\mathrm{c}}^{\mu}|\phi_{1\mathrm{h}}^{\nu}(\boldsymbol{k}_{\|})\rangle$ and so on are the overlap integrals of the envelop functions of the electron in the conduction band and of the hole in the valence band with respect to z in the quantum well width. As stated in Sect. 9.4.3, the momentum matrix element are related to P by the relations derived in the footnote 3 of p. 568,

$$\begin{aligned} \langle S|p_x|X\rangle^2 &= \langle S|p_y|Y\rangle^2 = \langle S|p_z|Z\rangle^2 \\ &= \frac{m^2}{\hbar^2}P^2 = \frac{m^2}{2m_0^*}\frac{\mathcal{E}_{\mathrm{G}}(\mathcal{E}_{\mathrm{G}} + \Delta_0)}{\mathcal{E}_{\mathrm{G}} + 2\Delta_0/3}\,, \end{aligned} \tag{9.172}$$

where P^2 given by (2.158) is used.

As mentioned in the case of GaAs, the spin–orbit split–off energy is appreciably large and we may neglect the contribution from the spin–orbit split–off band, the last terms of (9.170) and (9.171), except a case of a large strain. In this case the envelop function of the lowest subband arises from the heavy hole band $|3/2, \pm 3/2\rangle$ for lower values of $\boldsymbol{k}_{\|}$ and thus optical transition is governed by the conduction band to the heavy hole band. Therefore the TE mode oscillations dominates in the laser gain. The TM mode will be excited by the transition between the conduction band and the light hole band, and thus the TM mode oscillations will not take part in the GaAs based quantum well structures. On the other hand in a lattice mismatched GaInP–AlGaInP quantum well tensile strain results in a dramatic change in the subbands structures of the valence bands and the light hole subband ($|3/2, \pm 1/2\rangle$) is located at the band edge [36]. Then the TM mode gain overcomes the TE mode gain. We will not deal with the detail of the strained quantum wells here, but readers may analyze the laser gain in strained layer quantum wells by using the relations derived above and the strain effect described in Sect. 9.6.4.

9.6.3 Reduced Density of States and Gain

Here we will discuss the density of states in a quantum well structure. First we assume that the electron and hole dispersions are approximated by the parabolic functions as

$$\mathcal{E}_2 = \mathcal{E}_c + \mathcal{E}_\mu^c + \frac{\hbar^2 k_\parallel^2}{2m_e^*}, \tag{9.173a}$$

$$\mathcal{E}_1 = \mathcal{E}_v - \mathcal{E}_\nu^v - \frac{\hbar^2 k_\parallel^2}{2m_h^*}. \tag{9.173b}$$

The diagonal properties of the envelop functions give rise to the selection rule for the quantum number μ, ν,

$$\int \phi_c^\mu(z)\phi_v^\nu(z)\mathrm{d}z = 1 \quad \text{for} \quad \mu = \nu\,, \tag{9.174}$$

$$= 0 \quad \text{for} \quad \mu \neq \nu\,. \tag{9.175}$$

The wave vector conservation of $\boldsymbol{k}_\parallel$ and the energy conservation rule lead us to the following relations,

$$\mathcal{E}_2 = \mathcal{E}_c + \mathcal{E}_n^c + \frac{m_{\text{red}}}{m_e^*}(\hbar\omega - \mathcal{E}_{G,n})\,, \tag{9.176a}$$

$$\mathcal{E}_1 = \mathcal{E}_v - \mathcal{E}_n^v - \frac{m_{\text{red}}}{m_h^*}(\hbar\omega - \mathcal{E}_{G,n})\,, \tag{9.176b}$$

where

$$\frac{1}{m_{\text{red}}} = \frac{1}{m_e^*} + \frac{1}{m_h^*}, \tag{9.177a}$$

$$\mathcal{E}_{G,n} = \mathcal{E}_G + \mathcal{E}_n^c + \mathcal{E}_n^v = (\mathcal{E}_c - \mathcal{E}_v) + \mathcal{E}_n^c + \mathcal{E}_n^v\,, \tag{9.177b}$$

From these results we obtain the reduced density of states $\rho_{\text{2Dred},n}(\hbar\omega)$ for the nth subband,

$$\rho_{\text{2Dred},n}(\hbar\omega) = \frac{m_{\text{red}}}{2\pi\hbar^2} u\left(\hbar\omega - \mathcal{E}_{G,n}\right)\,, \tag{9.178}$$

where $u(x)$ is the unit–step function and $u(x \geq 0) = 1$, $u(x < 0) = 0$.

As shown in Sect. 9.6.2 the subband energy depends on the wave vector $\boldsymbol{k}_\parallel$ and thus the density of states is not given by the above relations. Here we will simplify the calculation of $\rho_{\rm 2Dred}(\boldsymbol{k}_\parallel)$ by taking one direction in the plane $(k_x, k_y, 0)$ as follows

$$\rho_{\rm 2Dred,\mathit{n}}(\boldsymbol{k}_\parallel) = \frac{1}{(2\pi)^2}\mathrm{d}^2\boldsymbol{k}_\parallel \simeq \frac{1}{(2\pi)^2}2\pi k_\parallel \mathrm{d}k_\parallel \,. \tag{9.179}$$

Then the density of states is expressed as

$$\rho_{\rm 2Dred,\mathit{n}}(\mathcal{E}) = \frac{1}{2\pi}\sum_n\sum_{k_\parallel} k_\parallel \left|\frac{\partial \mathcal{E}_n}{\partial k_\parallel}\right|^{-1} \quad \text{(per unit area)} \,, \tag{9.180a}$$

$$= \frac{1}{2\pi}\sum_n\sum_{k_\parallel} k_\parallel \left|\frac{\partial \mathcal{E}_n}{\partial k_\parallel}\right|^{-1} \times \frac{1}{L_z} \quad \text{(per unit volume)} \,, \tag{9.180b}$$

where the spin degeneracy factor 2 is excluded because it is included in the momentum matrix elements. When 6×6 or 4×4 matrix Hamiltonian is solved, we are able to calculated the density of states by summing up all the $\boldsymbol{k}_\parallel$ points, which is shown in Fig. 9.36.

When we define the subband energies $\mathcal{E}_n^{\rm e}$ for the conduction band, $\mathcal{E}_m^{\rm hh}$ for the heavy hole band, and $\mathcal{E}_{m'}^{\rm lh}$ for the light hole band $(n, m, m' = 0, 1, 2, \ldots)$, the orthogonality of the envelop functions gives rise to the selection rule of the optical transitions,

$$n = m, \; n = m' \,, \tag{9.181}$$

or in other words, interband transitions between the same quantum numbers are allowed. This is schematically shown in Fig. 9.37.

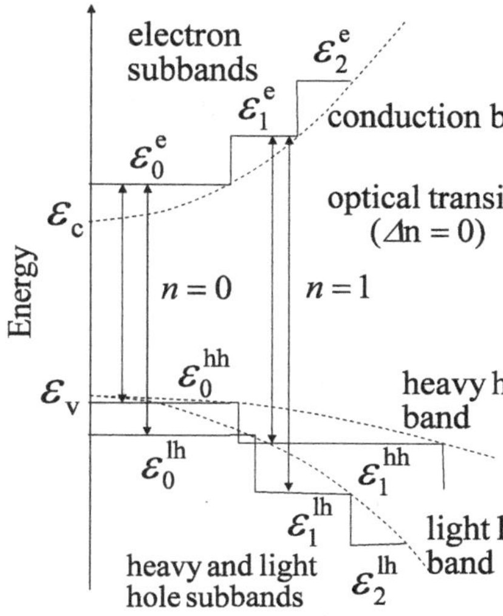

Fig. 9.37 Interband optical transition in a quantum well structure is shown, where the *dotted curves* are the conduction and valence bands in a bulk material and the step like curves are the subbands. Only transitions between the same quantum numbers are allowed

Assuming parabolic relations for the carrier energies *vs* wave vectors in the conduction and valence bands, the densities of electrons and holes in a quantum well of well width L_z are given by using (8.15)

$$n = \frac{m_c k_B T}{\pi \hbar^2 L_z} \sum_n \ln \left[\exp \left(\frac{\mathcal{E}_F^c - \mathcal{E}_n^c}{k_B T} \right) + 1 \right], \tag{9.182a}$$

$$p = \frac{m_v k_B T}{\pi \hbar^2 L_z} \sum_n \ln \left[\exp \left(\frac{\mathcal{E}_n^v - \mathcal{E}_F^v}{k_B T} \right) + 1 \right] \tag{9.182b}$$

where $\mathcal{E}_F^c$ and $\mathcal{E}_F^v$ are quasi–Fermi levels, and $\mathcal{E}_n^c$ and $\mathcal{E}_n^v$ are n–th subbands in the conduction and valence bands, respectively. The laser amplification gain under the carrier injection into a quantum well structure is given by,

$$\begin{aligned} g(\hbar\omega) &= -\alpha(\hbar\omega) \\ &= \frac{\pi e^2}{n_r c \epsilon_0 m^2 \omega} \times \sum_n |M|^2 (f_2 - f_1) \frac{m_{red}}{2\pi \hbar^2 L_z} u\left(\hbar\omega - \mathcal{E}_{G,n}\right), \end{aligned} \tag{9.183}$$

where m_{red} is the reduced effective mass, $\mathcal{E}_{G,n}$ is the gap energy of the subbands, and the summation is carried out with respect to the subband quantum number n. In Fig. 9.38 we show gain spectra in a simplified model of parabolic bands, where the upper and lower parts of the curves are for a bulk double heterostructure laser (DHL) and a quantum well laser (QWL), respectively. The left figures represent the density of states, the middle figures for the gains as a function of photon energy $\hbar\omega$, and the right figures for the maximum gains as a function of injection current. The gain spectra reflect the shape of the density of states curves.

In more general form we obtain the gain

$$\begin{aligned} g(\hbar\omega) = \frac{\pi e^2}{n_r c \epsilon_0 m^2 \omega} &\sum_{\mu,\nu} \sum_{\boldsymbol{k}_\parallel} \rho_{2Dred}(\boldsymbol{k}_\parallel) \frac{1}{L_z} \left| M_{cv}^{\mu\nu}(\boldsymbol{k}_\parallel) \right|^2 \\ &\times \left[f^c(\boldsymbol{k}_\parallel) - f^v(\boldsymbol{k}_\parallel) \right], \end{aligned} \tag{9.184}$$

where $f^c(\boldsymbol{k}_\parallel)$ and $f^v(\boldsymbol{k}_\parallel)$ are quasi–Fermi distribution functions of the electrons in the conduction band and holes in the valence bands. The density of states is given by

$$\sum_{\boldsymbol{k}_\parallel} \rho_{2Dred}(\boldsymbol{k}_\parallel) = \frac{1}{2\pi} \sum_{k_\parallel} k_\parallel \left| \frac{\partial \mathcal{E}_{\mu\nu}^{cv}(\boldsymbol{k}_\parallel)}{\partial k_\parallel} \right|^{-1} \times \frac{1}{L_z}, \tag{9.185a}$$

$$\mathcal{E}_{\mu\nu}^{cv}(\boldsymbol{k}_\parallel) = \mathcal{E}_\mu^c(\boldsymbol{k}_\parallel) - \mathcal{E}_\nu^v(\boldsymbol{k}_\parallel), \tag{9.185b}$$

where $\mathcal{E}_{\mu\nu}^{cv}(\boldsymbol{k}_\parallel)$ is the interband energy of the conduction band and valence band. In Fig. 9.36 we presented the subband energies of the valence bands, where the results are calculated by solving 4×4 Hamiltonian matrix of (9.148).

We have to mention here the effect of band tailing discussed in Sect. 9.3 for the optical transition in a bulk material. The band tail effect is also expected to play an important role in the laser oscillations in a quantum well. One of the most widely used methods to take account of the band tail effect in quantum well structures is the

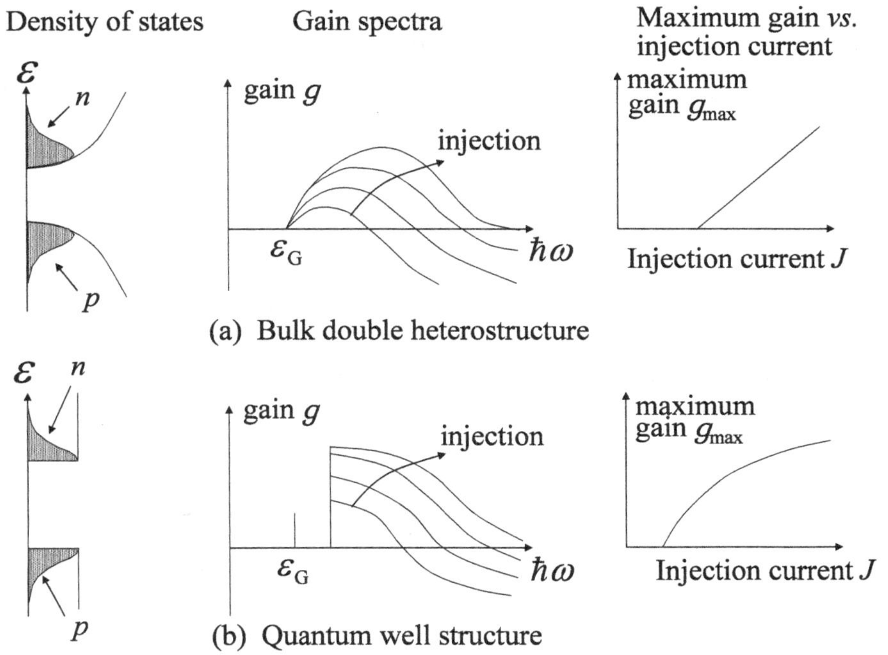

Fig. 9.38 The density of states, gain spectra and maximum gains are compared in **a** bulk double heterostructure laser (DHL) and **b** quantum well laser (QWL), where the gain reflects the shape of density of states. (after Suhara [24])

relaxation broadening model introduced by Asada and Suematsu [37]. They proposed to use Lorentzian broadening given by

$$L(\hbar\omega) = \frac{\hbar\Gamma/\pi}{(\mathcal{E}_{\mathrm{cv}}^{\mu\nu}(\boldsymbol{k}_{\parallel}) - \hbar\omega)^2 + (\hbar\Gamma)^2}, \tag{9.186}$$

where Γ is the intra–band relaxation time. Therefore the linear–gain is given by multiplying (9.184) by (9.186).

9.6.4 *Strain Effect*

Finally we discuss the strain effect of the energy bands. In Sect. 4.7.3 we dealt with the stress induced change in the energy bands and found that the valence band shift is given by (4.178). Figure 9.39 shows the change of the valence bands induced by uniaxial stress in GaAs, where we used Luttinger parameters $\gamma_1 = 6.85$, $\gamma_2 = 2.1$, deformation potentials $a_c = -7.17$, $a_v = -1.16$, $b = -1.7$ [eV], elastic

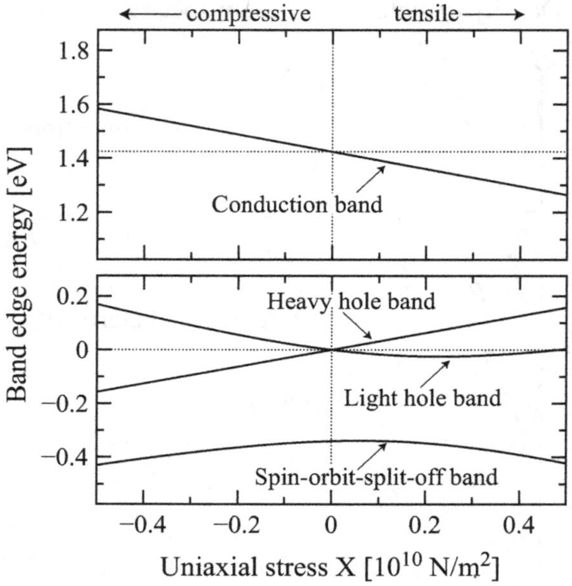

Fig. 9.39 The *upper curve* is the uniaxial stress effect of the conduction band edge. The lower curve represents the uniaxial stress effect of the heavy hole, light hole and spin–orbit split–off bands in GaAs, where the shift of the valence bands is calculated by (9.161) and the top of the valence bands at $X = 0$ is set to 0. The stress $X < 0$ is compressive and $X > 0$ is tensile. In the compressive stress $X < 0$ the light hole band is located higher than the heavy hole band

compliance constants $s_{11} = 0.1176 \times 10^{-10}$, $s_{12} = -0.0365 \times 10^{-10}$ [m^2/N], the lattice constant $a = 5.65$ A, the energy gap $\mathcal{E}_G = 1.425$ [eV], and the spin–orbit split–off energy $\Delta_0 = 0.34$ [eV]. The strain X is compressive for $X < 0$ and tensile for $X > 0$. We find that the compressive uniaxial stress results in the light hole band higher than the heavy hole band. In uniaxial stress case, the strain components are given by

$$e_{xx} = e_{yy} = s_{12}X\,, \tag{9.187a}$$
$$e_{zz} = s_{11}X\,, \tag{9.187b}$$
$$e_{xx} + e_{yy} + e_{zz} = (s_{11} + 2s_{12})X\,, \tag{9.187c}$$
$$e_{xx} + e_{yy} - 2e_{zz} = -2(s_{11} - s_{12})X\,, \tag{9.187d}$$

and thus we obtain

$$\Delta^{\text{hydro}} = -a_v(s_{11} + 2s_{12})X \; (= -\delta\mathcal{E}_H)\,, \tag{9.188a}$$
$$\Delta^{\text{shear}} = -b(s_{11} - s_{12})X \left(= -\frac{1}{2}\delta\mathcal{E}_{001}\right)\,. \tag{9.188b}$$

In Fig. 9.39 the effect of uniaxial stress on the conduction band edge calculated by

$$\mathcal{E}_c(X) + \mathcal{E}_G = a_c\,(s_{11} + 2s_{12})\,X + 1.425\,, \tag{9.189}$$

and the valence band edges in GaAs calculated from the 6×6 Hamiltonian given by (9.161) are shown, where we find that the heavy and light hole bands split into two

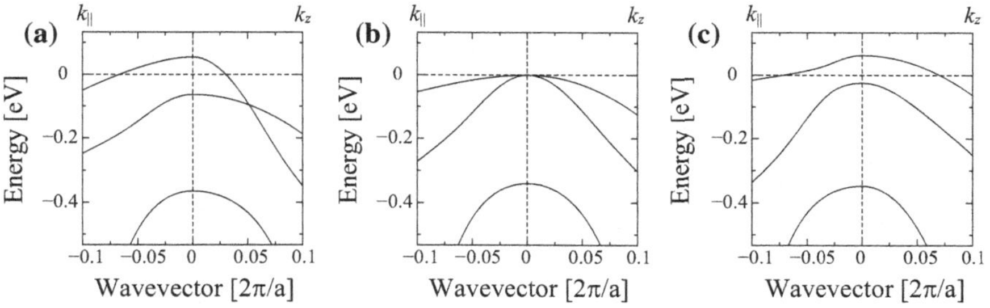

Fig. 9.40 Effect of the uniaxial stress on the valence bands of GaAs calculated by 6 × 6 Hamiltonian given by (9.161). Three valence bands are plotted as a function of the wave vectors $\boldsymbol{k}$ for **a** compressive stress of $X = -0.2 \times 10^{10}$ [N/m^2], **b** without stress ($X = 0$), and **c** tensile stress of $X = +0.2 \times 10^{10}$ [N/m^2]. The wave vectors $k_{\parallel}$ ($= \sqrt{2}k_x = \sqrt{2}k_y$) and k_z are in units of $(2\pi/a)$

bands and the compressive stress, $X < 0$, results in band edge locations of the light hole band higher than the heavy hole band. Typical examples of the three valence bands for the compressive stress $X = -0.2 \times 10^{10}$ [N/m^2] and for the tensile stress $X = +0.2 \times 10^{10}$ [N/m^2] are plotted as a function of $\boldsymbol{k}$, together with the bands for $X = 0$ in Fig. 9.40.

Let's consider the case of biaxial stress. In quantum well lasers AlGaAs/GaAs/AlGaAs, the well region of GaAs are sandwiched by the lattice matched AlGaAs barrier layers and the strain is quite small, enabling us to neglect the strain effect on the valence bands. In a lattice mismatched quantum well structures, however, the well region is subject to the strain effect. When the lattice constant of the well region is larger than the barrier layer (substrate), a compressive strain is induced. On the other hand, when the lattice constant of the well region is smaller than the barrier layer, a tensile strain is induced. Such a tensile biaxial strain results in a dramatic change in the subbands structures of the valence bands, and the light hole subband ($|3/2, \pm 1/2\rangle$) is located at the band edge for example in GaInP–AlGaInP [36]. In general, a lattice–mismatched well regions grown on the (001) surface of a substrate is subjected to a biaxial stress effect. In order to illustrate the biaxial effect on the valence band structure, we deal with a bulk GaAs under a biaxial stress effect. For the biaxial stress in the (001) plane, we obtain

$$e_{xx} = e_{yy} = (s_{11} + s_{12})X \,, \tag{9.190a}$$

$$e_{zz} = 2s_{12}X \,, \tag{9.190b}$$

$$e_{xx} + e_{yy} + e_{zz} = 2(s_{11} + 2s_{12})X \,, \tag{9.190c}$$

$$e_{xx} + e_{yy} - 2e_{zz} = 2(s_{11} - s_{12})X \,. \tag{9.190d}$$

Therefore the matrix elements of the strain Hamiltonian are given by

$$\Delta^{\text{hydro}} = -2a_{\text{v}}(s_{11} + 2s_{12})X \,, \tag{9.191a}$$

$$\Delta^{\text{shear}} = +b(s_{11} - s_{12})X \,. \tag{9.191b}$$

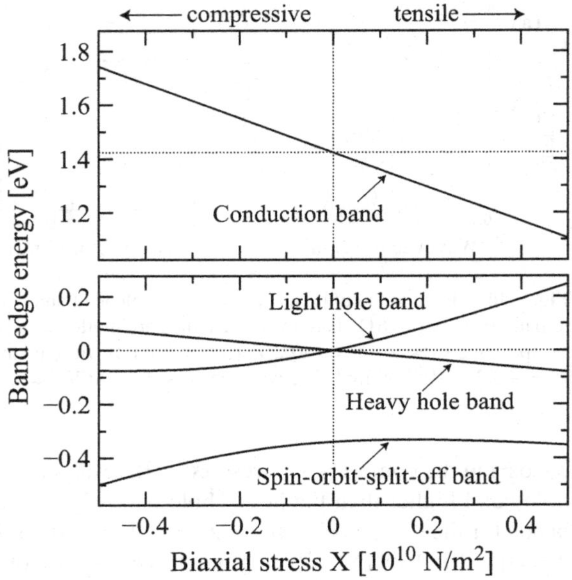

Fig. 9.41 Biaxial stress effect of the heavy hole, light hole and spin–orbit split–off bands in GaAs, where the shift of the valence bands is calculated by 6×6 matrix Hamiltonian, where the top of the valence bands at $X = 0$ is set to 0. The stress $X < 0$ is compressive and $X > 0$ is tensile. In the tensile stress $X > 0$ the light hole band is located higher than the heavy hole band

Defining the biaxial strain $e_{\parallel}$, we find the following relations

$$e_{xx} = e_{yy} \equiv e_{\parallel}\,, \tag{9.192a}$$

$$X = \frac{1}{s_{11} + s_{12}} e_{\parallel}\,. \tag{9.192b}$$

Figure 9.41 shows the biaxial stress effect of the valence band edges in GaAs calculated by solving 6×6 Hamiltonian of (9.161). In contrast to the case of uniaxial stress, we find that the band edge locations of the light hole band is higher than the heavy hole band in the tensile biaxial stress. Typical examples of the three valence bands in GaAs for the compressive biaxial stress $X = -0.2 \times 10^{10}$ [N/m^2] and for the tensile biaxial stress $X = +0.2 \times 10^{10}$ [N/m^2] are plotted as a function of $\boldsymbol{k}$, in Figs. 9.42 and 9.43, respectively. These results are obtained by solving the 6×6 matrix Hamiltonian of (9.161). Assuming a biaxial strain $e_{\parallel} = 0.5\%$, then the biaxial stress in GaAs is given by $X = +0.0617$ [10^{10}N/m^2], which gives the energy separation $\mathcal{E}_{\text{lh}}(0) - \mathcal{E}_{\text{hh}}(0) = 34$ [meV] and thus the strain effect cannot be neglected.

Since the strains is biaxial in a strained quantum well, the upper edge of the valence band is the edge of the heavy hole band for a compressive strain, and is the edge of the light hole band for a tensile strain, as seen in Fig. 9.41. Therefore the major transition is the electron–HH transition for the compressive strain and is the electron–LH transition for the tensile strain. The results indicate that the TE modes is excited in a quantum well with a compressive strain (or in un–strained quantum well) and that the TM mode oscillations will be observed in a quantum well laser with a tensile strain.

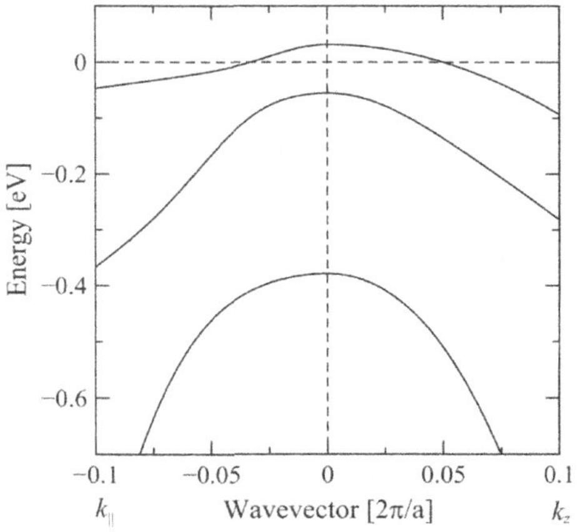

Fig. 9.42 Three valence bands of GaAs under the compressive biaxial stress of $X = -0.2 \times 10^{10}$ [N/m^2] as a function of the wave vector calculated from the 6 × 6 Hamiltonian given by (9.161). The wave vectors $k_{\parallel}$ $(= \sqrt{2}k_x = \sqrt{2}k_y)$ and k_z are in units of $(2\pi/a)$

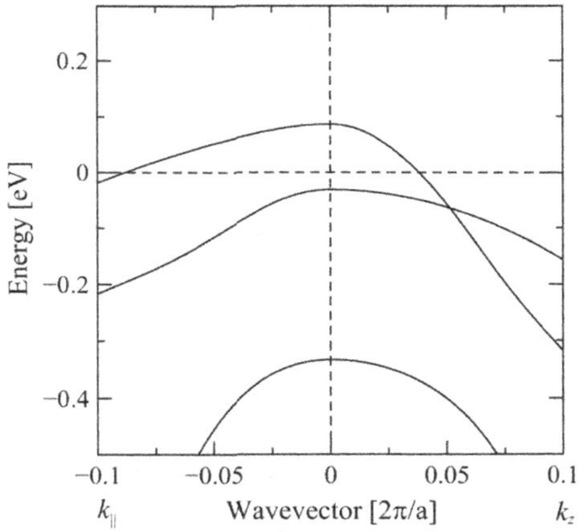

Fig. 9.43 Three valence bands of GaAs under the tensile biaxial stress of $X = +0.2 \times 10^{10}$ [N/m^2 as a function of the wave vector calculated from the 6 × 6 Hamiltonian given by (9.161). The wave vectors $k_{\parallel}$ $(= \sqrt{2}k_x = \sqrt{2}k_y)$ and k_z are in units of $(2\pi/a)$

In Table 9.1 we summarize the parameters of GaAs used in the present calculations together with some other materials. Detailed information of the band parameters in various semiconductors are well reviewed by Van de Walle [38] and Vurgaftman, Meyer and Ram–Mohan [39], and also listed by Chiao and Chuang [40] for GaAs, InAs, and InP which are used to analyze the valence band structures. Note here that the conduction and valence band deformation potentials of GaAs used in the present calculations are determined to match the present calculations with the piezoreflectance data of Pollak and Cardona [41]. For example, using $\mathcal{E}_G = 1.415$ [eV], the transition energies for the compressive uniaxial stress $X = -0.08 \times 10^{10}$ [N/m^2] are obtained as,

$$\mathcal{E}_c - \mathcal{E}_{LG} = 1.416, \qquad \mathcal{E}_c - \mathcal{E}_{HH} = 1.46, \qquad \mathcal{E}_c - \mathcal{E}_{SO} = 1.78\,, \tag{9.193}$$

Table 9.1 Material parameters of GaAs, InAs and InP at 300 K. Note that γ's are given by positive values and thus the valence band parameters A, B, and C (negative values) should be defined by the relations (9.150c). The value $a_c + a_v$ gives the total band gap dependence on the hydrostatic pressure. The listed parameters are estimated from the data reported by Van de Walle [38], Vurgaftman, Meyer and Ram–Mohan [39], and Chiao and Chuang [40]

Parameters	GaAs	InAs	InP
a [Å]	5.6533	6.0583	5.8688
$\mathcal{E}_G$ [eV]	1.425	0.36	1.344
Δ_0 [eV]	0.34	0.38	0.11
γ_1	6.85	20.0	5.08
γ_2	2.1	8.5	1.60
γ_3	2.9	9.2	2.10
c_{11} [10^{10} N/m^2]	11.879	8.329	10.11
c_{12} [10^{10} N/m^2]	5.376	4.526	5.61
$a_c + a_v$ [eV]	−8.33	−6.08	−6.31
a_c [eV]	−7.17	−5.08	−5.04
a_v [eV]	−1.16	−1.00	−1.27
b [eV]	−1.7	−1.8	−1.8
m_c^*/m_0	0.067	0.027	0.077

while the experimental values of Pollak and Cardona [41] are

$$\mathcal{E}_c - \mathcal{E}_{LH} = 1.41, \quad \mathcal{E}_c - \mathcal{E}_{HH} = 1.47, \quad \mathcal{E}_c - \mathcal{E}_{SO} = 1.78, \tag{9.194}$$

all in units of [eV].

The values of the deformation potentials reported so far vary in a wide range and adjustment is recommended to fit the experimental values. Also it should be reminded that the deformation potentials are determined by using the definition of the 6×6 Hamiltonian used in the text. The shear deformation energy $\Delta^{\text{shear}} = +(b/2)(e_{xx} + e_{yy} - 2e_{zz})$ of this textbook is sometimes cited as $\Delta^{\text{shear}} = -(b/2)(e_{xx} + e_{yy} - 2e_{zz})$.

In this section we discussed the strain effect of GaAs grown on the (001) surface. Recently epitaxial layers grown on the different surfaces have been reported to achieve higher mobility. The electronic properties of (110) surface under [110] uniaxial stress have been shown to exhibit a quite different behavior by Kajikawa [42].

9.7 Wurtzite Semiconductor Lasers

Wide-band-gap semiconductors are very important to produce laser diodes in blue ray regions. Most of the Blu-rays, LEDs and LDs are fabricated by utilizing GaN, AlN, and InN, and their ternary compounds, which are wurtzite. Nakamura et al. [43, 44] reported blue–green light emitting diodes based on $In_xGa_{1-x}N/Al_yGa_{1-y}N$

quantum well structures and Akasaki [45] demonstrated stimulated emission by AlGaN/GaN/GaInN quantum well. Then Nakamura et al. [46] have demonstrated room–temperature continuous–wave operation of strained $In_xGa_{1-x}N/Al_yGa_{1-y}N$ multiple quantum-well laser diodes with a long lifetime. Now LEDs and LDs based on such materials are on the market. Here we discuss electronic properties and lasers of wurtzite semiconductors.

9.7.1 Energy Band Structure of Wurtzite Crystals

We learned from Chaps. 1, 4 and 5 that the optical properties of semiconductors are characterized by the energy band structures. However, most of the results concern with semiconductors of face-centered-cubic lattices. In Chaps. 1 and 2 we discussed energy band calculations of such crystals by the pseudopotential theory and by $\boldsymbol{k} \cdot \boldsymbol{p}$ perturbation theory. Here we will concern with the energy band calculation of GaN, AlN, InN and their ternary alloys of wurtzite crystal using the pseudopotential theory [47]. Up to now various methods of the energy band calculations of GaN, AlN, InN, and their ternary alloys have been reported, by using first principle method, local density approximation, empirical pseudopotential methods and so on [47–52]. Experimental data of nitrides and their ternary compounds have been reported by many workers [53–55]. Also detailed experiments and analysis of the valence bands have been reported [56–58]. Controversy about the narrow gap InN and detailed discussions on the nitrides and their ternary alloys are reviewed by Furgaftman and Meyer [59], Wu [60], and Van de Walle [61].

As far as the reported band structure calculations are concerned, the empirical pseudopotential theory gives the most reliable results. Since this method is discussed in Chap. 1, we will describe the method in detail for the case of III-V(N) nitride compound semiconductors. Energy band structures of wurtzite crystals are quite different from the diamond and zinc blende crystals because of the crystal symmetries. Nitride semiconductors such as GaN, AlN and InN belong to hexagonal crystal and the hexagonal closed pack structure is shown in Fig. 9.44, where the lattice distance in the basal plane is a and the next equivalent plane is displaced by $c = \sqrt{8/3}a$, resulting in a closed pack structure with atoms of radius $a/2$ with the internal structural parameter $u = (a/c)^2 = 3/8 = 0.375$. The equilibrium lattice and internal structural parameters of nitride crystals GaN, InN and AlN are shown in Table 9.2, where we find the internal structural parameters are very close to that of the hexagonal closed pack structure.

To begin with, we deduce the reciprocal lattice vectors and the first Brillouin zone of the hexagonal crystal. When we choose the rectangular coordinates consisting unit vectors $[\boldsymbol{e}_x, \boldsymbol{e}_y, \boldsymbol{e}_z]$, an example of a set of the primitive vectors $[\boldsymbol{a}, \boldsymbol{b}, \boldsymbol{c}]$ of the hexagonal lattice is defined as

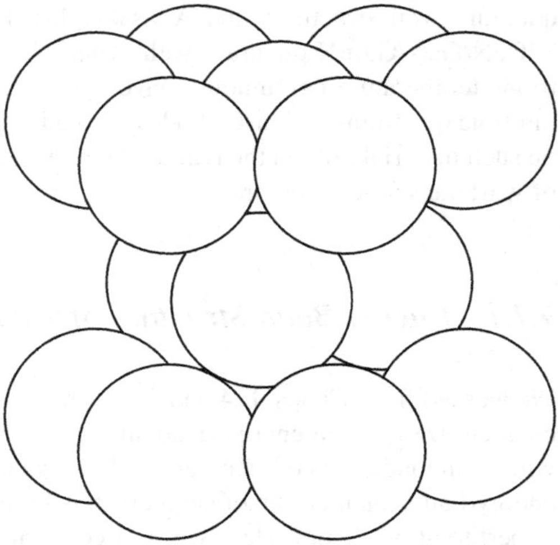

Fig. 9.44 Hexagonal closed pack crystal structure, where $c/a = \sqrt{8/3} = 1.63299$ with lattice constants a and c

Table 9.2 Equilibrium lattice and internal structural parameters of three nitride crystal GaN, InN and AlN after [48]. The sources of the experimental data are indicated by the reference numbers

	a [Å]	c [Å]	u
GaN	3.19 [62]	5.189 [62]	0.377 [63]
InN	3.544 [64]	5.718 [64]	0.38 [63]
AlN	3.11 [62]	4.98 [62]	0.38 [63]

$$\boldsymbol{a} = a\boldsymbol{e}_x, \qquad \boldsymbol{b} = a\left(\frac{1}{2}\boldsymbol{e}_x + \frac{\sqrt{3}}{2}\boldsymbol{e}_y\right), \qquad \boldsymbol{c} = c\boldsymbol{e}_z\,, \tag{9.195}$$

and then the reciprocal lattice vectors are given by

$$\boldsymbol{a}^* = \frac{1}{a}\left(\boldsymbol{e}_x - \frac{1}{\sqrt{3}}\boldsymbol{e}_y\right), \qquad \boldsymbol{b}^* = \frac{1}{a}\frac{2}{\sqrt{3}}\boldsymbol{e}_y, \qquad \boldsymbol{c}^* = \frac{1}{c}\boldsymbol{e}_z\,, \tag{9.196}$$

where the volume of the unit cell is $\boldsymbol{a} \cdot (\boldsymbol{b} \times \boldsymbol{c}) = 2(2\pi)^3/(\sqrt{3}a^2c)$. The reciprocal lattice vectors are then defined by using a set of integers (l, m, n),

$$\boldsymbol{G} = 2\pi(l\boldsymbol{a}^* + m\boldsymbol{b}^* + n\boldsymbol{c}^*) \tag{9.197}$$

which gives the squared magnitude of the reciprocal lattice vector

$$|\boldsymbol{G}|^2 = \left(\frac{2\pi}{a}\right)^2 \left[l^2 + \frac{(2m-l)^2}{3}\right] + \left(\frac{2\pi}{c}\right)^2 n^2 . \tag{9.198a}$$

For a hexagonal closed pack structure we put $c/a = \sqrt{8/3}$ and the reciprocal lattice vector is given by

$$|\boldsymbol{G}|^2 = \left(\frac{2\pi}{a}\right)^2 \left[l^2 + \frac{(2m-l)^2}{3} + \frac{3}{8}n^2\right] . \tag{9.198b}$$

Once we know the reciprocal lattice vectors, the first Brillouin zone is drawn by using electron wave vector $\boldsymbol{k}$ and reciprocal lattice vector $\boldsymbol{G}$. The boundaries of the first Brillouin zone is defined by

$$|\boldsymbol{k} + \boldsymbol{G}|^2 = \boldsymbol{G}^2, \qquad \boldsymbol{k} \cdot \boldsymbol{G} = |\boldsymbol{G}|/2 , \tag{9.199}$$

and therefore the distance between the Γ point ($\boldsymbol{k} = 0$) and the zone edge are $k_{\mathrm{M}} = 2\pi|\boldsymbol{a}^*|/2 = 2\pi|\boldsymbol{b}^*|/2 = (2\pi/a)/\sqrt{3}$ and the c plane is $k_{\mathrm{A}} = 2\pi|\boldsymbol{c}^*|/2 = (2\pi/c)/2$. Using these results we obtain the first Brillouin zone shown in Fig. 9.45. Since wurtzite crystal has six fold symmetry along the c axis, the box surrounded by the critical points is 1/24 of the first Brillouin zone as seen in Fig. 9.45.

When the electron wave vector $\boldsymbol{k} = (k_x, k_y, k_z)$ is defined as shown in Fig. 9.45, the free electron band is obtained from the following relation for the reciprocal lattice vector $\boldsymbol{G}$ given by (9.198b)

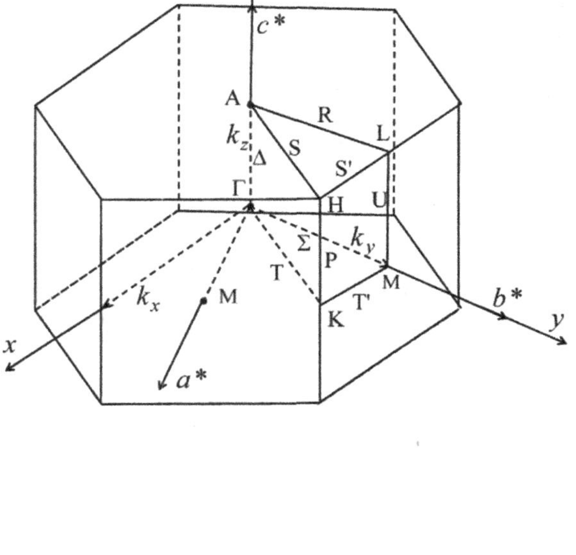

Fig. 9.45 The first Brillouin zone of a hexagonal crystal (C_{6v} = 6mm) is shown together with the critical points

$$\begin{aligned}\mathcal{E} &= \frac{\hbar^2}{2m}\nabla^2 = \frac{\hbar^2}{2m}\left[(k_x+G_x)^2+(k_y+G_y)^2+(k_z+G_z)^2\right] \\ &= \frac{\hbar^2}{2m}\left(\frac{2\pi}{a}\right)^2\left[(k_x+l)^2+\left(k_y+\frac{2m-l}{\sqrt{3}}\right)^2+\frac{3}{8}(k_z+n)^2\right], \end{aligned} \tag{9.200}$$

where (k_x, k_y, k_z) of the second line of (9.200) is redefined in units of $(2\pi/a)$, and the free electron bands are easily drawn by putting (k_x, k_y, k_z) in (9.200). Since the box surrounded by the critical points in Fig. 9.45 is one of the 24 equivalent boxes and thus we may rotate the box along the six fold symmetry axis c. The energy band calculations are carried out in a similar manner to the case of zinc blende crystals [47]. The pseudopotential for wurtzite may be written in a similar form to that of zinc blende crystal (see (1.92))

$$V_{\rm ps}(\boldsymbol{r}) = \sum_{\boldsymbol{G}}\left[S^{\rm S}(\boldsymbol{G})V^{\rm S}(|\boldsymbol{G}|)+{\rm i}S^{\rm A}(\boldsymbol{G})V^{\rm A}(|\boldsymbol{G}|)\right]{\rm e}^{-{\rm i}\boldsymbol{G}\cdot\boldsymbol{r}}, \tag{9.201}$$

where the symmetric and antisymmetric form factors $V^{\rm S}(\boldsymbol{G})$ and $V^{\rm A}(\boldsymbol{G})$ are given by the half sum and difference of the form factor for the two atoms in the unit cell, and the structure factor is written as (see (1.90c))

$$S^{\rm S}(\boldsymbol{G}) = \frac{1}{N_\alpha}\sum_j \exp(-{\rm i}\boldsymbol{G}\cdot\boldsymbol{\delta}_j), \tag{9.202a}$$

$$S^{\rm A}(\boldsymbol{G}) = -\frac{\rm i}{N_\alpha}\sum_j P_j \exp(-{\rm i}\boldsymbol{G}\cdot\boldsymbol{\delta}_j), \tag{9.202b}$$

where N_α is the number of atoms per unit cell (four for wurtzite), $\boldsymbol{\delta}_j$ is the basis vector of the jth atom in the unit cell, and the index j runs all over atoms in the cell. The operator P_j is $+1$ if j denotes one type of atom and -1 for the other type. The definition leads to the magnitude of the structure factor to be ≤ 1 as in the case of zinc blende crystal.

Explicit expressions for the wurtzite structure factors are obtained by choosing standard primitive translational vectors and place atoms of one type at the following two points,

$$\boldsymbol{r}_1 = \left[\frac{1}{6},\frac{1}{6},\frac{1}{2}\left(\frac{1}{2}+u\right)\right], \tag{9.203a}$$

$$-\boldsymbol{r}_2 = \left[-\frac{1}{6},-\frac{1}{6},-\frac{1}{2}\left(\frac{1}{2}-u\right)\right], \tag{9.203b}$$

and then atoms of the second type are located at $+\boldsymbol{r}_2$ and $-\boldsymbol{r}_1$. For the reciprocal lattice vector with (l, m, n) represents a relative translation of the primitive reciprocal lattice $[\boldsymbol{a}^*, \boldsymbol{b}^*, \boldsymbol{c}^*]$. The symmetric and antisymmetric structure factors are then given by

$$S^{\mathrm{S}} = \cos\left[2\pi\left(\frac{l}{6} + \frac{m}{6} + \frac{n}{4}\right)\right]\cos\left(\frac{2\pi n u}{2}\right), \tag{9.204a}$$

$$S^{\mathrm{A}} = \cos\left[2\pi\left(\frac{l}{6} + \frac{m}{6} + \frac{n}{4}\right)\right]\sin\left(\frac{2\pi n u}{2}\right). \tag{9.204b}$$

Once the pseudopotential form factors $V^{\mathrm{S}}(\boldsymbol{G})$ and $V^{\mathrm{A}}(\boldsymbol{G})$ are determined the energy band structure of wurtzite are easily calculated. We know from the energy band calculations of zinc blende crystals that the pseudopotential form factors $V^{\mathrm{S}}(|\boldsymbol{G}|^2)$ and $V^{\mathrm{A}}(|\boldsymbol{G}|^2)$ range form $V^{\mathrm{S,A}}(3)$ to $V^{\mathrm{S,A}}(11)$. Therefore it is very convenient to define the pseudopotential form factors in similar range. For this purpose we define the reciprocal lattice vector $|\boldsymbol{G}|$ of (9.198b) divided by the units of $\sqrt{2}\pi/a$ and thus we obtain

$$|\boldsymbol{G}|^2 = \left[\frac{8}{3}(l^2 + m^2 - lm) + \frac{3}{4}n^2\right]. \tag{9.205}$$

and then the free electron energy at the Γ point ($k_x = k_y = k_z = 0$) is given by the following relations

$$\begin{aligned} \mathcal{E}(\boldsymbol{k}=0) &= \frac{\hbar^2}{2m}\left(\frac{\sqrt{2}\pi}{a}\right)^2 2 \times \left[l^2 + \frac{(2m-l)^2}{3} + \frac{3}{8}n^2\right], \\ &= \frac{\hbar^2}{2m}\left(\frac{\sqrt{2}\pi}{a}\right)^2 |\boldsymbol{G}|^2, \end{aligned} \tag{9.206}$$

In the following we evaluate pseudopotential form factors by using newly defined $|\boldsymbol{G}|^2$ of (9.205), because many papers dealing with energy band calculations of wurtzite semiconductors use this definition. As described in Table 1.2 in Chap. 1 pseudopotentials of diamond and zinc blende type semiconductors $V(|\boldsymbol{G}|^2)$ becomes very small for $|\boldsymbol{G}|^2 > 11$ as illustrated in Fig. 6.10 for Si, and we may expect a similar behavior for the empirical pseudopotentials of wurtzite semiconductors. However, pseudopotentials of wurtzite are expected to be non-zero for many small values of $|\boldsymbol{G}|^2$ and thus we choose the new definition $|\boldsymbol{G}|^2$. Using this definition pseudopotentials of wurtzite semiconductors becomes very small beyond $|\boldsymbol{G}|^2 = 15$. As an example, pseudopotentials used for the energy band calculation of GaN are illustrated in Fig. 9.46, where $V^{\mathrm{S}}(|\boldsymbol{G}|^2)$ and $V^{\mathrm{A}}(|\boldsymbol{G}|^2)$ in units of [Ry] are plotted as a function of reciprocal lattice vector $|\boldsymbol{G}|^2$. Pseudopotentials are not continuous function of $|\boldsymbol{G}|$, but pseudopotential form factors are defined at typical values of $|\boldsymbol{G}|^2$. The reciprocal lattice vectors $|\boldsymbol{G}|^2$ and structure factors are also tabulated in Table 9.3 for $u = 3/8$. The pseudopotential form factors are estimated by the present author and N. Mori by modifying the reported values of Rezaei et al. [48] so that the calculated results agree with the recent experimental data, such as the band gap of InN (~0.7 eV) and the bowing of the direct band gaps for the ternary alloys as discussed below.

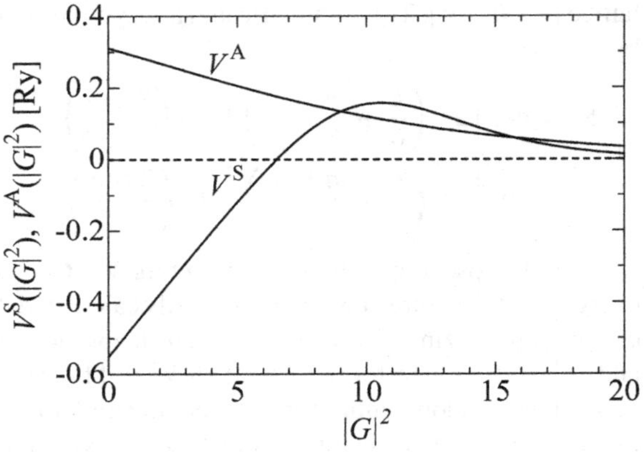

Fig. 9.46 Pseudopotentials V^s and V^A plotted as a function of reciprocal lattice vector $|\boldsymbol{G}|^2$ (defined in (9.205)) for wurtzite GaN, where parameters listed in Table 9.3 are used

Table 9.3 Reciprocal lattice vectors $\boldsymbol{G}$ and $|\boldsymbol{G}|^2$ (defined by (9.205)), structure factors $S^S(\boldsymbol{G})$ and $S^A(\boldsymbol{G})$, and pseudopotential form factors $V^S(|\boldsymbol{G}|^2)$ and $V^A(|\boldsymbol{G}|^2)$ in units of [Ry] of wurtzite semiconductors GaN, InN, and AlN. The pseudopotential form factors are estimated from the analytical equation (9.210) using the coefficients of the pseudopotentials form factors listed in Table 9.4

$\boldsymbol{G}$ (l, m, n)	$\|\boldsymbol{G}\|^2$	$\|S^S(\boldsymbol{G})\|$	$\|S^A(\boldsymbol{G})\|$	GaN		InN		AlN	
				V^S	V^A	V^S	V^A	V^S	V^A
000	0	1	0						
001	$\frac{3}{4}$	0	0						
100	$2\frac{2}{3}$	$\frac{1}{2}$	0	−0.319		−0.323		−0.236	
002	3	0.71	0.71	−0.290	0.247	−0.291	0.183	−0.212	0.296
101	$3\frac{5}{12}$	0.33	0.80	−0.254	0.238	−0.252	0.180	−0.181	0.277
102	$5\frac{2}{3}$	0.35	0.35	−0.066	0.192	−0.048	0.160	−0.021	0.192
003	$6\frac{3}{4}$	0	0						
210	8	1	0	0.091		0.099		0.119	
211	$8\frac{3}{4}$	0	0						
103	$9\frac{5}{12}$	0.80	0.33	0.144	0.128	0.120	0.128	0.172	0.101
200	$10\frac{2}{3}$	$\frac{1}{2}$	0	0.157		0.102		0.188	
212	11	0.71	0.71	0.156	0.106	0.094	0.115	0.187	0.077
201	$11\frac{5}{12}$	0.33	0.80	0.152	0.101	0.084	0.112	0.183	0.072
004	12	0.00	1.00		0.094		0.107		0.064
202	$13\frac{2}{3}$	0.35	0.35	0.105	0.077	0.036	0.095	0.130	0.048
104	$14\frac{2}{3}$	0.00	0.50		0.068		0.088		0.041
213	$14\frac{3}{4}$	0	0						

Table 9.4 The coefficients a_i of the pseudopotential form factor $V^S(q)$ and $V^A(q)$ functions used for the present calculations, which are readjusted to fit experimental data by modifying the form factors reported by Rezaei et al. The functions are defined by (9.210) after Rezaei et al. [48]

	$V^S(q)$				$V^A(q)$			
	a_1	a_2	a_3	a_4	a_1	a_2	a_3	a_4
GaN	0.04258	13.079	0.226	20.393	0.5114	−20.122	0.0835	−41.557
InN	0.0459	12.542	0.299	17.691	0.0221	−35.605	0.0574	−18.261
AlN	0.0363	11.960	0.234	22.233	0.0323	−145.212	0.0947	−19.160

Energy band calculations of wurtzite semiconductors are straight forward by solving the pseudopotential Hamiltonian

$$H_{\text{ps}} = -\frac{\hbar^2}{2m}\nabla^2 + V_{\text{ps}}(\boldsymbol{r}), \tag{9.207}$$

where $V_{\text{ps}}(\boldsymbol{r})$ is the pseudopotential that can be expanded in reciprocal lattice vectors $\boldsymbol{G}$ as:

$$V_{\text{ps}}(\boldsymbol{r}) = \sum_{\boldsymbol{G}} V_{\text{ps}}(\boldsymbol{G})\mathrm{e}^{-\mathrm{i}\boldsymbol{G}\cdot\boldsymbol{r}}. \tag{9.208}$$

In the present calculations the spin–orbit interactions are excluded. Using the definition of the structure factors derived above, the matrix elements of the pseudopotential Hamiltonian between the plane waves states $|\mathrm{i}(\boldsymbol{k}+\boldsymbol{G}_i)\rangle = \exp(-\mathrm{i}(\boldsymbol{k}+\boldsymbol{G}_i)\cdot\boldsymbol{r})$ and $|\mathrm{i}(\boldsymbol{k}+\boldsymbol{G}_j)\rangle = \exp(-\mathrm{i}(\boldsymbol{k}+\boldsymbol{G}_j)\cdot\boldsymbol{r})$ are given by

$$\begin{aligned} V_{\text{ps}}(\boldsymbol{G}_i - \boldsymbol{G}_j) = {} & S^{\text{S}}(\boldsymbol{G}_i - \boldsymbol{G}_j)V^{\text{S}}(|\boldsymbol{G}_i - \boldsymbol{G}_j|^2) \\ & + \mathrm{i}S^{\text{A}}(\boldsymbol{G}_i - \boldsymbol{G}_j)V^{\text{A}}(|\boldsymbol{G}_i - \boldsymbol{G}_j|^2) \end{aligned} \tag{9.209}$$

In the present calculation we adopt a suitable analytical expression for the pseudopotential form factors reported by Rezaei et al. [48]

$$V^{\text{S,A}}(q) = \frac{a_1(2q^2 - a_2)}{1 + \exp[a_3(2q^2 - a_4)]}, \tag{9.210}$$

where $q^2 = |\boldsymbol{G}|^2$ is squared reciprocal lattice vector defined by (9.205), and the coefficients a_j are determined from the fitting procedure of the energy bands (Table 9.4).

Calculated energy band structures of GaN, AlN and InN are shown in Figs. 9.47, 9.48 and 9.49, respectively, where 525 plane waves plane waves are used to form matrix elements. We find that all the nitrides AlN, GaN and InN with wurtzite structure have direct band gaps and agree well with the reported band gaps. Recent investigation reveals that the band gap of InN is quite small and ranges in the range 0.6 $\sim$ 0.9 eV as reported by the review articles of Furgaftman and Meyer [59], Wu [60], and Van de Walle [61]. The advance in the growth technology provided high quality nitrides and band parameters such as GaN [57], AlN [58] and GaInN [54]. The most

Table 9.5 Composition dependence of the direct band gaps of ternary nitrides $Al_xGa_{1-x}N$, $Al_xIn_{1-x}N$, and $Ga_xIn_{1-x}N$ calculated by the empirical pseudopotential method. The band gaps in the braces ( ) and the bowing parameter b are the recommended values of Vurgaftman and Meyer [59] based on various experimental data. The bowing parameter b_{ps} is introduced for the ternary alloys to estimate the empirical pseudopotentials (see (9.214))

$\mathcal{E}_G(x)$ [eV] $= \mathcal{E}_G(1)x + \mathcal{E}_G(0)(1-x) - bx(1-x)$

Ternary alloys	$\mathcal{E}_G(1)$	$\mathcal{E}_G(0)$	b	b_{ps}
$Al_xGa_{1-x}N$	6.183 (6.25)	3.510 (3.510)	0.7	0.10
$Al_xIn_{1-x}N$	6.183 (6.25)	0.771 (0.78)	2.5	0.86
$Ga_xIn_{1-x}N$	3.510 (3.510)	0.771 (0.78)	1.4	0.30

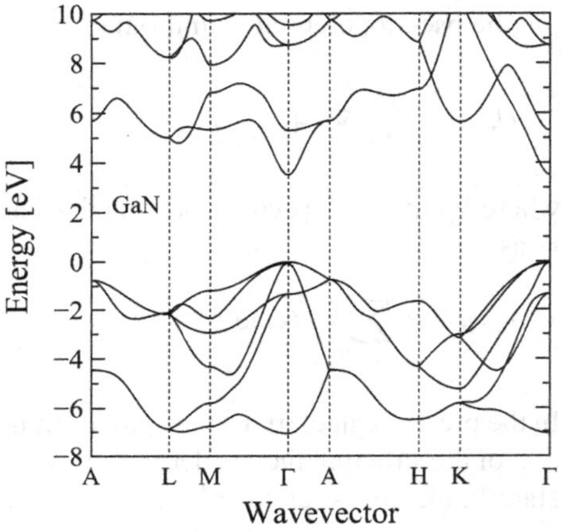

Fig. 9.47 Energy band structure of GaN calculated by empirical pseudopotential method

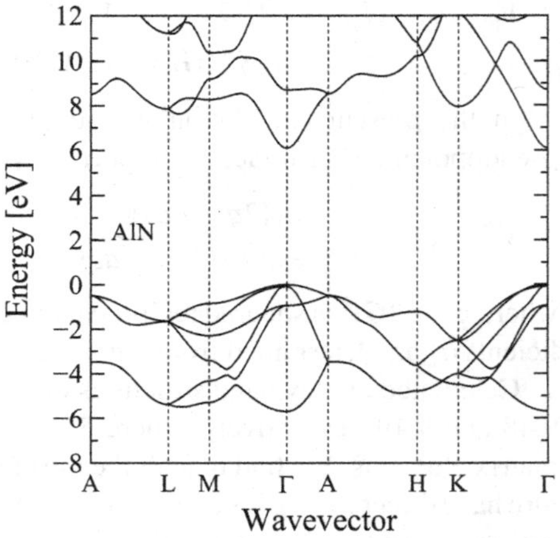

Fig. 9.48 Energy band structure of AlN calculated by empirical pseudopotential method

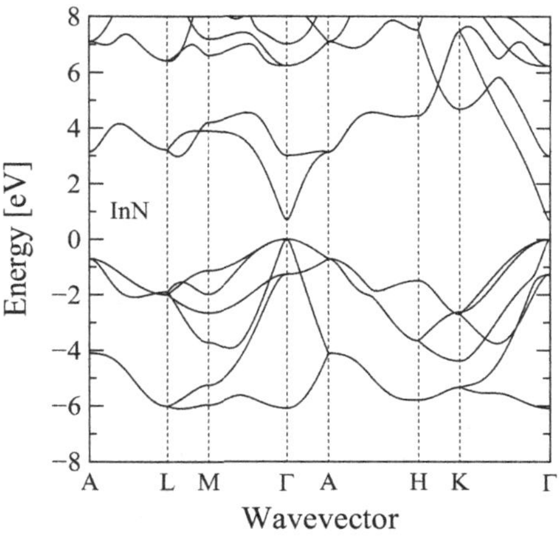

Fig. 9.49 Energy band structure of InN calculated by empirical pseudopotential method

important feature of the nitrides is the narrow band gap on InN which recalls revised calculations of the band structures. Energy band calculations are then carried out to explain the new results by empirical pseudopotential, *ab initio* and density functional method [50–52]. These results are reviewed by Furgaftman and Meyer [59], Wu [60], and Van de Walle [61].

9.7.2 Bowing of the Band Gaps and the Effective Masses in the Ternary Alloys

The composition dependence of the ternary alloys exhibit **bowing** (convex) and is well approximated by a parabolic relation for $A_xB_{1-x}N$ as

$$\mathcal{E}_G(A_xB_{1-x}N) = \mathcal{E}_G(AN) \cdot x + \mathcal{E}_G(BN) \cdot (1-x) - b \cdot x(1-x)\,, \tag{9.211}$$

where the linear relation between the composition x and the lattice constant a is assumed;

$$a(A_xB_{1-x}N) = a(AN) \cdot x + a(BN) \cdot (1-x)\,. \tag{9.212}$$

Above relation (9.212) is called Vegard's law [65]. The composition dependence of the direct band gaps is calculated by the pseudopotential method. When we assume the Vegard's law for the pseudopotentials;

$$V^S(A_xB_{1-x}N) = V^S(AN) \cdot x + V^S(BN) \cdot (1-x)\,, \tag{9.213}$$

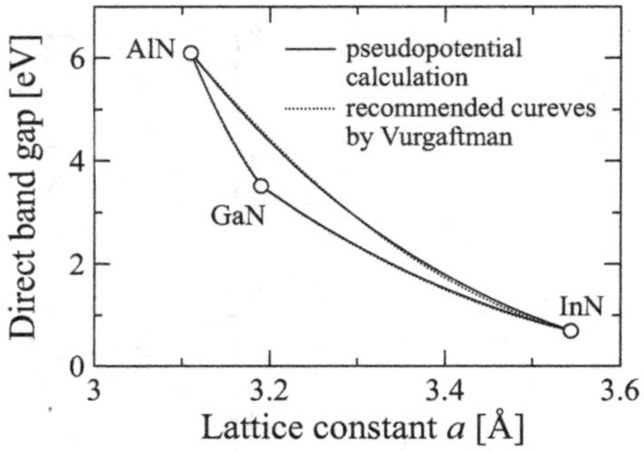

Fig. 9.50 Direct band gaps of ternary alloys as a function of the lattice constants, where the lattice constants of the ternary alloys are assumed to obey Vegard's law. The solid curves are calculated by the empirical pseudopotential method with one additional parameter b_{ps} to explain the bowing effect. The recommended curves for the ternary alloys by Vurgaftman and Meyer [59] are shown by *dotted lines*

and similar relation for V^A, the calculated results do not agree with the experimental data, but the curves are convex instead. We propose the following method to estimate pseudopotentials by introducing the bowing parameter b_{ps} for the pseudopotentials;

$$V^S(A_xB_{1-x}N) = V^S(AN) \cdot x - b_{ps} V^S_{ave} \cdot x(1-x) + V^S(BN) \cdot (1-x) \quad (9.214)$$

and similar relation for V^A. In (9.214), V^S_{ave} (similar relation for V^A_{ave}) is the average value of the binary crystals

$$V^S_{ave} = \frac{1}{2}\left[V^S(AN) + V^S(BN)\right] \quad (9.215)$$

The most important point of the introduction of the bowing parameter for the pseudopotentials b_{ps} is that only one additional parameter gives a good agreement with the experiment data. Calculated direct band gaps of ternary alloys are shown in in Fig. 9.50, where the present calculations are compared with the recommended curves of Vurgaftman and Meyer [59] and we find a very good agreement with each other. Theoretical calculations of the direct band gaps of the ternary nitrides have been carried out by using the first principles (ab initio) methods [49–52], where they adjust atomic positions of the cations (In or Ga) in the supercells so that the calculated results agree well with experimental observation. Cluster with different number of the cations and possible configurations are discussed in detail Caetano et al. [49].

The compositional dependence of the ternary alloys of nitrides are very useful for band gap engineering. We have to note here that the energy band structures of ternary alloys $A_xB_{1-x}N$ are calculated by assuming that the lattice constants follow Vegard's law. Vurgaftman and Meyer [59] proposed the following relation for the observed compositional dependence of the band gaps

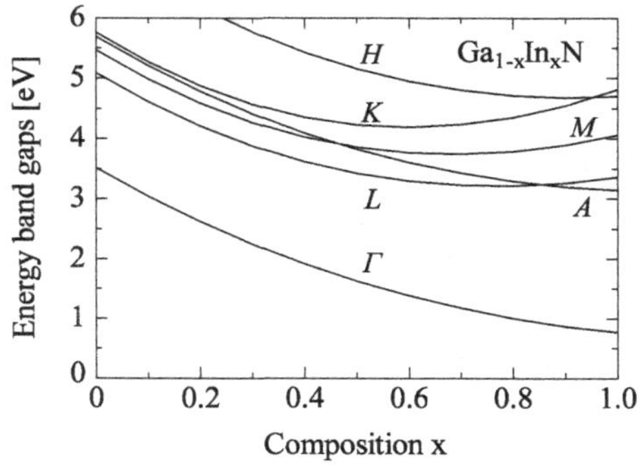

Fig. 9.51 Critical point energies at Γ, L, M, A, K, and H points as a function of the composition x for the ternary alloys $Ga_{1-x}In_xN$. The critical point energies are measured from the top of the valence band. The lowest conduction band is located at the Γ point and thus all the alloys are direct bang gap semiconductors

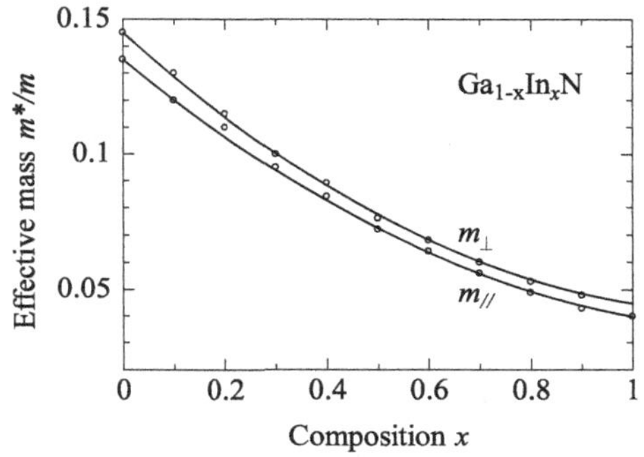

Fig. 9.52 Composition dependence of the band edge effective mass of $Ga_{1-x}In_xN$ obtained from the pseudopotential calculations of the energy bands of $Ga_{1-x}In_xN$. The solid curves are the best fitted empirical expressions with the bowing parameters as given by (9.218) and (9.219)

$$\mathcal{E}_G(A_xB_{1-x}N) = \mathcal{E}_G(AN) \cdot x + \mathcal{E}_G(BN) \cdot (1-x) - b \cdot x(1-x)\,, \tag{9.216}$$

where $-b \cdot x(1-x)$ is called "bowing term" and b is the bowing parameter. In Table 9.5 calculated energy gaps of AlN, GaN, InN, and the bowing parameters b of Vurgaftman and Meyer [59] and bowing parameters of pseudopotentials b_{ps} are summarized. Here we present the critical point energies with respect to the valence band top and the calculated band edge effective mass as a function of the composition x of the ternary alloys $Ga_{1-x}In_xN$ in Figs. 9.51 and 9.52, where we find that all of the ternary alloys have direct band gaps at the Γ point. The band edge effective masses are estimated from the energy band calculations based on the pseudopotential method as described below. As an example we present the conduction band dispersion of GaN near the Γ point in Fig. 9.53, where the conduction band energy is plotted in the vicinity of the Γ point in the direction $k_{\parallel}$ (parallel to the c–axis) and $k_{\perp}$ (in the basal plane). The solid lines shows the best fit curve of the non–parabolic band given by (2.97) of Chap. 2

$$\mathcal{E}\left(1 + \frac{\mathcal{E}}{\mathcal{E}_G}\right) = \frac{\hbar k^2}{2m^*} \tag{9.217}$$

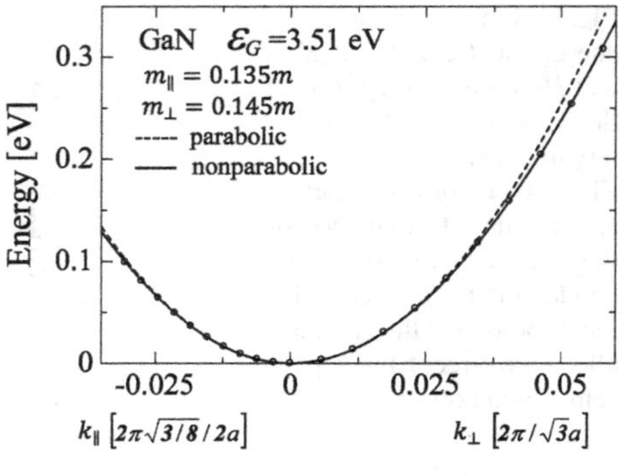

Fig. 9.53 Conduction band dispersion of GaN near the band edge. Data plotted by ∘ are the calculated result of the pseudopotential method and the solid curves are best fitted to the calculations with the masses $m_\parallel = 0.135$ m and $m_\perp = 0.145$ m by using the non–parabolic dispersion given by (2.97)

where $\mathcal{E}_G = 3.51$ eV is the energy gap and $m_\parallel$ and $m_\perp$ are the band edge effective masses in the direction $\parallel$ and $\perp$ (basal plane) to the c-axis. We find a very good agreement between the pseudopotential calculations and the non–parabolic band with $m_\parallel = 0.135$ m and $m_\perp = 0.145$ m. On the other hand the parabolic band dispersion deviates from the calculated energy band in the higher energy region. It is very interesting to point out that the effective mass anisotropy is quite small but the mass $m_\parallel < m_\perp$ in the whole range of the composition x. The composition dependence of the effective masses $m_\perp$ and $m_\parallel$ exhibits bowing characteristics and well approximated by the following relations. in the case of $Ga_(1-x)In_xN$

$$m_\perp/m = 0.145(1-x) - 0.070x(1-x) + 0.045x\,, \tag{9.218}$$

$$m_\parallel/m = 0.135(1-x) - 0.060x(1-x) + 0.040x\,. \tag{9.219}$$

We have to note here that the effective mass of the conduction band in GaN is $m^* = 0.20$ m reported so far by Vurgaftman et al. [59] as shown in Table 9.6, which is a little bit larger than the effective mass $m^* \simeq= 0.145$ m calculated by the pseudopotential energy band calculations.

In Fig. 9.54 valence band dispersions near the band edge at Γ point are plotted, where the pseudopotential calculations give the crystal field splitting $\Delta_{cr} = 0.0213$ eV and the heavy and light hole bands are slightly split in the $k_\perp$ direction. The solid curves are best fitted parabolic functions with $m_\parallel = m_\perp = 0.155m$, and the anisotropy of the heavy hole band is quite small, while the crystal field split valence band exhibits considerable anisotropy, $m_{ch\parallel} = 0.12m$ and $m_{ch\perp} = 0.16m$. Here we have to note that the spin–orbit interaction is not included in the present energy band calculations. The spin–orbit interaction results in additional valence band splitting of the heavy hole and light hole bands as discussed later. For comparison we estimate the valence band tops using the parameters of Vurgaftman et al. [59] in Table 9.6, $\Delta_{cr} = 0.010$ eV and $\Delta_{so} = 0.017$ eV, we obtain $\mathcal{E}_{hh} = 0$, $\mathcal{E}_{lh} = -0.0061$ eV, and $\mathcal{E}_{ch} = -0.0218$ eV, where we assumed $\Delta_2 = \Delta_3 = \Delta_{so}/3$.

Table 9.6 Energy band parameters of wurtzite nitrides at $T = 300\,\mathrm{K}$. (After Chuang and Park [71] and Vurgaftman and Meyer [59])

Parameters	GaN [71]	GaN [59]	AlN [59]	InN [59]
Lattice constant a [Å]	3.1892	3.189	3.112	3.545
Lattice constant c [Å]	5.1850	5.185	4.982	5.703
$\mathcal{E}_G$ [eV]	3.44	3.510	6.25	0.78
Δ_{cr} [eV]	0.016	0.010	−0.169	0.040
Δ_{so} [eV]	0.012	0.017	0.019	0.005
$m_e^{\parallel}/m_0$	0.20	0.20	0.32	0.07
$m_e^{\perp}/m_0$	0.18	0.20	0.30	0.07
A_1	−6.56	−7.21	−3.86	−8.21
A_2	−0.91	−0.44	−0.25	−0.68
A_3	5.65	6.68	3.58	7.57
A_4	5.65	−3.46	−1.32	−5.23
A_5	−3.13	−3.40	−1.47	−5.11
A_6	−4.86	−4.90	−1.64	−5.96
a_1 [eV]		−4.9	−3.4	−3.5
a_2 [eV]		−11.3	−11.8	−3.5
a [eV][a]	−8.16	−10.0	−9.0	−3.5
$a_c = 0.5a$ [eV][b]	−4.08	−5.0	−4.5	−1.75
D_1 [eV]	0.7	−3.7	−17.1	−3.7
D_2 [eV]	2.1	4.5	7.9	4.5
D_3 [eV]	1.4	8.2	8.8	8.2
D_4 [eV]	−0.7	−4.1	−3.9	−4.1
D_5 [eV]		−4.0	−3.4	−4.0
D_6 [eV]		−5.5	−3.4	−5.5
c_{11} [10^{10} N/m^2]	29.6	39.0	39.6	22.3
c_{12} [10^{10} N/m^2]	13.0	14.5	13.7	11.5
c_{13} [10^{10} N/m^2]	15.8	10.6	10.8	9.2
c_{33} [10^{10} N/m^2]	26.7	39.8	37.3	22.4
c_{44} [10^{10} N/m^2]	2.41	10.5	11.6	4.8
e_{31} [C/m^2]	(−1.7)[c]	−0.35	−0.50	−0.57
e_{33} [C/m^2]	(+3.4)[c]	1.27	1.79	0.97
P_{sp} [C/m^2][d]		−0.034	−0.090	−0.042

[a] Various values of the hydrostatic deformation potential a have been reported and the values listed above are not yet fixed.
[b] The hydrostatic deformation potential a_c is assumed to be $a_c = 0.5a$ (see [71]).
[c] The values in the parentheses are d_{31} and d_{33} [10^{-12} m/V] (from [66]).
[d] P_{sp} is the spontaneous polarization.

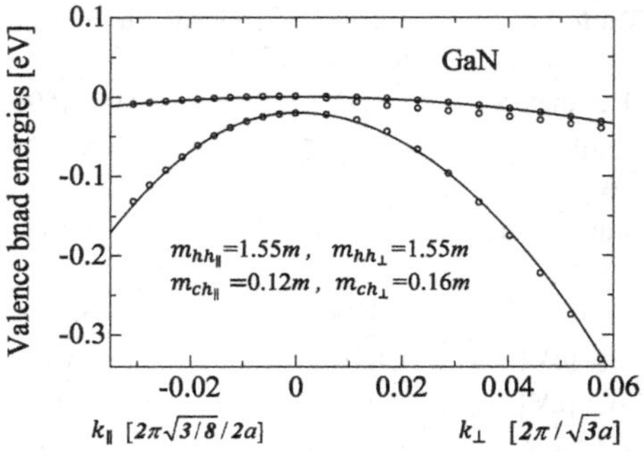

Fig. 9.54 Valence band dispersion of GaN near the band edge. The effective masses are determined by assuming a parabolic relation with the hole masses shown in the figure

9.7.3 Valence Band Structure in the Presence of Strain

In order to discuss the quantum well lasers based on the wurtzite crystals we have to note the following features. (1) The degeneracy of the valence bands of wurtzite is removed by the crystal field, and (2) strong piezoelectric field modifies the electronic and optical properties of the quantum well structures. However, the treatments of GaN based laser diodes are easily extended from the LDs based on the zinc blende type materials discussed in the previous sections. Strain Hamiltonian for wurtzite crystals are easily obtained from (4.168) by defining D_1, D_2, D_3, D_4, D_5, and D_6 using 4–th rank tensor for wurtzite (or D_{11}, D_{12}, D_{13}, D_{33}, D_{44}, and $D_{66} = (1/2)(D_{11} - D_{12})$), which are given by Park and Chuang [66] and also expressed by using different notation based on Picus–Bir strain Hamiltonian [67–70].

The valence band structure of strained wurtzite semiconductors with piezoelectric field is determined by the 6×6 Hamiltonian given by Park and Chuang [66]. In order to diagonalize the Hamiltonian we may use the same method used in the case of zinc blende semiconductors such as GaAs given in the previous sections. The energy states in strained quantum well of GaN are also calculated in a similar fashion as in the case of zinc blende semiconductors. The detailed treatment is given by Chuang [71].

The band structures of zinc blende and wurtzite crystals are illustrated in Fig. 9.55, where the effect of crystal field and the spin–orbit interactions in a wurtzite crystal are shown to result in the splitting of the valence bands, A, B and C. The crystal–field splitting leads to the band–edge energies (see the treatment of spin–orbit interaction given in Chap. 2):

$$\langle X|H_{\rm cr}|X\rangle = \langle Y|H_{\rm cr}|Y\rangle = \mathcal{E}_{\rm v} + \Delta_1\,, \tag{9.220a}$$

$$\langle Z|H_{\rm cr}|Z\rangle = \mathcal{E}_{\rm v}\,, \tag{9.220b}$$

where Δ_1 is the crystal–field–splitting energy between the $|X\rangle$, $|Y\rangle$ bands from the $|Z\rangle$ band. The spin–orbit splitting is parameterized by the following relations:

$$\langle X|H_{({\rm so})z}|Y\rangle = -{\rm i}\Delta_2\,, \tag{9.221a}$$

$$\langle Y|H_{({\rm so})x}|Z\rangle = \langle Z|H_{({\rm so})y}|X\rangle = -{\rm i}\Delta_3\,. \tag{9.221b}$$

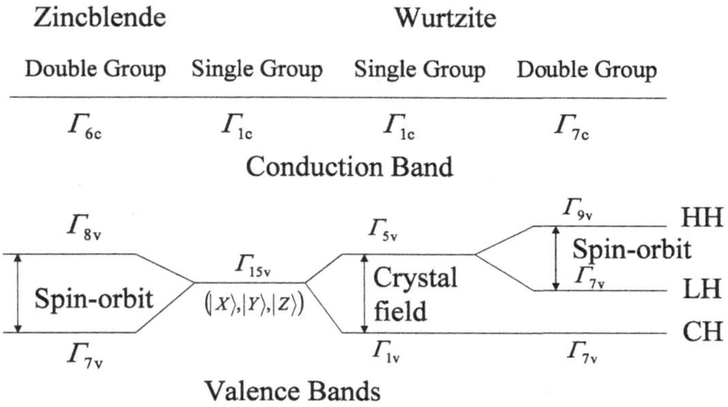

Fig. 9.55 Schematic illustration of the crystal–field splitting and spin–orbit splitting in wurtzite crystals as compared to zinc blende crystals

Here we will summarize the 6 × 6 Hamiltonian in the wurtzite symmetry, using the basis functions

$$|V_1\rangle = -\frac{1}{\sqrt{2}}|(X+\mathrm{i}Y)\uparrow\rangle\,,\ |V_2\rangle = \frac{1}{\sqrt{2}}|(X-\mathrm{i}Y)\uparrow\rangle\,,\ |V_3\rangle = |Z\uparrow\rangle\,,$$
$$|V_4\rangle = \frac{1}{\sqrt{2}}|(X-\mathrm{i}Y)\downarrow\rangle\,,\ |V_5\rangle = -\frac{1}{\sqrt{2}}|(X+\mathrm{i}Y)\downarrow\rangle\,,\ |V_6\rangle = |Z\downarrow\rangle\,.$$

The corresponding 6 × 6 matrix splits into two identical 3 × 3 matrices as shown by (2.59) of Chap. 2. The 3 × 3 matrix may be expressed by

$$\begin{vmatrix} \Delta_1+\Delta_2 & 0 & 0 \\ 0 & \Delta_1-\Delta_2 & \sqrt{2}\Delta_3 \\ 0 & \sqrt{2}\Delta_3 & 0 \end{vmatrix}\,. \tag{9.222}$$

The above equation gives the eigenstates for the valence–band edges, $\mathcal{E}_{\mathrm{HH}}$, $\mathcal{E}_{\mathrm{LH}}$, and $\mathcal{E}_{\mathrm{CH}}$,

$$\mathcal{E}_{\mathrm{HH}} = \Delta_1+\Delta_2\,, \tag{9.223a}$$
$$\mathcal{E}_{\mathrm{LH}} = \frac{1}{2}\left(\Delta_1-\Delta_2+\sqrt{(\Delta_2-\Delta_1)^2+8\Delta_3^2}\right)\,, \tag{9.223b}$$
$$\mathcal{E}_{\mathrm{CH}} = \frac{1}{2}\left(\Delta_1-\Delta_2-\sqrt{(\Delta_2-\Delta_1)^2+8\Delta_3^2}\right)\,. \tag{9.223c}$$

When we include the strain effect in wurtzite crystal, the 6 × 6 Hamiltonian is given by [66]

$$\begin{vmatrix} F & -K^* & -H^* & 0 & 0 & 0 \\ -K & G & H & 0 & 0 & \Delta \\ -H & H^* & \lambda & 0 & \Delta & 0 \\ 0 & 0 & 0 & F & -K & H \\ 0 & 0 & \Delta & -K^* & G & -H^* \\ 0 & \Delta & 0 & H^* & -H & \lambda \end{vmatrix}, \tag{9.224}$$

where

$$F = \Delta_1 + \Delta_2 + \lambda + \theta\,, \tag{9.225a}$$

$$G = \Delta_1 - \Delta_2 + \lambda + \theta\,, \tag{9.225b}$$

$$\lambda = \frac{\hbar^2}{2m_0}\left[A_1 k_z^2 + A_2(k_x^2 + k_y^2)\right] + \lambda_\epsilon\,, \tag{9.225c}$$

$$\theta = \frac{\hbar^2}{2m_0}\left[A_3 k_z^2 + A_4(k_x^2 + k_y^2)\right] + \theta_e\,, \tag{9.225d}$$

$$K = \frac{\hbar^2}{2m_0} A_5 (k_x + \mathrm{i}k_y)^2 + D_5 e_+\,, \tag{9.225e}$$

$$H = \frac{\hbar^2}{2m_0} A_6 (k_x + \mathrm{i}k_y)k_z + D_6 e_{z+}\,, \tag{9.225f}$$

$$\lambda_e = D_1 e_{zz} + D_2(e_{xx} + e_{yy})\,, \tag{9.225g}$$

$$\theta_e = D_3 e_{zz} + D_4(e_{xx} + e_{yy})\,, \tag{9.225h}$$

$$e_+ = e_{xx} - e_{yy} + 2\mathrm{i}e_{xy}\,, \tag{9.225i}$$

$$e_{z+} = e_{xz} + \mathrm{i}e_{yz}\,, \tag{9.225j}$$

$$\Delta = \sqrt{2}\Delta_3\,. \tag{9.225k}$$

Here the A_is are the valence–band effective–mass parameters which are similar to the Luttinger parameters in zinc blende crystals, and D_is are the deformation potentials for the wurtzite crystals. The crystal–field splitting energy is defined by $\Delta_{\mathrm{cr}} = \Delta_1$, and the spin–orbit splitting energy $\Delta_{\mathrm{so}} = 3\Delta_2 = 3\Delta_3$ is often used. The values of these parameters are tabulated for several wurtzite crystals by Vurgaftman [39], and by Park and Chuang [66]. The estimated values of the parameters are listed in Table 9.6.

The piezoelectric contribution to the total polarization is taken into account by the following relation for the piezoelectric polarization,

$$P_i^{\mathrm{pz}} = e_{ijk} e_{ij} = d_{ijk} T_{jk}\,, \tag{9.226}$$

where e_{ijk} and d_{ijk} are the piezoelectric coefficients and are often rewritten as $e_{i\alpha}$ (and $d_{i\alpha}$)

$$e_{ijk} = \begin{cases} e_{i\alpha}, & (i = 1, 2, 3;\ \alpha = 1, 2, 3)\,, \\ \frac{1}{2} e_{i\alpha}, & (i = 1, 2, 3;\ \alpha = 4, 5, 6)\,. \end{cases} \tag{9.227}$$

with the notations

tensor notation jk;	11	22	33	23, 32	13, 31	12, 21
matrix notation α;	1	2	3	4	5	6

The stress tensor T_{ij} is related to the strain tensor e_{kl} by

$$T_{ij} = c_{ijkl}e_{kl} \text{ or } e_{ij} = s_{ijkl}T_{kl} \tag{9.228}$$

where c_{ijkl} and s_{ijkl} are the elastic stiffness and elastic compliance constants, respectively (see also Appendix C for the stress–strain relations). Non–zero components of the piezoelectric constants $e_{i\alpha}$ (similarly for $d_{i\alpha}$) are $e_{14} = e_{15} = e_{16}$ for zinc blende crystals and $e_{15} = e_{24}$, $e_{31} = e_{32}$ and e_{33} for wurtzite crystals. The piezoelectric coefficients $e_{i\alpha}$ for zinc blende crystals are then given by

$$\begin{vmatrix} 0 & 0 & 0 & e_{14} & 0 & 0 \\ 0 & 0 & 0 & 0 & e_{14} & 0 \\ 0 & 0 & 0 & 0 & 0 & e_{14} \end{vmatrix}, \tag{9.229}$$

and for wurtzite crystals ($C_{6v} = 6mm$ symmetry)

$$\begin{vmatrix} 0 & 0 & 0 & 0 & e_{15} & 0 \\ 0 & 0 & 0 & e_{15} & 0 & 0 \\ e_{31} & e_{31} & e_{33} & 0 & 0 & 0 \end{vmatrix}. \tag{9.230}$$

Using the matrix notations for the strain tensors

$$\left|e_{xx}, e_{yy}, e_{zz}, \{e_{yz}, e_{zy}\}, \{e_{xz}, e_{zx}\}, \{e_{xy}, e_{yx}\}\right| = \left|e_1, e_2, e_3, e_4, e_5, e_6\right|,$$

ith component of the piezoelectric polarization is given by

$$P_i^{\text{pz}} = e_{i1}e_1 + e_{i2}e_2 + e_{i3}e_3 + e_{i4}e_4 + e_{i5}e_5 + e_{i6}e_6 . \tag{9.231}$$

Therefore the piezoelectric polarization components P_i^{pz} for zinc blende crystals are

$$P_i^{\text{pz}} = e_{14}(e_4 + e_5 + e_6) = e_{14}e_{jk}, \quad j \neq k, \tag{9.232}$$

and for wurtzite crystals are

$$\begin{aligned} P_x^{\text{pz}} &= e_{15}e_5 = e_{15}(e_{xz} + e_{zx}), \\ P_y^{\text{pz}} &= e_{15}e_4 = e_{15}\left(e_{yz} + e_{zy}\right), \\ P_z^{\text{pz}} &= e_{31}e_1 + e_{31}e_2 + e_{33}e_3 = e_{31}\left(e_{xx} + e_{yy}\right) + e_{33}e_{zz} . \end{aligned} \tag{9.233}$$

Calculating the electric field component $E^{\rm pz}$ due to the piezoelectric polarization[9] and rewriting the potential as $V(z)+eE^{\rm pz}z$, then we may solve the 6×6 Hamiltonian as described before in the case of (9.161). Various parameters of the wurtzite nitrides GaN, AlN, and InN are summarized in Table 9.6.

The strain effect on the GaN based lasers is treated as follows. Consider a quantum well structure of $\mathrm{Al}_x\mathrm{Ga}_{1-x}\mathrm{N}$ /GaN /$\mathrm{Al}_x\mathrm{Ga}_{1-x}\mathrm{N}$. The GaN well region is grown pseudomorphically along the c axis (z axis) on a thick $\mathrm{Al}_x\mathrm{Ga}_{1-x}\mathrm{N}$ layer and the strain tensor in the well region has the following components,

$$e_{xx} = e_{yy} = \frac{a_{\rm sub} - a_{\rm epi}}{a_{\rm epi}}\,, \tag{9.234a}$$

$$e_{zz} = -2\frac{c_{13}}{c_{33}}e_{xx}\,, \tag{9.234b}$$

$$e_{xy} = e_{yx} = e_{zx} = 0\,, \tag{9.234c}$$

where $a_{\rm sub}$ and $a_{\rm epi}$ are the lattice constants of the $\mathrm{Al}_x\mathrm{Ga}_{1-x}\mathrm{N}$ barrier (substrate) and the GaN well layers, respectively. The hydrostatic energy shift in the band–gap is written as

$$\mathcal{E}_{\rm G}(X) = a_1 e_{zz} + a_2\left(e_{xx} + e_{yy}\right) + \mathcal{E}_{\rm G}(0)\,, \tag{9.235}$$

where a_1 and a_2 are the overall deformation potentials parallel to the c axis and perpendicular to the c axis, respectively. Sometimes the anisotropy is neglected and approximation is made as $a_1 = a_2$ for simplicity. The values a and $a_{\rm c}$ shown in Table 9.6 are estimated by the following relation

$$\mathcal{E}_{\rm G}(X) \simeq a\left(e_{xx} + e_{yy} + e_{zz}\right) + \mathcal{E}_{\rm G}(0)\,, \tag{9.236a}$$

$$a \simeq \frac{a_1 + 2a_2}{3}\,, \tag{9.236b}$$

$$a_{\rm c} \simeq 0.5a\,. \tag{9.236c}$$

Here we have to note the validity of the reported values of the band parameters for wurtzite GaN, AlN, and InN listed in Table 9.6, where we revised the values of the data by Furgaftman et al. [39] (used in the second edition) by using new data by Furgaftman and Meyer [59]. These values are not yet well established for InN and all other data are subject to revision in the future. When we use the parameters reported by Furgaftman et al. [39], the strain dependence of the valence bands are too big, and the band edge of the valence band (HH) increases rapidly with the compressive strain, resulting in an anomalous behavior. Instead the parameters reported by Chuang [71] give a reasonable change in the valence bands labeled HH, LH, and CH, where HH, LH, and CH are heavy–hole–like, light–hole–like, and crystal–field (plus spin–orbit) split bands, respectively. Therefore the valence bands of GaN calculated by using the parameters of Chuang [71] are presented here. Figure 9.56 shows the conduction band edge $\mathcal{E}_{\rm c}$ and the three valence bands as a function of biaxial strain $e_{xx} = e_{yy}$ defined

[9]Electric displacement is given by $D_z = \epsilon_{zz}E_z + P_z^{\rm pz}$ with the dielectric constant ϵ_{zz}. No external charge gives rise to $D_z = 0$ and thus $E_z = -P_z^{\rm pz}/\epsilon_{zz}$.

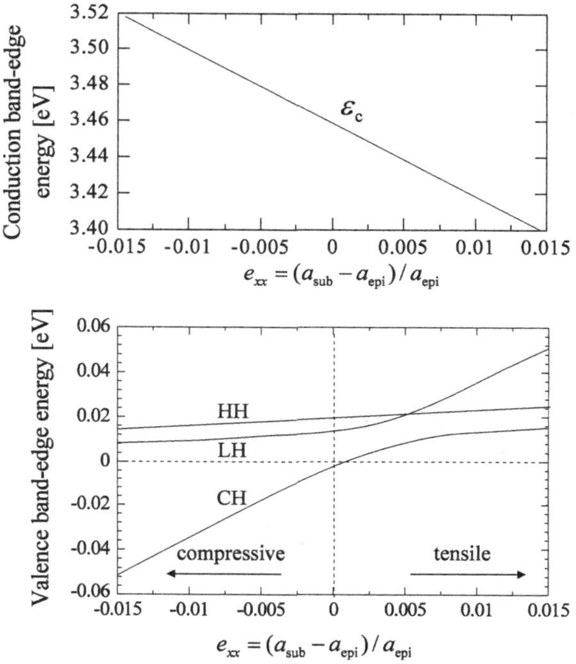

Fig. 9.56 Band–edge energies of wurtzite GaN as a function of biaxial strain $e_{xx} = e_{yy}$. The *upper figure* is for the conduction band edge and the *lower figure* shows the HH, LH, and CH bands as a function of biaxial strain

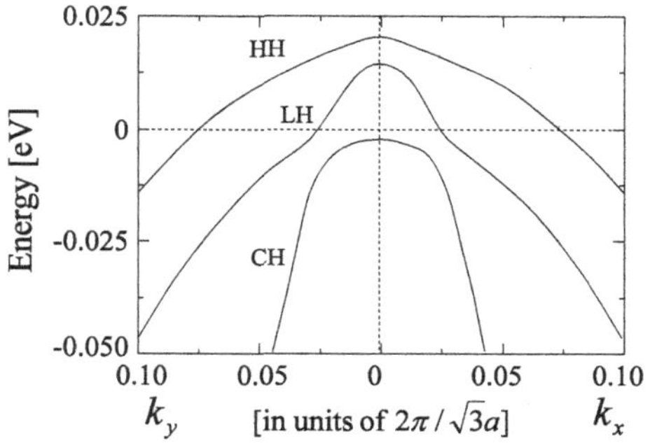

Fig. 9.57 Valence band dispersions of HH, LH, and CH bands without strain as a function of k_x and k_y in units of the zone boundary $\text{M} = (2\pi/a)[1/\sqrt{3}, 0, 0]$

by (9.234a)$\sim$ (9.234c). Here we find that the valence band ordering is unchanged for the compressive strain, but the ordering of HH and LH is changed at a higher tensile strain.

The first Brillouin zone of wurtzite crystal is shown in Fig. 9.45 with the critical point notations, where we have to note that the valence band dispersions in the case of the biaxial strain are the same for k_x and k_y. The valence band dispersions of HH, LH, and CH in GaN without strain are shown in Fig. 9.57. We find in Fig. 9.57 that the LH and CH bands exhibit anti-crossing due to the crystal–field and spin–orbit interactions. The valence band dispersions under biaxial strain given by (9.234a) $\sim$ (9.234c) are shown for a compressive biaxial strain of -0.005% ($e_{xx} = e_{yy} =$

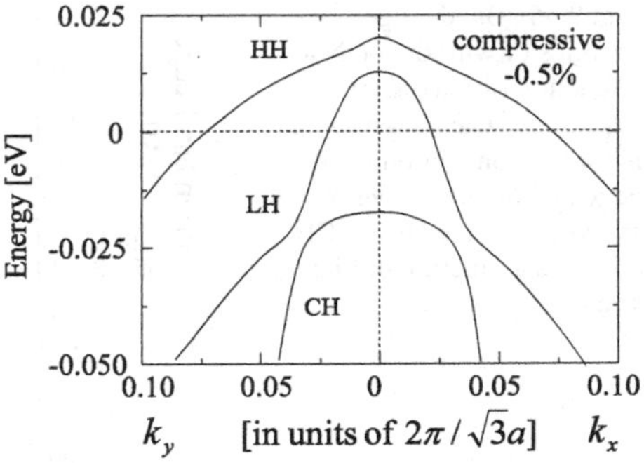

Fig. 9.58 Valence band dispersions of HH, LH, and CH bands with a compressive strain of -0.5% ($e_{xx} = e_{yy} = -0.005$) as a function of k_x and k_y in units of the zone boundary $(2\pi/a)[1/\sqrt{3}, 0, 0]$

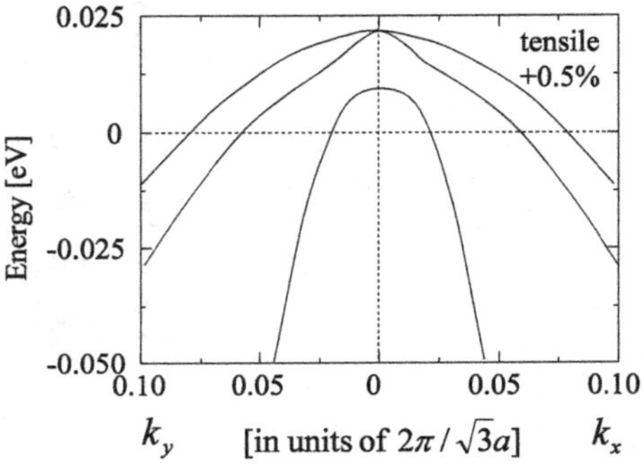

Fig. 9.59 Valence band dispersions of HH, LH, and CH bands with a tensile strain of $+0.5\%$ ($e_{xx} = e_{yy} = +0.005$) as a function of k_x and k_y in units of the zone boundary $(2\pi/a)[1/\sqrt{3}, 0, 0]$

-0.005) in Fig. 9.58 and for a tensile biaxial strain of $+0.005\%$ ($e_{xx} = e_{yy} = +0.005$) in Fig. 9.59. Here the energy bands are plotted as a function of k_x and k_y in the units of the zone boundary $\mathrm{M} = (2\pi/a)[1/\sqrt{3}, 0, 0]$. Chuang [71] reported the calculated results on the subband energies of the valence bands and the optical gain in a strained $Al_{0.3}Ga_{0.7}N$ /GaN /$Al_{0.3}Ga_{0.7}$ quantum–well structure. The band gaps of the nitrides calculated by the authors are shown in Figs. 9.47, 9.48, and 9.49, for GaN, AlN, and InN, respectively, and their ternary compounds are shown in Fig. 9.50 as a function of the lattice constants. Chuang [71] used a similar relation for a ternary compound and the band gap discontinuity is estimated by $\Delta\mathcal{E}_{\mathrm{v}} = 0.33\Delta\mathcal{E}_{\mathrm{G}}$. The lattice constants of GaN and AlN are $a = 3.19$ and 3.11 [Å], respectively, and the strain in $Al_xGa_{1-x}N/GaN/Al_xGa_{1-x}N$ ($x = 0.3$) is estimated by the linear interpolation of the lattice constants, giving rise to the compressive strain in GaN well region.

The subband energies in the valence bands are plotted in Fig. 9.60 for the well width $L_{\mathrm{w}} = 26$ [A] and Fig. 9.61 for the well width $L_{\mathrm{w}} = 50$ [A] as a function of k_{t}, where $k_{\mathrm{t}} = \sqrt{k_x^2 + k_y^2}$ is the magnitude of the wave vector in the (k_x, k_y) plane. From the results we find that HH1 and LH1 subbands are involved with the laser oscillations, and that the TE mode oscillations due to the transition between the electron subband and the HH1 subband dominate the laser oscillations.

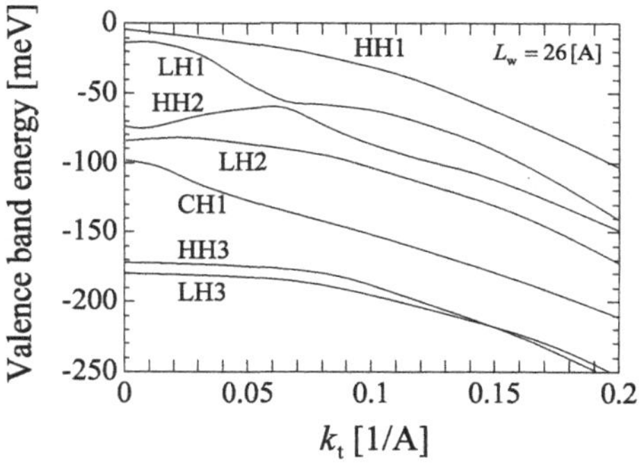

Fig. 9.60 The valence subband structures of a strained $Al_{0.3}Ga_{0.7}N$ /GaN/$Al_{0.3}Ga_{0.7}N$ quantum well with well width of $L_W = 26$ A. The reference energy is set at the valence band edge of unstrained GaN. After Chuang [71]

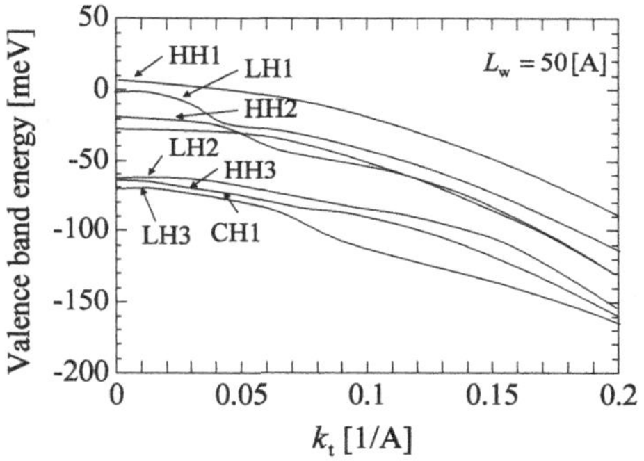

Fig. 9.61 The valence subband structures of a strained $Al_{0.3}Ga_{0.7}N$ /GaN/$Al_{0.3}Ga_{0.7}N$ quantum well with well width of $L_W = 50$ A. The reference energy is set at the valence band edge of unstrained GaN. After Chuang [71]

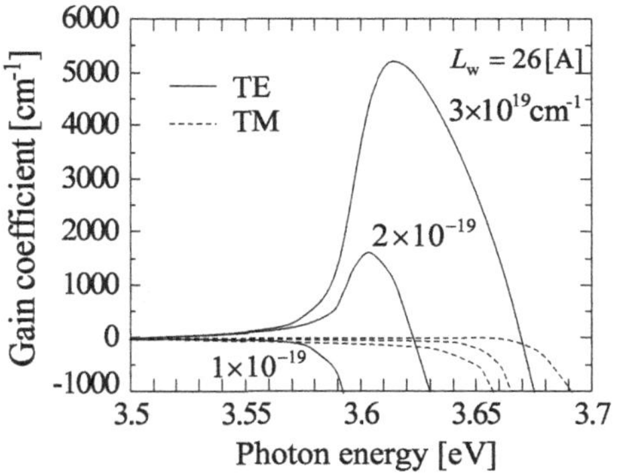

Fig. 9.62 Optical gain spectra for a strained $Al_{0.3}Ga_{0.7}N$ /GaN/$Al_{0.3}Ga_{0.7}N$ quantum well with well width of $L_W = 26$ A, at carrier concentrations $n = 1 \times 10^{19}$, 2×10^{19}, and 3×10^{19} cm^{-3}. The solid curves are for the TE mode polarization and the dashed curves are for the TM mode polarization. After Chuang [71]

9.7.4 Optical Gain of Nitride Quantum Well Structures

The calculations of the optical gain coefficients are straight forward. The results obtained by Chuang [71] are shown in Fig. 9.62 for the quantum well with well width $L_W = 26$ A and in Fig. 9.63 for the quantum well with well width $L_W = 50$ A for the carrier concentrations 1×10^{19}, 2×10^{19} and 3×10^{19} cm^{-3}, where solid curves are for TE mode and dashed curves are for TM mode. Here we see that the TE

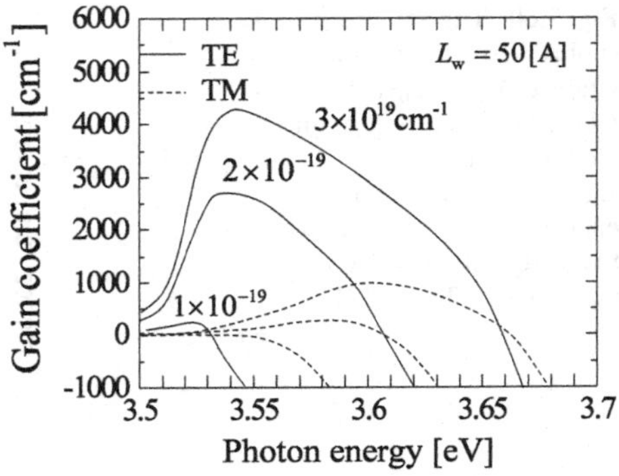

Fig. 9.63 Optical gain spectra for a strained $Al_{0.3}Ga_{0.7}N$ /$GaN/Al_{0.3}Ga_{0.7}N$ quantum well with well width of $L_w = 50$ A, at carrier concentrations $n = 1 \times 10^{19}$, 2×10^{19}, and 3×10^{19} cm^{-3}. The solid curves are for the TE mode polarization and the dashed curves are for the TM mode polarization. After Chuang [71]

mode gain is much higher than the TM mode gain, as expected from the valence band structures. The reason of the higher gain for TE mode is due the larger contribution from the HH subbands, while the TM mode oscillations arise from the LH subbands, as shown by (9.170) and (9.171).

9.8 Problems

(9.1) Luminescence experiments in direct gap semiconductors such as GaAs are used to evaluate hot electron distribution function. When a high electric filed is applied to a semiconductor, electrons are accelerated and resulting in hot electron state. Assuming the electron distribution is given by Maxwell–Boltzmann distribution function, plot luminescence spectra at $T_{\text{lattice}} = 77$ K and $T_{\text{electron}} = 150$ K for GaAs, where the energy band gap is assumed to be $\mathcal{E}_G = 1.50$ eV.

(9.2) Derive the relation (9.34) between the absorption coefficient α and stimulated emission rate r_{stim} for direct transition.

(9.3) Double heterostructure is very important for the quantum well lasers. Assume a double heterostructure laser consisting of the guiding region with higher refractive index $n_{\text{guide}} = 3.5$ sandwiched by symmetric cladding layers with lower refractive index $n_{\text{clad}} = 3.3$, and calculate the effective refractive index $\tilde{n}$.

(9.4) Using above result for the effective refractive index $\tilde{n}$, calculated the electric field distribution of TE_0 mode wave and discuss the confining of the waves in the guiding layer.

(9.5) Energy band calculations are carried out as follows: (1) calculate the reciprocal vectors $\boldsymbol{G}$ and then (2) plot the free electron bands in the full Brillouin zone.

References

1. G. Lasher, F. Stern, Spontaneous and atimulated recombination in semiconductors. Phys. Rev. **133**, A553–A563 (1964)
2. E.O. Kane, Phys. Rev. **131**, 79–88 (1963)
3. F. Stern, Phys. Rev. **148**, 186–194 (1966)
4. H.B. Bebb, E.W. Williams, in *Semiconductors and Semimetals*, ed. by R.K. Willardson, A.C. Beer. Photoluminescence I: Theory, vol. 8 (Academic Press, 1972) pp. 181–320
5. E.W. Williams, H.B. Bebb, , *Semiconductors and Semimetals*, ed. by R.K. Willardson, A.C. Beer Photoluminescence II: Gallium Arsenide, vol. 8 (Academic Press, 1972) pp. 321–392
6. N. Holonyak, M.H. Lee, *Semiconductors and Semimetals*, ed. by R.K. Willardson, A.C. Beer. Photopumped III-V Semiconductor Lasesr, vol. 14 (Academic Press, New York, 1979) Chapter 1, pp. 1–64
7. E.H. Bogardus, H.B. Bebb, Phys. Rev. **176**, 993 (1968)
8. W.P. Dumke, Phys. Rev. **132**, 1998 (1963)
9. D.M. Eagles, J. Phys. Chem. Solids **16**, 76 (1960)
10. E.W. Williams, H.B. Webb, J. Phys. Chem. Solids **30**, 1289 (1969)
11. P.J. Dean, J. Luminescence **1**(2), 398 (1970)
12. D.G. Thomas, J.J. Hopfield, C.J. Frosch, Phys. Rev. Lett. **15**, 857 (1965)
13. D.G. Thomas, J.J. Hopfield, Phys. Rev. **150**, 680 (1966)
14. J.S. Harris Jr., Semicond. Sci. Technol. **17**, 880 (2002)
15. J. Endicott, A. Patanè, J. Ibáñez, L. Eaves, M. Bissiri, M. Hopkinson, R. Airey, G. Hill, Phys. Rev. Lett. **91**, 126802 (2003)
16. H.P. Hjalmarson, P. Vogl, D.J. Worford, J.D. Daw, Phys. Rev. **44**, 810 (1980)
17. W. Shan, W. Wulukiewicz, J.W. Ager III, PHys. Rev. Lett. **82**, 1221 (1999)
18. P. Perlin, P. Wisniewski, C. Skiebiszewski, T. Suski, E. Kaminska, S.G. Subramanya, E.R. Weber, D. Mars, W. Walukiewcz, Appl. Phys. Lett. **76**, 1279 (2000)
19. M.I. Nathan, W.P. Dumke, G. Burns, F.H. Dill Jr., G. Lasher, Appl. Phys. Lett. **1**, 62 (1962)
20. R.N. Hall, G.E. Fenner, J.D. Kingsley, T.J. Soltys, R.O. Carlson, Phys. Rev Lett. **9**, 366 (1962)
21. T.M. Quist, R.H. Rediker, R.J. Keyes, W.E. Krag, B. Lax, A.L. McWhorter, H.J. Zeiger, Appl. Phys. Lett. **1**, 91 (1962)
22. N. Holonyak Jr., S.F. Bevacqua, Appl. Phys. Lett. **1**, 82 (1962)
23. I. Hayashi, M.B. Panish, P.W. Foy, S. Sumski, Appl. Phys. Leet **17**, 107 (1970)
24. T. Suhara: *Semiconductor Laser Fundamentals* (Marcel Dekker, New York, 2004) pp.1–308. This book deals with the mechanisms and device structures of semiconductor lasers in detail and is recommended to those readers who are interested in semiconductor lasers
25. H.C. Casey Jr., M.B. Panish, *Heterostructure Lasers* (Academic Press, Cambridge, 1978)
26. H.J. Lee, L.Y. Juravel, J.C. Wooley, A.J. Spring-Thorpe, Phys. Rev. B **21**, 695 (1980)
27. R.F. Kopf, M.H. Herman, M.L. Schnoes, J. Vac. Sci. Technol. **B11**, 813 (1993)
28. M.O. Watanabe, J. Yoshida, M. Mashita, T. Nakanisi, A. Hojo, J. Appl. Phys. **57**, 5340 (1985)
29. D. Arnold, A. Ketterson, T. Henderson, J. Klem, H. Morkoç, J. Appl. Phys. **57**, 2880 (1985)
30. H. Okumura, S. Misawa, S. Yoshida, S. Gonda, Appl. Phys. Lett. **46**, 377 (1985)
31. S.R. Forrst, P.H. Schmidt, R.B. Wilson, M.L. Kaplan, Appl. Phys. Lett. **45**, 1199 (1984)
32. J.M. Luttinger and W. Kohn: Phys. Rev. 97 (1955) 869
33. T. Ando, J. Phys. Soc. Jpn. **54**, 1528 (1985)
34. J.M. Luttinger, Phys. Rev. **102**, 1030 (1956)
35. J.M. Luttinger, W. Kohn, Phys. Rev. **97**, 869 (1955)
36. S. Kamiyama, T. Uenoyama, M. Mannoh, K. Ohnaka, IEEE J. Quantum Electron. **QE-31**, 1409 (1995)
37. M. Asada, Y. Suematsu, IEEE Quantum Electron. **QE-21**, 434 (1985)
38. C.G. Van de Walle, Phys. Rev. B **39**, 1871 (1989)
39. I. Vurgaftman, J.R. Meyer, L.R. Ram-Mohan, J. Appl. Phys. **89**, 5815 (2001)
40. C.Y.-P. Chiao, S.L. Chuang, Phys. Rev. B **46**, 4110 (1992)

41. F.H. Pollak, M. Cardona, Phys. Rev. **172**, 816 (1968)
42. Y. Kajikawa, Phys. Rev B **47**, 3649 (1993)
43. S. Nakamura, T. Mukai, M. Senoh, Jpn. J. Appl. Phys. **30**, L1998 (1991)
44. S. Nakamura, G. Fasol, *The Blue Green Diode* (Springer, Berlin, 1997)
45. I. Akasaki, H. Amano, S. Sota, H. Sakai, T. Tanaka, M. Koike, Jpn. J. Appl. Phys. **34**, L1517 (1995)
46. S. Nakamura, M. Senoh, S. Nagahama, N. Iwasa, T. Yamada, T. Matsushita, H. Kiyoku, Y. Sugimoto, T. Kozaki, H. Umemoto, M. Sano, K. Chocho, Appl. Phys. Leet. **72**, 2014 (1998)
47. M.L. Cohen, J.R. Chelikowsky, *Electronic Structure and Optical Properties of Semiconductors* (Springer, New York, 1989)
48. B. Rezaei, A. Asgari, M. Kalafi, Phys. B **371**, 107 (2006)
49. C. Caetano, L.K. Teles, M. Marques, A. Dal Pino, Jr., L.G. Ferreira, Phys. Rev. B **74**, 045215 (2006)
50. I. Gorczyca, S.P. Łepkowski, T. Suski, Phys. Rev. B **80**, 075202 (2009)
51. S. Schulz, M.A. Caro, L.-T. Tan, P.J. Parbrook, R.W. Martin, E.P. O'Reilly, Appl. Phys. Exp. **6**, 121001 (2013)
52. B.-T. Liou, S.-H. Yenb, Y.-K. Kuob, Semiconductor lasers and applications II, in *Proceedings of SPIE*, vol. 5628 (SPIE, Bellingham, WA, 2005), pp. 296–305
53. W. Terashima, S.-B. Che, Y. Ishitani, A. Yoshikawa, Jpn. J. Appl. Phys. **45**, L539 (2006)
54. M. Moret, B. Gil, S. Ruffenach, O. Briot, Ch. Giesen, M. Heuken, S. Rushworth, T. Leese, M. Succi, J. Cryst. Growth **311**, 2795 (2009)
55. T.V. Shubina, S.V. Ivanov, V.N. Jmerik, D.D. Solnyshkov, V.A. Vekshin, P.S. Kop'ev, A. Vasson, J. Leymarie, A. Kavokin, H. Amano, K. Shimono, A. Kasic, B. Monemar, Phys. Rev. Lett. **92**, 117407 (2004)
56. D.J. Dugdale, S. Brand, R.A. Abram, Phys. Rev. B **61**, 12933 (2000)
57. R. Ishii, A. Kaneta, M. Funato, Y. Kawakami, Phys. Rev. B **81**, 155202 (2010)
58. R. Ishii, A. Kaneta, M. Funato, Y. Kawakami, Phys. Rev. B **87**, 235201 (2013)
59. I. Vurgaftman, J.R. Meyer, Band parameters for nitrogen-containing semiconductors. J. Appl. Phys. **94**, 3675 (2003) (Applied Physics Reviews–Focused Review)
60. J. Wu, When group-III nitrides go infrared: New properties and perspectives. J. Appl. Phys. **106**, 011101 (2009) (Applied Physics Review)
61. C.G. Van de Walle, J. Neugebauer, First-principles calculations for defects and impurities: applications to III-nitrides. J. Appl. Phys. **95**, 3851 (2004) (Applied Physics Review)
62. H. Schulz, K.H. Thiemann, Solid State Commun. **23**, 815 (1977)
63. A.F. Wright, J.S. Nelson, Phys. Rev. B **51**, 7866 (1995)
64. K. Osamura, S. Naka, Y. Murakami, J. Appl. Phys. **46**, 3432 (1975)
65. L. Von Vegard, Z. Phys. **5**, 17 (1921)
66. S.-H. Park, S.-L. Chuang, Phys. Rev. B. **59**, 4725 (1999)
67. D.W. Langer, R.N. Euwema, K. Era, T. Koda, Phys. Rev. B. **2**, 4005 (1970)
68. G.E. Picus: Fiz. Tverd. Tela **6**, 324 (1964); Soviet Phys. Solid State, **6**, 261 (1964)
69. P.Y. Yu, M. Cardona, J. Phys. Chem. Solids **34**, 29 (1973)
70. K. Ando, C. Hamaguchi, Phys. Rev. B **11**, 3876 (1975)
71. S.L. Chuang, IEEE J. Quantum Electron. **32**, 1791 (1996)

Chapter 10
Answers for Problems

(1.1) The computer program to evaluate reciprocal vectors (in atomic units) is given below, where the program is written by using Octave (free software, compatible to MATLAB):
Efree=Gx^2+ Gy^2+ Gz^2; free electron energy
Plane waves 59 for Efree = 16 for example

```
% Start of program
R = 1;
for Nx = -4:4;
for Ny = -4:4;
for Nz = -4:4;
Efree = 16;
N2=(Nx - Ny + Nz)^2 + (Nx + Ny - Nz)^2 + (- Nx + Ny + Nz)^2;
if N2 ≤ = Efree;
Gx(R) = Nx - Ny + Nz;
Gy(R) = Nx + Ny - Nz;
Gz(R)= - Nx + Ny + Nz;
R = R + 1;
endif;
endfor;
endfor;
endfor;
Ncut = R - 1; % Number of plane waves for a given Efree
% End of plane waves
% A set of reciprocal lattice vectors is given by [Gx(R),Gy(R),Gz(R)]
% Compare the calculated results with Table 10.1.
% Use sort program to list from the lowest value to the highest value.
% Print [Gx(R),Gy(R),Gz(R)]
```


C. Hamaguchi, *Basic Semiconductor Physics*, Graduate Texts in Physics,
DOI 10.1007/978-3-319-66860-4_10

Table 10.1 Fundamental vectors $\boldsymbol{a}, \boldsymbol{b}, \boldsymbol{c}$, and unit cell volume v of single cubic, body centered, face centered and hexagonal closed pack crystals

	Simple cubic	Body centered cubic	Face centered cubic	Hexagonal c p
$\boldsymbol{a}$	$a\boldsymbol{e}_x$	$(a/2) - \boldsymbol{e}_x + \boldsymbol{e}_y + \boldsymbol{e}_z)$	$(a/2)(\boldsymbol{e}_y + \boldsymbol{e}_z)$	$a\boldsymbol{e}_x$
$\boldsymbol{b}$	$a\boldsymbol{e}_y$	$(a/2)(\boldsymbol{e}_x - \boldsymbol{e}_y + \boldsymbol{e}_z)$	$(a/2)(\boldsymbol{e}_x + \boldsymbol{e}_z)$	$(a/2)(\boldsymbol{e}_x + \sqrt{3}\boldsymbol{e}_y)$
$\boldsymbol{c}$	$a\boldsymbol{e}_z$	$(a/2)(\boldsymbol{e}_x + \boldsymbol{e}_y - \boldsymbol{e}_z)$	$(a/2)(\boldsymbol{e}_x + \boldsymbol{e}_y)$	$c\boldsymbol{e}_z$
v	a^3	$(1/2)a^3$	$(1/4)a^3$	$(\sqrt{3}/4)a^2c$

(1.2) Using Bohr radius $a_\mathrm{B} = 0.5.2917706\,[\text{Å}]$ and energy Rydberg Ry in atomic units for energy $\mathcal{E}$ and wave vector k in atomic units with a as lattice constant are given by

$$k = \frac{2\pi}{a} \to k \cdot a_\mathrm{B} = 2\pi \frac{a_\mathrm{B}}{a} \equiv D \text{ in [a.u]}$$

$$\mathcal{E} = \frac{\hbar^2}{2m * a_\mathrm{B}^2} = 2.1799 \times 10^{-18}\ [\mathrm{J}] = 1\,[\mathrm{Ry}] = 1[\mathrm{a.u}]$$

$$\mathrm{Ry}/e = 13.6077\ \ [\mathrm{eV}]$$

Ry = 13.6077; ! Rydberg in [eV]
See also (1.117a)–(1.117c) of the textbook.

(1.3) We define $D = 2\pi(a_\mathrm{B}/a)$ with the lattice constant a, and then wave vector $k_x = k$ in the first Brillouin zone for fcc crystal is given by $[-2\pi/a, +2\pi/a]$, which is rewritten as $[-1, +1]$ in atomic unit. Equation (1.35) is written as,

$$\mathcal{E}(k) = \frac{1}{2}\left\{D^2\left[k^2 + (k-1)^2 + 2\right]\right.$$
$$\left.\pm\sqrt{D^4[\boldsymbol{k}^2 - (\boldsymbol{k}-1)^2]^2 - 2|V(G_1)|^2}\right\}$$

Energy in [eV] is evaluated by $\mathcal{E} \times \mathrm{Ry}$ with $\mathrm{Ry} = 13.605\,[\mathrm{eV}]$.

(1.4) As shown in Fig. 1.7, the lower three free electron bands for $k_x = k$, $k_y = k_z = 0$ are

$$\mathcal{E}(k) = \mathrm{Ry} \times k^2,$$
$$\mathcal{E}(k) = \mathrm{Ry} \times (k \pm 1)^2 + 2,$$

which are plotted by dashed curves in Fig. 10.1.
As an example we calculate the energy band structure by nearly free electron approximation, taking account of the lowest two band. We use parameters of Si, where $a = 5.43\,[\text{Å}]$, $V(G_0) = 0$, $V(G_1) = V_3^\mathrm{s} = -0.21$ [a.u.], and $G_n = 1\pi n/a$ with $G_1 = \pm 1$ in [a.u.]. The calculated results are shown by solid curves in Fig. 10.1.

(1.5) The results are tabulated in Table 10.1. For hexagonal closed pack crystal refer Fig. 9.44 of Chap. 9.

(1.6) For hexagonal closed pack crystal refer Figs. 9.44 and 9.45 of Chap. 9. The results are summarized in Table 10.2.

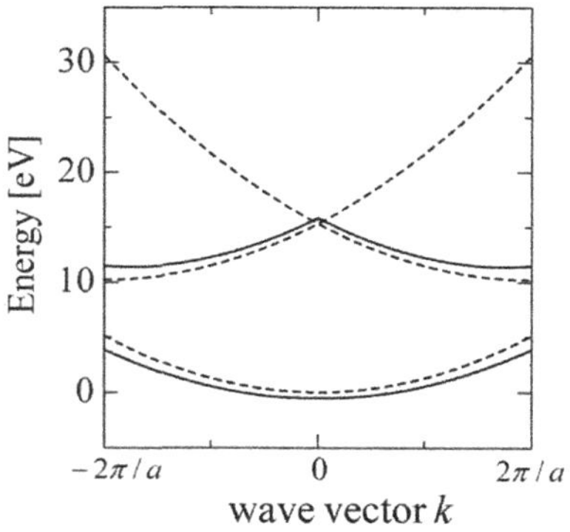

Fig. 10.1 Nearly free electron approximation based on 2 bands are shown by solid curves in the direction $[k_x, 0, 0]$. The energy is given by multiplying Ry $\times$ $D^2(k \pm 2)^2$

Table 10.2 Reciprocal lattice vectors $[\boldsymbol{a}^*, \boldsymbol{b}^*, \boldsymbol{c}^*]$ and the volume of the first Brillouin zone Ω_{BZ}

	Simple cubic	Body centered c	Face centered c	Hexagonal closed p
$\boldsymbol{a}^*$	$(1/a)\boldsymbol{e}_x$	$(1a)(\boldsymbol{e}_y + \boldsymbol{e}_z)$	$(1/a)(-\boldsymbol{e}_x + \boldsymbol{e}_y + \boldsymbol{e}_z)$	$(1/a)[\boldsymbol{e}_x - (1/\sqrt{3})\boldsymbol{e}_y]$
$\boldsymbol{b}^*$	$(1/a)\boldsymbol{e}_y$	$(1/a)(\boldsymbol{e}_x + \boldsymbol{e}_z)$	$(1/a)(\boldsymbol{e}_x - \boldsymbol{e}_y + \boldsymbol{e}_z)$	$(2/\sqrt{3}a)\boldsymbol{e}_y$
$\boldsymbol{c}^*$	$(1/a)\boldsymbol{e}_z$	$(1/a)(\boldsymbol{e}_x + \boldsymbol{e}_y)$	$(1/a)(\boldsymbol{e}_x + \boldsymbol{e}_y - \boldsymbol{e}_z)$	$(1/c)\boldsymbol{e}_z$
Ω_{BZ}	$(2\pi/a)^3$	$2\,(2\pi/a)^3$	$4\,(2\pi/a))^3$	$2(2\pi)^3/(\sqrt{3}a^2c)$

(1.7) Non–zero matrix element of the spin states is

$$\langle \uparrow |\sigma_z| \uparrow\rangle = 1\,,$$

and thus we obtain

$$-\mathrm{i}\hbar\langle X(\Gamma^l_{25'})\uparrow \left| \left(x\frac{\partial}{\mathrm{d}y} - y\frac{\partial}{\mathrm{d}x}\right)\sigma_z \right| Y(\Gamma^l_{25'})\uparrow\rangle$$
$$= -\mathrm{i}\hbar\langle X(\Gamma^l_{25'})\left| \left(x\frac{\partial}{\mathrm{d}y} - y\frac{\partial}{\mathrm{d}x}\right) \right| Y(\Gamma^l_{25'})\rangle\,.$$

Since the basis functions are given by $|X\rangle = |yz\rangle$ and $|Y\rangle = |zx\rangle$, non–zero component of the above equation results in

$$-\mathrm{i}\hbar\langle X(\Gamma^l_{25'})\left| \left(-y\frac{\partial}{\mathrm{d}x}\right) \right| Y(\Gamma^l_{25'})\rangle \equiv \mathrm{i}\Delta^l_{25'}/3\,,$$

where $\Delta^l_{25'}/3$ is defined to express the strength of the spin–orbit splitting. The result is just the same as (1.149a). In a similar fashion, (1.149b) and (1.149c) are derived.

(1.8) The symmetry points and the distances between the symmetry points are summarized in Table 10.3:

(2.1) Since cyclotron resonance condition is given by $\omega = \omega_c = eB/m^*$ with the microwave angular frequency $\omega = 2\pi \times 24 \times 10^9$, the required magnetic

Table 10.3 Symmetry points (critical points) and the lengths between the points for a face centered cubic crystal, where wave vector k is given in atomic units

Brillouin zone	Length	Symmetry points
Γ to X:	$\|k\| = 1$	[0,0,0] to [1,0,0]
X to W:	$\|k\| = 1/2$	[1,0,0] to [1,1/2,0]
W to K:	$\|k\| = \sqrt{2}/4$	[1,1/2,0] to [3/4,3/4,0]
K to Γ:	$\|k\| = 3/(2\sqrt{2})$	[3/4,3/4,0] to [0,0,0]
X to U (K):	$\|k\| = \sqrt{2}/4$	[1,0,0] to [1, 1/4, 1/4]
L to Γ:	$\|k\| = \sqrt{3}/2$	[1/2,1/2,1/2] to [0,0,0]

field is given by for
(1) $m^* = 0.3\,\text{m}$:

$$B = \frac{\omega \times m^*}{e} = \frac{2\pi 2.4 \times 10^{10} \times 0.3 \times 9.11 \times 10^{-31}}{1.602 \times 10^{-19}} = 0.257\,[\text{T}]$$

(2) $m^* = 0.067\,\text{m}$:

$$B = 0.0574\,[\text{T}]$$

(3) Cyclotron resonance condition $\omega\tau \gg 1$ is satisfied for Ge and Si at low temperature. As discussed in Chap. 6, electron mobility $\mu = e\tau/m^*$ and thus the condition is rewritten as $\omega_c\tau = \mu B \gg 1$. In the case of GaAs the resonance magnetic field is quite low and thus $\mu B \simeq 1$ ($\mu \simeq 1.0 \sim 10\,[\text{m}^2/\text{Vs}]$), not enough to observe clear resonance. Instead cyclotron resonance of GaAs is observed at higher frequency (infrared radiation) and at higher magnetic field at room temperature, where the condition $\mu B \gg 1$ is satisfied.

(2.2) We derive the following relation for an arbitrary scalar function $f(x)$:

$$[f(x), p_x] = i\hbar\frac{\partial f(x)}{\partial x} \equiv -p_x f(x)$$

We operate the commutation relation $[f(x), p_x]$ to an arbitrary scalar function $g(x)$, and we find

$$\begin{aligned}
[f(x), p_x]\, g(x) &= (f(x)p_x - p_x f(x))\, g(x) \\
&= -i\hbar\left(f(x)\frac{\partial g(x)}{\partial x} - \frac{\partial\,(f(x)g(x))}{\partial x}\right) \\
&= -i\hbar\left(f(x)\frac{\partial g(x)}{\partial x} - \frac{\partial f(x)}{\partial x}g(x) - f(x)\frac{\partial g(x)}{\partial x}\right) \\
&= +i\hbar\frac{\partial f(x)}{\partial x}g(x) \\
&\equiv -\,(p_x f(x))\, g(x)\,,
\end{aligned}$$

and then we obtain

$$[f(x), p_x] = i\hbar \frac{\partial f(x)}{\partial x} \equiv -p_x f(x) .$$

The commutation relation

$$[\boldsymbol{A}, \boldsymbol{p}] = \boldsymbol{A} \cdot \boldsymbol{p} - \boldsymbol{p} \cdot \boldsymbol{A} ,$$

is rewritten as follows by separating in xCyCz direction. Since $A_i \cdot p_j \equiv A_i p_j \delta_{i,j}$, only the case $i \neq j$ gives non–zero components and we obtain

$$[A_x, p_x] = -p_x A_x ,$$
$$\left[A_y, p_x\right] = -p_y A_y ,$$
$$\left[A_z, p_x\right] = -p_z A_z .$$

These relations give

$$[\boldsymbol{A}, \boldsymbol{p}] = -\boldsymbol{p} \cdot \boldsymbol{A} = \mathrm{i}\hbar\nabla \cdot \boldsymbol{A} .$$

Vector potential is defined as $\boldsymbol{A} = \nabla \times \boldsymbol{B}$ and $\nabla \cdot \boldsymbol{A} = \nabla \cdot \nabla \times \boldsymbol{B} = 0$. Therefore we obtain the following commutation relation

$$[\boldsymbol{A}, \boldsymbol{p}] = 0; \quad \boldsymbol{A} \cdot \boldsymbol{p} = \boldsymbol{p} \cdot \boldsymbol{A} .$$

(2.3) Non–parabolic conduction band is given by (2.97)

$$\frac{\hbar^2 k^2}{2m_0^*} \equiv \gamma(\mathcal{E}) = \mathcal{E}\left(1 + \frac{\mathcal{E}}{\mathcal{E}_\mathrm{G}}\right) .$$

In the limit $\mathcal{E}_\mathrm{G} \to 0$, we obtain

$$\mathcal{E} = \sqrt{\frac{\hbar^2}{2m_0^*}} k ,$$

and thus the energy band exhibits a linear dispersion relation for a narrow gap semiconductor.

(2.4) The constant energy contour of the valence band in Si shown in Fig. 2.13 exhibit convex surface (heavier effective mass) in the direction $< 1, 1, 1 >$ and concave surface (lighter effective mass) in the direction $< 0, 0, 1 >$. Cyclotron motion of a hole is in the plane perpendicular to the magnetic field. When a magnetic field is applied in the [1, 1, 1], heavy holes rotate crossing the convex region and exhibit heavier cyclotron mass. On the other hand, when a magnetic field is applied in the [0, 0, 1] direction, heavy holes rotate crossing convex regions, resulting in lighter cyclotron mass. As shown in Fig. 2.13, light hole band is almost spherical and the θ dependence of the cyclotron mass is almost constant. The warping of the light hole band is reversed as shown in Fig. 2.13b, c.

(2.5) Inserting given parameters into (2.158) and (2.159),

$$\frac{1}{m_0^*} = \frac{1}{m} + \frac{\mathcal{E}_{P0}}{3m} \frac{2\Delta_0 + 3\mathcal{E}_G}{\mathcal{E}_G(\Delta_0 + \mathcal{E}_G)},$$

$$\frac{g_0^*}{2} = 1 + \frac{\mathcal{E}_{P0}}{3}\left(\frac{1}{\mathcal{E}_G + \Delta_0} - \frac{1}{\mathcal{E}_G}\right) = 1 + \left(1 - \frac{m}{m_0^*}\right)\frac{\Delta_0}{2\Delta_0 + 3\mathcal{E}_G},$$

we obtain the following results.

$$m_0^*/m = 0.0227,$$
$$g_0^* = -14.5.$$

These values are in good agreement with the experiments $m_0^* = 0.024\,\mathrm{m}$ and $g_0^* = -14.7$. We may conclude that $\boldsymbol{k} \cdot \boldsymbol{p}$ perturbation theory is valid for the conduction band and valence band analysis.

(2.6) Following the procedure to derive (2.61) of Sect. 2.3 we obtain the following result:

$$\langle u_{-}\alpha | \boldsymbol{L} \cdot \boldsymbol{\sigma} | u_{-}\alpha \rangle = -\hbar,$$

(2.7) Density of states $\rho_c(\mathcal{E})\mathrm{d}\mathcal{E}$ is defined in terms of electronic states per volume in the energy range between $\mathcal{E}$ and $\mathcal{E} + \mathrm{d}\mathcal{E}$. Since the electronic states are defined by the wave vector $k_x = 2\pi n_x/L$, $ky = 2\pi n_y/L$, $k_z = 2\pi n_{=}z/L$ with $n_x,\ n_y,\ n_z = \pm 0, 1, 2, 3, \cdots$

$$\rho_c(\boldsymbol{k})\mathrm{D}^3\boldsymbol{k} = \frac{2}{L^3}\left(\frac{L}{2\pi}\right)^3 4\pi k^2 \mathrm{d}k,$$

where the factor 2 arises from the spin-degeneracy and $k = \sqrt{k_x^2 + k_y^2 + k_z^2}$.

(1) Using the relation between energy and wave vector for parabolic band:

$$\mathcal{E} = \frac{\hbar^2}{2m_0^*}k^2$$

and rewrite the density of states in energy $\mathcal{E}$ we obtain

$$\rho_c(\mathcal{E})\mathrm{d}\mathcal{E} = \frac{(2m_0^2)^{3/2}}{4\pi^2\hbar^3}\mathcal{E}^{1/2}\mathrm{d}\mathcal{E}$$

(2) For non–parabolic band the relation between energy and wave vector is given by (2.97):

$$\frac{\hbar^2 k^2}{2m_0^*} \equiv \gamma(\mathcal{E}) = \mathcal{E}\left(1 + \frac{\mathcal{E}}{\mathcal{E}_G}\right).$$

Therefore we obtain

$$\rho_c(\mathcal{E})\mathrm{d}\mathcal{E} = \frac{(2m_0^2)^{3/2}}{4\pi^2\hbar^3}\sqrt{\gamma(\mathcal{E})}\frac{\mathrm{d}\gamma(\mathcal{E})}{\mathrm{d}\mathcal{E}}\mathrm{d}\mathcal{E},$$

where

$$\frac{\mathrm{d}\gamma(\mathcal{E})}{\mathrm{d}\mathcal{E}}\mathrm{d}\mathcal{E} = \left(1 + 2\frac{\mathcal{E}}{\mathcal{E}_G}\right)\mathrm{d}\mathcal{E}.$$

(2.8) Magnetic field $\boldsymbol{B}$ is expressed by using vector potential $\boldsymbol{A}$ as $\boldsymbol{B} = \nabla \times \boldsymbol{A}$ D
(a) This relation is rewritten by separating into its components:

$$B_x = \frac{\partial A_z}{\partial y} - \frac{\partial A_y}{\partial z}$$
$$B_y = \frac{\partial A_x}{\partial z} - \frac{\partial A_z}{\partial x}$$
$$B_z = \frac{\partial A_y}{\partial x} - \frac{\partial A_x}{\partial y}$$

(b) When $\boldsymbol{A} = (0, B\,x, 0)$ used,

$$B_z = \frac{\partial B\,x}{\partial x} + 0 = B, \quad B_x = B_y = 0,$$

then magnetic field $B_z = B$ is derived and it is called Landau gauge). In this textbook, we use Landau gauge for Hamiltonian in a magnetic field.
(c) We $\boldsymbol{A} = (-B\,y/2, B\,x/2, 0)$ used, and insert it in the last equation of (a), we obtain

$$B_z = \frac{\partial(B\,x/2)}{\partial x} - \frac{\partial(-B\,y/2)}{\partial y} = (B/2)(1+1) = B, \quad B_x = B_y = 0,$$

and thus this representation is allowed.
(d) If $\boldsymbol{A} = (-By, 0, 0)$ is used, we find

$$B_z = 0 - \frac{\partial(-B\,y)}{\partial y} = B, \quad B_x = B_y = 0.$$

This representation is also called Landau gauge.

(3.1) J. M. Luttinger and W. Kohn: Phys. Rev. **97**, (1955) 869.

This paper is based on the $\boldsymbol{k} \cdot \boldsymbol{p}$ perturbation theory and the validity of the effective mass equation for $\boldsymbol{k} \simeq 0$ and very important to analyze the Landau levels in the valence bands.

(3.2) Electron motion in a magnetic field is solved by Newton's law.
(1) Since the Lorentz force acting to an electron with its effective mass m^* is given by

$$\boldsymbol{F} = -e\,[\boldsymbol{E} + \boldsymbol{v} \times \boldsymbol{B}]\ .$$

Without electric field, $\boldsymbol{E} = 0$, the electron motion is written as

$$m^* \frac{\boldsymbol{v}}{\mathrm{d}t} = -e\boldsymbol{v} \times \boldsymbol{B}$$

With the magnetic field $\boldsymbol{B} = (0, 0, B_z)$, the electron motion is written as

$$m^* \frac{v_x}{\mathrm{d}t} = -e v_y B_z\ ,$$
$$m^* \frac{v_y}{\mathrm{d}t} = e v_x B_z\ ,$$
$$m^* \frac{v_x}{\mathrm{d}t} = 0\ .$$

Take time derivative of the first equation and rewrite it by inserting the time derivative of the second equation $\mathrm{d}v_y/\mathrm{d}t$, and apply the same procedure to the second equation:

$$m^* \frac{\mathrm{d}^2 v_x}{\mathrm{d}t^2} = (eB_z)^2 v_x\ ,$$
$$m^* \frac{\mathrm{d}^2 v_y}{\mathrm{d}t^2} = (eB_z)^2 v_y\ .$$

We define

$$\omega_\mathrm{c} = \frac{eB_z}{m}\ ,$$

where ω_c is the cyclotron angular frequency. We obtain the following solutions:

$$v_x = v\cos(\omega_\mathrm{c} t)\ ,$$
$$v_y = v\sin(\omega_\mathrm{c} t)\ ,$$

and thus we obtain $v_x + \mathrm{i}v_y = v\exp(\mathrm{i}\omega_\mathrm{c} t)$. The result means that the electron makes cyclotron motion with angular frequency ω_c.
(2) Since Lorentz force is balanced with the centripetal force, we obtain

$$m^* \frac{v^2}{r} = evB$$

and then

$$r = \frac{mv}{eB} \equiv R_c .$$

(3) From Bohr's quantization condition $\int p = h$

$$\int p\mathrm{d}l = (m^* v) 2\pi R_c = h ,$$

is deduced. Therefore we find

$$R_c = \frac{1}{2\pi} \frac{h}{m^* v} \longrightarrow (m^* v)^2 = \frac{h}{2\pi} eB = \hbar eB .$$

Combining these relations we obtain the cyclotron radius given by

$$R_c = \sqrt{\frac{\hbar}{eB}} \equiv l .$$

This radius is exactly the cyclotron radius of the ground state of the electron motion, which is obtained as (2.108) of Chap. 2.

(4.1) Using reflectivity R given by (4.18), the results are plotted in Fig. 10.2. Note here that the reflectivity $R \simeq 0.2 \sim 0.3$ for many III-V compound semiconductors.

(4.2) The joint density of states for parabolic bands are calculated as follows.
(1) From (4.42)

$$J_{cv}(\hbar\omega) = \sum_{\boldsymbol{k}} \delta[\mathcal{E}_{cv}(\boldsymbol{k}) - \hbar\omega] = \frac{2}{(2\pi)^3} \int \mathrm{d}^3\boldsymbol{k} \cdot \delta[\mathcal{E}_{cv}(\boldsymbol{k}) - \hbar\omega] ,$$

$$= \frac{2}{(2\pi)^3} \int 4\pi k^2 \mathrm{d}k \cdot \delta[(\hbar^2/2\mu)k^2 + \mathcal{E}_G - \hbar\omega]$$

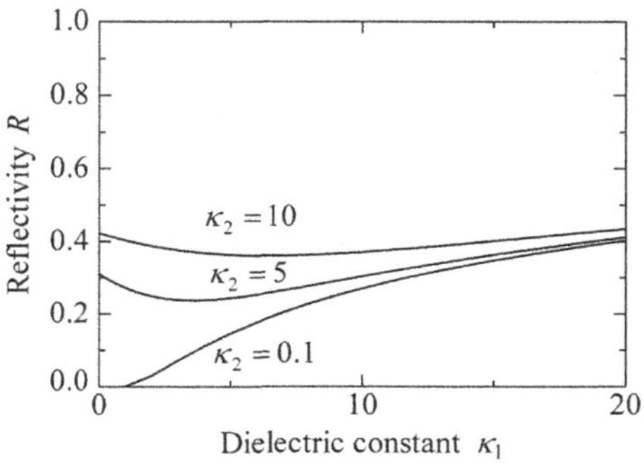

Fig. 10.2 Reflectivity R as a function of κ_1 for $\kappa_2 = 0.1$, 5.0 and 10

Here we use the formula (A.12) and (A.16) of Appendix A

$$\delta(ax) = \frac{1}{a}\delta(x), \quad a > 0$$

$$\delta(x^2 - a^2) = \frac{1}{2a}\left[\delta(x-a) + \delta(x+a)\right], \quad a > 0$$

and we find the following relation:

$$\delta(c_1 k^2 - c_2) = \frac{1}{c_1}\delta\left(k^2 - c_2/c_1\right) = \frac{1}{c_1}\delta\left(k^2 - c_0\right).$$

where $c_0 = c_2/c_1 = \left(\hbar\omega - \mathcal{E}_{\mathrm{G}}\right)/(\hbar^2/2\mu)$

$$\int k^2 \delta\left[\left(\frac{\hbar^2}{2\mu}k^2 + \mathcal{E}_{\mathrm{G}} - \hbar\omega\right)\right]\mathrm{d}k = \int k^2 \left(\frac{\hbar}{2\mu}\right)^{-1} \delta(k^2 - c_0)\mathrm{d}k$$

$$= \int k^2 \left(\frac{\hbar}{2\mu}\right)^{-1} \frac{1}{2\sqrt{c_0}}\left[\delta(k - \sqrt{c_0}) + \delta(k + \sqrt{c_0})\right]\mathrm{d}k$$

$$= \frac{2\mu}{\hbar^2}c_0 \frac{1}{2\sqrt{c_0}} = \frac{1}{2}\left(\frac{2\mu}{\hbar^2}\right)^{3/2}\sqrt{\hbar\omega - \mathcal{E}_{\mathrm{G}}}\,.$$

Finally we obtain

$$J_{\mathrm{cv}}(\omega) = \frac{4\pi}{(2\pi)^3}\left(\frac{2\mu}{\hbar^2}\right)^{3/2}\sqrt{\hbar\omega - \mathcal{E}_{\mathrm{G}}}\,.$$

(2) Next we evaluate the joint density of states for parabolic bands from (4.44)

$$J_{\mathrm{cv}}(\hbar\omega) = \frac{2}{(2\pi)^3}\int_{\hbar\omega = \mathcal{E}_{\mathrm{cv}}} \frac{\mathrm{d}S}{|\nabla_{\boldsymbol{k}}\mathcal{E}_{\mathrm{cv}}(\boldsymbol{k})|}\,.$$

We rewrite the combined bands as

$$\mathcal{E}_{\mathrm{cv}}(s) = s^2 + \mathcal{E}_{\mathrm{G}} - \hbar\omega$$

$$\nabla\mathcal{E}_{\mathrm{cv}}(s) = 2s$$

$$\int \mathrm{d}S = 4\pi s^2$$

and then the joint density of states is expressed as

$$J_{\mathrm{cv}}(\omega) = \frac{2}{(2\pi)^3}\left(\frac{2\mu}{\hbar^2}\right)^{3/2}\frac{4\pi s^2}{2s}, \quad \text{for} \quad \mathcal{E}_{\mathrm{cv}} = \hbar\omega$$

$$= \frac{4\pi}{(2\pi)^3}\left(\frac{2\mu}{\hbar^2}\right)^{3/2}\sqrt{\hbar\omega - \mathcal{E}_{\mathrm{G}}}\,.$$

Therefore two different methods give the same joint density of states.

(4.3) These two features are explained in terms of excited phonon number.
(1) The number of excited phonons is given by Bose–Einstein statistics and increases with increasing temperature, resulting in larger absorption coefficient at higher temperatures.
(2) At higher temperature, electron–phonon or hole–phonon interactions increase, leading stronger broadening of the density of states. Therefore the absorption coefficient looses its clearness at higher temperatures.

(4.4) Using (4.128) and (4.130a), we obtain the following relations:

$$\kappa_2(x_0) = AP_0^2\omega_0^{-3/2}\frac{1}{x_0^2}(x_0 - 1)^{1/2}$$

$$\kappa_1(x_0) = 1 + AP_0^2\omega_0^{-3/2}\frac{1}{x_0^2}\left[2 - (1 + x_0)^{1/2} - (1 - x_0)^{1/2}\right]$$

(4.5) The coefficient $C_0 = AP_0^2\omega_0^{-3/2}$ is given by

$$C_0 = \frac{e^2\hbar^{1/2}}{2\pi\epsilon_0 m^2}\frac{m\mathcal{E}_{P0}}{2}\left(\frac{2\mu}{\hbar^2}\right)^{3/2}\left(\frac{\mathcal{E}_G}{\hbar}\right)^{-3/2} = 4.31$$

where the value P_0^2 is evaluated from $P_0^2 = (m/2)\mathcal{E}_{P0}$.

(4.6) From the above result we may rewrite the dielectric functions:

$$\kappa_1(x_0) = 1 + C_0\frac{1}{x_0^2}\left[2 - (1 + x_0)^{1/2} - (1 - x_0)^{1/2}\right] \text{ for } x < x_0\,,$$

$$= 1 + C_0\frac{1}{x_0^2}\left[2 - (1 + x_0)^{1/2}\right] \text{ for } x > x_0$$

$$\kappa_2(x_0) = C_0\frac{1}{x_0^2}(x_0 - 1)^{1/2}$$

Using the parameters provided the dielectric functions are plotted in Fig. 10.3. The results are obtained for the direct E_0 transition from the heavy hole and light valence bands to conduction band, and the contributions from the

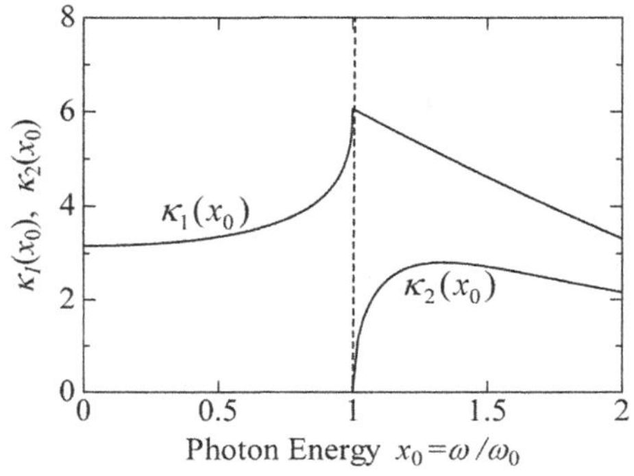

Fig. 10.3 Dielectric functions $\kappa_1(x_0)$ and $\kappa_2(x_0)$ for the transition between the heavy hole band and the conduction band

other interband transitions are neglected. Although the analysis is based on a simplified case, the obtained feature explains the experimental results quite well.

(4.7) We use (4.130a).

(1) Inserting the parameters in (4.130a) and putting $x_0 = 1$ we obtain

$$\begin{aligned}\kappa_1(\omega) &= 1 + AP_0^2\omega_0^{-3/2}\frac{1}{x_0^2}\left[2 - (1+x_0)^{1/2} - (1-x_0)^{1/2}\right] \\ &= 1 + C_0\frac{1}{x_0^2}\left[2 - (1+x_0)^{1/2} - (1-x_0)^{1/2}\right] \\ &\simeq 1 + C_0\frac{1}{x_0^2}\left[2 - \sqrt{2}\right] = 3.33\end{aligned}$$

(2) When we use (4.130a), we obtain $\kappa_1 = 7.37$. This value is very close to the experimental value $\kappa_1 = 10.7$ and the $\boldsymbol{k}\cdot\boldsymbol{p}$ theory gives a good explanation of the dielectric function. (3) Inserting $\mathcal{E}_G = 1.43\,\text{eV}$, we find (1) $\kappa_1(x_0) = 3.52$ and (2) $\kappa_1(x_0) = 7.88$.

(5.1) Two-dimensional band with reduced mass μ_y and μ_z is given by

$$\mathcal{E} = \frac{\hbar^2}{2\mu_y}k_y^2 + \frac{\hbar^2}{2\mu_z}k_z^2 + \mathcal{E}_G\,.$$

The density of states for this band is obtained applying the same procedure dealt in Chap. 8, where the density of states of two–dimensional electron gas is deduced from (8.12). First we define $k_i' = k_i/\sqrt{\mu_i/m}$ and rewrite the 2D electronic state as

$$\mathcal{E} = \frac{\hbar^2}{2m}\left(k_y'^2 + k_z'^2\right) + \mathcal{E}_G = \frac{\hbar^2}{2m}k'^2 + \mathcal{E}_G\,,$$

Since

$$\begin{aligned}\mathrm{d}k_y\mathrm{D}k_z &= \frac{\sqrt{\mu_y\mu_z}}{m}\mathrm{d}k_y'\mathrm{d}k_z' \\ \mathrm{d}k_y'\mathrm{d}k_z' &= 2\pi k'\mathrm{d}k' = \pi\,\mathrm{d}(k'^2) = \pi\frac{2\pi m}{\hbar^2}\mathrm{d}(\mathcal{E} - \mathcal{E}_G) \\ \mathrm{d}k_y\mathrm{D}k_z &= \frac{2}{(2\pi)^2}\mathrm{d}k_y\mathrm{D}k_z == \frac{2}{(2\pi)^2}\frac{\sqrt{\mu_y\mu_z}}{m}\frac{2\pi m}{\hbar^2}\mathrm{d}(\mathcal{E} - \mathcal{E}_G)\end{aligned}$$

Finally we obtain the following relation for the joint density of states:

$$J_{\mathrm{cv}}^{\mathrm{2D}}(\mathcal{E})\mathrm{d}(\mathcal{E} - \mathcal{E}_G) = \frac{\sqrt{\mu_y\mu_z}}{\pi\hbar^2}\mathrm{d}(\mathcal{E} - \mathcal{E}_G), \quad \mathcal{E} \geq \mathcal{E}_G\,,$$

where we included factor 2 for the spin degeneracy.

For optical transition we may rewrite this equation in the following form.

Fig. 10.4 Reflectivity calculated $R(\omega)$ for E_0 point, where only the direct transition form the heavy and light hole bands to the conduction band are included. Dielectric functions $\kappa_1(x_0)$ and $\kappa_2(x_0)$ for the transition between the heavy hole band and the conduction band

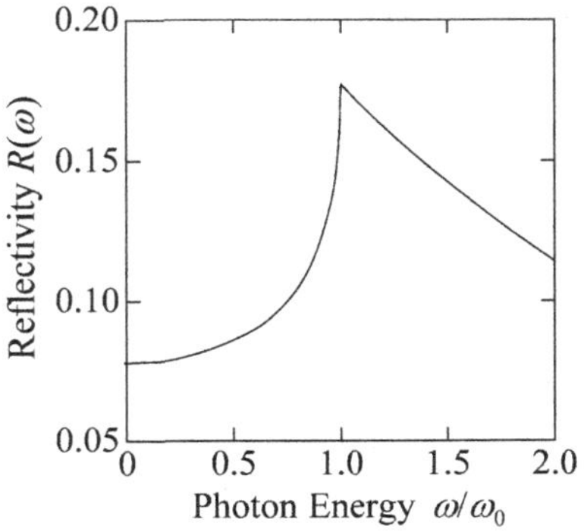

$$J_{\text{cv}}^{\text{2D}}(\hbar\omega)\text{d}(\hbar\omega - \mathcal{E}_{\text{G}}) = \frac{\sqrt{\mu_y \mu_z}}{\pi \hbar^2}\text{d}(\hbar\omega - \mathcal{E}_{\text{G}}), \quad \hbar\omega \geq \mathcal{E}_{\text{G}},$$

(5.2) Calculations are straight forward inserting the parameters. We use analytical expressions of the dielectric functions $\kappa_1(w)$ and $\kappa_2(\omega)$ and $2C_0$ which include the contribution from the heavy hole and light hole bands. The calculated result is shown in Fig. 10.4.

(5.3) The weak structure of observed in $R(\omega)$ spectra is due to the strong background of the real part of the dielectric constant. As shown in Fig. 10.3, Kramers–Kronig relation of the imaginary part κ_2 of the dielectric function give rise to a big background of the real part κ_1. There exist various critical points beyond E_0 point and these critical points give rise the strong background of the real part κ_1 and the structure of $R(\omega)$ near the E_0 critical points disappear.

(5.4) First we derive the relations (5.28d) $\sim$ (5.28d) using $n + \text{i}k = \sqrt{\kappa_1 + \text{i}\kappa_2}$,

$$n^2 = \frac{1}{2}\left(\kappa_1 + \sqrt{\kappa_1^2 + \kappa_2^2}\right),$$

$$k^2 = \frac{1}{2}\left(-\kappa_1 + \sqrt{\kappa_1^2 + \kappa_2^2}\right),$$

and insert analytical functions of dielectric functions for the critical point E_0 used for Problem **4.7**. Then we evaluate (5.27) and the result is given in Fig. 10.5. Note here that the coefficients are calculated for the model dielectric functions of the E_0 critical points and real coefficients are affected by the contribution from other critical points. Although the model calculations of Seraphin coefficients α and β are almost same order in the magnitude around the E_0 critical points, experimental data of $\Delta R/R$ is explained in terms of the derivative of the imaginary part of the dielectric constant κ_2. This is due to the fact that experimental data are obtained by modulation technique, where the derivative form of κ_2 is rich in fine structures as shown in the Problem **5.5**.

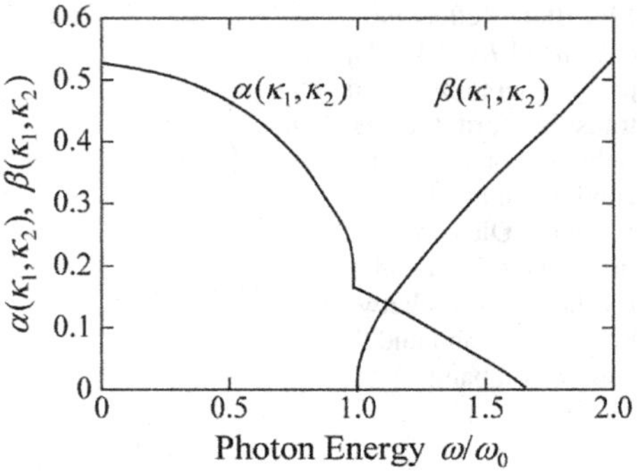

Fig. 10.5 Seraphin coefficients $\alpha(\kappa_1, \kappa_2)$ and $\beta(\kappa_1, \kappa_2)$ are plotted for the model dielectric functions of the E_0 critical point

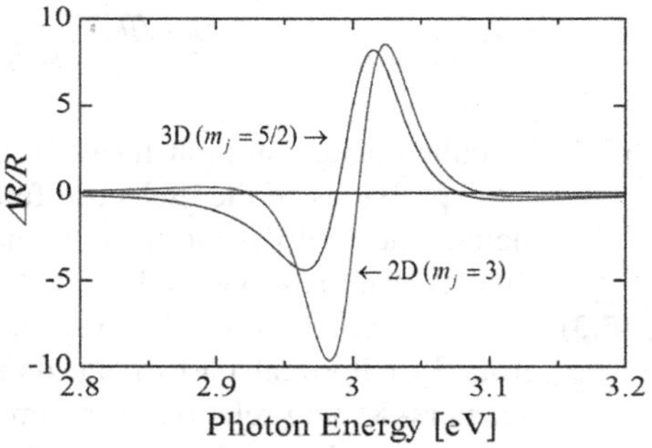

Fig. 10.6 Calculated electroreflectance spectra for (1) 3D and (2) 2D from (5.48)

(5.5) We use the analytical form of (5.48):

$$\frac{\Delta R}{R} = \Re\left[\sum_j C_j \mathrm{e}^{\mathrm{i}\theta_j}(\mathcal{E} - \mathcal{E}_j + \mathrm{i}\Gamma_j)^{-m_j}\right], \quad j = 3, 2\,.$$

Using the parameters we plot the results for (1) 3D ($m_i = 5/2$) and (2) 2D ($m_j = 3$) in Fig. 10.6

(5.6) For the free carrier absorption we obtain the following relations

$$\omega_\mathrm{p} = \sqrt{\frac{ne^2}{\kappa_\infty \epsilon_0 m^*}} = \kappa_\infty\left[1 - \frac{(\omega_\mathrm{p}\tau)^2}{(\omega\tau)^2 + 1}\right],$$

$$\kappa_1(\omega) = \kappa_\infty - \frac{ne^2\tau^2}{\epsilon_0 m^*(\omega^2\tau^2 + 1)} = \kappa_\infty \frac{\omega_\mathrm{p}^2}{\omega^2}\frac{1}{\omega^2\tau^2 + 1},$$

$$\kappa_2(\omega) = \frac{ne^2\tau}{\epsilon_0 m^* \omega(\omega^2\tau^2 + 1)}\,.$$

Inserting the dielectric functions for free carrier plasma into (5.27), and assuming $\omega_\mathrm{p}\tau = 10$ we obtain the result shown in Fig. 10.7, where calculated reflectivity R is plotted as a function of $\omega_\mathrm{p}/\omega = \lambda/\lambda_\mathrm{p}$ normalized by

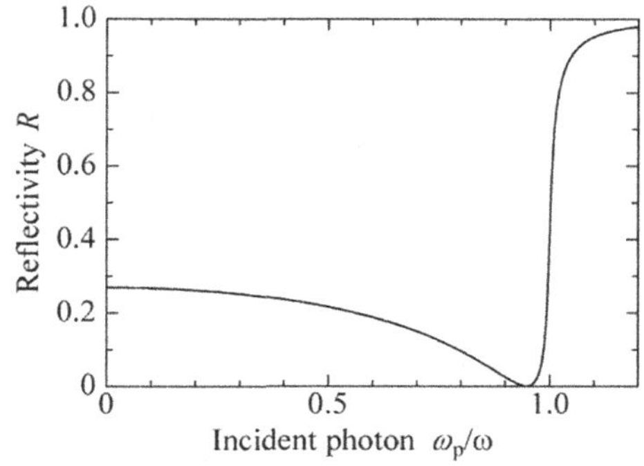

Fig. 10.7 Reflectivity due to free carrier plasma, where parameters $\omega_p\tau = 10$ and $\kappa_\infty = 10$ are used

the plasma wavelength $\lambda_p = 2\pi c/\omega_p$ corresponding to the plasma frequency ω_p.

We find here the calculated reflectivity R represents the observed characteristics in Fig. 5.33.

(6.1) The scattering probability is calculated according to quantum mechanics and given by the relation (6.129) for the interaction Hamiltonian H_{el}:

$$P(\boldsymbol{k}, \boldsymbol{k}') = \frac{2\pi}{\hbar}|\langle \boldsymbol{k}'|H_{el}|\boldsymbol{k}\rangle|^2 \delta\left[\mathcal{E}_{\boldsymbol{k}'} - \mathcal{E}_{\boldsymbol{k}}\right] ,$$

The scattering rate per second w_{ac} [1/s] is given by taking all the possible final states, as shown by (6.130):

$$\begin{aligned} w_{ac} &= \frac{2\pi}{\hbar}\sum_{\boldsymbol{k}'}|\langle \boldsymbol{k}'|H_{el}|\boldsymbol{k}\rangle|^2 \delta\left[\mathcal{E}_{\boldsymbol{k}'} - \mathcal{E}_{\boldsymbol{k}}\right] \\ &= \frac{2\pi}{\hbar}\frac{L^3}{(2\pi)^3}\int d^3\boldsymbol{k}'|\langle \boldsymbol{k}'|H_{el}|\boldsymbol{k}\rangle|^2 \delta\left[\mathcal{E}_{\boldsymbol{k}'} - \mathcal{E}_{\boldsymbol{k}}\right] . \end{aligned}$$

The matrix element for the acoustic phonon scattering is given by (6.156):

$$\begin{aligned} &\langle n_{\boldsymbol{q}'}|H_{el}(\boldsymbol{k}' - \boldsymbol{k})|n_{\boldsymbol{q}}\rangle \\ &\quad = \langle n_{\boldsymbol{q}'}|C(\boldsymbol{k}' - \boldsymbol{k})a_{\boldsymbol{k}'-\boldsymbol{k}} + C^\dagger(\boldsymbol{k} - \boldsymbol{k}')a^\dagger_{\boldsymbol{k}-\boldsymbol{k}'}|n_{\boldsymbol{q}}\rangle \\ &\quad = \begin{cases} C(\boldsymbol{q})\sqrt{n_{\boldsymbol{q}}} & (\boldsymbol{k}' = \boldsymbol{k} + \boldsymbol{q};\ \text{absorption}), \\ C^\dagger(\boldsymbol{q})\sqrt{n_{\boldsymbol{q}} + 1} & (\boldsymbol{k}' = \boldsymbol{k} - \boldsymbol{q};\ \text{emission}), \end{cases} \end{aligned}$$

where

$$C(\boldsymbol{q}) = D_{ac}\sqrt{\frac{\hbar}{2MN\omega_q}}(\mathrm{i}\boldsymbol{e}_q \cdot \boldsymbol{q}) .$$

Inserting these relations into the scattering rate w_{ac} (6.331), we obtain (6.332)

$$w_{\mathrm{ac}} = \frac{(2m^*)^{3/2} D_{\mathrm{ac}}^2 k_{\mathrm{B}} T}{2\pi \hbar^4 \rho v_{\mathrm{s}}^2} \mathcal{E}^{1/2} ,$$

where we assumed $n_q \simeq n_q + 1 \simeq k_{\mathrm{B}}T/\hbar\omega_q = k_{\mathrm{B}}T/\hbar v_{\mathrm{s}} q$. The assumptions used are (1) isotropic scattering and (2) purely elastic scattering: $q_{\max} = 2k(1 \pm m^* v_{\mathrm{s}}/\hbar k) \simeq 2k$.

(6.2) The relaxation time τ_{ac} is defined as

$$\frac{1}{\tau_{\mathrm{ac}}} = \frac{L^3}{(2\pi)^3} \int P(\boldsymbol{k}, \boldsymbol{k}') \left(1 - \frac{k' \cos\theta'}{k \cos\theta}\right) \mathrm{d}^3\boldsymbol{q}$$

which reduces to

$$\frac{1}{\tau_{\mathrm{ac}}} = \frac{L^3}{2\pi\hbar} \int_{q_{\min}}^{q_{\max}} \frac{1}{k} A(q) \left\{ (n_q + 1) \left(\frac{q}{2k} + \frac{m^* v_{\mathrm{s}}}{\hbar k} \right) \right. \\ \left. + n_q \left(\frac{q}{2k} - \frac{m^* v_{\mathrm{s}}}{\hbar k} \right) \right\} \frac{m^* q^2}{\hbar^2 k} \mathrm{d}q \, .$$

This equation is easily evaluated under the approximation $q_{\max} = 2k(1 \pm m^* v_{\mathrm{s}}/\hbar k) \simeq 2k$, and the two terms of the left hand side gives

$$2 \int_0^{2k} \int_0^{2k} \frac{q}{2k} \mathrm{d}q = 4k^3 \, .$$

Then we obtain the result (6.341) and we find the equality of $w_{\mathrm{ac}} = 1/\tau_{\mathrm{ac}}$.

(6.3) We define $q_\pm = 2k(1 \pm m^* v_{\mathrm{s}}/\hbar k)$ and the integral of the right hand side of $1/\tau_{\mathrm{ac}}$ is carried out:

$$\int_0^{q_+} \left(\frac{q}{2k} + \frac{m^* v_{\mathrm{s}}}{\hbar k} \right) \mathrm{d}q + \int_0^{q_-} \left(\frac{q}{2k} - \frac{m^* v_{\mathrm{s}}}{\hbar k} \right) \mathrm{d}q \\ = 4k^3 \left[1 + 10 \left(\frac{m^* v_{\mathrm{s}}}{\hbar k} \right)^2 + \frac{7}{3} \left(\frac{m^* v_{\mathrm{s}}}{\hbar k} \right)^4 \right]$$

The result indicate the anisotropy of scattering event (see Fig. 6.13 is given by the second and third terms of above equation and very small for $\pm m^* v_{\mathrm{s}}/\hbar k \ll 1$, and we may assume $1/\tau_{\mathrm{ac}} = w_{\mathrm{ac}}$

(6.4) The electron mobility due to polar optical phonon is evaluated by using the averaged relaxation time (6.116):

$$\langle \tau_{\mathrm{pop}} \rangle = \frac{\displaystyle\int_0^\infty \tau_{\mathrm{pop}} \mathcal{E}^{3/2} f_0 \mathrm{d}\mathcal{E}}{\displaystyle\int_0^\infty \mathcal{E}^{3/2} f_0 \mathrm{d}\mathcal{E}}$$

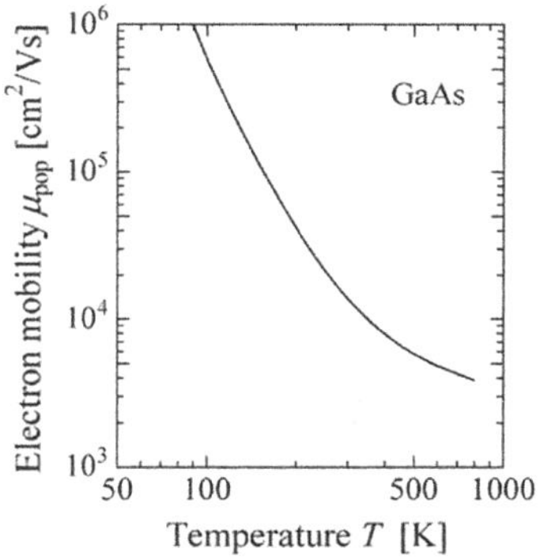

Fig. 10.8 Electron mobility of GaAs limited by polar optical phonon scattering which is calculated by numerical integration

Inserting (6.356) into above equation and carrying out the integration numerically we obtain the curve shown in Fig. 10.8, where numerical integration is carried out in the region $[0, \mathcal{E}_{\max}]$ with $\mathcal{E}_{\max} = 100 \times \hbar\omega_{LO}$. We see that the polar optical phonon scattering plays an important role in high temperature region.

(6.5) Electron mobility at low temperatures are limited by acoustic phonon scattering for lo impurity density and $\mu \propto m^{*-3/2}$ and we may estimate the electron mobility by the effective mass. We have to note that the electron effective mass is not well fixed by experiments and in this text we used the value $m^* = 0.22\,\mathrm{m}$ determined by the energy band calculations based on the empirical pseudopotential method as shown in Chap. 9. Using the relation (6.402) the following mobility values are estimated for the cases of $m^* = 0.22$ and $m^* = 0.10\,\mathrm{m}$.

$$\mu(100) = 1.75, \quad \mu(300) = 0.225, \quad \text{for} \quad m^* = 0.22\,\mathrm{m}$$
$$\mu(100) = 7.10, \quad \mu(300) = 0.967, \quad \text{for} \quad m^* = 0.10\,\mathrm{m}$$

(7.1) The answers are as follows.
(1) The type of carriers is determined by the direction of Hall field E_y. Since $E_y < 0$, carriers are electrons and the sample is n–type. Electrons move in the $-x$ direction and Lorentz force is $\boldsymbol{F} = (-e)\boldsymbol{v} \times \boldsymbol{B}$ and thus $F_y = -[-e(-v_x)B_z = -|ev_x B_z| < 0$. Since $F_y + (-e)E_y = 0$, we find $E_y = F_y/e = -|v_x B_z| < 0$. It confirms the type of carriers as electrons. (2) Using (7.6) we obtain

$$|R_{\mathrm{H}}| = \frac{V_{\mathrm{H}}t}{I_x B_z} = \frac{1.0 \times 10\ -3 \times 1.0 \times 10^{-3}}{1.0 \times 10^{-3} \times 0.4} = 2.5 \times 10^{-3}\,[\mathrm{m}^3/C]$$

Therefore the electron density is given by

$$n = \frac{1}{|R_{\mathrm{H}}|e} = \frac{1}{2.5 \times 10^{-3} \times 1.60 \times 10^{-19}} = 2.5 \times 10^{21}\,[\mathrm{m}^{-3}]$$

(3) Resistance of the sample is $R = V/Ix = 25.0 \times 10^{-3}/1 \times 10^{-3} = 25\,[\Omega]$ and then its resistivity is given by

$$\rho = \frac{Rwt}{l_x} = \frac{25 \times 4.0 \times 10^{-3} \times 1 \times 10^{-3}}{10 \times 10^{-3}} = 1.0 \times 10^{-2}\,[\Omega\text{m}],$$

$$\sigma = 1/\rho = 1.0 \times 10^{2}\,[1/\Omega\text{m}].$$

(4) Hall mobility is given by $\mu_H = |R_H|/\sigma = 0.25\,[\text{m}^2/\text{V s}]$.

(7.2) Using the parabolic band with isotropic effective mass, we obtain the relations (7.48) and (7.47b) and insert them in (7.51b). Then we obtain the relation given (7.51b).

(7.3) Using the parabolic band with isotropic effective mass, we obtain the relations (7.48) and (7.49) and insert them in (7.47c). Then we obtain the relation given (7.51c).

(7.4) (1) Inserting $m^* = 0.0135\,\text{m}$ and $n = 2 \times 10^{24}\,[\text{m}^{-3}]$ into equation (7.81) we find

$$\mathcal{E}_F = \frac{\hbar^w}{2m^*}\left(3\pi^2 n\right)^{2/3})\,/e = 0.429\,[\text{eV}]$$

(2) The oscillation period is given by (7.83)

$$\Delta\left(\frac{1}{B}\right) = \frac{2e}{\hbar}\left(3\pi^2 n\right)^{-2/3}\,. = 0.020\,[1/\text{T}]$$

(3) For GaSb we obtain $\mathcal{E}_F = 0.111\,[\text{eV}]$ and $\Delta(1/B) = 0.0267\,[1/\text{T}]$. The result is in good agreement of the oscillations for GaSb shown in Fig. 7.2.

(7.5) Inserting the given parameters in (7.136)

$$\sigma_{osc} \sim \exp(-0.5\omega_{LO}/\omega_c)\cos\left(2\pi\frac{\omega_{LO}}{\omega_c}\right),$$

we obtain the result shown in Fig. 10.9.

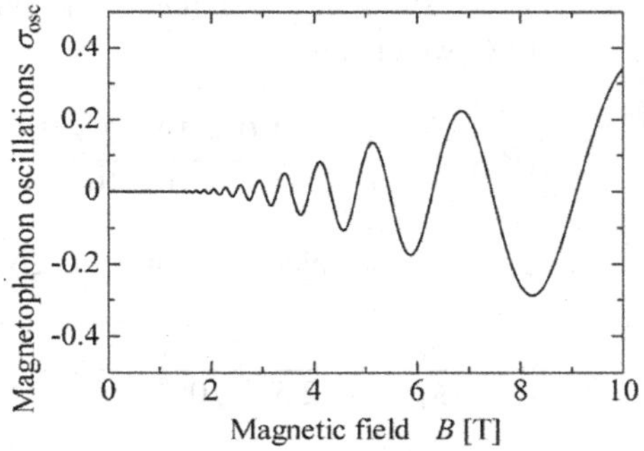

Fig. 10.9 Calculated oscillatory term of magnetophonon resonance as a function of applied magnetic field, where parameters are for GaAs

(7.6) Polaron effect is estimated by using equation (7.165) and the following parameters:

$$\alpha = \frac{e^2}{\hbar\epsilon_0}\left(\frac{1}{\kappa_\infty} - \frac{1}{\kappa_0}\right)\left(\frac{m^*}{2\hbar\omega_{\text{LO}}}\right)^{1/2}.$$

	GaAs	InAs
m^*/m	0.067	0.022
κ_0	12.90	15.15
κ_∞	10.92	12.25
$\hbar\omega_{\text{LO}}$	35.36	29.58

We obtain

$$\alpha(\text{GaAs}) = 0.02836, \quad \alpha(\text{InAs}) = 0.0197$$

Therefore we find that the polaron mass m^*_{pol} is

$$m^*_{\text{pol}}(\text{GaAs})/m = 0.06732, \quad m^*_{\text{pol}}(\text{InAs})/m = 0.02207$$

These results mean that the effective mass change due to the polaron effect is small, but effective mass correction is required in some case.

(8.1) Density of two–dimensional electron gas in subband $\mathcal{E}_i$ is given by

$$\frac{n_{\text{v}i} m_{\text{d}i}}{\pi\hbar^2}\int_0^\infty \frac{\mathrm{d}(\mathcal{E} - \mathcal{E}_i)}{\exp[(\mathcal{E} - \mathcal{E}_{\text{F}})/k_{\text{B}}T] + 1} = \frac{n_{\text{v}i} m_{\text{d}i}}{\pi\hbar^2}\int_{\mathcal{E}_i}^\infty \frac{\mathrm{d}\mathcal{E}}{\exp[(\mathcal{E} - \mathcal{E}_{\text{F}})/k_{\text{B}}T] + 1}.$$

We find the following relation:

$$\frac{1}{\mathrm{e}^{-x} + 1} = \frac{\mathrm{e}^x}{1 + \mathrm{e}^x} = \frac{\mathrm{d}}{\mathrm{d}x}\ln(1 + \mathrm{e}^x) \equiv \frac{\mathrm{d}}{\mathrm{d}x}F_0(x).$$

Defining $x = \mathcal{E}/k_{\text{B}}T$, $x_{\text{F}} = \mathcal{E}_{\text{F}}/k_{\text{B}}T$, and $x_i = \mathcal{E}_i/k_{\text{B}}T$, the above integration is easily carried out and leads to

$$\Big[F_0(x_{\text{F}} - x)\Big]_{x_i}^\infty = F_0\left[(\mathcal{E}_{\text{F}} - \mathcal{E}_i)/k_{\text{B}}T\right].$$

Then we obtain (8.15).

(8.2) Acoustic deformation potential scattering is dealt with (8.69) and for intra–subband scattering we obtain:

$$w_{\rm ac} = \frac{1}{b}\frac{m^*}{\hbar^3}\frac{k_{\rm B}T D_{\rm ac}^2}{2\rho v_{\rm s}^2},$$

$$\mu_{\rm ac} = \frac{b\, e\, \hbar^3}{m^{*2}}\frac{2\rho v_{\rm s}^2}{k_{\rm B}T D_{\rm ac}^2} \equiv C_{\rm ac}\frac{1}{T},$$

where we used the relation of (8.208) to derive the mobility;

$$\langle\tau\rangle = \frac{\int_0^\infty \mathcal{E}\tau f_0 \mathrm{d}\mathcal{E}}{\int_0^\infty \mathcal{E} f_0 \mathrm{d}\mathcal{E}}.$$

When we put $b = W/3$ and $W = 100\,\text{Å}$ and using the parameters in Table 6.3, we find

$$mu_{\rm ac} = 1.789 \times 10^4 \frac{1}{T}$$

(8.3) For non–polar optical phonon scattering (intra–subband scattering) from (8.71):

$$w_{\rm op} = \frac{1}{b}\frac{D_{\rm op}^2 m^*}{4\rho\hbar^2\omega_0}\left[n(\omega_0) + \frac{1}{2} \pm \frac{1}{2}\right] u(\mathcal{E} \mp \hbar\omega_0),$$

Here we introduce new parameter

$$x_0 = \frac{\hbar\omega_0}{k_{\rm B}T}, \quad n(x_0) = \frac{1}{\mathrm{e}^{x_0} - 1},$$

and we find that

$$A_{\rm op} = \left[n(x_0) + \frac{1}{2} \pm \frac{1}{2}\right]^{-1} u(\mathcal{E} \mp \hbar\omega_0) = \frac{\mathrm{e}^{x_0} - 1}{1 + \mathrm{e}^{x_0} u(x - x_0)}$$

and

$$A_{\rm op}^{\rm av} = \int_0^\infty A_{\rm op}\, x\, \mathrm{e}^{-x} \mathrm{d}x = \frac{(\mathrm{e}^{x_0} - 1)(\mathrm{e}^{x_0} - x_0)}{\mathrm{e}^{x_0} + 1}$$

$$\mu_{\rm op} = \frac{e\, b}{m^{*2}}\frac{4\rho\hbar^2\omega_0}{D_{\rm op}^2} A_{\rm op}^{\rm av} \equiv C_{\rm op} A_{\rm op}^{\rm av},$$

where we use the relation $D_{\rm op} = E_{\rm op}\omega_0/v_s$ and $E_{\rm op} = D_{\rm ac}/0.4$.

(8.4) For polar optical phonon scattering in a quantum well ($W \simeq 100\,\text{nm}$) the scattering rate and mobility are given by using (8.96) for $Wk_0 > 1$

$$w_{\rm pop}(W) \simeq \frac{3}{W}\frac{e^2}{8\hbar\epsilon_0}\left(\frac{1}{\kappa_\infty} - \frac{1}{\kappa_0}\right)\left[n(\omega_0) + \frac{1}{2} \pm \frac{1}{2}\right]u(\mathcal{E} \mp \hbar\omega_0)$$

$$\mu_{\rm pop}(W) \simeq \frac{W}{3m^*}\frac{8\hbar\epsilon_0}{e}\left(\frac{1}{\kappa_\infty} - \frac{1}{\kappa_0}\right)^{-1} A_{\rm op}^{\rm av} \equiv C_{\rm pop}A_{\rm op}^{\rm av} .$$

In a narrow quantum well, on the other hand, $Wk_0 < 1$ we have a approximated relation (8.97)

$$w_{\rm pop}(k) \simeq \left(\frac{2m^*\omega_0}{\hbar}\right)^{1/2}\frac{e^2}{8\hbar\epsilon_0}\left(\frac{1}{\kappa_\infty} - \frac{1}{\kappa_0}\right)\left[n(\omega_0) + \frac{1}{2} \pm \frac{1}{2}\right]u(\mathcal{E} \mp \hbar\omega_0)$$

$$\mu_{\rm pop}(k) \simeq \left(\frac{\hbar}{2m^*\omega_0}\right)^{1/2}\frac{1}{m^*}\frac{8\hbar\epsilon_0}{e}\left(\frac{1}{\kappa_\infty} - \frac{1}{\kappa_0}\right)^{-1} A_{\rm op}^{\rm av} \equiv C_{\rm pop}A_{\rm op}^{\rm av} .$$

(8.5) For piezoelectric potential scattering due to transverse mode of acoustic phonons with sound velocity $v_{\rm t}$ and piezoelectric coefficient e_{14}, the scattering rate and electron mobility are given by using (8.108)

$$\frac{1}{\tau_{\rm piez}} = \frac{\sqrt{m^*}}{\sqrt{2}\pi\hbar^2\epsilon^2}\left[\frac{13}{32}\frac{(e\cdot e_{14})^2}{\rho v_{\rm t}^2}\right]\frac{k_{\rm B}T}{\sqrt{\mathcal{E}}}$$

$$\mu_{\rm piez} = \frac{\sqrt{2}\pi e\hbar^2\epsilon^2}{m^{*3/2}}\left[\frac{32}{13}\frac{\rho v_{\rm t}^2}{(e\cdot e_{14})^2}\right]\frac{3\sqrt{\pi}}{4}(k_{\rm B}T)^{-1/2} \equiv C_{\rm piez}(k_{\rm B}T)^{-1/2}$$

(8.6) For ionized impurity scattering rate, we assume a quantum well $\zeta_1(z) = \sqrt{2/W}\sin(\pi z/W)$ with its width W (~100 Å) and the impurity is uniformly distributed along z direction with its volume density $n_{\rm ion}\,[{\rm m}^{-3}]$. The sheet density of the impurity is then given by $N_{\rm ion} = \int g_0(z_0){\rm d}z_0 = n_{\rm ion}W$. In addition we assume $Qz_0 \ll 1$, and therefore we get $I_{00}(Q) \simeq 1$, and then we have

$$J_{11}^{\rm ion}(Q) = AN_{\rm ion} .$$

Since $Q = \sqrt{2}k(1 - \cos\theta)$ we obtain from (8.125), we obtain

$$\frac{1}{\tau_{\rm ion}} = \sum_n\left(\frac{e^4}{2\pi\hbar(\kappa\epsilon_0)^2}\right)\int_0^{2\pi}\frac{N_{\rm ion}}{2k^2}{\rm d}\theta = \left(\frac{e^4}{2\pi\hbar(\kappa\epsilon_0)^2}\right)\frac{2\pi N_{\rm ion}}{2k^2} ,$$

$$= \left(\frac{e^4}{2\hbar(\kappa\epsilon_0)^2}\right)\frac{\hbar^2 N_{\rm ion}}{2m^*}\mathcal{E}^{-1} ,$$

where we neglect the screening effect and put $P = 0$. Then we obtain the electron mobility due to ionized impurity as;

$$\mu_{\rm ion} = \frac{e\langle\tau_{\rm ion}\rangle}{m^*} = \left(\frac{4\pi\hbar(\kappa\epsilon_0)^2}{m^*e^3}\right)\frac{k_{\rm B}T}{N_{\rm ion}} \equiv C_{\rm ion}k_{\rm B}T ,$$

When we use wave functions confined at the interface like MOSFETs and single heterostructures, we obtain $J_{11}^{\rm ion}/A \simeq n_{\rm ion}z_{\rm av}$, where $z_{\rm av}$ is estimated

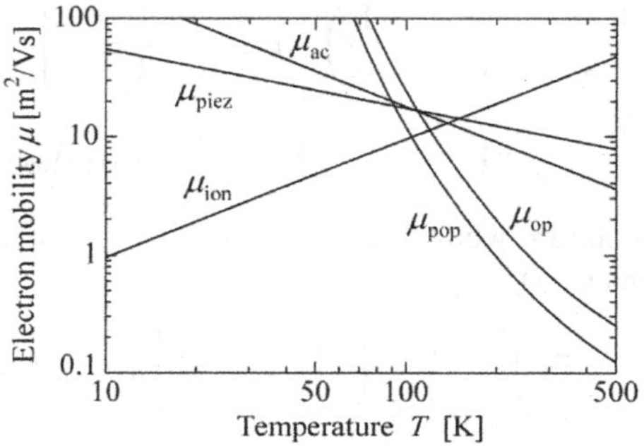

Fig. 10.10 Temperature dependence of two-dimensional electron gas mobilities (1) acoustic deformation potential scattering μ_{ac}, (2) non–polar optical phonon scattering μ_{op}, (3) polar optical phonon scattering $\mu_{pop}(k)$, (4) piezo–electric potential scattering μ_{piez}, and (5) ionized impurity scattering μ_{ion}

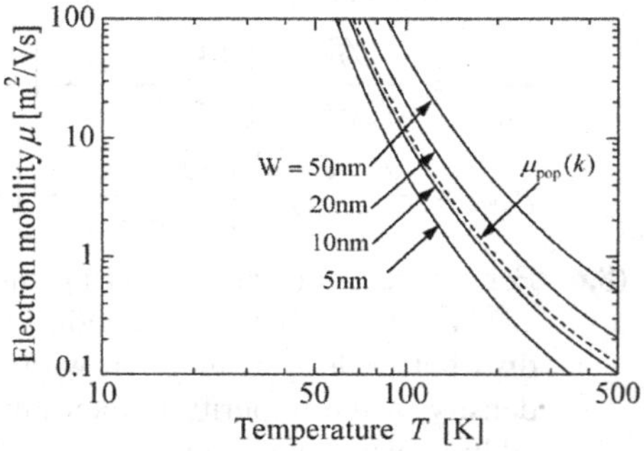

Fig. 10.11 Temperature dependence of two-dimensional electron gas mobilities due to polar optical phonon scattering, where the dashed curve (μ_{pop}) is obtained from (8.97) and the solid curves are obtained from (8.96) with the well width $W = 5, 10, 20, 50\,[\text{nm}]$ as a parameter

by using Stern–Howard empirical function, and the mobility μ_{ion} is evaluated. above relation for the simplified conditions.

(8.7) All the results are plotted in Fig. 10.10, where we find that ionized impurity scattering plays an important role in low temperature region and other scattering processes exhibit decrease in the mobility at higher temperatures.

(8.8) Using the results of previous problems, mobilities of two–dimensional electron gas are calculated, which are shown in Fig. 10.11. Temperature dependence of mobilities due to polar optical phonon scattering are plotted in Fig. 10.11, where the mobility expressions derived above are used. The dashed curve μ_{pop} is obtained from (8.97) and the solid curves are obtained from (8.96) with the well width $W = 5, 10, 20, 50\,[\text{nm}]$ as a parameter. It is evident that the electron mobility depends strongly on the well width and that the two models agree well for $W \simeq 10\,[\text{nm}]$ (200 Å).

(8.9) Using the expressions derived above the right hand side of the above equation is easily obtained. The calculated result is shown in Fig. 10.12.

(8.10) Since the density of states of Landau level for quantized electron gas in a magnetic field is given by a delta function $(eB/h)\delta(\mathcal{E} - \mathcal{E}_N)$ as shown in Fig. 10.13, where each straight bar has the areal density eB/h, and Fig. 10.13a

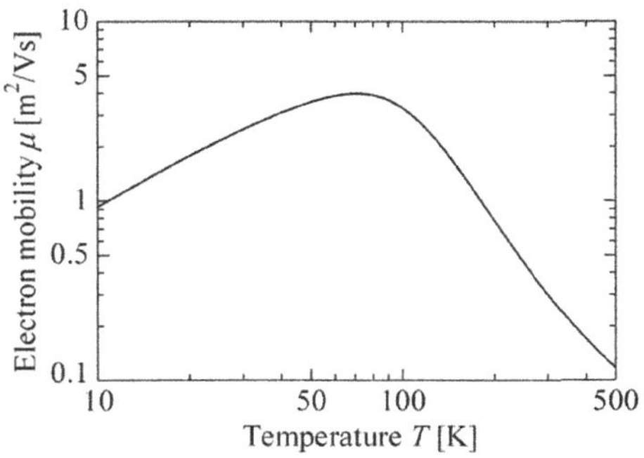

Fig. 10.12 Temperature dependence of two-dimensional electron gas mobility μ_{2D} given by $1/\mu_{2D} = \sum 1/\mu_j$ including the scattering processes of (1) acoustic deformation potential, (2) piezoelectric potential scattering, (3) ionized impurity scattering, and (4) polar optical phonon scattering

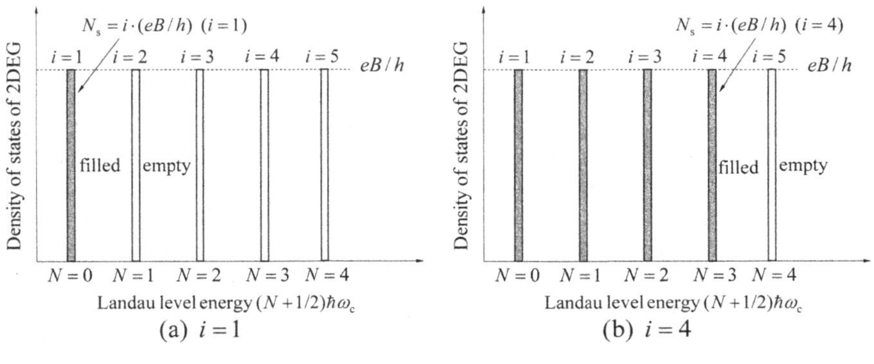

Fig. 10.13 Density of states for Landau levels in a high magnetic field. Occupied states of electron gas are shown for **a** filling factor $i = 1$ and **b** filling factor $i = 4$

is the case for N_s fills the lowest Landau level. Figure 10.13b shows the case where the sheet density N_s is increased and four Landau levels are filled. The plateaus observed by von Klitzing is explained later by introducing the broadening of the Landau levels and the mobility edges.

(8.11) Two methods are possible. (1) Solve the lattice vibrations numerically and obtain the modes of the phonons such as confined mode and interface mode in double heterostructures, and half–space mode and interface mode in single heterostructures. Then calculate the electron phonon interactions. (2) Using dielectric continuum model and calculate electron–optical phonon interaction. The latter method was used by Mori and Ando. They found the **sum rule** in which summation of the interactions are the same as the electron–bulk phonon interaction. See the detailed treatment of the following paper: N. Mori and T. Ando: "*Electron–optical-phonon interaction in single and double heterostructures*," Phys. Rev. B Vol.**40**, No.9, (1989) 6175–6188.

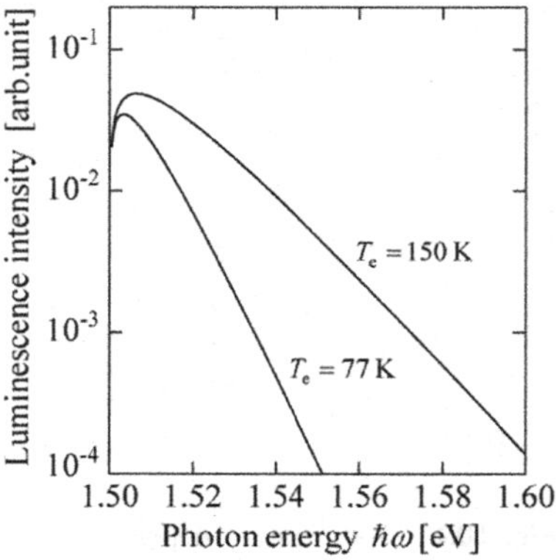

Fig. 10.14 Luminescence spectra in GaAs for electron temperature $T_e = 77\,\text{K}$ and $150\,\text{K}$, where we see the high energy tail is well expressed by $\exp[(\hbar\omega - \mathcal{E}_G)/(k_B T_e)]$

(9.1) Photoluminescence intensity as a function of photon energy is given by (9.55):

$$I(\hbar\omega) \propto r_{\text{spon}} \propto \sqrt{\hbar\omega - \mathcal{E}_G} \exp\left(-\frac{\hbar\omega - \mathcal{E}_G}{k_B T}\right).$$

Using this expression the calculated photoluminescence spectra at $T_e = T_{\text{lattice}} = 77\,\text{K}$, and $T_e = 150\,\text{K}$ are shown in Fig. 10.14. Hot electrons are produced by applying high electric fields to a semiconductor or by intense excitations of a semiconductor. The latter method leads to the heating the carriers and the generation of nonequilibrium phonons. The method provide a information of carrier–carrier interaction and dynamics of excited phonons by changing the excitation intensities. See the following paper for the detail. Jagdeep Shah: "*Hot Electrons and Phonons under High Intensity Photoexcitation of Semiconductors,*" Solis–State Electronics, Vol. 21, (1978) 43–50.

(9.2) Compare (4.58) and (9.15) or (9.18);

$$\alpha(\mathcal{E}) = \frac{e^2|M|^2}{2\pi\epsilon_0 m^2 c n_r \omega}(2\pi^2)\rho_{\text{red}}(\mathcal{E}),$$

$$r_{\text{stim}}(\mathcal{E}) = \frac{n_r e^2 \mathcal{E}}{\pi\epsilon_0 m^2 \hbar^2 c^3}|M|^2 \rho_{\text{red}}(\mathcal{E})(f_2 - f_1).$$

Using the relation $\hbar\omega = \mathcal{E}$ and putting $f_2 = 0$ and $f_1 = 1$, the above two equations lead to the relation (9.34).

$$\alpha(\mathcal{E}) = -\frac{\pi^2 c^2 \hbar^3}{n_r^2 \mathcal{E}^2} r_{\text{stim}}(\mathcal{E}),$$

(9.3) For a symmetric wave guide, the refractive indices for the cladding layers are the same and we put $n_{\text{UC}} = n_{\text{LC}} = n_{\text{C}}$ and

$$k_0 = \omega/c = 2\pi\nu/c = 2\pi/\lambda,$$

In the following we use the values normalized by λ and the boundary condition of (9.112) is rewritten as, (putting $d/\lambda \to 1$)

$$2\pi\kappa_G - 2\Phi_C = m\pi\,, \qquad \Phi_C = \tan^{-1}\left(\frac{\gamma_C}{\kappa_G}\right).$$

where

$$\kappa_G = (n_G^2 - \tilde{n}^2)^{1/2}, \quad \gamma_C = (\tilde{n}^2 - n_{UC}^2)^{1/2}.$$

This equation has no analytical solutions and we solve it by using numerical method, and the following results are obtained as given in the text.

$$\tilde{n} = 3.47808, \quad 3.41418, \quad \text{and} \quad 3.32125; \quad \text{for } TE_0,\ TE_1,\ \text{and } TE_2\,.$$

(9.4) From (9.111) we obtain the following solutions for the TE_0 mode ($m = 0$),

$$\begin{aligned} E_x(z) &= E_G \cos(\kappa_G/2) \exp\left[-\gamma_C(z - 1/2)\right], \quad &(z > 1/2) \\ E_x(z) &= E_G \cos(\kappa_G z), &(|z| < 1/2) \\ E_x(z) &= E_G \cos(\kappa_G/2) \exp\left[+\gamma_C(z + 1/2)\right], &(z < -1/2) \end{aligned}$$

The results calculated by using the refractive indices are plotted in Fig. 10.15, where we find the curve is continuously connected at the boundaries $z/\lambda = \pm 1/2$. We see the fundamental mode ($m = 0$) is well confined in the guiding region.

(9.5) The free electron bands are easily obtained by putting (k_x, k_y, k_z) into (9.200)

$$\mathcal{E}(k_x, k_y, k_z) = \frac{\hbar^2}{2m}\left(\frac{2\pi}{a}\right)^2 \left[(k_x + l)^2 + \left(k_y + \frac{2m - l}{\sqrt{3}}\right)^2 + \frac{3}{8}(k_z + n)^2\right].$$

or you may calculate using the following computer programs written in Octave (MATLAB).

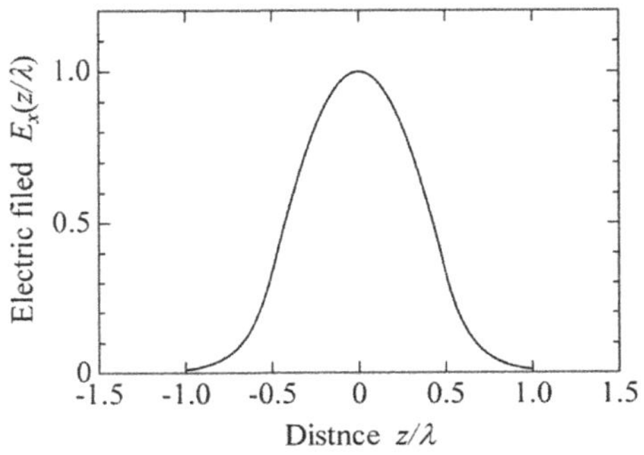

Fig. 10.15 Electric field distribution along the z direction, where we see the boundary conditions are well treated and the curve is smoothly connected at $z/\lambda = 1/2$

First, decide the lines to connect the critical points of the Brillouin zone. Next, choose (l, m, n) and put (k_x, k_y, k_z), then we obtain the free electron bands, which are obtained by the following programs:

```
% Plane Waves of hexagonal closed pack
% This program is a part of energy band calculations of GaN
% Nx=Ny=Nz=4 leads to 135 plane waves
Nx = 4;
Ny = 4;
Nz = 4;
% Wave vectors of plane waves at k=0
R=1;
for I = -Nx:Nx;
for J = -Ny:Ny;
for K = -Nz:Nz;
% G2=(8/3)*(I^2+J^2-I*J)+(6/8)*K^2
G2=2*I^2+(2/3)*(2*J-I)^2+(3/4)*K^2;
if (G2<15)
Gx(R)=I;
Gy(R)=J;
Gz(R)=K;
R=R+1;
endif
endfor
endfor
endfor
N_cut = R-1
% PRINT "number of plane waves N_cut
% Free electron bands are obtained by the following program;
```

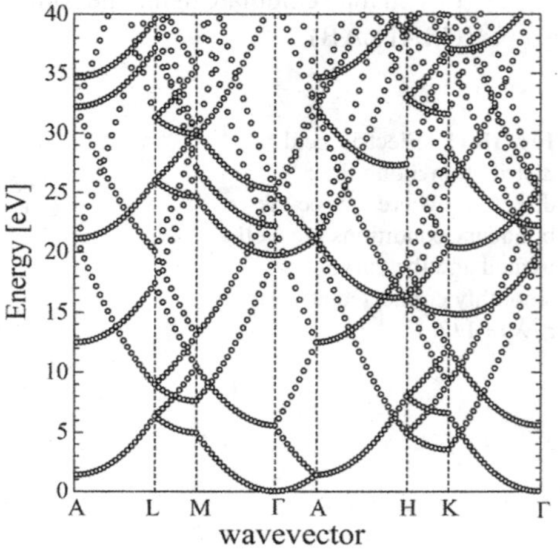

Fig. 10.16 Free electron full bands plot of GaN (wurtzite), where the critical points are shown

```
for I = 1:N_cut;
A(I,I)=EGA*((kx+Gx(I))^2+(ky+(2*Gy(I)-Gx(I))/sqrt(3))^2
+(3/8)*(kz+Gz(I))^2);
endfor
% End of Matrix Elements
% Plot A(I, I) as a function of (k_x, k_y, k_z) along the lines of the critical points
in the first Brillouin zone of Fig. 9.45
```

where Eg=2* π*Bohr/La in [a.u] with Bohr=5.29177 [Å] and lattice constant La. Energy in [eV] is obtained by Eg*Ry with Ry=13.6058. Obtained results are shown in Fig. 10.16.

Appendices

Abstract For better understandings of readers, several important mathematics and derivation of relations used in the text are given here. They are (A) Dirac delta function and Fourier transform, (B) uniaxial strain and strain components in cubic semiconductors, (C) boson operators, (D) random phase approximation, (E) density matrix, and (F) derivation of spontaneous and stimulated emission rates.

A Delta Function and Fourier Transform

A.1 Dirac Delta Function

The Delta function is very important for understanding semiconductor physics and some important relations will be discussed in this section. The **Dirac delta function** is defined by

$$\delta(\omega) = \frac{1}{2\pi}\int_{-\infty}^{+\infty} \mathrm{e}^{\mathrm{i}\omega t}\mathrm{d}t\,. \tag{A.1}$$

In the integral with respect to ω the following relation holds for $\varepsilon > 0$

$$\lim_{\varepsilon\to+0}\frac{1}{\omega - \mathrm{i}\varepsilon} = \mathcal{P}\frac{1}{\omega} + \mathrm{i}\pi\delta(\omega)\,, \tag{A.2}$$

where $\mathcal{P}[1/\omega]$ is the **Cauchy principal value** of $1/\omega$. This relation is called the **Dirac identity**.

First, we consider the integral

$$\int_0^{\infty} \mathrm{e}^{\mathrm{i}\omega t}\mathrm{d}t\,.$$

This integral does not converge when ω is real. Then we introduce an infinitesimal positive value ε, and replace ω by $\omega + \mathrm{i}\varepsilon$. The integral may be equivalently

C. Hamaguchi, *Basic Semiconductor Physics*, Graduate Texts in Physics,
DOI 10.1007/978-3-319-66860-4

written as[1]

$$\int_0^\infty e^{i\omega t}dt = \lim_{\varepsilon\to 0}\int_0^\infty e^{i(\omega+i\varepsilon)t}dt = \lim_{\varepsilon\to 0}\frac{i}{\omega+i\varepsilon}. \tag{A.3}$$

In a similar fashion, we obtain

$$\int_{-\infty}^0 e^{i\omega t}dt = \lim_{\varepsilon\to 0}\frac{-i}{\omega-i\varepsilon}. \tag{A.4}$$

These expressions lead to the following relation:

$$\int_{-\infty}^\infty e^{i\omega t}dt = \lim_{\varepsilon\to 0}\left[\frac{i}{\omega+i\varepsilon}-\frac{i}{\omega-i\varepsilon}\right] = \lim_{\varepsilon\to 0}\frac{2\varepsilon}{\omega^2+\varepsilon^2}. \tag{A.5}$$

Let us consider the following function

$$F_L(\omega) = \frac{\varepsilon/\pi}{\omega^2+\varepsilon^2},$$

which is called the **Lorentz function**. The Lorentz function has a peak at $\omega = 0$ with full width of the half maximum of 2ε and the integral with respect to ω is unity.[2] In the limit $\varepsilon \to 0$, the Lorentz function behaves like a delta function, or in other words

$$\int_{-\infty}^\infty F_L(\omega)d\omega = \int_{-\infty}^\infty \frac{\varepsilon/\pi}{\omega^2+\varepsilon^2}d\omega = \int_{-\infty}^\infty \delta(\omega)d\omega = 1.$$

Using this result we find the following expression for the Dirac delta function:

$$\frac{1}{2\pi}\int_{-\infty}^\infty e^{i\omega t}dt = \lim_{\varepsilon\to 0}\frac{\varepsilon/\pi}{\omega^2+\varepsilon^2} = \delta(\omega). \tag{A.6}$$

---

[1] Strictly speaking, we need to prove the validity of the interchange between the limit (lim) and the integral ($\int$).

[2] This is proved by the following relation.

$$\int_0^\infty \frac{dx}{a^2+b^2x^2} = \frac{1}{ab}\left[\arctan\frac{b}{a}x\right]_0^\infty = \frac{1}{ab}\cdot\frac{\pi}{2}.$$

The evaluation of the integral is made by introducing a new variable $bx/a = \tan z = \sin z/\cos z$. When the region of the integral is set to be $[-\infty, +\infty]$, we easily find the following result:

$$\frac{\varepsilon}{\pi}\int_{-\infty}^{+\infty}\frac{d\omega}{\omega^2+\varepsilon^2} = \frac{\varepsilon}{\pi}\left[\int_{-\infty}^{0}\frac{d\omega}{\omega^2+\varepsilon^2}+\int_0^{+\infty}\frac{d\omega}{\omega^2+\varepsilon^2}\right] = \frac{2\varepsilon}{\pi}\int_0^\infty\frac{d\omega}{\omega^2+\varepsilon^2}$$
$$= \frac{2\varepsilon}{\pi}\left[\frac{1}{\varepsilon}\frac{\pi}{2}\right] = 1.$$

Next, the second relation, the Dirac identity, is understood in the following way. When we rewrite $1/(\omega - \mathrm{i}\varepsilon)$ as

$$\frac{1}{\omega - \mathrm{i}\varepsilon} = \frac{\omega}{\omega^2 + \varepsilon^2} + \mathrm{i}\frac{\varepsilon}{\omega^2 + \varepsilon^2},$$

the first term on the right-hand side has the value $1/\omega$ in the limit $\varepsilon \to 0$. The second term may be replaced by a delta function. Therefore, we obtain

$$\lim_{\varepsilon \to 0} \frac{1}{\omega - \mathrm{i}\varepsilon} = \mathcal{P}\frac{1}{\omega} + \mathrm{i}\pi\delta(\omega), \tag{A.7}$$

where $\mathcal{P}$ is the Cauchy principal value and is defined by

$$\begin{aligned} \int_{-\infty}^{+\infty} f(\omega')\mathcal{P}\left[\frac{1}{\omega - \omega'}\right] \mathrm{d}\omega' &= \mathcal{P}\int_{-\infty}^{+\infty} \frac{f(\omega')}{\omega - \omega'}\mathrm{d}\omega' \\ &= \lim_{\varepsilon \to 0}\left(\int_{-\infty}^{\omega-\varepsilon} \frac{f(\omega')}{\omega - \omega'}\mathrm{d}\omega' + \int_{\omega+\varepsilon}^{+\infty} \frac{f(\omega')}{\omega - \omega'}\mathrm{d}\omega'\right). \end{aligned} \tag{A.8}$$

Several important relations of Dirac delta function are summarized in the following.

$$\delta(x) = \delta(-x) \tag{A.9}$$

$$\int f(x)\delta(x-a)\mathrm{d}x = f(a) \tag{A.10}$$

$$x\delta(x) = 0 \tag{A.11}$$

$$\delta(ax) = \frac{1}{a}\delta(x), \quad a > 0 \tag{A.12}$$

$$\int \delta(a-x)\delta(x-b)\mathrm{d}x = \delta(a-b) \tag{A.13}$$

$$\int \delta(x-a)\delta(a-b)\mathrm{d}x = \delta(x-b) \tag{A.14}$$

$$f(x)\delta(x-a) = f(a)\delta(x-a) \tag{A.15}$$

$$\delta(x^2 - a^2) = \frac{1}{2a}\Big[\delta(x-a) + \delta(x+a)\Big], \quad a > 0 \tag{A.16}$$

$$-\delta'(x) = \delta'(-x) \tag{A.17}$$

$$\int f(x)\delta'(x)\mathrm{d}x = -f'(a) \tag{A.18}$$

A.2 Cyclic Boundary Condition and Delta Function

In Appendix A.1 the Dirac delta function is defined in the region of integration $[-\infty, +\infty]$. In solid state physics, however, the dimension of a crystal of length L is defined by the region $[-L/2, L/2]$ and the cyclic boundary condition is adopted. For example, considering the one-dimensional case and letting the lattice constant be a, the wave vectors $q = 2\pi n/L$ ($n = 0, \pm 1, \pm 2, \cdots$) are those in the first Brillouin zone $[-\pi/a, +\pi/a]$, which correspond to the lattice points N ($n = -N/2, \ldots, N/2$). The calculations with respect to the wave vectors, therefore, can be carried out in the first Brillouin zone $[-\pi/a, +\pi/a]$ of the reciprocal lattice vector $2\pi/a$. Here we take account of the wave vector $\boldsymbol{q}$ of the lattice vibrations, but we may draw the same conclusion for the wave vectors of an electron in a crystal.

With this definition we find

$$\frac{1}{L}\int_{-L/2}^{L/2} \mathrm{e}^{\mathrm{i}(q-q')x}\mathrm{d}x = \delta_{q,q'}\,, \tag{A.19}$$

$$\sum_q \mathrm{e}^{\mathrm{i}q(x-x')} = L\delta(x-x')\,. \tag{A.20}$$

In the case of a crystal with d dimensions, we obtain

$$\frac{1}{L^d}\int_V \mathrm{e}^{\mathrm{i}(\boldsymbol{q}-\boldsymbol{q}')\cdot\boldsymbol{r}}\mathrm{d}^d\boldsymbol{r} = \delta_{\boldsymbol{q},\boldsymbol{q}'}\,, \tag{A.21}$$

$$\sum_{\boldsymbol{q}} \mathrm{e}^{\mathrm{i}\boldsymbol{q}\cdot(\boldsymbol{r}-\boldsymbol{r}')} = L^d\delta(\boldsymbol{r}-\boldsymbol{r}')\,. \tag{A.22}$$

First, we will prove the one-dimensional case. The cyclic boundary condition leads to the following relation

$$\mathrm{e}^{\mathrm{i}qL} = 1, \quad q = \frac{2\pi n}{L}\,.$$

For $q =\neq q'$ we find

$$\int_{-L/2}^{L/2} \mathrm{e}^{\mathrm{i}(q-q')x}\mathrm{d}x = \left[\frac{\mathrm{e}^{\mathrm{i}(q-q')x}}{\mathrm{i}(q-q')}\right]_{-L/2}^{L/2} = \frac{2\sin[(q-q')L/2]}{(q-q')}\,, \tag{A.23}$$

and putting $q = 2\pi n/L$ and $q' = 2\pi m/L$ ($m \neq n$), the above equation becomes equal to zero. On the other hand, for $q = q'$ we have

$$\int_{L/2}^{L/2} e^{i(q-q')x} dx = \int_{-L/2}^{L/2} 1 dx = L ,$$

and thus we obtain the following relation:

$$\frac{1}{L} \int_{-L/2}^{L/2} e^{i(q-q')x} dx = \delta_{q,q'} ,$$

and (A.19) is proved.

The relation given by (A.20) is just the inverse Fourier transform of (A.19), which will be understood from the following discussion of the Fourier transform. For simplicity we put $x' = 0$ and prove the following relation:

$$\sum_q e^{iqx} = L\delta(x) .$$

Multiplying both sides of this equation by $(1/L)\exp(-iq'x)$ and integrating over the region $[-L/2, L/2]$ leads to the following result with the help of (A.19):

$$\text{Left-hand side} = \sum_q \frac{1}{L} \int_{-L/2}^{L/2} e^{i(q-q')x} dx = \sum_q \delta_{q,q'} = 1 ,$$

$$\text{Right-hand side} = \int_{-L/2}^{L/2} e^{-iq'x} dx \delta(x) = 1 ,$$

and therefore the relation (A.20) is proved. The same result is obtained when the summation is replaced by the integral

$$\sum_q = \frac{L}{2\pi} \int dq ,$$

which leads to the following relation

$$\sum_q e^{iq(x-x')} = \frac{L}{2\pi} \int e^{iq(x-x')} dq = L\delta(x - x') .$$

From these results we may understand the relations between the delta function and the cyclic boundary condition.

In the case of the three-dimensional lattice we may deduce the same result. Consider a crystal with N lattice points and let the position vector be $\boldsymbol{R}_j$. Assuming the cyclic boundary condition, we rewrite the integral over the crystal as the sum of the integral over the unit cell Ω:

$$
\begin{aligned}
I &= \frac{1}{L^3}\int_V \exp[\mathrm{i}(\boldsymbol{q}-\boldsymbol{q}')\cdot\boldsymbol{r}]\mathrm{d}^3\boldsymbol{r} \\
&= \frac{1}{N\Omega}\sum_j^N \exp\left[\mathrm{i}(\boldsymbol{q}-\boldsymbol{q}')\cdot\boldsymbol{R}_j\right]\int_\Omega \exp[\mathrm{i}(\boldsymbol{q}-\boldsymbol{q}')\cdot\boldsymbol{r}]\mathrm{d}^3\boldsymbol{r} \\
&= \delta_{\boldsymbol{q},\boldsymbol{q}'}\frac{1}{\Omega}\int_\Omega \exp[\mathrm{i}(\boldsymbol{q}-\boldsymbol{q}')\cdot\boldsymbol{r}]\mathrm{d}^3\boldsymbol{r} \\
&= \begin{cases} 1 & (\text{for } \boldsymbol{q}=\boldsymbol{q}') \\ 0 & (\text{for } \boldsymbol{q}\neq\boldsymbol{q}'), \end{cases}
\end{aligned}
\tag{A.24}
$$

where

$$
\begin{aligned}
\sum_j^N \exp\left[\mathrm{i}(\boldsymbol{q}-\boldsymbol{q}')\cdot\boldsymbol{R}_j\right] &= \sum_{j=0}^{N-1} \exp\left[\mathrm{i}(\boldsymbol{q}-\boldsymbol{q}')\cdot\boldsymbol{R}_j\right] \\
&= \frac{1-\exp[\mathrm{i}(\boldsymbol{q}-\boldsymbol{q}')\cdot\boldsymbol{R}N]}{1-\exp[\mathrm{i}(\boldsymbol{q}-\boldsymbol{q}')\cdot\boldsymbol{R}]} \\
&= 0 \quad (\text{for } \boldsymbol{q}-\boldsymbol{q}'\neq 0),
\end{aligned}
\tag{A.25}
$$

which is shown in the following. Rewriting $\boldsymbol{q}$ and $\boldsymbol{R}$ in their vector components such as $q_x = (2\pi/L)n_x$ ($n_x = 0, \pm1, \pm2, \pm3, \cdots$) and $R_x = am_x$ ($m_x = 0, 1, 2, \ldots, N-1$), then we find

$$
\frac{2\pi}{L}n_x aN = 2\pi n_x\,. \tag{A.26}
$$

For $\boldsymbol{q}-\boldsymbol{q}'=0$, we obtain

$$
\sum_{j=0}^{N-1} \exp\left[\mathrm{i}(\boldsymbol{q}-\boldsymbol{q}')\cdot\boldsymbol{R}_j\right] = N\,. \tag{A.27}
$$

In general, therefore, the following relations hold for the wave vectors of electrons and phonons, $\boldsymbol{k}$ and $\boldsymbol{q}$:

$$
\sum_j \exp\left[\mathrm{i}(\boldsymbol{k}-\boldsymbol{k}')\cdot\boldsymbol{R}_j\right] = N\delta_{\boldsymbol{k},\boldsymbol{k}'}\,, \tag{A.28}
$$

$$
\sum_j \exp\left[\mathrm{i}(\boldsymbol{q}-\boldsymbol{q}')\cdot\boldsymbol{R}_j\right] = N\delta_{\boldsymbol{q},\boldsymbol{q}'}\,. \tag{A.29}
$$

In addition, we have the relation

$$\frac{1}{\Omega}\int_{\Omega}\exp[\mathrm{i}(\boldsymbol{q}-\boldsymbol{q}')\cdot\boldsymbol{r}]\mathrm{d}^3\boldsymbol{r}=1 \quad (\text{for } \boldsymbol{q}=\boldsymbol{q}'), \tag{A.30}$$

and therefore we obtain the final result.

$$\frac{1}{L^d}\int\exp[\mathrm{i}(\boldsymbol{q}-\boldsymbol{q}')\cdot\boldsymbol{r}]\mathrm{d}^d\boldsymbol{r}\equiv\delta_{\boldsymbol{q},\boldsymbol{q}'} = \begin{cases}1 & (\text{for } \boldsymbol{q}=\boldsymbol{q}')\\ 0 & (\text{for } \boldsymbol{q}\neq\boldsymbol{q}').\end{cases} \tag{A.31}$$

A.3 Fourier Transform

As is well known in mathematics, the Fourier transform is expressed as

$$f(x)=\frac{1}{\sqrt{2\pi}}\int_{-\infty}^{\infty}F(k)\exp(\mathrm{i}kx)|dk\,, \tag{A.32}$$

$$F(k)=\frac{1}{\sqrt{2\pi}}\int_{-\infty}^{\infty}f(x)\exp(-\mathrm{i}kx)\mathrm{d}x\,, \tag{A.33}$$

where the function $F(k)$ is called the **Fourier transform** of the function $f(x)$ and the function $f(x)$ is the Fourier transform of the function $F(k)$.

The Fourier transform given by (A.33) is realized when the function $f(x)$ satisfies the following condition

$$\int_{-\infty}^{\infty}|f(x)|^2\mathrm{d}x<\infty\,. \tag{A.34}$$

The Fourier transform discussed above is shown for the one-dimensional case. It is easy to extend it the three-dimensional case, which is written as

$$f(x,y,z)=\left(\frac{1}{2\pi}\right)^{3/2}\int_{-\infty}^{\infty}\int_{-\infty}^{\infty}\int_{-\infty}^{\infty}F(k_x,k_y,k_z) \times\exp\left[\mathrm{i}(k_xx,k_yy,k_zz)\right]\mathrm{d}k_x\mathrm{d}k_y\mathrm{d}k_z\,. \tag{A.35}$$

When we introduce the vector notation $(x,y,z)=\boldsymbol{r}$, $\mathrm{d}x\mathrm{d}y\mathrm{d}z=\mathrm{d}^3\boldsymbol{r}$, $(k_x,k_y,k_z)=\boldsymbol{k}$, and $\mathrm{d}k_x\mathrm{d}k_y\mathrm{d}k_z=\mathrm{d}^3\boldsymbol{k}$, the Fourier transform may be rewritten as

$$f(\boldsymbol{r})=\left(\frac{1}{2\pi}\right)^{3/2}\int_{-\infty}^{\infty}F(\boldsymbol{k})\exp(\mathrm{i}\boldsymbol{k}\cdot\boldsymbol{r})\mathrm{d}^3\boldsymbol{k}\,, \tag{A.36}$$

$$F(\boldsymbol{k})=\left(\frac{1}{2\pi}\right)^{3/2}\int_{-\infty}^{\infty}f(\boldsymbol{r})\exp(-\mathrm{i}\boldsymbol{k}\cdot\boldsymbol{r})\mathrm{d}^3\boldsymbol{r}\,. \tag{A.37}$$

It is very important in solid state physics to express the Fourier transform under the cyclic boundary condition. For one-dimensional space we find the following relations:

$$f(x) = \sum_q F(q)\mathrm{e}^{\mathrm{i}qx} , \tag{A.38}$$

$$F(q) = \frac{1}{L}\int f(x)\mathrm{e}^{-\mathrm{i}qx}\mathrm{d}x . \tag{A.39}$$

Here again (A.38) is called the Fourier transform of the function $f(x)$, and the coefficient $F(q)$ is the **Fourier coefficient**, or (A.39) is called the Fourier transform of $f(x)$. In general, the Fourier transform in d dimensional space is given by the following relations:

$$f(\boldsymbol{r}) = \sum_{\boldsymbol{q}} F(\boldsymbol{q})\mathrm{e}^{\mathrm{i}\boldsymbol{q}\cdot\boldsymbol{r}} , \tag{A.40}$$

$$F(\boldsymbol{q}) = \frac{1}{L^d}\int f(\boldsymbol{r})\mathrm{e}^{-\mathrm{i}\boldsymbol{q}\cdot\boldsymbol{r}}\mathrm{d}^d\boldsymbol{r} . \tag{A.41}$$

When we replace the summation in $\boldsymbol{q}$ space by an integral, (A.40) is written as follows.

$$f(\boldsymbol{r}) = \frac{L^d}{(2\pi)^d}\int F(\boldsymbol{q})\mathrm{e}^{\mathrm{i}\boldsymbol{q}\cdot\boldsymbol{r}}\mathrm{d}^d\boldsymbol{r} . \tag{A.42}$$

It may be proved as follows that the Fourier coefficient of (A.38) is given by (A.39). Inserting (A.38) into (A.39) and using (A.19), the following result follows:

$$F(q) = \sum_{q'} \frac{1}{L}\int F(q')\mathrm{e}^{\mathrm{i}(q'-q)x}\mathrm{d}x = \sum_{q'} F(q')\delta_{q',q} = F(q) .$$

In contrast, inserting (A.39) into (A.38), we find

$$f(x) = \sum_q \frac{1}{L}\int f(x')\mathrm{e}^{\mathrm{i}q(x-x')}\mathrm{d}x = \int f(x')\delta(x-x')\mathrm{d}x = f(x)$$

and thus (A.20)

$$\sum_q \mathrm{e}^{\mathrm{i}q(x-x')} = L\delta(x-x')$$

should hold.

B Gamma Function

Gamma function (Γ function) plays a very important role in the analysis of electron transport and electron statistics evaluated by using distribution function (exponential function). In Chap. 6, electron mobility is evaluated by averaging the relaxation time with the electron distribution and thus Γ function appears. Gamma function $\Gamma(z)$ is defined by the following relation

$$\Gamma(z) = \int_0^\infty x^{z-1} \mathrm{e}^{-x} \mathrm{d}x \,. \tag{B.1}$$

From this definition, $\Gamma(z+1)$ has the following relation:

$$\Gamma(z+1) = z\Gamma(z) \,, \tag{B.2}$$

which is proved by partial integration,

$$\Gamma(z+1) = \left[-x^z \mathrm{e}^{-x}\right]_0^\infty + \int_0^\infty x^{z-1} \mathrm{e}^{-x} \mathrm{d}x = z\Gamma(z) \,. \tag{B.3}$$

Next we show $\Gamma(1)$ and $\Gamma(1/2)$ have the following values.

$$\Gamma(1) = 1, \qquad \Gamma\left(\frac{1}{2}\right) = \sqrt{\pi} \,. \tag{B.4}$$

The first relation is obtained by using partial integration. When z=1, $\Gamma(1)$ is evaluated as

$$\Gamma(1) = \int_0^\infty x^{(1-1)} \mathrm{e}^{-x} \mathrm{d}x = \int_0^\infty \mathrm{e}^{-x} \mathrm{d}x = \left[-\mathrm{e}^{-x}\right]_0^\infty = 1 \,.$$

Then we obtain for all integer n,

$$\Gamma(n) = n\Gamma(n) = (n-1) \cdot (n-2) \cdots 3 \cdot 2 \cdot 1 = (n-1)! \tag{B.5}$$

Second relation for $\Gamma(1/2)$ is normally obtained by using β function in mathematical textbooks, but here we derive it by the following method, because we will not concern with β function in semiconductor physics. Now, we consider

$$I = \int_{-\infty}^\infty \mathrm{e}^{-x^2} \mathrm{d}x = 2\int_0^\infty \mathrm{e}^{-x^2} \mathrm{d}x$$

$$I = \int_{-\infty}^\infty \mathrm{e}^{-y^2} \mathrm{d}y = 2\int_0^\infty \mathrm{e}^{-y^2} \mathrm{d}y \,.$$

Putting $y^2 = t$, we find $2y\mathrm{d}y = \mathrm{d}t$, and thus we have the relations

$$dy = \frac{1}{2}\frac{1}{\sqrt{t}}dt\,,$$

and

$$I = 2\int_0^\infty \frac{1}{2}\frac{1}{\sqrt{t}}e^{-t}dt = \int_0^\infty t^{-1/2}e^{-t}dt = \Gamma\left(\frac{1}{2}\right). \tag{B.6}$$

Let's try the following manipulation

$$I^2 = \int_{-\infty}^{\infty} e^{-x^2}dx \int_{-\infty}^{\infty} e^{-y^2}dy = \int\int e^{-(x^2+y^2)}dxdy\,. \tag{B.7}$$

Integration over the space $[x, y]$ is changed into the spherical space integration over r and angle θ.

$$x^2 + y^2 = r^2$$

$$dxdy \rightarrow rd\theta dr \rightarrow 2\pi r dr$$

$$\int_{-\infty}^{\infty}\int_{-\infty}^{\infty} dxdy = \int_0^{2\pi}\int_0^\infty rd\theta dr = \int_0^\infty 2\pi r dr\,.$$

Since we have the relation

$$I^2 = \int_0^\infty 2\pi r e^{-r^2}dr = \pi\int_0^\infty e^{-t}dt = \pi\left[-e^{-t}\right]_0^\infty = \pi\,. \tag{B.8}$$

Here the relation $r^2 = t$, $2rdr = dt$ is used. From this relation C

$$I = \sqrt{\pi}$$

and thus we obtain the desired relation

$$\Gamma\left(\frac{1}{2}\right) = \sqrt{\pi}\,. \tag{B.9}$$

When n is a positive integer, then we have the following relations:

$$\Gamma(n+1) = (n-1)\cdot(n-2)\cdot\cdots 3\cdot 2\cdot 1 = (n-1)! \tag{B.10}$$

$$\Gamma\left(n+\frac{1}{2}\right) = \left(n-\frac{1}{2}\right)\Gamma\left(n-1+\frac{1}{2}\right) = \frac{(2n)!}{4^n n!}\sqrt{\pi} \tag{B.11}$$

C Uniaxial Stress and Strain Components in Cubic Crystals

Denoting the displacement of a crystal by $\boldsymbol{u}$ the strain tensor is defined by

$$e_{ij} = \left(\frac{\mathrm{d}u_i}{\mathrm{d}x_j} + \frac{\mathrm{d}u_j}{\mathrm{d}x_i} \right) . \tag{C.1}$$

When we define the force per unit area in the direction along the i axis in the plane perpendicular to the j axis by the stress tensor T_{ij}, Hooke's law is expressed as

$$T_{ij} = c_{ijkl} e_{kl} , \tag{C.2}$$

where c_{ijkl} is called the elastic constant. Let us define the notations

$$\begin{array}{lcccccc} ij: & xx & yy & zz & yz, zy & zx, xz & xy, yx \\ \alpha: & 1 & 2 & 3 & 4 & 5 & 6 \end{array} . \tag{C.3}$$

We may therefore rewrite (C.2) as

$$\begin{aligned} T_\alpha &= c_{\alpha\beta} e_\beta \\ & \qquad\qquad \alpha, \beta = 1, 2, 3, 4, 5, 6 \\ e_\alpha &= s_{\alpha\beta} T_\beta , \end{aligned} \tag{C.4}$$

where $s_{\alpha\beta}$ is the elastic compliance constant. When we define

$$c_{\alpha\beta} = c_{ijkl} , \tag{C.5}$$

we find for strain tensors the relations

$$\begin{aligned} &e_{xx} = e_1, && e_{yy} = e_2, && e_{zz} = e_3, \\ &2e_{yz} = 2e_{zy} = e_4, && 2e_{zx} = 2e_{xz} = e_5, && 2e_{xy} = 2e_{yx} = e_6, \end{aligned} \tag{C.6}$$

and for the elastic compliance constants the relations

$$\begin{aligned} &s_{xxxx} = s_{11}, && s_{xxyy} = s_{12}, && s_{xxzz} = s_{13}, \\ &2s_{xxyz} = s_{14}, && 2s_{xxzx} = s_{15}, && 2s_{xxxy} = s_{16}, \\ &4s_{yzyz} = s_{44}, && 4s_{yzzx} = s_{45} = s_{45}, && 4s_{yzxy} = s_{46} = s_{64}. \end{aligned} \tag{C.7}$$

Let us calculate the strain components under the application of a uniaxial stress in the (110) plane. We consider coordinates (x', y', z') such that a uniaxial stress X is applied in the z' direction, and the directions x', y' are perpendicular to the stress. Then the stress tensor is written as

$$||T'|| = \begin{vmatrix} 0 \\ 0 \\ X \\ 0 \\ 0 \\ 0 \end{vmatrix} . \tag{C.8}$$

We transform the stress tensor into the coordinates (x, y, z) of the crystal. Since the transform in general is expressed as

$$x_i = (a^{-1})_{ij} x'_j , \tag{C.9}$$

the transform matrix is given by

$$||a^{-1}|| = \begin{vmatrix} \frac{1}{\sqrt{2}} & \frac{1}{\sqrt{2}} \cos\theta & \frac{1}{\sqrt{2}} \sin\theta \\ -\frac{1}{\sqrt{2}} & \frac{1}{\sqrt{2}} \cos\theta & \frac{1}{\sqrt{2}} \sin\theta \\ 0 & -\sin\theta & \cos\theta \end{vmatrix} , \tag{C.10}$$

where θ is the angle between the z and z' axes. The transform of the stress is written as

$$T_{ik} = (a^{-1})_{ij} (a^{-1})_{kl} T'_{jl} , \tag{C.11}$$

and thus the stress in the coordinates (x, y, z) is given by the following relation:

$$||T|| = \begin{vmatrix} T_{xx} \\ T_{yy} \\ T_{zz} \\ T_{yz} \\ T_{zx} \\ T_{xy} \end{vmatrix} = X \begin{vmatrix} \frac{1}{2} \sin^2\theta \\ \frac{1}{2} \sin^2\theta \\ \cos^2\theta \\ \frac{1}{\sqrt{2}} \sin\theta\cos\theta \\ \frac{1}{\sqrt{2}} \sin\theta\cos\theta \\ \frac{1}{2} \sin^2\theta \end{vmatrix} . \tag{C.12}$$

From these results the strain tensor components are expressed as Table B.1

$$||e|| = \begin{vmatrix} e_1 \\ e_2 \\ e_3 \\ e_4 \\ e_5 \\ e_6 \end{vmatrix} = X \begin{vmatrix} s_{11} \frac{1}{2} \sin^2\theta + s_{12}(\frac{1}{2} \sin^2\theta + \cos^2\theta) \\ s_{11} \frac{1}{2} \sin^2\theta + s_{12}(\frac{1}{2} \sin^2\theta + \cos^2\theta) \\ s_{11} \cos^2\theta + s_{12} \sin^2\theta \\ \frac{1}{\sqrt{2}} s_{44} \cos\theta \sin\theta \\ \frac{1}{\sqrt{2}} s_{44} \cos\theta \sin\theta \\ \frac{1}{2} s_{44} \sin^2\theta \end{vmatrix} . \tag{C.13}$$

Table B.1 Character table of the T_d group and the basis functions

KST	BSW	MLC	E	$3C_2$	$6S_4$	6σ	$8C_3$	Basis functions
Γ_1	Γ_1	A_1	1	1	1	1	1	xyz
Γ_2	Γ_2	A_2	1	1	-1	-1	1	$x^4(y^2-z^2)+y^4(z^2-x^2)+z^4(x^2-y^2)$
Γ_3	Γ_{12}	E	2	2	0	0	-1	$(x^2-y^2), z^2-(x^2+y^2)/2$
Γ_5	Γ_{25}	T_1	3	-1	1	-1	0	$x(y^2-z^2), y(z^2-x^2), z(x^2-y^2)$
Γ_4	Γ_{15}	T_2	3	-1	-1	1	0	x, y, z

KST : Notation of Koster,
BSW : Notation of Bouckaert, Smoluchowski and Wigner,
MLC : Molecular notation.

Since the relation between the strain tensors e_α and e_{ij} is given by (C.6), we obtain the following result:

$$\begin{aligned}
e_{xx} &= e_{yy} = X\left[\frac{1}{2}s_{11}\sin^2\theta + s_{12}\left(\frac{1}{2}\sin^2\theta + \cos^2\theta\right)\right] \\
e_{zz} &= X[s_{11}\cos^2\theta + s_{12}\sin^2\theta] \\
e_{xy} &= \frac{X}{4}s_{44}\sin^2\theta \\
e_{zx} &= e_{yz} = \frac{X}{2\sqrt{2}}s_{44}\cos\theta\sin\theta\,.
\end{aligned} \tag{C.14}$$

As stated above, the strain tensor is a second-rank tensor and its six independent components are $e_{xx}, e_{yy}, e_{zz}, e_{yz} = e_{zy}, e_{zx} = e_{xz}, e_{xy} = e_{yx}$. The strain tensor is related to the symmetry of the crystal and the analysis of the Raman scattering tensor and the deformation potentials is classified with the help of the group theory analysis of the strain tensor. In this Appendix we briefly describe the irreducible representation of the strain tensor for a crystal with cubic symmetry, where we use the notation of group theory for the zinc blende-type crystals of the T_d group. In Table B.1 the character table for the T_d group is shown. From Table B.1 a symmetric strain tensor e_{ij} is classified into the irreducible representations of one-dimensional Γ_1, two-dimensional Γ_3 and three-dimensional Γ_4. In other words, we obtain

$$\begin{aligned}
&\Gamma_1 : e_{xx} + e_{yy} + e_{zz}\,, \\
&\Gamma_3 : e_{xx} - e_{yy}\,, \quad e_{zz} - (e_{xx} + e_{yy})/2\,, \\
&\Gamma_4 : e_{xy}\,, \quad e_{yz}\,, \quad e_{zx}\,.
\end{aligned} \tag{C.15}$$

It is also possible to express the strain tensor e_{ij} by the following three matrices:

$$[e_{ij}(\Gamma_1)] = \frac{1}{3}\begin{bmatrix} e_{xx}+e_{yy}+e_{zz} & 0 & 0 \\ 0 & e_{xx}+e_{yy}+e_{zz} & 0 \\ 0 & 0 & e_{xx}+e_{yy}+e_{zz} \end{bmatrix}$$

$$[e_{ij}(\Gamma_3)] = \frac{1}{3}\begin{bmatrix} 2e_{xx}-(e_{yy}+e_{zz}) & 0 & 0 \\ 0 & 2e_{yy}-(e_{zz}+e_{xx}) & 0 \\ 0 & 0 & 2e_{zz}-(e_{xx}+e_{yy}) \end{bmatrix}$$

$$[e_{ij}(\Gamma_4)] = \begin{bmatrix} 0 & e_{xy} & e_{xz} \\ e_{xy} & 0 & e_{yz} \\ e_{xz} & e_{yz} & 0 \end{bmatrix}.$$

D Boson Operators

In the main text we have discussed the quantization of the lattice vibrations, where boson operators appear. Here we will describe the boson operators to supplement the treatment. For simplicity we disregard the subscripts. The Hamiltonian for a simple harmonic oscillator is written as

$$H = \frac{1}{2M}\left(p^2 + M^2\omega^2 q^2\right), \tag{D.1}$$

where the following commutation relation holds:

$$[q, p] = i\hbar. \tag{D.2}$$

Defining new variables by

$$P = \sqrt{\frac{1}{M}} \cdot p, \tag{D.3}$$

$$Q = \sqrt{M} q, \tag{D.4}$$

the Hamiltonian is rewritten as

$$H = \frac{1}{2}\left(P^2 + \omega^2 Q^2\right) \tag{D.5}$$

and the commutation relation is expressed as

$$[Q, P] = \mathrm{i}\hbar. \tag{D.6}$$

As described in Sect. 6.1.2, we introduce new variables defined by

$$a = \left(\frac{1}{2\hbar\omega}\right)^{1/2} (\omega Q + \mathrm{i}P) , \tag{D.7}$$

$$a^{\dagger} = \left(\frac{1}{2\hbar\omega}\right)^{1/2} (\omega Q - \mathrm{i}P) , \tag{D.8}$$

where a and $a^{\dagger}$ are Hermite conjugates. Using these operators we may define Q and P, which are given by

$$Q = \left(\frac{\hbar}{2\omega}\right)^{1/2} \left(a + a^{\dagger}\right) , \tag{D.9}$$

$$P = -\mathrm{i}\left(\frac{\hbar\omega}{2}\right)^{1/2} \left(a - a^{\dagger}\right) . \tag{D.10}$$

From these results we easily find the following relations:

$$a^{\dagger}a = \frac{1}{\hbar\omega}\left(H - \frac{1}{2}\hbar\omega\right) , \tag{D.11}$$

$$aa^{\dagger} = \frac{1}{\hbar\omega}\left(H + \frac{1}{2}\hbar\omega\right) . \tag{D.12}$$

The commutation relation between a and $a^{\dagger}$ is written as

$$\left[a, a^{\dagger}\right] = aa^{\dagger} - a^{\dagger}a = 1 . \tag{D.13}$$

Finally, the Hamiltonian is rewritten with new operators as

$$H = \left(a^{\dagger}a + \frac{1}{2}\right) \hbar\omega . \tag{D.14}$$

We have to note here that the operators $a^{\dagger}$ and a are not observable, but that when the operators are applied to a state, they change the state. As shown in Sect. 6.1.2, we use the number operator $\hat{n}$ in the following.

Denoting the eigenstate of a simple harmonic oscillator by $|n\rangle$ and its eigenvalue by $\mathcal{E}_n$, we may write The Schrödinger equation as

$$H|n\rangle = \hbar\omega\left(a^{\dagger}a + \frac{1}{2}\right)|n\rangle = \mathcal{E}_n|n\rangle . \tag{D.15}$$

This may be rewritten as follows by using the commutation relation given by (D.13):

$$\hbar\omega\left(aa^{\dagger} - 1 + \frac{1}{2}\right)|n\rangle = \mathcal{E}_n|n\rangle . \tag{D.16}$$

Multiplication by $a^\dagger$ from the left to both sides of this equation leads to

$$\hbar\omega\left(a^\dagger a a^\dagger - a^\dagger + \frac{1}{2}a^\dagger\right)|n\rangle = \mathcal{E}_n a^\dagger |n\rangle \,. \tag{D.17}$$

Transposition of the second term on the left-hand side to the right-hand side results in

$$\hbar\omega\left(a^\dagger a + \frac{1}{2}\right)a^\dagger|n\rangle = H a^\dagger|n\rangle = (\mathcal{E}_n + \hbar\omega)\, a^\dagger|n\rangle \,. \tag{D.18}$$

This equation is understood to be an eigenequation with eigenstate $a^\dagger|n\rangle$ and eigenvalue $\mathcal{E}_n + \hbar\omega$. In other words, when $a^\dagger$ operates on the eigenstate $|n\rangle$, the eigenvalue is changed from $\mathcal{E}_n$ to $\mathcal{E}_n + \hbar\omega$. Considering this fact we may write new eigenstate and eigenvalue as

$$a^\dagger|n\rangle = c_n|n+1\rangle \tag{D.19}$$

$$\mathcal{E}_n + \hbar\omega = \mathcal{E}_{n+1} \,. \tag{D.20}$$

Here the constant c_n of (D.19) is introduced to normalize the state $|n+1\rangle$ and is determined later. Inserting these results in (D.18), we obtain

$$H|n+1\rangle = \mathcal{E}_{n+1}|n+1\rangle \,. \tag{D.21}$$

In a similar fashion, multiplying a from the left to both sides of (D.16) and applying a similar treatment, we obtain

$$Ha|n\rangle = (\mathcal{E}_n - \hbar\omega)\, a|n\rangle \,. \tag{D.22}$$

Here the above equation means that the eigenvalue for the eigenstate $a|n\rangle$ is given by $\mathcal{E}_n - \hbar\omega$. Therefore, in a similar fashion to the previous treatment we may rewrite as follows:

$$a|n\rangle = c'_n|n-1\rangle \,, \tag{D.23}$$

$$\mathcal{E}_n - \hbar\omega = \mathcal{E}_{n-1} \,, \tag{D.24}$$

where c'_n is a constant to normalize the eigenstate $|n-1\rangle$. These relations lead to

$$H|n-1\rangle = \mathcal{E}_{n-1}|n-1\rangle \,. \tag{D.25}$$

It is evident from the process of deriving (D.21) and (D.25) that starting with an eigenstate $|n\rangle$ and an eigenvalue $\mathcal{E}_n$ all other eigenstates and eigenvalues are calculated. In addition, these energy eigenvalues are equally spaced with the interval $\hbar\omega$. If $|n\rangle$ is not the ground state, the eigenstate $a|n\rangle$ exists and its energy eigenvalue lower than $\mathcal{E}_n$ by $\hbar\omega$. If $a|n\rangle$ is not the ground state, there exists the eigenstate $a^2|n\rangle$ and its energy is lower than that of $|n\rangle$ by $2\hbar\omega$. In this way we find the lowest eigenstate $|0\rangle$ and its eigenenergy $\mathcal{E}_0$, where the eigenenergy $\mathcal{E}_0$ should be positive. The lowest state is understood to be the ground state. When a operates on the ground state, we find

$$Ha|0\rangle = (\mathcal{E}_0 - \hbar\omega)a|0\rangle\,. \tag{D.26}$$

Since the eigenstate with eigenenergy lower than the ground state energy is not allowed, we find

$$a|0\rangle = 0\,. \tag{D.27}$$

Taking account of these results, the eigenequation of (D.25) for the ground state $|0\rangle$ is written as

$$H|0\rangle = \frac{1}{2}\hbar\omega|0\rangle = \mathcal{E}_0|0\rangle\,, \tag{D.28}$$

and thus the energy eigenvalue for the ground state is given by

$$\mathcal{E}_0 = \frac{1}{2}\hbar\omega\,. \tag{D.29}$$

From (D.20) or (D.24) the following relation is deduced.

$$\mathcal{E}_n = \left(n + \frac{1}{2}\right)\hbar\omega\,, \quad n = 0, 1, 2, \ldots\,. \tag{D.30}$$

Using (D.16) and (D.30) we may obtain

$$H|n\rangle = \hbar\omega\left(\hat{n} + \tfrac{1}{2}\right)|n\rangle = \hbar\omega\left(n + \tfrac{1}{2}\right)|n\rangle\,, \tag{D.31}$$

$$\hat{n}|n\rangle = a^\dagger a|n\rangle = n|n\rangle\,. \tag{D.32}$$

The above equation tells us that the eigenvalue for the operator $\hat{n} = a^\dagger a$ is n. In a similar fashion, using the commutation relation of (D.13)–(D.16), the following relation is derived:

$$aa^\dagger|n\rangle = (n+1)|n\rangle\,. \tag{D.33}$$

Let us determine the normalization constants c_n and c'_n. The normalization of the eigenstates $|n\rangle$, $|n+1\rangle$ and $|n-1\rangle$ is written as

$$\langle n|n\rangle = \langle n+1|n+1\rangle = \langle n-1|n-1\rangle\,. \tag{D.34}$$

Multiplying the Hermite conjugate of (D.19) to both sides and using (D.33) and (D.34), we obtain

$$\begin{aligned}\langle n+1|c_n^* c_n|n+1\rangle &= \langle n|aa^\dagger|n\rangle \\ &= (n+1)\langle n|n\rangle = n+1\,,\end{aligned} \tag{D.35}$$

which leads to

$$|c_n|^2 = n+1\,. \tag{D.36}$$

Assuming the phase factor of c_n to be zero, (D.19) can be rewritten as

$$a^\dagger|n\rangle = \sqrt{n+1}\,|n+1\rangle\,. \tag{D.37}$$

In a similar fashion, we obtain

$$a|n\rangle = \sqrt{n-1}\,|n-1\rangle\,. \tag{D.38}$$

Since the eigenfunctions are diagonal, i.e. $\langle n|n'\rangle = \delta_{n,n'}$, the non-zero matrix elements of the operators $a^\dagger$ and a are as follows:

$$\langle n+1|a^\dagger|n\rangle = \sqrt{n+1}\,, \tag{D.39}$$

$$\langle n-1|a|n\rangle = \sqrt{n}\,. \tag{D.40}$$

The operator $a^\dagger$ is called the **creation operator** and the operator a is called the **annihilation operator**.

We have mentioned that a known eigenstate will determine all the other eigenstates. An arbitrary eigenstate $|n\rangle$ is therefore derived from the ground state $|0\rangle$. Since (D.37) leads to $(n!)^{1/2}|n\rangle = (a^\dagger)^n|0\rangle$, we find the following relation:

$$|n\rangle = (n!)^{-1/2}(a^\dagger)^n|0\rangle\,. \tag{D.41}$$

As stated in Sect. 6.1.2, the lattice vibrations are expressed by a summation over the modes. Here we use the notation μ for the mode. The eigenstate of the lattice vibrations is given by $|n_1, n_2, \ldots, n_\mu, \ldots\rangle$ and therefore the following relations hold:

$$a_\mu|n_1, n_2, \ldots, n_\mu, \ldots\rangle = \sqrt{n_\mu}\,|n_1, n_2, \ldots, n_\mu - 1, \ldots\rangle\,, \tag{D.42}$$

$$a_\mu^\dagger|n_1, n_2, \ldots, n_\mu, \ldots\rangle = \sqrt{1+n_\mu}\,|n_1, n_2, \ldots, n_\mu + 1, \ldots\rangle\,, \tag{D.43}$$

$$a_\mu a_\nu - a_\nu a_\mu = a_\mu^\dagger a_\nu^\dagger - a_\nu^\dagger a_\mu^\dagger = 0\,, \tag{D.44}$$

$$a_\mu a_\nu^\dagger - a_\nu^\dagger a_\mu = \delta_{\mu\nu}\,, \tag{D.45}$$

or

$$[a_\mu, a_\nu]_- = [a_\mu^\dagger, a_\nu^\dagger]_- = 0\,, \tag{D.46}$$

$$[a_\mu, a_\nu^\dagger]_- = \delta_{\mu\nu}\,, \tag{D.47}$$

$$\begin{aligned} a_\mu a_\mu^\dagger |\ldots, n_\mu, \ldots\rangle &= \sqrt{n_\mu + 1}\, a_\mu |\ldots, n_\mu + 1, \ldots\rangle \\ &= (n_\mu + 1)\, |\ldots, n_\mu, \ldots\rangle\,, \end{aligned} \tag{D.48}$$

$$\begin{aligned} a_\mu^\dagger a_\mu |\ldots, n_\mu, \ldots\rangle &= \sqrt{n_\mu}\, a_\mu^\dagger |\ldots, n_\mu - 1, \ldots\rangle \\ &= (n_\mu)\, |\ldots, n_\mu, \ldots\rangle\,. \end{aligned} \tag{D.49}$$

E Random Phase Approximation and Lindhard Dielectric Function

In this section we follow the treatment of Haug and Koch[3] and discuss the plasma screening effect. The electron density operator $\langle \rho(\boldsymbol{q}) \rangle$ is defined, as shown in Appendix F by

$$\langle \rho(\boldsymbol{q}) \rangle = -\frac{e}{L^3} \sum_{\boldsymbol{k}} \langle c_{\boldsymbol{k}-\boldsymbol{q}}^\dagger c_{\boldsymbol{k}} \rangle\,. \tag{E.1}$$

Denoting the Coulomb potential $V(\boldsymbol{r})$ and the potential induced by electron fluctuations by $V_{\text{ind}}(\boldsymbol{r})$, the effective potential energy $V_{\text{eff}}(\boldsymbol{r})$ for an electron is written as

$$V_{\text{eff}}(\boldsymbol{r}) = V(\boldsymbol{r}) + V_{\text{ind}}(\boldsymbol{r})\,. \tag{E.2}$$

This effective potential energy should be determined self-consistently. The Fourier transform of the effective potential energy leads to

$$V_{\text{eff}}(\boldsymbol{q}) = V(\boldsymbol{q}) + V_{\text{ind}}(\boldsymbol{q})\,. \tag{E.3}$$

The electron Hamiltonian is then given by

$$\mathcal{H} = \sum_{\boldsymbol{k}} \mathcal{E}(\boldsymbol{k}) c_{\boldsymbol{k}}^\dagger c_{\boldsymbol{k}} + \sum_{\boldsymbol{k},\boldsymbol{q}'} V_{\text{eff}}(\boldsymbol{q}') c_{\boldsymbol{k}+\boldsymbol{q}'}^\dagger c_{\boldsymbol{k}}\,. \tag{E.4}$$

The Heisenberg equation for $c_{\boldsymbol{k}-\boldsymbol{q}}^\dagger c_{\boldsymbol{k}}$ gives the following relation:

[3] H. Haug and S.W. Koch: *Quantum Theory of the Optical and Electronic Properties of Semiconductors* (World Scientific, Singapore, 1993) Chaps. 7 and 8.

$$\begin{aligned}\frac{\mathrm{d}}{\mathrm{d}t}c^{\dagger}_{k-q}c_{k} &= \frac{\mathrm{i}}{\hbar}\left[\mathcal{H},\ c^{\dagger}_{k-q}c_{k}\right]\\ &= \frac{\mathrm{i}}{\hbar}\Big(\mathcal{E}(\boldsymbol{k}-\boldsymbol{q})-\mathcal{E}(\boldsymbol{k})\Big)c^{\dagger}_{k-q'}c_{k}\\ &\quad -\frac{\mathrm{i}}{\hbar}\sum_{q'}V_{\mathrm{eff}}(\boldsymbol{q}')\left(c^{\dagger}_{k-q}c_{k-q'}-c^{\dagger}_{k+q'-q}c_{k}\right).\end{aligned} \tag{E.5}$$

Here we use the random phase approximation to evaluate the above equation. The random phase approximation is based on the following assumption. We assume that the expectation value $\langle c^{\dagger}_{k}c_{k'}\rangle$ is approximated by $\langle c^{\dagger}_{k}c_{k'}\rangle \propto \exp[\mathrm{i}(\omega_{k}-\omega_{k'})t]$. In the summation $\sum_{k,k'}\exp[\mathrm{i}(\omega_{k}-\omega_{k'})t]$, the term $\boldsymbol{k}\neq\boldsymbol{k}'$ oscillates and the average contribution is assumed to vanish. Therefore, only the term for $\boldsymbol{k}=\boldsymbol{k}'$ will contribute to the summation. This assumption is called the **random phase approximation**. Applying the random phase approximation to the last two terms on the right hand side of (E.5) we obtain (see the reference in the footnote of p. 682)

$$\begin{aligned}\frac{\mathrm{d}}{\mathrm{d}t}\langle c^{\dagger}_{k-q}c_{k}\rangle &= \frac{\mathrm{i}}{\hbar}\Big(\mathcal{E}(\boldsymbol{k}-\boldsymbol{q})-\mathcal{E}(\boldsymbol{k})\Big)\langle c^{\dagger}_{k-q}c_{k}\rangle\\ &\quad -\frac{\mathrm{i}}{\hbar}V_{\mathrm{eff}}(\boldsymbol{q})\Big(f(\boldsymbol{k}-\boldsymbol{q})-f(\boldsymbol{k})\Big),\end{aligned} \tag{E.6}$$

where we use the following relation

$$f(\boldsymbol{k}) = \langle c^{\dagger}_{k}c_{k}\rangle\,. \tag{E.7}$$

When we assume that the electron density fluctuate as $\langle c^{\dagger}_{k-q}c_{k}\rangle \propto \exp[-\mathrm{i}(\omega + i\Gamma/\hbar)t$, the following relation may be obtained from (E.6)

$$\begin{aligned}&(\hbar\omega+\mathrm{i}\Gamma+\mathcal{E}(\boldsymbol{k}-\boldsymbol{q})-\mathcal{E}(\boldsymbol{k}))\,\langle c^{\dagger}_{k-q}c_{k}\rangle\\ &= V_{\mathrm{eff}}(\boldsymbol{q})\left[f(\boldsymbol{k}-\boldsymbol{q})-f(\boldsymbol{k})\right].\end{aligned} \tag{E.8}$$

Multiplying by $-e/L^3$ on both sides, summing up with respect to $\boldsymbol{k}$, and using the relation given by (E.1), we obtain

$$\langle\rho(\boldsymbol{q})\rangle = -\frac{e^2}{L^3}V_{\mathrm{eff}}(\boldsymbol{q})\sum_{k}\frac{f(\boldsymbol{k}-\boldsymbol{q})-f(\boldsymbol{k})}{\hbar\omega+\mathrm{i}\Gamma+\mathcal{E}(\boldsymbol{k}-\boldsymbol{q})-\mathcal{E}(\boldsymbol{k})}\,. \tag{E.9}$$

Since the potential due to the induced charge follows Poisson's equation, it is given by

$$\nabla^2 V_{\mathrm{ind}}(\boldsymbol{r}) = \frac{e\rho(\boldsymbol{r})}{\epsilon_0}\,. \tag{E.10}$$

The Fourier transform of this equation results in

$$V_{\text{ind}}(\boldsymbol{q}) = -\frac{e}{\epsilon_0 q^2}\rho(\boldsymbol{q}) = \frac{e^2}{\epsilon_0 q^2 L^3} V_{\text{eff}}(\boldsymbol{q}) \sum_{\boldsymbol{k}} \frac{f(\boldsymbol{k}-\boldsymbol{q}) - f(\boldsymbol{k})}{\hbar\omega + \mathrm{i}\Gamma + \mathcal{E}(\boldsymbol{k}-\boldsymbol{q}) - \mathcal{E}(\boldsymbol{k})}$$
$$= V(\boldsymbol{q}) V_{\text{eff}}(\boldsymbol{q}) \sum_{\boldsymbol{k}} \frac{f(\boldsymbol{k}-\boldsymbol{q}) - f(\boldsymbol{k})}{\hbar\omega + \mathrm{i}\Gamma + \mathcal{E}(\boldsymbol{k}-\boldsymbol{q}) - \mathcal{E}(\boldsymbol{k})} . \tag{E.11}$$

Inserting this into (E.3), (6.261) is derived as the following:

$$\kappa(q, \omega) = 1 - V(\boldsymbol{q}) \sum_{\boldsymbol{k}} \frac{f(\boldsymbol{k}-\boldsymbol{q}) - f(\boldsymbol{k})}{\hbar\omega + \mathrm{i}\Gamma + \mathcal{E}(\boldsymbol{k}-\boldsymbol{q}) - \mathcal{E}(\boldsymbol{k})}$$
$$= 1 - \frac{e^2}{\epsilon_0 q^2 L^3} \sum_{\boldsymbol{k}} \frac{f(\boldsymbol{k}-\boldsymbol{q}) - f(\boldsymbol{k})}{\hbar\omega + \mathrm{i}\Gamma + \mathcal{E}(\boldsymbol{k}-\boldsymbol{q}) - \mathcal{E}(\boldsymbol{k})} . \tag{E.12}$$

F Density Matrix

In this section the density matrix is summarized. A good introduction to the density matrix method is given in the text of Kittel,[4] which we shall follow here.

First, we assume a complete and orthonormal set of functions u_n. Any function may be expanded by using these functions and therefore an eigenstate for the Hamiltonian H is expressed as

$$\psi(x, t) = \sum_n c_n(t) u_n(x) , \tag{F.1}$$

where the orthonormality of the functions gives the following relation:

$$\langle u_n | u_m \rangle = \int u_n^* u_m dx = \delta_{nm} . \tag{F.2}$$

The density matrix is defined by

$$\rho_{nm} = \overline{c_m^* c_n} . \tag{F.3}$$

We have to note that the order of m and n on the two sides of (F.3) is interchangeable. The bar indicates the ensemble average over all the systems in the ensemble. Several important properties of the density matrix are summarized in the following.

1. $\sum_n \rho_{nn} = \mathrm{Tr}\{\rho\} = 1$.
   This property leads to the following relation:

[4] C. Kittel, *Elementary Statistical Mechanics* (John Wiley, New York, 1958).

$$\overline{\langle\psi|\psi\rangle} = \overline{\sum_n c_n^* c_n} = \sum_n \rho_{nn} = \mathrm{Tr}\,\{\rho\} = 1\,. \tag{F.4}$$

2. $\overline{\langle F\rangle} = \mathrm{Tr}\,\{F\rho\}$.
   Here $\overline{\langle F\rangle}$ represents the ensemble average of the expectation value of an observable F. This relation is derived as follows:

$$\overline{\langle F\rangle} = \overline{\langle\psi|F|\psi\rangle} = \sum_{m,n} F_{mn}\overline{c_m^* c_n} = \sum_{m,n} F_{mn}\rho_{nm} \tag{F.5}$$

   and thus

$$\overline{\langle F\rangle} = \sum_m (F\rho)_{mm} = \mathrm{Tr}\,\{F\rho\}\,. \tag{F.6}$$

   It is very important to point out that traces are independent of the representation and thus that the ensemble average $\overline{\langle F\rangle}$ is independent of the representation.
3. $\mathrm{i}\hbar\dfrac{\partial\rho}{\partial t} = -[\rho, H] = -(\rho H - H\rho)$.
   The above equation gives the time dependence of the density matrix ρ of the Hamiltonian H. In order to derive the equation, we begin with the wave function (F.1) and insert it into the Schrödinger equation

$$\mathrm{i}\hbar\frac{\partial\psi}{\partial t} = H\psi\,. \tag{F.7}$$

   First, we insert (F.1) into (F.7)

$$\mathrm{i}\hbar\frac{\partial\psi}{\partial t} = \mathrm{i}\hbar\sum_k \frac{\partial c_k}{\partial t}u_k(x) = H\psi = \sum_k c_k H u_k(x)\,, \tag{F.8}$$

   and then multiplying by $u_n(x)$ from the left and integrating over all space we obtain the equation

$$\mathrm{i}\hbar\frac{\partial c_n}{\partial t} = \sum_k H_{nk}c_k\,, \tag{F.9}$$

   where we have used the orthonormality property (F.2) and

$$H_{nk} = \langle u_n|H|u_k\rangle = \int u_n^*(x)Hu_k(x)\mathrm{d}x\,. \tag{F.10}$$

   In a similar fashion we obtain

$$-\mathrm{i}\hbar\frac{\partial c_m^*}{\partial t} = \sum_k H_{mk}^* c_k^*\,. \tag{F.11}$$

From (F.3) we obtain

$$i\hbar\frac{\partial \rho_{nm}}{\partial t} = i\hbar\frac{\partial}{\partial t}\overline{c_m^* c_n} = i\hbar\left(\overline{\frac{\partial c_m^*}{\partial t}c_n + c_m^*\frac{\partial c_n}{\partial t}}\right). \tag{F.12}$$

Inserting (F.9) and (F.11) into (F.12), we find

$$i\hbar\frac{\partial \rho_{nm}}{\partial t} = -\left(\rho H - H\rho\right)_{nm}. \tag{F.13}$$

4. $Z = \mathrm{Tr}\left\{e^{-\beta H}\right\}$

   Here, Z is the partition function. For a canonical ensemble (see the reference in the footnote of page 684)

$$\rho = e^{\beta(F - H)}, \tag{F.14}$$

where F is the Helmholtz free energy and H is the Hamiltonian. In the quantum mechanical representation the partition function Z is given by (see the reference of the footnote of p. 684)

$$Z = \sum_i e^{-\beta E_i}, \tag{F.15}$$

where $\beta = k_B T$. Using the relation between the Helmholtz free energy and the partition function $\log Z = -\beta F$, we find

$$Z = e^{-\beta F} = \sum e^{-\beta E_n} = \mathrm{Tr}\left\{e^{-\beta H}\right\}. \tag{F.16}$$

Since the trace is invariant under unitary transformations, the partition function may be calculated by taking the trace of $e^{-\beta H}$ in any representation. Using these results one may find

$$\rho = \frac{e^{-\beta H}}{\mathrm{Tr}\left\{e^{-\beta H}\right\}}, \tag{F.17}$$

which is a very important relation and is used to evaluate the ensemble average of an observable quantity.

G Spontaneous and Stimulated Emission Rates

Here we will derive the relations between absorption, spontaneous emission, and stimulated emission.

Let's define a vector potential by $\boldsymbol{A}$ and put $\boldsymbol{B} = \mathrm{rot}\,\boldsymbol{A} = \nabla \times \boldsymbol{A}$, then we obtain

$$\nabla \cdot \boldsymbol{B} = \nabla \cdot \nabla \times \boldsymbol{A} = 0\,, \tag{G.1}$$

and thus the vector potential satisfies the relation for the magnetic flux $\boldsymbol{B}$ of (9.90d), ($\nabla \cdot \boldsymbol{B} = 0$). We insert the relation $\boldsymbol{B} = \text{rot}\,\boldsymbol{A} = \nabla \times \boldsymbol{A}$ into Maxwell's equation (9.90b), we find

$$\nabla \times \boldsymbol{E} = -\frac{\partial}{\partial t} \nabla \times \boldsymbol{A}\,. \tag{G.2}$$

Therefore we obtain the following relation

$$\boldsymbol{E} = \mathrm{i}\omega \boldsymbol{A}\,. \tag{G.3}$$

Since we deal with the squared values of the vector potential and electric field, we put

$$E = \omega A\,. \tag{G.4}$$

The electromagnetic field interact with an electron of charge $-e$, and the interaction is given by the Hamiltonian

$$H = \frac{1}{2m}\left(\boldsymbol{p} + e\boldsymbol{A}\right)^2 + V(\boldsymbol{r})\,, \tag{G.5}$$

where $V(\boldsymbol{r})$ is the periodic potential of a crystal. This equation is rewritten as[5]

$$H = \frac{p^2}{2m} + V(\boldsymbol{r}) + \frac{e}{m}\left(\boldsymbol{A} \cdot \boldsymbol{p}\right) + \frac{1}{2m}\left(eA\right)^2 \tag{G.6}$$

and neglecting the last term because of its small contribution, we may rewrite

$$H = H_0 + H_1 \tag{G.7}$$

$$H_0 = \frac{p^2}{2m} + V(\boldsymbol{r})\,, \tag{G.8}$$

$$H_1 = \frac{e}{m}\left(\boldsymbol{A} \cdot \boldsymbol{p}\right). \tag{G.9}$$

Treating H_1 as a perturbation term, the transition probability between the initial state $|\mathrm{i}\rangle$ and the final state $|\mathrm{f}\rangle$ is given by

---

[5]In (G.5), we operate $\boldsymbol{A} \cdot \boldsymbol{p} + \boldsymbol{p} \cdot \boldsymbol{A}$ to a scalar function f and taking account of the vector potential of the electromagnetic field $\boldsymbol{A} = \boldsymbol{A}_0 \exp(\mathrm{i}\boldsymbol{k}_\mathrm{p} \cdot \boldsymbol{r})$ and of the momentum operator $\boldsymbol{p} = -\mathrm{i}\hbar\nabla$,

$$(\boldsymbol{A} \cdot \boldsymbol{p} + \boldsymbol{p} \cdot \boldsymbol{A})f = \boldsymbol{A} \cdot (\boldsymbol{p} + \boldsymbol{p} + \hbar\boldsymbol{k}_\mathrm{p})f = 2\boldsymbol{A} \cdot \boldsymbol{p}f$$

is obtained. The last relation was deduced from the fact that the electromagnetic field is transverse wave, resulting in $\boldsymbol{A} \cdot \boldsymbol{k}_\mathrm{p} = 0$.

$$w_{\rm if} = \frac{2\pi}{\hbar} \left|\langle {\rm f}|H_1|{\rm i}\rangle\right|^2 \delta\left[\mathcal{E}_{\rm f} - \mathcal{E}_{\rm i} - \hbar\omega\right] . \tag{G.10}$$

The transition rate between the valence band state $|c\boldsymbol{k}\rangle$ and the conduction band state $|v\boldsymbol{k}'\rangle$ is given by (4.31) in Sect. 4.2

$$\begin{aligned} w_{\rm cv} &= \frac{2\pi}{\hbar}\left|\langle c\boldsymbol{k}'|\frac{e}{m}\boldsymbol{A}\cdot\boldsymbol{p}|v\boldsymbol{k}\rangle\right|^2 \delta\left[\mathcal{E}_{\rm c}(\boldsymbol{k}') - \mathcal{E}_{\rm v}(\boldsymbol{k}) - \hbar\omega\right] \\ &= \frac{\pi e^2}{2\hbar m^2} A_0^2 \left|\langle c\boldsymbol{k}'|\exp({\rm i}\boldsymbol{k}_{\rm p}\cdot\boldsymbol{r})\boldsymbol{e}\cdot\boldsymbol{p}|v\boldsymbol{k}\rangle\right|^2 \delta\left[\mathcal{E}_{\rm c}(\boldsymbol{k}') - \mathcal{E}_{\rm v}(\boldsymbol{k}) - \hbar\omega\right] . \end{aligned}$$

Writing the vector potential $\boldsymbol{A}$ using the unit vector of the vector potential $\boldsymbol{e}$ as $\boldsymbol{A} = \boldsymbol{e}\cdot A$, the matrix element is given by

$$|M| = |\langle c\boldsymbol{k}'|\exp({\rm i}\boldsymbol{k}_{\rm p})\boldsymbol{e}\cdot\boldsymbol{p}|v\boldsymbol{k}\rangle| , \tag{G.11}$$

and we have

$$|M|^2 = \frac{1}{3}\left(|M_x|^2 + |M_y|^2 + |M_z|^2\right), , \tag{G.12a}$$

$$M_x = -{\rm i}\hbar\langle c\boldsymbol{k}'|\exp({\rm i}\boldsymbol{k}_{\rm p})\frac{\partial}{\partial x}|v\boldsymbol{k}\rangle . \tag{G.12b}$$

Since $\boldsymbol{k}_{\rm p}$ is very small, we put $\delta(\boldsymbol{k}' - \boldsymbol{k} - \boldsymbol{k}_{\rm p}) \equiv \delta(\boldsymbol{k}' - \boldsymbol{k}) = 0$ and thus the summation of the allowed $\boldsymbol{k}'$ and $\boldsymbol{k}$ give rise to Kronecker delta $\delta_{\boldsymbol{k}',v\boldsymbol{k}'}$.

Using (4.21) and (4.38) or (4.39), the absorption coefficient is given by

$$\begin{aligned} \alpha &= \frac{\omega\kappa_2}{n_{\rm r}c} = \frac{2\hbar\omega}{n_{\rm r}c\epsilon_0\omega^2 A_o^2} w_{\rm cv} \\ &= \frac{\pi e^2}{n_{\rm r}c\epsilon_0 m^2\omega}\sum_{\boldsymbol{k},\boldsymbol{k}'} |M|^2 \delta\left[\mathcal{E}_{\rm c}(\boldsymbol{k}') - \mathcal{E}_{\rm v}(\boldsymbol{k}) - \hbar\omega\right]\delta_{\boldsymbol{k},\boldsymbol{k}'} . \end{aligned} \tag{G.13}$$

In the case of semiconductor lasers, high densities of electrons and holes are injected in to the active region and occupy the conduction band and valence bands. Therefore the absorption coefficient depends on the occupation factors of the electron and holes. Let the occupation factor of the electrons in the upper and lower states as $f(\mathcal{E}_2)$ and $f(\mathcal{E}_1)$, respectively, and the net rate of the photon absorption and emission is proportional to $f(\mathcal{E}_1)[1 - f(\mathcal{E}_2)] - f(\mathcal{E}_2)[1 - f(\mathcal{E}_1)] = f(\mathcal{E}_1) - f(\mathcal{E}_2)$. Then the absorption coefficient is given by

$$\begin{aligned} \alpha &= \frac{\pi e^2}{n_{\rm r}c\epsilon_0 m^2\omega} \\ &\quad \times \sum_{\boldsymbol{k}} |M|^2 \delta\left[\mathcal{E}_2(\boldsymbol{k}) - \mathcal{E}_1(\boldsymbol{k}) - \hbar\omega\right]\delta_{\boldsymbol{k},\boldsymbol{k}'}\left[f(\mathcal{E}_1) - f(\mathcal{E}_2)\right] . \end{aligned} \tag{G.14}$$

In the above equation $\sum$ is carried out over the pair states of the valence band $|v\boldsymbol{k}\rangle$ and the conduction band $|c\boldsymbol{k}\rangle$, and thus we obtain

$$\sum_{\boldsymbol{k}} \delta[\mathcal{E}_{\mathrm{cv}}(\boldsymbol{k}) - \hbar\omega] = \frac{1}{(2\pi)^3} \int \mathrm{d}^3\boldsymbol{k} \cdot \delta[\mathcal{E}_{\mathrm{cv}}(\boldsymbol{k}) - \hbar\omega]$$

$$\equiv \int \rho_{\mathrm{red}}(\hbar\omega) \cdot \mathrm{d}(\hbar\omega)\,, \tag{G.15}$$

$$\mathcal{E}_{\mathrm{cv}} = \mathcal{E}_{\mathrm{c}}(\boldsymbol{k}) - \mathcal{E}_{\mathrm{v}}(\boldsymbol{k})\,, \tag{G.16}$$

where only one direction of spin orientation is considered. The last relation of (G.15) is obtained by putting the photon energy as $\hbar\omega$ and defining the density of states between the energies $\hbar\omega = \mathcal{E}$ and $\hbar\omega + \mathrm{d}(\hbar\omega) = \mathcal{E} + \mathrm{d}\mathcal{E}$ as $\rho_{\mathrm{red}}(\mathcal{E})\mathrm{d}\mathcal{E}$. When the effective masses of the conduction and valence bands are isotropic, putting the electron effective mass as m_{c} and the hole effective mass as m_{h}, we may write

$$\mathcal{E}_{\mathrm{cv}} = \frac{\hbar^2k^2}{2m_{\mathrm{c}}} + \frac{\hbar^2k^2}{2m_{\mathrm{h}}} + \mathcal{E}_{\mathrm{G}} = \frac{\hbar^2k^2}{2\mu} + \mathcal{E}_{\mathrm{G}}\,, \tag{G.17}$$

where $1/\mu = 1/m_{\mathrm{c}} + 1/m_{\mathrm{h}}$ and μ is called reduced mass. Then the density of states is given by

$$\rho_{\mathrm{red}} \cdot \mathrm{d}(\hbar\omega) = \frac{1}{(2\pi)^3} 4\pi k^2 \cdot \mathrm{d}k\,,$$

$$= \frac{2\pi}{(2\pi)^3}\left(\frac{2\mu}{\hbar^2}\right)^{3/2} \sqrt{\hbar\omega - \mathcal{E}_{\mathrm{G}}}\,\mathrm{d}\hbar\omega\,. \tag{G.18}$$

Writing the photon energy as $\hbar\omega = \mathcal{E}$, we find

$$\rho_{\mathrm{red}}\mathrm{d}(\mathcal{E}) = \frac{2\pi}{(2\pi)^3}\left(\frac{2\mu}{\hbar^2}\right)^{3/2} \sqrt{\mathcal{E} - \mathcal{E}_{\mathrm{G}}}\,\mathrm{d}\mathcal{E}\,. \tag{G.19}$$

Next we discuss quantum theory of spontaneous and stimulated emissions [1]. First, we deal with the radiation of the electromagnetic waves based on the classical theory. Using Poynting vector of the electromagnetic waves and vector potential $\boldsymbol{A}$ given by $\nabla \times \boldsymbol{A} = \boldsymbol{B}$, and (9.90b), the flux of the waves is given by the following relation

$$\langle \boldsymbol{S} \rangle = \frac{1}{2}\mathrm{Re}(\boldsymbol{E} \times \boldsymbol{H}) = \frac{1}{2}\mathrm{Re}\Big[(-\mathrm{i}\omega\boldsymbol{A}) \times (\mathrm{i}\boldsymbol{k}_{\mathrm{p}} \times \boldsymbol{A})/\mu_0\Big]$$

$$= \frac{\omega}{2\mu_0}\mathrm{Re}\Big[(\boldsymbol{A}\cdot\boldsymbol{A})\boldsymbol{k}_{\mathrm{p}} - (\boldsymbol{A}\cdot\boldsymbol{k}_{\mathrm{p}})\boldsymbol{A}\Big] = \frac{n_{\mathrm{r}}\omega^2}{2c\mu_0}|\boldsymbol{A}|^2\frac{\boldsymbol{k}_{\mathrm{p}}}{|\boldsymbol{k}_{\mathrm{p}}|}\,, \tag{G.20}$$

where μ_0 is the magnetic permeability in free space, $1/(\epsilon_0\mu_0) = c^2$ (c:the light speed in free space) and $\boldsymbol{k}_{\mathrm{p}}/|\boldsymbol{k}_{\mathrm{p}}|$ is a unit vector in the propagation direction of the vector potential $\boldsymbol{A}$. The last relation is derived for the transverse waves in uniform medium, and putting $\boldsymbol{k}_\mathrm{p} \cdot \boldsymbol{A} = 0$.

Next, we discuss Planck's radiation theory. Since the number of plane waves modes in a volume V in an element $\mathrm{d}k_x \mathrm{d}k_y \mathrm{d}k_z$ of the $\boldsymbol{k}_\mathrm{p}$ space is $V(2\pi)^{-3}\mathrm{d}k_x \mathrm{d}k_y \mathrm{d}k_z$, and is independent of the shape of the sample or of the boundary conditions provided the dimensions are large compared to the wavelength. Putting $k_\mathrm{p} = |\boldsymbol{k}_\mathrm{p}| = n_\mathrm{r}\omega/c$, and taking account of two independent polarization directions for each wave vector of the electromagnetic waves, the density of modes per unit volume between ω and $\omega + \mathrm{d}\omega$ is given by $G(\omega)\mathrm{d}\omega$ with

$$G(\omega) = \frac{2}{(2\pi)^3} 4\pi k_\mathrm{p}^2 \frac{\mathrm{d}k_\mathrm{p}}{\mathrm{d}\omega} = \frac{k_\mathrm{p}^2}{\pi^2 v_\mathrm{g}} = \frac{n_\mathrm{r}^2 \omega^2}{\pi^2 c^2 v_\mathrm{g}} . \tag{G.21}$$

Here the group velocity $v_\mathrm{g} = \mathrm{d}\omega/\mathrm{d}k_\mathrm{p}$ is used and assumed to be constant except strong absorption region with the anomalous dispersion.

The average energy of a mode with angular frequency ω at temperature T under the thermal equilibrium is given by

$$\langle \mathcal{E}(\omega) \rangle = \frac{\hbar\omega}{\exp(\hbar\omega/k_\mathrm{B}T) - 1} . \tag{G.22}$$

Thus the energy density of blackbody radiation $u(\omega)\mathrm{d}\omega = \langle \mathcal{E}(\omega) \rangle G(\omega)\mathrm{d}\omega$ in the range between ω and $\omega \mathrm{d}\omega$ is obtained as

$$u(\omega) = \frac{n_\mathrm{r}^2 \hbar\omega^3}{\pi^2 c^2 v_\mathrm{g}} \frac{1}{\exp(k_\mathrm{B}T) - 1} . \tag{G.23}$$

Since the velocity of the energy flow of the electromagnetic waves in a dielectric is the group velocity, the time average of the radiation with wave vector $\boldsymbol{k}_\mathrm{p}$ lying an element of solid angle $\mathrm{d}\Omega$, with polarization vector $\boldsymbol{e}$ lying in an angular interval $\mathrm{d}\theta$ in a plane perpendicular to $\boldsymbol{k}_\mathrm{p}$, and with an angular frequency in a range $\mathrm{d}\omega$, is

$$\begin{aligned} |\langle \boldsymbol{S} \rangle| &= u(\omega) v_\mathrm{g} \frac{\mathrm{d}\Omega}{4\pi} \frac{\mathrm{d}\theta}{2\pi} \mathrm{d}\omega \\ &= \frac{n_\mathrm{r}^2 \hbar\omega^3}{\pi^2 c^2} \frac{1}{\exp(k_\mathrm{B}T) - 1} \frac{\mathrm{d}\Omega}{4\pi} \frac{\mathrm{d}\theta}{2\pi} \mathrm{d}\omega . \end{aligned} \tag{G.24}$$

Averaging (G.24) in the solid angle Ω and polarization direction, we find

$$\int \frac{\mathrm{d}\omega}{4\pi} \int \frac{\mathrm{d}\theta}{2\pi} = 1 ,$$

and thus we obtain,

$$|\langle \boldsymbol{S} \rangle| = \frac{n_\mathrm{r}^2 \hbar\omega^3}{\pi^2 c^2} \frac{1}{\exp(k_\mathrm{B}T) - 1} \mathrm{d}\omega . \tag{G.25}$$

Using (G.11), the spontaneous emission rate from the upper state $\mathcal{E}_2$ to the lower state $\mathcal{E}_1$ is given by

$$r_{\text{spon}}(\mathcal{E}) = \frac{\pi e^2 |\boldsymbol{A}|^2}{2m^2 \hbar} |M|^2 \rho_{\text{red}}(\mathcal{E})[f(\mathcal{E}_2)(1 - f(\mathcal{E}_1)] . \tag{G.26}$$

Equating (G.20) and (G.25), and eliminating $|\boldsymbol{A}|$, we find

$$r_{\text{spon}}(\mathcal{E}) = \frac{n_{\text{r}} e^2 \mu_0 \omega}{\pi m^2 c} |M|^2 \rho_{\text{red}}(\mathcal{E}) f(\mathcal{E}_2)[1 - f(\mathcal{E}_1)] . \tag{G.27}$$

Using the relation $\epsilon_0 \mu_0 = 1/c^2$, and putting $\hbar\omega = \mathcal{E}$, the spontaneous emission rate between the energy separation $\mathcal{E}$ and $\mathcal{E} + \mathrm{d}\mathcal{E}$, is then given by

$$r_{\text{spon}}(\mathcal{E}) = \frac{n_{\text{r}} e^2 \mathcal{E}}{\pi \epsilon_0 m^2 \hbar^2 c^3} |M|^2 \rho_{\text{red}}(\mathcal{E}) f(\mathcal{E}_2)[1 - f(\mathcal{E}_1)] , \tag{G.28}$$

and thus we obtain the spontaneous emission rate given by (9.14). In a similar fashion, the stimulated emission rate (9.15) is given by

$$r_{\text{stim}}(\mathcal{E}) = \frac{n_{\text{r}} e^2 \mathcal{E}}{\pi \epsilon_0 m^2 \hbar^2 c^3} |M|^2 \rho_{\text{red}}(\mathcal{E})(f_2 - f_1) . \tag{G.29}$$

H Spin–Orbit Interaction

As stated in Chap. 2 energy bands are strongly affected by spin–orbit interaction. It is shown in Chap. 2 that the triply degenerate valence bands split into degenerate heavy hole and light hole bands and spin–orbit split-off band. Here an interpretation of spin–orbit interaction based on a semi-classical treatment is described. The spin–orbit interaction is given by solving the Dirac equation.[6] Here a semi-classical treatment of spin–orbit interaction is given for the purpose of introduction. Since the spin–orbit energy is usually given in units of [CGS] in many textbooks, it is derived here in units of [SI] and equations are also shown in units of [CGS].

Let's consider a simple system consisting of an electron $-e$ moving with the velocity $\boldsymbol{v}$ around the nucleus Ze with its relative position $\boldsymbol{r}$. The system is equivalently interpreted from the view point of the electron as that the nucleus is moving with the velocity $-\boldsymbol{v}$ around the electron as shown in Fig. H.1. The nucleus of charge Ze moving with velocity $-\boldsymbol{v}$ will produce a current

$$\boldsymbol{j} = -Ze\boldsymbol{v} \quad \text{[SI]}, \qquad \boldsymbol{j} = -\frac{Ze\boldsymbol{v}}{c} \quad \text{[CGS]} \tag{H.1}$$

[6]The spin–orbit energy is derived by the Dirac equation. See, for example, L. I. Schiff, *Quantum Mechanics*, 2nd edition (McGraw–Hill, New York, 1955) Chapter XII, Sect.44, p.332.

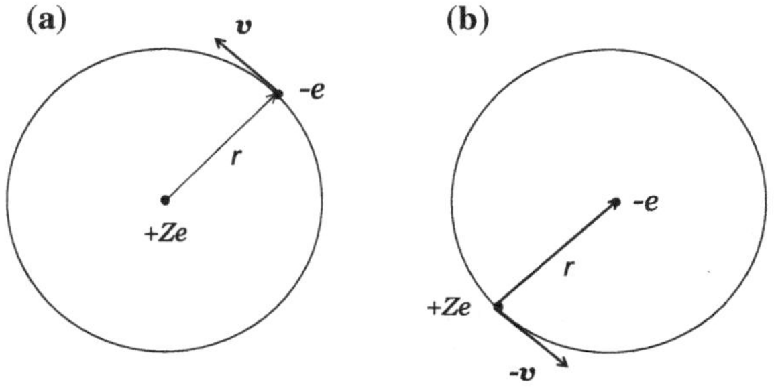

Fig. H.1 **a** Bohr orbital motion of an electron $-e$ with the velocity $\boldsymbol{v}$ as seen by the nucleus Ze is interpreted from the point of view of the electron as **b** the nucleus Ze is moving with velocity $-\boldsymbol{v}$ around the electron

The charge motion produces a magnetic field at the position of the electron, which is given by Biot-Sabart law,

$$\boldsymbol{B} = \frac{\mu_0}{4\pi}\frac{\boldsymbol{j}\times\boldsymbol{r}}{r^3} = -\frac{\mu_0 Ze}{4\pi}\frac{\boldsymbol{v}\times\boldsymbol{r}}{r^3} \ \text{[SI]}, \qquad \boldsymbol{H} = -\frac{Ze}{c}\frac{\boldsymbol{v}\times\boldsymbol{r}}{r^3} \ \text{[CGS]}. \tag{H.2}$$

Electric field $\boldsymbol{E}$ acting on the electron is given by Coulomb's law

$$\boldsymbol{E} = +\frac{Ze}{4\pi\epsilon_0}\frac{\boldsymbol{r}}{r^3} \ \text{[SI]}, \qquad \boldsymbol{E} = +Ze\frac{\boldsymbol{r}}{r^3} \ \text{[CGS]}. \tag{H.3}$$

Therefore the magnetic field is given by

$$\boldsymbol{B} = -\frac{1}{c^2}\boldsymbol{v}\times\boldsymbol{E} \ \text{[SI]}, \qquad \boldsymbol{B} = -\frac{1}{c}\boldsymbol{v}\times\boldsymbol{E} \ \text{[CGS]}, \tag{H.4}$$

where $c = 1/\sqrt{\epsilon_0\mu_0}$.

The magnetic field given by (H.2) interacts with the magnetic moment of the electron (the spin magnetic moment $\boldsymbol{\mu}_S$),

$$\boldsymbol{\mu}_S = -2\frac{\mu_B}{\hbar}\boldsymbol{S} = -\frac{e}{m}\boldsymbol{S} \ \text{[SI]}, \tag{H.5a}$$

$$\boldsymbol{\mu}_S = -2\frac{\mu_B}{\hbar}\boldsymbol{S} = -\frac{e}{mc}\boldsymbol{S} \ \text{[CGS]}, \tag{H.5b}$$

where the approximated factor 2 is used instead of the observed value $\simeq 2.002319$ and $\mu_B = e\hbar/2m$ ($= e\hbar/2mc$ in [CGS]) is called Bohr magneton. The interaction energy is evaluated as the scalar product of the magnetic moment and the magnetic field acting on the electron. A relativistic effect requires an additional factor of one-half [7] in the interaction energy, and thus using (H.4)

[7] See L. H. Thomas, *"The Motion of the Spinning Electron,"* Nature 117 (1926) 514.

$$H_{so} = -\frac{1}{2}\mu_S \cdot \left(\frac{1}{c^2}\boldsymbol{v} \times \boldsymbol{E}\right) = \frac{1}{2}\frac{\mu_S}{m} \cdot (\boldsymbol{E} \times \boldsymbol{p}) = -\frac{1}{2}\frac{1}{mc^2}\frac{1}{r}\frac{d\phi}{dr}(\boldsymbol{r} \times \boldsymbol{p}) \cdot \mu_S$$
$$= -\frac{e}{2m^2c^2}\frac{1}{r}\frac{d\phi}{dr}\boldsymbol{L} \cdot \boldsymbol{S} = \frac{1}{2m^2c^2}\frac{1}{r}\frac{dV}{dr}\boldsymbol{L} \cdot \boldsymbol{S}, \tag{H.6}$$

where we used the angular moment $\boldsymbol{L} = \boldsymbol{r} \times \boldsymbol{p}$, and the following relations. The final expression is exactly the same for the [CGS] unit. and related to the negative gradient of the potential, $\boldsymbol{E} = -d\phi/dr$. The potential energy is given by $V(r) = -e\phi(r)$.

Let's estimate the spin–orbit interaction energy of hydrogen in the state of $n = 2$, $l = 1$ state.The potential energy $V(r)$ for the electron is

$$V(r) = -\frac{e^2}{4\pi\epsilon_0 r} \tag{H.7}$$

and the gradient of the potential energy is

$$\frac{dV(r)}{dr} = \frac{e^2}{4\pi\epsilon_0 r^2}. \tag{H.8}$$

Therefore we obtain

$$\Delta_{so} = \frac{e^2}{4\pi\epsilon_0}\frac{1}{2m^2c^2}\frac{1}{r^3}\boldsymbol{S} \cdot \boldsymbol{L}. \tag{H.9}$$

The magnitude of the spin–orbit interaction energy is estimated by assuming $\boldsymbol{L} \cdot \boldsymbol{S} \simeq \hbar^2$, and the average value of $1/r^3 \simeq 1/2^3 a_B^3$ for the quantum number $n = 2$, where $a_B = 4\pi\epsilon_0\hbar^2/me^2$,

$$\Delta_{so} \sim \frac{(4\pi\epsilon_0 e)^2}{2m^2c^2}\frac{m^3e^6}{2^3\hbar^6} \sim 10^{-4}[\text{eV}], \tag{H.10}$$

and thus the spin–orbit energy is expected to be of the order of 0.1 meV.

From the treatment shown above we find that the spin–orbit interaction depends on the orbital wave function and on the angular momentum through the term $\boldsymbol{L} \cdot \boldsymbol{S}$. In order to get insight into a detailed interaction, we define the total momentum $\boldsymbol{J}$ by

$$\boldsymbol{J} = \boldsymbol{L} + \boldsymbol{S}. \tag{H.11}$$

From this relation we find

$$\boldsymbol{J}^2 = \boldsymbol{L}^2 + \boldsymbol{S}^2 + 2\boldsymbol{L} \cdot \boldsymbol{S} \tag{H.12a}$$
$$2\boldsymbol{L} \cdot \boldsymbol{S} = J^2 - L^2 - S^2. \tag{H.12b}$$

Then the spin–orbit interaction energy is given by

$$H_{\mathrm{so}} = \frac{1}{4m^2c^2}\frac{1}{r}\frac{\mathrm{d}V}{\mathrm{d}r}(J^2 - L^2 - S^2) \tag{H.13}$$

It is well known from quantum mechanics that the operators J^2, L^2, and S^2 commute with each other. Let define an atomic state by the quantum numbers, the principal quantum number n, the total angular momentum number j, the angular momentum number l and the spin quantum number s. Operating the first two operators in the parentheses to the eigenfunction, we obtain unique eigenvalues, and operation of the spin term to the spin state results in $(3/4)\hbar^2$. Therefore the energy of spin–orbit interaction is written as

$$\begin{aligned}\langle H_{\mathrm{so}}\rangle_{nl} &= \frac{1}{4m^2c^2}\left\langle\frac{1}{r}\frac{\mathrm{d}V}{\mathrm{d}r}\right\rangle_{nl}\Big[j(j+1) - l(l+1) - s(s+1)\Big]\hbar^2 \\ &= \frac{1}{4m^2c^2}\left\langle\frac{1}{r}\frac{\mathrm{d}V}{\mathrm{d}r}\right\rangle_{nl}\left[j(j+1) - l(l+1) - \frac{3}{4}\right]\hbar^2 .\end{aligned} \tag{H.14}$$

Using the results we may estimate the spin–orbit splitting energy. First we consider a case of the electronic state $l \neq 0$ of a free atom. Such a state is a doublet with $j = l \pm 1/2$, the spin–orbit splitting energy of the doublet is then given by

$$\Delta_{\mathrm{so}} = \frac{\hbar^2}{4m^2c^2}(2l+1)\left\langle\frac{1}{r}\frac{\mathrm{d}V}{\mathrm{d}r}\right\rangle_{\mathrm{nl}} . \tag{H.15}$$

Therefore the energy separation between the states of ($l = 1,\ j = 3/2$) and ($l = 1,\ j = 1/2$) is given by

$$\Delta_{\mathrm{so},3/2-1/2} = \frac{3\hbar^2}{4m^2c^2}\left\langle\frac{1}{r}\frac{\mathrm{d}V}{\mathrm{d}r}\right\rangle_{nl} . \tag{H.16}$$

The above results are well known as the separation of the D–lines of sodium atom ($P_{3/2}$– and $P_{1/2}$–states). An analysis of the spin–orbit interaction of the valence bands is dealt in Sect. 2.3 of Chap. 2, where the valence band states Γ'_{25} without spin–orbit interaction are six-fold degenerate with angular momentum $l_z = 1,\ 0,\ -1$ and spin $s_z = +1/2,\ -1/2$.

Next we will show how to calculate energy bands taking account of the spin–orbit interaction. Hamiltonian of an electron with the spin–orbit interaction is defined as

$$H = -\frac{\hbar^2}{2m}\nabla^2 + V(\boldsymbol{r}) + \frac{1}{2m^2c^2}\frac{1}{r}\frac{\mathrm{d}V}{\mathrm{d}r}\boldsymbol{L}\cdot\boldsymbol{S} . \tag{H.17}$$

The spin–orbit interaction term (spin–orbit Hamiltonian) is rewritten as

$$H_{\mathrm{so}} = \frac{\hbar}{4m^2c^2}\Big(\nabla V \times \boldsymbol{p}\Big)\cdot\boldsymbol{\sigma} , \tag{H.18}$$

where $\boldsymbol{\sigma}$ is Pauli spin matrix given by (2.50) of Sect. 2.3 in page 80 and we used the following relations

$$\boldsymbol{E} \times \boldsymbol{p} = -(\nabla \phi) \times \boldsymbol{p} = \frac{1}{e}\left(\nabla V\right) \times \boldsymbol{p}, \tag{H.19a}$$

$$\boldsymbol{S} = \frac{\hbar}{2}\boldsymbol{\sigma}. \tag{H.19b}$$

Now the energy band structure is calculated by utilizing the pseudopotential method. The matrix elements of the Hamiltonian is then written as by following Melz [2]

$$H(\boldsymbol{k})_{\boldsymbol{G},\boldsymbol{G}'} = \frac{\hbar^2}{2m}\left(\boldsymbol{k} + \boldsymbol{G}\right)^2 \delta_{\boldsymbol{G},\boldsymbol{G}'} + V_{\boldsymbol{G},\boldsymbol{G}'} + \Delta(\boldsymbol{k})_{\boldsymbol{G},\boldsymbol{G}'}, \tag{H.20}$$

where

$$V_{\boldsymbol{G},\boldsymbol{G}'} = \left[V^{\mathrm{S}}(\boldsymbol{Q})\cos(\boldsymbol{Q}\cdot\boldsymbol{\tau}) + \mathrm{i}V^{\mathrm{A}}(\boldsymbol{Q})\sin(\boldsymbol{Q}\cdot\boldsymbol{\tau})\right]_{\boldsymbol{Q}=\boldsymbol{G}-\boldsymbol{G}'} \tag{H.21}$$

$$\begin{aligned}\Delta(\boldsymbol{k})_{\boldsymbol{G},\boldsymbol{G}'} &= \mathrm{i}\boldsymbol{\sigma}\cdot\left[\boldsymbol{G}\times\boldsymbol{G}' - \boldsymbol{k}\times(\boldsymbol{G}-\boldsymbol{G}')\right] \\ &\quad\times\left[\lambda^{\mathrm{S}}\cos(\boldsymbol{Q}\cdot\boldsymbol{\tau}) + \mathrm{i}\lambda^{\mathrm{A}}\sin(\boldsymbol{Q}\cdot\boldsymbol{\tau})\right]_{\boldsymbol{Q}=\boldsymbol{G}-\boldsymbol{G}'}.\end{aligned} \tag{H.22}$$

The first two terms of the matrix elements given by (H.20) are the kinetic energy and the pseudopotential term $V_{\mathrm{ps}}(\boldsymbol{G})$ and the third term is the matrix element of the spin–orbit Hamiltonian, where $\boldsymbol{k}$ is the electron wave vector in the first Brillouin zone, $\boldsymbol{G}$ and $\boldsymbol{G}'$ are the reciprocal lattice vectors. $\boldsymbol{\tau}$ is half the vector separating the two atoms in the unit cell, and Ω_0 is half of the atomic volume Ω. The notation $V^{\mathrm{S}}(\boldsymbol{Q})$ or $V^{\mathrm{A}}(\boldsymbol{Q})$ refers to the symmetric and antisymmetric pseudopotential coefficient, corresponding to the wave vector $\boldsymbol{Q}$. The spin–orbit term is derived as follows,

$$\begin{aligned}\Delta(\boldsymbol{k})_{\boldsymbol{G},\boldsymbol{G}'} &= \frac{\hbar}{4m^2c^2\Omega}\langle\boldsymbol{k}+\boldsymbol{G}'|(\nabla V\times\boldsymbol{p})\cdot\boldsymbol{\sigma}|\boldsymbol{k}+\boldsymbol{G}\rangle \\ &= \frac{\hbar^2}{4m^2c^2\Omega}\boldsymbol{\sigma}\cdot\int \mathrm{e}^{\mathrm{i}(\boldsymbol{G}-\boldsymbol{G}')\cdot\boldsymbol{r}}[\nabla V\times(\boldsymbol{k}+\boldsymbol{G})]\mathrm{d}^3\boldsymbol{r}\end{aligned} \tag{H.23}$$

where Ω is the volume of the unit cell in the crystal.

In order to apply the spin–orbit interaction to a zinc blende crystal we define the potential by a superposition of the atomic potentials located at $\pm\boldsymbol{\tau}$ in the unit cell $V_{\mathrm{A}}(\boldsymbol{\tau}) + V_{\mathrm{B}}(-\boldsymbol{\tau})$, and the symmetric and antisymmetric potentials by

$$V^{\mathrm{S}} = \frac{1}{2}\left(V_{\mathrm{A}} + V_{\mathrm{B}}\right), \qquad V^{\mathrm{A}} = \frac{1}{2}\left(V_{\mathrm{A}} - V_{\mathrm{B}}\right), \tag{H.24}$$

and then (H.23) is rewritten as

$$\Delta(\boldsymbol{k})_{\boldsymbol{G},\boldsymbol{G}'} = \frac{\hbar^2}{4m^2c^2\Omega_0}\boldsymbol{\sigma}\cdot\left\{\cos[(\boldsymbol{G}-\boldsymbol{G}')\cdot\boldsymbol{\tau}]\int e^{i(\boldsymbol{G}-\boldsymbol{G}')\cdot\boldsymbol{r}} \times\left[\nabla V^{S}\times(\boldsymbol{k}+\boldsymbol{G})\right]d^3\boldsymbol{r} + i\sin\left[(\boldsymbol{G}-\boldsymbol{G}')\cdot\boldsymbol{\tau}\right] \times\int e^{i(\boldsymbol{G}-\boldsymbol{G}')\cdot\boldsymbol{r}}[\left[\nabla V^{A}\times(\boldsymbol{k}+\boldsymbol{G})\right]d^3\boldsymbol{r}\right\}, \tag{H.25}$$

where Ω_0 is half of the atomic volume Ω.

We approximate the integral of (H.25) by expanding the exponential terms,

$$e^{i(\boldsymbol{G}-\boldsymbol{G}')\cdot\boldsymbol{r}} \simeq 1 + i(\boldsymbol{G}-\boldsymbol{G}')\cdot\boldsymbol{r}. \tag{H.26}$$

Then we find that only the second term will give a non-vanishing value. Using the following relations [2]

$$(\boldsymbol{G}-\boldsymbol{G}')\cdot\boldsymbol{r}[\nabla V\times(\boldsymbol{k}+\boldsymbol{G})] = (\boldsymbol{G}-\boldsymbol{G}')\times(\boldsymbol{k}+\boldsymbol{G})\nabla V\cdot\boldsymbol{r}, \tag{H.27}$$

and

$$\lambda^{S} = \frac{\hbar^2}{4m^2c^2\Omega_0}\int\nabla V^{S}\cdot\boldsymbol{r}d^3\boldsymbol{r}, \quad \lambda^{A} = \frac{\hbar^2}{4m^2c^2\Omega_0}\int\nabla V^{A}\cdot\boldsymbol{r}d^3\boldsymbol{r}, \tag{H.28}$$

we obtain the spin–orbit interaction Hamiltonian given by (H.22)

$$\Delta(\boldsymbol{k})_{\boldsymbol{G},\boldsymbol{G}'} = i\boldsymbol{\sigma}\cdot\left[\boldsymbol{G}\times\boldsymbol{G}' - \boldsymbol{k}\times(\boldsymbol{G}-\boldsymbol{G}')\right] \times\left[\lambda^{S}\cos(\boldsymbol{Q}\cdot\boldsymbol{\tau}) + i\lambda^{A}\sin(\boldsymbol{Q}\cdot\boldsymbol{\tau})\right]_{\boldsymbol{Q}=\boldsymbol{G}-\boldsymbol{G}'}. \tag{H.29}$$

This is exactly the same as the expression (18) of Chelikowsky and Cohen [3].

$$H^{so}_{\boldsymbol{G},\boldsymbol{G}'}(\boldsymbol{k}) = (\boldsymbol{K}\times\boldsymbol{K}')\cdot\boldsymbol{\sigma}_{s,s'}\left\{-i\lambda^{S}\cos[(\boldsymbol{G}-\boldsymbol{G}')\cdot\boldsymbol{\tau}] + \lambda^{A}\sin[(\boldsymbol{G}-\boldsymbol{G}')\cdot\boldsymbol{\tau}]\right\}, \tag{H.30}$$

where $\boldsymbol{K} = \boldsymbol{k}+\boldsymbol{G}$, $\boldsymbol{K}' = \boldsymbol{k}+\boldsymbol{G}'$ and $\boldsymbol{k}\times\boldsymbol{k} = 0$.

When we calculate energy band structure including the spin–orbit interaction, above equation may be used by adjusting the parameters λ^{S} and λ^{A} so that the calculated energy bands fit to the experimental data. Here we consider how to manipulate the spin–orbit Hamiltonian. As an example we will show how to evaluate the matrix element of the term given below

$$(\boldsymbol{G}\times\boldsymbol{G}')\cdot\boldsymbol{\sigma}. \tag{H.31}$$

The above equation is given by the sum of the x, y, z components

$$(G_yG'_z - G_zG'_y)\sigma_x + (G_zG'_x - G_xG'_z)\sigma_y + (G_xG'_y - G_yG'_x)\sigma_z. \tag{H.32}$$

Since the wave function $|\chi_n\rangle$ used in the pseudopotential analysis has spin degeneracy $|\chi_n, \uparrow\rangle$ and $|\chi_n, \downarrow\rangle$ and the operation of the spin operators leads to non-vanishing terms shown as the following

$$\langle\uparrow|\sigma_x|\downarrow\rangle = \langle\downarrow|\sigma_x|\uparrow\rangle = 1\,, \tag{H.33a}$$

$$\langle\uparrow|\sigma_y|\downarrow\rangle = -\mathrm{i},\ \ \langle\downarrow|\sigma_y|\uparrow\rangle = \mathrm{i}\,, \tag{H.33b}$$

$$\langle\uparrow|\sigma_z|\uparrow\rangle = 1,\ \ \langle\downarrow|\sigma_z|\downarrow\rangle = -1\,. \tag{H.33c}$$

The above relations are easily obtained by using the relations (2.50) and (2.51) defined in Sect. 2.3. Then the evaluation of the spin–orbit matrix elements is straight forward.

Bibliography

1. F. Stern, in *Solid State Physics*, ed. by F. Setz, D. Turbull (Academic Press, New York, 1963)
2. P.J. Melz, J. Phys. Chem. Solids **32**, 209 (1971)
3. J.R. Chelikowsky, M.L. Cohen, Phys. Rev. B **14**, 556 (1976)
4. C. Kittel, *Quantum Theory of Solids* (John Wiley, New York, 1963)
5. J. Callaway, *Quantum Theory of the Solid State* (Academic Press, New York, 1974)
6. W.A. Harrison, *Electronic Structure and the Properties of Solids; The Physics of the Chemical Bond* (W.H. Freeman, San Francisco, 1980)
7. P. Yu, M. Cardona, *Fundamentals of Semiconductors* (Springer, Heidelberg, 1996)
8. G.F. Koster, J.O. Dimmock, R.G. Wheeler, H. Statz, *Properties of The Thirty-Two Point Groups* (M.I.T. Press, Cambridge, 1963)
9. D. Rideau, M. Feraille, L. Ciampolini, M. Minondo, C. Tavernier, Phys. Rev. B **74**, 195208 (2006)
10. M.E. Kurdi, G. Fishman, S. Sauvage, P. Boucaud, J. Appl. Phys. **107**, 013710 (2010)
11. F.H. Pollak, M. Cardona, C.W. Higginbotham, F. Herman, J.P. Van Dyke, Phys. Rev. B **2**, 352 (1970)
12. M. Willatzen, M. Cardona, N.E. Christensen, Phys. Rev. B **50**, 18054 (1994)
13. M. Cardona, N.E. Christensen, G. Fasol, Phys. Rev. B **38**, 1806 (1988)
14. S. Richard, F. Aniel, G. Fishman, Phys. Rev. B **70**, 235204–1 (2004)
15. P.Y. Yu, M. Cardona, *Fundamentals of Semiconductors* (Springer, Heidelberg, 1996)
16. J.O. Dimmock, in *Semiconductors and Semimetals*, ed. by R.K. Willardson, A.C. Beer (Academic Press, New York, 1967), pp. 259–319
17. F. Abelès (ed.), *Optical Properties of Solids* (North-Holland, Amsterdam, 1972)
18. D.L. Greenaway, G. Harbeke, *Optical Properties and Band Structure of Semiconductors* (Pergamon, New York, 1968)
19. M. Cardona, in *Modulation Spectroscopy*, ed. by F. Seitz, D. Turnbull, H. Ehrenreich (Academic Press, New York, 1969)
20. D.E. Aspnes, Modulation spectroscopy/electric field effects on the dielectric function of semiconductors, in *Handbook of Semiconductors*, vol. 2, ed. by M. Balkanski (North-Holland, Amsterdam, 1980), pp. 109–154
21. P. Yu, M. Cardona, *Fundamentals of Semiconductors: Physics and Materials Properties* (Springer, Heidelberg, 1996)
22. B.O. Seraphin, in *Semiconductors and Semimetals*, Electroreflectance, ed. by R.K. Willardson, A.C. Beer (Academic Press, New York, 1972), pp. 1–149
23. D.F. Blossey, P. Handler, Electreoabsorption, in *Semiconductors and Semimetals*, vol. 9, ed. by R.K. Willardson, A.C. Beer (Academic Press, New York, 1972), pp. 257–314

C. Hamaguchi, *Basic Semiconductor Physics*, Graduate Texts in Physics,
DOI 10.1007/978-3-319-66860-4

24. R. Loudon, *The Quantum Theory of Light* (Oxford University Press, Oxford, 1973)
25. C. Hamaguchi, Brillouin scattering in anisotropic crystals. Oyo Buturi. Jpn. Soc. Appl. Phys. **42**(9), 866 (1973)
26. J.M. Ziman, *Electrons and Phonons* (Oxford University Press, Oxford, 1963)
27. E.G.S. Paige, in *Progress in Semiconductors*, ed. by A.F. Gibson, R.E. Burgess (Temple Press Books, London, 1964)
28. B.R. Nag, *Theory of Electrical Transport* (Pergamon Press, New York, 1972)
29. A. Haug, *Theoretical Solid State Physics* (Pergamn Press, New York, 1972)
30. R.A. Smith, *Wave Mechanics of Crystalline Solids* (Chapman and Hall, London, 1969)
31. K. Seeger, *Semiconductor Physics* (Springer, Heidelberg, 1973)
32. H. Brooks, *Advances in Electronics and Electron Physics* (1955), p. 85
33. K. Seeger, *Semiconductor Physics* (Springer, New York, 1973)
34. R.V. Parfenev, G.J. Kharus, J.M. Tsidilkovskii, S.S. Shalyt, Sov. Phys.-USP **17**, 1 (1974)
35. R.L. Peterson: in *Semiconductors and Semimetals*, Vol. 10, ed. by R.K. Willardson and A.C. Beer (Academic Press, New York, 1975) p. 221
36. R.J. Nicholas, Prog. Quantum Electron. **10**, 1 (1985)
37. C. Hamaguchi, N. Mori, Physica B **164**, 85–96 (1990)
38. P.J. Price, F. Stern, Carrier confinement effects. Surf. Sci. **132**, 577–593 (1983)
39. B.K. Ridley, The electron-phonon interaction in quasi-two-dimensional semiconductor quantum-well structures. J. Phys. C: Solid State Phys. **15**, 5899–5917 (1982)
40. G. Bastard, J.A. Brum, Electronic states in semiconductor heterostructures. IEEE J. Quantum Electron. **QE–22**, 1625–1644 (1986)
41. G. Bastard, *Wave Mechanics Applied to Semiconductor Heterostructures* (Editions de Physique, Les Ulis, 1988)
42. G. Bastard, J.A. Brum, R. Ferreira, Electronic states in semiconductor heterostructures. Solid state phys. **44** (1991)
43. D.G. Austing, T. Honda, S. Tarucha, Semcond. Sci. Technol. **11**, 388 (1996)
44. M. Macucci, K. Hess, G.J. Iafrate, J. Appl. Phys. **77**, 3267 (1995)
45. M. Macucci, K. Hess, G.J. Iafrate, Phys. Rev. B **55**, R4879 (1997)
46. Y. Tanaka, H. Akera, J. Phys. Soc. Jpn. **66**, 15 (1997)
47. P.A. Maksym, T. Chakraborty, Phys. Rev. Lett. **65**, 108 (1990)
48. D. Pfannkuche, R.R. Gerhardts, Phys. Rev. B **44**, 13132 (1991)
49. T. Chakraborty, V. Halonen, P. Pietiläinen, Phys. Rev. B **43**, 14289 (1991)
50. P.A. Maksym, T. Chakraborty, Phys. Rev. B **45**, 1947 (1992)
51. U. Merkt, J. Huser, M. Wagner, Phys. Rev. B **43**, 7320 (1991)
52. M. Wagner, U. Merkt, A.V. Chaplik, Phis. Rev. B **45**, 1951 (1992)
53. D. Pfannkuche, V. Gudmundsson, P.A. Maksym, Phys. Rev. B **47**, 2244 (1993)
54. A.V. Madav, T. Chakraborty, Phys. Rev. B **49**, 8163 (1994)
55. S.M. Sze, Physics of semiconductor devices (Wiley-Interscience). New York. Chaps. **9–10**, 425–566 (1969)
56. Y. Suematsu, et al. *A Study on Long Wavelength Integrated Laser and Optical Integrated Circuits* (in Japanese) (Research Reports on Special Research Projects), (Ohmsha, 1984)
57. T. Kitatani, M. Kondow, K. Shinoda, Y. Yazawa, M. Okai, Jpn. J. Appl. Phys. **37**, 753 (1998)
58. J.D. Perkins, A. Mascarenhas, Y. Zhang, J.F. Geisz, D.J. Friedman, J.M. Olson, S.R. Kurz, Phys. Rev. Lett. **82**, 3312 (1999)
59. C. Skierbiszewski, P. Perlin, P. Wisniewski, T. Suski, Phys. Rev. B **65**, 035207 (2001)
60. M.Z. Huang, W.Y. Ching, J. Phys. Chem. Solids **46**, 977 (1985)
61. A. Rubio, J.L. Corkill, M.L. Cohen, E.L. Shirley, S.G. Louie, Phys. Rev. B **48**, 11810 (1993)
62. M. Suzuki, Takeshi Uenoyama, and A. Yanase. Phys. Rev. B **52**, 8132 (1995)
63. M.D. McCluskey, C.G.V. de Walle, L.T. Romano, B.S. Krusor, N.M. Johnson, J. Appl. Phys. **93**, 4340 (2003)
64. K. Kim, W.R.L. Lambrecht, B. Segall, M.V. Schilfgaarde, Phys. Rev. B **56**, 7363 (1997)
65. S.K. Pugh, D.J. Dugdale, S. Brand, R.A. Abram, Semicond. Sci. Technol. **14**, 23 (1999)
66. M. Goano, E. Bellotti, E. Ghillino, G. Ghione, K.F. Brennan, J. Appl. Phys. **88**, 6467 (2000)

67. C. Bulutay, C.M. Turgut, N.A. Zakhleniuk, Phys. Rev. B **81**, 155206 (2010)
68. E. Sakalauskas, *Optical Properties of Wurtzite InN and Related Alloys*, Dr. Thesis, (Technische Universitat Ilmenau, 2012)
69. R.E. Jones, R. Broesler, K.M. Yu, J.W. Ager, E.E. Haller, W. Walukiewicz, X. Chen, W.J. Schaff, J. Appl. Phys. **104**, 123501 (2008)
70. E. Iliopoulos, A. Adikimenakis, C. Giesen, M. Heuken, A. Georgakilas, Appl. Phys. Lett. **92**, 191907 (2008)
71. S. Nakamura, M. Senoh, S. Nagahama, N. Iwasa, T. Yamada, T. Matsushita, H. Kiyoku, Y. Sugimoto, Jpn. J. Appl. Phys. **35**, L74 (1996)

Index

C. Hamaguchi, *Basic Semiconductor Physics*, Graduate Texts in Physics,
DOI 10.1007/978-3-319-66860-4

Q

R

S

T

Printed in the United States
By Bookmasters